SYNTHETIC RUBBER

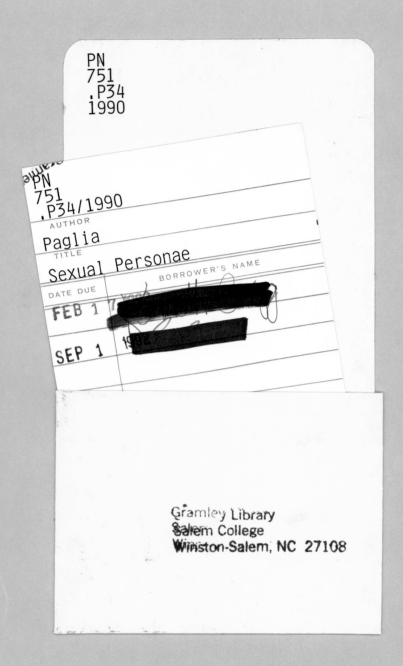

FRONTISPIECE.   Largest Plants in Government Synthetic Rubber Program
In the foreground a GR-S plant of capacity 120,000 tons a year
In the background a butadiene plant of capacity 100,000 tons a year
Courtesy B. F. Goodrich Chemical Co.

Elwood M. Payne, Houston, Texas

SYNTHETIC RUBBER

Prepared under the auspices of the Division of Rubber Chemistry, American Chemical Society

EDITORIAL BOARD

G. S. Whitby, Editor-in-Chief

C. C. Davis

R. F. Dunbrook

CONTRIBUTORS

Listed on the following pages

John Wiley & Sons, Inc., New York

Chapman & Hall, Limited, London

COPYRIGHT, 1954
BY
JOHN WILEY & SONS, INC.

———

All Rights Reserved

*This book or any part thereof must not
be reproduced in any form without
the written permission of the publisher.*

Library of Congress Catalog Card Number: 54–10308

PRINTED IN THE UNITED STATES OF AMERICA

CONTRIBUTORS

J. W. Adams, United States Rubber Company, Synthetic Rubber Division, Naugatuck, Conn. (present address: Marathon Corporation, Rothschild, Wis.)

J. Lawrence Amos, Physical Research Laboratory, Dow Chemical Company, Midland, Mich.

Henry E. Baumgarten, Department of Chemistry, University of Illinois, Urbana, Ill. (now Assistant Professor of Chemistry, University of Nebraska, Lincoln, Neb.)

E. M. Beavers, Rohm & Haas Company, Philadelphia, Pa.

R. L. Bebb, Research Laboratories, Firestone Tire & Rubber Company, Akron, Ohio

John T. Blake, Director of Research, Simplex Wire & Cable Company, Cambridge, Mass.

John O. Cole, Research Laboratories, Goodyear Tire & Rubber Company, Akron, Ohio

John T. Cox, Jr., Consulting Chemical Engineer, Washington, D. C. (formerly Deputy Director, Office of Rubber Reserve, Reconstruction Finance Corporation)

R. A. Crawford, Director of Research, B. F. Goodrich Company, Research Center, Brecksville, Ohio (deceased)

J. D. D'Ianni, Assistant to the Vice President, Research and Development, Goodyear Tire & Rubber Company, Akron, Ohio

R. P. Dinsmore, Vice President, Research and Development, Goodyear Tire & Rubber Company, Akron, Ohio

R. F. Dunbrook, Assistant Director, Research Laboratories, Firestone Tire & Rubber Company, Akron, Ohio

C. H. Fisher, Director, Southern Regional Research Laboratory, U. S. Department of Agriculture, New Orleans, La. (formerly Head, Carbohydrate Division, Eastern Regional Research Laboratory, Philadelphia, Pa.)

Harry L. Fisher, Professor of Chemical Engineering in Rubber Technology, University of Southern California, Los Angeles, Calif.

C. F. Fryling, Supervisor of Polymerization Research, Phillips Petroleum Company, Phillips, Texas (now Assistant Manager, Laboratory Branch, Verona Research Center, Koppers Company, Verona, Pa.)

L. H. Howland, Manager of Synthetic Rubber Development, Naugatuck Chemical Division of United States Rubber Company, Naugatuck, Conn.

A. E. Juve, Director of Compounding Research, B. F. Goodrich Company, Research Center, Brecksville, Ohio

R. D. Juve, Goodyear Tire & Rubber Company, Akron, Ohio (now Chief Chemist, Mohawk Rubber Company, Akron, Ohio)

F. L. Kilbourne, Jr., Xylos Rubber Company, Akron, Ohio (now Director of Research, Connecticut Hard Rubber Company, New Haven, Conn.)

John W. Livingston, Consultant, Manhasset, N. Y. (formerly Vice President, Rubber Reserve Company, Reconstruction Finance Corporation)

Frank S. Malm, Bell Telephone Laboratories, Murray Hill, N. J. (now retired, consultant, Newark, N. J.)

J. Lee Marsh, Carbide & Carbon Chemicals Corporation, New York, N. Y.

C. S. Marvel, Professor of Organic Chemistry, University of Illinois, Urbana, Ill.

LELAND R. MAYO, Rubber Laboratory, E. I. du Pont de Nemours & Company, Wilmington, Del.

J. E. MITCHELL, Dow Chemical Company, Midland, Mich.

C. E. MORRELL, Associate Director, Esso Laboratories, Chemical Division, Standard Oil Development Company, Linden, N. J.

ARTHUR M. NEAL, Assistant Director, Rubber Laboratory, E. I. du Pont de Nemours & Company, Wilmington, Del.

HENRY PETERS, In charge General Rubber Development, Bell Telephone Laboratories, Murray Hill, N. J.

WALDO L. SEMON, Director Pioneering Research, B. F. Goodrich Company, Research Center, Brecksville, Ohio

WILLIAM J. SPARKS, Director of Research, Esso Laboratories, Chemical Division, Standard Oil Development Company, Linden, N. J.

W. K. TAFT, General Manager, Government Laboratories, University of Akron, Akron 1, Ohio

ROBERT M. THOMAS, Head, New Products Development Laboratory, Chemical Division, Esso Laboratories, Standard Oil Development Company, Linden, N. J.

G. J. TIGER, Government Laboratories, University of Akron, Akron 1, Ohio (present address: Gladstone, N. J.)

W. J. TOUSSAINT, Research and Development Department, Carbide & Carbon Chemicals Corporation, South Charleston, W. Va.

L. B. WAKEFIELD, Research Laboratories, Firestone Tire & Rubber Company, Akron, Ohio

G. STAFFORD WHITBY, Professor of Rubber Chemistry, University of Akron, Akron 4, Ohio

LAWRENCE A. WOOD, Chief, Rubber Section, National Bureau of Standards, Washington, D. C.

PREFACE

This book has been prepared at the request and under the auspices of the Division of Rubber Chemistry of the American Chemical Society. The different aspects of the subject of synthetic rubber treated in the various chapters of the book are covered by authors each of whom has special familiarity with the aspects on which he writes. The time and effort that these authors have expended in contributing to the book is warmly acknowledged. The readiness of the Office of Rubber Reserve to allow reference in the book to certain unpublished reports is also acknowledged. And to Harry E. Outcault, former chairman of the Division of Rubber Chemistry, appreciation is expressed for his interest in the book at the time the volume was first projected.

Without doubt the not-distant future is destined to see many significant advances in synthetic rubber, in both its science and its technology. But already the development of synthetic rubber has reached such a point that the time is ripe for an attempt, which this volume represents, to review the whole subject. Such a review will serve, not only as a record of what has been accomplished so far, but also as a help in the further pursuit of research on the subject. The results of most of the intensive program of research that was initiated during World War II have now been published in papers, and the reader desirous of more detail than it is possible to convey in even a rather lengthy book may consult them for further information.

G. S. WHITBY

June 1954

CONTENTS

xi

ABBREVIATIONS

BIOS	British Intelligence Objectives Subcommittee
CIOS	Combined Intelligence Objectives Subcommittee
FD	Foreign Documents, United Kingdom Board of Trade
FIAT	Field Information Agency, Technical
Govt. Lab.	University of Akron Government Laboratories (Reconstruction Finance Corporation)
O.R.R.	Office of Rubber Reserve, Reconstruction Finance Corporation (also its predecessors, the Office of the Rubber Director, War Production Board and the Rubber Reserve Company and its successor, Office of Synthetic Rubber, Reconstruction Finance Corporation)
PB	Publication Board, Department of Commerce, Washington, D. C.
p.s.i.	Pounds per square inch
p.s.i.g.	Pounds per square inch, gage
TIIC	Technical Industrial Intelligence Committee

CHAPTER 1

INTRODUCTION

G. Stafford Whitby
University of Akron

That the rubber industry has undergone tremendous growth during the present century is generally realized.   How truly striking has been its development during a lifetime comes home with particular force to the present writer, who, as it happens, has been privileged to enjoy the unique experience of being associated in a technical capacity both with the early development of the plantation rubber-growing industry and, almost exactly a generation later, with the development of the large-scale manufacture of synthetic rubber in the United States and Canada.   In 1910, when, at the instance of his old professor, Tilden, who had made samples of synthetic rubber as far back as 1884, the writer went to work on rubber plantations in the East Indies, the world production of raw rubber was only 95,000 tons. None of the rubber was synthetic; most of it was from wild trees, the plantation rubber-growing industry at this early stage of its development contributing only 8000 tons of the year's supply.   By 1952 the world production of raw rubber had risen to about 2,600,000 tons, of which about 1,700,000 tons was natural rubber from plantations and about 900,000 tons was synthetic.   (Russian production is here left out of account, because of lack of authentic data concerning it.)   The growth of motor transport was the factor chiefly responsible for the rapid development of the plantation rubber industry.   During the decade from 1910 to 1920, the world production of raw rubber rose from 95,000 to 342,500 tons a year, and Akron, the largest center of rubber manufacturing, became the fastest-growing city in the United States and rose in population from 69,067 to 208,435.   The emergency of the Second World War was the factor chiefly responsible for the rapid development of the synthetic-rubber industry.   During the decade from 1942 to 1952 the production of synthetic rubber in the United States rose from 22,000 to about 800,000 tons a year.

Synthetic-Rubber Production Abroad.   Parallel with the establishment of a synthetic-rubber-producing industry in the United States, plants for the production of synthetic rubbers were built in Canada, and, as in the United States, production has continued since the war closed.   The Canadian output reached 75,000 tons in 1952.   As indicated in Chapter 2, the Canadian industry was initially established with the aid of American technologists on the basis of development work carried out in the United States.   In recent years Canadian scientists and technologists have themselves made worth-while contributions to the progress of synthetic-rubber research and technology.

1

In Germany synthetic rubber was first made commercially during World War I, but because of the high cost and the, at that time, inferior quality of the product the manufacture was discontinued at the close of the war. With the discovery of improved types of synthetic rubber, however, the manufacture of synthetic rubber was resumed in Germany at a later date and was conducted on a substantial scale. The history of the production of synthetic rubber in Germany is outlined in Chapter 2, and the important research carried out in Germany on synthetic rubber is reviewed in Chapter 26.

In Russia the manufacture of synthetic rubber was started in 1933, and by 1939 production had reached the level of 75,000 tons a year, obtained by the polymerization of butadiene by metallic sodium. In accord with the conspiratorial characteristics of the present Russian regime, no authentic information has been issued on postwar production in Russia or on any progress of first-class importance in research or in manufacturing technique that may have been made in that country since the war. An article by H. Schwartz in the New York Times, January 2, 1952, provides some estimates on the present scale and scope of synthetic-rubber manufacture in Russia. It appears that in 1951 Russian production probably reached a record level of 250,000 tons (tons in rubber production statistics are long tons), although one estimate places the figure at 300,000 tons. It appears likely that Russia is now making synthetic rubbers of other types in addition to sodium polybutadiene. During the war the United States provided Russia with information on the manufacture of Neoprene, and it appears that this type of rubber is now being made at Yerevan, Armenia, at a rate that one estimate places as high as 75,000 tons a year—a figure about which we may perhaps be permitted considerable scepticism. It appears too that emulsion polymerization is now being practiced in Russia for the production of rubber of the Buna type. In this connection it is to be noted that Schkopau and Auschwitz, at which places Germany built synthetic-rubber plants, are in the Russian zone, and it has been rumored that the latter plant has been transferred to Russia and is in operation there. (It has been stated that the Schkopau plant was in 1947 producing 3,500 tons of Buna a month.[1]) A further point of interest is that, whereas before the war all the butadiene used as starting material for the manufacture of synthetic rubber in Russia was made from alcohol obtained by the fermentation of potatoes or grain, some alcohol for butadiene production is now, it appears, being made from petroleum, at Sufgait, near Baku. It has been reported[30a] that in 1943 Neoprene (30 tons a month) was being made by Bata in Czechoslovakia.

In England, polysulfide rubbers and Vulcaprene (Chapter 25) have been manufactured to some extent. In Poland before the war and in Italy toward the end of the war, butadiene was manufactured in a relatively small way from alcohol and was presumably used to make synthetic rubber by treatment with sodium. In Japan small quantities of synthetic rubber have been manufactured. Before the war polysulfide rubbers and during the war Neoprene and nitrile rubber were, according to the writer's information, made in Japan. The rate of production of Neoprene was at one time 1 ton a day, according to a statement made by a Japanese chemist; 5 tons a day, according to a statement made by another Japanese chemist. It is interesting to note that in Japan the method adopted for the manufacture of the butadiene needed for the production of nitrile rubber was, after several other methods had been explored, the hydrogenation of vinylacetylene, an intermediate in the production of Neoprene, by treatment at room temperature with sodium hydroxide and zinc dust activated by mercury.

Postwar Position of Synthetic Rubber. The synthetic rubber for the production of which large plants were built in the United States and Canada during World War II has proved itself to be more than a mere temporary wartime substitute for natural rubber, so inferior to the latter as to fall into disuse after the war. Although it still has weaknesses and although, as is noted later, its improvement offers interesting and challenging problems to the research worker, synthetic rubber as now being manufactured has proved to be satisfactory in many classes of rubber goods to an extent such that the rubber is today being used in as large a quantity as it was during the war. In certain classes of rubber goods, synthetic rubbers have indeed in the United States replaced natural rubber almost completely.

PRESENT STATUS OF SYNTHETIC RUBBER IN THE MANUFACTURE OF RUBBER GOODS

Information on the actual usage of synthetic rubber in the manufacture of various classes of rubber goods is disclosed with a degree of detail not before publicly available in a survey issued on September 18, 1951, by the rubber division, National Production Authority, U.S. Department of Commerce. The survey shows the quantities and proportions of the various types of rubber (exclusive of reclaimed rubber) that manufacturers in the United States proposed to use in their operations during the third quarter of 1951. The survey was made at a time when the price of natural rubber was markedly higher than that of synthetic rubber but was tending downward. Natural rubber was at the time in the neighborhood of 50 cents a pound, while GR-S was 24.5 cents, Butyl 20.75 cents, and Neoprene GN 38 cents. The actual usage of the different types of rubber during the quarter to which the survey refers was, as data in the lower part of Table I show, quite close to the proposed usage, despite a drop in the price of natural rubber during the quarter. Further, in each of the 12 months succeeding the quarter in question, the ratio in which natural rubber and synthetic rubber were used by the rubber-manufacturing industry was closely similar to the ratio in the quarter of the survey, viz., about 35 per cent natural rubber, 65 per cent synthetic rubber. (During the 12 months' period mentioned the price of natural rubber was in the neighborhood of 30 cents a pound for a considerable part of the time.) Hence it is reasonable to consider that the information given in Tables I and II concerning the distribution between the various types of rubber in various classes of rubber goods still, at the present writing (early 1953), fairly represents industrial practice.

Among the points of interest regarding present practice in rubber manufacturing in the United States that may be read in or derived from the above statistics, the following may be mentioned.

1. Considering pneumatic tire casings for road service, the rubber in

small-size tires (tires for passenger cars and small trucks) is approximately 78 per cent GR-S and 22 per cent natural rubber, whereas in large-size tires (truck tires 8.25 and over in size) it is 77 per cent natural rubber and 23 per cent GR-S. In the size range of the tires used on most passenger automobiles (tires through 6.50 and 7.10) the proportion of GR-S is 83.3 per cent. That such tires give good service the average car owner can testify.

Table I.   Rubber Consumption in the United States, Third Quarter of 1951

Types of Rubber

Proposed Consumption	Dry NR	NR as Latex	GR-S	Butyl	Neo-prene	Other	All Types
1. As percentage of over-all consumption	31.6	4.0	53.2	6.4	3.7	1.1	100.0
2. As percentage of total consumption in							
(a) Transportation products	34.3	0.8	55.8	8.9	0.2	*	
(b) Nontransportation products	25.7	1.0	47.8	1.0	11.1	3.4	
3. Percentage of consumption of the type of rubber in question in							
(a) Transportation products	74.1	13.9	71.5	95.2	4.5	2.9	
(b) Nontransportation products	25.9	86.1	28.5	4.8	95.5	97.1	
4. Percentage of over-all consumption of all types in							
(a) Transportation products							68.2
(b) Nontransportation products							31.8

Actual Consumption

Over-all Actual Consumption: 306,343 Tons, i.e., at an Annual Rate of 1,225,372 Tons

	Dry NR	NR as Latex	GR-S	Butyl	Neo-prene	Other	All Types
1. As percentage of over-all consumption	32.7	3.1	53.5	5.8	3.8	1.0	100.0
2. As percentage of total consumption in							
(a) Transportation products	36.0		56.0		8.0		
(b) Nontransportation products	32.3		50.9		16.8		
4. Percentage of over-all consumption of all types in							
(a) Transportation products							69.7
(b) Nontransportation products							30.3

* Less than 0.05 per cent.

On the basis of the survey, of the total quantity of natural rubber used 39.8 per cent went into large-size and 18.7 per cent into small-size tires; of the total quantity of GR-S 44.8 per cent went into small tires and 7.9 per cent into large tires; of the total rubber the small-size tires consumed 30.6 per cent and the large-size tires 18.3 per cent.

Table II. Survey of Proposed Consumption in the United States of Various Types of Rubber in Various Classes of Products, Third Quarter of 1951

Product	Consumption by Products as Percentages of Total Consumption of Each Type							Consumption by Types as Percentages of Total Consumption in Each Product						
	Dry NR	NR as Latex	GR-S	Butyl	Neoprene	Other	All Types	Dry NR	NR as Latex	GR-S	Butyl	Neoprene	Other	All Types
Tires														
Transportation Products														
Airplane (large)	1.0	0.1	*	…	…	…	0.3	96.4	1.7	1.9	…	…	…	100.0
Airplane (small)	0.1	*	*	…	…	…	*	98.2	1.0	0.8	…	…	…	100.0
Bicycle	0.1	…	0.2	…	…	…	0.1	14.2	…	85.8	…	…	…	100.0
Motorcycle	*	*	*	…	…	…	*	20.4	0.1	79.5	…	…	…	100.0
Passenger through (6.50) and (7.10)	8.9	2.4	27.2	…	…	…	17.4	16.1	0.6	83.3	…	…	…	100.0
Passenger over (6.50) and (7.10)	5.0	0.3	9.4	…	…	…	6.7	24.1	0.2	75.7	…	…	…	100.0
Industrial pneumatic	0.1	*	0.2	…	…	…	0.1	20.0	0.1	79.9	…	…	…	100.0
Tractor implements (large)	1.0	*	3.8	…	…	…	2.3	13.3	*	86.7	…	…	…	100.0
Tractor implements (small)	0.3	*	1.4	…	…	…	0.8	12.1	0.1	87.8	…	…	…	100.0
Truck (7.50) and under	6.8	0.3	8.2	…	…	…	6.5	33.1	0.2	66.7	…	…	…	100.0
Truck (8.25) through (9.00)	15.3	0.3	4.7	…	…	…	7.4	65.7	0.2	34.1	…	…	…	100.0
Truck (10.00) through (12.00)	21.9	0.4	2.8	…	…	…	8.4	82.3	0.2	17.5	…	…	…	100.0
Truck over (12.00)	7.1	0.1	0.4	…	…	…	2.5	91.2	0.2	8.6	…	…	…	100.0
Solids—Airplane	*	…	…	…	*	…	*	70.2	…	…	…	29.8	…	100.0
Solids—Bogies, idlers and support rollers	0.1	…	1.1	…	*	…	0.6	4.9	…	95.1	…	*	…	100.0
Solids—Pressed and cured on	0.5	…	0.2	…	0.8	*	0.3	47.8	…	43.2	…	9.0	*	100.0
Tubes														
Airplane	0.2	…	…	*	…	…	0.1	99.8	…	…	0.2	…	…	100.0
Bicycle	…	…	…	1.7	…	…	0.1	…	…	…	100.0	…	…	100.0
Industrial	*	…	…	0.6	…	…	*	12.7	…	…	87.3	…	…	100.0
Passenger and motor cycle	0.3	…	…	48.4	…	…	3.2	3.0	…	…	97.0	…	…	100.0
Tractor implements	…	…	…	9.5	…	…	0.6	…	…	…	100.0	…	…	100.0

Table II—continued.

Product	Consumption by Products as Percentages of Total Consumption of Each Type							Consumption by Types as Percentages of Total Consumption in Each Product						
	Dry NR	NR as Latex	GR-S	Butyl	Neo-prene	Other	All Types	Dry NR	NR as Latex	GR-S	Butyl	Neo-prene	Other	All Types
Transportation Products														
Truck (8.25) and under	*	...	...	14.3	...	...	0.9	0.3	...	...	99.7	...	...	100.0
Truck (9.00) through (13.00)	0.1	...	...	16.8	...	...	1.1	2.3	...	...	97.7	...	...	100.0
Truck (14.00) and over	0.4	...	...	0.5	...	...	0.2	79.0	...	...	21.0	...	...	100.0
Other products														
Valves	*	...	*	0.8	*	*	0.1	10.3	...	2.9	86.3	0.2	0.3	100.0
Curing bags	0.6	...	*	2.3	0.1	...	0.3	56.3	...	*	42.8	0.9	...	100.0
Tire flaps	0.5	...	1.2	...	...	0.1	0.8	20.4	...	79.6	...	...	...	100.0
Camelback	1.6	...	8.4	...	...	0.1	5.0	10.3	...	89.7	...	...	*	100.0
Other retread materials	0.2	...	0.1	0.3	0.2	*	0.3	81.9	...	8.6	6.8	2.6	0.1	100.0
Tire and tube repair materials	0.6	...	0.1	*	*	*	0.2	87.4	...	12.1	0.4	...	0.1	100.0
Tank blocks, treads, and bend tracks	0.6	...	1.5	...	*	0.2	1.0	18.3	...	81.5	...	*	0.2	100.0
Masterbatches or compd's made for and/or sold to others	0.1	9.8	0.6	...	3.4	2.6	0.9	4.9	44.1	33.8	...	14.1	3.1	100.0
Nontransportation Products														
Belts and belting	2.1	*	0.9	*	4.2	0.7	1.3	52.2	0.1	35.2	*	11.9	0.6	100.0
Hose	1.2	...	1.5	0.4	17.5	10.7	1.9	19.1	...	40.2	1.3	33.3	6.1	100.0
Packing and gaskets	0.7	1.3	1.4	0.2	5.7	12.9	1.4	16.5	3.8	53.5	0.9	15.1	10.2	100.0
Aircraft equipment	0.2	*	*	*	1.6	2.2	0.2	40.8	0.2	8.7	0.1	33.7	16.5	100.0
Automotive equipment (including auto mats)	2.3	...	3.1	0.4	4.2	3.4	2.7	27.7	...	64.0	0.9	5.9	1.4	100.0
Household and appliance products	1.1	...	1.0	0.2	1.2	2.4	1.0	36.7	...	54.8	1.4	4.4	2.7	100.0
Mats and matting (excluding auto mats)	0.1	...	0.3	*	0.8	0.4	0.2	12.4	...	70.4	*	14.7	2.5	100.0
Hard rubber products (including auto battery containers)	0.6	0.4	1.5	*	0.1	0.4	1.0	19.0	1.8	78.5	*	0.2	0.5	100.0

Table II—continued.

Product	Consumption by Products as Percentages of Total Consumption of Each Type							Consumption by Types as Percentages of Total Consumption in Each Product						
	Dry NR	NR as Latex	GR-S	Butyl	Neoprene	Other	All Types	Dry NR	NR as Latex	GR-S	Butyl	Neoprene	Other	All Types
Nontransportation Products														
Other misc. mechanical goods	2.5	0.1	1.6	0.7	8.8	17.7	2.2	35.3	0.2	38.9	2.0	14.7	8.9	100.0
Wire and cable	0.7	0.5	2.4	1.0	28.8	1.6	2.7	7.9	0.7	48.7	2.4	39.7	0.6	100.0
Rubber footwear	4.0	1.5	1.0	*	0.3	...	1.9	68.9	3.3	27.3	*	0.5	...	100.0
Heels and soles	1.3	0.8	5.1	...	3.7	1.8	3.4	11.9	0.9	82.5	...	4.1	0.6	100.0
Inner shoe cushions and pads	*	0.3	0.4	*	0.8	*	0.2	3.0	4.4	80.6	0.1	11.7	0.2	100.0
Cements for shoes and shoe welding	0.2	3.5	*	*	2.1	0.7	0.3	18.4	47.9	5.6	0.3	26.2	2.6	100.0
Other shoe products	0.1	0.5	0.3	...	0.6	1.1	0.2	15.4	8.4	61.4	...	9.4	5.4	100.0
Cements	0.2	1.9	0.2	0.5	2.8	2.7	0.4	14.1	18.4	28.8	7.2	24.3	7.2	100.0
Proofing, coating or combining fabrics	0.5	0.4	1.0	0.9	3.2	6.7	1.0	17.0	1.8	55.2	6.1	12.2	7.7	100.0
Drug sundries	1.4	6.3	0.2	*	2.4	0.8	0.9	49.8	29.0	10.0	0.1	10.1	1.0	100.0
Flotation and lifesaving equipment	*	0.1	*	...	0.3	*	*	15.8	26.0	0.7	...	57.5	*	100.0
Bullet sealing fuel cells	0.4	...	*	*	0.2	8.6	0.3	48.5	...	12.3	*	2.2	37.0	100.0
Athletic goods	0.2	*	0.1	0.2	0.2	*	0.1	52.0	1.3	33.1	8.1	5.4	0.1	100.0
Toys and balloons	0.3	4.1	0.1	...	0.4	*	0.3	26.2	48.8	20.1	...	4.8	0.1	100.0
Sponge-rubber products and rubberized fiber and hair cushioning	1.9	2.7	1.8	*	2.8	*	1.8	35.0	6.1	52.9	*	6.0	*	100.0
Pressure-sensitive tape	1.0	0.1	0.3	*	*	6.2	0.5	59.2	0.5	26.5	0.2	0.3	13.3	100.0
Threads and related products	1.0	3.0	...	...	0.1	...	0.4	71.5	27.4	...	...	1.1	...	100.0
Rubber flooring and floor covering	*	0.6	1.8	...	0.2	2.2	1.0	0.1	2.3	94.7	...	0.5	2.4	100.0
Other misc. products	0.6	0.1	0.5	0.1	0.4	3.6	0.5	39.0	0.8	48.7	0.6	2.8	8.1	100.0
Latex foam	0.1	57.1	1.6	...	*	...	3.2	0.8	72.0	27.2	...	*	...	100.0
All other rubber products	1.2	0.8	0.4	0.2	2.1	9.9	0.8	45.9	4.2	24.4	2.0	9.7	13.8	100.0

* Less than 0.05 per cent.

2. Tires not designed primarily for road service are omitted from the calculations above. It may be noted that in tractor tires the rubber is chiefly GR-S. In large-size tractor tires the rubber is shown as being 86.5 per cent GR-S. These tires, although large, are not, of course, called upon to operate at high speeds.

3. Of the Butyl rubber 95 per cent is consumed in transportation products. Practically all tubes, except those for the largest-size truck tires (14.00 and over), are composed of Butyl rubber. More than 40 per cent of the rubber used for curing bags is Butyl rubber, the good aging properties of which enable it to serve for many heats.

4. Of the Neoprene used 95 per cent was in nontransportation products, the largest use being in wire and cable and the next largest in hose. Of the rubber used on wire and cable 88.4 per cent was either Neoprene or GR-S; only 7.9 per cent was natural rubber.

5. Of the rubber used for the production of hard rubber 78.5 per cent was GR-S and 20.8 per cent natural rubber.

6. The footwear manufacturing industry still finds it desirable to use natural rubber for the most part (72.2 per cent), but for heels and soles GR-S is mostly used (82.5 per cent).

7. Rubber flooring is made almost exclusively from GR-S.

8. Latex foam sponge rubber, use of which in upholstery and bedding has grown rapidly, consumed rubber in 1951, according to the survey, at the rate of about 39,000 tons a year, 72 per cent of this being natural rubber as latex and 27 per cent GR-S as latex.

RAW MATERIALS

Isoprene. Because the hydrocarbon of natural rubber is a polymer of isoprene, most of the early research work on the production of rubber by polymerization was based on isoprene as the monomer. Only after an interval was butadiene, the lower homolog of isoprene, used as the monomer in such work. It then appeared that isoprene had no obvious advantage over butadiene in the quality of the rubber obtained, and, because of this and because butadiene can be prepared more cheaply and by a greater variety of reactions than isoprene, the former has come to be used as the base monomer in most of today's synthetic-rubber production. A limited amount of isoprene finds use in the production of Butyl rubber and (Chapter 18) in the production of synthetic rubbers adapted to certain chemical reactions, expecially chlorination.

The isoprene used today is recovered from the C_5 cut derived from the deep cracking of petroleum. It is interesting that during World War II isoprene was manufactured also by the cracking of terpene hydrocarbons in equipment that was essentially an adaptation to large-scale operation of one of the early laboratory means of preparing isoprene, namely, the Harries-Gottlob isoprene lamp (Chapter 5).

Butadiene. It is interesting to note that, although isoprene has been known to be capable of undergoing polymerization to a rubbery product since at least 1874, and although butadiene was prepared at almost as early a date as isoprene, it was not until 1910 that butadiene was observed to be capable of polymerization, after programs of research on synthetic rubber had been started in several countries. This delay in examining the polymerizability of butadiene occurred presumably because butadiene, unlike isoprene, is a gas at ordinary temperature, and in the early days when it was discovered, chemists, lacking as they did the ready supplies of Dry Ice now available, worked with liquid gases only exceptionally.

Isoprene was first isolated definitively in 1860. Butadiene was first prepared in 1863—by the pyrolysis of amyl alcohol. Credit for this first preparation of butadiene belongs to Eugène Caventou, son of the Caventou (Joseph Bienaimé Caventou) famous for the isolation, in 1820 in conjunction with Pelletier, of quinine. The discovery of quinine was memorialized by the erection in Paris in 1900 of a statue to Pelletier and the elder Caventou. The discovery of butadiene by the younger Caventou has received no such signal notice!

The rapid development in the United States, under the pressure of the war emergency, of efficient methods for the manufacture of butadiene and the rapid erection and efficient operation of plants embodying these methods were truly impressive achievements of chemical research and chemical engineering. Two methods were chiefly used, viz.: (1) the catalytic dehydrogenation of straight-chain C_4 olefins derived from the gases produced in the deep cracking of petroleum, and (2) the catalytic dehydrogenation and dehydration of alcohol. The organizations largely responsible for the massive contributions to synthetic-rubber production which the successful development of these methods represented were respectively the Standard Oil Development Company and Carbide and Carbon Chemicals Corporation.

In discussions early in the emergency, it was contemplated that butadiene for synthetic rubber should be made largely from butane by dehydrogenation in two steps: the first to butene, the second to butadiene. Later it appeared, as indicated in the succeeding paragraph, that substantial supplies of butenes would become available through the aviation gasoline-production program. And in consequence, although certain of the petroleum butadiene plants were designed to start with butane and apply to it two-step dehydrogenation, most of the petroleum butadiene plants were designed to start with butene and apply to it one-step dehydrogenation. When the petroleum butadiene-production program was first discussed, the then known method of dehydrogenating butene, employing chromia-alumina as a catalyst, involved working under conditions of reduced pressure, which presented considerable operating difficulties. The Standard Oil Development Company, however, succeeded in developing a new catalyst, that, unlike previously known catalysts for the dehydrogenation, would work in the presence of steam, and would thus make it possible to bring the butene to the partial pressure necessary for successful dehydrogenation without working under reduced pressure, and would at the same time provide a means (by superheating the admixed steam) of bringing the reaction mixture to the necessary high temperature. Later, this same catalyst was found to be applicable to the dehydrogenation of ethylbenzene in the manufacture of styrene and to be an improvement on the bauxite previously used for that purpose.

The production of butadiene from petroleum was closely associated with developments in the oil industry to provide adequate supplies of military aviation gasoline. Highly important in this connection was the development of the "fluid" catalyst method of cracking oil, to operate which large plants were built during the war. By this technique, deep cracking was achieved and large quantities of C_4 gases were secured—quantities sufficient to supply the alkylate needed for high-antiknock gasoline and the butene needed for the manufacture of butadiene. In the fluid catalyst process, the high temperature required for deep cracking is secured by burning off (in a regenerator attached to the cracker) the carbon deposit which during cracking forms on the "fluid," i.e., suspended, finely divided catalyst.

In the meantime, the Carbide and Carbon Chemicals Corporation, thanks to an intensive 24-hour-a-day program of research and development, had worked out an efficient catalyst for the production of the 4-carbon compound butadiene from a mixture of the 2-carbon compounds ethyl alcohol and its dehydrogenation product acetaldehyde. The Carbide and Carbon process for the manufacture of butadiene from alcohol made a great contribution to synthetic-rubber production during the war, and the plants embodying it fortunately proved to have an output much in excess of their design capacity. However, alcohol as a raw material for butadiene manufacture is at present more expensive than are C_4 petroleum gases. The yield from alcohol approaches 2.5 lb. butadiene per U.S. gallon of alcohol, and, speaking very roughly, 2 gal. of molasses or 0.5 bushel of grain are needed to yield 1 gal. of alcohol by fermentation, the conversion cost being of the order of 15 cents per gallon of alcohol. Alcohol produced by the fermentation of vegetable materials, such as grains and potatoes, will probably always be a relatively expensive raw material for the manufacture of butadiene, although, at times of a glut of molasses, fermentation alcohol may be in a relatively more favorable position than normally. Recently the trend in regard to the manufacture of alcohol has been strongly toward synthetic alcohol, from ethylene derived from petroleum. And generally such alcohol is likely to be in a more favorable price position than fermentation alcohol in respect to butadiene manufacture.

Interesting types of fermentation studied during the war (Chapter 5) are those in which 2,3-butylene glycol is the main product and in which the yield of this glycol or of a mixture of this glycol and ethyl alcohol is such that the potential production of butadiene from a bushel of corn exceeds that obtainable by applying the ordinary alcoholic fermentation and using ethyl alcohol alone as the source of butadiene. However, the route to butadiene through 2,3-butylene glycol presents certain practical difficulties, viz.: (1) the cost and inconvenience of isolating the glycol from the dilute fermentation "beer," and (2) the fact that, unlike the corresponding 1,3- and 1,4-glycols, 2,3-butylene glycol cannot be dehydrated directly to butadiene but must first be converted to its diacetate.

Other methods of producing butadiene that have been shown, by actual plant or pilot-plant operation, to be industrially practical, but that are less economical than the method of butene dehydrogenation—now the standard method in the United States—are (1) from acetylene or acetaldehyde via acetaldol and 1,3-butylene glycol (the aldol process), (2) from acetylene via 1,4-butyndiol and 1,4-butylene glycol (the Reppe process), (3) from butene by chlorination and dehydrochlorination (the Shell process), and (4) from benzene via cyclohexane followed by cracking (the Koppers process).

The development of economical processes for the manufacture of butadiene, for which the synthetic-rubber industry is to be credited, can be regarded as a contribution, not to the rubber industry alone, but also to chemical industry generally.   Butadiene is now being used as a source of intermediates for the manufacture of nylon via butadiene dichloride, 1,4-dicyanobutane, and adiponitrile.   And it is not unreasonable to expect that, thanks to its great reactivity, butadiene will in time come to serve as the starting material for still other synthetic organic chemicals.

Styrene.   It had been found in Germany that synthetic rubber of a serviceable quality could be made more readily from a mixture of butadiene and a second, suitable monomer of the vinyl class than from butadiene alone. The second monomer decided upon in Germany for the production, by copolymerization with butadiene, of general-purpose synthetic rubber was styrene or vinybenzene.   In the United States too, when a wartime program of synthetic rubber was adopted, styrene was selected as the comonomer. It was known that other comonomers would yield serviceable rubber by copolymerization with butadiene, and in fact tires containing synthetic rubber involving a comonomer other than styrene were offered commercially in the United States 18 months before Pearl Harbor.   But the consensus was that, all things considered, styrene was the most practicable choice for a comonomer.

A process for the manufacture of styrene from cheap and abundant raw materials—benzene and ethylene—had shortly before been developed and had been operated on a modest scale.   As part of the synthetic-rubber program, large plants were built to manufacture styrene by this process. The plants proved to be highly efficient (Chapter 6), to be capable of an output much in excess of their design capacity, and to yield styrene at surprisingly low cost.   Thanks in considerable measure to this demonstration afforded by the styrene plants in the synthetic-rubber program that styrene of high purity can be manufactured at low cost, styrene since the war has, in addition to maintaining its importance in synthetic-rubber manufacture, become the most largely used source of thermoplastics in the United States. Polystyrene has taken the place of cellulose acetate as the most important thermoplastic molding material.

Although it has become an industrial chemical of outstanding importance only during the last decade, styrene has had a long prior history.   It occurs in small proportion in the balsam from a tree found in Asia Minor, *Liquidambar orientalis*.   This balsam, it has been stated, was used by the Greeks as incense, and later it was introduced into medicine under the name storax, now styrax in the U.S. Pharmacopoeia.   Its present major use is in soap and perfumery.   Styrax is now secured from the American species, *Liquidambar styraciflua*, as well as from the original, Levant tree.   It was demonstrated more than 100 years ago that Levantine styrax contains some free styrene, and later the presence of styrene in American styrax was established.[12]

Styrax contains, in addition, cinnamic acid and esters of cinnamic acid, as well as resins of unknown composition. The conjunction of styrene and cinnamic acid is interesting, because of the fact that the latter yields the former on loss of carbon dioxide.

It is pleasant to think of the sweetgum tree in the woods and forests, with its mellifluous name, *Liquidambar styraciflua*, its elegant star-shaped leaves, its brilliant fall colors, and its spiked seed balls, which remain hanging from it through the winter; and to reflect that, in addition to being the third largest source of hardwood lumber in the United States, it is able to produce, albeit in only small amount, the now important chemical, styrene.

How recent has been the attainment of industrial importance by styrene and how lately polymers have become a well-recognized, distinct field of study are indicated by the fact that, when in 1921 I first became interested in its polymers, styrene was available only as a purely laboratory chemical (from Germany), and, when in 1926 I published some observations on polystyrene, the latter was referred to as "meta-styrene" and—the general term "polymer" then being little used—was classed as an "organophilic colloid."[24]

POLYMERIZATION

The development of methods for the manufacture at low cost of the raw materials that enter into synthetic rubber represented a well-defined task in the field of standard organic chemistry—a task that has been executed in a highly meritorious manner The conversion of the raw materials into synthetic rubber represents a task, less well defined, in the field of polymer chemistry, and, in view of the newness of that field, it is not surprising that scope for improvement remains, and that, although the conversion can now be brought about under conditions that are practicable and that yield products of good serviceability, there are still many aspects of it in which improvements are desirable and conceivable, and in which further research will be of value.

Hitherto the conversion, i.e., the polymerization, has mostly been conducted in emulsion. The study of emulsion polymerization and rapid improvement in its practice have been factors of first-class importance in the development of synthetic-rubber manufacture. The story of emulsion polymerization as it concerns synthetic rubber is told in Chapter 8, and other references to the subject are to be found in Chapters 10 and 26.

If the polymerization of a mixture of butadiene and styrene in the GR-S ratio is brought about by merely heating the mixture in bulk without catalysts, a period of about 5 months at 55° C. is required to produce complete polymerization, and the product is insoluble and nonplastic and crumbles when it is worked on a rubber mill.[29] By working in emulsion, especially in the presence of suitable catalysts or initiators, much faster polymerization is obtained. In Germany, where the emulsion technique was first used commercially for the production of synthetic rubber, the period of reaction was 25 to 30 hours at temperatures in the range 45 to 50° C. In the recipe (the Mutual recipe) originally adopted in the United States for the manufacture of GR-S in emulsion, the time required for polymerization is only 12 hours at 50° C. A small proportion of a water-soluble persulfate is employed as a catalyst. Further, by including in the recipe a small proportion of a "modifier," e.g., *n*-dodecyl mercaptan, the product is plastic and can be compounded satisfactorily on a rubber mill.

Redox Polymerization. By the use of redox (reduction-oxidation) reactions for the initiation of polymerization, it has been found that rates of conversion much higher than those just instanced can be obtained—rates so high that it becomes practical to manufacture synthetic rubber at temperatures much lower than 50° C. In the manufacture of "cold GR-S," the temperature of polymerization is only 5° C., and yet, thanks to the use of redox polymerization recipes, the period of reaction is no longer than that in the Mutual recipe at 50° C. As is fully explained in Chapters 8 and 10 and elsewhere in this volume, the employment of a lower reaction temperature in its manufacture markedly improves the quality of GR-S, and, accordingly, the greater part of the manufacturing capacity for GR-S has now been or is being modified to produce cold rubber in place of the regular (50° C.) GR-S. Present plans call for the conversion of 11 of the 13 American GR-S polymerization plants to cold rubber before the middle of 1953. The growth in the production of cold GR-S to date is shown by the data in Table III.

Table III. *Government Production of GR-S in the United States*

Year	Tons			Per Cent Cold
	Total	Regular	Cold	
1949	288,881	221,369	67,512	23.4
1950	350,801	215,838	134,963	38.5
1951	694,583	431,088	263,495	37.9
1952	621,867	328,913	292,954	47.1
1953*	630,000	238,000	392,000	62.2

*Estimated on the basis of actual production in the first 11 months.

The redox principle of polymerization, i.e., the initiation of polymerization by free radicals generated by oxidation-reduction reactions, was discovered independently in the United States, Britain, and Germany. It was, however, the German discoveries in this connection, which had not reached the stage of plant application but about which information became available in America shortly after the war, that served to point the way to practical polymerization recipes capable of producing synthetic rubber at low temperatures, such as 5° C., in reasonably short periods of time. Such recipes were developed in the United States and made it possible for the manufacture of cold GR-S to be undertaken.

The British observations on redox polymerization were concerned with the use of redox pairs (typically alkali persulfates on the one hand and alkali bisulfites, thiosulfates, or dithionites on the other hand), both components of which are water-soluble.[2] These were applied chiefly to the polymerization and copolymerization of vinyl monomers, not to the production of synthetic rubber. It has not been shown that the redox systems here in question are suitable for the manufacture of synthetic rubber. Patents were applied for as early as June 4, 1941.

The first American observations on redox polymerization were apparently some on the basis of which patents were applied for by the B. F. Goodrich Company, first on February 19, 1941. In the redox systems here used, the oxidant was, in most examples quoted, hydrogen peroxide; and the other component or components consisted of a heavy metal complex (such as sodium iron pyrophosphates,[13] potassium cobaltic cyanide[19]) or a heavy metal (usually iron, but in some cases cobalt or copper, or a combination of these[22]) in conjunction with aliphatic acids, such as levulinic acid[14] or succinic acid,[14, 21] with reducing sugars or quebrachitol,[16] with sulfur compounds, such as cysteine or β-mercaptoethanol,[15] with cholesterol or bile,[17] with hydroxyamino acids, such as serine,[20] with dicyanodiamidine or hydantoin.[18] Such redox systems were applied to the emulsion polymerization of butadiene-acrylonitrile and butadiene-styrene, but the lowest temperature at which they are mentioned as being used is 30° C. These studies, although interesting as involving the redox principle of polymerization, were not in fact the basis on which cold GR-S was developed. Only when German data on redox polymerization became available at the close of the war, were there developed in the United States redox recipes capable of bringing about the polymerization and copolymerization of dienes in reasonably short periods of time at 5° C. and thus making the manufacture of cold GR-S practicable. In these recipes, instead of both members of the redox pair being water-soluble, or instead of at least the oxidant being water-soluble, the oxidant used is oil-soluble.

An important contribution to the original redox recipe for the production of cold GR-S and to the subsequent improvement of the recipe has been the development of new organic peroxides. In place of the benzoyl peroxide that figured in German recipes, cumene hydroperoxide was introduced and employed in the manufacture of cold GR-S. Since then many new hydroperoxides have been made and investigated. Cumene hydroperoxide has now been replaced by the more active hydroperoxides of isopropylcumene and p-menthane. It is not unlikely that, as investigation of the organic hydroperoxides is extended, still more active representatives will be discovered and will facilitate the application of redox polymerization to still lower temperatures and still faster reactions.

The Hercules Powder Company and the Phillips Petroleum Company have been particularly active in the investigation of new hydroperoxides. And it is interesting to note that the recent study of the chemistry of organic hydroperoxides, in the stimulation of which the synthetic-rubber program has been one of the factors, promises to bear fruit in the field of general organic synthesis; that the decomposition of appropriate hydroperoxides by suitable means promises to provide methods of making certain chemicals without recourse to drastic reactions, such as nitration and sulfonation. Already plants are in operation or under construction[1a] for the manufacture, by a process developed by the Hercules Powder Company and the Distillers'

Corporation, of phenol and acetone by the decomposition of the hydro-peroxide of cumene (isopropylbenzene). And there would seem to be many other synthetic possibilities via organic hydroperoxides.

The original recipe for the production of cold GR-S involves as an essential feature the decomposition of the hydroperoxide (to generate free radicals, which in turn initiate polymerization) by reaction with iron, the iron being introduced as a complex with an alkali pyrophosphate. Many variations of the original recipe, all dependent essentially on the decomposition of hydroperoxides by iron, have been devised. The "peroxamine" recipe, in which the oxidant is a hydroperoxide and the reductant a polyethylene polyamine, and in which no iron is added, appeared at first sight not to depend on iron, but it has been found experimentally that the recipe fails to produce polymerization if the reagents used are entirely free from iron[28] or if any iron in the reagents is sequestered by complexing it.[10] The level of iron needed in this recipe is very low, e.g., 1 part in 50 million on the system as a whole or 1 part on 40,000 parts of the hydroperoxide in the system, and this quantity is normally provided by the iron which the reagents carry as an impurity. The peroxamine polymerization recipe is in fact an interesting example of a highly turbulent reversible redox system, in which a minute amount of iron shuttles between the ferrous and ferric states, the ferrous iron decomposing peroxide and thus being oxidized to the ferric state, which in turn is reduced to the ferrous state by the polyamine. The fact that the polyamines used are capable of chelating iron is not improbably of significance in the mechanism.

The method of redox polymerization makes it possible, by the use in sufficient concentration of suitably chosen oxidants and reductants, to conduct the polymerization of butadiene-styrene in practicably short periods of time at temperatures much lower than 5° C., but it is not yet clear that the use of such temperatures will give an improvement in the quality of the rubber sufficient to justify the cost of the antifreeze and of the increased demand for refrigeration which polymerization at temperatures below the freezing point of water involves. Then too the redox principle makes possible polymerization at 5° C. in periods of time very much shorter than those now employed in the manufacture of cold GR-S; but, in order to control the heat of such fast reactions, drastic redesign of the present reactors or their replacement by tubular reactors or other means of providing better heat exchange would be necessary, as is pointed out in Chapter 8. And perhaps before the practical problems presented by very fast, low-temperature reactions have been worked out some very different mode of polymerization, e.g., continuous Alfin polymerization, may offer more attractions.

PROBLEMS OF POLYMER STRUCTURE

Possible Modes of Union of Dienes. In the polymerization of con-jugated dienes such as butadiene, the possible modes of union of the monomer units are more numerous than in the polymerization of vinyl monomers, and in view of this it is hardly surprising that, at the present early stage of the development of polymer chemistry, our understanding and control of diene polymerization is less advanced than our rapidly growing understanding and control of vinyl polymerization. While a vinyl monomer, $XCH : CH_2$, normally polmerizes by simple head-to-tail union, to yield a regular, linear structure of the form . . . $XCH \cdot CH_2 \cdot XCH \cdot CH_2$ . . . , a diene monomer such as butadiene, $CH_2 : CH \cdot CH : CH_2$, can and does unite in both the 1,4- and the 1,2-manner, thus: . . . $CH_2 \cdot CH : CH \cdot CH_2 \cdot CH \cdot CH_2$ . . . ;

$$| \\ CH : CH_2$$

and, as a further complication, in the 1,4-units both *cis* and *trans* configurations occur. The polymerization of butadiene is in fact a *co*polymerization.

At present no techniques are known capable of insuring the presence of only a single kind of structural unit in diene polymers; all known techniques result in polymers in which *cis* 1,4-, *trans* 1,4-, and 1,2-diene units occur. The discovery of a mode of polymerization that will yield from butadiene polymer in which all the units are joined 1,4- and possess the same sterical configuration is a desideratum, although it is not certain that such a polymer would have properties as good as those of a corresponding isoprene polymer, such as the hydrocarbon of natural rubber. It may prove that the bulking effect of the methyl substituents in poly-1,4-isoprene is advantageous.

Cross-Linking and Branching in Diene Polymerization. A further complication that enters into the polymerization of conjugated dienes, in comparison with the polymerization of vinyl compounds, is that, whereas in vinyl polymerization all the double bonds of the monomer disappear and the polymer is saturated, in diene polymerization the polymer retains one double bond per entering monomer unit. The unsaturation of diene polymers has both advantages and disadvantages. It is advantageous in rendering the polymers readily vulcanizable. (But even here difficulties enter, because, as is pointed out in a later paragraph, of the range of molecular weights in diene polymers.) The unsaturation is disadvantageous in that it facilitates the occurrence during polymerization of cross-linking and branching, with corresponding increase in the structural irregularity of the polymer molecules. Another disadvantage of the presence of unsaturation in diene polymers is that it makes the polymers far more susceptible than vinyl polymers to attack by oxygen. In fact, unless they are protected by an antioxidant, diene polymers change to inelastic, friable solids in the course of a few weeks' exposure to air (Chapter 13).

When polymerization is effected in emulsion, polymerization is initiated in the micelles of emulsifier, from which locus particles of polymer are ejected into the water phase; after about 13 per cent of the monomer has been converted to polymer, no new particles are formed, and all the subsequent polymerization takes place in the existing particles. The structure of the polymeric product is essentially determined by what takes place in the little world of the latex particle, within the boundaries of which the total number of molecules is relatively small. (A particle 500 A. in diameter in latex of polystyrene of molecular weight 1 million contains only about 40 molecules.) As the conversion of monomer to polymer proceeds, the ratio of polymer to monomer rises, and the chances increase of new polymer chains growing onto the molecules of already formed polymer and thus causing branching or cross-linking. If a growing chain, X·, terminates itself by extracting hydrogen from a molecule of already formed polymer, thus,

$$X· + \ldots CH_2 · CH : CH · CH_2 \ldots \rightarrow XH + \ldots \dot{C}H · CH : CH · CH_2 \ldots,$$

the polymer becomes a free radical, onto which a branch may grow or which may combine with another growing polymer chain in a termination reaction to form a branch. If a growing polymer chain terminates itself by addition to the double bond of a molecule of already existing polymer, thus,

$$X· + \ldots CH_2 · CH : CH · CH_2 \ldots \rightarrow \ldots CH_2 · CHX · \dot{C}H · CH_2 \ldots,$$

the polymer becomes a free radical, the formation of a branch onto which will produce a cross-link. Probably such a cross-linking reaction is more prone to occur than a mere branching reaction.

In the emulsion polymerization and copolymerization of dienes, the point at which such cross-linking reactions take place to a notable extent is known as the gel point, i.e., the point at which some of the polymer becomes insoluble. This is a critical point in the polymerization, as the rubber thereafter rapidly becomes increasingly insoluble and less plastic. In practice, because of the complications here outlined, the conversion of monomer to polymer is not carried to completion. In the manufacture of standard GR-S, the polymerization reaction is stopped at 72 per cent conversion, and in the manufacture of cold GR-S at 60 per cent conversion. Accordingly, it is necessary to incorporate in the GR-S plants equipment for the recovery of the butadiene and styrene which remain unreacted at these points.

Influence of Temperature on Polymer Structure in Emulsion Polymerization. In emulsion polymerization the only definitely known means of influencing the proportions in which the various diene forms (*cis* 1,4-, *trans* 1,4-, and 1,2-) enter into the polymer during the propagation step is by changing the temperature. By employing a lower temperature of polymerization, as is done in the manufacture of cold rubber, the proportion of 1,2- is somewhat reduced and—more important—the proportion of 1,4-butadiene units in the *trans* configuration is markedly raised, and, accordingly, the regularity of the polymer chains is increased. It appears that (with possible, doubtful exceptions at present under investigation) the structure of the polymer chains is not influenced by the nature of the free radicals that may be employed to initiate polymerization. It appears that, once a monomer free radical has been formed, no matter what the initiating radical, the manner in which the polymerization propagates itself is essentially the same, other conditions, such as temperature, being the same. As it has been phrased, once a monomer free radical has been formed, it is "on its own," and the nature of the initiating radical is without influence on the mode of propagation of the polymer chain; the resonance characteristics of the monomer determine the structure of the propagated chain.

Polymerization by Ionic Catalysts. In contrast to this state of affairs in polymerizations initiated by free radicals, when the polymerization of dienes is initiated by ionic catalysts, the structure of the polymer chain may be profoundly influenced by the nature of the catalyst. Thus, for example, sodium produces from butadiene a polymer in which the majority of the butadiene units have the 1,2-form; potassium, a polymer with a notably lower, and the Alfin catalyst, a polymer with a much lower proportion of such units. Further, the proportion of 1,2-units in the polymer is more susceptible to an influence of temperature in the ionic than in the free-radical polymerization of dienes; but the direction in which temperature influences the polymerization in this regard is opposite to that which prevails in emulsion polymerization; i.e., a reduction in the temperature of polymerization increases the proportion of 1,2-units.

In view of the possibility in the ionic polymerization of dienes of influencing markedly the structure of the polymer by the choice of the catalyst, and having regard to the striking and promising results achieved by the novel catalyst complexes known as Alfin catalysts (Chapter 21), it may be expected that the coming years will see an intensification of the study of ionic catalysts in the polymerization of dienes.

Influence of Molecular-Weight Distribution on Vulcanization. Products arising from addition polymerization normally consist of mixtures of molecules differing considerably in their degree of polymerization; that is to say, there is a rather wide distribution of molecular weights in addition polymers as ordinarily prepared. In this connection a complication presents itself in diene polymers destined to be vulcanized that is absent from vinyl polymers destined to be employed as thermoplastics. When polymers are vulcanized in order to unite the primary polymer chains into a network, the physical properties of the network depend on both the length of the primary chains and the number of cross-links between them; the longer the primary chains, the fewer the cross-links needed to produce a network possessing given physical properties, such as a given swelling capacity or a given resistance to extension, i.e., a given modulus. To instance a simple case on which quantitative measurements have been made: When a network is formed by the polymerization of styrene in the presence of small proportions of the cross-linking agent divinylbenzene, the proportion of cross-linking agent required to yield a product having a given swelling capacity is greater, the higher the temperature of polymerization, i.e., the shorter the primary (nonlinked) chains.[27] In the vulcanization of rubber, both natural and synthetic, the amount of cross-linking required to produce a given physical state of vulcanization is greater, the shorter the chains, i.e., the lower the molecular weight of the unvulcanized material.

Now, in rubber, both natural and synthetic, the individual molecules vary widely in size. The range of molecular weights in Hevea rubber is now known to be considerably wider than earlier work indicated. Bloomfield[4] finds that the hydrocarbon of Hevea rubber ranges in molecular weight from 50,000 to over 3,000,000, and that at least 60 per cent of it has molecular weights in excess of 1,300,000. But, when natural rubber is milled, in order to make it plastic for compounding, the molecular-weight range is leveled out to a very large extent. Indeed, according to Kemp and Peters,[8] after 15 minutes on a cold mill all the rubber is brought to practically the same molecular weight. In milled Hevea rubber, accordingly, the chains, being of approximately the same length, all require about the same amount of cross-linking to produce a vulcanizate having optimum physical properties; under any given conditions of vulcanization, the whole of the material is brought into the same state of vulcanization.

With synthetic diene polymers the situation is very different. The polymer originally embraces a wider range of molecular weights than does natural rubber, and, on milling, it suffers relatively less reduction in molecular weight than natural rubber and does not undergo a leveling of molecular weights to the extent that Hevea rubber does. The material subjected to vulcanization comprises a wide range of molecular sizes, which require different amounts of cross-linking to obtain optimum properties. Hence it is impossible (lacking a Maxwell demon which might portion out the sulfur in proper ratio to the different fractions) to bring all the material into the same relative state of cure. In vulcanization at a given temperature for a given period of time, the proportion of sulfur necessary to bring to an optimum state of vulcanization the fraction of lowest molecular weight is vastly greater

Table IV. Molecular Weight of Fractions of GR-S

| | | Unmilled | | Milled | |
	Per Cent of Total	Intrinsic Viscosity	Molecular Weight (Osmotic)	Intrinsic Viscosity	Molecular Weight (Osmotic)
Unfractionated GR-S	100.0	2.36	96,500	1.54	167,000
Fraction 1	14.4	6.57	1,652,000	2.24	303,000
2	14.3	3.85	723,900	1.95	246,000
3	12.0	2.86	482,000	1.92	240,000
4	10.0	1.68	193,000	1.61	184,000
5	11.2	1.07	103,400	1.16	112,000
6	10.5	0.90	65,800	0.90	75,000
7	7.9	0.54	23,600	0.54	35,000
8	11.5	0.30	12,400		
9	8.5	0.08	Too low for measurement		

than that necessary to bring to an optimum state the fraction of highest molecular weight. In the vulcanization of GR-S, the choice of a technically suitable proportion of sulfur is essentially a compromise between the different proportions that would be needed to secure optimum vulcanizates from the fractions of different molecular weight that compose GR-S.

These statements may be illustrated by data provided by Yanko[30] on the fractionation of GR-S. Table IV shows the molecular weights of fractions into which a sample of GR-S was separated, and also the molecular weights after the fractions had been subjected to a normal amount of milling preparatory to incorporating carbon black in them.

Vulcanization experiments in a black stock showed that the unfractionated rubber gave a technically satisfactory cure with 1.75 parts sulfur per 100

parts rubber.   But the fraction of highest molecular weight gave a similar
cure with as little as 0.75 part sulfur, and the fraction of lowest molecular
weight failed to cure with as much as 40 parts sulfur.   With 1.75 parts sulfur,
fractions 7, 8, and 9 failed to cure.   These three fractions together comprised
27.9 per cent of the rubber.   The fact that so large a proportion of standard
GR-S is of a molecular weight so low that under ordinary conditions of
vulcanization it remains in a substantially unvulcanized state and is in effect a
mere diluent or softener provides a rational basis for the practice of oil
masterbatching, according to which polymerization is conducted under
conditions of temperature and modifier concentration designed to produce
rubber high in molecular weight and lacking as far as possible low-molecular-
weight "diluent" fractions, and then blending with this rubber (by addition
as an emulsion to the latex) a nonvulcanizable oil softener.

PROCESSING, COMPOUNDING, AND TESTING GR-S

Although in respect to the conduct and control of processing, the composition of stocks, and the testing of vulcanizates, butadiene-styrene rubbers
are basically similar to natural rubber, they nevertheless differ from natural
rubber sufficiently to have made it necessary to conduct extensive programs of
investigation, in order to produce the best possible manufactured products
from them and in order to develop testing methods best adapted to evaluating
their vulcanizates.   Tree rubber is a "given" material, which, speaking by
and large, is relatively uniform in its properties and behavior.   Over the
years much experience has been accumulated concerning its behavior at the
various stages in the manufacture of rubber goods and concerning the
performance of the goods themselves.   Butadiene-styrene rubbers are
manufactured materials, the properties of which can be varied very considerably as a result of changes in the manufacturing procedure.   Much development work involving the study of processing, compounding, and testing was
devoted to the standardization of the rubbers, which are now being made in
a wide range of different grades (Chapter 7).   At the outset of their large-
scale manufacture, there was little in the way of experience to assist in
correlating the special properties of the butadiene-styrene rubbers with the
performance of products manufactured from them.   Such circumstances
have contributed to make the subject of rubber processing, compounding,
and testing more complicated today than it was in the dear, dead days when
only natural rubber was used in rubber manufacturing.   This increased
complexity is reflected, for example, in the fact that a review of recent studies
on rubber testing demands so much space as that devoted to it in this book
(Chapter 12).   Incidentally it may be remarked that the scientific and technical developments in the manufacture of rubber goods and the testing of
rubber vulcanizates that synthetic rubber has inspired have contributed to
improvements in the manufacture of goods from natural rubber.

An outstanding feature of the compounding of GR-S is the dependence of
the rubber for strength on carbon black.   Whereas natural rubber free from
reinforcing powders has, when vulcanized, a tensile strength up to 5000 p.s.i.,
GR-S without reinforcement has, when vulcanized, a tensile strength of only

200 to 300 p.s.i., although when it is reinforced by carbon black its strength is 3000 to 4000 p.s.i. The critical importance of carbon black in the manufacture of goods, especially tires, from GR-S has focused new attention on the manufacture of black for rubber compounding, has led to the development of many new grades of black adapted to secure the best possible products from GR-S, and, generally, has rendered the subject of rubber blacks decidedly more complicated than it was when the rubber industry used only natural rubber. A discussion of the various types and grades of black now available to the rubber compounder is given in Chapter 11. The carbon-black-producing industry has made valuable contributions to the technology of synthetic rubber and, as an incidental result of this, to the conservation of the country's resources of natural gas.

It would be unfair to say that the reinforcement of GR-S is no more than a black art, but it is true nevertheless that an understanding of the theory of the reinforcement of GR-S is less advanced than the practice. As stated above, carbon black vastly increases the breaking strength of GR-S. It does this, not only by increasing the stiffness (modulus) of the rubber, but also by increasing markedly its extensibility. In natural rubber, carbon black also increases the modulus, but it reduces the extensibility, and on balance it raises the breaking strength only moderately. However, as already stated, natural rubber possesses good tensile strength without reinforcement, whereas GR-S possesses little strength without it. Further, unreinforced natural rubber ("gum" rubber) retains a considerable part of its tensile strength when its temperature is raised to the neighborhood of 100° C.—a temperature frequently met in large tires—whereas GR-S, even when reinforced by carbon black, suffers a serious fall in tensile strength when its temperature is raised to the same extent (cf. Chapter 11, p. 419). In this connection, there is a marked contrast between the two rubbers in regard to the effect of temperature on extensibility. At elevated temperatures natural rubber becomes more extensible than it is at ordinary temperature, whereas the reverse is true for black-loaded GR-S. GR-S, when raised in temperature, tends as it were to revert toward the condition of low extensibility which it displays in the unreinforced state. The situation can perhaps be pictured in the following way. GR-S and similar polymers are composed of molecules so irregular in form that, as X-ray examination shows, when the polymer is stretched, crystallite lattices fail to form, and hence the rubbers lack gum strength and break before they can be stretched far; when they are blended with fine-particle carbon black, the interfacial forces between the black and the rubber confer strength, and enable the rubber to be stretched further before rupture occurs; when the temperature of the black-loaded rubber is raised, the interfacial forces fall, and accordingly the vulcanizates suffer rupture at a lower extension. Natural rubber, composed as it is of molecules regular in shape, forms crystallites on stretching, and this orientation of the originally higgledy-piggledy molecules is generally considered to have a reinforcing effect (self-reinforcement) and to give the material good gum strength. When natural rubber contains a particulate reinforcing agent, it may be considered that the strength of the material at ordinary temperatures derives partly from the crystallite effect and partly from the interfacial forces between rubber and pigment.   It is not, however, immediately obvious why black natural rubber retains good tensile strength at 100° C., because it is perhaps hardly to be expected that the thermal agitation of the rubber molecules at such a temperature will allow crystallites to form on stretching and because it is equally difficult to believe that the interfacial reinforcing effect does not fall off in natural rubber with rise in temperature, as it apparently does in GR-S.   The broad facts of reinforcement in natural rubber and the influence of temperature on the reinforcement remain a challenge to further investigation concerning their causes.

The resistance to abrasion of GR-S containing suitable reinforcing black is good.   And this fact too offers food for thought and scope for fundamental investigation.   The resistance to abrasion of natural rubber has usually been thought to be related to the resilient energy of the rubber, i.e., to the energy which the rubber when deformed is capable of absorbing before it ruptures. But this view may require modification when GR-S is brought into consideration, for the resistance of GR-S to abrasion, when compared with the actual magnitude of its resilient energy, is unexpectedly high.

Chapter 14, which concerns the use of GR-S in the·manufacture of certain specific types of rubber goods, is brief because of the fullness with which the present body of empirical knowledge about the general aspects of the processing and compounding of GR-S is reviewed in Chapter 11, and also because rubber manufacturers are naturally loath to disclose the details of their actual, working recipes.   (The compounding recipes included in Chapter 14 are quoted for illustrative purposes only.)   Separate chapters (Chapters 15 and 16) are devoted to the application of synthetic rubber in two specific fields, viz., in wire and cable manufacture and in hard-rubber manufacture respectively, largely because in these fields not only general-purpose synthetic rubber but also other types of synthetic rubber are employed.

COMONOMER VARIATIONS

In view of the fact that, with presently known techniques of preparation, rubber of better quality is obtainable by emulsion polymerization from a mixture consisting of a major proportion of butadiene and a minor proportion of the comonomer styrene than is obtainable from butadiene alone, it is not surprising that much effort has been devoted to the study of comonomers other than styrene, in order to determine whether any such have advantages over styrene in regard to the quality of their copolymers with butadiene.   Chapter 21 will serve to give some idea of the extent over which such study has ranged and of the manner in which it has been conducted. In general, the various comonomers have first been used in laboratory-scale polymerizations carried out in bottles; then copolymerizations have been conducted on a small pilot-plant scale; and, finally, the more promising copolymers have been made on a large pilot-plant scale, and tires have been built from them and subjected to road running tests.   The review of such studies here given, while amply sufficient to illustrate what has been done in this connection, covers only a portion of the work of this kind done in the

American synthetic-rubber program as a whole for the purpose of determining whether the substitution of the styrene of GR-S by some other comonomer might lead to a superior polymer. In the outcome, it has not appeared that for the production of general-purpose synthetic rubber any other comonomer has on balance—having regard to polymer quality, cost, and convenience—any or, at all events, any very substantial advantage over styrene, although for rubber uses in which some special quality, e.g., freeze resistance, is of outstanding importance a comonomer other than styrene may prove to be preferable to the latter.

NON-EMULSION POLYMERIZATION

Chapter 21 also provides an account of investigations, still in progress, on the production of synthetic rubber by reagents, especially sodium metal and the Alfin catalysts, which do not depend, as emulsion polymerization depends, on free radicals as polymerization initiators. The fact that metallic sodium will polymerize diene hydrocarbons was discovered in the early days of research on synthetic rubber—in 1910. It was almost inevitable that such a discovery should be made because of the fact that "drying over sodium" is a traditional laboratory technique in organic chemistry. In Germany polymerization by sodium was practiced on a moderately large manufacturing scale. In Russia polymerization by means of sodium has been employed exclusively, at all events until recently, for the manufacture of synthetic rubber. In both these countries polymerization by sodium has been applied, at all events on an industrial scale, only to butadiene unmixed with a comonomer. Work reported in Chapter 21, however, is concerned with polymerization by sodium not only of butadiene alone but also of butadiene-styrene mixtures. This work makes it appear that, as in emulsion polymerization so in sodium polymerization, copolymers are stronger than polymers of butadiene alone.

SPECIALTY RUBBERS

Of the synthetic rubbers made on a large scale in the United States, Neoprene (Chapter 22) was first on the scene. It has been commercially available since November 1932. Thanks presumably to the presence of chlorine, which confers polarity on the dienoid system of its monomer, $CH_2 : CH \cdot CCl : CH_2$, Neoprene, the polymer, in contrast to polymers of butadiene, has a regular 1,4-addition structure, gives an X-ray diffraction pattern when it is stretched, and has good strength in gum stocks. Its special applications were at first mostly centered around its property of resistance to petroleum oils, but more recently it has come to be particularly valued for its resistance to deterioration on aging, and many of its applications today depend on this property, as may be seen by an inspection of the data given earlier (Table II, pp. 5–7) for the consumption of Neoprene in various kinds of rubber goods. The total consumption of Neoprene by the rubber-manufacturing industry has grown continually. In the last four years especially a rapid expansion in the use of Neoprene has been witnessed (Tables I and II, pp. 51–2).

Copolymers of butadiene and acrylonitrile (Chapter 23) find most of their applications because of their special resistance to mineral oil. Their property of oil resistance derives from the presence in them of the highly polar cyano group, and it is presumably also the presence of this group which makes it in general advisable to incorporate in them fairly substantial proportions of plasticizers, in order to make it possible to process them readily and to secure adequate extensibility in their vulcanizates. In typical butadiene-nitrile rubbers the two monomers occur in a ratio of about 65 to 35. It may be noted that such a ratio by weight corresponds to a much higher molar ratio of comonomer to butadiene than is found in GR-S. Whereas in standard GR-S containing 23.5 per cent of styrene the molar ratio of butadiene to styrene is 6.25 to 1, in nitrile rubber containing 35 per cent acrylonitrile the molar ratio of butadiene to nitrile is only 1.8 to 1.

Butyl rubber, an account of which (Chapter 24) is contributed by its discoverers, is fundamentally polyisobutylene, represented by saturated units of the form $—CH_2 \cdot C(CH_3)_2—$, but it comprises also in its molecules just enough isoprene to make it vulcanizable. Like polyisobutylene, the manufacture of which preceded it, the preparation of Butyl rubber depends on the application of an ionic catalyst at a low temperature. But for the production of polymers of sufficient molecular weight from a mixture of isobutylene and a small proportion of isoprene an even lower temperature is needed than from isobutylene alone. Butyl rubber is in fact made at a temperature in the region of $—140°$ F. or $—100°$ C. When the catalyst, aluminum chloride, is added to the mixture of isobutylene and isoprene at this low temperature, polymer formation is almost instantaneous; indeed the reaction is almost explosive. To have succeeded in conducting a polymerization of this type on an industrial scale and in a semicontinuous manner; to have handled successfully the various problems, including that of temperature control, that it involves; and to have controlled the high-molecular reaction so effectively as to be able to manufacture by it a rubber uniform in regard to its processing and vulcanizing characteristics and in regard to the behavior of its vulcanizates is a noteworthy feat of chemical engineering.

Butyl rubber is here classed as a specialty rubber, because hitherto by far its largest application has been one, viz., inner tubes, dependent on its special property of air impermeability—a property that has been a boon to motorists. Active work on the application of Butyl rubber to other uses is under way, and it is not impossible that in the future Butyl rubber will approach the status of a general-purpose rubber.

MISCELLANEOUS ELASTOMERS

Under the heading "Miscellaneous Elastomers" (Chapter 25) are discussed a number of polymers that differ considerably in their chemical nature— polyesters, polyacrylates, polysulfides, polysilicones, polyesteramides. That these polymers, so various in their chemical constitution, are all elastic and vulcanizable serves to emphasize the fact that rubberlikeness is not confined to diene polymers and copolymers but may be found in any suitably bonded network of long molecules that are of suitable shape and have suitable interchain forces between them. This fact is likely to make itself even more conspicuous with the further development of the synthetic-rubber field, for undoubtedly many examples of elastic nondiene polymers remain for development in addition to those discussed in Chapter 25, which is restricted for the most part to such examples as have reached the point of commercial production.

Most of the polymers discussed are prepared by condensation—a process that, unlike the polymerization of conjugated dienes, is adapted to insuring linearity in the products. Some, however, viz., the polyacrylates, are made by addition polymerization, but here the monomer is of the vinyl type, and accordingly structural irregularities in the polymer chain, such as occur in the polymerization of dienes, do not enter.

Polyesters. Linear polyesters, prepared by the condensation of mixtures of dibasic acids and glycols, are, when vulcanized, elastic—indeed, under optimum conditions are remarkably snappy—provided that the mixture of reactants used in their preparation comprises suitable substituent groups, to exercise a bulking effect of a magnitude such as to reduce interchain forces to a suitable level and thus prevent the too ready occurrence of crystallization. By way of illustration it may be mentioned that, whereas the polyester from sebacic acid and ethylene glycol is a hard, crystalline material, with a melting point of 72° C., the polyester from sebacic acid and propylene glycol is, thanks to the effect of the side methyl groups in reducing interchain forces, a rubbery gum that is fairly soft and flexible down to 12° C. and that when compounded and cross-linked yields a strong, snappy vulcanizate.

An example of an elastic polyester has been encountered in nature. Kemp and Peters[9] found that the elastic skin of the berries of *Smilax rotundifolia* (the berries are sometimes known locally in Texas as "stretch berries"), unlike all rubber previously found in nature, is not a hydrocarbon, but consists of a polyester. On hydrolysis, it yields an ω-hydroxycarboxylic acid. (In passing, it is interesting to note that ω-hydroxycarboxylic acids, the structure of which is of course well adapted to the formation of linear polyesters by condensation, have been noted to occur in several places in nature. Aleuritic acid, derived from shellac, is an acid of this type. A waxy polyester derived from the needles of certain species of juniper and other conifers yields ω-hydroxylauric acid and ω-hydroxypalmitic acid on hydrolysis.[5] And an acid of the type of aleuritic acid has been isolated from cork.[31]) The present writer examined the material composing the elastic skins of the berries of *Smilax bona-nox* and *S. hispida,* and found it to be a polyester chemically similar to or identical with that of *S. rotundifolia,* and to have a tensile strength of about 2400 p.s.i. and a breaking elongation of about 650 per cent.[26] So far no other examples of naturally occurring elastic polyesters have been encountered. A cursory search among plants related to *Smilax* (e.g., asparagus, Solomon's seal, Butcher's broom) failed to disclose any others in which elastic seed skins occur.

Diisocyanate-Linked Elastomers. The elastomers described in Chapter 25 as "diisocyanate-linked condensation elastomers" involve in their preparation a device for the formation of high polymeric molecules of predetermined structure of which more is likely to be heard in the future development of polymer chemistry, namely, a two-step reaction in which primary polymer chains terminating in reactive groups are treated with a bifunctional reagent adapted to react with these terminal groups and thus, by uniting the primary chains, lead to molecules of a higher order of molecular weight. In the case of the elastomers involving this device in their preparation, the material of the first step consists of a polyester or polyesteramide, the molecular weight of which is only moderately high, and the molecules of which, thanks to the choice of an appropriate ratio between the dibasic acid and the glycol or aminoalcohol from which they are made by condensation, terminate in hydroxyl or amino groups. In the second step, such primary polymer chains are linked, with the consequent formation of much longer chains, by reaction of their terminal groups with a diisocyanate. Among the materials prepared on these lines are the Vulcollanes, which are now exciting considerable interest because of their outstanding resistance to abrasion and their good resistance to tearing.

Polysulfides. The device of building large linear polymers by uniting smaller polymer chains having reactive end groups is also encountered in some of the recent developments in the field of polysulfide rubbers. The chains of the polysulfides are characterized by the presence of disulfide linkages ($-S \cdot S-$). By controlled treatment of the polysulfide polymers with reducing agents or with thiuram disulfides, any desired proportion of the disulfide links in them can be ruptured and the polymer thus converted to material of lower molecular weight composed of chains shorter than the original chains and terminating in thiol groups. Then, in a further step, the shorter chains can be united into longer chains by the application of an oxidizing agent, which acts on the terminal thiol groups to re-form disulfide linkages. In the scission step, polysulfide rubbers, originally tough and low in plasticity, can if desired be brought down to materials of semifluid consistency (the liquid Thiokols), which may be applied, e.g., as putties, while of this consistency, and then cured in situ by oxidizing agents, to form oil-resistant seals.

The device of uniting primary polymer chains to form longer chains through reactive ends groups, of which examples in the field of elastomers have just been noted, has, it may be mentioned, been applied more recently in the preparation of the so-called block polymers.

Silicone Polymers. In the silicone polymers, the molecular chains are composed of a succession of silicon and oxygen atoms, thus:

$$-\overset{\cdot}{\underset{\cdot}{Si}} \cdot O-_{n}.$$

The absence in them of carbon atoms as chain elements gives the silicone polymers a unique position and unique properties. The polymers are outstanding among elastomers in their ability to withstand high temperatures. True, the polymers are not inorganic, but the carbon which their molecules comprises is not in the polymer chains but in side groups attached thereto, as in, e.g.,

$$\left(\begin{array}{c} CH_3 \\ | \\ -Si \cdot O- \\ | \\ CH_3 \end{array} \right)_n .$$

The silicone polymers and their applications to special conditions of service for which their unique properties fit them are in a rapid state of development. And in passing it may be remarked that the development of organic silicon polymers of practical usefulness illustrates well the dangers of prophecy in science and technology. In 1937, Kipping, near the close of his life, in a lecture devoted to a review of knowledge on organic silicon compounds, offered the pessimistic estimate of their future which the following words express: "as . . . the few (types of organic derivatives of silicon) which are known are very limited in their reactions, the prospects of any immediate and useful advance in this section of organic chemistry does not seem to be very hopeful."[11] Yet a period of only six or seven years was to elapse before there was begun the commercial preparation of a whole range of organic silicon products, among them the rubberlike polymers described in Chapter 25. Kipping had pioneered in the study of the chemistry of organic silicon compounds and had made to it contributions of great value, but, an organic chemist of the classical type, he had, in making the above estimate, reckoned without the rapidly growing activity in the field of polymers!

Hypalon. The recently introduced vulcanizable elastomer Hypalon (not described in Chapter 25) involves a manner of developing a polymer possessing special properties that has not been previously applied in the field of elastomers, viz., chemical treatment of an existing polymer.[23] Here the starting material is polyethylene, which lacks long-range elasticity. By treatment with chlorine and sulfur dioxide in suitable ratio, there are introduced into the polyethylene chain chlorine atoms, which serve to space the chains, and a small but sufficient number of chlorosulfonyl groups, $-SO_2Cl$, which, thanks to their chemical reactivity, permit cross-linking reactions to be applied readily and a vulcanized elastic product to be obtained.

REMARKS ON NOMENCLATURE

Objections have been made in some quarters to the designation of manufactured rubbers as *synthetic* rubbers, on the ground that none of manufactured products is chemically identical with natural rubber. And it has been suggested that they should be described as *chemical* rubbers. In the present volume, however, the prevailing practice of calling them synthetic rubbers has been generally followed, although in places they are, alternatively, referred to as elastomers—a term that, strictly speaking, is to be regarded as an abbreviation of "elastopolymers."

"Synrub" has perhaps something to commend it as an abbreviation for "synthetic rubber," and was suggested some time ago.[25] But it has never come into favor, although "syntan" found acceptance for "synthetic tanning agent," and although "syndet" has lately found some acceptance as an abbreviation for "synthetic detergent," and "surfactant" as an abbreviation for "surface-active agent."

In justification of the designation of manufactured rubber as synthetic rubber, two observations may perhaps be made:

1. It is not possible to define high polymeric materials in as clear-cut a way as that in which substances of low molecular weight can be defined. And this circumstance alone is sufficient to make it inappropriate to employ the adjective "synthetic" in connection with high polymers only in the rigid manner appropriate in connection with substances of low molecular weight. For example, methyl salicylate of natural origin, derived from oil of wintergreen, can be identified chemically with precision, and there can be no room for argument as to whether a given manufactured chemical is or is not synthetic methyl salicylate. But the hydrocarbon of natural rubber cannot be sharply defined. True, it can be defined to the extent of saying that it is a 1,4-addition polymer of isoprene, but, even if attention is confined to rubber from a single species, e.g., *Hevea brasilienesis*, the molecular weight cannot be strictly defined: the molecular weight and its distribution vary in samples from different trees and probably in samples from the same tree at different times, and, further, the nature of the end groups is unknown.

2. A more cogent consideration arises from the fact that polymeric materials are prized primarily for their useful mechanical properties, and that from a practical viewpoint their chemical nature is important only in so far as it determines their physical properties, i.e., is important only secondarily. On such grounds we are disposed today to accept as "rubber" any material, regardless of its chemical composition, that possesses mechanical properties similar to the special properties characteristic of natural rubber—high deformability, rapid recovery from deformation, good mechanical strength, etc.; we are disposed to take the view that, expressed epigrammatically, rubber is as rubber does.

Even among certain chemical products of low molecular weight there has been noticeable of late a tendency to apply the term "synthetic" to manufactured chemicals on the basis of their possession of certain physical properties rather than on the basis of their chemical constitution. Thus "synthetic detergents" are not synthesized naturally occurring detergents (saponins and soaps) but merely manufactured substances possessing detergent power.

In connection with the tendency just mentioned to place emphasis on physical behavior in the classification of polymeric materials, it may be noted that the description of the latter by chemical names that identify them precisely and state unambiguously their structure and molecular magnitude is hardly feasible and in fact, in our present state of knowledge, is often impossible. This applies especially to polymers and copolymers of diene hydrocarbons. The International Union of Pure and Applied Chemistry's Subcommittee on Nomenclature in the Field of Macromolecules has valiantly proposed a scheme for the systematic naming of polymers, but it is hardly feasible to employ it in ordinary discourse.[7] According to this scheme, a butadiene-styrene copolymer (which, because of the occurrence of 1,4- and 1,2-addition of the butadiene units, is a tripolymer rather than a dipolymer) would presumably be named poly-[butenamer-*co*-(vinyl)ethamer-*co*-(phenyl)ethamer]! And even such a name fails to take account of the facts that (1) some of the 1,4-butadiene units have a *cis* and some a *trans* configuration and (2) the ratio in which the butadiene and styrene enter into the copolymer molecules varies as the conversion of a butadiene-styrene mixture proceeds. Further, it gives no information regarding the widely different molecular weights of the many structurally different molecules of which the polymer is composed.

RESEARCH PROGRAM

At the basis of the remarkable accomplishments in the science and technology of synthetic rubber during the last ten years in the United States has been an intense program of research and development financed by the Reconstruction Finance Corporation, the Government agency responsible for the large-scale manufacture of general-purpose synthetic rubber. To date some 10,000 reports have been prepared on various aspects of this program—fundamental research, development, plant operation, standardization, etc.—and some 700 papers and articles arising from it have been published.

The Government-sponsored research and development program has been conducted in an enlightened spirit. On the strictly research aspects of the subject the workers have been allowed almost complete freedom in their choice of topics, always of course with the implicit understanding that the topics have a bearing on the advancement of knowledge on the subject of synthetic rubber. Taking part in the program have been groups in industrial firms and groups in universities. Naturally, only university workers and industrial firms possessing or ready to take an interest in some aspect of the subject have been invited to take part in the program. The special interests of the individual research workers have ranged from the most fundamental phases of the physical chemistry of macromolecules, through the fundamental organic chemistry of elastomers, to matters more immediately related to development problems. There has been a completely free exchange of information through the distribution of reports, through conferences held at intervals, and through personal contacts among the workers. The present writer cannot well imagine conditions more congenial and stimulating to the advancement of knowledge in a specified, broad field than the brotherhood of research workers that constitutes itself in circumstances such as those outlined. All the members are ipso facto interested in the same general subject, but different members approach it from different angles of interest and skill; all derive stimulus from the exchange of research results and ideas, and all find in it the urge to emulation. Even the most independent and free-ranging research mind derives stimulus of positive value from membership.

The author's ten-year experience of the excellent manner in which the Government-sponsored research program—sufficiently organized but not overorganized—has operated and of the technical successes that have marked it makes him feel that some contemporary writers, in their eagerness to defend freedom in research, have gone too far in deprecating any element whatever of planning when financial provision is made for research.[3] And certainly the author's experience makes him question the view of a recent writer who, after confessing that synthetic rubber in the United States "has progressed forward (*sic*) in its technology," adds dogmatically "but not as far as it would have had it been in the hands of private enterprise, where the stimulus of good, old-fashioned American competition would have taken it to even greater heights."[6] One may be strongly opposed to the Government ownership and operation of industrial plants in time of peace and yet feel unable to accept the ipse dixit just quoted, which seems to involve the unlikely assumption that research workers restricted in the exchange of information and ideas to the limits of the laboratory of a single industrial firm can be expected to make more rapid progress than workers free to exchange information and ideas with *all* others actively interested in the same subject.

REFERENCES

1. Anon., *Rubber Age N.Y.*, **62**, 81 (1947).
1a. Anon., *Chem. Eng. News*, **31**, 2298 (1953).
2. Bacon, R. G. R., Jarrett, S. G., and Morgan, L. B., Brit. Pat. 573,270; Bacon, R. G. R., and Morgan, L. B., Brit. Pat. 573,366. Other related patents are Bacon, R. G. R. et al., Brit. Pats. 573,369; 574,482; 575,616; 576,160; 577,317; 578,209; 583,166; 586,881; 586,988; 599,098; U.S. Pat. 2,391,218, Dec. 18, 1945. Cf. Brubaker, M. M., and Jacobson, R.A., U.S. Pat., 2,462,354, Feb. 22, 1949; Plambeck, L., U.S. Pat. 2,462,422, Feb. 22, 1949; Fellows, L., and Mellers, E. V., U.S. Pat. 2,403,788, Nov. 16, 1948; Jacobson, R. A., U.S. Pat. 2,519,135, Aug. 15, 1950.
3. Cf., for example, Baker, J. T., *Science and the Planned State*, Allen & Unwin, London, 1945.
4. Bloomfield, G. F., *J. Rubber Research Inst. Malaya*, **13**, Communication 271, 1–17, 18–24; Communication 272, 1–13, 14–25; Communication 273, 1–9, 10–23 (1951); *Rubber Chem. and Technol.*, **24**, 737–49 (1951).
5. Bougault, J., and Bourdier, L., *Compt. rend.*, **147**, 1311–4 (1908); *J. pharm. chim.* (6), **29**, 561–73 (1909); *ibid.*, **30**, 10–6 (1909); Bougault, J., *Compt. rend.*, **150**, 874–6 (1910); *J. pharm. chim.* (7), **1**, 425–32 (1910); *ibid.*, (7), **3** 101–3 (1911); Bougault, J., and Cattelain, E., *Compt. rend.*, **186**, 1746–8 (1928).
6. Cox, J. T., *Chem. Eng. News*, **31**, 35–6 (1953).
7. International Union of Pure and Applied Chemistry, Subcomm. on Nomenclature in the Field of Macromolecules, *J. Polymer Sci.*, **8**, 257–77 (1952).
8. Kemp, A. R., and Peters, H., *J. Phys. Chem.*, **43**, 1063–82 (1939).
9. Kemp, A. R., and Peters, H., *India Rubber World*, **110**, 639–41 (1944).
10. Kharasch, M. S., Nudenberg, W., White, E., and Glines, A., Private communication to O.R.R., May 28, 1950.
11. Kipping, F. S., *Proc. Roy. Soc.*, *London*, **159A**, 193 (1937).
12. Miller, W. von, *Arch. der Pharm.*, **220**, 648–51 (1882).
13. Stewart, W. D., U.S. Pat. 2,380,473, July 31, 1945.
14. Stewart, W. D., U.S. Pat. 2,380,474, July 31, 1945.
15. Stewart, W. D., U.S. Pat. 2,380,475, July 31, 1945.
16. Stewart, W. D., U.S. Pat. 2,380,476, July 31, 1945.
17. Stewart, W. D., U.S. Pat. 2,380,477, July 31, 1945.
18. Stewart, W. D., U.S. Pat. 2,380,710, July 31, 1945.

19. Stewart, W. D., U.S. Pat. 2,383,425, Aug. 21, 1945.
20. Stewart, W. D., U.S. Pat. 2,388,372, Nov. 6, 1945.
21. Stewart, W. D., U.S. Pat. 2,388,373, Nov. 6, 1945.
22. Stewart, W. D., and Zwicker, B. M. G., U.S. Pat. 2,380,617, July 31, 1945.
23. Warner, R. R., *Rubber Age N.Y.*, **71**, 205–21 (1952); Smook, M.A., Roche, I.D., Clark, W.B., and Youngquist, O.G., *India Rubber World*, **128**, 54–8 (1953); Busse, W.F., and Smook, M.A., *India Rubber World*, **128**, 348–51 (1953).
24. Whitby, G. S., *Colloid Symposium Monograph*, Vol. IV, Reinhold, New York, 1926, pp. 203–23.
25. Whitby, G. S., *Chem. and Ind.*, **20**, 277 (1942).
26. Whitby, G. S., in Mark, H., and Whitby, G. S., *Scientific Progress in the Field of Rubber and Synthetic Elastomers*, Interscience, N.Y., 1946, p. xxvi.
27. Whitby, G. S., and Wellman, N., Unpublished results.
28. Whitby, G. S., Wellman, N., and Stephens, H. L., Unpublished results.
29. Whitby, G. S., and Zomlefer, J., Unpublished results.
30. Yanko, J. A., *J. Polymer Sci.*, **4**, 576–601 (1948).
30a. Youker, M. A., and Copeland, N. A., *FIAT Rep.* **719,** Jan. 28, 1946.
31. Zetsche, F., and Weber, K., *J. prakt. Chem.*, **150**, 140 (1938).

CHAPTER 2

HISTORICAL REVIEW

R. F. Dunbrook
Firestone Tire & Rubber Company

EARLY INVESTIGATIONS

The synthesis of natural rubber has been the object of many researches during the last century and has received the attention of some of the most eminent chemists during that period. Yet the actual synthesis of a high-molecular-weight polymer duplicating all the properties of natural rubber has not thus far been accomplished. The term synthetic rubber therefore applies to any vulcanizable synthetic polymer having rubberlike properties and does not imply that natural rubber has been duplicated.

Michael Faraday[9] in 1826 first carried out the chemical analysis of rubber and established the empirical formula C_5H_8. Dumas[7] in 1838 confirmed the work of Faraday, and more recent investigators have firmly established this composition of the rubber hydrocarbon. After the elementary analysis of rubber attempts were made to learn more about the structure of the rubber hydrocarbon. In the early days of organic chemistry degradation by destructive distillation was commonly used to break complex molecules into simpler structural units. Himly[20] in 1835 resorted to this method to gain insight into the nature of rubber. He obtained a low-boiling fraction which he called "faradayine" and a high-boiling fraction, "caoutchine."[20] At about the same time A. Bouchardat,[2] Liebig,[31] Dalton,[6] and Gregory[17] also investigated the fractions obtained from the distillation of rubber. There is little reason to doubt that these investigators had isoprene in the low-boiling fraction from rubber. It was not, however, until 1860 that Greville Williams[50] succeeded in isolating a substance C_5H_8, which he named isoprene, from the distillation products of rubber. Williams noted that in the presence of air the isoprene became viscous, and that, when the oxidized (or peroxidized) isoprene thus formed was subjected to distillation, it changed at one stage of the distillation into a "white, spongy elastic mass," and he added of this mass that "when burnt it exhales a peculiar odor hitherto considered to be characteristic of caoutchouc itself." It seems reasonable to suppose that Williams thus observed for the first time that isoprene was capable of polymerizing to form an elastic solid. Analysis of his product, however, showed that considerable take-up of oxygen had occurred during the polymerization.

Williams had more than an inkling of the fact that isoprene and rubber were related as monomer and polymer, for he wrote "the action of heat on caoutchouc is considered to be merely the disruption of a polymeric body

32

into substances having a simple relationship to the parent hydrocarbon." But it remained for G. Bouchardat[3] in 1879 to postulate definitely that isoprene was the "mother substance" of rubber and to make attempts deliberately to convert the former into the latter. Bouchardat took isoprene obtained from natural rubber by distillation, treated it with hydrochloric acid, and obtained an elastic rubberlike solid. The first synthesis of rubber was thus accomplished.

Attention was then turned to other sources of isoprene. Tilden[44] in 1884 obtained isoprene by the pyrolysis of turpentine and showed that it could be converted into a tough elastic substance by treatment with hydrochloric acid as Bouchardat had done. Specimens of his isoprene were stored in bottles, and eight years later (1892) Tilden reported that the liquid had polymerized spontaneously to artificial rubber which was capable of uniting with sulfur in the same way as ordinary rubber, forming a tough, elastic insoluble product.

Tilden was unaware that Wallach[48] in 1887 had also reported the spontaneous polymerization of isoprene, when exposed to light in sealed tubes, to a product resembling rubber. Tilden[45] (1882) also proposed the structural formula $CH_2 : C(CH_3) \cdot CH : CH_2$ for isoprene, and Ipatieff and Wittorf[26] confirmed his findings. Final proof for the structure of isoprene was given by Euler,[8] who synthesized it from β-methyl pyrrolidine.

DEVELOPMENTS PRIOR TO WORLD WAR I

Although isoprene can be obtained from turpentine by pyrolysis, the yields are low, many by-products are formed and, furthermore, the supplies of turpentine are relatively limited. More economical methods starting with cheaper and more abundant raw materials were explored. Among the raw materials to which chief attention was given in the earliest attempts to develop the commercial production of synthetic rubber were coal tar, starch, and petroleum. Some of the early methods devised for the preparation of isoprene were complicated and, although made the subject of patents, were essentially of a laboratory nature. Thus, for example, Hofmann and Coutelle[21] started with p-cresol, converted it to methylcyclohexanol by catalytic hydrogenation, oxidized the latter to β-methyladipic acid, prepared the corresponding diamide, and then applied exhaustive methylation in order to secure isoprene.

Other methods devised, however, in the course of this early work were simpler and offered the possibility of forming the basis of practical procedures for manufacturing isoprene and butadiene. Thus, from starch, fermentation produces fusel oil from which amyl alcohol is obtainable. The amyl alcohol consisting of a mixture of iso- and active amyl alcohols is treated with hydrogen chloride to form monochloropentanes. These are chlorinated to dichloropentanes. The latter when passed over soda lime at 470° C. give isoprene in 40 per cent yield by loss of 2 molecules of hydrogen chloride. Similarly, n-butyl alcohol, obtained from starch by a special fermentation process introduced by Fernbach,[13] could be made to yield butadiene. Suitable petroleum fractions, by chlorination and subsequent dehydro-chlorination, could also be used to produce isoprene and butadiene.

During World War II isoprene was required for the production of Butyl rubber. The production of isoprene by cracking dipentene (from wood turpentine) was undertaken (cf. p. 134 et seq.), but with the large increase in the planned production of Butyl rubber it became apparent that this source would not be adequate. Certain streams of hydrocarbons resulting from high-temperature cracking of petroleum contain a mixture of C_5 hydrocarbons in which isoprene is present in small amounts. It was this source that supplied the necessary isoprene for the Butyl rubber plants. Isoprene was separated from the other C_5 hydrocarbons by a combination of close fractionation and extractive distillation with acetone (cf. p. 83).

Isoprene is a homolog of butadiene, $CH_2 : CH \cdot CH : CH_2$, known[5] since 1863. It was not until 1910, however, that Lebedev[30] reported the polymerization of butadiene to a rubbery polymer. In the meantime several other homologs of butadiene had been polymerized to rubberlike substances. Kondokow[27] in 1900 polymerized 2,3-dimethylbutadiene to a "leather-like elastic mass" by heating it with alcoholic potash and one year later reported the spontaneous polymerization of this diene to a rubberlike solid when left in sealed tubes. Thiele,[43] also in 1901, observed that 1-methylbutadiene or piperylene produces rubber by spontaneous polymerization.

The monumental work of Harries[18] between 1900 and 1910 on the structure of rubber stands out as a milestone in the synthesis of rubber. He carefully studied the products obtained by the ozonolysis of the hydrocarbon of natural rubber and identified levulinic acid and aldehyde as the principal products of ozonolysis. This established the existence of the recurring unit, $: CH \cdot CH_2 \cdot CH_2 \cdot C(CH_3) :$, and demonstrated that the isoprene units are joined in the rubber molecule principally by 1,4-addition, thus: $-CH_2 \cdot C(CH_3) : CH \cdot CH_2 \cdot CH_2 \cdot C(CH_3) : CH \cdot CH_2-$.

With the advent of the automobile at the beginning of the 20th century there came a greater demand for natural rubber and a renewed interest in the synthesis of rubber. The production of crude rubber increased from 44,131 long tons in 1900 to 94,013 long tons in 1910, and the price rose from $0.98 in 1900 to a peak of $3.06 in 1910. This situation led to intensive efforts in Germany, England, and the United States to produce a synthetic rubber.

In Germany two chemists Fritz Hofmann and Coutelle,[22] working for Bayer & Company, produced synthetic rubber by heating isoprene in closed tubes for 8 days at a temperature below 200° C. A variety of substances were explored as catalysts for the polymerization. Oxidation or ozonization products of rubber were found to be effective accelerators of polymerization. Harries[19] in 1910 developed a process for preparing synthetic rubber from isoprene by heating in the presence of acetic acid at 100° C. in a closed vessel for 8 days. In England Matthews and Strange[32] discovered that isoprene could be polymerized by sodium and applied for a British patent on the process on October 25, 1910. Three days later Harries, without knowledge of Matthews' work, announced his discovery of the same method to Bayer & Company and advised them to apply for a patent.[10] Later, in 1913, Holt[23] of the Badische Aniline and Sodafabrik modified the sodium process by conducting the polymerization in the presence of carbon dioxide. The resulting product is noticeably different from that produced by sodium in the presence of air in that it is insoluble in the usual solvents and does not swell in benzene.

These early synthetic rubbers differed from natural rubber in many respects, especially in the products they yielded on ozonolysis. Like natural rubber, the rubberlike materials obtained by heating isoprene alone or with acetic acid yielded on ozonolysis substantial amounts of levulinic aldehyde and levulinic acid. The sodium rubber of Harries, however, gave no levulinic aldehyde or acid on ozonolysis. The sodium rubber prepared by Holt in the presence of carbon dioxide also reacted normally on ozonolysis.

In 1911 Earle and Kyriakides,[29] at the Hood Rubber Company in Watertown, Mass., undertook a program of research with the object of developing a tire synthetic rubber. The early experiments were beset with many difficulties. Raw materials had to be synthesized in the laboratory by laborious methods, new techniques for the polymerization of the raw materials had to be worked out, and finally the preservation of synthetic rubber presented new problems.

Kyriakides emphasized the importance of using pure isoprene for polymerization in order to obtain a good rubber. He prepared isoprene by reacting ethyl magnesium chloride with chloroacetone and catalytically dehydrogenating the methyl-2-butylene oxide-1,2 which resulted. This pure isoprene was found to polymerize with sodium much more rapidly than had been reported by Harries. Butadiene prepared from acetaldehyde by a process similar to that developed and later used commercially in Germany was also polymerized with sodium.

The monomer regarded by Kyriakides as most suitable for commercial development was dimethylbutadiene obtained by using acetone as the starting material, which, it was known, could be converted to pinacol by reduction with magnesium amalgam or aluminum amalgam. Pinacol was catalytically dehydrated in the presence of traces of acids to dimethylbutadiene. Before the declaration of war in 1914, the work had progressed to the point where it was realized that it would be impossible to commercialize any of the processes. In addition the fall that had occurred in the price of natural rubber had a damping effect on the research. It is of interest that tires from the synthetic rubber were built and tested at that time.[15]

In Akron, Ohio, David Spence[40] and his associates at the Diamond Rubber Company were also active from 1910 to 1912 on the problem of synthesizing rubber. They worked with isoprene obtained from dipentene and turpentine. Acetic anhydride and pyridine were selected as catalysts for the polymerization of isoprene to rubber. Considerable rubber was made and evaluated in 1911. After further intensive studies of the relative merits of isoprene and dimethylbutadiene, it was decided that the latter offered most promise. Both sodium and acetic anhydride were studied as catalysts before the work was dropped in 1912 when rubber again became available at reduced prices.

PRODUCTION OF SYNTHETIC RUBBER DURING WORLD WAR I

At the outbreak of World War I in 1914 the Germans had only enough natural rubber for their immediate needs. The British blockade cut off Germany from further supplies, and the situation soon became critical. Attention was again directed to the work of Fritz Hofmann in 1910 on the production of synthetic rubber from dimethylbutadiene. Although this rubber was considered inferior to the natural product, it was decided to manufacture it commercially. The dimethylbutadiene monomer was produced from acetylene obtained from calcium carbide. Acetylene was converted to acetaldehyde and thence to acetic acid and calcium acetate. By heating calcium acetate, acetone was obtained which was reduced to pinacol with aluminum and sodium hydroxide. Pinacol was next subjected to distillation with a dehydrating agent to remove 2 molecules of water and form dimethylbutadiene. Since 2,3-dimethylbutadiene is the methyl homolog of isoprene, the monomer unit of natural rubber, the synthetic rubber was termed methyl rubber. Commercial methyl rubber was prepared by the Farbenfabriken vorm. Friedrich Bayer & Company by means of two fundamentally different methods, which produced two distinct final products, methyl rubber H, used especially for hard-rubber goods, and methyl rubber W, for soft-rubber goods.[15]

Methyl rubber H was produced by placing dimethylbutadiene in thin-walled metal vessels in such a way that an air space remained above the liquid. The drums were stored at approximately 30° C. for 6 to 10 weeks, during which time the container became filled with the polymer, which was formed as a white, solid mass having the appearance of cauliflower. A small amount of the product produced by the cold process of polymerization was added (as "seed") to the dimethylbutadiene to accelerate the conversion to polymer.[15] After the disappearance of the liquid by polymerization, the drums were cut open, and the mass, having little of the appearance of rubber, was converted to rubber sheets by prolonged milling, with or without the addition of softeners and preserving agents. The polymerization could be controlled more easily by the addition of catalysts, such as some of the final product.

For the production of methyl rubber W, dimethylbutadiene was placed in double-walled pressure vessels, and large charges were heated for 3 to 6 months[16] at about 70° C. The vessel then contained a slightly yellow, transparent, rather tough, rubberlike mass, which had to be cut from the containers in thin layers by special rotating knives. This product was then placed upon the mixing mill as described above. Methyl rubber W was used especially for the preparation of soft-rubber goods, either alone or mixed with a portion of natural rubber or of methyl rubber H.

Later on in the war another product of the Badische Company, designated as Marke B, was successfully used for the manufacture of cables. This was the so-called dimethylbutadiene-sodium-carbon dioxide rubber. Sodium wire was put in contact with dimethylbutadiene in the presence of carbon dioxide, and in a few weeks a white material was formed around the sodium.

The product was freed from sodium by washing with water. It was rather unstable but could be protected from oxidation by addition of organic bases and could be milled on hot rolls to very good, smooth sheets. Only small quantities of B rubber were produced.

Near the close of the war the Germans were producing rubber at the rate of 150 tons a month. A total of about 2350 tons was actually produced during World War I, and construction was started to produce 8000 tons per year.[15]

Methyl rubber H and methyl rubber W were very poor rubbers when compared to natural rubber in compounding formulas in use at that time. Whitby and Katz[49] re-examined the properties of vulcanizates of methyl rubber H and methyl rubber W. They found that the H rubber "gave in the unloaded stock at the best cures, vulcanized products less than one-third as strong and only one-third as extensible as products from natural rubber." Methyl rubber W "gave at the best cures in the unloaded stock, vulcanized products almost as extensible but not more than one-tenth as strong as the products from natural rubber." These investigators found that carbon black had a striking effect on the physical properties of vulcanizates of methyl rubber H and W. The effect of compounding with carbon black was greater in methyl rubber W than in methyl rubber H. The addition of 25 parts of carbon black to the gum stock raised the maximum tensile strength from 28.2 to 166.5 kg. per sq. cm. for W rubber, from 86.4 to 161 kg. per sq. cm. for H rubber. These results led the authors to remark that "had carbon black attained at the time of the war as it has today the position of a recognized rubber compounding ingredient, and had it been available in Germany, the story of synthetic rubber might have been different from what it was." In the postwar period the importance of carbon black to the physical properties of synthetic rubber was fully realized.

DEVELOPMENTS SUBSEQUENT TO WORLD WAR I

In the United States interest in synthetic rubber remained dormant until about 1921. In that year Ostromislensky and Maximoff began work on synthetic rubber for the U.S. Rubber Company. In 1922 they made butadiene synthetically from ethyl alcohol and acetaldehyde, and in 1923 they produced synthetic rubber from butadiene by the emulsion polymerization process.[47]

When natural rubber again became available in Germany after World War I, work in that country on synthetic rubber was dropped for a time. Experimental work was not resumed until about 1926. It then became necessary to decide whether dimethylbutadiene, isoprene, or butadiene was to be used as the principal raw material for the manufacture of synthetic rubber. Dimethylbutadiene was dropped from consideration because the product of its polymerization was of poor quality as a rubber. During 1927 and 1928 a great deal of work was conducted on both butadiene and isoprene, and it became clear that isoprene offered no advantage over butadiene. Furthermore, butadiene was more easily produced on a commercial scale from electric-furnace acetylene. The Germans were at this stage still using sodium as catalyst for the polymerization of butadiene and had developed a series of Buna polymers distinguished by numbers: e.g., Buna 85, Buna 115. The word Buna is derived by combining the first two letters of butadiene with the first two letters of Natrium, the German word for sodium. The numbered Bunas were never very satisfactory as general-purpose synthetic rubbers but were suitable for the manufacture of hard-rubber goods. Germany continued to manufacture and use the numbered Bunas on a limited scale for special applications throughout World War II.

Emulsion Polymerization. From 1926 to 1928 extensive experimental work was carried on in Germany to determine the best process of polymerization. It was at this time that the emulsion process was perfected and was adopted in place of the sodium process. The emulsion process was not new, for the first patent on this method of polymerization was granted to Bayer & Company[11] in 1912. Isoprene was polymerized in an aqueous solution of egg albumen, starch, or gelatin. A similar patent, also in 1912, covered the polymerization of butadiene.[12]

Copolymerization. In an effort to improve the physical properties of emulsion polybutadiene, the Germans tried the effect of polymerizing a second monomer with butadiene. Copolymers of butadiene with styrene or acrylonitrile were made and evaluated. A copolymer of butadiene with styrene produced in the emulsion process was found to have superior physical properties over polybutadiene, and copolymers of butadiene with acrylonitrile were found to have good oil-resistant properties. The term "Buna S" was used to designate copolymers of butadiene and styrene, and "Buna N" to designate copolymers of butadiene and acrylonitrile. The original patent on Buna S was issued in 1933 to Tschunkur and Bock[46] of the Leverkusen Works of I.G. Farbenindustrie and the patent on Buna N to Konrad and Tschunkur.[28] By the end of 1934 both Buna S and Buna N were being made on a pilot-plant scale and were available to the rubber industry in Germany.

SK Rubber. Although the Germans largely abandoned the sodium process in favor of the emulsion process, the Russians continued experiments with sodium polymerization of butadiene and selected sodium polybutadiene for commercial production. The Russian rubber was termed SK rubber, and in 1932 large-scale production is reported to have started. The production of Russian synthetic rubber in long tons from 1933 through 1939 was as follows:[25]

1933	1934	1935	1936	1937	1938	1939
2204	11,139	25,581	44,200	25,000	53,000	78,500[38]

Russian capacity for synthetic-rubber production in 1948 was estimated at 125,000 long tons[39] (based on reported production in 1940–41).

Neoprene. Meanwhile in the United States the du Pont Company had also become active in the field of synthetic rubber. Their efforts were based on the work of Nieuwland, who had studied the reactions of acetylene for years and had found that treatment of acetylene with cuprous ammonium chloride produced monovinylacetylene and divinylacetylene, the dimer and trimer of acetylene, respectively. In 1925 at a meeting of the American

Chemical Society in Rochester, N.Y., he mentioned that a soft, elastic solid was produced when divinylacetylene was treated with sulfur chloride.[51] This observation aroused the interest of du Pont chemists, and henceforth they worked with Nieuwland to produce a synthetic rubber. It was found more advantageous to work with monovinylacetylene, and a satisfactory method of synthesis from acetylene was worked out jointly. The du Pont chemists then developed the production of a new monomer, chloroprene, by treating monovinylacetylene with hydrochloric acid. Chloroprene or 2-chlorobutadiene, $CH_2 : CH \cdot CCl : CH_2$, differs from isoprene in having a chlorine atom substituted for the methyl group. When polymerized, it gives a rubber with unique properties.[4] In 1931 at a meeting of the Akron Section of the American Chemical Society the du Pont Company announced to the rubber industry that they had succeeded in developing a new synthetic rubber called Duprene, the name given to polychloroprene. The name was later changed to Neoprene. In 1932, 250 tons of this synthetic rubber was produced.

Thiokol. In 1924 Patrick attempted to utilize the large amounts of olefin gases, mostly ethylene and propylene, which were being burned at mid-continent refineries. He recovered the gases by chlorination and then attempted to prepare cheap and useful derivatives from the ethylene and propylene dichlorides. Attempts to prepare the glycols by treatment with alkali were not very encouraging, but treatment with sodium di- and polysulfides led to long-chain polymers having rubberlike properties. The name Thiokol was coined for these rubberlike materials. Patrick continued to work on this reaction and the product for about three years. The first patent application on Thiokol was filed in 1927 but did not issue until 1932.[33] Production of Thiokol was started in 1930 when 4000 lb. was manufactured. Patents were also issued to Jean Baer[1] of Switzerland, before the issuance of Patrick's patent. These patents claim rubberlike masses prepared by the action of sulfur or polysulfides on $C_2H_4Br_2$.

In passing, it may be mentioned that H. L. Fisher and Moskowitz[14] were among the first to observe the formation of rubbery masses by reaction of ethylene dichloride with alkali sulfides. During World War I, in the winter of 1917–18, while attempting to make mustard gas, they sought to shorten the reaction by treating ethylene dichloride with potassium sulfide or a mixture containing potassium sulfide. They were amazed to obtain rubbery masses. The reaction was repeated with the same result. Fisher continued work on the reaction at the B. F. Goodrich Company Laboratories in 1919, but the odor and poor rubbery qualities of the material dampened interest in it, and the work was dropped. It remained for Patrick to extend the reaction and to commercialize the products.

Butyl Rubber. On June 4, 1940, another synthetic rubber was announced, by the Standard Oil Company of New Jersey. It was called Butyl rubber and was an outgrowth of the work that had been done on polyisobutylene. Thomas and Sparks, working in the laboratories of the Standard Oil Company of New Jersey, had found in 1937 that small amounts of butadiene or isoprene could be copolymerized with isobutylene. Whereas polyisobutylene could not be vulcanized, the new copolymer of isobutylene

and butadiene could be vulcanized, because the molecule was still unsaturated. The polymerization leading to Butyl rubber is conducted at a
temperature of —150° F., and many chemical engineering problems had to
be solved before the rubber could be produced on a commercial scale. In
March 1943 the production of Butyl rubber was started.

SYNTHETIC-RUBBER PRODUCTION DURING WORLD WAR II

The shortages of natural rubber both in Germany and in the United States
during World War II served to accelerate the commercial production of the
synthetic rubbers. In Germany the production of Buna S, the general-
purpose rubber, was greatly increased. In the United States a large
synthetic-rubber industry was developed. The general-purpose rubber was
known as GR-S and was manufactured in largest volume. The production
of Neoprene was also greatly increased, and manufacture of Butyl rubber
was undertaken. (In subsequent chapters, the manufacture of these
synthetic rubbers will be described.)

EARLY INTEREST IN BUNA IN THE UNITED STATES

In 1929 Standard Oil of New Jersey[24] entered into an agreement with the
German I.G. Farbenindustrie for the use of the German process for the
hydrogenation of coal and oil. Under the contracts Standard Oil of New
Jersey had unrestricted access to the scientific work relating to the hydrogenation of coal and oil in Germany. In September 1930 Standard and
I.G. organized a Joint American Study Company (Jasco) for the purpose of
developing new chemical products from oil and commercializing and licensing these processes. Standard at this time was interested in the German
Buna rubber, but, since the rubber was not then made from raw materials
derived from oil, it did not come under the Standard-I.G. agreement.

I.G. had been working on a process for making acetylene from oil or
natural gas by the electric-arc process. It was decided that Standard should
undertake to develop the process for the conversion of oil and natural gas
into acetylene gas while I.G. continued its work on the production of Buna
rubber from acetylene derived from coal. An experimental electric-arc
acetylene plant was put into operation by Standard in 1932. By the end of
1935, however, Standard concluded that the cost of making butadiene from
acetylene produced by this method would be too high to permit synthetic
rubber to compete with natural rubber, and work was terminated.

In 1933 an agreement was reached with General Tire & Rubber Company
of Akron for evaluating Buna in tires, and a quantity of the rubber was shipped
to Akron. Also, Dr. Stoecklin of I.G. spent several months in the United
States conferring with various rubber technologists. In 1934 General Tire
& Rubber Company reported that Buna was unsuitable for handling in
standard factory equipment and the quality of the products made from it
definitely inferior to those made of natural rubber.

BUNA S AND GERMAN AUTARKY

Under the new National Socialist government's "Four-Year Plan" adopted in 1933 to make Germany self-sufficient, synthetic rubber received increased attention. By 1935 the production of Buna S in Germany had reached 25 tons per month, and was supposed to reach 200 tons a month within one year and 1000 tons a month within three years. Although Buna S was still considered inferior to natural rubber as a general-purpose rubber, the German rubber industry was forced to use it against its wishes. By the spring of 1937 production of the German Government-subsidized Buna S had reached 5000 tons per year, which was far behind the original schedule.

EXAMINATION OF BUNA N AND BUNA S IN THE UNITED STATES

In 1937 Buna N was introduced into the United States by an accidental circumstance; an explosion in the Neoprene plant of the du Pont Company had put the plant out of commission. The Company appealed to I.G. for some of their oil-resistant Buna N, and small shipments were soon received. The rubber immediately found favor with American manufacturers, and the demand for it increased. In February 1937 I.G. also furnished samples of Buna S totaling several hundred pounds to a number of American rubber companies. Up to this time, however, the American rubber companies were primarily interested in Buna N—a specialty rubber for oil-resistant applications.

In the fall of 1938 the German Government had decided to expand the use of the Buna rubbers in the United States. In November Dr. ter Meer of I.G. went thither to discuss the use of Buna S as a general-purpose rubber. Meetings were held with the larger rubber companies, and Dr. ter Meer presented data on tests made in Germany with Buna S in tire treads. The American rubber companies however wanted to run tests of their own, and it was agreed that necessary quantities of the latest-type Buna S should be provided to enable the manufacturers to conduct their own tests. In addition, I.G. early in 1939 sent Dr. Koch, who had experience in processing, compounding, fabrication, and vulcanization of Buna S to the United States to aid the manufacturers in the use of Buna S.

The availability of these small quantities of Buna S and Buna N from Germany in 1937 and 1938 and experiments with these rubbers by the rubber manufacturers served to increase interest in synthetic rubber. The major rubber companies and the Standard Oil Development Company were already carrying on research of their own on synthetic rubber and were producing pilot-plant quantities of these and similar rubbers at the outbreak of World War II in September 1939.

With the outbreak of war in Europe in September 1939, imports of Buna N and Buna S from Germany ceased. Standard Oil of New Jersey decided to build a plant for making butadiene, to erect a plant for producing Buna N, and to offer licenses to any of the rubber companies to manufacture the

Buna rubbers under the Standard-I.G. patents. In 1940 the Firestone Tire & Rubber Company, and the U.S. Rubber Company accepted licenses to manufacture the Buna rubbers under the Standard patents.

DEVELOPMENTS DURING WORLD WAR II

Establishment of Rubber Reserve Company. With the fall of France in May 1940, the possibility of a rubber shortage in the United States became more apparent. In June 1940 a committee within the Advisory Commission of the Council of National Defense was formed to handle the problem of rubber. The chairman was Clarence Francis. The production of synthetic rubber in this country became one of the main concerns of the Francis committee. After discussions with the leading manufacturers of rubber and other interested parties, a program for the production of 100,000 tons of synthetic rubber per annum was tentatively agreed upon, and plans were formulated for carrying on the design of plants to produce the rubber. In October 1940 the work of the Francis committee and the responsibility for the synthetic-rubber program was taken over by the Reconstruction Finance Corporation. On June 25, 1940, Congress amended the R.F.C. Act so as to authorize R.F.C. to create corporations for the purpose of acquiring strategic and critical materials as defined by the President. On June 28, 1940, the President designated rubber as a strategic and critical material, and on the same date Rubber Reserve Company was created by R.F.C.[36] Rubber Reserve Company henceforth took over the responsibility of laying the foundations of the synthetic-rubber industry and was charged with the responsibility of producing the raw materials necessary for the manufacture of synthetic rubber, the actual manufacture of the synthetic rubber, and the sale and use of the synthetic rubber.

Early Plans for Synthetic Rubber Production. At the time that the National Defense Advisory Committee turned over to R.F.C. its plans and data for a contemplated 100,000-ton synthetic-rubber industry, there remained a number of important questions to be answered with respect to the program. These included (*a*) the most satisfactory type of rubber to be manufactured, (*b*) the processes to be employed, (*c*) the size and location of the plants, (*d*) the availability of the necessary raw materials, (*e*) the critical construction materials required, and (*f*) the time to complete the plants.

The four major rubber-manufacturing companies and the Standard Oil Development Company were already producing or engaging in development work in the synthetic-rubber field, principally in relation to special-purpose synthetics, and their technique and "know-how" were adaptable to the production of "general-purpose" synthetics. Other major chemical and petroleum companies were engaged in the production of butadiene, styrene, and other ingredient materials, in some cases as by-products or coproducts, and it was believed that their experience and processes were adaptable to large-scale production of the same materials. Numerous conferences were held with these members of industry and with representatives of the Chemical Division of the National Defense Advisory Committee, and as a result requests were sent out by R.F.C. on December 5, 1940, to certain of these companies for the submission of definite plans for the construction of synthetic-rubber plants. The answers to this request were received during January 1941, and all of the proposals were discussed in detail with the companies submitting them.

After many such discussions with members of industry, Rubber Reserve Company, on March 28, 1941, asked the B. F. Goodrich Company, Goodyear Tire & Rubber Company, Firestone Tire & Rubber Company, and U.S. Rubber Company to submit proposals for a program involving the erection of four plants, each having the buildings laid out for a capacity of 10,000 tons but with initial equipment for only 2500 tons. These companies proceeded immediately with the development of plans for the construction and completion of the four plants.

On May 9, 1941, Rubber Reserve Company received a letter from Office of Production Management (O.P.M.) recommending the immediate construction of plants capable of producing 40,000 tons of synthetic rubber annually, and thereupon Rubber Reserve Company issued the necessary authorizations and concluded the necessary contracts with the interested companies covering the design, construction, and operation of such plants. Based on the recommendation of the rubber-manufacturing industry, such plants were designed to produce synthetic rubber of the butadiene-styrene-copolymer type, which was considered the most suitable for the manufacture of tires and tubes. As a result of technological developments during the ensuing few months, the designed capacity of the plants was increased from 10,000 to 15,000 tons per annum.

Exchange of Information and Patent Rights. The war, in necessitating the immediate production of synthetic rubber and constituent materials on a vastly accelerated and expanded scale, also made it necessary to insure that the best qualities of product should be secured and that the best processes suited to available raw materials should be rapidly advanced to an efficient and commercial state. It was necessary to accomplish in less than two years a swift expansion of the small synthetic-rubber industry to the status of a great and going essential industry, and to condense into that brief time technical and commercial developments which under normal circumstances might easily have taken twenty years or more. It was recognized that it was necessary to provide for a full and effective interchange of technical information among those private organizations having useful knowledge and experience in the various portions of the field, and also to provide a supporting framework of patent protection (i.e., an interchange of operating rights under patents, including both those already in existence and those arising out of developments made during the period of exchange). It was recognized that a series of agreements made under Government sponsorship would be necessary to accomplish these objectives. Accordingly, Rubber Reserve Company immediately undertook the negotiation with various groups of private organizations of a series of agreements relating to technical information and patent rights.

The earliest agreement was executed under date of December 19, 1941, and pertained to the manufacture of "Synthetic Rubber," which was therein broadly defined to include synthetic rubber of both the Buna-S and

Buna-N types and which provided for a full exchange of technical information through a "Technical Committee" comprised of representatives of Rubber Reserve Company and all the other parties, which included all of the companies operating copolymer plants in the program. This agreement also provided for a cross-grant of licenses under the patent rights relating to the various processes owned or controlled by the parties. The purpose of the agreement was to implement the operation of the copolymer plants in the Government's program and to secure a full exchange of information among companies which were naturally competitive and between which there existed conflicting patent claims. The agreement was successful from the start in composing such differences and engendering cooperative technical effort, which greatly benefited the production of synthetic rubber of the Buna-S type.

The "Cross-License Agreements (Buna Rubber)" likewise covered patent rights and technical information relating to the manufacture of Buna rubber but were limited to the field of "general-purpose" Buna rubber. These Cross-License Agreements were made possible by an offer from Standard Oil Company of New Jersey in April 1943 to donate to the Government, through Rubber Reserve Company, its general-purpose Buna rubber patent rights. The agreements were on a royalty-free basis and ran for the full life of all patents owned or controlled by the parties during the war period, thus relieving Rubber Reserve Company and other signatories of the obligation to pay royalties to other signatories for postwar operations in the field of "general-purpose" Buna rubber. As of January 1, 1945, Rubber Reserve Company had entered into this standard form of agreement with 36 private companies, including all of the parties to the agreement of December 19, 1941.

Expansion of Plans for Synthetic-Rubber Production. After the fall of Singapore, when the major supplies of crude rubber in the Far East were cut off, the necessity for an even greater expansion of the synthetic-rubber program became apparent. The first expansion step raised the amount of Buna-S-type rubber to be manufactured to 120,000 tons, 30,000 tons to be made by each of the four rubber companies by enlargement of the plants already under way. In the early part of January 1942 the program was increased to 400,000 tons. By successive stages the program was increased during the first half of 1942 to a total of 805,000 tons, to consist of 705,000 tons of GR-S (butadiene-styrene type), 60,000 tons of Butyl, and 40,000 tons of Neoprene. The War Production Board approved this expanded program and specifically provided that the program for 705,000 tons of GR-S be given all priority and allocation assistance needed to assure production of 350,000 tons in the calendar year 1943, this approval being in accordance with the recommendation submitted by Jesse Jones to the chairman of W.P.B. under date of April 6, 1942.

A Canadian synthetic-rubber program, with a rated capacity of 30,000 long tons per annum of butadiene-styrene rubber and 7000 tons of Butyl rubber, was adopted and thoroughly integrated with the U.S.A. program. Agreements similar to the Rubber Reserve agreements were made with U.S. companies, and complete cooperation was established between the two programs.

Baruch Committee. During the summer of 1942 the rubber situation became very critical. It became apparent that the stockpile of natural rubber was inadequate to meet military and essential civilian needs and that, unless new supplies of either natural or synthetic rubber could be obtained in time, the total military and export requirements would exhaust the stocks before the end of the summer of 1943. It became necessary to conserve every pound of rubber and to push forward as rapidly as possible the production of synthetic rubber. The synthetic-rubber program, however, had to compete with other military requirements for critical materials of construction and raw materials for producing butadiene, and as a result certain delays occurred in carrying through the program for synthetic rubber in this country.

On August 6, 1942, the President appointed a Rubber Survey Committee to study the rubber situation. The committee consisted of B. M. Baruch, chairman; J. B. Conant, president of Harvard University; Karl T. Compton, president of Massachusetts Institute of Technology; with a staff of experts. After holding a series of hearings and receiving various reports and other data, the committee, on September 10, 1942, submitted to the President a report containing recommendations. The committee specifically recommended a reorganization and consolidation of the governmental agencies then dealing with rubber through the appointment of a "Rubber Administrator" (later referred to as the Rubber Director). The Report of the Rubber Survey Committee considered all phases of the rubber program. The supply and demand for rubber with respect to military and civilian requirements was analyzed. Means for conserving rubber on hand were proposed, and the continued collection of scrap rubber was recommended. An increase in the country's reclaiming capacity was recommended. It was also suggested that provision be made for additional fabricating facilities necessary to handle the full volume of synthetic-rubber production expected in 1944. With regard to the synthetic-rubber program, an increase in the production of butadiene from alcohol and from petroleum and an increase in the proposed annual output of GR-S from 705,000 long tons to 845,000 long tons were urged. Additional capacity to produce Neoprene, Thiokol, and Butyl rubber was also recommended. It was found later that Thiokol was not needed, and this part of the program was canceled.

The report urged that inventors and research groups be encouraged to do work in the synthetic-rubber field, and accordingly the Executive Order of September 17, 1942, providing for the appointment of the Rubber Director, included among his duties full responsibility for technical research and development with respect to the Nation's rubber program in all of its phases.

The report of the Rubber Survey Committee stated *inter alia* that information on synthetic rubber should be obtained from Russia, and, as so recommended, the Rubber Director in the early part of December 1942 dispatched a mission of technical experts to Russia for the purpose of gathering information. The group returned in March 1943. The meager information obtained as a result of the trip has made no contribution whatever to the synthetic-rubber program in the United States.

Office of the Rubber Director. After the submission of the Rubber Survey Committee Report, the President, in September 1942, appointed

William Jeffers, president of Union Pacific Railroad, as Rubber Director, and vested in him broad powers in matters dealing with rubber. The organization established by him was highly effective in securing priority assistance and in expediting the delivery of construction materials necessary for completion of the synthetic-rubber plants. Close working relations were established among the Rubber Director's office, the Rubber Reserve Company, and the Petroleum Coordinator's office. The Rubber Director's office was to carry out the recommendations of the Baruch committee for the rapid upbuilding of the new synthetic-rubber industry.

On September 15, 1943, Mr. Jeffers resigned as Rubber Director, and Colonel Bradley Dewey, who had formerly been Deputy Director, was appointed to succeed him. Colonel Dewey resigned as of September 1, 1944, and in view of the progress that had been made in the program recommended termination of the special powers that had been given to his office. The resignation of the Rubber Director was accepted, and, pursuant to his recommendations, the President designated Rubber Reserve Company as the agency responsible for the administration of the synthetic-rubber program.

Rubber Reserve Company and Office of Rubber Reserve. As a result of the dissolution of the Office of the Rubber Director, certain functions which had been assumed by the Rubber Director, principally the direction of research pertaining to synthetic rubber, were returned to Rubber Reserve Company. The research program included operation of the large Government-owned pilot plant and laboratory facilities located in Akron, Ohio, and operated for the account of Rubber Reserve Company (later known as the Office of Rubber Reserve) by the University of Akron.

Rubber Reserve Company was dissolved as of June 30, 1945, pursuant to Public Law 109, and its functions, assets, and liabilities were transferred to the Reconstruction Finance Corporation. On July 1, 1945, R.F.C. established the Office of Rubber Reserve to carry on the Government synthetic-rubber program.[37]

The synthetic-rubber plant construction program had been carried on under the Defense Plant Corporation (D.P.C.). On December 31, 1945, the responsibility for the administration of the synthetic-rubber plants properties was transferred to the Office of Rubber Reserve.

POSTWAR SYNTHETIC-RUBBER POLICY

After V-J day, August 14, 1945, the Office of Rubber Reserve took steps to reduce the production of GR-S during the last quarter of 1945 to a level approaching 30,000 long tons per month. At this time reconversion problems and the experience of the war years indicated the need for a coordinated national rubber policy. On September 7, 1945, John W. Snyder, Director of the Office of War Mobilization and Reconversion (O.W.M.R.) announced the formation of an Inter-Agency Policy Committee on Rubber and appointed William Batt to represent him as chairman. The committee was instructed to survey all programs, plans, and problems of Federal agencies concerned with natural, synthetic, or reclaimed rubber, and to assemble the statistical, technical, and economic data necessary for the formulation of a coordinated national policy on rubber. It was further charged with the duty of making appropriate recommendations to O.W.M.R. on matters requiring action by the Director or by the President or the Congress of the United States.[37]

On February 19, 1946, the Inter-Agency Policy Committee issued its first report, and on July 22, 1946, its final report containing recommendations concerning the future of the synthetic-rubber program.[37] The chief recommendations were:[42]

1. Continued production and use, regardless of cost, of synthetic rubber in an amount equal to "at least one-third of our total rubber requirements."

2. Legislation that would assure the continuation of the industry at this level.

3. Private ownership and operation of the synthetic-rubber industry as a major objective, and immediate disposal negotiations, subject to certain considerations and in accordance with a plan outlined in the report.

4. Retention by the Government in stand-by condition of facilities not acquired by private industry, such that the total synthetic-rubber capacity either in operation or in stand-by condition would approach about 600,000 long tons yearly.

5. Accumulation and maintenance of a Government-owned strategic stockpile of natural rubber.

6. Continuation of broad research programs by Government and private industry in the synthetic-rubber field, and continued Government research on natural rubber capable of being cultivated in the United States and tropical America.

7. Continuous review of the rubber problem.

Since authority for Government control of the rubber industry was contained in the Second War Powers Act, due to expire on March 31, 1947, Congressional hearings on special rubber legislation were held early in 1947. No agreement, however, was reached on the solution of the many problems presented by the continued operation of the synthetic industry and the contemplated disposal of the facilities. The Congress, therefore, by a joint resolution in Public Law 24 (80th Congress), extended the specific wartime powers relating to rubber for one year, in order to permit further study of the situation; it terminated, however, exclusive Government purchase of natural rubber.

During extensive Congressional hearings in the latter part of 1947 and early 1948 no permanent or long-range solution to many of the problems was found. On March 31, 1948, Public Law 469 (80th Congress) known as the "Rubber Act of 1948" was approved. This Act became effective on April 1, 1948, and expired on June 30, 1950. It authorized the President to exercise controls to insure consumption of synthetic rubber at least in the amount of 221,667 long tons of which 200,000 long tons was to be general-purpose synthetic rubber (GR-S) and 21,667 long tons special-purpose synthetic rubber. Of the 21,667 tons of special-purpose rubber, at least 15,000 tons was to be of a type suitable for use in pneumatic tubes (Butyl rubber). The Act also provided that "there shall be maintained at all times within the

United States rubber-producing facilities having a rated production capacity of not less than 600,000 long tons per annum of general-purpose synthetic rubber and not less than 65,000 long tons per annum of special-purpose synthetic rubber," and, further, that "of the 65,000 long-ton rated production capacity for special-purpose synthetic rubber, at least 45,000 long tons shall be of a type suitable for use in pneumatic inner tubes."

The Act required that a report on the development of a disposal program of Government-owned rubber-producing facilities be made by April 1, 1949, and that "by January 15, 1950, the President shall recommend to the Congress legislation with respect to the disposal of such facilities together with other recommendations considered desirable."

In compliance with a requirement of the Act, the Reconstruction Finance Corporation submitted to the President and to the Congress on April 1, 1949, a "Report With Respect to the Development of a Program for Disposal of the Government-Owned Rubber-Producing Facilities,"[35] prepared with the assistance of a Government interagency committee and after consideration of suggested disposal procedures submitted by committees comprised of members from representative companies of the rubber, petroleum, and chemical industries.

Responsibility for administering certain provisions of the Rubber Act was assigned to the Secretary of Commerce. By Executive Order 9942 of April 1, 1948, the President delegated to the Secretary of Commerce the functions, power, authority, and discretions vested in the sections of the Act covering allocations, specifications, and inventory controls, regulation of imports and exports, and classification of types of rubber.

On January 3, 1950, John R. Steelman, Assistant to the President and acting chairman of the National Security Resources Board, at the request of the President, submitted a report[42] to the President covering the current status of the entire synthetic-rubber problem and recommending legislation to establish and implement the Government's rubber policy after June 30, 1950. The report is very complete and covers recommendations pertaining to the disposal of the Government-owned synthetic-rubber facilities. In transmitting the report to Congress on January 14, 1950, the President stated: "It is my earnest hope that controls on consumption of GR-S may be reduced or suspended over the next few years, as technological improvements result in increasing quantities of general-purpose synthetic rubber being consumed without Government support. This development should be stimulated by the disposal of the Government's plants to private owners. . . . I recommend the adoption of legislation of ten years' duration in order to provide adequate protection of the national security and to contribute to the development of a vigorous, competitive, and privately owned synthetic rubber industry in the United States."

On May 19, 1950, the House of Representatives, after extensive public hearings, passed a bill (HR-7579) extending the Rubber Act of 1948 for three years. On June 5 and 6 a subcommittee of the Senate Armed Services Committee headed by Senator Lyndon B. Johnson of Texas held public hearings on the legislation passed by the House of Representatives and on June 8 approved the bill passed by the House but favored limiting the law to two instead of three years. On June 19 the House of Representatives also approved the two-year extension of the Rubber Act. On June 24 President Truman signed the bill extending the Rubber Act for two more years to June 30, 1952. Senator Johnson, chairman of the Senate Armed Services Committee on rubber legislation, in his recommendations asked for another report from the Administration by April 1951 on the disposal of the synthetic-rubber plants. He also urged "continuing study" of the rubber problems by a Congressional committee.

Expansion of Production since 1950. The unprecedented demand for tires in the spring of 1950 caused a shortage of natural rubber and consequent rise in the price of this commodity. The Office of Rubber Reserve in the meantime increased production of GR-S synthetic rubber to 35,000 tons per month. Early in July it was decided to reactivate another GR-S plant which would increase the production to 40,000 tons per month. On July 28, in view of the unsettled world situation caused by Communist aggression, President Truman asked for the opening of additional stand-by GR-S plants which would bring the capacity to 600,000 long tons by the end of 1950.

In the fall of 1950 it became apparent that scheduled production of synthetic rubber would not meet the requirements of both civilian and military needs. Reactivation of all the GR-S synthetic plants to produce a maximum of 760,000 long tons was ordered. Approximately 250,000 tons was to be made by the low-temperature polymerization process, and the remaining 510,000 tons by the standard process. During 1951 plant modification and expansion brought the total capacity for GR-S production to 860,000 tons. Early in 1951 it was decided to install additional refrigeration equipment in the GR-S plants to produce 360,000 tons of cold rubber, and later it was decided to extend further the production of cold GR-S, according to a schedule calling for the conversion to cold rubber of 75 per cent of the total GR-S productive capacity before the end of 1952.

In 1951 the production of oil-extended rubber entered the picture. GR-S is made to a Mooney viscosity value which allows it to be processed on standard rubber machinery. GR-S polymerized to a higher Mooney value is tough and difficult to process. German Buna S3 was a tough polymer and was heat-softened with air to make it processible. The tough polymers have good physical properties and offer certain quality advantages, e.g., better treadwear. Chemical softeners have also been used to soften the tough polymers. Recent work has been directed toward other means of plasticizing tough polymers of the butadiene-styrene type.[41] Certain petroleum oils such as Circosol 2XH or Sundex 53 have been found to act as good plasticizers for tough polymers. The oils are masterbatched with the latex, and the rubber is coagulated in the usual manner. From 20 to 50 per cent of oil (based on the weight of the rubber) is added to latex containing high-Mooney-viscosity polymer. The amount of oil used depends on the toughness of the polymer; the higher the Mooney viscosity, the greater the amount of oil that is necessary to reduce the Mooney viscosity to that of standard GR-S (50 to 60). Oil-masterbatched high-Mooney-viscosity cold rubber containing 25 per cent of oil has properties at least equal to those of cold rubber of normal Mooney viscosity. The oils thus act not only as plasticizers for the tough polymers, but also as extenders for synthetic rubber, thereby increasing the supply of rubber and effecting a distinct saving since the oils are priced at around two cents per pound.

The production in the United States of latex-masterbatched, oil-extended high-Mooney GR-S was started in April 1951. In 1951 a total of 23,124 tons of GR-S was thus extended, and in 1952 the total was 55,839 tons or about 9 per cent of the GR-S produced in Government plants.

During 1953 the output of oil-extended GR-S rose further: in the first 8 months of the year 18 per cent of the GR-S produced in Government plants was extended with oil.

PRODUCTION STATISTICS

Through January 1, 1950, the Office of Rubber Reserve had authorized the expenditure of approximately $24,000,000 for research and development.[42] Fifty-one plants have been designed, constructed, and placed in operation, and 49 rubber, chemical, petroleum, and industrial companies participate in their operation under the supervision of the Office of Rubber Reserve. The estimated plant investment cost is in excess of $700,000,000.

Figures for world production of synthetic rubber and for the United States and Canada are given in Tables I, II, and III.

Germany continued to expand her synthetic-rubber industry before the start of World War II. There were four synthetic-rubber plants in Germany with the following rated capacities:

Ludwigshafen	2500 tons per month
Hüls	4000 tons per month
Schkopau	6000 tons per month
Auschwitz	3000 tons per month

Construction of the plant at Schkopau, the largest of the four, was started in 1937 and completed in 1939, at which time it was put into operation. In 1943 the Schkopau plant produced 68,000 tons of Buna S type. Construction of the second largest plant, at Hüls, was started in 1938: the plant was put into operation in August 1940. Construction of the Ludwigshafen plant was started in January 1941. It was designed to have a capacity of 30,000 tons per year. Production was started in March 1943. The highest output was 2000 tons per month in March 1944. Total production for 1944 was 12,000 tons. The plant at Auschwitz was designed to have a capacity of 3000 tons per month but never produced Buna S. At the time it was captured by the Russians the production of acetylene from calcium carbide had started. Thus the total production capacity of Buna S in Germany at the time the war ended was 150,000 tons per year. With the Auschwitz plant, the total planned production of Buna S types was 175,000 tons. The highest actual production of 110,000 tons was attained in 1944.

Total production of all types of synthetic rubber in Germany is given in Table IV.[34]

Table I.   World Production of Synthetic Rubber (Long Tons)

	1939	1940	1941	1942	1943	1944	1945
Buna S (GR-S)	20,251	36,550	65,075	96,399	293,603	798,283	756,042
Neoprene (GR-M)	1,738	2,469	5,423	8,998	33,648	58,102	45,651
Butyl (GR-I)				23	1,373	21,657	56,505
Buna N (GR-A) and other types	1,759	3,367	6,977	15,191	21,419	22,483	7,871
Total	23,748	42,386	77,475	120,661	350,043	900,525	866,069

	1946	1947	1948	1949	1950	1951†	1952
Buna S (GR-S)*	667,123	450,918	426,122	332,498	400,121	736,056	692,226‡
Neoprene (GR-M)	47,766	31,495	34,848	35,215	50,067	58,907	65,745
Butyl (GR-I)	85,325	69,899	60,555	59,284	68,753	88,859	96,267
Buna N (GR-A) and other types	6,350	7,012	10,661	13,334	15,683	24,559	23,531
Total	806,564	559,324	532,186	440,332	534,624	908,381	877,769

* For several years prior to March 1951 the output of high-styrene resins was included under this head.
† Includes 929 tons (all types) produced in the German Federal Republic.
‡ Included here is 4931 tons, which represents total West German production in 1952 and which may comprise a certain amount of N-type rubber.

Table II. Production of Synthetic Rubber in U.S.A.

	1939	1940	1941	1942	1943	1944	1945
GR-S	...	...	227	3,721	182,259	670,268	719,404
Neoprene (GR-M)	1,738	2,469	5,423	8,998	33,648	58,102	45,651
Butyl (GR-I)	...	...	...	23	1,373	18,890	47,426
Buna N (GR-A)	12	91	2,464	9,734	14,487	16,812	7,871
Total	1,750	2,560	8,114	22,476	231,767	764,072	820,352

	1946	1947	1948	1949	1950	1951	1952
GR-S	613,408	407,769	393,880	295,166	358,248	696,814*	637,225
Neoprene (GR-M)	47,766	31,495	34,848	35,215	50,067	58,907	65,745
Butyl (GR-I)	73,114	62,820	52,603	52,237	55,832	74,105	79,368
Buna N (GR-A)	5,738	6,618	7,012	11,072	12,037	15,333	16,228
Total	740,026	508,702	488,343	393,690	476,184	845,159	798,566

* Before March 1951 production figures for GR-S included high-styrene butadiene-styrene polymers.

Table III. *Production of Synthetic Rubber in Canada (Long Tons)*

Year	GR-S Type	Butyl Rubber	Other (Including Nitrile)	Total
1943	2,522	...	...	2,522
1944	32,062	2,769	...	34,829
1945	36,638	9,079	...	45,717
1946	38,770	12,211	...	50,981
1947	35,314	7,079	...	42,393
1948	32,242	7,952	261	40,455
1949	37,333	7,047	2263	46,642
1950	41,873	12,921	3646	58,440
1951	38,313	14,754	9226	62,293
1952	50,700	16,899	7303	74,272

Table IV. *Total Rubber Production in Germany (Metric Tons)*

	1937	1938	1939	1940	1941	1942	1943	1944
Buna S Types								
Schkopau	2110	3994	20,173	34,899	40,705	57,313	67,703	45,113
Hüls	...	...	...	2,045	25,020	36,680	34,693	39,105
Ludwigshafen	...	...	...	...	...	...	7,181	11,955
Leverkusen	*	*	403	193	164	173	992	1,320†
Total Buna S	2110	3994	20,576	37,137	65,889	94,166	110,569	97,493
Buna N	‡	‡						
Leverkusen	400	640	1,126	1,898	2,631	2,824	3,656	3,129†
Hüls	...	...	...	...	...	...	...	43
Total Buna N	400	640	1,126	1,898	2,631	2,824	3,656	3,172
Numbered Bunas								
Schkopau	637	848	649	1,431	1,955	2,721	3,388	2,590
Grand total synthetics	3147	5482	22,351	40,466	70,475	98,711	117,613	103,255

* Not available.
† To October 26, 1944.
‡ Approximate.

REFERENCES

1. Baer, J., Swiss Pat. 127,540, Oct. 20, 1926; also Brit. Pat. 279,406, Oct. 20, 1926; Swiss Pat. 132,505, Oct. 20, 1926; Ger. Pat. 526,121, June 2, 1931; Ger. Pat. 530,163, July 22, 1931.
2. Bouchardat, A., *Ann.*, **27,** 20 (1838).
3. Bouchardat, G., *Compt. rend.*, **89,** 1117 (1879); **80,** 1446 (1875); *Bull. soc. chim.*, **24,** 108 (1875).

4. Carothers, W. H., Williams, I., Collins, A. M., and Kirby, J. E., *J. Am. Chem. Soc.*, **53,** 4203–25 (1931).

5. Caventou, E., *Ann.*, **127,** 93 (1863).

6. Dalton, J., *Phil. Mag.*, **56,** 479 (1836); *J. prakt. Chem.*, **10,** 121 (1837).

7. Dumas, J., *Ann.*, **27,** 30 (1838).

8. Euler, W., *Ber.*, **30,** 1989 (1897); *J. prakt. Chem.*, **57** (2), 131 (1897).

9. Faraday, M., *Quart. J. Sci. and Arts*, **21,** 19 (1826).

10. Farbenfabriken vorm. Friedr. Bayer & Co., U.S. Pat. 1,058,056, Apr. 8, 1913.

11. Farbenfabriken vorm. Friedr. Bayer & Co., Ger. Pat. 254,672, Dec. 11, 1912.

12. Farbenfabriken vorm. Friedr. Bayer & Co., Ger. Pat. 255,129, Dec. 20, 1912.

13. Fernbach, A., Fr. Pat. 448,364, Sept. 16, 1912; Dubosc, A., and Luttringer, A., *Rubber: Its Production, Chemistry, and Synthesis*, Griffin, London, 1918, pp. 301–4.

14. Fisher, H. L., and Moskowitz, M., Private communication.

15. Gottlob, K., *Gummi-Ztg.*, **33,** 508, 534, 551, 576, 599 (1919).

16. Gottlob, K., *Technology of Rubber*, MacLaren, London, 1927, p. 221.

17. Gregory, Wilhelm J., *J. prakt. Chem.*, **9,** 387 (1836); *J. pharm. chim.*, **22,** 382 (1836); **23,** 454 (1837).

18. Harries, C. D., *Untersuchungen über die Einwirkung des Ozons auf organische Verbindungen,* 1903–1916, Springer, Berlin, 1916; *Untersuchungen über die natürlichen und künstlichen Kautschukarten*, Springer, Berlin, 1919.

19. Harries, C. D., *Ann.*, **383,** p. 190 (1911).

20. Himly, F. K., Dissertation, *De Caoutchouk ejusque destillationis siccae productis et ex his de Caoutchino, novo corpore ex hydrogenio et carboneo composito*, Dietrich, Göttingen, 1835; *Ann.*, **27,** 40 (1838).

21. Hofmann, F., and Coutelle, K., Ger. Pat. 231,806, Apr. 9, 1909.

22. Hofmann, F., and Coutelle, K., Ger. Pat. 250,690, Sept. 12, 1909.

23. Holt, A., *Chem.-Ztg.*, **38,** 188 (1914).

24. Howard, F. A., *Buna Rubber, the Birth of an Industry*, Van Nostrand, New York, 1947.

25. *India Rubber World*, **102,** 62 (1940).

26. Ipatieff, V. N., and Wittorf, N., *J. prakt. Chem.* (2), **55,** 1 (1896).

27. Kondakow, I., *J. prakt. Chem.*, **62,** 66 (1900).

28. Konrad, E., and Tschunkur, E., U.S. Pat. 1,973,000 (1934); Ger. Pat. 658,172 (1938).

29. Kyriakides, L. P., *Chem. Eng. News*, **23,** 531 (1945); *J. Am. Chem. Soc.*, **36,** 531, 657, 663, 980 (1914).

30. Lebedev, S. V., *J. Russ. Phys. Chem. Soc.*, **42,** 949 (1910).

31. Liebig, J. von, *Ann.*, **16,** 61 (1835).

32. Matthews, F. E., and Strange, E. H., Brit. Pat. 24,790, Oct. 25, 1910; *Chem. Absts.*, **6,** 1542 (1912).

33. Patrick, C. J., U.S. Pat. 1,890,191 (1932).

34. PB. No. 189.

35. Reconstruction Finance Corp., *Rubber Age N.Y.*, **65,** 207 (1949).

36. Rept. on the Rubber Program 1940–1945, Rubber Reserve Co., Feb. 24, 1945.

37. Rept. on the Rubber Program, Suppl. No. 1, Year 1945, O.R.R., Apr. 8, 1946; Rubber, First Rept. of the Inter-Agency Policy Comm. on Rubber, Washington, D.C., Feb. 19, 1946; Rubber, First and Second Repts. of the Inter-Agency Policy Comm., Office of War Mobilization and Reconversion, July 22, 1946.

38. *Rubber Statistical Bull.*, **1,** Nos. 5, 6 (1946).

39. *Rubber Statistical Bull.*, **2,** Nos. 8, 48 (1948).

40. Semon, W. L., *Chem. Eng. News*, **21,** 1613 (1943).

41. Swart, G. H., Pfau, E. S., and Weinstock, K. V., *India Rubber World*, **124,** 309–19 (1951); D'Ianni, J. D., Hoesly, J. J., and Greer, P. S., *Rubber Age N.Y.*, **69,** 317–21 (1951); Harrington, H. D., Weinstock, K. V., Legge, N. R., and Storey, E. B., *India Rubber*

World, **124,** 435–42, 571–5 (1951); Weinstock, K. V., Baker, L. M., and Jones, D. H., *Rubber Age N.Y.*, **70,** 333–8 (1951); Dinsmore, R. P., *Trans. Instn. Rubber Ind.*, **28,** 166–206 (1952); Weinstock, K. V., Storey, E. B., and Sweeley, J. S., *Ind. Eng. Chem.*, **45,** 1035–43 (1953); Taft, W. K., Duke, J., Snyder, A. D., Feldon, M., and Laundrie, R. W., *Ind. Eng. Chem.*, **45,** 1043–53 (1953).

42. Synthetic Rubber, Recommendations of the President to Congress, Jan., 1950, U.S. Govt. Printing Office, Washington, D.C.

43. Thiele, J., *Ann.*, **319,** 226 (1901).

44. Tilden, W. A., *J. Chem. Soc.*, **45,** 411 (1884).

45. Tilden, W. A., *Chem. News*, **46,** 120 (1882).

46. Tschunkur, E., and Bock, W., U.S. Pat. 1,938,731 (1933); Ger. Pat. 570,980 (1933).

47. U.S. Rubber Co., Private Communication.

48. Wallach, O., *Ann.*, **239,** 48 (1887).

49. Whitby, G. S., and Katz, M., *Ind. Eng. Chem.*, **25,** 1204–11 (1933).

50. Williams, Greville, *Proc. Roy. Soc.*, **10,** 516 (1860); *Phil. Trans.*, **1860,** 245.

51. Wolf, R. F., *Scientific Monthly*, **74,** 69–75 (1952).

CHAPTER 3

MANUFACTURE OF DIENES FROM PETROLEUM

C. E. Morrell

Standard Oil Development Company

HISTORICAL

A wide variety of dienes can be produced from petroleum hydrocarbons by pyrolytic or chemical transformations. (In this chapter the term "diene" is limited to conjugated diolefinic hydrocarbons.) However, only two hydrocarbons of this class, viz., butadiene and isoprene, are produced commercially for synthetic-rubber manufacture in the United States. Butadiene was first identified as a component of the products from cracking petroleum oils by Armstrong and Miller[4] in 1886. The presence of isoprene, as well as butadiene and other dienes, in the lower-boiling fractions of vapor-phase cracked gasoline was demonstrated by Birch and Scott[7] in 1932.

Extensive research work, both in the United States and abroad, during the two decades preceding World War II showed that butadiene results from the thermal treatment of a wide variety of hydrocarbons.[20] The yields of butadiene are markedly dependent on a number of factors, especially the type and molecular weight of the hydrocarbon pyrolyzed and the pyrolysis conditions employed. The lower paraffins from methane through the hexanes produce only small amounts of butadiene on thermal treatment.[15] Ethylene and propylene, at temperatures of 700 to 1100° C. and at subatmospheric pressures, produce butadiene in fair yields.[16] But, because of the difficulties attending operations under such conditions, pyrolysis of these olefins has not attained commercial significance for butadiene production. Under similar conditions, butene-2 is reported[60] to give yields as high as 45 per cent. Butene-1 gives decidedly lower yields than butene-2. In general, olefins of higher molecular weight than butene-2 are poorer than the latter as sources of butadiene. They yield higher-diene homologs, which do not crack readily to butadiene.

Early work showed that cycloparaffins (especially cyclohexane), in contrast to paraffins, can be thermally treated to produce high yields of butadiene. Yields as high as 60 per cent by weight on cyclohexane decomposed have been reported.[64] Among the cycloölefins, cyclohexene and its homologs crack readily to butadiene but are not available in sufficient amounts to be of commercial interest.[17] Yields of butadiene from benzene by pyrolysis appear to be very small.[50]

56

Following the identification of butadiene in cracked petroleum fractions by Armstrong and Miller, numerous researches relating to the production of butadiene by the thermal cracking of a wide variety of petroleum fractions appeared in the literature. A major portion of the work in this field was conducted by Russian investigators, especially Dobryanskii, Lebedev, and Buizov, although some work is reported by American and British investigators.[18, 66] Fractions boiling from the light naphtha to the fuel oil range were studied. A variety of thermal-cracking techniques was employed including the use of both tube furnaces and regenerative checkerbrick stoves. Butadiene yields were reported to range from 3 to 18 per cent by weight on the hydrocarbon cracked. In general, the yields were increased by the use of either reduced pressures or diluents such as steam or flue gas. In spite of their interest from a research viewpoint in hydrocarbon-cracking processes, the Russians apparently never used them on a significant commercial scale but instead relied, for the manufacture of butadiene, on the Lebedev alcohol process. As will be discussed later, however, thermal cracking of petroleum fractions became an important source of butadiene for American synthetic-rubber production. The Germans relied entirely on the acetylene routes for butadiene synthesis and hence did very little on the development of processes based on thermal cracking of petroleum. They displayed a research interest in the use of catalysts, especially graphite and magnesium oxide, for cracking naphtha and gas-oil fractions to olefinic products containing dienes but did not develop this approach for commercial use.[35]

Catalytic Dehydrogenation of n-Butenes and n-Butane. Since catalytic dehydrogenation of n-butenes and n-butane has become of major importance for butadiene production in the United States, a summary of the history of these processes is of interest. A number of early studies demonstrated that only small yields of n-butenes are obtained by thermal decomposition of n-butane, the primary products being largely methane, ethane, ethylene, and propylene. However, large quantities of n-butenes became available in refineries before and during the war as a result of extensive construction of catalytic cracking units. n-Butenes from this source constitute a large portion of the feed to plants producing butadiene by butene dehydrogenation. Other sources of n-butenes are the thermal cracking of naphtha and higher-boiling petroleum fractions and the catalytic dehydrogenation of n-butane recovered largely from natural-gas operations.

Interest in the catalytic dehydrogenation of n-butenes to butadiene developed during the decade preceding World War II. The earliest work in this field was of German origin, although Russian investigators became interested at a somewhat later date. The first German patents[35, 36] worthy of note claimed as catalysts such materials as lustrous carbon, graphite, magnesium oxide, and calcium oxide. The yields of butadiene obtained with these contact agents appear to have been rather poor however. A later patent of German origin[30] placed considerable emphasis on the use of catalysts containing zinc oxide as a major component. Minor components claimed include oxides of aluminum, calcium, chromium, vanadium, tungsten, and molybdenum. The passage of n-butenes over catalysts of this type with steam as a diluent was reported to give butadiene yields per pass of about 20 per cent by weight, and over-all yields, on recycling of undecomposed butenes, of about 70 per cent by weight. Early Russian work[9] on catalytic butene dehydrogenation (published in 1934 to 1942) was concerned largely with chromium oxide catalysts. Nitrogen and carbon dioxide were used as diluent gases. While the reported butadiene yields from butene-1 with these chromia catalysts are similar to those claimed for the German zinc oxide-based catalysts, yields from butene-2 with the chromia catalysts appear considerably lower. In 1940 Grosse, J. C. Morrell, and Mavity[29] reported 76 per cent over-all yields of butadiene from mixed n-butenes using an alumina-chromia catalyst without diluent gas at 0.022 atm. pressure. The same workers also demonstrated the effectiveness of alumina-supported oxides of tungsten, thorium, vanadium, tantalum, zirconium, and titanium as catalysts for n-butene dehydrogenation.[28, 45]

Production of butadiene by dehydrogenation of n-butane could, conceivably, be carried out in either of two ways. One of these, the so-called one-stage process, yields butadiene directly, the unconverted n-butane and the intermediate n-butenes being recycled together to the dehydrogenation step after removal of butadiene, hydrogen, and hydrocarbons boiling below and above the C_4 range. The other, the so-called two-stage process, employs relatively mild conditions in the first stage, producing largely n-butenes and little butadiene. In the second stage, n-butenes, after separation from unconverted n-butane and hydrogen, are subjected to further dehydrogenation to produce butadiene. Since n-butenes are the principal product of the first stage, the two-stage process must compete, in an economic sense, with the catalytic dehydrogenation of n-butenes from other sources, especially catalytic and thermal cracking of higher-boiling petroleum fractions.

A one-stage process employing a catalyst of undisclosed composition has been described by Komarewsky and Reisz.[39] Reduced pressures and short contact times are required to obtain selective conversion to butadiene. Yields of butadiene in a single pass over the catalyst are rather low (12 per cent), but, by recycling, a 74 weight per cent yield of butadiene may be obtained with little degradation to carbon. At temperatures in the range of 1100 to 1200° F., contact times of about 0.2 sec. and n-butane pressures of about 100 mm. of mercury appear to be optimum for maximum one-pass yields of butadiene. As later discussion will show, however, the one-stage type of process for producing butadiene from n-butane did not attain commercial status in the Synthetic Rubber Program.

A rather extensive background on the dehydrogenation of n-butane to n-butenes (first stage of the two-stage process outlined above) was available at the time intensive work on butadiene production started. Activated alumina[12] alone and alumina-supported oxides of chromium, molybdenum, thorium, vanadium, and a number of other metals[46, 24] had already been recognized as effective catalytic materials for this conversion. In general, temperatures of 1000 to 1150° F. are required for reasonable catalytic activity and conversions. The catalyst employed becomes coated with carbon during use, and this causes a decline in activity. Consequently the carbon deposit must be periodically removed, generally by burning with an oxygen-containing gas. This burning operation must generally be carried out under controlled temperature conditions to avoid injury to the catalyst.

GENERAL VIEW OF PETROLEUM DIENE PROCESSES

Following the realization of the need for large supplies of butadiene for synthetic-rubber production, extensive research and development work was undertaken within the petroleum industry to develop commercial production processes. This work quite naturally drew heavily on the background research (before about 1940) described in the preceding Historical Section. A wide variety of processes based on existing or potential petroleum raw materials received at least some degree of consideration for commercial use during the early development years (1940–1943). A general review of these is given in articles by Elder[21] and Frolich and Morrell,[25] which appeared during 1942 and 1943, respectively. Further information on general phases of the early development work is given in the Progress Reports issued during this period by Wm. M. Jeffers, Rubber Director.

The following processes based on raw materials of petroleum origin were considered:

Thermal Cracking of Hydrocarbons Containing More than Four Carbon Atoms. Among possible hydrocarbon raw materials are cyclohexane and naphtha and gas-oil fractions. Although cyclohexane gives good yields, it could be made available in significant quantities only by hydrogenating benzene. This did not prove feasible in view of other wartime demands for benzene. The thermal cracking of naphtha and gas-oil fractions was developed to the commercial stage both in privately owned plants and in the Government-sponsored "quickie" program. These developments will be discussed in more detail in a later section.

Aldol Condensation Process Based on Acetaldehyde Derived from the Lower Paraffins. The aldol condensation process, described in detail in Chapter 5, was utilized commercially by the Germans, who derived acetaldehyde from acetylene by catalytic hydration. The acetylene was produced from calcium carbide and also by the action of the electric arc on methane. A privately owned plant designed to operate on acetaldehyde from hydrocarbon oxidation was constructed at Bishop, Texas, by the Celanese Corporation of America, but was not placed in operation[26] by October 1944, and did not become a permanent part of the Synthetic Rubber Program.

Chlorination-Dehydrochlorination of n-Butenes. The literature shows that good yields of butadiene, viz., 60 per cent or more of theory, can be obtained by this method, which is described in detail in Chapter 5. Temperatures of 500° C. and higher are required to effect noncatalyzed conversion of the dichlorobutanes to butadiene. Short contact times at pyrolysis temperatures are also desirable to minimize loss of butadiene and formation of by-products. Predominant among the latter are monochloroölefins of the vinyl type which become degraded to lower-boiling materials on recycling. A plant privately owned by Shell Chemical Company started operations on this process before the inception of the Government-sponsored program of butadiene production.[26]

Table I. Petroleum Butadiene Plants in Actual or Contemplated Operation, October 1944

Plant Location	Operator	Rated Capacity, Tons per year	Ownership	Conversion Process
Baton Rouge, La.	Standard* Oil Co. of La.	6,800	Government	Thermal cracking of naphtha
Ingleside, Texas	Humble Oil and Refining Co.	7,000	Government	Thermal cracking of naphtha
El Dorado, Ark.	Lion Oil Refining Co.	6,700	Government	Thermal cracking†
Corpus Christi, Texas	Taylor Refining Co.	5,500	Government	Thermal cracking‡
Los Angeles, Calif.	Southern California Gas Co.	30,000	Government	Thermal cracking of naphtha§
Baton Rouge, La.	Standard Oil Co. of La.*	15,000	Government	Butene dehydrogenation
Baytown, Texas	Humble Oil and Refining Co.	30,000	Government	Butene dehydrogenation
Lake Charles, La.	Cities Service Refining Co.	55,000	Government	Butene dehydrogenation
Port Neches, Texas	Neches Butane Products Co.	100,000	Government	Butene dehydrogenation
Houston, Texas	Sinclair Rubber, Inc.	50,000	Government	Butene dehydrogenation
Los Angeles, Calif.	Shell Chemical Division	25,000	Government	Butene dehydrogenation
Borger, Texas	Phillips Petroleum Co.	45,000	Government	n-Butane dehydrogenation
El Segundo, Calif.	Standard Oil Co. of Calif.	15,000	Government	n-Butane dehydrogenation
Toledo, Ohio	Sun Oil Co.	15,000	Government	n-Butane dehydrogenation
Sarnia, Ontario	Imperial Oil, Ltd.	30,000	Canadian Government	Butene dehydrogenation
Charleston, W.Va.	Carbide & Carbon Chemicals Corp.‖	5,000	Private	Thermal cracking
Baton Rouge, La.	Standard Oil Co. of La.*	6,000	Private	Thermal cracking of gas oil
Bishop, Texas	Celanese Corp. of America	10,000	Private	Aldol process. Acetaldehyde derived from oxidation of propane and butane. (Cf. Chap. 5)

* Now Esso Standard Oil Co., Louisiana Division.
† Crude butadiene production only. Purification carried out at Standard Oil Co. of Louisiana plant at Baton Rouge, La.
‡ Production of crude butadiene only. Purification carried out at Humble Oil and Refining Co. plant at Ingleside, Texas.
§ Crude butadiene production only. Purification carried out at Shell Chemical Division plant at Los Angeles.
‖ Now Carbide and Carbon Chemicals Division of Union Carbide Corp.

Catalytic Treatment of Ethyl Alcohol or Ethyl Alcohol-Acetaldehyde Mixtures. These processes, known as the one-stage and two-stage butadiene-from-alcohol processes, respectively, are described in Chapter 4. They are mentioned here only because large volumes of ethyl alcohol are produced commercially by the hydration of petroleum-derived ethylene. However, it appears that essentially all alcohol raw material employed in the Synthetic Rubber Program was during World War II derived from fermentation sources.

Catalytic Dehydrogenation of n-Butenes. The basic process for dehydrogenating *n*-butenes was developed by the Standard Oil Development Company. It became the main process for the production of butadiene and will be discussed in more detail in subsequent sections.

Catalytic Dehydrogenation of n-Butane. Three plants utilizing *n*-butane as primary feed were constructed as part of the Synthetic Rubber Program. One of these was built to operate processes developed by the Phillips Petroleum Company. The other two, of lower rated capacities, employed processes contributed by the Catalytic Development Company (Houdry).

Petroleum Butadiene Plants. Table I lists petroleum butadiene plants known to be in operation, actual or contemplated, as of October 1944. This table also gives information on the rated capacities of the plants and the types of conversion processes employed. This list does not include plants privately owned by Shell Chemical Division of Shell Union Oil Company, Dow Chemical Company, United Gas Improvement Company, Phillips Petroleum Company, and Standard Oil Company of Louisiana (now Louisiana Division of Esso Standard Oil Company). These privately owned plants were placed in operation before the Government-sponsored program became effective. Most of the butadiene produced in these plants was derived from thermal cracking of petroleum hydrocarbons.

BUTADIENE FROM THERMAL CRACKING OF HYDROCARBONS

Especially during World War II, the production of butadiene by the thermal cracking of petroleum fractions became of some importance in the American Synthetic Rubber Program. The development of thermal-cracking processes was fostered to a large extent by the so-called "quickie" butadiene program sponsored by the Office of Petroleum Coordinator.[1] Before this program, however, butadiene was already being produced by thermal cracking in a number of small privately owned plants (Table I). The purpose of the "quickie" program and the related development work was to adapt existing petroleum refinery and gas-producing equipment to butadiene production. It was recognized that these expedients would not only save steel, which was critically needed at that time for other uses, but also would afford a badly needed supply of butadiene before production started from large plants specifically designed to produce the hydrocarbon.

In the development work a number of methods for heating petroleum fractions to the temperature range (1200 to 1400° F.) necessary for cracking to butadiene were considered. Among these were the use of tubular cracking furnaces and specially designed regenerative stoves of the type employed in the gas-generating industry. Methods such as partial combustion of the cracking stock with air and injection of hot flue gas or superheated steam were also considered, usually for auxiliary heating in combination with tubular heating.

At one time, participants in the "quickie" program were considering the production of as much as 200,000 tons a year of butadiene in approximately 30 plants.[2] [Also as part of this program considerable study was given to the conversion of existing refining equipment to dehydrogenating n-butane (see below).] For a number of reasons, however, the number of thermal-cracking plants actually placed in production and their total rated capacities were considerably less than the above figures. The plants actually placed in operating condition are listed in Table I. Included are those using converted refinery equipment and operated by Standard Oil Company of Louisiana (now Esso Standard Oil Company, Louisiana Division) at Baton Rouge, La.; by Humble Oil and Refining Company at Ingleside, Texas; by Lion Oil and Refining Company at El Dorado, Ark.; and by Taylor Refining Company at Corpus Christi, Texas. The Los Angeles plant operated by Southern California Gas Company employed thermal cracking in converted gas generating equipment and is discussed in more detail below. With the exception of the last, all these plants employed converted tubular cracking units and generally used heavy naphtha as charging stock and steam as diluent.

Very little has been published regarding the nature of the operations in the privately owned plants listed in Table I. Those operated by Esso Standard Oil Company at Baton Rouge employ tubular furnaces and use steam as a diluent in the cracking zone. These units also yield a crude isoprene fraction. The latter is processed by extractively distilling in the presence of aqueous acetone, yielding purified isoprene suitable for use in Butyl-rubber manufacture.

Another type of thermal-cracking process that found large-scale use in the Synthetic Rubber Program was that developed by the United Gas Improvement Company. This was employed not only in a small plant privately owned by this company but also in a large installation operated by the Southern California Gas Company in Los Angeles.[34] This process evolved from extensive research and development work on utility gas production and the utilization of by-products therefrom. The cracking equipment used consists of regenerative stoves or furnaces. These are large cylindrical shells containing baffles and liners fabricated from temperature-resistant brick. These cracking units or generators are operated in a cyclic manner, cracking proceeding for a portion of the cycle followed by a reheat period during which fuel is burned inside the generator. Temperature may be varied over wide ranges (1200 to 1900° F.). Total cycle time may also be varied, depending upon the products desired. At the Los Angeles plant a naphtha fraction was charged to cracking. The optimum butadiene yield was obtained at a cracking temperature of about 1300° F. and a hydrocarbon residence time of approximately one second. The total cycle time was about 6 minutes, during about 40 per cent of which naphtha was charged to cracking.   Gas or reduced crude was employed as fuel for heating the generators.   Also it was general practice to introduce an equal quantity by weight of superheated steam along with the naphtha charged to cracking, in order to lower the hydrocarbon partial pressure.

Besides butadiene, regenerative stove cracking, as described above, yields a number of other valuable hydrocarbons, including benzene, naphthalene, ethylene, propylene, butylenes, isoprene, and cyclopentadiene.   In the Los Angeles plant, a butadiene-butylene fraction was recovered from the product gases by compression, oil scrubbing, and distillation.   Butadiene was separated from the butenes by the copper solution process.   The n-butenes were then subjected to catalytic dehydrogenation in an adjoining plant to produce additional butadiene.   The naphtha-cracking plant, at peak production, charged about 17,000 barrels per day.   It had a rated butadiene production capacity of 30,000 tons a year.   The yield of butadiene amounted to about 3.5 weight per cent on the naphtha charged.[42]

CATALYTIC DEHYDROGENATION OF n-BUTANE

A general view of n-butane dehydrogenation has been given by Watson and co-workers[62] in 1944.   It was recognized quite early in the development that the catalyst used must meet a number of requirements to afford satisfactory results.   It must be highly active and selective and also must be stable under process conditions.   Relatively short contact times and low temperatures must be employed to prevent thermal-cracking reactions, which yield degradation products such as methane and ethane.   Catalysts that effectively promote paraffin dehydrogenation become coated during use with a carbonaceous deposit, which reduces catalyst activity and must be periodically removed by contacting the catalyst with an oxygen-containing gas.

The dehydrogenation of n-butane to n-butenes is endothermic, the heat of reaction amounting to about 1000 B.t.u. per lb. of butane.   Preferred temperatures for the dehydrogenation reaction are in the range 1000 to 1100° F.   At temperatures below 950° F., the thermodynamic equilibrium conversion is too low for practical application.   Large amounts of heat are evolved during the burning of carbonaceous deposits on the catalyst.   In order to prevent catalyst deactivation, it proved necessary to regulate heat evolution during the burning operation and avoid temperatures above 1300° F.

Dodd and Watson[14] have made a study of the kinetics of n-butane dehydrogenation, using a chromia-alumina catalyst.   They concluded that the rate of dehydrogenation is probably controlled by a surface reaction involving dual active centers.

Two processes for producing butadiene from n-butane were developed and placed in operation during the wartime rubber program.   One of these was developed by the Houdry Process Corporation and was utilized in two plants, one at El Segundo, Calif., operated by Standard Oil Company of California; the other at Toledo, Ohio, operated by Sun Oil Company.   Both plants had rated capacities of 15,000 tons a year.   The other process was developed by Phillips Petroleum Company.   A plant of 45,000 tons a year rated capacity using the Phillips process was installed at Borger, Texas, and operated by the Phillips Company.

Houdry Process. The Houdry process has been only briefly described in the literature.[59, 48] It was designed for smaller refineries and natural-gasoline plants having supplies of n-butane available. It was adapted to allow extensive use of existing refinery equipment. The plants were designed for two-stage catalytic dehydrogenation, using a reaction cycle consisting of a very short on-stream period followed by a regeneration period. During the latter, carbonaceous deposits that had formed on the catalyst during the on-stream period were removed by burning with air. The reactors or catalytic vessels were patterned after those used in the Houdry fixed-bed process for catalytically cracking gas oils to gasoline. In the first stage of the dehydrogenation, n-butane was partially dehydrogenated to n-butenes, hydrogen and lighter hydrocarbon gases also being formed. The butane-butene mixture was then separated from the other gases by absorption in oil under superatmospheric pressures and fed to the second dehydrogenation stage. In the latter, conversion of the butenes to butadiene was effected. The reactors were designed so that the heat required for the endothermic dehydrogenation reaction was furnished by burning the carbon formed on the catalyst. Sufficient heat-storage capacity was built into the reactors to prevent wide fluctuations in temperature during the dehydrogenation and oxidation phases of the process cycle. Pressures and flow rates were controlled to balance the heat evolved during catalyst regeneration with the heat required during the on-stream period. Butadiene yields of about 70 per cent on n-butane have been revealed, with 9.5 per cent of the feed converted to carbonaceous catalyst deposits. A catalyst life of 6 months was assumed in plant design.

Phillips Process. A flow diagram of the Phillips n-butane dehydrogenation process as installed in the Borger, Texas, plant has been published.[32] A more complete description of the integrated process is given in the patent literature.[52] Dehydrogenation is accomplished in two stages. In the first of these, n-butane is partially converted to butene-1, *cis*- and *trans*-butene-2. After separation of the gases boiling below the C_4 range, butene-1 is separated from the higher-boiling butenes-2 and n-butane by fractionation. The resulting mixture of butene-2 and n-butane is then subjected to solvent extraction or extractive distillation to recover the n-butane and butenes-2 separately. Aqueous furfural is described as a very satisfactory solvent for effecting this separation by extractive distillation. The Phillips Company also developed a butadiene purification process, using this solvent. This is described in more detail below, together with a short discussion of the principles of extractive distillation.

Butene-1 and butenes-2 recovered from the first-stage products as just described are mixed and fed to the second dehydrogenation stage, in which they are partially converted to butadiene. Catalytic dehydrogenation of n-butenes is discussed in considerable detail in subsequent paragraphs. Processing of the second-stage products includes removal of materials boiling both above and below the C_4 fraction, followed by fractionation of the latter to yield a lower-boiling butadiene-butene-1 mixture and a higher-boiling butene-2 concentrate, which is recycled to the second stage of dehydrogenation. Butadiene is then separated from butene-1 by extractive distillation with aqueous furfural (see below) and the butene-1 recycled to the second stage. n-Butane accumulating in the butene-2 recycle to the second stage may be removed by returning a portion of this stream to the extractive-distillation unit operating on the butane-butene-2 product from the first stage. n-Butane, after separation from butenes-2 and -1, is recycled to first-stage dehydrogenation. Isobutylene tends to accumulate in the recycle streams and may be removed selectively by polymerization with dilute sulfuric acid or a silica-alumina catalyst.

Pilot-plant studies of the first dehydrogenation stage (butane conversion stage) have been published along with results of special investigations conducted in the commercial plant.[31-2] Chromia-alumina and bauxite catalysts were studied. The former proved to be more satisfactory. One of the chromia-alumina catalysts investigated in considerable detail was employed in early commercial plant operations.

BUTENE DEHYDROGENATION

Fluid Catalytic Cracking. The introduction of the fluid catalytic cracking of oil, a process developed by the Standard Oil Development Company, played a very important role in the butadiene program. This process produces large quantities of C_4 olefins, including the n-butenes needed for one-step dehydrogenation to butadiene. The yield of butenes obtained in catalytic cracking increases as the temperature of cracking is raised, and the older methods of cracking were limited in cracking temperatures attainable and hence in butene yields.

In the fluid catalytic-cracking process, the catalyst, in powdered state, is maintained in a freely flowing condition and can be handled in much the same manner as a liquid. The operating equipment fulfills two basic functions in two separate primary zones. One of these functions is the contacting of oil vapors with the hot catalyst to effect cracking. The other is the treatment of the carbon-containing catalyst with an oxygen-containing gas to effect catalyst revivification. Each primary zone contains a catalyst standpipe which provides for gravity flow of the catalyst, a reaction vessel, and a system for separating the catalyst dust from the reactor effluent gases. In the cracking zone, oil vapors are contacted with regenerated catalyst flowing from a standpipe. A fluidlike suspension of the catalyst in the oil vapors forms in the reactor proper. Catalyst is continuously withdrawn from this reaction zone, separated from oil vapors, and transported to the revivification standpipe. At the bottom of the standpipe, the catalyst is injected into an air stream which flows to the revivification reactor, in which a fluidlike suspension of catalyst is also maintained. Catalyst is withdrawn continuously from the revivification reactor and returned to the cracking-zone standpipe, thereby completing the catalyst flow circuit.

An earlier catalytic-cracking process, the Houdry type, employs a multi-reactor system with the catalyst in stationary beds. Cracking and revivification are effected alternately in a given reactor, continuous flow of oil vapors being provided for by switching oil feed in a predetermined schedule from one reactor to another. Another catalytic-cracking process known as the Thermofor process provides separate cracking and revivification zones but circulates coarse catalyst between these by means of mechanical conveyors.

C_4 Cut from Catalytic Cracking. The yield of C_4 cut from the catalytic cracking of gas oils varies considerably with the cracking conditions, and especially with the temperature in the cracking zone. However, yields amounting to as much as 10 to 20 volume per cent on feed are quite generally obtained. The C_4 cut obtained from catalytic cracking is a mixture of n- and isobutanes, isobutylene, butene-1 and cis- and trans-butene-2. The ratios of these vary considerably with cracking conditions, especially with temperature, but the following analysis may be considered fairly typical:

Isobutane	26.0 to 38.5 mole per cent
n-Butane	6.5 to 9.5
Isobutylene	13–19
Butene-1	9–12
Butenes-2	27–36

The n-butenes are of value for the production of aviation alkylate as well as for butadiene production. The former is predominantly a mixture of highly branched octanes made by reacting the butenes with isobutane in the presence of strong acid catalysts, such as sulfuric and hydrofluoric acids.

In general, the C_4 fraction from catalytic cracking must be further processed to recover the n-butenes in a sufficiently concentrated form for feeding to catalytic dehydrogenation, for the purpose of manufacturing butadiene. The first step is to remove isobutylene by extraction with sulfuric acid (see Chapter 24). After removal of the isobutylene, the C_4 cut is processed to separate the n-butenes from the butanes. When C_4 cuts derived from catalytic cracking are used, this separation is quite generally carried out by extractive distillation with solvents such as aqueous acetone or aqueous furfural. This type of fractionation is discussed in more detail below under the section on Butadiene Purification. Details of some installed operations for concentrating n-butenes are also given there.

As indicated earlier, some dehydrogenation plants employed n-butenes derived from C_4 fractions produced by thermal cracking of gas oils or naphthas. In these instances, somewhat less elaborate processing of the C_4 fraction was necessary to obtain the n-butene concentrate needed for dehydrogenation plant feed.

Development of Standard Oil Development Process. A general review of the Standard Oil Development Company butene-dehydrogenation process employed for large-scale production of butadiene has been given by Russell, Murphree, and Asbury.[54] A description of the pilot-plant work affording design data for the large-scale plants has been given by Nicholson, Moise, Segura, and Kleiber.[47] As shown in Table I, seven plants, six in the United States and one in Canada, were constructed to produce butadiene by butene dehydrogenation. The total rated capacity of these plants was 300,000 short tons per year of butadiene.

The chief problems encountered in the evolution of this process were twofold in nature, viz.: (1) design of a suitable reactor, and (2) development of a catalyst capable of effecting selective dehydrogenation in the presence of a suitable diluent such as steam.

As pointed out in the Historical Section, there had been considerable interest in the oil industry in the catalytic dehydrogenation of C_4 hydrocarbons since the early 1930's. Reasonably satisfactory dehydrogenation to butenes had been obtained with a chromium oxide-alumina catalyst on the butanes at atmospheric pressure and temperatures around 1100° F. In order to convert butenes to butadiene, however, it is necessary to use subatmospheric hydrocarbon partial pressures at somewhat higher temperatures (1150 to 1240° F.). At these temperatures butadiene has a marked tendency to polymerize and cover catalytic surfaces with carbonaceous deposits. These degradation reactions are undesirable for two reasons: (1) They cause product losses, and (2) they deactivate the catalyst. Such reactions can be minimized, however, by the use of low residence times (few tenths of a second) and low hydrocarbon partial pressures. The latter are also desirable because of the thermodynamics of the dehydrogenation equilibrium, since they allow attainment of practical butene conversions at temperature levels below those producing excessive thermal cracking.

The butene dehydrogenation reaction is markedly endothermic, the heat of reaction amounting to about 22,500 calories per gram mole of butene converted to butadiene. Supplying this large amount of heat to the reactor was a problem of major consideration in reactor design. Another problem was of course the periodic revivification of the catalyst by removal of the carbonaceous deposits formed during butene reaction.

Early in the work it was recognized that the use of steam as a diluent to obtain low n-butene partial pressures would have a number of practical advantages. Necessary reactor temperatures could be maintained and the endothermic heat of reaction supplied by mixing the n-butene feed with superheated steam. In contrast to most other diluents which might conceivably be used, steam could be relatively easily removed from the product gases by simple cooling and condensation. Also, by virtue of the water-gas reaction, steam either would tend to keep the catalyst free of carbonaceous deposits or could be used to remove them from the catalyst, thereby avoiding the necessity for burning deposits off with oxygen. Practical realization of these potential advantages of steam dilution was not, however, possible with chromium oxide catalysts, since these perform very poorly in the presence of large amounts of diluent steam. The discovery of steam-insensitive catalysts of the 1707 type (described below) obviated these difficulties and gave impetus to full-scale development of the steam diluent process.

Outline of Standard Oil Development Process. The dehydrogenation reactors, as finally designed for the process based on the use of 1707 catalyst and steam diluent, are about 16 ft. in diameter and contain a fixed bed of catalyst 1.5 to 6.0 ft. in depth. The butene feed is preheated to about 1100° F. and is rapidly and thoroughly mixed at the top of the reactor with steam preheated to 1300° F. or even higher. Mixing is accomplished in jet-type mixers designed to give a minimum of gas swirl in the reactor. Products leaving the catalyst bed are rapidly quenched, i.e., cooled below 1000° F., by injecting water directly in the vapor stream. With catalysts requiring regeneration, reactor construction includes the necessary valving to allow periodic interruption of feeding the butene stream to the reactor.

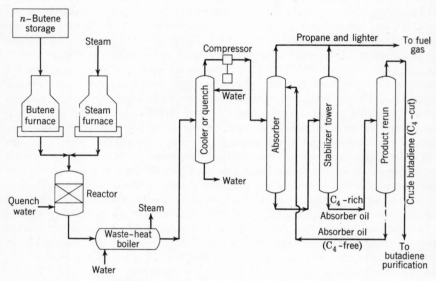

Fig. 1.   Simplified Flow Diagram of n-Butene Dehydrogenation Process

Representative operating data are given in the article by Nicholson and co-workers mentioned above, although actual operations in the different plants have varied somewhat. Usually catalyst-bed temperatures are maintained initially at about 1100° F., during both reaction and regeneration portions of the cycle, and are gradually raised to about 1250° F. as the catalyst becomes spent. Regeneration is carried out at slightly superatmospheric pressures (7 to 50 p.s.i.(g.)); the reaction phase of the cycle is essentially at atmospheric pressure. Duration of the reaction and regeneration periods can be varied somewhat, a schedule of one hour for each being satisfactory. Molal ratios of steam to n-butenes in the range 10 : 1 to 20 : 1 are suitable, although better selectivities and lower rates of accumulation of carbonaceous materials on the catalyst are experienced at the higher dilution levels. Practical n-butene space velocities are in the range of 200 to 500 volumes of hydrocarbon gas (at approximately standard conditions) per volume of catalyst per hour.

Figure 1 shows in simplified form the basic processing steps involved in n-butene dehydrogenation. Two of the larger dehydrogenation plants have been briefly described by Kunkel[40] and by Van Antwerpen.[61] A summary of the early operations of those located at Baton Rouge, La., and at Baytown, Texas, has been given by Russell, Murphree, and Asbury.[54] Operations of these plants have been eminently successful with only minor operating difficulties encountered. Some operating experience was necessary to obtain high selectivities and reasonably long catalyst lives. In some cases, nominal design capacities were exceeded relatively early in the operation.

Catalysts for Butene Dehydrogenation. The 1707 catalyst used initially in all the *n*-butene dehydrogenation plants has been described by Kearby.[37-8] Additional background is given in a patent to Sumerford.[58] The composition of this contact agent on a weight basis is 72.4 per cent magnesium oxide, 4.6 cupric oxide, 18.4 ferric oxide, and 4.6 potassium oxide. The potassium oxide component functions primarily to accelerate the water-gas reaction. It is gradually lost from the catalyst during use but can be replenished by adding a suitable potassium compound to the feed gases. Conversion products other than butadiene and hydrogen are oxides of carbon, methane, ethane, propane, and the lower olefins, ethylene and propylene. Carbon converted to oxides during the reaction is ordinarily 10 to 30 times greater than that deposited on the catalyst. Paraffins, such as *n*-butane, and also isobutylene are not appreciably converted by the catalyst. For this reason, these compounds accumulate in the recycle *n*-butene stream and must be periodically removed therefrom. The catalyst effectively shifts the double bond in the *n*-butene structure and hence gives the same yields from butene-1 and butenes-2. Selectivity to butadiene decreases with increasing *n*-butene conversion, ranging under fairly typical conditions from 85 mole per cent at 20 per cent conversion to 65 mole per cent at 50 per cent conversion. Very satisfactory catalyst life has been realized in plant operations—at least six months in a number of instances.

The components of the catalyst may be altered considerably without destroying dehydrogenating power. Numerous modified contact agents have been described in the patent literature. One of these, an unsupported 1707-type material (known as catalyst 105) containing ferric oxide, chromium oxide, and potassium oxide later came into extensive use in the dehydrogenation plants. This catalyst shows a long life and, at high ratios of steam to *n*-butene, operates without the need for frequent regeneration.

Beckberger and Watson[5] have studied the kinetics of *n*-butene dehydrogenation over the 1707 catalyst. They found the rate of the primary catalytic dehydrogenation reaction to be consistent with a mechanism involving the reaction of an adsorbed *n*-butene molecule with an adjacent vacant active center. They also found the more important secondary reactions, such as coke and water-gas formation, to originate from catalytic cracking of butadiene.

More recently, a catalyst of greatly improved selectivity has been developed by the Dow Chemical Company. This material has been described by Britton, Dietzler, and Noddings[10] as a chromium oxide-stabilized calcium-nickel phosphate of the approximate formula $Ca_8Ni(PO_4)_6$. In pilot-plant operation this catalyst gave selectivities to butadiene of 93 to 97 mole per cent at *n*-butene conversions ranging from 45 to 20 per cent. Butadiene selectivities in large-scale operation are said to be 86 to 88 mole per cent at 35 per cent conversion. Catalyst life is apparently quite long, exceeding six months. Steam to butene ratios around 20 : 1 are used during reaction. Periodic regeneration with a mixture of steam and air is necessary. Cycle times can be varied somewhat, but the pilot-plant work employed reaction and regeneration times of 0.5 to 1.0 hour each. n-Butene space velocities for reasonable conversions are in the range of 100 to 200 V/V/hour. Regeneration temperatures must be maintained below 1250° F., to prevent catalyst deactivation. Temperatures during n-butene reaction are in the range of 980 to 1200° F. The catalyst is said to be sensitive to nickel compounds and to alkali and alkaline earth oxides but is not poisoned by iron compounds, ammonia, acetone, and organic chlorides. The pilot-plant work referred to above established the feasibilty of using this contact agent in fixed beds of the type designed for the 1707 catalyst.

Reilly[52a] has summarized plant experience with the Dow catalyst, type B. High selectivities have been obtained, even at moderately high conversion levels, e.g., 90.8 mole per cent selectivity at 33.2 per cent conversion per pass. One disadvantage reported, however, is the production of considerable quantities of ketones—about 4 per cent (by weight, calculated as acetone) on the butadiene product. These ketones must be removed from the crude butadiene by water washing before fractionation. Plant experience with the 105 catalyst mentioned above is also given. Although this catalyst operates with less frequent regeneration than the Dow catalyst, it has the disadvantage of giving lower conversions and selectivities. Reilly gives diagrams showing the construction of the butene-dehydrogenation catalytic reactor and the gas flow to a pair of automatically switched reactors.

PURIFICATION OF BUTADIENE

In all the major processes for producing butadiene from petroleum, the products from the hydrocarbon conversion stage contain butadiene mixed with other closely boiling hydrocarbons. Fractionation alone proved to be incapable of separating butadiene of the desired purity from these other compounds. Table II lists closely boiling hydrocarbons that have been identified in crude butadiene fractions. The relative amounts of these hydrocarbons present in the crude butadiene depend on a number of factors, especially the type and temperature of the primary hydrocarbon conversion steps. Crude products from thermal cracking and butene dehydrogenation processes consist principally of unsaturated components with the monoölefins predominating. Nonconjugated dienes and acetylenes are ordinarily minor components except in the case of products from thermal cracking in the higher temperature range (above about 1350° F.).

It appears from the literature that a number of methods for purifying butadiene were considered in the early stages of the rubber program. The more important ones are discussed in the following sections.

Azeotropic Distillation of Crude Four-Carbon Hydrocarbon Fractions with Compounds Such as Ammonia. Azeotropic distillation was used in a small plant privately owned by Dow Chemical Company, to purify butadiene from oil cracking.[51] The azeotropic distillation column operated to recover the butenes-ammonia azeotropes as overhead. The tower bottoms was a mixture of butadiene and acetylenes. The latter were removed by chemical treatment, and pure butadiene was recovered by a final distillation.

Table II.   Compounds Present in Crude Butadiene from Petroleum Sources

Compound	Formula	Boiling Point, ° C. at 760 mm.
Propylene	$CH_3CH : CH_2$	—47.6
Propane	$CH_3CH_2CH_3$	—42.1
Propadiene (allene)	$CH_2 : C : CH_2$	—34.3
Methylacetylene	$CH_3C : CH$	—23.2
Isobutane	$(CH_3)_2CHCH_3$	—11.72
Isobutylene	$(CH_3)_2C : CH_2$	—6.93
1-Butene	$CH_2 : CHCH_2CH_3$	—6.32
1,3-Butadiene	$CH_2 : CHCH : CH_2$	—4.54
n-Butane	$CH_3CH_2CH_2CH_3$	—0.55
trans-2-Butene	$CH_3CH : CHCH_3$	0.86
cis-2-Butene	$CH_3CH : CHCH_3$	3.64
Vinylacetylene	$CH_2 : CHC : CH$	5.0
Ethylacetylene	$CH_3CH_2C : CH$	8.6
1,2-Butadiene (methylallene)	$CH_2 : C : CHCH_3$	10.3
1,3-Butadiyne (diacetylene)	$CH : C \cdot C : CH$	10.3
3-Methyl-1-butene	$CH_2 : CHCH(CH_3)_2$	18.8
1,4-Pentadiene	$CH_2 : CHCH_2CH : CH_2$	26.12
Dimethylacetylene	$CH_3 \cdot C : C \cdot CH_3$	27.1

Reversible Reaction of Butadiene with Sulfur Dioxide to Form a Cyclic Sulfone. The reaction involved is as follows:

$$CH_2 : CH \cdot CH : CH_2 + SO_2 \rightleftarrows \begin{array}{c} CH = CH \\ | \qquad | \\ H_2C \qquad CH_2 \\ \diagdown \quad \diagup \\ SO_2 \end{array}$$

The sulfone is formed by heating the components at elevated pressures. Unreacted hydrocarbons are removed by distillation, and the sulfone is decomposed at higher temperatures. Both selective solvents and chemical treatment were considered for separating sulfur dioxide from the butadiene-sulfur dioxide mixture resulting from decomposition of the sulfone. So far as is known, this method never came into commercial use. However, it was used extensively in the laboratory to obtain highly purified samples of butadiene and other dienes from crude hydrocarbon fractions. The reaction is quite selective for dienes and yields products of high purity.

Extractive Distillation with Selective Solvents. In extractive-distillation processes, a solvent is added to the distillation column, generally at a point close to the top of the column, in sufficient quantities to alter appreciably, from their normal values, the relative volatilities of the components of the mixture being distilled. Besides effecting the necessary changes in relative volatilities, a suitable solvent must be relatively stable and nonreactive toward butadiene and other unsaturated hydrocarbons. It should also be relatively nonvolatile compared to the hydrocarbons in the mixture to be separated. For a more detailed description of the principles and techniques involved in extractive distillation, the reader is referred to general articles on this subject.[6, 56, 65]

Extractive distillation is used extensively in petroleum butadiene manufacture. Table IIA lists the commercial plants constructed to use extractive distillation with aqueous furfural to purify butadiene. Not only is butadiene effectively purified by this technique, but also other C_4 hydrocarbons are separated in concentrated form for use in processing. This is especially true of the n-butenes, which are readily separated from the butanes by this method. Usually extractive distillation is used in combination with one or

Table IIA. Plants Constructed to Employ Extractive Distillation with Aqueous Furfural for Butadiene Purification

Operating Company	Plant Location	Hydrocarbon Conversion Process	Original Rated Annual Capacity, short tons a year
Neches Butane Products Co.	Port Neches, Texas	Butene dehydrogenation	100,000
Sinclair Rubber Co.	Houston, Texas	Butene dehydrogenation	50,000
Phillips Petroleum Co.	Borger, Texas	Butane dehydrogenation	45,000

more conventional fractionation steps. The latter produce relatively narrow-boiling fractions containing only a few of the many compounds normally present in the entire C_4 fraction. Thanks to the nature of the extractive-distillation process, these narrow-boiling fractions are easier to resolve into the desired products. For instance, one large plant (operated by Neches Butane Products Company) employs extractive distillation to process three different narrow-boiling portions of C_4 fractions.[33] (a) Separation of butene-1 from a mixture with isobutane, (b) separation of butenes-2 from a mixture with n-butane, (c) separation of butadiene from a mixture with butene-1.

Operations a and b are primarily for the preparation of n-butenes for dehydrogenation, and operation c is for butadiene purification. Furfural containing 4 to 6 per cent water is employed in this and a number of other plants for effecting the separations described.[11] This solvent has the general effect of increasing the volatilities of the more saturated hydrocarbons relative to those of the more unsaturated ones. This is shown in the following tabulation, in which the C_4 hydrocarbons are arranged in order of decreasing volatility, both with and without the presence of solvent.

Without Solvent	With Solvent
Isobutane	Isobutane
Isobutylene	n-Butane
Butene-1	Isobutylene
Butadiene	Butene-1
n-Butane	trans-Butene-2
trans-Butene-2	cis-Butene-2
cis-Butene-2	Butadiene

As in regular fractionation, the more volatile components of a mixture undergoing separation are taken overhead, and the less volatile ones are recovered at the tower bottom dissolved in the extractive-distillation solvent. The less volatile hydrocarbons are recovered from the solvent in a subsequent stripping operation, and the essentially hydrocarbon-free solvent is recycled to the extractive-distillation tower. When purifying butadiene from the butadiene–butene-1 mixture mentioned above, the butadiene is recovered as a solution in the aqueous furfural. The butadiene recovered from the solvent generally contains small amounts of butenes-2, from which it is separated to the desired degree of purity by regular fractionation.

In general, the changes produced in relative volatilities increase as the solvent-to-hydrocarbon ratio in the extractive-distillation tower increases. This ratio can be varied, depending upon the difficulty of the separation and the desired purities of the recovered products. The published information indicates that the butadiene recovery column ordinarily operates with about 78 to 88 volume per cent solvent in the tray liquid. For the isobutane–butene-1 and n-butane–butenes-2 separations, the tray solvent concentrations are 60 to 88 and 71 to 88 per cent, respectively.

Some difficulties arose in early operations with the aqueous furfural solvent, owing to the formation of polymeric and acidic materials. These have been overcome, however, largely by continuously removing these impurities by distillation. The use of aqueous furfural for separating C_4 hydrocarbons originated with the Phillips Petroleum Company. During the war, as part of the synthetic-rubber production program, a plant was built at Memphis, Tenn., to produce furfural from corncobs.

Another solvent also used for separating C_4 hydrocarbons is aqueous acetone. A number of plants use this solvent for separating butanes from n-butenes, the latter being utilized in dehydrogenation. However, it has not been applied in butadiene purification proper. The use of aqueous acetone originated with the Shell Development Company. In plant operation, butane-butene mixtures are fed to approximately the middle of a fractionating tower, the aqueous acetone solvent being added at a point near the top of the tower. Pressure is maintained at about 100 p.s.i.(g.). The butanes are recovered overhead and the butenes-solvent mixture as tower bottoms. After stripping from the solvent, which is recycled to the extractive distillation solvent, the butenes are generally redistilled to obtain a suitable feedstock for dehydrogenation.

Selective Absorption with Cuprous Salt Solutions. Butadiene purification processes based on the use of cuprous salt solutions were developed by the Standard Oil Development Company. These processes were employed in eleven plants constructed during the wartime rubber program. These plants produced crude butadiene by a variety of catalytic and thermal processes. Table III lists these plants along with information on hydrocarbon conversion processes employed and original rated capacities.[43]

Table III. Plants Constructed to Employ Copper Solutions for Butadiene Purification

Operating Company	Plant Location	Hydrocarbon Conversion Process	Original Rated Annual Capacity, short tons per year
Cities Service Refining Co.	Lake Charles, La.	Butene dehydrogenation	55,000
Humble Oil and Refining Co.	Baytown, Texas	Butene dehydrogenation	30,000
Humble Oil and Refining Co.	Ingleside, Texas	Thermal cracking	12,500
Publicker Comm. Alcohol Co.	Philadelphia, Pa.	*	10,000
Shell Union Oil Corp.	Los Angeles, Calif.	Thermal cracking and butene dehydrogenation	55,000
St. Clair Processing Corp.	Sarnia, Ontario	Butene dehydrogenation	30,000
Standard Oil Co. of Calif.	El Segundo, Calif.	Butane dehydrogenation	18,000
Esso Standard Oil Co.†	Baton Rouge, La.	Thermal cracking	22,000
Esso Standard Oil Co.	Baton Rouge, La.	Butene dehydrogenation	15,000
Sun Oil Co.	Toledo, Ohio	Butane dehydrogenation	15,000

* Catalytic conversion of ethyl alcohol.
† Includes three separate plants, two of which are privately owned.

The tendency of both solid cuprous salts and cuprous salt solutions to form additional compounds with monoölefinic hydrocarbons was first observed by Manchot and Brandt[41] in 1909. A patent issued to Feiler[22] in 1931 showed that diolefins also form complexes with certain cuprous salts.

The solutions used in the butadiene plants contain cuprous acetate solubilized with ammonia and ammonium acetate. These solvents are noncorrosive toward carbon steel and are completely stable at temperatures as high as 225° F. In normal practice the cuprous copper concentration of the solution is maintained in the range of 3.0 to 3.5 moles per liter. Ammonia and ammonium acetate concentrations must also be carefully controlled to obtain satisfactory results.[44] Either copper oxides or cupric acetate may be used to prepare these solutions. Reduction of cupric compounds to the cuprous state is effected by treatment with metallic copper.

The solubility of a particular unsaturated hydrocarbon in a cuprous salt solution decreases with increasing temperature and increases with increasing partial pressure of the hydrocarbon. Though all unsaturated hydrocarbons

Table IV. Solubility of Mono- and Diolefins in Ammoniacal Cuprous Acetate Solution (3 M/L cuprous copper) at 32° F. and 0.5 Atm. Hydrocarbon Partial Pressure

Hydrocarbon	Type	Solubility, moles per liter
Propylene	C_3 monoölefin	0.06
Isobutylene	C_4 monoölefin	0.02
1-Butene	C_4 monoölefin	0.068
trans-2-Butene	C_4 monoölefin	0.013
cis-2-Butene	C_4 monoölefin	0.028
Propadiene	C_3 diolefin	1.35
1,3-Butadiene	C_4 diolefin	0.70
1,2-Butadience	C_4 diolefin	1.15

boiling close to butadiene dissolve to some extent, solubility generally increases with degree of unsaturation and decreases with increasing hydrocarbon molecular weight. Typical solubility data illustrating these effects are shown in Table IV.

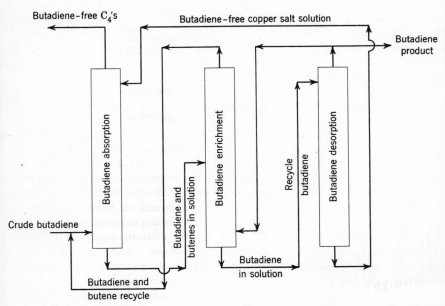

FIG. 2. Simplified Flow Diagram of Copper Salt Process for Purifying Butadiene

Plant operations using cuprous salt solvents vary somewhat in process details. However, all purification plants have the three following process steps, the relationship of which is shown in simplified form in Fig. 2: (*a*) an absorption step, in which butadiene together with smaller amounts of other unsaturated hydrocarbons are dissolved in the solvent; (*b*) an enrichment step, in which the solvent is stripped of essentially all unsaturated hydro-carbons other than butadiene; (c) a butadiene desorption step, in which this hydrocarbon is recovered from the solvent which is then returned to the absorption step.

The absorption step operates with countercurrent flow of the solvent and crude butadiene.   Equipment has been installed for absorbing butadiene from both gaseous and liquid C_4 fractions.   The enrichment step is based on the principle that a more soluble hydrocarbon such as butadiene tends to displace hydrocarbons of lower solubility such as the butenes.   For this reason, a portion of the butadiene product from the third or desorption stage is returned to the enrichment stage, generally in countercurrent flow to the solvent.   Enrichment is also aided by heating the solvent in this stage. Recovery of butadiene from the solvent in the third step is effected by heating. The butadiene vapors from the recovery step contain ammonia which is readily removed by water washing.

Acetylenes boiling close to butadiene, especially ethylacetylene and vinyl-acetylene, are completely extracted along with the butadiene by cuprous salt solutions.   These dissolved acetylenes are desorbed to a small extent along with butadiene in the third process step mentioned above.   However, for the most part they are polymerized in the solvent to higher-boiling materials which may be removed from the solvent by filtration and extraction with hydrocarbon solvents.   Polymerization of the acetylenes in this manner prevents accumulation of these in the solvent and hence prevents precipitation of copper acetylides.   In one plant using cuprous salt solutions for purifying butadiene, the acetylenes were removed from the crude butadiene by a selective hydrogenation reaction.[55]

PROPERTIES OF PETROLEUM BUTADIENE

Butadiene from petroleum has generally been produced at a purity of at least 98 per cent.   The nature of the impurities present has been a subject of considerable study, since they, for the most part, tend to build up in the recycle streams in the polymerization plants.[23]   The nature of the impurities present in butadiene purified by the copper salt process has been discussed in detail by Starr and Ratcliff.[57]   These impurities may be classified as follows: (a) hydrocarbons incompletely removed during the purification process and also hydrocarbons of higher-boiling points formed as a result of polymeriza-tion reactions, (b) oxygenated compounds such as carbonyl compounds and peroxides, (c) sulfur compounds, (d) water and other inorganic compounds and nonvolatile residues.

Because of the nature of the processes commonly employed for purifying petroleum butadiene, paraffins constitute only very small amounts of the impurities present.   Monoölefins, especially the butenes, constitute the bulk of the impurities, regardless of the concentration method employed.   For instance, the above authors report the monoölefin contents given in Table V for three samples of petroleum butadiene, all of which were above 98 per cent purity.   Although butenes are known to be the major contaminants in butadiene concentrated by procedures other than the copper salt method, it is to be expected that the distribution of them will vary somewhat with the nature of the purification process employed.

Table V.   Monoölefin Contents (Mole Per Cent) of Petroleum Butadiene
(Copper Salt Process)

Sample	A	B	C
Propylene	0.31	0.04	0.29
Isobutylene	0.27	0.11	0.14
1-Butene	0.62	1.07	0.92
2-Butenes	0.05	0.13	0.30
Total butenes	0.94	1.31	1.36
Pentenes	0.01	0.00	<0.01

Fortunately, the monoölefins are only very mild poisons in the polymerization reaction.   However, since they do not react extensively, they accumulate in the recycle streams in the rubber synthesis operations.

Nonconjugated dienes and acetylenes have been identified as contaminants in petroleum butadiene.   Generally speaking, all these hydrocarbons tend to retard the rubber polymerization reaction (see Chapter 21).   Hence, it has been necessary to control rather rigidly the amounts of these present. Among the nonconjugated dienes, 1,2-butadiene and 1,4-pentadiene are most objectionable from the standpoint of retarding polymerization. Propadiene has also been found in petroleum butadiene but appears to be less objectionable than the other two.   Starr and Ratcliff[57] report concentrations of 0.02 to 0.23, 0.06 to 0.24, and 0.00 to 0.02 mole per cent of propadiene, 1,2-butadiene, and 1,4-pentadiene, respectively, in samples of butadiene purified by the copper salt method.

Four acetylenic hydrocarbons, viz., methyl-, ethyl-, vinyl-, and dimethylacetylene, boil sufficiently close to butadiene-1,3 to be likely contaminants. Vinylacetylene is a powerful retardant in polymerization, but the others are considerably less objectionable.   General practice has been to produce butadiene of total acetylenes content below a specified maximum limit. Acetylene polymers are powerful poisons in the polymerization reaction but are generally excluded from product butadiene by fractionation.

Butadiene dimer (4-vinyl-1-cyclohexene) tends to accumulate in butadiene during storage.   The rate of dimerization of butadiene in the liquid phase has been studied by Robey, Wiese, and Morrell.[53]

ANALYSIS OF BUTADIENE

Considerable effort was expended during the Synthetic Rubber Program on the development of suitable analytical procedures for assaying butadiene. Methods for determining the nature and amounts of major and minor impurities were needed as well as improved methods for determining absolute purity, i.e., butadiene-1,3 content.   The necessary procedures were evolved as a result of an industry-wide effort coordinated by the Subcommittee on Butadiene Specifications and Methods of Analysis of the Butadiene Producers'

Technical Committee.   The procedures developed have been described in a form suitable for general use in a volume entitled *Light Hydrocarbon Analysis* edited by Burke, Starr, and Tuemmler.[13]

The industry-wide specifications established for polymerization butadiene, including that from petroleum sources, are listed in Table VI.

Table VI.   Specifications for Polymerization Grade 1,3-Butadiene

Specification	Limits
Sampling from containers	Must be sampled from liquid phase.
Appearance	Clear and free of entrained material.
Conjugated diene	98 wt. per cent, min.
Peroxides	10 p.p.m. max. as hydrogen peroxide
Carbonyl compounds	100 p.p.m. max. as acetaldehyde.
Acetylenes	0.10 wt. per cent max. as vinylacetylene.
Sulfur	100 p.p.m. max. as hydrogen sulfide.
C_5 Hydrocarbons—boiling-point rise	0.4° C. max. difference in boiling points between 2.0 and 0.5-ml. residual volumes.
Butadiene dimer	0.2 wt. per cent max.
Nonvolatile residue	0.1 wt. per cent max.
Inhibitor (*p-tert*-butyl catechol)	25–200 p.p.m.
Oxygen in vapor spaced in filled tank cars	0.3 vol. per cent max.

In addition to specification tests for the control of the quality of interplant shipments of butadiene, methods were developed for use in research work and also for the control of the compositions of streams in the polymer and butadiene-producing operations.   The methods finally developed are diverse in nature and are based not only on a variety of selective chemical reactions but also on measurements of a number of physical properties. Application of spectroscopic techniques, especially the use of infrared- and ultraviolet-absorption spectrometers, proved to be a very fruitful approach. To illustrate this point, Table VII lists the physical and chemical principles used as the bases for a number of the more important methods for analyzing purified butadiene and mixtures of this diolefin with other hydrocarbons. Generally speaking, the available analytical methods have proved to be sufficiently accurate and comprehensive to afford good control of product quality and process operations.

HANDLING OF BUTADIENE

The nature and rates of the reactions that butadiene undergoes during storage, transportation, and plant processes have been described by Robey, Wiese, and Morrell.[53]   Two mutually independent reactions were found to predominate, viz.: (1) dimerization to 4-vinyl-1-cyclohexene and (2) peroxide-catalyzed polymerization to plastic materials of high molecular weight. The first reaction is bimolecular and homogeneous and is unaffected by peroxides, steel surfaces, and antioxidants.   The second reaction proceeds at a rate directly proportional to the square root of the peroxide concentration

and is strongly inhibited by antioxidants.   Rates of both reactions increase
with increasing temperature.   The energies of activation appear to be almost

Table VII.   Methods for the Analysis of Butadiene

Property Measured	Principle of Measurement
Concentration of C_1-C_3 hydrocarbons	(a) Low-temperature distillation
	(b) Mass spectrometer
Total unsaturates in C_4 fractions	(a) Absorption in acid silver-mercuric nitrate solutions
	(b) Absorption in sulfuric acid
	(c) Catalytic hydrogenation
Isobutylene	(a) Absorption in 65 per cent sulfuric acid
	(b) Reaction with hydrogen chloride
	(c) Infrared absorption
	(d) Reaction with mercuric nitrate
n-Butenes	(a) Absorption in acid mercuric sulfate
	(b) Infrared absorption
1,3-Butadiene content	(a) Absorption in molten maleic anhydride
	(b) Freezing point
	(c) Ultraviolet absorption
	(d) Infrared absorption
	(e) Mass spectrometry
C_5 content	Cottrell boiling-point apparatus
Nonvolatile content	Controlled evaporation
Inhibitor (p-$tert$-butylcatechol) content	Light absorption by $FeCl_3$-inhibitor complex
Ammonia content	Acid-base titration
Total sulfur content	Turbidimetry ($BaSO_4$) on combustion products
Hydrogen sulfide and mercaptan contents	Reaction with cadmium sulfate—iodimetry
Chlorine content	Combustion and mercuric nitrate titration
Water content	Pyridine-iodine reagent
Peroxide content	Oxidimetry (ferrous-titanous system)
Carbonyl content	Hydroxylamine hydrochloride reagent
Acetaldehyde	Modified Nessler reagent
Oxygen content	Reaction with manganous hydroxide
Total acetylenes content	Ketalization and reaction of ketone with hydroxylamine hydrochloride
α-Acetylene content	(a) Reaction with alcoholic silver nitrate
	(b) Infrared absorption
Butadiene dimer (4-vinyl-1-cyclohexene) content	Bromination
Styrene content	Ultraviolet spectrometry
Cyclopentadiene content	Photometry on dimethylfulvene (acetone condensation product)
Isoprene content	Photometry on mercuric acetate reaction product

identical.   These authors also studied the action of oxygen on butadiene and
the nature of the resulting peroxides.   At temperatures ordinarily encountered
in storage and transportation, peroxides are formed as a result of contact of

butadiene with air. The presence of a phenolic antioxidant is necessary to prevent peroxide accumulation.

The distillation of butadiene containing small amounts of peroxides is accompanied by some hazards unless suitable precautions are taken. Because of their relatively lower volatility, peroxidic substances and decomposition products of these can accumulate in still bottoms in sufficient concentrations that explosions occur. These hazards can be eliminated, however, by distilling in the presence of a suitable antioxidant or by adding a high-boiling material to dilute the still bottoms.

Under certain conditions, especially in manufacturing operations, a so-called popcorn polymer of butadiene can be formed.[63] This material is an insoluble, infusible granular mass which shows a pronounced tendency to grow rapidly when in contact with butadiene. It has been established[27] that formation of this polymer is initiated by peroxides or other free radical-producing agents in butadiene under the accelerating influence of rusty iron and water. The phenomenon of growth of the popcorn polymer is attributed to the presence of peroxide linkages in the polymer network. The growth rate is not appreciably affected by ordinary antioxidants but is retarded by iodine, nitric oxide, vulcanization accelerators, and also by large amounts of air and soluble organic peroxides. Although popcorn polymer was frequently encountered in polymerization processing and at times became a nuisance, occurrence in petroleum butadiene production operations has fortunately been rather infrequent. Some troubles arising from this source were encountered in early operations of distillation towers at relatively high temperatures and pressures. For the most part, however, these were eliminated by modifying operating conditions.

The general experience in the handling and transportation of butadiene on a commercial scale has been discussed by Boyd.[9] Structural strengths necessary in storage and handling equipment have also been discussed.[3] Except for certain special precautions necessitated by the unsaturated character of this hydrocarbon, practices in bulk transportation of butadiene are essentially the same as those for other liquefied petroleum gases. Procedures for handling all tank cars, pipe lines, and storage equipment specify very low values for permissible oxygen contents. Recognizing that complete exclusion of oxygen from such systems is not possible, it is general practice to add an antioxidant, p-tert-butylcatechol, to all commercial butadiene. The presence of entrained water in butadiene is undesirable since it can cause freezing of lines and valves in cold weather. As a precautionary measure, small amounts of an antifreeze are generally added when loading tank cars. During the early stages of commercial development there was some fear that butadiene containers might be ruptured by high pressures developed as a result of exothermic, autocatalyzed polymerization reactions. This has never occurred, however, under conditions of commercial practice.

RECENT STATUS OF PLANTS AND ECONOMICS

The degree of success attained in the operation of the petroleum butadiene plants described above has varied considerably. Since the termination of

World War II a number of those operated for the Government have either ceased operation or have operated only on a part-time basis. Difficulties encountered in early operations of some of these were eliminated by appropriate changes in operations or by installation of additional equipment.

The three smaller plants employing thermal cracking in converted refinery equipment were shut down[49] in 1945. Another of this type has been transferred to private ownership. The large regenerative stove cracking plant at Los Angeles has subsequently ceased operation. For the most part, operation of these plants during the war was successful, considering the fact that they were not specifically designed for butadiene production. These five naphtha-cracking plants produced large quantities of olefins and aromatics which found use in the war effort; in fact, the recovery and utilization of products other than butadiene is necessary for this process to compete economically with dehydrogenation processes.

The plants employing the butene-dehydrogenation process have generally operated very successfully. In fact, some of these attained, and even notably exceeded, design production rates shortly after starting operations. During 1945, monthly production rates of those not limited by feed-stock availability reached values equivalent to 120 to 155 per cent of design capacity. Improvements resulted from general operating experience and from the use of new catalysts developed since the plants were placed in operation. Not all the butene-dehydrogenation plants have operated either continually or at full capacity since the war. Operations scheduled at the individual plants during this period have been largely influenced by over-all demands for butadiene and by the local availability of feed stocks. For instance,[8] in 1950, only three of the larger butene-dehydrogenation plants, viz., those operated by Cities Service Refining Corporation, Humble Oil and Refining Company, and Neches Butane Products Company, were in production. These together with the n-butane-dehydrogenation plant operated by Phillips Petroleum Company had at that time demonstrated a maximum annual productive capacity of 315,000 tons of purified butadiene as compared to an initial rated capacity of 241,000 tons.

The n-butane-dehydrogenation plants encountered numerous difficulties during initial operations. In fact, none of them was able to meet designed production rates during 1945. Since that time, however, operation of the larger of these, at Borger, Texas, has been substantially bettered by the use of an improved catalyst and by equipment additions.

Gilliland and Lavender[26] have summarized the original investments in the Government butadiene plants (Table VIII). These investment figures are probably mostly of historical interest, as they do not reflect technological advances since the start of the rubber program. Neither do they take into account the great increases in construction costs during the postwar era.

A detailed discussion of the over-all economics of butadiene production would be too complicated an undertaking to attempt here. The competitive status of the various petroleum processes depends upon a great number of factors. Among these are not only original investment costs but also such factors as feed-stock availability and costs and by-product values. It has been stated[8] that under proper conditions butane dehydrogenation can compete with the butylene route in spite of the lower investment for the latter as reflected in the figures in the table.   Also, in spite of the lower butadiene yields from the thermal-cracking processes, a number of privately owned

Table VIII

Investment,
Dollars per Annual Ton of Purified Butadiene

Type of Plant	Rated Capacity		Ultimate Capacity, Estimated in 1944	
	Average	Range	Average	Range
Alcohol	530	491–595	294	273–330
Butylene dehydrogenation	541	309–647	422	436–480
Naphtha and gas-oil cracking	393	294–586	553	294–1310
Butane dehydrogenation				
Permanent	...	778	778	...
Quickie	...	506	455	...

plants are carrying out such operations.   These plants, however, obtain substantial credits from other products such as olefins, aromatics, and other diolefins.

ISOPRENE AND OTHER DIOLEFINS

The literature records a number of chemical processes for making isoprene. The pyrolysis of terpene hydrocarbons, such as β-pinene and dipentene, gives significant yields of this diolefin.   Isoprene can also be made by the catalytic dehydrogenation of branched-chain pentenes, such as 2-methyl butenes-1 and -2.   Selectivities as high as 75 mole per cent to isoprene have been recorded using dehydrogenation catalysts of the older types, i.e., those not incorporating the recent improvements discussed previously under butadiene manufacture.   Thermal cracking of selected olefin fractions, e.g., polymers from the phosphoric acid-catalyzed polymerization of propylene and butenes, also yields isoprene, but these polymers are valuable gasoline components.

A number of other methods, employing such relatively well-known chemical reactions as aldol condensation, chlorination–dehydrochlorination, and addition of aldehydes to olefins, have been patented and described in the scientific literature.   Among these may be mentioned the condensation of formaldehyde and isobutylene, the condensation of formaldehyde and methyl ethyl ketone, and the chlorination–dehydrochlorination of branched-chain amyl alcohols and pentenes.   A review of such methods is given in Chapter 5.

Early in the synthetic-rubber program isoprene was produced by Newport Industries, Inc., by cracking terpene hydrocarbons recovered in the naval stores industry, as is described in Chapter 5.   Otherwise, so far as is known, none of the above-mentioned processes has been used for the commercial production of isoprene.

Isoprene used in Butyl-rubber manufacture and for the production of specialty emulsion polymers developed in recent years is obtained by the thermal cracking of gas oils and naphthas. The cracking processes employed are the same as those discussed under thermal cracking for butadiene, and hence isoprene is made today as a coproduct with butadiene.

Another conjugated diolefin recovered from thermal-cracking operations of this type is the cyclic compound cyclopentadiene. This hydrocarbon has a very pronounced tendency to dimerize and consequently is transported and marketed as the dimer, dicyclopentadiene. Cyclopentadiene is not used in synthetic-rubber manufacture but has attained considerable commercial importance as a raw material for insectide and resin manufacture and as a modifying agent for drying oils. Other conjugated diolefins, such as the piperylenes (cis- and trans-) and also those both acyclic and cyclic, boiling above the C_5 range, are present in the products from gas oil and naphtha cracking. However, none of these has yet attained commercial usefulness, either in synthetic rubber or elsewhere. Recovery of these hydrocarbons may well become commercialized in the future as uses for them develop.

The C_5 fraction as recovered from the products of thermal cracking contains both isoprene and piperylenes as well as large amounts of C_5 olefins. Separation of the isoprene from the other components of the C_5 fraction can be effected in a number of ways, e.g., sulfone formation and decomposition, extraction with cuprous salt solutions, and suitable combinations of conventional and extractive distillations. The large-scale installation for recovering isoprene for Butyl rubber is located at Baton Rouge, La., and is operated by Esso Standard Oil Company (Louisiana Division). It employs extractive distillation with an aqueous acetone solvent to recover isoprene from a crude feed comprising the lower-boiling portion of the total C_5 fraction. Enriched isoprene together with small amounts of other C_5 diolefins present in the crude feed are recovered in the bottoms from the extractive distillation, the monoölefins and paraffins passing overhead. Piperylene and cyclopentadiene are removed from the isoprene in a final distillation operation.

Specifications for Isoprene. Specifications on product isoprene for Butyl rubber manufacture are given in Table VIII. In a number of respects these specifications parallel quite closely those for butadiene for GR-S manufacture. Peroxides, sulfur, carbonyls, and oxygen in the gas phase are controlled to the specified levels in order to prevent deterioration reactions during storage and to minimize interferences in the Butyl-rubber polymerization reaction. Acetylene concentrations specified are larger than those for GR-S feed stocks. The specifications also allow for 1 to 2 per cent of the other C_5 diolefins. An antioxidant inhibitor is added to the purified isoprene to prevent deterioration reactions such as oxidation and polymerization. As for the latter, isoprene is quite similar to butadiene, generally showing no great difference from it in stability during processing and storage.

With a few exceptions, the analytical procedures used for determining impurities in commercial isoprene are quite similar to those for butadiene. Isoprene must, of course, be differentiated from the other C_5 diolefins, viz., cyclopentadiene and piperylene.   This can be accomplished in a number of ways, including the use of optical methods and freezing-point determination.

Table VIII.   Specifications for Isoprene for Butyl-Rubber Manufacture

Specification	Limits	
	Maximum	Minimum
Total diolefins, wt. per cent		95.0
Isoprene, wt. per cent		92.0
α-Acetylenes (as C_5 acetylene), wt. per cent	0.35	
Cyclopentadiene, wt. per cent	1.0	
Peroxides (as hydrogen peroxide), p.p.m.	10	
Sulfur as S, p.p.m.	100	
Carbonyl content (as acetone), p.p.m.	500	
Nonvolatile material at 60° C. wt. per cent	1.5	
Inhibitor, p.p.m.		50
Appearance	Clear and free of entrained material	
Oxygen in vapor space in filled tank cars, vol. per cent	0.3	

REFERENCES

1. Anon., *Natl. Petrol. News*, **34,** No. 28, 20 (1942).
2. Anon., *Natl. Petrol. News*, **34,** No. 31, 14–6, 18, 20, 38 (1942).
3. Anon., *Chem. and Met. Eng.*, **49,** No. 11, 117–9 (1942).
4. Armstrong, H. E., and Miller, A. K., *J. Chem. Soc.*, **49,** 80–5 (1886).
5. Beckberger, L. H., and Watson, D. M., *Chem. Eng. Prog.*, **44,** 229–48 (1948).
6. Benedict, M., and Rubin, L. C., *Trans. Am. Inst. Chem. Engrs.*, **41,** 353–70 (1945).
7. Birch, S. F., and Scott, W. D., *Ind. Eng. Chem.*, **24,** 49–50 (1932).
8. Bohmfalk, J. F., *Chem. Eng. News*, **28,** 2504–9 (1950).
9. Boyd, J. H., *Ind. Eng. Chem.*, **40,** 1703–7 (1948).
10. Britton, E. C., Dietzler, A. J., and Noddings, C. R., *Ind. Eng. Chem.*, **43,** 2871–4 (1951).
11. Buell, C. K., and Boatright, R. G., *Ind. Eng. Chem.*, **39,** 695–705 (1947).
12. Burgin, J., Groll, H., and Roberts, R. M., *Natl. Petrol. News*, **30,** R–432–6 (1938).
13. Burke, O. W., Starr, C. E., and Tuemmler, F. D., *Light Hydrocarbon Analysis*, Reinhold, New York, 1951.
14. Dodd, R. H., and Watson, K. M., *Trans. Am. Inst. Chem. Engrs.*, **42,** 263–91 (1946).
15. Egloff, G., and Hulla, G., *Oil and Gas J.*, **41,** No. 26, 40, 43 (1942).
16. Egloff, G., and Hulla, G., *Oil and Gas J.*, **41,** No. 27, 228–33 (1942).
17. Egloff, G., and Hulla, G., *Oil and Gas J.*, **41,** No. 28, 41–4, 46–7 (1942).
18. Egloff, G., and Hulla, G., *Oil and Gas J.*, **41,** No. 29, 124, 127–8, 130 (1942).
19. Egloff, G., and Hulla, G., *Oil and Gas J.*, **41,** No. 30, 36, 39–42, 45 (1942).
20. Egloff, G., and Hulla, G., *Chem. Rev.*, **35,** 279–333 (1944).
21. Elder, A. L., *Ind. Eng. Chem.*, **34,** 1260–6 (1942).
22. Feiler, P., U.S. Pat. 1,795,549, Mar. 10, 1931.
23. Frank, R. L., Blegen, J. R., Inskeep, G. E., and Smith, P. V., *Ind. Eng. Chem.*, **39,** 893–5 (1947).
24. Frey, F. E., and Huppke, W. F. (Phillips Petrol. Co.), U.S. Pat. 2,098,959, Nov. 16, 1937.
25. Frolich, P. K., and Morrell, C. E., *Chem. Eng. News*, **21,** 1138–45 (1943).
26. Gilliland, E. R., and Lavender, H. M., *India Rubber World*, **111,** 67–72 (1944).
27. Graham, W., and Winkler, C. A., *Can. J. Research*, **26B,** 564–80 (1948).
28. Grosse, A. V. (Universal Oil Products Co.), U.S. Pat. 2,178,584, Nov. 7, 1939.

29. Grosse, A. V., Morrell, J. C., and Mavity, J. M., *Ind. Eng. Chem.*, **32,** 309–11 (1940).
30. Grosskinsky, O., Roh, N., and Hoffman, G. (Jasco, Inc.), U.S. Pat. 2,265,641, Dec. 9, 1941.
31. Hachmuth, K., and Hanson, G. H., *Chem. Eng. Prog.*, **44,** 421–30 (1948).
32. Hanson, G. H., and Hays, H. L., *Chem. Eng. Prog.*, **44,** 431–42 (1948).
33. Happel, J., Cornell, P. W., Eastman, Du B., Fowle, M. J., Porter, C. A., and Schutte, A. H., *Trans. Am. Inst. Chem. Engrs.*, **42,** 189–214, 1001–7 (1946).
34. Heilman, H. H., *World Petroleum*, **18,** No. 12, 86–8 (1947).
35. I.G. Farb., Ger. Pats. 533,778 (1931) and 565,159 (1927).
36. I.G. Farb., Ger. Pat. 544,290 (1931).
37. Kearby, K. K. (Jasco, Inc.), U.S. Pat. 2,395,875, Mar. 5, 1946.
38. Kearby, K. K., *Ind. Eng. Chem.*, **42,** 295–30 (1950).
39. Komarewsky, V. I., and Reisz, C. H., *Oil and Gas J.*, **41,** No. 19, 33, 37, 39 (1942).
40. Kunkel, J. H., *Petrol. Eng.*, **15,** No. 11, 68 (1944).
41. Manchot, M., and Brandt, W., *Ann.*, **370,** 286–96 (1909).
42. Masser, H. L., *Gas*, **19,** No. 7, 21–4 (1943).
43. Morrell, C. E., Paltz, W. J., Packie, J. W., Asbury, W. C., and Brown, C. L., *Trans. Am. Inst. Chem. Engrs.*, **42,** 473–94 (1946).
44. Morrell, C. E., and Swaney, M. W. (Standard Oil), U.S. Pat. 2,429,134, Oct. 14, 1947.
45. Morrell, J. C. (Universal Oil Products Co.), U.S. Pat. 2,178,601, Nov. 7, 1939.
46. N.V. de Bataafsche Petrol. Mij., Fr. Pat. 805,690, Nov. 7, 1936.
47. Nicholson, E. W., Moise, J., Segura, M. A., and Kleiber, C. E., *Ind. Eng. Chem.*, **41,** 646–51 (1949).
48. O'Donnell, J. P., *Oil and Gas J.*, **41,** No. 29, 38–9, 122 (1942).
49. Office of Rubber Reserve, Report on the Rubber Program, Supplement No. 1, Year 1945.
50. Ostromislenski, I. I., *J. Russ. Phys. Chem. Soc.*, **47,** 1472 (1915).
51. Poffenberger, N., Horsley, L. H., Nutting, H. S., and Britton, E. C., *Trans. Am. Inst. Chem. Engrs.*, **42,** 814–26 (1946).
52. Phillips Petroleum Company, Brit. Pat. 602,499, May 27, 1948.
52a. Reilly, P. M., *Chem. in Canada*, **5,** 41–5 (1953).
53. Robey, R. F., Wiese, H. K., and Morrell, C. E., *Ind. Eng. Chem.*, **36,** 3–7 (1944).
54. Russell, R. P., Murphree, E. V., and Asbury, W. C., *Trans. Am. Inst. Chem. Engrs.*, **42,** 1–14 (1946).
55. Sawdon, W. A., *Petrol. Eng.*, **15,** No. 8, 162, 164, 166, 168 (1944).
56. Scheibel, E. G., *Chem. Eng. Prog.*, **44,** 927–31 (1948).
57. Starr, C. E., and Ratcliff, W. F., *Ind. Eng. Chem.*, **38,** 1020–5 (1946).
58. Sumerford, S. D. (Standard Oil Development Company), U.S. Pat. 2,436,616, Feb 24, 1948.
59. Thayer, C. H., Lassiat, R. C., and Lederer, E. R., *Chem. and Met. Eng.*, **49,** No. 11, 116–7 (1942); *Natl. Petrol. News*, **34.** No. 39, R-305–6 (1942).
60. Tropsch, H., Parrish, C. I., and Egloff, G., *Ind. Eng. Chem.*, **28,** 581–6 (1936).
61. Van Antwerpen, F. J., *Chem. Eng. News*, **22,** 316–9, 398 (1944).
62. Watson, C. C., Newton, F., McCausland, J. W., McGrew, E. H., and Kassel, L. S., *Trans. Am. Inst. Chem. Engrs.*, **40,** 309–15 (1944).
63. Welch, L. M., Swaney, M. W., Gleason, A. H., and Beckwith, R. K., *Ind. Eng. Chem.*, **39,** 826–9 (1947).
64. Whitby, G. S., and Katz, M., *Ind. Eng. Chem.*, **25,** 1338–48 (1933).
65. York, R., and Holmes, R. C., *Ind. Eng. Chem.*, **34,** 345–50 (1952).
66. Zelinski, N. D., Mikhailov, B. M., and Arbuzov, Y. M., *J. Gen. Chem. U.S.S.R.*, **4,** 856 (1934); *Compt. rend. acad. sci. U.S.S.R.*, **5,** 208 (1934).

CHAPTER 4

MANUFACTURE OF BUTADIENE FROM ALCOHOL

W. J. Toussaint and J. Lee Marsh

Carbide & Carbon Chemicals Corporation

The first significant preparation of butadiene from alcohol as raw material was reported in 1915 by Ostromislensky.[15]   In these experiments, ethanol and acetaldehyde (which latter can be obtained readily from ethanol) were passed over certain clays at 300 to 450° C.   Subsequent tests of the Ostromislensky procedure in 1922 under official Russian auspices[25] showed yields of 6 per cent of the reactants to butadiene.   To relieve the dependence of Russia on natural sources of rubber, the Russian Government endeavored to obtain an industrially feasible process for the production of butadiene. One answer to this problem was found in 1928 in the Lebedev process, which has since continued in use in Russia on a growing scale.

The two processes—Ostromislensky's and Lebedev's—represent the two lines of development that have been followed in the production of butadiene from alcohol.   In the Ostromislensky process, alcohol and acetaldehyde are fed over a dehydrating type of catalyst to produce butadiene, the acetaldehyde being supplied by the dehydrogenation of ethanol in a separate, prior operation.   These steps may be represented broadly by the following equations:

$$CH_3CH_2OH \rightarrow CH_3CHO + H_2$$
$$CH_3CH_2OH + CH_3CHO \rightarrow CH_2 : CH \cdot CH : CH_2 + 2H_2O$$

In the Lebedev process, alcohol alone is fed over a mixture of dehydrogenating and dehydrating catalysts, and the required acetaldehyde is thus supplied in situ.   The over-all steps in this process may be represented by the equation:

$$2CH_3CH_2OH \rightarrow CH_2 : CH \cdot CH : CH_2 + H_2 + 2H_2O$$

At first glance the Lebedev process would seem to be the simpler of the two processes.   It requires, however, the selection of catalysts of two quite different types which will give good results at the same temperature. Further, it presents the problem of adjusting the relative amounts of the two components in the catalyst mixture and maintaining within reasonable limits a constant relative activity of the components during use between reactivations and over the life of the catalysts.

THE LEBEDEV PROCESS

Russian Practice. Many of the details concerning this process and the laboratory and pilot-plant work involved in its development are given in a monograph by Talalay and Magat,[24] but very important specific information regarding it, especially concerning the catalyst, only recently became available in German documents.[16, 26] These documents throw a new light on the nature of the catalyst actually used in Russia in the operation of the Lebedev process. Russian publications had left the impression that the catalyst consisted essentially of alumina (dehydrating) and zinc oxide (dehydrogenating). At Efremov, according to one German source,[26] the alcohol feed (of 80 per cent concentration) containing 2 to 3.5 per cent of acetaldehyde was passed upward through the catalyst bed. The catalyst was contained in rectangular iron boxes, approximately 4 by 10 in. cross section and 16 ft. long. Sixteen such containers were grouped circumferentially in an oil-fired furnace. Operation was at reduced pressures of 50 mm. of mercury inlet and 10 mm. outlet. The reaction temperature was 385 to 390° C. After 12 hours the catalyst was regenerated by burning off carbonaceous deposits; after 1000 hours it was discarded.

The percentage composition of the new and used catalyst at Efremov was given as follows:

	SiO_2	Al_2O_3	Fe_2O_3	TiO_2	ZnO	MgO	Na_2O	CO_2	SO_3	Loss at 100°	Red Heat
Used	10.5	2.3	0.4	0.1	0.2	40.2	0.9	3.5	0.6	25.0	20.3
Fresh	10.1	2.3	0.5	0.1	0.2	44.6	0.9	10.9	0.8	12.8	27.5

A thorough study of the Lebedev catalyst and its composition was reported by von Susich.[16] Unused samples of the catalyst contained chiefly magnesium hydroxide. The other components were largely kaolin and silicic acid. Von Susich concluded that the Lebedev catalyst consisted of 100 parts by weight of magnesium hydroxide, 15 to 18 parts of kaolin, and an equal amount of hydrous silica. The latter (21 to 26 parts by weight total SiO_2 per 100 parts of MgO) was thought to be present, except for that in the kaolin, as silica gel or diatomaceous earth. The smaller amounts of iron, titania, and zinc oxides present may have acted as promoters. A more recent catalyst allegedly consisted of five components.

The Germans tested a Lebedev catalyst which they prepared in the following manner. A solution of a magnesium salt was precipitated with aqueous caustic soda containing sodium carbonate. The precipitate was washed with water, homogenized with kaolin and silica gel, and formed to suitable shape under hydraulic pressure. It was activated by heating slowly first to 400° C., to dehydrate the magnesium hydroxide, and then to 570° C., to dehydrate the kaolin. Such a catalyst was tested at Ludwigshafen, probably as reported in the other German document. The results were in accord with the Russian claims, except that the catalyst life was somewhat shorter.

There is some question about the actual results that were obtained in the Russian plants operating the Lebedev process. The theoretical weight yield of butadiene from absolute alcohol is 58.7 per cent. A 20 per cent weight yield from ethanol was claimed in 1934. Later, in 1939, the yield on the alcohol converted (chemical efficiency) was alleged to be 70 per cent, equivalent to a weight yield of about 41 per cent through the use of an improved catalyst and by recycling acetaldehyde.[25] According to the German report,[26] the Lebedev catalyst apparently provided a yield of 55 to 72 per cent of crude butadiene of about 80 per cent purity; at the higher figure this is 57.6 per cent actual chemical efficiency to butadiene from alcohol. The available data are confusing, but it seems likely that the efficiency to butadiene is 50 to 60 per cent (30 to 35 per cent weight yield) without credit for butylene, which amounts to 5 to 10 per cent of the refined butadiene.

The Lebedev Process in Other Countries. Another catalyst for the Lebedev process was devised by Szukiewicz. It[23] consisted principally of magnesia and silica, and the addition of chromium oxide was said to increase the yield of butadiene by about 5 per cent. (Others[14] state that chromium oxide represses the formation of magnesium silicate.) The inventor had built a plant in Poland[29] in 1937, which by 1939 was producing about 1000 tons per year. In Italy an adaptation of this catalyst was presumably employed[16] at Ferrara during the latter part of World War II by SAICS (Società Anonima Gomma Sintetica, Milano, Pirelli Group). The catalyst, by analysis, contained 33.3 per cent MgO, 19.7 SiO_2, 17.9 CO_2, 1.3 Na_2O, and 21.1 chemically bound and 5.9 free water. X-ray examination revealed the catalyst as a basic magnesium carbonate, $5MgO \cdot 4CO_2 \cdot 5H_2O$, and hydrous silica. During use it lost its activity in 2 to 3 weeks, because of the formation of forsterite (Mg_2SiO_4). The operating cycle comprised 4 to 6 hours on stream and 3 to 4 hours' regeneration with steam and air. The operating temperature was 410 to 430° C. The "yield" (chemical efficiency) was stated to be 55 per cent, but information is lacking concerning the purity of the butadiene.

In early 1942 Szukiewicz concluded an agreement with the then Publicker Commercial Alcohol Company for operation of the process in the United States. Probably an improved catalyst was employed, perhaps containing chromium oxide as indicated in the patent.[23] Laboratory and pilot-plant operation of the process was investigated and reported on by Weiss.[30, 31] Alcohol and a small amount of acetaldehyde were passed through the catalyst at about 400° C. A molten mixture of inorganic salts was used as a heating medium. The on-stream time was 4 hours, and this was followed by a reactivation period of about 2 hours. A production ratio of about 3 lb. of butadiene per cu. ft. of catalyst per hour was obtained. The efficiency from alcohol was given as 57 per cent, and the butadiene contained about 20 per cent butylene.

DEVELOPMENT OF THE OSTROMISLENSKY PROCESS

Meanwhile little had been published on the development of the Ostromislensky process. Maximoff[11] disclosed the preparation of butadiene from a mixture of alcohol and aldol or crotonaldehyde over a dehydrating catalyst (alumina).   In the research laboratories of the Carbide & Carbon Chemicals Corporation an experimental program was carried on intermittently from about 1930, and the published results of the Lebedev process, which carefully avoided mention of specific, effective catalysts,[3] were thoroughly examined.   This effort resulted in finding new catalysts[27-8] for the Lebedev process.   Zirconia or tantulum oxide, both on silica gel, may be used as the dehydrating component, and cadmium oxide or copper as the dehydrogenating component.   With copper, the feed must contain acetaldehyde to repress hydrogenation of butadiene to butylene, and under these circumstances an efficiency above 60 per cent to butadiene can be obtained, with 5 per cent or less of butylene as by-product.   Presumably the necessary acetaldehyde could be provided by using in the catalyst mixture the proper ratio of the dehydrogenating to the dehydrating component.   How all the various catalysts would compare is a moot question, and the complexity of the problem is thoroughly presented by Corson and co-workers.[3]

During the early stages of experimental work by the Carbide & Carbon Chemicals Corporation, it became evident that crotonaldehyde was an intermediate,[18] even in the Lebedev process.   Attention was then focused on the preparation of butadiene from ethanol and crotonaldehyde.   Alumina and aluminum silicate catalysts were quite active, but tended to give more or less butylene.   Silica gel of high purity was also active and gave very little butylene.   Since established processes could provide crotonaldehyde from acetaldehyde, and acetaldehyde from ethanol, it was possible to envisage a practical process[17, 28] of making butadiene from ethanol.   The ultimate efficiency from ethanol was estimated to be 60 to 65 per cent. Through improvement in the silica catalyst, it was considered possible to eliminate the step of separately forming crotonaldehyde, and this was accomplished without sacrifice in the purity of the butadiene or in the ultimate yield of alcohol through the use of silica gel promoted by tantalum oxide.

MECHANISM OF BUTADIENE FORMATION

Various mechanisms have been proposed for the reaction of ethanol and acetaldehyde to form butadiene.   A good digest of these speculations is to be found in an article by Egloff and Hulla.[5]   The experimental evidence,[18] however, supports the following course for the reaction:

$$2CH_3CHO \rightarrow CH_3CH : CHCHO + H_2O$$
$$CH_3CH_2OH + CH_3CH : CHCHO \rightarrow CH_2 : CH \cdot CH : CH_2 + CH_3CHO + H_2O$$

All of the active catalysts for the formation of butadiene from ethanol and acetaldehyde are dehydrating in character and catalyze the condensation of acetaldehyde to crotonaldehyde.   Unpromoted silica gel is a dehydrating catalyst, and, although unsatisfactory for the preparation of butadiene from ethanol and acetaldehyde, it is an effective catalyst for each of the above reactions separately.[7, 17, 28]   In the second reaction, with a 3-to-1 molar ratio of ethanol to crotonaldehyde, hydrolysis of the crotonaldehyde—the reverse of the first reaction—has been observed[28] over silica gel. Apparently the reason for the failure of silica gel to catalyze satisfactorily the reaction of acetaldehyde and ethanol to butadiene is its inability to deoxygenate relatively small concentrations of crotonaldehyde, in ethanol, to butadiene.

Certain catalysts such as promoted silica gel and activated alumina[3, 7, 18, 27] catalyze both of the above reactions and also convert ethanol and acetaldehyde to butadiene. In the presence of sufficient ethanol, crotonaldehyde[2, 7, 8, 18] does not appear in the product. Evidently these catalysts are especially active in deoxygenation and can convert even small concentrations of crotonaldehyde, in ethanol, to butadiene.

The above mechanism is believed to apply also to the Lebedev process. In this process, the reported[24] by-products are similar in kind to those in the Ostromislensky process. The two processes differ essentially only in that in the one the acetaldehyde is formed from ethanol by a dehydrogenation catalyst in actual contact with a butadiene-forming catalyst, whereas in the other the acetaldehyde is formed in a separate converter. The larger amounts of butanol and butylene in the Lebedev process would be expected from the alumina component of the catalyst. Other possible mechanisms, as for example the formation of 1,3-butylene glycol as intermediate, suggested by Ostromislensky, are not in accord with experimental evidence.[18] The formation of radicals as a part of the principal course of the reaction, which was suggested by Lebedev, seems highly improbable. And in fact the more recent views of Russian chemists on the mechanism of the reaction involve the idea that crotonaldehyde is an essential intermediate. Thus Gorin[5a] postulates crotyl alcohol as an intermediate in the series of steps, and Kagan[7a] considers that the crotonaldehyde intermediate is reduced by activated, adsorbed hydrogen from the conjugate dehydrogenation of alcohol.

THE PROCESS OF THE CARBIDE & CARBON CHEMICALS CORPORATION

At the time the war-emergency demand for butadiene became evident, the best process available in the United States for the manufacture of butadiene from alcohol seemed to be that of Ostromislensky, modified by the newly invented catalysts. In separating the two steps of dissimilar nature, this process, as compared with the one-step Lebedev process, offered greater flexibility of operation, more certainty of accomplishment, and also more opportunity for improvement during plant operation. In the process, alcohol is dehydrogenated to acetaldehyde, and alcohol and acetaldehyde are converted to butadiene in separate converters containing, respectively, a dehydrogenating and a dehydrating catalyst. For the best results the conditions of operation are different. The required temperatures are more moderate than for the Lebedev catalyst and can be secured by means of an organic liquid heating medium. The equations for the steps are as follows:

$$CH_3CH_2OH \rightarrow CH_3CHO + H_2$$
$$CH_3CH_2OH + CH_3CHO \rightarrow CH_2 : CH \cdot CH : CH_2 + 2H_2O$$

DEHYDROGENATION OF ETHANOL

The dehydrogenation of ethanol is an old established process which can be conducted with relatively good chemical efficiency under suitable conditions.[33] The catalyst is copper on an inert support, promoted with a few per cent of chromium oxide. In a temperature range of 250 to 300° C., single-pass yields of 30 to 40 per cent of acetaldehyde can be obtained with a chemical efficiency (yield on consumed alcohol) of 92 per cent and higher. The production ratio is upward from 1.25 lb. of acetaldehyde per cu. ft. of catalyst per hour. The chief by-products are butyraldehyde, ethyl acetate, and acetic acid. Pressure has an unfavorable effect on the reaction. The catalyst can operate for an extended period before reactivation is necessary. The reactivation consists in burning carbonaceous deposits and reducing the copper oxide to the metal.

FACTORS IN THE PRODUCTION OF BUTADIENE

In the preparation of butadiene from ethanol and acetaldehyde, the most significant variables[2, 4, 7, 28] seem to be the catalyst, the molar ratio of ethanol to acetaldehyde in the feed, the feed rate, and the temperature. At a given temperature in the range of about 300 to 375° C., the single-pass yield decreases with increase in feed rate, but, within the limits of feed rate employed, the production rate increases. The chemical efficiency is dependent on the catalyst, the molar ratio of ethanol to acetaldehyde in the feed, and the temperature. At the lower molar ratios of ethanol to acetaldehyde, however, the single-pass yield also seems to be a factor in securing good efficiency.

Catalyst. The production of butadiene from ethanol and acetaldehyde can be carried out with a variety of dehydrating catalysts. The catalyst of superior demonstrated value is tantalum oxide on silica gel.[2-4, 18, 28] Others[3] have reported hafnia on silica gel to be about as effective. Zirconia on silica gel gives nearly as good results, but extensive testing has indicated that the tantalum catalyst is somewhat superior in providing a greater ultimate yield of butadiene from ethanol. Alumina catalysts are quite active, but tend to give also butylene[2, 18] to the extent of at least 10 per cent (approximately) of the butadiene. Many details on these catalysts and their preparation may be found in the article by Corson and co-workers.[3]

The concentration of tantalum oxide on the silica gel is an important factor in the activity of the catalyst.[4, 28] The single-pass yield increases with concentration of the oxide, as indicated in Table I, but the efficiency is somewhat more favorable at 2.0 per cent than at substantially higher concentrations, even when the feed rate is adjusted to give about the same single-pass yield for the catalyst of higher tantalum oxide content. The apparent decrease in efficiency observed at the lowest tantalum oxide concentration is thought to be attributable in part to the analytical difficulties encountered in determining ethanol and acetaldehyde in the recovered feed material.

The 2 per cent catalyst, as compared with one of low concentration, also seems to offer some advantage in retaining a practical activity in long-term use. Comparative data covering the use of the catalysts for extended periods of time are not available from laboratory work, and the data from the plant units are vitiated by the difficulty of maintaining exactly the same conditions of temperature in the reactivation step, during which the catalyst is subjected

*Table I. Effect of Concentration of Tantalum Oxide**

Temperature, 320° C.; Feed Ratio Ethanol/Acetaldehyde, 3 to 1

Experiment no.	1†	2	3	4‡	5	6
Tantalum oxide, %	0.0	0.6	2.0	2.5	5.5	5.5
L.h.s.v. §	0.58	0.58	0.58	0.33	0.58	1.0
Production ratio, g. per liter of catalyst per hr.	6	37	66	40	96	160
Single-pass yield to butadiene from ethanol and acetaldehyde, % ‖	2.3	15.0	27.3	30	35.0	27.9
Efficiency % to butadiene from:						
Ethanol ¶	...	60.9	61.2	60	52.1	50.3
Acetaldehyde ¶	...	68.8	80.4	80	80.0	79.3
Ethanol, ultimate ¶	...	62	67	63	61	60

* Reference 28, p. 124, except as noted.   Reprinted by permission.
† Temperature, 365° C.
‡ Reference 2.
§ L.h.s.v. = liquid hourly space velocity (volume of liquid feed/volume of catalyst per hr.).
‖ Single-pass yield, % = (moles C_4H_6 × 200) ÷ (moles C_2H_5OH + CH_3CHO fed).
¶ Efficiencies: Ethanol = (moles C_4H_6 × 100) ÷ (moles C_2H_5OH consumed);
         Acetaldehyde = (moles C_4H_6 × 100) ÷ (moles CH_3CHO consumed);
         Ethanol, ultimate = (moles C_4H_6 × 200) ÷ (consumed moles C_2H_5OH
                + consumed moles CH_3CHO 0.92);
    92% assumed efficiency from ethanol to acetaldehyde.

to the maximum temperature. In the laboratory, heating[2] the catalyst to temperatures of 500° C. for a period of 40 hours resulted in a loss of about one-third its activity for the catalyst containing 2 per cent of tantalum oxide, and the effect was somewhat greater for catalysts containing 1 per cent. It was also found that the volume of the catalyst decreased, as indicated by an increase in its bulk density—presumably a result of general shrinkage of the particles. X-ray examinations of catalysts treated more drastically showed that crystalline tantalum oxide was formed. Various devices were tried to mitigate these effects, at least in so far as they affected the decrease in activity. Preheating the silica gel, the addition of various substances to the catalyst, and purification of the catalyst components were without definite beneficial effect. The reactivation procedure actually adopted is described in a later section.

Feed Ratio and Temperature. The over-all equation for the reaction of ethanol and acetaldehyde to form butadiene shows a theoretical requirement of one mole of ethanol to one of acetaldehyde, but somewhat higher ratios give better results.[2, 4, 7, 28] It is convenient to state the composition of the feed material in terms of the ratio of the reactants, rather than in terms of its percentage composition, because of the variable composition of the feed and the generally lesser importance of the diluents, which exert their effect chiefly on single-pass yield and production ratio.   Between the ratios of about 2 to 1 and 5 to 1 similar results with respect to efficiency from ethanol and acetaldehyde are obtained, provided a suitable temperature and feed rate are used.   Thus, with a 3-to-1 feed ratio at about 325° C. the ethanol

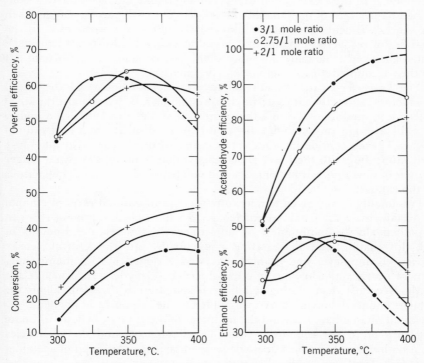

Fig. 1.   Effect of Feed Ratio and Temperature on Formation of Butadiene from Alcohol and Acetaldehyde*

*Reference 4, p. 1017. Reprinted by permission

efficiency was 55 to 60 per cent, the acetaldehyde efficiency 75 to 80 per cent, and the ultimate ethanol efficiency about 63 to 65 per cent in a large number of experiments.[2]   A feed containing a 2–2.5-to-1 ratio requires a temperature of 350 to 375° C. for like results.   The general trend is an increase in efficiency from acetaldehyde, with rise in temperature, to a maximum determined by temperature and feed rate.   Similar effects are evident in the ethanol efficiency.

A comprehensive survey of the effect of feed ratio and temperature was made by Corson, Stahly, Jones, and Bishop[4] (Fig. 1).   In general, their results agree with those outlined above.   For the 3-to-1 ratio a maximal "overall efficiency" (ultimate efficiency from ethanol) was obtained at a little above 325° C. For the 2.75- and 2.0-to-1 ratios the maxima were, respectively, at about 350 and 375° C.

A more exact and precise value for the efficiency could be obtained by repassing the recovered ethanol and acetaldehyde with additional quantities of these reactants to compensate for the amounts consumed. In a series of experiments[28] at 320° C., in which approximately a 3-to-1 molar feed ratio was maintained, an ultimate efficiency of 67 per cent from ethanol was obtained. In these experiments recycling of ethanol and acetaldehyde consumed 65 per cent of the ethanol and 90 per cent of the acetaldehyde taken for the experiments and thus minimized the effect of uncertainties in the analyses of recovered ethanol and acetaldehyde. It also tended to make butadiene from some materials other than these reactants, which would not be thus accounted for in single-pass experiments.

The lower feed ratios of ethanol to acetaldehyde increased the initial production ratio of the catalyst; but the production ratio fell off more rapidly[28] with a feed mixture of 2.5 to 1 than with one of 3 to 1. Actually, a higher over-all average production ratio could be predicated for a feed ratio of 2.75 to 1, even considering time spent in reactivation, than with the 3-to-1 feed ratio; but with the lower ratio reactivation required somewhat longer periods of time than with the higher and was believed to be more deleterious to the catalyst.

Presumably at any given temperature there is an optimal ratio of ethanol to acetaldehyde for maximal efficiency. In the Lebedev process this can hypothetically be maintained by adjustment in the relative amounts of dehydrogenating and dehydrating components. In the Ostromislensky process, Kampmeyer and Stahly[8] suggested various ways for maintaining a more nearly uniform ratio of ethanol to acetaldehyde. In principle the methods consisted in adding auxiliary feed materials at points along the catalyst bed while the main feed was added at the beginning of the bed. By the addition of ethanol beyond the midpoint of the catalyst bed, an ultimate efficiency of 76 per cent from ethanol was reported, and the efficiencies from both ethanol and acetaldehyde were increased. Addition of acetaldehyde-rich feed was made above the midpoint; it gave less advantage, and the improvement noted came solely from increase in the efficiency from ethanol to butadiene. To apply these findings in practice, it was proposed that two converters should be operated in series and auxiliary ethanol added to effluent of the first converter before it entered the second converter. For this type of operation, an ultimate efficiency of 67 per cent from ethanol was reported on the basis of limited experimental work (curtailed by a change in the program).

Kampmeyer and Stahly based their explanation of these results on the mechanism of the reaction and also, as regards the high efficiency with auxiliary ethanol feed, on the increase in partial pressure of ethanol in the feed and the decrease in contact time. However, it is not apparent that these results should materially differ from those obtained at corresponding over-all feed ratios and space velocities with all of the feed added at the top of the converter, inasmuch as the only apparent intermediate is crotonaldehyde and it is present in minute concentration.[7, 8, 18]

PILQT-UNIT OPERATION

In the course of its development, the process was operated on a pilot-plant scale.[2, 31] Chief attention was directed to the performance of the butadiene catalyst, and only occasional experiments were made in a separate installation with the dehydrogenating catalyst. The butadiene converter was a multitube vessel with steel tubes, 3 in. in diameter and 16 ft. long, and it held about 40 cu. ft. of catalyst. The catalyst contained 1.4 per cent of tantalum oxide. Provision was made for recovery of the products and (for recycling) the unreacted materials.

Twenty-six runs were made for a total of about 3100 hours. During these runs recovered ethanol and acetaldehyde were used, containing about 20 per cent of water and by-products as diluents. Significant data from certain groups of these runs are given in Table II. The variation in feed ratio does not seem to be large, but it had considerable effect on the results. After runs 14–15, with a 2-to-1 feed ratio, considerable difficulty was experienced during reactivation of the catalyst in maintaining a satisfactorily low temperature. Comparison of runs 9–11 and 19–23 showed a significant deterioration in the activity of the catalyst with age.

Table II. Production of Butadiene, Pilot Unit

Run No.	6–7	9–11	12–15	14–15	16–18	19–23	24–26
Temperature, ° C.	325	325	325	325	335	325	325
Molar ratio, C_2H_5OH/CH_3CHO	2.7	3.0	2.9	2.0	3.0	3.0	3.4
L.h.s.v.*	0.34	0.34	0.45	0.34	0.36	0.36	0.43
Production ratio (avg.), lb. C_4H_6 per cu. ft. catalyst per hr.	1.8	1.7	1.9	1.7	1.7	1.5	1.3

* L.h.s.v.: Liquid hourly space velocity (volume of liquid feed per volume catalyst per hr.).

Examination of the catalyst upon its removal from the converter showed that the physical deterioration was minor, although laboratory tests tended to confirm the decreased activity shown in Table II. During the on-stream and reactivation periods, the catalyst had been subjected to severe conditions, sometimes inadvertently, which probably accelerated its deterioration. The experiments demonstrated that the catalyst would perform satisfactorily over a reasonably wide range of selected conditions. Further, this experience demonstrated the desirability of employing a feed of 3 moles of ethanol to 1 of acetaldehyde and a temperature of 325° C. for the initial period of operation of the large plants.

Reactivation of the Catalyst. During use of the butadiene catalyst, fouling by carbonaceous deposits occurred, causing a decrease in activity. The activity could be restored by burning off the deposit with air containing nitric acid at a temperature of 400° C. In the plant this would have required corrosion-resistant equipment. Since temperatures above 400° C. were regarded as undesirable in their effect on the catalyst activity, the use of air under pressure was indicated, at least during the later stages of the burn-off.

In 1-in. tubes no difficulties were met, but with the 3-in. tubes excessive temperatures were encountered during the initial stages of the burn-off and particularly during the later stages when air under pressure was used. Even with multiple-point and movable thermocouples, complete assurance could not be secured regarding the temperature of the individual catalyst particles. After repeated experiments,[2] in which reactivation of fouled catalyst was studied, it was concluded that a satisfactory reactivation procedure could be achieved by diluting the air initially with steam.

After the catalyst had been steamed for about 1 hour, air was gradually admitted, and the effluent gas was analyzed for carbon dioxide and carbon monoxide. Any sharp rise in the concentration of these oxides in the effluent was regarded as undesirable, necessitating a cutback in the air flow. When the volume of air attained a flow of 60 cu. ft. per cu. ft. of catalyst per hour, the steam was gradually cut off. Then air alone was used until the concentration of carbon dioxide fell to 1 per cent in the effluent. Pressure was gradually increased until a final pressure of 100 p.s.i. was reached. This was continued until the carbon dioxide concentration fell to a few tenths of 1 per cent, at which point the reactivation was considered to be complete.

PLANT DESIGN

Large-scale production of butadiene became essential with the war emergency of 1941. Design of the butadiene-from-alcohol plants was developed near the end of that year, at a time when the research and development program had not been fully completed. The data clearly defined the critical ranges in the operating variables, but were not sufficient to predict the exact values that would yield the optimum results. The data showed, for example, that the best operating temperature for the butadiene reaction would probably be about 330° C., but that it might be desirable to operate at temperatures up to 360 to 370° C. Equipment was selected that was suitable for a maximum temperature of 385° C. The research data likewise indicated that the preferred molar ratio of ethanol to acetaldehyde in the converter feed mixture would be about 3 to 1 but that some variation from this ratio might be necessary. For plant design, a range in molar ratio from 2 to 1 to 3.5 to 1 was allowed for. Pressure seemed to have little or no effect on the reaction; the conversion system was designed for operation at minimum pressure.

The nature of the butadiene reaction determined the converter design. From research data it appeared that catalyst activity would be reduced substantially after 24 hours of operation because of the deposition of carbon. To remove the carbon by careful oxidation with air would take about 8 hours, during which it would be necessary to raise the air pressure to 100 p.s.i. In the reactivation period the heat of combustion would have to be removed and the catalyst temperature kept below 400° C. In the reaction period, however, heat would have to be supplied. Experiments had shown that heat transfer with a 3-in.-diameter tube would be adequate for the necessary control of temperature in both the reactivation and the reaction periods. The converters were designed, therefore, as large heat exchangers, fitted with 3-in. tubes 20 ft. long. Steel was used for both tube and shell side, since research had shown steel to be a satisfactory material. For heating and cooling the butadiene converters, a circulating, liquid Dowtherm system was selected.

Design of the separate acetaldehyde reaction unit, which was based on data from large-scale operations over a period of years, offered no problem. For convenience in the arrangement and construction of the plant, tubular converters identical with those for the butadiene reaction were selected.

The complex recovery and purification system[12-3] developed for the butadiene operaton is illustrated diagrammatically in Fig. 2. The product from the acetaldehyde converters is a mixture of acetaldehyde, ethanol, hydrogen, and small quantities of by-products (acetic acid, butyraldehyde, ethyl acetate). After the mixture has been cooled and condensed partially, the gas fraction is compressed, cooled, and scrubbed with water. The scrubber liquid is delivered to an acetaldehyde distillation column, but the condensate, rich in ethanol, is used in the recovery of butadiene. Butadiene converter product is cooled and condensed partially, to make a rough separation between, on the one hand, butadiene, butylenes, and other fixed gases, and, on the other hand, a mixture of ethanol, acetaldehyde, and by-products. The gas fraction is compressed and is scrubbed with the condensate from the acetaldehyde reaction-product condenser, to remove butadiene. The gas is then scrubbed with water, to recover ethanol and acetaldehyde, before it is discharged as a by-product stream containing ethylene and propylene. The scrubber liquid from the butadiene removal scrubber is combined with the butadiene condensate and delivered to a butadiene distillation column. Scrubber liquid from the final water scrubber flows to the acetaldehyde distillation column.

Separation and purification of the two reaction products is accomplished in a series of distillation and absorption columns. The first in the series is the butadiene distillation column, to which the condensates and scrubber liquids containing butadiene are delivered. The distillate is a mixture of 88 to 90 per cent butadiene, 5 to 6 acetaldehyde, and 5 to 6 butylenes, butane, and fixed gases. The distillate is scrubbed with water to remove acetaldehyde and is then passed into an absorption unit[34] for removal of the 4-carbon impurities and the fixed gases, "Chlorex" ($\beta\beta'$-dichloroethyl ether) being the selective solvent. Butadiene can be purified in the absorption system to meet the specifications, but usually undergoes a final distillation.

The residue from the butadiene distillation is a mixture of alcohol, acetaldehyde, ethyl ether, and other by-products. This mixture, combined with condensates and scrubber liquids from the acetaldehyde reaction system, enters the acetaldehyde distillation column. The composition of the distillate is approximately that of the constant-boiling mixture of acetaldehyde and ethyl ether, containing 75 to 80 per cent acetaldehyde. Part of the distillate is condensed for reflux. Product is withdrawn as vapor to supply acetaldehyde to the butadiene reaction system.

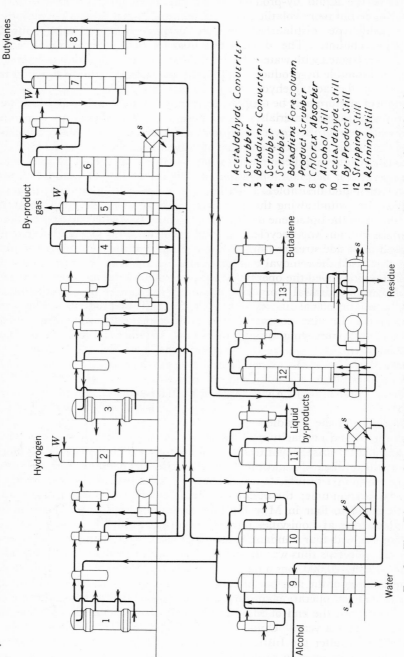

FIG. 2. Flow Diagram of the Carbide and Carbon process for the Production of Butadiene from Alcohol

1 Acetaldehyde Converter
2 Scrubber
3 Butadiene Converter
4 Scrubber
5 Scrubber
6 Butadiene Fore column
7 Product Scrubber
8 Chlorex Absorber
9 Alcohol Still
10 Acetaldehyde Still
11 By-Product Still
12 Stripping Still
13 Refining Still

Most of the liquid by-products of the reactions are less volatile than acetaldehyde but more volatile than alcohol.   They are removed by passing the acetaldehyde distillation residue through a by-products-removal distillation column.   The distillate, containing some acetaldehyde and alcohol, is extracted with water to recover the useful materials and to separate the water-insoluble by-products.   The water fraction from the extraction is returned to the acetaldehyde distillation column.   The water-insoluble by-products mixture may be used as fuel or may be treated separately for the recovery of valuable materials.

The final step is the recovery of alcohol, which is accomplished by distillation.   Product is withdrawn as vapor to form the alcohol feed for both the acetaldehyde and butadiene reactions.

In the course of plant operation several refinements in the recovery system were found desirable.   Separation of butadiene from fixed gases was simplified[1] by withdrawing the butadiene distillate as a liquid from a tray near the top of the butadiene distillation column; uncondensed gas from the top of the column was recycled through the primary butadiene scrubber. Withdrawal of side streams from the butadiene distillation column near the base[10] permitted the removal of substantial amounts of ethyl ether from the system, and improved the operation of the acetaldehyde distillation column. Side-stream withdrawal from the acetaldehyde distillation column[6] was also found beneficial in eliminating troublesome accumulations of by-products.   In the course of the operation, other schemes[9] were developed for removing minor amounts of by-products from the recovery system.

PLANT CONSTRUCTION

The design of the butadiene-from-alcohol process was established at the end of 1941, when a capacity of 20,000 tons a year was selected for the standard production unit.   As the synthetic-rubber program expanded in the following months, the number of standard units was increased.   The butadiene-from-alcohol plant at Institute, W. Va., was originally planned to consist of a single production unit; the number of units was doubled in December 1941 and increased to four in May 1942.   In July two additional plants were authorized, one at Louisville, Ky., with three standard units, and one at Kobuta, Pa., with four standard units.   The rated capacity of the butadiene-from-alcohol plants thus was increased in a period of 7 months from 20,000 tons a year to 220,000 tons a year.

Construction of the plant at Institute, W. Va., began in April 1942.   The first unit was in operation during January 1943, and all four units were operating before the end of May 1943.   Production was then at a rate of over 85,000 tons a year.   Units at the Louisville and Kobuta plants began operation soon after the Institute plant was completed, and by December 1943 all 11 units were in production.   Almost 130,000 tons of butadiene, representing 77 per cent of all the butadiene available for synthetic rubber during 1943, was produced from alcohol.[19]

RESULTS IN PLANT OPERATION

The results obtained in the butadiene-from-alcohol plants closely confirmed the research data. From the experimental work it had been predicted that an ultimate yield (efficiency) of 61 per cent of theory could be achieved. After two years of plant operation, during which ranges of temperature, feed rate, and feed composition were thoroughly explored, the best operating procedure gave an efficiency that was consistently about 64 per cent. Many

FIG. 3. Unit for Production of Butadiene from Alcohol by Carbide and Carbon Process
Nominal capacity, 20,000 short tons butadiene a year. Some of the butadiene convertors are to be seen in the foreground; fractionating columns, in the rear
Courtesy Carbide and Carbon Chemicals Corp.

of the by-products had been identified in the laboratory, and the yields of them to be expected had been predicted, despite the extreme difficulty of quantitative determination in small-scale work. When typical by-products from plant operation were carefully analyzed,[28] the agreement with the predicted yields was remarkably good, as is shown in Table III.

The flexibility that had been allowed in plant design, in order to permit the adjustment of feed composition and temperature to the values shown best by operating experience, made it possible to exceed the rated plant capacities by substantial margins. After extensive investigation of the variables, it was generally found that a molar ratio of 2.75 parts of ethanol to 1 part of acetaldehyde in the converter feed gave optimum results; the best operating temperature was about 350° C. Because these optimum values were so close to those recommended by research data, the reserve capacity which had been provided to permit flexibility in operation could be used to increase capacity. The converter feed rate was raised gradually from the original design value of 0.23 cu. ft. of liquid feed per cu. ft. of catalyst per hour to an average value of about 0.4. Plant production rates increased during 1944 and 1945, when maximum production was essential, to about 180 per cent of the original rated capacity.

Another reason for a high production rate in the plants was that catalyst activity fell off less rapidly than had been expected.   Removal of carbon by

Table III.   Products of the Butadiene-from-Alcohol Process

Product	Per cent of Ethanol Consumed	
	Predicted	Actual
Butadiene	61.0	63.9
Butylenes	1.5	1.4
Olefins		
Ethylene		3.9
Propylene		1.7
	6.0	5.6
Other gaseous by-products		
Carbon oxides		0.12
Butane		0.14
Ethane		0.09
Propane		0.12
	1.0	0.5
Low-boiling by-products		
Methyl ethyl ether		0.3
Ethyl ether		8.0
Ethyl vinyl ether		0.1
Butyraldehyde and methyl ethyl ketone		0.3
Ethyl acetate		1.2
Pentenes		0.4
Pentadienes		0.5
Hexenes and hexadienes		4.0
	11.0	14.8
High-boiling by-products		
Ethyl acetal		1.3
Crotonaldehyde		0.1
Butanol and crotyl alcohol		1.2
Butyl acetate		0.1
Vinylcyclohexene		0.1
Hexanol		0.2
Hexaldehyde		0.5
Acetic acid		1.4
Unidentified		3.8
	13.0	8.7
Carbon deposit	1.5	...
Loss	5.0	5.1
	100.0	100.0

controlled combustion was necessary only once a week rather than once a day.   Although the time for reactivation was increased from the anticipated 8 hours to about 24 hours, the catalyst was available for production 87 per cent of the time instead of 75 per cent as had been expected.

Among the other variables explored in plant operation was pressure, which the laboratory had predicted would have little effect on efficiency or average production rate. Trials in the plants confirmed the prediction; they showed that the gain in productivity during operation was offset by the much longer time required for reactivation.

Utilization of By-Products. The recovery and utilization of by-products, which was undertaken after the plants were in operation, materially increased the total yield from alcohol. Before the end of the program, large-scale recovery of ethylene, ethyl ether, butylenes, and butanol was being effected, and a scheme for the recovery and utilization of ethyl acetate had been developed.

As shown in Table III, nearly 4 per cent of the alcohol consumed in the operation was converted to ethylene, which appeared in the by-product gas stream mixed with propylene and other fixed gases. By removal of the propylene, the gas stream was made suitable for use in the styrene plants at Institute and Kobuta.

Approximately 8 per cent of the alcohol used in the operation was dehydrated to ethyl ether, which was removed from the process in a mixture of low-boiling by-products. Methods were developed for concentrating the ether fraction, so that it could be hydrated. Substantial quantities of alcohol were recovered by this operation, for use in the butadiene process; ethylene formed in the hydration step was useful in the production of styrene. Still later, a process for recovery of a purified grade of ethyl ether was developed.[32]

A small butylene fraction was isolated in the final purification of butadiene. This fraction was used as raw material in butadiene-from-petroleum plants.

Since butanol was in great demand during World War II, a process was developed for hydrogenating and distilling the high-boiling by-product fraction, and substantial amounts of butanol were obtained.

Experimental work near the end of the program showed that the utilization of alcohol could be increased by recycling ethyl acetate recovered from the low-boiling by-products.[21-2]

The net result of recovering and using the by-products at the several plants was an increase in alcohol utilization from 64 to over 72 per cent.[20] Further development of recovery projects for these by-products would have raised the utilization to a still higher figure. When the program ended, recovery of other materials from the by-product fractions was being explored with promising results.

STATISTICS OF BUTADIENE-FROM-ALCOHOL PRODUCTION

The three butadiene-from-alcohol plants in the United States, which began operation during 1943, produced about 130,000 tons of butadiene in that year. In 1944, with all units in operation for the full year, production was about 362,000 tons. Butadiene-from-alcohol provided 64 per cent of all butadiene for synthetic rubber in 1944. Production continued in 1945 until the end of the war, after which the butadiene-from-alcohol plants were progressively shut down.   About 233,000 tons of butadiene, or 39 per cent of the total used for synthetic rubber, was made from alcohol in 1945.   The

FIG. 4.   Butadiene-Storage Spheres of 250,000 Gallons Capacity
Courtesy Carbide and Carbon Chemicals Corp.

importance of the process in supplying butadiene for synthetic rubber during the war emergency is shown by the production figures in Table IV.[19, 20]

Table IV.   Production of Butadiene-from-Alcohol

	Tons a Year	% of Total Butadiene Available for Synthetic Rubber
Rated capacity	220,000	34
Actual production		
1943	130,000	77
1944	362,000	64
1945	233,000	39
Total 1943–45	725,000	54 (weighted average)

REFERENCES

1. Cannon, E. J., and Stuewe, H. A., U.S. Pat. 2,409,250 (1946).
2. Carbide & Carbon Chemicals Corp., Unpublished Data.
3. Corson, B. B., Jones, H. E., Welling, C. E., Hinckley, J. A., and Stahly, E. E., *Ind. Eng. Chem.*, **42**, 359–73 (1950).
4. Corson, B. B., Stahly, E. E., Jones, H. E., and Bishop, H. D., *Ind. Eng. Chem.*, **41**, 1012–7 (1949).
5. Egloff, G., and Hulla, G., *Chem. Revs.*, **36**, 63–141 (1945).
5a. Gorin, Yu. A., *J. Gen. Chem.*, *U.R.S.S.*, **19**, 877–83 (1949) ; *Chem. Absts.*, **44**, 1006 (1950). Cf. Gorin, Yu. A., *J. Gen. Chem.*, *U.R.S.S.*, **16**, 283–94 (1946) ; *Chem. Absts.*, **41**, 685 (1947); Gorin, Yu. A., and Gorn, I. K., *J. Gen. Chem.*, *U.R.S.S.*, **18**, 645–55 (1948); *Chem. Absts.*, **43**, 991 (1949).
6. Hitchcock, M. E., and Field, J. A., U.S. Pat. 2,403,743 (1946).
7. Jones, H. E., Stahly, E. E., and Corson, B. B., *J. Am. Chem. Soc.*, **71**, 1822–8 (1949).
7a. Kagan, M. Ya., Lyubarskii, G. D., and Podurovskaya, O. M., *Bull. acad. sci.*, *U.R.S.S.*, *Classe sci. chim.*, **1947**, 173–81; *Chem. Absts.*, **42**, 4515.
8. Kampmeyer, P. M., and Stahly, E. E., *Ind. Eng. Chem.*, **41**, 550–5 (1949).
9. Kinsey, H. D., Kelly, T. H., and Ferrara, P. J., U.S. Pat. 2,393,381 (1946).
10. Marsh, J. L., and Hitchcock, M. E., U.S. Pat. 2,395,057 (1946).
11. Maximoff, A. T., U.S. Pat. 1,682,919 (1928).
12. Murray, I. L., Marsh, J. L., and Smith, S. P., U.S. Pat. 2,403,741 (1946).
13. Murray, I. L., Marsh, J. L., and Smith, S. P., U.S. Pat. 2,403,742 (1946).
14. Natta, G., and Rigamonti, R., *Chimica e industria*, **29**, 195–200 (1947).
15. Ostromislensky, I. I., *J. Russ. Phys. Chem. Soc.*, **47**, 1472–1506 (1915).
16. Publication Board Rept. No. 85,172, Vol. III, Dyestuff Research, U.S. Dept. Commerce, Washington, D.C., pp. 427–33, 457.
17. Quattlebaum, W. M., and Toussaint, W. J., U.S. Pat. 2,407,291 (1946).
18. Quattlebaum, W. M., Toussaint, W. J., and Dunn, J. T., *J. Am. Chem. Soc.*, **69**, 593–9 (1947).
19. Reconstruction Finance Corp., O.R.R., Rept. on the Rubber Program, Suppl. No. 1, Year 1945, Washington, D.C., Apr. 8, 1946.
20. Rubber Reserve Co., Rept. on the Rubber Program 1940–1945, Washington, D.C., Feb. 24, 1945.
21. Stahly, E. E., U.S. Pat. 2,439,587 (1948).
22. Stahly, E. E., Jones, H. E., and Corson, B. B., *Ind. Eng. Chem.*, **40**, 2301–3 (1948).
23. Szukiewicz, W., U.S. Pat. 2,357,855 (1944).
24. Talalay, A., and Magat, M., *Synthetic Rubber from Alcohol*, Interscience, New York, 1945, pp. 1–89.
25. Talalay, A., and Talalay, L., *Rubber Chem. and Technol.*, **15**, 403–29 (1942).
26. Technical Oil Mission Rept., Reel 2, Section 12, Library of Congress, Washington, D.C.
27. Toussaint, W. J., and Dunn, J. T., U.S. Pat. 2,421,361 (1947).
28. Toussaint, W. J., Dunn, J. T., and Jackson, D. R., *Ind. Eng. Chem.*, **39**, 120–5 (1947).
29. U.S. Senate Hearing, 77th Cong., 2d Sess., S. Res. 224, Utilization of Farm Crops, U.S. Govt. Printing Office, Washington, D.C., 1942, pp. 259–60.
30. U.S. Senate Hearing, *ibid.*, pp. 864–7.
31. U.S. Senate Hearing, *ibid.*, pp. 1257–61.
32. Van der Hoeven, B. J. C., and Ghoul, W., U.S. Pat. 2,474,874 (1949).
33. Young, C. O., U.S. Pat. 1,977,750 (1934).
34. Young, C. O., and Perkins, G. A., U.S. Pat. 1,948,777 (1934).

CHAPTER 5

PREPARATION AND PRODUCTION OF DIENES BY OTHER METHODS

Harry L. Fisher

University of Southern California

EARLY HISTORICAL AND GENERAL METHODS

The methods of preparing conjugated diene hydrocarbons are numerous, and there are many variations of practically all of them. In this chapter a survey of the methods is given, and many examples are included. The chronological order of their discovery is followed only in part. It is very interesting that many of the important methods were discovered, tested, and even used in small operations in the short period from about 1909 to 1914. The outburst of interest in the synthesis of rubber which marked this period is understandable in view of the fact[56] that the price of natural rubber in 1910 attained its highest annual average of $2.07 per pound on the New York market and at one time during the year actually reached $3.12.

BUTADIENE AND ISOPRENE

The methods of preparing butadiene and isoprene are discussed more or less together because these two dienes were frequently prepared concurrently by the same general methods. The methods of preparation of piperylene, dimethylbutadiene, methylpentadienes, and myrcene are given at the end of this section. When the term butadiene is used it is understood that 1,3-butadiene is always meant.

1,3-Butadiene	Isoprene (2-Methyl-1,3-butadiene)
$CH_2 : CH \cdot CH : CH_2$	$CH_2 : C(CH_3) \cdot CH : CH_2$
b.p., $-4.41°$ C., 760 mm. Hg^{124}	b.p., $34.076°$ C., 760 mm. Hg^{18}
$d_{15.56°C.}^{15.56°C.}$, 0.6274^{46}	$d_{20°C.}^{20°C.}$, 0.6805^{18}
$n_D^{-25°C.}$, 1.4293^{66}	$n_D^{20°C.}$, 1.42160^{18}

Diene hydrocarbons having conjugated double bonds and capable of polymerization to rubberlike products have been known for a little less than a century, but commercial products have been made from them for only about the past twenty years. The chief dienes of commercial importance are the simplest of the series and contain only 4 and 5 carbon atoms. Butadiene is manufactured in the greatest amount, and isoprene is next although in much smaller quantity. Dimethylbutadiene was used by the Germans to make "methyl rubber" during World War I, and piperylene, an isomer of isoprene, has never been of any commercial consequence. How much this picture of relative amounts will change in the next decade or two is, of course, open to conjecture.

Preparation by Pyrolysis of Natural Products. Isoprene, the first member of the series to be isolated, was obtained by C. Greville Williams in 1860 by the destructive distillation or pyrolysis of natural rubber and of gutta-percha.[179]

$$(C_5H_8)_x \xrightarrow{\text{heat}} CH_2 : C(CH_3) \cdot CH : CH_2 \qquad (1)*$$

Natural-rubber hydrocarbon   Isoprene

In 1884 isoprene was obtained by Tilden[173] by the passage of redistilled American turpentine through a hot iron tube. The yield was about 7.5 per cent, and the product was identified by analysis and molecular-weight determination.

$$C_{10}H_{16} \xrightarrow{\text{hot iron tube}} CH_2 : C(CH_3) \cdot CH : CH_2 \qquad (2)$$

Pinene (chief constituent of turpentine)   Isoprene

Crude pyrolytic methods usually give low yields. A careful study of the conditions made it possible to raise the yield of isoprene from natural rubber up to 10 per cent,[116] then to 16.7 per cent,[11] and recently to 58 per cent.[29] Conditions necessary for the highest yield of isoprene appear to be (a) to bring the rubber to a high temperature as rapidly as possible, and (b) by working under reduced pressure to remove the cracking products from the hot zone as quickly as possible.

Later work with turpentine and other terpenes, especially dl-limonene (dipentene), produced isoprene in yields up to 60 per cent. Myrcene on pyrolysis also yields isoprene[45] (see also p. 121). Harries[73] in 1911 devised his "isoprene lamp," by means of which the vapors of turpentine are conducted over an electrically heated coil of wire in the neck of a flask, to which is attached a reflux condenser maintained at a temperature such that uncracked terpene will condense while isoprene will pass through. Staudinger,[169] using this device and also reduced pressure, obtained the high yield mentioned above. Gottlob immersed an electrically heated filament in the terpene liquid.[62] The method of turpentine to dl-limonene to isoprene was developed commercially just before World War II and is described later (p. 134).

Butadiene Not Found in Nature. Butadiene apparently is not related directly to any natural product, except through its homolog isoprene. Isoprene itself has never been found free in nature, but many substances, such as rubber, the terpenes, and the carotenoids, contain its skeleton and bear a definite relation to it.

Exhaustive Methylation of Cyclic Imines and 1,4-Diamines. The correct structural formula for isoprene was first proposed in 1882, by Tilden,[172] but it was not until 1897 that this structure was proved, by Euler,[53] who synthesized isoprene from 3-methylpyrrolidine by exhaustive methylation—a

*A number in parentheses is used to identify the reaction.

general method for the preparation of dienes adapted from a similar method of preparing monenes.

$$CH_3CH\!-\!CH_2 \qquad CH_3CH\!-\!CH_2$$

(3-Methylpyrrolidine) $\xrightarrow{CH_3I}$ $\xrightarrow[\text{AgOH}]{\text{KOH or}}$

$$CH_3CH\!-\!CH_2 \xrightarrow{\text{heat}} CH_3CH\!=\!CH \xrightarrow[\text{KOH, and heat}]{\text{repetition with } CH_3I,}$$

$$\begin{array}{c} CH_3C\!-\!-\!CH \\ \parallel \qquad \parallel \\ H_2C \qquad CH_2 \end{array} + N(CH_3)_3 + H_2O \qquad (3)$$

Isoprene

Exhaustive methylation had been used twelve years earlier by Ciamician and Magnaghi[40] to prepare butadiene in a similar manner from pyrrolidine itself, and to establish its structure. Exhaustive methylation with methyl chloride as well as methyl iodide was also used on tetramethylenediamine and its β-methyl derivative to prepare butadiene and isoprene, respectively (see also reaction 17).[72, 78, 162]

From Halogen Derivatives. Treatment of dichlorobutanes and their β-methyl derivatives with the time-honored agent, alcoholic potash, or with heat over barium chloride[10] and other substances as catalysts, yields butadiene and isoprene, respectively, by dehydrohalogenation.

$$CH_3C(CH_3)Br \cdot CH_2CH_2Br \xrightarrow{\text{alc. KOH}} CH_2 : C(CH_3) \cdot CH : CH_2 \quad (4)$$

The 2,4-dibromo-2-methylbutane in reaction 4 above was obtained by the action of 73 per cent hydrobromic acid on isoprene by Ipatieff[85] who then reconverted it to isoprene as indicated. The isoprene used by him was obtained originally by the dry distillation of Para rubber. The 2,4-dibromo-2-methylbutane (α,α-dimethyltrimethylene dibromide) was also obtained by the addition of 2HBr to unsym.-dimethylallene which also was converted into isoprene in the early experiment.[86]

1-Bromo-2-butene distilled with potassium formate[38] and, more recently,

2-chloro-2-butene over alumina containing chromium or iron at 475° C.[54] yield butadiene.

$$BrCH_2CH : CHCH_3 \xrightarrow{HCOOK} CH_2 : CH \cdot CH : CH_2 \qquad (5)$$

$$CH_3CCl : CHCH_3 \xrightarrow{Al_2O_3 + Cr, 475° C.} CH_2 : CH \cdot CH : CH_2 \qquad (6)$$

Recrystallized 1,2,3,4-tetrabromobutane (butadiene tetrabromide) treated with zinc in alcohol gives butadiene of high purity,[91, 171] isoprene tetrabromide treated similarly gives isoprene;[49] and 1,4-dibromo-2-butene heated under the same general conditions also gives butadiene of high purity.[171]

$$BrCH_2 \cdot CHBr \cdot CHBr \cdot CH_2Br \xrightarrow{Zn + alcohol} CH_2 : CH \cdot CH : CH_2 \qquad (7)$$

$$BrCH_2 \cdot CH : CH \cdot CH_2Br \xrightarrow{Zn + alcohol} CH_2 : CH \cdot CH : CH_2 \qquad (8)$$

By Pyrolysis of Alcohols. Butadiene was obtained first in 1863 by E. Caventou[35] by passing the vapors of fusel oil—a mixture chiefly of isomeric branched amyl alcohols—through a hot tube. In this pyrolysis the amyl alcohols are, no doubt, converted into isoamylenes or isopentenes. Ostromislensky[135] showed later that isoamylene can in fact be cracked to produce butadiene. Very recently this work was resubstantiated, trimethylethylene (isopentene), easily obtained from isoamyl alcohol by dehydration and the shifting of the double bond, giving both butadiene and isoprene.[75] The fusel oil used by Caventou may also have contained some butyl alcohol.

$$\left.\begin{array}{l} CH_3CH(CH_3)CH_2CH_2OH \xrightarrow{heat} CH_3CH(CH_3)CH : CH_2 \\ \quad\text{Isoamyl alcohol} \\[1em] CH_3CH_2CH(CH_3)CH_2OH \xrightarrow{heat} CH_3CHC(CH_3) : CH_2 \\ \quad\text{Active amyl alcohol} \end{array}\right\} \rightarrow \begin{array}{l} CH_2 : CH \cdot CH : CH_2 + \\ \quad\text{Butadiene} \\ CH_2 : C(CH_3) \cdot CH : CH_2 \\ \quad\text{Isoprene} \\ \text{and other products} \end{array} \quad (9)$$

From Acetylene and Ethylene. In 1866 Berthelot[21] obtained some butadiene by passing a mixture of acetylene and ethylene through a red-hot iron tube.

$$HC : CH + CH_2 : CH_2 \rightarrow CH_2 : CH \cdot CH : CH_2 \qquad (10)$$

This intriguing reaction has often been studied since that time, even very recently.[123]

By Pyrolysis of Oils and Hydrocarbons. In 1873 Caventou[36] isolated butadiene from compressed illuminating gas, and in 1886 Armstrong and Miller[4] verified his report. In 1886 butadiene was obtained by Norton and Noyes[132] by the pyrolysis of ethylene, and by Norton and Andrews[131] by the pyrolysis of pentane and hexane. In fact, butadiene is obtained in some amount by the pyrolysis of nearly all simple and complex hydrocarbons, for example:

$$C_4H_8RR' \rightarrow C_4H_6 + RH + R'H \qquad (11)$$

Butadiene is easily identified in pyrolytic gases because it readily forms a crystalline tetrabromide having a sharp melting point, 118° C. It also forms a stereoisomeric tetrabromide that melts at 38° C., but usually is a liquid. Isoprene apparently forms only a liquid tetrabromide, which boils at 153 to 155° C. at 12 mm.[170] Piperylene forms a solid tetrabromide (m.p. 114.5° C.) and a liquid isomer (b.p. 115 to 118° C. at 4 mm.); 2,3-dimethylbutadiene tetrabromide melts 139 to 140° C. If pyrolytic gases are passed into a solution of bromine in accordance with normal laboratory practice, the crystalline butadiene tetrabromide, from the butadiene that usually forms the greater part of the dienes present, is easily isolated and may be the only substance isolated and identified. Therefore, there are more data on the occurrence of butadiene in these pyrolytic gases than of any other substance.

Butadiene can be separated from monenes and paraffins and can be purified by forming the crystalline sulfur dioxide addition product, butadiene sulfone or 2,5-dihydrothiophene-1-dioxide (1-thia-3-cyclopentene-1-dioxide), heating the latter to 125° C. to decompose it,[44] and finally washing the gas with water and drying it.

$$CH_2 : CH \cdot CH : CH_2 + SO_2 \rightarrow CH_2 \cdot CH : CH \cdot CH_2 \qquad (12)$$
$$\underline{SO_2}$$

1,2-Butadiene or methylallene (b.p. 10.3° C.) and 1-butyne or ethylacetylene (b.p. 8.5° C.) are also found along with 1,3-butadiene among the gaseous products of cracked petroleum. Neither of them on further pyrolysis is isomerized to 1,3-butadiene[82] but, when passed over heated impure clays ("Argilla pura"),[163] over floridin[167] at 275 to 300° C., or over hot activated fuller's earth,[27] they are converted in part to 1,3-butadiene. 3-Methyl-1,2-butadiene or *asym*-dimethylallene gives a 50 to 55 per cent yield of isoprene when heated with quinoline hydrobromide in quinoline solution[100] at 130 to 135° C. The mechanism of the reaction in this method may follow the interesting conversion of 3-methyl-1,2-butadiene into isoprene in the early long stepwise method that starts with methyl ethyl ketone:

$$CH_3COCH_2CH_3 + CH_3I + Mg \xrightarrow{+H_2O} CH_3C(CH_3)(OH) \cdot CH_2CH_3 \xrightarrow{-H_2O}$$

$$\underset{\text{Trimethylethylene}}{CH_3C(CH_3) : CHCH_3} \xrightarrow{Br_2} CH_3C(CH_3)Br \cdot CHBrCH_3 \xrightarrow{\text{alc. KOH}^{52}}$$

$$\underset{\substack{\text{3-Methyl-1,2-butadiene} \\ (asym\text{-dimethylallene})}}{CH_3C(CH_3) : C : CH_2} \xrightarrow{\text{HBr in AcOH}^{84}} CH_3C(CH_3)Br \cdot CH_2CH_2Br \xrightarrow{\text{alc. KOH}^{87}}$$

$$\underset{\text{Isoprene}}{CH_2 : C(CH_3) \cdot CH : CH_2} \qquad (13)$$

From Esters of Polyhydric Alcohols. In 1886 Henninger[74] obtained butadiene by distilling erythritol (1,2,3,4-butanetetrol) with concentrated formic acid. In this reaction esters of formic acid are produced which then are decomposed, but the decomposition mechanism is not simple. One of the early names of butadiene, "erythrene," comes from this experiment.

The reaction is a general one, especially the simple decomposition of the esters of those polyhydroxyl compounds that have the hydroxyl groups on adjacent carbons (see reaction 32).

$$HOCH_2 \cdot CH(OH) \cdot CH(OH) \cdot CH_2OH \xrightarrow{HCOOH} CH_2 : CH \cdot CH : CH_2 \quad (14)$$
$$\underset{\text{Erythritol}}{} \qquad\qquad\qquad\qquad \underset{\text{Butadiene}}{}$$

Butadiene from Ethyl Alcohol in Early Reactions. Ipatieff[88] in 1903 passed the vapor of ethyl alcohol through a hot tube filled with aluminum powder and obtained a small yield of butadiene along with ethylene and acetaldehyde. In 1915 Ostromislensky[134, 137] published his classic work on the preparation of butadiene by the pyrolysis of a mixture of ethyl alcohol and acetaldehyde over alumina at 360 to 460° C.

$$(\text{Ipatieff}) \quad 2CH_3CH_2OH \xrightarrow{\text{heat and Al powder}} CH_2 : CH \cdot CH : CH_2 \quad (15)$$

(Ostromislensky)
$$CH_3CH_2OH + CH_3CHO \xrightarrow{\text{heat and catalyst}} CH_2 : CH \cdot CH : CH_2 \quad (16)$$

The pyrolysis of alcohol was studied further and the method used in the U.S.S.R. by Lebedev,[103] who received a special (Soviet) Government prize. During World War II alcohol was used extensively in the United States in the manufacture of butadiene for GR-S (see Chapter 4).

It should be added that in 1910, under the same conditions used by Ipatieff, Filippov[55] found that diethyl ether—a simple derivative of ethyl alcohol—gives three times more butadiene than ethyl alcohol itself.

International Race for Synthetic Rubber. With the increase in the use of automobiles early in the twentieth century and the rise in the price of natural rubber an international race for the discovery of methods of preparation and polymerization of dienes and for the economical manufacture of synthetic rubber took place amid feverish activity especially among chemists in England, Germany, and Russia. Cheap sources of butadiene and isoprene were sought, chiefly coal tar, fermentation products of starch and sugars, and calcium carbide. These are now described and later methods also included in the description of methods of preparing butadiene and isoprene from these three general sources.

The chief substances studied were: benzene, phenol, and the cresols, from coal tar; ethyl, butyl and amyl alcohols, and acetone, from starch and sugars; and acetaldehyde and, later, butynediol, from calcium carbide. Turpentine was also considered. All of these, even of late years, have played their parts although their pathways were somewhat different. The earlier methods were the more direct definite methods by steps according to strictly organic chemical procedures, whereas the more modern methods depend almost entirely on controlled pyrolysis and the subsequently necessary purification of the products.

From Coal Tar. Phenol is converted to butadiene and *p*-cresol to isoprene by a similar series of reactions developed by F. Hofmann and

Coutelle[78] of the Elberfelder Farbenfabriken of Friedrich Bayer & Company.[12, 14, 108, 162]   For example:

$$CH_3\text{-}\underset{p\text{-Cresol}}{\boxed{}}\text{-OH} \xrightarrow{H_2}$$

4-Methylcyclohexanol

$$\xrightarrow{oxidation}$$

$$\underset{\beta\text{-Methyladipic acid}}{HOOCCH_2CH(CH_3)CH_2CH_2COOH} \xrightarrow{NH_3,\ heat}$$

$$\underset{\beta\text{-Methyladipamide}}{H_2NOCCH_2CH(CH_3)CH_2CH_2CONH_2} \xrightarrow[\text{reaction (NaClO)}]{Hofmann}$$

$$\underset{\text{2-Methyltetramethylenediamine}}{H_2NCH_2CH(CH_3)CH_2CH_2NH_2} \xrightarrow[\text{methylation}]{\text{exhaustive}} \underset{\text{Isoprene}}{CH_2:C(CH_3)\cdot CH:CH_2}\ (17)$$

Benzene is hydrogenated to cyclohexane, which is pyrolyzed to butadiene, ethylene, and hydrogen,[15, 89] and phenol is hydrogenated to cyclohexanol, which is dehydrated to cyclohexene, and the cyclohexene is then pyrolyzed to give a high yield of butadiene and ethylene, practically quantitative at 700 to 800° C. and 7 to 15 mm.[8, 9, 23, 99, 151, 180]

$$\boxed{} \xrightarrow{H_2} \quad \xrightarrow{\text{pyrolysis}} CH_2:CH\cdot CH:CH_2 + CH_2:CH_2 + H_2$$

$$(18)$$

$$\boxed{}\text{-OH} \xrightarrow{H_2} \quad \xrightarrow{-H_2O} \quad \xrightarrow{\text{pyrolysis}}$$

$$CH_2:CH\cdot CH:CH_2 + CH_2:CH_2 \quad\quad (19)$$

The pyrolysis of cyclohexane to produce commercial quantities of butadiene is discussed in some detail later (p. 126).

Cyclohexane and also methylcyclopentane are found in certain petroleum oils. Thermal decomposition of them at 1500° C. and contact time of 0.4 second give yields of 22 and 7 per cent, respectively, of butadiene.[19]

From Fermentation Products. Perkin, Matthews, Strange, Weizmann, and co-workers developed methods of preparing butadiene and isoprene from *n*-butyl alcohol and isoamyl alcohol, respectively. Both these alcohols are products of fermentation although the availability of isoamyl alcohol is low. In 1913–14 the Synthetic Products Company, founded in 1912, built a plant at King's Lynn, England, for making 15 tons of butyl alcohol a week[164] by the Fernbach and Weizmann fermentation process. Butyl alcohol is also made from acetylene through acetaldehyde, acetaldol, and crotonaldehyde, as follows:

$$HC : CH \xrightarrow{\text{hydration}} CH_3CHO \xrightarrow{\text{NaOH}} CH_3CHOHCH_2CHO \xrightarrow{-H_2O}$$

$$CH_3CH : CHCHO \xrightarrow{H_2} CH_3CH_2CH_2CH_2OH \quad (20)$$

n-Butyl alcohol is converted to *n*-butyl chloride by means of dry hydrogen chloride, and the butyl chloride is chlorinated to give a mixture of dichlorobutanes, from which there is obtained by distillation a fraction containing three isomeric dichlorides. These latter when passed over heated soda lime yield butadiene directly, an intramolecular change taking place in one instance.[141]

$$CH_3CH_2CH_2CH_2OH \xrightarrow{HCl} CH_3CH_2CH_2CH_2Cl \xrightarrow{Cl_2}$$
n-Butyl alcohol   *n*-Butyl chloride

1,2-$CH_3CH_2CHClCH_2Cl$ (b.p. 125° C.)
1,3-$CH_3CHClCH_2CH_2Cl$ (b.p. 134–7° C.) $\left.\right\} \xrightarrow[\text{soda lime}]{\text{heated}} CH_2 : CH \cdot CH : CH_2$
1,4-$CH_2ClCH_2CH_2CH_2Cl$ (b.p. 161–3° C.)
Dichlorobutanes
(by fractionation)

$$(21)$$

The 1,3-compound gives the best yield of butadiene, 29.6 per cent at 700 to 730° C., according to a rather recent report.[121]

Similarly, isoprene is obtained from isoamyl alcohol.[111, 141]

$$CH_3CH(CH_3)CH_2CH_2OH \xrightarrow{HCl} CH_3CH(CH_3)CH_2CH_2Cl \xrightarrow{Cl_2}$$
Isoamyl alcohol   Isoamyl chloride

$$\left[\begin{array}{l} CH_3CH(CH_3)CHClCH_2Cl \\ \quad \text{(b.p. 142° C.)} \\ \qquad \text{2-Methyl-3,4-dichlorobutane} \\ CH_3C(CH_3)Cl \cdot CH_2CH_2Cl \\ \quad \text{(b.p. 152–154° C.)} \\ \qquad \text{2-Methyl-2,4-dichlorobutane} \\ ClCH_2CH(CH_3)CH_2CH_2Cl \\ \quad \text{(b.p. 170–172° C.)} \\ \qquad \text{2-Methyl-1,4-dichlorobutane} \end{array} \right\} \xrightarrow[470° \text{C.}]{\text{soda lime}} CH_2 : C(CH_3) \cdot CH : CH_2 \quad (22)$$

Fraction boiling 140–175° C.
Isoprene

Dichlorobutanes can also be obtained by the action of chlorine on butenes and butane; direct thermal decomposition of them with steam, and with or without catalysts, at temperatures of 450 to 700° C., produces butadiene.[65, 110]

$$CH : CHCH_2CH_3 \xrightarrow{2Cl} CH_2ClCHClCH_2CH_3 \longrightarrow \qquad (23)$$
1-Butene   1,2-Dichlorobutane

$$CH_3CH_2CH_2CH_3 \xrightarrow{2Cl_2}$$
n-Butane

$$\rightarrow CH_2ClCH_2CHClCH_3 \longrightarrow$$
1,3-Dichlorobutane

$$\rightarrow CH_2ClCH_2CH_2CH_2Cl \longrightarrow$$
1,4-Dichlorobutane

thermal decomp'n or soda lime $\rightarrow CH_2 : CH \cdot CH : CH_2$
Butadiene   (24)

$$CH_3CH : CHCH_3 \xrightarrow{2Cl} CH_3CHClCHClCH_3 \longrightarrow \qquad (25)$$
2-Butene   2,3-Dichlorobutane

1-Butene is prepared by the dehydration of n-butyl alcohol, by the dehydrochlorination of n-butyl chloride—the butyl chloride coming from n-butyl alcohol and hydrogen chloride or by the chlorination of butane—and by the cracking of petroleum. 2-Butene is prepared similarly from sec-butyl alcohol and by the cracking of butane or petroleum.

The process actually adopted by Strange and Graham consisted in preparing n-butyl alcohol by fermentation.[164] The butyl alcohol was then converted into 2-butene by dehydration over pumice saturated with phosphoric acid at 300 to 400° C., by which treatment the 1-butene first formed was isomerized to 2-butene, all in practically quantitative yield. Treatment of the 2-butene with chlorine gave 2,3-dichlorobutane which was pyrolyzed over a red-hot spiral of Nichrome wire at about 600° C. in a silica tube to produce butadiene and hydrogen chloride. The butadiene was purified by reacting it with sulfur dioxide in the presence of strong acids or acetyl chloride to give butadiene sulfone in high yield and then decomposing the sulfone (compare reaction 12).

$$CH_3CH_2CH_2CH_2OH \rightarrow [CH_3CH_2CH : CH_2] \rightarrow CH_3CH : CHCH_3 \xrightarrow{2Cl}$$

$$CH_3CHClCHClCH_3 \xrightarrow{\text{Nichrome coil, 600° C.}} CH_2 : CH \cdot CH : CH_2 + 2HCl \quad (26)$$

This general process, using butenes from cracked petroleum, was developed commercially just before World War II, and is described later (p. 124).

1- and 2-Butenes when treated with hypochlorous acid give the corresponding chlorohydrins which with alkali yield epoxides. These epoxides on being passed over heated pumice saturated with syrupy phosphoric acid[113] or over heated kaolin in vacuo produce butadiene. In the following equations only the reactions starting with 2-butene are shown.

$$CH_3CH : CHCH_3 \xrightarrow{HOCl} CH_3CHCl \cdot CH(OH)CH_3 \xrightarrow{alkali}$$
2-Butene   Butylene chlorohydrin

$$CH_3CH \cdot CHCH_3 \xrightarrow{H_3PO_4 \text{ or kaolin}} CH_2 : CH \cdot CH : CH_2 \quad (27)$$
$\diagdown O \diagup$ Butadiene
2,3-Butylene oxide

Similarly isoprene is obtained from 2-methyl-1-chloro-2-butanol.[101]

From Lactic Acid. Another product of fermentation is lactic acid which which can be converted into isoprene by a rather long series of steps:[141]

$$CH_3CHOHCOOH \xrightarrow{HBr} CH_3CHBrCOOH \xrightarrow{esterification}$$
$$\underset{\text{Lactic acid}}{}$$

$$CH_3CHBrCOOC_2H_5 + CH_3COCH_3 + Zn \rightarrow$$

$$CH_3C(CH_3)(OH) \cdot CH(CH_3)COOC_2H_5 \xrightarrow{-H_2O}$$

$$CH_3C(CH_3) : C(CH_3)COOC_2H_5 \xrightarrow[\text{then} + 2HBr]{\text{hydrolysis,}}$$

$$CH_3C(CH_3)Br \cdot C(CH_3)Br \cdot COOH \xrightarrow{\text{pyridine}}$$

$$\underset{\text{Isoprene}}{CH_2 : C(CH_3) \cdot CH : CH_2} \xleftarrow[\text{lime}]{\text{soda}} CH_3C(CH_3) : CBrCH_3 \quad (28)$$

From Ethyl Alcohol. Ethyl alcohol—another product of fermentation—is used for the production of butadiene, as already mentioned. Also it can be converted by dehydrogenation to acetaldehyde, which is the basis of important methods of producing butadiene as described later (p. 121).

From 2,3-Butylene Glycol. 2,3-Butylene glycol—still another product of fermentation—can be converted into butadiene by direct dehydration in the presence of the proper catalyst[184] and by the pyrolysis of its esters. Ordinary dehydration methods convert 1,2-glycols into aldehydes or ketones, with or without isomerization, and they convert 2,3-butylene glycol almost entirely into methyl ethyl ketone through the loss of one mole of water followed probably by the pinacol rearrangement of the first product or by the inter-

$$\underset{\text{2,3-Butylene glycol}}{CH_3CH(OH) \cdot CH(OH)CH_3} \xrightarrow{-1H_2O} \begin{matrix} \nearrow [CH_3CH \cdot CH(OH)CH_3]^- \\ + \\ \searrow [CH_3CH : C(OH)CH_3]^- \end{matrix} \rightarrow \underset{\text{Methyl ethyl ketone}}{CH_3CH_2COCH_3}$$

$$(29)$$

mediate formation of an enol. Sometimes the cyclic acetal of 2,3-butylene glycol and methyl ethyl ketone is formed in addition.[7, 127] Pyrolysis of this compound gives only a small yield of butadiene.

$$\begin{matrix} CH_3CHO|H & | & C_2H_5 \\ | & & O|C & \\ | & & | & \\ CH_3CHO|H & | & CH_3 \end{matrix} \longrightarrow \begin{matrix} CH_3CH \cdot O & C_2H_5 \\ | & C & \\ | & & \\ CH_3CH \cdot O & CH_3 \end{matrix} \quad (30)$$

However, as stated, by means of the proper catalyst, 2,3-butylene glycol has been dehydrated directly to butadiene:[184]

$$CH_3CH(OH) \cdot CH(OH)CH_3 \xrightarrow{ThO_2} CH_2 : CH \cdot CH : CH_2 \quad (31)$$

This direct dehydration method was discovered rather late in World War II. Success in transforming 2,3-butylene glycol into butadiene was attained earlier by the well-known method of forming the di-ester and pyrolyzing it, since this method amounts to the removal of 2 moles of water without the occurrence of any structural change.[76]   The butadiene prepared in this

$$CH_3CH(OH) \cdot CH(OH)CH_3 \xrightarrow[\substack{\text{or AcOH} \\ H_2SO_4}]{Ac_2O}$$

$$CH_3CH(OCOCH_3) \cdot CH(OCOCH_3)CH_3 \xrightarrow{\text{pyrolysis}}$$
$$CH_2 : CH \cdot CH : CH_2 + 2CH_3COOH \qquad (32)$$

manner is very pure and can be used without further purification for making synthetic rubbers.   The yield of butadiene and the recovery of acetic acid are high, but the extra steps and the use of important strategic materials of construction required elsewhere made the process more costly and less adequate than those actually used on a large scale during the war (see p. 128 *et seq*).

Another method that involves a di-ester as an intermediate and also one that produces no structural change is the xanthate method of Chugaev (Tschugaeff).   Ordinarily the alcohol or glycol is treated with sodium or sodium methoxide to make the sodium salt which is reacted with carbon disulfide and methyl iodide to form the methyl xanthate of the alcohol or glycol.   It happens that this reaction does not go with the 1,2-glycols, but a reversed dixanthate can be made from the corresponding dibromide by reaction with sodium ethyl xanthate.   Pyrolysis of the dixanthate thus

$$\begin{array}{l} CH_3CHBr \\ \quad | \qquad\qquad + 2NaSCSOEt \rightarrow \\ CH_3CHBr \end{array} \quad \begin{array}{l} CH_3CHSCSOEt \\ \quad | \qquad \xrightarrow{230\text{–}240° \text{ C.}} \\ CH_3CHSCSOEt \end{array}$$
2,3-Butylene bis(ethylxanthate)

$$\begin{array}{l} CH_2 : CH \\ \quad | \qquad + 2COS + 2EtSH \qquad (33) \\ CH_2 : CH \end{array}$$

formed gives butadiene.[97, 174]   This method is included here because of its relation to the glycol route although it really belongs in the section on butadiene from halogen derivatives.

The inner carbonate of 2,3-butylene glycol,

$$\begin{array}{l} CH_3CH \cdot O \\ \quad | \qquad\qquad \diagdown CO \\ CH_3CH \cdot O \diagup \end{array}$$

when pyrolyzed gives only about 1 per cent butadiene.[94]   The monomethyl ether of the glycol on pyrolysis gives 13 per cent butadiene.[94]

From Acetone.   Acetone, also a product of fermentation, has been used as a starting material for the production of isoprene and dimethylbutadiene (see reactions 34 and 42).

From Calcium Carbide. Both butadiene and isoprene have been produced in quantity from calcium carbide as the starting material, and a great deal of work in this connection has been done in Germany, especially on the production of butadiene from carbide. However, only outlines are given here, and the more important methods are described in greater detail later.

Calcium carbide is first decomposed by water to give acetylene which is then converted by different methods into butadiene and isoprene. One of the first uses of acetylene in synthetic-rubber work was its conversion into isoprene by reacting it with acetone—a product also obtainable from it—in the presence of a sodium catalyst such as sodamide. The 5-carbon acetylenic alcohol thus formed was partially reduced to the 5-carbon olefinic alcohol, which in turn was dehydrated to isoprene.[13, 115] Acetone is, of course, obtainable by cheaper methods than from acetylene.

$$HC : CH \xrightarrow[\text{Hg catalyst}]{\text{hydration}} CH_3CHO \xrightarrow{O_2} CH_3COOH \xrightarrow[\text{or catalytically}]{Ca(OH)_2 + \text{heat}}$$

$$CH_3COCH_3 + HC : CH + NaNH_2 \rightarrow CH_3C(CH_3)(OH) \cdot C : CH \xrightarrow{H_2}$$
2-Methyl-2-butyn-2-ol

$$CH_3C(CH_3)(OH) \cdot CH : CH_2 \xrightarrow{-H_2O} CH_2 : C(CH_3) \cdot CH : CH_2 \quad (34)$$
2-Methyl-3-buten-2-ol

Butadiene also can be made by this method by substituting acetaldehyde for acetone.[89, 165]

Another early method also started with the catalytic hydration of acetylene to acetaldehyde. The acetaldehyde on treatment with alkali becomes acetaldol and the acetaldol is reduced to 1,3-butylene glycol. Perkin and co-workers[141] converted this glycol with hydrogen chloride into 1,3-dichlorobutane which they passed over heated soda lime to produce butadiene.

$$HC : CH \xrightarrow[\text{hydration}]{\text{Hg catalyst}} CH_3CHO \xrightarrow{NaOH} CH_3CHOHCH_2CHO \xrightarrow{H_2}$$
Acetaldol

$$CH_3CHOHCH_2CH_2OH \xrightarrow{2HCl} CH_3CHClCH_2CH_2Cl \xrightarrow{\text{heated soda lime}}$$
1,3-Butylene glycol   1,3-Dichlorobutane

$$CH_2 : CH \cdot CH : CH_2 \quad (35)$$
Butadiene

In a later method, which has been much used in Germany, the 1,3-butylene glycol is dehydrated directly to butadiene,[112] as described in some detail later.

1,3-Butylene glycol can be converted into esters, such as the diacetate and the ester pyrolyzed to give butadiene[134] (compare reaction 32). This reaction has been used commercially on a small scale.[32] The glycol can probably also be converted into butadiene through the formation of the dixanthate followed by thermal decomposition (compare reaction 33). Acetals of 1,3-butylene glycol on pyrolysis give butadiene[58] (compare reaction 30). Thus, for example, 1,3-butanediol formal (4-methyl-1,3-dioxolane or 4-methyl-*m*-dioxane) gives butadiene, when passed over phosphoric acid on graphite at 270° C.

Similarly, 4,4-dimethyl-*m*-dioxane treated with dilute sulfuric acid yields isoprene.[117]

From Dimerized Acetylene. Acetylene can be dimerized to vinylacetylene[129] which is then selectively hydrogenated to butadiene.[20, 28, 81]

$$2HC:CH \xrightarrow{\quad NH_4Cl + CuCl \quad} \underset{\substack{\text{Vinylacetylene} \\ \text{(diacetylene)}}}{CH_2:CH \cdot C:CH} \xrightarrow[\substack{KHg + H_2O, \text{ Pd-diatomaceous} \\ \text{earth, or Pd-black}}]{\substack{\text{hydrogenation in the} \\ \text{presence of Ni, with}}}$$

$$CH_2:CH \cdot CH:CH_2 \qquad (36)$$

Reppe's Process. Another process for preparing butadiene from acetylene was developed by Reppe[70, 144] in Germany in the period just before and early in World War II. Acetylene and formaldehyde, in the presence of copper acetylide, react to form 2-butyne-1,4-diol; the latter is reduced catalytically with hydrogen to 1,4-butanediol, and the diol is dehydrated catalytically to yield butadiene. In the dehydration tetrahydrofuran is usually an intermediate. The yields are high and so is the purity. The commercial method is described later.

$$HC:CH + 2HCHO \xrightarrow{\quad Cu_2C_2 \quad} \underset{\text{2-Butyne-1,4-diol}}{HOCH_2C:CCH_2OH} \xrightarrow{\quad H_2 \quad}$$

$$\underset{\text{1,4-Butanediol}}{HOCH_2CH_2CH_2CH_2OH} \xrightarrow{\quad -H_2O \quad} \underset{\underset{\text{Tetrahydrofuran}}{\underline{\qquad O \qquad}}}{CH_2CH_2CH_2CH_2} \xrightarrow{\quad -H_2O \quad}$$

$$CH_2:CH \cdot CH:CH_2 \qquad (37)$$

Tetrahydrofuran (tetramethylene oxide) itself, isolated in the above reaction or prepared by other methods, can be used directly, and is dehydrated by passing it over heated alumina (80 to 90 per cent yield, as shown by Ostromislensky in 1915)[134] or by mixing it with at least 30 per cent of water and heating above 180° C. over phosphoric acid[149] or over a composite phosphate catalyst prepared by heating a mixture of monocalcium phosphate, monosodium phosphate, butylammonium phosphate, and phosphoric acid.[26]

Tetrahydrofuran is also produced from furfural: (1) by hydrogenation to tetrahydrofurfuryl alcohol and then removing the side chain by passing the alcohol over a nickel catalyst[183] at 240 to 300° C., (2) by decarbonylation of furfural to furan in the presence of steam and manganese chromite containing potassium chromate[50] at 400° C. and then hydrogenating the furan, and (3) by converting the furfural into furoic acid by a Cannizzaro reaction, hydrogenating the furoic acid to tetrahydrofuroic acid, and decarboxylating the latter to tetrahydrofuran.[41]

Miscellaneous Methods. Methyl ethyl ketone gives on pyrolysis only traces of butadiene, but, when water vapor is added as a diluent in a molar ratio of 1 to 40, and the mixture is heated at 700° C., a yield of 44.8 per cent of butadiene is obtained.[30] Methyl ethyl ketone can be transformed into isoprene by aldoling formaldehyde with it, reducing the hydroxy ketone formed, and then dehydrating the glycol.[83]

$$CH_3COCH_2CH_3 + HCHO \rightarrow CH_3COCH(CH_3)CH_2OH \rightarrow$$
$$CH_3CH(OH)CH(CH_3)CH_2OH \rightarrow CH_2 : CH \cdot C(CH_3) : CH_2 \qquad (38)$$

Dry distillation of calcium dihydromuconate, $(OOCCH_2CH : CHCH_2COO)Ca$, gives butadiene together with carbon monoxide and a residue of calcium carbonate.

Catechol heated at 550° C. with nitrogen yields butadiene and carbon monoxide.[68] Perhaps there is a prior tautomerization into 2,4-cyclo-hexadien-1-on-2-ol or 4-cyclohexene-1,2-dione.

Butadiene dimerizes, slowly at room temperature, forming in 30 days 0.3 per cent of its weight of 4-vinyl-1-cyclohexene,[59] more rapidly at higher temperatures.[90, 150] Heating the dimer with benzene vapor over a hot platinum gauze[134] or in a heated tube[152] reconverts it to butadiene.

$$\begin{array}{c} CH \\ H_2C \quad\quad CH \\ | \quad\quad\quad\quad | \\ CH_2 : CH \cdot CH \quad CH_2 \\ CH_2 \end{array} \rightarrow 2CH_2 : CH \cdot CH : CH_2 \qquad (39)$$

Butadiene dimer

Then there are the cyclobutane derivatives. Cyclobutyl bromide heated with an excess of quinoline[136] at 110 to 120° C. and 1,2-dibromocyclo-butane[182] under similar conditions are converted into butadiene. Butadiene can be obtained from cyclobutylamine by exhaustive methylation or by heating with phosphoric acid,[181] and from cyclobutanol by passage over alumina at 300 to 350° C. (93 per cent yield).[136] Ordinarily in these cases cyclobutene should be obtained. It probably is formed first but rearranges, at least in part, to butadiene since butadiene is more stable thermally than cyclobutene.

PIPERYLENE

cis-Piperylene,   b.p. 43.8° C. at 750 mm., $d_{20° C.}^{20° C.}$ 0.6916, $n_D^{20° C.}$ 1.4360[44]

trans-Piperylene   b.p. 41.7° C. at 745 mm., $d_{20° C.}^{20° C.}$ 0.6771, $n_D^{20° C.}$ 1.4300[44]

Piperylene is 1,3-pentadiene, $CH_3CH : CH \cdot CH : CH_2$; it has also been called α-methyldivinyl and α-methylbutadiene. It is an isomer of isoprene, which is β-methylbutadiene.

It was first prepared by Hofmann by his method of exhaustive methylation starting with piperidine[77] (compare reaction 3). The method should give 1,4-pentadiene but a shift of a double bond takes place to form the more stable conjugated system. Piperylene has also been prepared by heating pentamethylenediamine nitrite.[47]

$$HNO_2 \cdot NH_2CH_2CH_2CH_2CH_2CH_2NH_2 \cdot HNO_2 \xrightarrow{\text{heat}} CH_3CH:CH \cdot CH:CH_2$$
$$(40)$$

It has been prepared by other standard methods including the two following methods from furfural:[33, 67]

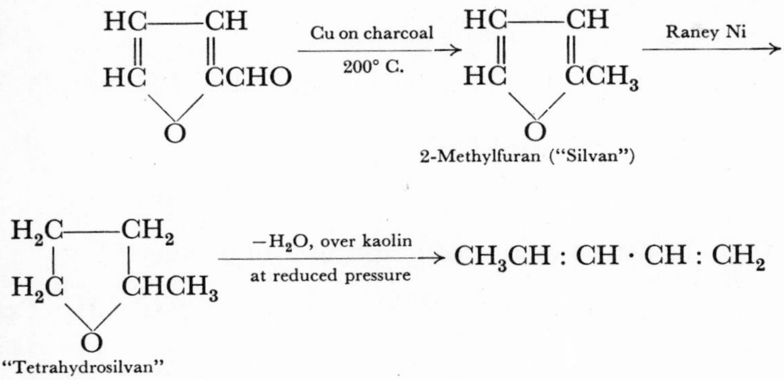

$$\begin{array}{c} HC\!\!-\!\!-\!\!-\!\!CH \\ \| \quad\quad \| \\ HC \quad\quad CCHO \\ \diagdown\!O\!\diagup \end{array} \xrightarrow[\text{200° C.}]{\text{Cu on charcoal}} \begin{array}{c} HC\!\!-\!\!-\!\!-\!\!CH \\ \| \quad\quad \| \\ HC \quad\quad CCH_3 \\ \diagdown\!O\!\diagup \end{array} \xrightarrow{\text{Raney Ni}}$$

2-Methylfuran ("Silvan")

$$\begin{array}{c} H_2C\!\!-\!\!-\!\!-\!\!CH_2 \\ | \quad\quad\quad | \\ H_2C \quad\quad CHCH_3 \\ \diagdown\!O\!\diagup \end{array} \xrightarrow[\text{at reduced pressure}]{-H_2O,\ \text{over kaolin}} CH_3CH:CH\cdot CH:CH_2 \qquad (41)$$

"Tetrahydrosilvan"

From furfural on hydrogenation there can also be obtained 1,2- and 1,5-pentanediols. Dehydration of the 1,2-diol over kaolin, alumina, and phosphorus gives valeraldehyde, but the 1,5-diol over phosphorus gives some piperylene.[16] It is also of interest that pyrolysis of the diacetate of 1,2-pentanediol gives good yields (60 to 65 per cent) of piperylene.[161] Evidently 1,2-pentadiene is formed first and rearranges to the more stable 1,3-pentadiene. Generally, double-bond displacement does not normally occur when acetates of mono- and dihydroxy compounds are converted to olefins by pyrolysis.

Piperylene was found in condensed oil-gas[5] in 1886 and is obtained commercially by the cracking of C_5 hydrocarbons from natural gas and petroleum.

2,3-DIMETHYL-1,3-BUTADIENE (DIISOPROPENYL)

B.p. 69–70° C.; $d_{20°\,C.}^{20°\,C.}$ 0.7262; $n_D^{22°\,C.}$ 1,4370[17]

This diene hydrocarbon, $CH_2:C(CH_3)\cdot C(CH_3):CH_2$, was manufactured by the Germans during World War I to make "Methylrubber," a name given to the polymerized product because each structural unit had one more methyl group than the corresponding unit of natural rubber.[63] The commercial method of preparing this hydrocarbon involved reducing acetone with aluminum and sodium hydroxide to pinacol (pinacone) and distilling this under pressure to dehydrate it.[130] Other methods of dehydration use dilute sulfuric acid,[43, 95] distillation with a small proportion of concentrated hydrobromic acid,[102] and aluminum oxide,[2, 128] at 424 to 470° C.

$$2(CH_3)_2CO \xrightarrow{Al + NaOH} CH_3C(CH_3)(OH)\cdot(OH)(CH_3)CCH_3 \xrightarrow{-2H_2O}$$
$$\text{Pinacol}$$
$$CH_2:C(CH_3)\cdot(CH_3)C:CH_2 \qquad (42)$$

Three other methods are as follows: (a) Heat tetramethylethylene dichloride with alcoholic KOH.[95]

$$CH_3C(CH_3)Cl \cdot Cl(CH_3)CCH_3 \xrightarrow{\text{alc. KOH, 130° C.}} CH_2 : C(CH_3) \cdot (CH_3)C : CH_2 \tag{43}$$

(b) Heat dimethylisopropenyl carbinol with dilute hydrochloric acid:[39, 109]

$$(CH_3)_2C : C(CH_3)_2 \xrightarrow{Cl_2} C_6H_{11}Cl \xrightarrow{H_2O} \underset{\text{Dimethylisopropenylcarbinol}}{(CH_3)_2C(OH) \cdot C(CH_3) : CH_2}$$

$$\xrightarrow{\text{0.1\% HCl, 100° C.}} CH_2 : C(CH_3) \cdot C(CH_3) : CH_2 \tag{44}$$

(c) Heat 2,2,3-trimethyl-3,4-dibromobutyric acid with pyridine:[42]

$$BrCH_2 \cdot C(CH_3)Br \cdot C(CH_3)_2COOH \xrightarrow{\text{pyridine}} CH_2 : C(CH_3) \cdot C(CH_3) : CH_2 \tag{45}$$

METHYLPENTADIENES

There are three branched-chain conjugated methylpentadienes; their structural formulas, names, and physical constants are:

(a) $CH_2 : C(CH_3) \cdot CH : CHCH_3$ b.p. 75.8° C. at 763 mm.; $d_{4° C.}^{20° C.}$ 0.7326;
2-Methyl-1,3-pentadiene $n_D^{20° C.}$ 1.4479[120]

(b) $CH_3C(CH_3) : CH \cdot CH : CH_2$ b.p. 76.3° C. at 760 mm.; $d_{4° C.}^{20° C.}$ 0.7189;
4-Methyl-1,3-pentadiene   b.p. 76.7° C. at 763.1 mm.;
                                                  $n_D^{20° C.}$ 1.4505[6]
                                                  $n_D^{20° C.}$ 1.4520[120]

(c) $CH_2 : CH \cdot C(CH_3) : CHCH_3$ b.p. 77–78° C.; $n_D^{21° C.}$ 1.4561[57]
3-Methyl-1,3-pentadiene

The first two compounds above are easily obtained by the dehydration of 2-methyl-2,4-pentanediol ("diacetone glycol"), which is prepared by the hydrogenation of 4-hydroxy-4-methyl-2-pentanone ("diacetone alcohol"), which is a simple aldol product of 2 molecules of acetone. Both isomers are obtained at the same time, the mixture ordinarily consisting of 85 per cent of 2-methyl-1,3-pentadiene and 15 per cent of its isomer, 4-methyl-1,3-pentadiene.[102, 120] Their boiling points are too close to allow separation by

$$2CH_3COCH_3 \xrightarrow{Ba(OH)_2} (CH_3)_2C(OH) \cdot CH_2COCH_3 \xrightarrow{H_2}$$

$$(CH_3)_2C(OH) \cdot CH_2 \cdot CH(OH)CH_3 \xrightarrow[\text{aniline} \cdot HBr \text{ in } HBr]{-2H_2O,}$$

$$\underset{(85\%)}{CH_2 : C(CH_3) \cdot CH : CHCH_3} + \underset{(15\%)}{CH_3C(CH_3) : CH \cdot CH : CH_2} \tag{46}$$

fractionation. 4-Methyl-1,3-pentadiene can be isolated by treatment of the mixture with maleic anhydride in dioxane at 0° C., in the presence of an antipolymerization agent. It does not react with maleic anhydride while 2-methyl-1,3-pentadiene forms an addition product.[6] 4-Methyl-1,3-pentadiene is isomerized to 2-methyl-1,3-pentadiene by heating with sulfur dioxide.[120]   Treatment of the mixture of isomers with sulfur dioxide gives a product consisting of 2-methyl-1,3-pentadiene only.

3-Methyl-1,3-pentadiene is prepared by a series of standard reactions starting with methyl ethyl ketone and acetaldehyde.[57]   No isomers are produced in this reaction, since only one compound is formed.

$$CH_3COCH_2CH_3 + CH_3CHO \xrightarrow{\text{alkali}} CH_3COCH(CH_3)CH(OH)CH_3 \xrightarrow{H_2}$$

$$CH_3CH(OH)CH(CH_3)CH(OH)CH_3 \xrightarrow[\text{aniline} \cdot \text{HBr in HBr}]{-2H_2O}$$

$$CH_2 : CH \cdot C(CH_3) : CHCH_3 \tag{47}$$

All three methylpentadienes can be obtained by other standard organic reactions.

MYRCENE

Myrcene, $C_{10}H_{16}$ or $CH_3C(CH_3) : CHCH_2CH_2C(: CH_2) \cdot CH : CH_2$ (b.p. 67° C. at 20 mm.) has three double bonds, two of which are conjugated. It occurs in bayberry and hop oils.   Pyrolysis of myrcene gives isoprene.[45] p-tert-Butylcatechol (0.1 per cent) is effective in inhibiting the polymerization of myrcene at room temperature.[155]

Myrcene is prepared by the dehydration of linalool, $CH_3C(CH_3) :$ $CHCH_2CH_2C(CH_3)(OH) \cdot CH : CH_2$, over Japanese acid earth[133] at 159° C., or with potassium hydrogen sulfate or iodine.[31]   Thermal isomerization of β-pinene (formula on p. 135) at above 700° C. gives yields of myrcene up to 85 per cent.[61, 156]

INDUSTRIAL METHODS

BUTADIENE BY THE ALDOL PROCESS

The aldol or "four-step" process was the favored method of producing butadiene in Germany before and during World War II.   As already noted, acetylene is hydrated in the presence of a mercury catalyst to acetaldehyde, the aldehyde treated with dilute potassium hydroxide to convert it into acetaldol, the latter hydrogenated to 1,3-butylene glycol, and the glycol dehydrated to butadiene.   The following description is chiefly from two postwar reports.[69, 105]

Acetylene was manufactured by the usual calcium carbide process at Ludwigshafen and Schkopau and by the electric-arc cracking of hydrocarbon gases at Hüls.

Acetylene by the Electric-Arc Process.   Acetylene is produced at Hüls by the electric-arc cracking of methane (Bentheim natural gas) or methane-ethane mixtures (by-product from coal hydrogenation) by the use of direct current at 7000 volts.   Each of the 15 reactor sets consists of a mercury-arc rectifier and two arc-reaction tube units.   The reaction tube is 1 meter long, 9.5 cm. inside diameter, and has a wall thickness of 9 mm.   It is made of mild carbon steel and is jacketed for water cooling.   Operation of the arc is at 1.5 atm. pressure.   The terminal temperature in the reaction tube is 1600° C., and the reaction gases are immediately quenched by water to 150° C. Conversion per pass through the arc is approximately 50 per cent. One hundred kilograms of fresh gas yield 45 kg. acetylene, 9.2 kg ethylene, 5.3 kg. carbon black, and 143 cubic meters hydrogen. The carbon black is removed first in cyclones and then in bag filters, higher-boiling constituents are removed with oil, and the acetylene is compressed to 19 atm. and absorbed in water. It is stripped from the water by flashing to a final pressure of 0.05 atm., recompressed and recycled until it reaches 90 per cent purity, and then, by low-temperature condensation and evaporation with the use of liquid ammonia, and scrubbing with a petroleum distillate, it is brought to 97 to 98 per cent purity and ready for conversion into acetaldehyde.

Hydration of Acetylene to Acetaldehyde. Acetaldehyde is produced by continuous hydration of acetylene in the presence of sulfuric acid, mercuric sulfate, and iron sulfate over metallic mercury. Conversion per pass is about 55 per cent, and the yield is 93 to 95 per cent of the theoretical. The reaction is conducted in vertical metal towers, lined with rubber, 1.3 meters in diameter and 15 meters high. Acetylene and steam pass up through the mercury and the solution of salts which is constantly being withdrawn for oxidation of the ferrous iron to ferric with nitric acid. The nitrous oxide liberated in this reaction is then oxidized with air to nitric oxide and converted into nitric acid and the cycle is repeated. Mercury usage is about 0.1 per cent based on acetaldehyde. The bottom of the reactor operates at 97° C. and the top at 94° C. The acetylene in the vapors is removed while under a little pressure, the steam condensed and re-used, entrained mercury recovered, and the aldehyde, after having been washed and purified, obtained first as a 7 per cent solution and then, after redistillation, is obtained at about 99.9 per cent purity. Acetone formed in the reaction is recovered, but the diacetyl and crotonaldehyde are discarded.

Production of Aldol. Acetaldol is made by treatment of acetaldehyde with 0.02 to 0.1 per cent of dilute potassium hydroxide, based on the aldehyde, at 20 to 30° C., by rapid circulation through long heat exchangers cooled with water. The conversion is kept at about 46 per cent by controlling the alkali concentration and the withdrawal rate, and thus avoiding the formation of too much crotonaldehyde and other higher aldehydes. The yield is 88 per cent of theory. The average time of contact is 2 to 3 hours. The reactor tubes are of iron, about 20 mm. in diameter, 6 meters long, 12 tubes connected in horizontal runs making up a reactor. The heat production is 300 kg.-cal. per kg. of acetaldehyde reacted.

The aldehyde-aldol mixture is neutralized with phosphoric acid, and the crystals of potassium phosphate formed are removed, partly by centrifuging and then by a leaf filter. The aldehyde is distilled off at atmospheric pressure at a reflux ratio of 1 to 5, surplus water is flashed off, and the crude aldol cooled for storage. A typical analysis shows aldol, 72.82 per cent; acetaldehyde, 4.65, crotonaldehyde, 1.59, water, 18.34; residue, 2.60.

Butylene Glycol from Aldol. 1,3-Butylene glycol (known in Germany as "Butol") is produced by the continuous hydrogenation of the crude acetaldol described above, at 300 atm. pressure and 50 to 150° C., over a catalyst made of 17 to 20 per cent copper and 0.7 to 1.0 per cent chromium on calcined silica gel.   One reactor working at 700 atm. pressure produces twice the amount of glycol in the same time, but it is not known which pressure is the more advantageous.   The reactors are vertical, 18 meters high, 0.8 meters in diameter, and have a volume of 9 cubic meters each; they are constructed of chromium-molybdenum steel and lined with copper.

The catalyst granules are prepared by impregnating the silica gel with the nitrates of copper and chromium, heating them in nitrogen to 200° C., and then reducing the oxides that are formed to the active metals with hydrogen, which is added to the nitrogen in the ratio 80 to 20 at 300 atm. The temperature is then reduced to 50° C.   Aldol heated to 50 to 70° C. enters at the top of the reactor, and a large excess of hydrogen is used to help remove the heat of the reaction, which raises the temperature 50 to 90° C., depending on the age of the catalyst, the lower when the catalyst is new. The reaction is carried to completion in one pass, to only 0.1 per cent aldehyde content.

The crude glycol is heated in a distillation column, and ethanol, butanol, and other low boilers are removed at atmospheric pressure.   It is then heated in a large column first at 51 mm. pressure to remove water, and finally distilled in a third column at 29 mm. pressure, 120° C. top temperature and 174° C. base temperature.   The product is 98.5 to 99.5 per cent pure, and the yield is 93 per cent of the theoretical.

Butadiene from Butylene Glycol.   1,3-Butylene glycol is dehydrated over a sodium phosphate-coke catalyst in the presence of steam at 280° C. and 1 atm. pressure.   The conversion per pass is 100 per cent and the over-all yield is 81 per cent of the theoretical.   The catalyst consists of 5 to 8 mm. granules of coke impregnated with 56 parts of disodium hydrogen phosphate, 8.5 of phosphoric acid, and sufficient butylamine to neutralize the free acid, dried and heated.   The dehydration reactor is a vertical cast-iron cylinder, 3 meters in diameter by 5 meters high, made in segments, and containing 12 cubic meters of catalyst (12 tons).   Between each segment is a pancake coil of steel tubing through which hot water under high pressure is circulated to maintain the reaction temperature.

1,3-Butylene glycol is heated to 210° C., mixed with an equal quantity of steam at 400° C. and 1 atm. pressure, and allowed to flow upward through the reactor, while the temperature is maintained at 280° C., the pressure continuing at 1 atm.   The gases leaving the top of the reactor are cooled to 25° C., and water and oil are separated.   The organic products consist of butadiene 80 per cent, propylene 2, butyraldehyde 2, allylcarbinol (3-buten-1-ol) 10, 2-ethylhexanol 1, heavier oils 4, and butenes 1.   The gas is compressed to about 200 mm. mercury above atmospheric pressure and scrubbed with water.   By a series of compressions, liquefactions, and distillations, butadiene of 99.5 per cent purity is obtained.   The by-products are largely partially dehydrated butylene glycol consisting of the three possible butenols, $CH_2 : CHCH_2CH_2OH$, $CH_3CH : CHCH_2OH$, $CH_3CHOHCH : CH_2$, chiefly the first one, 3-buten-1-ol.

Use of the Aldol Process outside Germany.   The aldol process was used in France by the Usines de Melle[3, 32] in 1936.   The aldol was hydrogenated in the liquid phase under a pressure of 20 kg. per sq. cm. in the presence of Raney nickel. Difficulties caused by incomplete dehydration of the glycol were overcome in 1939 by the use of monocalcium phosphate with amine phosphates as the catalyst. The yield of butadiene reached 87 per cent. During the war the process was improved by preparing 1,3-diacetoxy-butane from butylene glycol and pyrolyzing it (see p. 130, and compare reaction 32).

In the United States during World War II the aldol process was operated on a pilot-plant scale by at least three chemical companies which had been investigating different methods of making butadiene. As soon as the war started, Air Reduction Company and U.S. Industrial Chemicals abandoned their other butadiene experimental work (see p. 129) and in their combined laboratories began small pilot work on the aldol process. The reactions in general were standard except that the dehydration was done in the presence of a suspended catalyst of ammonium phosphate on Filtercel in an inert liquid medium which was heated to about 275° C. and agitated vigorously.[106-7, 166, 175] At the end of the first year these two companies joined with the Celanese Corporation of America in a united effort to build a large pilot plant and perfect the complete process, with the result that something like two tank cars of butadiene eventually were shipped for synthetic-rubber manufacture. Since alcohol was then already being used in a different process and necessary materials of construction were required elsewhere in the manufacture of war products, the work was discontinued.

Work on the aldol process was also carried on in Japan as evidenced by a series of articles in a Japanese journal in 1942.[122] A mixed catalyst of kaolin-ferric oxide-potash at 300° C. and 710 mm. pressure gave only a 40.1 per cent yield of butadiene from 1,3-butylene glycol. A kaolin-phosphoric acid catalyst below 350° C. caused neither polymerization nor decomposition of butadiene.

BUTADIENE BY DEHYDROCHLORINATION OF DICHLOROBUTANE (SHELL PROCESS)*

In the international race, in 1910–14, for the development of methods for the preparation of simple diene hydrocarbons and, from them, of synthetic rubbers, the method selected for the production of butadiene by the English group (Perkin, Matthews, Strange, Graham, and Weizmann) was that from n-butyl alcohol (obtained by fermentation) involving simultaneous dehydration and isomerization to 2-butene, addition of chlorine to the latter, and dehydrochlorination of the 2,3-dichlorobutane to butadiene[141, 164] (see reaction 26). World War I put an end to the operation, but in 1938, just before the beginning of World War II, the Shell Development Company, Emeryville, Calif., commenced large pilot-scale production of butadiene by this same general method, but using as the starting material butene obtained by the cracking of petroleum. The butenes were not separated from the un-cracked butane in the C_4 fraction; the entire mixture was treated with chlorine, and the dichloro compounds were separated, purified, and then pyrolyzed.

* Nearly all the information in this section was kindly provided by the Shell Chemical Corporation, New York.

The production of butadiene was up to one ton daily during the years 1939, 1940, and 1941.   During the latter part of this period a large Shell plant designed to produce 15 tons of butadiene per calendar day was erected at Houston, Texas, and initial operation was begun in August 1941.   The butadiene thus produced was used chiefly by the B. F. Goodrich Company and the Goodyear Tire & Rubber Company in their synthetic-rubber developments.

The process involved the following steps: (a) Vapor-phase chlorination of a C_4-hydrocarbon fraction containing approximately 33 per cent of butenes, (b) separation and pyrolysis of the butylene dichlorides, (c) separation of the products of pyrolysis into hydrogen chloride as 20° Bé. hydrochloric acid, unsaturated C_4-monochlorides (chlorobutenes), and a hydrocarbon fraction containing butadiene, (d) purification of the C_4-hydrocarbon fraction from step c by fractional distillation, to yield a butadiene product of 98.5 per cent purity.   These steps are considered in somewhat greater detail in the following description of the process.

Chlorination.   Chlorination was effected[64] in the vapor phase at a temperature of approximately 70° C.   Under these conditions, substitution in the saturated hydrocarbon (butane) portion of the C_4 feed is slight, and the main reaction consists of addition of chlorine to the double bond to form mainly 2,3-dichlorobutane.   The residual normal butane was washed free of acidic constituents and returned to the refinery whence it came.   The chlorine-to-olefin ratio was so adjusted that substantially all of the olefin portion of the C_4 feed was reacted.   Small amounts of hydrogen chloride were formed by substitution and were removed along with the residual butane as an overhead product in a butane recovery column.   The bottom product of the column was substantially free of hydrogen chloride and hydrocarbon and consisted mainly of 2,3-dichlorobutane.   This material was further distilled to separate high-boiling residues, and the overhead product, which was neutral and water-white in color, was used as feed to the pyrolysis section.

Pyrolysis of Dichlorobutane.   Pyrolysis of the dichlorobutane was carried out in a tubular heater operating at 580° C. without the aid of a catalyst.[65]   Examination of the structure of 2,3-dichlorobutane indicates that on pyrolysis the molecule may be expected to decompose to form 75 mole per cent of butadiene and 25 mole per cent of 2-chloro-2-butene. Hydrogen chloride equivalent to the loss in chlorine is, of course, generated in this reaction.   In actual practice it is not possible to decompose the dichlorobutane completely without encountering severe carbonization. Accordingly, conversion levels were maintained somewhat below the theoretical, and provision was made to recover the unreacted dichlorobutane from the pyrolysis products.

The first step in the recovery of the pyrolysis products after quenching was the absorption of hydrogen chloride in water to form 20° Bé. hydrochloric acid which, during the operation of the plant, was largely sold to the Tin Processing Corporation at Texas City.   This hydrochloric acid amounted to one of the largest single sources of this material in the United States.   After the absorption of the hydrogen chloride the residual products were scrubbed with caustic and separated by distillation into a hydrocarbon fraction and a residual chloride fraction. The residual chloride fraction consisted largely of 2-chloro-2-butene and some unreacted dichlorobutane. The dichlorobutane was recovered by distillation and recycled through the pyrolysis step. The 2-chloro-2-butene was disposed of by being burned in the refinery boilers.

Purification of Butadiene. The hydrocarbon portion of the pyrolysis products boiling in the C_4 range and containing approximately 90 per cent of butadiene was fractionated to remove low-boiling and high-boiling impurities. The high-boiling impurities, consisting of 2-butyne (dimethylacetylene) (b.p. 27° C.) and 1,2-butadiene (b.p. 10.3° C.), were removed as a bottom product in the first of three columns. The butadiene-rich overhead from the first column was fed to a second column operating as an extractive distillation column, using an acetone-water mixture as an entraining liquid.[119] Impurities which consisted mainly of butenes were removed as a top cut from the extractive distillation column. The acetone-water-butadiene stream from the bottom of the column was fed to a third column which acted as a stripping column to separate butadiene of specification grade as a top product and return the acetone-water solvent to the extractive column.

Although initial operating difficulties prevented attainment of the design output for some months, slight changes in operating technique and improvements in equipment permitted attainment of this output early in 1942; and further improvements in equipment led to a production of approximately 22 tons per calendar day of finished butadiene by the end of 1942. Though this amount of butadiene is small compared with the productions from the alcohol and catalytic dehydrogenation processes which were developed in the later stages of the rubber program, the timeliness of the product of this plant was very important in the early stages of the development of synthetic rubber.

A serious drawback to the use of the process on a larger scale was the difficulty of obtaining adequate supplies of chlorine. Further expansion in capacity would have necessitated construction of facilities for the recovery of chlorine from the hydrogen chloride by-product and this was one of the factors that influenced the decision to use either the alcohol or the catalytic dehydrogenation processes in the war program.

BUTADIENE BY PYROLYSIS OF CYCLOHEXANE (KOPPERS PROCESS)*

Introduction. In the international race for synthetic rubber around 1910 two of the earliest methods considered and patented for preparing butadiene were the pyrolysis of cyclohexane and of cyclohexene. The cyclohexane was obtained by the then rather new catalytic hydrogenation of benzene, and the cyclohexene came from phenol by hydrogenation followed by dehydration (see reactions 18 and 19). Both the benzene and phenol were, of course, produced from coal tar, of which a fairly large amount was available.

In the beginning of World War II the two chief processes proposed by the

* Nearly all the information in this section was kindly provided by Koppers Company, Pittsburgh, Pa.

chemical industry to supply the butadiene for synthetic rubber were from alcohol (see Chapter 4) and from benzene through cyclohexane.[80] The cyclohexane process was proposed and studied by the Koppers Company, producers and refiners of coal tar.   This process had the advantage of also supplying the ethylene, formed as a by-product, which was required for the manufacture of the second ingredient necessary for synthetic rubber (Buna S, later GR-S), namely, styrene.   Although the process was piloted and found reliable, and although full-scale plans for a commercial unit that would produce 20,000 short tons of butadiene a year were drawn up, the process was abandoned[154] during the summer of 1942 because the method used more critical materials of construction than other methods and because there was not enough benzene for the process since benzene was also required for the production of styrene and of cumene and other components of aviation fuels. The pilot plant was built and operated at the Koppers Seaboard plant, Kearney, N.J., and had an output of 80 lb. of butadiene of 93 to 94 per cent purity a day.   The benzene hydrogenation section of the pilot plant was started in September 1941 and the cyclohexane cracking section in March 1942.

Hydrogenation of Benzene. The process of hydrogenation of benzene required specially purified benzene, since it involved the use of a sulfur-sensitive catalyst and only low pressure.   With ordinary commercial benzene, a catalyst less sensitive to sulfur and a high pressure would have been required. It was necessary to reduce the amount of sulfur in the benzene to less than 1 p.p.m.   Nitration-quality benzene was purified by freezing; treatment with sodium hydroxide in methanol would also have been suitable.

Impure hydrogen from pitch cooking, which contained 80 per cent hydrogen, the remainder being methane, carbon monoxide, and nitrogen, was purified in oxide boxes in the usual way, compressed, and passed over the nickel catalyst at 200° C. to convert the carbon monoxide into methane. The catalyst was prepared from nickel oxide on kieselguhr in 1/8-in. pellets.   It was contained in 1-in.-diameter tubes cooled with water and kept at 338° F. (170° C.) under accurate temperature control.   The catalyst consumption was 1 lb. for every 300 gal. of cyclohexane formed.   The benzene was preheated and together with the hydrogen was passed through two converters in series.   The first converter contained the partially spent catalyst and acted as a guard for removing any sulfur.   The reaction pressure was 100 p.s.i., and the temperature, as mentioned, was 338° F. Practically complete conversion of the benzene to cyclohexane was obtained in one pass, and the yield was almost quantitative.

Pyrolysis (Cracking) of Cyclohexane. Cyclohexane heated to 1000° F. (538° C.) was mixed with 2.5 times its weight of steam that had been super-heated to 1950 to 2000° F. (1066 to 1093° C.), the mixture thus being brought to a temperature of about 1450° F. (788° C.), under which the cyclohexane was cracked to produce butadiene and equivalent proportions of ethylene and hydrogen, with some acetylene, propylene, and butylenes as by-products.[89]   The procedure was as follows:

Cyclohexane and water were pumped separately at the desired rate through each of two parallel coils in a pipe still and heated to 1000° F., which is the maximum temperature for the carbon-steel equipment used and the maximum to which cyclohexane could be heated without being prematurely decomposed. The steam, after having gone through one coil of the pipe still and having reached the temperature of 1000° F., was passed over bricks in a stove heated beforehand by a downdraft burner to a temperature above 2000° F., and then brought to the constant temperature of 1950 to 2000° F. by the addition of the proper amount of cool steam by means of an automatic controller. The proportion of cool steam added was, of course, diminished continually until the temperature of the bricks fell to 2000° F., when the operation was stopped and the bricks reheated (about an 8-hour cycle).

The hot cyclohexane at 1000° F. and the steam at 2000° F. were mixed, and the mixed gases, now at a mean temperature of 1450° F., were passed directly into the reactor, which was a horizontal cylindrical vessel lined with fire brick and containing a nest of Carborundum tubes. The cracking took place with a reaction time of 0.1 second. In order to recover butadiene[61a] the gases then went directly to a quenching tower in which hot water was circulated, and from this into a condenser. The water in the condensate was separated and returned to the quenching tower. The uncondensed gas was scrubbed with cyclohexane to absorb the butadiene; this solution was mixed with the hydrocarbons that were separated from the water in the condensate and the mixture fed to a rectifier. The tops from the rectifier were sent back through the absorbers, and the main portion was fed to a column working at 45 p.s.i. pressure, from which butadiene was obtained overhead and cyclohexane as bottoms. All columns were 8 in. in diameter, about 40 ft. long, and packed with 1.5-in. Raschig rings. The butadiene was 93 to 94 per cent pure, the remainder consisting chiefly of other C_4 hydrocarbons. In the projected full-scale plant the butadiene was to be purified by azeotropic distillation with ammonia according to the Dow process.[143]

During the operation 2.5 gal. (17 lb.) of cyclohexane were consumed per hour, giving 7 lb. of butadiene. Since the conversion was 30 per cent, the yield was about 62 per cent of the theoretical. From the cyclohexane solution there were obtained 2 to 2.5 gal. of dimerized butadiene each week. The gas from the absorber consisted approximately of hydrogen and inert gases 44 per cent; methane 6.1, ethylene 36.7, other C_2 hydrocarbons 4.0, C_3 hydrocarbons 8.0, C_4 hydrocarbons 1.2.

BUTADIENE FROM FERMENTATION
2,3-BUTYLENE GLYCOL

General Discussion. 2,3-Butylene glycol or 2,3-butanediol was first prepared almost a century ago,[185] in 1859; that was by a standard organic reaction, and it was not until almost half a century later,[71] in 1906, that it was prepared by a fermentation method which was to be the most interesting method of all. Production of the glycol by the fermentation of starches and sugars led a number of people early in World War II to consider it as a probable cheap source of butadiene. However, pyrolysis of the glycol with the use of the common dehydration catalysts, alumina and kaolin, ordinarily gives only very poor yields of butadiene,[118] since the glycol is converted largely into methyl ethyl ketone (see reaction 29), except in the one instance reported late in the war,[185] in which thoria was used as the catalyst and a yield of 70 to 80 per cent of butadiene obtained (reaction 31).

It was known that α- or 1,2-glycols, that is, glycols with the hydroxyl groups on adjacent carbons, can be converted into diene hydrocarbons by first forming a di-ester, such as the diacetate, and pyrolyzing it. The acid obtained as a by-product is recovered and can be recycled to form more ester, and so on. There are several steps in the process, but the yield of the diene is high, and the recovery of the acid is almost quantitative. In the case of 2,3-butylene glycol the method was described in 1938 in a patent[76] and in 1939 in a scientific article.[48] Furthermore, a workable method of preparation of the glycol by fermentation had been described also in a patent,[92] in 1933, and in a scientific article,[37] in 1936.

More Recent History. 2,3-Butylene glycol was prepared in large pilot-plant lots by U.S. Industrial Chemicals, in 1939, and its dehydration to butadiene was studied. The results were similar to those of other investigators and not encouraging. The inner carbonate of the glycol was prepared, but pyrolysis of it gave very low yields of butadiene, as was reported later from independent work.[94] As soon as World War II began, the investigation was discontinued in favor of the aldol process (see p. 124).

Early in World War II other companies and institutions became interested in producing butadiene from 2,3-butylene glycol. Preliminary studies had shown that from 1 bushel of corn there could be obtained 7 to 7.5 lb. of butadiene by going through the 2,3-butylene glycol diacetate process,[51, 93] whereas the yield of butadiene by the way of fermentation ethyl alcohol and either the aldol or the catalytic alcohol-acetaldehyde process was only 5.6 lb.[59, 60, 93] Alcohol could, of course, be obtained more cheaply from molasses and from ethylene than from corn or other grains.

On August 3, 1942, a conference was held at the Northern Regional Research Laboratory of the U.S. Department of Agriculture at Peoria, Ill., to disclose the data already obtained and discuss the commercial possibilities of obtaining butadiene through the 2,3-butylene glycol diacetate reaction, and as a result of the conference a cooperative research program was organized under the general sponsorship of the Office of Rubber Director, War Production Board.[118] The following agencies collaborated: Columbia Brewing Company (Doane Agricultural Service), Commercial Solvents Corporation, Heyden Chemical Corporation, Iowa State College, Lucidol Corporation, Merck & Company, National Research Council of Canada, Northern Regional Research Laboratory, Pennsylvania Sugar Company, Polytechnic Institute of Brooklyn, Schenley Research Institute, Joseph E. Seagram & Sons, and University of Wisconsin. The work reached large pilot-plant scale but never attained commercial production because to handle the acetic acid the process required important strategic materials necessary elsewhere and because the difficulties involved in separating the glycol from the dilute fermentation "beer" were never completely overcome. The following description of the general process is from published articles on laboratory and especially pilot-plant work.

Fermentation. Bacteria of the type of *Bacillus polymyxa* convert unsaccharified starchy substrates to a mixture of *l*-2,3-butanediol and ethyl alcohol, whereas bacteria of the genus *Aerobacter* transform most mono- and disaccharides principally to *meso*-2,3-butanediol, accompanied by only small quantities of *d*-2,3-butanediol and ethyl alcohol.[104, 126, 168, 177] The use of *Aerobacter* requires that starchy substrates be converted first to reducing sugar by either acid or enzymes before the diol fermentation. In the first case, using *Aerobacillus polymyxa*, the molar ratio of diol to ethyl alcohol hovers[96] around 1 to 1 but may vary considerably. The fermentation takes 40 to 60 hours and the yields are approximately 27 per cent 2,3-butylene glycol, 15 per cent ethyl alcohol, and 1 per cent acetoin.[96] In the second case, using *Aerobacter aerogenes*, there is obtained from acid-hydrolyzed starches 13.5 to 14 lb. 2,3-butanediol per bushel of corn. The higher yield represents 41 per cent on the starch (34 lb.). When converted to butadiene, this in turn represents (at a yield of 85 per cent) 7.1 lb. butadiene per bushel of corn—a yield noticeably higher than that obtainable by first fermenting the corn to ethyl alcohol and then converting the latter to butadiene by either the aldol process or the Carbide and Carbon process.[59, 60, 93]

Recovery of the glycol is difficult because of its comparatively high boiling point (the *meso*-form boils at 183° C. at 760 mm.) and the presence in the liquor of 2 parts of nonvolatile solids to 1 part of glycol.[25] The glycol can be isolated by filtration of the mixture followed by distillation or by solvent extraction with, for example, butyl alcohol.[24, 96, 138-9] Losses in the recovery process sometimes amount to as much as 50 per cent.[153]

Esterification. The esterification of 2,3-butylene glycol to the diacetate has been worked out satisfactorily by the use of glacial acetic acid with sulfuric acid as the catalyst in a continuous process with and without entraining agents for the water formed in the reaction.[34, 158, 160] The over-all yield is 97 per cent. 2,3-Butylene glycol and a catalytic amount of sulfuric acid are fed into the top of a reaction column while a continuous stream of glacial acetic acid is introduced into the bottom. Temperatures of 140 to 150° C. are maintained in the zone between the feeds. The column distillate, consisting of acetic acid, water, and traces of methyl ethyl ketone, is dehydrated in an auxiliary column and the acetic acid returned to the first column. The base product from the reaction column is a mixture of butylene glycol diacetate, acetic acid, and sulfuric acid esters of the glycol and is separated by vacuum distillation, the acetic and sulfuric esters being returned to the esterification column as feed and catalyst, respectively.

The side reactions to methyl ethyl ketone and butadiene occur to a certain extent under all conditions although minimized by conducting the esterification as a continuous rather than as a batch operation. These products probably result chiefly from the decomposition of the mono- and disulfuric acid esters.

Pyrolysis. Pyrolysis of 2,3-butylene glycol diacetate to butadiene is best accomplished at a temperature of 595° C. and a contact time of about 1.06 seconds.[34, 118, 125, 157, 159] The conversion is 83 per cent to butadiene of 99+ per cent purity, and the yield can be increased to 88 per cent by the repyrolysis of recovered intermediate butenol acetates. About 8 per cent of useful by-products are obtained, consisting of methyl ethyl ketone, methyl ethyl ketone enol-acetate (2-acetoxy-2-butene), and "methyl acetyl acetone," the last two being hydrolyzable to methyl ethyl ketone and acetic acid. The over-all recovery of acetic acid is 99 per cent. About 0.8 per cent of the initial diacetate is converted to vent gases.

The pyrolysis appears to be a noncatalyzed, homogeneous, gas-phase reaction involving two stages: (a) elimination of one molecule of acetic acid to form a mixture of three butenol acetates,

(I) $CH_3CH(OCOCH_3) \cdot CH : CH_2$
Methyl vinyl carbinol acetate (3-acetoxy-1-butene)

(II) $CH_3C(OCOCH_3) : CHCH_3$
Methyl ethyl ketone enol-acetate (2-acetoxy-2-butene)

(III) $CH_3CH : CHCH_2OCOCH_3$
Crotyl acetate (1-acetoxy-2-butene)

and (b) elimination of another molecule of acetic acid from I and III to form butadiene. In the conversion of compound III to butadiene a rearrangement takes place.[134] The action of heat on compound II transforms it into "methyl acetyl acetone," or 3-methyl-2,4-pentanedione, $CH_3COCH(CH_3)COCH_3$, which, as mentioned above, can be hydrolyzed to methyl ethyl ketone and acetic acid. The over-all course of the pyrolysis reactions—that is, the conversion to butadiene, the composition and yield of unsaturated acetate intermediates, and the formation of by-products—is determined essentially by the temperature and time of contact. The quality of butadiene produced by this method is exceptionally high, 99 per cent pure, under all conditions of operation.

The pyrolysis unit consisted of a gas-heated lead bath in which the pyrolysis coil was immersed. The glycol diacetate was pumped from a weigh tank to a vaporizer which was connected to the inlet of the coil; the outlet of the coil was connected to the base of a jacketed column where the butadiene was separated from condensed acetic acid and other liquid products. The butadiene from the head of this column was washed with water, dried in a tower packed with calcium chloride, compressed and condensed for storage. All parts of the unit that came in contact with liquid or gaseous acetic acid were constructed of stainless steel.

BUTADIENE FROM ACETYLENE BY THE REPPE PROCESS

In recent years the Germans built their chemical economy on acetylene which they obtained mostly from calcium carbide but partly by the electric-arc cracking of methane from natural gas and by-product methane-ethane mixtures from the hydrogenation of coal. A description of the manufacture of butadiene from acetylene through the aldol process has already been given. During the 1930's and the war years Walter Reppe[145-148] of the I.G. Farbenindustrie A.-G. developed several important methods of making some old and some new commercial chemicals by the bold use of acetylene under high temperatures and high pressures. Copper acetylide—an

explosive compound—was harnessed as a catalyst. By diluting the acetylene
with an inert gas, nitrogen, or by constructing suitable apparatus with pipes
filled with bundles of small tubes and bends filled with Raschig rings, Reppe
and his co-workers[70] found it possible to handle acetylene at 200 to 300 p.s.i.
pressures at temperatures up to 200° C. A full-scale plant for the production of butadiene by the Reppe process was built in Germany during World
War II and operated successfully, although when the war came to an end
experience with the plant had not been sufficient to determine conclusively
whether the process was cheaper than the aldol process.

The Reactions. The reactions and a brief discussion of the organic
chemistry of the process are found on page 117. Suffice it to repeat that
(*a*) 2 moles of formaldehyde are reacted with 1 mole of acetylene at 4.5 atm.
and 100 to 120° C. in the presence of copper acetylide as the catalyst, with
release of 55 kg.-cal. per gram-mole; (*b*) the 2-butyne-1,4-diol thus formed
is hydrogenated to 1,4-butanediol at 300 atm. and 120° C. in the presence of
a copper nickel catalyst, with generation of 60 kg.-cal.; and (*c*) the 1,4-
butanediol is dehydrated to butadiene at 1 atm. and 280° C. over a sodium
phosphate catalyst, with absorption of 23.5 kg.-cal. In the dehydration
step, with the loss of the first molecule of water there is formed tetrahydrofuran (tetramethylene oxide) as an intermediate, and butadiene by further
dehydration with the loss of the second molecule of water. The reaction
can be carried out in two separate steps, and this is done in a special plant at
Ludwigshafen. Apparently it has not yet been decided which is the more
economic and advantageous process. Besides, tetrahydrofuran is becoming
an important intermediate for making other chemicals.

The practical over-all yield of butadiene by the Reppe process is 71 per
cent of the theoretical, based on formaldehyde; earlier research had indicated
80 to 87 per cent. For comparison it is mentioned that the over-all yield of
butadiene by the aldol process is 60 per cent of the theoretical, based on
acetylene, and 64 per cent, based on acetaldehyde.

The general description of the process which follows is chiefly from a
postwar report.[79]

The Process—First Step to Butynediol. Methanol diluted to 60 per
cent with water is mixed with air, and the mixture is heated to 600° C. and
passed over an appropriate catalyst to oxidize it to formaldehyde. The
formaldehyde as a 30 per cent solution and an excess of acetylene are circulated over a copper acetylide catalyst at 100 to 120° C. and 5 atm.
pressure to react all the formaldehyde and produce 2-butyne-1,4-diol.
If only equimolecular proportions are used, propargyl alcohol (2-propyn-
1-ol), $HC : C \cdot CH_2OH$, is obtained. Some of this substance is always
formed as a by-product in the first reaction. Each reactor is 1.5 meters in
diameter, 18 meters high, and 30 cubic meters in volume, is fabricated out
of V_2A metal to withstand 100 atm. pressure and is housed in a separate
explosion bay. Usually two sets of two reactors are used in series to prevent
too high a rise in temperature, to keep the formaldehyde solution dilute, and
to use up all the formaldehyde. The over-all conversion of formaldehyde is
95 to 98 per cent, and the ultimate yield based on the formaldehyde is
90 per cent.

The reaction mixture is filtered to remove any copper acetylide that may be present and passed into a separator. The gases are washed and cooled to recover the acetylene, which is recycled. The crude butynediol is fractionated, methanol, propargyl alcohol, and unreacted formaldehyde going overhead, and a 35 to 40 per cent aqueous solution of butynediol obtained. The latter is hydrogenated to butanediol.

Copper Acetylide Catalyst. The catalyst is prepared by impregnating pellets of silica, 2 to 6 mm., with a solution of copper and bismuth nitrates, drying, and heating in a muffle to 500° C. The mixture is placed in a reactor and treated with formaldehyde and acetylene at a temperature of about 70° C. By being careful to saturate the impregnated silica with formaldehyde solution first, adding acetylene slowly, and bringing up the temperature slowly, the active copper acetylide catalyst is formed. The catalyst is probably a complex mixture of CuC_2, H_2O, C_2H_2.

Hydrogenation to Butanediol. The hydrogenation of the butynediol as a 35 to 40 per cent solution takes place with hydrogen at 300 atm. pressure in the presence of a copper-nickel catalyst on silica. There are seven high-pressure reactors in parallel, two of which are used for hydrogen purification. Each reactor is 0.8 meter in diameter and 18 meters high. The feed enters the reactors at 20° C. and leaves at 120° C. Since the reaction is highly exothermic (60 kg.-cal. per gram-mole), 80 times the theoretical amount of hydrogen is circulated through the reactors. The liquid product is a solution of mainly 35 to 40 per cent butanediol and 1 to 2 per cent butanol. The proportion of butanol gradually increases as the catalyst gets older. The butanediol solution is purified in seven bubble columns.

The catalyst is basically copper-nickel on silica, and contains 16 parts of nickel, 5 of copper, and, as an activator, 0.7 of manganese. Sodium silicate is put into a stirred vessel and dilute sulfuric acid run in below the stirrer. The precipitate is filtered in a filter press, dried in a drum drier, ground and made into a paste in a pug mill. The paste is extruded in cylindrical form, dried and heated in a rotary kiln. The calcined pellets are impregnated with the salts of the desired metals, dried and heated in a kiln to 450° C. The dried product is placed in a reactor and treated for 24 hours at 320° C. and 50 to 100 atm. with hydrogen to reduce the metals to the active form. The use of nickel and copper, instead of copper only as in the hydrogenation of aldol, prevents extensive formation of butanol but requires purer hydrogen.

Butanediol to Butadiene. The dehydration takes place at 280° C. and 1 atm. pressure over sodium phosphate deposited on coke as catalyst. Tetrahydrofuran is formed in the first step very readily with the release of only a small amount of heat, 3.2 kg.-cal. per gram-mole. The heat absorption in the second step, 26.5 kg.-cal. per gram-mole, constitutes a difficulty because of the use of large granules of coke catalyst. The vertical cast-iron cylinder reactors are similar to those used for dehydrating 1,3-butylene glycol to butadiene. The reaction is carried to 20 per cent conversion per pass. The catalyst life is 3 weeks and is limited by the increase in pressure drop.

The butanediol and tetrahydrofuran from other runs are vaporized separately, mixed together and then with steam, and preheated to 330° C. with steam at 100 atm. in an electrical-resistance superheater. The gases enter the base of the reactor at 320° C., and the pancake coils between the segments keep the temperature at about 280° C. The space velocity is about 0.5 kg. butadiene per liter of catalyst per day. From the composition of the feed and the product, it is believed that the reaction is in effect one of dehydration of butanediol in the presence of steam with tetrahydrofuran acting as an intermediate product and diluent. The butadiene is separated from the other products by several distillations and dried and compressed for storage.

For the preparation of the catalyst, coke is broken into 8 to 15 mm. granules and treated with solutions of disodium hydrogen phosphate and butylammonium phosphate. The impregnated coke is sprayed in a rotary drum with sodium hydroxide solution and the water then evaporated by direct contact with flue gases. The particles are heated to 260° C. in a tunnel kiln to transform the Na_2HPO_4 first to $Na_2H_2P_2O_7$ and finally to 80 per cent $(NaPO_3)_2$ and 20 per cent $Na_2H_2P_2O_7$. This catalyst is the same as for the dehydration of 1,3-butylene glycol to butadiene except that it is more completely converted to $(NaPO_3)_2$.

ISOPRENE FROM TURPENTINE (NEWPORT INDUSTRIES)*

Isoprene was first isolated from among the products of the dry distillation or pyrolysis of natural rubber and gutta-percha by C. G. Williams[179] in 1860, and later this method was frequently used for the preparation of small laboratory quantities of it—even as late as 1930, by Midgley and co-workers for synthetic-rubber studies.[116] The yield, however, was small, sometimes only up to 3 to 5 per cent. Tilden[173] obtained isoprene by the pyrolysis of turpentine, the yield being only about 7.5 per cent. Harries[73] invented an "isoprene lamp," with which he obtained fairly high yields of isoprene by passing the vapors of dl-limonene (dipentene) over an electrically heated platinum spiral; this method was also used by Gottlob,[62] a student of Harries; by Staudinger and Klever,[169] under reduced pressure (yield 67 per cent); and by Whitby and Crozier.[178] F. Hoffman[78] in 1909 synthesized pilot-scale lots of isoprene by starting with p-cresol (see reaction 17). Perkin[141] started with isoamyl alcohol (reaction 22), and during World War I Merling's [115] process from acetylene and acetone (reaction 34) was used on a small commercial scale in Germany.

The pyrolytic method came to the fore again when, just before and during World War II, Newport Industries, Inc., produced isoprene on a large scale by the pyrolysis of dl-limonene with electrically heated Nichrome coils, in accordance with Bibb's process.[22, 140]

The source of terpene hydrocarbon for conversion was largely determined by the availability of raw material, but in any event the process was primarily dependent on the presence of a substantial concentration of dl-limonene (dipentene) in the feed. The commercial dipentene obtained by the fractional distillation of oils extracted from stump pine wood was employed in part, but the synthetic dl-limonene produced by the isomerization of pinene

* Nearly all the information in this section was kindly provided by Newport Industries, Inc., Pensacola, Fla.

was preferred because of its greater purity which resulted in higher yields of isoprene.

Freshly distilled pinene from wood turpentine, which consists chiefly of α-pinene (b.p. 155° C.) and practically no β-pinene (b.p. 165° C.), was heated for 2 hours at 250° C. and thus converted into a mixture of *dl*-limonene, alloöcimene, and polymers of alloöcimene, with only a small proportion of unconverted pinene remaining. The major portion of the conversion products was *dl*-limonene. For the formation of isoprene it is desirable to split the butane ring in pinene to make limonene without at the same time allowing the double bond that is formed in the external isopropylene group to wander into the ring, since terpene hydrocarbons having two double bonds in the ring do not yield commercial quantities of isoprene.

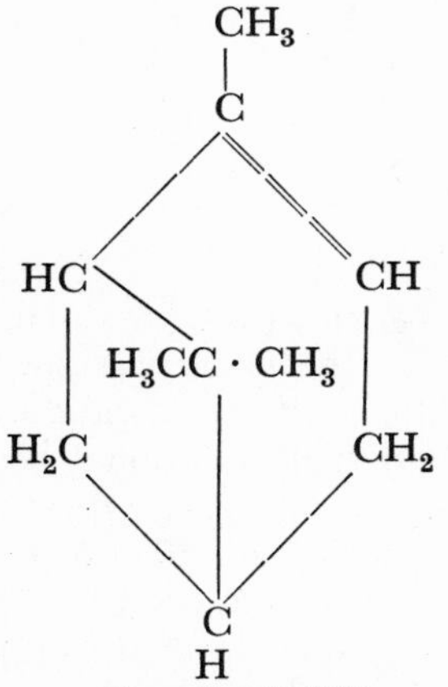

α-Pinene (b.p. 155° C.)

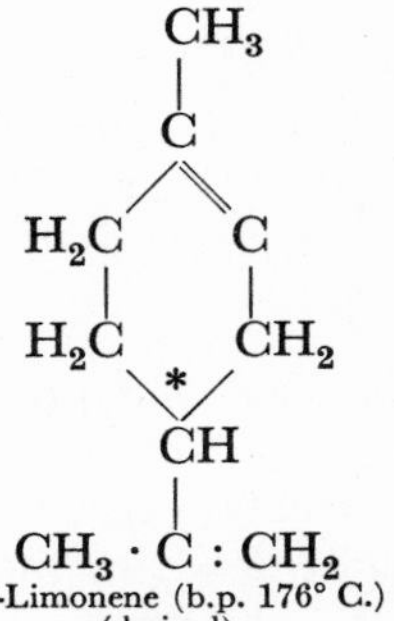

dl-Limonene (b.p. 176° C.)
(desired)

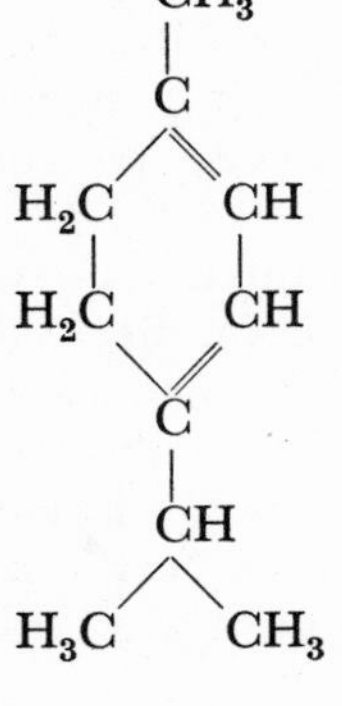

α-Terpinene (b.p. 174° C. at 755 mm.)
(not desired)

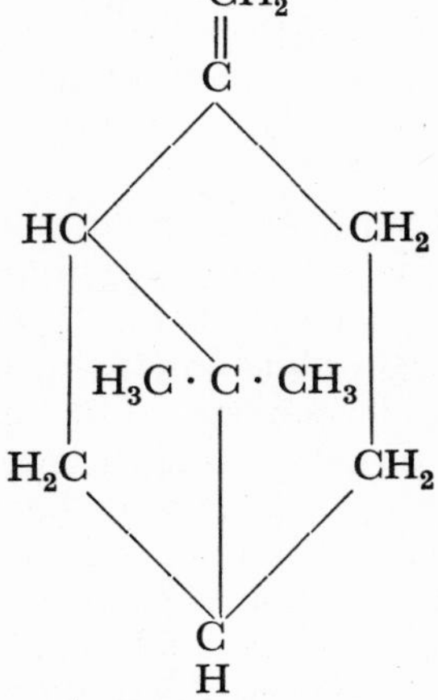

β-Pinene (b.p. 165° C.)

The reaction that forms alloöcimene results from a split of the hexene ring in addition to a split of the butane ring but apparently does not occur after the butane ring is broken.   By fractional distillation it is relatively easy to isolate the unreacted pinene, *dl*-limonene, alloöcimene, and its polymers as separate cuts.

$$
\begin{array}{c}
\text{CH}_3 \\
| \\
\text{C} \\
\diagup\!\!\diagdown \\
\text{HC} \quad\quad \text{CH} \\
\| \quad\quad\quad | \\
\text{HC} \quad\quad \text{CH}_3 \\
\diagdown \\
\text{CH} \\
\| \\
\text{H}_3\text{C}\!-\!\text{C}\!-\!\text{CH}_3
\end{array}
\quad\text{or}\quad (\text{CH}_3)_2\text{C} : \text{CH} \cdot \text{CH} : \text{CH} \cdot \text{C(CH}_3) : \text{CHCH}_3
$$

Alloöcimene (2,6-dimethyl-2,4,6-octatriene,
b.p. 85–86° C. at 16 mm.)

Stainless-steel-clad pressure vessels heated with Dowtherm and holding 700 to 1000 gal. of charge were employed for the isomerization.   At the end of the reaction period, they were vented into condensers, and the pressure was brought down to that of the atmosphere, such additional heat being supplied as was necessary to distil over all the oil except the polymeric residue.   The distillate was the feed stock for the vacuum fractionating stills.

Twelve hundred gallons of *dl*-limonene were charged into horizontal iron pyrolyzers.   Electrically heated Nichrome coils were located well down in the charge.   Regulated current was turned on, and the coils were brought to a red heat which decomposed some of the limonene into isoprene.   The vapors which contained a considerable proportion of *dl*-limonene were conducted to a 12-plate fractionating column with dephlegmator.   The limonene refluxed back to the pyrolyzer, and the isoprene passed the dephlegmator and was condensed in brine-cooled condensers.   This treatment was continued until the rate of isoprene formation reached an uneconomic level.

The crude isoprene so produced was refined by distillation at atmospheric pressure.   The columns used were packed with 1/2-in. Berl saddles and operated at a high reflux ratio.   Materials boiling below and above isoprene were separated.   One hundred gallons of *dl*-limonene would deliver about 85 gal. of crude isoprene.   One hundred gallons of crude isoprene would yield about 80 gal. of refined isoprene, of 96 per cent purity, an over-all yield of about 65 gal. from the limonene.   The isoprene was inhibited and stored in insulated brine-cooled tanks.

The isolation and purification of isoprene from the C_5-hydrocarbon stream from cracked petroleum and from isopentane is a cheaper method and from more available materials than the method described above, and superseded it as World War II wore on[143] (see p. 83).

REFERENCES

1. Adams, G. A., and Stanier, R. Y., *Can. J. Research*, **23B**, 1–9 (1945).
2. Allen, C. F. H., and Bell, A., *Organic Syntheses*, Vol. 22, Wiley, New York, 1942, pp. 39–40.
3. Anon., *India Rubber World*, **121**, 599 (1950).
4. Armstrong, H. E., and Miller, A. K., *J. Chem. Soc.*, **49**, 80 (1886).
5. Armstrong, H. E., and Miller, A. K., *J. Chem. Soc.*, **49**, 83 (1886).
6. Bachman, G. B., and Goebel, C. G., *J. Am. Chem. Soc.*, **64**, 787–90 (1942).
7. Backer, H. J., *Rec. trav. chim.*, **55**, 1036–9 (1936).
8. Badische Anilin u. Soda-Fabrik, Brit. Pat. 27,387, Dec. 6, 1911.
9. Badische Anilin u. Soda-Fabrik, Ger. Pat. 252,499, Oct. 23, 1912.
10. Badische Anilin u. Soda-Fabrik, U.S. Pat. 1,026,418 (1912).
11. Bassett, H. L., and Williams, H. G., *J. Chem. Soc.*, 1932, 2324.
12. Bayer, Friedr. & Co., Brit. Pat. 8100, Apr. 4, 1910.
13. Bayer, Friedr. & Co., Ger. Pat. 246,241 (1910).
14. Bayer, Friedr. & Co., U.S. Pat. 1,005,217 (1911).
15. Bayer, Friedr. & Co., Ger. Pat. 248,738, June 29, 1912.
16. Beati, E., and Mattei, G., *Ann. chim. applicata*, **30**, 21–28 (1940).
17. Beilstein's *Handbuch der Organischen Chemie*, edited by B. Prager and P. Jacobson, Vol. I, Sec. Suppl., Springer, Berlin, 1918, p. 232.
18. Bekkedahl, N., Wood, L. A., and Wojciechowski, M., *J. Research Natl. Bur. Standards*, **17**, 883–94 (1936).
19. Berg, L., Sumner, G. L., Montgomery, C. W., and Coull, J., *Ind. Eng. Chem.*, **37**, 352–5 (1945).
20. Berndt, W., and Wulff, O., to I.G. Farbenindustrie A.-G., U.S. Pat. 2,145,387, Jan. 31, 1939.
21. Berthelot, M., *Ann. chim. et phys.* (4), **9**, 445, 466 (1866).
22. Bibb, C. H. (Newport Industries, Inc.), U.S. Pat. 2,386,537, Oct. 9, 1945.
23. Blatt, A. H., editor, *Organic Syntheses*, Coll. Vol. II, Wiley, New York, 1943, pp. 102–5.
24. Blom, R. H., Mustakas, G. C., Efron, A., and Reed, D. L., *Ind. Eng. Chem.*, **37**, 870–2 (1945).
25. Blom, R. H., Reed, D. L., Efron, A., and Mustakas, G. C., *Ind. Eng. Chem.*, **37**, 865–70 (1945).
26. Bludworth, J. E., and Robeson, M. O. (Celanese Corp. of America), U.S. Pat. 2,427,704, Sept. 23, 1947.
27. Blumer, D. R. (Phillips Petrol. Co.), U.S. Pat. 2,389,231, Nov. 20, 1945.
28. Bolton, E. K., and Downing, F. B. (du Pont), U.S. Pat. 1,777,600, Oct. 7, 1930.
29. Boonstra, B. B. S. T., and van Amerongen, G. J., *Ind. Eng. Chem.*, **41**, 161 (1949).
30. Bourns, A. N., and Nicholls, R. V. V., *Can. J. Research*, **25B**, 80–9 (1947).
31. Brooks, B. T., *J. Am. Chem. Soc.*, **40**, 845 (1918).
32. Buret, R., *Rev. gén. caoutchouc*, **26**, 738–9 (1949).
33. Burnette, L. W., *Iowa State Coll. J. Sci.*, **19**, 9–10 (1944).
34. Callaham, J. R., *Chem. Met. Eng.*, **51**, No. 11, 94–8 (1944).
35. Caventou, E., *Ann.*, **127**, 93 (1863).
36. Caventou, E., *Ber.*, **6**, 70 (1873).
37. Chappell, C. H., *Iowa State Coll. J. Sci.*, **11**, 45–7 (1936).
38. Charon, E., *Ann. chim. et phys.* (7), **17**, 234 (1899).
39. Chupotsky, A., and Mariutza, N., *J. Russ. Phys. Chem. Soc.*, **21**, 431–4 (1889).
40. Ciamician, G., and Magnaghi, P., *Gazz. chim. ital.*, **15**, 504 (1885); *Ber.*, **18**, 2079–85 (1885).
41. Codignola, F., and Piacenza, M., Ital. Pat. 421,628, May 28, 1947.
42. Courtot, M. A., *Bull. soc. chim. France* (3), **35**, 969–88 (1906).

43. Couturier, F., *ibid.* (3), **33**, 454 (1880); *Ann. chim. et phys.* (6), **26**, 485 (1892).
44. Craig, D., *J. Am. Chem. Soc.*, **65**, 1006–13 (1943).
45. Davis, B. L., Goldblatt, L. A., and Palkin, S., *Ind. Eng. Chem.*, **38**, 53–7 (1946).
46. Dean, M. R., and Legatski, T. W., *Ind. Eng. Chem., Anal. Ed.*, **16**, 7–8 (1944).
47. Demjanow, N., and Dojarenko, M., *Ber.*, **40**, 2589–94 (1907).
48. Denivelle, L., *Compt. rend.*, **208**, 1024–5 (1939).
49. Doyarenko, M., *Ber.*, **59**, p. 2940 (1926).
50. Du Pont, E. I., de Nemours & Co., Brit. Pat. 575,362, Feb. 14, 1946.
51. Elder, A. L., *Ind. Eng. Chem.*, **34**, 1260–6 (1942).
52. Eltekow, A., *Ber.*, **10**, 2059 (1877).
53. Euler, W., *Ber.*, **30**, 1989–91 (1897); *J. prakt. Chem.* (2), **57**, 131–59 (1898).
54. Evans, T. W., Morris, R. C., and Melchior, N. C. (Shell Develop. Co.), U.S. Pat. 2,379,697, July 3, 1945.
55. Filippov, O. G., *J. Russ. Phys. Chem. Soc.*, **42**, 364–5 (1910).
56. Fisher, H. L., *Rubber and Its Use*, Chemical Publishing Co., Brooklyn, 1941, pp. 22–4.
57. Fisher, H. L., and Chittenden, F. D., *Ind. Eng. Chem.*, **22**, 869–71 (1930).
58. Friedrichsen, W., and Fitzky (General Aniline & Film Corp.), U.S. Pat. 2,218,640, Oct. 22, 1940.
59. Frolich, P. K., and Morrell, C. E., *Chem. Eng. News*, **21**, 1138–45 (1943).
60. Gabriel, C. L., *Chem. Eng. News*, **21**, 490 (1943).
61. Goldblatt, L. A., and Palkin, S., *J. Am. Chem. Soc.*, **63**, 3517–22 (1941).
61*a.* Gollman, H. A. (Koppers Co.), Process for the recovery of butadiene from cyclohexane pyrolate, U.S. Pat. 2,575,341, Nov. 20, 1951.
62. Gottlob, K. (Farbenfabriken vorm. Friedr. Bayer & Co.), U.S. Pat. 1,065,522, June 24, 1913.
63. Gottlob, K., *India Rubber J.*, **58**, 305, 348, 391, 433 (1919).
64. Groll, H. P. A., Hearne, G., and LaFrance, D. S. (Shell Devel. Co.), U.S. Pat. 2,245,776, June 17, 1941.
65. Groll, H. P. A., Hearne, G. W., and von Stietz, G. E. G. (Shell Devel. Co.), U.S. Pat. 2,310,523, Feb. 9, 1943.
66. Grosse, A. V., *J. Am. Chem. Soc.*, **59**, 2739–41 (1937).
67. Guinot, H. M. E. (Les Usines de Melle), U.S. Pat. 2,273,484, Feb. 17, 1941.
68. Hagemann, A., *Z. angew. Chem.*, **42**, 355–61 (1929).
69. Handley, E. T., Rowzee, R. W., Fennebresque, J. D., Garvey, B. S., Juve, R. D., Monrad, C. C., and Troyan, J. E., U.S. Dept. Commerce OTS, PB 189 (1945).
70. Hanford, W. E., and Fuller, D. L., *Ind. Eng. Chem.*, **40**, 1171–7 (1948). This article has many excellent references.
71. Harden, A., and Walpole, G. S., *Proc. Roy. Soc. London*, **77**, B, 399–405 (1906).
72. Harries, C., *Untersuchungen über die Natürlichen und Künstlichen Kautschukarten*, Springer, Berlin, 1919.
73. Harries, C., and Gottlob, K., *Ann.*, **383**, 228 (1911).
74. Henninger, A. M., *Ann. chim. et phys.* (6), **7**, pp. 211, 216 (1886).
75. Hepp, H. J., and Frey, F. E., *Ind. Eng. Chem.*, **41**, 827–30 (1949).
76. Hill, R., and Isaacs, E. (Imperial Chemical Industries Ltd.), Brit. Pat. 483,989, Apr. 28, 1938; U.S. Pat. 2,224,912, Dec. 17, 1940.
77. Hofmann, A. W., *Ber.*, **14**, 664 (1881).
78. Hofmann, F., and Coutelle, C. (Farbenfabriken vorm. Friedr. Bayer & Co.), Ger. Pat. 231,806 (1911); U.S. Pat. 1,030,239, June 18, 1912.
79. Hopkinson, R., Davey, W. C., Patterson, P. B., Monrad, C. C., and Glenn, L., U.S. Dept. Commerce OTS, PB 1763, Feb. 15, 1946. Cf. Reppe, W., Acetylene Chemistry, PB 18852-S.
80. Howard, Frank A., *Buna Rubber—the Birth of an Industry*, Van Nostrand, New York, 1947, pp. 163–4.

81. Hurakawa, Z., and Nakaguti, K., *J. Soc. Chem. Ind. Japan*, **43**, No. 5; Suppl. binding, 142–4B (1940); *Rubber Chem. and Technol.*, **13**, 856–60 (1940).

82. Hurd, C. D., and Meinert, R. W., *J. Am. Chem. Soc.*, **53**, 289–300 (1931).

83. I.G. Farbenindustrie A.-G., Brit. Pat. 320,362, Mar. 30, 1928.

84. Ipatieff, V. N., *J. prakt. Chem.* (2), **53**, 145–68 (1896).

85. Ipatieff, V. N., and Wittorf, N., *J. prakt. Chem.* (2), **55**, 1–4 (1897).

86. Ipatieff, V. N., *J. prakt. Chem.* (2), **55**, 4–13 (1897).

87. Ipatieff, V. N., *J. prakt. Chem.* (2), **59**, 522, 525 (1899).

88. Ipatieff, V. N., *Chem. Ztg.*, **26**, 530 (1902); *J. prakt. Chem.* (2), **67**, 420–2 (1903).

89. Keeling, W. O. (Koppers Co.), Brit. Pat. 595,879, Jan. 13, 1948.

90. Kistiakowsky, G. B., and Ransom, W. W., *J. Chem. Phys.*, **7**, 725–35 (1939).

91. Kistiakowsky, G. B., Ruhoff, J. R., Smith, H. A., and Vaughan, W. E., *J. Am. Chem. Soc.*, **58**, 146–53 (1936).

92. Kluyver, A. J., and Scheffer, M. A. (Thomas H. Verhave), U.S. Pat. 1,899,156, Feb. 28, 1933.

93. Kolachov, P. J., *Chem. Eng. News*, **21**, 488 (1943).

94. Kolfenbach, J. J., *Iowa State Coll. J. Sci.*, **19**, 35–7 (1944).

95. Kondakow, I., *J. prakt. Chem.* (2), **62**, 170–2 (1900).

96. Kooi, E. R., Fulmer, E. I., and Underkofler, L. A., *Ind. Eng. Chem.*, **40**, 1440–5 (1948).

97. Kosternaya, A. F., *Uchenye Zapiski Leningrad. Gosudarst. Univ. Ser. Khim. Nauk.*, **3**, 126–56 (1938).

98. Kreimeier, O. R. (du Pont), U.S. Pat. 2,106,181, Jan. 25, 1938.

99. Küchler, L., *Trans. Faraday Soc.*, **35**, 874–80 (1939).

100. Kutscherow, L., *J. Russ. Phys. Chem. Soc.*, **45**, 1634–54 (1913).

101. Kyriakides, L. P., *J. Am. Chem. Soc.*, **36**, 663–70 (1914).

102. Kyriakides, L. P., *J. Am. Chem. Soc.*, **36**, 987–1005 (1914).

103. Lebedev, S. V., Brit. Pat. 331,482 (1930); cf. *Rubber Chem. and Technol.*, **15**, 403–29 (1942),

104. Ledingham, G. A., Adams, G. A., and Stanier, R. Y., *Can. J. Research*, **23F**, 48–71 (1945).

105. Livingston, J. W., U.S. Dept. Commerce OTS, PB 517 (1945).

106. Lorch, A. E. (Air Reduction Co.), U.S. Pat. 2,371,634, Mar. 20, 1945.

107. Manninen, T. H. (U.S. Industrial Chemicals, Inc.), U.S. Pat. 2,399,049, Apr. 23, 1946.

108. Marchionna, F., *Butalastic Polymers*, Reinhold, New York, 1946, p. 76.

109. Mariutza, A., *J. Russ. Phys. Chem. Soc.*, **21**, 434–6 (1889).

110. Matthews, F. E., Bliss, H. J. W., and Elder, H. M., Brit. Pat. 17,234, July 24, 1912.

111. Matthews, F. E., and Strange, E. H., Brit. Pat. 4,572, Feb. 23, 1910.

112. Matthews, F. E., Strange, E. H., and Bliss, H. J. W., Brit. Pat. 3873 (1912).

113. Matthews, F. E., Strange, E. H., and Bliss, H. J. W., Brit. Pat. 12,771, May 30, 1912.

114. Merezhkovskii, B. K., *Bull. soc. chim. France* (4), **37**, 1174–87 (1925).

115. Merling, G., and Kochler, H. (Friedr. Bayer & Co.), U.S. Pat. 1,026,691 (1912).

116. Midgley, T., and Henne, A. L., *J. Am. Chem. Soc.*, **51**, 1215–26 (1929).

117. Mikeska, L. A., and Arundale, E. (Jasco., Inc.), U.S. Pat. 2,350,517, June 6, 1944.

118. Morell, S. A., Geller, H. H., and Lathrop, E. C., *Ind. Eng. Chem.*, **37**, 877–84 (1945).

119. Morris, R. C., and Evans, T. W. (Shell Development Co.), U.S. Pat. 2,371,908, Mar. 20, 1945.

120. Morris, R. C., and Van Winkle, J. L. (Shell Development Co.), U.S. Pat. 2,375,024, May 1, 1945.

121. Muskat, I. E., and Northrup, H. E., *J. Am. Chem. Soc.*, **52**, 4043–55 (1930).

122. Nagai, H., *J. Soc. Chem. Ind. Japan*, **44**, Suppl. binding, 64–6; No. 1, Suppl. binding, 41B–3B (1941); *C. A.*, **35**, 3960, 5096 (1941); *J. Soc. Rubber Ind. Japan*, **15**, 350–65, 591–601 (1942); *C. A.*, **43**, 2105, 4045 (1949).

123. Naragon, E. A., Burk, R. E., and Lankelma, H. P., *Ind. Eng. Chem.*, **34**, 355–8 (1942).

140 SYNTHETIC RUBBER

124. Nat. Bur. Standards, Cir. C-461, Nov. 1947.
125. Neish, A. C., *Can. Chem. Process Inds.*, **28,** 862, 864, 866 (1944).
126. Neish, A. C., *Can. J. Research*, **23B,** 10–6 (1945).
127. Neish, A. C., Haskell, V. C., and Macdonald, F. J., *Can. J. Research*, **25B,** 266–71 (1947).
128. Newton, L. W., and Coburn, E. R., in *Organic Syntheses*, Vol. 22, Wiley, New York, 1942, pp. 40–2.
129. Nieuwland, J. A., Calcott, W. S., Downing, F. B., and Carter, A. S., *J. Am. Chem. Soc.*, **53,** 4197–202 (1931).
130. Norris, J. F., *Ind. Eng. Chem.*, **11,** 819 (1919).
131. Norton, L. M., and Andrews, C. W., *Am. Chem. J.*, **8,** 4, 8 (1886).
132. Norton, L. M., and Noyes, A. A., *Am. Chem. J.*, **8,** 362–4 (1886).
133. Ono, K., and Takedo, Z., *Bull. Chem. Soc. Japan*, **2,** 16–9 (1927).
134. Ostromislensky, I., *J. Russ. Phys. Chem. Soc.*, **47,** 1472–94 (1915).
135. Ostromislensky, I., *J. Russ. Phys. Chem. Soc.*, **47,** 1974 (1915).
136. Ostromislensky, I., *J. Russ. Phys. Chem. Soc.*, **47,** 1978–82 (1915).
137. Ostromislensky, I., and Kelbasinsky, S. S., *J. Russ. Phys. Chem. Soc.*, **47,** 1509–28 (1915).
138. Othmer, D. F., Bergen, W. S., Schlechter, N., and Bruins, P. F., *Ind. Eng. Chem.*, **37,** 890–4 (1945).
139. Othmer, D. F., Schlechter, N., and Koszalka, W. A., *Ind. Eng. Chem.*, **37,** 895–900 (1945).
140. Palmer, R. C., *Ind. Eng. Chem.*, **34,** 1028–34 (1942).
141. Perkin, W. H., Jr., *J. Soc. Chem. Ind.*, **31,** 616–24 (1912).
142. Perkin, W. H., Jr., Matthews, F. E., and Strange, E. H., Brit. Pat. 5931 (1910).
143. Poffenberger, N., Horsley, L. H., Nutting, H. S., and Britton, E. C., *Trans. Am. Inst. Chem. Engrs.*, **42,** 815–26 (1946).
144. Reppe, W., Reichsamt Wirtschaftsaufbau Chem. Ber. Prüf-Nr. 38 (PB 52007), 37–68 (1940).
145. Reppe, W., U.S. Dept. Commerce OTS, PB 2437, Feb. 8, 1946.
146. Reppe, W., and Keyssner, E. (General Aniline & Film Corp.), U.S. Pat. 2,232,867, Feb. 25, 1941.
147. Reppe, W., Schmidt, W., Schulz, A., and Wenderlein, H. (General Aniline & Film Corp.), U.S. Pat. 2,319,707, May 18, 1943.
148. Reppe, W., Steinhofer, A., Spaenig, H., and Lockei, K. (General Aniline & Film Corp.), U.S. Pat. 2,300,969, Nov. 3, 1942.
149. Reppe, W., and Trieschmann, H. G. (General Aniline & Film Corp.), U.S. Pat. 2,251,835, Aug. 5, 1941.
150. Rice, F. O., and Murphy, M. T., *J. Am. Chem. Soc.*, **66,** 765–7 (1944).
151. Rice, F. O., Ruoff, P. M., and Rodowskas, E. L., *J. Am. Chem. Soc.*, **60,** 955–61 (1938).
152. Robey, R. F., Wiese, H. K., and Morrell, C. E., *Ind. Eng. Chem.*, **36,** 3–7 (1944).
153. Rose, D., and King, W. S., *Can. J. Research*, **23F,** 79–89 (1945).
154. Rubber Reserve Co., Rept. on the Rubber Program, 1940–45, Feb. 24, 1945, p. 22.
155. Runckel, W. J., and Goldblatt, L. A., *Ind. Eng. Chem.*, **38,** 749–51 (1946).
156. Savich, T. R., and Goldblatt, L. A. (U.S. of America as represented by the Secretary of Agriculture), U.S. Pat. 2,507,546, May 16, 1950.
157. Schlechter, N., Othmer, D. F., and Brand, R., *Ind. Eng. Chem.*, **37,** 905–8 (1945).
158. Schlechter, N., Othmer, D. F., and Marshak, S., *Ind. Eng. Chem.*, **37,** 900–5 (1945).
159. Schniepp, L. E., Dunning, J. W., Geller, H. H., Morell, S. A., and Lathrop, E. C., *Ind. Eng. Chem.*, **37,** 884–9 (1945).
160. Schniepp, L. E., Dunning, J. W., and Lathrop, E. C., *Ind. Eng. Chem.*, **37,** 872–7 (1945).
161. Schniepp, L. E., and Geller, H. H., *J. Am. Chem. Soc.*, **67,** 54–6 (1945).
162. Schotz, S. P., *Synthetic Rubber*, Benn, London, 1926, p. 44.
163. Schotz, S. P., *Synthetic Rubber*, Benn, London, 1926, p. 49.
164. Schotz, S. P., *Synthetic Rubber*, Benn, London, 1926, pp. 124–5.

165. Sequin, P., *Bull. soc. chim.*, **12,** 948–9 (1945).
166. Seymour, G. W., and Fortress, F. (Celanese Corp. of America), U.S. Pat. 2,444,538, July 6, 1948.
167. Slobodin, Y. M., *J. Gen. Chem. U.S.S.R.*, **7,** 2376–80 (1937).
168. Stanier, R. Y., Adams, G. A., and Ledingham, G. A., *Can. J. Research,* **23F,** 72–8 (1945).
169. Staudinger, H., and Klever, H. W., *Ber.*, **44,** 2212–5 (1911).
170. Staudinger, H., Muntwyler, O., and Kupfer, O., *Helv. Chim. Acta*, **5,** 756–67 (1922).
171. Thiele, J., *Ann.*, **308,** 339 (1899).
172. Tilden, W. A., *Chem. News*, **46,** 129 (1882).
173. Tilden, W. A., *J. Chem. Soc.*, **45,** 410–20 (1884).
174. Tishchenko, V. S., and Kosternaya, A. F., *J. Gen. Chem. U.S.S.R.*, **7,** 1366–77 (1937).
175. Tollefson, R. C. (Air Reduction Co.), U.S. Pat. 2,373,153, Apr. 10, 1945.
176. Ward, G. E., Pettijohn, O. G., and Coghill, R. D., *Ind. Eng. Chem.*, **37,** 1189–94 (1945).
177. Ward, G. E., Pettijohn, O. G., Lockwood, L. B., and Coghill, R. D., *J. Am. Chem. Soc.*, **66,** 541–2 (1944).
178. Whitby, G. S., and Crozier, R. N., *Can. J. Research*, **6,** 203–25 (1932); *Rubber Chem. and Technol.*, **5,** 546–65 (1932).
179. Williams, C. G., *Trans. Roy. Soc. London*, **150,** 241 (1860); *Proc. Roy. Soc. London*, **10,** 516 (1860); *J. Chem. Soc.*, **15,** 110 (1862).
180. Williams, J. W., and Hurd, C. D., *J. Org. Chem.*, **5,** 122–5 (1940).
181. Willstätter, R., and Bruce, J., *Ber.*, **40,** 3979–99 (1907).
182. Willstätter, R., and von Schmädel, W., *Ber.*, **38,** 1992–9 (1905).
183. Wilson, C. L. (Imperial Chemical Industries, Ltd.), Brit. Pat. 550,105, Dec. 23, 1942; U.S. Pat. 2,366,464, Jan. 2, 1945.
184. Winfield, M. E., *J. Council Sci. Ind. Research*, **18,** 412–23 (1945); *C. A.*, **40,** 3719 (1946).
185. Wurtz, A., *Ann. chim. et phys.* (3), **55,** 452 (1859).

CHAPTER 6

MANUFACTURE OF STYRENE MONOMER*

J. E. Mitchell and J. Lawrence Amos

The Dow Chemical Company

Out of the classic researches of Berthelot on pyrolysis has come the most important technical means of producing synthetic styrene commercially today. In 1851 Berthelot showed that certain hydrocarbons when passed through a red-hot tube gave styrene and indene.[6, 7] His discovery of the dehydrogenation of ethylbenzene in 1869 by pyrolysis to form styrene as the principal product is the basis of the most widely used present-day commercial styrene process.[8] Later, Ostromislensky and Shepard in 1925 disclosed the most important information on obtaining styrene from hydrocarbons. They found that some hydrocarbons, instead of splitting off hydrogen, yield styrene by the elimination of methane or other hydrocarbons of low molecular weight. Cymene (*p*-methylisopropylbenzene) was claimed to yield well over 50 per cent of *p*-methylstyrene.[54] Methyl-, dimethyl-, or ethylstyrenes were produced by dehydrogenating the corresponding alkylbenzenes[53] at 450 to 700° C. Ostromislensky and Shepard did their most notable work in proving that the process involving the dehydrogenation of ethylbenzene could be used for producing styrene;[52] however, their process was not commercialized.

Research on the present Dow method for the manufacture of styrene began in 1931. (Independently, work on the German process began about the same time.[18]) The two features of these two processes that distinguish them from methods previously described for the manufacture of styrene are (*a*) the use of catalysts for the dehydrogenation of ethylbenzene and (*b*) the use of polymerization inhibitors in the styrene during its purification in low-pressure continuous stills.

When Japanese attacks in the Pacific area in late 1941 cut off the supply of natural rubber, the United States decided quickly on emergency production of a styrene-butadiene type of synthetic rubber designated as GR-S. Since styrene makes up about one-quarter of this synthetic, enormous expansion of styrene-manufacturing facilities became necessary.[72] Fortunately sufficient knowledge of the manufacture of styrene had been developed in the United States to make it possible to meet this emergency. The following companies were called on to share technical details on styrene manufacture: Carbide and Carbon Chemicals Corporation, the Dow

* This chapter is closely similar to chapters on the same subject in *ACS Monograph* 115, Styrene, Its Polymers, Copolymers, and Derivatives, by R. H. Boundy, R. F. Boyer, and S. M. Stoesser, published by Reinhold Publishing Corporation, New York (1952), whose permission for its use is hereby acknowledged.

Chemical Company, Jasco, Inc., the Koppers Company, the Lummus Company, Monsanto Chemical Company, Phillips Petroleum Company, Standard Oil Development Company, and Universal Oil Products Company. Each of these had done laboratory work on the production of ethylbenzene or styrene. The Dow Company had been in commercial production of styrene monomer since 1937 and consequently was relied on most heavily for the styrene part of the rubber program.[18, 56] Carbide had developed a process based on the ethylation of benzene to ethylbenzene and an oxidation step for producing styrene from ethylbenzene. Since semiplant work had proved this method, construction of a plant at Institute, W. Va., was authorized early in the program. The process will be described later in this chapter.

The Dow Chemical Company was asked to take over styrene-manufacturing projects at Los Angeles, Calif.; Port Neches, Texas; Gary, Ind.; and Sarnia, Ontario. In addition, an expansion of the existing privately owned facilities at Midland, Mich., was requested. A few weeks later the location of the Texas plant was switched to Velasco adjacent to the Government-financed Dow Magnesium Corporation seawater magnesium plant. This move made possible the joint use of service and powerhouse facilities. The Gary project was canceled when other scheduled plants appeared adequate.

After considerable study by industry and Government representatives of the various processes proposed, it was decided that Monsanto should employ a procedure for the production of ethylbenzene which they themselves had developed, a dehydrogenation process used by Dow, and a styrene-finishing process to be designed jointly by Monsanto and the Lummus Company. The plant for this was built at Texas City, Texas.

The Koppers Company agreed to construct a plant at Kobuta, Pa. This project used an ethylbenzene unit contributed to by Koppers, Phillips, and Universal Oil Products. It is an aklylation unit operating at high temperature and pressure. At first phosphoric acid was used as a catalytic agent in the alkylation,[16, 33] but this was later replaced by a silica-alumina catalyst.[51, 58-60] The dehydrogenation and finishing steps were of Dow design, patterned after the Midland unit.

The sharing of process information on styrene was an outstanding example of the way private industry, in time of emergency, can put down the normal barriers of competitive business and work unselfishly toward a common goal. Improvements made by one producer were quickly made available to others, with consequent economy in manpower and raw-material requirements. This exchange of technical information was centered in J. W. Livingston of Rubber Reserve Company, whose capable guidance was largely responsible for the successful operation of the technical exchange program.

Until 1943 essentially all of the high-purity synthetic styrene in the United States was made by the catalytic dehydrogenation process. During World War II the process was used in whole or in part to make about 90 per cent of the total supply. Production of synthetic styrene in the United States is shown in Fig. 1. Since the close of the war, Monsanto, Koppers, and Dow have employed the method to make almost all of the monomer needed by the United States.   In April 1947 an explosion in the harbor at Texas City, Texas, completely destroyed Monsanto's styrene plant.   The Monsanto Company, however, has rebuilt this plant.

A plant at Sarnia, Ontario, was operated by Dow for the Canadian Government from 1943 to 1951.   This is the first synthetic-styrene producer in the British Commonwealth.

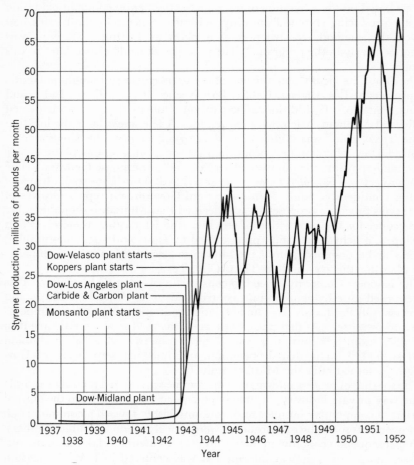

FIG. 1.   Synthetic-Styrene Production in the United States

As the data in Fig. 1 imply, expansion in the uses of styrene has proceeded at a very rapid rate; and stable peacetime uses promise to absorb the output of the tremendous expansions in styrene-productive capacity made during wartime.[27]   And new uses for styrene, its polymers, and its copolymers are constantly appearing.[27, 65]   During World War II styrene was the largest consumer of benzene, and even during peacetime it has gone beyond phenol and assumed top place in benzene consumption.

DOW PROCESS

The basic chemistry of the Dow styrene process is outlined by two simple equations. Ethylbenzene is produced by the alkylation of benzene with ethylene.

$$C_2H_4 + C_6H_6 \xrightarrow{95° C. AlCl_3} C_6H_5 \cdot C_2H_5$$

Purified ethylbenzene is then dehydrogenated catalytically in the presence of steam to give styrene.

$$C_6H_5 \cdot CH_2 \cdot CH_3 \xrightarrow{630° C.} C_6H_5 \cdot CH : CH_2 + H_2$$

The simplified flow diagram (Fig. 2) points out the three essential steps in making styrene and gives a picture of the relative flow quantities involved.[14]

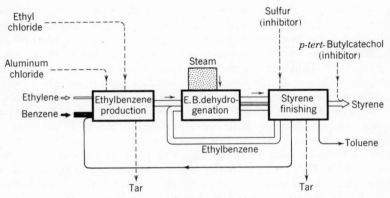

Fig. 2. Simplified Flow Diagram of Dow Process

The Velasco plant was laid out in four independent parallel trains, each train consisting of an ethylbenzene block, an ethylbenzene dehydrogenation block, and a styrene-finishing block. Thus, there are 12 separate operating units. Such separation has proved to be a desirable safety feature when flammable hydrocarbons are handled.

Raw Materials, By-Products, and Yields. The basic raw materials used in the manufacture of styrene are benzene and ethylene. Both coal-tar and petroleum by-product benzene have been used successfully. The petroleum industry is producing benzene and has become an important source of this raw material since the supply of coal-tar benzene became inadequate. The product has been found satisfactory from a chemical standpoint and is used interchangeably with coal-tar benzene. The ethylene used is ordinarily obtained from the cracking of crude oil or propane. Ethylene made by the dehydration of ethyl alcohol, however, has been used successfully. Other chemicals are aluminum chloride, ethyl chloride, liquid 50 per cent caustic, flake caustic, dehydrogenation catalyst, sulfur, and *p-tert*-butylcatechol. Toluene is made as a by-product.

Table I gives a summary of average yields at different stages in the process. Higher efficiencies have been realized for single months.

Table I.   Representative Yields in Dow Styrene Process

	Per cent
Benzene to ethylbenzene	95.5
Ethylene to ethylbenzene	96.8
Ethylbenzene to crude styrene	90.1
Crude styrene to finished styrene	99.4
Over-all—Benzene to styrene	86.5
Over-all—Ethylene to styrene	86.5

Ethylbenzene Production Step

Raw Materials.   When ethylene and benzene react in the presence of aluminum chloride and hydrogen chloride in an anhydrous system, alkylation of the benzene ring occurs to produce ethylbenzene and higher ethylated benzenes.[1, 2]   Considerable work has been done on this reaction.[24, 55]   Ethylene purity is not critical so long as the impurities are hydrogen, light paraffin hydrocarbons, or other inert materials.   Acetylene is to be avoided.   Its presence materially increases catalyst consumption. The purity of ethylene ordinarily used averages 95 per cent, but the process has operated satisfactorily with concentrations as low as 38 per cent.   The benzene used is known in the industry as "Styrene Grade."   This term defines a benzene with a boiling range of 1° C. and a minimum freezing point of 4.85° C.   This ordinarily corresponds to a purity slightly above 99 per cent.   Total sulfur should be below 0.10 per cent.   It has been found that normal variations in the benzene supply have little effect on plant operation.

Aluminum chloride with a minimum purity of 97.5 per cent is used as an alkylation catalyst.   A granular material has proved most desirable for the best operation of mechanical feeders.   In order to achieve high catalyst efficiencies, hydrogen chloride must be added as a promoter.   This is accomplished by furnishing the reaction mixture with ethyl chloride, which in turn provides the desired hydrogen chloride as well as ethyl groups.

Formation of Polyethylbenzenes.   If one mole of ethylene comes to equilibrium with one mole of benzene at 95° C., the resulting product contains by weight 18 per cent benzene, 51 per cent ethylbenzene, and 31 per cent polyethylbenzenes.   Obviously, satisfactory yields of ethylbenzene can be obtained only when the quantity of polyethyls is cut down.   This is accomplished in two ways: (1) by reducing the ratio of ethyl groups to benzene rings in the mixture, thus creating a more favorable equilibrium, and (2) by recycling the polyethyls formed.   Two simultaneous reactions come into consideration—the forward one of alkylation, the backward one of dealkylation.[40]   In Fig. 3 the equilibrium conditions at various ratios of ethylene to benzene groups indicate the optimum operating conditions. The percentage of polyethyls rises steadily, while the ethylbenzene concentration levels out.   Although plant alkylators do not operate exactly at equilibrium conditions, production experience substantiates the optimum operating point as predicted by theory. The theoretical equilibrium equation then gives for the ethylene-to-benzene ratio used commercially

$$0.58C_2H_4 + 1.00C_6H_6 \xrightarrow[\text{95}^\circ \text{ C.}]{\text{Al Cl}_3} 0.51C_6H_6 + 0.41 \text{ ethylbenzene} + 0.08 \text{ polyethylbenzenes}$$

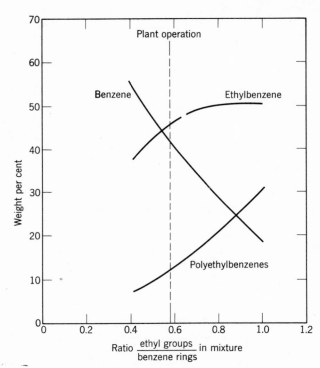

Fig. 3. Alkylation Equilibrium Conditions at Various Ratios of Ethylene to Benzene

Benzene in the alkylated product is recycled. Polyethylbenzenes are returned to the reactor for dealkylation.

Plant Operation. A flow diagram of an alkylation unit is shown in Fig. 4. Benzene is received either by barge or by railroad tank car and consequently is often saturated with water. Successful operation of an aluminum chloride alkylation depends on maintaining strictly anhydrous conditions, since the presence of water reflects itself in increased aluminum chloride consumption, sludge formation, and a greatly increased rate of corrosion in the highly acidic system. Therefore, an azeotropic drying column is employed on benzene fed to the alkylators. Overhead containing the constant-boiling mixture of benzene and water is returned to a decanting tank where water is withdrawn from the system. Benzene containing less than 30 p.p.m. water is cooled and sent to the alkylator.

Ethylene is used to vaporize ethyl chloride and carry it to the reaction zone for use as a catalyst promoter.

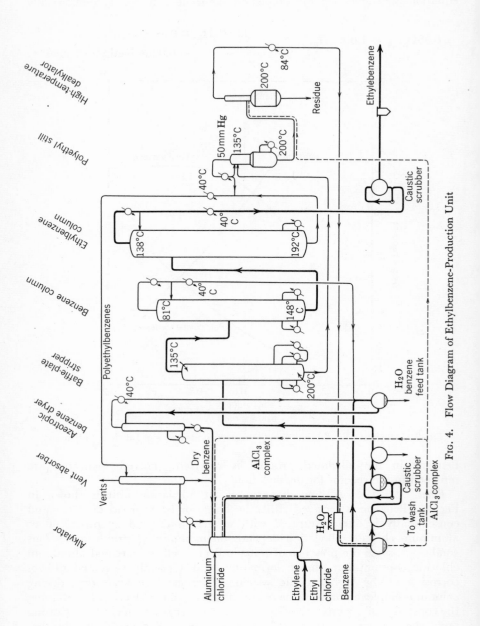

FIG. 4.   Flow Diagram of Ethylbenzene-Production Unit

Aluminum chloride, which is received in drums, is handled carefully to prevent its picking up moisture and is dumped into the hopper of a screw conveyor.   The screw feeds the granular catalyst at a constant rate.

The alkylators[3] operate with a liquid depth of about 34 ft.   Incoming ethylene, fed at the bottom of the reactor, provides circulation by a gas-lift effect.   In the alkylation, aluminum chloride combines with hydrocarbons present to form a complex which is practically insoluble in the hydrocarbon layer.   This complex appears to play an important part in the reaction mechanism.   The complex is a reddish-brown oil with a specific gravity greater than 1.   A representative composition would be:

$AlCl_3$ (combined)	26 per cent
$AlCl_3$ (free)	1 per cent
High-mol.-wt. hydrocarbons	25 per cent
Benzene and ethylbenzene	48 per cent

These proportions, however, may vary considerably.

The reaction is almost quantitative.   During smooth operation, ethylene lost in the vents is negligible.   The exothermic alkylation which liberates 1740 B.t.u. per lb. of ethylene reacted, is held at 95° C. by means of an overhead condenser and cooling water running down the outside of the alkylation shell.   Top reactor pressure is held at 5 lb. gage.   Aromatic hydrocarbons are removed from the condenser vents in a system to be described later.   The crude alkylate, along with entrained complex, is cooled in drip-type exchangers to about 40° C.   This cooling is necessary to provide proper settling.   The stream passes to a horizontal cylindrical tank, where aluminum chloride complex settles out and is pumped to the dealkylator or back to the alkylator.   The crude ethylbenzene is decanted once more in a second tank and sent to a caustic scrubbing system.   Here a greatly oversize centrifugal pump, altered mechanically to run at low hydraulic efficiency, mixes 50 per cent caustic solution and crude ethylbenzene.   This method of contacting two liquid phases has proved quite effective.   A third tank serves as a settler in this system.   Fifty per cent caustic is charged to the tank periodically.   When the caustic concentration has dropped to 30 per cent, the charge is renewed.   Another horizontal process tank serves as a final settler to remove any caustic solution carried over.   Crude ethylbenzene, after thus being sweetened, is charged to a baffle-plate stripping column.   This type of column is used because of possible fouling of a conventional plate column with heavy materials in the crude alkylate.   After being heated to 200° C. by steam at 400 lb. pressure, the heavy materials pass from the bottom of the stripper to the polyethyl still.   The overhead from the stripper at 135° C. enters the benzene column as a vapor feed.   A reflux ratio of 3 to 1 is maintained with a column of 20 actual plates.   Benzene product at 81° C. and 1/2 lb. gage is cooled to 40° C. and returned to the wet benzene feed tank.   Thus, any water entering the stripper or benzene column from leaks is removed before the stream is returned to the alkylation zone.   Bottoms from the benzene column, consisting of ethylbenzene and material of higher-boiling point, are fed to the ethylbenzene column.   This column operates with a top pressure of 1/2 lb.

gage. It uses a 3-to-1 reflux ratio and contains 58 actual plates. The ethylbenzene product, at a purity of over 99 per cent, is cooled to 40° C. and sent to a caustic scrubbing system similar to the one previously described. In this case, however, 20 per cent aqueous caustic is the scrubbing agent. Finished ethylbenzene is then dried by passing it through a bed of flake caustic. The chloride content of the product ethylbenzene is only 10 p.p.m.

Bottoms from the baffle-plate stripper enter the polyethyl still, which operates at 50 mm. mercury absolute pressure and a top temperature of 135° C. It contains ten baffle plates and runs at a 0.5-to-1 reflux ratio. The overhead product combines with the bottoms from the ethyl benzene column, is cooled to 40° C., and then is used as absorption oil in the alkylator vent recovery system. These polyethylbenzenes pass through a small absorption column, picking up aromatics and some of the ethylene and hydrogen chloride in the vents and returning them to the alkylator. Vents from this absorber are scrubbed with water to remove the hydrogen chloride left in the stream. The polyethylbenzenes themselves are dealkylated to ethylbenzene in the alkylator. Bottoms from the polyethyl still go to a high-temperature dealkylator.

In the high-temperature dealkylator[22] a stream of aluminum chloride complex joins the hydrocarbons fed to the reactor. Here the highly alkylated hydrocarbons and the complex are broken down at a temperature of 200° C. to benzene, ethylbenzene, and diethylbenzene, which pass overhead, and a tar-aluminum chloride residue. Rotating scraper knives keep the walls of the unit clean. Heat is provided electrically—both by resistance strips and by induction heating of the shell. The condensed overhead product is returned to the alkylator product stream.

Ethylbenzene Dehydrogenation Step

Dilution with Steam. The second important step in the manufacture of styrene is the dehydrogenation of ethylbenzene. In this endothermic reaction, a volume increase accompanies dehydrogenation, and hence decreased pressure favors its progress.[33] Instead of conducting this high-temperature reaction under a partial vacuum, steam is used to reduce the partial pressure of the reactants and accomplish the desired purpose. Steam fed to the reactor in the proportion of 2.6 lb. per lb. ethylbenzene reduces the partial pressure of the reaction products to about 0.1 atm. At total operating pressure this theoretical equilibrium at 630° C. would give only 25 to 30 per cent styrene in the liquid product. The addition of steam shifts the equilibrium and makes possible a theoretical styrene concentration of 80 to 85 per cent. Although the equilibrium relations are favorable at this temperature, the reaction rate, without catalysis, is too slow for practical application.

Catalysts. The dehydrogenation, which consists in subjecting the ethylbenzene vapors to the action of heat at a high temperature ranging from 500 to 800° C., appears to be a simple and convenient process. Actually it is very complex. Direct conversion gives only a 45 per cent yield.[51] Simple pyrolytic dehydrogenation gives, in addition to styrene, numerous other products, such as benzene, ethylene, methane, toluene, and colored materials.

The patent literature shows, however, that appreciably better results are obtainable through the application of selective catalysts which accelerate the dehydrogenation and allow operation with less by-product formation. Suida suggested such catalysts as copper, iron, and nickel on pumice and difficultly reducible oxides of metals and their compounds.[62] Mark and Wulff[46] prepared styrene from ethylbenzene by passing the vapor of the latter over difficultly reducible oxides of metals and their compounds (i.e., calcium and magnesium oxides) in admixture with steam or carbon dioxide as a diluent at a temperature of 500 to 700° C. Wulff and Roell[75] carried out this reaction in a similar manner with a catalyst comprising ingredients described as readily reducible oxides of metals and their compounds (i.e., copper and iron oxides). Graves[25] claims the use of a dehydrogenation catalyst difficultly reducible under the conditions of working (without use of diluents). The patents just cited have the broadest coverage concerning ethylbenzene dehydrogenation catalysts.

During World War II, and for several years thereafter, the catalysts most widely employed by those styrene producers using the dehydrogenation process were Standard Oil 1707-W catalyst, Shell Development Company 105 catalyst, and (used in Germany) the German high zinc oxide catalyst. All these catalysts contain, as a minor component, a potassium compound, and are used with steam as the diluent while dehydrogenating ethylbenzene. Employed as a fixed bed, catalysts of this type allow completely continuous operation without periodic regeneration in order to remove carbon. This is quite unique for fixed-bed catalyst practice, which normally requires periodic regeneration to reactivate the catalyst, and entails considerably greater complexity of operation.

Hansgirg[30–32] and later Grinevich[26] have described how the presence of potassium salts (i.e., potassium carbonate) with magnesium oxide and iron oxide accelerates reaction between water and carbon to form carbon monoxide (water-gas reaction) and enables satisfactory results to be obtained at temperatures as low as 400° C. Normally, when conducted noncatalytically, this reaction requires a temperature of about 900° C.[23] for satisfactory results. The information disclosed by Hansgirg and Grinevich makes understandable the role that potassium compounds play in ethylbenzene dehydrogenation catalysts that are self-regenerative when steam is used and that therefore allow continuous operation.

The Standard Oil 1707-W dehydrogenation catalyst[38, 63] contains a major portion of magnesia; minor amounts of iron oxide, copper oxide, potassium dichromate; and a small amount of potassium oxide. It is supplied in $3/16$-in. pellets, and was first used for butadiene production and later adapted for ethylbenzene dehydrogenation (cf. Chap. 3).

The Shell Development Company 105 catalyst[29] contains a major amount of iron oxide, a minor amount of an alkaline compound of potassium, and a minor amount of chromium oxide. It is supplied in $3/16$-in. pellets and has been used only by Dow for styrene production.

The catalyst used in the German styrene process had the following percentage composition: zinc oxide 80, calcium oxide 5 to 7, aluminum oxide 10, potassium hydroxide 2 to 3, and chromic oxide 0.5 to 0.7. This catalyst was supplied as $1/_8$-in. pellets and used, in a heated catalyst case, for the continuous dehydrogenation of ethylbenzene, with steam present in equimolecular ratios. The German process used as a diluent only about $1/_{20}$th of the steam employed in the Dow process. The difference arises from the fact that the necessary heat for the dehydrogenation reaction is provided in the one process by the heated catalyst case and in the other by the sensible heat within the steam.

Recent patents by Kearby disclose other compositions of styrene catalysts having three features in common. They are all used with steam, comprise potassium oxide as a minor component, and comprise one ingredient as a major component. In different patent compositions the major component is zinc oxide,[34] beryllium oxide,[35] iron oxide,[36] or copper oxide.[37]

For several other reasons the use of steam in the dehydrogenation has proved desirable. By superheating steam to the point where it provides the heat of dehydrogenation and mixing it with ethylbenzene immediately before contacting the catalyst, side reactions are kept at a minimum.[19] Direct heating of ethylbenzene in a furnace is to be avoided. By superheating steam alone, furnace operation is greatly simplified, no decoking operation is necessary, and nickel alloys can be used in furnace tubes. Nickel, however, is not used in contact with ethylbenzene at elevated temperatures because of carbon formation. In addition, as previously pointed out, steam continuously removes carbon deposits on the catalyst by the water-gas reaction. The catalyst operates continuously throughout its life, which is a year or more. The dehydrogenation step is now a very smooth-running continuous operation giving ultimate yields of 90 to 92 per cent at conversions of 35 to 40 per cent per pass.

Divinylbenzene. Ethylbenzene fed to the dehydrogenation step must have a diethylbenzene content below 0.04 per cent, since this material is partially converted to divinylbenzene. Divinylbenzene polymerizes very rapidly to form insoluble residues in the purification system and may be very troublesome. Small amounts of benzene in the ethylbenzene fed act only as a diluent. In fact dehydrogenation has been carried on using benzene rather than steam as the inert vapor.[47-8]

Plant Operation. A flow sheet of the dehydrogenation step is given in Fig. 5. The temperature level in the system varies considerably during the life of the catalyst. Fresh catalyst gives 37 per cent conversion per pass at about 600° C. As the catalyst becomes less active, the steam temperature is raised to provide a constant conversion. The reaction temperature may thus be as high as 660° C. when old catalyst is used, and other temperatures are correspondingly higher. For discussion purposes, a temperature of 630° C. at the bottom of the catalyst bed has been assumed.

Steam is fed at a rate of 2.6 lb. per lb. ethylbenzene. About 90 per cent of this steam quantity is raised to a temperature of 385° C. by heat exchange before entering the superheating furnace which produces steam at 710° C. The remaining 10 per cent of the steam is mixed with ethylbenzene before being passed through a vaporizer, attaining a temperature of 160° C. before heat exchange with hot reactor product. At 520° C. the ethylbenzene vapors meet superheated steam at 710° C. The resulting temperature at the bottom of the catalyst bed is 630° C. This temperature point is used for controlling the reaction system. Reactor product at 565° C. is cooled by heat exchange, first with incoming ethylbenzene, and then with steam. From this point a spray-type desuperheater lowers the crude-product temperature to 105° C. and condenses out tars formed at high temperature. Vapors then enter the final condenser, where steam, styrene, ethylbenzene

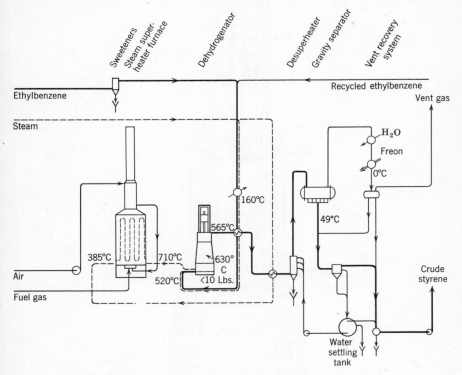

FIG. 5. Flow Diagram of Dehydrogenation Unit

benzene, toluene, and small amounts of tar are liquefied. Vent gases containing hydrogen, carbon monoxide, carbon dioxide, methane, ethane, plus some of the aromatics pass overhead to a refrigerated recovery system.

The condensed materials go to a gravity separator, where the hydrocarbons are decanted from the water phase. After further settling, part of the water is recycled to the sprays of the desuperheater, and the rest is discharged to a disposal system. The hydrocarbons pass to another settler, where insoluble tar can drop out along with entrained water. Crude styrene is then pumped to storage. An average composition by weight for this stream is 37.0 per cent styrene, 1.1 toluene, 0.2 tar, 61.1 ethylbenzene, 0.6 benzene.

The superheaters ordinarily used are Petro-Chem Iso-Flow furnaces, with combustion chambers 9 ft. in diameter by 20 ft. in height. An air preheater is mounted between furnace and stack. Since each furnace feeds two reactors, it has parallel superheating coils, each consisting of 14 vertical alloy steel tubes in series, connected by return bends. The furnace is fired by gas burners located in the flat bottom of the combustion chamber. One large stationary burner is placed at the center with smaller movable ones placed symmetrically around it.

Each dehydrogenation block in the Government-built plants contains eight dehydrogenation reactors. Figure 6 illustrates the inside construction

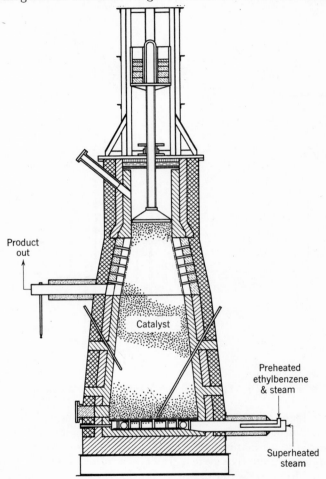

Product
out

Catalyst

Preheated
ethylbenzene
& steam

Superheated
steam

Fig. 6.   Ethylbenzene-Dehydrogenation Reactor

of a reactor. The reactor consists of a refractory-lined catalyst case employing upflow from a distributing ring below the catalyst. A screen in the top part of the units allows passage of outgoing vapors. Catalyst is retained in place by a weighted ram. Filling is accomplished through the tube shown. Thermocouple wells are indicated. Since the reaction is endothermic and the heat of reaction is furnished mostly by superheated steam, the catalyst case must be insulated to minimize heat loss. An inside layer of insulating cement backed up by two layers of insulating brick serves this purpose. A 1/4-in. steel shell makes the chamber gas-tight. Thorough mixing of steam and ethylbenzene takes place in the concentric-tube inlet arrangement shown.

Styrene-Finishing Step

Problems in Styrene Distillation. The third and final step is the purification of the crude dehydrogenated material containing 35 to 40 per cent styrene.

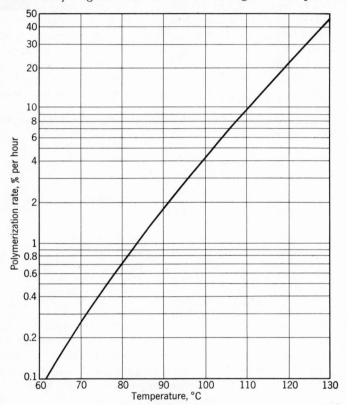

FIG. 7.   Polymerization Rate of Styrene at Various Temperatures

The fractionation requirements are rather strict. Not only the styrene product but also recycled benzene and toluene must be of high purity. The normal boiling point of ethylbenzene is only 9° C. lower than that of styrene.[51] Other factors complicate the distillation step to an even greater extent. It is evident from Fig. 7 that, because of concurrent polymerization, distillation of styrene at its atmospheric boiling point of 145.2° C. is out of the question. At 30 mm. mercury absolute, the boiling point of styrene is 54° C., and the polymerization rate has dropped to less than 0.1 per cent per hour. Running under a vacuum, then, appears necessary. However, the ethylbenzene-styrene separation requires 70 actual plates. With a pressure drop of

4 mm. mercury per plate throughout such a column, the bottom temperature at 310 mm. mercury could cause excessive polymer formation. Even though a low distillation temperature is achieved, the normal irregularities of operation make further precautions necessary. Very small quantities of certain foreign materials catalyze the polymerization reaction greatly. To guard against such possibilities, elemental sulfur is dissolved in styrene to act as a polymerization inhibitor.[11, 21]   Only by a combination of vacuum operation, suitable inhibition of styrene polymerization, and special column design can styrene be distilled successfully.   Now, after years of experience, the problem seems straightforward, but this purification step was actually the major stumbling block in the commercial production of styrene.[72] Relatively trouble-free operation can be attained by distilling inhibited styrene in a system which never allows the concentrated monomer to exceed 90° C. in temperature.   The penalty for mistakes is rather severe, since cleaning a thoroughly plugged column would require hundreds of man hours and infinite patience.

Quality of Styrene Recovered.   A representative analysis of finished styrene is given in Table II.

Table II.   Representative Analysis of Styrene Product

	Per cent
Styrene	99.65
Aldehydes (as acetaldehyde)	0.0039
Peroxides (as diethyl peroxide)	0.0008
Sulfur	0.0015
Chlorides	0.0059
Polymer	nil
Ethylbenzene and isopropylbenzene	0.35
p-tert-Butylcatechol	10 p.p.m.
(added as an inhibitor)	

Since the sulfur which is added as a polymerization inhibitor must be very low in the final product, a different inhibitor is added to prevent polymer formation during shipment.   The material commonly used is *p-tert*-butylcatechol (TBC) in a concentration of 10 to 15 p.p.m.[66]   If necessary, this inhibitor can be removed at the polymer plant by a caustic wash.

In making monomeric materials to be used for polymerization, it is often found that a few parts per million of certain materials can influence the polymerization tremendously.   For this reason styrene production has its tricky aspects, and the manufacturer must be constantly alert for trace materials which may act as inhibitors or catalysts.   Even a purity of 99.95 per cent does not always guarantee an acceptable product.

A styrene plant operates with an average styrene recovery of 99.4 per cent in the purification step.   Nearly all of the 0.6 per cent loss goes to polymer or compounds formed by the reaction of styrene and sulfur.

Plant Operation.   A flow-sheet of the styrene-finishing system is shown in Fig. 8.   Crude styrene passes through a pot containing sulfur.   Enough sulfur is dissolved to act as a polymerization inhibitor.[21]   After preheating, the feed enters the 30-plate benzene-toluene column, where benzene and

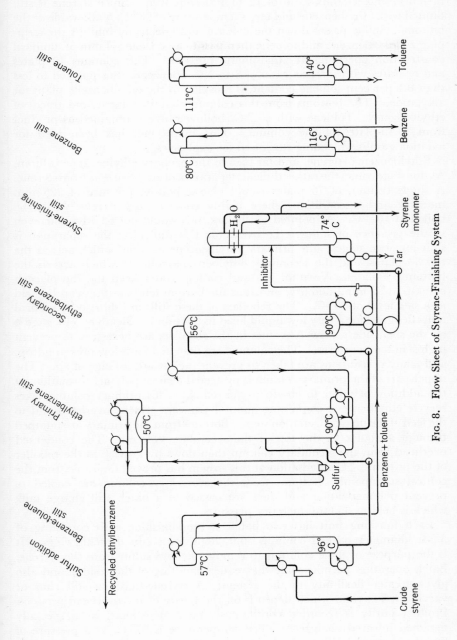

FIG. 8.   Flow Sheet of Styrene-Finishing System

toluene pass overhead at 57° C.   The top of this column operates at 175 mm. mercury, and a reflux ratio of 12 to 1 is employed.   Since styrene is still diluted with ethylbenzene and tar, a temperature of 96° C. is allowable at the bottom.   Sulfur passes down the column and effectively inhibits the stripping zone.   Benzene and toluene then pass to a benzene column of standard construction operating at atmospheric pressure.   This contains 40 plates and operates with a reflux ratio of 3 to 1.[28]   Benzene, thus purified to less than 0.1 per cent toluene content, is recycled to the ethylbenzene plant for alkylation.   The bottoms from this column contain toluene and traces of ethylbenzene.   Toluene with a 1° C. boiling range is obtained as product from a 35-plate toluene column.   This by-product has been used for aviation gasoline blending stock and for explosives.

Ethylbenzene, styrene, and tar pass to the primary ethylbenzene column. At this stage comes the rather difficult separation of ethylbenzene from styrene. A single column of 70 plates would give a bottom pressure of 225 mm. mercury and a condition where highly concentrated styrene would be boiled at 104° C.   As pointed out before, this would not be advisable even in the presence of sulfur.   To overcome this difficulty, the separation is split into two stages: the primary ethylbenzene column, which acts as the top 38 plates, and the secondary ethylbenzene column, which acts as the bottom 32 plates.   Vacuum jets and cooling water keep the top of each column at 35 mm. mercury, and thus the bottom temperatures can be held at a safe level of 90° C.   The reboilers of these stills are short-tube vertical units designed for a very low liquid head in the tubes.   Steam at 5 lb. gage is on the shell side.   Low Δt values in the reboilers are necessary to prevent high skin temperatures.   The columns are designed for a low pressure drop, and many precautions are taken to prevent the inward leakage of air.   The upper part of the primary column is protected against polymer formation by the addition of sulfur to ethylbenzene reflux.   This primary column takes recycle ethylbenzene containing about 1 per cent styrene overhead and to the feed of the dehydrogenation step.   Bottoms from the primary are pumped through a cooler to the top of the secondary column.   The condensed overhead from the secondary column then flows by gravity to the reboiler of the primary.   Condensation of this stream has proved desirable from the control standpoint.   Bottoms from the secondary column are cooled to prevent polymerization and then are passed to a batch still charge tank which is held at 35 mm. mercury pressure.

Each finishing unit has two batch styrene-finishing stills consisting of 3.5-ft. diameter packed columns 36 ft. high, with 2-in. Raschig rings.   It is the purpose of these columns to remove tar and sulfur from the styrene. Batch operation permits somewhat easier handling of the residue and also gives greater flexibility to the system.   A styrene-finishing still runs at varying reflux ratio which depends on the purity of the overhead product. Product purity is recorded continuously on a strip chart by a specially designed infrared analyzer.[74]   Top temperature is 57° C. at a pressure of 35 mm. mercury.   Since the upper part of the column operates with little sulfur present, a solution of *p-tert*-butylcatechol in styrene is pumped into the reflux to this still by means of a proportioning pump.   Five parts per million of inhibitor are effective in preventing excessive polymer formation. The residue from these columns is drained while hot into dump buckets. This heavy material, consisting of tars formed in the dehydrogenation step, styrene-sulfur compounds, elemental sulfur compounds, elemental sulfur, polystyrene, and styrene, is then burned.

The product styrene which passes overhead goes to horizontal receivers. Here *p-tert*-butylcatechol is added to bring the concentration to a minimum of 10 p.p.m. Inventories of this high-purity styrene are held at a minimum for obvious reasons. Shipment of the finished material is ordinarily made in insulated tank cars, although standard cars have been used in moderately cool weather. Styrene is pumped through a refrigerated cooler to these cars to provide a loading temperature below 20° C. With these precautions, the probability of polymer formation in normal tank car shipment is exceedingly small.

A point requiring further emphasis is the continuity of operation of the process just described. Its on-stream efficiency is more than 95 per cent. A factor of importance in maintaining continuity of operation has been the highly instrumented nature of the process. In all, about 960 indicating and recording instruments, not counting gages and thermometers, are used at the Velasco styrene plant described.[13]

Variations in Plant Operation. In the Monsanto plant at Texas City, the steps employed are alkylation with aluminum chloride as catalyst, dehydrogenation as practiced by Dow, and a distillation purification system developed by Monsanto and the Lummus Company.[15] The distillation method differs from that used by Dow in the order of separation of components of the crude styrene. Benzene, toluene, and ethylbenzene are taken overhead in the first column, with styrene and tars passing out as bottoms. The styrene is then separated from the residue by continuous distillation. Benzene, toluene, and ethylbenzene are then separated for recycle and sale. This method has an advantage over the Dow system in that styrene passes through the reboiler of one column rather than two. In actual operation, the distillation temperatures in the Texas City plant ran considerably higher than those in Dow plants, with consequently higher losses of styrene. After the disaster occasioned by the explosion of a cargo of ammonium nitrate in the Texas City harbor, which resulted in the destruction of a large portion of the dock area of Texas City where the Monsanto Chemical Company is located, the plant was erected anew without any substantial departure from the process previously in use.

CARBIDE AND CARBON PROCESS

For several years prior to 1942, Carbide and Carbon Chemicals Corporation had been working on a styrene monomer process which resembled the Dow Company's method only in the alkylation step. In manufacturing ethylbenzene, Carbide also used an aluminum chloride ethylation, but engineering layout and equipment details differed considerably. The alkylation step was carried out under pressure. In other respects the process itself was quite similar to the method already described. It is interesting to note that Monsanto, Carbide and Carbon, Dow, and I.G. Farbenindustrie in Germany independently arrived at processes for the production of ethylbenzene that were essentially the same.

In working out the styrene process to be used at Institute, Carbide recognized the difficulty of the ethylbenzene-styrene distillation step and chose to avoid it by employing a chemical method of separation.[43] Whereas most of the development work on the dehydrogenation process centered on the specialized distillation step, Carbide concentrated on the development of selective chemical reactions, followed by simple distillations. The process arrived at employed oxidation, hydrogenation, and dehydrogenation steps to convert ethylbenzene to styrene. A generalized flow sheet is given in Fig. 9.

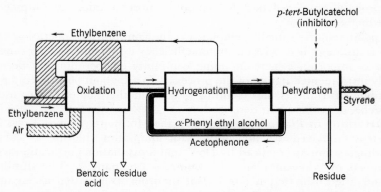

FIG. 9. Simplified Flow Diagram of Carbide and Carbon Chemicals Corporation's Styrene Process

Ethylbenzene is oxidized with air in the presence of manganese acetate catalyst. This operation is conducted in the liquid phase. Products formed are α-phenylethyl alcohol, acetophenone, and small quantities of benzoic acid and other by-products.

$$(C_6H_5 \cdot CH_2 \cdot CH_3 \rightarrow C_6H_5 \cdot CH(OH) \cdot CH_3 + C_6H_5COCH_3)$$

Benzoic acid is then removed by a caustic wash, yielding sodium benzoate which may be recovered. After a water wash, unreacted ethylbenzene is removed from the product by vacuum distillation. In another vacuum still acetophenone and phenylethyl alcohol are separated from high-boiling residue. This distillate is then hydrogenated in the liquid phase, a copper-chrome-iron catalyst being used. In this operation most of the acetophenone is converted to phenylethyl alcohol. A small amount of the phenylethyl alcohol is also hydrogenated to ethylbenzene in this step, and the hydrogenator product is vacuum-distilled to remove such ethylbenzene, which is returned to the feed of the oxidation step. It is essential at this point in the process to remove the ethylbenzene completely, as it would otherwise appear in the finished styrene. The residue from this distillation, consisting of phenylethyl alcohol and unconverted acetophenone, is then passed in the vapor phase over alumina catalyst to dehydrate the alcohol to styrene. The reaction product is distilled under vacuum for the recovery of refined styrene. The unconverted alcohol and acetophenone are redistilled to separate them from residue and are then returned to the feed of the hydrogenation step.[57a, 76]

Efficiencies are:	Conversion per Pass, %	Over-all Yield, %
Oxidation step	25	90
Hydrogenation step	80	98
Dehydration step	80	91

The economics of this process depends on utilizing the intermediates and by-products. Although the over-all yield of styrene is lower than that of the dehydrogenation method, several chemical by-products formed have quite substantial value. In fact, acetophenone is now being manufactured from intermediates which the Carbide and Carbon styrene process makes available.[57a]

As previously mentioned, the reaction steps involved in this process are relatively complicated when compared with the dehydrogenation process, but the distillation steps are easy. The maximum number of trays required in any of the stills[12] is only 24.

GERMAN STYRENE-MANUFACTURING PRACTICE[17]

The I.G. Farbenindustrie A.-G. at Ludwigshafen was responsible for the original development of the German styrene-manufacturing process. Here styrene was first intended for the manufacture of polystyrene. Later it was made in modern well-engineered plants at the locations listed in Table III. Much of the styrene produced in Germany went into the manufacture of Buna S rubber.

Table III. Approximate Rated Productive Capacity of German Plants[17]

	Long Tons per Annum		
	Styrene	Polystyrene	Buna S
Ludwigshafen	12,000	6000	24,000
Schkopau	20,400	600	72,000
Hüls	21,600	...	48,000
Auschitz	12,000	...	36,000

The process used at Schkopau for the production of styrene monomer will be described in some detail. It was based on brown coal as the fundamental raw material. Benzene was obtained from the coal directly. Acetylene from calcium carbide, made from coal and limestone, was hydrogenated to ethylene. The ethylene and benzene were combined to form ethylbenzene which was dehydrogenated to styrene.

Ethylbenzene Production. The ethylbenzene plant at Schkopau had a capacity of 2700 metric tons per month. The raw materials consisted of (a) benzene refined from coal distillation, (b) 98.0 per cent ethylene from

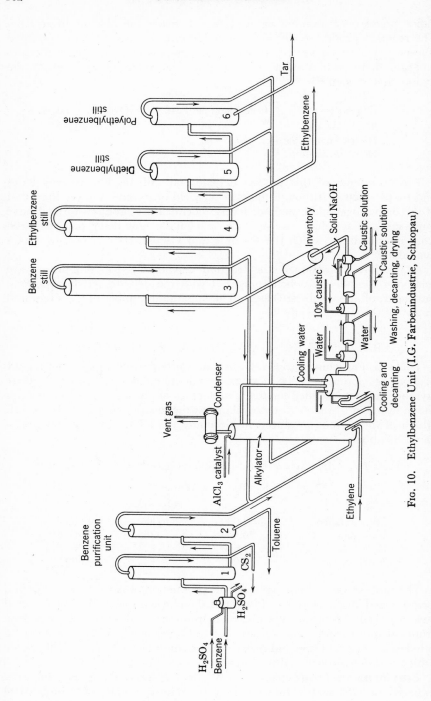

Fig. 10. Ethylbenzene Unit (I.G. Farbenindustrie, Schkopau)

the hydrogenation of acetylene, and (c) powdered aluminum chloride containing less than 0.1 per cent iron.   A flow sheet of a plant unit is shown in Fig. 10.

The ethylene and benzene were reacted in an enameled steel tower in the presence of aluminum chloride as catalyst.   Here the reactor composition was 45 to 50 per cent benzene, 35 per cent ethylbenzene, and 15 to 17 per cent higher alkylbenzenes.   The reactor operated about 80 per cent full at a temperature of 90° C. and was controlled by reflux.   No separate dealky-lators were necessary, as the polyalkylbenzenes were recycled to the alkylating reactor, where an equilibrium was established.   The reaction mixture over-flowed from the reactor and was cooled to about 40° C. in a combined separator and cooler.   The heavy catalyst layer was recycled to the reactor, and the top ethylbenzene layer was washed, first with water and then with 10 per cent sodium hydroxide solution.   After drying with solid caustic, the top layer was pumped to a 200-cubic meter iron storage tank.   This acted as an inventory between the reaction and distillation steps.   The mixture was pumped to a series of continuous stills.

Information regarding the steps followed and the equipment used in a unit for the production of ethylbenzene is given[9] in Table IV.

The yield based on benzene was 95 to 96 per cent; based on ethylene, 92 to 94 per cent.   The ratio of ethylbenzene to aluminum chloride was approximately 35 to 1.   Ethylbenzene was maintained at high purity, the diethylbenzene and toluene content being less than 0.002 per cent.   The boiling range was 0.5° C. to the dry point.

Dehydrogenation of Ethylbenzene.   The final major step in the process (diagrammed in Fig. 11) consisted in dehydrogenating the ethyl-benzene to styrene and distilling to obtain pure styrene.   Here the 15 dehydrogenation furnaces had a total capacity of 1700 tons per month. The raw materials consisted of (a) ethylbenzene having a boiling range of 5° C. to dryness and free from chlorides, toluene, and diethylbenzene; (b) low-pressure steam, and (c) a special catalyst, of the following percentage composition: zinc oxide 80, calcium oxide 5 to 7, aluminum oxide 10, potassium hydroxide 2 to 3, chromic oxide 0.5 to 0.7.

When the catalyst was new, the reaction temperature was 580° C.; when it was a year old, a temperature of 600 to 610° C. was required.   For the manufacture of the catalyst, the ingredients were mixed with a small proportion of water, extruded as 1/8-in. rod, dried at 150° C., broken into 1/4-in. lengths, and heated to reaction temperature.   With this catalyst, an over-all yield of 90 per cent was obtained.   If the chromium and potassium were omitted, the yield dropped to 84 per cent.

The dehydrogenation reaction furnace was about 4 ft. in diameter and consisted of bundles (22 tubes per furnace) of verticle tubes 8 ft. long filled with catalyst and set into a header sheet covered with graphite Raschig rings to aid gas distribution.   The tubes were copper-manganese-coated steel, heated by direct gas firing from front and back with combustion gases baffled around the tube bank.   Newer furnaces had 6-in. cylindrical tubes; the older ones had elliptical tubes.   The use of many thermocouples made possible good control of the temperature.   Steam and ethylbenzene in

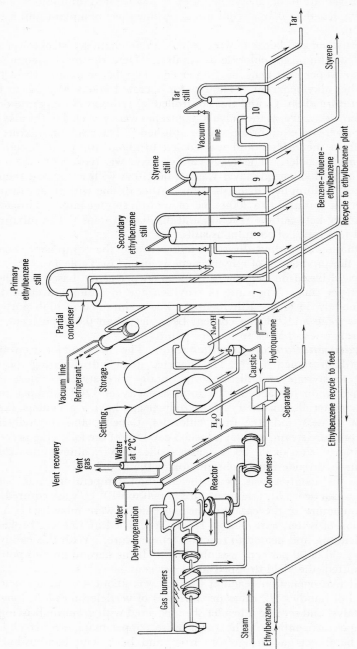

FIG. 11. Styrene Unit (I.G. Farbenindustrie)

Table IV. Ethylbenzene Unit at Schkopau

Process Step	Equipment Size	Material of Construction	Stills Type	Plates	Reflux Ratio	Pressure	No. of Units
Benzene Purification							
1. 96% sulfuric acid wash		Iron					1
2. Separator	100 cu. ft.	Iron					1
3. 5% caustic wash		Iron					1
4. Separator	100 cu. ft.	Iron					1
5. Carbon disulfide removal	40 in.	Iron	Cont.	30		Atm.	1
6. Toluene removal	40 in.	Iron	Cont.	30		Atm.	1
Alkylation							
1. Alkylators	45 in. × 40 ft.	Enameled Steel					4
2. Separators	6 ft. × 8 ft.	Phenolic Resin-Lined					4
3. Alkylate coolers	6 ft. × 8 ft.	Enameled Steel					4
4. Product washers and separators	30 cu. ft.	Steel					4
5. 10% caustic wash	100 cu. ft.	Iron					4
6. Caustic drier		Iron					4
Distillation							
1. Benzene	5.23 ft.	Iron	Cont.	50	1:1 to 3:1	Atm.	1
2. Ethylbenzene	5.23 ft.	Iron	Cont.	50	1:1 to 3:1	Atm.	1
3. Diethylbenzene still	5 ft. × 25 ft.	Iron	Cont.		1:1 to 3:1	100 mm.	1
4. Polyethylbenzene still	3 ft. × 25 ft.	Iron	Cont.		1:1 to 3:1	20 to 50 mm.	1

equimolecular ratios were preheated in a two-stage heat exchanger and passed through the catalyst at the temperatures ranging from 520 to 610° C., with a pressure drop of 20 mm. to about 1 atm. The exchangers, heated by the combustion gases from the furnace and the reaction gases, brought the reactants nearly to reaction temperature. The reaction gases cooled by heat exchanger were condensed to a liquid and entered the condenser at 280° C. The liquid contained 40 per cent styrene, 0.5 per cent benzene, 1 per cent toluene, 0.3 per cent tar, and the remainder ethylbenzene. The mixture was separated from water by a gravity separator and then stabilized with 10 grams per ton of hydroquinone and passed to storage tanks.

The product was then ready for distillation. The distillation system consisted of two fractionating columns, a rectifying column, and a tar still. Two trains were used in the distillation plant. The first consisted of two plate columns and two ring-packed stills (for styrene and tar). The second used ring packing for the tar still only. All stills and condensers were tin-lined, but later work showed that such lining was unnecessary and that iron was satisfactory.

The mixture of ethylbenzene, styrene, and "heavies" was fed to a 50-plate column, 6 ft. in diameter. Ethylbenzene almost pure was taken off at the base of the column condenser and returned to the reactor system. The benzene, toluene, and ethylbenzene mixture went overhead out of the condenser and was returned to the ethylbenzene plant. The styrene–ethylbenzene (60 : 40) and heavies mixture from the bottom of the still was fed to the second column. The second column was of about 25 plates and 5 ft. in diameter. The overhead from this column contained about 40 per cent styrene and 60 per cent ethylbenzene and was fed back to the first column. The bottom, containing impure styrene with some heavies, was fed to the third still. The bottom had no components of lower boiling point than styrene. The third column in the first train was about 4 ft. in diameter and consisted of 20 plates. In the second train this column was ring-packed. The top of this still delivered styrene of 99.5 to 99.8 per cent purity to storage. The bottoms were heavy oils, a small amount of polystyrene, stilbene, and diphenyl. These bottoms were delivered to the tar still. This still was ring-packed and delivered tar out at the bottom and a styrene-tar mixture out at the top. This mixture was returned to the third column.

All columns were continuous and operated under vacuum. The plate spacing was about 14 in., and the bubble caps were well designed with deep, square-cut bubble slots. The bubble-plate wall faces were accurately machined with small annular grooves for a rubber-asbestos gasket. Leaks were rare but when they happened led to rapid polymer accumulation. That they seldom occurred was indicated by the relatively small vacuum jet used for the entire distillation train. Only occasionally was it necessary to remove polymer from the condensers. It was possible to remove the lower section of the stills for cleaning. This was necessary about once a year to remove the polymer caused by divinylbenzene or air in the system. Some details regarding the construction and operation of the styrene distillation unit diagrammed in Fig. 11 are shown in Table V.

Table V.   Distillation System Details

Still	Plates	Diameter	Plate Spacing	Pressure mm. Hg. Top	Pressure mm. Hg. Bottom	Temp., °C. Top	Temp., °C. Bottom	Reflux Ratio	Feed Plate
7	45	72 in.	15 in.	25	175	45	100	4.1	18
8	25	60 in.	15 in.	15	75	35	85	8.1	15
9	20	48 in.	12 in.	10	50	35	85	5.1	10
10	Ring packing	20 in. × 12 ft. packed column							

The inventory of liquid in the system was kept to a minimum through rapid circulation of the liquid in the small vertical thermal circulation reboilers on the still, where liquid levels were accurately controlled.

The purity of the final product was controlled by continuous measurement of the refractive index of the liquid. The pure styrene fraction was stored unstabilized under nitrogen in aluminum tanks.

The styrene-production process and equipment used at Schkopau was quite typical of the operations in Germany, and only minor equipment changes and slight temperature, pressure, and feed differences were noted in other German plants. Mitchell,[49] reporting on "The Dow Process for Styrene Production," made the following comparison of German and American styrene processes. "Recently, information has been released on details of the styrene process employed by I.G. Farbenindustrie A.-G. in Germany. American technical experts found a process very much like the one just described (Dow process). In fact, the only variations are minor ones—mostly in materials of construction. Yields on raw materials used are almost identical . . . . In making a general comparison between the two processes, we find little to choose. Yields are comparable and product purities are almost identical."

PHYSICAL PROPERTIES OF STYRENE MONOMER[71]

Styrene is a colorless, transparent, mobile liquid with a sweet, pleasant odor. On exposure to air, oxidation rapidly forms certain aldehydes and ketones, which give to styrene the sharp, penetrating disagreeable odor sometimes associated with it. Odor in fact furnishes an extremely sensitive test for the presence of aldehydes. Styrene containing these and other oxidation products does not behave like styrene of high purity; it shows a marked difference in rate of polymerization and in the induction period which precedes polymerization. Some of the oxidation products have quinoid structures and hence give rise to color—in its more advanced stages to yellowing. These oxidation products have dipole moments, so that the dielectric constant of monomer containing them (and of polymer formed from such monomer) is not so low as that of material of higher purity. The proper handling of styrene will prevent the troubles that can arise from impurities caused by oxidation of this monomer.[69]

The general properties of styrene monomer as a chemical substance are shown in several tables. Table VI lists some standard physical properties

Table VI.   Physical Properties of Pure Styrene

Formula	C_8H_8
Molecular weight	104.14
Density at 25° C.	0.9019
Pounds per gallon at 25° C.	7.5
Refractive index at 25° C.	1.5439
Viscosity, centipoises at 25° C.	0.730
Surface tension, dynes per cm. at 25° C.	31.7
Coefficient of expansion, cc. per ° at 25° C.	0.0009719
Boiling point, °C.	145.2
Heat of vaporization, cal. per gram. at 25° C.	102.65
Freezing point, °C., in air at 1 atm. pressure	−30.628
Heat of fusion, cal. per gram.	25.4
Specific heat, cal. per gram. per °C. at 25° C.	0.416
Heat of combustion, kg.-cal. per gram. at 25° C.	10.086
Critical temperature, °C.	373.0
Critical pressure, atm.	40.0
Critical density, grams per ml.	0.30
Flash point, °C.	31
Fire point, °C.	34
Explosive limits, volume per cent in air	1.1 to 6.1
Heat of polymerization, cal. per gram.	160.2
Volume shrinkage on polymerization, per cent	17

Table VII.   Typical Analysis of a Commercial Styrene Monomer

Color	Saybolt 25
Specific gravity at 25/25° C.	0.9044
Refractive index (D line) at 25° C.	1.5437
% Styrene from refractive index	99.65
Freezing point, °C.	−30.64
Styrene from freezing point, %	99.98
Polymer in monomer, %	None
Phenylacetylene, %	0.0002
Aldehydes as acetaldehyde, %	0.0039
Aldehydes as benzaldehyde, %	0.0094
Peroxides as diethyl peroxide, %	0.0008
Chlorine, %	0.0059
Sulfur, %	0.0015

Table VIII.   Specific Heat of Styrene Vapor[67]

Temperature °C.	Specific Heat Cal. per Gram.	Temperature °C.	Specific Heat Cal. per Gram.
50	0.289	500	0.567
100	0.326	550	0.586
150	0.366	600	0.605
200	0.405	700	0.639
250	0.439	800	0.666
300	0.470	900	0.688
350	0.497	1000	0.708
400	0.521		
450	0.545		

Table IX. Properties of Styrene Monomer as a Function of Temperature[68]

T° C.	Density, Grams per Ml.		Specific Volume, Ml. per Gram		Viscosity, Centipoise	Surface Tension, Dynes per Cm.	Vapor Pressure, mm. of Hg.	Heat of Vaporization, Kg.-cal. per Mole
	Liquid	Vapor (Calc.)	Liquid	Vapor (Calc.)				
0.0	0.9238	0.00000819	1.0824	122,000	0.976	34.5	1.13	11,250
10.0	0.9150	0.0000156	1.0928	63,700	0.877	33.4	2.34	11,010
20.0	0.9063	0.0000283	1.1033	35,200	0.781	32.3	4.53	10,790
25.0	0.9019	0.0000374	1.1087	28,700	0.730	31.7	6.17	10,690
30.0	0.8975	0.0000489	1.1142	20,400	0.694	31.2	8.31	10,590
40.0	0.8887	0.0000812	1.1252	12,300	0.621	30.0	14.51	10,410
50.0	0.8800	0.000129	1.1363	7,750	0.552	28.9	24.26	10,230
60.0	0.8712	0.000200	1.1478	5,000	0.490	27.8	39.07	10,070
70.0	(0.8624)	0.000300	(1.1595)	3,330	(0.438)	(26.7)	60.78	9,920
80.0	(0.8535)	0.000438	(1.1716)	2,280	(0.392)	(25.6)	91.74	9,770
90.0	(0.8446)	0.000622	(1.1839)	1,610	(0.348)	(24.6)	134.7	9,630
100.0	(0.8356)	0.000871	(1.1967)	1,150	(0.312)	(23.5)	192.9	9,490
110.0	(0.8265)	0.00119	(1.2099)	840	(0.278)	(22.5)	270.2	9,350
120.0	(0.8174)	0.00160	(1.2234)	625	(0.248)	(21.5)	370.7	9,220
130.0	(0.8081)	0.00211	(1.2374)	474	(0.221)	(20.4)	500.4	9,080
140.0	(0.7988)	0.00276	(1.2518)	362	(0.196)	(19.4)	660.9	8,940
145.2	(0.7939)	0.00311	(1.2596)	321	(0.184)	(18.9)	760.0	8,870
150.0	(0.7893)	0.00346	(1.2669)	289	(0.174)	(18.4)	861.5	8,800

Values in parenthesis are extrapolated.

of styrene monomer. Table VII represents an analysis of a commercial styrene monomer and illustrates the types of impurities that might be expected. Tables VIII to XI list in greater detail some common physical properties of styrene monomer as a function of temperature. The numerical values are believed to be typical of styrene monomer of a purity equal to or greater than 99.87 per cent, unless otherwise stated in the individual tables. More detailed information concerning the physical properties of styrene monomer is presented in an American Chemical Society monograph on styrene.[65]

Solubility Characteristics. Styrene monomer is miscible with a large variety of organic liquids. It is miscible in all proportions with ether, methyl alcohol, ethyl alcohol, carbon disulfide, acetone, benzene, toluene, and carbon tetrachloride. Hydroquinone is soluble in styrene to only a limited extent; quinone is very soluble; p-tert-butylcatechol is soluble in all proportions, and benzoyl peroxide, lauroyl peroxide, and diacetyl peroxide are all very soluble. Hydrogen peroxide is only very slightly soluble, and urea peroxide is practically insoluble in styrene. 2,4-Dichloro-6-nitrophenol[10] is soluble to the extent of several per cent. Styrene is also a good solvent for synthetic and natural rubbers and plastics as well as many other organic compounds. Table X, taken from Lane,[42] shows the limited solubility of styrene in water, and of water in styrene.

Table X.   Solubility of Styrene in Water and Vice-Versa, Weight %

Temperature, °C.	Water in Styrene	Styrene in Water
0	0.020	0.018
10	0.040	0.023
20	0.060	0.029
30	0.080	0.034
40	0.100	0.040
50	0.120	0.045
60	0.140	0.051
70	0.160	0.056
80	0.180	0.062

Electrical Properties of Styrene Monomer. The electrical properties of styrene monomer depend very much on the history of the sample. Freshly distilled material may show a volume resistivity of 10^{14} ohm-cm. and a power factor of less than 0.01 per cent. Contact with oxygen or moisture quickly impairs these values. Table XI shows a typical set of measurements on styrene monomer of 99.87 per cent purity (by freezing point), treated with silica gel and containing 5 p.p.m. of p-tert-butylcatechol to prevent polymerization during the heating cycle.

The breakdown strength of styrene monomer is in excess of 30,000 volts for a 0.1-in. gap, thus corresponding to a dielectric strength greater than 300 volts per mil at room temperature.

Table XI. Electrical Properties of 99.87 per cent Styrene Monomer Treated with Silica Gel and Stabilized with 5 p.p.m. of Inhibitor

Temperature, °C.	1000-Cycle Dielectric Constant	1000-Cycle % Power Factor	Specific D.C. Resistivity
10	2.43	...	5.6×10^{13} ohm-cm.
20	2.415	0.011	4.0×10^{13}
25	2.41	0.014	3.5×10^{13}
30	2.40	0.018	3.0×10^{13}
40	2.38	0.028	2.1×10^{13}
50	2.365	0.040	1.5×10^{13}
60	2.35	0.058	1.1×10^{13}
70	2.335	0.083	8.0×10^{12}
80	2.32	0.115	5.8×10^{12}
90	2.305	0.150	4.2×10^{12}
100	2.29	0.210	3.0×10^{12}
110	2.275	0.280	2.2×10^{12}

ANALYTICAL TECHNIQUES FOR STYRENE MONOMER

During the commercial development of styrene monomer, a number of analytical procedures especially applicable to styrene were worked out and perfected. The most important ones concern tests for percentage styrene and tests for minor impurities. Tests for the latter are included because even traces of certain impurities may markedly influence the polymerization of styrene. Information on the following has been published.[39, 65, 70, 73a]

A. Assay methods for styrene monomer.
    1. By freezing point:
        (a) Hand-operated Rubber Reserve method.
        (b) Automatic platinum resistance thermometer freezing-point recorder.
    2. By refractive index.[4]
    3. By titration.[44, 45]
B. Aldehydes.
C. Peroxides.
D. Phenylacetylene.
E. Water.
F. p-tert-Butylcatechol.
G. Total sulfur.
H. Total chlorides.
I. Polymer in monomer:
    1. Rapid qualitative method.
    2. Precipitation—gravimetric method.
    3. Distillation—gravimetric method.
J. Solubility of polymer prepared from styrene monomer.

K. Viscosity of styrene monomer.
L. Determination of color in styrene monomer.
M. Specific gravity of styrene monomer.
N. Rate of polymerization of styrene monomer.
O. Determination of monomeric styrene in air [57]
P. Infrared transmission of styrene monomer.
Q. Ultraviolet absorption of styrene monomer.
R. Polarographic analysis.[41]

STANDARD PRACTICE IN HANDLING STYRENE MONOMER

The chemical structure and reactivity of styrene which account for its ready polymerizability to a valuable end product also create certain problems in handling and using it. These problems are not in fact difficult to meet, as is evidenced by the tremendous quantities of styrene that are being successfully made and transported today. Over a period of years, experience has been accumulated in handling styrene monomer from a few ounces in the laboratory to tank-car quantities. For the benefit of those groups that are learning to use styrene, perhaps on a relatively large scale from the first, it seems desirable to emphasize what can be done and what must be done with styrene. Information on the following has been published.[39, 50, 65, 69, 73a]

A. Inhibitors for storage and shipment.
B. Heat developed during polymerization.
C. Flammability.[73]
D. Storage of flammable liquids.
E. Toxicity.[61]
F. Purification:
   1. Removal of moisture.
   2. Removal of inhibitors.
   3. Removal of polymer.
   4. Removal of dissolved gases.
   5. Filtering.
   6. Distillation.
G. Effect of impurities.

REFERENCES

1. Amos, J. L., Dreisbach, R. R., and Williams, J. L. (Dow), U.S. Pat. 2,198,595 (1940).
2. Amos, J. L., Schwegler, C. C., and Bezenah, W. H. (Dow), U.S. Pat. 2,443,758 (1948).
3. Amos, J. L., Williams, J. L., and Winnicki, H. S. (Dow), U.S. Pat. 2,222,012 (1940).
4. Barstow, O. E., A New Recording Refractometer, paper No. 49–4–2, presented before Conf. of the Instrument Soc. of America, Sept. 12–16, 1949, St. Louis, Mo.
5. Berg, L., Harrison, J. M., and Montgomery, C. W., *Ind. Eng. Chem.*, **38**, 1149–52 (1946).
6. Berthelot, M., *Les Carbures d'hydrogène*, p. 901 (1851).
7. Berthelot, M., *Ann.*, **142**, 257 (1867).
8. Berthelot, M., *Ann. chim. et phys.* (4), **16**, 153–62 (1869).
9. Boundy, R. H., and Hasche, R. R., Technical Report 1069, *Manufacturing of Thermoplastics in Plants of I.G. Farbenindustrie A.-G.*, U.S. Tech. Ind. Intelligence Comm. (1945).
10. Boyer, R. F., and Rubens, L. C. (Dow), U.S. Pat. 2,304,728 (1942).

11. Britton, J. W., Prescott, R. F., and Dosser, R. C. (Dow), U.S. Pat. 2,166,125 (1939).
12. Carbide and Carbon Chemicals Corp., Private Communication, July 10, 1947.
13. Cermak, R. W., *Instrumentation*, **2**, Sept.–Oct., 7, 1946.
14. Chemical and Metallurgical Engineering Pictured Flowsheet, *Chem. Met. Eng.*, **51**, 160–3 (1944).
15. *Chem. Met. Eng.*, **50**, 133 (1943); also Private Communication from Monsanto Chemical Co.'s Texas Division.
16. Corson, B. B., and Brady, L. J., U.S. Pat. 2,417,454 (1947).
17. De Bell, J. M., Goggin, W. C., and Gloor, W. E., *German Plastics Practice*, De Bell and Richardson, Springfield, Mass., 1946.
18. Dow, W. H., *Ind. Eng. Chem.*, **34**, 1267–8 (1942).
18a. Dow Chemical Co., The, *see* The Dow Chemical Co.
19. Dreisbach, R. R. (Dow), U.S. Pat. 2,110,829 (1938).
20. Dreisbach, R. R., and Pierce, J. E. (Dow), U.S. Pat. 2,188,772 (1940).
21. Dreisbach, R. R., and Pierce, J. E. (Dow), U.S. Pat. 2,240,764 (1941).
22. Dreisbach, R. R. (Dow), U.S. Pats. 2,282,327 (1942), 2,308,415 (1943).
23. Foster, J. F., *Ind. Eng. Chem.*, **40**, 586–92 (1948).
24. Francis, A. W., and Reid, E. E., *Ind. Eng. Chem.*, **38**, 1194–1203 (1946).
25. Graves, G. D. (du Pont), U.S. Pat. 2,036,410 (1936).
26. Grinevich, V. M., *J. Applied Chem. U.S.S.R.*, **13**, 831–40 (1940).
27. Goggin, W. C., *Petroleum Refiner*, **23**, 87–92, Mar. 1942.
28. Griswold, J., Andres, D., and Klein, V. A., *Trans. Am. Inst. Chem. Engrs.*, **39**, 223–40 (1943).
29. Gutzeit, C. L. (Shell Devel. Co.), U.S. Pat. 2,408,140 (1946).
30. Hansgirg, F. (Austro-American Magnesite Co.), Aus. Pat. 142,219 (1935).
31. Hansgirg, F. (Austro-American Magnesite Co.), Ger. Pat. 614,507 (1935).
32. Hansgirg, F. (Austro-American Magnesite Co.), Fr. Pat. 811,736 (1937).
33. Ipatieff, V. N., and Schmerling, L., *Ind. Eng. Chem.*, **38**, 400–2 (1946).
34. Kearby, K. K. (Standard Oil), U.S. Pat. 2,370,797 (1945).
35. Kearby, K. K. (Standard Oil), U.S. Pat. 2,370,798 (1945).
36. Kearby, K. K. (Standard Oil), U.S. Pat. 2,426,829 (1947).
37. Kearby, K. K. (Standard Oil), U.S. Pat. 2,407,373 (1946).
38. Kearby, K. K. (Jasco, Inc.), U.S. Pat. 2,395,875 (1946).
39. Koppers Co., Inc., Bull. No. C-9-119-1.
40. Kutz, W. M., and Corson, B. B., *Ind. Eng. Chem.*, **38**, 761–4 (1946).
41. Laitinen, H. A., and Wawzonek, S., *J. Am. Chem. Soc.*, **64**, 1765–6 (1942).
42. Lane, W. H., *Ind. Eng. Chem.*, *Anal. Ed.*, **18**, 295–6 (1946).
43. Lee, J. A., *Chem. Met. Eng.*, **50**, 98–102, June 1943.
44. Marquardt, R. P., and Luce, E. N., *Ind. Eng. Chem.*, *Anal. Ed.*, **21**, 1194–6 (1949).
45. Marquardt, R. P., and Luce, E. N., *Ind. Eng. Chem.*, *Anal. Ed.*, **20**, 751–3 (1948).
46. Mark, H., and Wulff, C. (I.G. Farbenindustrie), U.S. Pat. 2,110,833 (1938).
47. Mavity, J. M., Zetterholm, E. E., and Hervert, G. L., *Trans. Am. Inst. Chem. Engrs.*, **41**, 519–28 (1945).
48. Mavity, J. M., Zetterholm, E. C., and Hervert, G. L., *Ind. Eng. Chem.*, **38**, 829–32 (1946).
49. Mitchell, J. E., *Trans. Am. Inst. Chem. Engrs.*, **42**, 293–307 (1946).
50. Monsanto Chemical Co., Bull. No. 5M-12-48, 71.
51. O'Kelly, A. A., Kellett, J., and Plucker, J., *Ind. Eng. Chem.*, **39**, 154–8 (1947).
52. Ostromislensky, I., and Shepard, M. G. (Naugatuck Chemical Co.), U.S. Pat. 1,541,175 (1925).
53. Ostromislensky, I., and Shepard, M. G. (Naugatuck Chemical Co.), U.S. Pat. 1,552,874 (1925).
54. Ostromislensky, I., and Shepard, M. G. (Naugatuck Chemical Co.), U.S. Pat. 1,552,875 (1925).

55. Pardee, W. A., and Dodge, B. F., *Ind. Eng. Chem.*, **35,** 273–8 (1943).
56. Rept. on the Rubber Program, 1940–5, Rubber Reserve Co., Feb. 24, 1945, p. 25.
57. Rowe, V. K., Atchison, G. J., Luce, E. N., and Adams, E. M., *J. Ind. Hyg. Toxicol.*, **25,** No. 8, Oct. 1943.
57a. Sanders, W. A., Keag, H. F., and McCullough, H. S., *Ind. Eng. Chem.*, **45,** 2–14 (1953).
58. Schulze, W. A., and Lyon, J. P. (Phillips Petrol. Co.), U.S. Pat. 2,416,022 (1947).
59. Schulze, W. A. (Phillips Petrol. Co.), U.S. Pat. 2,419,599 (1947).
60. Schulze, W. A. (Phillips Petrol. Co.), U.S. Pat. 2,419,796 (1947).
61. Spencer, H. C., Irish, D. D., Adams, E. M., and Rowe, V. K., *J. Ind. Hyg. Toxicol.*, **24,** No. 10, 295–301 (1942).
62. Suida, H., Austrian Pat. 132,642 (1933).
63. Sumerford, S. D. (Standard Oil), U.S. Pat. 2,436,616 (1948).
64. Stanley, A. M., *Chem. and Ind.*, **57,** 93–8 (1938).
65. Styrene, A. C. S. monograph, edited by R. H. Boundy, R. F. Boyer, and S. M. Stoesser, Reinhold, New York, 1952.
66. Stoesser, S. M., and Stoesser, W. C. (Dow), U.S. Pat. 2,181,102 (1939).
67. Stull, D. R., and Mayfield, F. D., *Ind. Eng. Chem.*, **35,** 639–45, 1303–4 (1943).
68. Stull, D. R., Physical Research Lab., The Dow Chemical Co.
69. The Dow Chemical Co., Bull. Form No. PL 44-2M-346.
70. The Dow Chemical Co., Bull. Form No. PL 43-15C-346.
71. The Dow Chemical Co., Bull. Form. No. PL 31-2M-346.
72. The Rubber Situation (Baruch Comm. Rept.), House Doc. No. 836, U.S. Govt. Printing Office, Sept. 10, 1942.
73. U.S. Bureau of Mines, R. I. 3630.
73a. Ward, A. L., and Roberts, W. J., *Styrene*, Interscience, New York, 1951.
74. Wright, N., and Herscher, L. W., *J. Optical Soc.*, **36,** 195–202 (1946).
75. Wulff, C., and Roell, E. (I.G. Farbenindustrie A.-G.), U.S. Pat. 1,986,241 (1935).
76. Young, D. M., Young, F. G., and Guest, H. R. (Carbide & Carbon Chemicals Corp.), Brit. Pat. 587,181 (1947).

CHAPTER 7

THE MANUFACTURE OF GR-S

John W. Livingston and John T. Cox, Jr.

Formerly Vice President, Rubber Reserve Company, and Deputy Director,
Office Of Rubber Reserve, Respectively

Rubber is indisputably one of the most important basic raw materials of American industry. Before 1940 up to 600,000 long tons of natural rubber a year had been used in the United States. Of the world's supply of natural rubber, approximately 97 per cent came from the Far East, the major sources being British Malaya and the Dutch East Indies. During World War II not only did the supply of natural rubber largely fail but also the tonnage of rubber required mounted. And, as described in Chapter 2, the establishment of a vast industry for the manufacture of synthetic rubber was undertaken, the major type of rubber produced being GR-S (Government Rubber-Styrene). The plants in which GR-S was made by the copolymerization of butadiene and styrene were mostly built adjacent to large butadiene producing plants, which in turn were located nearby the feedstock producers, which for petroleum-derived butadiene were the large oil refineries. GR-S plants using butadiene derived from petroleum were located in Louisiana, Texas, and California; those using butadiene made from alcohol were located in West Virginia, Pennsylvania, and Kentucky. The location and the designed capacities of the polymerization plants are given in Table I. These plants cost a total of about $164,340,000, and, since their rated capacity was 705,000 long tons a year, their initial cost was $233 per ton of rated capacity.

The policy of Rubber Reserve was to select those companies believed to be best qualified to carry out the different parts of the Government Synthetic Rubber Program and rely on these companies for the processes and plant designs needed for the accomplishment of their part of the work. This policy was followed in the manufacture of the copolymer GR-S, and the four principal rubber companies, all of whom had some experience with synthetic-rubber polymers, were selected to supervise the building of the copolymer plants and to operate them after they were built. Since the rubber companies were responsible for the results obtained from the copolymer plants, their recommendations regarding process and plant design were followed by Rubber Reserve.

Rubber Reserve's three main responsibilities were (1) over-all management of the program, (2) making effective arrangements covering information to be exchanged among the companies engaged in each phase or section of the program, (3) approval of expenditures of Government money.

The synthetic-rubber plants were actually built by the Defense Plant Corporation, a sister corporation to Rubber Reserve in the Reconstruction

Finance Corporation.   In the R.F.C., however, Defense Plant Corporation required Rubber Reserve's approval on all phases of plant design, materials of construction, and all other matters of consequence in the erection of the plants.   After the plants were built, Rubber Reserve took over the operation, and later took over all phases of Governmental responsibility regarding additional construction, such as an expansion in plant capacity in 1944 and other additions and improvements.

It should be noted that the name of the Government agency that supervised the erection and operation of the Government owned synthetic rubber plants was first Rubber Reserve Company, a subsidiary of the R.F.C.; then Office of Rubber Reserve, R.F.C., and is now Office of Synthetic Rubber, R.F.C.

EXCHANGE OF INFORMATION

To facilitate the exchange of information among all companies operating copolymer plants, Rubber Reserve set up an operating committee in which all problems relating to the production of GR-S were periodically discussed and reports on all phases of the work were exchanged between the operating companies and Rubber Reserve.   Operators of copolymer plants were also members of other Rubber Reserve committees, such as those concerned with safety, the handling of butadiene and styrene, etc., the development of methods of analysis for monomers and other chemical raw materials.   Visits of company personnel from one plant to another were arranged and encouraged, and in these and other ways the experiences of the plants were effectively exchanged.   There is no doubt that this interchange of information was of great aid to synthetic-rubber plant operations and output, since improvements developed at one plant very quickly spread to the other copolymer plants.   With the agreement of the operating companies, cost sheets and other operating information were exchanged each month, this being preceded by a study and an agreement between cost accountants of the companies and Rubber Reserve concerning a uniform basis of making up costs of all plants.   Rubber Reserve circulated each month a list of all complaints about GR-S quality, and later monthly distributions were made of graphs showing the relative uniformity of production from each plant, developed by statistical methods, from the quality tests on GR-S production.

Although the fee of the companies operating the copolymer plants was in no way affected by the relative performance of the plants with regard to either quality or cost, nevertheless intense competition prevailed both between copolymer plants operated by different rubber companies and between those operated by the same company.   The reason primarily for this no doubt was the desire to maintain the best possible performance in the production of a material vital to the war effort, and secondarily the natural rivalry existing between companies and plants operating in the same field.   Rubber Reserve encouraged this rivalry while insisting on complete exchange of information. Scoreboard figures were furnished to each copolymer operating unit as well as to the general offices of each agent company, giving plant performance on all major matters and showing in each category the relative standing of the 15 plants engaged in the production of GR-S.

Agreements made by Rubber Reserve in December 1941 with the four major rubber companies and Standard Oil Company of New Jersey established the basis by which they undertook to furnish Rubber Reserve with all information they then had on butadiene-styrene rubbers, as well as other diene rubbers and all information they might acquire or develop in the future from research and experimental work.

In order to increase the number of rubber companies participating in the synthetic-rubber program, two additional companies (the National Synthetic Rubber Corporation and the Copolymer Corporation) were formed to operate synthetic-rubber plants. The ownership of these corporations is shown on Table I. The General Tire and Rubber Company along with its associate, the General Latex and Chemical Company of Cambridge, Mass., was also brought in as an operating company. Rubber Reserve Company arranged for these new cooperating companies to receive technical information under the related operator's agreement. The operating personnel of each of the new groups received intensive preliminary training from the original operating companies in existing copolymer plants, thus making it possible for these group companies to assume the operation of the plants assigned to them within a comparatively short time after their completion.

STANDARDIZATION OF PLANTS, PROCESSES, AND PRODUCTS

Undoubtedly the greatest problem in the production of fabricated rubber articles, including tires, from synthetic rubber was learning how to use this new material. Since the only knowledge of synthetic-rubber use was the relatively limited experience of the small experimental groups of four rubber companies and Standard Oil of New Jersey, and since no group had any real production experience with the actual polymer, GR-S, which was to be the only general-purpose synthetic rubber produced, it was apparent that a tremendous amount of experimentation was necessary on the part of users to determine the optimum conditions of compounding, mixing, tire design and fabrication, and all other operations involved in the production of rubber articles. It was obvious that, unless GR-S was produced with reasonably uniform qualities, this necessary experimentation on the part of the rubber companies would be at least greatly complicated and perhaps even made valueless by simultaneous changes in GR-S properties. Even changes that gave improved polymer quality might under these conditions actually retard progress in improving tire quality and output. For these reasons it was deemed necessary to standardize the GR-S process, its operation and the testing of the product, and much time and effort were spent in improving the precision of the necessary process controls.

It was decided to make a butadiene-styrene copolymer of approximately the same composition as the German Buna S rubber. However, the polymer, GR-S, was to have a low viscosity and be capable of being used readily by the rubber-fabricating industry in existing rubber-mixing equipment without the preliminary softening by heat which the high-viscosity German polymer required.   Heat softening would have required extensive additional equipment at all rubber-fabricating plants, as it was not thought feasible to conduct heat softening at the synthetic-rubber-producing plants on account of the "reversion" that would occur within the normal time of storage and shipment, to say nothing of a reasonable stockpiling period.

As stated, it was considered necessary initially to standardize the process and equipment for the butadiene-styrene copolymer production, to concentrate on only one grade of polymer, GR-S, and to bend all efforts to insure that the product made by all synthetic-rubber plants was the same.   To accomplish this and to facilitate the building of the plants, it was decided to build the large majority of the synthetic-rubber production facilities as identical "standard plants."   This standardization, which speeded up the fabrication of materials and provided flexibility in shipping materials between different plant locations, was agreed upon after plans for the first four plants had been made and erection was already under way, so that these plants differed in several respects from the standard plants.   In a few instances deviations from the standard plant design were made at specific plant locations.   These differences included tank stripping of latex at the Institute plant, alum coagulation at the Lake Charles plant, and special handling of solutions and salt recovery from the rubber-coagulation operations at Borger.   With these exceptions all of the standard plants were practically identical.

The "standard plant" design, although containing some novelties and to some extent representing a compromise among the ideas of the chemists and engineers of different companies, proved quite satisfactory, and in general the performance of these plants was superior to that of any of the first four plants built from the designs of the separate companies.   In fact, some changes were afterward made in the first four plants to obtain the improved performance secured by the methods used in the standard plants. Other and more extensive changes were proposed for the nonstandard plants such as the installation of stripping columns and standard plant three-pass driers, but these came too late in the program to be justified for the war emergency and were not needed for lower postwar output.

EXPERIMENTAL X POLYMERS

For uniform production of GR-S it was necessary not only to standardize a single process to be used at all plants but also to require identical methods of operation in all steps where the quality and uniformity of the polymer could be affected.   Either because the effect of many variables could not be accurately evaluated or because under war conditions their evaluation was not possible or profitable, it was considered necessary to "freeze"the process arbitrarily and not permit any changes unless these were shown to be thoroughly justified and could be applied to all plants, in which case the standard process would be changed accordingly.   Though this standardization of process and operations to some extent limited efforts of chemists and engineers to improve plant operations and output, it was deemed to be essential to turn out a uniform product.   Other means of developing improvements both in the polymer and in the process and equipment were later provided in the experimental polymer program under an X-number designation. This program allowed polymers made by various process modifications to be produced on a plant scale and made available for experimental factory use by all interested rubber companies. The X-number polymers were announced by Rubber Reserve as they were produced; their properties were described, and all rubber fabricators were urged to try them for any application that appeared suitable. Those companies that obtained lots of the experimental polymer were obligated to render a report to Rubber Reserve on their experience with it in any new applications tried. Any rubber company operating a copolymer plant could introduce any change in process desired under this program, provided it agreed to take the entire production at the standard GR-S price, if so required by Rubber Reserve. Experimental polymers from other available monomers were also made, for example, polyisoprene (used for chlorinated rubber) and isoprene-styrene copolymers. Consumers of rubber could also request the production of polymers with special properties such as, for example, the experimental polymers which later became the special GR-S types used by the wire and cable industry.

Under this system more than 750 different experimental rubbers and master batches with carbon black and other pigments have been produced and tested by rubber companies. Those that found favor with customers were turned out in increasing quantities and finally established as GR-S types, of which there are now around 65. All experimental polymers were sold to consumers at the standard price of regular GR-S. This was done chiefly to encourage the tryout of new polymers by consumers in order that new types of synthetic rubbers which better fitted specific end uses might be discovered. It was realized that, since no cost penalty attached to the use of any type of experimental polymer, eventually an excessive number of different GR-S types would very likely be put into production, because many experimental polymers differed only very slightly in their qualities and the selection of one type or another might be largely a personal preference. Since the production of a large variety of product types increases production costs, such a situation has in many lines of synthetic products, e.g. plastics, resulted in the establishment of price differentials between "standard" grades, which are large production items, and "special" grades, the latter selling at a higher price than the former. But in a new industry just starting, with a vital need for better synthetic rubbers for specific end-product uses, many of them war necessities, it was deemed necessary to provide every incentive to insure the fullest cooperation between the user and the producer of synthetic rubbers, to the end that new types with superior properties for specific end-product uses might be developed.

TIRE TESTS

It is particularly necessary in the rubber industry that the closest cooperation be effected between the producer of the primary material, the raw rubber, and the rubber companies which manufacture end products

Table I. Authorization and Oth

Name of Company	Location of Plant	Rated Capacity,* Long Tons per Year	Date of First Authorization
Firestone Tire & Rubber Co.	Akron, Ohio	15,000	5-15-41
		15,000	1- 9-42
Firestone Tire & Rubber Co.	Lake Charles, La.	30,000	3-26-42
		30,000	3-26-42
Firestone Tire & Rubber Co.	Port Neches, Texas	30,000	8- 5-42§
		30,000	8- 5-42§
B. F. Goodrich Co.	Louisville, Ky.	22,500	5-15-41
		22,500	1- 9-42
		15,000	6-24-42
B. F. Goodrich Co.	Borger, Texas	30,000	2-14-42
		15,000	2-14-42
B. F. Goodrich Co.	Port Neches, Texas	30,000	3-26-42
		30,000	3-26-42
Goodyear Synthetic Rubber Corp.	Akron, Ohio	15,000	5-15-41
		15,000	1- 9-42
Goodyear Synthetic Rubber Corp.	Houston, Texas	30,000	3-26-42
		30,000	3-26-42
Goodyear Synthetic Rubber Corp.	Los Angeles, Calif.	30,000	2-14-42
		30,000	3-26-42
U.S. Rubber Co.	Naugatuck, Conn.	10,000	5-15-41
		20,000	1- 9-42
U.S. Rubber Co.	Institute, W. Va.	30,000	3-26-42
		30,000	3-26-42
		30,000	5-20-42
U.S. Rubber Co.	Los Angeles, Calif.	30,000	8- 5-42
Copolymer Corp.‖	Baton Rouge, La.	30,000	7-17-42
General Tire & Rubber Co.¶	Baytown, Texas	30,000	7-17-42
National Synthetic Rubber Corp.**	Louisville, Ky.	30,000	7-17-42
		705,000	

* Plants have demonstrated over-all average capacity of approximately 125 per cent rating. Production was restricted because of limited butadiene supplies and inability of the rubber industry to consume additional material.

† Plant investment includes contemplated expenditures for changes and improvements. Investment includes funds for utility facilities at some projects, while other projects obtain utilities from outside sources.

‡ Excludes materials blended into the latex.

§ Four 30,000-ton units constructed by B. F. Goodrick Co., at Port Neches, Texas. Operation of two units assigned to Firestone Tire & Rubber Co. on August 5, 1942.

‖ Operated by Copolymer Corp. comprised of: Armstrong Rubber Co., West Haven, Conn.; Dayton Rubber Manufacturing Co., Dayton, Ohio; Gates Rubber Co., Denver, *ata Pertinent to Copolymer Plants*

Date Construction Started	Date Operation Started	Estimated Plant Investment† $	$ per Annual Ton of Product	Production‡ 1942	1943	1944	1945
7-28-41	4-26-42	...	...	944	24,196	39,379	30,976
...	1-21-43	7,300,000	243				
9-14-42	9- 1-43	...	...	...	5,466	46,326	66,306
...	4-19-44	14,000,000	233				
6- 4-42	11-28-43	...	...	...	668	50,396	63,505
...	3-15-44	16,400,000	273				
2-15-42	11-27-42	...	...				
...	2-19-43	...	...	102	29,695	57,952	62,213
...	6 -4-43	11,200,000	187				
9-15-42	7-27-43	...	...	...	6,735	39,633	48,524
...	11-18-43	8,900,000	198				
6- 4-42	8-22-43	...	...	...	5,667	50,245	63,430
...	3- 3-44	16,400,000	273				
6- 1-41	5-18-42	...	...	868	17,040	37,158	29,697
...	6- 8-43	8,400,000	280				
8- 8-42	10-26-43	...	...	...	1,581	45,569	63,997
...	4- 9-44	13,500,000	225				
9- 1-42	6-15-43	...	...	...	7,047	29,305	29,730
...	10-25-43	11,340,000	189				
9-31-41	9- 4-42	...	...	327	11,270	27,598	30,357
...	6- 9-43	9,000,000	300				
6-18-42	3-31-43	...	...	...	36,816	112,948	97,845
...	7-28-43	...	...				
...	9-15-43	18,750,000	208				
9- 1-42	10-13-43	5,670,000	189	...	1,303	21,399	33,805
10-29-42	3-31-43	7,850,000	262	...	19,741	38,090	32,849
9-24-42	7-21-43	8,000,000	267	...	8,367	37,124	32,660
12-15-42	9-30-43	7,630,000	254	...	5,428	35,512	31,794
		164,340,000	233	2,241	181,470	668,834	717,688

Colo.; Lake Shore Tire & Rubber Co., Des Moines, Ia.; Sears, Roebuck & Co., Chicago, Ill. (including Armstrong Tire & Rubber Co., Natchez, Miss., in which Sears, Roebuck & Co. owns 50 per cent of the voting stock); Mansfield Tire & Rubber Co., Mansfield, Ohio; Pennsylvania Rubber Co., Jeannette, Pa.

¶ Operated by General Tire & Rubber Co. (General Latex & Chemical Co., Cambridge, Mass. Associates).

** Operated by National Synthetic Rubber Corp. comprised of: Goodall Rubber Co., Trenton, N.J.; Hewitt Rubber Co., Buffalo, N.Y.; Lee Rubber & Tire Co., Conshohocken, Pa.; Minnesota Mining & Manufacturing Co., St. Paul, Minn. (including Inland Rubber Corp., a subsidiary).

from it. The precision of laboratory and experimental scale testing of synthetic rubbers or compounded rubber articles such as tires, does not allow accurate prediction of commercial performance. This is, of course, true of all components of rubber articles and has made the improvement of such products as tires, for instance, unusually difficult. For these reasons, laboratory tests on synthetic rubbers or on other materials going into rubber articles are considered only to *indicate* commercial value. If the tests appear promising, wheel tests are made, then tires are fabricated for actual road test. If these latter tests do not indicate serious inferiority in any respect and the over-all results are good, then thousands of tires are fabricated and given mass testing on taxicabs and mileage accounts. It is only when these last test results are tabulated that the relative performance of tires made with different synthetic rubbers or different methods of compounding, or a different tire design, etc., can be accurately evaluated.

In order to increase the facilities for testing tires made from synthetic rubbers beyond those maintained by the principal rubber companies and U.S. Army Ordnance, Rubber Reserve established its own tire-testing fleet, which was managed and maintained by the Copolymer Corporation. These additional testing facilities were very helpful in the development of new GR-S-type polymers.

PERFORMANCE OF THE GR-S PLANTS

Reference to Table I will show the output of the copolymer plants in the U.S.A. program for the years 1942, '43, '44, and '45. By the end of 1944 the demonstrated capacity of the GR-S copolymer plants had substantially exceeded their rated capacity of 705,000 long tons per year. Owing to a shortage of butadiene it had not been possible to operate the polymerization plants at full capacity, which was thought to be around 840,000 long tons per year, or about 30 per cent over rating. At the end of 1944 a further increase in capacity for both rubber tires and synthetic rubber was projected by the War Production Board. GR-S capacity was to be increased to around 1,054,000 long tons per year for the year 1946, chiefly by converting the copolymer plants from batch to continuous polymerization and by converting monomer recovery operations in the nonstandard plants from batch to continuous. The butadiene and styrene sections of the program were to be expanded correspondingly to give an over-all increase in total synthetic-rubber capacity of approximately 25 per cent at a cost of approximately $22,500,000 or about 3 per cent of the cost of the original synthetic-rubber program as a whole.

The production peak of GR-S was reached in May 1945, when 72,306 long tons of GR-S was produced, corresponding to an annual production of over 860,000 long tons per year. Since this production rate exceeded the processing capacity of the rubber factories at that time, GR-S production was cut back to release certain raw materials of the synthetic-rubber program, such as butylenes and benzene, for use in producing more high-octane gasoline which was then in very short supply.

With the end of the war, the synthetic-rubber expansion program was partially suspended, since additional capacity was no longer needed, and only those construction projects justified for other reasons were continued. By the end of 1945 sufficient equipment for the production of 300,000 long tons of GR-S by the continuous polymerization process had been installed.

The actual demonstrated capacity of the synthetic-rubber plants, which amounted to 1,000,000 long tons per year, including 860,000 long tons of GR-S, exceeded by far the prewar consumption of rubber in the United States and was actually in excess of the average world production and consumption of natural rubber in the ten years preceding the war. In addition the Canadian synthetic-rubber plant demonstrated in 1945 a production output of 36,654 long tons of GR-S and 9080 long tons of GR-I. Creation of this large industry in the short space of time required, from relatively new processes, many not operated at all previously except on a small experimental scale, was an accomplishment of the highest order and testified to the ingenuity, skill, and technical and managerial ability of the companies and research groups participating in the Rubber Reserve Program.

At the end of 1944 the total number of employees in the Government Synthetic Rubber Program was 21,713, operating 51 plants. Of those employed 8105 were in the production of GR-S and GR-S types in 15 copolymer plants. Rubber Reserve Company at this time had a total personnel of 284 engaged in the planning and supervision of the program and the necessary office functions including the sale and distribution of the synthetic rubbers, and the sale, storage, and distribution of natural rubber. This figure is slightly less than 1.2 per cent of the total number of people engaged in the synthetic-rubber production program. In addition there were others employed on rubber company and university research staffs in Rubber Reserve research work, engineers in the employ of engineering firms who were consultants to Rubber Reserve, and those in other specialized groups of the R.F.C. carrying out accounting, auditing, and legal functions.

Plant Safety. In spite of the hazardous nature of some of the products used in the production of synthetic rubber, such as butadiene, which must be handled as a liquid under pressure and which forms an explosive mixture when its vapor is mixed with air, and styrene, which also is inflammable and capable of forming explosive mixtures with air, and in spite of the relatively small experience of the copolymer-plant operators and others in handling these materials, an outstanding safety record was established by the synthetic-rubber plants. This safety record was substantially lower in both frequency and severity of accidents than the records of either the petroleum industry or the chemical industry (whose operations are most similar to those of the synthetic-rubber industry) in the years 1944 and 1945, the record for the latter year being particularly outstanding. Table II gives the data for the synthetic-rubber plants.

Plant Wastes. When the copolymerization plants were built, relatively little information was available concerning the treatment and handling of the trade wastes from the operations. Furthermore the plants were constructed at a time of serious emergency with all personnel doing their utmost to get the plants built and in operation at the earliest possible time. At all plants the sanitary wastes were kept separate from the plant effluent and treated according to well-established and satisfactory methods. It was believed

Table II. Safety Record of the Government Synthetic Rubber Program

Frequency and Severity Rates

Year	Man-Hours	Disabling Injuries	Days Charged	Frequency*	Severity
1943 (4 mo.)	13,408,000	160	19,083	11.9	1.42
1944	51,814,000	424	29,930	8.5	0.58
1945	48,533,000	303	32,395	6.3	0.67
1946	37,126,000	168	34,446	4.5	0.93
1947	24,289,000	63	14,802	2.6	0.61
1948	18,541,000	46	2,119	2.5	0.11
1949	14,515,000	21	2,230	1.4	0.15

Annual Injury Frequency Rates

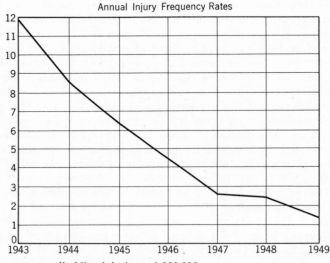

$$* \text{ Frequency} = \frac{\text{no. disabling injuries} \times 1,000,000}{\text{no. employee hours}}$$

that the trade waste problem was not so serious that it could not be **deferred** until a satisfactory production of synthetic rubber had been attained.

Facilities necessary to accomplish a satisfactory cleanup of plant wastes were considered by Rubber Reserve to be an essential part of the plants, and as a matter of policy it was desired that the effluent from all plants in the Government Synthetic Rubber Program be freed from oils, other immiscible liquids, suspended solids, and harmful chemicals, which could cause trouble when discharged into waste channels, streams, or bodies of water adjacent to the plants. After a thorough study of the problems involved in treating copolymer-plant wastes with the help of engineering firms specializing in this field, the plants were all equipped with treatment and disposal facilities.[10] Since the degree of treatment of plant wastes varies greatly with the plant locations and the volume of the rivers or tidal movement of bays, etc., into which the waste is discharged, the type of installation required varied considerably at the different plants. Solid rubber from spills and washing operations as well as latex from the same sources is collected in settling basins. The latex is coagulated, when sufficient amounts have been collected, the crumb is separated from the waste liquor, and the waste liquor is then discarded. At plants producing carbon-black masterbatches, special facilities had to be installed to prevent discharge of small amounts of finely divided black into disposal waters. The primary control of plant wastes, particularly GR-S crumb and latex, is the satisfactory operation of the plant, but the collection facilities provided give immediate notice of any accidental spillage and also prevent stream pollution by any process material.

GR-S MANUFACTURE IN OUTLINE

Standard GR-S is manufactured by reacting an aqueous emulsion of butadiene and styrene in the presence of soap solution, polymerization

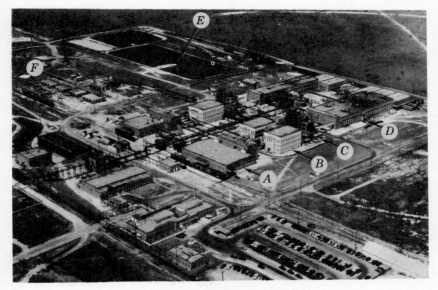

Fig. 1.   Air View of a Standard GR-S Plant

A.   Raw-material storage and pigment-preparation area
B.   Reactor building
C.   Recovery
D.   Finishing building
E.   Industrial-waste treatment
F.   Tank farm

Courtesy U.S. Rubber Co.

initiators, and regulators to form a high-molecular-weight polymer with elastic properties. This reaction is carried out at a temperature of 50° C. and a pressure of about 60 to 45 p.s.i.(g.). The reaction is carried to approximately 72 per cent of completion at which point it is terminated by the addition of chemicals known generally as shortstoppers. The polymer, in the form of latex, is then stripped of unreacted monomers after which it is coagulated, washed, dried, and packaged. The recipe for the material used in GR-S, which has remained essentially unchanged since the start of the synthetic-rubber program, is given in Table III. The flow diagram (Fig. 2; see also Fig. 1) shows that the standard copolymer plant is divided into four basic areas:

1. Tank form and pigment preparation.
2. Reactor.
3. Recovery.
4. Coagulation and finishing.

In addition, each plant has complete maintenance facilities, raw-material and finished-products storage, shipping docks, laboratories, and offices. In

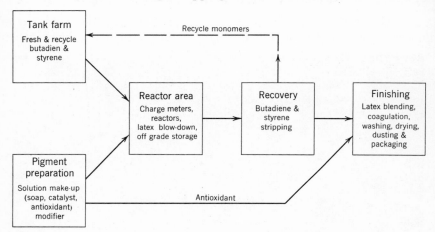

FIG. 2.   Condensed Flow Diagram of GR-S Process

the standard plant, pigment preparation and reactor operation are batchwise processes; recovery, coagulation, and finishing are continuous operations. The general steps in the copolymerization process are best followed by a review of the operations in each major plant area.

Tank Farm and Pigment Preparation Area.   The monomers (butadiene and styrene) are received by pipe line or tank car and are stored in the tank farm. Small quantities of polymerization inhibitors are added to the monomers at the time of manufacture, the purpose of which is to prevent polymerization during transportation and storage. Fresh liquid butadiene is caustic-washed to remove the inhibitor and then is blended with recovered lower-purity butadiene, the blend purity being adjusted to the desired final concentration.   The inhibitor in styrene is not removed since the concentration is so low that it has a negligible effect on the rate of copolymerization with butadiene.   Fresh styrene is blended with recovered lower-purity styrene and the mixture adjusted for the specification concentration. Aqueous solutions of soap, caustic, and shortstopper as well as emulsions and dispersions of antioxidant are prepared in the pigment preparation area.

Reactor Area.   Polymerization is carried out in glass-lined, water-jacketed steel reactors built to operate at pressures up to 125 p.s.i.(g.).   The unstripped latex is held in intermediate storage in glass-lined blow-down tanks. Also included in the reactor area are meters and weight scales for the measurement of the monomers and other ingredients which are charged to the reactors in accordance with the formulation given in Table III. The mixture is reacted at a temperature of approximately 50° C. with constant agitation. When the copolymerization reaction has proceeded until the desired conversion of monomers is obtained, the reaction is terminated by the addition of the shortstop chemicals which terminate the polymerization reaction leaving unreacted butadiene and styrene present. The charge of "unvented" latex is dropped to the blow-down tanks.

Table III.   GR-S Polymerization Formula

	Parts per 100 Parts Total Monomers	Reactor Charge, Lb.
Butadiene (100% basis)	71 ± 0.5*	5,680 ± 40
Styrene (100% basis)	29 ± 0.5*	2,320 ± 40
Dodecyl mercaptan	0.5†	40
Potassium persulfate	0.23	18.4
Soap (anhydrous basis)	4.3	344
Water	180	14,400
Total	285.03	22,802
Short-stop Solution‡		
Hydroquinone	0.08	6.40
Sodium sulfite	0.008	0.64
Water	1.60 to 3.20	128 to 256

* The monomer charge ratio is to be adjusted to give a product having a bound styrene content as close as possible to 23.5 per cent of the hydrocarbon.

† Modifier figures are only approximate. This amount is adjusted to give a product of the desired viscosity at the desired conversion.

‡ Added to the reactor at the end of polymerization.

Recovery Area. The unreacted monomers are recovered from the GR-S latex; the butadiene is taken off by a flash distillation while the unreacted styrene is removed by low-pressure steam stripping in a perforated plate column operating at approximately 50 mm. mercury absolute.

Coagulation and Finishing Area. The stripped latex is stored in large tile-lined concrete tanks. Antioxidant emulsion or dispersion is added to the latex to stabilize the product during the drying process. The latex is "creamed" by the addition of sodium chloride brine and coagulated by the addition of dilute sulfuric acid. The resulting slurry is discharged into a holding tank where the pH is such that most of the soap contained in the rubber is converted into fatty acid. The crumb is then thoroughly washed on an Oliver rotary vacuum filter to remove the soluble salts, after which the product is shredded, dried, baled, dusted with talc, and packaged.

BATCH PRODUCTION OF GR-S

Butadiene and Styrene. The manufacture of GR-S[22, 27, 29] is essentially the conversion of butadiene and styrene into an elastic high polymer. Both of these monomers are highly reactive, and their handling and storage necessitate precautionary measures to insure safe and efficient operation of the plants.   Since it was necessary in some of the GR-S plants to transport the monomers in large quantities over long distances, their effective stabilization (inhibition) was of great importance.   Butadiene is capable of polymerization with the formation of high-molecular-weight hyrocarbons varying in consistency from a tacky fluid to a rubbery mass and in some cases to a resinous solid.   Peroxides have been demonstrated to be active catalysts for this type of reaction, and it should be noted that butadiene is readily peroxidized by air.[24]   However, this peroxidation and accompanying polymerization may be inhibited by the addition of certain stabilizing agents.

The inhibitor used for the stabilization of both butadiene and styrene in the GR-S program is *p-tert*-butylcatechol (TBC).   It is added (approximately 200 p.p.m. concentration) at the butadiene-producing plants before shipment by tank car.   Where the butadiene and GR-S plants are connected by pipe line, the inhibitor content has been reduced to 50 p.p.m. without any difficulties being experienced and has even been omitted entirely over substantial periods of time.

Since butadiene is a gas at normal temperatures and pressures (b.p. —4.5° C.), it must be stored and shipped in pressure vessels.   The copolymer-plant tank farm has steel pressure tanks for the storage of fresh, recycled, and blended butadiene and can receive this material either by pipe line (when located adjacent to the butadiene plant) or by insulated steel pressure tank cars (10,000 gal. capacity).

Styrene has a boiling point of 145.1° C.; consequently its storage at ordinary temperatures offers no particular problem from a volatilization standpoint.   Uninhibited styrene polymerizes slowly at summer temperatures to form resinous polymers, this polymerization also being catalyzed by peroxides.   Only about 10 p.p.m. of the inhibitor TBC is required for the stabilization of styrene.   Styrene is conveyed to the copolymer plants by steel tank car (not insulated) or by pipe line.

The monomers used by the copolymer plants are purchased under specifications that were set up and agreed to by the producers of butadiene and styrene and the copolymer plants and were approved by Rubber Reserve. Butadiene must have a minimum conjugated diene content of at least 98 per cent by weight and must be clear and free of entrained material.   There are also limits on peroxides, acetylenes, aldehydes, sulfides, butadiene dimer, and other materials which may be present in the butadiene and which may affect the rate of polymerization or the quality of the polymer.[7] (Cf. p. 78.) Fresh styrene must have a monomer content of 99 per cent minimum by weight, although actual purity usually exceeds 99.5 per cent.   The styrene specifications limit the amount of aldehydes, peroxides, and sulfides that may be present. (p. 168.)

Inhibited butadiene and styrene received at the copolymer plants are stored in 30,000-gal. horizontal steel tanks or sometimes, in the case of butadiene, in Horton spheres.   Fresh butadiene is mixed with recovered butadiene to give a blend containing the desired proportion of reactive monomer.   Thorough mixing of fresh and recycle stocks is accomplished by removing the butadiene vapor from the upper portion of the blending tank and discharging it under the surface of the liquid in the tank.   The use of recovered butadiene represents an important phase of the economic operation of the copolymer plant.   Inhibitor is removed from the butadiene blend before use, by washing with caustic solution in a contactor packed with Raschig rings, followed by decantation.   Styrene blends are prepared by mixing fresh and recycle styrene and adjusting the purity to the desired level.   The blends are agitated by pump circulation.   Since the fresh styrene is stabilized with only a very small quantity of TBC, it is unnecessary, as has been noted previously, to remove this inhibitor before the production of GR-S.   The normal practice, in the standard copolymer plants, is to use butadiene and styrene blends which analyze 95.5 per cent pure.   As indicated in Table III, the polymerization formula is based on pure monomers and correction must be made in charging, therefore, for the $4^1/_2$ per cent accumulated impurities (chiefly butylenes) normally present.

The control of quality in GR-S manufacture starts with careful analysis of incoming shipments of raw materials.   To establish and maintain a high standard of precision in testing, frequent cross-checks are run on exchange samples between the laboratories of the copolymer plants and the monomer suppliers.   The use of standard samples has played an important part in developing and maintaining a high level of precision in all the copolymer-plant laboratories.   All butadiene blends are checked for purity by the Koppers-Hinckley rapid analytical test method[11] which was adopted after careful comparison of the results obtained by this method with those obtained by freezing-point and mass-spectrometer determinations in the laboratories of the National Bureau of Standards.   Styrene purity is determined by freezing point.[3]

The storage tanks for both butadiene and styrene are provided with auxiliary equipment for the addition of extra inhibitor, should this become necessary.   The elevation of the tanks provides a positive liquid pressure head on the transfer pumps.   All equipment containing butadiene is provided with safety valves which discharge through a collecting system of pipes leading to a burning stack.   A perpetual gas flame burns at the top of this stack to ignite any combustible materials which may issue from the safety valves throughout the plant, thereby preventing the accumulation of inflammable vapors which are explosive when mixed with air.

Soap.   The quality of the soaps used exerts a profound effect upon the initiation and rate of polymerization.   With the GR-S-type recipe the rate of polymerization is roughly proportional to the concentration of fatty acid soap.   A wide variety of fatty acids can be employed to make soaps for use in the emulsion polymerization process.   Table IV lists the comparative values of the sodium soaps of common fatty acids in making GR-S at 50° C. Special attention should be directed to the harmful effects of linoleic, linolenic, and other unsaturated nonconjugated acids if these are present in soaps to be used in emulsion polymerization.   Because of such harmful effects the great bulk of the soap used in the GR-S program has been manufactured from mildly hydrogenated tallow.

At the outset of the GR-S production program, the properties of the soaps from different manufacturers and sometimes from the same manufacturer varied considerably, causing variations in polymerization time. To overcome this, a cooperative research program was carried out by the copolymer plants and the soap manufacturers under the joint direction of Rubber Reserve and the War Production Board. This resulted in determining the effect of various soap impurities on the polymerization and the establishment of

Table IV.   Comparative Conversions Obtained with Sodium Soaps[26]

Sodium Soap of	% Conversion of Monomers in 12 Hrs. at 50° C.
Lauric acid	71
Myristic acid	75
Palmitic acid	83
Stearic acid	82
Elaidic acid	81
Palmitic acid 90% } Linoleic acid 10% }	72
Palmitic acid 90% } Linolenic acid 10% }	41
Mixed hydrogenated tallow acids	80

satisfactory specifications.   All soaps used in the manufacture of GR-S must meet these rigid specifications, among the most important of which are the stipulations that there shall be no more than 1 per cent saturated acids below C_{12} and that there shall be less than 2 per cent unsaturated acids above C_{18} in the dry soap.   The unsaturated materials are determined by the ultraviolet spectrometer which is used by both the soap producers and the copolymer plants for testing all soap shipments.   Each lot of soap must carry a producer's guarantee of uniformity, and samples of each lot must pass a copolymer-plant laboratory polymerization test of polymerization rate and product quality.

In the GR-S plant, soap solution is prepared from soap chips and zeolite-treated (softened) water or water (condensate) from the styrene decanter. The solution is made up in wooden tanks provided with propeller-type agitators.   The soap solution is heated before use, but the temperature must not exceed 75° C.   By the addition of caustic soda or fatty acid, as necessary, the solution is adjusted to pH 10, and laboratory checks are made on all batches of soap solution before they are released for use.   Because of the danger of spontaneous heating at summer temperatures, bags of flake soap should be stored on pallets at least 1 ft. from each other and from walls or partitions.   Care should also be taken to avoid storing soap in damp or heated areas.

Although this chapter is mainly concerned with the production of standard GR-S, which requires the use of tallow soaps, mention must be made of another major polymerization emulsifier that made its appearance in 1944. The use of rosin acid soap as a polymerization emulsifier had received active attention previous to the inception of the Government program.   However, inhibitors present in the rosin soap which was then available slowed down the polymerization in the GR-S process to such an extent that this soap could not be used except in the manufacture of a special latex (type 3). As a result of research carried out by the Hercules Powder Company in collaboration with the rubber companies working in the Rubber Reserve program, Hercules developed a disproportionated rosin soap of uniform quality. It gave a constant polymerization rate closely approximating the normal rate with tallow soap. The first major polymer to emerge using this soap was GR-S-10 which will be discussed later.

Other Chemicals. Potassium persulfate is used as the catalyst in the GR-S formula. A solution is usually prepared with zeolite-treated water in a no. 316 stainless-steel vessel provided with a propeller-type agitator. An alternate procedure is to use a single soap-catalyst solution, in which case styrene decanter water may not be used for making up or diluting the solution and the temperature must not exceed 60° C.

The shortstop solution (see Table III) is also made up in stainless-steel equipment. In some plants it has been found desirable to include in this solution up to 0.04 parts (on the basis of 100 parts of monomers charged) of sodium nitrite to help prevent the formation of "popcorn polymer" in the recovery equipment. It should be noted, however, that sodium nitrite plays no part in the actual shortstopping of the GR-S reaction.

When butadiene and styrene are copolymerized in an aqueous emulsion containing only the monomers, an emulsifying agent and a peroxy catalyst, the product contains a considerable amount of cross-linked polymer which is difficult to process and has a low value as a rubber for use in American fabricating plants. Processible copolymers are obtained by adding to the polymerization system a substance known as a "modifier," which also increases the rate of polymerization and renders the copolymer benzene-soluble. In the preparation of standard GR-S, dodecyl mercaptan is used as the modifier. It functions probably by a chain-transfer mechanism which diminishes the molecular weight of the polymer.[28] Small changes in the amount of dodecyl mercaptan charged have a profound effect upon the properties, especially the Mooney viscosity, of the finished rubber. Since it is vitally necessary that extreme care must be taken in its measurement, all modifier is measured into the reactors with calibrated weigh scales.

Emulsions and dispersions of rubber antioxidants are also made up in the pigment preparation area. These emulsions are added to the unstripped latex to prevent the deterioration of the polymer during the styrene stripping and drying operations. Because of the physically unstable nature of most antioxidant suspensions, which causes agglomeration of the particles, it is necessary to provide mild agitation in the antioxidant suspension storage tanks. The antioxidants used in standard GR-S are phenyl-β-naphthylamine (PBNA) and a reaction product of acetone and diphenylamine (known to the trade as BLE). In special nonstaining grades of GR-S for use in making white or light-colored stock other stabilizers are used.

Operation of Reactors. The reactors in the standard GR-S plants are glass-lined steel vessels with a capacity of 3750 gal. and a working pressure of 125 p.s.i. and equipped with modified Brumagin-type agitators. Individual pressure-sealing units are installed on each reactor agitator shaft, oil or other liquid pressure of 100 p.s.i. being maintained on the rotating ring in contact with a stationary gland to prevent leakage.   Twenty-four autoclaves are provided for each 30,000-ton unit (Fig. 3).

As originally installed in the plants, the agitators rotated at 235 r.p.m. and had three blades pitched at 5° 21'.   A modification of the reactor stirrers was instituted which resulted in a considerable reduction in the amount of modifier required in the production of GR-S.   The agitation in all the

FIG. 3.   Top View of GR-S Reactors
Courtesy Firestone Tire & Rubber Co.

standard plant reactors was changed to that provided by a modified Brumagin-type rotating at 105 r.p.m.   The revised agitators employ three blades with surface dimensions 5 in. by 13 in. and a pitch of 45°.

To obtain a uniform product, it is necessary to control the reaction temperature within narrow limits.   The polymerization of butadiene and styrene evolves considerable reaction heat which is removed through the water jacket of the reactors.   The correct water temperature is maintained by automatic charging of water, refrigerated water, or steam, as required, and a circulating pump is provided for recirculation of water, if needed, to increase the heat-transfer rate.   Each reactor is provided with a recording temperature controller and pressure recorder, the temperature control bulb being located in the bottom section of the vessel.

The butadiene, styrene, soap solution, and water are charged through separate meters to the reactor.   These meters are calibrated at frequent intervals by the use of calibrating tanks of known capacity.   Since slight changes in the amounts of modifier and catalyst added to the reaction greatly affect the rate of reaction and the properties of the finished rubber, these solutions are weighed before charging, the respective weigh scales also being calibrated once each day by the use of standard weights.

All of the ingredients, except the catalyst solution and flush water, are charged into the reactor simultaneously while the agitator is running.   The catalyst solution is then added, followed by the remainder of the water to flush the charge header.   In order that the initial mixture will approach the proper polymerization temperature, the temperature of the water and pigment solutions is adjusted before charging them.   The emulsified monomers are polymerized with constant agitation at a temperature between 48.4 and

FIG. 4.   Blowdown Tanks and Fresh Styrene-Storage Reactor Area
Courtesy Copolymer Corp.

50° C. until proper conversion and viscosity have been obtained.   The polymerization temperature is automatically controlled within plus or minus 0.25° C. of the desired temperature, and the reaction is permitted to proceed until the desired conversion of the monomers to polymer has been attained.   The butadiene-styrene ratio is adjusted to result in 23.5 per cent bound styrene content in the polymer as determined by refractive index (see Chapter 10).   The primary control of the polymerization process is based on the shortstop viscosity (rapid Mooney), which indicates the potential Mooney viscosity (plasticity) of the finished polymer.   This property of the polymer is controlled by adjusting the amount of dodecyl mercaptan in the formulation or by varying the conversion of the monomers, within the process specification limits.

The production of a uniform polymer requires that the reaction be terminated when the conversion and shortstop viscosity of the product are both within certain narrow limits.   This control is achieved by the analysis of a series of samples taken during the course of the reaction.   The monomer conversion is measured by determining the total solids present in each sample of the latex and reading the results from a predetermined curve (in this test, provision is made for weathering off the unreacted butadiene and driving off the other volatile constituents by heating). The viscosity of the polymer is measured by the Mooney or Williams technique on samples of the latex which have been rapidly coagulated and dried. Most of the troubles encountered in these quick-viscosity measurements are due to irregularities

FIG. 5.   Tower Stripping Columns for Recovery of Unreacted Styrene from GR-S Latex
Courtesy Goodyear Synthetic Rubber Corp.

in the drying of the samples. By plotting these data and extrapolating the curves, it is possible to predict the time at which each reaction should be terminated. A great deal of time and effort have been expended in the refinement of both conversion and viscosity testing, but it will probably be necessary to improve the precision of both measurements before any marked improvement can be obtained in the control of the GR-S polymerization, since present production variability is of the same order of magnitude as the test errors.

The pressure at the start of the reaction is approximately 60 p.s.i.(g.) but, because of the conversion of butadiene and styrene to polymer during the course of the reaction, the pressure drops gradually until it reaches about 45 p.s.i.(g.) at the desired conversion. When the specified level of reaction has been reached, the shortstop solution is added to the reactor to destroy the residual catalyst remaining and stop the polymerization. At this time the monomer conversion should be approximately 72 per cent, the predicted viscosity approximately 50 Mooney points; the average size of the emulsified polymer particles will then be between 700 and 1000 A.

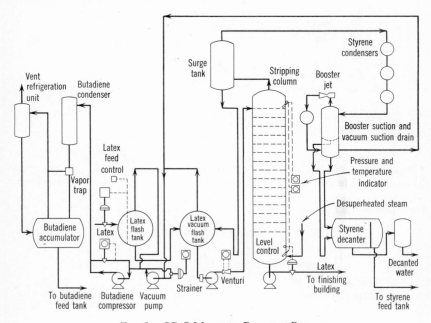

Fig. 6. GR-S Monomer-Recovery Process

The unstripped latex is dropped from the reactor to one of six blow-down tanks provided in each 30,000-ton unit. These tanks are glass-lined, have turbine agitators, and serve as intermediate storage vessels for the unstripped latex before delivery to the recovery area. Each blow-down tank is of sufficient size to hold two reactor batches with vapor space enough to allow for normal working pressures. There are also additional tanks provided to allow for the storage of off-grade batches—batches that do not meet the requirements of conversion or viscosity or are in some other manner abnormal.

Recovery of Unreacted Monomers.[13] (See Fig. 6.) The unstripped latex is transferred from the blow-down tanks to the recovery area by differential pressure. Since the charged monomers are reacted to only 72 per cent conversion, it is necessary to recover the unreacted monomers before the latex is coagulated to form GR-S rubber. The composition of the latex before recovering the unreacted butadiene and styrene is shown in Table V. Monomer properties are listed in Table VI which shows that there is a wide difference between the boiling points and vapor pressure of butadiene and styrene and that the monomers are only slightly soluble in water.

Table V.   Composition of Latex before Monomer Recovery

72% Conversion

	Parts per 100 Parts Total Monomers Charged
Water	185
Butadiene (in polymer), 76.5%	55.1
Styrene (in polymer), 23.5%	16.9
Butadiene (unreacted)	15.9
Styrene (unreacted)	12.1
Butadiene (impurities)*	3.4
Styrene (impurities)†	1.4
Soap	4.3
Plasticity modifier, catalyst, and reaction stopping agent‡	0.8
Total	294.9

* Principal impurities are 1-butene, 2-butene, propylene, *n*-butane.

† Principal impurities are ethylbenzene, 4-vinyl-l-cyclohexene, isopropylbenzene, 1, 3-butadiene.

‡ Parts given are the quantities charged.

Table VI.   Physical Constants of Monomers

	Butadiene	Styrene
Formulas	$CH_2 : CH \cdot CH : CH_2$	$C_6H_5CH : CH_2$
Molecular weight	54.088	104.14
Physical appearance	Clear, colorless liquid under pressure; gas at normal atmospheric conditions	Colorless liquid, sharp aromatic odor
Boiling point at 760 mm. Hg.	−4.6° C. (23.7° F.)	145.2° C. (293.4° F.)
Vapor pressure	11.2 mm. Hg − 78.52° C.	1.3 mm. Hg 0° C.
	760 mm. Hg − 4.6° C.	6.9 mm. Hg 25° C.
	2144 mm. Hg 25° C.	20.7 mm. Hg 50° C.
	3338 mm. Hg 40° C.	166 mm. Hg 100° C.
Melting point	−108.9° C.	−30.60° C.
Density at 60° F.	5.229 lb./gal.	7.582 lb./gal.
Solubility in water at 15° C.	0.13 g./100 g. water	0.0063 g./100 g. water
Heat of vaporization	99.8 cal./g. at −4.6° C.	85.0 cal./g.-mole
	179.6 B.t.u./lb.	153.0° B.t.u./lb. at 146° C.
Specific heat of of liquid	0.157 cal./(g.) (°C.) at 25° C.	0.413 cal./(g.) (°C.)
Heat of combustion of liquid	11,158 cal./g. or 607.4 kg.-cal./g.-mole	1.046 kg.-cal./g.-mole
Explosive limits in air	2.0 to 11.5% by vol.	1.1 to 6.1% by vol.

Several of the first copolymer plants built were provided with a batchwise recovery process, but the standard plants have a continuous recovery system which is much more satisfactory.   The recovery system of a standard plant is divided into two sections, one for butadiene and one for styrene.   The continuous recovery of the monomers is accomplished in two stages by first passing the unstripped latex at a rate of 40 to 60 g.p.m. through two flash tanks where the unreacted butadiene is vaporized by first reducing the pressure from 30 p.s.i.(g.) to 3 p.s.i.(g.) in the first tank and then to 220 mm. mercury absolute in the second tank.   Horizontal cylindrical vessels of 10,000-gal. capacity are used for this purpose.   They are glass-lined and designed for a working pressure up to 85 p.s.i.(g.).   The latex level in the flash tanks is controlled by a bubble-type differential controller using butadiene vapor to determine the static head pressure of the liquid level in the vessels.   The latex intake pipe is fitted with a deflector so that incoming fluid will hit the side wall of the vessel and run down into the liquid in the flash tank with little turbulence, which helps to hold foaming to a minimum. The butadiene vapors taken from the flash tanks are compressed by water-sealed Nash-type rotary vacuum pumps and compressors and are discharged into a refrigerated vertical-tube condenser at 60 p.s.i.(g.) pressure.   The condensed butadiene flows by gravity to a receiver from which it is pumped to feed tanks and returned to the tank farm for the preparation of blends. From the second flash tank the latex goes to the styrene-recovery section of the area.

To recover the unreacted styrene, the partially stripped latex is pumped to the top of a stripping column, the latex passing downward over perforated plates, countercurrent to an upward flow of steam.   The stripping columns are 9 ft. in diameter and $49^1/_2$ ft. high; they contain 12 perforated porcelain-enameled no.-11 gage Armco iron trays which have $^3/_{16}$-in. holes on a $^5/_8$-in. triangular pitch.   Adjustable weirs are used to maintain 3 in. of latex on the trays.   The columns operate at 60 to 100 mm. mercury absolute pressure at the top and at 100 to 150 mm. mercury pressure at the bottom. The residual styrene content of the latex coming from the stripping column runs about 0.2 per cent.

The vapors from the stripping column contain water, styrene, small amounts of butadiene, and entrained particles of latex.   These vapors pass through a foam trap where the entrained particles of latex are removed and returned to the low-pressure flash tank.   The stream then flows through a triple bank of water-cooled condensers (in series) where most of the styrene and water are condensed.   The liquid styrene accompanied by water and vapors of styrene, butadiene, and water flows from this condenser bank to a separation vessel where the vapors are removed and compressed by a steam jet from approximately 40 mm. mercury absolute to 220 mm. mercury absolute.   The condensed liquids flow from the separator drum to the decanter.   The compressed vapors are discharged into a refrigerated booster condenser where most of the remaining styrene and water are condensed. Liquid styrene, water, and butadiene vapor enter the vacuum suction drum for separation and the vapors are sent to the vacuum-pump suction line where the butadiene is recovered as previously described.   The liquid outlets of both the booster and vacuum suction drums are connected to the styrene decanter. The recovered styrene overflows from the decanter into an accumulator and on to the feed tanks, and is then pumped to the tank farm for re-use.

Butadiene recovery efficiency is kept high by passing the mixed vapors above the butadiene received through a refrigerated vent trap and then through an oil-absorption column where 90 to 95 per cent of the butadiene vapors are removed before the residual noncondensables are vented through the burning stack. Butadiene is desorbed from the oil on heating and is recovered. After cooling, the oil is continuously recirculated to the absorption column. The water which condenses with the styrene is decanted but still contains an appreciable quantity (up to 0.1 per cent) of dissolved and entrained styrene. This styrene is recovered by steam stripping the decanter water in a packed column which is tied into the vacuum system.

Recovery efficiencies of unreacted monomers has been unusually high in plants equipped with continuous recovery equipment and using columns for styrene stripping. Butadiene recovery averages 98.5 to 99.0 per cent, and styrene recovery is 98.0 to 98.5 per cent. Batch recovery of unreacted monomers has averaged 95 per cent.

The greatest difficulties encountered in the operation of the styrene stripping columns arise from foaming and from prefloc formation. If foaming becomes excessive, small quantities of silicone defoamer are added to the top of the column. Prefloc formation, which plugs the holes in the plates and limits the operating life of a column between cleanings, is greater on the top three plates and the bottom two plates than throughout the rest of the column. Bottom fouling may be largely overcome by desuperheating the stripping steam which then enters the column at approximately 65° C. and 150 mm. mercury absolute pressure.

Impurities accumulate in recycle butadiene as a result of its continued re-usage in the blending of fresh monomers. When the purity of the recycle has dropped to about 88 per cent, it is pumped into pressure tank cars and sent for purification to a butadiene plant, or it is sent thither through pipe lines. This practice permits a relatively constant purity of the butadiene blend charged to the reactors to be maintained and also prevents the accumulation of inhibitors or other impurities in the copolymer system. The impurities, particularly vinyl acetylene and butadiene dimer, if present in large enough quantities, slow down reactions and may give inferior polymers, although there is some controversy on this latter point. The removal of low-purity recycle material also diminishes the accumulation of butadiene peroxides, which represent a hazard, since they will explode spontaneously under certain conditions. Recycle styrene of about 88 per cent purity was returned for purification to the Kobuta styrene plant, operated by Koppers, which handled all the recycle styrene from the GR-S program.

Since the maintenance of low pressure is the basis of all operations in the recovery system, special attention must be given to the operation of the equipment involved, namely, the rotary vacuum pumps, compressors, and their related seal-water system. The vacuum pump operates at a suction pressure of 220 mm. mercury, receiving butadiene vapors from the vacuum flash tank and vacuum suction drum and discharging them at 3 p.s.i.(g.) and not over 40° C. to the compressor pump suction   The compressor takes the butadiene vapors from the pressure flash tank at 3 p.s.i.(g.) and combines these vapors with those of the vacuum-pump discharge.   Vapor at 60 p.s.i.(g.) and 55° C. is discharged from the compressor and sent to the butadiene condenser for condensation.   Operation of both the vacuum pump and compressor is dependent on the use of water to compress the vapors within the pump body.   Water used for this purpose is called seal water, and the system employed for handling it in the standard GR-S plants is shown in Fig. 7.   This is a closed system whereby a solution of 0.5 per cent sodium

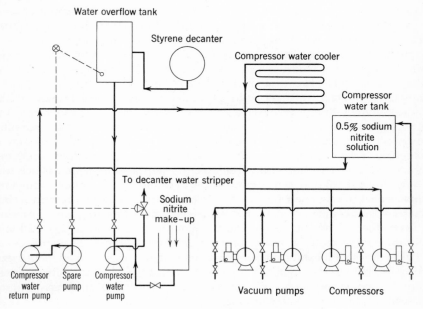

FIG. 7.   Revised Seal-Water System

nitrite, used to inhibit the formation of cross-linked "popcorn" polymer, is pumped from the compressor water tank, through the refrigerated cooler to the vacuum pump and compressor, and back to the compressor water tank. Seal water is also used on pump packing glands to prevent dilution of the sodium nitrite solution.

It was originally thought that mechanical agitation of latex during pumping would cause fine particles of coagulum, known as prefloc, to be formed in the latex before coagulation.   Diaphragm pumps controlled by variable-speed drives were therefore installed in the copolymer plants since these pumps had sufficient capacity to handle large volumes of latex flowing through the recovery system with a minimum amount of agitation.   The pumps performed satisfactorily except that diaphragm breakage proved excessive, and the capacity of one pump was strictly limited to the design capacity of 38.0 g.p.m. Open impeller-type centrifugal pumps were later tried and found to operate satisfactorily.   In addition the centrifugal pumps had the ability to deliver up to 70 g.p.m. each without formation of appreciable amounts of prefloc. Consequently centrifugal pumps have now become standard through the plants for latex service.

It was also believed that positive-displacement piston-type pumps would be required to handle recovered butadiene and styrene. However, the small amount of water entrained in the recovered monomers tended to cause valve sticking, and also it was difficult to prevent piston leakage.   It was found that centrifugal pumps equipped with metallic seals would handle both butadiene and styrene satisfactorily.   Further, since centrifugal pumps discharge at constant pressure, the breakage of rupture disks on storage vessels due to the pulsation of the reciprocating pumps was eliminated.

Popcorn Polymer. At the outset of the GR-S program, the autopolymerization of styrene was recognized as an important factor in the fouling of condensers and other equipment in the monomer recovery system.   Provisions were made for the addition of *p-tert*-butylcatechol to the vapors entering the vacuum and booster condenser in the recovery system so that the inhibitor would be present when the styrene condensed and would remain in the recycle styrene during recovery and storage operations.   By this means the formation of polystyrene in the recovery equipment was held to a minimum. After 6 to 9 months of operation, however, each of the plants experienced a plugging of the vacuum and booster condensers, vapor lines, flash tanks, and other equipment with a white, insoluble or "popcorn" polymer, which was shown to consist of polystyrene cross-linked by butadiene.[34]   Once this polymer had formed, some vessels and lines required cleaning every 6 to 12 weeks, thus increasing maintenance cost, reducing monomer efficiency, and lowering the productive capacity of the plants.   Popcorn polymer is usually a hard, porous, opaque material and its most remarkable property is the phenomenon of growth which it exhibits when in contact with liquid or gaseous monomers. A large increase in apparent volume accompanies the transformation of liquid monomer into popcorn polymer, and the polymerization may produce pressures sufficient to bulge or crack steel equipment.

Two separate and distinct methods may be employed for the prevention of popcorn growth in a plant already contaminated with active seed.[14] The first, which may be called "seed destruction," requires shutting down the plant and introducing into the system (from which the monomers have been removed) an agent which in one treatment greatly reduces or destroys the activity of the seed.   The second method, which may be called "seed inhibition," consists of maintaining constantly in the system a specific inhibitor in concentration sufficient to minimize seed activity.   The best procedure for controlling popcorn polymer in industrial equipment consists of a combination of both methods.   Deactivation treatment should first be undertaken to kill all active seeds present in the system and to decontaminate the equipment.   Since the initial formation of popcorn polymer in clean equipment is slow and much more easily controlled than seed growth, it should, after deactivation, be possible to prevent formation of new popcorn seeds by constantly maintaining a small concentration of inhibitor in the system.

Coagulation and Drying. The stripped latex, meeting in-process specifications of residual styrene content, bound styrene content, polymer viscosity, and concentration, is blended and stored in 30,000-gal. rectangular concrete tanks under mild agitation. The finishing area for each 30,000-ton unit of the standard plant is provided with four of these blend tanks. Each finishing area is also equipped with four lines for continuous coagulation, filtration, washing, shredding, drying, and baling. In addition, each 30,000-ton unit has provision for talc dusting and packaging the finished rubber.

The finishing areas of the copolymer plants have undergone more changes than any other section. Some of these changes have been made to gain greater flexibility in the production of special types of polymers, to improve polymer quality, and to reduce operating costs. The finishing operation described in what follows is a combination of the most satisfactory features of several different plants.

Synthetic-rubber latex is coagulated by the addition of acid or salts or by a combination of the two. The coagulation product thus obtained requires thorough washing to remove residual amounts of these agents and soluble polymerization chemicals followed by drying to produce practically moisture-free rubber which can be put into packages convenient for handling in shipment, storage, and use at consumer's plants. The GR-S plants conduct a "creaming" operation by the addition of brine to the latex. This is actually a partial flocculation or agglomeration of the rubber particles, and changes the latex consistency from a mobile liquid to a heavy cream. The addition of dilute acid to the creamed latex converts the layer of soap molecules on the surface of each particle to fatty acid. Thus the stability of the system is destroyed, and the GR-S globules agglomerate to form porous particles of substantial size. These crumbs are readily washed free of salt and acid because of the large surface exposed to the washing action. Since the drying operation subjects the rubber crumbs to conditions that are favorable for oxidative degradation and also cross-linking, a quantity of antioxidant (1.25 parts phenyl-β-naphthylamine or BLE per 100 parts of finished rubber) is incorporated in the latex before creaming and coagulation. Practically all of the antioxidant remains in the finished GR-S and permits a corresponding reduction in the antioxidant added when the rubber is compounded.

A flow diagram of the finishing area is shown in Fig. 8. The latex is pumped from the blend tank to the coagulation equipment by means of an open impeller centrifugal pump. Antioxidant emulsion or dispersion is metered continuously into this latex stream, thorough mixing being achieved in the pipe line. The creaming operation is also carried out in the pipe, brine being introduced into the flow a few feet before the acid addition. Dilute sulfuric acid is added to the creamed latex at the intake of a centrifugal pump which is oversized for the volume of slurry handled. The violent agitation in the pump body produces a porous crumb which has excellent filtering and drying properties.

Brine and acid solutions are both made up continuously and automatically. Fresh saturated (26 per cent) sodium chloride brine and fresh concentrated (66° Bé.) sulfuric acid are used as make-up to the serum recycled from the dewatering screen. The brine in the creaming operation is prepared in a separate unit by dissolving rock salt to give a saturated solution, treated with soda ash and lime followed by filtration to remove calcium and magnesium salts. The brine concentration for creaming is controlled by specific gravity, and the coagulating acid solution is controlled by pH.

The crumb and serum are discharged from the coagulation pump into a holding tank, where coagulation is completed. This tank is equipped

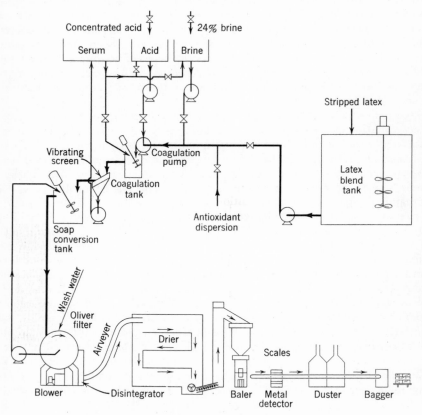

FIG. 8.   Flow Sheet of GR-S Finishing

with marine-type propellers which provide agitation sufficient to maintain the density of the crumb slurry. The overflow from this tank leads to a vibrating screen where the crumb is separated from the serum. The crumb is then reslurried in the receiving tank with wash water recovered from the Oliver rotary filter. Caustic soda solution, if required, may be added to the reslurry tank in sufficient quantities to raise the pH to a point where a small quantity of the fatty acid in the polymer is converted back into soap (a soap content of 0.3 per cent on the finished GR-S is considered to be near optimum). Agitation in the reslurry tank must be sufficient to maintain the density of the crumb slurry. The discharge from the reslurry tank is fed to an Oliver rotary vacuum-drum filter covered with stainless-steel wire screen. The filter is provided with two 8-in. smoothing rolls and three 12-in. press rolls, all covered with stainless-steel sheet. Wash water is sprayed on the GR-S filter blanket to further remove the soluble impurities. The dilute serum filtrate obtained is discarded, while the wash water is recovered and delivered to the reslurrying tank.

The compacted GR-S blanket coming off the filter should not contain more than about 30 per cent moisture. This crumb sheet is dropped into a high-speed Jeffrey hammer disintegrator, and the resulting crumbs are blown to the top of the drier with a pneumatic conveyor. The standard finishing areas are equipped with three-pass, perforated apron-type hot-air

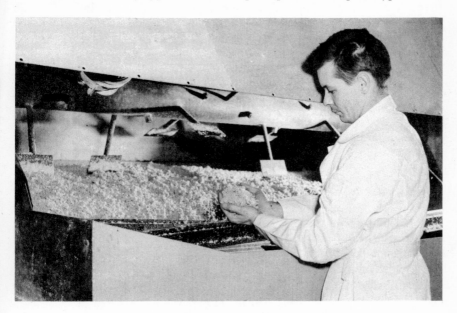

FIG. 9. GR-S Crumb being removed from a Drier Shelf for Examination
Courtesy Firestone Tire & Rubber Co.

continuous driers, having an effective drying area of approximately 400 sq. ft. in each pass. The drier is compartmented so that temperatures may be controlled at any stage of the drying process. The drying temperature for GR-S is set at 82° C. maximum. The passage of time in normal drying is slightly under 2 hours. The finished-product specifications require that standard GR-S shall contain less than 0.50 per cent volatile matter (mostly water). (An alternative method of drying is, after coagulating the latex in a manner designed to give a suitable slurry of rubber, to form a thin sheet of rubber on a Fourdrinier paper machine, and pass the sheet through a drier at 280 to 300° F., the drying time being 7 to 8 minutes. The feasibility of this method has been demonstrated in plant operation, but the method has not been adopted in standard GR-S manufacture.[2])

The dried GR-S is transferred to automatic scales by means of a screw conveyor.   Approximately 75 lb. of rubber crumb is delivered by the scales to the hydraulic baler where it is formed into a block approximately 28 in. by 14 in. by 7 in. in size.   This size bale has been found to be the most satisfactory weight for handling in the American consuming plants.   Since there is a slight variation in weight between bales, all GR-S is sold on the basis of the actual weight of each shipment and not by bale count.

The baled GR-S from each of the four units passes to a conveyor belt provided with photoelectric control to insure minimum spacing between bales.   The bales pass through an electronic metal-detecting device (designed to detect and reject any bales containing any metallic contamination) and then into an automatic dusting machine, consisting of a dust chamber through which the baled rubber is passed on a conveyor belt.   Dusting is accomplished by jets of air introduced at the bottom of a talc hopper, the equipment being designed to minimize the amount of free dust outside the chamber.   The bales are thus coated uniformly with talc and are dropped into multilayer kraft paper bags, provided with a light clay inner coating to prevent adhesion of the baled rubber.   The bags are closed by sewing, a code identification is stenciled on the face, and they are then stored on pallets until released by the laboratory.   The finished product is sampled at regular intervals, and this material is sent to the laboratory for testing.   Each lot of material must meet all specifications before it is released as GR-S.   The pallets of specification GR-S are handled by lift trucks into boxcars for shipment.   Manual operation in this area has been cut to a minimum through the use of automatic machinery.   Actually, the first hand operation on GR-S takes place as it emerges from a dusting machine and is directed by the operator into a kraft paper bag, run through a sewing machine, and put on a storage pallet as a finished product.   A crew of three men handles the output from four driers, about 100 tons of GR-S per day.

CONTROL OF THE MANUFACTURE BY STANDARDIZED RUBBER TESTING

The Government Synthetic Rubber Program was from its inception organized by men trained in chemical process control, and from the start the plant laboratories have acted as the nerve center of copolymer-plant operations.   As has already been pointed out, all raw materials are purchased under specifications accompanied by official methods of analysis.   Table VII summarizes the salient factors of the GR-S process and the corresponding controls.   It is evident from this that tremendous masses of testing data are accumulated which would be of little value unless recorded and correlated to reveal the pertinent facts.   This is done effectively by the application of modern statistical methods.[31]   The statistical information thus secured is used for better control of plant operations, and also for historical background in the long-range development of more efficient operating and testing methods and products.   The crux of the problem is to organize the data in such manner that significant trends or correlations are clearly indicated.   Normal variations must be clearly segregated from abnormal ones which originate from some phase of the process or equipment that has undergone significant

Table VII.   Summary of the Salient Features of the GR-S Process and its Concomitant Control

Operation	Points of Control	Methods of Control
I. Raw materials preparation	(1) Purity of raw materials	Analysis of sample
	(2) Concentration of solution	Analysis of sample
II. Reactor charging	(1) Charging meters	Daily calibration
	(2) Weigh scales	Daily calibration
III. Polymerization	(1) Temperature	Instrumentation
	(2) Time	Reaction cycle
	(3) Yield	Solids content of sample
	(4) Viscosity	Quick Mooney viscosity of flocculated sample
IV. Recovery of unreacted hydrocarbons	(1) Vacuum	Instrumentation
	(2) Steam flow	Instrumentation
	(3) Hold-up time	Material flow rate
	(4) Residual unstripped hydrocarbons	Analysis of stripped sample
V. Flocculation	(1) Composition of brine	Analysis of sample
	(2) Quantity of brine	Instrumentation
	(3) pH of acid	Instrumentation
	(4) Temperature	Instrumentation
	(5) Hold-up time	Material flow rate
	(6) Porosity and size of	Visual inspection
	(7) Composition of latex	Analysis
	(8) Composition of antiox't	Analysis
VI. Washing	(1) Hardness of wash water	Analysis of sample
	(2) Quantity of wash water	Valve setting
	(3) pH of wash water	Analysis of sample
	(4) Temperature of wash water	Instrumentation
	(5) Time of washing	Speed of filter rotation
	(6) Removal of wash water	Vacuum and pressure on filter
VII. Drying		
VIII. Baling and Packaging	(1) Weight	Automatic weigh scales
	(2) Talc to prevent adhesion to package	Visual inspection
IX. Testing of finished product	(1) Viscosity (plasticity)	Mooney viscometer
	(2) Chemical properties	Analysis of sample
	(3) Stress-strain characteristics	Rubber testing

change. Control charts, correlation studies, and the technique of variance analysis have been used for this purpose.

One of the first steps in the improvement of testing took the form of numerous cross-checks among the different laboratories. Before 1943 the accuracy of the physical testing of vulcanized rubber by the rubber industry left much to be desired; systematic errors were found in most physical testing laboratories, and there was considerable variation between different laboratories. Standardization of these physical test methods was advanced by the daily testing of a reference lot of standard GR-S by each copolymer-plant laboratory. It was also found necessary to establish uniform lots of compounding ingredients in order to eliminate variability from this source. When these lots were used up, they had to be replaced with new lots of identical characteristics. The suppliers of the compounding ingredients were called upon to meet tolerances not previously attained in their industries. New lots of these materials were standardized at the National Bureau of Standards and taken to the Government Evaluation Laboratories in Akron, which makes shipments to the copolymer-plant laboratories and others. The Office of Rubber Reserve standard materials were also made available to other rubber-testing laboratories desiring to use them.

It was also found necessary to standardize the testing equipment and correct defects in equipment which gave variable results. An example of the latter is the modification in the design of the Mooney viscometer by the National Bureau of Standards which was adopted early in the program.

A second step in the improvement of laboratory operations consisted of an educational program. Technical representatives from the National Bureau of Standards and the Office of Rubber Reserve (previously the Rubber Reserve Company) visited each laboratory in turn and assisted in an instruction program to insure that all technicians performed each test using the same prescribed procedure. In addition, field representatives or inspectors, whose prime responsibility was to check on the testing techniques and quality of production, were assigned to each plant by Rubber Reserve.

Good results were obtained from this intensive drive to improve the precision of laboratory testing, and the standard deviations of test results obtained at the beginning of the program (1943) have been substantially reduced each year thereafter. Depending on the particular copolymer plant, the results in 1948 compared with those in 1943 showed that the standard deviations were from one-half to one-sixth as great in 1948 as in 1943.[19, 30] In most of the plant laboratories the only technically trained personnel are the shift supervisors and the chief chemist; the actual tests are made by nontechnical personnel, chiefly girls with high school education. The improvement in testing was also made without any material increase in the cost of laboratory operation. Papers which have appeared in technical journals give all the technical information necessary for any industrial laboratory to duplicate the accuracy of testing that is now attained by the copolymer laboratories.

The need for a standard reference rubber became apparent when the refinement of testing procedures was undertaken. Such a rubber would have widespread use in the copolymer plants, and the quantities needed would be quite large, since it was contemplated that the standard rubber should be made available to the consuming industry should they desire to use it for calibration of their testing instruments. A procedure was adopted whereby a copolymer plant was authorized to manufacture, under closely controlled standard process conditions, a lot (150,000 to 250,000 lb.) of rubber that was segregated, sampled, and tested for variation. If its variability was found to fit into very close limits, it was then numbered and designated as the "Standard Bale." This bale is used *daily* in copolymer plants to calibrate testing instruments and procedures, and to act as a referee sample. The testing of the Standard Bale is also done by the National Bureau of Standards, and the latter's test figures are considered official and are published. Usually a given lot of Standard Bale rubber lasts 3 months, at the end of which time another lot is manufactured, usually by another plant selected on the basis of a good record in production of uniform specification rubber.

SPECIFICATIONS FOR GR-S

It has already been pointed out that the Government Synthetic Rubber Program was based primarily on the production of a single general-purpose rubber, GR-S. The goal of the program was to produce at each plant GR-S of such uniform quality and processing characteristics that it could be interchanged with the production of any other plant without affecting the operation of any rubber-fabricating plant or the quality of their products. It was necessary, of course, to establish specifications for GR-S, defining its important properties and setting limits to physical and chemical tests. A committee on specifications was established by joint action of the Rubber Director and Rubber Reserve with representatives of rubber compounders, synthetic-rubber plant operators, the National Bureau of Standards, and representatives of Rubber Reserve and the Office of Rubber Director. Later a representative of the Govenment Evaluation Laboratory was added to this committee. This group was charged with the responsibility for establishing specifications covering the quality of all the synthetic rubbers produced in Government-owned plants, including GR-S, GR-I, and GR-M.

Typical of the specifications established by this committee is that for GR-S which is given in Table VIII. As new types of Government synthetic rubbers were developed and passed beyond the experimental stage, separate specifications for each new rubber type were set up by the committee. All material produced by the copolymer plants that tests showed to be outside of the limits established is graded as off-specification material and is disposed of separately.

Since 1944 customers using GR-S types have been supplied with copies of the copolymer laboratory reports on all lots shipped to them. These reports give the results of all specification tests, both chemical and physical. A large majority of rubber companies no longer themselves test the lots of GR-S types received but use the copolymer-plant laboratory tests as a basis for their compounding.

Table VIII. Specification for GR-S*

GR-S is a copolymer of butadiene and styrene to which approximately 1.25 per cent of a standard rubber antioxidant has been added during manufacture.

General

All GR-S shall be free of foreign or extraneous material which is objectionable in normal rubber practice.

Chemical

Volatile matter	0.50%	maximum
Ash	1.50%	maximum
Fatty acid (as stearic acid)	3.75%	minimum
	6.00%	maximum
Soap (soluble; as sodium stearate)	0.75%	maximum
Stabilizer	1.00%	minimum
	1.75%	maximum
Bound styrene	22.5%	minimum
	24.5%	maximum

Physical

Viscosity of GR-S	46	minimum
ML 212° F. at 4 minutes	54	maximum
Viscosity of compounded stock,		
ML 212° F. at 4 minutes	73	maximum
Properties of vulcanizate at 77° F.		
Tensile strength,		
50-minute cure at 292° F.	2700 p.s.i.	minimum
Ultimate elongation,		
50-minute cure at 292° F.	550%	minimum
Modulus at 300% elongation,		
25-minute cure at 292° F.	350 p.s.i.	minimum
	650 p.s.i.	maximum
50-minute cure at 292° F.	850 p.s.i.	minimum
	1200 p.s.i.	maximum
100-minute cure at 292° F	1300 p.s.i.	minimum
	1700 p.s.i.	maximum

*Now GR-S 1000, having slightly modified specifications.

AGING OF RAW GR-S IN STORAGE

One of the advantages demonstrated by GR-S has been its excellent aging qualities. Extensive tests made indicate that GR-S stored under normal conditions undergoes little change and may still be used with complete confidence after several years of storage. In the spring of 1943 three batches of GR-S produced in the Goodyear-operated Akron plant were blended in a Banbury mixer. Portions of this blend were stored in open bins while other portions were sealed in metal containers. These blends were tested in 1945 and again in 1948, and no significant differences could be detected between the processing or stress-strain properties of the material whether stored in bins or containers. Normal practice in the Government Synthetic Rubber Program has been to rotate warehoused polymers at regular intervals. All tests made, however, indicate that GR-S can withstand long periods of storage, which makes it possible to maintain a strategic stockpile of GR-S, without deterioration in quality.

THE CONTINUOUS POLYMERIZATION OF GR-S [6, 8, 21]

At the time of the emergency in late 1941 and early 1942 not enough was known about the butadiene-styrene polymerization and the effect of different factors on quality and output to make a continuous polymerization process practicable.   For this reason the synthetic-rubber plants were designed for batch polymerization, the process becoming continuous after the latex blow-down tanks and continuing so throughout the rest of the procedure including continuous monomer stripping, coagulation, and drying.   It was realized that an entirely continuous process would have many advantages, and pilot-plant studies on continuous polymerization with several small reactor units in series were carried out by Goodyear.   However, the processing characteristics of the polymer were not entirely satisfactory, owing, it was thought, to "short-circuiting" in the reactor setup.   Since it seemed possible that more GR-S would be needed in the war than the copolymer plants could turn out by the standard batch process, it seemed desirable to carry out further experimentation with multiunit continuous polymerization. It was apparent that a continuous polymerization setup, if it operated satisfactorily, would have substantially more capacity than a batch-operated unit, since the full volumetric capacity of the reactors could be utilized and downtime required for charging and discharging reactors eliminated, as well as the inhibition period at the beginning of polymerization. Consequently it was decided in late 1943 to convert 12 reactors of the Goodyear-operated Houston plant to continuous polymerization, and operations were started in the spring of 1944.   It was hoped that by considerably increasing the number of reactors in the continuous polymerization chain (from 4 to 12) over the pilot-plant setup previously used and by installing horizontal baffles in each reactor to convert it in effect into two units the "short-circuiting" action could be largely eliminated and a hoped-for material improvement in the processing characteristics of the polymer obtained.   To reduce further the short circuiting and to provide additional polymerizing capacity, ten displacement reactor pipes, operating in series, were added after the twelfth polymerizer.

The GR-S polymer made in the continuous polymerization plant described in outline above was carefully and thoroughly tested for quality and processing characteristics and proved to be entirely satisfactory.   A substantial improvement in reactor output was demonstrated, and those sections of the plants that succeeded the reactors, viz. the strippers and the coagulation and drying equipment, were shown to have capacity to handle the increased output. Consequently, when far greater production of GR-S was called for at the end of 1944, it was evident that this could be best obtained not by building additional copolymer plants but by changing practically all the existing plants over from batch to continuous polymerization.   The war ended before this program was completed, and some of the projects were canceled, but a substantial proportion of the plants were converted.

Advantages claimed for continuous polymerization over the batch technique are as follows: 1. The inhibition periods encountered in batch polymerization are eliminated through the maintenance of a closed, oxygen-free system.   2. Monomer recovery operations are smoothed out through the application of a constant load on the system as compared to the surges in batchwise operation. 3. Peak demand for cooling water levels off in continuous polymerization. 4. The full volumetric utilization of the reactors, combined with the elimination of downtime for charging and discharging

Fig. 10.   Flow-Control Mechanism (Rotameters) for Continuous Polymerization of GR-S
Courtesy B. F. Goodrich Chemical Co.

batchwise, yields a 20 per cent increase in effective reactor capacity. 5. A polymer equal in quality to that given by batch operation and of improved uniformity is produced.

Charging the Reagents. The heart of the continuous polymerization system is the equipment controlling charge rates, since the uniformity of the polymer produced is largely determined by the accuracy of charging.   In the original plant scale continuous installation (Goodyear-Houston), reciprocating proportioning pumps were used.   They were operated in parallel from one variable-speed drive shaft with each pump in the series sized for the normal flow of material.   The variable-speed drive permitted changes in production rate while a constant ratio of charge materials was maintained.   Adjustments of individual flow rates were made by changes in setting of the microscrew adjusting arms.   The accuracy of these pumps was satisfactory and, aside from their massiveness, the chief objection to them arose from the pulsating flow from the reciprocating pumps.   In later installations centrifugal or turbine pumps with flow controllers were used, the material to be metered being pumped through the flow controller and control valve at a pressure exceeding the reactor pressure.

One system of flow-control equipment employs rotameter controllers. With this type of volumetric controller a pneumatic controlling mechanism containing proportional response and automatic reset operates the control valve.   By proper sizing of the float and the bore of the tapered tube, good accuracy of flow measurement may be obtained.   Another system of flow measurement utilizes orifice meters.   Since the material measured must be kept out of the meter body, seal pots are used.   For butadiene and styrene, water is used as a sealing fluid, as these materials have a density below unity and are immiscible with water.   Carbon tetrachloride is used for sealing the soap-catalyst solution.   Orifices are so located as to prevent swirling or helical flow in the upstream channel.   The simplicity and resultant low maintenance cost on orifice meters are their primary advantage over other types used.

The equipment so far described is for the addition of butadiene, styrene, soap-catalyst solution, and shortstop solution.   The addition of modifier presents a difficult problem since the amount charged determines to a large extent the properties of the finished polymer.   Small changes in the amount of mercaptan used make appreciable changes in the viscosity of GR-S. Extremely accurate control of the addition has been achieved by the use of reciprocating proportioning pumps.   Although the average rate obtainable with a rotamatic controller is comparable with that obtained with a proportioning pump, the standard deviation is greater with the former than with the latter equipment.   In order to obtain a continuous record of the amount of modifier charged, the pump is fed from a tank mounted on a weight scale equipped with a recorder.   The accuracy of the charging rate for all ingredients to the continuous polymerizer is largely dependent on the accuracy of calibration of the flow meters or the proportioning pumps. Both volumetric tanks and weigh tanks are used for this purpose.

The Reactors.   The reactors used in the continuous process of polymerizing GR-S are those originally installed in the standard plants for batchwise operation.   The reactors are connected in series, the outlet being at the top, and the inlet at the bottom.   In order to reduce short circuiting in the reactors, a horizontal baffle is inserted in each with an annular opening around the agitator shaft.   This, in effect, doubles the number of vessels in any given reactor chain.   The agitation employed is the modified Brumagin type already described.

Displacement reactors consisting of steel pipes 18 in. in diameter and 30 ft. in length, operating in series, are provided at the end of each reactor chain. Ten such pipes are used for 12 standard reactors. These displacement reactors afford a convenient means of controlling the viscosity since the shortstopper may be added to any unit. The primary control of the continuous process is based on the shortstop viscosity, as in the case of the batchwise reaction. Control of the reaction is maintained by slight changes of the temperature in the last several reactors, by varying the location of shortstop addition, and by changing the amount of modifier added to the system. One of the weaknesses of the continuous system is that the effect of the change in modifier charge does not show up in the finished product for approximately 18 hours. A considerable amount of work has been conducted attempting to predict these changes on the basis of samples taken early in the reaction.

Shutdowns. Occasionally it becomes necessary to suspend operations of a continuous system. For prolonged shutdowns, the polymerization of the materials in the reactors is completed in a batchwise manner. Under these conditions, the hydrocarbon conversion may be expected to fall off, depending on the position of the polymerizer in the continuous chain. To prevent this, the correct amount of additional modifier is added to each reactor at the time the process is changed from continuous to batchwise. If the shutdown can be anticipated, the modifier addition by the proportioning pump may be increased increment-wise until the requirements for batch polymerization have been reached before the continuous charging is stopped. Since considerable time is lost in completely shutting down a continuous unit, a technique has been devised whereby the reaction may be retarded to permit the discontinuation of charging. This is accomplished by dropping the temperature over a period of 3 hours to approximately 25° C. This is called a "chilled shutdown." During the chilled period the conversion of the system remains approximately constant; however, the viscosity continues to rise at a slow rate.

PROGRESS IN THE PRODUCTION AND UTILIZATION OF GR-S

It has already been shown that the American synthetic-rubber program was closely integrated with the existing rubber-fabricating industry. The basic philosophy has always been to try to produce elastomers that will do the best job in ‧the fabricating equipment available, working in close co-operation with the rubber companies. This is in sharp contrast to German practice where the polymer was produced by a single chemical company to suit its own operations. Although the I.G. undoubtedly tried to turn out those synthetic rubbers which they believed to be the best that could economically be produced, nevertheless the consumer had to install expensive heat-softening and other equipment to be able to use the I.G. polymer, and, though theoretically he could heat soften the polymer to the desired viscosity, this does not seem to have worked out very well in practice.

It was learned after the war that the German rubber companies complained often and bitterly of the toughness and other undesirable properties of the Buna rubber, and apparently these characteristics had something to do with the relatively poor performance of German-made tires tested on American test fleets after the war. There were also other reasons, such as inferior rayon cord, for the limited mileages obtained. Satisfactory results were obtained from tires built from Buna S3 by a rubber company in the United States, but much care and time were required to handle the Buna rubber, which was tough indeed, even after heat softening. The special treatment given to the experimental tires could not have been readily applied on a production-line basis.

It was probably fortunate that the development of butadiene-styrene-type synthetic rubbers in the United States was undertaken by the large rubber companies since this insured that throughout the development and manu-facture of GR-S types the rubber consumers' interests were given first con-sideration. More important still was the fact that a number of completely separate rubber companies, each with well-equipped experimental facilities and an experienced staff, were engaged in the development and manufacture of the synthetic rubbers. Naturally there were many different ideas regarding the development of new and superior polymers and their optimum compounding, as well as regarding the tire design and construction methods, etc., best suited to the new materials. No attempt was made by Rubber Reserve or the Rubber Section of the War Production Board to insist on any single program for developing the best end products from synthetic rubbers, although full exchange of information was provided for and insisted upon. Undoubtedly this combination of competitive experimentation, together with full exchange of information was largely responsible both for the develop-ment of new types of synthetic rubbers better adapted to some specific end-product use and for the rapid progress in learning how to compound and fabricate superior-quality products from GR-S type polymers. Although the quality of the polymer did not greatly improve from its first production in 1942 and 1943, except for greater uniformity, the quality of the end products such as tires showed outstanding improvements each year over the year before from 1943 to 1947 and later.

The system employed in the Government synthetic-rubber program permitted and encouraged the production of experimental polymers. As previously mentioned, experimental types produced in the large-scale plants have been designated as X polymers (e.g., X-430 GR-S). New types of polymer continue to be manufactured under an X designation until all the production difficulties have been ironed out. When it becomes possible to produce a uniform material of satisfactory quality, and provided there is sufficient demand for the new polymer, specifications are established and a regulation-type designation is assigned to the material (e.g., GR-S 10).

LOW-TEMPERATURE POLYMERIZATION

Most of the changes in the GR-S process and equipment made during the war did not involve radical alterations in the fundamental polymerization recipe. It had earlier been thought that some improvement in polymer quality could be obtained by reducing the temperature of polymerization. However, the reaction rate dropped off very sharply with lower temperature, cutting the production rate to a low figure. In 1943–44 a number of large-scale polymerizations were carried out at 86° F. (instead of 122° F., the standard GR-S process temperature). These runs required several times the normal polymerization time, and no significant improvement in quality was detected in the polymer either from laboratory physical tests or from actual road tests on tires made from the polymer. Although some further means of speeding up the polymerization were later developed, no practical method was known for carrying out polymerization in a reasonably short period of time at fairly low temperatures such as around 40° F. until information was obtained immediately after the war regarding the German redox formulas and their application to polymerization at 5° C. (41° F.).[18, 33] This information greatly stimulated interest in low-temperature polymerizations at the laboratories of several universities and companies doing research work in the Rubber Reserve program. The German redox low-temperature formulas were greatly improved and a number of new low-temperature polymerization systems discovered. Pilot-plant runs in 1946 and 1947 at 41° F. made sufficient amounts of polymer available for construction of tires, and the road tests of these tires showed marked superiority over tires made from GR-S. New furnace blacks developed about the same time contributed further notable improvements when compounded with the low-temperature polymers and used for tire treads. Low-temperature polymer thus compounded gave results that were not only substantially superior to regular GR-S, but in tread stocks on passenger tires were better than natural rubber. Pilot-plant polymerizations were made at still lower temperatures (14 and 0° F.), and some further improvement in rubber quality was obtained.

In view of the outstanding improvement in GR-S properties obtained by polymerization at low temperatures the Office of Rubber Reserve in October 1947 authorized the Copolymer Corporation to convert one-half of the Office of Rubber Reserve Copolymer plant at Baton Rouge to the production of 41° F. rubber. This plant, using a recipe which had been demonstrated on a pilot-plant scale, started operations in late February 1948 and made "cold rubber" available for extensive industry testing and for the manufacture of tires made by the rubber companies which formed the Copolymer Corporation. (Under the Office of Rubber Reserve regulations the sponsors of a new experimental rubber are entitled to the major share of the first production.) The new product was so well received that in the fall of 1948 the Office of Rubber Reserve decided to convert a substantial part of its capacity, 188,000 long tons annually, to cold rubber.[9] (Later, a still larger proportion of the capacity was converted to cold rubber.) Both batch- and continuous-polymerization copolymer plants were converted to the new material, with operations starting batchwise in the continuous-polymerization plants, then changing to continuous after experience had been gained with the new process.

Batch Low-Temperature Polymerization.[20, 27] The adaptation of standard GR-S plants to cold-rubber polymerization involved several areas of the plant. The composition of an applicable polymerization recipe

Table IX.   Cold-Rubber Polymerization Formula "1949"

	Parts
Butadiene	72
Styrene	28
Sulfole (technical *tert*-dodecyl mercaptan)	0.17
Water	200
Dresinate 214 (potassium soap of disproportionated rosin, anhydrous basis)	4.5
Sodium hydroxide	0.10
Trisodium phosphate	0.5
Daxad 11 (formaldehyde condensation product of sodium alkylnaphthalene sulfonate)	0.1
Potassium pyrophosphate	0.177
Dextrose	1.0
Ferrous sulfate heptahydrate	0.14
Cumene hydroperoxide	0.01

is given in Table IX.   In the pigment-preparation area the available space had to be reorganized to provide for the activator solution storage (sodium pyrophosphate, glucose, ferrous sulfate).   Additional tankage was added to prepare the soap solution, the recipe calling for rosin soaps along with small amounts of caustic soda and sodium pyrophosphate.   A considerable number of changes had to be made in the reactor area, where the reactors were insulated with 4 in. of cork and provisions made for the circulation in the reactor jackets of cold brine from a new refrigerating system.   A major addition required for cold-rubber production was the extra refrigeration plant.   For one-half of a standard plant of capacity 30,000 long tons per year, about 400 tons refrigeration capacity at 0° F. is required or 800 tons for a 30,000-ton-per-year plant.   Besides this it is desirable to have a spare refrigerating unit available when needed.   In addition heat exchangers had to be provided for cooling the charging stocks previous to their entry into the reactor.

Since the latex emerging from the reactors was at polymerization temperature, it had to be heated before entering the stripping columns.   Though it would have been desirable to use this cold latex for cooling the incoming monomers using heat exchangers, and thereby effecting a substantial economy in refrigeration, this operation is complicated by the physical instability of the latex from cold polymerization, and consequent fouling of heat-exchange surfaces—a difficulty that has not yet been solved on a production scale. The latex is now heated by direct-steam sparge lines inserted in the receiving vessel, and the latex is brought up to stripping temperature by open steam addition.   The shortstopping agent first used, viz., 2,4-dinitrochlorobenzene, has now been abandoned, and water-soluble dithiocarbamates have come into use as shortstops in cold GR-S production.[12]   Other substances have been studied and advocated as shortstops from time to time.[32]

Normal procedures are followed in stripping and coagulating cold rubber, with the exception that greater vacuum pump and compressor capacity is needed to handle the larger amount of vapors occasioned by the. lower conversion used (60 per cent instead of 72 per cent) in making the cold polymer.

Batch-charging procedure for cold-rubber follows that prescribed for standard GR-S, with the exception that, as previously noted, all reactants must be cooled to the reaction temperature (41° F.) before entering the reactor. Catalyst and modifier are added separately through individual pipe lines.

When low-temperature polymerization was first put into plant operation long reaction times (20 to 35 hours compared with 14 hours for regular GR-S) were experienced, but as the plant chemists and operators became more familiar with the process and its variables the reaction time was substantially reduced, it is now 14 hours or less, with occasional batches running longer. One of the most important factors in improving reaction rate was the elimination of oxygen from the system as far as possible, including the use of water with an oxygen content reduced to less than 0.1 p.p.m. by vacuum deaeration, avoidance of vortex formation by the agitators in the tanks used for making up solutions, evacuation of the reactors before introducing the charge, etc.

Improved technique in making up the activator solution, the choice of a suitable electrolyte as a component of the recipe, and the use of peroxides more active than cumene hydroperoxide have all contributed to improve the rate of polymerization in the manufacture[23] of GR-S at 41° F. Another contributing factor has been the installation in the reactors of auxiliary refrigeration in the form of bundles of tubes in which coolant is circulated. The installation makes it feasible to conduct polymerization more quickly without loss of temperature control than in reactors of the original design.[12, 15, 23]

The development of polymerization recipes adapted to the low-temperature production of GR-S is fully discussed in Chapter 8.

Developments in Low-Temperature Polymerization. With the advent of more powerful oxidants, such as diisopropylbenzene and *p*-menthane hydroperoxides, it has become possible to polymerize cold (5° C.) rubber to 60 per cent conversion in 7 to 10 hours compared to a former average time of 14 hours. Such fast reactions have been made feasible by the installation of better heat transfer equipment in the copolymer plants. In most of the plants, cooling coils mounted internally in the reactors either supplement or have replaced the less efficient external cooling jackets. In these coils the direct evaporation of ammonia is used as the cooling medium, the latent heat of vaporization cooling the polymerization emulsion. With this arrangement, reactions as fast as 5 hours to 60 per cent conversion can be controlled at 5° C.

Although disproportionated rosin soap continues to be the emulsifier in many cold GR-S polymerizations, other emulsion systems have gained favor. For products in which rosin must be avoided, cold polymers are made with potassium fatty acid soap as the sole emulsifier. In this instance the peroxamine activation system is usually employed instead of the usual iron-pyrophosphate system. Many cold polymers are made in which the emulsifier is a 50/50 mixture of the potassium soap of disproportionated rosin and the sodium soap of fatty acids. The ratio of sodium and potassium ions is dictated by the fluidity desired at the temperature of polymerization. Commercially prepared potassium soaps of 50/50 mixtures of disproportionated rosin and fatty acids derived from tall oil have recently become available and are in use.

Peroxamine Recipe. When iron is to be kept at a minimum in the polymerization of GR-S, peroxamine formulations are used. A typical peroxamine recipe is as follows.

Water	200
Butadiene	72
Styrene	28
Potassium fatty acid soap	4.7
Trisodium phosphate 12H_2O	0.8
Tamol N	0.15
Diethylenetriamine	0.15
p-Menthane hydroperoxide	0.12
Ferrous sulfate 7H_2O	0.002
Versene Fe-3	0.02
tert-Dodecyl mercaptan	0.20

Sulfoxylate Recipe. The most recent formulation to come into commercial use in the production of cold rubber is the sulfoxylate recipe, in which the activator is composed of sodium formaldehyde sulfoxylate, ferrous sulfate, and Versene Fe-3 or other chelating agent. The sulfoxylate recipe offers the following favorable features in regard to ease of plant operation: (1) the activator solution is simple to prepare and does not involve critical time-temperature factors; (2) the activator is a clear solution, not a slurry; (3) polymerization rates are more uniform from batch to batch; (4) polymerizations that die or slow down can be revived by the addition of small quantities of activator. In addition, the cost of chemicals for the sulfoxylate recipe is substantially less than that for the iron-pyrophosphate or the peroxamine recipe. No improvement in quality can be detected in polymers prepared in the sulfoxylate system, but latexes prepared by it are much lighter in color than latexes prepared in the usual recipes. A typical sulfoxylate recipe is as follows.

Water	200
Butadiene	70
Styrene	30
Potassium soap of disproportionated rosin	4.5
Trisodium phosphate 12H_2O	0.8
Tamol N	0.15
Sodium formaldehyde sulfoxylate 2H_2O	0.15
Ferrous sulfate 7H_2O	0.05
Versene Fe-3	0.07
p-Menthane hydroperoxide	0.10
tert-Dodecyl mercaptan	0.20

Superfast Polymerization. Extremely fast polymerizations have been conducted using jacketed pipe as the polymerizer instead of the conventional autoclave. Reactions to 60 per cent conversion in as short a time as 10 to 20 minutes have been achieved, with phenylcyclohexyl hydroperoxide as the oxidant. This process is not yet commercially practicable, primarily because, in order to maintain latex stability during the reaction, large quantities of soap (ca. 20 parts) must be used and excessive amounts of water (300 to 500 parts) are required.

Continuous Low-Temperature Polymerization. Since the polymerization initiation reagents in some of the cold-rubber polymerization recipes are sensitive to oxygen, the leakage of air into vessels and pipes was a constant worry to the operators. The production of cold rubber by continuous polymerization has eliminated the entrance of oxygen, because the reactors are under pressure at all times. The application of continuous polymerization in the production of cold rubber has shown all the favorable features that attended its application to the production of standard GR-S. In continuous cold polymerization, the reagents constituting the charge must be cooled before being metered to the first reactor in the line, and provision must be made for continuously shortstopping the reaction as the latex leaves the displacement tubes, and also for raising the temperature of the latex before it enters the butadiene flash tanks in the recovery area. The operational techniques are otherwise essentially the same as those described for the continuous production of standard GR-S. The shortstop may be added at the top of any one of the series of displacement tubes, thus giving a measure of control over the period of polymerization without altering the rate of charging. The coolant circulated in the reactor jackets may be, for example, isopropanol–water or calcium chloride brine containing corrosion inhibitor.

A number of plants are now operating cold polymerization as a continuous process and are very well satisfied with it. The chief advantage of the continuous over the batch process of polymerization is, of course, a higher plant capacity because (*a*) the reactors are run full instead of, as in batch operation, only 90 per cent filled, (*b*) the displacement tubes add extra capacity equal about to that of an extra reactor, (*c*) elimination of the necessity for charging and discharging individual reactors involved in batch operation represents a saving of more than 1 hour in the over-all time of polymerization. When first the continuous polymerization of cold GR-S was put into plant operation, the saving of time over batch polymerization was in fact only 4 per cent because of slower reaction rates, but with greater experience and improved recipes and techniques, the theoretically expected saving has been attained.

Another advantage of the continuous process is that the refrigeration load is more uniform than in batch operation. About 25 per cent of the total load comes in cooling the charge before its entry to the reactors, and in batch operation the ingredients are charged in a 20-minute period or at a rate about four times as fast as in continuous operation. A surge demand on refrigeration is thus avoided by continuous operation, and along with it a surge demand on steam for the turbines. Other advantages are that the continuous as compared with the batch process requires smaller feedstock coolers, pumps, and meters, makes smaller labor demands (a saving of one operator per shift per 30,000-ton unit); involves lower power costs; and turns out a product of more uniform quality.[1, 16, 17, 20]

OIL-EXTENDED GR-S

A growing recent practice in the manufacture of GR-S involves extending GR-S of high Mooney viscosity by means of oil, the oil being added as an emulsion to the latex before coagulation of the latter. Petroleum oils used as extenders in this connection have been divided into three classes: naphthenic, aromatic, and highly aromatic. The composition of the oils appears to influence the characteristics of the masterbatch with regard to processing into finished products. Masterbatches containing naphthenic or highly saturated oils tend to be poorer processing than those containing highly aromatic oils. But the highly saturated or naphthenic oils have better resistance to staining and discoloration than the highly aromatic oils. Circosol 2XH is an example of a naphthenic oil; Sundex 53 and Shell SPX97 are classed as aromatic oils, and Dutrex 20 and Califlux TT are highly aromatic oils.

At present in the United States a masterbatch containing 37.5 parts of processing oil per 100 parts of GR-S represents the highest oil loading commercially available. Masterbatches containing 50 parts of oil have not yet become practicable because of physical difficulties encountered in handling the coagulated masterbatch in the copolymer plants.

Almost all the oil masterbatches have been made with cold rubber as the base polymer. Several masterbatches of hot (50° C.) GR-S with 25 parts of aromatic oil have been made, but have been poor processing and have not enjoyed wide popularity. Poor processing may be on account of large proportions of gel in the base polymer. In the hot polymerization system, large proportions of insoluble gel (up to 50 per cent) are formed when the Mooney is high enough to support the oil loading. In the cold polymerization system, polymers of a Mooney viscosity high enough to support moderately large quantities of oil are either gel-free or contain only small amounts of insoluble gel. Even in this system it is necessary to keep the hydrocarbon conversion below 72 per cent in order to minimize the effect of gel on processing.

OIL-BLACK MASTERBATCH

It has been found that masterbatches of cold GR-S and HAF carbon black can be extended with petroleum processing oils just as cold GR-S of high Mooney viscosity can itself be extended with oil. The principle of operation is the same in that the base polymer is made to a sufficiently high viscosity that the compounded Mooney viscosity of the oil—GR-S—black masterbatch will be equivalent to the compounded Mooney viscosity of cold GR-S-black masterbatches normally used. The oil, as a water emulsion, may be incorporated into the masterbatch in one of two ways. The emulsion may be added to the latex and the resultant mixture co-precipitated with carbon black slurry in the usual way (Chapter 20) or the oil emulsion may be added to the carbon black slurry and this mixture co-precipitated with latex in the normal manner. Some work has been done on mixing processing oil with carbon black at the carbon black plant but this operation has not yet proved economically practicable.

The only oil-black masterbatch at present commercially available as a regular grade of GR-S, viz., GR-S 1801, is composed of 100 parts cold GR-S, 25 parts processing oil, and 50 parts HAF black. It is technically possible to load 37.5 parts of oil into this kind of masterbatch but this type of material presents problems of plant operation.

In the matter of carbon black loading, in order to achieve true equivalence with the non-oil-extended masterbatch counterpart, the oil plus polymer must be considered as "apparent" polymer or usable rubber. Since almost all black masterbatches are loaded with 50 parts of black to 100 parts of polymer, in the oil-black masterbatch 50 parts of black should be loaded on 100 parts of apparent polymer. Thus, for a loading of 37.5 parts of oil, the composition of the masterbatch should be 100 parts cold GR-S, 37.5 parts oil, and 68.75 parts black. It is apparent that more than half of the material passing through the driers in this case is not the result of polymerization. Consequently, in order to utilize fully the polymerization equipment in a copolymer plant devoted to making such a product, it would be necessary to double the drier capacity. Oil-black masterbatches containing high loadings of black have been produced experimentally. In X-668 there is 62.5 parts HAF black and 25 parts oil per 100 parts polymer; in X-691, 77 parts HAF black and 40 parts oil per 100 parts polymer.

At present HAF carbon blacks are, because of their quality advantages, the only blacks used in oil-black masterbatch. The newer ISAF blacks show considerable improvement in quality and may eventually replace HAF blacks in the production of oil-black masterbatch.

GRADES OF GR-S

The initial Government Synthetic Rubber Program emphasized the development and production of a single general-purpose polymer, GR-S, and efforts were concentrated on producing this material by exactly the same process in each copolymer plant. Mention has been made of the policy of producing special-purpose rubbers which has led to the general development of a variety of rubbers that have afforded the rubber-consuming industry a source of new materials tailored to meet specific conditions of service and utility. The first deviation from basic GR-S was the use of alum instead of salt and acid as a coagulant for the standard GR-S latex. The polymer made in this fashion is known as GR-S-AC (now GR-S 1004) and has been produced in substantial quantities for a number of years. GR-S-AC is a slightly slower-curing polymer than GR-S, but some segments of industry feel that by its use they are able to produce products with superior aging properties.

Another variation of the standard polymer is GR-S 10 (now GR-S 1002), made by the substitution of disproportionated rosin soap for fatty acid soap

Table X.   *Grades of GR-S Offered as at August* 1, 1951*

Grade*	Distinguishing Features
GR-S (1000)	Standard product
GR-S Wire and Cable	Low water-soluble ash
GR-S-SP	Has been strained and milled after drying
GR-S 10 (1002)	Contains rosin acids in place of fatty acids
GR-S 17	Similar to GR-S 10 but has stain-resistant antioxidant
GR-S 18 (1003)	Antioxidant is heptylated diphenylamine; contains rosin acids
GR-S 20 (1000)	Mooney viscosity and combined styrene slightly lower than for standard GR-S
GR-S 21 (1006)	Antioxidant is hydroquinone†
GR-S 25 (1006)	Has strain-resistant antioxidant†
GR-S 26 (1006)	Antioxidant is Wingstay-S†
GR-S 50 (1001)	Antioxidant is heptylated diphenylamine
GR-S 60 (1008)	Polymer is cross-linked by use of a small amount of divinylbenzene as a comonomer
GR-S 60-SP (1017)	GR-S 60 strained and milled after drying
GR-S 61-SP (1018)	Similar to GR-S 60-SP but with UBUB as the antioxidant§
GR-S 62 (1009)	Similar to GR-S 60 but with nonstaining IRUN as the antioxidant§
GR-S 65 (1007)	Lower ash and better electrical properties than standard GR-S
GR-S 65-SP (1016)	GR-S 65 strained and milled after drying
GR-S 66-SP (1019)	Has stain-resistant antioxidant†; lower ash than GR-S-SP; strained and milled after drying
GR-S 85-SP	Mooney viscosity (90-110) much higher than that of GR-S
GR-S 86-SP (1012)	Similar to GR-S-SP but has high Mooney (95-115)
GR-S 100 (1500)‡	Cold GR-S made at 41° F. by a sugar-hydroperoxide recipe
GR-S 101 (1500)	Cold GR-S made at 41° F. by a sugar-free recipe
GR-S 117	Similar to GR-S 17 but darker in color
GR-S-AC (1004)	Similar to standard GR-S but coagulated by alum
GR-S-AC for Wire Cable	Similar to GR-S Wire and Cable but coagulated by alum
GR-S 10-AC (1005)	Similar to GR-S 10 but coagulated by alum
GR-S 20-AC (1004)	Lower Mooney viscosity (40-50) than GR-S-AC (46-54)
GR-S 30-AC (1010)	Considerably lower Mooney viscosity (25-35) than GR-S-AC; has UBUB as stabilizer§; coagulated by alum
GR-S 40-AC (1013)	Combined styrene is 43 per cent; stain-resistant antioxidant§; alum coagulated
GR-S 45-AC	Antioxidant is EFED; alum coagulated

* Revised code designations as of July 1, 1952, are shown in parentheses.

† Antioxidant is now alkylated phenol type.

‡ Now made in a sugar-free recipe.

§ Antioxidant is now tris-nonyl phenyl phosphite type.

as the emulsifier in polymerization. This polymer is somewhat more tacky than standard GR-S, thus making it easier to fabricate tires and other products.[4] It is preferred by many customers to regular GR-S. GR-S 25 (now GR-S 1006) and GR-S 17 are specialty rubbers that have been developed using relatively nonstaining antioxidants in place of PBNA and BLE. These polymers have been used in the manufacture of articles which come in contact with enameled surfaces and in similar applications where migratory staining might make the article unacceptable. Several grades of polymers embodying any or all of the above modifications have been manufactured over a considerable range of viscosities (polymers have been regularly prepared as low as 25 Mooney and as high as 135 Mooney).

A special group of copolymers containing a small amount of divinylbenzene as a cross-linking agent have been made. Members of this group of elastomers when compounded with natural rubber or other GR-S types impart improved extrusion and processing properties to the mixtures.

A group of copolymers coagulated by special techniques resulting in extremely low water-soluble ash contents have been developed for the use of the wire and cable industry. Some of the polymers in this class are prepared in a special finishing line in which the wet polymer is strained, dried in a one-pass drier, strained again, and sheeted out on a rubber mill before baling. The additional mechanical breakdown imparted by this treatment produces an easy processing polymer. These are but a few of the types of special rubbers which have been made in the Government Synthetic Rubber Program.

The August 1951 edition of the "Specifications for Government Synthetic Rubber" listed 32 different types of GR-S (most of which are shown in Table X), 4 GR-S black masterbatches, and 7 types of GR-S latex. These masterbatches and latexes are discussed in Chapters 19 and 20. In addition 754 different experimental modifications had been authorized for production up to November 1953.

As the technology of GR-S progresses, changes are not infrequently made in the grades of GR-S produced. A revised list of grades effective February 15, 1954, shows 28 types of GR-S, 6 GR-S black masterbatches, 11 GR-S oil (or oil/black) masterbatches, and 7 types of GR-S latex. (Cf. ref. 5.)

REFERENCES

1. Anon., *Chem. Inds.*, **66**, 836–7 (1950).
2. Bixby, W. F., *Chem. Eng. Prog.*, **45**, 81–6 (1949).
3. Burke, O. W., Starr, C. E., and Tuemmler, F. D., *Light Hydrocarbon Analysis*, Reinhold, New York, 1952.
4. Cuthbertson, G. R., Coe, W. S., and Brady, J. L., *Ind. Eng. Chem.*, **38**, 975–6 (1946).
5. Drogin, I., *India Rubber World*, **127**, 505–10 (1953).
6. Francis, D. H., and Sontag, H. R., *Rubber Age N.Y.*, **65**, 183–7 (1948); *Chem. Eng. Prog.*, **45**, 402–6 (1949).
7. Frank, R. L., Blegen, J. R., Inskeep, G. E., and Smith, P. V., *Ind. Eng. Chem.*, **39**, 893–5 (1947).
8. Gracia, A. J., *Chem. Inds.*, **57**, 628–30 (1945).
9. Hadlock, G. B., Cox, J. T., and Burke, O. W., *Chem. Eng. News*, **27**, 27–8 (1949).

10. Hebbard, G. M., Powell, S. T., and Rostenbach, R. E., *Ind. Eng. Chem.*, **39**, 589–95 (1947).

11. Hillyer, J. C., *Proc. Am. Soc. Testing Materials*, **47**, 283–8, 300–6 (1947).   Cf. M. Sheperd, R. Thomas, S. Schumann, and V. Dibeler, *J. Research Natl. Bur. Standards*, **39**, 435–51 (1947).

12. Howland, L. H., Neklutin, V. C., Provost, R. L., and Mauger, F. A., *Ind. Eng. Chem.*, **45**, 1304–11 (1953).

13. Johnson, C. R., and Otto, W. M., *Chem. Eng. Prog.*, **45**, 407–14 (1949).

14. Kharasch, M. S., Nudenberg, W., Jensen, E. V., Fischer, P. E., and Mayfield, D. L., *Ind. Eng. Chem.*, **39**, 830–7 (1947).

15. Kirkpatrick, S. D., *Chem. Eng.*, **59**, Nov., 148–52 (1951).

16. Larson, M. W., *Chem. Eng. Prog.*, **47**, 270–4 (1951).

17. Laundrie, R. W., Rowland, E. E., Snyder, A. D., Taft, W. K., and Tiger, G. J., *Ind. Eng. Chem.*, **42**, 1439–42 (1950).

18. Livingston, J. W., *Chem. Eng. News*, **27**, 2444 (1949).

19. Meuser, L., Stiehler, R. D., and Hackett, R. W., Symposium on Rubber Testing, June 1947, Am. Soc. Testing Materials, Spec. Technol., Pub. No. 74; *India Rubber World*, **117**, 57–61 (1947).

20. Mitchelson, J. B., and Francis, D. H., *Chem. Eng.*, **57**, 102–5, Apr. 1950.

21. Owen, J. J., Steel, C. T., Parker, P. T., and Carrier, E. W., *Ind. Eng. Chem.*, **39**, 110–3 (1947).

22. Pahl, W. H., *Chem. Eng. Prog.*, **43**, 515–22 (1947).

23. Pryor, B. C., Harrington, E. W., and Druesedow, D., *Ind. Eng. Chem.*, **45**, 1311–5 (1953).

24. Robey, R. F., Wiese, H. K., and Morrell, C. E., *Ind. Eng. Chem.*, **36**, 3–7 (1944).

25. Schulze, W. A., Reynolds, W. B., Fryling, C. F., Sperberg, L. R., and Troyan, J. E., *India Rubber World*, **117**, 739–42 (1948).

26. Semon, W. L., *J. Am. Oil Chemists' Soc.*, **24**, 33–6 (1947); *India Rubber World*, **116**, 63–5, 132 (1947).

27. Shearon, W. H., McKenzie, J. P., and Samuels, M. E., *Ind. Eng. Chem.*, **40**, 769–77 (1948).

28. Snyder, H. R., Stewart, J. M., Allen, R. E., and Dearborn, R. J., *J. Am. Chem. Soc.*, **68**, 1422–8 (1946).

29. Soday, F. J., *Trans. Am. Inst. Chem. Engrs.*, **42**, 647–64 (1946).

30. Stiehler, R. D., and Hackett, R. W., *Anal. Chem.*, **20**, 292–6 (1948).

31. Vila, G. R., and Gross, M. D., *Rubber Age N.Y.*, **57**, 551–8 (1945).

32. Wakefield, L. B., and Bebb, R. L., *Ind. Eng. Chem.*, **42**, 838–41 (1950).

33. Weidlein, E. R., Jr., *Chem. Eng. News*, **24**, 771–4 (1946).

34. Whitby, G. S., and Zomlefer, J., in Dunbrook, R. F., *India Rubber World*, **117**, 745 (1948).

CHAPTER 8

EMULSION POLYMERIZATION SYSTEMS

C. F. Fryling

Phillips Petroleum Company

INTRODUCTION

The two fundamental contributions to modern synthetic-rubber technique are polymerization in aqueous dispersed systems and copolymerization, i.e., the use of a mixture of polymerizable monomers. The more recent application of modifiers for controlling plastic properties ranks in importance with the fundamental discoveries. Subsequent developments have been largely concerned with increasing the rate of emulsion copolymerization and improving the properties of the product.

Emulsion polymerization owes its origin to attempts at duplicating conditions under which rubber is formed in nature. Patents embodying this idea were issued to F. Hofmann and co-workers[6] in 1912. Two unverified hypotheses appear to have influenced the thinking of these early investigators. No evidence has been obtained to indicate that natural rubber is produced by polymerization of isoprene in the cambium of the parent tree. From the behavior of extracted rubber it was concluded that elastomeric properties resulted from a colloidal complex of polymer and naturally occurring colloids—an error arising from ignorance of the role played by antioxidants in preserving the polymer. As a result, Hofmann[83] and others proposed the use in polymerization of various colloidal materials, such as casein, glue, dried blood, and albumins. More recent experience[54] has shown that such colloidal additives frequently exert deleterious effects in emulsion polymerization. Nevertheless the positive benefits derived from emulsion polymerization have been sufficiently great to warrant its retention and use in large-scale manufacturing operations.[129] A review with extensive bibliography of this early work was prepared by Whitby and Katz.[216]

The first to use a mixture of monomers, a procedure now designated as copolymerization or interpolymerization, appears to have been Kondakow.[120] Another early example was reported from Traun's research laboratory.[201] These matters and subsequent developments have been reviewed by Wood,[223] H. L. Fisher,[37] Breuer,[10] and Hauser.[75]

Dispersed Polymerization Systems. Polymerizations conducted as aqueous dispersions can be classified in three categories depending on the amount and to a lesser extent the type of dispersing agent employed. The designation of these as emulsifier-free, suspension, and emulsion polymerizations is somewhat ambiguous. Suspension polymerizations start as a dispersion of liquid in liquid and in a strict sense can only be considered suspensions if the final product contains discrete solid particles. Emulsion polymerizations are characterized by high concentrations of emulsifying agents. The final particles may be solid or liquid.

Emulsifier-free systems are as yet of limited practical utility, but they may be very useful for demonstrating phenomena of fundamental significance. As an example, the initiation of polymerization of acrylonitrile was shown to take place in the aqueous phase by allowing the monomer to stand quietly over a dilute solution of potassium persulfate at room temperature.[56] Others have used such systems at various times.[5] Whitby and co-workers have shown that, by using a small proportion of a water-soluble monomer along with a major proportion of water-insoluble monomers, it is possible to obtain stable latexes in emulsifier-free systems.[213, 214, 220]

Of greater utility, although not hitherto in the field of synthetic rubber, are polymerizations conducted in suspension. These were described in detail by Trommsdorff[202] and more recently by Hohenstein and Mark,[79] who have emphasized the more fundamental aspects of the subject. Suspension polymerization requires in general minimal amounts of emulsifying or dispersing agents. The process is essentially a bulk reaction carried out in a finely subdivided state. The monomer or mixture of monomers is dispersed in water by mechanical agitation in the presence of a catalyst soluble in the organic phase and of a suspension stabilizer to keep the globules dispersed during reaction. The final size of the particles or "pearls" is determined by the several reaction variables. In general, suspension polymerization may be employed when a product with a minimum of contamination from the emulsifying agent is desired.

Emulsion polymerization requires what by contrast at first appears to be an excessive amount of emulsifying agent. In general, the rate of polymerization increases with the emulsifier content, and the emulsifier must be a solubilizing agent.[56] The path followed by the reaction is very different from that of a suspension polymerization. With the emulsion system, polymerization is initiated in the micelles; the polymer particles are much smaller than, and grow independently of, the monomer droplets, which serve as reactant reservoirs, and at about 60 per cent conversion the monomer droplets disappear as a separate phase.

Depending on circumstances, the differences between these three types of aqueous dispersions may become indistinct. Thus surface-active agents might be generated in the emulsifier-free systems. An example of such would be afforded by the formation of 2-phenylethanesulfonate by a bisulfite-initiated polymerization of styrene.[95] It is also conceivable that conditions could be so adjusted as to bring about simultaneous polymerizations by both the suspension and emulsion mechanisms.

Against this background the advantages of emulsion polymerization can be specifically enumerated. First and foremost, emulsion polymerization provides a convenient method of dissipating the heat of polymerization. The same applies also to suspension systems. Mass polymerizations, however, must be conducted slowly, as otherwise overheating, due to low heat conductivity, leads to the formation of low molecular weight and at times porous and even charred products. Emulsion systems can be polymerized at very high rates with steady, rapid removal of the heat of reaction. Further, emulsion polymerizations are faster and lead to products of higher molecular weight than other methods of polymerization. One reason is undoubtedly the catalytic effect exerted by soap micelles on the rate of chain initiation. Another factor, pointed out by Hohenstein and Mark,[79] is the decreased accessibility of the growing chains to chain terminators and chain-transfer agents.

Emulsion Polymerization Recipes. The results of emulsion polymerization research can be concisely summarized in the form of a recipe. Such a formulation is convenient for transmittal and classification of information or as a basis on which systematic research can be conducted. Moreover, the number of emulsion polymerization recipes has become so great that their discussion may lead to an almost endless repetition of details which are not pertinent to the matter under consideration. Therefore, it is proposed to introduce a general recipe, based on the Mutual recipe,[46] in terms of seven functional items so chosen that specific recipes can be concisely formulated in terms of their distinctive components.

General Recipe

I. Monomers:	100
(*a*) Diene (butadiene)	75 (70-80)
(*b*) Comonomer (styrene)	25 (20-30)
II. Dispersion medium:	
(*a*) Water	180
(*b*) Antifreeze	
III. Emulsifier:	
(*a*) Soap flakes	5.0
IV. Modifier	0.10-1.0
V. Initiator:	
(*a*) Oxidant	
(*b*) Reductant	
(*c*) Catalyst	
VI. Electrolyte	
VII. Additive	
VIII. Performance	
(*a*) Temperature, °C.	
(*b*) Time, hours	
(*c*) Conversion, %	

In accordance with adopted practice, the total monomer charged, item I, is arbitrarily set at 100 parts by weight. All other constituents, unless otherwise specified, must then be given in the same units, i.e., pounds or grams. Although a 75/25 mixture of monomers is specified in the General recipe, probably more investigations have been conducted with a 70/30 ratio. Only very slight differences in behavior and properties result from this change, and, in the case of the butadiene-styrene system, all work with initial comonomer ratios between 70/30 and 80/20 will be classified together without reference to the absolute value of the ratio.

The dispersion medium, item II, may vary from 100 parts or less to 250, with 180 parts being considered a practical mean which has been used in a great number of investigations and in commercial production. This item, plus the soluble soaps, initiators, and catalysts included in the subsequent items, comprises the aqueous phase.

Although the emulsifier content, item III, may vary, five parts is commonly used. Mixed soaps, such as a 3.5-to-1.5 mixture of potassium disproportionated resinate and potassium fatty acid soap, are frequently advantageous. The pH of the soap solution is an important variable and, when known, should be included in the recipe. It may be expressed as such or, in the case of fatty acids, indirectly as the percentage neutralization. Sometimes the amount of sodium or potassium hydroxide added in excess to a completely neutralized soap is specified. It is the consensus of opinion that polymerization rates obtained when the system is adjusted by the latter two methods can be more readily duplicated than by direct adjustment of the initial pH of the soap solution with the aid of a meter.

An important ingredient of the polymerization recipe is the modifier, item IV, or chain-transfer agent, sometimes called the regulator. In most cases small amounts of an aliphatic mercaptan are employed. The function of this ingredient is to control the average molecular weight, which determines to a great extent the processibility of the elastomeric product.

Item V, the initiator, is frequently termed the polymerization catalyst. However, since it is consumed during the reaction, initiator appears to be a better designation. In the case of redox initiator systems, the description frequently becomes complicated, because of the inclusion of both oxidant and reductant initiators and oxidation-reduction catalysts. A redox initiator can be defined as an oxidation-reduction system which generates free radicals. It may or may not contain complexing agents and variable-valence metallic ions added to increase the rate of polymerization. The amounts of initiator ingredients employed are sometimes given in terms of millimoles per 100 grams of monomers. This method of formulation has led to the discovery of interesting stoichiometric relationships.

Electrolytes, item VI, must be present to prevent gelation of the latex. Generally the amounts required are small, viz., 0.1 to 0.5 part. Earlier proposals to include buffer systems[216] in emulsion polymerizations have not materialized for the reason that salts in the high concentrations required bring about flocculation of the latex. Furthermore, the soaps and cationic emulsifiers employed are themselves buffering agents. However, use is frequently made of sodium and potassium orthophosphates to increase the pH of alkaline systems.

Under additives, item VII, can be included a great variety of substances not normally present in a polymerization system. For an example, variable amounts of albumin might be included in order to study its effect on pre-coagulation of the latex.

Item VIII, performance, is, with the exception of temperature, an addendum to the recipe which is included for the purpose of comparison between recipes. This item can be expanded to include a considerable number of variables and properties.

Certain of the ingredients of the recipe may exercise a dual function. Thus in the Mutual recipe, which follows, lorol mercaptan is both a modifier, item

IV, and a reductant initiator, item V*b*. The potassium persulfate functions as an oxidant, item V*a*, and as an electrolyte, item VI. Many investigators consider it advisable for each ingredient to exercise a single function, as more satisfactory control is thereby obtainable. However this desideratum is not always possible.

Mutual Recipe

As Adopted by the Technical Advisory Committee of Rubber Reserve,[26, 46]
March 26, 1942

Butadiene, parts	75
Styrene, parts	25
Water, parts	180
Soap, parts	5.0
Lorol mercaptan, parts	0.50
Potassium persulfate, parts	0.30
Temperature, °C	50
Time, hours	12
Conversion, %	75

The abbreviated formulation of the Mutual recipe in terms of the General recipe given on page 226 follows. Any item not specified in the general formulation or in the abbreviated formulation is assumed to be zero.

Mutual Recipe

Abbreviated Formulation

IV, V*b*.	Lorol mercaptan	0.50
V*a*, VI.	$K_2S_2O_8$	0.30
VIII.	*a.* 50° C.; *b.* 12 hr.; *c.* 75%	

Certain terms, commonly used in describing emulsion polymerizations, require definition: *Inhibitor*: A material which when present prevents polymerization for a certain length of time, thus giving rise to an induction period. Oxygen is an inhibitor of the Mutual recipe. This is indicated by curve *B* of Fig. 1. *Retarder*: A material which slows down the reaction (curve *C*). *Shortstop*: A material which when added terminates the reaction as, for example, hydroquinone (curve *D*). *Promoter*: A term used by Carlin[14] for the reductant initiator. Promoter or activator is loosely used to designate any additive which increases the rate of polymerization. *Activator solution*: An aqueous solution containing initiator ingredients. When it is added to the system as the last ingredient of the recipe, polymerization starts, provided no inhibitors are present.

Before terminating this consideration of the mode of recording polymerization recipes, it is desirable to state the limitations of such records. They convey no information regarding the manner or order in which the system is best put together; yet these are frequently determining factors. Further, the recipes convey no information as to whether the reaction is considered batchwise, continuing, or semicontinuing. Consequently detailed instructions are required to accompany a research recipe when submitted for pilot plant or production tests. In addition, all ingredients are assumed to be as highly purified as possible. Impurities frequently cause considerable trouble and disappointment. However, with the high-grade polymerization ingredients now available this difficulty has been greatly alleviated.

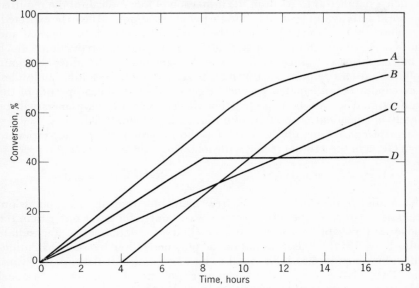

FIG. 1. Schematic Reaction Curves in the Mutual Recipe

A. Normal polymerization   C. Retarded polymerization

B. Inhibited polymerization   D. Short-stopped at 8 hours

THE MECHANISM OF EMULSION POLYMERIZATION

General Considerations. Logic requires that a polymerization reaction exhibit a beginning, a progression, and an end. The technical terms for these stages are initiation, chain propagation, and termination. Polymerization phenomena have been considered in these terms by many investigators, among whom may be mentioned Mark[130] and Flory[38]. Kharasch and Nudenberg[94] have applied these concepts to the preparation of GR-S, the product obtained by conducting the Mutual recipe.

Acceptance of a free-radical-chain mechanism to account for polymerization in emulsion appears at the present time to be unanimous. Although incontrovertible proof of it is certainly impossible, this mechanism is highly useful in explaining phenomena of abnormal temperature coefficients, initiation, propagation, chain transfer as revealed by modification, and effects of trace catalysts and inhibitors. Copolymerization in a polyphase system introduces complexities that are absent in most fundamental investigations. These will be considered in a general manner here and in greater detail later.

Initiation by Free Radicals. The production of free radicals, i.e., fragments containing tervalent carbon, bivalent nitrogen, or monovalent oxygen, is endothermic to a greater or lesser degree, depending on the extent of resonance stabilization of the radical. Such radicals usually exhibit very short half-life periods, but again this is influenced by resonance. Any process by which energy can be supplied to suitable energy acceptor molecules in such a manner as to result in the formation of free radicals of the required degree of activity can be used to initiate polymerizations. It is to be expected, therefore, that polymerizations can be initiated by reactions of thermal and catalytic decomposition, photoactivation, and oxidation–reduction. Such is indeed the case. Since the free radical fragments which induce the chain polymerization comprise such a small part of the total resulting molecules, they cannot be anticipated to exert a great effect on the properties of the product, and it is to be expected that the properties of elastomers will be largely independent of the type of initiating agent employed.

The reaction generally assumed for the initiation of polymerization to form GR-S in the Mutual recipe is the following oxidation-reduction process:

$$K_2S_2O_8 + 2C_{12}H_{25}SH \longrightarrow 2KHSO_4 + 2C_{12}H_{25}S^{\cdot}$$

This leads to the formation of free mercapto radicals and potassium bisulfate. However the initial process is certainly of a lower order than the above and possibly involves the momentary formation of a free sulfate radical ion (SO_4^{-}). The initiation of polymerization by such oxidation-reduction reactions has been found to be so general that Baxendale, Evans, and Park[5] have proposed to consider polymerizations as free-radical indicators. Considerable experience with the vagaries of polymerization indicates that, although positive results establish the presence of free radicals, negative results do not necessarily prove their absence.

The importance of initiation cannot be overemphasized. Early in the Government-sponsored research program, Carlin[14] came to the conclusion that the initiation step is the rate-determining step. Consequently recent investigations which have led to the formulation of active low-temperature recipes have been predominantly concerned with the development of new and more potent initiator systems.

Chain Propagation. The first step in the creation of a polymer molecule is the reaction between a free radical and a monomer molecule.

$$C_{12}H_{25}S^{\cdot} + C_6H_5CH : CH_2 \longrightarrow C_6H_5\overset{\cdot}{C}HCH_2SC_{12}H_{25}$$

The product of this reaction is a new free radical, which in turn reacts with another monomer molecule, thus (disregarding the mercaptan fragment) giving in succession dimer, trimer, etc. chains, up to the final high-molecular-weight elastomeric product. Experiments conducted on the termination and re-initiation of polymerization indicate that the growth of the chains is, despite their length, a fast process. Swain and Bartlett[197] have shown that the average lifetime of growing chains in the homogenous photopolymerization of vinyl acetate is about one second. In the example given above, the dodecyl mercaptan free radical is depicted as reacting with styrene; its initial reaction with butadiene would presumably be almost as probable.

Neglecting for the time the matter of molecular weight, consider briefly the variations of structure and composition inherent in the formation of the elastomer GR-S. Butadiene can polymerize in two ways, giving 1,2- and 1,4-structures represented, respectively, as follows:

$$—CH_2 \cdot CH— \atop \underset{CH : CH_2}{|} \qquad \text{and} \quad —CH_2 \cdot CH : CH \cdot CH_2—$$

With more complex dienes, such as isoprene, 3,4-addition also occurs. After the original addition, free radical attack at the double bond or, possibly, at the methylenic groups alpha to them, or, again, possibly at the vinyl side groups, may produce cross-linked or branched structures. Further, the double bond resulting from conjugated (1,4-) reaction may exist in either *cis* or *trans* form. That these forms have different effects on physical properties is evident from the difference between the two polyisoprenes, natural rubber (*cis*) and gutta-percha (*trans*). In addition, the residual unsaturation can saturate itself by reaction with similar unsaturated bonds in adjacent molecules, thereby producing cross-linked structures. Finally, poorly defined phenomena of cyclization are a possibility. Thus polybutadiene can exhibit the following seven structural variants; 1,2-addition, 1,4-*cis* and 1,4-*trans* addition, linear, branched, and cross-linked chains, and possibly cyclization. When styrene is present, additional complications ensue. The amount of styrene entering the growing chain depends on its specific reactivity and its concentration in the growing polymer-monomer particle, which latter in turn is determined by the total amount present, i.e., the initial comonomer ratio, and the rate of transfer from the comonomer reservoir particles to the polymer-monomer particles. The styrene units may enter the growing chain in clumped groups or be rather evenly distributed. Mayo and Walling[138, 211] have proposed two extreme types of copolymerization, the "ideal" and the "alternating." The majority of copolymerizations exhibit features of both to a greater or lesser extent. With the ideal type, the relative reactivities of the two monomers are the same, and a random copolymer is formed. The butadiene-styrene system appears to undergo random addition rather than alternation. However, whatever may in fact be the exact order of succession of butadiene and styrene units in the copolymer, there can be no doubt regarding the beneficial effects that result from the addition of styrene to butadiene, though the fundamental reason for this remains unknown. It has been suggested that the comonomer exerts a directive influence on subsequent addition affecting either the *cis-trans* arrangement or the tendency toward branching.

The contributions of the free-radical theory to an understanding of chain propagation in the polymerization and copolymerization of dienes have been outlined by Kharasch and Nudenberg.[94] The addition of a free radical to styrene results in a rather unreactive type of free radical stabilized by resonance.

$$C_6H_5CH : CH_2 + X^{\cdot} \longrightarrow C_6H_5\overset{\cdot}{C}HCH_2X$$

Consequently the product does not react with soap and other inert materials present. Likewise butadiene reacts to form a pair of resonance-stabilized hybrids.

$$CH_2 : CHCH : CH_2 + X \cdot \Big\langle \begin{array}{l} XCH_2\dot{C}HCH_2 : CH_2 \\[4pt] XCH_2CH : CH_2C\dot{H}_2 \end{array}$$

These butadiene free radicals are somewhat more reactive than those derived from styrene. The new free radicals from styrene can react in one or more of several possible ways:

(1) $C_6H_5\dot{C}HCH_2X + RSH$ $\longrightarrow C_6H_5CH_2CH_2X + RS\cdot$

(2) $C_6H_5\dot{C}HCH_2X + C_6H_5CH : CH_2 \longrightarrow C_6H_5\dot{C}HCH_2CH(C_6H_5)$
$$\hspace{9cm} CH_2X$$

(3) $2C_6H_5\dot{C}HCH_2X$ $\longrightarrow C_6H_5CH_2CH_2X$
$$\hspace{6cm} + C_6H_5CH : CHX$$

(4) $2C_6H_5\dot{C}HCH_2X$ $\longrightarrow C_6H_5CHCH_2X$
$$\hspace{6.5cm} | $$
$$\hspace{6cm} C_6H_5CHCH_2X$$

Similar equations may be written for radicals derived from butadiene. Of the four possible reactions that the initial activated-monomer radical might undergo, (2) is most probable. Experimental support for this view was obtained by examining the nature of the reaction products from equimolar mixtures of mercaptans and the monomers in the presence of about 1 mole per cent of ascaridole.

Flory has postulated that all additions to butadiene take place at a terminal carbon and that resonance of the resulting free radical between two forms leads to 1,4- or 1,2-addition.[39] The similarity of the propagation reaction and the abnormal addition of hydrogen bromide to olefins led Sivertz and his co-workers to postulate the addition of a free radical to either a terminal or a central carbon of the butadiene molecule.[124] Later, however, Longfield, Jones, and Sivertz found that the initial attack of n-butyl mercaptan on butadiene in the presence of persulfate led exclusively to the formation of crotyl n-butyl thioether.[125] Thus only the terminal carbon atom of butadiene is susceptible to attack by free radicals. These authors were forced to the conclusion that the initiation step and the propagation step do not necessarily proceed by identical mechanisms and that resonance does determine the relative amounts of 1,4- and 1,2-addition reactions during propagation of the chains.

Chain propagation is of practical importance because the final properties of the product are determined by what occurs during this process. The outstanding excellence of emulsion-polymerized products can be attributed in no small measure to segregation from outside influences of the growing chain in the polymer-monomer particles. At the same time this makes it difficult to change the polymerization environment in such a way as to secure improvement.

Chain Termination. At least five possibilities for stopping chain growth are recognized. Of greatest importance are the processes of mutual termination and chain transfer. However the possibility of reactions of the growing chains with free radicals derived from the initiator system should not be overlooked. Disproportionation and reactions that lead to the production of unreactive, resonance-stabilized free radicals[169] must also be considered.

It is common experience that increasing the initiator content increases the rate of polymerization through a maximum. At low initiator concentrations certain relationships of theoretical significance may hold, such as the $\frac{1}{2}$ power expression shown by Kolthoff and Dale[103] for the influence of potassium persulfate on the rate of polymerization of styrene. At higher initiator concentrations, however, the decrease in rate can be attributed to reactions of the following type, where $R^{\cdot}$ is a growing polymer chain.

$$R^{\cdot} + {}^{\cdot}SC_{12}H_{25} \longrightarrow RSC_{12}H_{25} \tag{1}$$

Fortunately in emulsion systems the growing chains are segregated from the initiator free radicals in the aqueous phase.

The reactions leading to the formation of unreactive resonance-stabilized free radicals are of importance in connection with the use of shortstops or reaction terminators, which will be described later.

Mutual termination or coupling of polymer free radicals takes place when two free-radical chains react with each other. The resemblance to dimerization is evident.

$$R_1^{\cdot} + R_2^{\cdot} \longrightarrow R_1R_2 \tag{2}$$

Chain transfer takes place in accord with the following equation:

$$R^{\cdot} + C_{12}H_{25}SH \longrightarrow RH + C_{12}H_{25}S^{\cdot} \tag{3}$$

In the absence of mercaptan modifiers, reaction 2, leading to the formation of high-molecular-weight products, is undoubtedly the most prevalent mode of termination. Reaction 3 indicates that, if modifiers are not involved in initiation, their presence will not change the rate of polymerization but will, however, greatly influence the molecular weight and hence the plastic properties of the product. Thus chain-transfer agents are employed to obtain synthetic elastomers which can be readily processed in the manufacture of vulcanized products.

Disproportionation is formulated as follows:

$$2RCH_2\overset{\cdot}{C}H(C_6H_5) \longrightarrow RCH:CHC_6H_5 + RCH_2CH_2C_6H_5 \tag{4}$$

Because of the high heat of activation, disproportionation does not appear to be of much significance as a mechanism for chain termination.

The Locus of Reaction. Early investigators naturally assumed that the polymerization was initiated at the water-monomer interface and that the chain grew into the monomer phase. According to Barron,[4] Staudinger specifically proposed this idea. However, experimentation conducted in several laboratories yielded results that could not be reconciled with this simple theory.   Fikentscher[35] was the first to suggest in a publication that emulsion polymerizations took place in the aqueous phase.   A few years later Fryling and Harrington,[56] after a discussion with J. W. McBain of their experimental data on the emulsion copolymerization of butadiene and acrylonitrile, suggested that the initiation occurred in soap micelles and pointed out that different mechanisms would be required for suspension and emulsion polymerization systems.   At the same time Harkins was working on the problem, and soon he began to contribute extensively to the subject.   There can be no doubt that his work will constitute a classic contribution both to the theory of colloid chemistry and to emulsion polymerization.   Since it is impossible to do justice to the work in the space allotted to this review, it is fortunate that a considerable portion of it has been published in current periodicals.[70-1]   The theory of reaction locus presented here in abbreviated form is based on summaries prepared by Harkins.[69]

Harkin's conclusions have been derived from a wealth of experimental evidence and highly accurate measurements.   A method of determining X-ray diffraction values at extremely small angles was developed.   This resulted in determinations of micellar magnitudes of soap systems, alone and in the presence of solubilized hydrocarbons.   Methods of determining the critical concentrations at which emulsifying agents exhibit the formation of micelles were also developed.

Four possible loci of reaction are recognized by Harkins.   Two of these, viz., the soap micelles, where the formation of polymer particles occurs, and the polymer-monomer particles, where growth of the polymer takes place, are of outstanding importance.   Initiation and propagation in aqueous solution, as distinguished from micelles,[227] and within or on the surface of dispersed monomer droplets do not result in the formation of significant amounts of product.   The various loci are illustrated diagrammatically in Fig. 2.

In emulsion polymerization soap functions in two ways.   It stabilizes the monomer and polymer-monomer particles, giving electric charges to all of the particles.   In addition, soap micelles solubilize a small amount of monomer, thereby increasing in principal dimension from about 40 to approximately 50 A. and providing a larger target than a single monomer molecule with which a free radical can contact by diffusion.   With relatively insoluble monomers, most of the polymer nuclei are formed within the micelles.   In the Mutual recipe there are approximately 10^{18} micelles per milliliter of aqueous phase at the start of the process and the ultimate number of polymer particles is of the order of 10^{15} per milliliter.   Harkins calculates that only about one in 700 micelles initiates a latex particle.   The polymer molecule acquires a certain amount of monomer by diffusion from the monomer phase and a surface layer of oriented soap; it thus becomes a polymer-monomer particle.   At about 13 per cent conversion the uptake of soap by polymer-monomer particles is so great that all of the micelles have disappeared.   Therefore quantitatively the most important locus of reaction is the aggregate of polymer-monomer particles wherein more than 83 per cent of the total polymerization process takes place.   That polymerization in the monomer phase (the droplets) contributes little to the yield is evident from centrifugation of the latex. The monomers separate as a clear polymer-free phase which gradually decreases in volume and disappears at from 50 to 60 per cent conversion. The monomer droplets serve merely as a reservoir to supply monomer first to the micelles and then to the polymer-monomer particles. The reason for the disparity in conversion between these two loci can be attributed largely to the ratio of their relative surfaces. At an indeterminate conversion, higher than 13 per cent, such that, as per Fig. 2, half of the initial monomer phase can be considered as in the form of

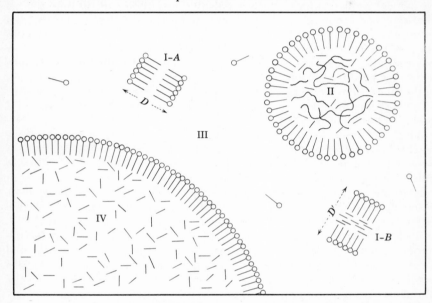

FIG. 2. Diagrammatic Representation of Various Loci of Polymerization in Emulsion

Locus I-*A*. Soap micelle. $d =$ ca. 40 A.
Locus I-*B*. Soap micelle with solubilized monomer. $d' =$ ca. 50 A.
Locus II. Polymer-monomer particle. $r =$ ca. 110 A.
Locus III. Aqueous phase.
Locus IV. Dispersed monomer particle. $r =$ ca. 5,000 A.

droplets of monomer phase with a radius of 5,000 A. and such that the polymer-monomer particles have radii of 110 A., the ratio of total polymer-monomer surface to total monomer-phase surface is approximately 46.

Harkins emphasizes the point that so little polymer is produced outside the polymer-monomer particles that the properties of the polymer depend almost entirely on what occurs in these particles. At a 10 per cent yield the polymer-monomer particle contains 2.5 times more monomer than polymer. This proportion varies with the initial soap concentration and with conversion. At high conversions there is very little monomer present. Experiments conducted with styrene show that about one-half of all the polymer-monomer particles contain a growing chain at any given time during the process. A completed polymer particle with a diameter of 400 A. contains approximately 250 polymer molecules of average molecular weight, 150,000. It is evident that any given polymer-monomer particle acquires a great number of free radicals from the initiator system during the process of polymerization.

Little need be said regarding the polymer molecules generated in loci III and IV because of their relatively small number. Those initiated in the aqueous phase acquire both soap and monomer by diffusion and become indistinguishable from the other monomer-polymer particles. The large monomer droplets which acquire polymer lose monomer as the process continues and eventually become almost identical with the particles initiated in micelles. Since it cannot be expected that the modifier concentration in the large monomer droplets will be the same as in the polymer-monomer particles, the rubber produced in this locus should differ in molecular weight from that produced in the polymer-monomer particles. In consideration of the fact that in most experimental techniques the mercaptan is charged in solution in the monomer phase, such rubber as may be formed in locus IV probably has a lower-than-average molecular weight.

The number of polymer-monomer particles per milliliter of aqueous phase reaches a value of about 2.8×10^{15} at 20 per cent conversion in GR-S latex and remains quite constant up to 80 per cent conversion. Certain other systems show a decrease in the number of particles per milliliter at conversions above 60 per cent, while in some cases the number has been observed to increase slightly after all micelles have disappeared. This latter case can be explained by initiation in locus III, the aqueous phase.

Based on the theory that micelles constitute the principal locus for the initiation of polymer particles (this should not be confused with the principal locus of polymerization, viz., the polymer-monomer particles), Harkins predicted that the rate of polymerization should be proportional to the initial amount of micellar soap. For example, in one experiment a conversion of 4.4 per cent per hour was obtained. When the aqueous phase was increased four times and the soap only twice, to yield half the concentration of soap in four times the volume of water, the conversion was 9.0 per cent per hour, or twice as great. So important is this generalization, which has been observed by a number of other investigators, that it might serve as a criterion of emulsion polymerization.

On the basis of Harkins' theory of reaction locus, Reynolds[171, 173] deduced that low-molecular-weight soaps, such as sodium caprylate, would behave as satisfactory emulsifying agents for polymerization systems, provided sufficient electrolyte were present to bring about the formation of the micellar state. This deduction was verified, and it established one of the positive electrolyte effects which have played an important role in the development of highly active recipes.[60] A concise summary of the most pertinent features of emulsion polymerization has been presented by Reynolds.[172]

When soaps of rosin acids are used as emulsifiers, the rate of polymerization is considerably less than when soaps of fatty acids are used. This is especially true in polymerizations at subfreezing temperatures, which require the inclusion of an antifreeze component. The antifreeze most generally employed is methanol. It has been reported by Howland and co-workers that, contrary to its relatively small effect with fatty acid soaps, methanol raises greatly the critical micelle concentration of rosin acid soaps and hence reduces the effectiveness of the soaps in promoting polymerization.[82a]   These authors find that replacement of methanol in part by ethylene glycol markedly improves the rate of polymerization when a rosin acid soap is the emulsifier.

The Kinetics of Emulsion Polymerizations.   The difficulties involved in deducing a mathematical expression for the rate of emulsion polymerization should now be evident.   Such an expression would be required to take account of the generation of free radicals, since this is basically the rate-determining step, and the efficiency of transfer of free-radical characteristics to the monomers.   Other factors which influence the rate are the type and amount of emulsifier, the relative surface of polymer-monomer particles compared with the monomer droplets, the rates of diffusion of the various reactants since these determine the concentrations at the locus of reaction, the rates of chain termination and transfer, the specific reactivities of the monomers, and the influence of various inhibitors.   Knowledge of the heats of activation of several simultaneous processes would likewise be required. It is not surprising, therefore, that no one has managed to solve this baffling problem completely.   However, many experimental studies bearing on the problem have been conducted,[14, 25, 96, 140, 146, 168] and several attempts have been made to treat its theory.   One of the more successful efforts was that of Corrin,[19] who simplified the problem by considering only styrene and its reaction in locus II, the polymer-monomer particles, during the time of coexistence of a separate monomer phase.   Under these conditions he obtained agreement between experiment and theory when the reaction in the monomer-polymer particle is considered three-halves order with respect to the monomer.   The concentration of monomer in the polymer particles when a separate monomer phase exists is correctly predicted, and the energy of activation of styrene is calculated as 22,600 calories per mole.

A comprehensive treatment of reaction kinetics based on Harkins' theory of locus is that of W. V. Smith and Ewart.[190–1]   Here again the system was simplified by confining the treatment to styrene.   High soap and persulfate concentrations and high temperature favor the production of a large number of polymer-monomer particles.   The number of particles was given by the expression:

$$\mathcal{N} = k(\rho/\mu)^{2/5} \, (a_s S)^{3/5}$$

where $\mathcal{N}$ is the number of polymer-monomer particles formed per milliliter of water up to the time of complete absorption of soap on particles (13 to 20 per cent conversion), k is a numerical constant with values between 0.37 and 0.53, ρ is the rate of formation of free radicals per milliliter of water, μ is the rate of increase in volume of a polymerizing polymer-monomer particle, a_s is the interfacial area occupied by one gram of absorbed soap when micelles are present, and S is the initial concentration of the soap in the aqueous phase.   A number of such theoretical efforts is considered in the review by Hohenstein and Mark.[79]

Recently, M. Morton, by considering the rate of polymerization in the interval between the time at which new particles cease to be formed and micellar soap is no longer present and the time at which the free monomer phase disappears, i.e., in the interval during which no new particles are generated and the existing particles are in equilibrium with free monomer, has been successful in deriving an absolute rate for the propagation reaction in the emulsion polymerization of butadiene.[153]

LABORATORY POLYMERIZATION TECHNIQUES

The experimental techniques employed in conducting emulsion polymerizations are of great importance. When polymerization ingredients were difficult to obtain, small-scale sealed-tube techniques, such as described by Fryling,[50] were of considerable utility. However, inability to prevent inhibition by atmospheric oxygen sealed within the tubes proved to be a serious limitation to this technique. Despite this, some interesting work was performed in sealed tubes. The correlation between decrease in volume and percentage conversion of a single monomer proved to be more accurate than was at first evident. Oxygen inhibition could be avoided by filling the tubes from high-vacuum systems, and highly purified fatty acid soaps were readily prepared in situ. Much of the precise fundamental research of Harkins, Kharasch, and others was conducted by similar dilatometric techniques. One difficulty which could not be overcome was that filling from high-vacuum systems is always laborious and time-consuming.

Stirred glass reactor techniques have been used at various times. In these a stream of purified nitrogen is passed continuously through the system, and escape of reactants is prevented by a reflux condenser. It is evident that such a technique cannot be employed with butadiene at and above $-3°$ C. at atmospheric pressure, but satisfactory operation has been secured with isoprene. Corrin[18] has described a stirred, sealed dilatometric technique in which the change of volume is read on a calibrated side arm. Such techniques are feasible with butadiene when subfreezing temperatures are employed, and Uraneck[206] has employed a stirred reactor for butadiene in which one or more of the reaction components, such as modifier or initiator, can be added in very small concentrations, either continuously or intermittently, over considerable periods of time. All such techniques, however, suffer the disadvantage that a number of bottle experiments can be conducted in the time required for one experiment in a stirred reactor.

The bottle technique was introduced to the rubber program with the Mutual recipe in December 1941. It was an adaptation of the procedure developed in Germany. At a time when it was considered possible that any experimental variant might conceivably lead to a radical improvement in the quality of the product, large samples of rubber were desired for evaluation after compounding. By that time butadiene, styrene, and acrylonitrile could be readily obtained in any desired amount, and the practice of conducting emulsion polymerizations in 32-oz. crown-capped bottles containing up to 200 grams of monomers was widely adopted. However, this advantage alone would not have led to the complete abandonment of the simple sealed-tube technique, which, in conjunction with Garvey's[65] microtesting procedure, is considerably faster and more economical than the bottle technique, had it not been for the discovery of oxygen inhibition. This can be readily overcome in bottles by the simple expedient of adding a slight excess of butadiene and allowing the excess to evaporate, thereby removing sufficient oxygen from the system to reduce the induction period to a negligible value.

The bottle technique, as originally introduced, had several objectionable features. A fire hazard was created by handling large amounts of volatile hydrocarbon, particularly during the venting process. The high pressure generated by butadiene at 50° C., 60 lb. gage pressure and above, resulted in rupture of defective bottles, and various methods of enclosing the bottles in metal sleeves were devised. No method was at first available for adding or subtracting reaction ingredients, and, in order to carry the reaction to a definite conversion, the use of several bottles that had to be opened at successive partial conversions was required. In addition the large constant-temperature water baths required were both expensive to construct and difficult to maintain.

Houston[80] described a procedure which made use of self-sealing gaskets and a hypodermic syringe equipped with a safety guard to prevent accidental expulsion of the piston. The progress of a polymerization could be followed by removing samples and determining their total solids content. Harrison and Meincke[74] developed a similar method by which an instantaneous determination of conversion could be obtained by inserting a hypodermic needle attached to a pressure gage through a self-sealing gasket. This latter method, though rapid, requires a separate calibration for each individual temperature and monomeric system, and the gage must be maintained at the temperature of the polymerization reaction. Figure 3 illustrates data obtained by Houston's total solids method in regard to oxygen inhibition with the Mutual recipe. Harrison and Meincke's pressure-conversion relation is illustrated by Fig. 4.

It gradually became evident that large samples of rubber were not usually required, and a trend toward the use of smaller bottles developed. Safety, convenience, and the desirability of using standardized samples of reaction ingredients over considerable periods of time were responsible for this reversion of technique. Medalia[141] described the use of four-oz. screw-capped bottles with self-sealing gaskets and a device for removing samples periodically with a hypodermic needle.

By way of example, a recent method of conducting emulsion polymerization reactions is described in some detail. Although the procedure is that used with low-temperature, redox-initiated recipes, in most cases the changes required for other types of recipes should usually be readily evident.

Butadiene is bled from a 28-gallon cylinder, passed through a short packed column with reflux induced by a Dry Ice-isopropanol mixture, and condensed by surrounding the glass receiver by a similar cooling mixture. To avoid a measurable retardation frequently observed when original cylinders of butadiene are opened, the first two or three liters of condensate should be discarded. Styrene is distilled through a 20-in. Vigreux column at 10 mm. pressure, to remove inhibitor. A very small stream of air passed into the

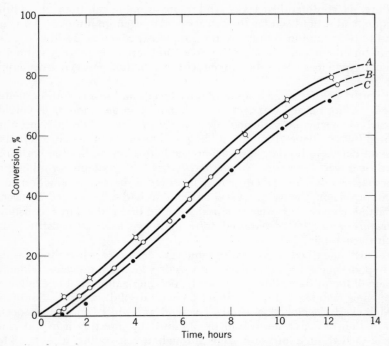

FIG. 3.   Effect of Oxygen on Polymerization Rate of Butadiene-Styrene in the
Mutual Recipe
   A.   Control (no oxygen)
   B.   25 ml. of oxygen per 100 grams of monomers
   C.   50 ml. of oxygen per 100 grams of monomers

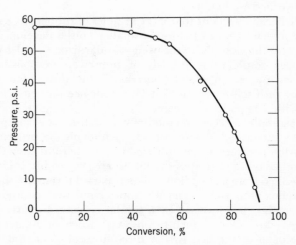

FIG. 4.   Pressure-Conversion Relation in Emulsion Polymerization

boiling flask prevents bumping and inhibits bulk polymerization. Both styrene and butadiene can be readily stored in crown-cap bottles at $-20°$ C., but it is considered preferable to use the butadiene shortly after distillation.

Beverage bottles of 6-oz. capacity are favored. These are washed in a bottle-washing machine with a good grade of washing compound and rinsed several times with distilled water. It is more economical to discard the bottles after an experiment than to attempt to clean them with concentrated nitric acid, which is needed to remove the surface-adherent layer of rubber. Crown caps can be purchased without liners and with a 3/32-in.-diameter hole punched through the center. The self-sealing inserts and gaskets are cut by circular dies from sheets of Butyl and nitrile (Hycar) rubber, 2 mm. thick, compounded and cured in accordance with the formulas presented by Harrison and Meincke.[74] Before being used, the Hycar gaskets are extracted with acetone until the solvent is practically colorless. About 50 hours' extraction is required per batch of 200 gaskets. The soft Butyl rubber insert is centered over the hole on the inner side of the crown cap and retained in place with rubber cement. This is covered by the nitrile gasket, similarly held by rubber cement.

The soap for a series of bottle experiments is prepared in large beakers. In preparing soap solution, it is brought to a boil, care being exercised to prevent foaming over. If desired, a 20 per cent soap solution can be made in large volume for future use. Such soap solutions have been stored in tightly stoppered bottles for 6 months without any detectable decrease in activity. The required amount of soap solution is weighed into the bottles on a platform balance. Styrene, containing dissolved mercaptan and hydroperoxide in correct proportions, is added to the bottles from a burette. Methanol, if used as an antifreeze, is then added.

Since it is generally desirable to start the reaction in the morning, the bottles can be made up thus far on the previous day, then stoppered temporarily with tinfoil-covered corks, and stored in an icebox at $0°$ C. overnight. If this procedure is followed, fastest rates are obtained when, by refraining from shaking, the aqueous soap, styrene solution, and methanol are allowed to stand in the bottles as supernatant layers. For some unaccountable reason, certain redox systems so prepared react at rates approximately 10 per cent faster than systems prepared immediately before initiation of reaction. The bottles, brought to room temperature, are placed on a suitably tared torsion balance, and butadiene is added. From 1 to 2 grams excess is included, and the prepared caps are placed loosely on top of the bottles while venting takes place in a ventilated hood well removed from any possible source of ignition. When the bottles come to the desired weight by evaporation, the caps are quickly fastened.

Experiments to be conducted at subfreezing temperatures require pressuring so that samples can be removed with a syringe. Oxygen-free nitrogen to 30 lb. pressure is added through a hypodermic needle from a cylinder.

The bottles are rotated in a constant-temperature water-glycol reaction bath from 15 to 30 minutes before the final operation is performed. The reaction is started by the addition of either a definite amount of an aqueous activator solution or a polyamine. The activator solution consists of a mixture of water, potassium chloride, ferrous sulfate, and sodium pyro-phosphate, heated to 60° C. and then cooled to room temperature. Such a system is heterogeneous and requires thorough shaking before a sample is removed in a 10-ml. syringe equipped with a number-19 needle. Zero reaction time is taken as the moment the contents of the syringe are injected into the bottle.

The amounts of the various ingredients used in the above procedure can be calculated from one of the polymerization recipes presented on pages 263 and 268. Tetraethylenepentamine would also form a satisfactory reductant initiator to use with this procedure.

To follow the course of the polymerization by the "total solids" method, samples are removed in a 5-ml. syringe with a number-19 needle. The needle should be inserted, the bottle inverted, the contents of the bottle vigorously shaken, and, before phase separation can occur, a sample of approximately 2.5 ml. allowed to enter the syringe. The syringe cock should then be closed, the needle removed from the self-sealing gasket, and the total weight of the syringe and its contents determined on an analytical balance. The contents are then discharged into a weighed aluminum-foil moisture dish (58 mm. diameter, 15 mm. depth) containing 2 drops of a 1 per cent alcoholic solution of di-*tert*-butyl-hydroquinone to act as a shortstop. A weight correction of 1 mg. per drop of shortstop solution is applied. The empty syringe is weighed to obtain the weight of sample by difference. (After the syringe has been used, it should be immediately washed out with clean soap solution and filled with distilled water. With this care the syringe can be used for months if the plungers and cocks are periodically lubricated.) The sample is evaporated to dryness on a hot plate, cooled to room temperature, and weighed. The percentage conversion can then be calculated on the basis of the weight of solids contained in the recipe or of a blank obtained by withdrawing a sample of the bottle contents immediately after initiation.

When the reaction system in the bottle has reached the desired conversion, as indicated by the reaction curve constructed from the several conversion data obtained, the polymerization is stopped by the addition of 0.1 per cent of di-*tert*-butylhydroquinone based on the weight of monomers charged. After the bottles have been warmed to room temperature, unreacted buta-diene is vented with a hypodermic needle inserted through the crown cap. A foam suppressor might be a convenient additive at this point. The cap is then removed and the contents of the bottle are poured into a 1-liter stainless-steel beaker. Based on the yield of polymer, 2 per cent of phenyl-β-naph-thylamine is added to the latex. This is prepared as a 2 per cent solution in ethyl alcohol diluted with water to incipient precipitation just before addition. The latex is then stirred with a propeller at about 1700 r.p.m., operated by compressed air to avoid ignition of residual butadiene fumes, and concentrated brine is added to bring about flocculation. Then, depending on whether a fatty-acid-free or an acid-containing rubber is desired (in general, research technique favors the acid-free product), isopropanol or dilute sulfuric acid is added to obtain complete coagulation. The rubber is washed well with hot distilled or softened water, to remove soap and electrolytes, and is spread on stainless-steel screens and dried overnight at 60° C. in a circulating air oven. If the operations of coagulation, washing, and drying are carefully conducted, the weight of the product will agree to within 1 per cent with the conversion determined by the total solids method.

The resemblances of the bottle technique to large-scale procedure are evident. However, there are also certain limitations. The agitation is constant, being fixed by the size of bottles, the distance from the center of the shaft, and the rate of rotation. Temperatures cannot be increased temporarily to start a slowly reacting, inhibited system. On the other hand, the removal of unreacted monomers is quickly accomplished by venting, washing with hot water, and drying. Consequently it is never necessary to hold large batches of latex at high temperatures for considerable periods of time, and gelation or cross-linking of the rubber is thereby avoided. Bottle reactions conducted in accordance with low-temperature recipes are usually faster than large-scale reactions, but exact reproduction of the over-all rate of reaction has been attained with the Mutual recipe in both bottles and reactors.

Numerous variations of the foregoing procedure are possible. The order in which reactants are added to the bottles can be varied and various changes can be made in the preparation of raw materials and in the treatment of the product. Certain recipes will not work unless the polymerization system is formulated in a very exact manner both with respect to the preparation of the ingredients and the order in which they are put together.

THE MUTUAL RECIPE

The emergency of December 1941 dictated the pooling of information on the manufacture of tire-tread synthetic rubber by all parties involved. Three processes were submitted for consideration. Each exhibited certain advantages. At the time practical manufacturing experience was so meager that accurate evaluation of the numerous factors involved was difficult. The problem was further complicated by a shortage of materials of construction. Nevertheless, by March of 1942 all concerned were able to agree on the adoption of what was shortly thereafter named the Mutual recipe. The product name, GR-S, was an abbreviation for Government rubber containing styrene.

The Mutual recipe, presented on page 228, was based on prior German experience. Negotiations by which important relevant information was obtained by the Standard Oil Company are interestingly described by Howard.[82] However, credit for the inclusion of a modifier in the recipe should be given to the research organizations of Standard Oil Development Company and U.S. Rubber Company. The addition of lorol mercaptan made the process practical both in rate of conversion and quality of the product. It is somewhat strange that up to the conclusion of the war the Germans, who had discovered the function of modifying agents, continued to produce essentially unmodified rubber.

The Mutual recipe was adopted because all concerned could duplicate each other's results and because the product proved to be satisfactory as a replacement for natural rubber.   The principal disadvantage of the process as it appeared at the time of its adoption was that termination of reaction at 75 per cent conversion required the provision of large equipment for monomer recovery.   In retrospect this is seen to have been an advantage because it made possible the attainment of a highly uniform product.

Four years of research effort was devoted to the Mutual recipe after its adoption in 1942.   Despite this, little change was made in the procedure. The monomer ratio was changed to 72/28, principally to balance the recipe with respect to the production of monomers.   A slight improvement was effected by decreasing the conversion from 75 to 72 per cent.   The modifier content was found to depend on the degree of agitation, and it was possible to obtain a minor economy by decreasing the persulfate content from 0.30 to 0.25 part.   Various soaps, including those from disproportionated rosin acids, were substituted for sodium oleate, which had been first preferred. The recipe as originally formulated was almost perfectly balanced and therefore susceptible to only slight improvement.   Nevertheless the extensive research conducted on this recipe furnished a substantial foundation of knowledge of emulsion polymerization, it enlarged and refined the theory of the process, it resulted in the development of a convenient experimental technique, and it eventually prepared the way for the successful formulation of improved low-temperature procedures.

Comonomer Ratio.   A series of copolymers having widely differing properties can be prepared by varying the monomer charge ratio from that of pure butadiene to pure styrene.   This is accompanied by an increase in the rate of polymerization as the styrene content is increased.   At each end of the scale the properties of the polymer are progressively modified by increasing amounts of comonomer.   Throughout the whole range of compositions there is a continuous variation of properties, such as density, rebound, residual unsaturation, and freezing point.   It is convenient to consider arbitrarily four regions of composition.   These can be designated as the polybutadiene range with amounts of styrene up to 10 per cent, the optimum elastomer range varying from a butadiene-styrene comonomer ratio of 80/20 to one of 70/30, the subelastomer range from approximately 60/40 to 50/50, and the polystyrene range from 10/90 to 0/100.   The properties of products made with the Mutual recipe in the polybutadiene range are unpromising except when made at low conversions.   The subelastomers, because of their low residual unsaturation, are resistant toward oxidation, and they exhibit promise for uses in which high rebound and extensibility are not important, as, for example, cable coverings.   Most attention will be devoted to compositions in and adjacent to the optimum elastomer range.

W. C. Smith[189] observed that increasing the ratio of butadiene to styrene from 75/25 to 100/0 improved the rebound but reduced the stress-strain properties and plasticity, as shown in Table I.   He recommended charging an 80/20 mixture, thereby obtaining the advantages of a higher rebound without serious loss of tensile properties.   Zwicker[225] investigated copolymers with butadiene-styrene ratios varying from 75/25 to 50/50.   Increasing the styrene content led to improved plasticity, easier processing, and lower modifier requirements.   The flex life and hysteresis of compounded stocks increased greatly.   However, rebound, abrasion resistance, and rate of cure fell with increasing styrene contents.   When the copolymers were measured at 100° C., hysteresis and rebound were nearly independent of the composition.   This behavior correlates with the rubbery condition assumed by polystyrene at elevated temperatures.   A series of copolymers containing from 0 to 40 per cent initial styrene content was investigaged by Vila.[208] With increase in styrene content, the change in Mooney viscosity on compounding (Mooney difference), the tensile strength, elongation, and resistance to flex crack growth became progressively better.   Hysteresis, abrasion resistance, and tear resistance were reported to be best at the 85/15 ratio and to deteriorate at higher styrene contents.   A careful comparison of 70/30 and 75/25 copolymers by R. D. Juve[92] revealed only slight differences in the rate of copolymerization, curing characteristics, tensile properties, rebound, and tear resistance.   Such differences as were observed were in the directions previously indicated.

Table I.   Variation of Properties at Approximately 76 per cent Conversion as a Function of Initial Comonomer Ratio

Butadiene–styrene	Scott Plasticity	Rebound %	$T/300$ Kg. per Cm.^{2}	T, Kg. per Cm.^{2}	E, %
75/25	35	52	72	204	614
80/20	50	54	79	200	562
85/15	59	57	80	164	487
90/10	58	58	72	147	488
100/0	89	60	99	112	316

Conversion. *Progressive Change in Properties of the Polymer.*   Conversion may be treated as an independent reaction variable for the reason that the process can be terminated at will by the addition of an inhibitor such as hydroquinone, commonly termed a shortstop.   From theory it is evident that the properties of the product vary continuously as the reaction proceeds toward completion.   In the polybutadiene and optimum elastomer ranges of composition, the copolymer formed first is highly modified.   As modifier is consumed, the molecular weight of the copolymer increases.   After the disappearance of the monomer phase, the monomer content of the polymer-monomer particles is progressively consumed, thus bringing the polymer molecules into closer contact and increasing the probability of cross-linking reactions.   Accordingly, it is to be expected that properties such as Mooney viscosity will increase with conversion.   This behavior is discussed in greater detail under the heading of modifiers.   The conversion at which the Mutual recipe is usually terminated represents a compromise.   Conversions lower than 72 to 75 per cent result in better products at the expense of higher monomer recovery costs.   Upper limits of conversion are set by the product becoming difficult to process.   This, of course, is greatly influenced by the modifier content.   Another factor affecting the conversion compromise is the amount of fatty acid, generated from the soap during coagulation of the latex, that can be tolerated in the rubber.

Mazur and Coe[139] investigated the influence of conversion on a number of physical properties at two dodecyl mercaptan levels, 0.50 and 0.60 parts. A portion of their data is presented in Table II. With increasing conversion a regular increase ensues in raw Mooney viscosity, compounded Mooney, and modulus. The tensile data are irregular. Compression sets and elongations decrease at high conversions. Messer and Howland[150] conducted a similar but more extended investigation with both the Mutual and a hydrogen peroxide-ferric pyrophosphate recipe at several temperatures. With the Mutual recipe maximum tensile strengths were exhibited at approximately 75 per cent conversion; the modulus, resilience, and torsional hysteresis increased with conversion, while the tear resistance decreased. Similar results were obtained later by Messer,[147] who investigated the variation of conversion and the properties obtained on rubber made by following five recipes.

Table II. Influence of Variable Conversion on the Properties of GR-S Modified by 0.50 Part Dodecyl Mercaptan

Time, hours	10	13	16	18
Conversion, %	64	76	84	86
Raw Mooney (ML-4)	32	53	63	71
Compounded Mooney (ML-4)	59	81	109	125
$T/300$, 60' cure, p.s.i.	780	710	900	980
T, 60' cure, p.s.i.	3090	2750	3200	2690
E, 60' cure, %	683	606	605	546
Set at 45 lb., 60' cure	30	21	20	18

Progressive Change in Composition of the Polymer. The composition of GR-S varies with conversion in a regular manner. To determine composition, two methods have been developed. The more easily applied method requires only a determination of refractive index[2] and comparison of this value with the refractive indexes of polystyrene and polybutadiene prepared under conditions simulating those used for the sample (see Chapter 10). The copolymer sample must first be purified in a manner similar to that of Kolthoff, Carr, and Carr,[102] who proposed the use of an azeotropic (64/27/10) mixture of ethanol, toluene, and water as an extractant. More accurate determinations of composition can be obtained by the spectrophotometric method described by Meehan.[143] This depends on the ultraviolet absorption at 262 mμ by the phenyl residues in the copolymer.

Meehan[144] derived a set of equations that permits the calculation of F_s, the fraction of styrene in the entire copolymer formed up to any conversion, and C, the degree of conversion of monomers to copolymer, for any initial comonomer ratio R. It is thus possible to construct curves showing the variation of composition as a function of conversion for any initial comonomer ratio. The equations follow:

$$F_s = \left(1 + R\frac{1 - x^\alpha}{1 - x}\right)^{-1} \tag{1}$$

$$C = \frac{1 - x}{1 + R}\frac{1}{F_s} \tag{2}$$

in which x is defined as a parameter equal to S/S_0, where S is the amount of styrene unpolymerized at any conversion, and S_0 is the original amount of styrene. x can vary between 0 and 1; i.e., $(0 \leq x \leq 1)$. α can be evaluated from the composition of the copolymer increment formed at zero conversion as determined by extrapolation of an experimentally determined composition versus conversion curve.

$$\alpha = \frac{1}{R}\left(\frac{1}{F_s} - 1\right) \quad \text{for} \quad C = 0 \tag{3}$$

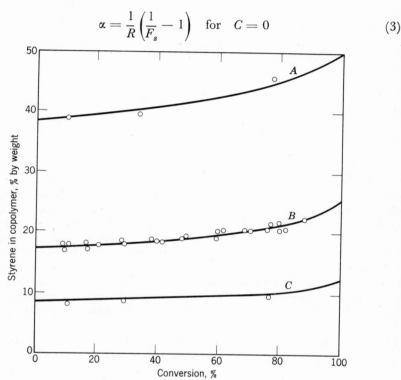

FIG. 5. Composition-Conversion Curves of Butadiene-Styrene Copolymers

The circles are experimental points; the curves are calculated with
$$\alpha = 1.60$$
Initial styrene charge contents: A, 50; B, 25; and C, 12.5

In the example used by Meehan, the initial (extrapolated) composition obtained with a $75/25(R = 3)$ monomer charge ratio was 17.2 per cent styrene. The calculated value was $\alpha = 1.60$. With this value of α, complete composition-conversion curves can be calculated for the whole range of initial comonomer ratios. To make the calculations a series of values of x between 0 and 1 is assumed, and corresponding values of F_s and C are then calculated by means of equations 1 and 2.

The agreement between the calculated and experimental determinations is excellent. This is illustrated by Fig. 5.

Table III gives data from Meehan's paper showing calculated values of composition for three monomer ratios. Extensive investigations have shown that the composition of the copolymer is unaffected by variations of modifier content, amount and kind of emulsifier, initiator system, and temperature within reasonable limits. There is evidence that Meehan's relationships do not hold at subfreezing temperatures with recipes containing a freezing-point depressant.[156, 166]

Table III. Copolymer Composition at Various Conversions

	Styrene in copolymer, % Butadiene-Styrene Charge Ratios		
Conversion %	87.5/12.5	75/25	50/50
0	8.2	17.2	38.4
20	8.5	17.9	39.7
40	9.0	18.7	41.1
60	9.5	19.7	42.8
80	10.3	21.2	45.4
90	11.0	22.5	47.2
100	(12.5)	(25.0)	(50.0)

It is evident that there is an increasing difference between the average composition of the copolymer and the composition of copolymer formed at any instant as the conversion increases. These compositions can be designated as integral and differential. Meehan also derived an expression for differential compositions. It will suffice to illustrate the difference in tabular form. Table IV shows that the limiting integral composition at 100 per cent conversion is equal to the initial monomer ratio, whereas the limiting differential composition is that of polystyrene. Therefore, one of the minor contributing factors to the production of poor quality at high conversions is the inclusion of subelastomers and polystyrene in the product.

Table IV. Integral and Differential Copolymer Compositions at Various Conversions Obtained with a 75/25 Butadiene-Styrene Initial Charge Ratio

	Styrene in Copolymer, %	
Conversion, %	Integral	Differential
0	17.2	17.2
20	17.9	18.8
40	18.7	20.6
60	19.7	23.3
80	21.2	29.5
90	22.5	36.4
100	(25.0)	(100)

Change in Vapor Pressure. Another property that changes with conversion is the vapor pressure of the system. Harrison and Meincke[74] have used pressure as a means of following the course of the reaction. Meehan[145] has pointed out that, if butadiene followed Raoult's law, the mole fraction of monomeric butadiene at 99 per cent polymerization is 0.80, assuming a number-average molecular weight of 40,000, and high pressure would be expected almost to the end of the process. However, the deviation from

Raoult's law by polymer-solvent solution leads to a perceptible decrease in pressure in the case of the Mutual recipe system at the lowest conversion. At about 52 per cent conversion when the free monomer phase has disappeared (cf. Fig. 4), the rate of pressure drop increases markedly. Absolute pressures, corrected by subtraction of 92 mm. (corresponding to the vapor pressure of water at 50° C.), are presented in Table V. These values were obtained by interpolation of the data of Meehan for a 75/25 initial monomer mixture. They cover a range of conversions up to 60 per cent over which range the conversion is linear with respect to time.

Table V. *Variation of Pressure with Conversion of a 75/25 Initial Butadiene-Styrene Mixture*

Conversion, %	Pressure,* mm.
0	3650
10	3640
20	3630
30	3610
40	3575
45	3545
50	3500
55	3415
60	3290

* Adjusted for the theoretical initial pressure of 3650 mm. mercury.

Emulsifiers. *Cationic Emulsifiers.* Although many different types of emulsifiers can be used in the Mutual recipe,[16, 64, 148] the two found to be most useful are soaps of fatty acids and sodium disproportionated-resinates. Satisfactory experimental results have been obtained by using cationic emulsifying agents, such as dodecylamine hydrochloride.[81, 147, 215] The adoption of a low pH recipe would have required extensive research and development on equipment and compounding techniques because of the difficulty of removing the free amine from the product. The advantages foreseen did not warrant such a course.

Fatty Acid Soaps. Extensive tests have failed to disclose significant differences in the rates of polymerizations or the properties of the product obtained with soaps prepared from purified samples of lauric, myristic, palmitic, and oleic acids. The use of sodium stearate frequently results in slow polymerizations (cf., however, ref. 15), but satisfactory rates are obtained with this and higher-molecular-weight soaps when prepared in situ, i.e., by dissolving stearic acid in the monomers and dispersing the monomers in an aqueous solution of sodium hydroxide. This behavior was investigated by Osterhof and his co-workers[161] and confirmed by Kolthoff and Guss.[105] Soaps of lower molecular weight than lauric can be employed, if, as Reynolds[171] has shown, the soap is thrown out of isotropic solution into the micellar state by the addition of an electrolyte to the system. Mixtures of soaps in general appear to be more satisfactory than the pure components. Early experience indicated that the use of sodium oleate resulted in slow polymerization and the production of more highly modified products. This behavior was traced to the presence of linoleic[165, 221] acid containing two double bonds separated by a methylene group. When the double-bond system is conjugated by heating the soap, the retardation and modification disappear. Commercial soaps used in production of GR-S were sold on the basis of performance rather than of composition. This enabled the soap manufacturers to exercise freedom of choice with respect to raw materials. It appears probable that fatty acids of both animal and vegetable origin were used and that the reactions of partial hydrogenation and conjugation of double bonds were employed when necessary.

Soaps and fatty acids deteriorate on storage. The first stage consists in the formation of soap or fatty acid peroxides. Potter, Borders, and Osterhof[167] proposed the use of such peroxidized emulsifying agents for the initiation of the Mutual recipe without potassium persulfate. Although satisfactory rates of polymerization could be realized, it was the consensus of opinion that greater uniformity of product and reproducibility of reaction rate could be obtained with unoxidized soaps and potassium persulfate. Consequently the producers of GR-S are restricted to the use of rather freshly made fatty acids for the preparation of soaps. When the oxidation of soaps and fatty acids has proceeded too far, retarding and inhibiting substances are formed. Potter, Borders, and Osterhof[167] also reported that naturally occurring inhibitols such as α-tocopherol retard the reaction. Wilson and Pfau,[221] however, did not find α-tocopherol to cause retardation, and Carr and co-workers[15] found only slight retardation. Marvel et al.[133] found that the presence of lecithin or of choline in amounts up to 0.4 and 1 per cent of the soap content, respectively, exercised no inhibiting influence.

In accordance with theory, doubling the soap content should approximately double the rate of polymerization. Exact linearity is not, however, observed. As W. V. Smith has remarked in connection with a study of the emulsion polymerization of styrene, "a given weight of soap is more effective in forming particles in a dilute than in a concentrated solution."[190] Brucksch, Messer, and Howland[12] found that, in the polymerization of butadiene-styrene in the Mutual recipe, increasing the content of a standard soap from 4 to 7.3 parts increased the conversion approximtely 73 per cent. At soap concentrations higher than 11 parts, the conversions leveled off. A more extensive investigation of the influence of soap concentration was conducted by Kolthoff and Meehan.[115] With the standard soap designated as SF flakes, it was found that sigmoidal curves were obtained at soap concentrations lower than 5 parts (Fig. 6). At higher soap contents typical linear reaction curves are obtained. The initial (below 10 per cent conversion) rate of polymerization, tabulated in Table VI, is roughly proportional to the soap concentration, while the maximum rate is independent of the concentration below 5 parts.

Kolthoff and his co-workers[105, 118] found also that, unlike the situation with some other recipes, the addition of excess sodium hydroxide to the Mutual recipe did not affect the rate of conversion at several soap contents. The addition of excess free fatty acid, however, considerably increased the rate. Replacement of soap by fatty acid did not change the rate. In other words the rate of polymerization with the Mutual recipe is not influenced greatly by the percentage neutralization of the fatty acid within fairly wide limits, and within these limits the addition of free fatty acid is equivalent in its effect to the addition of soap.

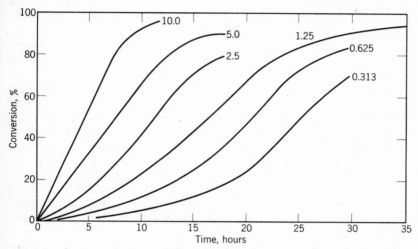

FIG. 6. Conversion Curves with Various Soap Concentrations in the Mutual Recipe S.F. Flakes in parts per 100 of monomers

Table VI. *The Effect of Soap Concentration on the Rate of Copolymerization with the Mutual Recipe*

SF Flakes Parts	Rate of Polymerization, % per hour	
	Initial (below 10% Conversions)	Maximum
10	11.0	12.8
5	6.2	6.5
2.5	3.3	6.3
1.25	2.0	6.2
0.625	1.0	6.5
0.313	0.7	6.3

Rosin Acid Soaps. The possibility that fatty acid soaps might be required in larger amounts than were commercially available led to an investigation of sodium disproportionated resinates. After the first promising results,[46] a considerable number of investigations were conducted, among which may be mentioned those of Zwicker and Garvey,[226] Fryling,[47-8] and Tierney and Coe.[200] Despite all this effort it proved impossible to develop a rosin soap recipe which exhibited a rate of conversion equal to that obtained with fatty acid soap until cumene hydroperoxide was introduced as an oxidizing initiator. The slow rates of reaction were eventually shown by Shepherd and his co-workers[186] and by Hays, Drake, and Pratt[76] to be the result of the presence of unconverted abietic acid and phenolic impurities. However the improved building tack conferred on GR-S by the rosin acids generated from rosin soaps in coagulation[21, 226] led to the widespread use of so-called "rosin rubber," or GR-S 10, after aging tests had shown that the material was as resistant to oxidative degradation as GR-S. The superiority of rosin rubber was found by Juve, Schroeder, Ludwig, and Helms[91] to be attributable to the fact that it contained rosin acid in place of the fatty acids in standard GR-S and not to any difference in the quality of the copolymers. For accounts of later investigations of rosin soap emulsification, the reader is referred to publications by Azorlosa[1] and by Fryling.[53]

Modifiers. Viewed from a practical standpoint, a modifier is a substance which, when included in a polymerization recipe, makes possible the production of a plastic, workable, soluble polymer under conditions of polymerization which in its absence would lead to a tough, unworkable product.

Modification of GR-S was first accomplished by the addition of lorol mercaptan, later known as DDM, a mixture of mercaptans prepared from coconut oil and consisting mostly of n-dodecyl mercaptan. Subsequent investigation showed that this material exercised a dual function: It was both a reducing initiator and a chain-transfer agent. By its use the persulfate-initiated system was made sufficiently fast to be practical, and the plasticity of the product could be controlled and adjusted to a range suitable for standard rubber-processing equipment.

Three motives appear to have inspired research on modification of the Mutual recipe. Before the function of modifiers as chain-transfer agents was understood, the possibility of preparing elastomers exhibiting improved properties with new and as yet undiscovered modifiers appeared promising. When large-scale rubber production came into being it appeared that proper control of plastic properties in the product was something of an art. Among the factors involved were the purity of raw materials, rate of reaction, intensity of agitation, percentage conversion, initial comonomer ratio, pH of the system, rate and temperature of drying, amount and type of antioxidant employed and its method of addition, and the specific activity of various lots of mercaptan. Intensive investigation of all these variables provided the knowledge on the basis of which it was possible to produce synthetic elastomer of a high degree of uniformity. Finally, the necessity of having available mercaptans derived from materials other than coconut oil, to forestall a possible shortage of this raw material, led to the introduction of tertiary mercaptans produced by the addition of hydrogen sulfide to unsaturated petroleum hydrocarbons.[180]

A large number of materials were investigated for possible modifier activity. Messer and Howland[148] reported tests on more than 63 substances. Likewise Stewart[194] investigated materials from many sources. The early work of these and others showed that highest modifier activity was exhibited by n-dodecyl and tetradecyl mercaptans, that n-hexadecyl mercaptan was an efficient initiator but that it lacked modifier activity, that normal mercaptans of molecular weight lower than heptyl, though exhibiting modifier activity, were potent inhibitors. Modification produced by n-alkyl mercaptans was influenced by the pH of the system, an increase in pH increasing the rate of disappearance of the mercaptan.[106] Among substances that produced modification were trithiocarbamate, isopropylxanthogen disulfide, α-thio-naphthol, *p*-alkylphenoxymonomercaptodiethylene glycol, 2-mercapto-4-phenylthiazole, and di-(4-phenylthiazolyl-2) disulfide.

Increasing the concentration of *n*-dodecyl mercaptan from 0 to 0.10 part greatly increased the rate of copolymerization; further increase had no effect on the rate but did increase the plasticity of the product.[26, 68] McCleary, Messer, and Howland[127] reported as follows on the effect of increased *n*-dodecyl mercaptan content on the properties of GR-S: Variation of mercaptan content above 0.05 part has no significant effect on the rate of polymerization; uncompounded Mooney viscosity and the Mooney difference decrease; maximum tensile strengths are obtained with 0.4 to 0.6 part; resilience or rebound and durometer hardness decrease; and torsional hysteresis together with the tendency toward undercure increase. Kolthoff and Harris[107, 108] showed that, though, with increase in mercaptan concentration, the rate of copolymerization is constant over a wide range in the case of normal mercaptans, the rate goes through a maximum in the case of tertiary mercaptans.

Increment and Continuous Addition of Modifier. When the Mutual recipe was proposed in December of 1941, the statement was made that less modifier was required if a portion were left out at the start and added toward the end of the process. Eventually this observation led to a considerable interest in the possibilities of increment and continuous addition of modifier throughout the duration of the process. A great number of investigations were conducted,[11, 67, 77, 163–4, 170, 175] and economies in mercaptan requirements up to 17 per cent were demonstrated. But more than this was expected. Common sense indicated that no benefit is to be derived from having high concentrations of modifier present at the start of the process when low-molecular-weight polymers are normally produced, since at high conversions, when the tendency is toward the production of high-molecular-weight products, the modifier content is depleted to a low value. In addition, low-molecular-weight copolymers cure at a slower rate than those of high molecular weight. The best that can be expected from a mixture of high- and low-molecular-weight products is a compromise cure with both undercured and overcured materials present. Such being the case, efforts to produce a "constant-viscosity" product, i.e., a product of uniform molecular weight throughout the whole range of conversions, were considered very promising. Interest in this possibility dropped rapidly after Laurence, Hobson, and Borders[121] reported the surprising observation that, although continuous modifier addition produced rubber of constant intrinsic viscosity with increase of conversion throughout the course of the polymerization, the molecular-weight distribution of the final product was only slightly different from that of normal GR-S, in the production of which all the mercaptan is added in the original charge.

Rate of Disappearance of Mercaptans. The determination of the mercaptan content of partially reacted latexes by the amperometric titration method developed by Kolthoff and Harris[106] proved to be valuable for following the modifier concentration during the reaction. A rather complete account of these investigations was given by Dunbrook.[26] Figures 7 and 8 illustrate

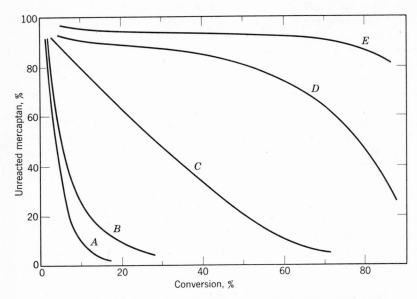

FIG. 7. Depletion of Normal Mercaptans in the Mutual Recipe
A, Octyl; *B*, decyl; *C*, dodecyl; *D*, tetradecyl; *E*, hexadecyl

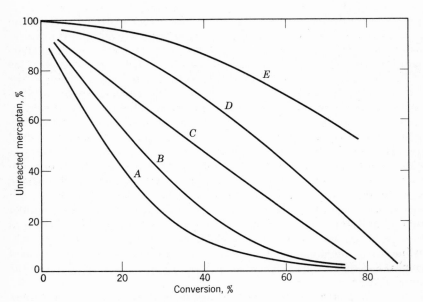

FIG. 8. Depletion of Tertiary Mercaptans in the Mutual Recipe
A, Octyl; *B*, decyl; *C*, dodecyl; *D*, tetradecyl; *E*, hexadecyl

data obtained by Kolthoff and Harris[108] on the disappearance of primary and tertiary mercaptans as a function of conversion. The *n*- and *tert*-dodecyl mercaptans exhibit very similar disappearance curves. Even though the *n*-dodecyl mercaptan disappears at a faster rate with respect to conversion than the *tert*-dodecyl mercaptan, a somewhat lower concentration of the tertiary mercaptan suffices to attain a given Mooney viscosity at a definite conversion. Data such as these indicate why mixtures of mercaptans of different molecular weights constitute more effective modifying agents than do pure compounds. The greater efficiency of mixtures has been repeatedly demonstrated by experience obtained in pilot plants and production.

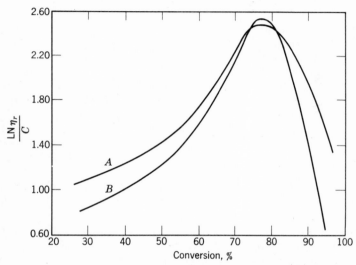

FIG. 9. Inherent Viscosity versus Conversion of GR-S Modified by
(*A*) Primary and (*B*) Tertiary Dodecyl Mercaptan

Influence of Mercaptan on Change of Inherent Viscosity with Conversion. As conversion increases, the Mooney viscosity of the product rises. Although modifiers reduce the Mooney viscosity over the whole range of conversions, nevertheless the same trend of increasing Mooney is observed in their presence. When, however, inherent viscosities are investigated, by measurements on solutions of the polymers, a different situation exists. Schulze and Crouch[179] have shown that the solution viscosity versus conversion curves depend on the amount and type of various mercaptans charged and correlate with the rates of depletion of the mercaptans. The viscosity increases with conversion and reaches a maximum at monomer conversions of approximately 75 per cent. Insoluble gel forms at higher conversions, and the viscosity of the residual soluble copolymer decreases thereafter with increasing conversion. Curves illustrating the viscosities obtained with *n*- and *tert*-dodecyl mercaptans are shown in Fig. 9.

The statement has been made that tertiary mercaptan modified rubbers are not equal in processibility and have properties inferior to rubbers made with DDM.[26]   However, careful comparative tests have failed to reveal any basis for such supposed inferiority, and in fact tertiary mercaptans are at present widely used in the manufacture of cold GR-S.   It is evident that modifiers exercise one predominant function, that of reducing the average molecular weight of the product by acting as a chain-transfer agent.   Experimental evidence for this was furnished by Snyder[192] and Wall,[209] who have shown that there is one atom of sulfur in each molecule of copolymer produced in mercaptan-persulfate systems.   Were it not for the many other factors that influence modification, it should be possible to use any efficient chain-transfer agent to accomplish the same result.

   Influence of Agitation on Modifier Consumption.   The degree of agitation is one . of the variables that influence the modifier activity of normal mercaptans. Insufficient agitation results in low rates of polymerization and inadequate modification.[63]   If, however, the agitation is excessive, mercaptan depletion is accelerated, and the process of modification becomes inefficient.   Accordingly, less modifier can be used to obtain the desired plasticity if agitation is decreased.[33]   Optimum agitation depends on a number of factors, such as the size and shape of the reactor, the design of the propeller, and its speed. Kolthoff and Harris[108] reported that the rate of disappearance of *tert*-dodecyl mercaptan, in contrast to that of normal mercaptan, does not change when the speed of agitation is varied.   It was subsequently found that this behavior is characteristic of all tertiary mercaptans except hexadecyl, the depletion of which is influenced by agitation.

   Addition of Mercaptans to Polymer.   In addition to acting as chain-transfer agents during polymerization, mercaptans possess the ability to add to double bonds.   The addition of mercaptans to polymer is accompanied by an induced oxidation leading to the formation of products of lower inherent viscosity.   Kolthoff and Stenberg[117] were in fact able to prepare syrupy materials by the slow addition of *n*-dodecyl mercaptan to GR-S in the presence of air and an iron-cobalt oxidation catalyst.

   Concentration of Mercaptan and Rate of Polymerization.   Above 0.05 to 0.10 part the concentration of *n*-dodecyl mercaptan has little influence on the rate of copolymerization.   The corresponding concentration for tertiary mercaptans is about 0.30 part.   With amounts of tertiary mercaptans equivalent in modifying effectiveness to 0.50 part of dodecyl mercaptan, the rates of polymerization were found to be about equal when disproportionated rosin soap was used as the emulsifying agent.[53]   However when ammonium salts are present, rosin soap-tertiary mercaptan recipes exhibit a characteristic retardation which was first observed by McCleary.[126]   This behavior was also found with *n*-dodecyl mercaptan when its concentration was in the range of 0.0 to 0.10.   The effect with tertiary mercaptans can be suppressed by increasing the *p*H of the system.

   Factors in Mercaptan Consumption.   Reynolds and Canterino[26, 173] explain the behavior shown in emulsion polymerization by mercaptans of different molecular weights and structures on the basis of solubility, rates of diffusion, and relative rates of oxidation.   Generally speaking, rates of diffusion and rates of oxidation in solutions of soap and persulfate increase with decrease in the molecular weight of normal mercaptans.   To generate free radicals by an oxidation-reduction system appears to require something akin to a balance in the relative reactivities of the oxidant and reductant. For example, if the more powerful reductant hydroquinone is substituted for mercaptan in the Mutual recipe, inhibition ensues. Consequently the inhibitory effects of the lower mercaptans can be attributed to their greater tendency to be oxidized and at the same time to their higher effective concentration by virtue of their solubility in the aqueous phase. With n-hexadecyl mercaptan the concentration in the aqueous phase is high enough to enable it to act as an efficient initiator, but its rate of diffusion through the aqueous medium is so low that its concentration in the polymer-monomer particles never reaches a value high enough for it to be an effective chain-transfer agent. By measuring the rates of diffusion of various mercaptans as a function of pH, Reynolds and his co-workers[44, 173] succeeded in presenting a well-integrated theoretical explanation of the differences exhibited by various normal and tertiary mercaptan modifiers.

Calculation of Modifier Requirements. Harris and Kolthoff[73] proposed a method of calculating the theoretical minimum modifier requirement. This value, divided by the actual amount used, constitutes the modifier efficiency. The efficiency is low at low conversions and increases through a maximum with increased conversion. For a 75/25 butadiene-styrene copolymerization conducted in accordance with the Mutual recipe, E, the efficiency, is calculated from the equation

$$E = \frac{P(0.2 + P)}{R_O(M_V \times 10^{-5})}$$

where M_V is the viscosity molecular weight of the polymer at conversion P, with an initial mercaptan content of R_O. The theoretical minimum modifier requirement at 75 per cent conversion of polymer of intrinsic viscosity 2.0 is about 60 per cent of the amount of primary dodecyl mercaptan actually used in large commercial reactors.

Persulfate Initiation. A certain similarity exists between the behavior in polymerization of dodecyl mercaptan and potassium persulfate. Both materials perform a dual function; both are consumed during the process, and with both an effect of concentration on the rate of emulsion copolymerization is evident only at very low values, while the rate is essentially constant over the range of concentrations considered as practical. Potassium persulfate, however, is a water-soluble oxidizing agent, whereas dodecyl mercaptan is a reducing agent which is predominantly soluble in the organic phase. On this account it might be concluded that the interaction between these materials takes place at the water-organic interface. However, as has been shown, all pertinent evidence indicates that the initiating reaction between persulfate and mercaptan takes place in the aqueous phase, and that the mercaptan must migrate to the polymer-monomer particles to exercise its function as a chain-transfer agent.

One aspect of the dual function exercised by potassium persulfate is not revealed by any investigation of the performance of the Mutual recipe. Studies conducted with other recipes, however, show that unless a certain concentration of electrolyte is present in addition to the sodium or potassium hydroxide required to neutralize exactly the fatty acid, a soft pasty gel, which cannot be stirred and which retards the transfer of the heat of reaction, is obtained instead of a free-flowing latex. Fortunately potassium persulfate is converted to potassium acid sulfate which is as effective as the former in its action as an electrolyte.

Data obtained by Kolthoff and Meehan[115] by varying the persulfate content of the Mutual recipe are given in Table VII. When the persulfate concentration is varied in the range of 0.15 to 3.0 parts, the rate of polymerization in the 3- to 6-hour period is almost constant. As the persulfate content is further lowered, the rate decreases from 7.7 to 5.3 per cent per hour.

Table VII.   Variation of Potassium Persulfate Content of the Mutual Recipe

Persulfate Parts	2	Conversion (%) in 3	6	12 hours
3.0	8	'14	37	82
1.5	9	15	37	80
0.3	11	16	37	78
0.15	...	17	37	77
0.075	...	14	32	73
0.030	...	11	27	62

The variation of persulfate content influences the length of the induction period caused by oxygen. Kolthoff and Meehan's[115] data, presented in Table VIII, indicate that the induction period is inversely proportional to the square root of the persulfate concentration. If the induction period I is expressed in minutes and potassium persulfate in parts, the following relation is obtained:

$$I = K_b(K_2S_2O_8)^{-1/2} - 36$$

Table VIII.   Influence of Potassium Persulfate Concentration on the Induction Period with 20 ml. of Oxygen in the Charge (Mutual Recipe)

$K_2S_2O_8$, Parts	$(K_2S_2O_8)^{-1/2}$	Induction Period, Minutes	Kb
0.60	1.28	60	47
0.30	1.82	105	58
0.15	2.57	160	62
0.075	3.44	220	60

The data tabulated in Tables VII and VIII are illustrative only of what can be expected with the Mutual recipe as originally formulated. With butadiene alone quite similar results are obtained. With styrene alone the results are very different.[9, 103] Differences are also introduced by changes in the type and concentration of soaps and mercaptans. Other relationships are exhibited when mercaptan is absent from the system.

Temperature. With free-radical-chain reactions, abnormal temperature effects on the rate of reaction are to be expected. Emulsion polymerizations are no exception in this respect. In at least one case it has been definitely shown that raising the temperature greatly decreases the over-all rate of reaction. However, it appears possible to deduce a simple rule on the basis of experience with a number of systems. With a well-balanced emulsion polymerization formula, the temperature coefficient of reaction rate is normal and equal to approximately 2 per 10° C. in the temperature range immediately adjacent to the temperature at which the recipe is balanced. Thus one might expect a doubling of the rate of reaction with the Mutual recipe by raising the temperature to 60° C. and a halving of the rate by lowering the temperature to 40° C. Errors can be expected if calculations are made over a greater range of temperatures, and, conversely, higher accuracy can be obtained if the temperature range is restricted to, say, $\pm 5°$ C. How well this rule is obeyed with the Mutual recipe is evident from a consideration of data obtained by Messer and Howland.[150] In Table IX the times required for 60 per cent conversion at three temperatures, obtained by interpolation of Messer and Howland's reaction curves, are tabulated.

Table IX. Influence of Temperature on Time Required to Obtain 60 per cent Conversions with the Mutual Recipe

60% Conversion		Temperature Coefficients	
Time, Hours	Temp., °C.	Temp. Range, °C.	Coefficient
4.0	60	50-60	2.1
12.0	45	45-55	2.0
57.2	30	40-50	2.3
		30-40	3.1

·That the temperature of polymerization affected the physical properties of the product can be seen from Messer and Howland's data presented in Table X. It is evident that lowering the temperature of polymerization improves the product. However, the extent of improvement within the temperature range studied was considered insufficient to warrant the longer time required for production in the polymerization recipes then (1943) available, and in the opinion of some the improvement could be attributed to the slower reaction rather than to the lower temperature. Subsequent developments have demonstrated the fallacy of this latter point of view.

Table X. Physical Properties of Elastomers Prepared with the Mutual Recipe at 30 and 60° C.

	30	60
Temperature of polymerization, °C.	30	60
Polymer time, hours	65	6
Conversion, %	76	75
Mooney viscosity at 212° F.		
Uncompounded	36	62
Compounded	61	127
Difference	25	65
Modulus, 300%, 60′ cure; p.s.i.	840	1070
Tensile strength, 60′ cure, p.s.i.	3740	2840
Elongation, 60′ cure, %	737	567
Tensile strength at 212° F., 60′ cure, p.s.i.	1200	987
Elongation at 212° F., 60′ cure, %	610	480

Inhibitors, Retarders, and Impurities. Inhibitors and retarders, once the bugbears of polymerization research, have to a great extent become materials of theoretical interest only. The shift of practical interest away from these substances is the result of the present availability of highly purified "polymerization-grade" raw materials.

Common sense indicates that those materials used in greatest amounts should be most suspect as sources of inhibiting impurities. For example, a low concentration of inhibitor in feed water can be of greater importance than a considerable concentration in a catalyst used in mere traces. Nevertheless, because decanter water is at times recycled through the reactors, small concentrations of ammonium persulfate in the potassium persulfate can lead to serious retardation (the ammonia effect) when tertiary mercaptans are used. Marvel and his co-workers have investigated the inhibitory and retarding activity of many substances likely to be present in butadiene, isoprene, and emulsifying agents. These results are summarized in Chapter 21. Only a few matters of outstanding importance need be considered here.

Vinylacetylene, a common impurity in butadiene produced by certain processes, does not retard the reaction but acts as a potent cross-linking agent. The effect of vinylacetylene on decreasing the solubility of the elastomeric product is well known.[26] The presence of 2 per cent of vinylacetylene in butadiene can bring about a decrease of 75 per cent in tensile strength of the cured rubber.[45]

Phenolic compounds, aniline compounds, and, in general, substances possessing strong reducing activity are inhibitors. One class of materials exhibiting high retardation, particularly when present in soaps, contains two double bonds separated by a methylene group.[221] Such groups are found in linoleic and linolenic acids.

Inhibition by atmospheric oxygen may be a serious matter when reactions are conducted in accordance with the Mutual recipe. Since its first observation,[61, 97, 98] oxygen inhibition has been subjected to considerable investigation. The influence of this behavior on the development of experimental technique has been noted (p. 239). The variation of the oxygen induction period with persulfate content has been discussed (p. 258). Particularly interesting in this connection is Bovey and Kolthoff's[9] observation that a copolymer of styrene and oxygen is formed during the induction period preceding the emulsion polymerization of styrene. During an induction period in the Mutual recipe, the oxygen present is used up, and a portion of the mercaptan is consumed.[108]

Frank[43] and his co-workers attribute retardation to inefficient chain transfer. Examples of retarders are substances containing an allylic methinyl group $> C : \dot{C}—CH <$ which lose the tertiary hydrogen atom and form a free radical insufficiently reactive to initiate a new polymer chain. However, retardation and inhibition can arise from so many sources that one theory is unable to account for all the facts. Electrolytes in sufficient concentration to precipitate the soap bring about retardation by decreasing the initial micelle content of the system. It is conceivable that certain substances may generate free radicals so rapidly that dimerization of them occurs rather than chain initiation. Reaction shortstops such as hydroquinone, which when added to the Mutual system in concentrations[26] of 0.05 to 0.10 per cent are effective, probably act by destroying the persulfate. Dinitrochlorobenzene, another shortstop, on the other hand, is believed to act by termination of growing chains. These last examples are cited to indicate that the subject of inhibitors of emulsion polymerization may be very complex.

LOW-TEMPERATURE PYROPHOSPHATE ACTIVATED REDOX RECIPES

The Hydrogen Peroxide Prototype. A recipe submitted for consideration to the Government in December 1941 was the following, presented in abbreviated form:[66]

IIa.	Water	250
IIIa.	Soap flakes, 95% neutralized	5.0
IV.	Isopropylxanthogen disulfide	0.45
Va.	H_2O_2, 100%	0.25
c.	$Na_4P_2O_7 \cdot 10H_2O$ (VI)	1.00
	$Fe_2(SO_4)_3$ (VI)	0.055
	$CoCl_2 \cdot 6H_2O$ (VI)	0.0025
VIII.	a. 40° C.; b. 20 hr.; c. 90%	

Though this system was of empirical origin, its development was influenced by ideas derived from biological oxidation-reduction phenomena.[196] An oxidant initiator, hydrogen peroxide, was specified. Subsequently it was found that the polymerization did not start until, as indicated by titration with ceric sulfate, all the hydrogen peroxide had disappeared, a process requiring about 5 minutes. Marvel and his co-workers[23, 135] reported that peroxide depletion required approximately one hour and that during this period there was formed a quantity of organic peroxide which might conceivably be the cause of the subsequent polymerization.

The use of dodecyl mercaptan as a modifier in the Mutual recipe suggested its substitution for isopropylxanthogen disulfide. A faster formulation was thereby obtained, and with its use an interesting and suggestive study of the effect of temperature on elastomer properties was conducted by Messer and Howland[149] (Table XI).

Further acceleration of the hydrogen peroxide recipe was obtained by including in the system an additional reductant initiator, e.g., sodium sulfite, *tert*-butylcatechol, or acetaldehyde. The best reductant was hydroquinone in extremely low concentrations.[72] Data illustrating its effect are given in Table XII. The water content was adjusted to that of the Mutual recipe. A mixed mercaptan-xanthogen modifier was employed. The maximum rate of polymerization was obtained with 1 mg. of hydroquinone per 100 grams of monomers. This constitutes an interesting example of trace catalysis. The dodecyl mercaptan in this and in the Messer and Howland formulation is considered as functioning as a reductant initiator.

Sugar-Containing Organic Hydroperoxide Recipes. The next advance in the development of pyrophosphate activated recipes followed the introduction of cumene hydroperoxide as an oxidant initiator. Marvel and

Table XI.   Polymerization in Type-H Recipe at Various Temperatures

IIa.	Water				200	
IIIa.	Ivory flakes				7.3	
IV.	n-Dodecyl mercaptan (Vb?)				0.55	
Va.	H_2O_2				0.056	
c.	$Na_4P_2O_7 \cdot 10H_2O$				1.25	
	$Fe_2(SO_4)_3$				0.041	
	$CoCl_2 \cdot 6H_2O$				0.0018	
VIII.	a. Temp. °C.	b. Time, Hr.		c. Convn., %		
	15	50		65		
	30	13		73		
	45	4.7		79		
	60	2.5		69		

d.	Physical Properties				
	Temperature	15	30	45	60
	Mooney	24	45	67	36
	Tensile strength				
	60' cure, p.s.i.	3220	3800	3440	2960
	Elongation				
	60' cure, %	846	760	627	600

Table XII.   The Addition of a Reductant Initiator to the Hydrogen Peroxide Recipe

IVa.	Isopropylxanthogen disulfide		0.3
b.	n-Dodecyl mercaptan (Vb?)		0.3
Va.	H_2O_2		0.35
b.	Hydroquinone		variable
c.	$Na_4P_2O_7 \cdot 10H_2O$	(VI)	0.80
	$Fe_2(SO_4)_3$	(VI)	0.044
	$CoCl_2 \cdot 6H_2O$	(VI)	0.0016
VIII.	a. 30° C.; b. 10 hr.		

Hydroquinone, (Vb.)	Convn. % (VIIIc.)
0.000	42
0.003	55
0.005	60
0.008	64
0.010	68
0.015	66
0.020	63

his co-workers,[135-6] following the recommendations of the German investigator, Becker,[162] had developed a promising formulation containing sorbose and ferrous pyrophosphate.   A number of reductants and oxidants were evaluated.   Difficulty was encountered at the start of this investigation because of inability to duplicate the poorly characterized emulsifying agents proposed by the German investigators.   Kolthoff and Meehan reported the use of cumene hydroperoxide in a recipe containing ferrous sulfate, sodium pyrophosphate, and glucose.[112-3]   Even with disproportionated rosin soap as the emulsifier, the at that time surprisingly high rate of conversion of 17 per cent per hour at 30° C. was obtained.   Their recipe follows:

IIIa.	Rosin soap (Dresinate 731)	5.0
IV.	n-Dodecyl mercaptan	0.50
Va.	Cumene hydroperoxide	0.17
b.	Glucose	1.0
c.	$Na_4P_2O_7 \cdot 10H_2O$ (VI)	1.0
	$FeSO_4 \cdot 7H_2O$ (VI)	0.10
VIII.	a. 30° C.; b. 7 hr.; c. 75%	

This development commanded immediate attention for several reasons. It indicated for the first time a method, employing hydroperoxides, whereby fast reactions could be secured with rosin acid soaps as emulsifiers. The amount of iron required for activation was sufficiently low to promise freedom from iron-catalyzed oxidative degradation of the elastomer. In sharp contrast to information received shortly before this development from German sources,[100, 122, 131, 212] the advantages of modification could be retained in a low-temperature process. Finally the ingredients were inexpensive.

Improvements in this formulation were briefly described by Fryling and his co-workers.[58] Of interest was the stoichiometric equivalence observed between the ferrous iron content and the cumene hydroperoxide. With molar ratios of Fe^{++} to hydroperoxide greater than 1, complete inhibition resulted, while maximum rates of copolymerization were observed at ratios only slightly lower than unity. The rate was increased by heating the sugar reductant in a dilute alkaline solution. Sugar formulations were in practical use for the large-scale production of "cold rubber" at 41° F. in 1948 and 1949.[185] By that time a number of similar recipes involving the use of reducing sugars had been developed. As an example of such, the abbreviated form of the so-called Custom recipe[181, 204] is presented:

IIIa.	Rosin soap, Dresinate 214	4.7
b.	Daxad 11 (optional)	0.1
IV.	Mixed tertiary mercaptans (MTM)	0.24
Va.	Cumene hydroperoxide	0.10
b.	Dextrose	1.00
c.	$K_4P_2O_7$	0.177
	$FeSO_4 \cdot 7H_2O$	0.14
VI.	KCl	0.50
VII.	KOH	0.10
VIII.	a. 5° C.; b. 13 hr.; c. 60%	

The sugar recipe, thanks to the control over the rate of polymerization which it provided, immediately afforded the means of conducting a crucial experiment on the influence of temperature of polymerization on the physical properties of the product. Though many previous investigations had shown that rubbers prepared at relatively low temperatures exhibited improved properties, the improvements were not so great as to command attention and objection could always be raised that the improvements were due to slower reaction, decreased modification, or a change of recipe. Now, with rosin soap as the emulsifier, rubber was prepared at two temperatures, 15 and 50° C., to identical Mooney viscosities, to the same conversion, in approximately the same period of time.[59] The products were processed in an identical manner. Results are presented in Table XIII. The improvement in physical properties is evident. Nevertheless it appeared certain that, to obtain a degree of improvement that would be commercially significant, it would be necessary to employ temperatures of polymerization still lower than 15° C. The only objection that might be raised against the procedure adopted was that the difference in concentrations of initiating ingredients might have influenced the properties of the product. However, when experiments were conducted at approximately the same yield, time, temperature, and Mooney viscosity of product with the sugar-free recipe and with the diazothioether recipe described later, identical properties were realized, thus showing that, in accordance with theory, differences in the initiating system do not influence the properties of the product to any significant extent.

Table XIII. The Influence of Temperature of Polymerization on the Properties of Rosin Rubber[51]

		15	50
	Temperature, °C.	15	50
I.	Butadiene/styrene	72/28	72/28
III.	Dresinate 731 soap*	3.5/1.2	3.5/1.2
IV.	Mixed tertiary mercaptans	0.40	0.35
Va.	Cumene hydroperoxide, 100%	0.15	0.15
b.	Levulose	1.00	...
	Glucose	...	0.01
c.	$Na_4P_2O_7 \cdot 10H_2O$	1.00	0.02
	$FeSO_4 \cdot 7H_2O$	0.10	0.002
VI.	Na_2SO_4	...	0.20
VIIIb.	Hr.	10.0	10.5
c.	%	73.4	73.2
d.	Mooney, (ML-4)	75	73
e.	Inherent viscosity, (η)	2.34	2.39
f.	Tensile str., 45' cure at 80 F., p.s.i.	4250	3940
g.	Elong., 45' cure at 80 F., p.s.i.	640	630
h.	Tensile str., 45' cure at 200 F., p.s.i.	1730	1500
i.	Elong., 45' cure at 200 F., p.s.i.	540	300
j.	Hysteresis, $\Delta T°$ F., 45' cure	82	72
k.	Flex, M at 210° F.	72	37
l.	Shore hardness	60	64

* Sodium disproportionated resinate.

Further Study of Sugar-Containing Recipes. It has been possible to increase the rate of polymerization of the sugar-reductant systems by careful balancing of the recipe and by the discovery of more efficient oxidant and reductant initiators. Kolthoff and his co-workers[109, 142] conducted a series of investigations of this recipe. They confirmed the observation that heating glucose solution with a low concentration of sodium hydroxide increased its activity to an extent that made it comparable to that of fructose. In addition, two reductants, dihydroxyacetone, first used by the Germans, and *scyllo*inosose, the latter suggested by Marvel,[135] were found to give extremely rapid rates at concentrations much lower than was required for dextrose. These reductants made it possible to use a sugar-reductant recipe at subfreezing temperatures with the attainment of high rates of polymerization. Practical substitutes for dihydroxyacetone and inosose could be provided by subjecting glycerol and inositol, respectively, to oxidation with Fenton reagent.[110] The fast reactions obtained with these new reductants used in conjunction with a new oxidant, the hydroperoxide prepared from *tert*-butylisopropylbenzene,[55] at $-10°$ C., are illustrated in Table XIV.[116] It is evident that the sugar recipe has been developed to a point where its potential rate of reaction at subfreezing temperatures cannot be utilized in plant practice until production equipment is redesigned to handle the resulting high rate of evolution of heat.

Table XIV. Conversions Obtained with Improved Initiators at $-10°$ C.

II*a*.	Water	140
b.	Methanol	40
III.	Potassium myristate	5.0
IV.	*tert*-Dodecyl mercaptan	0.35
V*a*.	See below	
b.	See below	
c.	Pyrophosphate (see below)	1.0
	$Fe(NO_3)_3 \cdot 9H_2O$	0.036
VIII.	*a*. $-10°$ C.; *b*. See below; *c*. See below	

Pyrophosphate (V*c*.)	Oxidant (V*a*.)	Reductant (V*b*.)	Conversion (%) in 2.5	4.5	8.5	12.5	25 Hr.
1 Na pyro.	0.037 CHP*	3 glucose	0	0	1	3	0
1 K pyro.	0.066 *t*-BIBHP*	3 fructose	0	1	2	5	16
1 K pyro.	0.05 CHP	0.07 DHA*	2	8	20	26	38
1 K pyro.	0.066 *t*-BIBHP	0.07 DHA	12	24	30	45	76
1 K pyro.	0.05 CHP	0.14 inosose	6	16	25	29	47
1 K pyro.	0.066 *t*-BIBHP	0.14 inosose	41	62	74	...	...

* CHP cumene hydroperoxide; *t*-BIBHP, *tert*-butylisopropylbenzene hydroperoxide; DHA, dihydroxyacetone.

According to Kolthoff and Youse[119] the initiation reaction in emulsion systems such as those now under consideration occurs in the aqueous phase. In the original sugar recipes, ferric salts are reduced to ferrous salts, which in turn react with peroxides. The new reductants, such as inosose, appear to be capable of reacting directly with hydroperoxides. The peroxides, even though as insoluble in water as benzoyl peroxide, react with the reductant in the water phase. In this connection, solubilization of the peroxide by the soap is a determining factor. The polymer-monomer particles, which owe their origin to the soap micelles, remain the principal locus of polymerization. Thus, despite a considerable advance in ability to accelerate emulsion polymerization processes, the theory of the process remains essentially unchanged. It seems fair to say that fundamental theory has furnished a firm foundation for ensuing developments.

The Sugar-Free Hydroperoxide-Ferrous Pyrophosphate Recipes. The direct oxidation of ferrous ions by peroxides for the initiation of polymerization was employed in Germany[100, 122, 131, 212] and considered theoretically by Baxendale, Evans, and Park[5] and by Wall and Swaboda.[210] The presence of a pyrophosphate or a similar metal-complexing agent appears necessary to obtain satisfactory rates of polymerization when ferrous salts are used as reductant initiators in butadiene-styrene polymerizations.

When formulations containing reducing sugars, such as glucose, were investigated at subfreezing temperatures, low, erratic rates of reaction were obtained. Removal of the sugar from the system led to immediate improvement. However, the concentration of ferrous ion required when a reductant was lacking was considered too high, and the latexes exhibited a tendency toward gelation and instability.[58] Subsequently, considerable improvement was effected as the result of investigations conducted on balancing the initiator ingredients; studying the influence of electrolytes, water content, and freezing-point depressants; and introducing new, more effective oxidant initiators. The following information was secured:

1. Faster rates of reaction at subfreezing temperatures can be obtained with sodium pyrophosphate than with the potassium salt, whereas the latter is more effective at temperatures above 0° C. With other recipes the substitution of potassium pyrophosphate for sodium may result in a different behavior.

2. Methanol is a satisfactory freezing-point depressant. Latex fluidity and rate of reaction can be increased by increasing the aqueous phase to about 240 parts.

3. The optimum molar ratios of the initiator ingredients, i.e., hydroperoxide, ferrous ion, and pyrophosphate, are one to one to one.

4. Satisfactory performance can be obtained with as little as 0.3 millimole of initiators per 100 grams of monomers. This corresponds to 0.083 part of crystalline ferrous sulfate.

5. To prevent cessation of reaction (dying out) at low conversions when low initiator concentrations are employed, the addition of an electrolyte, preferably potassium chloride, to the activator solution before "heat ripening" is effective. This also prevents gelation of the latex. The activator solution is prepared by dissolving ferrous sulfate, sodium pyrophosphate, and potassium chloride in distilled water in the absence of air, heating to 60° C., and cooling to room temperature. A thin slurry of ferrous pyrophosphate is thereby attained.

6. Potassium soaps, because of their high solubilities at low temperatures, are used.

7. The introduction of new hydroperoxides, including isopropylcumene hydroperoxide, tert-butylcumene hydroperoxide, and p-menthane hydroperoxide has greatly improved the performance of these systems.[36, 55, 219]

Two recipes are presented in Table XV. The results shown are indicative of what may be expected from sugar-free ferrous pyrophosphate formulations. The first is an example of a very fast formulation obtained by using high concentrations of initiator ingredients (1 millimole per 100 grams of monomers) and 3 parts each of potassium laurate and potassium myristate as the emulsifying agent. The conversion was slightly better than 60 per cent in 2 hours at −10° C. The second is a more practical formulation at a 0.3-millimole level of initiator with a rosin soap at the same temperature. That the high rates of reaction did not adversely affect the properties of the product is evident from the data of Table XVI where the physical properties of cured samples are compared with those of a sample of rosin rubber prepared at −10° C. to 69 per cent conversion in 16.3 hours in accordance with a similar recipe initiated with cumene hydroperoxide.[55] The fatty acid and accelerator contents were adjusted to obtain comparable states of

Table XV.   Two Sugar-Free Formulations for Polymerization at −10° C.

		I	II
II*a*.	Water	192	192
b.	Methanol	48	48
III*a*.	Potassium laurate	3.0	...
b.	Potassium myristate	3.0	...
c.	Dresinate 214*	...	3.5
d.	Potassium soap flakes	...	1.5
IV.	Mixed tertiary mercaptan	0.25	0.25
V*a*.	*t*-BIBHP	0.156	0.047
b.	$FeSO_4 \cdot 7H_2O$	0.278	0.084
c.	$Na_4P_2O_7 \cdot 10H_2O$	0.446	0.134
VI.	KCl	0.25	0.25
VIII*a*.	Temperature, °C.	−10	−10
b.	Time, hr.	1.9	9.4
c.	Conversion, %	59.5	60.0

* Potassium disproportionated resinate.

Table XVI.   Physical Evaluation of Butadiene-Styrene Copolymers Prepared at −10° C.

Compounding Recipe

Ingredients	Parts by Weight		
−10° C. copolymer (1.9 hr.)	100	...	...
−10° C. copolymer (16.3 hr.)	...	100	...
GR-S (X-452)	...	...	100
HAF black	50	50	50
Zinc oxide	3	3	3
Asphalt no. 6	6	6	6
Stearic acid	2	1	...
Sulfur	1.75	1.75	1.75
Benzothiazolyl cyclohexyl sulfenamide	0.60	0.95	0.80

	Cure (Min.) at 307° F.	$T/300$, p.s.i.	T, p.s.i.	E, %	Hysteresis,* T° F.	Resilience,† %	Flex,‡ M
−10° C. copolymer	30	1440	4120	606	71.6	64.5	71.5
(1.9) hr.	45	1510	3910	565	70.3	65.0	84.3
−10° C. copolymer	30	2000	4130	560	63.2	63.4	29.5
(16.3 hr.)	45	2040	4210	545	61.4	64.7	20.0
GR-S (X-452)	45	1840	3075	430	70.9	61.2	3.8
	45	1910	2860	400	68.6	60.9	2.2

Oven-Aged 24 Hours at 212° F

	Cure						
−10° C. copolymer	30	2170	3610	445	61.5	69.4	11.6
(1.9 hr.)	75	1860	3360	415	...	...	...
−10° C. copolymer	30	2980	4080	410	67.4	67.8	6.5
(16.3 hr.)	75	2680	4100	450	...	...	...
GR-S (X-452)	30	1450	2400	275	58.8	67.5	3.1
	75	1380	2900	330	...	...	...

* 100° F. oven, 143 p.s.i. load, 0.175-in. stroke.

† Yerzley oscillograph, 12 to 13 weights inertia load, 16 to 20% deflection.

‡ DeMattia, 210° F., 3-in. stroke, 500 r.p.m.

cure. Such differences as are evident between the cured properties of the −10° C. samples are attributable to the higher conversion of the rosin rubber samples.

In the actual manufacture of GR-S at 41° F., sugar-free iron-pyrophosphate recipes have now[169a] displaced the sugar-containing iron-pyrophosphate recipes first used, and, further, the more active peroxides, isopropylcumene hydroperoxide and *p*-menthane hydroperoxide, have taken the place of cumene hydroperoxide. A recipe typical at the present writing is as follows[82a]

A Sugar-Free Ferrous-Pyrophosphate Recipe for use at 5° C.

II.	Water	200
IIIa.	Potassium disproportionated rosin soap	4.0
b.	Daxad (latex stabilizer)	0.15
IV.	*tert*-Dodecyl mercaptan	0.21
Va.	Diisopropylbenzene hydroperoxide	0.08
b.	FeSO$_4$·7H$_2$O	0.20 max.
c.	K$_4$P$_2$O$_7$	0.22 max.
VIIa.	Potassium hydroxide	0.5
b.	Na$_3$PO$_4$·12H$_2$O	0.3
VIII.	*a.* 5° C.; *b.* 14 hr.; *c.* 60%	

Very fast recipes are now being developed for polymerization at 41° F., the plant application of which will require the use of equipment such as tubular reactors or heat exchangers for the purpose of controlling the temperature. There has been described an iron-pyrophosphate recipe that employs normal amounts of initiating components and gives 60 per cent conversion in 1.67 hours,[82a] and an iron-pyrophosphate recipe that employs very high levels of soap and of iron together with a new hydroperoxide (phenylcyclohexane hydroperoxide) and gives 60 per cent conversion in 11 minutes.[169a] The application of redox recipes to continuous polymerization at 122° F. in a tubular reactor, with a residence time of 15 to 20 minutes, has been described.[32a]

SYSTEMS INITIATED BY COMPOUNDS OF NITROGEN

Sapamine Recipes. A number of early patents, mostly of German origin, describe the copolymerization of diene hydrocarbons with styrene, acrylonitrile, and various acrylates. Simple polymerization systems are described which consist of comonomers, water, and nitrogen-containing emulsifiers, such as diethylaminoethoxyoleylanilide hydrochloride, diethylaminoetholeylamide hydrochloride, N-cetyl-α-betaine, etc.[8, 13, 84–6] These cationic emulsifying agents are generally obtained as free amines and require the addition of acids to render them water-soluble. Such recipes perform under the conditions stipulated if and only if oxidizing agents are present. The active oxygen required can be so readily supplied by allowing the free amines to stand exposed to the atmosphere for a few days that the significance of the phenomenon was at first apparently overlooked. However, the first redox-initiated system, which contained an oxidant, reductant, and an oxidation-reduction catalyst intentionally added to the recipe, employed such an oxidant.[51] The reductant proposed was sodium bisulfite and the catalyst was cupric sulfate in concentrations of 5×10^{-6} to 3×10^{-5} per cent of copper based on the monomers. At 50° C. the time required for 90 per cent conversion was reduced from 117 to 26 hours by the use of such a combination with air-oxidized diethylaminoetholeylamide trichloracetate.

Diazoaminobenzene Recipes. The decomposition of diazoaminobenzene, $C_6H_5N : NNHC_6H_5$, with the production of free radicals, was employed to initiate emulsion polymerization of butadiene in Russia before 1936.[3] A perusal of the original publications shows that the Russians possessed a good comprehension of the free-radical-chain mechanism of polymerization and applied an intelligent systematic technique to their investigation. Had the quality of their butadiene been better and had they made use of copolymerization, their efforts would have received more recognition. When in 1937 Wolfe[222] and Fryling[49] employed diazoaminobenzene with soap-emulsified systems, the reactions were considerably faster and could be more readily duplicated than with any emulsion polymerization system previously investigated. Later Semon and Fryling[183, 184] proposed and demonstrated the possibilities as initiators of polymerization of several water-soluble and oil-soluble aliphatic diazo compounds.

The Ferricyanide Activated Mutual Recipe. In May 1943 the du Pont Company[27] presented to the Office of the Rubber Director a complete account of investigations of emulsion copolymerization conducted for the production of synthetic elastomers from 1935 to 1942. Much of this material has since been published by Starkweather and others.[193] Several types of recipes were investigated, and 214 monomers were copolymerized with butadiene. The recipe indicated as most practical was initiated by potassium persulfate, potassium ferricyanide, and dodecyl mercaptan.[224] This recipe is as follows:

II.	Water	150
IIIa.	Sodium oleate	4.3
b.	Daxad 11	1.0
IV.	Dodecyl mercaptan	0.75
Va.	$K_2S_2O_8$ (VI)	0.75
c.	$K_3Fe(CN)_6$	0.15
VII.	Sodium hydroxide, excess	0.2
VIII.	a. 40° C.; b. 12 hr.; c. 70%	

The contribution of Starkweather and his co-workers is recommended for study because of its coverage of a great number of factors that influence the rate of emulsion polymerization, despite the fact that its conclusions and inferences cannot be applied indiscriminately to other types of initiator systems. An excellent bibliography is appended.

Considerable investigation, largely on a pilot-plant scale, was conducted under the auspices of the Rubber Director, and later Rubber Reserve Company, in an effort to achieve practical results with the above formulation. A commercial mixture of sodium and potassium ferricyanides, called Redsol, was employed as the activator. Edwards[28] found that aging a solution of Redsol and persulfate in the presence of caustic at 50° C. increased its activity. The reproducibility of a Redsol-activated rosin soap formula was found to be as good at 40 as at 50° C.   Silicates in the water were reported as increasing the rate of reaction slightly.[29]   Costanza[20] reported that copper and silicate retarded the reaction at +10° C. and that n-octyl mercaptan and potassium cyanide increased the rate when dodecyl mercaptan was present.

Since the ferricyanide-activated process has been superseded by more reactive systems, the results of numerous investigations of it will not be considered.   Troyan[203] prepared a review of this work and conducted an extensive investigation of the chemical engineering aspects of its application to a large-scale production program at 40° C. undertaken in California in 1944.   The general conclusion appeared to be that, although the presence of ferricyanide undoubtedly resulted in a considerable increase in the rate of reaction as compared with the Mutual recipe, its use was attended with difficulties arising from the sensitivity of the system to accidental retardation of various kinds.

Diazothioether Recipes.   In 1944 Reynolds and Cotten,[174] in an attempt to combine the functions of initiator and modifier in one molecule, discovered the extraordinary initiating activity of diazothioethers (also termed diazo-thiolic esters), prepared by coupling diazotized aromatic amines with aliphatic or aromatic mercaptans.   A great number of such compounds can be synthesized.   They can be made water-soluble or oil-soluble, depending on the substituent groups.   Solubilizing groups can be added to the thio component, the diazo component, or both.   Compounds varying widely in activity and able to function in different ranges of pH and at different temperatures can be prepared.   Such being the case, great versatility is to be expected on the part of diazothioether recipes, and it is not surprising that polymerizations can be conducted with widely differing emulsifiers.   Soaps, synthetic emulsifiers such as Aersol AY, and even casein have given interesting results.   Because of the involved nomenclature required to designate the diazothioethers (DTE's), RN : NSR', a series of abbreviations has been adopted.   Two compounds that have been especially important are MDN or 2-(4-methoxybenzenediazomercapto)naphthalene, prepared by coupling diazotized p-anisidine with β-naphthyl mercaptan, and α-XDN or (1-(2,4-dimethylbenzenediazomercapto)naphthalene, prepared from diazotized 2,4-dimethylaniline and α-naphthyl mercaptan.

When MDN was tested by Kolthoff[99] in the several formulations that were under investigation at the time, a number of interesting facts were revealed.   Thus MDN (0.20 part) could be substituted for potassium persulfate in the Mutual recipe.   When MDN, ferricyanide, and mercaptan were used together, a very fast polymerization rate was obtained.   The combination of 0.20 part MDN and 0.35 part $tert$-dodecyl mercaptan resulted in a 62 per cent conversion in 2 hours.   These observations led to the development of the ferricyanide-activated MDN formulations described by Kolthoff and Dale.[104]

All three substances, ferricyanide, mercaptan, and diazothioether, are required for satisfactory rates of reaction.   The elimination of any one results in extreme retardation.   The rate of reaction is greatly influenced by the pH of the system.   Increasing the alkali content increases the reaction velocity. However, highly alkaline systems tend to die out at low conversions. Tertiary mercaptans of carbon content as low as hexyl are satisfactory. Rosin soaps and disproportionated rosin soaps can be used with equal facility. The following recipe was proposed by Kolthoff and Dale as exhibiting an attractive rate:

IV.	tert-Dodecyl mercaptan (Vb)	0.2
Va.	Potassium ferricyanide	0.3
b.	MDN	0.2
VI.	Na$_2$SO$_4$	0.3
VIIIa.	50° C.	

b. Time, Hr.	c. Conversion, %
1	50
2	64
3	72

When it became urgent to conduct polymerizations on a large scale at near-zero and subfreezing temperatures, the DTE-ferricyanide recipe was selected for further investigation. Experiments conducted by Thompson in cooperation with Reynolds[62, 199] were discussed by Fryling[58] and his co-workers. Various freezing-point depressants were investigated in the following recipe:

II.	Water	250
III.	Dresinate 731 (pH = 10.3)	5.0
IV.	Mixed tertiary mercaptan (Vb)	0.40
Va.	Potassium ferricyanide	0.30
b.	MDN	0.30
VI.	Na$_3$PO$_4$ · 12H$_2$O	0.50
VIII.	a. 5° C.; b. 3.5 hr.; c. 60%	

Kinetic data on the influence of freezing-point depressants on the above recipe at $+5°$ C. are tabulated in Table XVII. Glycerol and glycol retarded only slightly. Methanol was selected for further investigation

Table XVII. Conversions with Various Freezing-Point Depressants at $+5°$ C. with the MDN-Ferricyanide Recipe[58]

Substance	Water-Antifreeze Ratio, Parts	Conversion (%) in 4	7.5	24 Hr.
Control	250 : 0	68	...	...
Isopropanol	200 : 50	1	...	3
Ethanol	200 : 50	0	...	4
Methanol	200 : 50	16	30	62
Acetone	200 : 50	4	...	25
Acetonitrile	200 : 50	24	36	38
Dioxane	200 : 50	13	30	60
Ethylene glycol	200 : 50	47	71	88
Glycerol	175 : 75	61	70	72

and for use on a large scale. At lower temperatures of polymerization (-10 and $-18°$ C.), closely similar rates of polymerization were subsequently obtained with methanol and with ethylene glycol.

A number of changes in the above formulation were required for satisfactory operation at subfreezing temperatures. Substitution of potassium oleate for Dresinate 731 increased polymerization rates from 8 to 12 times. For fast reactions, rather large amounts of mercaptans were required. The copolymers were consequently overmodified. This led to the use of the nonmodifying, n-hexadecyl mercaptan.[52] Further increase in rate was obtained by using α-XDN. The recipe follows:

IIa.	Water	175
b.	Methanol	75
III.	Potassium oleate	5.0
IV.	See Vb.	
Va.	Potassium ferricyanide	0.30
b_1.	α-XDN	0.30
b_2.	n-Hexadecyl mercaptan	1.00
VI.	$Na_3PO_4 \cdot 12H_2O$	0.50
VII.	KOH (excess)	0.10
VIII	a_1. $-10°$ C.; b_1. 20 hr.; c_1. 60%	
	a_2. $-18°$ C.; b_2. 31 hr.; c_2. 58%	

The use of diazothioether recipes for production of low-temperature rubber on a large scale has been superseded by the sugar and sugar-free hydroperoxide-iron pyrophosphate formulations. The principal reason for this has been the bother and slightly greater cost of preparing the DTE as compared with pyrophosphate activators. However, it is believed that further progress can still be made by investigation of diazothioether initiation.

Peroxamine Recipes. "Peroxamine" is suggested to designate a low-temperature initiator system developed by Whitby.[217] The addition of one or two drops of certain complex amines, such as tetraethylenepentamine, to approximately one gram of benzoyl peroxide contained in a test tube affords a spectacular demonstration of a free-radical reaction. Instantaneous decomposition occurs, and a large cloud of smoke containing biphenyl is vigorously expelled from the tube. This behavior suggested to Whitby the possibility of using a combination of hydroperoxide and amines for the initiation of polymerization at low temperatures.

Preliminary investigations by Whitby showed that the combination of cumene hydroperoxide (CHP) and tetraethylenepentamine (Tepa) was the most active available and that the polymerization product could be readily modified with mercaptans. Soaps prepared by reacting fatty acids with amine were without activity. In other words, it is the free amine that is the reductant initiator. The use of an amine with a peroxide oxidant is not an entirely novel idea.[123] However, nothing of this type so active as the system proposed by Whitby has previously been known. The combination of persulfate and dimethylamine was used at Leverkusen for effecting the copolymerization of butadiene and acrylonitrile.[32, 93] A similar system, consisting of hydrogen peroxide and dicyandiamidine, with an iron-cobalt catalyst, was proposed by Stewart and Fryling.[196] Whitby mentions a patent[205] on the use of alkyl polyamines and benzoyl peroxide to promote the gelation of an unsaturated alkyd-styrene mixture.

An early recipe for peroxamine polymerization, taken from a report by Hobson and D'Ianni,[78] is given in Table XVIII.

Table XVIII.   Example of a Peroxamine Recipe

III.	Potassium O.R.R. soap	4.5
IV.	Mixed tert. mercaptans	0.2
Va.	Cumene hydroperoxide	0.2
b.	Tetraethylenepentamine	0.2
VI.	Potassium chloride	0.5
VII.	Triton R-100	0.1
VIII.	a. 5° C.; b. 8 hr.; c. 60%	

The greater activity of hydroperoxides of higher molecular weight, as compared with CHP, in a Peroxamine recipe was reported was Schulze.[177] The data are given in Table XIX.   It is evident that the relative reactivities of hydroperoxides in the Peroxamine recipe are similar to those exhibited in the sugar and sugar-free pyrophosphate formulations.

Table XIX.   Active Hydroperoxides in the Peroxamine Recipe

IV.	Mixed tertiary mercaptans	0.1
Va.	See below	
b.	Tetraethylenepentamine	1.50
VI.	KCl	0.40
VIIIa.	Temperature, °C.	+5

(Va) Hydroperoxide of (2.27 millimoles)	Conversion (%) in	
	2.5	13.7 Hr.
Cumene	33	92
Diisopropylbenzene	43	100
Triisopropylbenzene	66	...
Cyclohexylbenzene	84	...
tert-Butylisopropylbenzene	66	...

After disclosure of the Peroxamine recipe, active development work on it was carried out,[22, 30, 82a, 154, 188] and Peroxamine recipes are now employed in the manufacture by polymerization at 5° C. of certain types of GR-S and of GR-S latex.   In this connection peroxides used are isopropylcumene, and p-menthane hydroperoxides; polyamines used are tetraethylenepentamine, triethylenetetramine, and diethylenetriamine, the last of these amines, although the least active, being preferred in plant operation because of its economy, uniformity, and availability.

Kolthoff and Meehan[116] have presented data on the influence of iron salts and phosphates on the rate of reaction in Peroxamine-type recipes.   Potassium orthophosphate was somewhat more active than potassium pyrophosphate.   That high rates of polymerization are attainable at −10° C. is evident from the data of Table XX.   The ferric nitrate content required to obtain the highest rate is about 0.01 part.

Table XX.   Conversions at $-10°$ C. in a Peroxamine Recipe
with Variable Iron Content

IIa.	Water	140
b.	Methanol	40
IV.	tert-Dodecyl mercaptan	0.20
Va.	tert-Butylisopropylbenzene hydroperoxide	0.53
b.	Tetraethylenepentamine	0.75
c.	See below	
VIIa.	K_3PO_4	0.50
b.	KOH	0.01 N

| (Vc) $Fe(NO_3)_3 \cdot 9H_2O$, | Conversion (%) in | | | |
Parts	1	2	4	8 Hr.
0.0000	...	10	21	53
0.0005	...	17	30	67
0.0010	6	24	34	70
0.0036	...	24	35	43 ?
0.0050	20	32	51	100
0.009	17	37	62	...
0.036	22	38	61	...

The rate of polymerization in Peroxamine recipes is influenced by the presence of traces of iron. The proportion of iron giving an optimum rate depends on the particular polyamine used and the temperature. As just indicated, at $-10°$ C. in the presence of the pentamine, a suitable proportion is 0.01 part ferric nitrate (16 p.p.m. iron on the monomers). At 5° C., with the same polyamine, the very minute quantity of iron present as impurity in the reagents (especially in the soap) is sufficient.

Peroxamine recipes have proved to be less readily adaptable than iron-pyrophosphate recipes to polymerizations in which the emulsifier is wholly or in large part not fatty acid soap but disproportionated rosin soap. But recently Howland and his co-workers have described a recipe, containing diethylenetetriamine along with a little iron (12 p.p.m. on the monomers) and a supplementary reducing agent in the form of sodium hydrosulfite, that is capable of bringing about polymerization at a reasonable rate at $-18°$ C. in the presence of Dresinate soap as the major emulsifying agent.[82a] The Government Laboratories have developed a Peroxamine recipe giving 60 per cent conversion in 15 hours at $-29°$ C.[67a]

Bearing on the theory of Peroxamine polymerization are some interesting studies made by Williams and his associates on the kinetics of the action of iron on cumene hydroperoxide,[40–42] isopropylcumene hydroperoxide,[157] tert-butylcumene hydroperoxide,[158] and p-menthane hydroperoxide,[159] and on the kinetics of the reactions between polyethylene polyamines on the one hand and cumene hydroperoxide[160] and hydrogen peroxide on the other.

MISCELLANEOUS EMULSION POLYMERIZATION FORMULATIONS

The number of possible emulsion polymerization recipes differing in regard to the emulsifiers and initiators used in them is great. In general the behavior of monomers and comonomeric mixtures is similar in various formulations. Yet differences must be anticipated. For example, isoprene, which polymerizes more slowly than butadiene in the Mutual recipe, may react faster with a redox system. Acrylonitrile increases the rate of polymerization when admixed with butadiene in the hydrogen peroxide-ferric pyrophosphate recipe, but in the absence of butadiene its polymerization is totally inhibited by the system. The reasons for such vagaries are almost completely unrecognized. It is fortunate therefore that a variety of different emulsion polymerization systems is available. The systems discussed briefly in what follows may be of value under exceptional circumstances, or they may serve as the starting point for further investigation.

Differences in the behavior of different monomers in various systems can probably be attributed to differences in the free radicals involved in chain initiation. The possibility that hydroxyl free radicals may be the initiating agents in certain formulations while mercaptan free radicals perform similarly in other systems has received consideration by a number of investigators. However the average life of free radicals is so brief and the actual number required for initiation of chain processes is, relatively speaking, so small that identification of such entities is always difficult. One factor that may account for variable monomer behavior and that has received inadequate consideration is hydrolysis. The pH ranges in which emulsion polymerizations are generally conducted are conducive to this reaction.

Initiation by Radiant Energy. Although both theory and experience support the thesis that the properties of the product of emulsion polymerization depend not on the mode of initiation but on the environment in which chain growth occurs, the evidence afforded by photochemical initiation appeared when first considered to be in conflict with this conclusion. Reference is made to a formulation (given in Table XXI) of Nudenberg and Kharasch,[155] who conducted the polymerization of butadiene in the presence of the photosensitizer, uranyl acetate. The light from an ordinary Mazda lamp at room temperature afforded sufficient energy to induce a rapid polymerization.

Table XXI. Uranyl Acetate-Photosensitized Initiator System

I.	Butadiene	100
II.	Water	240
III.	Potassium stearate ("peroxide-free")	5.2
IV.	Modifier	None
V*c*.	Uranyl acetate	0.053
d.	100 watt lamp at 12–18 in.	...
VIII.	*a*. 30° C.; *b*. 3 hr.; *c*. 60%	

The reaction did not proceed in the dark. Oxygen in a concentration of 0.1 mole per cent on the monomer completely inhibited the reaction during the time required for a control run. The reaction did not occur when a dodecylammonium soap was the emulsifier and only to a slight extent when Nekal BXG was the emulsifier. Modifiers were without effect on the properties. Finally, the product resembled cured rubber. However, Carr, Bebb, and B. L. Johnson[17] have reported results with a modification of uranyl-photoactivated emulsion polymerizations which are in better agreement with those obtained with other recipes. Their work indicates that photoactivated initiation does not influence the properties of the product to any significant extent.

Redox Recipes With Synthetic Anionic Emulsifiers. Wartime shortage of fatty acids was responsible for the utilization of synthetic emulsifying agents by the Germans in their efforts to develop low-temperature emulsion polymerization recipes. News of these developments after the war created widespread interest in this type of recipe, and a number of investigations were undertaken. Two difficulties were encountered: The emulsifying agents were either inadequately characterized or impossible to duplicate, and the unmodified rubber proved to be ill-suited to American manufacturing methods. The former of these made it difficult to correlate results between various laboratories. However P. H. Johnson and Bebb,[88] starting with a German recipe, eventually developed a satisfactory formulation. Their optimum formulation for polymerization at $+10°$ C. is presented in Table XXII.

Table XXII. Recipe for Copolymerization at 10° C.

II.	Water	200
IIIa.	Sodium oleate ($pH = 8.6$)	1.5
b.	Sodium hydrocarbon sulfonate (MP-189-S)*	3.5
IV.	Modifier	none
Va.	Benzoyl peroxide	0.5
b.	(See below)	0.6
c.	$Na_4P_2O_7$ (anhydrous)	0.2
	$Fe(NH_4)_2(SO_4)_2.6H_2O$	0.35
VIII.	a. 10° C.; b. 6 hr.	

Reductant (Vb)	Conversion, % (VIIIc)
None†	44
Sorbose	61
Dextrose	33
Fructose	46
Invert sugar	45
Dihydroxyacetone	37

* Supplied by du Pont.
† In this case the ferrous sulfate is the reductant.

To obtain the results in Table XXII the use of a portion of sodium oleate was required. Without this, and also at a higher temperature, the addition of organic reductants had no effect on the rate of reaction. The system was retarded by oxygen. As with many redox systems, aging the activator system improved the rate. Aging was carried out on the total aqueous phase after addition of ferrous ammonium sulfate at room temperature. Twenty minutes was found to be optimum. Unmodified copolymers prepared at 10° C. broke down rapidly on the mill. The addition of small amounts of mercaptan modifiers led to the production of polymers which were easier to process but the physical properties of which were reported as being no better than those of GR-S. The substitution of isoprene for a mixture of butadiene and styrene required that the recipe be "rebalanced." Decreases were made in the content of benzoyl peroxide, sodium oleate, and ferrous ammonium sulfate, while that of MP-189-S was increased. Under these conditions the polymerization of isoprene was somewhat faster than that of butadiene–styrene.

Copolymerizations were conducted at −40° C. by the inclusion of glycerol as an antifreeze.[90] Table XXIII illustrates the alteration of formula required to obtain polymerization at this remarkably low temperature.

Table XXIII. Recipe for Copolymerization at −40° C.

IIa.	Water	113
b.	Glycerol	263
IIIa.	Potassium oleate	5.0
b.	MP-189-S	3.5
IV.		None
Va.	Benzoyl peroxide	0.5
b.	Sorbose	0.6
c.	$Na_4P_2O_7$ (anhydrous)	0.35
	$Fe(NH_4)_2(SO_4)_2.6H_2O$	0.60
VIII.	a. −40° C.; b. 96 hr.; c. 60%	

It has been shown by Schulze, Tucker, and Crouch[182] that results substantially similar to those of P. H. Johnson and Bebb are obtained when synthetic anionic emulsifiers are substituted for fatty and rosin acid soaps in the formulations initiated by hydroperoxides.

Activation by Cobalt and Other Variable Valence Metallic Compounds. Mention only will be accorded this possibility. The reader is referred to published material by Stewart,[195] Marvel,[24, 137] and Mitchell.[152] As yet, none of these systems has assumed any degree of commercial importance. However compounds such as sodium cobaltinitrite and the iron complexes formed with sodium ethylene–dinitrilotetracetate and many other similar substances will continue to interest those engaged in the development of emulsion polymerization systems. Kolthoff and Meehan have described a redox recipe in which potassium dichromate is the oxidant and arsenious oxide the reductant.[116a]

The Ferrous Sulfide Recipe. Theoretical considerations indicate that the efficiency of utilization of peroxide initiators can be increased by maintaining an extremely low concentration of ferrous ions.[5] Kolthoff and his co-workers[111, 114] proposed to utilize the low solubility of ferrous sulfide to accomplish this. The resulting ferrous sulfide formulation has assumed some importance for the preparation of low-temperature latexes for use as such.[187] Description of this recipe is based on a summary report prepared by Kolthoff, Medalia, and Held.[111]

A precipitate of ferrous sulfide is formed on mixing ferrous sulfate and sodium sulfide in the charge. Excess of either reagent reacts rapidly with cumene hydroperoxide and results in an initial decrease of hydroperoxide content. The soluble ferrous ion content from solution equilibrium of the ferrous sulfide reacts with the CHP to form free radicals. The sulfide ion may react with CHP or with ferric iron, thereby regenerating ferrous sulfide. When first mixed the system exhibits a blue-black color, which disappears during the course of the polymerization or on admission of oxygen. Pyrophosphate appears to exert only a neutral electrolyte effect in maintaining the latex in a fluid state since it can be replaced by sodium sulfate without decrease in rate of reaction.

According to Neklutin, Westerhoff, and Howland,[154] the optimum rate of polymerization in the ferrous sulfide recipe is obtained when the hydroperoxide, ferrous sulfate, and sodium hydrosulfide charged (the last to generate ferrous sulfide in situ) are in a 1.5 : 1 : 1.5 molar ratio. The rate of reaction between oxidant and reductant is such that both are present in the latex up to high conversion.[111] The composition of a recipe recommended for use at 5° C. is given[154] in Table XXIV.

Table XXIV. A Ferrous Sulfide Recipe

III.	K.O.R.R. soap	4.7
IV.	Mixed tertiary mercaptans	0.25
Va.	Cumene hydroperoxide	0.10
b_1.	Sodium hydrosulfide	0.022
b_2.	Ferrous sulfate · $7H_2O$	0.10
VII.	Potassium hydroxide	0.02
VIII.	a. 5° C.; b. 11 hrs.; c. 60%	

If the sodium salt of disproportionated rosin acid (Dresinate 731) is used as the emulsifier in place of potash fatty acid soap, it is necessary, in order to secure a similar rate of reaction to that indicated in the table, to increase somewhat the concentrations of peroxide and ferrous sulfide.

Many other variations of the iron-peroxide recipe can be devised. A recipe in which the iron is present as ferric Versenate and in which hydrazine is included as a reducing agent may be instanced.[116b]

GENERAL CONSIDERATIONS PERTAINING TO THE INFLUENCE OF POLYMERIZATION VARIABLES ON EMULSION POLYMERIZATION

With so many variables to be manipulated and with so many properties to be considered, successful investigation of emulsion polymerization requires the development of an exacting technique. All procedures must be conducted in an identical manner both in preparing the reaction mixture and in handling the product. Finally the investigator must be ever observant to detect any unforeseen aberrations of procedure.

The Rate of Polymerization. There are three valid reasons for research directed toward increasing the rate of emulsion polymerization. Faster reactions reduce overhead expenses and make possible the use of smaller, less expensive equipment. If the rate of polymerization can be speeded up, less pure and consequently cheaper reaction ingredients can be used. Finally, the more vigorous processes can be conducted at lower temperatures, thereby securing products exhibiting improved properties.

Almost every polymerization variable influences the rate of polymerization either adversely or favorably. Progress has been made in increasing the rate of polymerization by following four more or less interrelated lines of attack, viz.: (1) the provision of more highly purified and therefore less retarded ingredients, (2) the discovery of new types of initiator systems, (3) the discovery of more efficient initiator ingredients such as the alkyl aromatic hydroperoxides and the polyamines, and (4) "balancing" the recipe, i.e., adjusting all the concentrations to give optimum rates of reaction. This last procedure requires meticulous attention to details and is generally conditioned by the necessity of making compromises of a practical nature. One cannot use all the iron-containing catalyst or all the soap that might seem desirable without conferring harmful incidental properties on the product and unduly increasing the cost of the process. In balancing the recipe, peculiar and unexpected polymerization variables are at times encountered, examples of such being the effect of aging one or more of the initiator ingredients or the effect of small concentrations of electrolytes in preventing premature cessation of reaction. Attention must be devoted to induction periods, termination of reaction, elimination of the tendency frequently encountered for the reaction to stop of its own accord, and the average rate of polymerization. The advances made during recent years in the technique of conducting emulsion polymerization are evident from the high rate of reaction now attained at subfreezing temperatures. On the assumption of a normal temperature coefficient of 2 per 10° C., the reaction rate has been increased by a factor of approximately 300.

The Behavior and Properties of Synthetic Latex. The preparation of a satisfactory latex is essential to a successful investigation of emulsion polymerization. A partially precoagulated product introduces errors in measuring the rate of polymerization because of inability to obtain representative samples. Gelation of latex, by decreasing the transfer of heat and thereby raising the temperature, may raise the rate of reaction and bring about premature cessation of the process. The difficulties encountered in coagulating an unsatisfactory latex frequently confer deleterious incidental properties on the product. Poor latexes, when encountered in commercial production, can be the cause of serious economic losses. When it is considered that the stability of synthetic latex is impaired by the inclusion of modifiers and freezing-point depressants, it is evident that insufficient fundamental research has yet been devoted to this important problem.

The fluidity of latex varies with the amount of water, conversion, and particle size. Since the number of latex particles is practically established at 20 per cent conversion, at which point the soap is no longer in aqueous solution, the soap interface must be expanded as the system reacts up to 60 per cent where the monomer phase disappears. It is significant that gelation and precoagulum formation, when encountered, become evident at conversions of approximately 50 per cent and higher. Although it may appear anomalous that decreasing the soap content can correct such conditions to a certain extent, this observation is in exact accord with theory. By decreasing the concentration of soap, the number of polymer-monomer particles is decreased, the average particle size increased, and less viscous latex is obtained.   Of course, the rate of reaction is thereby decreased.   This is one of the expedients employed in making type-III latex with a solids content of 41 per cent compared with 29 per cent usually obtained in following the Mutual recipe[87] (cf. Chapter 19).

Gelation of latex is especially marked with electrolyte-free initiator systems, such as diazoaminobenzene.   Gelation can be usually suppressed by the addition of small amounts of electrolyte in the concentration range of 0.05 to 0.30 part.[60]   Greater concentrations bring about flocculation of the polymer-monomer particles.

Precoagulation of latex may become a very serious impediment to successful polymerization.   It is characterized by the formation of large masses of elastomer plasticized by unreacted comonomer which gives an objectionable degree of stickiness to the material.   When analyzed, the precoagulum is found to be practically free of soap or fatty acid, indicating that the emulsifying agent has transferred to some other portion of the system.   A clue to one cause for precoagulum formation was afforded by investigation of substances similar to those first proposed by Hofmann.[6]   Organic colloids such as gelatin and albumin at very low concentrations were found to promote the formation of precoagulum and to increase the particle size of the latex.[54] The range of concentration of albumin which brought about precoagulum formation was 0.01 to 0.10 part per 100 of monomers.   Higher concentrations appeared to exercise a stabilizing effect.   One particularly annoying example of precoagulum formation on a large scale was traced by McKenzie[128] to an infusion of gallinippers which had accidentally gotten into the soap solution.

Inherent Elastomer Properties. Theory, in agreement with experience, indicates what reaction variables influence the inherent properties of elastomers prepared by emulsion polymerization.   The variables that determine composition, structure, and molecular weight are those that influence the environment within the polymer-monomer particles during the process of chain growth.   Therefore one would not expect changes in the emulsifier, pH, or initiator system to affect the inherent properties of the resulting elastomer.   The variables with which the investigator must operate in his efforts to produce a tailor-made molecule are the following: (1) monomers and comonomers, (2) the ratio of comonomers, (3) the concentration of the modifier, which in turn is a function of the specific properties of the modifying substance, (4) certain cross-linking agents, such as divinylbenzene, which may be deliberately or accidentally added to the systems, (5) percentage conversion, and (6) the temperature of reaction.

Considerable information concerning these variables is to be found in the section devoted to the Mutual recipe.   Intensive study of the Mutual system showed it to be well balanced within a framework of certain practical limitations which could not be profitably exceeded.   The conviction that an improvement in properties could be obtained by operating at lower reaction temperatures gradually became a certainty as the development of fast low-temperature recipes progressed.   The improvement was first shown by laboratory tests, [34, 66, 89, 151, 158, 218] and has been substantiated by road tests on tires (see Chapter 21).   Not only have service tests shown the wear resistance of rubber made at low temperatures to be better than that of regular GR-S, but also the absence of cracking and chipping in cold-rubber tires has been noteworthy. There is a good prospect that in time it will no longer be necessary to confine the manufacture of large bus and truck tires to natural rubber.

In addition to its influence per se, the lower temperature of reaction induced certain variations in the behavior of four of the other five variables which govern the properties. These changes will be discussed in some detail. In passing, mention should be made of the possibility that future investigation may show pressure to be another such variable.

Influence of Reaction Temperature on Properties Determined by Monomers and Comonomer Ratios. Elastomers prepared by emulsion copolymerization from butadiene, styrene, and acrylonitrile were introduced commercially by the Germans[37] in 1937. Why these three monomeric materials have remained pre-eminent is a matter of interest. Butadiene, being the lowest-molecular-weight conjugated diene, possesses certain unique properties. However other comonomeric materials are available, and their investigation also offers a sphere of opportunity for future research. To date, however, unique advantages attaching to them, other than that of resistance to swelling by oil exhibited by the nitrile copolymers, have not been demonstrated (cf. Chapter 21). This indicates that structure and molecular weight determine the properties of an elastomer to a greater extent than composition.

Evidence for the influence of structure on elastomer properties is afforded by the marked difference between polybutadiene prepared at subfreezing temperatures and at temperatures higher than 30° C. The low-temperature polymer possesses a more regular chain structure. X-ray diffraction patterns demonstrate crystallinity with double bonds in the *trans* configuration.[7] This improved structure of low-temperature rubber extends the optimum elastomer range of compositions well toward the polybutadiene limit. Copolymers containing 5 to 10 per cent initial styrene content are promising for use as general-purpose rubbers. They exhibit higher rebound, better flexibility, and satisfactory tensile properties. Similar materials prepared to reasonable conversions at 50° C. show such poor working and tensile properties that their value is definitely limited. A copolymer made at −18° C. from a butadiene-styrene charge of 90/10 was superior in vulcanization tests to standard GR-S (made at 50° C. from a 71/29 charge).[58, 62]

Influence of Reaction Temperature on Modifier Requirements. Since modifiers are chain-transfer agents, it is evident that their function is simply that of reducing the average molecular weight and possibly influencing the distribution of molecular weights. They apparently have little or no influence on the degree of branching, cross-linkage, regularity of structure, 1,2- versus 1,4-addition, or *cis-trans* configuration. If the temperature of polymerization is lowered, the modification of the product may be decreased or increased, depending on which of the above-mentioned factors is influenced to the greatest extent. When recipes that could be employed satisfactorily at from 15 to 30° C. were first investigated, the indications were that the products exhibited higher Mooney viscosities and that the modifier requirements were high. This was in accord with the well-known principle that reduction of the temperature of polymerization increases the molecular weight of polymers. However, the products, while exhibiting high inherent viscosities, were frequently obtained in a gel-free condition, thus indicating greater linearity of polymer structure. It now appears that copolymerizations conducted at subfreezing temperatures require considerably less modifier to attain a standard Mooney viscosity than does GR-S. Apparently the greater regularity of structure obtained at such temperatures has a more pronounced effect in lowering Mooney viscosity than the increased molecular weight has in the opposite direction. The product obtained under these conditions would undergo less degradation than usual in milling, and this may account in part for the better mechanical properties exhibited by low-temperature rubbers.

One additional consideration pertinent to the utilization of mercaptans at low temperatures requires attention. The degree of modification is influenced by the type of freezing-point depressant employed.[57] Recipes containing glycerol need two to three times the quantity of mercaptan required by recipes containing methanol. This behavior is in accord with views advanced by Reynolds[173] to the effect that the solubility and rate of diffusion of mercaptans in the aqueous medium are among the determining factors in modification.

Influence of Reaction Temperature on Properties that Vary with Conversion. Although it is apparent that the effect of percentage conversion on the properties of the elastomer can be attributed in great measure to an excess of modifier and monomer in the polymer-monomer particles at the start of the reaction and to a deficiency at high conversions, this does not appear to account for the regular variation of composition of the product. At subfreezing temperatures of polymerization it has been found that compositions vary less with conversion than is indicated by Meehan's formulation.[144] A series of copolymerizations was conducted at −20° C., and the products were analyzed by measuring their refractive indexes.[166] To insure that the copolymers should be soluble in benzene, i.e., free from gel, the initial concentrations of modifier were varied in such a way that all the samples were of approximately the same Mooney and inherent viscosity.

Table XXV.   Variation of Composition with Conversion of a 70/30 Initial Butadiene-Styrene Mixture Polymerized at −20° C.[166]

Conversion, %	Percentage Styrene in Copolymer	
	Observed	Calculated
10	24.8	21.4
20	24.4	21.8
30	24.0	22.4
40	23.7	22.9
50	23.5	23.5
60	23.5	24.1
70	23.7	24.9
80	24.3	25.9
90	25.4	27.3
100	(30.0)	(30.0)

A low-temperature, diazothioether-glycerol recipe was employed. The results were checked by Meehan[116] with a cumene hydroperoxide-dihydroxy-acetone recipe at 0° C. Meehan's values were in agreement with those obtained previously at −20° C. by refractive index. The data presented in Table XXV were obtained from a smooth curve drawn through the experimental values obtained by the refractive index method. For comparison, the results calculated according to Meehan's formulas are also presented. The variation of composition with conversion of the elastomer produced at −20° C. is seen to be less than that observed for GR-S. These data indicate the possibility of obtaining satisfactory products by polymerization at subfreezing temperatures at conversions up to 90 per cent.

REFERENCES

1. Azorlosa, J. L., *Ind. Eng. Chem.*, **41**, 1626–9 (1949).
2. Baker, W. O., and Heiss, J. H., Private Communication to O.R.R., May 11, 1944.
3. Balandina, V., Berezan, K., Dobromyslova, A., Dogadkin, B., and Lapuk, M., *Bull. Acad. Sci. U.S.S.R.*, **1936**, 423–33.
4. Barron, H., *Modern Synthetic Rubbers*, 2d. ed., Van Nostrand, New York, 1943.
5. Baxendale, J. H., Evans, M. G., and Park, G. S., *Trans. Faraday Soc.*, **42**, 155–69 (1946).
6. Bayer and Co., Brit. Pat. 14,566 (1912); Ger. Pats. 254,672 and 255,129 (1912).
7. Beu, K. E., Reynolds, W. B., Fryling, C. F., and McMurray, H. L., *J. Polymer Sci.*, **3**, 465–80 (1948).
8. Bock, W., and Tschunkur, E., U.S. Pats. 1,898,522 (1933); 1,938,730 (1933).
9. Bovey, F. A., and Kolthoff, I. M., *J. Am. Chem. Soc.*, **69**, 2143–53 (1947).
10. Breuer, F. W., *India Rubber World*, **109**, 585–6, 590; **110**, 55–6, 63, 172–3, 301–3 (1944).
11. Bristol, K. E., Borders, A. M., and Osterhof, H. J., Private Communication to O.R.R., Sept. 5, 1943.
12. Brucksch, W. F., Messer, W. E., and Howland, L. H., Private Communication to O.R.R., Mar. 14, 1943.
13. Calcott, W. S., U.S. Pat. 2,161,949 (1939).
14. Carlin, R. B., Private Communication to O.R.R., July 16, 1943.
15. Carr, C. W., Kolthoff, I. M., Meehan, E. J., and Stenberg, R. J., *J. Polymer Sci.*, **5**, 191–200 (1950).
16. Carr, C. W., Kolthoff, I. M., Meehan, E. J., and Williams, D. E., *J. Polymer Sci.*, **5**, 201–6 (1950).
17. Carr, E. L., Bebb, R. L., and Johnson, B. L., Private Communication to O.R.R., Oct. 19, 1949.
18. Corrin, M. L., Private Communication to O.R.R., Dec. 30, 1946.
19. Corrin, M. L., *J. Polymer Sci.*, **2**, 257–62 (1947).
20. Costanza, A. J., Private Communication to O.R.R., Jan. 13, 1949.
21. Cuthbertson, G. R., Coe, W. S., and Brady, W. L., *Ind. Eng. Chem.*, **38**, 975–6 (1946).
22. Davidson, M. J. G., Embree, W. H., and Williams, H. L., Presented to Div. High Polymer Chem., Chemical Inst. Canada, Montreal, June, 1952.
23. Deanin, R., Claus, C. J., Wyld, M. B., and Seitz, R. L., Private Communication to O.R.R., June 3, 1947.
24. Deanin, R., Lindsay, R. D., and Leventer, S. E., *J. Polymer Sci.*, **3**, 421–32 (1948).
25. Dornte, R. W., Private Communication to O.R.R., Sept. 1943.
26. Dunbrook, R. F., *India Rubber World*, **117**, 203–7 (1947).
27. Du Pont de Nemours and Co., E. I., Private Communication to O.R.R., May 25, 1943.
28. Edwards, B. C., Private Communication to O.R.R., Feb. 13, 1946.

29. Edwards, B. C., Private Communication to O.R.R., June 13, 1946.
30. Embree, W. H., Spolsky, R., and Williams, H. L., *Ind. Eng. Chem.*, **43**, 2553–9 (1951).
31. Ewart, R. H. and Hulse, D. E., Private Communication to O.R.R., Aug. 26, 1942.
32. Falk, Wissenschaftliche Kautschuk Kommission, July 1944.
32a. Feldon, M., McCann, R. F., and Laundrie, R. W., *India Rubber World*, **128**, 51–3, 63 (1953).
33. Fennebresque, J. D., Private Communication to O.R.R., Jan. 18, 1944.
34. Fielding, J. H., *Ind. Eng. Chem.*, **41**, 1560–4 (1949).
35. Fikentscher, H., *Angew. Chem.*, **51**, 433 (1938).
36. Fisher, G. S., Goldblatt, L. A., Kniel, I., and Snyder, A. D., *Ind. Eng. Chem.*, **43**, 671–4 (1951).
37. Fisher, H. L., The Origin and Development of Synthetic Rubbers, Symposium Am. Soc. Testing Materials, Cincinnati, Mar. 2, 1944.
38. Flory, P. J., *J. Am. Chem. Soc.*, **59**, 241–53 (1937).
39. Flory, P. J., Private Communication to O.R.R., Apr. 20, 1945.
40. Fordham, J. W. L., and Williams, H. L., *J. Am. Chem. Soc.*, **72**, 4465–9 (1950).
41. Fordham, J. W. L., and Williams, H. L., *Can. J. Research*, **28B**, 551–5 (1950).
42. Fordham, J. W. L., and Williams, H. L., *J. Am. Chem. Soc.*, **73**, 1643–7 (1951).
43. Frank, R. L., et al., Private Communication to O.R.R., May 2, 1946.
44. Frank, R. L., Smith, P. V., Woodward, F. E., Reynolds, W. B., and Canterino, P. J., *J. Polymer Sci.*, **3**, 39–49 (1948).
45. Fryling, C. F., Private Communication to O.R.R., Mar. 12, 1942.
46. Fryling, C. F., Private Communication to O.R.R., Mar. 26, 1942.
47. Fryling, C. F., Private Communication to O.R.R., May 21, 1942.
48. Fryling, C. F., Private Communication to O.R.R., Sept. 12, 1943.
49. Fryling, C. F., U.S. Pat. 2,313,233 (1943).
50. Fryling, C. F., *Ind. Eng. Chem., Anal. Ed.*, **16**, 1–4 (1944).
51. Fryling, C. F., U.S. Pat. 2,379,431, July 3, 1945.
52. Fryling, C. F., U.S. Pat. 2,416,440, Feb. 25, 1947.
53. Fryling, C. F., *Ind. Eng. Chem.*, **40**, 928–32 (1948).
54. Fryling, C. F., Fauske, S. C., and Burleigh, J. E., Private Communication to O.R.R., Aug. 4, 1949.
55. Fryling, C. F., and Follett, A. E., *J. Polymer Sci.*, **6**, 59–72 (1951).
56. Fryling, C. F., and Harrington, E. W., *Ind. Eng. Chem.*, **36**, 114–7 (1944).
57. Fryling, C. F., and Landes, S. H., Private Communication to O.R.R., Sept. 30, 1947.
58. Fryling, C. F., Landes, S. H., St. John, W. M., and Uraneck, C. A., *Ind. Eng. Chem.*, **41**, 986–91 (1949).
59. Fryling, C. F., and St. John, W. M., Private Communication to O.R.R., Feb., 12, 1947.
60. Fryling, C. F., and St. John, W. M., *Ind. Eng. Chem.*, **42**, 2164–70 (1950).
61. Fryling, C. F., Sundet, S. A., Pfau, E. S., and Houston, R. J., Private Communication to O.R.R., Dec. 1943.
62. Fryling, C. F., Thompson, R. D., and Landes, S. H., Private Communication to O.R.R., Sept. 30, 1947.
63. Fryling, C. F., and Zwicker, B. M. G., Private Communication to O.R.R., Mar. 23, 1943.
64. Gander, R., Private Communication to O.R.R., July 11, 1944.
65. Garvey, B. S., *Ind. Eng. Chem.*, **34**, 1320—3 (1942).
66. Goodrich Co., The B. F., Private Communication to O.R.R., Dec. 1941.
67. Goodrich Co., The B. F., Private Communication to O.R.R., Oct. 29, 1943.
67a. Government Laboratories, Private Communication, Jan. 1953.
68. Haden, R. L., Private Communication to O.R.R., May 28, 1943.
69. Harkins, W. D., Private Communication to O.R.R., Dec. 6, 1945; Apr. 11, 1946; Dec. 30, 1946, and June 15, 1948.

70. Harkins, W. D., *J. Chem. Phys.*, **13**, 381 (1945); **14**, 47–48 (1946); **16**, 156–7 (1948); *J. Am. Chem. Soc.*, **69**, 1428–44 (1947); *J. Polymer Sci.*, **5**, 217–51 (1950); *The Physical Chemistry of Surface Films*, Reinhold, New York, 1952, pp. 298–354.

71. Harkins, W. D., *et al.*, *J. Chem. Phys.*, **15**, 209–11 (1947); **15**, 763–6 (1947); *J. Am. Chem. Soc.*, **68**, 221–8 (1946).

72. Harrington, E. W., and Fryling, C. F., Private Communication to O.R.R., Aug. 13, 1942.

73. Harris, W. E., and Kolthoff, I. M., *J. Polymer Sci.*, **2**, 82–89 (1947).

74. Harrison, S. A., and Meincke, E. R., *Anal. Chem.*, **20**, 47–48 (1948).

75. Hauser, E. A., and le Beau, D. S., in *Colloid Chemistry*, edited by J. Alexander, Reinhold, New York, 1946, Vol. VI, pp. 356–407.

76. Hays, J. T., Drake, A. E., and Pratt, Y. T., *Ind. Eng. Chem.*, **39**, 1129–32 (1947).

77. Hobson, R. W., Borders, A. M., and Osterhof, H. J., Private Communication to O.R.R., Oct. 29, 1943.

78. Hobson, R. W., and D'Ianni, J. D., *Ind. Eng. Chem.*, **42**, 1572–7 (1950).

79. Hohenstein, W. P., and Mark, H., Polymerization in Suspension and Emulsion, in *Frontiers in Chemistry*, Interscience, New York, 1949, Vol. VI, pp. 1–74.

80. Houston, R. J., *Anal. Chem.*, **20**, 49–51 (1948).

81. Houston, R. J., and Fryling, C. F., Private Communication to O.R.R., Mar. 13, 1944.

82. Howard, F. A., *Buna Rubber*, Van Nostrand, New York, 1947.

82a. Howland, L. H., Neklutin, V. C., Provost, R. L., and Mauger, F. A., *Ind. Eng. Chem.*, **45**, 1304–11 (1953).

83. I.G. Farbenindustrie, Brit. Pat. 283,840 (1929).

84. I.G. Farbenindustrie, Brit. Pats. 339,255 (1930); 360,822 (1931).

85. I.G. Farbenindustrie, Brit. Pat. 349,499 (1930); Fr. Pat. 699,154 (1931); Ger. Pat. 588,785 (1933).

86. I.G. Farbenindustrie, Fr. Pat. 715,982 (1931).

87. Jackson, D. L., Private Communication to O.R.R., July 31, 1947.

88. Johnson, P. H., and Bebb, R. L., *J. Polymer Sci.*, **3**, 389–99 (1948).

89. Johnson, P. H., and Bebb, R. L., *Ind. Eng. Chem.*, **41**, 1577–80 (1949).

90. Johnson, P. H., Brown, R. R., and Bebb, R. L., *Ind. Eng. Chem.*, **41**, 1617–21 (1949).

91. Juve, A. E., Schroder, C. H., Ludwig, L. E., and Helms, J. R., Private Communication to O.R.R., Sept. 4, 1944.

92. Juve, R. D., Private Communication to O.R.R., Apr. 30, 1943.

93. Kern, W., *Makromol. Chem.*, **B1**, 209 (1948).

94. Kharasch, M. S., and Nudenberg, W., Private Communication to O.R.R., May 10, 1945.

95. Kharasch, M. S., Schenck, R. T. E., and Mayo, F. R., *J. Am. Chem. Soc.*, **61**, 3092–8 (1939).

96. Kharasch, M. S., and Westheimer, F. H., Private Communication to O.R.R., June 24, 1943.

97. Kharasch, M. S., and Westheimer, F. H., Private Communication to O.R.R., Dec. 1943.

98. Kolthoff, I. M., Private Communication to O.R.R., Dec. 1943.

99. Kolthoff, I. M., Private Communication to O.R.R., Oct. 1944.

100. Kolthoff, I. M., Private Communication to O.R.R., Aug. 28, 1946.

101. Kolthoff, I. M., and Bovey, F. A., *J. Am. Chem. Soc.*, **70**, 791–9 (1948).

102. Kolthoff, I. M., Carr, C. W., and Carr, B. J., *J. Polymer Sci.*, **2**, 637–42 (1947).

103. Kolthoff, I. M., and Dale, W. J., *J. Am. Chem. Soc.*, **67**, 1672–5 (1945); **69**, 441–6 (1947); Dale, W. J., and Miller, I. K., *J. Polymer Sci.*, **5**, 667–72 (1950).

104. Kolthoff, I. M., and Dale, W. J., *J. Polymer Sci.*, **3**, 400–9 (1948); **5**, 301–6 (1950); Kolthoff, I. M., Dale, W. J., and Miller, I. K., *J. Polymer Sci.*, **5**, 667–72 (1950).

105. Kolthoff, I. M., and Guss, L. S., Private Communication to O.R.R., Dec. 6, 1943.

106. Kolthoff, I. M., and Harris, W. E., *Ind. Eng. Chem., Anal. Ed.*, **18**, 161–2 (1946); *J. Polymer Sci.*, **2**, 49–71 (1947).
107. Kolthoff, I. M., and Harris, W. E., *J. Polymer Sci.*, **2**, 41–48 (1947).
108. Kolthoff, I. M., and Harris, W. E., *J. Polymer Sci.*, **2**, 49–71 (1947).
109. Kolthoff, I. M., and Medalia, A. I., *J. Polymer Sci.*, **5**, 391–427 (1950).
110. Kolthoff, I. M., and Medalia, A. I., *J. Polymer Sci.*, **6**, 189–207 (1951).
111. Kolthoff, I. M., Medalia, A. I., and Held, R., Private Communication to O.R.R., Mar. 27, 1947; Kolthoff, I. M., and Medalia, A. I., *J. Polymer Sci.*, **6**, 209–23 (1951).
112. Kolthoff, I. M., and Meehan, E. J., Private Communication to O.R.R., Apr. 1946.
113. Kolthoff, I. M., and Meehan, E. J., Private Communication to O.R.R., July, 1946.
114. Kolthoff, I. M., and Meehan, E. J., Private Communication to O.R.R., Dec. 1946; Kolthoff, I. M., and Medalia, A. I., *J. Polymer Sci.*, **6**, 209–23 (1951).
115. Kolthoff, I. M., and Meehan, E. J., Private Communication to O.R.R., Jan. 1947.
116. Kolthoff, I. M., and Meehan, E. J., Private Communication to O.R.R., Aug. 1949.
116a. Kolthoff, I. M., and Meehan, E. J., *J. Polymer Sci.*, **9**, 327–42 (1952).
116b. Kolthoff, I. M., and Meehan, E. J., *J. Polymer Sci.*, **9**, 343–67 (1952).
117. Kolthoff, I. M., and Stenberg, R. J., Private Communication to O.R.R., Dec. 11, 1945.
118. Kolthoff, I. M., Williams, D. E., and Carr, C. W., Private Communication to O.R.R., Sept. 30, 1944; Dec. 5, 1944.
119. Kolthoff, I. M., and Youse, M., *J. Am. Chem. Soc.*, **72**, 3431–5 (1950).
120. Kondakow, I., *Rev. Gen. Chem.*, **15**, 408 (1912).
121. Lawrence, J., Hobson, R. W., and Borders, A. M., Private Communication to O.R.R., Dec. 1, 1943.
122. Livingston, J. W., U.S. Dept. Commerce OTS Rept. PB 13356.
123. Logemann, H., U.S. Dept. Commerce OTS Rept. PB 4670.
124. Longfield, J., Blades, H., and Sivertz, C., Private Communication to O.R.R., May 20, 1948.
125. Longfield, J., Jones, R., and Sivertz, C., *Can. J. Research*, **B28**, 373–82 (1950).
126. McCleary, C. D., Private Communication to O.R.R., June 12, 1944.
127. McCleary, C. D., Messer, W. E., and Howland, L. H., Private Communication to O.R.R., June 11, 1943.
128. McKenzie, J. P., Private Communication to O.R.R., Feb. 1949.
129. Marchionna, F., *Butalastic Polymers*, Reinhold, New York, 1946.
130. Mark, H., and Raff, R., *High Polymeric Reactions*, Interscience, New York, 1941.
131. Marvel, C. S., Private Communications to O.R.R., Feb. 21 and Mar. 8, 1946.
132. Marvel, C. S., Private Communication to O.R.R., Apr. 4, 1947.
133. Marvel, C. S., Blackburn, W. E., Sheperd, D. A., and Dammon, J. A., Private Communication to O.R.R., Mar. 10, 1944.
134. Marvel, C. S., Deanin, R., Claus, C. J., Wyld, M. B., and Seitz, R. L., *J. Polymer Sci.*, **3**, 350–3 (1948).
135. Marvel, C. S., Deanin, R., Kuhn, B. M., and Landes, G. B., *J. Polymer Sci.*, **3**, 433–7 (1948).
136. Marvel, C. S., Deanin, R., Overberger, C. G., and Kuhn, B. M., *J. Polymer Sci.*, **3**, 128–37 (1948).
137. Marvel, C. S., *et al.*, *J. Polymer Sci.*, **3**, 181–94 (1948); Marvel, C. S., and Keplinger, O., *J. Polymer Sci.*, **6**, 83–91 (1951).
138. Mayo, F. R., Lewis, F. M., and Walling, C., *Discussions Faraday Soc.*, **2**, 285–95 (1947).
139. Mazur, J. W., and Coe, W. S., Private Communication to O.R.R., Feb. 12, 1943.
140. McBain, J. W., and Dean, H. B., Private Communication to O.R.R., Mar. 1944.
141. Medalia, A. I., *J. Polymer Sci.*, **1**, 245–6 (1946).
142. Medalia, A. I., and Kolthoff, I. M., *J. Polymer Sci.*, **4**, 377–98 (1949).
143. Meehan, E. J., *J. Polymer Sci.*, **1**, 175–82 (1946).

144. Meehan, E. J., *J. Polymer Sci.*, **1**, 318–28 (1946).
145. Meehan, E. J., *J. Am. Chem. Soc.*, **71**, 628–33 (1949).
146. Meehan, E. J., and Kolthoff, I. M., Private Communication to O.R.R., Apr. 1945.
147. Messer, W. E., Private Communication to O.R.R., June 12, 1944.
148. Messer, W. E., and Howland, L. H., Private Communication to O.R.R., Dec. 28, 1942.
149. Messer, W. E., and Howland, L. H., Private Communication to O.R.R., June 9, 1943.
150. Messer, W. E., and Howland, L. H., Private Communication to O.R.R., June 24, 1943.
151. Meyer, A. W., *Ind. Eng. Chem.*, **41**, 1570–7 (1949).
152. Mitchell, J. M., Spolsky, R., and Williams, H. L., *Ind. Eng. Chem.*, **41**, 1592–1603 (1949).
153. Morton, M., Salatiello, P. P., and Landfield, H., *J. Polymer Sci.*, **8**, 215–24 (1952).
154. Neklutin, V. C., Westerhoff, C. B., and Howland, L. H., *Ind. Eng. Chem.*, **43**, 1246–52 (1951).
155. Nudenberg, W., and Kharasch, M. S., Private Communication to O.R.R., Oct. 24, 1947.
156. Orr, R. J., and Williams, H. L., *Can. J. Chem.*, **29**, 270–83 (1951).
157. Orr, R. J., and Williams, H. L., *Can. J. Chem.*, **30**, 985–93 (1952).
158. Orr, R. J., and Williams, H. L., *Can. J. Chem.*, **30**, 985–93 (1952).
159. Orr, R. J., and Williams, H. L., Paper presented to Div. High Polymer Chem., Canadian Inst. Chem., Montreal, June 1952.
160. Orr, R. J., and Williams, H. L., *Faraday Soc. Discussion*, No. 14, 170–81 (1953).
161. Osterhof, H. J., Private Communication to O.R.R., Mar. 1943.
162. PB Rept. 5521, Dept. Commerce, Washington, D.C.
163. Perloff, J. W., Private Communication to O.R.R., Dec. 12, 1943.
164. Perloff, J. W., and Childress, V. R., Private Communication to O.R.R., Mar. 13, 1944.
165. Pfau, E. S., and Wilson, J. W., Private Communication to O.R.R., Mar. 13, 1944; Wilson, J. W., and Pfau, E. S., *Ind. Eng. Chem.*, **40**, 530–4 (1948).
166. Phillips Petroleum Co., Private Communication to O.R.R., July 10, 1948.
167. Potter, W. J., Borders, A. M., and Osterhof, H. J., Private Communication to O.R.R., June 23, 1943.
168. Price, C. C., and Adams, C. E., *J. Am. Chem. Soc.*, **67**, 1674–80 (1945).
169. Price, C. C., and Durham, D. A., *J. Am. Chem. Soc.*, **65**, 757–9 (1943).
169a. Pryor, B. C., Harrington, E. W., and Druesedow, D., *Ind. Eng. Chem.*, **45**, 1311–5 (1953).
170. Renfro, R. W., Private Communication to O.R.R., Nov. 26, 1943.
171. Reynolds, W. B., Private Communication to O.R.R., July 25, 1943.
172. Reynolds, W. B., *J. Chem. Education*, **26**, 135–8 (1949).
173. Reynolds, W. B., and Canterino, P. J., Private Communication to O.R.R., Apr. 25, 1946. Also High Polymer Section, A.A.A.S., Gibson Island, Md., July 4, 1946.
174. Reynolds, W. B., and Cotten, E. W., Private Communication to O.R.R., June 4, 1944; *Ind. Eng. Chem.*, **42**, 1905–10 (1950).
175. Roberts, H. P., Private Communication to O.R.R., Oct. 27, 1943.
176. St. John, W. M., Uraneck, C. A., and Fryling, C. F., *J. Polymer Sci.*, **7**, 159–73 (1951).
177. Schulze, W. A., Private Communication to O.R.R., May 1949.
178. Schulze, W. A., Private Communication to O.R.R., Oct. 1949.
179. Schulze, W. A., and Crouch, W. W., *Ind. Eng. Chem.*, **40**, 151–4 (1948).
180. Schulze, W. A., Lyon, J. P., and Short, G. H., *Ind. Eng. Chem.*, **40**, 2308–13 (1948).
181. Schulze, W. A., Troyan, J. E., Fryling, C. F., and Reynolds, W. B., Private Communication to O.R.R., July 9, 1948.
182. Schulze, W. A., Tucker, C. M., and Crouch, W. W., *Ind. Eng. Chem.*, **41**, 1599–1603 (1949).
183. Semon, W. L., U.S. Pat. 2,376,015 (1945).
184. Semon, W. L., and Fryling, C. F., U.S. Pat. 2,376,014 (1945).
185. Shearon, W. H., McKenzie, J. P., and Samuels, M. E., *Ind. Eng. Chem.*, **40**, 769–77 (1948).
186. Shepherd, D. A., Higgins, N. A., and Runge, W. F., Private Communication to O.R.R., June 15, 1946.

187. Smith, H. S., Werner, H. G., Madigan, J. C., and Howland, L. H., *Ind. Eng. Chem.*, **41**, 1584–7 (1949).
188. Smith, H. S., Werner, H. G., Westerhoff, C. B., and Howland, L. H., *Ind. Eng. Chem.*, **43**, 212–6 (1951).
189. Smith, W. C., Private Communication, Goodyear Tire & Rubber Co. to O.R.R., July 23, 1942.
190. Smith, W. V., *J. Am. Chem. Soc.*, **70**, 3695–3702 (1948).
191. Smith, W. V., and Ewart, R. H., *J. Chem. Phys.*, **16**, 592–9 (1948).
192. Snyder, H. R., Stewart, J. M., Allen, R. E., and Dearborn, R. J., *J. Am. Chem. Soc.*, **68**, 1422–8 (1946).
193. Starkweather, H. W., *et al.*, *Ind. Eng. Chem.*, **39**, 210–22 (1947).
194. Stewart, W. D., Private Communication to O.R.R., Mar. 20, 1943.
195. Stewart, W. D., U.S. Pat. 2,383,425 (1945).
196. Stewart, W. D., and Fryling, C. F., Private Communication to O.R.R., Mar. 23, 1943.
197. Swain, C. G., and Bartlett, P. D., *J. Am. Chem. Soc.*, **68**, 2381–6 (1946).
198. Taft, W. K., Private Communication to O.R.R., Apr. 15, 1946.
199. Thompson, R. D., Private Communication to O.R.R., Aug. 26, 1947.
200. Tierney, M. J., and Coe, W. S., Private Communication to O.R.R., June 23, 1943.
201. Traun's Forschungslaboratorium, Ger. Pat. 329,593 (1918).
202. Trommsdorff, E., Kunststoffe aus Polymerizaten von Athylenderivaten, in *Chemie und Technologie der Kunststoffe*, edited by R. Houwink, 1939, pp. 304–60.
203. Troyan, J. E., Private Communication to O.R.R., July 5, 1944.
204. Troyan, J. E., and Tucker, C. M., *India Rubber World*, **121**, 67–70, 190–2 (1949).
205. U.S. Rubber Co., Brit. Pat. 598,871 (1948).
206. Uraneck, C. A., Private Communication.
207. Vandenberg, E. J., and Hulse, G. E., *Ind. Eng. Chem.*, **40**, 932–7 (1948).
208. Vila, G. R., Private Communication to O.R.R., June 16, 1943.
209. Wall, F. T., Banes, F. W., and Sands, G. D., *J. Am. Chem. Soc.*, **68**, 1429–31 (1946).
210. Wall, F. T., and Swaboda, J., *J. Am. Chem. Soc.*, **71**, 919–24 (1949).
211. Walling, C., and Mayo, F. R., *Discussions Faraday Soc.*, **2**, 295–303 (1947).
212. Weidlein, E. R., Jr., *Chem. Eng. News*, **24**, 774 (1946).
213. Whitby, G. S., and Gross, M. D., Private Communication to O.R.R., Dec. 2, 1946.
214. Whitby, G. S., Gross, M. D., Miller, J. R., and Constanza, A. J., *Chem. Eng. News*, **29**, 3952 (1951).
215. Whitby, G. S., and Kaplan, S. J., Private Communication to O.R.R., Dec. 6–7, 1943.
216. Whitby, G. S., and Katz, M., *Ind. Eng. Chem.*, **25**, 1204–11, 1338–48 (1933).
217. Whitby, G. S., Wellman, N., Floutz, V. W., and Stephens, H. L., *Ind. Eng. Chem.*, **42**, 445–56 (1950).
218. White, L. M., *Ind. Eng. Chem.*, **41**, 1554–60 (1949).
219. Wicklatz, J. E., Kennedy, T. J., and Reynolds, W. B., *J. Polymer Sci.*, **6**, 45–58 (1951).
220. Willis, J. M., *Ind. Eng. Chem.*, **41**, 2272–6 (1949).
221. Wilson, J. W., and Pfau, E. S., *Ind. Eng. Chem.*, **40**, 530–4 (1948).
222. Wolfe, W. D., U.S. Pat. 2,235,625 (1941).
223. Wood, L. A., Synthetic Rubbers, *Natl. Bur. Standards Circ.* **C427**, U.S. Dept. Commerce, 1940.
224. Youker, M. A., U.S. Pat. 2,417,038, Mar. 4, 1947.
225. Zwicker, B. M. G., Private Communication to O.R.R., Sept. 9, 1943.
226. Zwicker, B. M. G., and Garvey, B. S., Private Communication to O.R.R., Sept. 9, 1943.
227. There has been considerable discussion as to whether soap micelles are part of the aqueous phase or exist as a separate phase. Fryling and Harrington[56] considered the solubilized micelles as being in the aqueous phase, similar in this respect to water-soluble complex compounds. Professor Harkins neatly sidestepped this question by considering the water phase and the micelles as two separate loci of reaction, as indeed they are.

CHAPTER 9

CHEMICAL STUDY OF THE STRUCTURE OF DIENE POLYMERS AND COPOLYMERS

C. S. Marvel and Henry E. Baumgarten

University of Illinois

INTRODUCTION

The major structural problems concerning synthetic diene polymers and copolymers that have been studied by chemical methods are: (1) the total unsaturation, (2) the ratio of 1,2- to 1,4-diene units in the chain, (3) the comonomer ratio, (4) the arrangement of the various units in the chain, and (5) the nature of the branching reaction which accompanies polymerization.

In a butadiene polymer the two units that may occur in the chain are the 1,4-unit (I) and the 1,2-unit (II).   When a monosubstituted butadiene such

$$-CH_2CH : CH \cdot CH_2-$$
I

$$\begin{array}{c} -CHCH_2- \\ | \\ CH \\ \| \\ CH_2 \end{array}$$
II

as isoprene is polymerized the possibilities are greater, for the 1,4-units may be united "head-to-head and tail-to-tail" (III) and head-to-tail (IV), and, further, addition may occur 1,2 at the substituted double bond to produce the 1,2-unit (V) or 3,4 at the unsubstituted double bond to produce the 3,4-unit (VI).

$$\begin{array}{c} -CH_2\cdot CH:C\cdot CH_2\cdot CH_2\cdot C:CH\cdot CH_2\cdot CH_2\cdot CH:C\cdot CH_2\cdot CH_2\cdot C:CH\cdot CH_2- \\ | \qquad\qquad | \qquad\qquad\qquad | \qquad\qquad | \\ CH_3 \qquad\quad CH_3 \qquad\qquad\quad CH_3 \qquad\quad CH_3 \end{array}$$
III

$$\begin{array}{c} -CH_2 \cdot C : CH \cdot CH_2 \cdot CH_2 \cdot C : CH \cdot CH_2- \\ | \qquad\qquad\qquad\qquad | \\ CH_3 \qquad\qquad\qquad CH_3 \end{array}$$
IV

$$\begin{array}{c} CH_3 \\ | \\ -CH_2 \cdot C- \\ | \\ CH \\ \| \\ CH_2 \end{array} \qquad \begin{array}{c} -CH_2 \cdot CH- \\ | \\ C \cdot CH_3 \\ \| \\ CH_2 \end{array}$$
V   VI

It should be noted that structures I, III, and IV are capable of existing in *cis* and *trans* configurations at each double union, and that structures II, V, and VI have asymmetric carbon atoms which make optical isomerism possible. The recurring styrene unit in GR-S also has an asymmetric carbon atom and therefore can exist in *d* and *l* configurations. Chemical reactions have not helped in characterizing these stereoisomeric units, and they will not be discussed further in this chapter.

The extent to which the various possible diene units will occur in any polymer is influenced by the method used for polymerization. Natural rubber has a regular isoprene 1,4, all *cis*, head-to-tail structure. In the synthetic diene polymers such regularity is missing, and the ratio of different structural units is dependent on the method of polymerization.

TOTAL UNSATURATION

In a linear polymer molecule from a 1,3-diene there should be one olefinic double bond (either internal or external) for each diene unit in the molecule. Deviations from regularity or linearity of the molecule such as would be occasioned by branching, cross-linking, or cyclization might result in a decrease in the number of double bonds in the polymer chain if these inter- or intramolecular secondary reactions involved either the internal or external double bonds. The addition of fragments of catalysts or impurities to either the internal or external double bonds would also lower the number of double bonds present in the molecule. However, these reactions occur to such a limited extent that none is likely to effect a measurable reduction in the total unsaturation.

From the standpoint of structural studies it would be desirable for any determination of the total unsaturation to be sufficiently accurate to indicate the loss of every double bond that is saturated by one of the above secondary reactions. Such a determination would be of great assistance in elucidating the mechanism of and in ascertaining the extent of branching or cross-linking; however, in a polybutadiene molecule having a molecular weight of about 216,000 there are approximately 4000 double bonds, and a method capable of accounting for every double bond would have to have an accuracy of about 0.025 per cent. No such method is available at present. There are methods currently available accurate to within about 1 per cent (absolute), so that, although minute deviations from linearity may not be detected, appreciable saturation of the double bonds in diene polymers and copolymers can be determined with reasonable accuracy.

A number of methods for the determination of the total unsaturation of diene polymers have appeared in the literature. Most of these involve the addition of an active reagent to the double bonds and the determination of the excess reagent or of the amount of actual addition. Among the reagents that have been used in these determinations are bromine,[33, 40, 63, 65–6, 70] dithiocyanogen,[6, 28] iodine monobromide,[14, 31] perbenzoic acid,[56] and iodine monochloride.[5, 9, 29–32, 35, 39, 56–7, 60, 71]

The procedure generally used at present is based on the experiments of Kemp and Peters[32] and involves the use of iodine monochloride. The reaction at any double bond in the polymer chain can be represented as follows.

$$
\begin{array}{ccc}
& I & Cl \\
& | & | \\
P \cdot CH : CH \cdot P' + ICl \longrightarrow & P \cdot CH \cdot CH \cdot P'
\end{array}
$$

In the equation P and P′ represent the remainder of the polymer chain. Although the reagent is applicable to the analysis of a wide variety of unsaturated hydrocarbons, the interpretation of results obtained with high polymers is complicated by the occurrence of two side reactions (substitution and splitting out),[32, 39, 60] both of which involve the liberation of hydrogen iodide.

The unsaturation values are generally reported as iodine numbers or as percentages of unsaturation, where percentages of unsaturation for both homopolymers and copolymers are weight percentages of unsaturation calculated on the assumption that the polymer or copolymer is made up entirely of diene units. Thus, in copolymers where the unsaturation is equal to the theoretical unsaturation, the percentage of unsaturation is equal to the percentage of polymerized diene in the copolymer. Iodine numbers for butadiene polymers and for copolymers of butadiene with vinyl monomers may be converted to percentage unsaturation by dividing by 4.696; for isoprene polymers, by 3.728.

Although a considerable number of investigations of the unsaturation of diene polymers and copolymers have been reported, most of these are concerned with the structure of natural rubber[14, 28–31, 33, 40, 56–7, 60, 65–6, 69] and the mechanism of the vulcanization process.[7, 11–2, 23–4, 40] Early reports on butadiene-styrene copolymers showed conflicting results, possibly because of the wide variation in experimental conditions during the determination. Cheyney and Kelley[9] found that their unsaturation values for 75/25 butadiene-styrene copolymers varied with the solvent, but in general the values were in the range of 85 to 90 per cent of theoretical values for the linear polymers. The difference was attributed by them to cross-linkage in the polymer. Kemp and Peters[32] reported that the unsaturation values of a series of 75/25 butadiene-styrene copolymers examined by them were nearly the same as the theoretical values. German chemists[71] during World War II, using the iodine monochloride titration, could account for only 90 per cent of the theoretical number of double bonds in Buna S, and so they utilized an empirical correction factor to bring the values up to the theoretical figures. Using the same correction factor with Buna N, they obtained unsaturation values that agreed with the results of nitrogen analysis. Lee, Kolthoff, and Mairs[39] made an extensive study of the application of iodine monochloride to various diene polymers and copolymers and developed general procedures for both routine analyses and determinations on new polymers. Their procedures incorporate determination of the extent of both substitution and "splitting out" and allow for correction for both. Tables I and II give some of the data obtained by them for several types of diene polymers and for GR-S at various conversions.

Table I.   Percentage Unsaturation of 1,3-Diene Polymers

Type of Polymer	Corrected Unsaturation, %
GR-S (64% conversion)	80.9
Emulsion polybutadiene	97.7
Emulsion polyisoprene	96.8
Sodium butadiene-styrene (75/25)	72.1
Sodium polybutadiene	92.1
Sodium polyisoprene	83.9
Natural rubber	96.7

Table II.   Unsaturation of GR-S at Different Conversions

Conversion,* %	Unsaturation, Corrected, %	Styrene Content from Unsatn., %	Styrene Content[49] by U.V. Spectrophotometry, %
21	82.5	17.5	17.6
41	81.1	18.9	18.4
64	80.9	19.1	20.0
77	80.5	19.5	20.7

* Temperature of polymerization was 50° C.

From the data in Table II it is concluded that the unsaturation of GR-S at various conversions corresponds to the spectrophotometrically determined styrene content, i.e., that GR-S has the theoretical unsaturation.   Emulsion polybutadiene exhibits 97 to 98 per cent of the theoretical unsaturation, emulsion polyisoprene 97 per cent, and natural rubber 97 per cent.   The amount of unsaturation in the sodium polymers and copolymers is distinctly less than theoretical.

The implications of the foregoing results are most important to the complete determination of the structure of the GR-S copolymer.   Apparently for conversions of 77 per cent or less, the GR-S molecule has one double bond for each butadiene unit (or nearly so); thus, no extensive secondary reactions involving the double bonds can have occurred during polymerization.   In other words any extensive branching (since the polymers examined were soluble, presumably no cross-linking had occurred) during the polymerization reaction could not have involved addition to the double bonds.   The slightly less than theoretical unsaturations of emulsion polybutadiene and polyisoprene are attributed by Lee, Kolthoff, and Mairs[39] to intramolecular cyclization.

Application of the iodine monochloride technique to the determination of the total unsaturation of polybutadiene and butadiene-styrene copolymers prepared by means of catalysts of the Friedel-Crafts type indicated that such polymers have considerably less than the theoretical unsaturation.[43]   The values for the total unsaturation vary widely among individual batches of polymer and appear to depend to a large extent on the choice of catalyst, solvent (usually an alkyl halide), and experimental conditions.   Values as low as 60 per cent of the theoretical unsaturation have been obtained for these polymers. The loss of unsaturation may occur through Friedel-Crafts-type addition of the catalyst solvent to the double bonds or through intramolecular cyclization.

RATIO OF 1,2- TO 1,4-DIENE UNITS

By Ozonolysis. The classical researches of Harries[22] (1905–19) established that there are at least two types of synthetic polydienes. The first, or "normal" type, which was prepared by the bulk polymerization of butadiene or isoprene, was considered to be analogous to natural rubber; it yielded 1,4-dicarbonyl compounds on ozonolysis and was thought to be formed by 1,4-addition of the dienes. That the "abnormal" type of polymer, which was prepared by the polymerization of butadiene or isoprene with sodium, was formed by the combination of 1,2- and 1,4-addition was demonstrated later by Pummerer[53] and by Ziegler, Dersch, and Wolthan.[74] Pummerer[53] found that the ozonization of sodium polybutadiene and decomposition of the ozonide yielded a small amount of succinic acid and a larger amount of a white polymeric aldehyde–acid, which accounted for approximately 80 per cent of the carbon skeleton. The succinic acid was derived from adjacent 1,4-units in the polymer chain, and the polymeric substance was thought to have been formed from the oxidation of sections of the chain in which there were a number of contiguous 1,2-units. This work indicated that 1,2-addition could predominate over 1,4-addition in the formation of sodium polybutadienes.

From the ozonolysis products of polyisoprene prepared with a potassium dimethylphenylmethyl catalyst, Ziegler and his students[74] isolated formaldehyde, formic acid, succinic acid, levulinic acid, and acetonylacetone, all of the degradation products to be expected from the oxidative cleavage of a polymer formed by a combination of 1,2-, 1,4-, and 3,4-addition. Ziegler, Grimm, and Willer[75] found that the ratio of 1,2-butadiene units to 1,4-butadiene units could be varied markedly by changing the temperature of polymerization when carried out with alkali metal alkyls. Thus in polybutadiene formed at 25 to 30° C. both 1,2- and 1,4-units were present; at 100 to 115° C. the polymer was formed principally from 1,4-units (approximately 80 per cent), and at −50° C. the 1,2-units predominated (approximately 95 per cent).

The observations of the previous workers have been extended to emulsion polybutadiene by Hill, Lewis, and Simonsen.[25] By ozonolysis of an insoluble polybutadiene, permanganate oxidation of the intermediate carbonyl compounds, esterification of the acids mixture, and fractionation of the methyl esters, they obtained two pure esters, dimethyl succinate and trimethyl 1,2,4-butanetricarboxylate. The former represented adjacent 1,4-units in the polymer molecule and the latter a 1,2-unit sandwiched in between two 1,4-units. In addition to these products they obtained others which could not be definitely characterized but which indicated that there were sections of the chain in which two or more adjacent 1,2-units occurred between 1,4-units. Unfortunately their data were not sufficiently quantitative to permit estimation of the number of 1,2-units.

Most of the early estimates of amounts of 1,2- and 1,4-addition occurring during polymerization were based on qualitative rather than quantitative data, and no particular effort was made to develop quantitative techniques for the determination of the 1,2–1,4-ratio. Later, however, during the early period of the development of GR-S, many workers thought that the less desirable properties of the synthetic rubber could be attributed to the 1,2-butadiene units (or vinyl side chains) present in the polymer chain, and that, the greater the regularity of the polymer molecule through 1,4-polymerization of butadiene, all other variables being unchanged, the more desirable would be the properties of the rubber. Therefore, much interest was evinced in the determination of the ratio of 1,2-butadiene units to 1,4-butadiene units in polymers and copolymers prepared under varied conditions, and several chemical analytical methods were developed for this determination: ozonolysis,[2, 58, 71, 73] potassium permanganate oxidation,[71] perbenzoic acid oxidation,[37, 62, 71] and addition of aliphatic thiols.[64]

The ozonolysis of a 1,3-diene polymer and decomposition of the resultant ozonide result, of course, in the scission of the polymer at the double bonds. Each external double bond (vinyl side chain, VII) derived from the 1,2-polymerization of a diene molecule gives rise to a molecule of formaldehyde

$$
\begin{array}{ccc}
\begin{array}{c}
-CH \cdot CH_2- \\
| \\
CH \\
\| \\
CH_2 \\
\text{VII}
\end{array}
& \longrightarrow &
\begin{array}{c}
-CH \cdot CH- \\
| \\
CO_2H \\
+ \\
HCO_2H
\end{array}
\end{array}
$$

or formic acid (depending on the degree of oxidation occurring during the decomposition of the ozonide), and these should be the only volatile compounds that appear in measurable quantities, provided there is no oxidative scission beyond the expected cleavage of the double bonds. The products resulting from the cleavage of the internal double bonds should be nonvolatile di- or polyfunctional compounds. The determination of the volatile products, therefore, gives a measure of the extent of 1,2-addition. In the procedure used by Rabjohn and co-workers,[58] chloroform solutions of the purified polymers were treated with ozonized oxygen, and the ozonides were decomposed by steam distillation. The results, although quite reproducible, were not necessarily accurate in the absolute sense, and for this reason the apparent percentage of 1,2-polymerization was given the less specific title of "ozonization number."

The results obtained by Rabjohn et al.[58] are summarized in Table III. The method is inaccurate as an absolute method because ozone attacks the polymer molecule at points other than the double bonds.[55, 67] However, the results of this work gave the limits of 1,2-addition as between 18 and 22 per cent for emulsion polybutadiene and for emulsion copolymers of butadiene with styrene and offered the first indications that the amount of 1,2-addition is not greatly affected by changes in the emulsion system.

The ozonolysis method of Yakubchik, Vasiliev, and Zhabina[73] differed from the method of Rabjohn et al. principally in the analytical procedure used for the determination of the formaldehyde and formic acid obtained from the decomposition of the ozonide. Their data on various butadiene polymers are included in Table III.

Table III. Percentage 1,2-Addition as Determined by Ozonolysis

	1,2-Addition, %		
Description of Polymer	Rabjohn Ozonization Number[58]	Yakubchik Ozonization Method[73]	Perbenzoic Acid Method
Natural rubber	11.8	...	0
Polybutadiene			
Emulsion at 15°	17.9*	...	...
Emulsion at 50°	19.4; 20.7	31	23
Emulsion at 110°	12.0; 11.8	...	...
Bulk	...	37.4	...
Sodium at 10°		...	73.5
Sodium at 22°	...	59.4	...
Sodium at 30°		...	61; 67
Sodium at 50°		...	58; 59
Sodium-rod	...	43.8	...
Sodium-rodless	...	49.0	...
Sodium at 60°	...	43.8	...
Sodium (Buna-115)	38.5; 42.2	...	...
Sodium, Russian	42.8; 40.5	...	...
Lithium at 60°	...	23.9	...
Potassium at 60°	...	37.0	...
Butadiene/styrene copolymer			
GR-S, 75/25, 71% conversion	19.1; 19.3	...	20.6
GR-S, 75/25, at 110–130°	17.1; 18.0	...	...
GR-S, 75/25, at 50°	21.1; 22.3*	...	20.6
GR-S, 65/35, 81% conversion	20.9	...	...
GR-S, 0.533 parts DDM	16.5; 18.7*	...	...
GR-S, large amount cetyl mercaptan	18.7	...	...
GR-S, ferricyanide initiated	19.1	...	...
GR-S, milled 10 minutes at 300° F.	19.2*	...	...
GR-S, cold-milled 45 minutes	22.4	...	...
GR-S, acid side (pH 4.5)	13.4; 14.3	...	14
Latex treated with DDM	8.1; 10.1	...	...
Latex treated with butyl mercaptan	11.7; 12.7	...	...
Bulk	...	34.6	...
Diazoaminobenzene recipe	...	31.6	...

* Unpublished results.

By Perbenzoic Acid. The reaction of perbenzoic acid with diene polymers was first described by Pummerer and his students,[54, 55] who found that the reagent reacted rapidly and quantitatively with the double bonds in natural rubber. The use of perbenzoic acid for determining the ratio of 1,2- to 1,4-addition was developed by German chemists during World War II.[71] The German procedure was modified somewhat by Kolthoff, Lee, and Mairs[37] and by Saffer and Johnson,[62] who utilized it in extensive examinations of the variation of the 1,2–1,4-ratio in butadiene-type polymers with variations in the conditions of polymerization.

The method is based on the reaction of perbenzoic acid with many olefinic double bonds to form an epoxide and benzoic acid.[52] The rate at which

$$-C:C- + C_6H_5COOOH \longrightarrow -\overset{\displaystyle O}{\overset{\displaystyle \triangle}{C-C}}- + C_6H_5COOH$$

perbenzoic acid reacts with an unsaturated compound depends on the nature of the substituents on the carbon atoms joined by the double bond. In the 1,3-diene polymers, perbenzoic acid reacts more rapidly with the internal double bonds (1,4-units) than with the external double bonds (1,2-units). Fortunately the difference in the rates of reaction with the perbenzoic acid is sufficiently great to permit the separation of the contributions of the two

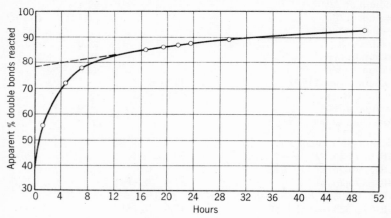

FIG. 1. Typical Rate Curve for Addition of Perbenzoic Acid to GR-S
Industrial and Engineering Chemistry. Reprinted by permission

species of double bonds to the total reaction. The reliability of the method has been well demonstrated by Eimers[10] in experiments with model substances of known constitution.

Slightly different procedures for this determination have been described by Kolthoff, Lee, and Mairs[37] and by Saffer and Johnson.[62] Both, however, are based on the differences in the rate of addition of perbenzoic acid to different types of olefinic bonds. A typical GR-S addition rate curve, obtained by Saffer and Johnson[62] under the standard conditions of 6° C. and 25 per cent excess of perbenzoic acid, is shown in Fig. 1. The data obtained by Kolthoff, Lee, and Mairs[37] and by Saffer and Johnson[62] are given in Table IV.

A sample of sodium polyisoprene was found to contain 36 per cent internal double bonds and 64 per cent external double bonds. From an analysis of the reaction-rate curve for this polymer, Kolthoff, Lee, and Mairs[37] estimated that approximately 40 per cent (absolute) of the external double bonds were formed by 3,4-addition and 24 per cent by 1,2-addition. The same workers found that natural rubber (*cis*) reacts at a slightly faster rate than balata (*trans*) with perbenzoic acid, but the rates were so nearly equal that the *cis* and *trans* configurations of double bonds could not be distinguished by the perbenzoic acid method.

Table IV. Percentage 1,2-Addition as Determined by Perbenzoic Acid Method

Description of Polymer	Conversion %	Unsaturation %	1,2-Addition (%) Perbenzoic Acid	Reference
Natural Rubber (*cis*)		98.5	0	37
Balata (*trans*)		98	0	37
Natural rubber				
Pale crepe (Hevea)			5	62
Smoked sheet (Hevea)			3	62
Red Congo (Landolphia)			5	62
Benguella (Landolphia)			3	62
Black Congo (Funtumia)			6	62
Polybutadiene				
Emulsion at 50°	74	96	23	37
Emulsion at 50°	44	98	22	37
Emulsion, alkaline			25	62
Sodium at 50°		94	58.5	37
Sodium at 50°		96	59	37
Sodium at 30°		94	67	37
Sodium at 30°		98	61	37
Sodium at 10°		96	73.5	37
Polyisoprene				
Emulsion at 50°	58	99	13	37
Emulsion, acid system			12	62
Emulsion, alkaline			14	62
Sodium at 50°		94	50	37
Sodium at 50°		98	56.5	37
Sodium at 43°		96	53	37
Sodium at 30°		95	53.5	37
Sodium at 30°		90	55	37
Poly-2,3-dimethyl-1,3-butadiene				
Emulsion (0.6 part DDM)			12	62
Emulsion (0.1 part DDM)			11	62
Butadiene-styrene copolymer				
GR-S (50°)	20	80	21.1	37
GR-S	37	80	21.6	37
GR-S	60	78.5	20.6	37
GR-S	80	78	20.6	37
GR-S			22	62
GR-S	21.0		22	62
GR-S	42.6		24	62
GR-S	60.6		22	62

Table IV (continued)

Description of Polymer	Con-version %	Unsatur-ation %	1,2-Addition (%) Per-benzoic Acid	Refer-ence
Butadiene-styrene copolymer				
GR-S, acid system, 50°, pH 4.3			17	62
GR-S, acid system, 10°, pH 4.5			15	62
Buna S-3			22	62
Buna S			27	71
Buna S-3			25	71
Buna SS			15	71
75/25 copolymer				
Emulsion at 100°, Mutual recipe	75	74	19.9	37
Emulsion at 50°, Mutual recipe	80	78	20.5	37
Emulsion at 30°, Mutual recipe	84	75	19.5	37
Emulsion at 30°, Mutual recipe	60	77	19.0	37
Emulsion at 10°, Mutual recipe	60	77	18.5	37
Emulsion at 5°, MDN recipe	68	76	18.5	37
Emulsion at 30°, Redox I	61	78.5	20	37
Emulsion at 30°, Redox II	61	78.5	24	37
Emulsion at 50°, DDM	65	80	22	37
95/5 copolymer	49	95	21	37
75/25 copolymer	77	78	20	37
50/50 copolymer	8î	54	19	37
75/25 copolymer				
Sodium at 50°		72	58	37
Sodium at 30°		72	59	37
Sodium at 30°		72.5	55	37
Isoprene–styrene, 75/25, emulsion, at 50°	60	79	5	37
Emulsion at 50°				
Butadiene–styrene–acrylonitrile, 73/25/2	76	77	22.5	37
Butadiene–p-chlorostyrene, 73/27		73	20.5	37
Butadiene–2,5-dichlorostyrene, 73/27	76	67	17	37
Butadiene–o- and p-chlorostyrene, 73/27	67	73	20	37

Other Methods. Thioglycolic acid adds exothermally to butadiene polymers and copolymers in benzene solution under mild conditions to give apparent double-bond saturation values of 38 to 47 per cent.[64] When the same polymers react with aliphatic mercaptans of C_2 to C_{16} chain length, in mass or latex reactions, saturation values are obtained that are in accord with those found by thioglycolic acid addition. The lack of correlation of the results with determinations made by the perbenzoic acid method and the infrared method indicate that the mercaptan addition method in its present form is not sufficiently accurate for analytical use.

Conclusions. There are sufficient data available at present to draw some tentative general conclusions concerning the relative amounts of 1,2- and 1,4-diene units in diene polymers and copolymers. Emulsion polybutadiene and emulsion copolymers of butadiene and styrene contain in respect to the butadiene in them about 20 per cent external double bonds (1,2-units). Polymerization variables such as temperature, percentage conversion, type of catalyst, type of emulsifier, type of modifier, and styrene content have little effect on the ratio of 1,2- to 1,4-units. Even a change of the comonomer does not alter the ratio in some cases.

Emulsion polyisoprene and emulsion poly-2,3-dimethyl-1,3-butadiene contain about 12 to 13 per cent 1,2-diene units. The 75/25 emulsion copolymer of isoprene and styrene has only 5 per cent 1,2-units. Other poly-3-alkyl-1,3-butadienes show approximately the same ratio of 1,2- to 1,4-units[48] as polyisoprene and polydimethylbutadiene.

The amount of 1,2-addition in polymers prepared with alkali metal catalysts is dependent on the temperature, more 1,2-units being formed at low temperatures than at high temperatures. Polybutadiene, butadiene-styrene copolymers, and polyisoprene prepared by sodium catalysis at 50° C. contain 50 to 60 per cent 1,2-diene units.

Emulsion polymers prepared in media of low pH tend to contain a slightly smaller percentage of 1,2-units than polymers prepared in the standard alkaline emulsion system. Mill treatment does not appear to affect the 1,2-unit content, but treatment of the polymer latex with mercaptans appears to decrease the number of 1,2-units, as might be expected from the experiments of Serniuk, Banes, and Swaney.[64]

COMONOMER RATIO

One of the important considerations in determining the structure of the diene copolymers is the actual ratio of diene units to vinyl monomer units in the copolymer. The ratio of comonomer units in any copolymer depends on the ratio of the comonomers in the polymerization recipe, the relative rate at which each monomer enters the polymer molecule, the percentage conversion to which the polymerization is carried, and perhaps other factors. In practice the variables in the polymerization are adjusted so that the ratio of the comonomers in the polymer produces a material of desirable physical characteristics. For example, the optimum butadiene-styrene ratio for GR-S appears to be near 75 parts of butadiene to 25 parts of styrene. These optimum values are average values, and some deviation in individual batches of copolymer is inevitable. Thus, to know the precise composition of a particular batch of copolymer it is necessary to be able to determine accurately the number of each monomer unit present.

With the exception of butadiene-styrene and similar copolymers, the comonomer compositions of most polymers are determined by combustion analysis. The butadiene-styrene ratio in butadiene-styrene copolymers may be determined from index of refraction measurements on the solid polymer or on benzene solutions of the polymer[4, 27] or by ultraviolet spectrophotometry.[49] These methods are described and the leading references to them given in Chapter 10. The chemical methods of determining the ratio include the determination of the carbon-hydrogen ratio by combustion analysis[32, 41] and determination of the theoretical butadiene content by the use of the iodine monochloride unsaturation determination.[32, 39]

The only absolute method of determination of the composition of butadiene-styrene (and many other) copolymers is the determination of the carbon-hydrogen ratio by precise elementary combustion analysis. However, since the elementary compositions of butadiene (carbon, 88.8 per cent) and styrene (carbon, 92.3 per cent) do not differ greatly, the determinations must be unusually accurate if significant results are to be obtained for the composition of the copolymer. Kemp and Peters[32] reported that their results using combustion analysis for the determination of the compositions of various butadiene-styrene copolymers were accurate to within 1 or 2 per cent. The method of elementary analysis is particularly applicable to copolymers in which one of the comonomers contains an analyzable element not contained by the other; e.g., nitrogen, halogen. Copolymers for which this method might be suitable are those containing monomers such as chlorostyrene, acrylonitrile, methacrylonitrile.

As indicated in the section on the determination of total unsaturation, GR-S has, within the limits of experimental error, the theoretical unsaturation; therefore, the total number of double bonds present is very nearly equal to the number of butadiene units in the copolymer molecule. Because of this relationship, it has been suggested that the determination of unsaturation by the iodine monochloride method be used also as a method for determination of the butadiene-styrene ratio.[32, 39] Meehan[49] has estimated that the accuracy of the method for butadiene polymers is about 3 per cent; however, Lee, Kolthoff, and Mairs[39] obtained results (Table I) with GR-S that show that the correspondence of values obtained with iodine monochloride and ultraviolet spectrophotometry is within about 1 per cent.

The combustion analysis method, the refractive index method, and the iodine monochloride titration may be used with gelled or insoluble polymers, whereas the use of ultraviolet spectrophotometry is limited to soluble polymers.

ARRANGEMENT OF STRUCTURAL UNITS IN THE DIENE POLYMER CHAIN

Products of Oxidative Scission. After determining the amount of 1,2- and 1,4-addition of the diene unit and of the ratio of the comonomers present, the major problem of monomer distribution along the polymer chain remains. The chemical method almost universally used for studying the arrangement of the structural units in diene polymer and copolymer molecules involves the oxidative scission of the double bonds in the polymer and the separation and identification of the resulting fragments. By varying the nature of the oxidizing agent and the conditions of the oxidation, each double bond may be cleaved to form either two carbonyl groups or two carboxyl groups. Procedures involving both types of degradation have been used with some success; however, with synthetic polymers and copolymers the acidic degradation products are frequently more easily handled than the carbonyl products.

If there is random orientation of the structural units in butadiene polymer and copolymer molecules, oxidative degradation of the polymers at the double bonds to form acid products will lead, theoretically, to the formation of a large number of polybasic acids. Fortunately, the majority of these acids will appear in negligible amounts. It is useful, however, to calculate the relative amounts of each significant degradation product that should result from the quantitative scission of the double bonds in butadiene polymers and copolymers having the random arrangement of structural units and to compare these theoretical values with experimental values for actual polymers to determine whether or not there is any unusual distribution of units in the latter.

The manner in which the three structural units expected to be present in butadiene polymers and copolymers break down on oxidation is shown in the following formulas. Scission of the double bond in a 1,4-unit results in

$$-CH_2 \cdot CH : CH \cdot CH_2- \qquad -CH_2 \cdot CH- \qquad -CH_2 \cdot CH-$$

$$VIII \qquad\qquad\qquad\qquad CH \qquad\qquad\qquad C_6H_5$$

$$\downarrow \qquad\qquad\qquad\qquad \|$$

$$-CH_2 \cdot CO_2H + HO_2C \cdot CH_2- \qquad CH_2 \qquad\qquad\qquad X$$

$$A \qquad\qquad A \qquad\qquad\qquad IX$$

$$\downarrow \qquad\qquad\qquad\qquad\qquad \downarrow$$

$$-CH_2 \cdot CH- + HCO_2H \qquad -CH_2 \cdot CH-$$

$$CO_2H \qquad\qquad\qquad\qquad\qquad C_6H_5$$

$$B \qquad\qquad\qquad\qquad\qquad\qquad C$$

cleavage of the polymer chain; oxidation of the double bond in a 1,2-unit leads to the formation of the volatile acid, formic acid, and a carboxyl group attached to the polymer chain, but does not break the chain. The vinyl comonomer unit is normally unaffected by oxidation. Thus, three types of acid structural units will be formed: from the 1,4-diene unit, an acidic terminal unit, A; from the 1,2-diene unit, an acidic internal unit, B, and from the vinyl monomer unit, a neutral internal unit, C. Therefore, the nonvolatile polybasic acids derived from the cleavage of the polymer will have the general formula, AB_mC_nA (XI), without regard to the arrangement of the internal units. Formic acid should be the only volatile acid formed

$$HO_2C \cdot CH_2- \left[\begin{array}{c} -CH_2 \cdot CH- \\ | \\ CO_2H \end{array}\right]_m - \left[\begin{array}{c} -CH_2 \cdot CH- \\ | \\ C_6H_5 \end{array}\right]_n -CH_2 \cdot CO_2H$$

$$XI$$

on oxidation of the double bonds, and it should be formed in amounts that can be determined directly from the number of 1,2-diene units in the polymer molecule.

Since the 1,4-diene unit is the only unit with a double bond in the polymer chain, it represents the end of one potential acid molecule and the beginning of another. Thus, the number of acid molecules formed on oxidation of the double bonds in the chain is equal to the number of 1,4-diene units in the chain.* The 75/25 butadiene-styrene copolymer may be taken as an example. In this copolymer there are 14.8 styrene units and 85.2 butadiene units in a section of polymer chain 100 units long. If it is assumed that there is 20 per cent 1,2-addition, there will be 17.0 1,2-units and 68.2 1,4-units in this length of polymer chain, and on oxidation 68.2 molecules of nonvolatile acids will be formed. Furthermore, 68.2 per cent of the 1,4-units will be followed in the chain by other 1,4-units; thus, 68.2 per cent of all the acid molecules will be succinic acid (AA) molecules. Similarly, 14.8 per cent of the 1,4-units will be followed by styrene units and 68.2 per cent of these styrene units will be followed in turn by other 1,4-units; therefore, 0.148×68.2 per cent or 10.1 per cent of all the acid molecules will be β-phenyladipic acid (ACA) molecules. To find the number of such molecules formed from a 100-unit section of the chain, it is necessary only to form the product $0.101 \times 68.2 = 6.88$ molecules, since 68.2 molecules are formed from this length of chain. As the number of 1,2-units and styrene units occurring between two 1,4-units increases, the method of calculation remains the same, but the possibilities of various combinations of the 1,2-units and styrene units must be taken into consideration. For example, the number of acid molecules formed when a 1,4-unit is followed successively by two styrene units, a 1,2-unit, and another 1,4-unit (ACCBA) is $0.148 \times 0.148 \times 0.170 \times 0.682 \times 68.2 = 0.198$. This arrangement of units in the polymer chain is, however, only one of three ways in which two styrene units and one 1,2-unit can be combined; therefore, there will be formed $3 \times 0.198 = 0.594$ acid molecules (ABCCA, ACBCA, ACCBA) containing the proper units. Calculations for the expected acids in any butadiene polymer or copolymer are exactly the same in principle as those outlined above; in each case the chance of obtaining any given configuration depends on the relative number of 1,2-butadiene units, 1,4-butadiene units, and vinyl monomer units available.

Tables V and VI give the results of the calculation of the amounts of the various acids to be expected from polybutadiene, emulsion 75/25 butadiene-styrene copolymer (20 per cent 1,2-addition), and emulsion 69/31 butadiene-chlorostyrene copolymer (20 per cent 1,2-addition), assuming that there is random distribution of the structural units in each of the polymers.

In practice it is found increasingly difficult to separate and identify the acids from the oxidative degradation of 1,3-diene polymers and copolymers as the molecular weight of the acids increases. Among the reasons for this difficulty are: (a) the amount of the individual higher-molecular-weight acids present is usually small; (b) the differences in physical properties of the acids and their derivatives usually helpful in isolation work decreases as molecular weight increases; (c) the number of isomeric acids for any m and/or n values increases, and (d) comparison with authentic specimens of the higher-molecular-weight acids is usually not possible, as most of these have never been prepared and appear to be difficult to prepare.

---

* Actually the number of 1,4-units is one less than the number of acid molecules formed, but, as there are approximately 2000 to 3000 1,4-units in an emulsion diene polymer or copolymer molecule of average size (200,000 molecular weight), the extra molecule of acid may be neglected to simplify calculations.

Table V. Amounts of Acids, AB_mA, Expected from the Oxidation of Polybutadiene

A. *Per Cent Butadiene Accounted for*

m	1,2-Addition, %						
	0	10	20	30	40	50	60
0	100	81.0	64.0	49.0	36.0	25.0	16.0
1	0	16.2	25.6	29.4	28.8	25.0	19.2
2	0	2.43	7.68	13.2	17.3	18.8	17.3
3	0	0.32	2.05	5.3	9.2	12.5	13.8
4	0	0.04	0.51	2.0	4.6	7.8	10.5
Other	0	0.01	0.16	1.1	4.1	10.9	23.2

B. *Grams of Acid per 100 Grams Polymer*

m	1,2-Addition, %						
	0	10	20	30	40	50	60
0	218.3	176.7	139.6	107.0	78.5	54.6	35.0
1	0	28.5	45.0	51.6	50.6	43.9	33.8
2	0	3.9	12.4	21.4	27.9	30.3	27.9
3	0	0.5	3.2	8.2	14.2	19.3	21.3
4	0	0.1	0.8	2.5	6.9	11.7	15.7

Table VI. Amounts of Acids, AB_mC_nA, Expected from 75/25 Butadiene-Styrene and 69/31 Butadiene-o-Chlorostyrene Copolymers

AB_mC_nA		Butadiene Accounted for, %	Styrene or o-Chlorostyrene Accounted for %	Grams of Acid per 100 Grams Polymer	
m	n			Styrene Copolymer	o-Chlorostyrene Copolymer
0	0	54.6		82.5	82.5
1	0	18.6		22.6	22.6
0	1	8.08	46.5	23.0	26.6
1	1	5.50	15.8	10.4	11.6
2	0	4.75		5.3	5.3
0	2	1.19	13.7	5.0	5.7

Ozonolysis and Other Methods of Scission. Ozonolysis has been the most widely used method for the cleavage of the diene polymers at the double bonds. Since its introduction by Harries[15–6, 21] the method has been applied by Pummerer, Ebermayer, and Gerlach[55] to natural rubber, by Alekseeva and Belitskaya[2] to a 50/50 butadiene-styrene colpoymer, by Hill, Lewis, and Simonsen[25, 26] to polybutadiene and to a butadiene-methyl-methacrylate copolymer, by Alekseeva[1] to a copolymer of butadiene and acrylonitrile and a mixed polymer of butadiene and methacrylonitrile, by Klebanskii and Vasil'eva[34] to polychloroprene, by Rabjohn, Bryan, Inskeep, Johnson, and Lawson[58] to GR-S, and by Marvel and co-workers[44–5] to polybutadiene, GR-S, and other polymers. As stated previously, the method is not entirely satisfactory because ozone attacks the polymer molecule at points other than the double bonds and may give rise to degradation products that are not directly relevant to the structure of the molecule. Nevertheless, ozonolysis remains at present the simplest and cleanest procedure available for the degradation of diene polymers and copolymers.

German workers[71] have claimed that better results may be obtained by oxidizing the diene polymers in nitrobenzene solution with aqueous potassium permanganate. Using this procedure, they were able to account for 93 per cent of the carbon in emulsion polybutadiene and 80 per cent of the carbon in sodium-polymerized polybutadiene. Similarly, from 100 grams of Buna S they obtained 30 grams of β-phenyladipic acid (ACA) instead of 9 grams of the acid obtained by them using ozonolysis.

It has been suggested by Kolthoff, Lee, and Carr[36] that the use of tert-butylhydroperoxide with a trace of osmium tetroxide might give better results than ozonolysis for the degradation of diene polymer molecules. These authors utilized the reagent to oxidize butadiene-styrene copolymers to determine the polystyrene content. The reagent cleaved the polymer molecules at the double bonds to form a mixture of low-molecular-weight aldehydes and possibly other oxygenated products. The low-molecular-weight material was dissolved in alcohol, and any polystyrene present was removed by filtration. Ordinary GR-S was found to be free from polystyrene. Suitable procedures for the use of the reagent to cleave polymer molecules to useful fragments have been developed by Laitinen and Lukes[38] for polybutadiene and GR-S; however, these procedures are necessarily somewhat involved. It was found that acetals, esters, glycols, etc., as well as aldehydes, are formed on oxidation of the polymers with the reagent and that it is necessary to convert these products into separable and identifiable compounds. In its present form the method appears to offer no advantages over ozonolysis and is considerably less attractive from a manipulative viewpoint.

Other methods of oxidative degradation which have been used but which have not been proved to be superior to ozonolysis include the use of alkaline,[61] neutral,[6, 17, 61] and acidic[51] aqueous potassium permanganate[68] to oxidize solutions of the polymer and the use of hydrogen peroxide.[6, 61] Of these methods only the use of acidic permanganate has been successful in yielding ultimate oxidation products similar to those produced by ozonolysis, and only this method has been applied to synthetic rubbers (see p. 313).

That ozonolysis, despite its advantages, is not always so clear-cut and free from uncertainties as might be hoped is shown by the fact that, among the products of the oxidative hydrolysis (by 3 per cent hydrogen peroxide) of the ozonide of polybutadiene, Marvel and his co-workers[42, 47] have identified not only the major fragments expected from 1,4- and 1,2-union of the diene units, viz., succinic acid, 1,2,4-butanetricarboxylic acid, and formic acid, but also small amounts of the following products, viz.: (1) 1,2,3-propanetricarboxylic acid, (2) the half aldehyde of succinic acid, (3) β-formyladipic acid, (4) acetic acid, (5) acetaldehyde, (6) acetone, (7) levulinic acid. The significance of product 1 is discussed later. The products 2 and 3 clearly represent partially oxidized fragments corresponding to succinic acid and 1,2,4-butanetricarboxylic acid, respectively. But the other products are less readily explicable: They may arise as a result of the rearrangement of double bonds during the course of polymerization, or they may be due to some at present obscure, "abnormal" mode of scission of the ozonide.

A study bearing on the question of "abnormal" ozonolysis products has also been published by Ziegler.[76] It involved the ozonization of model molecules and the application to the ozonides of two methods of oxidative scission, viz.: (a) the peracetic acid method of Wilms,[72] (b) the sodium hydroxide-silver oxide method of Asinger.[3] The former method is more drastic and liable to produce more extensive oxidative degradation than the latter. Thus, when treated with peracetic acid, the ozonide of 5-vinyldecane yielded not only the expected α-amylcaproic acid but also decanol-5 and probably decanone-5, and the ozonide of 4-vinylcyclohexene yielded not only the expected 1,2,4-butanetricarboxylic acid but also small amounts of levulinic and succinic acids.

Ziegler considers one possible reason for abnormalities in the process of ozonolysis to be the presence in the ozonization product of hydroperoxides resulting from the attack of oxygen at the α-methylene group. . He reports that the ozonide of cyclohexene hydroperoxide, when subjected to oxidative hydrolysis, yielded no adipic acid but only succinic and glutaric acids.

Separation of Scission Products. A number of methods are available for the separation of the degradation products. Where the oxidation products are carbonyl compounds (from ozonolysis generally), they may be separated by fractional crystallization of carbonyl derivatives, such as phenylhydrazones and semicarbazones; where the oxidation products are acids, they may be separated by fractional crystallization or fractional extraction, or the acids may be converted into the methyl esters and the esters separated by fractional distillation. None of these methods is particularly adapted to handling quantitatively the complex mixture of products that usually results, and most of them have the common disadvantage that they require a relatively large percentage of a given component to be present before that component can be isolated. Improvement in some of the methods of separation has been reported in recent years, but none of these methods is truly quantitative. Thus, Marvel and Richards[46] have reported some improvement in the use of fractional extraction as a method for the separation of the acids from the ozonolysis of GR-S. By using diazomethane for the esterification of the acids, German workers eliminated most of the loss in that step of the separation, but the difficulty in separating by distillation the complex mixture of high-molecular-weight esters remains.[71] A chromatographic separation procedure is described in a later section.

Results of Ozonolysis of GR-S. The difficulties encountered in diene polymer structure studies made using conventional methods for the separation of the acid mixtures are amply illustrated by the results of Rabjohn, Bryan, Inskeep, Johnson, and Lawson.[58] Following the procedure of Alekseeva and Belitskaya,[2] these workers ozonized two samples of GR-S (a standard soluble polymer and an insoluble polymer), converted the ozonide into di- and polybasic acids by hydrolysis with hydrogen peroxide, and separated these into water-soluble and water-insoluble fractions. Succinic acid was separated from the water-soluble acids by fractional crystallization. Then the remaining water-soluble acids were converted into the methyl esters, and the esters were fractionally distilled. A number of ester fractions were obtained, as well as a higher-boiling residue which could not be satisfactorily fractionated. Esterification of the water-insoluble acids and subsequent distillation of the esters yielded fractions that were satisfactory only for partial characterization. Identification of the esters in these fractions was only tentative. The results obtained for the two polymers are indicated in Table VII.

Table VII. Acids from Ozonolysis of 77/23 Butadiene-Styrene (GR-S)

		Soluble GR-S			Insoluble GR-S		
AB_mC_nA		Butadiene	Styrene	% of Theory for	Butadiene	Styrene	% of Theory for
		Accounted	Accounted	Random	Accounted	Accounted	Random
m	n	for, %	for, %	Pattern*	for, %	for, %	Pattern*
0	0	40.2		79	36.4		75
1	0	17.3		98	15.3		91
0	1	3.2	20.7	47	2.5	15.9	38
2	0	4.3		94	4.6		104
1	1	2.0	6.5	43	2.2	7.0	48
(0	2)†		(14.5)	(121)		(26.2)	(230)

  * 20% 1,2-addition assumed, based on total amount of di- and polybasic acids obtained from ozonolysis.
  † Not definitely characterized.

All of the acids in the table, except the last, were isolated from the water-soluble acids fraction. In addition to these acids (or esters) isolated from the water-soluble acids, a supposed fraction of dimethyl diphenylsuberate (derived from a portion of the polymer chain in which two styrene units occur in immediate succession) was obtained from the esters of the water-insoluble acids (last entry in the table). A fraction of volatile acids (assumed to be formic acid) was obtained also for each polymer; for the soluble polymer this fraction was equivalent to about 29.9 per cent of the butadiene present, for the insoluble polymer 25.7 per cent. About 92 per cent of the expected amount of di- and polybasic acids was obtained from the soluble polymer and about 88 per cent from the insoluble polymer.

The percentages of the calculated amounts for random distribution of units for the various acids are shown in Table VII. These values in the table are based on the total amounts of di- and polybasic acids actually obtained from ozonolysis but are not corrected for the indeterminate losses incurred in separating the individual acids. Considering that these losses may be rather high, the amounts of acids found correspond well with the values calculated for random distribution of the units in the chain, with the exceptions of the phenyl-containing acids and formic acid. β-Phenyladipic acid (ACA) was found in only about one-half the predicted amounts, while the amounts of the supposed diphenylsuberic acid (ACCA) were larger than predicted. Assuming that the diphenylsuberic acid was properly characterized, only 40 to 50 per cent of the styrene in the polymers was accounted for. Presumably the remainder of the styrene units were lost in the high-boiling residue, which could not be distilled.

The high values for formic acid apparently indicate a greater percentage of 1,2-addition than is normally expected for emulsion GR-S. However, it has been pointed out by Pummerer[55] that the formation of excessive formaldehyde or formic acid is not necessarily significant, for even hexane will yield a small amount of both of these products on treatment with ozone.

Ozonolysis of Copolymer of Butadiene and o-Chlorostyrene. Chromatographic Separation of Scission Products. A butadiene-*o*-chlorostyrene (69/31) copolymer which corresponds in molar ratio to the 75/25 butadiene-styrene copolymer has been ozonized in a further study of the distribution of the styrene units in the GR-S-type polymer chain.[44] The chlorine atoms in the ring were a convenient tag to follow the benzene ring through the oxidation process and the acid-separation system. The polymer was purified carefully and ozonized in chloroform solution. The chloroform was evaporated under reduced pressure and the ozonide decomposed with 3 to 10 per cent hydrogen peroxide. The excess hydrogen peroxide was destroyed by treatment of the solution with a small amount of platinum oxide. The water-insoluble acids were separated and the water-soluble acids obtained by careful evaporation of the solution under reduced pressure.

The water-soluble acids were then separated by a partition chromatographic method developed by Marvel and Rands.[45] This method consists of adsorbing the mixed water-soluble acids on a column of moist silicic acid gel and separating the acids by eluting them with chloroform-ethanol mixtures which are gradually increased in percentage of ethanol. The water-insoluble acid mixture was separated by use of adsorption chromatography, using silicic acid as adsorbent and chloroform as the eluant. The total weight of acids obtained was 73 per cent of the calculated amount and the chlorine content of the acids was at least 95 per cent of the theoretical, which showed that the benzene rings were essentially intact after ozonolysis. The three major acids isolated were succinic, 1,2,4-butanetricarboxylic, and β-(o-chlorophenyl)adipic. Small amounts of unidentified acids were also obtained. The quantities of these acids from 100 grams of polymer found in the water-soluble and water-insoluble portions are given in Table VIII.

It may be significant that the amount of 1,2,4-butanetricarboxylic acid is

Table VIII. Ozonolysis of 69/31 Butadiene-o-Chlorostyrene Copolymer

	Succinic Acid	1,2,4-Butane-tricarboxylic Acid	β-(o-Chloro-phenyl)adipic Acid
Water-soluble acids	69.2 g.	10.6	16.2
Water-insoluble acids	0.16	0.11	2.27
Total found	69.36	10.71	18.47
Calculated weight that should be found, assuming 20% 1,2-addition of butadiene*	82.5	22.6	26.6
Per cent recovery	84.1	47.3	69.6

* See Table VI.

low. Perhaps there is less than 20 per cent 1,2-addition of the butadiene in this polymer. The yield of β-(o-chlorophenyl)adipic acid is nearly up to the over-all average yield of acids obtained in the oxidation (73 per cent); hence there is no evidence that the styrene units tend to bunch together in the chain.

Further Results on Butadiene-Styrene Copolymers. German workers,[71] using both ozonolysis and potassium permanagnate oxidation, separated β-phenyladipic acid from the polyphenyl acids by benzene solubility, the monophenyldibasic acid being insoluble. Oxidation of Buna S is reported to give 25 to 30 per cent polyphenyl and 65 to 70 per cent monophenyl acids, Buna SS to give 70 per cent polyphenyl and 30 per cent monophenyl acids. Although detailed data are not available on these investigations, the above results indicate a random distribution of styrene units throughout the polymer chain.

Somewhat similar results were obtained by Alekseeva and Belitzkaya[2] by the ozonolysis of a 50/50 butadiene-styrene copolymer (prepared in bulk). From the amounts of acids isolated, these workers calculated that 31.2 per cent of the styrene occurred between 1,4-butadiene units, 40 per cent occurred in pairs between 1,4-units, and 29 per cent was found in conjunction with 1,2-units as well as other styrene units. About 23.2 per cent 1,2-addition was indicated from the amount of formic acid isolated. These values correspond roughly to those predicted for a random distribution of structural units in a 50/50 copolymer of butadiene and styrene. The higher percentage of styrene present in the copolymer (as compared with GR-S) increased the amount of β-phenyladipic acid to be expected on oxidative ozonolysis and made separation of this degradation product somewhat easier; however, difficulty was experienced by Alekseeva and Belitzkaya[2] in the isolation of the higher-molecular-weight acids, and the figures derived from the amounts of these acids are only approximate.

Random and Alternating Distribution of Structural Units. The best evidence thus indicates that in polybutadienes or butadiene-styrene copolymers prepared by free-radical catalysis the three common structural units (VIII, IX, and X) are distributed along the chain according to chance. Not all of the copolymers of butadiene display the random arrangement of structural units. Among those that have been shown to have the alternating pattern of monomer units are the copolymers of butadiene with methacrylonitrile and with methyl methacrylate.

Copolymers of Butadiene and Unsaturated Nitriles. The arrangement of the structural units in the copolymer of butadiene and methacrylonitrile was studied by Alekseeva.[1] Ozonolysis of the copolymer and separation through the methyl esters of the acidic degradation products gave 62 weight per cent of 2-methyl-1,2,4-butanetricarboxylic acid (one carboxyl group coming from the hydrolysis of the nitrile group) and 8 weight per cent succinic acid. The conclusion drawn from these data was that the copolymer consisted mostly of alternate units of each monomer, the addition to butadiene being almost entirely in the 1,4-manner. A small amount of 1,4-addition of butadiene to butadiene also occurred. The oxidative degradation products from a copolymer of equimolecular parts of butadiene and acrylonitrile indicated that about one-half of the units in the polymer occurred alternately in the chain.[1]

Copolymer of Butadiene and Methyl Methacrylate. Utilizing a similar procedure, Hill, Lewis, and Simonsen[26] studied the structure of a 1 : 1 (molar) butadiene-methyl methacrylate copolymer. They found that 51 per cent of the butadiene was accounted for by the acid in which only one methyl methacrylate unit appears, 4.1 per cent by the acid in which two methyl methacrylate units appear, and 10.4 per cent (approximate) by the acid in which three methyl methacrylate units appear, and that only 9.1 per cent of the butadiene appeared as succinic acid. About 76 per cent of the butadiene and 89 per cent of the methyl methacrylate were accounted for. The conclusion was that more than half of the copolymer consisted of alternate butadiene and methyl methacrylate units, with addition to butadiene occurring principally in the 1,4-manner.

Isoprene Polymers and Copolymers. Comparatively little work has been done on the structures of synthetic isoprene polymers and copolymers. Most of our present knowledge on this subject comes from the early ozonolysis studies of Harries[18-9, 21-2] and Ziegler[74] and the more recent investigations of the relative amounts of 1,2-, 1,4-, and 3,4-addition. The principal products isolated by Harries[18-9, 21-2] from the ozonolysis of polyisoprene prepared in bulk were levulinic aldehyde and levulinic acid. Traces of acetonylacetone and succinaldehyde also were obtained. These data indicated that the polymer was formed predominantly by 1,4-addition and that, although there were some head-to-head and tail-to-tail linkages of 1,4-units, most of the 1,4-units were linked head-to-tail. From the ozonolysis of sodium polyisoprene Harries isolated only a trace of levulinic aldehyde. Although Harries stated only that this result indicated that sodium isoprene was an "abnormal" polymer' (i.e., different from natural rubber), it can be seen that the result is in agreement with more recent experiments showing that sodium polyisoprene contains a large percentage of 1,2- (and 3,4-) units.

It is difficult to draw a satisfactory picture of the synthetic isoprene polymer molecules from the rather fragmentary data available from the researches just described. What are needed for a general representation of these molecules are more data of a semiquantitative nature on the relative amounts of 1,2- and 3,4-addition and of head-to-tail and head-to-head (tail-to-tail) union of 1,4-units.

Copolymers of Isobutylene and Isoprene. The structure of the isobutylene-isoprene copolymer has been investigated by Rehner[59] by use of the ozonolysis technique. Several polymers, in which the percentage of isoprene varied from 0.6 to 2.4 per cent, were ozonized and the degradation products examined for the presence of formaldehyde, formic acid, levulinic acid, succinic acid, and acetonylacetone, the compounds that might be expected from the scission of sections of the polymer chain in which at least two isoprene units were joined in any of the possible arrangements. None of these compounds was found in any of the ozonolysis products. The conclusion was that the isoprene units enter the polymer by 1,4-addition and that there is no bunching of isoprene units; i.e., the polymer is formed by the random entrance of the monomer units into the polymer chain.

Polydimethylbutadiene. The oxidative scission of the double bonds in the two diene structural units expected to be present in poly-2,3-dimethyl-1,3-butadiene, the 1,2-unit and the 1,4-unit, would give rise to the structural units in the degradation products shown in the formulas. The only acid

$$
\begin{matrix}
& CH_3 \ CH_3 & & O & & O \\
& | \quad | & & \parallel & & \parallel \\
-CH_2 \cdot C{=}C \cdot CH_2{-} & \longrightarrow & -CH_2 \cdot C \cdot CH_3 + CH_3 \cdot C \cdot CH_2{-}
\end{matrix}
$$

XII

$$
\begin{matrix}
& CH_3 & & CH_3 \\
& | & & | \\
-CH_2 \cdot C{-} & \longrightarrow & -CH_2 \cdot C{-} & + \quad HCO_2H \\
& | & & | \\
& C \cdot CH_3 & & C : O \\
& \parallel & & | \\
& CH_2 & & CH_3
\end{matrix}
$$

XIII

to be formed by normal oxidative cleavage would be formic acid; the other degradation products would be ketones. The simplest ketones expected are acetonylacetone, from the oxidation of adjacent 1,4-units, and 4-methyl-4-acetyl-2,7-octandione, from a single 1,2-unit sandwiched in between 1,4-units.

Harries[18, 20, 22] is responsible for the only available information on the structures of the 2,3-dimethyl-1,3-butadiene polymers other than that already described in the section on 1,2- and 1,4-addition in diene polymers. From the ozonolysis of bulk poly-2,3-dimethyl-1,3-butadiene, Harries[20, 22] obtained acetonylacetone in sufficient quantity to account for about 80 per cent of the total yield of degradation products. This result indicates a structure in which 1,4-units predominate, an indication that agrees very well with the more recent determinations of 1,2- to 1,4-addition ratios. From the ozonolysis products of sodium poly-2,3-dimethyl-1,3-butadiene, Harries[18, 22] isolated only a small amount of acetonylacetone, indicating a low percentage of 1,4-addition in the polymer—again in agreement with recent determinations of 1,4-addition.

Polychloroprene. The oxidative cleavage of the double bonds in polychloroprene might be expected to result in the formation of the structural units in the degradation products shown in the formulas. Probably the intermediate acyl halides would be hydrolyzed to the corresponding acids during the oxidation process. Thus, the acid to be expected from the oxidation of adjacent 1,4-units is succinic acid and from a single 1,2-unit between 1,4-units is either 1,2,4-butanetricarboxylic acid or 2-chloro-1,2,4-butane-tricarboxylic acid. Since both halves of the 1,4-unit may yield the acid structure unit A, on oxidative cleavage (and subsequent hydrolysis of the acyl halide), it is not possible to distinguish conclusively, by identification of the degradation products, between 1,4-units linked head-to-head (tail-to-tail) and those linked head-to-tail.

Klebanskii and co-workers[34] degraded polychloroprene by the ozonolysis method and found that 82.3 per cent of the carbon chain in the α-polymer,[8]

81.6 per cent in the μ-polymer,[8] and 87.1 per cent in the ω-polymer were accounted for by the succinic acid isolated and only 2.2 per cent of the carbon chain in the α-polymer, 5.2 per cent in the μ-polymer, and 0.3 per cent in the ω-polymer were accounted for by formic acid. These data indicate

$$\underset{\text{XIV}}{-CH_2 \cdot \overset{\overset{\displaystyle Cl}{|}}{C} : CH \cdot CH_2 -} \longrightarrow -CH_2 \cdot COCl + HO_2C \cdot CH_2 - \quad A$$

$$\underset{A}{\overset{\downarrow}{-CH_2 \cdot CO_2H}}$$

$$\underset{\text{XV}}{-CH_2 \cdot \overset{\overset{\displaystyle Cl}{|}}{\underset{\overset{\displaystyle |}{\underset{\overset{\displaystyle CH}{\|}}{CH}}}{C}}-} \longrightarrow -CH_2 \cdot \overset{\overset{\displaystyle Cl}{|}}{\underset{\overset{\displaystyle |}{CO_2H}}{C}}- + HCO_2H$$

$$\underset{\text{XVI}}{-CH_2 \cdot \overset{\overset{\displaystyle |}{\underset{\overset{\displaystyle |}{\underset{\overset{\displaystyle CH_2}{\|}}{C \cdot Cl}}}{CH}}}-} \qquad -CH_2 \cdot \overset{\overset{\displaystyle |}{\underset{\overset{\displaystyle |}{COCl}}{CH}}}- + HCO_2H$$

$$\underset{\overset{\displaystyle |}{CO_2H}}{\overset{\downarrow}{-CH_2 \cdot CH-}}$$

that the polymer molecules are formed predominantly from 1,4-units. There was no evidence for head-to-head structural units. Somewhat similar experimental results were obtained by Carothers, Williams, Collins, and Kirby[8] from the oxidation of polychloroprene with nitric acid. Although these workers did not examine the oxidative degradation extensively, they reported that the principal product isolated was succinic acid. Walker and Mochel[70] have obtained 93, 95, and 94 per cent, respectively, of the expected amount of succinic acid by ozonolysis and oxidative cleavage of 10°, 40°, and popcorn polymers of chloroprene, and subsequently Mochel and Nichols[50] obtained by the ozonolysis of Neoprene W more than 98 per cent of the amount expected from a linear, 1,4-polymer. Further, by working under mild conditions to avoid secondary oxidation products, very much less formic acid and/or formaldehyde was obtained than the quantity reported by Klebanskii; in fact, so little that Mochel and Walker conclude that it probably is not significant or indicative of the presence of side vinyl groups. From highly cross-linked ω-polychloroprene they obtained evidence (from an infrared analysis of the esters of an acid fraction higher than succinic acid) of the presence of a small amount of 1,2,4-butanetricarboxylic acid—1 mole per 15,000 molecular weight.

BRANCHING

Diene polymers prepared at high conversions or with too little modifier present give gels which are undoubtedly cross-linked structures. Physical evidence has indicated that most of the diene polymers are not entirely linear but are branched to a considerable extent.

Flory[13] has suggested that the cross-linking reaction is the result of a growing polymer free-radical chain (P·) adding to a double bond in a section of preformed polymer molecule to give a new free radical which then grows, thus producing a cross-linked, insoluble structure thus:

$$
\begin{array}{ccc}
\quad|\quad & \quad|\quad & \quad|\quad \\
CH_2 & CH_2 & CH_2 \\
| & | & | \\
CH & P\cdot CH & P\cdot CH \\
P\cdot + \| \longrightarrow & | \quad\xrightarrow[\text{butadiene}]{} & | \\
CH & CH\cdot & CH\cdot CH_2CH:CH\cdot CH_2- \\
| & | & | \\
CH_2 & CH_2 & CH_2 \\
| & | & | \\
\text{XVII} & \text{XVIII} & \text{XIX}
\end{array}
$$

$$
\begin{array}{ccc}
-CH\cdot CH_2- & -CH\cdot CH_2- & -CH\cdot CH_2- \\
| & | & | \\
CH & CH\cdot & CH\cdot CH_2CH:CH\cdot CH_2- \\
P\cdot + \| \longrightarrow & | \longrightarrow & | \\
CH_2 & P\cdot CH_2 & P\cdot CH_2 \\
\text{XX} & \text{XXI} & \text{XXII}
\end{array}
$$

This type of reaction should reduce the total unsaturation of the final diene polymer, but the extent of reduction of unsaturation will undoubtedly be too little to estimate by the titration methods available for this determination.

A type of branching that may occur involves a chain-transfer reaction between a growing polymer chain and the active allylic hydrogens on a preformed polymer chain.[13]   The free radical produced as an intermediate

$$
\begin{array}{ccc}
CH_2CH:CHCH_2- & CH_2CH:CHCH_2- & CH_2CH:CHCH_2- \\
| & | & | \\
CH_2 & CH\cdot & CHCH_2CH:CHCH_2- \\
P\cdot + | \longrightarrow PH+ & | \quad\xrightarrow[\text{butadiene}]{} & | \\
CH & CH & CH \\
\| & \| & \| \\
CH & CH & CH \\
| & | & | \\
CH_2 & CH_2 & CH_2 \\
| & | & | \\
\text{XXIII} & \text{XXIV} & \text{XXV}
\end{array}
$$

in this chain transfer is allylic in character and could readily undergo a 1,3-shift to produce a different intermediate.   It is also possible in a GR-S polymerization that such a chain-transfer reaction might involve the benzyl hydrogens of the styrene residues.   Free radicals such as these could dimerize to produce cross-links or add monomer units to produce branching (XXV).

At this writing the evidence regarding the occurrence of such cross-linked and branched structures in diene polymers is unfortunately insufficient and inconclusive, but it is interesting enough to be worth noting.   A branched structure such as XXV would be expected on ozonolysis to yield 1,2,3-propanetricarboxylic acid (tricarballylic acid).   And Marvel and co-workers[47] actually isolated this acid from the products of the oxidative ozonolysis of insoluble, highly gelled polybutadiene.   However, the acid was also obtainable from soluble samples of polybutadiene.   Although under mild conditions highly gelled polybutadiene gave more 1,2,3-propanetricarboxylic acid than did soluble polymer, the application to the ozonides of more severe methods of oxidative hydrolysis might reverse this situation. Further, the ozonide of 4-vinylcyclohexene, which would be expected to yield only 1,2,4-butanetricarboxylic acid and formic acid, gave in fact a small amount of the propanetricarboxylic acid.

It remains uncertain how far the propanetricarboxylic acid is a primary oxidative scission product, indicative of branching, and how far it is a secondary product.   The possibility that a portion of it is a primary product finds some support in the observation that reoxidation with silver oxide of an acid mixture formed by the mild oxidative hydrolysis of an ozonide produced only a small amount of the propanetricarboxylic acid.   It may further be noted that, when the butanetricarboxylic acid was subjected to the conditions applied in the oxidative hydrolysis of the ozonides, no oxidative degradation to 1,2,3-propanetricarboxylic acid took place.

Naples and D'Ianni[51] obtained 1,2,3-propanetricarboxylic acid by the oxidation, with aqueous, acidic permanganate, of solutions in carbon tetrachloride of polybutadiene samples prepared in emulsion, by an Alfin catalyst and by sodium.   They think that this acid is veritably a primary oxidation fragment and not derived from the secondary oxidation of 1,2,4-butanetricarboxylic acid, since the latter was found to be more resistant than the former to destruction by acidic permanganate solution.   It may however be remarked that the quantities of tricarballylic acid produced from the polybutadiene samples were relatively high and, if accepted as a true measure of the branching at the α-methylenic carbon, indicate an unexpectedly large amount of such branching.   Thus, for example, from a sample of polybutadiene made in emulsion to 60 per cent conversion at 50° C. and free from gel the following oxidation products were identified, the amounts shown being based on a recovery of 62.2 per cent of the theoretical amount of solid acids: succinic acid 57.4 per cent, 1,2,4-butanetricarboxylic acid 11.5 per cent, tricarballylic acid 3.0 per cent.   Such a quantity of tricarballylic acid would correspond to as many as 46 branches in a polybutadiene of 216,000 molecular weight having 20 per cent of 1,2-butadiene units.   (It is assumed that the above percentages are weight percentages.)

It may be noted that Ziegler[76] found that the action of acid permanganate on a benzene solution of 4-vinylcyclohexene yielded not only the expected 1,2,4-butanetricarboxylic acid but also succinic acid.

REFERENCES

1. Alekseeva, E. N., *J. Gen. Chem. U.S.S.R.*, **9**, 1426–30 (1939); **11**, 353–7 (1941); *Rubber Chem. and Technol.*, **15**, 698–703 (1942).
2. Alekseeva, E. N., and Belitzkaya, R. N., *J. Gen. Chem. U.S.S.R.*, **11**, 358–62 (1941); *Rubber Chem. and Technol.*, **15**, 693–7 (1942).
3. Asinger, F., *Ber.*, **75**, 656–60 (1942).
4. Baker, W. O., and Heiss, J. H., Private Communication to O.R.R., Mar. 23, 1943.
5. Bloomfield, G. F., *J. Soc. Chem. Ind.*, **64**, 274–8 (1945).
6. Boswell, M. C., Hambleton, A., Parker, R. R., and MacLaughlin, R. R., *Trans. Roy. Soc. Can.*, Section III, **16**, 27–47 (1922); *India-Rubber J.*, **64**, 981–7 (1922).
7. Brown, J. R., and Hauser, E. A., *Ind. Eng. Chem.*, **30**, 1291–6 (1938).
8. Carothers, W. H., Williams, I., Collins, A. M., and Kirby, J. E., *J. Am. Chem. Soc.*, **53**, 4203–25 (1931).
9. Cheyney, L. E., and Kelley, E. J., *Ind. Eng. Chem.*, **34**, 1323–6 (1942).
10. Eimers, E., *Ann.*, **567**, 116–23 (1950).
11. Fisher, H. L., *Ind. Eng. Chem.*, **19**, 1325–8 (1927).
12. Fisher, H. L., and McColm, E. M., *Ind. Eng. Chem.*, **19**, 1328–33 (1927).
13. Flory, P. J., *J. Am. Chem. Soc.*, **69**, 2893–9 (1947).
14. Gorgas, A., *Kautschuk*, **4**, 253–4 (1928).
15. Harries, C. D. (a) *Ber.*, **37**, 2708–11 (1904); (b) *Ber.*, **38**, 1195–1203 (1905); (c) *Ber.*, **38**, 3985–9 (1905).
16. Harries, C. D., (d) *Ann.*, **383**, 157–227 (1911); (e) *Ber.*, **45**, 936–44 (1912); (f) *Ann.*, **395**, 211–64 (1913); (g) *Ann.*, **395**, 264–72 (1913); (h) *Ber.*, **47**, 573–7 (1914); (i) *Ber.*, **48**, 863–8 (1915).
17. Harries, C. D., *Ber.*, **37**, 2708–11 (1904).
18. Harries, C. D., *Ann.*, **383**, 157–227 (1911).
19. Harries, C. D., *Ann.*, **395**, 211–64 (1913).
20. Harries, C. D., *Ann.*, **395**, 264–72 (1913).
21. Harries, C. D., *Ber.*, **48**, 863–8 (1915).
22. Harries, C. D., *Untersuchungen über die natürlichen und künstlichen Kautschukarten*, Springer, Berlin, 1919.
23. Hauser, E. A., and Brown, J. R., *Ind. Eng. Chem.*, **31**, 1225–9 (1939).
24. Hauser, E. A., and Sze, M. C., *J. Phys. Chem.*, **46**, 118–31 (1942).
25. Hill, R., Lewis, J. R., and Simonsen, J. L., *Trans. Faraday Soc.*, **35**, 1067–73 (1939).
26. Hill, R., Lewis, J. R., and Simonsen, J. L., *Trans. Faraday Soc.*, **35**, 1073–9 (1939).
27. Johnson, B. L., Private Communication to O.R.R., Jan. 13, 1943.
28. Kaufmann, H. P., and Gaertner, P., *Ber.*, **57**, 928–34 (1924).
29. Kemp, A. R., *Ind. Eng. Chem.*, **19**, 531–3 (1927).
30. Kemp, A. R., Bishop, W. S., and Lackner, T. J., *Ind. Eng. Chem.*, **20**, 427–9 (1928).
31. Kemp, A. R., and Mueller, G. S., *Ind. Eng. Chem., Anal. Ed.*, **6**, 52–6 (1934).
32. Kemp, A. R., and Peters, H., *Ind. Eng. Chem., Anal. Ed.*, **15**, 453–9 (1943).
33. Kirchhof, F., *Gummi-Ztg.*, **27**, 9 (1913).
34. Klebanskii, A. L., and Vasil'eva, V. G., *J. prakt. Chem.*, **144**, 251–64 (1936); *Rubber Chem. and Technol.*, **10**, 126–34 (1937); Klebanskii, A. L., and Chevychalova, K. K., *J. Gen. Chem. U.S.S.R.*, **17**, 941–56 (1947).
35. Kobeko, P. P., and Moskvina, E. K., *Zhur. Priklad. Khim.*, **19**, 1143–8 (1946); *Rubber Chem. and Technol.*, **21**, 830–4 (1948).

36. Kolthoff, I. M., Lee, T. S., and Carr, C. W., *J. Polymer Sci.*, **1**, 429–33 (1946).
37. Kolthoff, I. M., Lee, T. S., and Mairs, M. A., *J. Polymer Sci.*, **2**, 199–205 (1947); Kolthoff, I. M., and Lee, T. S., *J. Polymer Sci.*, **2**, 206–19 (1947); Kolthoff, I. M., Lee, T. S., and Mairs, M. A., *J. Polymer Sci.*, **2**, 220–8 (1947).
38. Laitinen, H. A., and Lukes, G. E., Private Communication to O.R.R., Mar. 22, 1948.
39. Lee, T. S., Kolthoff, I. M., and Mairs, M. A., *J. Polymer Sci.*, **3**, 66–84 (1948).
40. Lewis, W. K., and McAdams, W. H., *Ind. Eng. Chem.*, **12**, 673–6 (1920).
41. Madorsky, S. L., Wagman, D. D., and Rossini, F. D., Private Communication to O.R.R.
42. Marvel, C. S., Bluestein, C., Schilling, W. M., and Sheth, P. G., *J. Org. Chem.*, **16**, 854–9 (1951).
43. Marvel, C. S., Gilkey, R., Morgan, C. R., Noth, J. R., Rands, R. D., and Young, C. H., Unpublished Data. Cf. same authors, *J. Polymer Sci.*, **6**, 483–502 (1951).
44. Marvel, C. S., and Light, R. E., *J. Am. Chem. Soc.*, **72**, 3887–91 (1950).
45. Marvel, C. S., and Rands, R. D., *J. Am. Chem. Soc.*, **72**, 2642–6 (1950).
46. Marvel, C. S., and Richards, J. C., *Anal. Chem.*, **21**, 1480–3 (1949).
47. Marvel, C. S., Schilling, W. M., Shields, J. D., Bluestein, C., Irwin, O. R., Sheth, P. G., and Honig, J., *J. Org. Chem.*, **16**, 838–53 (1951).
48. Marvel, C. S., Williams, J. L. R., and Baumgarten, H. E., *J. Polymer Sci.*, **4**, 583–95 (1949).
49. Meehan, E. J., *J. Polymer Sci.*, **1**, 175–82 (1946).
50. Mochel, W. E., and Nichols, J. B., *Ind. Eng. Chem.*, **43**, 154–7 (1951).
51. Naples, F. J., and D'Ianni, J. D., *Ind. Eng. Chem.*, **43**, 471–6 (1951).
52. Prilezhaev, N., *Ber.*, **42**, 4811–5 (1909).
53. Pummerer, R., *Kautschuk*, **10**, 149–51 (1934).
54. Pummerer, R., and Burkard, P. A., *Ber.*, **55**, 3458–72 (1922).
55. Pummerer, R., Ebermayer, G., and Gerlach, K., *Ber.*, **64**, 809–25 (1931).
56. Pummerer, R., and Mann, P. J., *Ber.*, **62**, 2636–47 (1929).
57. Pummerer, R., and Stärk, H., *Ber.*, **64**, 825–30 (1931).
58. Rabjohn, N., Bryan, C. E., Inskeep, G. E., Johnson, H. W., and Lawson, J. K., *J. Am. Chem. Soc.*, **69**, 314–9 (1947).
59. Rehner, J., *Ind. Eng. Chem.*, **36**, 46–51 (1944).
60. Rehner, J., *Ind. Eng. Chem.*, **36**, 118–24 (1944).
61. Robertson, J. M., and Mair, J. A., *J. Soc. Chem. Ind.*, **46**, 41–9T (1927).
62. Saffer, A., and Johnson, B. L., *Ind. Eng. Chem.*, **40**, 538–41 (1948).
63. Schmitz, W., *Gummi-Ztg.*, **27**, 1342–8 (1913).
64. Serniuk, G. E., Banes, F. W., and Swaney, M. W., *J. Am. Chem. Soc.*, **70**, 1804–8 (1948).
65. Staudinger, H., and Geiger, E., *Helv. Chim. Acta*, **9**, 549–57 (1926).
66. Staudinger, H., and Widmer, W., *Helv. Chim. Acta*, **9**, 529–49 (1926).
67. Stoll, M., and Rouve, A., *Helv. Chim. Acta*, **27**, 950–61 (1944).
68. Van Rossem, A., *Kolloid-Beih.*, **10**, 9 (1918).
69. Vaubel, W., *Gummi-Ztg.*, **26**, 1879–80 (1912).
70. Walker, H. W., and Mochel, W. E., *Proc. Second Rubber Technol. Conf.*, London, 1948, 69–78.
71. Weidlein, E. R., *Chem. Eng. News*, **24**, 771–4 (1946).
72. Wilms, H., *Ann.*, **567**, 96–9 (1950).
73. Yakubchik, A. I., Visiliev, A. A., and Zhabina, V. M., *J. Appl. Chem. U.S.S.R.*, **17**, 107–13 (1944); *Rubber Chem. and Technol.*, **18**, 780–4 (1945).
74. Ziegler, K., Dersch, F., and Wolthan, H. V., *Ann.*, **511**, 13–44 (1934).
75. Ziegler, K., Grimm, H., and Willer, R., *Ann.*, **542**, 90–122 (1939).
76. Ziegler, K., Hechelhammer, W., Wagner, H. D., and Wilms, H., *Ann.*, **567**, 99–115 (1950).

CHAPTER 10

PHYSICAL CHEMISTRY OF SYNTHETIC RUBBERS

Lawrence A. Wood

National Bureau of Standards

INTRODUCTION

The principles of physical chemistry have been invaluable in the development of synthetic rubbers. It is the object of this chapter to show how these principles have been applied, particularly in the American synthetic-rubber industry in the period since 1940, for the purpose of obtaining information about the structure and properties of the rubbers and of altering if possible the structure to produce new rubbers with desired properties.

Attention here will necessarily be specially concentrated on GR-S, the Government-produced butadiene-styrene copolymer, since it is the type which has been in greatest production and on which by far the largest amount of research has been done during the decade.[100] A certain amount of attention will be given to GR-I (Butyl rubber), Neoprene, and the nitrile rubbers, which are also being produced on a large scale. Other polymers will also be treated where they may be expected to cast light on the synthetic rubbers already mentioned. The strong predominance of references to work in the United States reflects the relative research activity in this field since 1940, not any consciously nationalistic bias of the author's.

A considerable amount of the work done in the Government Synthetic Rubber Program was first presented in unpublished reports to the Office of the Rubber Director and the Office of Rubber Reserve. Where the results have now appeared in published form reference will be made only to the publication rather than to the report. The reports, although useful in showing when the work was done, are often rather preliminary in nature. Because of conditions during and after the war an interval of several years is noted in many instances between the report and the publication. Many significant books on high-polymer physics and chemistry have been published since 1940, and the reader is referred to the list on page 361 for more detailed presentations of many of the points covered here.[1-19]

MECHANISM AND KINETICS OF POLYMERIZATION AS RELATED TO COMPOSITION, MOLECULAR WEIGHT, AND STRUCTURE

The mechanism and kinetics of the polymerization reaction are dominant factors in determining the composition, molecular weight, and structure of the product, as will be discussed in detail in the following sections of this chapter. However, many of the practices in polymerization technique are of empirical origin, and it is only recently that the mechanism and kinetics have been explored with some thoroughness. Copolymerization, for example, was of empirical origin but is now beginning to receive adequate theoretical consideration.[235]

Condensation Polymerization

In condensation polymerization two monomeric molecules react in such a fashion that (1) a by-product is eliminated, usually water but sometimes another simple molecule like $NaCl$, HCl, or NH_3, and (2) the condensation product can itself undergo a similar reaction with additional monomer molecules. Flory[127] has published an excellent summary of this subject. The reaction is usually a familiar chemical process like esterification, and it is found that the rate constant is always essentially equal to that for the monomer and independent of the degree of polymerization. Bifunctional reactants give rise to linear polymers while reactants of higher functionality can grow in more than one dimension and form three-dimensional networks.

Condensation polymerization proceeds by a series of independent condensation reactions extended over the whole time of polymerization. The molecular weight increases continuously, and the distribution of molecular weights can be calculated from the conversion. The final molecular weight of a linear polymer is limited only by the attainable completeness of the condensation reaction. In these respects condensation polymerization differs markedly from addition polymerization, to be discussed later. Because of the relative simplicity and slowness of the chemical reactions involved, condensation polymerization furnishes a good method of obtaining polymers of known structure and of controlled molecular weight. The phenomenon of gelation is more easily studied in condensation polymerization,[121-2] although it is by no means limited to this type of polymerization.

Two types of commercial synthetic rubber are made by condensation polymerization. They are the organic polysulfide rubbers and the silicone rubbers. The former, usually known under the trade name of Thiokols, are condensation products of organic dichlorides with a sodium polysulfide, and sodium chloride is the by-product of the reaction.[111]

The rank of the polysulfide, defined as the average number of sulfur atoms in the anion, varies from about 1.8 to 4 in the commercial production of different varieties. Some varieties (Thiokol A and Thiokol FA) are made processible by the use of a chemical plasticizer which causes scission of disulfide linkages and thus reduction of the molecular weight. These are vulcanized with zinc oxide and the broken linkages re-formed without cross-linking. Other varieties (Thiokol ST and Thiokol PR-1) are cross-linked because of the use of small amounts of trifunctional halides, have thiol end groups, are processible without breakdown, and can be vulcanized with oxidizing agents.

The simplest silicone rubbers[39] are made by the reaction of dimethyl dichlorosilane with water and subsequent condensation to polymethyl silicones of the dimethyl silicol thus formed. Others can be made by varying the initial materials used.

Ionic Addition Polymerization

Polymerization with Friedel-Crafts Type Catalysts. In addition polymerization no by-product is eliminated, and the composition of a simple polymer is the same as that of the monomer. Isobutylene is polymerized commercially by the aid of ionic catalysts of the Friedel-Crafts type such as boron trifluoride or aluminum trichloride. This type of reaction, known as ionic polymerization,[103, 178, 343] has received much less attention than other types of addition polymerization to be discussed later. The presence of a cocatalyst, usually water, has been found essential.[103] The reaction is extremely rapid, often being completed in a time estimated as less than one second. The average molecular weight is determined chiefly by the temperature and increases with decreasing temperature. Molecular weights in the millions are obtainable. An inert diluent such as methyl chloride is often used. Higher molecular weights are thereby obtained up to diluent concentrations of about 80 per cent and more. Within limits the molecular weight obtained does not depend on the conversion.[343] The structure is considered to be chiefly head-to-tail addition.

Several mechanisms of initiation, propagation, and termination have been proposed and are given in a summary article by Heiligmann.[178] None has as yet gained general acceptance.

Copolymerization of isobutylene with a diolefin (isoprene in commercial practice) yields a Butyl rubber,[342] known as GR-I when produced in Government-owned plants. The reaction appears to be somewhat slower and a somewhat narrower range of molecular weights is obtained than in the polymerization of isobutylene alone.

Marvel and his co-workers have shown how to bring about ionic polymerization of butadiene and ionic copolymerization of butadiene and isoprene.[233] Styrene has also been polymerized by this mechanism.[384]

Alkali Metal-Catalyzed Polymerization. The alkali metals, sodium and potassium, were found many years ago to catalyze the polymerization of isoprene and butadiene. Most of the commercial rubbers produced by bulk polymerization have been made by their means. The chief examples of this type of polybutadienes are the German Buna 85 and Buna 115, and the Russian SK-B. The mechanism of catalytic action is not very well understood but is thought to be ionic in nature.

When polymerization is occurring at a reasonable rate in bulk, there is often local overheating because the solid polymer prevents the convection and circulation necessary to dissipate the heat of polymerization. For similar reasons it is difficult to reach high values of conversion. Likewise there are inhomogeneities in composition arising from this lack of circulation. For these and other reasons, diluents are used extensively in present-day alkali-catalyzed polymerizations. Some diluents exert comparatively little effect on the structure of the polymer, whereas the reverse is true for others. In the Russian process most completely described,[341] butadiene is produced by a process that leaves almost 25 per cent of butene-1 as an impurity.   In the polymerization of this mixture with an alkali catalyst the butene-1 very conveniently serves as a diluent, and the effective maximum conversion is limited to 75 per cent.   The solubility of the polymer in the diluent is, of course, a factor to be considered.

The sodium-catalyzed copolymerization of butadiene and styrene has been carried out by Marvel, Bailey, and Inskeep[232] and by Schulze and Crouch.[303]

Alfin-Catalyzed Polymerization.   More recently, considerable attention has been given to a new class of catalysts for bulk polymerization.   These are the "alfin" catalysts, which are double organic salts of sodium, one from an alcohol and the other from an olefin.   The polymerization is very rapid, and products of very high molecular weight are obtained.[255]   Gel formation proved troublesome at first, but later developments have led to the production of polymers containing less than 5 per cent gel and having molecular weights of one to two million.   Diluents are usually used in alfin-catalyzed polymerizations for reasons given in the previous section.

Free Radical Addition Polymerization

Bulk Polymerization and Solution Polymerization.   The earliest synthetic rubbers were made directly from liquid monomers in which, for one reason or another, free radicals were being produced.   The role of the free radicals was usually not well understood and the rate of their formation not under very good control.   In a few instances they can be formed by light, especially ultraviolet.   Often impurities serve as a source of free radicals.   The presence or absence of oxygen usually has an important bearing on the rate of formation of free radicals, since peroxides are often involved.[62, 63]

Suspension Polymerization.   A monomer dispersed in a nonsolvent medium (usually water) and maintained in suspension by vigorous stirring or shaking can be polymerized by a mechanism almost the same as that for free-radical bulk polymerization.[182]   Each liquid globule yields a corresponding sphere of polymer, sometimes called a "pearl" or "bead."   The size of the globule is controlled chiefly by the intensity of agitation and can be varied from about 10 microns to about 10 mm.   At conversions between 20 and 80 per cent the proportions of polymer and monomer in the spheres are such that the spheres are rather sticky and agglomeration almost invariably occurs unless steps are taken to prevent it.   These steps include any one or preferably several of the following: the use of suspension stabilizers, such as for example carbonates, silicates, talc, gelatin, pectin, or starch in concentrations not over 0.5 per cent; increase of interfacial tension by the solution of electrolytes; equalization of the density of the sticky globules and the carrier liquid; or increase of viscosity of the carrier liquid, for example by the addition of glycerol.   The added materials are readily removed from the final polymer, and polymers of relatively high purity can be prepared by suspension polymerization.

The mechanism and kinetics of polymerization in each small globule are considered to be essentially the same as those in the bulk polymerization. The presence of the water phase makes possible the circulation necessary for the homogeneity of the polymer and the dissipation of the heat of polymerization.   Suspension polymerization is a relatively new development and has not yet been utilized in the commercial production of synthetic rubbers. It would appear to deserve further investigation for this purpose.

Emulsion Polymerization. Emulsion polymerization is used in the commercial production of GR-S,[82, 328] Neoprene,[74] the nitrile rubbers,[266] and many other polymers. Although empirical studies have led to a good knowledge of the effects of many of the variables, the mechanism and kinetics of polymerization are not so well understood as for the simpler systems previously mentioned. Emulsion polymerization, unlike suspension polymerization, differs in many respects from bulk polymerization or solution polymerization, although in all of them polymerization can be initiated by free-radicals. The mechanism of emulsion polymerization is treated in detail in Chapter 8. Consequently this section will treat mainly those aspects that relate directly to composition, structure, and molecular weight.

In emulsion polymerization, a monomer or mixture of monomers is emulsified by the addition of, e.g., soap and water. A catalyst, such as a persulfate or a peroxide, and a modifier or regulator, such as a mercaptan, are present in small amounts.[152] The modifier, acting as a chain-transfer agent, limits the growth of the chains.[321-3]

The complexity of emulsion polymerization is related to the fact that there are many phases present. There are a monomer phase, a polymer phase, a water phase, and soap "micelles." The monomer is soluble in the polymer, is present in the soap micelles, and is slightly soluble in the water phase. At the beginning of polymerization the soap coats the monomer particles with a monomolecular layer and any excess soap is dispersed as micelles in the water. At this stage there is of course no polymer phase. There is a distribution of monomer particle sizes with a peak normally between 5000 and 10,000 A.[171] The size of the particles is determined to quite a large extent by the amount of soap present; the less the amount of soap, the larger the particles.

Polymerization is thought to be initiated chiefly by free radicals from the catalyst and modifier, formed at first in the soap micelles. Thus polymer particles are formed, coated with a layer of soap and capable of dissolving considerable quantities of monomer. The polymer particles are much smaller than the monomer droplets, and the total surface area of monomer and polymer particles is increased. At a certain conversion, normally between 13 and 20 per cent,[75, 171] the amount of soap becomes insufficient to cover the surfaces completely, and the soap micelles disappear. Beyond this point it is considered that very few additional polymer particles are formed. The monomer is so soluble in the polymer that the monomer phase disappears completely after a certain later conversion, near 52 per cent for GR-S or polybutadiene.[242] By far the major portion of the polymer is formed from the monomer dissolved in the polymer particles. At any given time during polymerization, there is evidence[324, 326] that about half the particles contain one free radical and a growing chain, while there are none in the other half. The reaction is normally stopped at 60 to 75 per cent conversion. At this time the number of polymer molecules in each particle is of the order of several hundred. The particle has been growing steadily in size since its formation but is still only about a tenth the diameter of the original monomer droplets. The mass median particle size of ordinary GR-S latex (after polymerization) has been reported to be about 600 to 800 A., with a weight-average particle diameter of 800 to 1000 A. With latexes of higher styrene content, the values are somewhat higher.[231, 300] Willson and co-workers[385] have shown how to produce latexes of different particle sizes. For polystyrene produced in emulsion the peak in the distribution curve occurs at about 550 A. at 15 per cent conversion, increasing to about 800 A. at 74 per cent conversion.[171-2] On the average each emulsified droplet of monomer has thus yielded about a thousand latex particles.

The rate of growth of a polymer chain, once initiated, is relatively rapid, and it is thought that each molecule is formed in a period of the order of seconds or less.[38, 324, 336] Since this step is so rapid, the over-all rate of formation of polymer is controlled by the rate of formation of free radicals and their diffusion to the locus of the polymerization reaction.[325] During the formation of most of the polymer in the GR-S reaction this over-all rate is nearly constant.[71-2, 256, 258]

The rate of formation of free radicals in the normal GR-S polymerization recipe is a sensitive function of temperature, and the rate of polymerization becomes inordinately small at temperatures much below the normal 50° C. Systems in which reduction and oxidation occur simultaneously ("redox systems") have been developed which will furnish an adequate supply of free radicals at temperatures as low as about −20° C. For temperatures below the ice point, antifreeze solutions must be used, of course.[153, 270, 350, 359] The name "cold rubber" is given to polymers made at temperatures below room temperature. GR-S polymerized at 5° C. (41° F.) is being made on a very large scale at present.

Flory[130] has pointed out that the cross-linking of polymer chains occurs when a growing free-radical chain adds to an unsaturated carbon of a previously polymerized diene. This reaction is incapable of producing highly branched molecules of finite size, since gelation occurs before this state is reached. (Cross-linking can also occur after polymerization is complete, as in vulcanization.)   A branching reaction occurs when there is a chain transfer between the free-radical terminus of a growing chain and a previously formed polymer molecule. Subsequent growth of the latter molecule by the addition of fresh monomer at the point of transfer gives rise to a branch in the structure.   Such a process, unlike cross-linking, by itself does not produce gelation or infinite networks.   However, it may produce conditions favorable for cross-linking and gelation to occur.   Morton and Salatiello,[257] studying the cross-linking reaction of butadiene, conclude that the ratio of the rate of cross-linking to the rate of propagation is constant up to the point where the monomer phase disappears or where gelation occurs, whichever is the earlier.   This ratio decreases with decreasing temperature so that a structure with fewer cross-links is to be found in low-temperature polymerization.[247]

Since cross-linking requires unsaturation in the chain, it is clear that no cross-linking can occur in polystyrene.   Controlled amounts of cross-linking can be introduced by copolymerizing styrene with known amounts of divinylbenzene, which is effectively a tetrafunctional monomer.

THERMODYNAMICS OF POLYMERIZATION

Heats of Polymerization. A summary article by Roberts[288] lists values of the heats of polymerization of many polymers and discusses the relation of these values to the structure of the polymers. For 1,3-butadiene a value of 18 kcal per mole of monomer, derived from heats of combustion, is given for a polymerization from liquid monomer to solid polymer. More recently new experimental results have been obtained for butadiene polymers and copolymers.[267] The value obtained for butadiene polymerized at either 50 or at 5° C. is found to be 17.6 kcal per mole of C_4H_6. For styrene the value of 16.7 kcal per mole of monomer is taken.[288-9] The copolymers are found to have values lying above the straight line joining these two values in accordance with the theoretical predictions of Alfrey and Lewis.[22] For a copolymer containing 23.9 per cent bound styrene the value found was about 17.1 kcal per mole. Since, as is discussed later, the styrene content of copolymer formed during any interval increases somewhat as polymerization proceeds, the heat of polymerization may be expected to decrease slightly as a function of conversion.

Isobutylene has been found[343] to have a heat of polymerization in the neighborhood of 10 kcal per mole. Roberts[288] points out that this is considerably lower than the values just given because of the steric hindrance of the two methyl groups in isobutylene. Isoprene, the minor constituent of GR-I, has a heat of polymerization of 17.9 kcal per mole when the polymer is natural rubber.[46]

Free Energy of Polymerization. The free energy of polymerization can be calculated from heats of combustion and entropy values. The latter can be obtained from measurements of specific heats or from spectroscopic data. Bekkedahl[46] has carried out such calculations for the polymerization of isoprene to form natural rubber. Similar calculations have not yet been made for the synthetic rubbers. The heats of combustion for butadiene,[278] styrene,[289] and polystyrene[289] are available in the published literature. The heat of combustion of polybutadiene has recently been found[267] to be 17.6 kcal per mole of C_4H_6. Specific heat data are available for butadiene,[307] styrene,[276] polybutadiene,[156] polystyrene,[262] and GR-S containing about 25.5 per cent bound styrene.[281] There is considerable uncertainty regarding the entropy of the polymer at 0° K., but this is probably only a small portion of the total entropy at room temperature.

PHYSICOCHEMICAL METHODS FOR DETERMINING POLYMER COMPOSITION

The methods of chemical analysis developed in the Government Synthetic Rubber Program have been given in summary articles by Tyler and Higuchi,[354] by Bekkedahl and Stiehler,[53] and by Bekkedahl.[48, 49] Attention is concentrated here on those that involve most clearly the principles of physical chemistry. No discussion is given regarding the more conventional chemical methods such as the determination of unsaturation by iodine chloride,[202, 217] the determination of external double bonds by perbenzoic acid,[208, 296] or the determination of structure by ozonolysis,[279, 286] since these are treated in Chapter 9.   However, some of the results obtained by these methods will be discussed.   The determination of the composition of nitrile rubbers by measurement of the nitrogen content by Kjeldahl analysis[52, 93, 382] is likewise to be omitted, as is also the determination of polystyrene, mixed with or attached to the copolymer GR-S, by degradation by *tert*-butyl hydroperoxide in the presence of osmium tetroxide.[209]

Carbon-Hydrogen Ratio.   The measurement of the carbon-hydrogen ratio by combustion analysis is probably the most fundamental and unambiguous method of determining the composition of a hydrocarbon.   It is useful in determining the relative amounts of the two constituents of a copolymer if the constituents differ sufficiently in carbon-hydrogen ratio. Kemp and Peters[202-3] were the first to publish results obtained by this method when applied to copolymers of butadiene and styrene.   I. Madorsky and Wood[224] have developed the application of the method to determining the bound styrene content of GR-S.   It has been applied chiefly to establishing the composition of samples intended for standards in developing other procedures which are easier to apply.   Some of those other procedures are described in the following sections.   It will be noted that each of them depends on the use of copolymer samples of known composition for calibration or on the assumption of a linear change in some property with the bound styrene content.   It is useful to note the oxygen content,[372] which is normally 0.1 to 0.2 per cent.

Refractive Index of Solid Polymers.   The high refractivity of the benzene ring as compared with straight-chain hydrocarbons suggested the measurement of refractive index as a means of determining the bound styrene in GR-S.   The first experimental investigation of this possibility was given in a preliminary unpublished report by B. L. Johnson[191] in early 1943.   In the following year work on the method was begun independently in three different laboratories at about the same time.[31, 221, 223]   The procedure actually put into operation in control testing in the synthetic-rubber plants[246, 333] was that of I. Madorsky and Wood.[223]   Since 1945 this test has been applied to the whole production of GR-S, not because the properties of the polymer are very sensitive to small changes in styrene content, but more particularly because a varying styrene content is indicative of non-uniform polymerization conditions.

After careful extraction of stabilizer, fatty acid, soap, and other minor ingredients the sample is pressed into a sheet about 0.010 to 0.020 in. thick. The refractive index of the sheet is then determined by the conventional Abbé refractometer in much the same way as that of a glass test slab or other solid specimen.   The details of the procedure for measurement of index have been the subject of a publication by Arnold, I. Madorsky, and Wood.[23]

The relation between refractive index and styrene content, where the latter has been determined by measurement of carbon-hydrogen ratio, has been found in recent work at the National Bureau of Standards[260-1] to be

$$S = 23.50 + 1164\ (n - 1.53456) - 3497\ (n - 1.53456)^2 \qquad (1)$$

Here S is the bound styrene content by weight and n the refractive index at 25° C. of a copolymer of butadiene and styrene prepared in the normal GR-S emulsion polymerization system at 50° C. Variations in bound mercaptan content over the range normally encountered do not affect the value of index. If the index is measured at a temperature other than 25°, the value at 25° is computed. The rate of change of refractive index of GR-S with temperature has been found[23-4] to be -3.7×10^{-4} per degree centigrade.

This relation differs slightly from that developed in 1944 on the basis of less satisfactory measurements. For the same refractive index the earlier relation gave styrene contents about 0.2 per cent higher than those given by equation 1 in the range 20 to 25 per cent bound styrene, and about 1.7 per cent higher, when the polymer contains 55 per cent bound styrene.[260] The earlier relation was used from 1945 to 1950 in the synthetic-rubber plants and applied in published work by several other investigators.[168, 177, 246-7, 249, 303, 354, 397]

The refractive index n_D^{25} of polybutadiene polymerized in emulsion at 50° C. (122° F.) was found to be 1.5154 in the work establishing this relation. Polybutadiene made in emulsion at $-19°$ C. is reported[177, 247] to have an index of 1.5143, while that made at 97° C. has a value of 1.5159. Alkali metal-catalyzed polymers of butadiene[351] have indexes ranging from 1.5074 to 1.5117. Unpublished observations at the National Bureau of Standards on German and Russian sodium-catalyzed polybutadienes produced commercially have yielded values of refractive index near 1.5090. These rubbers are reported[341] to be produced in a reactor where the temperature rises during polymerization and usually includes the range from 40 to 65° C. Similar observations on potassium-catalyzed polybutadienes yield values of refractive index near 1.5118.

These observations indicate that the refractive index of a polymer is affected by its structure as well as by its composition; consequently the relation given above between refractive index and styrene content may not be valid if the polymerization system differs from that of normal GR-S polymerized at 50° C. An investigation is under way at the National Bureau of Standards to determine the applicability of the relationship to GR-S polymerized at 5° C. (41° F.). Schulze and Crouch[303] present data to show that the index for sodium-catalyzed copolymers is less by a constant amount than that given by the above relation over a range of styrene contents from 0 to 40 per cent. They made index measurements on sodium-catalyzed polymers run to 100 per cent conversion in order to determine styrene content from the charging ratio.

Measurements of the refractive index at 25° C. of copolymers containing more than 70 per cent styrene are difficult because of the stiffness of the polymers. Furthermore, for these polymers the second-order transition is near or above 25° C. and a change in the relation between index and styrene content would be expected above the second-order transition. Measurements of index at higher temperatures[249] offer a possibility of extending the usefulness of the method to polymers of higher styrene content.

Refractive Index of Solutions. A method for determining styrene content by measurements of the refractive index of a solution of a copolymer was developed at the Bell Telephone Laboratories,[27, 30, 154] and used there in the control of the uniformity of polymerization of GR-S in 1943 and 1944. Since the solution was relatively dilute, the change of index with styrene content was much smaller than with the bulk polymer. Consequently a Zeiss interferometric refractometer was employed. A linear relation between index and styrene content was assumed, making use of measurements of the index of a polybutadiene solution and that of a polystyrene solution. The necessity for the use of interferometry and for precise temperature control has prevented the method from finding extensive use in other laboratories.

Ultraviolet Absorption. The fact that styrene, polystyrene, and copolymers of styrene possess an absorption band in the ultraviolet near 260 mμ makes it possible to base a method for styrene content on a determination of ultraviolet absorption. After preliminary work at the U.S. Rubber Company,[189] the method was developed by Meehan at the University of Minnesota.[239, 243] The Beckman ultraviolet spectrophotometer is used to measure the specific extinction coefficient of a chloroform solution of the copolymer, usually at a wavelength of 262 mμ. The styrene content is determined by a linear interpolation of the measured extinction coefficient between the values 0.042 for polybutadiene and 2.16 for polystyrene.

Meehan points out that the assumption of simple additivity of the absorption spectra of polystyrene and polybutadiene to produce that of the copolymer is not exact, since there is a change in the shape of the absorption curve and a shift in the wavelength of maximum absorption from below 260 mμ for polymers of low styrene content to 262 mμ for polystyrene. The stabilizers usually used in GR-S absorb strongly in the ultraviolet, and it is recommended that they be removed by extraction or by solution and precipitation of the polymer until the concentration of stabilizer is below 0.004 per cent. Polymers that have stood for many months cannot be readily purified to this extent. An alternative approximate method involving measurements at two wavelengths is suggested for such polymers. The accuracy of the normal method is considered to be about 3 per cent.

Meehan, Parks, and Laitinen[243] found that the results obtained by applying a similar method to copolymers of butadiene and p-chlorostyrene were in agreement with those calculated from the measured chlorine contents of three copolymers.

Crippen and Bonilla[83] have proposed a method by which the copolymer is broken down by superheated steam at 300° C. to yield monomeric styrene, which is then removed by a steam-distillation method, nitrated with a mixture of nitric and sulfuric acids, and determined by measuring the transmission of a solution at a wavelength near 365 mμ, where there is a maximum in the absorption of the nitrated styrene. When applied to a series of GR-S samples, the method gave results in reasonably good agreement with those obtained from carbon-hydrogen determinations.

Infrared Absorption. The benzene ring has an absorption band at a wavelength of about 14.3 microns (699 cm.$^{-1}$), which has been used as the basis for determining the styrene content of butadiene-styrene copolymers. Hampton[168] has published a description of such a method using carbon disulfide solutions. Simultaneous determinations made at other wavelengths permit the determination of 1,2-addition, *cis* 1,4-addition, and *trans* 1,4-addition. These will be discussed in a later section.

The cyanide group, $C \equiv N$, has an absorption band at a wavelength of 4.47 microns (2237 cm.$^{-1}$) which has been used as the basis for the determination of the nitrile content of nitrile rubbers. The method, described in detail by H. L. Dinsmore and D. C. Smith,[93] uses film specimens and is capable of giving results with an average deviation of ± 1 per cent.

Infrared measurements can also be used as the basis for determining the composition of mixtures of rubbers in either the vulcanized or the unvulcanized states and for identification and qualitative analysis.[229] Dinsmore and Smith[93] have worked out complete procedures.

Analysis for Minor Ingredients. About 9 per cent of the weight of normal GR-S is not derived from butadiene or styrene. Fatty acid and soap can be extracted from the rubber and titrated by conventional chemical methods. The stabilizer, however, is usually determined by the measurement of its ultraviolet absorption.[34] The wavelengths used are those of maximum absorption for the particular stabilizer, and are as follows: phenyl-β-naphthylamine, 309 mμ; BLE (an acetonediphenylamine reaction product), 288 mμ; Stalite (a heptylated diphenylamine), 288 mμ. The solvents most commonly used at present are ethylene dichloride, toluene, methylcyclohexane, and mixtures of methylcyclohexane and ethanol.[352]

Analysis for Bound Isoprene in GR-I Butyl Rubber. The amount of bound isoprene in GR-I Butyl rubber (about 1 to 4 per cent) is so small that the procedures of analysis already mentioned have not been very effective. A method developed by Rehner and Gray,[286] however, has proved satisfactory. It involves ozonizing the polymer, so that each chain is broken at all its double bonds, and then determining the molecular weight of the degraded polymer by viscosity measurements. The viscosity–average molecular weights are in the range 2,000 to 10,000. This type of measurement is discussed in the next section.

METHODS FOR DETERMINING MOLECULAR WEIGHT

A single sample of a synthetic rubber is usually made up of molecules with molecular weights ranging from a few thousand into the millions. This section describes the methods employed to measure such molecular weights, defines the different types of average values obtained, and quotes results obtained on typical synthetic rubbers. The scope of the present chapter does not permit more than a brief survey of these topics, limiting their application largely to the synthetic rubbers. For more detailed discussions particularly including applications to other polymers one should consult some of the texts listed in the references, for example Bawn's book.[2]

The wide range of molecular weights usually found in a polymer makes it necessary to express results in terms of average values and the distribution of values around the average. In order to postpone discussion of the various kinds of average values of molecular weight, let us consider, for a time, only measurements on a sample containing molecules of a single molecular weight or, more realistically, a range of molecular weights sufficiently narrow that no distinction between kinds of averages need be made.

Osmotic Pressure. The osmotic pressure developed by an ideal solution is proportional to the number of molecules of solute per unit volume. This number divided by Avogadro's number is equal to the number of moles of solute per unit volume and this is the concentration divided by the molecular weight. One can readily see in principle how to base a method for measuring molecular weight on measurements of osmotic pressure and concentration. High-polymer solutions become more nearly ideal as the concentration is reduced, and so actual measurements involve observations at several concentrations and an extrapolation to zero concentration. A plot of the ratio of osmotic pressure to concentration as a function of concentration is usually employed for this purpose. This plot is often almost linear, and the different equations derived from theory agree in relating the intercept of this plot to the reciprocal of the molecular weight.[161]

The measurement of osmotic pressure has usually involved the measurement of the liquid head arising from the influx of solvent through a semipermeable membrane of denitrated nitrocellulose film. At least two new osmometers[151, 297] were developed in the course of the synthetic-rubber program.

The membranes normally used permit the measurement of molecular weights between about 10,000 and 1,000,000, the most favorable conditions occurring[2] between 40,000 and 500,000. The lower-molecular-weight polymers diffuse through the membrane, and the higher ones give values of osmotic pressure that are too small for precise measurement. The deswelling of a gel has been proposed[64] as a variant on the usual osmotic-pressure method for determining molecular weights.

Light Scattering. The intensity of the light scattered by a polymer solution may be made the basis of a method of determining the size of the dissolved molecules. The development of this method by Debye[85, 87] was one of the major contributions of the synthetic-rubber program to polymer research generally. The applicability was extended in further work by Doty, Zimm, and Mark,[97–8, 405] and by others.[108, 166] Measurements are made of the turbidity at a series of concentrations and of the dependence of refractive index on concentration. The latter quantity can usually be determined from a measurement at a single concentration and a measurement on the solvent. The turbidity is usually measured at some fixed angle with the incident beam. A graph of an ordinate involving the reciprocal of the turbidity plotted against the concentration is found to be linear, in accordance with the theoretical considerations. The intercept of this graph at zero concentration is the reciprocal of the molecular weight.

The theory on which the method is based assumes that the scattering molecules are small compared with the wavelength of the light used. Consequently there is in principle no lower limit of molecular weights to which the method may be applied, and results have been reported for sucrose molecules, where the molecular weight is only a few hundred.[87]

When the largest dimension of the solute molecule exceeds about onetwentieth of the wavelength of the light used, there are significant interference effects between light scattered from different parts of the molecule. The result is an all-round decrease in the intensity of the scattered light with a greater reduction in the backward scattering than in the forward scattering. A spherical particle of the diameter mentioned would have a molecular weight of several million, but, of course, with nonspherical particles the interference effects become perceptible at lower molecular weights. The asymmetry of scattering can be measured and used[403-4] to give information about the size and shape of molecules too large to be treated by the simpler theory. Thus the applicability of the method has been extended to particles as large as the wavelength of the light used.

In practice the light-scattering method requires that the solutions be exceptionally free from particles other than those of the polymer. Recent work[95, 347] has shown differences in molecular weights above 700,000 when values obtained by light scattering are compared with those from osmometry. The discrepancy is ascribed to solvent effects rather than to failure of the method.

Solution Viscosity. The viscosity of a dilute solution of a polymer is related to its molecular weight. Because of the simplicity of the experimental observations this method has been applied more often than any other to measuring the molecular weights of synthetic rubbers. Although many attempts have been made, it has not been possible to derive the form of the relation and the constants independently of other molecular-weight measurements. Consequently this method, unlike the others discussed here, is at present a relative method and requires the evaluation of constants for a particular polymer-solvent system by the use of a series of polymers with known values of molecular weight, as determined by one of the other methods.

The time of flow of a given volume of polymer solution through a capillary viscometer is compared with the corresponding time for the pure solvent. This is often repeated at several concentrations of polymer. No additional observations are required.

The actual viscosity η of the solution and the actual viscosity η_0 of the solvent are seldom calculated since they are each proportional to the observed times of flow, and it is possible to deal with their ratio η/η_0, called the "relative viscosity," η_r. The fractional increase of viscosity caused by the presence of the solute molecules is called the specific viscosity, $\eta_{sp} = (\eta - \eta_0)/\eta_0 = \eta_r - 1$. The reduced viscosity, η_{sp}/c (where c is the concentration), and the inherent viscosity,[76] $(\ln \eta_r)/c$, each approach the same limit as the concentration approaches zero. This limit is written in square brackets as $[\eta]$ and is called the intrinsic viscosity. Some workers in the past have failed to make a clear distinction between reduced viscosity and the limiting value, the intrinsic viscosity. Others have failed to distinguish between inherent viscosity and intrinsic viscosity. The reduced viscosity at a low concentration has been called the dilute solution viscosity (DSV) by some authors who did not wish to determine the actual intrinsic viscosity.

Most polymer solutions show non-Newtonian behavior, and the viscosities mentioned are functions of the rate of shear. In the past, very few calculations have taken account of this fact. Fox, Fox, and Flory[150] have now shown the magnitude of the effect in polyisobutylene fractions and how to extrapolate observations to zero rate of shear. GR-S fractions show a rate of change of the logarithm of viscosity with rate of shear which is two to three times as great as that for polyisobutylene fractions. The intrinsic viscosity is, as shown below, the quantity related to molecular weight. It should be noted that, of the quantities just discussed, only η and η_0 are true viscosities, some of the others being dimensionless ratios and some having the dimensions of an inverse concentration. It has become conventional to express concentrations in grams of polymer per 100 ml. of solution.

The observer who has measured relative viscosities at different concentrations finds that plots of reduced viscosity and inherent viscosity are linear at very low concentrations. The plot of reduced viscosity has a positive slope and that of inherent viscosity a negative slope of smaller magnitude. Both of these have been extensively used to determine the intrinsic viscosity by graphical extrapolation.

Cragg, Rogers, and Henderson[80] have studied the applicability of the Baker-Philippoff function $8(\eta_r^{1/8} - 1)/c$ to the system GR-S in benzene. They find that a plot of this function when extrapolated to zero concentration gives the intrinsic viscosity. Furthermore and more important they observe that the plot has zero slope up to concentrations of about 0.5 gram per 100 ml. Under these conditions observations at only a single concentration are required. The advantages of the use of this plot, where applicable, are obvious. The observations of Cragg, Rogers, and Henderson relate to five different GR-S-type unfractionated polymers. They report that they have found some evidence for a change in the coefficient and exponent in the function when a fractionated polymer is used. It is known that other polymer-solvent systems require different numerical values in the Baker-Philippoff function.[294]

As discussed in more detail in Chapter 26, German workers have made considerable use of a quantity called the Fikentscher k value, appearing as a constant in an equation connecting relative viscosity and concentration.[11] From it the intrinsic viscosity can be calculated by the following relation: $[\eta] = 2.303 \ (75k^2 + k)$. For convenience, workers have commonly quoted values of K defined as 1000 k. American researchers have not found that it offered any advantage over intrinsic viscosity as a parameter for the characterization of a polymer.

Observations of solution viscosity are commonly made at some well-controlled temperature between 15 and 45° C. The values of intrinsic viscosity obtained for most systems are not greatly dependent on the temperature used. Rogers, Henderson, and Cragg[292] could find no variation in the intrinsic viscosity of GR-S in benzene over the range 10 to 55° C. when the concentrations were measured at the temperatures investigated. Later work[79] showed small increases of intrinsic viscosity of GR-S with temperature for some solvents and small decreases for other solvents. The change with temperature is greater for poor solvents and for fractions of high molecular weight.[81]

Vistex Technique. The solvents used for the synthetic rubbers are usually the conventional aromatic or chlorinated hydrocarbons. However, where speed of preparation is a factor and the solution is to be prepared from a latex, Baker, Mullen, and Heiss[32] have suggested the use of a mixture of a hydrophobic and a hydrophilic solvent, such as xylene and pyridine. With a 75–25 mixture of xylene and pyridine, clear solutions of 1 ml. of GR-S latex per 100 ml. of solution can be made and measured at 40° C. They coined the term "vistex" to describe viscosity measurements on a solution prepared directly from latex without coagulation. Harris and Kolthoff[174] studied the use of benzene-isopropanol (80–20) as a vistex solvent applied at room temperature. The inherent viscosities of GR-S polymers in dilute benzene solution were found to be 16 to 42 per cent greater than the inherent vistex viscosities, when the latter ranged from 0.3 to 1.8.

Henderson and Legge[181] in a subsequent complete publication have shown how the intrinsic viscosity of a GR-S polymer in toluene may be calculated from observations of the intrinsic vistex viscosity in an 80–20 toluene-isopropanol mixture and have evaluated the constants in the linear relation between inherent viscosity and concentration. Others[141] have made similar studies. The vistex method, in addition to saving time, is found to avoid variations associated with coagulation and drying of the polymer.

Intrinsic Viscosity–Molecular Weight Relation. The relation between intrinsic viscosity $[\eta]$ and molecular weight M most commonly used for rubber polymers in recent years is the equation proposed by Kuhn[213] from theoretical considerations and suggested by Mark[230] for empirical reasons

$$[\eta] = KM^a \tag{2}$$

where K and a are constants to be determined for the particular polymer-solvent system in question. Table I shows values of these constants for the most common synthetic rubbers, polyisobutylene, and natural rubber. It will be noted that the exponent a is quite near to the value 2/3 in most cases. There is very good agreement in the values for GR-S reported by independent observers, and a remarkably close approach to the same constants for GR-S and natural rubber.

The idea, proposed by Staudinger, that the exponent a for a linear polymer has the value unity dominated the thinking of many of the German research workers. The data in Table I relating to elastomers do not support this idea. The European literature contains a great many values of "Staudinger molecular weights," calculated as proportional to intrinsic viscosity with a proportionality factor based on measurements at very low molecular weights. The actual molecular weights are frequently as much as ten or more times the values calculated in this manner.

Other Methods. Three other methods of determining molecular weights have not been much applied to synthetic rubbers. They are the ultracentrifuge, the measurement of freezing-point depression, and the determination of end groups.

The ultracentrifuge,[268] requiring expensive equipment and considerable skill in its operation, has thus far shown its chief utility with proteins and viruses, where the molecules are more nearly spheroidal or ellipsoidal in shape than those of the long-chain rubber polymers. The difficulties of

Table I. Molecular Weight—Intrinsic Viscosity Relation

$$[\eta] = KM^a$$

Material	Temperature of Polymerization	Solvent	Molecular Weight Range	a	K	Reference
Polyisobutylene	−85 to −27° C.	Diisobutylene	5,620–1,300,000	0.64	3.6×10^{-4}	124
		Cyclohexane		0.69	2.65	144
		Carbon tetrachloride		0.68	2.9	144
		Carbon tetrachloride	233,000–1,300,000	0.64	4.52	286
		Toluene		0.67	2.0	144
		Toluene		0.64	2.6	312
		Benzene		0.56	6.1	144
Butyl rubber	35 to 60° (Na)	Diisobutylene		0.64	3.6	128
Polybutadiene	−20°	Toluene		0.62	11.0	312
	5° (emulsion)	Toluene		0.63	10.6	174
		Toluene		0.55	26.4	174
	50°	Toluene		0.45	72.5	174
	50° (Nitrazole-CF system)	Toluene		0.70	4.93	118
GR-S	27° (Photo-initiated)	Toluene		0.75	2.36	117
	97° (emulsion)	Toluene		0.18	1093.0	115
	−10° C.	Toluene		0.68	5.16	116
	50°	Toluene	25,000–500,000	0.67 ±0.05	5.25	312
Nitrile rubber (Buna N)	50°	Toluene	10,000–920,000	0.66	5.4	151
	50°	Toluene		0.67	4.9	151
	50°	Benzene	12,400–1,650,000	0.67	5.25	107
	50°	Benzene		0.66	5.4	397
		Toluene		0.64	4.9	312
		Benzene		0.55	1.3	312
		Chloroform		0.68	5.4	312
		Acetone		0.64	5.0	312
Neoprene CG	10°	Benzene	61,400–1,450,000	0.89	2.02	251
Neoprene GN	40°	Benzene	20,500–959,000	0.73	1.46	253
Neoprene GN	40°	Toluene		0.615	5.0	312
Neoprene W		Benzene	47,000–1,050,000	0.71	1.55	252
Silicone rubber		Toluene	2,500–200,000	0.66	2.0	37
Natural rubber		Toluene	420–1,500,000	0.667 ±0.007	5.02 ±0.04	73

extrapolation to infinite dilution in order to eliminate interaction between individual molecules are greater with the rubber polymers. In spite of this, some measurements have been carried out on natural rubber[211] and on polystyrene.[360]

The measurement of the freezing point of a solution of a polymer as compared with that of the pure solvent is useful only for molecular weights below 5000. Above this limit the depression is too small to be readily measured unless the concentration is excessively high. As in the other methods, extrapolation to infinite dilution is necessary. This extrapolation is extremely difficult if the solution fails to obey Raoult's law. Kemp and Peters have shown that such failure makes the cryoscopic method unsatisfactory for molecular weights much above 2200 for polyisobutylene,[201] 1360 for natural rubber,[199] or 1200 for polystyrene.[200]

It often happens that a polymer molecule contains an atom or group of atoms different from the atoms and groups in the rest of the molecule. The atom usually originates from a fragment of the catalyst or modifier used in the polymerization or from an unintentional impurity. If it can be shown that there is exactly one (or any other known constant number) of such atoms per molecule, an accurate analysis of the polymer for the particular atom or group makes possible a simple calculation of the molecular weight. The method loses sensitivity as the molecular weight increases, since analysis of increasingly higher accuracy are required. It has been little used with synthetic rubbers. Wall and co-workers[327, 364] have concluded that each molecule of GR-S contains one sulfur atom. Their measurements were in the range of molecular weights below 50,000. However, Mochel and Peterson,[254] using radioactive sulfur as a tracer, conclude that 1.1 to 4 sulfur atoms per molecule are combined when chloroprene is polymerized under conditions similar to those used in the polymerization of GR-S. In unpublished work at the National Bureau of Standards[224] the bound sulfur content of a number of unfractionated normal GR-S polymers has been found to be near 0.1 per cent. For a number-average molecular weight of about 100,000, as given below, this is equivalent to 2 to 3 sulfur atoms per molecule. Further studies of greater precision on different polymers are required before the end-group method can be established as a satisfactory method of determining molecular weights of synthetic rubbers.

The direct measurement of molecular sizes by electron microscopy followed by calculations of molecular weights from density considerations is an intriguing possibility under exploration. Rochow and Rochow[290-1] have ascribed observed discontinuities in the fractured surfaces of silicone resins and rubbers to individual molecules and have calculated molecular weights of the order of 300,000, in agreement with those obtained by osmometry.[306]

Molecular-Weight Averages. Each method just described when applied to a polymer comprising only a narrow range of molecular weights yields a value in reasonable agreement with that obtained by any other method. It can be readily demonstrated experimentally that such agreement is not obtained for a mixture of polymer fractions or for an unfractionated polymer. It becomes evident on further consideration that an average value will be obtained in each case but that there are different kinds of average values corresponding to the different types of measurement. For a fraction containing a narrow range of molecular weights all kinds of average values agree sufficiently well but the difference between them becomes greater as the range of molecular weights is increased.

Number Average. The most familiar type of average, obtained by dividing the total weight of the polymer by the number of molecules, is called the number-average molecular weight. It is quite sensitive to the number of molecules of low molecular weight, which may contribute little to the total weight of the polymer. The number-average molecular weight is the average obtained from measurements of osmotic pressure, freezing-point depression, and end groups.

Weight Average. In computing the weight-average molecular weight, the weight of each group of molecules of similar weight is multiplied by itself and the sum of all the products is divided by the sum of the products of the weight and number in each group. . This process is simply one of "weighting," giving to each group a "statistical weight" equal to the actual weight of the group. The weight average, consequently, is not nearly so sensitive to the number of molecules of low molecular weight as the number average. The relation between the two averages depends on the particular distribution of molecular weights in the sample under consideration. The weight average is always greater than the number average and is often several times as large. The weight-average molecular weight is the average obtained from measurements of light scattering and some types of ultracentrifuge measurements.

Viscosity Average. Other weighting factors can be used in computing other kinds of average values. The average molecular weight computed from measurements of intrinsic viscosity requires the weighting factor to be M^a where M is the molecular weight of a group of molecules and a is the exponent in the relation between intrinsic viscosity and molecular weight discussed in the section dealing with intrinsic viscosity. This average is called the viscosity average. For a polymer where $a = 1$ the viscosity average becomes the same as the weight average. Where a is less than one, as it usually is, this average is somewhat less than the weight average. It is usually much nearer to it than to the number average. The Fikentscher K values mentioned earlier correspond to viscosity-average molecular weights. This fact, generally overlooked, vitiates some of the conclusions drawn by those who have used these values.

METHODS FOR DETERMINING STRUCTURE

1,4- and 1,2-Addition, cis- and trans-Addition, and Head-to-Tail and Head-to-Head Addition

Natural rubber consists, as far as can be determined, of *cis* units of the polyisoprene type, joined 1,4 in head-to-tail fashion. The gutta hydrocarbon, found in gutta-percha and balata, is the corresponding *trans* isomer. It has not been found possible to polymerize a synthetic rubber as uniform in primary structure as natural rubber or the gutta hydrocarbon. The synthetic diene rubbers possess structures that are mixtures of *cis* and *trans* units joined by 1,4-addition or of units joined by 1,2-addition. There may be head-to-tail or head-to-head addition in those cases where the monomer is unsymmetrical, like isoprene, isobutylene, or chloroprene; this possibility does not arise with butadiene. GR-S, like other copolymers, has an additional possibility for irregularity, in that there can be variation in the number of units derived from butadiene between each unit derived from styrene.

Infrared Absorption. An estimation of the amount of 1,2-addition in polymers can be made by measurements of infrared absorption.[229] First applied by Swaney, White, and Flory,[338, 377] this method has been further studied by Rassmussen and Brattain[282] and published in detail by Field, Woodford, and Gehman.[112] More recently Hampton[168] and Treumann and Wall[349] have published the details of the procedures they follow. It has been applied to Neoprene by Mochel and Hall.[250] The absorption bands in the silicone rubbers have been investigated by Richards and Thompson,[287] and a number of other polymers have been studied by Thompson and Torkington.[344] No significant changes in the position or intensity of the infrared absorption of polymers are observed when the temperature is lowered to that of liquid helium.[205]

Hampton[168] considers the absorption of GR-S at the following wave numbers to be a measure of the amount of the corresponding structure:

911 cm.$^{-1}$	1,2-addition
967 cm.$^{-1}$	*trans* 1,4-addition
724 cm.$^{-1}$	*cis* 1,4-addition
699 cm.$^{-1}$	Styrene

By measurements at these wave numbers and the solution of four simultaneous equations, the relative amounts of the four types of structure can be calculated. The other observers used only the first two of these absorption bands or a band at 996 cm.$^{-1}$ of less certain origin and based the values on comparisons with hydrocarbons containing 8 to 12 carbon atoms.

X-Ray Diffraction. X-ray diffraction furnishes useful information about the structure of certain polymers, especially those that can be induced to crystallize, as discussed in a later section of this chapter. The identity period along the fiber axis can be related to the structure. The *cis* isomer for example has an identity period approximately twice that of the *trans* isomer. The method has been applied to low-temperature polymers by Beu, Reynolds, Fryling, and McMurray.[59, 60] The method is not so sensitive as the infrared and is not well suited to dealing with mixtures or polymers that do not crystallize.

Pyrolysis. A study of the products of pyrolysis can also be made to furnish information about the structure of a polymer, especially the relative strengths of the chemical bonds. The presence of free radicals causes the decomposition temperature of hydrocarbons of low molecular weight to be lowered from 800° C. or more to temperatures as low as 300° C. Since the hydrocarbon synthetic rubbers decompose on heating in the absence of oxygen at temperatures from 300 to 480° C., it can be assumed[225, 317] that decomposition in them is due to the presence of free radicals. Decomposition occurs in air at lower temperatures.[317]

L. A. Wall[370-1] developed a simple pyrolysis apparatus in which the volatile components formed at 400° C. in a closed system are condensed with liquid air. These are then analyzed by a mass spectrometer. S. L. Madorsky and his co-workers[225-7] pyrolyze a polymer film on a platinum tray in a system connected through traps to a pump. The pressure can thus be kept very low, and the residue can be measured and analyzed. The volatile products are analyzed by the use of a mass spectrometer. Temperatures of 300 to 480° C. are employed.

Branching and Cross-Linking.   Gel and Microgel

Formation.   Some high polymers are adequately characterized as perfectly linear. The atoms of a single molecule are joined by primary valence bonds and except for side groups of limited extent (already present in the monomer) lie along a single chain. In other words, a linear polymer is one that is free of either cross-linking or branching, as already mentioned in the section on polymerization. The chain, of course, need not be fully extended and is in fact usually folded or coiled up considerably. The polymerization of bifunctional units yields linear polymers. Linear polymers can be dissolved in suitable solvents to form homogeneous solutions.

In a branched polymer, on the other hand, there are a number of forks at intervals. These may originate during the polymerization of trifunctional or tetrafunctional units or by the activation of a chain already formed, as discussed in the section on the mechanism of polymerization (p. 321).

In condensation polymers, the amount of branching can be expressed quantitatively in terms of a branching coefficient α which is the probability that on moving along a given chain one will find a branch rather than a terminal group. Flory[121-2] has presented theoretical considerations to show that α increases with conversion and that at or above a critical value of α (equal to $1/2$ for trifunctional and $1/3$ for tetrafunctional units) the formation of an infinite network should occur. It is "infinite" in size in that it extends to the boundary of the phase in which polymerization is proceeding, and often exceeds by many orders of magnitude the dimensions of linear high-polymer molecules. It is a network because of the junction, by cross-linking, of different branches as the number of branches increases without limit. The infinite network, which has been observed experimentally, is commonly called a gel. Gels are also formed in addition polymers if cross-linking is possible, as for example by the activation of polymer atoms other than those at the ends of a chain. A gel is effectively a single molecule, and does not disperse in any solvent that does not break down its primary chain structure.[28, 36] In contact with a solvent, it swells to a definite limit while retaining its shape. The portion of a polymer that is not included in the gel network is called the sol fraction.

The conversion at which the formation of the infinite network occurs is usually observed to be quite sharply defined. At conversions just above this gel point the gel includes only a portion of the total polymer formed, but it increases rapidly by the addition of the remaining polymer to its structure. The gel point of GR-S, as currently produced, is just above 80 per cent conversion. However, the polymerization is stopped at an earlier stage, so that normal GR-S is free from gel. Since polystyrene lacks unsaturation in its main chain, cross-linking and gel formation are not found in this polymer. There is even evidence[272] that polystyrene is not branched to any significant extent. The same statements may be made of polyisobutylene.[283, 399]

Microgel. In the latter stages of the normal GR-S polymerization the reaction is proceeding in swollen latex particles. Under these conditions, Baker[28] has pointed out, it is obvious that when the gel point is reached the "infinite network" can extend no farther than the boundaries of the particle. Consequently the size of a given gel particle cannot exceed the size of the latex particle in which it is formed. The latex particle[300] usually ranges from 1000 to 2000 A. in diameter, corresponding to a molecular weight of the order of 60×10^6. Baker,[28] who has devoted particular attention to gel particles of this size, has called them "microgel." He has made direct measurements on microgel by light-scattering techniques. However, it is quite difficult to measure and study microgel by most ordinary techniques, because of the smallness of the particle size. Microgel can be agglomerated to form macrogel, and is most readily studied in this form. Medalia and Kolthoff[238] have recommended a standard agglomeration procedure for GR-S, in which the polymer (preferably coagulated from the latex by ethanol) is milled for about 10 passes with the addition of a stabilizer, and the milled polymer is heated for 12 hours in a vacuum oven at 80° C. After this treatment the sol portion can be leached out of the gel by a suitable solvent. They find that this procedure agglomerates any microgel particles present but does not produce gel in GR-S which is free from microgel. However, gel can be formed, even in polymers free from microgel, by heating under certain conditions. It appears likely that this is due to cross-linking of the polymer chains by oxygen. Though the infinite network is produced in this case by a different mechanism, the resultant structure is equivalent to that arising from a tetrafunctional polymerization unit. Likewise, rubber free from gel can form an insoluble complex with carbon black[339] often called "bound rubber."

Medalia[237] has shown that microgel can be dispersed in solvents and precipitated by the addition of nonsolvent. Thus it is, thermodynamically speaking, soluble in a suitable solvent in the same sense as are straight-chain polymers of the same molecular weight.

Measurement. There are unfortunately no satisfactory methods for the quantitative determination of the amount of branching of a polymer. B. L. Johnson and Wolfangel[193-4] consider it likely that different amounts of branching are reflected in the exponent *a* of the relation between intrinsic viscosity and molecular weight. The lower values of the exponent shown in their data in Table I for polybutadiene prepared at the higher temperatures may then connote greater amounts of branching. A recent report by Holdsworth[183] tends to confirm this view since the exponent for polystyrene was found to be lowered markedly by the incorporation of 0.05 or 0.10 per cent of divinylbenzene in the polymer. Similar work[120] showed a reduction in the value of a from 0.92 for polystyrene to 0.78 for styrene copolymerized with 0.025 per cent divinylbenzene. Staudinger and Fischer[332] consider that branching is reflected in any deviations of the exponent from the value of unity. The treatment of this subject has more recently been carried forward by Flory and Fox.[135]

A method of detecting branching by measurements of the constant k' in the Huggins equation (equation 5, page 346) for the concentration dependence of viscosity has been investigated by Cragg and co-workers. They have applied it to branching in GR-S as a function of conversion[76a] and as a function of polymerization temperature.[77a] They conclude that branching is negligible at conversions below 30 per cent for GR-S made at 50° C. and that it is negligible even up to conversions of 65 per cent for polymerization temperatures of 15° C. and below.

Wall and Zelikoff[368] found an increase in the minimum thickness of films of nitrile rubber from 6 A. for low-conversion rubber to 30 A. for high-conversion rubber in which branching had become pronounced. This method offers promise as a tool for the study of branching.

German workers,[157, 332, 373] as mentioned in Chapter 26, have sometimes taken as an index of branching the ratio of molecular weight as determined by osmotic pressure to the "Staudinger molecular weight" calculated as proportional to intrinsic viscosity. Since the latter quantity, as already mentioned, is based on a proportionality factor derived from measurements at very low molecular weights, the ratio is usually considerably greater than unity. American workers have not regarded this ratio as significant, because the lack of proportionality between molecular weight and intrinsic viscosity has been shown to be not primarily due to branching, and because in polymers comprising a range of molecular weights the osmotic-pressure method gives a "number average" whereas the viscosity method gives a "viscosity average." Consequently different values for this ratio would be obtained from polymers having the same branching but differing in distribution of molecular weights. Garten and Becker[157] could find no quantitative relationship between this ratio and 1,2-addition as determined by degradation methods.

The degree of cross-linking has been estimated from swelling measurements or from measurements of modulus.[64, 128, 130, 313, 314] The theory involved has been reviewed and developed by Flory.[125]

Methods for measuring the amount of gel in GR-S were described in unpublished reports in 1943 by Mullen and Baker and by Hulse and others. They have summarized by Back.[25] The sol portion is characterized by a measurement of its dilute solution viscosity and the gel by a measurement of the amount it swells. The visual method of measuring swelling as proposed by Berueffy and Wood[54] has the advantage of eliminating any manipulation or weighing of the swollen gel.

The swelling of a GR-S gel reaches a limiting value after a period of the order of 48 hours in most instances. The ratio of the volume of the swollen gel to the original volume of the specimen has been called the swelling volume ratio.[54] This quantity, or its reciprocal, has been utilized by several earlier workers[26, 64, 96, 139] and seems the best parameter by which to characterize a gel. A "tight gel" is one with a low swelling volume ratio and a "loose gel" one with a high ratio.

Another quantity, the swelling index S, defined as the ratio of the weight of the swollen gel to the weight of the dry gel, has found use where the weight of the swollen gel is measured. Berueffy and Wood[54] have shown that the swelling index S is related to the swelling volume ratio K and the weight fraction of gel G by the following equation, applicable to normal GR-S and benzene,

$$S = 0.937K/G \qquad (3)$$

The numerical factor is the ratio of the densities of benzene and GR-S. Equations applicable to other systems are obtained by altering this factor.

Because the weight of the swollen gel is made up partly of gel and partly of solution and for other reasons the swelling index seems to be of less fundamental significance than the swelling volume ratio.

GR-S 60, a Cross-Linked Polymer. The presence of gel has been regarded as undesirable in American synthetic-rubber production, and, as already mentioned, polymerization is normally stopped before the gel point is reached. However, in the polymerization of the particular variety designated as GR-S 60, gel is intentionally produced by the replacement of about 0.5 per cent styrene by divinylbenzene.[301] The divinylbenzene, by furnishing a tetrafunctional unit, produces a controlled amount of cross-linking which is determined by the quantity used. GR-S 60 has better tubing and molding properties and less nerve and shrinkage than standard GR-S, but has lower tensile strength when vulcanized and is used for only certain purposes.

COMPOSITION, MOLECULAR WEIGHT, AND STRUCTURE OF SYNTHETIC RUBBERS

The methods given in the three preceding sections, when applied to synthetic rubbers, have yielded results which are summarized in this section. There is particular interest in the effects of varying the polymerization conditions, including especially the temperature of polymerization, and in varying the conversion.

Composition as a Function of Conversion. The bound styrene content of a butadiene-styrene copolymer increases with conversion in emulsion polymerization. In a normal GR-S system with a charge ratio of 25 parts of styrene to 75 of butadiene the polymer formed at the beginning contains about 17 per cent styrene whereas that being formed at 70 per cent conversion contains about 26 per cent styrene.[240] The polymer formed during the reaction up to 70 per cent conversion has an average styrene content of about 20.5 per cent. Raising the charge ratio to 30 parts of styrene to 70 butadiene increases each of these figures by 4 to 5 per cent.[228, 247] High-styrene copolymers also show a rise in styrene content with conversion.[249] In a $-18°$ C. emulsion polymerization system the bound styrene content of the copolymer has been found to be relatively constant over a considerable range of conversion, rising at the higher conversions to conform to the charge ratio.[271] The bound styrene content in this case shows the greatest difference from the ratio charged when the latter is about 50 to 50. At any temperature, copolymers of greater compositional homogeneity will be obtained if conversion is kept below 10 per cent and modifiers of low reactivity are used.[222]

In sodium-catalyzed copolymerization, as contrasted with emulsion copolymerization, the styrene content for a charge of 25 parts of styrene to 75 of butadiene decreases from about 32 per cent at the beginning to an average value of about 28 per cent for the polymer formed up to 70 per cent conversion.[303] In alfin-catalyzed copolymers also the styrene appears to enter the copolymer more rapidly than the butadiene.[91] Several different modes of variation of composition with conversion are found in butadiene-acrylonitrile copolymers, depending on the conditions of polymerization.[102, 315, 367] Here there seems to be a much stronger tendency toward alternation of the two components than in butadiene-styrene systems. The theory of copolymer composition has been treated by F. T. Wall[362-3, 367] and by L. A. Wall.[369]

Molecular Weight. In those cases where a synthetic rubber has been fractionated and the distribution of molecular weights measured, the properties of the polymer may be regarded as determined by the distribution. The various average values can be computed and each property correlated with its respective properly weighted average molecular weight. It should be recognized, however, that a given average value of molecular weight may have been derived from any one of many different distributions and that there can be no correlation between the given average molecular weight and a property that depends on one of the other averages.

Since fractionation requires considerable time, care, and effort, relatively few polymers have been subjected to careful studies of molecular-weight distribution. The average values obtained in the more significant fractionation studies are given in Table II, together with a number of other average values observed on unfractionated polymers. It should be recognized of course that these values refer strictly only to rubbers polymerized under exactly the same conditions as those investigated. The temperature of polymerization is obviously one of the important variables controlling the molecular weight.

The intrinsic viscosity of normal GR-S in toluene or benzene is usually found to be about 2.09, corresponding to a viscosity-average molecular weight of about 270,000. The inherent viscosity at low concentrations for alfin-catalyzed polybutadiene may be as high[113] as the range 6 to 16, or even 29 according to another source.[91] The lower value corresponds to a molecular weight of the order of 1,400,000 and an average greater than 2,000,000 according to the results of measurements of osmotic pressure.[113]

A fractionation of a silicone rubber[306] yielded fractions with osmotic molecular weights ranging from 290,000 to 2,800,000. Ultracentrifuge studies have been made on fractions of polychloroprene by Svedberg and Kinell.[334] The molecular weights of the fractions ranged from 40,500 to 657,000 for a bulk polymer and from 62,000 to 1,200,000 for an emulsion polymer.

The molecular weight of the polymer being formed in the normal GR-S

Table II. Molecular Weights of Unfractionated Polymers

Material	Temperature of Polymerization, °C.	Number Average, $\overline{M}_n$	Weight Average, $\overline{M}_w$	Viscosity Average, $\overline{M}_v$	Reference
Natural rubber		400,000		1.74 to 2.8 × 10^6	373
			400,000		192
					211
			435,000		211
Polybutadiene emulsion	−20	242,000			193
	5	138,000			193
	50	119,000			193
Butadiene-styrene copolymers					
GR-S	50	92,000		330,000	151
	50	96,500		265,000	397
	50			270,000	79
	50			277,000	81
	50			400,000	192
	50			213,000	247, 337
	50	27,000		154,000	242
	50	31,000		282,000	242
	5	113,000			252
Polyisobutylene	−85			700,000	124
	−78			381,000	124
	−47			34,500	124
	−27			42,200	124
Isobutylene-isoprene copolymers					
Butyl rubber	−101			293,000	399
	−101			320,000	400
Polychloroprenes					
Neoprene CG	10	168,000	316,000	290,000	251, 252
Neoprene GN	40	114,000	257,000	233,000	253
Neoprene W		206,000		336,000	252

system increases as conversion progresses, reaches a maximum near the gel point, and then decreases rapidly at higher conversions.[28, 238, 249, 257, 304, 365-6] Low-temperature polymerizations can be arranged to give little change in molecular weight with conversion until just before the gel point.[270-1] The molecular weight distribution is narrower if conversion is held below 10 per cent and a modifier of low reactivity is used.[222]   A more complicated behavior is observed in butadiene-acrylonitrile copolymers.[102]

Bardwell and Winkler[36] have shown how to calculate the weight-average molecular weight from the amount of modifier consumed and the conversion. Meehan[242] has gone even further in showing how to calculate the approximate distribution of molecular weights from measurements of mercaptan consumption (yielding the number-average molecular weight) and intrinsic viscosity (yielding viscosity-average molecular weight).

Yanko[397] found the bound styrene content of copolymer fractions to increase from about 22.2 to 25.1 per cent as the number-average molecular weight increased from 12,400 to 1,652,000.   Kemp and Straitiff[203] could find no systematic variation with molecular weight.

The change in distribution of molecular weights as a function of temperature of polymerization is shown in Table III taken from a report of P. H. Johnson.[195]   The number-average molecular weight, determined osmometrically, increased linearly from 200,000 to 500,000 as the polymerization temperature was lowered from 10 to $-40°$ C.   Others[99, 185] have noted the same effect.   Rabjohn and co-workers[280] found the viscosity-average molecular weights of polymers prepared in emulsion at temperatures up to $160°$ C. to be considerably lower than those for polymers prepared at $50°$ C.

Table III.   Distribution of Intrinsic Viscosity[195]

Intrinsic Viscosity Range	0–1	1–2	2–3	3–4	4–5	Over 5
GR-S	29.8%	19.7%	26.5%	14.4%		
Redox (modified)						
at 10° C.	39.9	30.8	20.2	9.1		
Redox						
10° C. (unmodified)	10.1	19.8	21.0	18.3	15.9	13.8
−10° C. (unmodified)	11.3	18.9	25.9	24.1	19.8	
−25° C. (unmodified)	3.5	28.2	24.8	24.0	19.5	
−40° C. (unmodified)	5.0	13.5	17.4	17.4	17.4	29.0

Structure. The structure of a polymer depends on the type of polymerization mechanism and on the temperature of polymerization.   It does not vary much with the amount of modifier or the conversion, except possibly in the later stages beyond the conversions in commercial use.

For polymers made from unsymmetrical monomer units a head-to-tail sequence has been found almost universally.[131]   There are a few exceptions outside the field of rubber, but the evidence for them has been stated to be open to question.   There is evidence of this head-to-tail type of sequence in polystyrene and in polyisobutylene.   With a symmetrical monomer, like 1,3-butadiene, of course the question does not arise.

Neoprene. The infrared spectra of Neoprenes[250] polymerized at 40 and at 10° C. show no evidence of any 1,2- or 3,4-addition, and ozonolysis[207] also indicates at least 95 per cent 1,4-addition. There is X-ray evidence[70] that the polymers are largely *trans* isomers. The chlorine atom is readily removed by pyrolysis[370] to form hydrogen chloride on combination with hydrogen.

Alkali Metal- and Alfin-Catalyzed Polybutadiene. It has long been recognized that polybutadiene prepared by an alkali metal catalyst has a preponderance of 1,2-addition. Hampton[168] found 68 per cent 1,2-addition, 14 per cent *cis* 1,4-addition, and 20 per cent *trans* 1,4-addition in a polymer made at 30° C. Treumann and Wall[349] found 71 per cent 1,2-addition for a similar polymer. Kolthoff, Lee, and Mairs,[210] using the perbenzoic acid technique,[208] found 61 to 67 per cent 1,2-addition in sodium polymers prepared at 30° C. Such addition was found to increase from about 59 to 73.5 per cent as the temperature of polymerization was lowered from 50 to 10° C. Saffer and B. L. Johnson,[169, 296] using the same method, obtained similar values, as did Meyer and collaborators by using infrared absorption,[248] but Schulze and Crouch[303] report somewhat lower values.

Alfin-catalyzed polybutadiene is considered to have a larger amount of 1,2-addition than the emulsion polymer and a larger amount of *trans* 1,4-addition than any emulsion polymers, even those[91] prepared at −20° C. Apparently the amount of *cis* 1,4-addition is negligible. The amount of external double bonds, corresponding to 1,2-addition, was measured by the perbenzoic acid method as 24 to 27.5 per cent.

Emulsion Polybutadiene. Hampton[168] used the infrared technique to measure the structure of a series of polybutadienes prepared in emulsion at various temperatures. All of the results he obtained are plotted in Fig. 1, which includes more values than a similar figure given in Hampton's paper. It can readily be seen that the amount of 1,2-addition is much lower than in alkali metal-catalyzed polybutadiene and that it shows relatively little change with polymerization temperature. As this temperature is lowered, the striking feature is the decrease in the amount of *cis* isomer and the corresponding increase in *trans* isomer. Hampton made some revisions in values reported earlier[177, 247] based on the same measurements. Infrared analyses of GR-S at the B. F. Goodrich Laboratory[165, 406-7] differ from these only in indicating substantially lower amounts of *cis* 1,4-addition at all temperatures of polymerization. The perbenzoic acid method yields values[91] of 17 to 22.5 per cent or 11 to 25 per cent for the amount of 1,2-addition in emulsion polybutadiene.[169, 296] Beu and his co-workers[59, 60] report about 14 per cent 1,2-addition in a polybutadiene prepared in emulsion at −20° C. They concluded from X-ray evidence that the major portion of the polymer could be crystallized and that the identity period of these crystals corresponds to the *trans* configuration. Condon[74a] has collated data from a number of different laboratories on the structure, as indicated by infrared analyses, of polybutadiene samples prepared at various temperatures in a variety of polymerization recipes.

Polybutadiene, along with polyisoprene and GR-S, is found by pyrolysis experiments[225] to be intermediate between those polymers (polystyrene and polyisobutylene) that decompose at relatively low temperatures and yield large amounts of monomer and those polymers (like polyethylene) that require higher temperatures for decomposition and yield relatively little monomer. With the first polymers, fragments break away most readily at

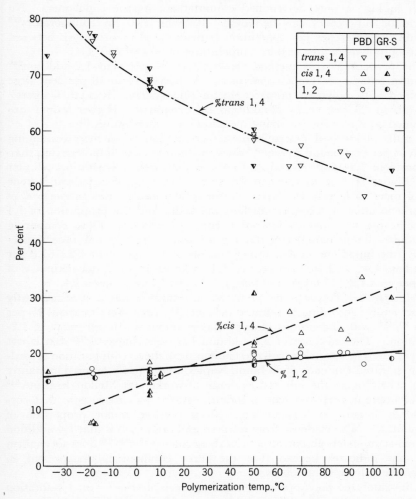

	PBD	GR-S
trans 1, 4	▽	▼
cis 1, 4	△	▲
1, 2	○	◐

FIG. 1.   Structure of Butadiene Units as a Function of Polymerization Temperature for Emulsion Polymerization

Data from a paper by Hampton[168]

the ends of the molecules, with the last type, random breaks at any point along the chain predominate, and, with polybutadiene, there is a combination of both mechanisms. These observations permit conclusions as to the relative bond strengths in different positions along the chain.

Butadiene-Styrene Copolymers. It is found from both infrared studies on butadiene-styrene copolymers and analysis by the perbenzoic acid method[210] that the presence of the styrene does not alter appreciably the distribution of the different structures in the butadiene units. This result makes it possible to regard Fig. 1 also as a representation of the structure of the butadiene units in the copolymers. In fact it will be noted that a number of the values given in Fig. 1 were determined on butadiene-styrene copolymers. No systematic differences can be noted. Rabjohn and his collaborators[279] found little difference in 1,2-addition as determined by ozonolysis between emulsion polymers prepared at temperatures of 50 and 130° C.

The perbenzoic acid method shows 22 to 24 per cent 1,2-addition[296] for most butadiene-styrene copolymers. Values of 14 to 18 per cent were obtained with polymers prepared in an acid system. Russian workers[21] find about 23 per cent 1,2-addition by ozonolysis. Higher values are obtained by a reaction involving the addition of mercaptans.[316]

Results of infrared determinations on emulsion copolymers containing 45 to 63 per cent bound styrene[269] show, in regard to the butadiene, less than 15 per cent 1,2-addition and 21 to 48 per cent *trans* 1,4-addition. Foster and Binder[141a] find that, when the styrene content of butadiene-styrene copolymers is largely in excess of the GR-S range, the proportion of butadiene units of 1,2 structure falls markedly and the proportion of 1,4 units having the *trans* configuration rises appreciably. Thus, comparing butadiene-styrene samples containing 23 and 87 per cent styrene respectively, the butadiene in the former consists of 17.6 per cent 1,2; 65.8 per cent *trans* 1,4; and 16.7 per cent *cis* 1,4; whereas in the latter it consists of 9.2 per cent 1,2; 77.6 per cent *trans* 1,4; and 13.4 per cent *cis* 1,4.

Polyisoprenes. Polyisoprene prepared in emulsion has a predominantly 1,4-structure, according to infrared evidence,[112] amounting to about 88 per cent,[89, 296] with indications of appreciable amounts (10 per cent) of 1,2-addition. The existence of 3,4-addition has been suggested[112] but is not certain.[131] The 1,4-structure is predominantly of the *cis* configuration according to infrared evidence of D'Ianni, Naples, and Field[91] or predominantly *trans* according to the infrared evidence of workers at Mellon Institute.[244]

Polyisoprene prepared with a sodium catalyst has a structure that corresponds largely to 3,4-addition with a possible minor proportion of 1,2-units.[112] The evidence from infrared and ozonolysis is that 1,4-addition is almost completely absent, although other evidence[89, 169, 296] does not confirm this but indicates instead that the three components are present in approximately equal amounts.

Alfin-catalyzed polyisoprene contains appreciably more *trans* 1,4-addition than the emulsion polymer[91] and an appreciable amount of 3,4-addition.

Isobutylene-Isoprene Copolymers. The copolymers of isobutylene and isoprene were found by Rehner[283] to be formed by 1,4-addition of the isoprene at random intervals along the chain. Branching and gel formation are not observed here.

RUBBERS IN SOLUTION AND SWOLLEN RUBBERS

Physical chemistry as related to the properties of polymers in solution will be briefly discussed in this section. Specific applications of some of its principles have already been discussed in connection with methods for determining molecular weight and structure.

Properties of Dilute Solutions

Thermodynamic Considerations. The thermodynamics of a polymer solution includes the formulation of expressions for its volume, free energy, and heat content in terms of composition, temperature, and pressure. The first point to be noted is that the concentrations often studied are extremely small when expressed in mole per cent, since the molecular weight of the polymer is so very large compared with that of the solvent. For example, a solution containing 1 per cent by weight of a polymer having a molecular weight of 100,000 in a solvent with a molecular weight of 100 has a concentration of approximately 0.001 mole per cent. The second point to be noted is that many processes in high-polymer systems take place so slowly that it is often difficult to attain a state of equilibrium or to determine whether or not the system is in a state of equilibrium. Thermodynamic reasoning is of course normally applicable only to equilibrium states.[161]

A knowledge of the Gibbs free energy makes possible the calculation of many solution phenomena. For example, when a polymer and a liquid of low molecular weight are brought into contact, if the Gibbs free energy of mixing is negative, a solution will be formed, and the liquid will be a solvent for the polymer. The free energy is the difference between the heat of solution and the product of the absolute temperature and the entropy change. The entropy term is relatively large because the number of possible configurations for polymer molecules in the swollen or dissolved state is much larger than the number in the solid state. From the rate of change of free energy with temperature the magnitude of the heat of solution and the entropy change may be calculated.

Observations of the vapor pressure and osmotic pressure of solutions of different concentrations furnish the most useful method of determining experimentally the free energy and related thermodynamic functions. In a few instances direct measurements of the heat of solution have also been made. The free energy is equal to the product of the gas constant R, the absolute temperature, and the logarithm of the ratio of the vapor pressures of the polymer solution (or swollen polymer) and of the pure solvent, corrected, if necessary, for any deviation of the vapor from an ideal gas. Alternatively, the free energy is the product of the osmotic pressure by the partial molar volume of the solvent liquid.

Vapor Pressure and Osmotic Pressure. The rate of evaporation of molecules from the surface of a liquid obviously depends on the number of molecules in a layer near the surface. If molecules of a volatile solvent liquid are partially replaced by molecules of an almost completely nonvolatile polymer, it is clear that the vapor pressure will be lowered. If there were no interaction forces between polymer and solvent, and if the vapor behaved as an ideal gas, a linear relation of negative slope would exist between vapor pressure and concentration of polymer. Deviations from such a simple relation (known as Raoult's law) are very common among solutes of low molecular weight and are universal among high polymers. Studies of the vapor pressure of a polymer-solvent system as a function of concentration yield significant information about the polymer molecules and their interaction with the solvent by furnishing the necessary data for the calculations of free energy.

At low concentrations the vapor pressures to be measured are too near the vapor pressure of the solvent to furnish useful information. Osmotic-pressure measurements are used to determine the free energy in this region of concentration. The use of osmotic-pressure measurements to determine molecular weights has been discussed earlier. It was mentioned that the molecular weight was related to the intercept of the plot of the ratio of osmotic pressure to concentration against concentration. The slope of this plot and its curvature, if any, are related to structural factors in the molecule and its interaction with the solvent.

This relation has received theoretical consideration by Huggins,[186, 188] Flory,[123, 126] and Zimm[402] and experimental study by Bawn and co-workers,[41–2] and many others.[272, 290] The Flory-Huggins equation is

$$\frac{\pi}{c} = \frac{RT}{M_2} + \frac{RTd_1}{M_1 d_2^2}\left(\frac{1}{2} - \mu\right)c \tag{4}$$

where π is the osmotic pressure, c the concentration, R the gas constant, T the absolute temperature, M_2 the molecular weight of the polymer, M_1 the molecular weight of the solvent, d_1 and d_2 the densities of solvent and polymer, respectively, and μ the "solvent-solute interaction constant," to be evaluated for each polymer-solvent system. The constant μ approaches the value 0.5 as progressively poorer solvents are used. For GR-S in benzene a value of about 0.35 is found.[397]

Heats of Solution and Dilution. Heats of solution and dilution have usually been calculated from the rate of change with temperature of the vapor pressure or the osmotic pressure.

Tager and Sanatina[340] have made direct calorimetric measurements of the heats of solution of several of the synthetic rubbers in benzene and find them to be independent of the concentration in the range 0.2 to 4 per cent. Roberts, Walton, and Jessup,[289] by direct calorimetric measurements, found the heat of forming an 8 per cent solution of polystyrene in monomeric styrene to be -0.86 kcal per mole of C_8H_8. The negative sign indicates that solution is exothermic in this case. More recent results[264] show the heat of forming a 3 per cent solution of polystyrene in toluene to be also negative and equal to about -0.54 kcal per mole of C_8H_8, in agreement with the indirect measurements of some observers[299] but not of others.[42]

Viscosity. The viscosity of dilute polymer solutions is a readily measured quantity yielding information about the polymer. The use of viscosity measurements to determine molecular weights has already been discussed. The concentration dependence is represented by the following equation:[187]

$$\eta_{sp}/c = [\eta] + k'[\eta]^2 c \tag{5}$$

where k' for many polymers in good solvents has a value of about 0.34–0.40.[146–7] GR-S shows this value in benzene and in bromoform solutions, as does Neoprene W in benzene;[252] polyisobutylene in diisobutylene, cyclohexane, toluene, n-heptane, triptane, and n-hexadecane;[146] and polystyrene in benzene and toluene.[147]   In poor solvents, as the precipitation point is approached, k' increases and sometimes becomes dependent on the concentration.[146]   Bawn[40] finds the value to increase as a nonsolvent is added to polystyrene solutions.   Simha[319] and Eirich and Riseman[101] have discussed the theory of the concentration dependence.

The relation $[\eta] = KM^a$, already discussed in connection with the determination of molecular weights, is of empirical origin.   Debye and Bueche[88] show that with a coiled structure like a polymer molecule the outer elements disturb the flow inside the coil so that the inner elements are partially shielded from the solvent.   They and Brinkman[67] independently showed that the exponent a should decrease from 1.0 to 0.5 as the molecular weight increases.   Kirkwood and Riseman[206] reached a similar conclusion from a different mathematical approach.   Flory and Fox[134-5] have presented theoretical considerations to show why the value normally lies between 0.5 and 0.8 and have developed methods for determining the parameters involved.

On the addition of a nonsolvent to a polymer solution, the intrinsic viscosity is reduced.   Cragg and Rogers[79] find that the value of the intrinsic viscosity at the precipitation point of GR-S fractions is independent of the nature of the nonsolvent and independent of the temperature.   Bawn and co-workers[44] find that at the precipitation point of polystyrene fractions the exponent a has the value 0.5.   Both of these facts are indicative of a highly coiled molecule at the precipitation point.

Size and Shape of Molecules in Dilute Solution.   The molecules of the synthetic rubbers are flexible long chains, and the distance between the ends is always very much less than the sum of the interatomic distances along the chain. Entropy considerations readily demonstrate the extreme improbability of such an extended form of the molecule.   However, the distance between ends is appreciably greater than that which would be obtained if there were free rotation about the interatomic bonds.   Fox and Flory[146] have found values of the order of 800 A. for a randomly coiled polyisobutylene chain unperturbed by thermodynamic interactions and having a molecular weight of 10^6.   The fully extended length would be of the order of 50,000 A., and the value corresponding to free rotation would be 412 A.   In an actual solvent there may be thermodynamic interactions; in cyclohexane, for example, the value is increased to 1220 A.   Fox and Flory[147] found similar results for polystyrene, but concluded from the values obtained that the molecular chain is somewhat stiffer than that of polyisobutylene.   Similar results for these two polymers were obtained by Kunst[214] on the basis of light scattering and viscosity.   Heller and A. C. Thompson[180] find the apparent specific volume of polystyrene in chlorobenzene solution to be about 3 per cent lower than that of bulk polystyrene.

In general, it may be said that the shape of the molecule is more nearly spherical, the poorer the solvent, since the number of polymer-solvent contacts is thus a minimum.   In a good solvent the number of these contacts is increased and the axial ratio of the polymer molecule deviates more widely from unity.

The viscosity of a solution of given concentration of polymer in the form of spheres is less than that of a solution of the same concentration in the form of ellipsoids. The exact form of the relationship has been investigated by Simha,[318-9] who shows how to relate such measurements to the shape factors.

It is likewise apparent from the values obtained for the viscosity of solutions that a high-polymer molecule in motion through a solvent carries along with it a considerable amount of solvent. In other words it departs considerably from the concept of a "freely draining" molecule. The extent of this deviation depends on both the shape of the molecule and the nature of molecule and solvent. The effective volume of the domain of the molecule often is from ten to several hundred times its actual volume.

The value of the exponent a in the relation between intrinsic viscosity and molecular weight previously discussed is considered to be related to the shape of the molecule. The shape is probably influenced by the amount of branching, as already mentioned, but is also dependent on the stiffness of the molecular chain and on the nature of the solvent. For a linear rodlike molecule the exponent would be expected to be about 1.7 whereas for a compact sphere of large molecular weight[318] it would be 0.5. The significance of values lower than 0.5, which have been occasionally reported,[115, 183, 193-4] is not clear. The most recent theories[135] do not predict such values.

Properties of Concentrated Solutions and Swollen Rubbers

Viscosity. In the limiting case of infinite dilution there is of course no interaction between polymer molecules. As the concentration is increased, the interactions become of increasingly greater importance. At high concentrations the solution may be regarded as a swollen polymer. Viscosity studies[55-7, 77, 109, 302] over the whole range of concentrations show no discontinuities that could be interpreted as indicative of any definite concentration at which the interaction first becomes evident or of any other concentration that could be taken as a limit to distinguish between solution and swollen polymer. The viscosity is found to be dependent on the rate of shear (non-Newtonian flow) for solutions of elastomers. At a constant rate of shear the logarithm of the viscosity increases with concentration, without discontinuities in the logarithm or its slope. The slope of a graph of the logarithm against the concentration decreases by a factor of about one hundred between the concentration limits of 0 and 100 per cent for the system GR-S in α-methylnaphthalene.[57]

Cohesive Energy Density. The concept of cohesive energy density (C.E.D.) is a useful one in dealing with the swelling of polymers by liquids, since the swelling is greatest when the values of this quantity for liquid and polymer are equal.[158-9] For a liquid the cohesive energy density is simply defined as the energy required to evaporate unit volume at constant pressure. This involves merely a determination of the latent heat of evaporation. The cohesive energy density of a polymer is obtained by cross-linking it somewhat and observing the swelling in a series of liquids of varying cohesive energy densities. Varying amounts of cross-linking, if not too large, are found not to alter the observed values of the cohesive energy density. The latter can also be computed as the product of the absolute temperature by the ratio of volume expansivity to compressibility,[160] but not much use has been made of this relation. The "interaction constant" α is defined as the square root of the C.E.D.[106] Table IV shows values of these two quantities for several rubbers. Gee[158-9] and Ewart[106] have given values for various liquids used in the swelling measurements.

Table IV. Cohesive Energy Densities

Material	C.E.D.	Interaction Constant	
	Cal. per cc.	α (Cal. per cc.)$^{1/2}$	Reference
Natural rubber	63.7	7.98	158, 159
Natural rubber	69.7	8.35	308, 314
Polybutadiene, emulsion	70.6	8.40	308
Polybutadiene, sodium-catalyzed	74.0	8.60	308
Butadiene-styrene copolymers			
GR-S	65.6	8.10	114
Buna S	65.4	8.09	158, 159
15% styrene	72.2	8.50	308
25% styrene	73.1	8.55	308
40% styrene	75.2	8.67	308
Polystyrene	83	9.1	308
Butadiene-acrylonitrile copolymers			
75/25 ratio	90	9.5	308, 314
Buna N	88.0	9.38	158, 159
Polyisobutylene	64.8	8.05	308
Polychloroprene			
Neoprene GN	67.1	8.19	158, 159
Neoprene GN	85.6	9.25	314
Polysulfide rubber			
Thiokol F and FA	88.0	9.38	158, 159

Mechanical Properties of Swollen Elastomers. Swelling affects the stress-strain-time relations of rubbers.[35, 94, 128, 136-7, 159-60] Conversely elastomers under strain show a swelling behavior dependent on the kind and amount of the strain.[138-40, 346]

PROPERTIES OF SOLID POLYMERS

Some of the properties of solid polymers of interest in physical chemistry are discussed in this section. Numerical values are given wherever possible.

Densities and Expansivities. Density values for hydrocarbon elastomers other than those containing large amounts of bound styrene are found to be near 0.93 gram per cubic centimeter as shown in Table V. It is not surprising that elastomers containing large amounts of bound styrene or those containing chlorine or nitrogen should have higher values. Expansivity values are given in Table VI. Procedures for determining density[393] and expansivity[47] with good precision have been developed at the National Bureau of Standards. A correction for ash, corresponding to an ash content of 0.1 per cent, would reduce the values given for the butadiene-styrene copolymers by about 0.0005 to 0.0006 gram per cubic centimeter in each case.

Table V. Densities of Unvulcanized Polymers

Material	Polymer-ization Tempera-ture, °C.	Density at 25° C. g. per cc.	Reference
Natural rubber		0.911	388
Polybutadiene			
Emulsion	5	0.892	219
	50	0.8936	219
Butadiene-styrene copolymers			
Emulsion			
0% bound styrene	5	0.892	219
7.9	5	0.9045	219
22.6	5	0.9288	219
36.3	5	0.9526	219
53.1	5	0.9837	219
0% bound styrene	50	0.8936	219
8.6	50	0.9065	219
23.9	50	0.9326	219
43.0	50	0.9667	219
55.7	50	0.9929	219
Polystyrene		1.053	90
Polystyrene		1.054	145
Butadiene-acrylonitrile copolymers			
27% bound acrylonitrile		0.968	393
33%		0.978	391
40%		0.999	393
Polyisobutylene		0.913	148, 389
Isobutylene-isoprene copolymer (Butyl rubber)		0.92	47
Polychloroprene (Neoprene GN)		1.229	393
Silicone rubber		0.974	375
Polysulfide rubber (Thiokol FA)		1.330	393

Table VI. Expansivities $(1/V)(dV/dT)$ of Unvulcanized Materials

Material	Expansivity at 25° C.	Reference
Natural rubber	67×10^{-5}	388
Polybutadiene	70×10^{-5}	395
Butadiene-styrene copolymers		
0–56% styrene	70×10^{-5}	395
GR-S		
23.5% bound styrene, 50° C. polymer	66×10^{-5}	51
Polystyrene	27×10^{-5}	145
Polyisobutylene	62×10^{-5}	148
Isobutylene-isoprene copolymers		
Butyl rubber	57×10^{-5}	47
Polychloroprene (Neoprene GN)	61×10^{-5}	51
Silicone rubber	120×10^{-5}	375
Thiokols		
FA	74×10^{-5}	395
ST	68×10^{-5}	395
PR-1	61×10^{-5}	395

Table VII. *Refractive Indexes*

Material	Temperature of Polymerization °C.	Refractive Index at 25° C.	Reference
Natural rubber		1.5191	394
Polybutadiene			
Emulsion	−19	1.5143	177, 247
	−10	1.5144	177, 247
	5	1.5148	177, 247
			263
	50	1.5154	259
	50	1.5156	177, 247
	65	1.5156	177, 247
	97	1.5159	177, 247
Sodium-catalyzed	5	1.5074	351
	20	1.5080	303
	30	1.5085	351
	40	1.5090	303
	45	1.5099	351
	60	1.5110	303
Potassium-catalyzed	5	1.5102	351
	30	1.5118	351
	45	1.5117	351
Butadiene-styrene copolymers			
0% bound styrene	5	1.5148	263
2.1	5	1.5166	351
7.9	5	1.5217	351
22.6	5	1.5342	351
23.5	5	1.5350	351
36.3	5	1.5456	351
53.1	5	1.5611	351
0% bound styrene	50	1.5154	259
8.6	50	1.5222	259
23.5	50	1.5346	259
23.9	50	1.5350	259
43.0	50	1.5523	259
55.7	50	1.5654	259
Polystyrene		1.5935	90
		1.5929	218
Butadiene-acrylonitrile copolymers			
27% bound acrylonitrile		1.5213	197
Polyisobutylene			
Mol. wt. 4,300		1.5018	383
Mol. wt. 11,000		1.5060	383
Mol. wt. 233,000		1.5077	383
Vistanex		1.5089	383
Isobutylene-isoprene copolymers*			
GR-I R-2 (2% isoprene)		1.5078	383
GR-I Y-15 (2.5% isoprene)		1.5080	383
GR-I Y-25 (3% isoprene)		1.5081	383
Polychloroprene			
Neoprene		1.5578	383
Neoprene GN		1.5580	197
Silicone rubber		1.4040	375

* Correspond apparently to present GR-I grades 35, 15, and 25, respectively.

Refractive Index. Values of the refractive index of unvulcanized synthetic rubbers are given in Table VII along with the index of natural rubber for comparison. The molecular refraction, calculated from these values and the densities, is in most cases very nearly that which is calculated from the sum of atomic refractivities. However, this has been studied carefully only for natural rubber.[394]

Thermal Properties

Heat of Combustion. Heats of combustion of some polymers are listed in Table VIII. These values are useful in indirect determinations of heats of polymerization as previously discussed.

Table VIII

Material		Heat of Combustion		Reference
Natural rubber		10.767	kcal. per g.	388
Polybutadiene		13,876	j./g. CO_2	267
Butadiene-styrene copolymers				
Bound Styrene Wt. %	Temperature of Polymerization			
8.58	50° C. polymer	13,744.4	j./g. CO_2	267
8.58	5° C.	13,667.5		267
22.6	5° C.	13,498.4		267
23.89	50° C.	13,493.2		267
36.26	5° C.	13,283.0		267
42.98	50° C.	13,179.7		267
53.09	5° C.	13,011.3		267
55.73	50° C.	12,961.2		
Buna S		10.396	kcal/g.	293
Polystyrene		1,033.89	kcal/C_8H_8 unit	289
Polyisobutylene		628.2	kcal/C_4H_8 unit	274

Specific Heat. Calorimetric measurements of specific heats over very extended temperature ranges have been reported for polystyrene and several elastomers. Table IX lists the values at 25° C. only. Reference should be made to the original publications for values at other temperatures.

Hamill, Mrowca, and Anthony[167] have shown that the specific heat of a rubber compound at room temperature can usually be calculated by the addition of the weight-fractional values of its components.

Thermal Conductivity. The thermal conductivity of pure gum vulcanizates of GR-S and natural rubber has been studied by Dauphinee, Ivey, and Smith.[84] Values at 0° C. for GR-S were 3.8 to 4.0 $\times$ 10^{-4} cal. sec.$^{-1}$ cm.$^{-1}$ deg.$^{-1}$; two values for natural rubber were found, 3.1 $\times$ 10^{-4} or 3.5 $\times$ 10^{-4} cal. sec.$^{-1}$ cm.$^{-1}$ deg.$^{-1}$, depending on the immediate previous treatment of the specimen. The values for GR-S decreased considerably on stretching, whereas those for natural rubber increased very slightly.

Rehner[285] measured the thermal diffusivity of a number of different synthetic rubbers at 60 and 140° C. The thermal conductivity is the product of the diffusivity by the specific heat and the density, but the latter values were not measured.   Rehner gives a comprehensive discussion of the relation of conductivity to structure.

Table IX

Material	Specific Heat at 25° C. cal gr^{-1} deg^{-1}	Reference
Natural Rubber	0.449	388
Polybutadiene		
5° C. polymer	0.471	155
50° C. polymer	0.467	155
Butadiene-styrene copolymers		
8.58% bound styrene, 5° C. polymer	0.462	156
8.58% bound styrene, 50° C. polymer	0.463	155
25.5% bound styrene, 50° C. polymer	0.45	281
Buna S	0.47	293
43.0% bound styrene, 50° C. polymer	0.435	155
Polystyrene	0.292	262
Butadiene-acrylonitrile copolymer		
39% bound acrylonitrile	0.471	52
Isobutylene-isoprene copolymer		
GR-I (Butyl rubber)	0.464	167

Permeability to Gases

The most extensive measurements of permeability of different elastomers to various gases have been made by van Amerongen.[357-8]   The results (cf. p. 499) have led to the conclusion that permeation is a process in which gas molecules dissolve in the elastomer on one side of a membrane, diffuse through to the other side, and there evaporate.   The permeability can be calculated from measurements of solubility and rate of diffusion.   The solubility of a gas in a given polymer is closely related to its tendency to condense, as manifested by its critical temperature, and is also related to interactions between the gas molecules and polymer molecules.   The rate of diffusion in a given polymer is found to be related chiefly to the size of the gas molecule.   It is observed that the presence of polar groups or methyl groups in the polymer molecule reduces the permeability to a given gas.   Consequently, the nitrile rubbers and Butyl rubber have low values of permeability.   Van Amerongen[358] has discussed the results in terms of activation energies.

Viscosity

Viscosity-Molecular Weight Relation.   An empirical equation applied with success by Flory to polyesters and polyamides calls for a linear relation between the logarithm of the viscosity and the square root of the molecular weight. In spite of the fact that this gives a satisfactory representation over limited ranges for polyisobutylene,[284] Butyl rubber,[399] dimethylsiloxane polymers,[37] and polyethylene,[92] it has been shown to be definitely inadequate for polystyrene.[33, 143, 145, 148, 330-1]   The more general relation found valid in these cases is a linear relation between the logarithm of the viscosity and the logarithm of the weight-average molecular weight. Baldwin[33] shows a linear relation between the logarithm of viscosity and the logarithm of the intrinsic viscosity obtained from solution measurements on polyisobutylene and Butyl rubber. This would be expected from the more general equation if the distinction between viscosity average and weight average does not introduce complications. A more complicated behavior is shown at 30° C. by polyisobutylene fractions[148] of molecular weight less than 17,000.

When a definite relationship between viscosity and molecular weight is established for a given polymer at a given temperature, viscosity measurements can be used to determine molecular weights. Rehner,[284] Zapp and Baldwin,[399] and also Baldwin[33] indicate the convenience of such measurements, the viscosity being determined from measurements with a parallel-plate plastometer according to the method of Dienes and Klemm.[92] Presumably the Mooney viscometer or the worker consistometer could also be used for this purpose, as would appear from the work of Bestul, Decker, and White.[58]

Viscosity-Temperature Relation. The variation of viscosity with temperature has been most extensively studied in polyisobutylene and in polystyrene. A linear relation between the logarithm of the viscosity and the inverse square of the absolute temperature is found for polyisobutylene by Leaderman and co-workers,[215-6] and by Fox and Flory.[146] For polystyrene a linear relation is found between the logarithm of the viscosity and the inverse sixth power of the absolute temperature.[145]

Viscosity and Rate of Shear. The viscosity of a polymer is a function of the rate of shear. It has been tacitly assumed that measurements previously compared were made at the same rate of shear. Investigations have been conducted on GR-S by Treloar[345] and by Bestul, Decker, and White;[58] on Butyl rubber by Bestul, Decker, and White;[58] and on natural rubber by Saunders and Treloar.[298] The rubbers are found to be highly non-Newtonian. Bestul, Decker, and White found the logarithm of the viscosity usually to be linearly related to the logarithm of the rate of shear, with occasional rubbers showing a discontinuity of slope at an intermediate rate of shear, and a few rubbers requiring a line with a small curvature. They found the viscosity to decrease from a value of the order of 5,000,000 poises to one of the order of 10,000 poises as the rate of shear was increased from 0.1 to 500 sec.$^{-1}$.

Crystallization

Some of the synthetic elastomers under the proper conditions show the phenomena associated with crystallization and fusion of crystals; in others these phenomena are completely absent. Crystallizable synthetic polymers include polybutadiene, polyisobutylene, Butyl rubber, Neoprene, the silicone rubbers, and the polysulfide rubbers. Crystallization effects are not normally observed in copolymer rubbers containing substantial amounts of the secondary monomer. The crystallization of natural rubber has been investigated far more thoroughly than that of any of the synthetic elastomers. A review by the present author[390] of the field of crystallization phenomena in natural and synthetic rubbers was published in 1946, and the reader is referred to it for a more comprehensive discussion.

The tensile strength of pure gum compounds of the rubbers which crystallize on stretching is notably higher than that of those which do not.   It is considered that the crystallites produce effects on stress-strain relations similar to those produced by reinforcing filler particles.[20, 162]   A comprehensive theory of crystallization in high polymers has been proposed by Flory and his co-workers.[104-5, 132-3, 149]   James has discussed the size of crystallites.[190]

Polybutadiene crystallizes to some extent, as shown by the evidence of calorimetric measurements,[155]   volume changes,[220, 247, 374]   X-ray diffraction,[59, 60, 196]   birefringence,[69]   and mechanical properties.[163, 320, 335]   Because of the increasing amount of *trans* 1,4-structure with decreasing temperature of polymerization (mentioned in an earlier section; see Fig. 1), a larger fraction of crystalline material is found in emulsion polybutadiene prepared at low temperatures than in that prepared at high temperatures. Calorimetric measurements[155] and volume changes[247] show crystallization in polymers prepared at 50° C., although it could not be detected by the less sensitive methods of X-ray diffraction.[60]   By extrapolation of the resultene Meyer,[247] it appears that no crystallization would be found in polybutadis of prepared at a temperature above 60° C.

The introduction of styrene decreases the fraction of crystallizable material, and it appears from the results of Meyer[247] that 5, 15, or 30 per cent styrene in the polymerization recipe suffices to prevent detectable volume changes for copolymers made at 50, 5, or −20° C., respectively.   No evidence of crystallization, of course, is found in GR-S polymerized at 50° and containing about 23.5 per cent bound styrene.[179]

Polyisobutylene and Butyl rubber crystallize readily on stretching.[390] At low temperatures in the region favorable for crystallization the modulus has been reported to increase slowly with time, suggesting the formation of crystals,[163-4, 387] and the retraction of a vulcanizate stretched to a low elongation also is indicative of crystallization.[335]

Chloroprene polymers[70, 142] are found to crystallize more rapidly, the lower the temperature of polymerization.[361, 376]   No crystallization was observable in a polymer prepared at 100° C.   Crystallization in Neoprene CG made at 10° C. is much more rapid than that in Neoprene GN prepared at 40° C.[251]   A chloroprene-styrene copolymer of low styrene content (Neoprene RT) is found to crystallize more slowly than a simple chloroprene polymer made at the same temperature.[236]   Information about crystallinity in polychloroprene, as with other polymers, can also be obtained from stress-time studies.[204]

Most varieties of the silicone rubbers[375] crystallize more rapidly than any of the other synthetic rubbers.   The rate of crystallization is so great that these rubbers are crystallizable for the most part at a definite temperature, rather than at any temperature in a range like other types of rubber studied. This temperature was found to be near −60° C. for one variety and near −75° C. for another.   A third variety did not crystallize at all.

Crystallization effects are clearly evident in many of the polysulfide rubbers in both the unstretched and the stretched condition.[348, 390]   Shorter repeating units are more favorable for the formation of crystals than longer ones.

Crystallization effects are pronounced in polyethylene and other polymers[29] not so closely related to the synthetic rubbers.

The determination of the fraction of crystalline material present in elastomers has been largely limited to work on natural rubber,[390] where, however, the X-ray evidence has been somewhat conflicting. Recent calculations based on expansivity measurements[273, 396] indicate a maximum of 30 to 40 per cent crystalline material in natural rubber under the most favorable conditions. From changes in the specific volume on crystallization, it would appear that the synthetic rubbers show considerably lower degrees of crystallinity than natural rubber.

Glass Transitions (Second-Order Transitions)

As the temperature is lowered, the expansivity of all elastomers decreases from a value about that of a normal liquid to a value near that of a normal crystalline solid. The change, which usually occurs over quite a narrow range of temperature, is evident also in other properties, such as specific heat, which are derivatives of primary thermodynamic variables. Consequently the term "second-order transition" has usually been applied to the phenomenon. However, there appears to be growing sentiment for the use of a phrase like "glass transition" which does not necessarily imply a similarity to certain well-recognized transitions in materials of low molecular weight or imply that equilibrium in the true thermodynamic sense has been reached. The type of molecular motion necessary for rubber elasticity ceases as the temperature is reduced below that of the glass transition. Consequently the glass-transition temperature characteristic of each elastomer is closely related to its molecular structure and has an important bearing on the use of the elastomer at low temperatures. A comprehensive review by Boyer and Spencer[65] should be consulted for a summary of work on this subject up to 1946. Kauzmann[198] has published an excellent review of the nature of the glassy state, and Buchdahl and Nielsen[68] have discussed the nature of the transition.

The results of Fox and Flory[145] on fractionated polystyrene have contributed considerably to an understanding of the phenomenon, including an explanation of the asymptotic increase of transition temperature with increasing molecular weight. Krimm and Tobolsky[212] have found changes in X-ray diffraction in the region of the glass transition. The glass-transition temperature of polystyrene is a function somewhat of the rate of cooling,[329] but similar effects do not seem to be so pronounced in the elastomers. Nuclear magnetic resonance studies[184] are beginning to throw more light on the nature of the transition. The addition of a plasticizer depresses the glass-transition temperature.[66]

The brittle point, defined as ·the maximum temperature at which an elastomer undergoes brittle fracture under some arbitrary test condition, is related to the glass-transition temperature. It is higher than the latter for a given elastomer, decreases with increasing molecular weight, and depends much more than the latter on the exact conditions of test, particularly the rate of deformation. A correlation between tensile strengths and brittle points has been noted by Borders and Juve.[61]

Values reported for the glass-transition temperatures of elastomers are summarized in Table X. They were determined for the most part as the intersection of the linear portions of graphs of volume or length against temperature. Some of the data were obtained in a corresponding manner from graphs of refractive index against temperature by the method of Wiley and Brauer.[379–83] In a few instances[50, 52, 110, 155, 281] they were obtained by measurements of specific heat. Good agreement is found among these methods when they are applied to the same sample. Stress-temperature studies[386] can also serve to locate the transition temperature.

From an examination of Table X it can be concluded that butadiene polymerized in emulsion at 50° C. has a lower glass-transition temperature than any other polymer reported except the silicone rubbers. A lowering of the polymerization temperature produces a moderate rise in transition temperature. The alkali metal-catalyzed polybutadienes have markedly higher transition temperatures than polybutadienes prepared in emulsion. Copolymerization with styrene raises the transition temperature at the rate of about 1.2° C. for each weight per cent of bound styrene in the region of low styrene content. In the higher region a higher rate is observed. Values in the region of high styrene contents approach 100° C., the transition temperature given for carefully purified polystyrene samples. (Commercial samples of polystyrene usually contain enough monomeric styrene, (1 to 3 per cent), to lower the transition[65, 145, 245, 275] to values between 75 and 95° C.) A plot of the data given in the table for butadiene-acrylonitrile copolymers shows a linear increase of transition temperature with bound acrylonitrile content from a value of about −85° C. for polybutadiene, with a slope of 1.4° C. for each weight per cent of acrylonitrile. Data are not available appreciably above 50 per cent bound acrylonitrile. The equation valid below 50 per cent for the transition temperature T_g in degrees centigrade is thus

$$T_g = -85 + 1.4A \tag{6}$$

where A is the bound acrylonitrile content in weight per cent.

By examining the series polybutadiene, polyisoprene, polydimethylbutadiene it may be concluded that successive added methyl groups raise the transition temperature. In polychloroprene the rise caused by a heavy atom like chlorine may be noted; polar effects also may be important here.

Properties of Fractionated Solid Polymers

Relatively few studies have been made of the physical properties of fractionated solid polymers. Early work by Kemp and Straitiff[203] furnished indications that the low-molecular-weight portions are less readily cross-linked on vulcanization. This was confirmed in more detail by Yanko,[397] who made perhaps the most complete study hitherto of GR-S fractions. Yanko was able to vulcanize a fraction having a number-average molecular weight of 65,800, but could not vulcanize fractions having a number-average molecular weight of 23,600 or less. B. L. Johnson[192] compared the properties of GR-S fractions with those of natural-rubber fractions. Flory[128] made an exhaustive study of the properties of fractionated Butyl rubber, and Zapp and Baldwin[399]

Table X. Glass-Transition (Second-Order Transition) Temperatures

Key to methods: ED, Expansivity by volume dilatometer
EI, Expansivity by interferometer
ER, Expansivity by refractive index
H, Heat capacity

Material	Type of Polymer	Polymerization Temperature, °C.	Transition Temperature, °C.	Reference	Method
Natural rubber			−72	45	ED
			−72 to −78	50	H
			−70	374	EI
			−75	383	ER
			−67 to −71	392	EI
Synthetic polyisoprene	Emulsion	50	−60	374	EI
Polybutadiene	Emulsion	−20	−75	356	ED
		5	−79	395	EI
		5	−78	155	H
		50	−86	374	EI
		50	−85	383	ER
		50	−86	155	H
	Na-catalyzed	30	−45.5	248	ED
			−46	395	EI
		40	−48.5	248	ED
		50	−54	248	ED
		60	−55	248	ED
		75	−63.5	248	ED
	K-catalyzed	30	−60	395	EI
Butadiene-styrene copolymers	Emulsion	−20			
	0% bound styrene		−75	356	ED
	6.8		−71.5	356	ED
	16.0		−68.5	356	ED
	23.0		−62.5	356	ED

Emulsion	5			
0% bound styrene		−79	395	EI
		−78	155	H
2.5		−74	395	EI
8.6		−70	395	EI
		−73	156	H
22.6		−52	395	EI
36.3		−38	395	EI
53.1		−14	395	EI
Emulsion	50			
0% bound styrene		−86	395	EI
		−86	155	H
8.6		−74	395	EI
		−80	155	H
24.0		−56	395	EI
43.0		−34	395	EI
		−36	155	H
55.7		−13	395	EI
c. 44		−28	395	EI
c. 65		−2	395	EI
c. 75		+13	395	EI
c. 80		+21	395	EI
c. 85		+36	395	EI
25.5		−59 to −64	281	H
Polystyrene Solution	150	100	145	ED
Bulk	100	100	262	H
Butadiene-acrylonitrile copolymers Emulsion				
20% bound acrylonitrile		−56	382	ER
22		−52	382	ER
26		−52	382	ER
29		−46	382	ER
30		−41	382	ER
31		−43	382	ER
33		−39	382	ER

Table X—continued

Material	Type of Polymer	Polymerization Temperature, °C.	Transition Temperature, °C.	Reference	Method
Butadiene-acrylonitrile copolymers	33%		−37	382	ER
	37		−34	382	ER
	39		−33	382	ER
	39		−32	382	ER
	39		−26	382	H, ED
	39		−23	52	EI
	40		−22	374	EI
	52		−16	382	ER
Dimethylbutadiene polymers	H-rubber		−12 to +5	295	H
	W-rubber		−22 to −5	295	H
Polyisobutylene	Mol. wt. 4,900		−73 to −83	110	H
	4,900		−68 to −88	110	ED
	4,300		−80.5	383	ER
	11,000		−77	383	ER
	233,000		−77	383	ER
	$1.4 \times 10^6\,(\overline{M}_v)$		−71	395	EI
	Vistanex		−84	383	ER
	Oppanol B-3		−85	355	ED
	Oppanol B15 and B100		−65	355	ED
Isobutylene-isoprene copolymers	2% isoprene		−80	383	ER
	2.5%		−79	383	ER
	3		−75	383	ER
	GR-I (Butyl rubber)		−69	374	EI
Polychloroprene	Neoprene GN	40	−40	374	EI
			−40	395	EI
	Neoprene		−50	383	ER
Silicone rubbers	Silastic 6–160		−118 to −128	375	EI
			−113	395	EI
Polysulfide rubbers	Thiokol FA		−50	395	EI
	Thiokol ST		−56	395	EI
	Thiokol PR–1		−50	395	EI

paid particular attention to the elastic and plastic properties of different fractions.   Methods and techniques of fractionation are described by Cragg and Hammerschlag.[78]   There is general agreement that the modulus and tensile strength increase with increasing molecular weight.   However, the increase is asymptotic, and there is little gain in increasing the molecular weight beyond about a few hundred thousand.   Dynamic properties[192] are somewhat better for the fractions of higher molecular weight.   On the other hand the higher-molecular-weight fractions had such a high Mooney viscosity that processing them under normal factory conditions would be impossible.   There are some indications that a mixture of high- and low-molecular-weight material furnishes a rubber that is more satisfactory than a more homogeneous rubber of the same average molecular weight.   Yanko[397] found the properties of a heterogeneous polymer to be somewhat better defined by calculations from the number-average molecular weight than by other types of average.

REFERENCES

BOOKS

1. Alfrey, T., *Mechanical Behavior of High Polymers*, Interscience, New York, 1948.
2. Bawn, C. E. H., *The Chemistry of High Polymers*, Interscience, New York, 1948.
3. Burk, R. E., and Grummit, O., editors, *Frontiers in Chemistry*, Vol. I, *The Chemistry of Large Molecules*, Interscience, New York, 1943.
4. Burk, R. E., and Grummit, O., editors, *Frontiers in Chemistry*, Vol. VI, *High Molecular Weight Organic Compounds*, Interscience, New York, 1949.
5. Dawson, T. R., Blow, C. M., and Gee, G., *Proc. Second Rubber Technol. Conf.*, London, 1948, Heffer, Cambridge, England.
5a. Flory, P. J., *Principles of Polymer Chemistry*, Cornell Univ. Press, Ithaca, N.Y., 1953.
6. Houwink, R., editor, *Elastomers and Plastomers*, Vol. I, *General Theory*, Elsevier, New York, Amsterdam, 1950.
7. Leaderman, H., *Elastic and Creep Properties of Filamentous Materials and other High Polymers*, Textile Foundation, Washington, D.C., 1943.
8. Mark, H., and Raff, R., *High Polymeric Reactions, Their Theory and Practice*, Interscience, New York, 1941.
9. Mark, H., and Tobolsky, A. V., *Physical Chemistry of High Polymeric Systems*, Interscience, New York, 1950.
10. Mark, H., and Whitby, G. S., *Advances in Colloid Science*, Vol. II, *Scientific Progress in the Field of Rubber and Synthetic Elastomers*, Interscience, New York, 1946.
11. Meyer, K. H., *Natural and Synthetic High Polymers*, Interscience, New York, 1950.
12. Miller, A. R., *The Theory of Solutions of High Polymers*, Oxford Univ. Press, London, 1948.
13. N.Y. Acad. Sci., High Polymers, *Ann. N.Y. Acad. Sci.*, **44**, 263–443 (1943).
14. *Proc. Internatl. Colloq. Macromolecules*, Amsterdam, 1949.
15. *Proc. Internatl. Rheology Congr.*, Holland, 1948, North-Holland Publishing Co. (Interscience), Amsterdam, 1949.
16. Robinson, H. A., editor, *High Polymer Physics*, Remsen Press Div., Chemical Publishing Co., Brooklyn, 1948.
17. Treloar, L. R. G., *The Physics of Rubber Elasticity*, Oxford Univ. Press, London, 1949.
18. Twiss, S. B., editor, *Advancing Fronts in Chemistry*, Vol. I, *High Polymers*, Reinhold, New York, 1945.
19. Wildschut, A. J., *Technological and Physical Investigations on Natural and Synthetic Rubbers*, Elsevier, New York and Amsterdam, 1946.

ARTICLES

20. Aleksandrov, A. P., and Lazurkin, J. S., *Compt. rend. acad. sci. U.R.S.S.*, **45**, 291–4 (1944); *Rubber Chem. and Technol.*, **19**, 42–5 (1946).

21. Alekseeva, E. N., and Belitzkaya, R. N., *J. Gen. Chem. U.S.S.R.*, **11**, 358–62 (1941); *Rubber Chem. and Tech.*, **15**, 693–7 (1942).

22. Alfrey, T., and Lewis, C., *J. Polymer Sci.*, **4**, 221–3 (1949).

22a. Amerongen, G. J. van, *see* Van Amerongen, G. J.

23. Arnold, A., Madorsky, I., and Wood, L. A., *Anal. Chem.*, **23**, 1656–9 (1951); *Rubber Chem. and Technol.*, **25**, 693–9 (1952).

24. Arnold, A., and Wood, L. A., Private Communication to O.R.R., Feb. 28, 1949.

25. Back, A. L., *Ind. Eng. Chem.*, **39**, 1339–43 (1947).

26. Baker, W. O., Private Communication to O.R.R., Apr. 20, 1945.

27. Baker, W. O., *Bell Labs. Record*, **25**, 447–51 (1947).

28. Baker, W. O., *Ind. Eng. Chem.*, **41**, 511–20 (1949); *Rubber Chem. and Technol.*, **22**, 935–55 (1949).

29. Baker, W. O., and Fuller, C. S., *Ind. Eng. Chem.*, **38**, 272–7 (1946).

30. Baker, W. O., and Heiss, J. H., Private Communication to O.R.R., Mar. 23, 1943.

31. Baker, W. O., and Heiss, J. H., Private Communication to O.R.R., May 23, 1944.

32. Baker, W. O., Mullen, J. W., and Heiss, J. H., Private Communication to O.R.R. Feb. 11, 1944.

33. Baldwin, F. P., *J. Am. Chem. Soc.*, **72**, 1833–4 (1950).

34. Banes, F. W., and Eby, L. T., *Ind. Eng. Chem., Anal Ed.*, **18**, 535–8 (1946); *Rubber Chem. and Technol.*, **20**, 55–62 (1947).

35. Bardwell, J., and Winkler, C. A., *India Rubber World*, **118**, 509–12 (1948); *Rubber Chem. and Technol.*, **22**, 96–104 (1949).

36. Bardwell, J., and Winkler, C. A., *Can. J. Research*, **B27**, 116–27, 128–38, 139–50 (1949).

37. Barry, A. J., *J. Applied Phys.*, **17**, 1020–4 (1946); also Ref. 16, p. 144.

38. Bartlett, P. D., and Swain, C. G., *J. Am. Chem. Soc.*, **67**, 2273–4 (1945).

39. Bass, S. L., *Proc. Second Rubber Technol. Conf.*, Heffer, Cambridge, England, 1948, 17–33.

40. Bawn, C. E. H., *Trans. Faraday Soc.*, **47**, 97–100 (1951).

41. Bawn, C. E. H., Freeman, R. F. J., and Kamaliddin, A. R., *Trans. Faraday Soc.*, **46**, 677–84 (1950).

42. Bawn, C. E. H., Freeman, R. F. J., and Kamaliddin, A. R., *Trans. Faraday Soc.*, **46**, 862–72 (1950).

43. Bawn, C. E. H., Freeman, R. F. J., and Kamaliddin, A. R., *Trans. Faraday Soc.*, **46**, 1107–12 (1950).

44. Bawn, C. E. H., Grimley, T. B., and Wajid, M. A., *Trans. Faraday Soc.*, **46**, 1112–20 (1950).

45. Bekkedahl, N., *J. Research Natl. Bur. Standards*, **13**, 411–31 (1934); *Rubber Chem. and Technol*, **8**, 5–22 (1935).

46. Bekkedahl, N., *Proc. Rubber Technol. Conf.*, Heffer, Cambridge, England, 1938; *Rubber Chem. and Technol.*, **12**, 150–62 (1939).

47. Bekkedahl, N., *J. Research Natl. Bur. Standards*, **43**, 145–56 (1949).

48. Bekkedahl, N., *Anal. Chem.*, **22**, 253–64 (1950).

49. Bekkedahl, N., *Anal. Chem.*, **23**, 243–53 (1951).

50. Bekkedahl, N., and Matheson, H., *J. Research Natl. Bur. Standards*, **15**, 503–15 (1935); *Rubber Chem. and Technol.*, **9**, 264–274 (1936).

51. Bekkedahl, N., and Roth, F. L., Unpublished Observations.

52. Bekkedahl, N., and Scott, R. B., *J. Research Natl. Bur. Standards*, **29**, 87–95 (1942); *Rubber Chem. and Technol.*, **16**, 310–317 (1943).

53. Bekkedahl, N., and Stiehler, R. D., *Anal. Chem.*, **21**, 266–78 (1949).

54. Berueffy, R. R., and Wood, L. A., Private Communication to O.R.R., Oct. 14, 1946.

55. Bestul, A. B., and Belcher, H. V., *J. Colloid Sci.*, **5**, 303–14 (1950).

56. Bestul, A. B., and Belcher, H. V., *J. Applied Phys.*, **24**, 1011–4 (1953).
57. Bestul, A. B., Belcher, H. V., Quinn, F. A., and Bryant, C. B., *J. Phys. Chem.*, **56**, 432–9 (1952).
58. Bestul, A. B., Decker, G. E., and White, H. S., *J. Research Natl. Bur. Standards*, **46**, 283–7 (1951).
59. Beu, K. E., *J. Polymer Sci.*, **3**, 801–3 (1948).
60. Beu, K. E., Reynolds, W. B., Fryling, C. F., and McMurry, H. L., *J. Polymer Sci.*, **3**, 465–79 (1948); *Rubber Chem. and Technol.*, **22**, 356–369 (1949).
61. Borders, A. M., and Juve, R. D., *Ind. Eng. Chem.*, **38**, 1066–70 (1946); *Rubber Chem. and Technol.*, **20**, 515–524 (1947).
62. Bovey, F. A., and Kolthoff, I. M., *Chem. Revs.*, **42**, 491–525 (1948).
63. Bovey, F. A., and Kolthoff, I. M., *J. Polymer Sci.*, **5**, 569–86 (1950).
64. Boyer, R. F., *J. Chem. Phys.*, **13**, 363–72 (1945).
65. Boyer, R. F., and Spencer, R. S., Chap. I in Ref. 10.
66. Boyer, R. F., and Spencer, R. S., *J. Polymer Sci.*, **2**, 157–77 (1947).
67. Brinkman, H. C., *Proc. Amsterdam Acad.*, **50**, No. 6 (1947); *Applied Sci. Research*, **A1**, 27 (1947); *Physica*, **13**, 447 (1947).
68. Buchdahl, R., and Nielsen, L. E., *J. Applied Phys.*, **21**, 482–7 (1950).
69. Campbell, H. N., and Allen, M. D., *Ind. Eng. Chem.*, **43**, 413–5 (1951); *Rubber Chem. and Technol.*, **24**, 550–556 (1951).
70. Carothers, W. H., Williams, I., Collins, A. M., and Kirby, J. E., *J. Am. Chem. Soc.*, **53**, 4203–25 (1931); *Rubber Chem. and Technol.*, **5**, 7 (1932).
71. Carr, C. W., Kolthoff, I. M., Meehan, E. J., and Stenberg, R. J., *J. Polymer Sci.*, **5**, 191–200 (1950).
72. Carr, C. W., Kolthoff, I. M., Meehan, E. J., and Williams, D. E., *J. Polymer Sci.*, **5**, 201–6 (1950).
73. Carter, W. C., Scott, R. L., and Magat, M., *J. Am. Chem. Soc.*, **68**, 1480–3 (1946); *Rubber Chem. and Technol.*, **20**, 78 (1947)
74. Collins, A. M., U.S. Pat. 2,264,173, Nov. 25, 1941.
74a. Condon, F. E., *J. Polymer Sci.*, **11**, 139–49 (1953).
75. Corrin, M. L., *J. Polymer Sci.*, **2**, 257–62 (1947).
76. Cragg, L. H., *J. Colloid Sci.*, **1**, 261–9 (1946); *Rubber Chem. and Technol.*, **19**, 1092–8 (1946).
76a. Cragg, L. H., and Brown, A. T., *Canadian J. Chem.*, **30**, 1033–43 (1952); *Rubber Chem. and Technol.*, **26**, 764–74 (1953).
77. Cragg, L. H., Faichney, L. M., and Olds, H. F., *Can. J. Research*, **B26**, 551–63 (1948).
77a. Cragg, L. H., and Fern, G. R. H., *J. Polymer Sci.*, **10**, 185–99 (1953); *Rubber Chem. and Technol.*, **26**, 775–86 (1953).
78. Cragg, L. H., and Hammerschlag, H., *Chem. Revs.*, **39**, 79–135 (1946).
79. Cragg, L. H., and Rogers, T. M., *Can. J. Research*, **B26**, 230–47 (1948).
80. Cragg, L. H., Rogers, T. M., and Henderson, D. A., *Can. J. Research*, **B25**, 333–50 (1947).
81. Cragg, L. H., and Simkins, J. E., *Can. J. Research*, **B27**, 961–71 (1949).
82. Craig, D., U.S. Pat. 2,362,052 (1944).
83. Crippen, R. C., and Bonilla, C. F., *Anal. Chem.*, **21**, 927–30 (1949).
84. Dauphinee, T. M., Ivey, D. G., and Smith, H. D., *Can. J. Research*, **28A**, 596–615 (1950).
85. Debye, P., *J. Applied Phys.*, **15**, 338–42 (1944).
86. Debye, P., *J. Chem. Phys.*, **14**, 636–9 (1946).
87. Debye, P., *J. Phys. and Colloid Chem.*, **51**, 18–32 (1947).
88. Debye, P. and Bueche, *J. Chem. Phys.*, **16**, 573–9 (1948).
89. D'Ianni, J. D., *Ind. Eng. Chem.*, **40**, 253–6 (1948); *Rubber Chem. and Technol.*, **21**, 596–604 (1948).
90. D'Ianni, J. D., Hess, L. D., and Mast, W. C., *Ind. Eng. Chem.*, **43**, 319–24 (1951); *Rubber Chem. and Technol.*, **24**, 697–708 (1951).
91. D'Ianni, J. D., Naples, F. J., and Field, J. E., *Ind. Eng. Chem.*, **42**, 95–102 (1950).

92. Dienes, G. J., and Klemm, H. F., *J. Applied Phys.*, **17**, 458–71 (1946).
93. Dinsmore, H. L., and Smith, D. C., *Anal. Chem.*, **20**, 11–24 (1948); *Rubber Chem. and Technol.*, **22**, 572 (1949).
94. Dogadkin, B. A., and Gul, V. E., *Repts. Acad. Sci. U.S.S.R.*, **70**, 1017–9 (1950); *Rubber Chem. and Technol.*, **24**, 140–3 (1951).
95. Dogadkin, B. A., Soboleva, I., and Arkhangel'skaya, M., *Kolloidnii Zhurnal*, **11**, 143–50 (1949); *Rubber Chem. and Technol.*, **23**, 89–97 (1950).
96. Doty, P., and Zable, H. S., *J. Polymer Sci.*, **1**, 90–101 (1946); Ref. 16, 440–56.
97. Doty, P. M., Zimm, B. H., and Mark, H., *J. Chem. Phys*, **12**, 144–5 (1944).
98. Doty, P. M., Zimm, B. H., and Mark, H., *J. Chem. Phys.*, **13**, 159–66 (1945).
99. Duke, J., Private Communication to O.R.R., July 11, 1950.
100. Dunbrook, R. F., *India Rubber World*, **117**, 203–7, 355–9, 486, 552, 617–9, 745–8 (1947–48).
101. Eirich, F., and Riseman, J., *J. Polymer Sci.*, **4**, 417–34 (1949).
102. Embree, W. H., Mitchell, J. W., and Williams, H. L., *Can. J. Chem.*, **29**, 253–69 (1951).
103. Evans, A. G., and Meadows, G. W., *J. Polymer Sci.*, **4**, 359–76 (1949).
104. Evans, R. D., Mighton, H. R., and Flory, P. J., *J. Chem. Phys.*, **15**, 685 (1947).
105. Evans, R. D., Mighton, H. R., and Flory, P. J., *J. Am. Chem. Soc.*, **72**, 2018–28 (1950).
106. Ewart, R. H., chapter in Ref. 10.
107. Ewart, R. H., and co-workers, quoted in Ref. 312; *Rubber Chem. and Technol.*, **22**, 660–6 (1949).
108. Ewart, R. H., Roe, C. P., Debye, P., and McCartney, J. R., *J. Chem. Phys.*, **14**, 687–95 (1946).
109. Ferry, J. D., Fitzgerald, E. R., Johnson, M. F., and Grandine, L. D., *J. Applied Phys.*, **22**, 717–22 (1951).
110. Ferry, J. D., and Parks, G. S., *J. Chem. Phys.*, **4**, 70–5 (1936).
111. Fettes, E. M., and Jorczak, J. S., *Ind. Eng. Chem.*, **42**, 2217–23 (1950); *Rubber Chem. and Technol.*, **24**, 709–723 (1951).
112. Field, J. E., Woodford, D. E., and Gehman, S. D., *J. Applied Phys.*, **17**, 386–92 (1946); also Ref. 16; *Rubber Chem. and Technol.*, **19**, 1113–1123 (1946).
113. Firestone Tire & Rubber Co., Private Communication to O.R.R., Sept. 10, 1948.
114. Firestone Tire & Rubber Co., Private Communication to O.R.R., Apr. 10, 1949.
115. Firestone Tire & Rubber Co., Private Communication to O.R.R., July 15, 1949.
116. Firestone Tire & Rubber Co., Private Communication to O.R.R., Aug. 15, 1949.
117. Firestone Tire & Rubber Co., Private Communication to O.R.R., Sept. 10, 1949.
118. Firestone Tire & Rubber Co., Private Communication to O.R.R., Oct. 1949.
119. Firestone Tire & Rubber Co., Private Communication to O.R.R., Feb. 15, 1951.
120. Firestone Tire & Rubber Co., Private Communication to O.R.R., May 15, 1951.
121. Flory, P. J., *J. Am. Chem. Soc.*, **63**, 3083–3100 (1941).
122. Flory, P. J., *J. Phys. Chem.*, **46**, 132–40 (1942).
123. Flory, P. J., *J. Chem. Phys.*, **10**, 51–61 (1942).
124. Flory, P. J., *J. Amer. Chem. Soc.*, **65**, 372–82 (1943); *Rubber Chem. and Technol.*, **16**, 493 (1943).
125. Flory, P. J., *Chem. Revs.*, **35**, 51–75 (1944).
126. Flory, P. J., *J. Chem. Phys.*, **13**, 453–65 (1945).
127. Flory, P. J., *Chem Revs.*, **39**, 137–97 (1946); Ref. 4.
128. Flory, P. J., *Ind. Eng. Chem.*, **38**, 417–36 (1946); *Rubber Chem. and Technol.*, **19**, 552–598 (1946).
129. Flory, P. J., *J. Am. Chem. Soc.*, **69**, 30–35 (1947).
130. Flory, P. J., *J. Am. Chem. Soc.*, **69**, 2893–9 (1947); *Rubber Chem. and Technol.*, **21**, 461–470 (1948).
131. Flory, P. J., *J. Polymer Sci.*, **2**, 36–40 (1947).
132. Flory, P. J., *J. Chem. Phys.*, **15**, 397–408, 684 (1947).
133. Flory, P. J., *J. Chem. Phys.*, **17**, 223–40 (1949).

134. Flory, P. J., *J. Chem. Phys.*, **17**, 303–10 (1949).

135. Flory, P. J., and Fox, T. G., *J. Am. Chem. Soc.*, **73**, 1904–8 (1951).

136. Flory, P. J., Rabjohn, N., and Shaffer, M. C., *J. Polymer Sci.*, **4**, 225–44 (1949); *Rubber Chem. and Technol.*, **23**, 9–26 (1950).

137. Flory, P. J., Rabjohn, N., and Shaffer, M. C., *J. Polymer Sci.*, **4**, 435–53 (1949); *Rubber Chem. and Technol.*, **23**, 27–43 (1950).

138. Flory, P. J., and Rehner, J., *J. Chem. Phys.*, **11**, 512–20 (1943).

139. Flory, P. J., and Rehner, J., *J. Chem. Phys.*, **11**, 521–6 (1943).

140. Flory, P. J., and Rehner, J., *J. Chem. Phys.*, **12**, 412–4 (1944).

141. Fordham, J. W. L., O'Neill, A. N., and Williams, H. L., *Can. J. Research*, **F27,** 119–42 (1949).

141a. Foster, F. C., and Binder, J. L., *J. Am. Chem. Soc.*, **75**, 2910–3 (1953); *Rubber Chem. and Technol.*, **26**, 832–9 (1953).

142. Fournier, H., and Trillat, J. J., *Compt. rend.*, **219**, 447–8 (1944); *Rubber Chem. and Technol.*, **19**, 1088–9 (1946).

143. Fox, T. G., and Flory, P. J., *J. Am. Chem. Soc.*, **70**, 2384–95 (1948).

144. Fox, T. G., and Flory, P. J., *J. Phys. and Colloid Chem.*, **53**, 197–212 (1949).

145. Fox, T. G., and Flory, P. J., *J. Applied Phys.*, **21**, 581–91 (1950).

146. Fox, T. G., and Flory, P. J., *J. Am. Chem. Soc.*, **73**, 1909–15 (1951).

147. Fox, T. G., and Flory, P. J., *J. Am. Chem. Soc.*, **73**, 1915–20 (1951).

148. Fox, T. G., and Flory, P. J., *J. Phys. and Colloid Chem.*, **55**, 221–34 (1951).

149. Fox, T. G., Flory, P. J., and Marshall, R. E., *J. Chem. Phys.*, **17**, 704–6 (1949); *Rubber Chem. and Technol.*, **23**, 576–580 (1950).

150. Fox, T. G., Fox, J. C., and Flory, P. J., *J. Am. Chem. Soc.*, **73**, 1901–4 (1951).

151. French, D. M., and Ewart, R. H., *Anal. Chem.*, **19**, 165–7 (1947); *Rubber Chem. and Technol.*, **20**, 984–989 (1947).

152. Fryling, C. F., *Ind. Eng. Chem.*, **40**, 928–32 (1948).

153. Fryling, C. F., Landes, S. H., St. John, W. M., and Uraneck, C. A., *Ind. Eng. Chem.*, **41**, 986–91 (1949).

154. Fuller, C. S., *Bell System Tech. J.*, **25**, 351–84 (1946).

155. Furukawa, G. T., National Bureau of Standards, Unpublished Work, 1950.

156. Furukawa, G. T., McCoskey, R. E., and King, G. J., *J. Research Natl. Bur. Standards*, **50**, 357–65 (1953); **51**, 321–6 (1953).

157. Garten, V., and Becker, W., *Makromol. Chem.*, **3**, 78–110 (1949).

158. Gee, G., *Trans. Instn. Rubber Ind.*, **18**, 266–81 (1943); *Rubber Chem. and Technol.*, **16**, 818–833 (1943).

159. Gee, G., chapter in Ref. 10.

160. Gee, G., *Trans. Faraday Soc.*, **42**, 585–98 (1946); *Rubber Chem. and Technol.*, **20**, 442–456 (1947).

161. Gee, G., *Quart. Rev., Chem. Soc.*, **1**, 265–98 (1947); *Rubber Chem. and Technol.*, **21**, 564–95 (1948).

162. Gee, G., *J. Polymer Sci.*, **2**, 451–62 (1947); *Rubber Chem. and Technol.*, **21**, 301 (1948).

163. Gehman, S. D., Jones, P. J., Wilkinson, C. S., and Woodford, D. E., *Ind. Eng. Chem.*, **42**, 475–82 (1950); *Rubber Chem. and Technol.*, **23**, 770–785 (1950).

164. Gehman, S. D., Woodford, D. E., and Wilkinson, C. E., *Ind. Eng. Chem.*, **39**, 1108–15 (1947); *Rubber Chem. and Technol.*, **21**, 94–111 (1948).

165. Goodrich Chem. Co., B. F., Private Communication to O.R.R., Aug. 1949.

166. Hadow, H. J., Sheffer, H., and Hyde, J. C., *Can. J. Research*, **B27**, 791–806 (1949).

167. Hamill, W. H., Mrowca, B. A., and Anthony, R. L., *Ind. Eng. Chem.*, **38**, 106–10 (1946); *Rubber Chem. and Technol.*, **19**, 622–631 (1946).

168. Hampton, R. R., *Anal. Chem.*, **21**, 923–6 (1949).
169. Hanson, E. E., and Halverson, G., *J. Am. Chem. Soc.*, **70**, 779–83 (1948); *Rubber Chem. and Technol.*, **21**, 627–638 (1948).
170. Harkins, W. D., *J. Chem. Phys.*, **14**, 47–8 (1946).
171. Harkins, W. D., *J. Am. Chem. Soc.*, **69**, 1428–44 (1947).
172. Harkins, W. D., *J. Polymer Sci.*, **5**, 217–52 (1950).
173. Harkins, W. D., and Stearns, R. S., *J. Chem. Phys.*, **14**, 215–6 (1945).
174. Harris, W. E., and Kolthoff, I. M., Private Communication to O.R.R., July 16, 1945.
175. Harris, W. E., and Kolthoff, I. M., *J. Polymer Sci.*, **2**, 72–81 (1947).
176. Harris, W. E., and Kolthoff, I. M., *J. Polymer Sci.*, **2**, 82–9 (1947).
177. Hart, E. J., and Meyer, A. W., *J. Am. Chem. Soc.*, **71**, 1980–5 (1949); *Rubber Chem. and Technol.*, **23**, 98–106 (1950).
178. Heiligmann, R. G., *J. Polymer Sci.*, **4**, 183–202 (1949).
179. Heller, W., and Oppenheimer, H., *J. Colloid Sci.*, **3**, 33–43 (1948); *Rubber Chem. and Technol.*, **21**, 790–798 (1948).
180. Heller, W., and Thompson, A. C., *J. Colloid Sci.*, **6**, 57–74 (1951).
181. Henderson, D. A., and Legge, N. R., *Can. J. Research*, **B27**, 666–81 (1949).
182. Hohenstein, W. P., and Mark, H., chapter 1 in Ref. 3.
183. Holdsworth, R. S., Private Communication to O.R.R., Apr. 9, 1951.
184. Holroyd, L. V., Codrington, R. S., Mrowca, B. A., and Guth, E., *J. Applied Phys.*, **22**, 696–705 (1951); *Rubber Chem. and Technol.*, **25**, 767–83 (1952).
185. Huddleston, G. R., Private Communication to O.R.R., June 21, 1950.
186. Huggins, M. L., *J. Am. Chem. Soc.*, **64**, 1712–9 (1942).
187. Huggins, M. L., *J. Am. Chem. Soc.*, **64**, 2716–8 (1942).
188. Huggins, M. L., *J. Phys. Chem.*, **46**, 151–8 (1942).
189. Hulse, G. E., Hart, E. J., French, D. M., and Matheson, M. S., Private Communication to O.R.R., June, 1943.
190. James, H. M., *Am. Assoc. Advanc. Sci. Centennial 1950*, 263–9.
191. Johnson, B. L., Private Communication to O.R.R., Jan. 13, 1943.
192. Johnson, B. L., *Ind. Eng. Chem.*, **40**, 351–6 (1948); *Rubber Chem. and Technol.*, **21**, 654 (1948).
193. Johnson, B. L., and Wolfangel, R. D., *Ind. Eng. Chem.*, **41**, 1580–4 (1949).
194. Johnson, B. L., and Wolfangel, R. D., *Ind. Eng. Chem.*, **44**, 752–6 (1952).
195. Johnson, P. H., Private Communication to O.R.R., June 14, 1948.
196. Johnson, P. H., and Bebb, R. L., *Ind. Eng. Chem.*, **41**, 1577–80 (1949).
197. Jones, H. C., *Ind. Eng. Chem.*, **32**, 331–4 (1940); *Rubber Chem. and Technol.*, **13**, 649–654 (1940).
198. Kauzmann, W., *Chem. Revs.*, **43**, 219–56 (1948).
199. Kemp, A. R., and Peters, H., *Ind. Eng. Chem.*, **33**, 1263–9 (1941); *Rubber Chem. and Technol.*, **15**, 60–71 (1942).
200. Kemp, A. R., and Peters, H., *Ind. Eng. Chem.*, **34**, 1097–1102 (1942); *Rubber Chem. and Technol.*, **16**, 58–68 (1943).
201. Kemp, A. R., and Peters, H., *Ind. Eng. Chem.*, **34**, 1192–9 (1942); *Rubber Chem. and Technol.*, **16**, 69–84 (1943).
202. Kemp, A. R., and Peters, H., *Ind. Eng. Chem., Anal. Ed.*, **15**, 453–9 (1943); *Rubber Chem. and Technol.*, **17**, 61–75 (1944).
203. Kemp, A. R., and Straitiff, W. G., *Ind. Eng. Chem.*, **36**, 707–15 (1944); *Rubber Chem. and Technol.*, **18**, 41–61 (1945).
204. Kinell, P. O., *J. Phys. and Colloid Chem.*, **51**, 70–9 (1947).
205. King, G. W., Hainer, R. M., and McMahon, H. O., *J. Applied Phys.*, **20**, 559–63 (1949).
206. Kirkwood, J. G., and Riseman, J., *J. Chem. Phys.*, **16**, 565–73 (1948).

207. Klebanskii, A. L., and Vasil'eva, W. G., *J. prakt. Chem.*, **144,** 251–64 (1936); *Rubber Chem. and Technol.*, **10,** 126–34 (1937).
208. Kolthoff, I. M., and Lee, T. S., *J. Polymer Sci.*, **2,** 206–19 (1947).
209. Kolthoff, I. M., Lee, T. S., and Carr, C. W., *J. Polymer Sci.*, **1,** 429–33 (1946); *Rubber Chem. and Technol,* **20,** 546 (1947).
210. Kolthoff, I. M., Lee, T. S., and Mairs, M. A., *J. Polymer Sci.*, **2,** 220–8 (1947).
211. Kraemer, E. O., and Nichols, J. B., in Svedberg, T., and Pedersen, K. O., *The Ultracentrifuge*, Oxford Univ. Press, London, 1940, pp. 416–31.
212. Krimm, S., and Tobolsky, A. V., *J. Polymer Sci.*, **6,** 667–8 (1951).
213. Kuhn, W., *Kolloid-Z.*, **68,** 2–15 (1934).
214. Kunst, E. D., *Rec. trav. chim.*, **69,** 125–40 (1950); Ref. 14, 145–60.
215. Leaderman, H. and Smith, R. G., *Phys. Rev.*, **81,** 303 (1951) (Abstract).
216. Leaderman, H., Smith, R. G., and Jones, R. W., Private Communication to Office of Naval Research, Dec. 1951; *J. Polymer Sci.*, **13,** 371–84 (1954).
217. Lee, T. S., Kolthoff, I. M., and Mairs, M. A., *J. Polymer Sci.*, **3,** 66–84 (1948); *Rubber Chem. and Technol.*, **21,** 835–852 (1948).
218. Lewis, F. M., Walling, C., Cummings, W., Briggs, E. R., and Wenisch, W. J., *J. Am. Chem. Soc.*, **70,** 1527–9 (1948).
219. Linnig, F. J., and co-workers, Unpublished Data.
220. Lucas, V. E., Johnson, P. H., Wakefield, L. B., and Johnson, B. L., *Ind. Eng. Chem.*, **41,** 1629–32 (1949).
221. Lundstedt, O. W., and Hampton, R. R., Private Communication to O.R.R., June 7, 1944.
222. MacLean, D. B., Morton, M., and Nicholls, R. V. V., *Ind. Eng. Chem.*, **41,** 1622–5 (1949).
223. Madorsky, I., and Wood, L. A., Private Communications to O.R.R., Sept. 13, Nov. 30, 1944, and Jan. 1, 1945.
224. Madorsky, I., and Wood, L. A., Unpublished Rept., 1950.
225. Madorsky, S. L., *Science*, **111,** 360 (1950).
226. Madorsky, S. L., and Strauss, S., *Ind. Eng. Chem.*, **40,** 848–52 (1948); *J. Research Natl. Bur. Standards*, **40,** 417–25 (1948); Madorsky, S. L., *J. Polymer Sci.*, **9,** 133–56 (1952).
227. Madorsky, S. L., Strauss, S., Thompson, D., and Williamson, L., *J. Polymer Sci.*, **4,** 639–64 (1949); *J. Research Natl. Bur. Standards*, **42,** 499–514 (1949); Strauss, S., and Madorsky, S. L., *J. Research Natl. Bur. Standards*, **50,** 165–76 (1953).
228. Maher, E. D., and Davies, T. L., *Rubber Age N.Y.*, **59,** 557–62 (1946).
229. Mann, J., *J. Rubber Research*, **18,** 79–88, 89–97 (1949).
230. Mark, H., *Die Feste Korper*, Hirzel, Leipzig, 1938, p. 103.
231. Maron, S. H., and Ulevitch, I. N., Private Communication to O.R.R., Oct. 5, 1950; Maron, S. H., Moore, C., and Ulevitch, I. N., Private Communication to O.R.R., Feb. 26, 1952.
232. Marvel, C. S., Bailey, W. J., and Inskeep, G. E., *J. Polymer Sci.*, **1,** 275–88 (1946); *Rubber Chem. and Technol.*, **20,** 1–13 (1947).
233. Marvel, C. S., Gilkey, R., Morgan, C. R., Noth, J. F., Rands, R. D., and Young, C. H., *J. Polymer Sci.*, **6,** 483–502 (1951).
234. Marvel, C. S., and Lewis, C. D., *J. Polymer Sci.*, **3,** 354–7 (1948).
235. Mayo, F. R., and Walling, C., *Chem. Revs.*, **46,** 191–287 (1950). Cf. T. Alfrey, J. J. Bohrer, and H. Mark, *Copolymerization*, Interscience, New York, 1952.
236. Mayo, L. R., *Ind. Eng. Chem.*, **42,** 696–700 (1950).
237. Medalia, A. I., *J. Polymer Sci.*, **6,** 423–31 (1951).
238. Medalia, A. I., and Kolthoff, I. M., *J. Polymer Sci.*, **6,** 433–55 (1951).
239. Meehan, E. J., *J. Polymer Sci.*, **1,** 175–82 (1946); *Rubber Chem. and Technol.*, **19,** 1077 (1946).

240. Meehan, E. J., *J. Polymer Sci.*, **1**, 318–28 (1946); *Rubber Chem. and Technol.*, **20**, 14 (1947).
241. Meehan, E. J., *J. Am. Chem. Soc.*, **71**, 628–33 (1949).
242. Meehan, E. J., *J. Polymer Sci.*, **6**, 255–60 (1951).
243. Meehan, E. J., Parks, T. D., and Laitinen, H. A., *J. Polymer Sci.*, **1**, 247–8 (1946); *Rubber Chem. and Technol.*, **20**, 313 (1947).
244. Mellon Inst. Ind. Research, Private Communication to O.R.R., Apr. 15, 1951; Richardson, W. S., and Sacher, A., *J. Polymer Sci.*, **10**, 353–70 (1953).
245. Merz, E. H., Nielsen, L. E., and Buchdahl, R., *Ind. Eng. Chem.*, **43**, 1396–401 (1951).
246. Meuser, L., Stiehler, R. D., and Hackett, R. W., "Symposium on Rubber Testing," Special Tech. Pub. No. 74, Am. Soc. Testing Materials, 1947; *India Rubber World*, **117**, 57–61 (1947).
247. Meyer, A. W., *Ind. Eng. Chem.*, **41**, 1570–7 (1949).
248. Meyer, A. W., Hampton, R. R., and Davison, J. A., *J. Am. Chem. Soc.*, **74**, 2294–6 (1952); *Rubber Chem. and Technol.*, **26**, 522–7 (1953).
249. Mitchell, J. M., and Williams, H. L., *Can. J. Research*, **F27**, 35–46 (1949).
250. Mochel, W. E., and Hall, M. B., *J. Am. Chem. Soc.*, **71**, 4082–8 (1949).
251. Mochel, W. E., and Nichols, J. B., *J. Am. Chem. Soc.*, **71**, 3435–8 (1949).
252. Mochel, W. E., and Nichols, J. B., *Ind. Eng. Chem.*, **43**, 154–7 (1951).
253. Mochel, W. E., Nichols, J. B., and Mighton, C. J., *J. Am. Chem. Soc.*, **70**, 2185–90 (1948); *Rubber Chem. and Technol.*, **22**, 680 (1949).
254. Mochel, W. E., and Peterson, J. H., *J. Am. Chem. Soc.*, **71**, 1426–32 (1949); *Rubber Chem. and Technol.*, **22**, 1092–1102 (1949).
255. Morton, A. A., *Ind. Eng. Chem.*, **42**, 1488–96 (1950); *Rubber Chem. and Technol.*, **24**, 35–53 (1951).
256. Morton, M., and Nicholls, R. V. V., *Can. J. Research*, **25B**, 159–182 (1947).
257. Morton, M., and Salatiello, P. P., *J. Polymer Sci.*, **6**, 225–37 (1951).
258. Morton, M., Salatiello, P. P., and Landfield, H., *J. Polymer Sci.*, **8**, 111–21, 215–24, 279–87 (1952).
259. Nat. Bur. Standards, Private Communication to O.R.R., May 15, 1950.
260. Nat. Bur. Standards, Private Communication to O.R.R., Aug. 14, 1950.
261. Nat. Bur. Standards, Private Communication to O.R.R., Dec. 15, 1950.
262. Nat. Bur. Standards, Data given in *Styrene*, edited by R. H. Boundy and R. F. Boyer, Reinhold, New York, 1952.
263. Nat. Bur. Standards, Private Communication to O.R.R., Jan. 15, 1951.
264. Nat. Bur. Standards, Private Communication to O.R.R., Apr. 15, 1951.
265. Nat. Bur. Standards, Unpublished Data.
266. Nelson, J. F., and Vanderbilt, B. M., *Proc. Second Rubber Technol. Conf.*, London, 1948, 49–60.
267. Nelson, R. S., Jessup, R. S., and Roberts, D. E., *J. Research Natl. Bur. Standards*, **48**, 275–80 (1952).
268. Nichols, J. B., and Bailey, E. D., Determinations with the Ultracentrifuge, in *Physical Methods of Organic Chemistry*, edited by A. Weissberger, 2d. ed., Interscience, New York, 1949, pp. 621–730.
269. Nielsen, L. E., Buchdahl, R., and Claver, G. C., *Ind. Eng. Chem.*, **43**, 341–5 (1951); *Rubber Chem. and Technol.*, **24**, 574–584 (1951).
270. Orr, R. J., and Williams, H. L., *Can. J. Technol.*, **29**, 29–42 (1951).
271. Orr, R. J., and Williams, H. L., *Can. J. Chem.*, **29**, 270–83 (1951).
272. Outer, P., Carr, C. I., and Zimm, B. H., *J. Chem. Phys.*, **18**, 830–9 (1950).
273. Parks, G. S., *J. Chem. Phys.*, **4**, 459 (1936); *Rubber Chem. and Technol.*, **10**, 135–136 (1937).
274. Parks, G. S., and Mosley, J. R., *J. Chem. Phys.*, **17**, 691–4 (1949).
275. Patnode, W., and Scheiber, W. J., *J. Am. Chem. Soc.*, **61**, 3449–51 (1939).

276. Pitzer, K. S., Guttman, L., and Westrum, E. F., *J. Am. Chem. Soc.*, **68**, 2209–12 (1946).
277. Price, C. C., and Adams, C. E., *J. Am. Chem. Soc.*, **67**, 1674–80 (1945).
278. Prosen, E. J., and Rossini, F. D., *J. Research Natl. Bur. Standards*, **34**, 59–63 (1945).
279. Rabjohn, N., Bryan, C. E., Inskeep, G. E., Johnson, H. W., and Lawson, J. K., *J. Am. Chem. Soc.*, **69**, 314–9 (1947); *Rubber Chem. and Technol.*, **20**, 916–926 (1947).
280. Rabjohn, N., Dearborn, R. J., Blackburn, W. E., Inskeep, G. E., Snyder, H. R., and Marvel, C. S., *J. Polymer Sci.*, **2**, 488–502 (1947).
281. Rands, R. D., Ferguson, W. J., and Prather, J. L., *J. Research Natl. Bur. Standards*, **33**, 63–70 (1944).
282. Rasmussen, R. S., and Brattain, R. R., Private Communication to O.R.R., May 1945.
283. Rehner, J., *Ind. Eng. Chem.*, **36**, 46–51 (1944); *Rubber Chem. and Technol.*, **17**, 346 (1944).
284. Rehner, J., *J. Polymer Sci.*, **1**, 225–8 (1946).
285. Rehner, J., *J. Polymer Sci.*, **2**, 263–74 (1947); *Rubber Chem. and Technol.*, **21**, 82–93 (1948).
286. Rehner, J., and Gray, P., *Ind. Eng. Chem., Anal. Ed.*, **17**, 367–70 (1945); *Rubber Chem. and Technol.*, **18**, 887–895 (1945).
287. Richards, R. E., and Thompson, H. W., *J. Chem. Soc.*, **1949**, 124–32.
288. Roberts, D. E., *J. Research Natl. Bur. Standards*, **44**, 221–32 (1950).
289. Roberts, D. E., Walton, W. W., and Jessup, R. S., *J. Polymer Sci.*, **2**, 420–31 (1947); *J. Research Natl. Bur. Standards*, **38**, 627–35 (1947).
290. Rochow, E. G., and Rochow, T. G., *J. Phys. and Colloid Chem.*, **55**, 9–16 (1951).
291. Rochow, T. G., and Rochow, E. G., *Science*, **111**, 271–5 (1950).
292. Rogers, T. M., Henderson, D. A., and Cragg, L. H., *Can. J. Research*, **B25**, 351–6 (1947).
293. Roth, W. A., Wirths, G., and Berendt, H., *Kautschuk*, **17**, 31–33 (1941); *Rubber Chem. and Technol.*, **15**, 874–8 (1942).
294. Rothman, S., Simha, R., and Weissberg, S. G., *J. Polymer Sci.*, **5**, 141–2 (1950).
295. Ruhemann, M., and Simon, F., *Zeit. phys. Chem.*, **138A**, 1–20 (1928).
296. Saffer, A., and Johnson, B. L., *Ind. Eng. Chem.*, **40**, 538–41 (1948); *Rubber Chem. and Technol.*, **21**, 821–829 (1948).
297. Sands, G. D., and Johnson, B. L., *Ind. Eng. Chem., Anal. Ed.*, **19**, 261–4 (1947).
298. Saunders, D. W., and Treloar, L. R. G., *Trans. Instn. Rubber Ind.*, **24**, 92–100 (1948); *Rubber Chem. and Technol.*, **22**, 333–341 (1949).
299. Schick, M. J., Doty, P., and Zimm, B. H., *J. Am. Chem. Soc.*, **72**, 530–4 (1950).
300. Schmidt, E., and Kelsey, R. H., *Ind. Eng. Chem.*, **43**, 406–12 (1951).
301. Schoene, D. L., Green, A. J., Burns, E. R., and Vila, G. R., *Ind. Eng. Chem.*, **38**, 1246–9 (1946).
302. Schremp, F. W., Ferry, J. D., and Evans, W. W., *J. Applied Phys.*, **22**, 711–7 (1951).
303. Schulze, W. A., and Crouch, W. W., *J. Am. Chem. Soc.*, **70**, 3891–3 (1948).
304. Schulze, W. A., and Crouch, W. W., *Ind. Eng. Chem.*, **40**, 151–4 (1948).
305. Schulze, W. A., Tucker, C. M., and Crouch, W. W., *Ind. Eng. Chem.*, **41**, 1599–603 (1949).
306. Scott, D. W., *J. Am. Chem. Soc.*, **68**, 1877–9 (1946).
307. Scott, R. B., Meyers, C. H., Rands, R. D., Brickwedde, F. G., and Bekkedahl, N., *J. Research Natl. Bur. Standards*, **35**, 39–85 (1945).
308. Scott, R. L., Thesis, Princeton University, 1945, quoted in Ref. 9.
309. Scott, R. L., *J. Chem. Phys.*, **13**, 178–87 (1945).
310. Scott, R. L., *J. Chem. Phys.*, **17**, 268–79 (1949).
311. Scott, R. L., *J. Chem. Phys.*, **17**, 279–84 (1949).
312. Scott, R. L., Carter, W. C., and Magat, M., *J. Am. Chem. Soc.*, **71**, 220–3 (1949); *Rubber Chem. and Technol.*, **22**, 660 (1949).
313. Scott, R. L., and Magat, M., *J. Chem. Phys.*, **13**, 172–7 (1945).
314. Scott, R. L., and Magat, M., *J. Polymer Sci.*, **4**, 555–71 (1949).

315. Semon, W. L., *Chem. Eng. News*, **24,** 2900–5 (1946); *India Rubber World*, **115,** 364–9 (1946).
316. Serniuk, G. E., Banes, F. W., and Swaney, M. A., *J. Am. Chem. Soc.*, **70,** 1804–8 (1948); *Rubber Chem. and Technol.*, **22,** 148–154 (1949).
317. Seymour, R. B., *Ind. Eng. Chem.*, **40,** 524–7 (1948).
318. Simha, R., *J. Chem. Phys.*, **13,** 188–95 (1945).
319. Simha, R., *J. Research Natl. Bur. Standards*, **42,** 409–18 (1949).
320. Smith, O. H., Hermonat, W. A., Haxo, H. E., and Meyer, A. W., *Anal. Chem.*, **23,** 322–7 (1951); *Rubber Chem. and Technol.*, **24,** 684–696 (1951).
321. Smith, W. V., *J. Am. Chem. Soc.*, **68,** 2059–64 (1946).
322. Smith, W. V., *J. Am. Chem. Soc.*, **68,** 2064–9 (1946).
323. Smith, W. V., *J. Am. Chem. Soc.*, **68,** 2069–71 (1946).
324. Smith, W. V., *J. Am. Chem. Soc.*, **70,** 3695–3702 (1948).
325. Smith, W. V., *J. Am. Chem. Soc.*, **71,** 4077–82 (1949).
326. Smith, W. V., and Ewart, R. H., *J. Chem. Phys.*, **16,** 592–9 (1948).
327. Snyder, H. R., Stewart, J. M., Allen, R. E., and Dearborn, R. J., *J. Am. Chem. Soc.*, **68,** 1422–8 (1946); *Rubber Chem. and Technol.*, **20,** 29–40 (1947).
328. Soday, F. J., *Trans. Am. Inst. Chem. Engrs.*, **42,** 647–64 (1946).
329. Spencer, R. S., *J. Colloid Sci.*, **4,** 229–40 (1949).
330. Spencer, R. S., and Dillon, R. E., *J. Colloid Sci.*, **3,** 163–80 (1948).
331. Spencer, R. S., and Dillon, R. E., *J. Colloid Sci.*, **4,** 241–51 (1949).
332. Staudinger, H., and Fischer, K., *J. prakt. Chem.*, **157,** 158–76 (1941); *Rubber Chem. and Technol.*, **15,** 523–34 (1942).
333. Stiehler, R. D., and Hackett, R. W., *Anal. Chem.*, **20,** 292–6 (1948).
334. Svedberg, T., and Kinell, P. O., *Harald Nordenson 60th Anniv. Vol.*, 321–39 (1946); *Rubber Chem. and Technol.*, **21,** 436–51 (1948).
335. Svetlik, J. F., and Sperberg, L. R., *India Rubber World*, **124,** 182–7 (1951); *Rubber Chem. and Technol.*, **25,** 140–151 (1952).
336. Swain, C. G., and Bartlett, P. D., *J. Am. Chem. Soc.*, **68,** 2381–6 (1946).
337. Swaney, M. W., and Baldwin, F. P., Private Communication to O.R.R., Aug. 28, 1943.
338. Swaney, M. W., and White, J. U., Private Communication to O.R.R., Sept. 1943.
339. Sweitzer, C. W., Goodrich, W. C., and Burgess, K. A., *Rubber Age N.Y.*, **65,** 651–62 (1949).
340. Tager, A., and Sanatina, V., *Kolloid Zhur.*, **12,** 474–77 (1950); *Rubber Chem. and Technol.*, **24,** 773–776 (1951).
341. Talaly, A., and Magat, M., *Synthetic Rubber from Alcohol*, Interscience, New York, 1945, pp. 189–97.
342. Thomas, R. M., Lightbown, I. E., Sparks, W. J., Frolich, P. K., and Murphree, E. V., *Ind. Eng. Chem.*, **32,** 1283–92 (1940).
343. Thomas, R. M., Sparks, W. J., Frolich, P. K., Otto, M., and Mueller-Cunradi, N., *J. Am. Chem. Soc.*, **62,** 276–80 (1940).
344. Thompson, H. W., and Torkington, P., *Trans. Faraday Soc.*, **41,** 246–60 (1945); *Rubber Chem. and Technol.*, **19,** 46–62 (1946).
345. Treloar, L. R. G., *Trans. Instn. Rubber Ind.*, **25,** 167–71 (1949); *Rubber Chem. and Technol.*, **23,** 347–351 (1950).
346. Treloar, L. R. G., *Trans. Faraday Soc.*, **46,** 783–9 (1950).
347. Trementozzi, Q. A., *J. Phys. and Colloid Chem.*, **54,** 1227–39 (1950).
348. Trillat, J. J., and Tertian, R., *Compt. rend.*, **219,** 395–7 (1944); *Rubber Chem. and Technol.*, **19,** 1090–1 (1946).
349. Treuman, W. B., and Wall, F. T., *Anal. Chem.*, **21,** 1161–5 (1949).
350. Troyan, J. E., *Rubber Age N.Y.*, **63,** 585–95 (1948); *Rubber Chem. and Technol.*, **22,** 405–426 (1949).

351. Tryon, M., and Arnold, A., Unpublished Data.
352. Tryon, M., and Mandel, J., Private Communication to O.R.R., Apr. 15, 1949.
353. Tunnicliff, D. D., Rasmussen, R. S., and Brattain, R. R., Private Communication to O.R.R., May 5, 1945.
354. Tyler, W. P., and Higuchi, T., Symposium on Rubber Testing, Am. Soc. Testing Materials Spec. Tech. Pub. 74, 1947, p. 18; *India Rubber World*, **116**, 635–8, 640 (1947).
355. Überreiter, K., *Z. phys. Chem.*, **B45**, 361–73 (1940).
356. U.S. Rubber Co., Private Communication.
357. Van Amerongen, G. J. *J. Applied Phys.*, **17**, 972–85 (1946); *Rubber Chem. and Technol.*, **20**, 494–514 (1947).
358. Van Amerongen, G. J., *J. Polymer Sci.*, **5**, 307–32 (1950); *Rubber Chem. and Technol.*, **24**, 109–131 (1951).
359. Vandenberg, E. J., and Hulse, G. E., *Ind. Eng. Chem.*, **40**, 932–7 (1948).
360. Wales, M., Williams, J. W., Thompson, J. O., and Ewart, R. H., *J. Phys. and Colloid Chem.*, **52**, 983–8 (1948).
361. Walker, H. W., and Mochel, W. E., *Proc. Second Rubber Technol. Conf.*, London, 1948, pp. 69–78, Heffer, Cambridge, England, 1949.
362. Wall, F. T., *J. Am. Chem. Soc.*, **66**, 2050–7 (1944).
363. Wall, F. T., *J. Am. Chem. Soc.*, **67**, 1929–31 (1945).
364. Wall, F. T., Banes, F. W., and Sands, G. D., *J. Am. Chem. Soc.*, **68**, 1429–31 (1946); *Rubber Chem. and Technol.*, **20**, 41–44 (1947).
365. Wall, F. T., and Beste, L. F., *J. Am. Chem. Soc.*, **69**, 1761–4 (1947).
366. Wall, F. T., Powers, R. W., Sands, G. D., and Stent, G. S., *J. Am. Chem. Soc.*, **69**, 904–7 (1947).
367. Wall, F. T., Powers, R. W., Sands, G. D., and Stent, G. S., *J. Am. Chem. Soc.*, **70**, 1031–7 (1948).
368. Wall, F. T., and Zelikoff, M., *J. Am. Chem. Soc.*, **68**, 726 (1946); *Rubber Chem. and Technol.*, **19**, 551 (1946).
369. Wall, L. A., *J. Polymer Sci.*, **2**, 542–3 (1947).
370. Wall, L. A., *J. Research Natl. Bur. Standards*, **41**, 315–22 (1948).
371. Wall, L. A., *India Rubber World*, **119**, 615 (1949).
372. Walton, W. W., McCulloch, F. W., and Smith, W. H., *J. Research Natl. Bur. Standards*, **40**, 443–7 (1948).
373. Weidlein, E. R., Jr., *Chem. Eng. News*, **24**, 771–4 (1946).
374. Weir, C. E., Nat. Bur. Standards, Unpublished Results, 1949.
375. Weir, C. E., Leser, W. H., and Wood, L. A., *J. Research Natl. Bur. Standards*, **44**, 367–72 (1950); *Rubber Chem. and Technol.*, **24**, 366–373 (1951).
376. Whitby, G. S., Private Communication to O.R.R., Apr. 10, 1947.
377. White, J. U., and Flory, P. J., Private Communication to O.R.R., Dec. 16, 1942.
378. White, L. M., Ebers, E. S., Shriver, G. E., and Breck, S., *Ind. Eng. Chem.*, **37**, 770–5 (1945); *Rubber Chem. and Technol.*, **18**, 833 (1945).
379. Wiley, R. H., *J. Polymer Sci.*, **2**, 10–1 (1947).
380. Wiley, R. H., and Brauer, G. M., *J. Polymer Sci.*, **3**, 455–61 (1948).
381. Wiley, R. H., and Brauer, G. M., *J. Polymer Sci.*, **3**, 647–51 (1948).
382. Wiley, R. H., and Brauer, G. M., *J. Polymer Sci.*, **3**, 704–7 (1948); *Rubber Chem. and Technol.*, **22**, 402 (1949).
383. Wiley, R. H., Brauer, G. M., and Bennett, A. R., *J. Polymer Sci.*, **5**, 609–14 (1950); *Rubber Chem. and Technol.*, **24**, 585–590 (1951).
384. Williams, G., *J. Chem. Soc.*, **1940**, 775–84 (1940).
385. Willson, E. A., Miller, J. R., and Rowe, E. H., *J. Phys. and Colloid Chem.*, **53**, 357–74 (1949).
386. Witte, R. S., and Anthony, R. L., *J. Applied Phys.*, **22**, 689–95 (1951); *Rubber Chem. and Technol.*, **25**, 468–79 (1952).

387. Wolstenholme, W. E., Mooney, M., Div. Rubber Chem., Am. Chem. Soc., Mar. 2, 1951.
388. Wood, L. A., *Proc. Rubber Technol. Conf.*, London, 1938, pp. 933–54; *Rubber Chem. and Technol.*, **12**, 130–149 (1939).
389. Wood, L. A., *Natl. Bur. Standards Circ.*, **C427**; *India Rubber World*, **102**, No. 4, 33–43, 51 (1940); *Rubber Chem. and Technol.*, **13**, 861–885 (1940).
390. Wood, L. A., Chapter in Ref. 10.
391. Wood, L. A., Unpublished Observations.
392. Wood, L. A., Bekkedahl, N., and Peters, C. G., *J. Research Natl. Bur. Standards*, **23**, 571–83 (1939); *Rubber Chem. and Technol.*, **13**, 290–301 (1940).
393. Wood, L. A., Bekkedahl, N., and Roth, F. L., *J. Research Natl. Bur. Standards*, **29**, 391–6 (1942); *Ind. Eng. Chem.*, **34**, 1291–3 (1942); *Rubber Chem. and Technol.*, **16**, 244 (1943).
394. Wood, L. A., and Tilton, L. W., *Proc. Second Rubber Technol. Conf.*, London, 1948, Heffer, Cambridge, England; *J. Research Natl. Bur. Standards*, **43**, 57–64 (1949); *Rubber Chem. and Technol.*, **23**, 661–669 (1950).
395. Work, R. N., Unpublished Observations, 1951.
396. Work, R. N., *Rubber Age N.Y.*, **69**, 59 (1951); *Phys. Rev.*, **83**, 204 (1951).
397. Yanko, J. A., *J. Polymer Sci.*, **3**, 576–600 (1948); *Rubber Chem. and Technol.*, **22**, 494–517 (1949).
398. Zapp, R. L., *Rubber Age N.Y.*, **59**, 574–5 (1946).
399. Zapp, R. L., and Baldwin, F. P., *Ind. Eng. Chem.*, **38**, 948–55 (1946); *Rubber Chem. and Technol.*, **20**, 84–98 (1947).
400. Zapp, R. L., Decker, R. H., Dyroff, M. S., and Rayner, H. A., *J. Polymer Sci.*, **6**, 331–49 (1951).
401. Zapp, R. L., and Guth, E., *Ind. Eng. Chem.*, **43**, 430–8 (1951); *Rubber Chem. and Technol.*, **24**, 894–913 (1951).
402. Zimm, B. H., *J. Chem. Phys.*, **14**, 164–79 (1946).
403. Zimm, B. H., *J. Phys. and Colloid Chem.*, **52**, 260–7 (1948).
404. Zimm, B. H., *J. Chem. Phys.*, **16**, 1099–1116 (1948).
405. Zimm, H. B., Stein, R. S., and Doty, P. M., *Polymer Bull.*, **1**, 90 (1945).
406. Zwicker, B. M. G., Private Communication to O.R.R., June 5, 1950. See also Beatty, J. R., and Zwicker, B. M. G., *Ind. Eng. Chem.*, **44**, 742–52 (1952).
407. Zwicker, B. M. G., *India Rubber World*, **121**, 431–2 (1950).

CHAPTER 11

THE PROCESSING AND COMPOUNDING OF GR-S

R. P. Dinsmore and R. D. Juve

Goodyear Tire & Rubber Company

INTRODUCTION

At the outset of the wartime development of GR-S as a general-purpose replacement for Hevea the rubber produced was of comparatively poor quality. With production improvements, quality and uniformity have improved, so that GR-S can now be considered to be a satisfactory raw material. The GR-S-type polymer exhibits general similarity to natural rubber in that it can be processed on the same factory equipment as natural rubber, the same compounding materials can be used with it as with natural rubber, and similar finished products are produced by vulcanization. Although it is generally similar to Hevea in many respects, its differences have presented not a few problems in its utilization. Pronounced dissimilarities exist in processing behavior, in curative and accelerator requirements, in fabrication techniques, and in ultimate physical properties. In the development of suitable compounds of GR-S, the background of information and experience that had been accumulated with natural rubber was useful but required certain modifications. Although similar types of softeners, accelerators, pigments, etc., are useful in both GR-S and Hevea, the same specific materials or types of materials that have proved best for natural rubber are not necessarily the best for GR-S.

A great variety of GR-S-type polymers have been produced, each with its own peculiarities and its own particular compounding problems. In this chapter the compounding of only the following general-purpose polymers will be considered; they represent the greatest share of the production of general-purpose synthetic rubber: GR-S, GR-S 10, GR-S-AC, cold rubber (GR-S 100). These grades are now designated as GR-S 1000, 1002, 1004, and 1500, respectively.

Types of GR-S. As distinguished from the German Buna S3, which requires thermal treatment before it can be processed on conventional equipment, GR-S is made soft enough in the polymerization process so that it can be directly mixed and processed by moderate mastication and the use of softeners.

Standard GR-S is a copolymer of butadiene and styrene in the ratio of approximately 76.5 to 23.5 per cent. The mixture of monomers is polymerized at 50° C. to a Mooney viscosity of 46 to 54. In addition to the hydrocarbon, standard GR-S contains roughly 1.25 per cent of an antioxidant, 3.75 to 6.0 per cent fatty acid, and small amounts of impurities.

Of the other types of GR-S in most general use, GR-S 10 is the same as standard GR-S except that it contains rosin acid derivatives instead of fatty acid. GR-S-AC similarly is the same as standard GR-S except that its coagulation from latex with alum in place of salt and acid causes aluminum soaps and salts to be present in the dried polymer in place of fatty acid. So-called cold rubber is produced by polymerization at a low temperature instead of the 50° C. employed in the manufacture of standard GR-S. As used in this chapter, the name cold rubber refers to GR-S 100 (now GR-S 1500) and those closely related modifications most widely used in 1950. In each of these rubbers, butadiene and styrene have been copolymerized in the same ratio. These types of GR-S are distinguished by differences that influence the methods of processing, the compounding, and some of the physical properties of the finished goods, but the basic characteristics of all of them are essentially the same.

In addition to these general-purpose polymers, specialty polymers are made for wire and cable insulation, low-temperature service, blown sponge, and many other applications. Many of these variations require special compounding and processing considerations, but these are beyond the scope of this chapter.

STORAGE STABILITY OF RAW GR-S

When protected with a sufficient amount of a good antioxidant, for example 1.0 per cent of phenyl-β-naphthylamine, GR-S undergoes very slight change in viscosity, and only slight decomposition of the antioxidant occurs during five years' storage under normal conditions.[1] This bespeaks of satisfactory stability in storage. In general, GR-S aged up to one year has been found to suffer only very minor changes in processing characteristics and cured physical properties.[103]

It has been found that standard GR-S absorbs approximately 2.5 per cent by weight of oxygen when exposed to oxygen at 70° C. for 60 days.[63] At the end of 60 days the oxygen absorption rose abruptly, indicating rather severe decomposition of the polymer. Evidence has been reported[32, 90] that raw cold rubber made by an iron-containing polymerization recipe is less stable than standard GR-S in the presence of an antioxidant.

MILL BREAKDOWN BEHAVIOR

When GR-S was first introduced on a large scale to the rubber-manufacturing industry, it caused factory processing difficulties and increased labor costs. For the most part the difficulties were encountered because the rubber was different from, not fundamentally inferior to, rubber previously used. Most of the difficulties were overcome as the polymer was improved and as skill in factory handling grew. After the war the return of natural rubber in larger quantities caused similar processing difficulties in the factory because many of the workers at that time either had never seen natural rubber or had forgotten how it should be handled. An effort will be made to describe the most important differences between GR-S and Hevea with respect to mill breakdown.

For the proper dispersion of pigments during the mixing operation, it is required that the rubber shall be neither too soft nor too tough. Although crude natural rubber is initially very tough, it softens rather rapidly when masticated. The mastication also reduces the tendency for the rubber to recover from distortion. GR-S, although initially more plastic than crude Hevea, does not break down so rapidly as the latter during milling. In addition, the relationship between the nerve or elasticity and the degree of plasticity for GR-S is different from that for natural rubber.

Heat Softening. GR-S can be softened by heating in air, but continued heating tends to cause it to resinify.[75] Tobolsky[140] and others are of the opinion that the autooxidation of synthetic rubbers produces two effects: chain scission and cross-linking. Chain scission results in a reduced viscosity, whereas cross-linking results in a gradual toughening which ultimately leads to a brittle, resinified condition. The relative rates of these two reactions depend, among other things, on temperature. Piper and Scott[101] have shown that GR-S is not softened by heat in the absence of air. Natural rubber behaves similarly at moderate temperatures, but it will soften in the absence of air when heated at 200° C. or a higher temperature.

Influence of Processing Temperature. More commonly GR-S is softened by mastication on an open mill or in a Banbury mixer or a Gordon plasticator. The manufacture of most GR-S products requires some degree of premastication for practical processing. Some slight deterioration of physical properties in the vulcanized product usually occurs as a result. Wiegand and Braendle[146] have emphasized the importance, in breaking down and mixing GR-S, of avoiding elevated temperatures because of the deterioration in the physical properties of the vulcanizate which results from such temperatures. In factory practice, on open mills smaller-size batches generally are used for GR-S than for natural rubber in order to provide more efficient cooling. In Banbury mixing, however, full-size loads are charged to obtain adequate ram pressure. Much of the Banbury mixing of GR-S is done at very high temperatures with good results, in contradiction to laboratory evidence that high temperatures should be avoided throughout the processing. The literature on the effects of processing and mixing temperature on processibility and product quality has been reviewed by Drogin, Bishop, and Wiseman.[47]

Influence of Gel. Investigations by White and co-workers[145] have correlated the effect of benzene-insoluble fractions (gel, formed either during polymerization or during plastication and mixing) with the processing characteristics and the ultimate physical properties of GR-S. They have shown that increased proportions of tight gel make mixing operations more difficult, reduce tensile strength, increase modulus, and decrease cut-growth resistance. On the other hand, the presence of tight gel reduces shrinkage and roughness, thereby aiding in calendering and extrusion operations. White and co-workers concluded that processing conditions favoring the formation of tight gel should be avoided, where maximum quality in the final vulcanizate is desired. If quality can be sacrificed for ease of processing, gel formation during mastication can be used to advantage. Similar results[114] are achieved, as in GR-S 60, now GR-S 1008, by introducing a small amount of divinylbenzene during polymerization and thus forming a controlled amount of gel in the preparation of the polymer. An attempt to prepare a stock combining the excellent processing properties of the divinylbenzene cross-linked GR-S 60 with the superior physical properties of standard GR-S was made by Crawford and Tiger.[34] These investigators found that the subjection of GR-S to slight precuring by means of Banbury treatment at 193° C. for 5 minutes in the presence of 0.5 part sulfur, 1.5 parts mercaptobenzothiazole, and 50 parts carbon black prior to the addition of other ingredients gave an easy processing stock somewhat superior to Banbury-mixed GR-S 60 in both processing properties and tensile strength.

The Mooney Viscosimeter. The changes in plasticity that occur during mill or Banbury mastication can be studied with the Mooney viscosimeter.[91] This instrument measures the torque required to revolve a rotor at constant speed in a sample of the polymer at a constant temperature. The shearing action in the Mooney viscosimeter has been shown to be analogous to mill mastication.[71] Mooney viscosity values are usually described in terms of rotor size, test temperature, and the time of running before the reading is taken. Thus "ML/212-4" means the Mooney viscosity of a raw, uncompounded polymer read after 4 minutes' operation of the large rotor at 212° F.

Internal Mixers. Because the cost, labor, time, and power are lower with internal mixers than they are for open mill operation, such mixers are used rather generally. The Gordon plasticator is an extrusion type of machine often used to soften rubbers. After the softening, mixing is accomplished in a Banbury or on a mill. Similar results can be achieved by both masticating and mixing in the Banbury. By premastication of the raw polymer in the Banbury the viscosity can be markedly reduced. However, when black is added to the broken-down GR-S in the subsequent Banbury mixing, the compounded Mooney viscosity rises to almost the same value as that reached when the untreated GR-S is Banbury mixed with black. Consequently, an efficient procedure has been to mix in the black and soften the GR-S in the first operation. If further breakdown is required, the batch is remilled in a later Banbury operation. The age between remillings seems to have only a minor effect on the changes in processibility that take place in factory scale work.

Peptizers. In some instances it is necessary to soften GR-S to a high degree. In order to do this without prolonging the factory mastication cycle, chemical plasticizers, called peptizers, are employed. Typical peptizers for GR-S are naphthyl-β-mercaptan, xylyl mercaptan, and di-o-benzamidophenyl disulfide. Small quantities of such chemicals facilitate the attack of oxygen and consequent scission of the polymer chains.[15] In general 0.5 to 3.0 parts per 100 parts GR-S are used. Furfural phenylhydrazone has been shown[4] to be one of the most active peptizers for GR-S. The same investigators found furfural-p-bromophenylhydrazone and furfural-1-naphthylhydrazone to be active and furfural-p-nitrophenylhydrazone to be inactive. GR-S softened by peptizers tends to recover stiffness more slowly during subsequent storage than GR-S softened by mastication in the absence of peptizers.

Although useful in peptizing standard GR-S, small amounts of xylyl mercaptan and di-o-benzamidophenyl disulfide are relatively ineffective as peptizers for GR-S polymerized at 5° C. (cold rubber).[112] Stangor and Radcliff[128] have reported that, if the amount of aromatic mercaptan is increased to approximately 1 per cent, cold rubber will soften when masticated for 5 minutes at a temperature lower than 177° C.

Polymerization Variables and Processibility. Variations in the polymerization recipe and conditions of the reaction may markedly influence the processing properties of the elastomer. A few such variables which might be considered are the degree of monomer conversion, the reaction temperature, and the type and amount of modifier. In general, increases in monomer conversion cause decreases in processing efficiency, even though modifier adjustment be made so that the plasticity of the polymer is unchanged. Within the range of polymerization temperatures from 30 to 50° C. the lower the temperature, the smoother the processing seems to be. The type of modifier has a pronounced effect on processibility. For example, it has been found that GR-S modified with tertiary dodecyl mercaptan is more difficult to process than that modified with the standard primary dodecyl mercaptan, at a similar Mooney viscosity level. Since increased concentrations of modifier result in polymer of greater plasticity, the amount of modifier has a critical effect on the ease of processing. At the outset GR-S was tough and difficult to process.[118] In the interest of producing a processible GR-S, industry agreed on a mean Mooney viscosity of 50 (large rotor at 100° C.). With rubber of this plasticity, existing factory machinery could be used with reasonable efficiency. By improved control of the many polymerization variables, the uniformity and the processing efficiency of GR-S have become quite satisfactory.

Some discussion of the influence of polymerization variables on the Mooney viscosity and the solution viscosity (molecular weight) of GR-S has already been offered in Chapter 8.

Cold Rubber. GR-S X-478 cold rubber has been reported to be more difficult to process than standard GR-S.[129] In laboratory Banbury tests, Zwicker and co-workers[157] found cold rubber to be softened at an appreciably slower rate than GR-S. The same investigator reported that at Banbury temperatures in excess of 157° C. stiffening occurred, probably due to gel formation. Tiger, Reich, and Taft[138] found that in the Banbury mastication of cold rubber the temperature rose to a higher point than with standard GR-S. They found that cold rubber broke down more slowly than standard GR-S under normal conditions of operation, and, further, that, if the Banbury was fully loaded (to 95 per cent of capacity) or if its speed was higher than normal, cold rubber underwent actual stiffening (with gel formation) instead of breakdown. However, when air was fed to the mixer during mastication, the breakdown of cold rubber was greatly facilitated and rendered similar to that of standard GR-S, both when the Banbury was heavily loaded and when it was less heavily loaded. These findings are illustrated by data in Table I.

PROCESSING CHARACTERISTICS

Despite the inherent processing deficiencies of GR-S, large volumes have been handled in factories with little or no loss in gross output. This has been achieved by adjustment of processing conditions and by the increased use of reinforcing furnace and high-modulus furnace blacks.

In common practice GR-S is either softened as a raw polymer, for example in a Gordon plasticator, or in the form of a black masterbatch in a Banbury. In Banbury mastication the presence of carbon black causes more rapid breakdown of the polymer. Mixing in a Banbury at a slow speed and for a proportionately longer time yields essentially the same plasticity as mixing at a high speed.

Table I. *Mastication of Standard and Cold GR-S for 6 Minutes in Laboratory-Size-B Banbury Mixer Loaded to 95 Per Cent Capacity*[138]

Polymer	Original Mooney	Conditions of Operation	Final Mooney	Temp. of Rubber when Dumped, °C.	Gel. (%)	Inher. Visc. of Soluble
GR-S 10	56	Normal	35	200	9	1.65
GR-S 10		Air stream	30	155	2	1.79
Cold GR-S (X-478)	59	Normal	64	252	34	1.40
Cold GR-S (X-478)		Air stream	36	163	2	1.80

Reprinted by permission of *Ind. Eng. Chem.*

Influence of Mooney Viscosity on Processing, Efficiency. Using the procedure of masticating in the presence of black, Dinsmore and Fielding[42] compared the mixing and extrusion efficiencies of eight GR-S polymers of different Mooney viscosities. The GR-S was mixed with carbon black and softener, in some cases being returned to the Banbury for a remill after cooling, and finally it was mixed with the other tread stock ingredients in the Banbury. Data for this comparison in EPC black stocks are shown in Table II.

Banbury efficiencies were calculated on the total time in minutes that a Banbury would be occupied in producing 1000 lb. of finished stock, based on 100 as the efficiency for a natural-rubber tread stock. In analzying the extrusion data, it can be considered that a natural rubber tread would be smooth and that in its case the maximum extrusion temperature would be 116° C. The data indicate 100 per cent mixing efficiency, and acceptable extrusion could be obtained with GR-S of 45 Mooney viscosity.

The comparison just described was made in 1942. Subsequent improvements in the control of polymerization variables and in the uniformity of

Table II. The Effect of Viscosity on the Processing of GR-S Tread Stock[42]

Raw-polymer Mooney viscosity*	16	31	32	43	47	52	60	60
Black masterbatch								
30 r.p.m., Banbury min.	3½	4¾	3½	4	4½	4½	4½	4½
Discharge temperature, °C.	111	147	129	134	147	146	151	147
Mooney viscosity†	16	51	43	48	50	53	57	74
Remill								
30 r.p.m., Banbury min.		2½			2½	2½	2½	2½
Discharge temperature, °C.		138			146	147	149	129
Mooney viscosity†		33			48	51	59	69
Complete Tread Stock								
30 r.p.m., Banbury min.	2¼	2¼	2¼	2¼	2¼	2¼	2	2¼
Discharge temperature, °C.	113	117	120	123	123	129	127	129
Mooney viscosity†	17	26	39	47	34	39	52	66
Banbury Efficiency, %	148	92	148	139	100	100	102	96
Extrusion								
Extrusion temperature, °C.	96	102	109	115	113	116	121	120
Appearance	Smooth	Very smooth	Rough	Rough	Smooth	Slightly rough	Very rough	Very rough
Order of acceptability	2	1	4	5	3	4	6	6

*Mooney viscosity, large rotor, 100° C.
†Mooney viscosity, small rotor, 100° C.

Reprinted by permission of *India Rubber World*.

GR-S have improved mixing efficiency. Also, the trend to reinforcing furnace and high-modulus furnace blacks has aided processing to the extent that GR-S now is usually considered to be processed more efficiently than natural rubber. Dinsmore and Fielding have estimated that 100 per cent efficiency and acceptable tubing can be obtained with a 50 Mooney viscosity GR-S when the common mixture of EPC and HMF blacks is used. When a latex-black masterbatch is used, an additional advantage in processing is obtained because of the reduced Banbury mixing time. GR-S can be processed more efficiently than natural rubber by combining the use of lower-viscosity polymer, furnace blacks, and latex masterbatches of polymer and black.

Other Aspects of the Processibility of GR-S. The mixing of GR-S consumes more power and generates more heat than the mixing of natural rubber. When hot, GR-S stock is tender, and the edges of extruded stock tend to tear unless the stock is compounded and handled properly. The shrinkage is greater, and the extruded and calendered stock is rougher in the case of GR-S than in that of natural rubber. Most GR-S compounds extrude at a slower speed than natural-rubber compounds and extrusion temperatures are higher for GR-S compounds. If compounded satisfactorily, GR-S can be calendered and frictioned with ease. Although more calender crown is necessary for GR-S stocks, the calender crown is usually designed to be as nearly correct as possible for the greater majority of stocks run in a plant, and both natural-rubber and GR-S stocks can be run on the same calender. The lower sensitivity of GR-S to roll temperatures is an advantage in calendering.

The plasticity and the flow characteristics of any particular compound determine its molding characteristics. In respect to molding there are no general differences between GR-S and Hevea, except that GR-S stocks, which fall off in tensile strength more rapidly as their temperature is raised (p. 419), tear more readily than Hevea stocks when removed from a hot mold.

Cold rubber requires approximately the same power as GR-S in the mixing of black masterbatches, and pigments are incorporated in it at approximately the same rate as in standard GR-S. Extruder screw speed sometimes is increased in the case of cold rubber to obtain the same extrusion rate as with GR-S, but scorch may become a problem with the resulting higher temperature. Cold rubber extrudes more smoothly, swells less after leaving the die, and shrinks less on cooling than standard GR-S.

Since GR-S is less subject to precure or scorch at processing temperature, it has in this respect a distinct advantage over natural rubber. Nevertheless, the balance between the cure rate and the tendency to scorch is a factor that always must be considered.

GR-S can be used for the spread coating of fabrics in a manner similar to natural rubber. The same solvents can be used, and the technique is essentially the same for both rubbers. However, because of its low degree of tackiness, GR-S finds only limited application, particularly where more than one ply is required. The absence of tackiness is a serious handicap of GR-S for cements and results in very little practical application of GR-S in this field.

PROCESSIBILITY EVALUATION METHODS

In order to predict the factory processing characteristics of various GR-S elastomers, a number of laboratory processibility tests have been developed. Production processing operations are so complex and so varied that the selection of processibility test methods must be based on the conditions employed in each specific factory operation. In some cases a single test may be satisfactory for production control, whereas a combination of a number of tests may be necessary for evaluation of the processing characteristics of a new GR-S polymer or a new compound. Although the Mooney plastometer has found extensive use in this connection, it is often insufficient by itself, for two GR-S polymers of identical plasticity may not behave alike during processing. Vila[142] has shown that the plasticity is related to the power required and the heat generated during extrusion but is not correlated with the rate of extrusion.

By the addition of a recovery device, the Mooney plastometer has been adapted to the measurement of the elastic recovery. A general relationship between the elastic recovery of GR-S stocks and their processibility has been found to exist.[134]

The Brabender plastograph[74] has been used to study on a reduced scale the factors involved in the mastication breakdown of GR-S. It consists of a jacketed mixing chamber with two rotors which revolve in opposite directions, one at 90 r.p.m. and the other at 60 r.p.m. A continuous recording is made of the consistency.

In addition to consideration of the work required to break down GR-S to the desired workability, the extrusion and calendering behavior of the polymer is an important factor in processing.

The tendency for uncured elastomer to recover from distortion is referred to as nerve. In calendering or extrusion, nerve is objectionable. Lack of cohesion, resulting in holes in milled sheets, is also objectionable. These and other processing characteristics have been evaluated by numerous methods.

Laboratory-Scale Evaluation of Processibility. A laboratory-mill method of evaluating processibility was proposed by White, Ebers, and Shriver[144] and was further modified by Schade and Labbe.[113] The method consists essentially of noting (1) the time required to form a band on the mill rolls, (2) the time required to mill until the band is free of holes, (3) the Mooney viscosity increase caused by the incorporation of the black, (4) the roughness of the surface of the compounded stock, (5) the shrinkage of the sheet after removal from the mill, and then calculating a numerical processibility rating from the values assigned for each individual stage of the test. In a somewhat similar test a laboratory Banbury is used to mix the stocks, after which measurements are made of surface roughness and shrinkage after sheeting out on a mill.[106] In summarizing the utility of the mill and Banbury tests, Taylor, Fielding, and Mooney[136] concluded that these tests show reasonably good correlation with factory processing but fail to discriminate between polymers with minor differences in processibility.

Laboratory extrusion tests have been described by Dillon,[41] Garvey,[57] and Vila.[142] A laboratory extrusion test[136] that has shown good correlation with factory processibility may be described briefly as follows. Two batches of black masterbatch are prepared. Then half portions of each of the two batches are blended together in a laboratory Banbury. This provides three samples, two of them half batches that have had a simple Banbury mix and the third that has been returned to the Banbury for a second mix. These stocks are extruded through a straight, tread-shaped die $1\frac{3}{4}$ in. wide and 3/32 in. thick at the center. Part of the extruded stock is weighed, and part is extruded through a triangular-shaped die. In evaluating the processibility, most importance is attached to the Mooney viscosity of the Banbury mixed stocks, the weight of the straight die extruded stock per unit of length, and the surface and edge appearance of the triangular extruded sections.

Factory-Scale Evaluation of Processibility. A factory-scale evaluation procedure[136] has been devised to compare the processibility of GR-S polymers. The method involves the mixing of a standard tread stock by a standard Banbury schedule. Half of the Banbury batch is then submitted to the normal amount of Banbury remilling, and the other half is given one more or one less remill. Next the batches are extruded through standard dies under controlled conditions. The smoothness of the extruded stock, the occurrence of torn edges, the percentage of acceptable treads, and the extrusion temperature are observed. The observations allow a comparison to be made with a control stock run in the same program.

Rugosity. The surface roughness or rugosity of a milled or calendered sheet can be measured by an instrument called a Rugosimeter.[92] This device measures the resistance to air flow between the surface of the experimental sheet and a plane test surface which rests on it.

Scorching. Another problem encountered during processing is the precuring or scorching of compounded stocks which may be sensitive to the rather high temperatures involved in mixing, calendering, or extrusion. The scorching tendencies of stocks can be studied by a number of methods, according to which samples are heated, and the changes in various physical properties determined. The use of the Mooney plastometer for studying scorch tendencies by measuring the time required to reach a viscosity of 100 at 121° C. was proposed by Weaver.[143] Other investigators[121] later suggested that the time in the plastometer necessary to reach the last minimum viscosity prior to the increase in viscosity caused by incipient curing is a more satisfactory criterion of scorching. Factors appreciably affecting the scorch test results were found by the National Bureau of Standards[136] to be the control of the temperature of the test specimen and the amount of moisture in the sample. Since moisture accelerates the cure of GR-S, conditioning the samples to uniform moisture control before testing improved duplicability. The Bureau of Standards defines the "cure factor" as the time required for the viscosity to increase 40 units above the minimum value.

Scorching tests have found considerable use in assisting in the choice of accelerators and fillers and combinations of accelerators and fillers such as will offer a sufficient margin of safety from prevulcanization in factory operations.[57a, 85a, 90a, 137a]

THE ROLE OF COMPOUNDING INGREDIENTS
IN PROCESSING

Softeners. In addition to the plasticity arising from breakdown of the GR-S itself, further increase in plasticity is achieved with softeners. Softeners or plasticizers may reduce the mixing and processing temperature, aid in incorporating and dispersing dry compounds, reduce nerve at the extruder or calender, and improve flow during molding. More softener is generally necessary in GR-S than in natural rubber to get similar improvement in processing. Since the larger softener content causes a greater effect on the physical properties of the vulcanizate, the proper softener or blend of softeners is carefully selected. Within the class of softeners come both materials that soften the rubber and materials that improve processing without softening. As the plasticity is increased by greater softener content, less power is consumed and less heat is generated during milling. Also, increased plasticity may improve extrusion and calendering. There are, however, some materials that will improve extrusion and calendering operations without markedly increasing the plasticity.

Ludwig, Sarbach, Garvey, and A. E. Juve[82] examined some 650 materials as softeners in a GR-S black stock, noting their effects on the processibility and tack of the uncured stock and on the properties of the vulcanized stocks. From the viewpoint of the mutual solubility relationship between softener and rubber, they classified softeners as follows:

Solvents. Aromatic hydrocarbons, chlorinated hydrocarbons, aliphatic hydrocarbons, and terpenes and related compounds such as gum turpentine and rosin.

Partial Solvents. Esters, high-molecular-weight ketones, and naphthalenes.

Non-solvents. Alcohols, phenols, low-molecular-weight ketones, branched-chain aliphatic hydrocarbons, amines, and ether–alcohols.

The relative effectiveness of the softeners in any one class is dependent on their chemical structure and physical properties. The solvent type of softeners give, other things being equal, soft stocks which retain considerable nerve. With softeners of less solubility, a lubricating effect comes into play. When the solubility is good but not excessive, a flat, smooth running stock is obtained. With still lower solubility on the part of the softener, the lubricating effect becomes more marked, and it becomes permissible to use only a small amount of the softener. The general effect of lubrication is to make extrusions more smooth, reduce sticking to mill rolls, and reduce building tack. Softener solubility may be greater during hot milling than at room temperature, resulting in bleeding to the surface when stock is cooled. Ludwig and co-workers found improved tubing characteristics in the presence of cottonseed fatty acid, ricinoleic acid, oleic acid, soft coal tar, dimethoxy-ethyl phthalate, polybutene, and lauric acid. On the other hand, materials that impart tack or stickiness will tend to reduce extrusion speeds.

Among the softeners used most widely in GR-S compounding are refined, heavy coal-tar fractions, moderately heavy petroleum distillates (process oils), and petroleum asphalts.

Softeners usually are added late in the mixing cycle. The lower-viscosity liquid softeners generally require a longer time to incorporate than higher-viscosity materials which can be kneaded more readily.

A miscellaneous group of materials classed as extenders also impart beneficial processing characteristics to GR-S. The results have been published[83] of a survey of 110 materials which might serve as extenders of GR-S, i.e., might be capable of replacing part of the rubber without serious sacrifice of quality in the final vulcanized product.

The addition of factices, solid oil, sulfur-reactive mixtures of selected unsaturated hydrocarbons, coal-tar pitch, mineral rubber, and asphaltic hydrocarbon (gilsonite) gives smooth stocks with good extrudability. Asphaltenes, the principal components of gilsonite and mineral rubber, are stated by Rostler and Sternberg[107] to aid in processibility by reducing nerve without exerting more than a slight effect on the plasticity.

Other Blending Agents. The addition of natural rubber to GR-S causes roughness in extruded and calendered stock. Cold rubber, however, appears to be more compatible with natural rubber.

Resins such as high-styrene/low-butadiene copolymers[16] promote smoothness during processing and reduce shrinkage of GR-S. At elevated temperatures these resins act as plasticizers. If the GR-S is peptized to reduce nerve, resin-rubber blends can be made to extrude smoothly even in the absence of any compound loading. Somewhat similar results can be obtained with copolymers of styrene and isobutylene.[35]

Carbon Blacks

Comparison of Blacks. Since by far the greatest amount of GR-S is consumed in combination with reinforcing pigment, it is important to consider the effect of pigments on the processing behavior. Differences in the inherent characteristics of the various types of black cause marked differences in the manner in which they affect processing. The effect of particle size was pointed out early by Garvey and Freese,[56] when they showed a progressive processing advantage in going from ink black to standard channel to easy-processing channel to large particle reinforcing black. Softer blacks were all reported to be easier processing. Also, they observed acetylene black to give in extruded stock an unusually smooth, shiny surface with sharp corners and no porosity. In a comparison of channel blacks, Brown[21] found easy-processing channel (EPC) to give the lowest mixing temperature, the lowest power consumption, the lowest Mooney viscosity, and the smoothest and coolest extrusion. The same author reported medium-processing channel (MPC) to be nearly equivalent to EPC but hard-processing channel (HPC) to require extra remills if the stock was to extrude as smoothly as an EPC stock. Conductive channel (CC) produced a higher viscosity, but extruded almost as smoothly as EPC.

The processing characteristics of a wide variety of carbon blacks have been reviewed by Drogin,[44, 46] with the following findings. More power is required to mix the same black in GR-S than in natural rubber. The greatest difference between power consumption in natural rubber and in GR-S occurs with CC black and the least with high-modulus furnace (HMF) and fine thermal (FT) blacks. The temperature rise during mixing is greatest for channel and least for thermal blacks. Channel blacks cause hotter mixing in GR-S than in natural rubber, whereas blacks with reticular chain structure and blacks from aromatic hydrocarbons cause less heat generation in GR-S than in natural rubber. Blacks cause higher Mooney viscosity in GR-S than in natural rubber, and the viscosity increases as the reinforcing power of the black increases. Speaking broadly, the extrusion rate of GR-S is increased as the loading of black is increased, in contradistinction to natural rubber, where increased black usually slows down the extrusion rate. Carbon blacks are less scorchy in GR-S, but the differences between the scorch tendencies of semireinforcing furnace (SRF), HMF, and fine furnace (FF) are greater than in natural rubber.

The data in Table III will serve to illustrate most of the general statements made in the preceding paragraph. In addition to data showing the effect of the various types of black on the processing qualities of GR-S and Hevea, the table also includes data on their effect on the mechanical properties of the vulcanized products. It will be seen that for all types of black the tensile strength, elongation, hardness, and resilience are lower and the laboratory abrasion resistance and the heat buildup higher with GR-S than with Hevea.

Structure Blacks. Structure carbons, i.e., those types of carbon black in which the particles, instead of being discrete, tend to form chainlike aggregates, are known to impart smoothness in processing and to confer stiffness (high modulus) on the vulcanized stock. Dobbin and Rossman[43] subjected a high-structure black to densification by progressive mechanical work which altered the secondary aggregates, and found that a progressive decrease in the smoothness of the stock occurred when the black was mixed with GR-S and the mixture calendered. There also followed a decrease in the stiffness of the stock after vulcanization, as indicated by a markedly lower modulus and a somewhat lower hardness.

Dannenberg and Stokes[38] likewise attribute the low shrinkage and smoothness obtained with certain furnace blacks to their structure. The effects of structure on processibility are greater in GR-S than in natural rubber. With regard to the shrinkage characteristics of GR-S with blacks, Dannenberg and Stokes found the following. Those blacks that cause the largest reduction in shrinkage vary greatly in particle size, but they are generally distinguished by high oil absorption and the ability to cause smooth processing. As the loading is increased, the smoothness increases. The structure blacks cause smooth processing at lower loadings than nonstructure blacks. A possible explanation of the reduction in shrinkage caused by structure blacks is that nonuniform dispersion resulting from the original aggregated state of the black causes a partial restriction of elasticity in the regions of high black concentration, somewhat similar to the effect of tight gel.

Blacks and Rate of Cure. Blacks affect the scorchiness of GR-S and cold rubber by virtue of their effect on the rate of cure and the generation of heat. The low pH of channel blacks causes slow curing and freedom from scorch, whereas the high pH of furnace blacks tends to cause scorchiness.

Table III. Comparison of Carbon Blacks in Natural Rubber and in GR-S[46]

Test Stocks

Smoked sheet	100	...
GR-S	...	100
Zinc oxide	3	5
Sulfur	3	2
Mercaptobenzothiazole	1	1.5
Stearic acid	4	...
Pine tar	1.5	...
Coal tar softener	...	5
Antioxidant	...	1
Carbon black	50	50

Test details. Mixings were made in a Banbury mixer, size *B*. The Mooney scorch value was determined at 121° C. Rebound was measured by the Lupke pendulum. Abrasion resistance was measured as the cc. lost per hp.-hour on a du Pont abrader. Heat buildup was determined by the Goodrich flexometer, with a stroke of 0.175 in., a load of 175 lb., and a temperature of 38° C., using a 30-minute cycle at 1800 r.p.m.

The Reinforcement Data refer to cures of 90 minutes at 127° C. with natural rubber and 90 minutes at 138° C. with GR-S, except as follows: The tensile strength refers to the optimum cure; the abrasion loss and the heat buildup refer to cures of 120 minutes at 127° C. with natural rubber and 120 minutes at 138° C. with GR-S.

The data for high-modulus furnace (HMF), semireinforcing furnace (SRF), and fine furnace (FF) blacks are in each case the average results for 3 commercial brands. CK-4 is a German HAF black made from coal-tar raw material.

NR means natural rubber.

Black	MT		FT		SRF		HMF		FF		RF		CK-4		EPC	
Rubber	NR	GRS	NR	GRS	NR	GRS	NR	GRS	NR	GRS	NR	GRS	NR	GRS	NR	GRS
Processing Data																
Banbury temp, max., °C.	117	128	117	132	130	141	150	148	124	146	162	161	143	164	142	167
Power consumption, watt-hr.	1315	1540	1314	1550	1340	1613	1480	1687	1353	1640	1570	1770	1630	1980	1458	1830
Mooney viscosity	37	48	38	48	45	54	62	70	51	60	69	88	56	102	57	112
Mooney scorch, minutes	21	29	12	21	10	23	7	17	9	18	8		8		14	31
Tubing rate, g. per minute	20.9	5.1	21.4	4.0	19.5	9.0	17.1	7.8	18.8	4.1	15.5	2.3	17.6	0.9	19.5	1.1
Reinforcement Data																
Modulus at 300%	900	660	640	550	1640	1130	2540	2030	1680	1330	2800	2610	2280	3000	1600	1960
Tensile strength, p.s.i.	3510	1280	4200	1560	3640	1890	3850	2530	4250	2740	4080	2950	4640	3520	4580	3140
Elongation, %	550	410	625	590	510	500	430	370	545	505	430	340	525	325	585	430
Hardness, Shore	51	51	67	53	63	59	67	64	63	59	69	68	65	69	66	67
Rebound, %	73	57	67	51	64	50	61	47	59	45	53	43	53	38	47	41
Abrasion loss	610	555	564	581	414	312	251	184	340	223	247	120	294	108	294	138
Heat buildup, C.	51	94	54	103	63	121	66	118	70	127	76	127	75	107	75	124

Quoted by permission from *Today's Furnace Blacks.*

The scorch tendencies caused by heat generation during mixing increase as the particle size of the black becomes smaller. A study of the influence of certain accelerators and combination of accelerators in controlling the liability of HAF black to cause scorching has been published by Moakes.[90a]

Blends of Blacks. To achieve a balance between processibility and good physical properties after vulcanization or for reasons of availability or costs, various types of blacks are often blended in GR-S. Blends of EPC and HMF have been used extensively. Cohan[29] obtained essentially the same processibility, flex life, and hysteresis loss with a straight EPC stock and with one containing a 90/10 MPC–SRF blend.

Balance between Black and Softener. Higher than normal concentrations of carbon black, balanced with appropriate increases in softener concentrations, have been shown by McMillan and co-workers[87] to behave as extenders for GR-S, reducing nerve and giving extruded products with a smoother surface and sharper edges. In similar experiments with cold rubber, McMillan, Winkler, and Anderson[88] have shown that in a tread-type stock the properties of the vulcanizates remain substantially constant when the content of black is raised, provided that at the same time there is added an extra quantity of softener sufficient to balance the black. Thus, by using a HAF black and, as the softener, an unsaturated petroleum oil, it was found that if, for each 0.94 part of black added beyond 40 parts, there was also added 1 part of softener, a "balance" was obtained and little change occurred in the properties of the vulcanizate. The loadings listed in Table IV all gave vulcanizates with similar properties.

Table IV. Balanced Loadings of Black and Softener[88]

Cold GR-S	100	100	100	100	100	100
HAF black	40	45	50	55	60	65
Unsaturated petroleum softener	0	5	10.3	15.7	21.0	26.5

Reprinted by permission of *Rubber Age.*

In such experiments the combination of extra black and softener may perhaps be regarded as an extender for the GR-S. In a more recent development it has been found that cold GR-S having a Mooney viscosity much higher than usual can be extended by 25 per cent or more petroleum oil on the weight of the rubber, the latter being introduced as a dispersion to the latex before coagulation. Such oil-extended rubber, without extra black beyond the loading normal in tread stocks, is stated to have superior resilience and to behave similarly in other properties to unextended cold rubber of normal Mooney viscosity both in tests in the laboratory and in tire tests on the road.[6, 40, 41a, 66, 106a, 135a, 143a]

Tack. The building tack of GR-S is notoriously poor. Piper and Scott[101] have attributed the lack of tack, at least partly, to the resistance to flow under small stresses. Freshening GR-S stock surfaces with solvent yields poor response in the matter of tackiness. What little tack is present is destroyed by such lubricants as stearic acid, and therefore lubricants are omitted from compounds when tack is important. Softeners or resins which are themselves sticky or tacky tend to give some tack to GR-S, but in most cases their action is nullified by the drying action of carbon black.

Of a number of synthetic resinous materials which have been examined as tackifiers[24, 85, 123, 156] for GR-S, the condensation product of acetylene and p-tert-butylphenol (Koresin) which was used in Germany in Buna S-3 stocks is perhaps the most effective resin in making GR-S sticky, but the extra building tack which it confers is only moderate when it is used in amounts that do not severely harm the physical properties of the rubber. To obtain adequate tack most success is achieved by replacing part of the GR-S with natural rubber. As little as 5 per cent of natural rubber imparts some tack, and greater amounts yield more tack. Natural-rubber cements have been used extensively to obtain adhesion of GR-S to itself; they have a toughness or "nerve" which retards lateral slippage. GR-S 10 which contains rosin acid instead of fatty acid has more tack than standard GR-S.[36]

Adhesion of GR-S. No particular problem occurs in uniting GR-S to cord fabric in tires.[62, 65, 84, 97] The best adhesives comprise substantial quantities of GR-S latex. The currently used resorcinol formaldehyde and casein base adhesives can be employed with GR-S equally as well as with natural rubber.

The prevention of the sticking together of uncured slabs or sheets of GR-S is less of a problem than with natural rubber. The same dusting materials such as soapstone, talc, and mica or the same aqueous dispersions of powders or solutions of soaps can be used for both rubbers.

Most of the lubricants added to improve processing also function somewhat as mold-release agents. Mold release is usually further facilitated either by a surface treatment of the stock or a surface treatment of the mold, similar materials being used with both GR-S and natural rubber and similar results being obtained.

CURATIVES

Comparative Vulcanization Behavior of GR-S and Hevea. The differences between natural rubber and GR-S in practical vulcanization characteristics are a matter of degree rather than kind. GR-S requires less sulfur to reach maximum properties, and the vulcanization proceeds at a slower rate. Generally, for soft-rubber goods only 1.5 to 2.5 parts by weight of sulfur is used per 100 parts of GR-S, in comparison to 2.0 to 3.5 parts of sulfur per 100 parts of natural rubber. The data in Table V, published by Dinsmore and Fielding,[42] illustrate the differences in the curing characteristics of GR-S and Hevea. The change from natural rubber to GR-S necessitated either an increase of approximately 14° C. in curing temperature or the use of more active acceleration.

Sulfur. Kemp and his co-workers[79] found the solubility of sulfur to be less in GR-S than it is in natural rubber at room temperature, although greater above 55° C., and the rate of diffusion of sulfur at 86° C. to be lower in GR-S than in Hevea. They concluded that the possibility of heterogeneous vulcanization as a result of incomplete solution or local high concentration of sulfur is greater in GR-S than in natural rubber. Mastication

Table V. Comparative Vulcanization Behavior of GR-S and Hevea[42]

Hevea		100		
GR-S			100	100
EPC black		50	50	50
Stearic acid		3		
Paraflux		3	6	6
Zinc oxide		3	3	3
Phenyl-β-naphthylamine		1		
Sulfur		2.8	2.8	1.6
Mercaptobenzothiazole		0.9	0.9	0.5
Diphenylguanidine				0.6
		163.7	162.7	161.7
Vulcanization temperature, °C.		135	149	135
Stress-strain	Time of Cure,			
Properties	Min.			
Modulus				
at 300%, p.s.i.	20	810	460	240
	30	1200	720	490
	40	1300	1020	800
	60	1520	1300	980
	80	1620	1580	1080
Tensile strength,	20	2850	1100	750
p.s.i.	30	3900	2350	1850
	40	4030	2750	2350
	60	4050	3000	2900
	80	4050	3000	2700
Elongation, %	20	600	600	705
	30	620	680	740
	40	590	600	630
	60	565	530	610
	80	560	480	545
Flex tests				
Cut strip, cold, min.*	40	456	92	150
	80	327	40	89
Cut strip, hot, min.*	40	600	40	58
	80	354	29	39
Tear tests, cold†	40	296	340	550
	80	198	203	259
Tear tests, hot†	40	373	184	335
	80	206	55	131
Resilience tests				
Rebound, cold, %	55	71.8	54.5	52.5
	95	70.7	55.0	54.0
Rebound, hot, %	55	81.8	64.7	61.5
	95	81.8	67.6	63.6

* Method 10 reported by Holt and Knox[68] run at 360 cycles per minute.   Overall stretch is 25% for cold (27° C.) test and 20% for hot (93° C.) test.

† Tear test uses the piece illustrated in Fig. 13 of Reinsmith,[105] but no weight is used on the pendulum.   Each scale unit represents 0.35 lb per in. of thickness.

Reprinted by permission of India Rubber World.

of GR-S increases the solubility of sulfur in it. The same investigators found the particle size of the sulfur to have marked influence on the rate of solution. In adding three parts of sulfur to 100 parts of GR-S it was observed that micronized sulfur of an average particle size of 3 to 4 microns had a much more rapid rate of solution than commercial grades of ground sulfur. However, the commercial grades of ground sulfur could be dispersed satisfactorily by proper mixing procedures.

The effect of sulfur content on the stress-strain properties of simple vulcanizates of GR-S and sulfur, in the absence of accelerator, has been studied by Cheyney and Robinson.[27] As the sulfur was increased, the tensile strength increased from very low values for soft vulcanizates to high values for hard rubber, and the elongation decreased slowly to low values for the hard vulcanizates.

In practical soft-rubber stocks, which contain accelerator, increases in sulfur content usually result in higher modulus, lower elongation, lower permanent set, greater tear resistance to an optimum point and thereafter reduced tear resistance, lower cut-growth resistance, higher rebound, and shorter age life. Carlton and Reinbold[25] found the maximum tensile strength to be developed at a sulfur level of 2.5 to 3.0 parts.

Sulfur is the most commonly used vulcanizing agent for GR-S. For certain special properties, such as superior heat age life, vulcanization with sulfur-liberating materials either without other sulfur or in combination with small quantities of free sulfur can be employed. Examples of such sulfur-liberating materials are thiuram disulfides[73, 141] and polysulfides,[137] the alkylphenol sulfides,[153] and N,N'-dithioamines.[137a]

Excellent heat resistance and flex life were found by Beaver and Throdahl[10] to result from the use of 4.0 per cent of ethylxanthogen disulfide as a source of sulfur and either 2.0 per cent of N-cyclohexylbenzothiazole-2-sulfenamide or 0.5 per cent of tetramethylthiuram disulfide as an accelerator. Small additions of *tert*-butyl-, xylyl-, and *p*-methylphenylthiols caused significant improvement in the flex-cracking resistance of vulcanizates containing ethylxanthogen disulfide. This method of securing vulcanization and the succeeding methods of vulcanization described in the present section have not, so far as is known, been used in industrial practice.

Special Curatives. The Peachey process of subjecting rubber alternately to sulfur dioxide and hydrogen sulfide gases has been found[11, 89] to vulcanize GR-S in similar manner to natural rubber but at a slower rate.

As in the case of natural rubber, materials other than sulfur can be used. Nonsulfur vulcanizing agents have been classified[2] as follows: (1) compounds that decompose thermally at vulcanization temperatures to yield free radicals, e.g., benzoyl peroxide, diazoaminobenzenes, and dichloroazodicarbonamidine; (2) oxidants of appropriate resonance structure, e.g., quinones, quinone oximes and imines, and polynitrobenzenes; (3) agents that yield free radicals on oxidation, e.g., aromatic amines and mercaptans, phenols, and dihydric phenols in combination with certain oxidizing agents. Some of these materials are relatively ineffective vulcanizing agents, whereas others, such as polynitrobenzenes,[132] produce rather good vulcanizates.

A unique class reported[133] to be primary vulcanizing agents for GR-S is composed of halogenated organic compounds falling into three groups as follows:

1. Chlorinated arylmethyl compounds containing at least one chlorine atom substituted in the methyl group, e.g., benzal chloride and o-chlorobenzotrichloride. Such compounds alone are capable of vulcanizing GR-S, but their effectiveness is greater in the presence of various metallic oxides. An example is as follows: A stock consisting of GR-S 100, MPC black 50, o-chlorobenzotrichloride 2.5, litharge 10, when heated for 30 minutes at 153° C. gave a cured product having a 300 per cent modulus of 1710, a tensile strength of 3140, and an elongation of 430 per cent. The heat buildup of the vulcanizate was similar to that of a corresponding GR-S stock cured with sulfur and accelerators.

2. Halogenated aliphatic hydrocarbons containing at least one CX_3 group, where X is any halogen other than fluorine, e.g., hexachloroethane and 1,1,1,3-tetrachloropropane. These agents require the presence of litharge in order to produce vulcanization.

3. Aliphatic compounds containing a trichloromethyl group attached to a strongly polar group, e.g., trichloroacetic acid and trichloromethane sulfochloride. These agents alone will produce vulcanization but are markedly activated by a small proportion of zinc oxide and also by sulfur.

Vulcanizates produced by polyhalogen compound of the above classes have in general good resistance to heat aging. For example, whereas, after heat aging for 2 days at 100° C., a GR-S black stock cured with sulfur and accelerators rose in 200 per cent modulus from 575 to 1600 p.s.i. and fell in elongation from 460 to 260 per cent, a similar stock cured with 1,1,1,5-tetrachloropentane and litharge suffered corresponding changes of only 400 to 475 p.s.i. and 530 to 520 per cent. It is supposed that vulcanization by polyhalogen compounds follows a free radical mechanism. In various examples the chlorine content of the vulcanizates was 0.1 to 0.5 per cent.

ACCELERATORS

The best accelerators for the vulcanization of GR-S are also good accelerators in natural-rubber vulcanization. However, their specific behavior in GR-S is different from their behavior in natural rubber. The relatively slower rate of vulcanization requires greater acceleration in GR-S. In general, mercaptobenzothiazole and its derivatives, thiuram sulfides, and dithiocarbamates are considered the best accelerators for GR-S. The combination of a thiazole and an auxiliary accelerator has become popular. Vila[141] in experiments with a variety of accelerators, found that, with a fixed quantity of 2 parts of sulfur, there was for each accelerator an optimum proportion at which the tensile strength was maximal, and that this strength was not significantly different with different accelerators. Increase in the proportion of accelerator beyond the optimum proportion led to reduction in the tensile strength and elongation and to increase in the modulus.

Classes of Accelerators. In a review of the use of accelerators in GR-S-type polymers, Neal[99] made the following classification.

1. *Basic-Type Accelerators.* Guanidines, aldehyde-amines, etc. generally are very weak accelerators for GR-S. Even the more active aldehyde-amine accelerators, such as butyraldehyde–aniline, must be used in amounts of more than 2 per cent in order to obtain reasonably rapid curing. Several, however, have found extended use as secondary accelerators in combination with acidic-type accelerators and with thiuram accelerators.

2. *Acidic-Type Accelerators and Their Derivatives.* Members of this group, including thiazoles, thiazolines, and their derivatives, are very useful in GR-S. Some can be used successfully alone, and some are used principally in combination with a secondary accelerator.

3. *Thiuram Sulfides.* Thiuram monosulfides, thiuram disulfides, and thiuram polysulfides are very good accelerators for GR-S and can be used in GR-S with less danger of scorching than in natural rubber.

4. *Dithiocarbamates.* These accelerators, such as, for example, zinc diethyldithiocarbamate, are extremely active. Since stocks containing them are scorchy, they are seldom used except where curing at relatively low temperature is desired.

Table VI shows an estimate of the consumption of accelerators of various classes by the American rubber industry in 1948, a year in which the total usage of raw rubber was 1,069,404 long tons, of which 32.2 per cent was GR-S. It also shows data for the production of accelerators in America in 1951, a year in which the total usage of raw rubber was 1,214,298 tons, of which 51.6 per cent was GR-S. (Production of accelerators, although probably somewhat higher than actual consumption, is probably closer to the latter than are sales figures, because some companies make accelerators for their own use.)

Table VI.   Accelerators in the United States[55]

In Millions of Pounds

	1948 Consumption (Estimated)		1951 Production
Aldehyde–amines	1.6	Butyraldehyde–aniline	1.025
Guanidines	5.2	Guanidines	6.373
Thiazoles	35.6	Mercaptobenzothiazole	18.160
Dithiocarbamates	1.2	Benzothiazolyl disulfide	17.118
Thiuram sulfides	3.2	Other thiazoles	11.905
		(Total thiazoles, 47.183)	
Total	46.8	Other cyclic compounds	1.400
		Zinc diethyldithiocarbamate	1.354
		Tetramethylthiuram disulfide	4.577
		Tetramethylthiuram mono- and tetrasulfides	0.814
		Other acyclic compounds	4.850
		Total	67.576

In addition to cost considerations the selection of the type and the amount of curative and accelerator depends on the temperature and speed of processing, the period of cure allowed, the influence of other compounding ingredients, and the physical properties and aging characteristics desired in the vulcanizate. Neal[99] has pointed out that GR-S is more sensitive than natural rubber to the proper balance between sulfur and accelerator concentrations. The fact that vulcanization is less readily induced in GR-S than in natural rubber allows greater freedom from scorch during processing and necessitates the use of somewhat greater acceleration.

Basic Accelerators. Vila showed that, as primary accelerators for GR-S, aldehyde–amines and di-o-tolylguanidine are not only comparatively weak but also are more prone to cause prevulcanization (scorching) than the thiazoles and dithiocarbamates.[141] Although the basic-type accelerators are generally weak primary accelerators for GR-S, certain combinations of organic bases with magnesia or sodium hydroxide were reported by Breckley[20] to be powerful. An example of such a combination which yields good physical properties in the vulcanizate is 2.0 parts of magnesia, 2.0 parts triethanolamine, and 2.0 parts of sulfur. Other strong bases stated by the same investigator to produce similar results are long-chain aliphatic amines, benzylamine, piperidine, and morpholine.

Breckley found the weak bases, pyridine, aniline, and hexamethylene tetramine to have little activity. Urea, acetamide, and long-chain aliphatic amine acetates are inactive alone but are rather active in the presence of magnesia.

Thiazoles. The acidic-type accelerators and their derivatives, as for example mercaptobenzothiazole and N-cyclohexylbenzothiazole-2-sulfenamide, can be used to produce good physical properties in GR-S stocks. These accelerators are generally considered to be persistent, that is, to continue to develop higher modulus and lower elongation during extended cure or high-temperature aging.

A study by Sperberg[124] indicated that varying the accelerator N-cyclohexylbenzothiazole-2-sulfenamide concentration from 0.75 to 3.0 parts per 100 parts of polymer has negligible effect on the scorch characteristics. This author presents a systematic study of the effect on the properties of cold GR-S tread stocks of the 16 combinations of sulfur and accelerator possible with sulfur levels of 0.5, 1.0, 2.0, and 3.0 and accelerator levels of 0.75, 1.37, 2.0, and 2.75. When the level of the curative combination is lowered, in general, resistance to abrasion is affected very little, cut growth resistance improves, and resilience falls off.

Thiuram Sulfides. In comparison to the relatively slow development of physical properties and the continued change in physical properties beyond the optimum cure as obtained with the persistent accelerators, the thiuram sulfides cause a rapid development of physical properties and a fairly steady retention of physical properties beyond optimum cure. A typical curing combination of this type consists of 2.0 parts by weight of sulfur and 0.4 part of tetramethylthiuram monosulfide.[73] An evaluation of a homologous series of tetraalkylthiuram sulfides by Wolf and deHilster[154] showed in general that an increase in the alkyl group decreases the activity. The ethyl derivatives are less scorchy and give better bin age life than the corresponding methyl derivatives. The tetraalkylthiuram disulfides are more active than the monosulfides.

Dithiocarbamates. When used alone as the primary accelerator, some of the dithiocarbamate accelerators produce relatively nonpersistent cures and good physical properties in GR-S. The order of decreasing accelerator activity for the metal dialkyldithiocarbamates is copper, selenium, zinc, and lead.[154] The selenium dialkyldithiocarbamates promote more rapid vulcanization than the corresponding tetraalkylthiuram disulfides.[154] The scorchiness of the dithiocarbamates limits their use.

Secondary Accelerators. The popular usage of the basic type of accelerators as secondary accelerators is derived primarily from efforts to modify the rate of cure, while at the same time maintaining desirable physical properties. The persistence of the thiazole acceleration is lessened by reducing the concentration of the thiazole and adding a secondary accelerator. Basic accelerators such as di-*o*-tolylguanidine and diphenylguanidine are often employed as the secondaries. Such combinations are relatively nonscorchy, yet produce tight cures. The economy in cost of certain accelerator combinations is sometimes a factor in their selection.

Also, the thiuram sulfides and the dithiocarbamates find use as secondary accelerators. A typical curing combination of this sort given by Neal[99] is 3.0 parts by weight of sulfur, 0.5 part tetramethylthiuram monosulfide, and 0.5 part benzothiazolyl disulfide. Breckley[20] states that a combination of triethanolamine and tetramethylthiuram monosulfide will cause vulcanization of GR-S at room temperature.

Inorganic Accelerators. Certain inorganic substances can be used to accelerate the sulfur vulcanization of GR-S. Fisher and Davis[51] studied a number of inorganic oxides and oxidizing agents in GR-S–sulfur mixtures and found that magnesia, litharge, and calcium hydroxide led to curing. The further addition of inorganic oxidizing agents, such as red lead, lead dioxide, ferric oxide, and lead chromate, accelerated the vulcanization rate and increased tensile strength. For example, a stock consisting of GR-S 100, sulfur 2, magnesia 10, red lead 5 had a tensile strength of 715 p.s.i. after curing for 60 minutes at 144° C. In comparison with organic accelerators the inorganic accelerators are relatively weak; hence they find very little commercial application.

Activators and Retarders

Metallic Oxides. With few exceptions the action of accelerator activators is similar in GR-S to their action in natural rubber. Most of the activators used in GR-S are basic in character or decompose to yield bases. Typical activators are zinc oxide, litharge, magnesium oxide, organic amines, alkali carbonates, and alkali hydroxides. It is common practice to include 1 to 5 per cent of zinc oxide in GR-S stocks, particularly when thiazole acceleration is used. With most commonly used accelerators, as the zinc oxide concentration is increased up to approximately 3.0 per cent, a tighter cure is obtained. Beyond this concentration only minor additional activation occurs. Litharge is a powerful activator of mercaptobenzothiazole acceleration. Svetlik, Cooper, and Railsback[134] concluded that, in the vulcanization of cold GR-S in black stocks with 1.75 parts of sulfur and 0.95

part of N-cyclohexylbenzothiazole-2-sulfenamide as the curatives, the quality of the vulcanizates was at an optimum with 1.0 part of zinc oxide. Beyond this level of zinc oxide, they found the abrasion resistance and the flex life after aging to fall appreciably. Other metallic oxides, such as mercuric oxide, are even stronger than zinc oxide as activators of mercapto-benzothiazole and cause stocks to scorch during milling. As stated under accelerators, magnesuim oxide markedly activates many basic accelerators.

The selection of activators or combination of activators has considerable influence on the rate of vulcanization, and on the physical properties and aging characteristics of the vulcanizates. Henricks[67] has cited a GR-S vulcanization mixture of 1.0 per cent benzothiazolyl disulfide, 1.5 per cent litharge, and 3.5 per cent zinc oxide as being capable of curing a 0.5 per cent sulfur stock at a practical rate without a sacrifice in the vulcanizate of the good heat stability characteristic of low sulfur stocks.

The persistence of accelerators can be modified by proper selection of the type and the amount of activator. For example, with magnesium oxide less accelerator persistence is evidenced than with zinc oxide.

Fatty and Rosin Acids. In addition to the curative, the accelerator, and the inorganic activator, the nonrubber constituents in the various GR-S polymers can be considered as being part of the vulcanization system. In standard GR-S the fatty acid present tends to retard the cure slightly in the early stages; it then behaves as a mild activator after the initial stages of cure with mercaptobenzothiazole acceleration. The addition of more fatty acid than that present in the raw GR-S is seldom necessary for activating the accelerator.

The rosin acid derivative present[36] in GR-S 10 and cold rubber and the aluminum salts and soaps in GR-S-AC cause a slower rate of cure with acid-type accelerator systems than the fatty acids in GR-S. Consequently, organic activators are often added to stocks of GR-S 10, cold rubber, and GR-S-AC stocks. These may include organic acids and their salts, organic amines, and amine soaps. Stearic acid, zinc stearate, and triethanolamine are typical materials used.

Retarders. In very rapid curing combinations it is sometimes necessary to use accelerator–retarders to allow mixing and processing without scorching. Benzoic acid, salicyclic acid, N-nitrosodiphenylamine, 1-chloro-1-nitro-propane, and phthalic anhydride are recommended for this use. Their relative effectiveness varies with the type of accelerator. Although they retard incipient cure at processing temperatures, some retarders exert a mild activation of accelerators at curing temperature. Because GR-S is less susceptible to scorch than natural rubber, retarders are not generally used in normal stocks. Their use is mainly in rapid-curing wire insulation, cement, or other specialty stocks.

VULCANIZATION

Temperature Coefficient of Vulcanization. The temperature coefficient of vulcanization, defined as the ratio of the time at a given temperature to the time at a 10° C. higher temperature to produce a similar state of vulcanization, has been found to be approximately the same for GR-S and natural rubber.[26] A value of 1.95 for a 10° C. change was obtained from physical test data over the range of 132 to 149° C. using sulfur contents varying from 1 to 5 per cent in a compound accelerated with benzothiazolyl disulfide, both with and without carbon black reinforcement. From combined sulfur analyses a coefficient of 2.06 was calculated for the same stocks. These results were essentially substantiated by another investigation[73] in which it was found that the rate of cure, determined on the basis of physical test data, doubled for an increase of 10° C. over the range of 127 to 149° C. using various accelerator systems. No significant difference between the physical properties at equivalent states of cure was found over the range of vulcanization temperatures 127 to 149° C.

Heat of Vulcanization. The heat evolved during vulcanization is nearly the same for both GR-S and natural rubber.[22] When less than 2 per cent sulfur is present, little or no heat is evolved. With greater concentrations of sulfur, however, the reaction is more exothermic. A study by A. E. Juve and Garvey[73] showed that with increase in the thickness of the rubber less additional time is required for GR-S than for natural rubber to reach a given state of cure. They found, for example, that a tread-type GR-S stock, as a slab 2 in. thick, when cured at 138° C. required heating for about 50 minutes longer than as a slab $1/_{16}$ in. thick, in order to cure it in the center, whereas, they calculated, the corresponding increase in the period of cure necessary to cure a 2 in. slab of natural-rubber tread stock was about 80 minutes. These workers found insufficient differences between tread stocks of GR-S and of Hevea in thermal conductivity and specific heat and attributed the difference in the curing of thick sections to an exothermic reaction. Morris and co-workers,[95] however, believe the difference to be due to the 16 per cent greater heat diffusivity that they observed for GR-S.

An interesting difference between GR-S and natural rubber in the effect of pressure during vulcanization has been observed by Wilkinson and Gehman.[147] At the high molding pressure of 100,000 p.s.i. the rate of cure of Hevea appears to be retarded, whereas GR-S seems to cure more rapidly than at conventional pressures.

Change of Physical Properties with Progress of Vulcanization. Vulcanization produces marked alteration of physical properties. As the extent of vulcanization progresses, as a result of a higher curing temperature, a longer curing time, or changes in the curative-accelerator system, the physical properties of GR-S usually are affected as follows:

Increase	Decrease	Increase to a Maximum and Then Decrease
Modulus	Ultimate elongation	Tensile strength
Resilience	Permanent set	Tear resistance
Abrasion resistance	Flex life and cut growth	
Speed of retraction	Swelling in solvents	

When vulcanization is extended beyond the optimum point, GR-S usually continues to increase in modulus and hardness instead of undergoing the reversion or softening phenomenon which frequently takes place with Hevea. In curing with certain dithiocarbamates and other ultra-accelerators, however, a reversion in these properties takes place even with GR-S.

Choice of State of Cure. The optimum cure, from the practical standpoint, is that state of cure at which the best balance of all physical properties is obtained, and not necessarily merely a time at which the various chemical processes have reached a certain point. In determining the optimum state of cure of a stock, the balance of physical properties most desirable for the particular application in view for the product must be considered. Further, in comparing various experimental polymers or various compounds, consideration must be given to the peculiarities of the stocks in question. For example, comparison of the physical properties at equal moduli may be satisfactory when GR-S polymers that differ only in soap content are being compared. However, modulus would not be a satisfactory criterion of state of cure in the comparison of GR-S polymers of widely varied Mooney viscosity.

Since the tear resistance of GR-S decreases rapidly beyond an optimum state of cure, the evaluation of this property can be used as a means of determining the extent of cure.

Permanent set is often used as a criterion of cure. For example, Sperberg[124, 125] used a relaxed compression test, in which, employing the apparatus specified in ASTM compression set method B, the sample is subjected to 35 per cent deflection for 2 hours at 100° C. and then allowed to relax for 1 hour at 100° C. The permanent set value for GR-S products varies considerably with the state of cure. In many products reasonably low plastic flow or permanent set is required. Since, for example, the permanent set is of much greater consequence in a tire carcass stock than it is in a tread stock, the interpretation of test results requires an intimate knowledge of service requirements. The determination of permanent set under dynamic vibration, as in the Goodrich Flexometer, is useful in studying the state of vulcanization.

Cut-growth resistance is sensitive to the extent of cure in GR-S stocks, and particular care must be exerted in the interpretation of cut growth data. An inverse relationship between cut-growth resistance and hysteresis at various levels of vulcanization has been shown to hold.[72] The balance between flex life or cut growth and the hysteresis characteristics in GR-S is of great importance for many applications. The hysteresis characteristics of synthetic rubbers are less dependent on the extent of cure than are their flex properties. In general, the results of resilience and hysteresis tests do not vary nearly so much as cut-growth results within normal ranges of cure. Hence, qualitative comparisons of the dynamic properties of different rubbers are likely to be more significant than comparisons of the cut-growth resistance unless the degrees of cure are considered carefully.

A method of measuring the speed of retraction by an electronic timing circuit with photocell input can be utilized to determine the state of cure.[127]

The analysis of combined or free sulfur offers interesting information regarding the course of the vulcanization of GR-S. Used as a tool to determine state of cure, combined sulfur analyses must be interpreted with a knowledge of the course of sulfur combination over a wide range of cures in the particular stock being studied.

EFFECT OF POLYMER IMPURITIES ON RATE OF CURE

The amount and type of impurities present in the various GR-S polymers influence the rate of vulcanization. GR-S 10, with its rosin acid, has a slower rate of cure than standard GR-S, with its fatty acid. This, of course, can be counteracted by using a stronger accelerator-activator system.

Available data are somewhat conflicting, but various soaps may be expected to have some effect on rate of cure. One study[139] indicated that, as the soap in standard GR-S was decreased below 0.35 per cent, the rate of cure increased. The maximum soap content permitted under the specifications for GR-S is 0.75 per cent. Other nonrubber constituents, such as sodium chloride, also may affect the cure rate.

Moisture concentration has been found to affect the rate of vulcanization[19, 109–10] of GR-S. Rush and Kilbank[111] found that, as the amount of moisture retained by a GR-S stock is increased, a retarding effect on cure occurs up to a moisture content of 0.25 per cent, that thereafter the rate of cure increases, but that even at 0.5 per cent moisture (the maximum permitted under the specifications for GR-S) the rate is not so high as that of bone-dry stock.

THEORETICAL CONSIDERATIONS REGARDING VULCANIZATION

The mechanism of vulcanization of GR-S is generally believed to be similar to that of natural rubber, which has been reviewed by Williams.[148] The lack of structural symmetry has been pointed to by Kemp and Straitiff[80] as probably responsible for the inability of GR-S to cross-link effectively, during vulcanization with sulfur, and its consequent low tensile strength.

Flory[52–3] and Gee[58] have described the structure of vulcanized polymer as a network of primary molecules of given molecular weight and molecular-weight distribution randomly interlinked by cross-linkages. Some of the cross-links may be formed during polymerization or during processing; the remainder during vulcanization. It is the total number of cross-linkages that is significant from the viewpoint of the physical properties of the vulcanizate. The relatively large concentrations of cross-linkages required for the formation of a network structure in polymers of low molecular weight is believed by Bardwell and Winkler[8] to handicap their reversible elasticity. The influence of the molecular weight of GR-S on its vulcanization characteristics was made clear by Yanko[155] by vulcanizing fractions of GR-S. With a sulfur content of 1.75 per cent, the tensile strength increased linearly with increase in molecular weight until a limiting value appeared to be reached at a number-average molecular weight of approximately 400,000. The percentage of sulfur necessary to give, from the lower-molecular-weight fractions, vulcanizates having a 300 per cent modulus of 1000 p.s.i. increased as an inverse function of the molecular weight. During processing and compounding, GR-S may undergo mechanical breakdown, oxidative breakdown, and gel formation. Therefore, in correlating molecular structure with physical properties, Yanko stated that the vulcanized structure of GR-S is determined by the average molecular weight or the molecular-weight distribution immediately before vulcanization and the amount of effective cross-linkage that occurs during vulcanization.

CARBON BLACK REINFORCEMENT*

Comparative Effect of Blacks in GR-S and Hevea. Good physical properties in GR-S vulcanizates are obtained only if reinforcing agents are included. In a gum stock containing only the ingredients essential for vulcanization, GR-S has a tensile strength of no better than approximately 300 to 400 p.s.i. By reinforcement, with, for example, carbon black, the tensile strength can be raised to slightly over 3000 p.s.i. In comparison the tensile strength of a natural-rubber gum stock containing only the curative essentials may be in the vicinity of 4000 p.s.i., and carbon black reinforcement will cause an increase to perhaps 4900 p.s.i.[108] A comparison of the behavior of carbon blacks in natural and synthetic rubbers was made by Braendle.[17] He concluded that the behavior of carbon black is a function of its pH, specific surface, and structure as follows: Increased pH accelerates the cure of normal GR-S as well as of natural-rubber compounds. The extent of the carbon black surface is the most important factor in relation to tensile strength of both GR-S and Hevea. Although the surface is only of minor importance in relation to modulus in natural rubber, it has important bearing on the modulus of GR-S. Increased carbon surface decreases the rebound of GR-S as well as that of Hevea. The rate of cure of GR-S is accelerated slightly by increased carbon surface. The carbon structure has somewhat similar effect on both GR-S and Hevea, and a higher modulus (i.e., greater stiffness) is developed by the structure carbons than is to be expected from their surface area.

In a discussion of pigment reinforcement, Parkinson[100] has expressed the opinion that the increase in the abrasion resistance of both GR-S and Hevea produced by such reinforcement depends on the same mechanism, because the distortions involved in abrasive wear are in both rubbers lower than those that give diffraction patterns, but that the increase in tensile strength produced by carbon black possibly depends on a different mechanism in the two rubbers, because crystallization of the rubber chains occurs when Hevea is stretched to the point of rupture but not when GR-S is so stretched.

Characteristics of Various Types of Black. In view of the importance of carbon blacks in GR-S, unpublished data from Drogin[45] have been tabulated in Table VII to describe the major types.

---

* For a description of the manufacture of carbon blacks and of the grades used in rubber, consult W. R. Smith, *Encyclopedia of Chemical Technology*, 3, pp. 34–65, Interscience Encyclopedia, Inc., N.Y., c. 1949. For a briefer account of modern carbon blacks, consult C. W. Sweitzer, *J. Chem. Ed.*, **29**, 493–502 (1952).

Table VII. Physical and Chemical Characteristics

Type		Representative Brand	Mfr.	Surface Area, Sq. Meters per Gram.	Calculated Average Diameter, Milli-microns	Color, Nigro-meter
EPC	(Easy-processing channel)	Kosmobile 77*	1	106.8	30.0	85.4
MPC	(Medium-processing channel)	Kosmobile HM*	1	120.4	26.0	85.0
HPC	(Hard-processing channel)	Kosmobile*	1	165.0	18.9	82.7
SRF	(Semireinforcing furnace)	Kosmos 20*	1	19.5	160.0	103.3
		Lampblack†		16.1	193.9	106.8
HMF	(High-modulus furnace)	Kosmos 40*	1	33.0	95.0	99.3
		Statex-93*	2	38.6	80.9	101.2
MAF	(Medium-abrasion furnace)	Kosmos 50†	1	34.5	93.0	99.1
	or	Philblack A†	3	29.0	80.0	95.7
FEF	(Fast-extrusion furnace)	Statex M	2	No information		
		Sterling SO†	4	42.1	74.2	101.5
FF	(Fine furnace)	Kosmos 80*	1	43.7	71.5	94.6
		Sterling 99*	4	68.0	44.5	95.3
		Statex-B*	2	76.1	41.0	94.6
		Continex FF*	5	35.3	88.5	97.8
VFF	(Very fine furnace)	Statex-K*	2	63.4	47.8	90.8
RF	(Reinforcing furnace)	Kosmos 60†	1	81.4	38.0	90.7
		Sterling 105*	4	95.7	32.7	92.0
HAF	(High-abrasion furnace)	Philblack O†	3	69.2	45.2	92.2
		Vulcan 3†	4	80.6	38.8	94.3
		Aromex†	6			95.1
CC	(Conductive channel)	Voltex*	1	387.0	8.1	62.3
CF	(Conductive furnace)	Acetylene*	7	56.3	55.5	94.1
		Statex-A*	2	62.8	49.8	92.0
MT	(Medium thermal)	Thermax*	8	6.6	472.8	116.2
FT	(Fine thermal)	P-33*	8	18.1	172.4	106.0
German		CK-4		85.1	37.0	85.9
		P-1250		74.1	42.0	85.7

* From natural gas.
† From oil.
‡ Brown when compared with furnace black; blue when compared with channel black.
§ Brown when compared with both furnace and channel blacks.
‖ Very brown when compared with furnace black; same undertone as channel black.

f Various Types of Rubber-Grade Carbon Blacks[45]

Tint Value	Tint Under-tone	pH	Mois-ture %	Volatile Matter, %	Ash, %	Benzol Extract, %	Iodine Absorp-tion, %	Oil Absorp-tion Factor	Electrical Resis-tance, Ohms
364	Very brown	4.3	1.41	5.76	0.05	0.05	15.6	112	173.0
375	Very brown	4.2	2.30	5.52	0.06	0.03	17.4	116	52.0
383	Brown	4.3	1.09	6.10	0.15	0.05	34.1	119	9.0
240	Blue	9.4	0.13	1.08	0.26	0.13	2.2	87	2.9
173	Blue	6.3	0.17	0.93	0.06	0.09	2.3	169	0.9
287	Blue	9.0	0.20	1.27	0.53	0.13	3.1	83	3.0
324	‡	8.7	0.26	0.90	0.15	0.09	4.5	123	2.0
286	Blue	8.9	0.20	1.78	0.09	0.08	3.8	170	1.0
301	‡	8.5	0.54	1.09	0.13	0.12	4.5	172	1.8
321	Blue	7.6	0.21	0.67	0.08	0.04	12.9	155	0.3
352	‡	8.9	0.21	1.41	1.43	0.11	7.0	84	1.8
347	Blue	10.2	0.50	1.35	1.38	0.04	8.3	95	1.5
348	Blue	9.2	0.33	1.23	0.69	0.15	5.7	107	2.5
320	‡	8.6	0.16	0.69	0.28	0.34	4.5	90	3.0
375	‡	8.9	0.63	1.70	0.74	0.28	9.0	122	3.0
347	Blue	8.8	0.18	1.33	0.29	0.09	13.2	187	1.1
362	‡	8.6	1.05	1.07	0.35	0.17	13.6	119	1.0
358	‡	8.6	0.93	1.24	0.12	0.12	9.5	150	1.2
364	Blue	8.0	1.76	1.69	0.66	0.27	21.8	130	0.5
379	Blue	8.8	0.64	1.26	0.23	0.12	22.2	133	0.3
394	Brown	4.6	1.30	5.05	0.03	0.01	41.1	163	1.9
300	Blue	5.1	0.14	1.81	0.03	0.03	8.9	290	0.5
365	§	8.1	0.87	2.71	1.50	0.02	13.1	109	1.5
100	Blue	7.4	0.13	0.96	0.06	0.11	0.4	39	2.7
241	Blue	7.2	0.26	1.12	0.12	1.31	2.1	59	950.0
377	‖	3.4	2.68	5.08	0.05	0.53	14.1	127	42.0
342	‡	5.4	0.27	0.72	0.03	0.03	10.1	176	1.3

. United Carbon Co.

. Columbian Carbon Co., Binney and Smith Co.

. Phillips Chemical Co.

. Godfrey L. Cabot, Inc.

5. Continental Carbon Co., Witco Chemical Co.

6. J. M. Huber Corp.

7. Shawinigan Chemicals, Ltd.

8. Thermatomic Carbon Co., R. T. Vanderbilt Co.

In addition to the representative brands listed in Table VII, the following typical blacks also fall in the various classes:

EPC (*Easy-Processing Channel*)
Continental AA*
Micronex W6†
Spheron no. 9‡
Texas E§
Wyex‖

HPC (*Hard-Processing Channel*)
Continental F*
HX‖
Micronex MK II†
Spheron no. 4‡

HMF (*High-Modulus Furnace*)
Continex HMF*
Modulex‖
Sterling L‡

CC (*Conductive Channel*)
Continental R-40*
Spheron N or C‡

MPC (*Medium-Processing Channel*)
Arrow TX‖
Continental A*
Micronex Standard†
Spheron no. 6‡
Texas M§

SRF (*Semireinforcing Furnace*)
Continex SRF*
Essex‖
Furnex†
Pelletex¶
Sterling NS and S‡

HAF (*High-Abrasion Furnace*)
Continex HAF*
Kosmos 60 Improved**
Statex R†

CF (*Conductive Furnace*)
FB-200‖
Sterling I‡

* Continental Carbon Co., Witco Chemical Co.
† Columbian Carbon Co., Binney and Smith Co.
‡ Godfrey L. Cabot, Inc.
§ Sid Richardson Carbon Co.
‖ J. M. Huber Corp.
¶ General Atlas Carbon Co.
** United Carbon Co.

In Table VIII are described the particle configurations of some of the common blacks as observed from an electron microscopic examination. The data in Tables VII and VIII together provide a good general description of the particle size and shape and the surface characteristics of the various types of rubber blacks.

New Types of Blacks. Although MPC and HPC blacks were used to a large extent in natural rubber, their poor processing characteristics in uncured stock and the high heat generation in vulcanizates to which they lead are highly undesirable in GR-S, which is already deficient in both these respects. The relative advantages of EPC black in processibility and heat generation have promoted its use. With 50 parts black, EPC treads have been reported[49] to run approximately 8° C. cooler than GR-S treads containing MPC black. To improve processibility and achieve a better balance of physical properties, HMF or SRF blacks are often blended with EPC. Recently there has been a strong trend toward certain types of furnace blacks. The newer types of furnace blacks include HMF, FF, VFF, RF, HAF, and FEF. Two still newer furnace blacks, rather similar to HAF but of finer particle size, have been introduced under the designations ISAF and SAF. The use of SAF (Super Abrasion Furnace) black results in a significant improvement in road wear. ISAF (Intermediate SAF) is

Table VIII. Microscopic Classification and Particle Configuration of Some Common Carbon Blacks[149]

Trade Name and Type of Black		W-6 (EPC)	Gastex (SRF)	Philblack A (HMF)	Statex A (CF)	Shawinigan (CF)	P-33 (FT)	Thermax (MT)
Particle Shape — Nonlinear	Spheres	Present	...	...	...	...	C	C
	Approximate spheres	Present	Present	...	Present	...	C	C
	Compact clusters	Present	Some	Present	Present	...	C	C
Particle Shape — Linear	Nonbranching	Many (C)	Present	...	Present	Present	...	...
	Simple branching	Many (C)	Present	Present	Present	C	...	...
	Multiple branching	Present	Some	Many	Present	Present	...	...
	Flat clusters	Present	Some (C)	...	Present	Some	...	...
	Rangy	Present	...	C	...	...	...	...
	Stocky	Present	C	...	...	...	...	...
Size	Small	Mostly	...	...	Present	Tiny (C)	Present	...
	Medium	...	...	Mostly	Present	...	Mostly	...
	Large	...	Mostly	...	Present	...	...	Very large
Size Variation		Some	Little	Some	Marked	Little	Some	Some
Subunits — Rounded — Indentations	Prominent	...	C	...	Present	...	...	...
	Moderate	...	...	C	Present	Mostly	...	...
	Slight	Mostly	...	...	Present	Mostly	Present	Present
	Ofiset	Mostly	Mostly	Mostly	Mostly	Mostly	...	...
	Single rows	Mostly	Some	Mostly	Present	Present	...	...
	Multiple rows	Present	Some	Some	Present	Mostly	...	...
	Doublets	...	Some C	...	...	...	Some	Some
Crystallinity (Bragg)		...	...	...	...	C	...	...
Geometric, flat		...	...	...	...	Some (C)	...	...
Size range of subunit diameter		10–40*	16–150	7–90	8–100	5–30	6–600	–1500

C. Characteristic.

* Units in millimicrons. Does not mean particle diameter, particle length, particle volume, or particle size.

intermediate between HAF and SAF in particle size and in the improvement it produces in abrasion resistance.

The continuous furnace process allows the introduction of wide variations in the properties of the blacks by virtue of its flexibility of operation and the fact that different types of raw materials can be used. Many of these new furnace blacks possess rubber reinforcing abilities in the general range of those of channel blacks and at the same time have the advantage of superior processing properties. In general, the fine furnace blacks are fast curing. Dannenberg and Collyer[37] have attributed at least part of the superior physical properties obtained with HAF black to the fact that, unlike the channel blacks, it exerts no retardation of cure. A similar statement probably could be made for RF and VFF blacks.

The same investigators have shown that calcination of channel blacks modifies their effect on the curing characteristics of stocks containing them. Although HAF, RF, and VFF blacks are superior to channel blacks of approximately equivalent particle size in the abrasion resistance that they produce, Dannenberg and Collyer found that MPC which has been calcined is approximately equal to HAF black in respect of abrasion resistance. The calcination of channel black removes volatile matter from the surface of the particles, changes the pH from acid to alkaline, and, presumably by destroying its polar character, renders the black more easily wettable by rubber. Table IX shows the behavior of the calcined channel black in a cold GR-S tread stock in comparison with the original black and with an HAF black.

Table IX. Effect of Calcining Channel Black on its Behavior in Cold GR-S[37]

	Original MPC Black	Calcined MPC Black	Original HAF Black
Volatile, %	4.9	1.6	0.9
pH of water slurry	5.2	9.6	9.5
Modulus at 300%, p.s.i.			
30 min. at 144° C.	740	1600	1550
90 min. at 144° C.	1590	2020	2180
Tensile strength, p.s.i.	4120	4200	3950
Elongation, %	540	500	470
Abrasion resistance, % of HAF	90	100	100
Rebound, %	55.2	54.2	60
Torsional hysteresis	0.243	0.204	0.134
Electrical resistance, megohms-cm.	100.0	0.300	0.006
Mooney viscosity (ML/212-4)	68	71	73

Reprinted by permission of *Ind. Eng. Chem.*

Behavior of Various Blacks in GR-S. Some data showing inter alia the effect of various grades of carbon black on the physical properties of vulcanized GR-S have already been given in Table III. Drogin, Bishop, and Wiseman[48] examined the behavior of 12 commercial blacks, representing 8 types of black, in regular GR-S, in GR-S made at three different low temperatures (5, 10, and −18° C.), and in natural rubber. The data obtained are so extensive that references must be made to the original paper for a full assessment of them.   Here only some of the highlights can be mentioned.   In all four types of GR-S the following relations were observed. RF and HAF blacks led to fast curing, EPC black to slow curing, and VFF to an intermediate rate.   RF black imparted high modulus, and HAF (Philblack 0), VVF, HAF (Vulcan 3), and EPC progressively lower moduli. In tensile strength VFF, HAF (Philblack 0), RF, and HAF (Vulcan 3) were progressively lower than EPC.   RF and HAF (Philblack 0) gave the best resistance to abrasion.

The resistance to road wear of the FF blacks was stated by Stokes and Dannenberg[130] to be 80 to 90 per cent of that of channel black, and the RF blacks were said to be equal or, in some cases, superior to channel blacks in the same property.   In a road test on 6.00–16 tires run for 22,320 miles at 60 m.p.h. on paved roads at a mean prevailing temperature of 14° C. these authors found the average mileage per 0.001 in. of treadwear to be 112.8 for treads containing EPC black and 105.6 for treads containing RF black.   In a laboratory comparison of the two blacks in GR-S, RF black caused the more rapid vulcanization, but in general the properties of the vulcanizates at optimum cure were not very different for the two blacks.   The greatest difference between stocks containing the two blacks was in electrical resistance, which was 1×10^5 megohm-cm. less for RF than for EPC black. Correspondingly, the furnace black gave better processing than the channel black, as judged by the smoothness of extruded treads.

Braendle and co-workers[18] pointed out that the high-pH VFF black Statex K produces outstanding age life and cut-growth resistance in GR-S, but introduces a problem of scorching and rapid cure rate.   The suggestion was made by these authors that scorching and cure rate can be modified by using softeners of retarding action in addition to making a minor adjustment in sulfur and accelerator.   Sperberg, Svetlik, and Bliss,[126] in five road tests with passenger-car size tires found 8 to 19 per cent greater resistance to road wear with the HAF black Philblack 0 than with EPC black in GR-S.   The percentage improvement in tread wear obtained with HAF black in comparison to EPC was found by Fielding[50] to be approximately the same in GR-S, cold rubber, and natural rubber.   Similarly, Sjothun and Cole[122] found two fine furnace blacks, HAF and VFF, to give significant improvement in treadwear in both standard GR-S and cold rubber.

The electric conductivity of GR-S can be increased by the use of conductive blacks.   Benson[12] has shown acetylene black to give, as compared with other blacks high electric conductivity, improved heat conductivity, and physical properties intermediate between HMF and EPC blacks.   Channel black of unusually small particle size (estimated average diameter 10 mμ) has been reported[31] to produce a semiconductive vulcanizate, but the high-structure furnace blacks impart greater electric conductivity than the channel blacks.

Mode of Incorporation of Black in GR-S.   In regard to the method of addition of blacks to GR-S and more particularly to cold rubber, there is considerable diversity of opinion on the merits and the disadvantages of high-temperature mixing, as has been discussed under processing.   In some cases high-temperature mixing is reported to give superior physical properties, such as improved treadwear.[129, 17a]

Dannenberg[36a] reports experiments on the influence of the conditions of Banbury mixing (speed, temperature, time) on the effect of blacks in GR-S. In general, more severe milling conditions lead to lower Mooney viscosity, and, after vulcanization, higher modulus and higher rebound. Abrasion resistance is highest when the amount of milling is least; when it is barely sufficient to incorporate the black.

To develop the maximum physical properties of GR-S stocks it is necessary, of course, to get uniform dispersion of pigment. Softeners, in addition to their effect on subsequent processing, facilitate the dispersion of pigments. Morris and Hollister[94] believe, contrary to a widely held opinion, that fatty acids have little effect on the dispersion of black in rubbers. They base this belief on experiments on the dispersion of conducting furnace black in Butyl rubber, in the course of which no important differences could be found between stearic acid and paraffin in their effect on the rate of incorporation of the black and on the Mooney viscosity, the Goodrich plasticity, the electrical resistivity, the bound rubber content, and the tensile strength of Banbury-mixed stocks. The rate of incorporation of carbon black in GR-S was shown by Amberg and Elliott[5] to be increased by rosin and terpene softeners.

In general the effect of increased black loading in GR-S is similar to its effect in natural rubber, with the exception that the increased reinforcement is more pronounced than in natural rubber. In many cases the incorporation of increased loadings of black requires the use of increased softener. The selection of the black loading and the ratio of softener to black depends on the processing and service requirements. The volume of GR-S stock can be extended by using increased black and increased softener, such as the combinations, mentioned earlier, of black and unsaturated petroleum hydrocarbon described by McMillan and co-workers.[87]

Masterbatches[61] prepared by adding carbon black to GR-S latexes and coprecipitating the rubber and black have the advantages of reducing subsequent processing time, making mill-room operations cleaner, giving more assurance of uniform dispersion, and allowing the manufacture of polymer of lower Mooney viscosity. The last item is possible because the carbon black increases viscosity and reduces stickiness so that polymer which otherwise could not be handled in the driers and other equipment can be made with ease. (Cf. Chapter 20.)

NONBLACK PIGMENTS

Silica. Although carbon black is the most outstanding reinforcing pigment, certain other materials are effective to varying degrees. Special fine-particle silicas reinforce GR-S to a greater extent than other commercially available inorganic pigments. A hydrated silicon dioxide of particle size approximately the same as that of EPC black was reported by Allen, Gage, and Wolf[3] to give GR-S stocks with good physical properties. The authors stated that the hydrated silica causes a slow cure rate and thus necessitates the use of increased acceleration. The use of 5 to 10 per cent of diethylene glycol on the weight of the pigment is recommended to increase the rate of cure and improve physical properties. To develop maximum physical properties, it is said to be necessary to use approximately 10 parts of a hard coumarone-indene resin. Table X offers a comparison at optimum cures of (A) fine particle silica, (B) extra fine precipitated calcium silicate, and (C) EPC black.

Table X. Fine Particle Silica as a Pigment in GR-S[3]

	A	B	C
GR-S	100	100	100
Pigment: By volume	30	30	30
By weight	58.5	63	54.6
Zinc oxide	5	5	5
Antioxidant	1	1	1
Sulfur	3	3	1.75
Coal-tar softener	...	...	5
Pine tar	...	...	3
Coumarone resin, 100° C. softening point	10	10	...
Diethylene glycol	3.5	3.5	...
N-Cyclohexylbenzothiazole-2-sulfenamide	...	1.5	1.25
Benzothiazolyl disulfide	1.2	...	...
Tetramethylthiuram disulfide	0.15	...	...
Cure, min. at 138° C.	15	15	120
Modulus at 300%	620	890	1360
Tensile strength, p.s.i.	2550	2030	3400
Elongation, %	620	490	560
Hardness	55	65	61
Tear, lb. per in.	200	160	300
Rebound, %	46	45	40
Heat buildup, °C.	38	27	44

Reprinted by permission of *India Rubber World.*

A fine-particle-size silica[7] having a particle size and shape similar to those of the fine carbon blacks has been obtained from ethyl silicate by a combustion process. Good tensile strength, tear resistance, and abrasion resistance were obtained in light-colored stocks with this silica.

Other Nonblack Pigments. The tensile strength of GR-S is increased to varying relatively moderate degrees by calcium silicate, titanium dioxide, zinc sulfide, hydrated alumina, fine calcium carbonate, zinc oxide, and extra light calcined magnesia. Some clays increase tensile strength slightly. A study by Morris, James, and Evans[96] shows that higher modulus is obtained with clays than with calcium carbonates, zinc oxides, titanium dioxides, zinc sulfides, and hydrated alumina. Fine-particle-size calcium silicate was shown to give higher modulus than the clays.

According to Gage,[54] fine-particle-size hydrated calcium silicate is somewhat reinforcing in GR-S and results in age life superior to that obtained with carbon black. It is hydrophilic and increases the water absorption of the stock. The high hysteresis of GR-S containing calcium silicate, which results in a high heat buildup such as to make the stocks unsuitable for application involving subjection to dynamic flexing, can be reduced by the inclusion in the stock of polyhydroxy compounds, especially diethylene glycol.

In a classification of clays and a description of their properties, Gongwer[64] showed that a lower content of grit gives better properties in clay-containing vulcanizates. Particle size is also shown to be a factor in the properties obtained in vulcanizates. To illustrate this and to show the reinforcing effect of fine clay in comparison with certain other noncarbon pigments, Gongwer gives the data in Table XI for GR-S loaded with 40 volumes of pigment and cured to maximum tensile strength.

Table XI. Comparison of Nonblack Pigments[64] in GR-S

Pigment	Tensile Strength, P.S.I.	Relative Resistance to Abrasion
Hard clay	2000	100
Soft clay	1250	92
Treated precipitated calcium carbonate	1200	61
Water-ground whiting	630	70
Calcium silicate	1390	66
Zinc oxide	1060	110

Reprinted by permission of *India Rubber World*.

In a so-called hard clay the minimum proportion of particles 2 μ or less in size is 82 per cent, whereas in a soft clay it is 56 per cent.

Moakes[90a] has published a compounding study of certain clays which shows that the treatment of clays with an organic base may improve their behavior in rubber. The compounding material Franclay is a fine-particle kaolin, normally having a pH of 3.5, which has been rendered alkaline and "activated" by treatment with dibutylamine.

In a study of the use of zinc oxide as a filler for GR-S, Jones[70] made the interesting observation that, in GR-S freed from fatty acid by acetone extraction, zinc oxide produces much higher stiffness and tensile strength than in regular GR-S. For example, whereas in regular GR-S 150 parts of fine zinc oxide produced a 300 per cent modulus of 285 p.s.i., a tensile strength of 890 p.s.i. and an elongation of 510 per cent, in acetone-extracted GR-S the corresponding figures were 935 modulus, 2160 tensile strength, and 510 per cent elongation. This author reported further that the effectiveness of zinc oxide as a filler in regular GR-S is markedly improved by the inclusion in the stock of a small proportion of magnesia along with coumarone resin. Barnett and Jones[9] concluded that the special effectiveness of such magnesia depends on its power to absorb fatty acid and soap from the GR-S.

The last-mentioned authors demonstrated a relation between pigment-reinforcing powers and surface activity in GR-S by experiments with basic zinc carbonate calcined over a range of temperatures. Calcination at 350° C., which is considered, as a result of the elimination of carbon dioxide, to render the pigment porous and surface-active while still retaining its crystalline configuration, markedly alters its behavior in GR-S, as is shown by Table XII, referring to stocks containing 20 volumes of pigment and cured to maximum tensile strength.

Table XII. Comparison of Calcined and Uncalcined Zinc Carbonate in GR-S[9]

	Basic Zinc Carbonate, Uncalcined, (81.5 Parts)	Zinc Carbonate, Calcined 1 Hour at 350° C. (120.4 Parts)
Modulus at 300%, p.s.i.	365	1360
Tensile Strength, p.s.i.	815	2060
Elongation, %	595	400

Reprinted by permission of *Ind. Eng. Chem.*

The tensile strength, tear resistance, and abrasion resistance were found by Cohan and Spielman[30] to increase as the particle size of calcium carbonate pigments of fairly symmetrical shape, similar crystalline structure, and similar surface nature was reduced. This is shown by the data in Table XIII, which refer to GR-S stocks loaded with 40 volumes of various forms of calcium carbonate and containing also 30 parts by weight of hard coumarone resin.

Table XIII. Effect of Particle Size of Calcium Carbonate on the Properties of GR-S Vulcanizates[30]

	Whiting	Finely Ground (Micro-) pulverized) Whiting	Precipitated Calcium Carbonate	Ultrafine Precipitated Calcium Carbonate
Mean particle size, mμ	3900	1500	145	50
Tensile strength, p.s.i.	1100	1400	1600	2300
Tear resistance, lb. per in.	20	40	50	190
Abrasion loss, cc. per hp.-hr.	2100	1500	1000	750

Reprinted by permission of *Ind. Eng. Chem.*

It is concluded that with pigments, such as calcium carbonate, having fairly symmetrical shape, as with carbon blacks, the predominant factor determining reinforcing properties is particle size. Clay, however, which has an unsymmetrical shape (its particles are platelets), leads to greater stiffening than would be expected from its surface area alone. A comparison of the effect on the physical properties (including flex life) of GR-S produced by channel black and by precipitated calcium carbonate of very fine particle size has been offered by Vodra.[142a]

The inorganic pigments find use in GR-S products where a light color, low cost, high hardness, and stiffness, or special properties such as good dielectric characteristics are desired.

Dispersion of Pigments. To aid in the dispersion of pigments fatty acid soaps, sulfonated petroleum and vegetable oils, and other materials are sometimes used. When added to the rubber mix, these materials reduce the time required for dispersing pigments and improve the degree of dispersion. In many cases the pigments are coated by the manufacturers with coconut oil, tall oil, and similar materials which are miscible with GR-S, in order to improve their dispersion during kneading operations.

The incorporation of inorganic pigments as well as carbon black can be facilitated by dispersion in GR-S latex, followed by coprecipitation of the pigment and polymer. In some cases[84] higher tensile strength and modulus and otherwise improved physical properties, explained by improved dispersion, have been observed when inorganic pigments were latex-masterbatched.

ORGANIC REINFORCING AGENTS OTHER THAN CARBON BLACK

Coumarone-Indene Resins. A number of organic materials other than carbon black which reinforce GR-S have been introduced. The higher-melting-point coumarone-indene and other coal-tar resins function both as plasticizers during processing and as reinforcers[60] in the vulcanizate. The harder resins contain a relatively large proportion of styrene polymers. They reduce nerve and cause smooth extrusion. Vulcanizates containing some of these resins are high in hardness and modulus, low in elongation, and have fairly good tensile strength. The light-colored types can be used to produce light-colored stocks with only very slight discoloration and only slight staining tendency. Rather high loadings of hard or medium-hard coumarone-indene resins are often used in combination with mineral pigments to aid in pigment dispersion and to increase tensile strength, tear resistance, and cut-growth resistance. In GR-S stocks highly loaded with mineral pigments, the resins produce fairly good physical properties at low cost. The resins can also be used in black stocks. Under "Softeners" the coumarone-indene resins are discussed more completely.

The favorable effects of a coumarone resin of a softening point of 115° C. in a nonblack GR-S stock may be illustrated by the data in Table XIV.

High-Styrene Resins. The introduction of high-styrene/low-butadiene copolymers[16] has allowed the manufacture of smooth-processing GR-S stocks of high strength and low gravity in a wide variety of colors. Resins of this type vary in heat distortion point according to the ratio of butadiene to styrene and other factors, but the ones most used currently are in the approximate range of 50 to 60° C. In addition to having a beneficial effect on processing, they increase tensile strength, hardness, stiffness, and flex life. They have low moisture absorption and good dielectric properties. They find use either alone or in combination with carbon black or mineral pigments. Only very slight adjustment in curative and accelerator is necessary for these materials. Vulcanization reduces their solubility and the tendency to flow. In moderate loadings in GR-S the styrene-butadiene resins have been used extensively in such products as flooring, shoe soles, and wire insulation. In higher loadings,[120] where the amount of resin exceeds the amount of GR-S, stocks of high impact resistance are produced

Table XIV. Influence of Coumarone Resin in GR-S Loaded with a Nonblack Pigment[60]

Test Stock

GR-S		100	
Zinc oxide		10	
Lithopone (50 volumes)		207.5	
Coumarone resin		Various	
Sulfur		2.5	
Benzothiazolyl disulfide		0.5	
Tetramethylthiuram monosulfide		0.5	

Cure: 15 minutes at 158° C.

Coumarone resin	0	10	20	40
Modulus at 300%, p.s.i.	275	150	150	75
Modulus at 500%, p.s.i.	...	450	275	150
Tensile strength, p.s.i.	475	625	1525	1775
Elongation, %	400	540	715	870
Hardness	58	55	49	39

Reprinted by permission of *India Rubber World.*

which find useful application in athletic protective wear and other fields. The resins are discussed more fully in Chapter 18.

Lignin. Special lignins[39, 78] from the black liquor of the sulfate wood pulp process reinforce the tensile strength and tear resistance of GR-S. In comparison with carbon black reinforcement, their use results in low modulus but inferior abrasion resistance. The lignins are low in gravity. They can be used in making bright-colored GR-S goods but cannot be used to produce white or very light colors. Although the dry lignins can be milled into GR-S, the most successful results are obtained by coprecipitating an aqueous alkaline solution of lignin and GR-S latex. A fairly good dispersion of lignin in the polymer is thus obtained, requiring only a moderate amount of mill mixing. The lignins can be used in combination with mineral pigments. Oxidation of the lignin from the alkali pulping process has been shown[104] to increase its ability to reinforce GR-S. Vulcanizates prepared from GR-S coprecipitated with oxidized lignin are reported to have physical properties comparable to those of GR-S reinforced with EPC black.[77]

Styrene-Isobutylene Resins. Resins of styrene–isobutylene[35] also are compatible with GR-S. Like the styrene-butadiene resins they serve as aids in processing. They are recommended primarily to improve light stability and to lower the rate of gas or vapor diffusion. Since these polymers contain no unsaturation, vulcanization should have no effect on their physical properties.

Phenolic Resins. Some phenolic resins[116] plasticize GR-S and improve processibility. However, the plasticization does not improve extrusion characteristics. The phenol-aldehyde resins increase hardness, abrasion resistance, and heat resistance but do not increase tensile strength significantly. Because of their limited compatibility the phenolics are not used in high proportions in GR-S.

ORGANIC LOADING MATERIALS. EXTENDERS AND DILUENTS

In addition to organic reinforcing agents, such as have just been discussed, there is a class of organic materials which, although they are without marked reinforcing action, either impart some special properties or merely dilute the volume of GR-S. These materials include both extenders and diluents. Extenders can be used to replace rather large proportions of the rubber hydrocarbon without a very great loss in certain physical properties of the vulcanized product. Diluents, on the other hand, are used to dilute the rubber hydrocarbon, a loss in the physical properties of the vulcanizate being expected. A number of organic extenders and diluents can also be regarded as softeners. Organic loading materials have been classified by Ludwig and co-workers[83] as follows: vulcanized vegetable oils, mineral rubber and pitches, processed vegetable oils, natural rubbers, synthetic elastomers, petroleum derivatives, resinous materials, and miscellaneous other materials.

The factices and other vulcanized vegetable oils[83] decrease plasticity but somewhat improve extrusion behavior. They cause a slight decrease in tensile strength, an increase in modulus, and a decrease in ultimate elongation. They reduce the rebound and flex life and have little effect on tear resistance.

Asphaltic materials such as mineral rubber, sometimes called hard hydrocarbon or hard asphalt, decrease the nerve, promote smoother processing, and reduce the tendency of stocks to swell when extruded. Although not strictly reinforcing in behavior, they have a useful influence on the physical properties of cured GR-S stocks. At a loading of 10 parts in GR-S black stocks, their effect on tensile strength is very slight, but their presence in GR-S produces increased resistance to tear.[115] Rebound and modulus are reduced. The asphaltic materials and also coal tar and hardwood pitch have no effect on the rate of cure. The use of these materials is limited to applications where rather high cured hardness is useful or can be tolerated.

The processed vegetable oils other than the vulcanized oils contain varying amounts of reactive unsaturation, which necessitates the use of additional sulfur in compounding. The gel and solid types of linseed and soybean oils have been reported[83] to have little effect on plasticity, to decrease tensile strength, modulus, and rebound, and to improve tear resistance slightly.

Admixtures of GR-S with other synthetic rubbers and with natural rubber are discussed later in this chapter.

The unsaturated hydrocarbon materials derived from petroleum, such as for example Naftolen or Dutrex, require the use of additional sulfur because of the unsaturation. It has been asserted that these hydrocarbons can be used in substantial quantities without excessive deterioration in stress-strain characteristics. Ludwig and co-workers[83] found these extenders to cause in uncured stocks appreciable increase in plasticity, improved processing, and extruding characteristics, and in cured stocks lower hardness, moderately lower rebound, and somewhat better tear resistance and flex life. By properly balancing the amounts of the unsaturated petroleum derivatives and pigment, rather high loadings can be used to obtain smooth extrusion and calendering with satisfactory physical properties.

Under resinous and miscellaneous materials are or have been offered a wide variety of extenders and diluents having a correspondingly wide variety of effects on processibility and physical properties. For the most part the chemical nature of the materials has not been disclosed. Polymers of styrene and its homologs were stated by Winkelmann[150] to improve the processibility and tack of GR-S in cements, pressure-sensitive adhesives, and sponge rubber and to increase the hardness and stiffness of shoe-sole stock. The same author showed that 10 parts of polyvinylbutyral reduces the shrinkage of GR-S, increases the hardness and modulus, and slightly increases the tensile strength. Polyethylene improves processing characteristics and increases the toughness of GR-S. Shellac or glue can be used where a high degree of hardness is desired.

Various flocks of cut or ground fibers are used in GR-S to increase the stiffness and hardness while at the same time maintaining low gravity.

SOFTENERS

Effect on Properties of Vulcanizates. The principal use of softeners in GR-S, discussed earlier (p. 383 et seq.), is to make the practical processing of GR-S stocks possible. In addition to their effect on processing, softeners exert strong effects on the physical properties of the cured products. Not all of them increase the softness of the cured stocks; some give low and some high hardness in the final products. The characteristics of individual softeners are also reflected in other physical properties of the products. And, since more softener is generally used in GR-S than Hevea, more attention must be given to their effect on the physical properties of the vulcanizates.

Softeners of strong solvent power tend to produce snappy stocks of low durometer hardness. With less solubility in the polymer, processing smoothness increases, but bleeding to the surface of the cured stock may occur. In addition to solvent power other physical properties in a softener are important. A relatively high boiling point and low vapor pressure are desirable features from the standpoint of volatility during processing and during service. The melting point and viscosity are important because of their effect on the ease of incorporation and their relationship to the physical properties after vulcanization. The melting point and the viscosity have a very strong influence on the flexibility of the cured stock at low temperatures. Further, a lower viscosity in a softener, other things being the same, results in a lower hardness and a higher resilience in the final product. The rate of extraction of a softener by water from a vulcanizate is another property usually of importance.

Types of Softeners. Ludwig and co-workers[82] classified softeners according to their chemical nature as follows: (1) esters, (2) aromatic hydrocarbons, (3) chlorinated hydrocarbons, (4) ethers, ketones, (5) alcohols, phenols, (6) amines, (7) vegetable oils, (8) fatty acids, (9) petroleum products, (10) pine products, (11) resins other than rosin, (12) coal-tar products, (13) waxes, and (14) miscellaneous other materials. If we take these groups

in the order given and note the results produced in the vulcanizate by the
addition of 10 parts of softener per 100 parts of rubber to a GR-S black
stock, the following points may be noted.

1. The *esters* include a variety of materials, many having a solvent
action in GR-S.   In general the esters reduce tensile strength, modulus, and
hardness.   Most do not impair rebound but have adverse effect on tear
resistance and flex life.   With few exceptions no pronounced retardation or
acceleration of vulcanization occurs.   In this group may be found many
softeners which improve flexibility at low temperature.   Typical of the most
effective from the standpoint of low-temperature flexibility[75] are butyl
cellosolve pelargonate, di-*n*-hexyl adipate, and trioctyl phosphate.

2. *Aromatic hydrocarbons* such as Decalin and diphenyl were said by Ludwig
and co-workers to cause minor decreases in tensile strength, modulus, and
hardness.   These materials have no commercial use as softeners.

3. Most of the *chlorinated hydrocarbons* are solvent-type plasticizers, and their
effect on physical properties is similar to that of the aromatic hydrocarbons.

4. *Ethers and ketones* of higher molecular weight are fairly good solvents for
GR-S.   Ludwig and co-workers found them, like other solvents, to cause
only relatively small changes in physical properties.

5. *Alcohols and phenols* are poor solvents for GR-S and of little value as
softeners.

6. Although not usually regarded as softeners, certain *amines* in concentrations of 0.5 to 1 part are shown by Ludwig and co-workers to be useful in that
they form amine salts of the fatty acid present in GR-S, thereby acting as
wetting and dispersing agents for pigment addition.   Amines accelerate the
cure rate.

7. Unpolymerized or raw *vegetable oils*, such as raw linseed, cottonseed,
and palm oils, reduce tensile strength, modulus, and hardness.   They do not
reduce the rebound or tear resistance significantly.   These oils have only a
small effect on cure rate.   Their use is principally in hard GR-S stocks.
Polymerized oils, considered as a group distinct from factices or vulcanized
oils, reduce tensile strength and modulus with little change in hardness.   In
general, rebound is reduced, but flex life and tear resistance are improved.
The rate of cure is retarded somewhat by the polymerized oils.

8. *Fatty acids* behave rather similarly to the unpolymerized vegetable oils
in GR-S.

9. An illuminating study of *petroleum softeners*, embracing 84 examples,
was made by Rostler and Sternberg.[107]   This study makes it apparent that
physical properties alone, such as viscosity and distillation range, are
insufficient to characterize such softeners which vary widely in the chemical
nature of their components.   These components may be divided as follows:
(*a*) asphaltenes—black, pulverable solid material insoluble in cold petroleum
ether, (*b*) nitrogen bases—dark-brown, medium- to heavy-viscous, sticky
liquids, precipitable by hydrogen chloride gas or extractable by 85 per cent
sulfuric acid, (*c*) unsaturated hydrocarbons—extractable by cold 97 per cent
sulfuric acid (group I) or by cold 30 per cent oleum (group II), (*d*) saturated,
liquid hydrocarbons.   Samples of the naturally occurring bitumen gilsonite
were included in the survey for comparison with the petroleum products.

Among the softeners that contained asphaltenes the content of the latter ranged from about 60 per cent in gilsonite and 50 per cent in mineral rubber (blown petroleum bitumen) to quite low values in other commercial softeners. The mineral rubbers differ from gilsonite in having (1) little or no nitrogen-base components, (2) a higher content of unsaturated hydrocarbons and of saturated oils. Among the softener samples (65 in number) containing no asphaltenes, the content of saturated oil ranged from 1.4 to 84.4 per cent, and that of nitrogen bases from 0.0 to 57.9 per cent. Certain additional data on the composition of petroleum softeners is to be found in other references.[81a, 106b]

That the influence on the properties of GR-S stocks of the various types of petroleum softener components may be noticeably different is illustrated by the following data, in which a comparison is made of Gilsonite (with a high content of asphaltenes), nitrogenous bases, and unsaturated hydrocarbons isolated from a California crude oil, and (as a representative of saturated hydrocarbons) medicinal white oil.

Table XV.   Effect of Petroleum Components in GR-S[107]

Stock

GR-S	100	White oil	2
Stearic acid	1	Benzothiazolyl	
Sulfur	2.3*	disulfide	1.5
Zinc oxide	5	Component	
MPC black	30	tested	15

* 2.0 in stocks 1 and 6.

Predominant Component Tested	1 None	2 Asphal-tenes	3 Nitro-gen Bases	4 Unsat-urated Hydro-carbons Group I	5 Unsat-urated Hydro-carbons Group II	6 White Oil
Composition of softeners						
Asphaltenes	...	53.4	0.0	0.0	0.0	0.0
Nitrogen bases	...	32.2	84.5	8.9	1.6	0.0
Unsaturated hydrocarbons, I	...	4.5	7.0	85.5	3.7	0.0
Unsaturated hydrocarbons, II	...	9.9	7.5	5.6	94.7	0.0
Saturated hydrocarbons	...	0.0	0.0	0.0	0.0	100.0
Scott plasticity, compression in 0.001 in.	160	165	235	249	260	235
Optimum cure at 45 lb., min.	80	80	30	80	80	80
Hardness, Shore	54	61	52	44	44	43
Modulus, p.s.i.						
At 300%	810	710	460	320	370	360
At 500%	1940	1210	1100	870	950	960
Tensile strength, p.s.i.	2170	2300	2100	1860	1730	1140
Elongation, %	550	640	660	740	710	570
Set at break, %	13	34	17	17	13	14
Resilience, Luepke						
First impact	71	67	70	69	69	69
Fourth impact	26	21	24	22	24	24

Reprinted by permission of *Ind. Eng. Chem.*

Asphaltenes have practically no effect on the plasticity, but nevertheless aid processing by reducing the nerve of a rubber stock. They increase hardness and set markedly and reduce the resilience slightly. They have less effect in reducing the modulus than the other components have. In a gum stock they show a marked reinforcing effect and raise the tensile strength. The nitrogen bases markedly increase the rate of cure. The unsaturated hydrocarbons, especially those of type I, consume sulfur and hence demand an upward adjustment of the sulfur in the compound. The oily components especially the saturated hydrocarbons, markedly reduce the tensile strength.

10. *Pine Products* include terpenes, gum turpentine, rosins, and pine tars. The terpenes are excellent solvents for GR-S. Amberg and Elliott[5] report on the examination of 65 softeners, mostly consisting of terpenes, wood rosins, and their derivatives. They conclude that of these materials hydrogenated and dehydrogenated rosin give the best all-round result. These reduce the modulus of loaded GR-S to a somewhat less extent than coal-tar and petroleum softeners and much less than wood rosin.

The mixtures of rosin and terpenes called crude gum turpentine impart good tensile strength, low modulus, rather low hardness, lowered rebound, poor tear resistance, and good flex life. Crude gum turpentine aids pigment dispersion. Rosins and rosin derivatives are readily admixed with GR-S. In vulcanizates they give good physical properties. In comparison with hydrogenated rosin, the same workers found the pale and dark wood rosins and commercial abietic acid to cause slower cure rate. In the same work it was found that esters derived from rosins had effects similar to the resin acids on the properties of the vulcanizates but that the esters had a less pronounced retarding effect on cure. Pine tars aid in pigment dispersion and produce GR-S stocks of good physical properties. Like most pine products the pine tars impart some tack to GR-S and retard the rate of cure.

11. *Resins* other than rosins belong rather in the categories of reinforcing or loading materials and are discussed under those subjects.

12. *Coal-tar products* vary widely in their properties and in their effects on the physical characteristics of GR-S. Geiger[60] divided these products into three classes: liquids from distillates, including alkylnaphthalenes and polynuclear benzenes; semisolids from coal tar, including low-molecular-weight polymers of coumarone-indene and related resins; and solids from coal-tar pitch, with which are also classed medium-hard and hard coumarone-indene resins. The last group was included in this chapter under organic reinforcing agents. Some of the liquid oils are widely used either alone or in combination with more resinous plasticizers to improve processibility and certain physical properties, such as resilience, in the cured products. A number of grades, varying in softening point from 10 to about 125° C. and in consistency from viscous fluids to brittle resins, are available. The hardness, rebound, and set are affected only slightly by coumarone-indene resins. They have a solvent action on sulfur and promote uniform distribution of sulfur. A slight sulfur adjustment in GR-S stocks is necessary to compensate for the mild retarding action of coumarone-indene resins. According to Bulifant[23] the resins result in flat-curing, good-aging stocks of good tear resistance. The variety of grades available, of varied softening points, makes it possible to

select a coumarone resin softener on the basis of either the processing characteristics or the cured physical properties desired.

13. *Waxes* are of limited solubility in GR-S and tend to bleed to the surface. The tendency to bloom is utilized to form a protective wax film to retard sun checking during service.   The waxes have little effect on cure, but they lower tensile and modulus.

14. *Low-molecular-weight polybutadiene* such as was used extensively to soften Buna S3 in Germany has found little acceptance to date in GR-S compounding.

ANTIOXIDANTS

Approximately 1.25 per cent of a standard antioxidant or 1.5 per cent of a nonstaining antioxidant is added to GR-S during its manufacture.   The standard antioxidants are phenyl-β-naphthylamine and a reaction product of diphenylamine and acetone.   A heptylated diphenylamine is less staining than the standard antioxidants.   Typical of the nonstaining antioxidants are triphenyl phosphite,[69] di-*tert*-butylhydroquinone, alklyated phenols[81, 151] and their sulfides, and cresol-styrene condensation products.   Unoxidized soda pulp lignin has been found[98] to be a good age resistor of a relatively nonstaining character.

The incorporation of antioxidant in the raw GR-S protects it from deterioration during the drying operation in its manufacture, during storage, during processing and protects its vulcanizates during storage and service. Although marked differences between various antioxidants have been observed[1] in the natural and accelerated aging of raw polymer, the differences between antioxidants are of smaller magnitude in the aging characteristics of vulcanizates.   It has been pointed out by Winn and Shelton[152] that the best antioxidant for the raw polymer is not necessarily the best for the vulcanizate.   Thus, for example, although 2 per cent of phenyl-β-naphthylamine protects raw GR-S better than does 2 per cent of 2,2,4-trimethyl-6-phenyl-1,2-dihydroquinoline, the latter protects vulcanized GR-S better than the former.[1]

Except for special applications, it is not considered necessary to add additional antioxidants to GR-S during compounding.   The subject of antioxidant action is treated in Chapter 13.

MIXTURES OF GR-S AND OTHER ELASTOMERS

The addition of small amounts of Hevea rubber to GR-S imparts tack. Fairly good building tack is obtained when 20 to 30 per cent of the GR-S in a stock is replaced by Hevea.   Prettyman[102] found a linear relation between the proportion of Hevea in tread stocks made from mixtures of GR-S and Hevea on the one hand and the resistance to cut growth on the other; the replacement of as little as 10 per cent of GR-S by Hevea gave a noticeable improvement.   Morris and co-workers[93] reported that, when 20 per cent of the GR-S was replaced with Hevea, cryptostegia, or deresinated guayule, the tack was improved; the ultimate elongation and the tensile strength at room temperature were enhanced slightly and their retention at 93° C. was improved; the tear resistance was improved and its retention at 93° C. was greatly improved. Modulus, rebound and heat buildup were little affected. The use of an amine soap as an activator of thiazole acceleration provided a better balance between the degrees of vulcanization of the two rubbers in a blend of GR-S and Hevea than benzothiazolyl disulfide alone, according to Stubbs and C. R. Johnson.[131] In blending guayule and GR-S, Clark and Place[28] observed that up to 20 per cent of guayule improved GR-S in some respects but impaired abrasion resistance. The same workers found that guayule of low resin content was more desirable for quality in vulcanizates of the blends, but that guayule of relatively high resin content was more desirable from the standpoint of processibility and tack in uncured stock.

The use of natural rubbers in admixture with GR-S and the proportion of natural rubber employed depend on the processing and fabricating requirements and the physical properties desired. For example, a study by Vodra and Jarvis[142b] indicates how stocks having specified properties (in their case a tensile strength of 1500 p.s.i. and a Shore hardness of 65) can be prepared, not only from natural rubber and from GR-S, but also from blends of these rubbers and from blends of natural rubber, GR-S, and reclaim by suitable adjustment of the loading and acceleration.

Alkali reclaims speed up the rate of cure, but after a small amount has been added less increase in cure rate is caused by further additions. The use of reclaim in GR-S or in mixtures of GR-S and natural rubber retards degradation on overcuring and improves processing. In GR-S its use as a means of reducing cost is frequently secondary to its advantageous effect on the rate of cure and on processing.

Ludwig and co-workers[83] stated that blends of GR-S with nitrile rubbers, Thiokol, or Neoprene have greater solvent resistance than straight GR-S stocks. The nitrile rubbers were found by the same workers to increase the tensile strength, tear resistance, and elongation but to reduce the rebound of GR-S. Thiokol reduced tensile strength, modulus, rebound, and tear resistance. Neoprene GN caused no appreciable change in the tensile strength or elongation but lowered the modulus of GR-S. In the same work it was reported that polyisobutylene decreased the tensile strength, elongation, rebound, and tear resistance but markedly improved the flex life of GR-S.

In mixing GR-S with other elastomers it is good general practice to make separate masterbatches of the rubbers with carbon black before blending. In mixing with natural rubber, it may be desirable to remill the GR-S masterbatch before blending the individual masterbatches. If natural rubber and GR-S are blended before black is added, greater roughness and more broken edges may occur during extrusion. The tubing characteristics of most types of GR-S are widely different from those of natural rubber, and therefore mixtures of these rubbers regardless of the manner in which they are made, will exhibit unsatisfactory extrusion. On the other hand, since cold rubber is quite similar to natural rubber with respect to extrusion, its admixture with natural rubber can be accomplished in any desired manner, and good extrusion can be expected in the product.

GENERAL COMMENTS ON THE PHYSICAL PROPERTIES OF VULCANIZED GR-S

Tread-type vulcanizates of GR-S have lower tensile strength than those of Hevea. Cold-rubber treads have higher tensile strength than standard GR-S and at room temperature have almost the tensile strength of natural rubber. Data published by Fielding[50] emphasize the severe loss in strength suffered by GR-S and cold rubber at elevated temperature. This effect is illustrated by the data in Table XVI.

Table XVI. Influence of Elevated Temperature on Properties of GR-S and Natural Rubber in Tread-Type Stock[50]

	Hevea	GR-S	Cold GR-S
Tensile strength, p.s.i.			
At room temperature	4100	2900	3800
At 93° C.	3100	1100	1500
Elongation, %			
At room temperature	600	600	650
At 93° C.	680	480	550
Resistance to cut growth (DeMattia pierced groove test)			
At room temperature	1200	200	400
At 93° C.	700	21	30

Reprinted by permission of *Ind. Eng. Chem.*

Whereas Hevea loses approximately 25 per cent of its tensile strength when the test temperature is increased from 27 to 93° C., GR-S and cold rubber lose almost 66 per cent. The elongation at break decreases for GR-S and cold rubber, but increases for Hevea, when tested at the higher temperature. X-ray diffraction studies[119] have shown standard GR-S to be amorphous under all conditions, whereas natural rubber develops a crystalline pattern when stretched. Beu and co-workers[13] have observed a tendency for polybutadiene and 90/10 butadiene-styrene copolymers made at low temperature to crystallize. However, these workers found that, when the amount of styrene was increased to 20 per cent, crystallization was prevented. The inability of GR-S to crystallize results in low tensile strength in unpigmented stocks and poor tear resistance.

It has been reported by Sebrell[117] that in the range 25 to 90° C. the dynamic modulus of GR-S tread stocks has a greater dependence on temperature than that of Hevea.

Since properly compounded GR-S is superior to Hevea in ozone resistance,[33] exposed surfaces are less likely to start to crack. However, once cracks are started, they grow at a greater rate in GR-S than in Hevea. Dinsmore and Fielding[42] have pointed out that the flex life of GR-S is better when it is flexed at low elongations than in higher elongation cycles. With Hevea, however, thanks to the effect of crystallization, flex life is superior when Hevea is flexed from 200 to 500 per cent elongation than when flexed from zero to 300 per cent elongation. Cold rubber is superior to standard GR-S but inferior to Hevea in cut-growth resistance. The inferiority of GR-S and cold rubber in cut-growth resistance is more pronounced at elevated temperature than at room temperature.

Perhaps the most serious weakness of GR-S in comparison with Hevea is its markedly lower resilience. Cold rubber is slightly better than GR-S in resilience but definitely inferior to Hevea. In tire carcasses and other products that are flexed rapidly in service the higher heat generation of the synthetics is a serious handicap, and is the chief reason why the percentage of GR-S used in truck tires is lower than in other tire uses of the rubber.

GR-S and cold rubber exhibit good resistance to tread wear. In two separate tests[122, 50] cold rubber was shown to have 15 and 23 per cent improved treadwear in comparison to GR-S. In the second test the treadwear obtained with cold rubber was somewhat better than that of the Hevea control tires.

The traction[122] of GR-S treads on wet roads is slightly better than that of Hevea, but on snow and ice GR-S is inferior to Hevea in traction.

The creep of GR-S vulcanizates has been shown by Gehman[59] to be greater than that of Hevea vulcanizates at $35°$ C.

GR-S which has been specially coagulated to reduce its water-absorptive impurities has excellent electrical properties which are in some cases in the same range as those of deproteinized Hevea.

At normal temperatures GR-S vulcanizates have good age life. In air at high temperatures they suffer a rise in modulus and a fall in elongation, but they can be so compounded as to possess good heat aging characteristics.

Blake[14] has indicated that GR-S vulcanizates are more susceptible to atmospheric cracking than Hevea in the absence of wax. However, GR-S is generally more resistant than Hevea when it is compounded properly with wax that may bloom to the surface. In a review article Blake states that dilute ozone is the active cause of cracking in stretched rubber, and that light has little effect except as it may produce ozone.

Despite its definite inferiority in strength and cut-growth resistance when hot and its high hysteresis characteristics, standard GR-S has proved useful in many applications. For special applications, many modifications of GR-S having special properties such as superior cut-growth resistance, resilience, electrical properties, or flexibility at low temperature are being made. Cold rubber is superior to GR-S for tire treads and numerous other products, possibly because it contains less low-molecular-weight (nonreinforcible) polymer, less chain branching and cross-linking, and a greater uniformity in regard to the sterical configuration about the double bond. Further advances, via the proper choice of polymerization conditions, in achieving a still more desirable molecular structure undoubtedly will result in butadiene-styrene copolymers superior to those already made.

REFERENCES

1. Albert, H. E., *Ind. Eng. Chem.*, **40**, 1746–50 (1948).
2. Alfrey, T., Hendricks, J. G., Hershey, R. M., and Mark, H., *India Rubber World*, **112** 577–81 (1945).

3. Allen, E. M., Gage, F. W., and Wolf, R. F., *India Rubber World*, **120**, 577–81, 586 (1949).
4. Ambelang, J. C., Smith, G. E. P., and Gottschalk, G. W., *Ind. Eng. Chem.*, **40,** 2186–92 (1948).
5. Amberg, L. O., and Elliott, J. H., *India Rubber World*, **112**, 309–12 (1945).
6. Anon., *Chem. and Eng. News*, **29**, 979 (1951).
7. Anon., *Rubber Age N.Y.*, **59**, 197 (1946).
8. Bardwell, J., and Winkler, C. A., *India Rubber World*, **118**, 509–12, 520 (1948).
9. Barnett, C. E., and Jones, H. C., *Ind. Eng. Chem.*, **41**, 1518–22 (1949).
10. Beaver, D. J., and Throdahl, M. C., *Rubber Chem. and Technol.*, **17**, 896–902 (1944).
11. Bekkedahl, N., Quinn, F. A., and Zimmerman, E. W., *J. Research Natl. Bur. Standards*, **40**, 1–7 (1948).
12. Benson, C., *Rubber Age N.Y.*, **58**, 461–5 (1946).
13. Beu, K. E., Reynolds, W. B., Fryling, C. F., and McMurry, H. L., *J. Polymer Sci.*, **3,** 465–80 (1948).
14. Blake, J. T., Symposium on Aging of Rubbers, Am. Soc. Testing Materials, Philadelphia, 1949, pp. 48–55.
15. Blow, C. M., and Wood, R. I., *Trans. Instn. Rubber Ind.*, **25**, 309–27 (1950).
16. Borders, A. M., Juve, R. D., and Hess, L. D., *Ind. Eng. Chem.*, **38**, 955–8 (1946).
17. Braendle, H. A., in J. Alexander, *Colloid Chemistry*, Reinhold, New York, 1946, Vol. VI, pp. 408–35.
17a. Braendle, H. A., *Rubber Age N.Y.*, **72**, 205–10 (1952).
18. Braendle, H. A., Steffen, H. C., and Sheppard, J. R., *India Rubber World*, **119**, 57–62 (1948).
19. Braendle, H. A., and Wiegand, W. B., *Ind. Eng. Chem.*, **36**, 724–7 (1944).
20. Breckley, J., *India Rubber World*, **114**, 663–5 (1946).
21. Brown, G. L., *India Rubber World*, **116**, 787–8 (1947).
22. Bruce, P. L., Lyle, R., and Blake, J. T., *Ind. Eng. Chem.*, **36**, 37–9 (1944).
23. Bulifant, T. A., *Rubber Age N.Y.*, **62**, 300–2 (1947).
24. Busse, W. F., Lambert, J. M., and Verdery, R. B., *J. Applied Phys.*, **17,** 376–85 (1946).
25. Carlton, C. A., and Reinbold, E. B., *Rubber Age N.Y.*, **52**, 29–34 (1942).
26. Cheyney, L. E., and Duncan, R. W., *Ind. Eng. Chem.*, **36**, 33–6 (1944).
27. Cheyney, L. E., and Robinson, A. L., *Ind. Eng. Chem.*, **35,** 976–9 (1943).
28. Clark, F. E., and Place, W. F. L., *Ind. Eng. Chem.*, **38**, 1026–33 (1946).
29. Cohan, L. H., *Rubber Age N.Y.*, **55**, 263–6 (1944).
30. Cohan, L. H., and Spielman, R., *Ind. Eng. Chem.*, **40**, 2204–10 (1948).
31. Cohan, L. H., and Steinberg, M., *Ind. Eng. Chem.*, **36**, 7–15 (1944).
32. Cole, J. O., Parks, C. R., and D'Ianni, J. D., Private Communication to O.R.R., Oct. 7, 1949.
33. Crabtree, J., and Kemp, A. R., *Ind. Eng. Chem., Anal. Ed.*, **18**, 769–74 (1946).
34. Crawford, R. A., and Tiger, G. J., *Ind. Eng. Chem.*, **41**, 592–6 (1949).
35. Cunningham, E. N., *Rubber Age N.Y.*, **62**, 187–90 (1947).
36. Cuthbertson, G. R., Coe, W. S., and Brady, J. L., *Ind. Eng. Chem.*, **38**, 975–6 (1946).
36a. Dannenberg, E. M., *Ind. Eng. Chem.*, **44**, 813–8 (1952).
37. Dannenberg, E. M., and Collyer, H. J., *Ind. Eng. Chem.*, **41**, 1607–16 (1949).
38. Dannenberg, E. M., and Stokes, C. A., *Ind. Eng. Chem.*, **41**, 812–7 (1949).
39. Dawson, T. R., *Trans. Instn. Rubber Ind.*, **24**, 227–38 (1949).
40. D'Ianni, J. D., Hosely, J. J., and Greer, P. S., *Rubber Age N.Y.*, **69**, 317–21 (1951).
41. Dillon, J. H., *Physics*, **7**, 73 (1936); *Rubber Chem. and Technol.*, **9**, 496–501 (1936).
41a. Dinsmore, R. P., *Trans. Instn. Rubber Ind.*, **28**, 166–206 (1952).
42. Dinsmore, R. P., and Fielding, J. H., *India Rubber World*, **119**, 457–61 (1949).
43. Dobbin, R. E., and Rossman, R. P., *Ind. Eng. Chem.*, **38**, 1145–8 (1946).
44. Drogin, I., *Developments and Status of Carbon Black*, United Carbon Co., Maryland, 1945.

45. Drogin, I. (United Carbon Co.), Private Communication, Nov. 14, 1949.
46. Drogin, I., and Bishop, H. R., *Today's Furnace Black*, United Carbon Co., Maryland, 1948.
47. Drogin, I., Bishop, H. R., and Wiseman, P., *India Rubber World*, **120**, 693–7 (1949).
48. Drogin, I., Bishop, H. R., and Wiseman, P., *Rubber Age N.Y.*, **64**, 309–50 (1948).
49. Eagles, R. H., and Carlton, C. A., *India Rubber World*, **111**, 693 (1945).
50. Fielding, J. H., *Ind. Eng. Chem.*, **41**, 1560–3 (1949).
51. Fisher, H. L., and Davis, A. R., *Ind. Eng. Chem.*, **40**, 143–50 (1948).
52. Flory, P. J., *Chem. Rev.*, **35**, 51–75 (1944).
53. Flory, P. J., *J. Am. Chem. Soc.*, **69**, 2893–9 (1947).
54. Gage, F. W., *Rubber Age N.Y.*, **58**, 343–6, 351 (1945).
55. Garvey, B. S., *Chem. Eng. News*, **29**, 2468 (1951); Anon., *Rubber Age N.Y.*, **71**, 645 (1952).
56. Garvey, B. S., and Freese, J. A., *Ind. Eng. Chem.*, **34**, 1277–83 (1942).
57. Garvey, B. S., Whitlock, M. H., and Freese, J. A., *Ind. Eng. Chem.*, **34**, 1309–12 (1942).
57a. Garvey, B. S., Yochum, D. W., and Morschausen, C. A., *Rubber Age N.Y.*, **73**, 361–8 (1953).
58. Gee, G., *J. Polymer Sci.*, **2**, 451–62 (1947).
59. Gehman, S. D., *J. Applied Phys.*, **19**, 456–63 (1948); *Rubber Chem. and Technol.*, **22**, 105–17 (1949).
60. Geiger, L. M., *India Rubber World*, **111**, 312–7 (1944).
61. General Tire & Rubber Co., *India Rubber World*, **118**, 660–2 (1948).
62. Gillman, H. H., and Thoman, R., *Ind. Eng. Chem.*, **40**, 1237–42 (1948).
63. Glazer, E. J., Parks, C. R., Cole, J. O., and D'Ianni, J. D., *Ind. Eng. Chem.*, **41**, 2270–2 (1949).
64. Gongwer, L. F., *India Rubber World*, **118**, 793–5 (1948).
65. Hammond, G. L., and Moakes, R. C. W., *Trans. Instn. Rubber Ind.*, **25**, 172–89 (1949).
66. Harrington, H. D., Weinstock, K. V., Legge, N. R., and Storey, E. B., *India Rubber World*, **124**, 435–42, 571–5 (1951).
67. Hendricks, J. G., *Rubber Age N.Y.*, **54**, 521–5 (1944).
68. Holt, W. L., and Knox, E. O., *Rubber Age N.Y.*, **60**, 689–92 (1947).
69. Howland, L. H., and Hunter, B. A., U.S. Pat. 2,419,354, Apr. 22, 1947.
70. Jones, H. C., *Ind. Eng. Chem.*, **36**, 641–8 (1944).
71. Juve, A. E., Symposium on the Application of Synthetic Rubbers, Am. Soc. Testing Materials, Philadelphia, 1944, pp. 50–58.
72. Juve, A. E., *Ind. Eng. Chem.*, **39**, 1494–8 (1947).
73. Juve, A. E., and Garvey, B. S., *Ind. Eng. Chem.*, **36**, 212–8 (1944).
74. Juve, A. E., and Hay, D. C., *India Rubber World*, **117**, 62–4 (1947).
75. Juve, R. D., *India Rubber World*, **115**, 657–8 (1947).
76. Juve, R. D., and Marsh, J. W., *Ind. Eng. Chem.*, **41**, 2535–8 (1949).
77. Keilen, J. J., Dougherty, W. K., and Cook, W. K., *India Rubber World*. **142**, 178–81 (1951).
78. Keilen, J. J., and Pollak, A., *Ind. Eng. Chem.*, **39**, 480–3 (1947).
79. Kemp, A. R., Malm, F. S., and Stiratelli, B., *Ind. Eng. Chem.*, **36**, 109–13 (1944).
80. Kemp, A. R., and Straitiff, W. G., *Ind. Eng. Chem.*, **36**, 707–15 (1944).
81. Kitchen, L. J., Albert, H. E., and Smith, G. E. P., *Ind. Eng. Chem.*, **42**, 675–85 (1950).
81a. Kurtz, S. S., and Martin, C. C., *India Rubber World*, **126**, 495–6 (1952).
82. Ludwig, L. E., Sarbach, D. V., Garvey, B. S., and Juve, A. E., *India Rubber World*, **111**, 55–62, 180–6 (1944).
83. Ludwig, L. E., Sarbach, D. V., Garvey, B. S., and Juve, A. E., *India Rubber World*, **112**, 731–7 (1945).
84. Lyons, W. J., Nelson, M. L., and Conrad, C. M., *India Rubber World*, **114**, 213–7, 219 (1946).
85. Marvel, C. S., Gander, R. J., and Chambers, R. R., *J. Polymer Sci.*, **4**, 689–702 (1949).
85a. Mason, J., *Trans. Instn. Rubber Ind.*, **29**, 148–59 (1953).

86. McMahon, W., and Kemp, A. R., *Ind. Eng. Chem.*, **36,** 735–8 (1944).

87. McMillan, F. M., Wheeler, V. V., Blackburn, B. A., *Rubber Age N.Y.*, **61,** 555–62 (1947).

88. McMillan, F. M., Winkler, D. E., and Anderson, M. D., *Rubber Age N.Y.*, **66,** 663–6 (1950).

89. McPherson, A. T., *Rubber Age N.Y.*, **59,** 323 (1946).

90. Mitchell, G. R., and Lufter, C. H., Private Communication to O.R.R., May 15, 1949.

90a. Moakes, R. C. W., in *Rubber Technology*, edited by R. C. W. Moakes and W. C. Wake, Butterworths Scientific Publications, London, 1951, pp. 26–67.

91. Mooney, M., *Ind. Eng. Chem., Anal. Ed.*, **6,** 147–51 (1934).

92. Mooney, M., *Ind. Eng. Chem., Anal. Ed.*, **17,** 514–7 (1945).

93. Morris, R. E., Barrett, A. E., Harmon, R. E., and Werkenthin, T. A., *Ind. Eng. Chem.*, **36,** 60–3 (1944).

94. Morris, R. E., and Hollister, J. W., *Ind. Eng. Chem.*, **40,** 2325–33 (1948).

95. Morris, R. E., Hollister, J. W., and Mallard, P. A., *Ind. Eng. Chem.*, **36,** 649–53 (1944).

96. Morris, R. E., James, R. R., and Evans, E. R., *Rubber Age N.Y.*, **58,** 331–7 (1945).

97. Murphy, R. T., Baker, L. M., and Reinhardt, R., *Ind. Eng. Chem.*, **40,** 2292–5 (1948).

98. Murray, G. S., and Watson, W. H., *India Rubber World*, **118,** 667–9 (1948).

99. Neal, A. M., *Rubber Age N.Y.*, **53,** 31–4 (1943).

100. Parkinson, D., in Mark and Whitby, *Advances in Colloid Science*, Vol. II, Interscience, New York, 1946, pp. 389–429.

101. Piper, G. H., and Scott, J. R., *J. Rubber Research*, **16,** 151–60 (1947); *Rubber Chem. and Technol.*, **21,** 149–63 (1948).

102. Prettyman, I. B., *Ind. Eng. Chem.*, **36,** 29–33 (1944).

103. Private Communication, Goodyear to the O.R.R., Mar. 15, 1945.

104. Raff, R. A. V., Tomlinson, G. H., II, Davies, T. L., and Watson, W. H., *Rubber Age N.Y.*, **64,** 197–200 (1948).

105. Reinsmith, G., *India Rubber World*, **116,** 499–503, 507 (1947).

106. Riddel, C. F., and Woltz, F. E , Private Communication to O.R.R., Mar. 8, 1946.

106a. Rostler, F. S., *Rubber Age N.Y.*, **69,** 559–78 (1951); **71,** 223–8 (1952); Rostler, F. S., and White, R. M., *Rubber Age N.Y.*, **70,** 735–47 (1952).

106b. Rostler, F. S., *Rubber Age N.Y.*, **71,** 223–8 (1952).

107. Rostler, F. S., and Sternberg, H. W., *Ind. Eng. Chem.*, **41,** 598–608 (1949).

108. R. T. Vanderbilt Co., *The Vanderbilt Rubber Handbook*, 9th ed., R. T. Vanderbilt Co., New York, 1948, p. 114.

109. Rupert, F. E., and Gage, F. W., *Ind. Eng. Chem.*, **37,** 378–82 (1945).

110. Rush, I. C., *Ind. Eng. Chem.*, **38,** 58–61 (1946).

111. Rush, I. C., and Kilbank, S. C., *Ind. Eng. Chem.*, **41,** 167–71 (1949).

112. Samuels, M. E., and P'Pool, W. B., Private Comm. to O.R.R., Oct. 5, 1948.

113. Schade, J. W., and Labbe, B. G., Private Communication to O.R.R., June 20, 1946.

114. Schoene, D. L., Green, A. J., Burns, E. R., and Vila, G. R., *Ind. Eng. Chem.*, **38,** 1246–9 (1946).

115. Schwarz, H. F., *Ind. Eng. Chem.*, **36,** 51–4 (1944).

116. Searer, J. C., *Rubber Age N.Y.*, **62,** 191–3 (1947).

117. Sebrell, L. B., *Ind. Eng. Chem.*, **35,** 736–50 (1943).

118. Sebrell, L. B., and Dinsmore, R. P., *India Rubber World.*, **103,** 37–40 (1941); **104,** 45–50 (1941).

119. Sebrell, L. B., and Dinsmore, R. P., *SAE Journal*, **49,** 368 (1941).

120. Sell, H. S., and McCutcheon, R. J., *India Rubber World*, **119,** 66–8, 116 (1948).

121. Shearer, R., Juve, A. E., and Musch, J. H., *India Rubber World*, **117,** 216–9 (1947).

122. Sjothun, I. J., and Cole, O. D., *Ind. Eng. Chem.*, **41,** 1564–7 (1949).

123. Smith, G. E. P., Amberlang, J. C., and Gottschalk, G. W., *Ind. Eng. Chem.*, **38,** 1166–70 (1946).

124. Sperberg, L. R., *Ind. Eng. Chem.*, **42,** 1412–7 (1950).

125. Sperberg, L. R., Bliss, L. A., and Svetlik, J. F., *Ind. Eng. Chem.*, **39**, 511–4 (1947).
126. Sperberg, L. R., Svetlik, J. F., and Bliss, L. A., *Ind. Eng. Chem.*, **41**, 1641–6 (1949).
127. Stambaugh, R. B., Rohner, M., and Gehman, S. D., *J. Applied Phys.*, **15**, 740–8 (1944).
128. Stangor, E. L., and Radcliff, R. R., *Ind. Eng. Chem.*, **41**, 1603–7 (1949).
129. Steffen, H. C., *India Rubber World*, **120**, 60–2 (1949).
130. Stokes, C. A., and Dannenberg, E. M., *Ind. Eng. Chem.*, **41**, 381–9 (1949).
131. Stubbs, A. P., and Johnson, C. R., *Rubber Age N.Y.*, **59**, 567–9 (1946).
132. Sturgis, B. M., Baum, A. A., and Vincent, J. R., *Ind. Eng. Chem.*, **36**, 348–51 (1944).
133. Sturgis, B. M., Baum, A. A., and Trepagnier, J. H., *Ind. Eng. Chem.*, **39**, 64–8 (1947).
134. Svetlik, J. F., Cooper, W. T., and Railsback, H. E., Private Communication to O.R.R., Oct. 3, 1949.
135. Swart, G. H., Pfau, E. S., and Weinstock, K. V., *India Rubber World*, **124**, 309–19 (1951).
135a. Taft, W. K., Duke, J., Snyder, A. D., Feldon, M., and Laundrie, R. W., *Ind. Eng. Chem.*, **45**, 1043–53 (1953).
136. Taylor, R. H., Fielding, J. H., and Mooney, M., *Rubber Age N.Y.*, **61**, 567–73, 705–10 (1947).
137. Throdahl, M. C., and Beaver, D. J., *Rubber Chem. and Technol.*, **18**, 110–5 (1945).
137a. Throdahl, M. C., and Harman, M. W., *Ind. Eng. Chem.*, **43**, 421–9 (1951).
138. Tiger, G. J., Reich, M. H., and Taft, W. K., *Ind. Eng. Chem.*, **42**, 2562–9 (1950); Reich, M. H., and Taft, W. K., *Rubber Age N.Y.*, **72**, 619–24 (1953).
139. Tiger, G. J., Snyder, A. D., and Taft, W. K., Private Communication to O.R.R., Oct. 20, 1947.
140. Tobolsky, R. V., Prettyman, I. B., and Dillon, J. H., *J. Applied Phys.*, **15**, 380–95 (1944).
141. Vila, G. R., *Ind. Eng. Chem.*, **34**, 1269–76 (1942).
142. Vila, G. R., *Ind. Eng. Chem.*, **36**, 1113–9 (1944).
142a. Vodra, V. H., *Rubber Age N.Y.*, **71**, 507–14 (1952).
142b. Vodra, V. H., and Jarvis, L. A., *India Rubber World*, **127**, 633–9 (1953).
143. Weaver, J. V., *Rubber Age N.Y.*, **48**, 89–95 (1940).
143a. Weinstock, K. V., Baker, L. M., and Jones, D. H., *Rubber Age N.Y.*, **70**, 333–8 (1951); Weinstock, K. V., Storey, E. B., and Sweeley, J. S., *Ind. Eng. Chem.*, **45**, 1035–43 (1953).
144. White, L. M., Ebers, E. S., and Shriver, G. E., *Ind. Eng. Chem.*, **37**, 767–9 (1945).
145. White, L. M., Ebers, E. S., Shriver, G. E., and Breck, S., *Ind. Eng. Chem.*, **37**, 770–5 (1945).
146. Wiegand, W. B., and Braendle, H. A., *Ind. Eng. Chem.*, **36**, 699–702 (1944).
147. Wilkinson, C. S., and Gehman, S. D., *Ind. Eng. Chem.*, **41**, 841–6 (1949).
148. Williams, I., in Davis and Blake, *Chemistry and Technology of Rubber*, Reinhold, New York, 1937, pp. 237–68.
149. Willisford, L. H., Private Communication, Nov. 18, 1949.
150. Winkeimann, H. A., *India Rubber World*, **113**, 799–804 (1946).
151. Winkler, D. E., and McMillan, F. M., *Rubber Age N.Y.*, **66**, 299–304 (1949).
152. Winn, H., and Shelton, J. R., *Ind. Eng. Chem.*, **40**, 2081–5 (1948).
153. Wolf, G. M., Deger, T. E., Cramer, H. I., and deHilster, C. C., *Ind. Eng. Chem.*, **38**, 1157–66 (1946).
154. Wolf, G. M., and deHilster, C. C., *India Rubber World*, **120**, 191–8 (1949).
155. Yanko, J. A., *J. Polymer Science*, **3**, 576–600 (1948).
156. Zoss, A., Hanford, W., and Schildknecht, C. E., *Ind. Eng. Chem.*, **41**, 73–7 (1949).
157. Zwicker, B. M. G., Private Communication to O.R.R., Apr. 1949; *India Rubber World*, **121**, 431–2 (1950).

CHAPTER 12

PHYSICAL TEST METHODS AND POLYMER EVALUATION

A. E. Juve

The B. F. Goodrich Company Research Center

INTRODUCTION

The development of new test methods and improvements in the methods in existence at the time of Carpenter's[48] review of 1937 have been numerous and important. A large proportion of these developments resulted directly from the stimulus of the synthetic-rubber program. This chapter will consider these new and improved test methods and will also discuss the problem of the evaluation of experimental synthetic rubbers, omitting only tests for processibility, which are discussed in Chapter 11.

As was to have been expected, more emphasis has been given to test methods relating to those properties in which the synthetic rubbers are inferior to natural rubber than to those in which natural rubber is inferior to the synthetics. For example, the hysteresis loss in natural-rubber vulcanizates is less than that of any of the commercially available synthetics when equitably compared. There has therefore been a marked interest in testing methods for measuring this property. On the other hand, the stability of synthetic-rubber vulcanizates during "normal" aging is generally superior to that of natural-rubber vulcanizates, and there have accordingly been few significant advances in laboratory aging procedures for studying this property with greater speed or accuracy. An account of the behavior of GR-S vulcanizates on aging and a discussion of factors influencing the aging and of its mechanism are given in Chapter 13.

In addition to those properties in which the synthetics were inferior to natural rubber, there were certain properties in which the special-purpose synthetics were greatly superior to natural rubber. These include the oil resistance of the nitrile rubbers, the Neoprenes and the Thiokols, the resistance to ozone cracking of the Neoprenes and the Thiokols and the lower permeability to gases of GR-I and the Thiokols. Though test methods had been used to measure these properties in natural-rubber vulcanizates, the appearance of the synthetics stimulated the development and improvement of the methods.

Further, the necessity of controlling the uniformity of the synthetic product made it desirable, for specification purposes, to establish acceptable values for certain of the physical properties, both of the raw rubber and its vulcanizates. And in order that all the production plants should operate uniformly, it was necessary to improve and refine the existing test methods.

METHODS FOR MEASURING RATE OF CURE

Standard Procedure. For many years the standard procedure for estimating the rate of cure of a vulcanizable mixture has been to cure a series of sheets for increasing periods of time at a particular temperature, measure their stress-strain properties, and plot these values according to one or other of the methods described by Carpenter.[48] The rate of cure was estimated either by the rate at which the stress-strain properties developed with increasing cure in the early stages of the cure or by the time required to reach the optimum cure. But since the optimum cure for one desirable property in a synthetic rubber (e.g., high tensile) may be quite different from that for another property (e.g., resistance to tear), there has been no general agreement on which optimum cure to select as the basis for measuring the rate of cure. The criterion most frequently used has been tensile strength, though tensile strength is admittedly only one of several criteria to be considered in ascertaining the best technical cure for a particular product. The method commonly employed has been to measure the time that elapses between the initiation of the cure and the earliest attainment of an approximately maximum strength. This procedure has never been considered very satisfactory, from the standpoint of precision and reproducibility. Errors in the stress-strain measurements, the difficulty of weighting the usually scattered points when drawing the curves, and the different interpretation of different individuals as to what constituted the earliest approach to a maximum tensile strength meant that the lack of agreement between laboratories and between individuals was appreciable and that small differences in curing rates could not be readily recognized.

Other Procedures. The proposal of Cohan and Steinberg[56] that the ratio of tensile strength at an appreciable undercure to the maximum tensile strength be employed as a measure of rate of cure seems to be a sounder approach and more precise than the time to reach maximum tensile strength.

Other proposals that have been suggested for measuring or estimating the rate of cure include the following:

1. *The T-50 Test,*[280] in which temperature changes are used to measure the tendency for crystallization resulting from stretching to be suppressed as the state of cure is increased. This method is not very satisfactory for most of the synthetics because with these rubbers the T-50 temperature changes but slightly with increasing cure.

2. *Combined Sulfur.* Measurement of combined sulfur as a function of time of cure is sometimes used but is somewhat cumbersome, and the rate of sulfur combination does not necessarily parallel the rate of development of physical properties.

3. *Modulus Corner.* The time required to reach the break in the curve of modulus versus time of cure, sometimes referred to as the "modulus corner" was suggested by Juve and Garvey[130] for GR-S stocks as the basis for selection of equivalent states of cure when comparing different compounds. The break is fairly sharp and reproducible for most compositions of GR-S.

4. The time required to reach a cure that imparts to the vulcanizate a "reasonable snap with substantially unimpaired tear" (both based on hand tests) has also been suggested.[58]

5. *Strain Test.* The test method described by Roth and Stiehler[220] for measuring the elongation at constant stress, which will be discussed in more detail in a subsequent section, has been applied to the problem of the measurement of rate of cure. It is capable of giving more precise data on rate of cure than any of the methods mentioned above. The data obtained for a particular material on a series of cures, when plotted as a function of curing time, gives curves of the form of a rectangular hyperbola. From the equation for a particular curve the cure rate constant, the time for incipient cure,

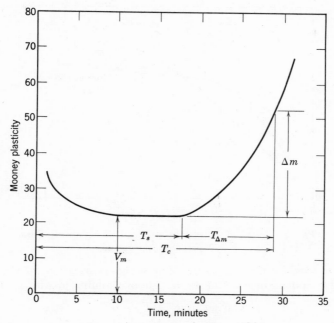

FIG. 1. Typical Mooney Cure Curve

and the elongation at infinite cure (if no reversion is assumed) can be calculated. Schade[227] has shown how these data may be used to calculate equivalent cures for two materials based either on the time at which the rate of change of strain is equal or the time at which the acceleration of the above rate is equal.

6. *Mooney Curing Test.* Another method, first suggested by Weaver[313] and later extended by Shearer and co-workers,[243] involves the use of the Mooney plastometer for measuring both the time for scorching or incipient vulcanization to occur and the rate at which vulcanization occurs after its inception. Figure 1 is a typical curing curve obtained in this way and shows the measurements used to characterize cure curves obtained by this method. The time $T_{\Delta m}$, is used as a measure of the cure rate. Later work has shown that, if the scorching time is taken as the time required to increase the viscosity a few specified units above the minimum, somewhat better reproducibility of this measurement is attained. This method has the advantage that it is fast

and reproducible and that all measurements are made with a single small specimen.   It has the disadvantage that cure can be carried only to a relatively low state, after which crumbling occurs.

The present author has suggested[125] that the relative over-all curing rates of vulcanizable materials in terms of the time required to reach the approximate optimum could be calculated from the Mooney cure data by assuming that this time is made up of the scorch time and the product of a constant and the time required for the viscosity to increase above the minimum

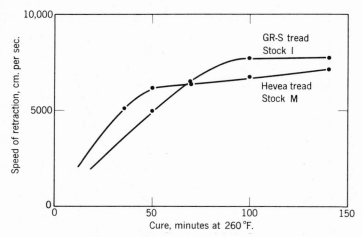

FIG. 2.   Speed of Retraction of Tread Stocks versus Cure,
250 per cent Elongation

viscosity a fixed number of units.   The suggested relation for use with the small rotor and an increase of 30 units above minimum was

$$\text{Curing time} = T_s + 6T_{\Delta 30}$$

A more precise relation can be worked out if the times obtained in the Mooney cure test are corrected for the time lag in heating the specimen to the impressed temperature.

7. *Speed of Retraction.*   The speed of retraction of specimens cut from sheets cured for increasing time periods has been suggested as a method for selecting the optimum cure to serve as a basis for measuring the curing rate. The method described by Stambaugh and co-workers[263] employs an electronic timing circuit which measures the time interval during which the free end of a retracting strip passes between two sharply focused light beams.   Plots of speed of retraction versus curing time exhibit fairly sharp bends which correspond fairly well to the "best" cure from other data, as judged by experienced technical men.   Figure 2 shows two of these curves taken from the report of Stambaugh et al.

Definition of Terms.   There has been much confusion concerning the precise meaning of some of the terms used in this section, particularly optimum cure, state of cure, and rate of cure.   Cohan and Steinberg[56] offered definitions of these terms based both on their own experience and on the expressed opinion of other workers. The following definitions and comments are from their report:

Optimum cure. The optimum cure for a particular property is the time of cure that yields the maximum value for that property.

State of cure. This has been defined by Whitby[317] as the position of the cure in question in a series of cures. Wiegand[320] has suggested that a more precise definition would be that fraction of the maximum value for the particular property under consideration exhibited by the cure in question.

Rate of cure. The rate of cure is measured by the time required to reach a given state of cure compared to the time required by some standard stock.

In view of the fact that most present-day compositions exhibit induction periods of varying lengths of time depending on the degree of delayed action imparted by the accelerator used, it might be desirable to modify the rate of cure definition to read as follows: Rate of cure is measured by the time from the onset of vulcanization to the time required to reach a given state of cure compared to similar data obtained on a standard stock.

TESTS RELATING TO STRESS-STRAIN MEASUREMENTS IN TENSION

The stress-strain test in tension continues to be the test most used for rubber vulcanizates. Its utility is unquestioned for a variety of purposes, such as following the changes occurring with advancing cure or with artificial or natural aging or the changes taking place during the service life of an article, or studying the effects of changes in composition, or for specification and control testing. However, many workers have suggested that its value in assessing product quality is overrated since few rubber products are strained in service to the extent that a tensile break occurs. Most rubber products are subjected in service to deformations of small amplitude, and failures occur for a variety of other reasons. For this reason it is frequently stated that little correlation exists between stress-strain characteristics and product performance.

However, the enormous background of data on the stress-strain properties of various compositions and its value in rubber technology cannot be minimized. The reported poor correlation with service may be a hasty conclusion resulting from the sparseness of adequate data, particularly on service performance. The lack of good data on the service performance of products and on the relation between service performance and the usual laboratory-measured properties is one of the more important blanks in the field of rubber testing which it is hoped will be filled in the future.

Improvements in Standard Stress-Strain Testing Procedure. A number of modifications in the procedure for running stress-strain tests as outlined[267] in the ASTM method D 412–41 have been suggested as the result of the experience in the control testing of the synthetics. Some of these modifications have been adopted by ASTM and are now incorporated in the tentative version of the method, designated as D 412–51T.

Specimen width. One potential source of error is variation in the width of the specimen. Although the method D 412–41 specifies that the width of the restricted section shall not vary more than ±0.001 in. from the specified dimensions and that the minimum measured width shall be used in the calculations, in actual practice it has been found that the die width usually varies more than ±0.001 in. from the specified dimension at the restricted section, that specimens prepared from stocks of different consistencies vary

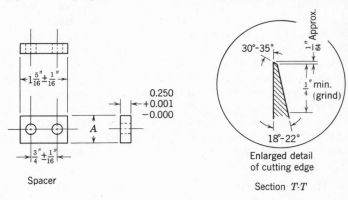

Spacer

Enlarged detail
of cutting edge

Section T-T

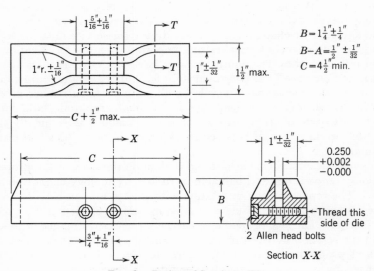

FIG. 3. Design of Specimen Die

appreciably from the die width, and that because of difficulties in measuring the specimen width with any accuracy the nominal die width has been used as the basis for calculations of the unit stresses.

The tentative method suggests die designs that permit a better control of the spacing of the cutting edges and specifies that the die width, in the restricted section, shall be used as the specimen width. The suggested die design is illustrated in Fig. 3.[258] A procedure,[257] which is not a part of the tentative method, has been developed which makes possible the measurement of variations in the width of specimens cut with a die arising from differences in the consistency of the material being tested. This involves cutting a specimen from the sheet, replacing it in the position from which it was cut, and making three transverse cuts with the die across the restricted section of the specimen in such a way that each square so produced is cut in both directions in the same portion of the die. This gives three $\frac{1}{4}$ in. squares when die C is used. From the weight, thickness, and density of the material the width of the specimen can be accurately calculated and can be subsequently applied in calculations for specimens tested from the same material.

Specimen Length. One source of error was found to be the inaccurate placement of the bench marks. Because of the shortness of the restricted section, one of the marks is occasionally placed on the flared portion of the specimen. This possibility of error has been eliminated by increasing the length of the restricted section by $^{5}/_{16}$ in. The method of applying the bench marks has also been specified in greater detail.

Averaging Tensile-Strength Data. The requirement of the ASTM method D 412–41 with respect to the selection of typical specimens for ultimate tensile strength was that three specimens be pulled and the average reported, provided that the breaks occurred in the restricted portion of the specimen and provided that the results agreed within 10 per cent of the highest value. In the event that two results failed to check within 10 per cent, additional specimens were to be tested until two or more check results were obtained. Results from obviously defective specimens were also to be discarded. This procedure was based on the assumption that the highest values obtained are most likely to approach the "true" single value which it was presumed was characteristic of the material. The present view, which is supported by statistical studies,[294, 201, 117, 171, 34] is that the strength of a material such as rubber can best be represented by a statistical distribution function and that any portion of the variation in a set of results not attributable to testing error is due to inherent variations in the material. As a result of these studies it has been concluded that the soundest procedure that can be used when only a limited number of breaks can be made is to employ the median or middlemost value. This eliminates nearly all calculations and does not reject any breaks as atypical.

Other Details of Testing. In addition to the changes discussed above, certain of the operations involved in the testing procedure are now more rigorously specified so that different laboratories will operate on a more standard basis than in the past. For example, the calibration procedure has been modified to require the interposition of a dumbell specimen between the weight and the upper grip. This procedure permits a more gradual application of the load, more nearly approaching actual testing conditions. Also the calibration tolerances are more precisely defined. Daily honing of the die with hard Arkansas honing stones is suggested. It is recommended that the pad supporting the sheet during the dieing-out operation be made up of a solid foundation (hard wood or steel), a semihard pad (vinyl plastic, gutta-percha, rubber or Masonite) and a smooth cutting surface (hard paper board or Manila folder).

Improvement Achieved. The degree of improvement that can be effected by the adoption of these changes, along with the use of an improved modification of the tensile tester and the use of a temperature correction table (both to be described later) is appreciable. Schade and Roth[228] have reported that the standard deviation in tests for stress at 300 per cent was reduced from 75 to 30 p.s.i. when a Standard Reference Bale of GR-S was tested by 16 different laboratories.

Effect of Temperature of Test. Since the stress-strain properties of rubber and synthetic-rubber vulcanizates are affected by the temperature at which the test is conducted and since many laboratories are not equipped with constant-temperature rooms, errors in testing will be introduced from this source. To permit correction of stress-strain values to a constant temperature in the control testing of production lots of synthetic rubbers, correction tables were prepared for the test specification compounds[259, 260] of GR-S, GR-I, and GR-M. The values given in these tables have been recalculated and plotted in Fig. 4, to show the changes in stress-strain properties that occur over a range of temperatures higher and lower than the "standard" test temperature. The GR-S and GR-I values are based on channel black, tread-type stocks, whereas the GR-M values are based on a pure gum stock. These correction values do not necessarily apply to stocks compounded differently. It is frequently desirable to test at temperatures higher or lower than room temperature, since products are actually used under these conditions. Because of the different effects of temperature on the stress-strain properties of synthetic vulcanizates as compared to natural-rubber vulcanizates, it has become common in many laboratories to make routine tests on experimental materials at temperatures other than room temperature (usually higher).

Boonstra[20] has published data on the variation of tensile strength and elongation with test temperature over a wide range of temperatures for both gum and tread-type vulcanizates of the principal rubbers. Figures 5 and 6 are taken from his work and show the effects of test temperature on tread-type vulcanizates. It will be noted that for all the rubbers the tensile strength decreases with an increase in temperature but at different rates. Elongation generally increases up to about room temperature, after which it decreases rapidly. However, the natural-rubber stocks behave differently in that the breaking elongation either increases steadily with testing temperature (stock without softener) or decreases slightly up to about 50° C. and then increases to a maximum at about 100° C. (stock with softener). Data have also been reported by Morron and co-workers[181] on the effect of temperature ranging from −60 to +260° F. on the tensile strength of vulcanizates of natural rubber, GR-I, GR-M, GR-S, and Perbunan.

Since most rubber products do perform in service at temperatures considerably higher than room temperature, it may be anticipated that in the future more emphasis will be placed on stress-strain tests at temperatures more nearly comparable to those encountered in service than on room-temperature tests.

Apparatus for Tests at Elevated and Reduced Temperatures. Special equipment or a modification of the standard Scott machine is required for running tests at temperatures other than room temperature. Probably the simplest and most satisfactory modification permitting a fair degree of accuracy is to build an enclosure having suitable temperature-control facilities around the space traversed by the grips. One such design has been described by Scheu and Schade.[229] Scott Testers, Inc., have

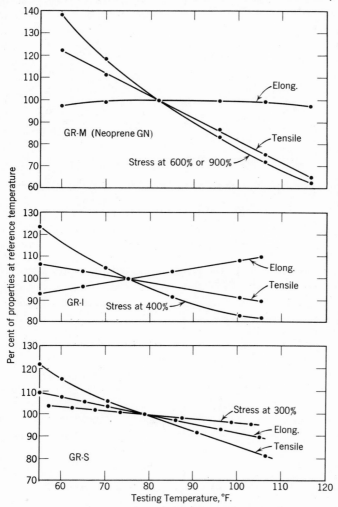

Fig. 4. Effect of Testing Temperature

developed a tensile tester capable of testing over the range —70 to +300° F. This machine is also designed to permit rates of jaw separation up to 200 in. a minute.

A simple method for obtaining a high-temperature tensile test is one described by Braendle, Valden, and Wiegand.[25] It involves the use of an electrically heated patch heater having a face grooved to fit the restricted section of a dumbell. The heater is held against the specimen during extension until the break occurs. It was found that by regulating the temperature

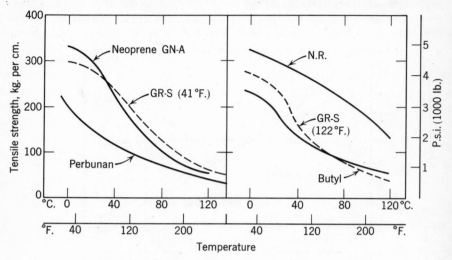

Fig. 5. Tensile Strength as a Function of Temperature or Carbon Black Vulcanizates[20]

of the patch heater to 230° F. the results obtained were very close to those obtained by other methods at 212° F.

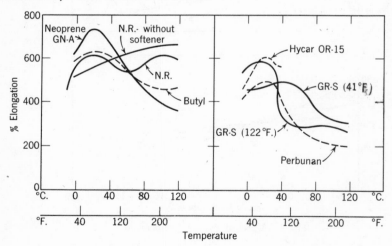

Fig. 6. Elongation at Break as a Function of Temperature for Carbon Black Vulcanizates[20]

Effect of Specimen Size. As with other materials of construction, the ultimate tensile strength and elongation of rubber vulcanizates depend on the dimensions of the test piece. For this reason, the standard procedures for testing specify the shape of the dumbbell and the permissible range of thicknesses. However, for products that are thinner than this or are of such a size that a standard specimen cannot be cut, nonstandard specimens must be used. Also in several of the research laboratories involved in the synthetic-rubber program, it was found desirable (in order to conserve material) to use a "micro" scale test procedure, which utilizes small-scale test pieces considerably thinner than standard specimens.[90, 92] These specimens gave test results appreciably higher for tensile strength than standard specimens, but equal moduli. The relationship between these higher values and those obtained on standard specimens was found not to be constant for different materials.

A study by Higuchi and co-workers[117] showed that if ultimate tensile strength is plotted against the logarithm of the volume of the test piece (between bench marks) a straight line is obtained, the slope of which varies with different materials. On the basis of a relationship that had been developed in the textile field for the effect of the length of the specimen on tensile strength it was deduced that the slope of tensile strength versus log volume curve was a measure of the heterogeneity of the material. Since this slope differs with different materials, it is obvious that no correction factor can be utilized for converting results on "micro" scale specimens to standard specimens.

Effect of Prestretching. It is well known that if a rubber specimen is prestretched before being tested for its stress-strain characteristics the resultant curve will be different from that for the unworked specimen. An interesting study of this phenomenon has been reported by Mullins,[186] whose conclusions are summarized below:

Prestretching results in a softening of the material. The greater the stiffening effect of the pigment used in the compound, the greater the degree of softening. For unloaded stocks the softening is nearly negligible. Softening of the loaded stocks is apparent only at elongations below that of the prestretch; beyond this point the stress-strain curve is practically identical with that of the original unstressed material. With successive prestretches the stress-strain curve at elongations less than that used in prestretching approaches that of the unloaded stock.

Figures 7 and 8 are illustrations from Mullin's work. Figure 7 shows the effect of a single prestretch to various elongations up to 400 per cent followed by a determination of tensile stress up to 280 per cent stretch for both a pure gum mix (A) and a channel black mix (B). Figure 8 illustrates the effect of the number of prestretches to 420 per cent (approximately three fourths of the breaking elongation) for the channel black mix (B).

Some recovery of the stiffness destroyed during the prestretching occurs on standing, and the recovery is accelerated by raising the temperature. However, the recovery at room temperature is so slow that the effect, for practical purposes, can be considered as permanent. The ultimate tensile strength is but slightly affected by prestretching.

Effect of Speed of Extension. The standard rate of jaw separation prescribed[267] by the method D 412–41 is 20 in. per minute. A proposal by the committee responsible for this method has been made that the rate may be

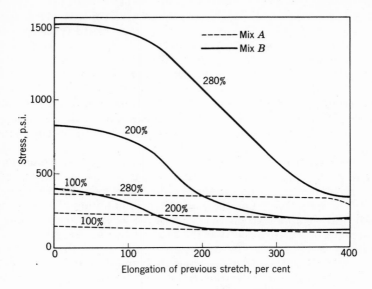

FIG. 7.   Effect of Prestretching[186]

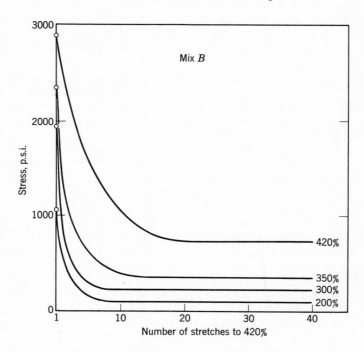

FIG. 8.   Effect of Prestretching[186]

increased up to 40 in. per minute but that for referee purposes 20 in. per minute shall be standard. The R. T. Vanderbilt Company has reported[307] that no differences were found with five natural-rubber compounds when the rate of jaw separation was increased from 20 to 40 in. per minute. However, materials exhibiting high set and low resilience, as, for example, plasticized polyvinyl chloride, show a marked dependence of modulus on rate of elongation over this range.

With a dumbbell-shaped specimen, the rate of elongation of the restricted portion is not identical with the jaw-opening rate and is not constant throughout the test. A rough average for the rate of elongation is about 10 per cent per second.

Roth and Holt[219] reported data on a series of natural-rubber compounds using ring specimens and a falling-weight device designed to give a rate of elongation of the order of 100 per cent per second. In general the stresses at elongations somewhat short of the break were increased as the rate of elongation was increased. More recently Villars[310] has reported data at rates of elongation up to 270,000 per cent per second. These data show marked increases in modulus, increases in tensile strength, and decreases in elongation at rates above 10,000 per cent per second. Apparently Villars based his calculations of rate of extension on the rate of grip separation and not on the rate of elongation of the restricted section of the dumbbell specimen. This is indicated by the points on his graph for the 20 in. per minute speed for the standard test which he calculated at 20 per cent per second. If the same calculations were applied to the high-speed tests, the values quoted above might be too high by a factor of approximately 2. Dart and co-workers[70] have described a fast stress-strain tester which can be operated at speeds up to 200 per cent extension per second. However, no data were reported on the effects found as compared with tests at the standard speed of extension of 20 in. per minute.

It would be desirable to have more data on the effect of variation in test rate on materials widely different in composition and including in particular a variety of synthetic rubbers.

Importance of Variable Test Conditions. The effects of testing temperature, speed of extension, specimen size, and preworking of the test piece discussed in the preceding pages should be considered in characterizing a rubbery material instead of relying solely on the results of tests under "standard" conditions on a "standard" test piece. This is particularly important when attempts at correlation with service performance are being made, since the rubber in products during use is seldom subjected to standard temperature conditions, stretched at a rate corresponding to that used in standard stress-strain tests, or used in the unworked condition or in the form of press-cured slabs.

Improvement in Tensile Testers. In the search for sources of error in the specification and control testing of the output of synthetic rubber at the producing plants, it was found that some errors were introduced by certain features of the standard Scott machines in use. The manufacturers, in cooperation with a Rubber Reserve test methods committee and the National Bureau of Standards, announced[122] a new model designated as

ORR L-5 which embodies improvements designed to overcome these errors. The changes involve an added adjustment for compensating for width variations of the test specimens, a solenoid which holds the pawls up during the course of the test but engages them at the instant the specimen breaks, and a stylus motion which permits the use of a chart having equal spacings representing equal increments of stress, and thus more accurate readings of modulus values. Older machines may be revised to incorporate these improvements.

One of the sources of inaccuracy in modulus determinations on the Scott machine is the difficulty of following accurately the separation of bench marks. For improving accuracy in this regard, a suggestion by Shearer,[242] which can be adapted to the trammel-point method of following the bench marks, involves the use of a steel tape preforated at intervals corresponding to elongations of 100, 200, 300 per cent, etc. One end of the tape is attached to the lower trammel point, and the other trammel point has a fixture through which the tape is drawn during elongation of the specimen. As the holes in the tape pass through this fixture, an air-actuated microswitch in a suitable circuit fires a spark of controlled duration. The reported advantages of this device are that the operator need not be distracted from the problem of following the bench marks by the necessity of operating a switch periodically and that the intensity and duration of the spark is uniform at all times. Both features contribute to greater accuracy.

New Machines for Tension Testing. Several new machines have been developed which are reputed to be useful for testing materials (such as rubber) of high extensibility. Some of them are refinements of standard laboratory testing machines adapted for a high rate of extension and low load as required for rubber testing. Among these are the Baldwin Tate-Emery machine[53] and the Tinius Olson Plastiversal machine. Both are very versatile, permitting tests in shear and compression as well as tension and having a wide range of rates of extension. However, neither machine is designed specifically for testing dumbbell-shaped samples, nor is either equipped with facilities for following the separation of bench marks so that a complete stress-strain curve up to the break can be obtained.

The Instron machine[43] was developed primarily for use with plastics but may also be used for rubbers. Like the two machines mentioned above it does not have facilities for following the bench marks on dumbbell-shaped specimens. This machine utilizes strain gages and an electronic circuit for measuring and recording the load and an electrical system for following strains. A variety of tests including stress relaxation and hysteresis loops can be run with it. The Amsler horizontal tensile-testing machine[6] is an apparatus designed for testing paper and yarns as well as rubber. Variable rates of jaw separation from 0.1 to 20 in. per minute are provided, as well as five load scales from 5 to 200 kg. This is a pendulum-type machine.

New Test Methods Relating to Tension Testing. One of the more important properties brought out in a stress-strain test is stress at relatively low elongations. This is an important measurement both for research and development purposes and for estimating service performance. However, by the conventional methods of testing, the accuracy of its determination is quite poor. The substitution of a test in which the elongation at some fixed stress is measured instead of the stress at a fixed elongation has occurred to a number of workers (see, e.g., Messenger[162]) as a means of improving accuracy. The National Bureau of Standards undertook a study of this method and has developed an apparatus which is now commercially available.[44] Holt, Knox, and Roth[120] have described the details of construction of the apparatus; Roth and Stiehler[220] have described the testing procedure followed and the results obtained with it. The specimen used is approximately $1/4$ in. wide and 6 in. long. Bench marks 10 cm. apart are marked on the specimen, and the distance between bench marks measured 60 seconds after the application of the load. For tread-type compounds a stress of 400 p.s.i. is used; for gum stocks this is reduced to 100. Since creep is involved in the test, particularly for low cures, it is necessary that the timing be done accurately and that the time of load application be sufficiently long to minimize errors from this source. The 60 second interval appears to be adequate. Roth and Stiehler[220] report that the standard deviation in this test is approximately 40 per cent of that of the 300 per cent modulus determination by the conventional test method.

If desired, the permanent set of the specimen may be measured at some fixed time after the completion of the strain test.

ACCELERATED AGING TESTS

It has previously been indicated that, since all the synthetic rubbers are more stable than natural rubber to ordinary static exposure to air or oxygen, relatively little recent work has been done in this field directed toward the development of improved test methods. The oxygen bomb test of Bierer and Davis[14, 271] and the Geer oven test[96, 272] remain the standard accelerated aging test methods. The air-pressure heat test[269] continues to be used, but less frequently than the air-oven or oxygen-bomb methods.

Test-Tube Technique. One new method, the test-tube method of high-temperature aging,[287] in which the specimens are suspended in large test tubes provided with vent tubes and heated by immersion in an oil bath, was developed for high-temperature exposures up to 150° C. The anticipated advantages of this method over the oven method were as follows: (1) The specimens of each material are isolated from other materials being aged so that contamination of one material by volatile products from another is eliminated. (2) The use of an oil bath as the source of heat permits better control of temperature. (3) The air circulation (by convection) is moderate and constant for a particular exposure temperature. In a more recent development of this method a heated aluminum block is substituted for the oil bath.

One report[127] on the test-tube method showed improved reproducibility between laboratories over the standard oven method at a temperature of 150° C.; a later report[232] showed no marked improvement at a temperature range of 70 to 150° C. However, one of the materials tested in the latter temperature range, a nitrile-rubber stock vulcanized with tetramethyl-thiuram disulfide, gave very erratic results when oven-aged, particularly at

70° C., but consistent results when aged by the test-tube method. Subsequent tests conducted by the same committee[225] that conducted the above-mentioned tests demonstrated that the reason for the erratic results was contamination of the nitrile-rubber stock by the other stocks that were aged with it or by the use of contaminated ovens. Thus for materials sensitive to contaminants during aging, the test-tube method does give more reliable and consistent results.

It is generally recognized that, during accelerated oven- or bomb-aging tests in which more than one material is aged simultaneously, contamination of one material by volatile products from another may occur and affect the results obtained. The ASTM aging methods contain warnings on this point, but in practice they are frequently ignored. Some good data on this point have been reported by Fackler[85] on the extent of the effect in the oxygen-bomb method. In oven-aging experiments at 78° C. by Eccher and Oberto,[79a] in which stocks originally free from antioxidant were aged alongside stocks containing 2 per cent of antioxidant (phenyl-β-naphthylamine), it was found that, if the air in the oven was agitated, the former stocks acquired about 0.1 per cent of antioxidant after 1 day and more than 0.3 per cent after 4 days, whereas, if the air in the oven was still, substantially no migration of antioxidant occurred. These authors describe an aging oven of large capacity (60 liters) in which, it is asserted, a variety of stocks can be tested together without the danger of contamination of one stocks by volatile products from another. The oven is heated by means of a vapor jacket supplied by a boiler containing a liquid of suitable boiling point. The air in the oven is not subjected to agitation.

Oxygen Absorption in Aging. Methods other than those mentioned above have been suggested for evaluating the stability of rubber vulcanizates. One of these is the measurement of oxygen absorption under controlled conditions. Shelton[244] has reviewed the methods that have been used and discussed the advantages and disadvantages of oxygen-absorption methods in general.

In the gravimetric method the gain in weight of the specimen is followed during exposure to the aging conditions. The manometric method involves measurement of the drop in pressure as oxygen is absorbed; the volumetric method, the measurement of the change in volume at constant pressure. In all cases the gas used is oxygen, and the exposure temperature is relatively high. Shelton[244] prefers the volumetric method because of fewer complications and better reproducibility. The correlation between the quantity of oxygen absorbed and the change in physical properties for a particular material is good,[210, 246] and so is the correlation between natural aging and oxygen absorption.[3] However, the effect of the absorption of equal quantities of oxygen by compositions based on different polymer systems is quite different, so that the application of the method for some purposes, as for example product specifications, would be limited.

The work of Tobolsky and co-workers[161, 297] on stress relaxation, creep, and permanent set, which is discussed in a later section of this chapter, suggests that the measurement of these can be used for the evaluation of the stability of rubbers to oxidative degradation. Throdahl[296] has reported on the use of creep and stress-relaxation tests as a means of evaluating the effectiveness of antioxidants and claims improved ability to discriminate between materials over the conventional oven methods.

Object of Aging Tests. The principal object of accelerated aging tests is to permit an estimate to be made of the stability of a material under some conditions of storage or use that are different from the conditions employed in the test. This requires a knowledge of the effect of temperature, oxygen pressure, time, and other factors that may be involved. Accelerated aging tests are also run for control purposes (for checking the uniformity of products being produced on a routine basis), for specification purposes (to be certain that certain minimum requirements are being met), and for research and development purposes (for testing the comparative performance of different materials).

Influence of Temperature of Test. The effect of the temperature in air-aging tests has been discussed by Schoch and Juve,[232] who summarized prior published data in terms of the temperature coefficient. The reported values varied from 1.97 to 4.04 per 10° C., depending on the properties measured, the criteria used to judge equal degrees of deterioration, and the kind of rubber used. The data reported in this reference on vulcanizates of natural rubber, Neoprene, nitrile rubber, GR-I, and GR-S gave temperature coefficients from 1.80 to 3.05 per 10° C., also depending on the property and the composition. At very high temperatures nonhomogeneous deterioration occurs, as a result of a reaction rate of the rubber with oxygen exceeding the diffusion rate of the oxygen through the rubber. This phenomenon occurs more readily, the higher the temperature, the more readily oxidizable the material, and the thicker the test specimen. With most of the synthetic rubbers it does not occur below 150° C. The temperature coefficient for oxygen bomb aging as reported in the literature[191, 123-4, 134, 308, 247] varies from 1.63 to 3.88.

Neal, Bimmerman, and Vincent[191] studied the effects of variation in the oxygen pressure by the bomb method. Over the range of 15.2 to 314.7 p.s.i. the effect of pressure varied according to the composition of the material, but in no case was the deterioration rate proportional to the oxygen pressure.

Estimation of Degree of Deterioration. The usual procedure for judging the degree of deterioration after the aging interval is to compare the stress-strain properties, particularly tensile strength, with those of the unaged material. However, when two or more materials are being compared, it is sometimes difficult to know whether comparison should be made on the basis of the time required to reach a particular stage of deterioration or on the basis of the deterioration at a particular time. Figure 9, taken from a report by Neal and Vincent,[194] was used by these authors as an illustration of this point. The curves for compounds *A* and *B* are similar in shape, and there is no doubt that compound *B* is superior to compound *A* (the control, without added antioxidant) at all times of aging. Compound *C* is superior to compound *B* during the early stages of aging but inferior in the later stages.

For the synthetic rubbers, tensile strength is a less satisfactory criterion of resistance to aging than modulus or elongation.[193, 245] Buist and Welding[40] suggest that elongation is the best single property to consider except in those cases in which reversion occurs, when tensile strength is to be preferred. Other tests, such as swelling, hardness and resilience, are occasionally used but appear to have no advantage over stress-strain tests.

Correlation of Accelerated with Natural and Service Aging. The general conclusion of investigators is that the correlation between accelerated aging tests and natural aging is poor and that the principal reason is that more than one reaction occurs during aging, one of which may be more dependent on temperature or oxygen pressure than another. However, the

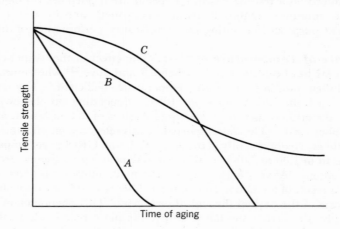

FIG. 9.   Typical Types of Aging Curves[194]

published data on which the conclusion concerning poor correlation is based are not altogether convincing because of the lack of detail concerning the precise conditions for shelf aging and the criteria used for the selection of equal degrees of deterioration.   Practically, the reported poor correlation between accelerated aging tests and shelf aging is of less importance than the correlation between accelerated aging tests and performance in a product exposed to temperatures appreciably higher than room temperature.   In this case the correlation is probably quite good.   An excellent correlation would be expected when the ultimate failure of a product which is operated at the same temperature as the aging test can be traced to those changes in properties that are observed in the laboratory test.   However, as Buist and Welding[40] have pointed out, the stress-strain tests run both before and after aging should preferably be conducted at the same temperature as the aging and not at room temperature.

It should be apparent that no correlation between accelerated aging tests and product performance can be expected if the ultimate failure of the product is due to conditions unrelated to the oxidative changes occurring in the laboratory test.   Failures due to flex cracking, ozone cracking, light catalyzed oxidation, fatigue, etc., would not be expected to correlate with the results of the standard laboratory aging tests.

TESTS FOR HARDNESS

Hardness is one of the more important properties of rubber vulcanizates and should rank in importance with stress-strain and dynamic properties. It is difficult to conceive of an application for rubber in which it is not an important characteristic, and in many it is of vital importance. This section will be concerned with static measurements as distinct from dynamic measurements, which are discussed later.

The real objective in making hardness tests is to measure the elastic modulus under conditions of small strain. With but one exception, all hardness-measuring devices record the depth of penetration of an indentor under either a fixed weight or a spring load. The one exception measures the force required to produce a given indentation. The penetration may be reported on an arbitrary scale or in terms of the actual depth of penetration. By either method the reading obtained is related to the elastic modulus. For small deformations and if rubber is assumed to be an ideal elastic isotropic medium, the indentation obtained on a specimen of reasonable thickness can be shown to depend entirely on the elastic modulus, the load applied, and the dimensions of the indentor.[251, 238] For spherical indentors,

$$H = K \frac{L^{0.75}}{D^{0.5}}$$

where H is the depth of penetration, L is the load, and D is the diameter of the indentor. For cylindrical indentors,

$$H = k \frac{L}{D}$$

where K and k are constants incorporating Young's modulus. In spite of the assumptions that the rubber is ideally elastic and isotropic, the above equations have been found to agree well with experimental data on ordinary rubbers. Scott[238] has shown that for indentations greater than about 0.8 of the diameter of a ball indentor the first equation does not hold. At these depths the ball indentor behaves like a cylindrical indentor, and the relations conform to the second equation. Scott[235] has also considered the use of a conical indentor which would have the advantages that the shape of the indentation would be independent of the depth of penetration and that the penetration of a perfectly elastic material at a constant force would be inversely proportional to the square root of the modulus.

In all hardness-measuring instruments (except one which employs a maximum indicator) the time at which the reading is taken is important. Since for a particular set of conditions the usual hardness test imposes a constant load on the specimen, the test becomes a creep test if readings are taken at various time intervals after the application of the load. This factor is responsible for the failure of exact correlations between different instruments which require different time intervals for reading. The correlation is particularly poor when materials having varying creep rates are compared.

Hardness-Measuring Instruments

Portable Instruments. The most widely used instrument of this type in the United States is the Shore A durometer. In spite of its disadvantages, it has become popular because of its portability and its applicability to odd-shaped specimens. Its disadvantages include the necessity of frequent adjustments, the ease with which the indentor becomes worn and damaged, and the personal element involved in operating the instrument and reading the scale.

This instrument uses a calibrated spring to provide the indenting force, and the load imposed by the spring varies with the indentation. Readings are normally made at the instant when firm contact has been established with the specimen, although some specifications require readings at 15 or 30 seconds thereafter. Taylor[277] has reported the effect of the time of reading on the value of the readings for a variety of materials. His data gave nearly a linear relation between the hardness readings and the logarithm of the time. The differences between the slopes of the curves obtained for different materials was appreciable.

Another portable instrument, which gives hardness values on the same scale as the Shore instrument, is the Rex hardness gage.[189] This has been introduced to the industry relatively recently but has gained ready acceptance. Its shape is that of a small cylinder which can be carried like a fountain pen. The gage or hardness indicator is ejected from the top of the gage when the indentor is applied to the specimen and does not retract until pressed back by hand. Thus the reading given is the maximum or "instantaneous" value. Several reports have indicated that some practice is required to read the vernier scale accurately. This instrument cannot be adjusted by the user but must be returned to the factory for alteration.

The ASTM method D 676–49T covers[282] the use of the two instruments described above. The R.A.B.R.M. (Research Association of British Rubber Manufacturers) hardness meter[202] is another portable tester and employs a $^3/_{32}$-in. ball indentor. The Wallace pocket meter recently placed on the market is thought to be an improvement over most if not all existing hand-operated meters[252] using a calibrated spring. It reads directly in B.S. (British Standard) degrees and has a $^1/_{16}$-in. spherical-ended indentor which is not so subject to wear or damage as the truncated-cone indentor of the Shore A. This meter is easily calibrated by the use of a dead-weight device supplied with it and may also be checked against the Wallace dead-weight tester, which reads in the same units.

Constant-Dead-Load Instruments. A number of instruments employing a constant dead load and indentors with spherical tips of varying sizes are designated below.

The Pusey and Jones plastometer[270] is the subject of the ASTM method D 531–49 and consists essentially of (1) a ball-point indentor 0.1250 in. in diameter (or 0.250 in. in diameter for extremely soft stocks) acting under a dead load of 1000 grams and (2) a device for measuring the indentation in hundredths of millimeters.

The Admiralty rubber meter[2] of British origin resembles the Pusey and Jones plastometer in operation. A ball-point indentor of $^3/_{32}$-in. diameter is loaded first with 30 grams for 5 seconds and then with an additional 535 grams. The difference in indentation after 30 seconds may be converted to B.S. degrees by means of a conversion chart or graph, or the dial may be calibrated in B.S. degrees, which bear a fixed relationship to Young's modulus and are roughly comparable to Shore A units. This method is the subject[27] of the British Standard 903–1950. The device also employs a buzzer or vibrator in the base to overcome frictional effects.

The ASTM tester described in detail[265] in method D 314–39 employs a hemispherical indentor 0.0938 in. in diameter acting under a dead weight load of 3 lb. and with an annular presser foot around the indentor subject to a load of 5 lb. The indentation in thousandths of an inch is measured 30 seconds after the application of the load.

The R.A.B.R.M. tester[74] is similar to the ASTM tester in that it employs a foot which is in light contact with the upper surface of the specimen. The indentation, using either a 1/8- or 1/4-in. ball indentor under a 1-kg. load, is measured with respect to the position of the foot. This instrument has been modified to conform[27] to B.S. 903–1950 by use of the prescribed loads and indentor size. In this case the indentation is readily converted to B.S. degrees.

Constant-Indentation Tester. Buist and Kennedy[37] have developed a hardness tester that measures the force required to secure a given indentation of the rubber. The force on the indentor is measured hydraulically. This type of tester would have at least one advantage over the dead-load instruments in that the effect of a limiting indentation would not be encountered with extremely soft rubbers.

Scales of Hardness. It will be noted that the scale used on the Shore and Rex instruments assigns a value of 0 for an infinitely soft material and 100 for an infinitely hard one. The dead-load devices usually report the depth of penetration as a hardness number obtained by multiplying the actual depth by 100. On these scales a larger number denotes a soft rubber and a small number a hard one. This reversed scale is somewhat awkward and is less acceptable to most people than the Shore scale. Scott[239–40] has proposed a new method of expressing hardness which has been included[27] in B.S. 903–1950. This method provides a scale reading from 0 to 100 B.S. degrees. This proposal does not change the present form of test, but the proposed scale can be obtained either by means of a conversion chart or by application of the new scale directly to the micrometer gage.

An excellent review of the use, adjustment, and calibration of all the more commonly used hardness testers has been prepared by Soden.[252]

Factors Affecting Tests for Hardness. The hardness of rubber vulcanizates changes with temperature, and it therefore is important that tests for specification purposes or in different laboratories should be made at the same temperature. Dawson and Porritt[71] have reported that natural-rubber vulcanizates become softer with an increase in the temperature of test up to 100° C. Taylor[277] reported some tests at various temperatures up to 100° F. and showed that softening occurred with a variety of vulcanizates of different rubbers. The effect of low temperatures will be discussed in a subsequent section.

The condition of the surface of the rubber has been shown[78] to have an appreciable effect on indentation with spherical indentors. Factors that lower the coefficient of friction between the indentor and the rubber such as dust or a lubricant will increase the indentation.

The thickness of the specimen must be sufficiently great that the rigid supporting surface does not influence the results. Scott[236] has studied this factor carefully with a variety of instruments. Figures 10 and 11 are re-

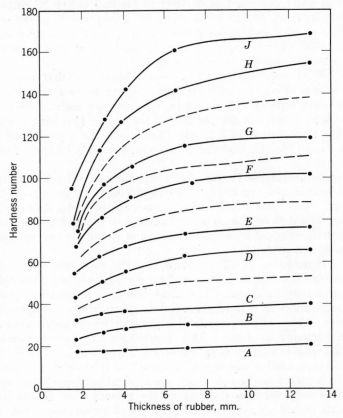

FIG. 10.   Effect of Varying Thickness—0.075-in. Ball Indentor[236]

produced from his report and show the effects of thickness variations from 2 to 12 mm. on the hardness as measured in a dead-weight apparatus using a 0.075-in.-diameter ball indentor and the Shore durometer. The limiting thickness becomes greater, the softer the rubber, and is greater for the 0.075-in. ball indentor, except for the very hard rubbers, than for the Shore durometer. Stocks H and J are gum Hevea; stocks A to G contain MPC black diminishing in amount from 100 to 12 parts per 100 parts Hevea.

Newton[200] has pointed out that, of six common sources of error in hardness measurements, errors due to variations between nominally identical samples of rubber are the greatest whereas the personal error in repeated observations is small. He also points out that results from the variable-load-type instruments (such as the Shore) are most easily influenced by the observer. Frequent calibration of the spring and checking of the dimensions of the indentor are particularly necessary with the Shore instrument.[144]

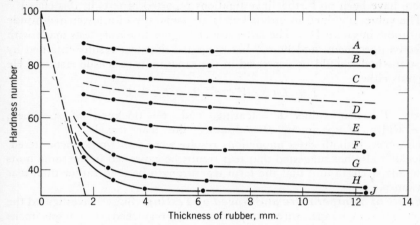

FIG. 11. Effect of Varying Thickness—Shore Durometer[236]

TEAR TESTING

Tests for tear resistance continue to occupy the attention of many workers. New methods of testing continue to be developed, and new analyses of the mechanisms of tearing continue to appear. But in spite of all this effort tear resistance remains an elusive property, and there is reasonable doubt whether any of the currently used methods really measure the property they are intended to measure.

The Mechanism of Tearing. The basic requirement for continuous tearing from an initial cut is that the local deformation be sufficiently great to cause consecutive rupture of the structural elements in the immediate proximity of the cut. Nijveld[205] states that observation of the process of tearing under a microscope shows it to be like the tearing of a bunch of fibers one by one. At a particular moment only one fiber needs to reach the ultimate elongation. Most investigators have concluded that the local deformation in a tear test is always less than that to which the material can be stretched in a conventional stress-strain test. However, Nijveld presents convincing evidence that it approaches very closely the latter value.

Types of Tear Tests. Patrikeev and Melnikov[207] classify tearing tests as follows:

Group I. Direct tearing methods. For example, the methods proposed by Zimmerman,[331] Evans,[82] and Lefcaditis and Cotton.[147]

Group II. Tearing perpendicular to the direction of stretch. For example, the tests with the crescent and angle specimens as described[273] in the ASTM method D 624–48.

Group III. Tearing of a sample sliced in the direction of stretching. (A test is suggested by Patrikeev and Melnikov[207] that utilizes a specimen

180 × 10 × 1 mm. A cut 80 mm. long is made in the center of the specimen. The top of the uncut end is held in the upper grip of the testing machine. The two legs are separated and held by a fixture attached to the lower grip. On separation, the load at failure and the angle between the legs are noted. There have been no further investigations reported of tests in this class.)

The values obtained in group I tests are generally of a lower magnitude than those in group II. The latter are analogous to tensile tests made with defective specimens, and Buist[30] has suggested that tear results obtained by these methods should be expressed in the same units as tensile results. He suggests either

$$T = L/t \times W \quad \text{or} \quad T = L/t \times D$$

where $T =$ tear value, $L =$ tearing load, $t =$ thickness of specimen, $W =$ width of specimen, $D =$ distance the tear travels.

However, in both cases anomalous results were sometimes encountered. Nijveld[205] also has suggested that tear results be expressed in the same units as tensile strength and that the term tear strength be used rather than tear resistance.

Effect of Temperature and Speed of Testing. Buist[30] investigated the effects of testing temperature and speed of testing on nicked crescent specimens of gum and tread-type stocks of the principal rubbers. Each kind of rubber exhibited a characteristic behavior as the testing temperature was increased. In all cases except the GR-S tread, the tear resistance was decreased. In several cases a knotty tear developed with an increase in test temperature. This was particularly noticed in the Butyl stocks. Figures 12 to 16, inclusive, showing the effect of cut depth and testing temperature for the various rubbers, are taken from the above report.

Table I shows the effect of varying the speed of testing. The speed of testing specified by the ASTM method D 624–48 is 508 mm. per minute.

Table I. Tear Resistance, Kg. per Cm.

Speed, mm. per min.	400	200	100
Natural rubber tread	124.5	132.6	130
GR-S tread	54.6	59.8	61.4
Butyl tread	80.6	88.2	85.3

Patrikeev and Melnikov[182] also reported little effect of variations in testing speed over the range of 200 to 600 mm. per minute. Buist[30] concludes that the effect of an increase in speed is the same as a decrease in testing temperature and that in those materials that have a tendency toward knotty tears, this tendency can be accentuated by either lowering the speed of testing or raising the temperature. Both Patrikeev and Melnikov[207] and Nijveld[205] report that tear strength is not strictly proportional to thickness.

Crescent and Angle Methods. Judging from the published literature, the most popular of the tear tests (excluding hand tests, which will be discussed in a later section) is that which requires the use of the crescent-shaped specimen. Probably the most important of the reasons for its popularity is that the loads required for failure of a standard specimen are such that they can be readily measured on a standard tensile-testing machine. The

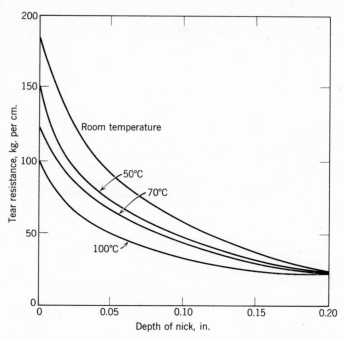

FIG. 12. Effect of Nick Depth on Tear Resistance—Natural-Rubber Tread[30]

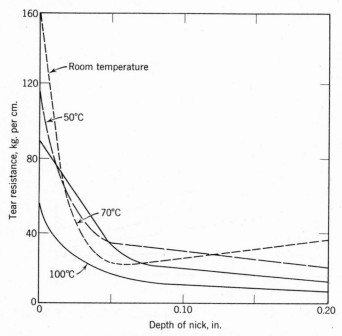

FIG. 13. Effect of Nick Depth on Tear Resistance—GR-S Tread[30]

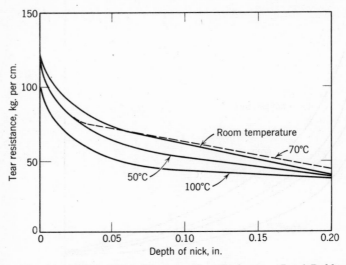

FIG. 14.   Effect of Nick Depth on Tear Resistance—Butyl Rubber[30]

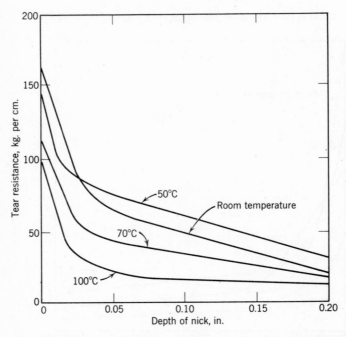

FIG. 15.   Effect of Nick Depth on Tear Resistance—Neoprene-GN Tread[30]

method for conducting the test is described in detail[273] in the ASTM method D 624–48. Die *A* gives the peanut-shaped specimen, whereas die *B* is a modification of die *A*, providing tabs on the two ends of the specimen to permit gripping the latter in the same grips as would be used for stress-strain testing. In 1948 a new shape designated as die *C* was added to the method. This is the angle specimen proposed by Graves.[103]

The description of the ASTM method mentions that the results obtained with the three dies for a particular material are not the same. Morris and Bonnar[173] report that die *A* gives slightly higher results than die *B*, whereas the Graves angle die gives appreciably lower results (except in two cases) which could not be directly correlated with the results obtained with either die *A* or die *B*. This comparison did not include materials exhibiting good tear resistance.

The ASTM method requires that the crescent specimen dies *A* and *B* be nicked to a depth of 0.020 in. by a suitable means before testing. Before its adoption as an ASTM standard, the method as described by Carpenter and Sargisson[49] required five nicks. Poules[211] reported that a single nick gave more reproducible results than five nicks, and his suggestion was adopted in the first issue of the method.

Buist and Kennedy[36] object to the omission of tolerances on the depth of the nick on the ground that the results obtained on pulling the nicked specimen are very sensitive to the depth of the nick and that variations in the depth of the nick are therefore a source of error in the method. The data of Buist and Kennedy,[36] Busse,[45] and Patrikeev and Melnikov[207] showing the marked dependence of tear strength on depth of nick, particularly in the region of 0.020 in., amply support this objection.

Buist and Kennedy[35] describe a cutter which was designed for the purpose of controlling the depth of the cut within the limits of ±0.003 in. But Newman and Taylor[195] point out that, when test specimens are cut with a die from a flat slab, the cut edges are not normal to the surfaces, and thus, if the base of the nick is normal to the surfaces, it may make an appreciable difference from which side the depth of the nick is measured. In their measurements, which covered a variety of compositions, the variation of the cut edge from perpendicularity varied from 0.0012 to 0.0161 in.

One of the advantages claimed for the Graves[103] angle die is that nicking of the specimen is not required and that the errors that may be introduced by this operation are eliminated, thus improving the reproducibility of the test. The latter part of the claim has been substantiated by Morris and Bonnar,[173] Buist,[32] and by the ASTM committee[213] which investigated this die before its adoption as a part of D 624–48.

Comparison of Crescent and Angle Methods. The fact that the angle-die specimen gives different tear values from the crescent specimen and that these values are not a constant fraction of the values obtained with the crescent specimen is evidence that the property or combination of properties being measured is different for the two specimens. Graves[103] reasoned that the load reported as the measurement of tear strength for a particular material is made up of stresses at the apex of the tear and of other stresses not involved in the process of tearing. An ideal test piece would exclude the latter, and a test piece designed to be an improvement over the crescent would minimize the unwanted stresses. This was accomplished in the angle specimen, according to Graves, by effecting (as examination of transparent specimens by polarized light demonstrates) a higher stress concentration at the point of tearing than can be accomplished with the nicked crescent specimen.

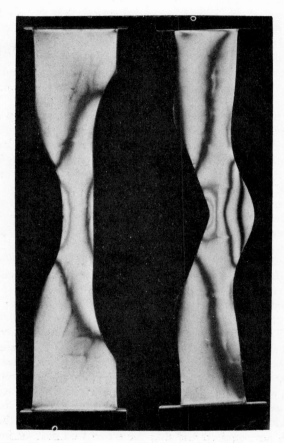

FIG. 16a.   Stress Patterns in Crescent and Angle Specimens[103]

Figure 16a shows photographs, taken from Graves's report, of transparent rubber specimens stressed and illuminated by polarized light.

Buist and co-workers[36, 30, 32] in several publications have maintained that the results obtained with the angle specimen measure the force required to initiate a tear (or in their latest publication a combination of initiation and propagation), while the nicked crescent specimen gives a measure of the force required for propagating an initiated tear.   Thus it is reasoned that the angle specimen should not be considered as a replacement for the crescent specimen but rather as a supplement to provide information not obtainable with the latter.   Buist[32] also criticizes the angle specimen on the ground that the stress concentrations are too high, so that for nonblack filled compounds the test fails to discriminate between materials.

Nijveld[205] disagrees with Buist on the initiation versus propagation point and states that all tests of this type, i.e., tearing in a direction perpendicular to the direction of pulling, have a tear initiated. To determine tear initiation would require a test such as the wire-cutting test described by Werkenthin.[314]

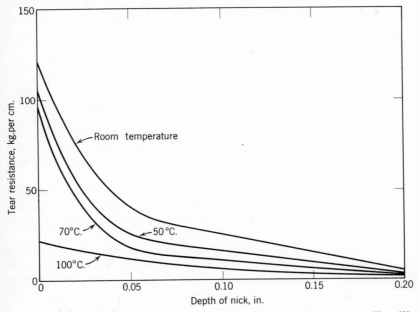

FIG. 16b. Effect of Nick Depth on Tear Resistance—Vulcaprene Tread[30]

Another view of this matter may be taken, namely that the nicking of a crescent specimen does not initiate the tear. Observations of the tearing of nicked specimens during testing show that, more frequently than not, the ultimate tear is not a continuation of the initial cut. The indications are that the function of the cut (besides reducing the untorn area) is to increase further the stress gradient. With the angle specimen also, a preliminary tearing occurs at the base of the angle because it is not normal to the faces of the specimen. This preliminary tearing occurs well in advance of the main and final tear and serves to normalize the base of the angle. The essential difference between the unnicked crescent, the nicked crescent, and the angle specimen is stress concentration; with increasing stress concentration the test tends to approach a group-I type of test, with a lower contribution due to tensile strength and a higher contribution due to "true tear." For these reasons there seems to be little justification for considering tear propagation a characteristic of one specimen and not of another.

The point made by Buist[31-2] that the angle test specimen has poor discriminatory power is based on tests on a variety of nonblack compounds of varying loadings and times of cure. These results showed the angle test specimen to give very small differences, regardless of the kind of pigment used, its concentration, or the time of cure. On the other hand, the nicked crescent specimen rated the pigments, according to Buist, in approximately the order in which they are rated by practical compounders.

New Tear Test Methods. Aside from the development of the Graves right-angle specimen described above, there have been several new methods of testing proposed. One method, described by Cooper,[65] required the preparation of a special test piece consisting of two slabs of the rubber being tested spaced some distance apart and connected by several fins formed when the specimen was molded. The two slabs were then separated by tearing of the fins. No data were reported comparing the results obtained by this method with those obtained by conventional methods.

A tester described by Bashore,[8] sometimes referred to as a tear tester, is a device that can be adapted to a standard Scott machine and that punctures a disk of $^3/_4$-in. diameter cut from a standard slab and held in a suitable fixture. The load required for the indentor point to pierce the specimen is recorded. Leeper[146] reports that the reproducibility of the test is good when the puncturing device is operated at 20 in. per minute. The data obtained do not correlate with conventional tear tests but show some correlation with tensile strength and 300 per cent modulus.

A device patented by Martin[158] is apparently an attempt to mechanize one of the old hand tear procedures. It involves stretching a strip and then making a lateral cut into the material sufficiently deep to permit tearing to occur.

A tear test described by Werkinthin[314] uses a plain rectangular strip and a novel set of grips which stress one side of the strip earlier than the other side during extension. According to the author, the load required to start tearing and the load required to sustain tearing may be recorded. No data were reported.

The method used by Nijveld[205] belongs to the group-II classification of Patrikeev and Melnikov.[207] The specimen consists of a rectangular strip 60 × 9 mm., cut by a die from a molded slab. The die used is provided with a cutter 5 mm. wide which cuts an incision crosswise in the specimen and midway between the two ends. The strip is pulled in a suitable testing machine and the load calculated in terms of kilograms per square centimeter on the basis of the uncut area.

ABRASION TESTING

Laboratory abrasion tests belong to a class of tests that are expected to have a direct relation to product performance and, if they do not correlate with service, are of little value. In spite of the voluminous quantity of publications on this subject over the past several years, it would be fairly safe to say that the reliability of laboratory abrasion tests for predicting service performance is little better today than it has ever been. Within certain very narrow limits of composition, fairly satisfactory correlations have been reported, but outside these limits laboratory results are more likely to be misleading than informative.

Abrasion resistance is obviously a composite property to which a number of other basic properties contribute in varying degrees, depending on the precise details of the service. The principal conditions that influence the rate of abrasive wear are the rapidity of relative movement between the abrasive and the rubber, the sharpness of the abrasive, the pressure with which the rubber is held against the abrasive or alternatively the pressure with which the abrasive strikes the rubber, and the temperature at which the process takes place. One composition that is superior to another under one particular set of conditions may be inferior under another set of conditions. Thus a statement that one material is superior to another in abrasion resistance is meaningless unless the precise conditions of abrasive wear are stipulated.

Because of the different uses to which they are put, different rubber products are subjected to widely different service conditions (cf. heels and soles, conveyor-belt covers, sand-blast hose lines, and hose covers). Even a single product, such as an automobile tire, may be operated under entirely different conditions of climate or road service. For these reasons it could not be anticipated that a single laboratory test would be capable of measuring the abrasion resistance of a material in a variety of abrasive services or of indicating the relative excellence of two materials in the same product subjected to widely different service conditions. One view of the problem is that the properties that contribute to abrasion resistance are so numerous (and not known with any certainty) and the service conditions so varied that the problem of analysis that would be required to elucidate the contribution of the various properties is almost insoluble. To avoid the necessity of such an analysis, laboratory abrasion tests have been devised in great numbers which attempt to duplicate the essential features of the abrasive services.

Laboratory Abrasion Testers. With several unimportant exceptions, all laboratory abrasion testers operate in such a way that the rubber specimen is made to move relative to the movement of an abrading surface. The methods for accomplishing this differ according to the type of abrasive used, the speed of the relative movement, the load on the specimen, the temperature developed at the abrading surface, or other details of operation. In all cases except one, the loss in volume of the specimen, after one or more periods of test, is measured and usually compared to the volume loss of a control run under the same conditions. The exception is the du Pont or Williams method,[266] in which the loss in volume may be calculated either in terms of volume loss per unit time or as volume loss per unit of power expended.

Probably the most popular of the abrasion test methods is this last method, first described by Williams,[324] although the Lambourn abrader is preferred by Adams and co-workers.[1a] In this method a pair of samples mounted on a bar are pressed against opposite sides of the center of a flat abrasive disk rotating in a vertical plane. The bar on which the specimens are mounted is arranged to operate as a Prony brake so that the work expended during the test may be measured.

Variation in Abrasive Paper. The question of how uniform the abrasive paper is and how successfully the use of a standard control rubber cancels out the differences between different lots or kinds of paper is a constantly recurring one. Morley and Scott[172] reported a maximum difference of

13 per cent between sheets from a single shipment of 500 sheets supplied by the instrument manufacturer. When different grades of abrasives (10 types of abrasive paper and seven types of bonded abrasive) were used, the calculated indexes of different compositions varied by a factor of four.[203] These differences, according to Morley and Scott,[172] are not due exclusively to differences in the degree of abrasiveness of the paper and indicate the necessity of careful control of the quality of the abrasive paper if reproducible results are to be expected.

Along the same line, if different abrasive papers do not rate two compositions in the same order, it would not be anticipated that two different testing machines would, since in addition to differences in the abrasive used on the two machines, other differences—in loads, speeds, and the mechanics of the instruments—are involved. In a comparison of the du Pont and Akron[72] machines Scott[237] reports that this is indeed the case. Also, on the basis of tests of 31 commercial compositions and two standard mixes, it was not found possible to convert the results obtained on one of these instruments to those on the other with any degree of exactness.

Contamination of Abrasive. One of the constantly recurring complaints with respect to abrasion tests is that of the sample gumming the abrasive and thus reducing its abrasiveness. Morley,[170] in a study of the effect of the addition of stearic acid and mixtures of stearic acid and pine tar to a compound tested on both the du Pont and the Akron-Croydon[72] machines, found that softeners influenced the results obtained to a considerably greater extent than would be anticipated from the slight effects they had on the stress-strain properties. He suggests that the effect of the softeners is to lubricate the abrasive surface, thus reducing the abrasiveness and giving results that are greatly different from those that would be expected from corresponding road tests. The fact that this effect was more pronounced with the du Pont machine than with the Akron-Croydon machine was attributed to the lower rate of wear per unit area of abrasive in the test with the latter. The use of a continuous-strip abrasive was suggested to overcome this difficulty.

An interesting suggestion by Griffith and co-workers[111] involves extraction of the rubber in ethanol-toluene azeotrope before testing. For a series of eight GR-S recapping compounds the correlation with road wear was found to be very poor when based on the abrasion results with unextracted specimens but very good when based on the results with extracted specimens. For a series of 15 commercial footwear compounds, the results on extracted specimens appeared to be more nearly in line with service performance, as reported by the manufacturers, than the results on the unextracted specimens. It was observed that those specimens that gave a fictitiously high abrasion index or low abrasion loss left visible contaminant on the abrasive paper. In an effort to determine the source of the contaminant, it was found that added softeners, the extractable material present in the crude GR-S, and the extract from carbon black, though contributing to the contamination, were not exclusively responsible. A natural-rubber tread stock behaved much like the GR-S stocks. Different accelerator combinations had a profound effect on the degree of contamination, with variations from 3 to 2275 units difference between the extracted and unextracted index. It was concluded that at least a contributing factor in the contamination was the formation of rubber of low molecular weight during vulcanization and that the proportion of this material produced depended on the curing agents used. It was also observed that the horsepower required for the unextracted samples was higher than for the extracted samples. The contaminant thus increased the coefficient of friction between the sample and the abrasive and at the same time reduced its abrasiveness. It was calculated in one instance that 93 per cent of the power used in a test was due to the contamination effect.

New Abrasion Testers. Two new abrasion-testing machines have been described which incorporate the frequently made suggestion that a continuous-strip abrasive be employed to prevent contamination of the abrasive by the specimen. The apparatus described by Gavan, Eby, and Schrader[94] was designed primarily for testing flooring, resins, painted surfaces, etc. The other apparatus, described by Lewis,[150] was designed primarily for testing shoe-sole compositions. A roll of abrasive which is slowly drawn over the bed of the machine and a specimen 1 in. square by 0.20 to 0.25 in. thick is pressed against the abrasive with a load of 8 lb. (approximately equivalent to the load on a shoe sole during use) and moved forward and back over the slowly moving abrasive. By appropriate means the thickness of the specimen may be measured during the course of the test. No data on the correlation of the test results with service performance were reported.

A new type of abrasion tester developed by Schiefer at the National Bureau of Standards[230] is designed with the object of providing a uniform abrasive action from every azimuthal direction over the surface of the specimen. The specimen is held with a constant pressure against an abrading surface consisting of hard metal disks having a suitably serrated surface presented to the specimen. Although this apparatus was designed primarily for testing textiles, it has also been used to test rubber-coated fabrics and may be applicable to unsupported rubber specimens as well. The feature of a uniform abrasive action over the entire surface would seem to be of less importance in most rubber applications (where this condition does not exist) than in the textile field.

Laboratory Tests and Road Wear. As implied above, laboratory abrasion tests can only be justified if they correlate with service performance. Within certain ill-defined limits they have some utility. However, it would be most advantageous if some method were developed which did not suffer from the present limitations.

In a report by an ASTM section[128] on the question of correlation of laboratory and service abrasion tests, reasons are suggested for the failure of laboratory tests to predict service performance over more than a limited range of compounding variations. These reasons with amplifications are repeated here:

1. Road tests on the same pair of compositions run at different times frequently reverse the relative rating of the two materials. This might be due to relatively large differences in the conditions under which the two tests are run. If actual performance tests fail to rate two materials in the same way both qualitatively and quantitatively, it would obviously not be possible for a single laboratory test to give a precise correlation with all road tests.

The suggestion of Scott[237] that road testing of tires represents a particular type of abrasion test which may differ as much from a particular laboratory test as one laboratory test differs from another is, it appears, too limited a view. It is more likely that the differences between a test run at high speed, high temperature, and on a highly abrasive road surface and one at low speed, at low temperature, and on a less abrasive road surface are as great as the difference between two different laboratory machines.

2. Since the state of cure affects the performance of a material both in a laboratory test and in a service test, it is important that the state of cure of the laboratory specimen should be the same as that of the product. It is believed that too little care has been exercised on this point.

3. Laboratory-prepared specimens cannot be expected to duplicate factory-prepared products in degree of pigment dispersion, amount of breakdown of the rubber, or other respects and thus would not be expected to give the same performance in a laboratory test as the product in service will give, even if a perfect correlation is assumed. This objection, of course, can be obviated by utilizing compounded stock from the same supply as that used for building the tire, preferably a portion of the extruded tread. The misleading results that can be obtained by ignoring this point are well illustrated by Braendle and co-workers.[24]

4. The rate of wear of a tire tread is not necessarily constant during its life. Changes in stiffness of the tread due to continued cure or oxidation may result in reversing the relative rating of two compositions during the course of a test. This probably does not occur in laboratory abrasion tests.

5. In line with the above point, the time required to complete a laboratory test is so much shorter than that of the usual road test that the factors that may affect the stiffness of the composition in the latter test have insufficient time to be operative in the former test.

6. *Effect of Temperature on Road Wear.* It is generally considered that high operating temperatures have an adverse effect on the rate of abrasive wear. For example, Evans[83] estimates that treadwear is increased approximately 10 per cent for each 16° F. increase in the ambient air temperature. This view appears to be corroborated by the reports of Lambourn[141] and H. V. Carpenter.[50] The latter reported that rates of wear in the summer months were as much as six times the rates during the winter months. The data in Table II from a Rubber Reserve report[222] of a test in which the rate of wear of natural-rubber treads and GR-S treads were compared in the winter season and in the summer season also show the same trend. On the

Tire Size	Polymer	Avg. Winter Temperature, °F.	Avg. Summer Temperature, °F.	Summer Treadwear Index (Winter Index = 100)
6.00 × 16	Natural rubber	64	81	81
6.00 × 16	GR-S	64	75	91
7.00 × 15	Natural rubber	65	82	61
7.00 × 15	GR-S	63	83	73

Table II.

basis of a winter index of 100, the corresponding values for the summer test, along with the average temperatures, are given.

Additional data have been reported by Moyer and Tesdall[184] on the effect of temperature on rate of wear. Figure 17 is taken from this report. The rate of wear on gravel roads is increased approximately 10 per cent for a 10° F. increase in air temperature whereas on concrete pavements the rate of wear is reduced by 5 per cent. This latter effect might be explained on the basis

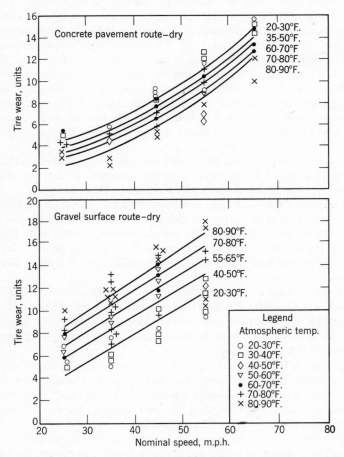

Fig. 17. Effect of Variations in Atmospheric Temperature on Tire Wear on the Iowa Test Routes

of the contamination effect discussed by Griffith and co-workers[111] in connection with tests on the du Pont machine. During the summer months the surface of concrete roads may become less abrasive as the result of the accumulation of oil and grease.

A report by Mandel, Steel, and Stiehler[156] on a series of road tests run during the four seasons of the year (with an additional check test of the first season) on tread materials formulated from various polymers and a variety of carbon blacks showed that the effect of variations in the ambient air temperatures was relatively small and that in certain cases the treadwear improved slightly with an increase in temperature.

A completely satisfactory study of this effect is lacking and is difficult to make for the following reasons: (a) The temperature of concern is the road-surface temperature and not the air temperature, since it is the former that affects the tread surface directly, and its value will vary independently of the air temperature, depending on whether or not the sun is shining. (b) The length of time required to produce sufficient wear for an adequate measurement in any season is relatively long, and wide variations in temperature may occur in the course of the test. (c) To obtain significant temperature differences during a pair of tests requires an interval of 3 to 6 months, during which the test tires may change owing to aging, or, more important, the abrasive character of the road surface may change. (d) During the low-temperature tests the proportion of wet miles is usually high, which complicates the interpretation of the test data since the rate of wear under wet condition is no doubt lower than under dry condition at the same temperature.

The various laboratory tests carry out the abrasion process at different temperatures, but there are no reports in the literature to indicate that the temperature factor was taken into consideration in their design.

7. The removal of material from any portion of a tread surface occurs only at the moment when that portion of the tread is in contact with the road; during the rest of the revolution of the tire the rubber is recovering from the distortion to which it was subjected during contact with the road. Continuous abrasion tests in the laboratory do not reproduce this feature of alternate distortion and recovery.

8. Test methods such as the du Pont method and the Bureau of Standards method tend to overemphasize the effect of stiffness of the rubber. Tests of tire-tread stocks, for example, fail to reproduce the cushioning effect of a pneumatic tire in service. Those methods that utilize a rolling-ring specimen are somewhat better in this respect.

9. The gumming of the abrasive (discussed at some length above) is one of the most frequently mentioned objections to laboratory abrasion tests. The suggestions that a continuous-strip abrasive be used and that the test specimens be extracted before testing are two methods for overcoming this difficulty which will bear further investigation.

It is believed that it will be necessary to give careful consideration to the above points in the design of better test methods before the present limitations of laboratory abrasion tests can be overcome.

LIGHT AND OZONE TESTS

Ozone and light are responsible for characteristic types of deterioration in rubber vulcanizates in service. A better understanding of their effects has resulted from the work of Williams,[323] Van Rossem and Talen,[303] Crabtree and Kemp,[68] Newton,[197] and Buist and Welding.[40] In testing methods designed to evaluate these effects it is important to have the deteriorating agencies segregated as effectively as possible in order to avoid confusion and to aid in the interpretation of the results.

On exposure to outdoor weathering, rubber is subjected to a variety of conditions, but the resulting effects (excluding normal aging which goes on concurrently) can be separated into two distinct processes, light-energized oxidation and atmospheric or ozone cracking. The light-energized oxidation produces a hard, inelastic skin which, on leaching with water or on flexing, produces a checked or crazed appearance on the surface. This can be most readily observed by the exposure of a nonblack, unstretched stock. Black stocks are less affected since the penetration of the light is greatly reduced.

There are several difficulties involved in attempting to duplicate the effects of outdoor exposure in a laboratory apparatus using an artificial light source. Radiation of too great intensity may overheat the specimens so that temperature effects are accelerated. The spectral energy distribution of the light source may be quite different from that of natural sunlight,[22] so that different photochemical reactions occur. In addition to these difficulties, ozone may be generated by the light source and cause unwanted reactions.

It is difficult to estimate the extent of skin formation by visual observation. Crabtree and Kemp[68] have used the increase in weight due to combination of oxygen as a measure of its extent, but this is complicated, particularly for outdoor exposure or laboratory tests in which the specimens are periodically sprayed with water, by the leaching of water-soluble oxidation products. Changes in tensile strength and elongation[178, 285] are commonly used as measures of the extent of the light-energized oxidation.

This phenomenon has been studied in some detail by Newton and Wake[204] who suggest the term *light stiffening* for the effect and define it as "the noticeable increase in the bending modulus of thin sheets of rubber which occurs with exposure to light and air." The method used by them to measure the change in stiffness is very simple and effective. Proofed fabric samples were used, and the bending modulus was determined before and after sunlight exposure. It was established by these workers that ozone was not a factor in this process, that oxygen was necessary, that in the absence of light no effect was observed, and that waxes did not have a beneficial effect.

Chalking of the surface of the specimen frequently occurs during weathering, particularly with pigmented stocks. This is due to oxidation of the rubber at the surface accompanied by release of the pigment. It should not be confused with the phenomenon of frosting, which is discussed below. Chalking occurs fairly late in the weathering process, whereas frosting occurs very rapidly.

Ozone Cracking. Atmospheric cracking has been shown[323, 303, 197, 68] to be due exclusively to ozone and to occur only when the specimen is stretched. A more precise designation for this effect is "ozone cracking," which avoids confusion with the crazing resulting from weathering. The ozone concentration at the earth's surface varies widely from day to day, seasonally and geographically. Crabtree and Kemp[68] published the results of analyses covering a period of one year at the location of their laboratory and showed a spread of values from approximately 0.3 to 6 parts per 100,000,000. Thus outdoor exposures cannot be expected to give a uniform or reproducible test for ozone cracking. The usual procedure of exposure under tension in sunlight is subject to these variations and is also complicated by the presence of the light-energized oxidation. The effect of the latter is to retard the development of ozone cracks. The rate at which cracking occurs for a particular material and method of exposure depends on the ozone concentration. Very high concentrations cause cracking so rapidly that the degree of discrimination between materials is poor, with a consequent reduction in the degree of correlation with outdoor exposure cracking.[93, 69] The preferred procedure is to utilize a somewhat higher concentration of ozone than that normally encountered in service and to conduct the exposure in the absence of light. Crabtree and Kemp[69] suggest a concentration of 25 parts per 100,000,000 and also a method of analysis adaptable to concentrations as low as this. An ASTM method of test for susceptibility to ozone cracking[293] has been developed, based largely on the suggestions of Crabtree and Kemp.[69]

Concentrations of ozone considerably higher than those occurring naturally are frequently encountered around high-voltage electric equipment. For applications of rubber under these conditions, testing is done at much higher concentrations. The ASTM method D 470–52T[278] for testing insulated wire for ozone resistance specifies an ozone concentration of 0.01 to 0.015 per cent.

The effect of an increase in the temperature of the test is to decrease the time required for the first cracks to appear, although Crabtree and Kemp[69] report one instance of a compound containing a particular wax which gave better performance at high temperatures than at room temperature. Figure 18 from this report shows the effect of temperature on the time required for the first cracks to appear in a series of compounds containing increasing quantities of wax. For a particular degree of stretch, an increase in temperature increases the number of cracks found but decreases their size.[68]

Ozone reacts with unstressed rubber but does not cause cracking, nor does cracking occur with subsequent stretching in the absence of ozone. With increasing strains there is an increase in the rate of uptake of ozone[197] and in the number of cracks per unit area. Most investigators who have studied the effect of strain have concluded that there is a critical elongation, which varies somewhat according to the composition of the stock, at which cracking is most severe. This conclusion is based on the fact that the disruptive character of the cracks, their length and depth, is greatest at this elongation, although the number of cracks may be relatively few. Further, as the time of exposure is increased, the number of cracks for a particular strain changes, because of the growth of existing cracks, their coalescence with smaller neighboring cracks, and the local relief of surface strain resulting in the collapse of some of the early cracks. Figure 19 from Newton's work[197] illustrates the effects of both strain and time of exposure on the number of cracks.

Nearly all tests for ozone cracking involve stretching the specimen to a specified elongation before exposure. It is apparent that the stresses in the specimen will decay at a rate dependent on the stress-relaxation characteristics of the material. Apparently good resistance to ozone cracking may be obtained with materials that relax so rapidly that stresses reach zero before the test is started or shortly thereafter. Use is made of this fact in compounding materials to withstand high ozone concentration.[17]

One of the standard devices to improve the performance of a rubber subject to early ozone cracking during static exposure under strain is to provide a wax bloom on the surface or to apply a surface film such as a lacquer or a film of an elastomer which is unaffected by ozone. These

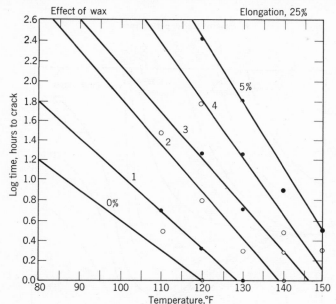

FIG. 18. Effect of Temperature on Time Taken to Crack a GR-S Cable Jacket Compound Containing Different Amounts of Protective Wax.[69] Ozone concentration 25 parts per 10^8, elongation 25%

devices are usually unsatisfactory for dynamic exposures either because the film becomes broken during flexing or because it has inadequate adhesion or unsatisfactory dynamic properties.

Dynamic Exposure Tests. Although flexing tests are usually not considered as ozone cracking tests, numerous investigators have suggested that ozone is the agency responsible for the initiation of flex cracking because of the marked similarity in appearance of the cracks to those that occur in a static test. Eccher[79] has demonstrated that ozone does indeed play an important part in conventional flexing tests.

Throdahl[295] and Fielding[88] have described devices for dynamic-exposure cracking tests. The former used an artificial light source, the latter natural sunlight. Fielding reported good correlation between dynamic sunlight exposure results on wax-containing compounds and static exposure on the same compounds from which the wax film had been removed.

Aside from the report of Eccher,[79] no reports have been made of flexing tests in an atmosphere of controlled ozone concentration. It seems likely that a part of the notoriously poor reproducibility of flexing tests can be attributed to variations in the ozone concentration of the atmosphere in which they are conducted.

The evaluation of the resistance to either static or dynamic cracking due to ozone is a difficult problem. Crabtree and Kemp[69] used the length of time for the first appearance of cracking as observed under a magnification of 7 to 10 times as a measure of performance. Newton[199] used a series of

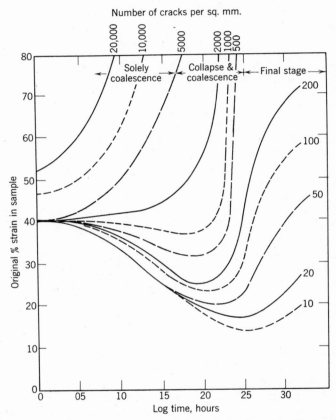

FIG. 19. Relation between Number of Cracks per Sq. Mm., Surface Strain, and Logarithm of the Period of Exposure[197]

graded photographs to assess the degree of cracking. Both these methods are dependent on the experience and reliability of the operator. Werkenthin and co-workers[315] reported that the change in physical properties was a more satisfactory means of judging deterioration than the appearance of the exposed surface. But several investigators[93, 73] have shown that tensile tests are not entirely satisfactory for evaluating surface cracking, since the tensile strength is also affected by other deteriorating influences. A novel method has been suggested by Werkenthin,[315] which involves filling the cracks with powder, removing the powder, and calculating from its weight the corresponding space occupied by the cracks. Errors are caused by sticking and packing of the powder.

Frosting. Frosting is a common phenomenon which is probably merely a variant of ozone cracking. A smooth, shiny surface of a freshly vulcanized material may, under the right conditions, become dull and frosted in appearance on account of the formation of a maze of minute cracks. The action is very rapid, requiring only minutes, or a few hours at the most. Tuley[300] has shown that frosting can be produced in the laboratory by exposure to ozone in moist air. He suggests that residual strains in the material are responsible for the effect.

Corona Cracking. Corona cracking, which occurs when a voltage sufficiently high to produce a corona is applied to an insulated conductor bent into an arc, has generally been attributed to ozone cracking. However, Blake[17] suggests that this cannot be the only factor involved, since materials that are relatively inert to ozone also exhibit this effect. He suggests that electron bombardment may be at least partly responsible.

DYNAMIC TESTS FOR CRACK INITIATION AND CRACK GROWTH

Frequently rubber products which are subjected to flexing during their service life fail because of the appearance and growth of cracks. This type of failure should be distinguished from fatigue failure during flexing, which results from a gradual reduction in the stress-strain properties such that eventually the strength or elongation of the material is inadequate to withstand the stress or strain required by the service. The tread and sidewall of a tire are examples of applications subject to cracking. A strain cycle of about 20 per cent has been measured in the tread grooves[11] as the tire revolves. The growth of cracks, whether initiated by ozone or by mechanical means, results from the strain cycle imposing high stress concentrations at the base and ends of the crack.

A considerable volume of work has been done to investigate the causes of cracking and to develop test methods which can be used in the laboratory for estimating product performance. An excellent review of this subject has been published recently by Buist and G. E. Williams,[41] who discuss the various test methods in use, the effect of test condition, the mechanism of crack initiation and crack growth, the analysis of the test data, and definition of the terms in general use.

Tests for Crack Initiation. Under ASTM designation D 430–51T, methods B and C describe procedures[268] by which the initiation of cracks and their subsequent growth may be measured. Method B employs a DeMattia flexing machine for testing either by extending a standard dumbbell (from 0 per cent to a maximum of 25 per cent of the breaking elongation) or by bend-flexing a standard specimen $6 \times 1 \times {}^{1}/_{4}$ in. having a half-round groove molded transversely in the center of the strip. Method C requires a specially molded specimen with a fabric backing and with eight transverse V-shaped grooves on the surface. A number of these are fastened together to form an endless belt which is then driven over pulleys which give both extension and compression to any one element of the specimen.

Throdahl[295] has described a cracking test conducted between strain limits of 0 and 35 per cent on dumbbell specimens under an artificial light source which presumably produced some ozone. Results correlated well with outdoor tests using the same apparatus.

Fielding[88] has described an outdoor dynamic flexer which, he reports, gave results that correlate well with static ozone tests on materials from which the wax bloom had been removed. Eccher[79] also reported the close similarity between static ozone cracks and flex cracks and showed that ozone was the agency responsible for spontaneous crack initiation.

The results of tests of this type are usually reported in terms of the length of time or number of cycles required to produce a degree of cracking equal to that of a standard specimen, although sometimes the cracking is carried to complete failure, and the time required to reach various stages of cracking as represented by a series of standards is reported. It would probably be more precise if the time or number of cycles required for the first appearance of cracking were reported as the crack-initiation time and the subsequent time as the crack-growth time.

Tests for Crack Growth. The ASTM designation D 813–52T is an adaptation of the bend-flex method[268] of designation D 430–51T (*supra*), in which a cut is introduced[286] through the bottom of the groove with a spear-shaped tool to initiate the crack. On flexing, the rate of growth of the initiated crack is measured. A wide variety of conditions, differing in details as regards frequency, amplitude, temperature, and specimen dimensions, are used by laboratories reporting flexing results from machines that are essentially like the DeMattia machine.

Many other machines and methods for crack-growth measurements have been described in the literature. The machine described by Prettyman[212] uses a specimen resembling a cross section through a tread groove which is subjected to varying degrees of extension and compression. The speed of flexing, 1160 cycles per minute, is quite high and results in a temperature rise in the specimen of 45 to 100° C. Fielding[87] describes a machine that flexes ring specimens at amplitudes as low as 25 per cent or as high as 400 per cent at a speed of 550 cycles per minute. Trepp and co-workers[299] used a novel static test which they report correlates well with crack growth in tires during service. A half cylinder with a nick parallel to the axis of the cylinder and in the center of the curved surface is compressed between heated platens and the growth of the initial cut measured.

The results obtained in crack-growth tests are somewhat erratic, partly on account of the different methods used to initiate the crack. Ranier and Gerke[215] used ozone, which has the disadvantage that a multiplicity of cracks is initiated. Breckley[26] used a needle having a 0.025 in. diameter. Examination of the character of the crack initiated by a needle puncture shows it to be a ragged tear rather than a smooth hole. Gray and co-workers[106] and Trepp et al.[299] used knife cuts to initiate the crack, while Schwarz,[233] Vila,[309] and Carlton and Reinbold[46] used razor blade nicks. The Columbian Carbon Company[57] and Winn and Shelton[326] employed special tools to obtain uniform nicks. Buist and Powell[39] have reported data on the improvement (of the order of 40 per cent) in the reproducibility of results by the use of a spear-shaped puncturing tool instead of a round cross-section needle.

Factors Affecting Cracking Tests. All cracking tests are run under conditions of constant amplitude, and hence the modulus of the material for a particular set of conditions determines the stresses imposed during flexing. For the bending type of test, variations in thickness likewise affect the stresses and consequently the rate of crack growth.

There is some disagreement concerning the effect of strain, some authors maintaining that there is a critical strain, while others maintain that cracking tendency increases with strain. It is probable that, as in static ozone cracking, the cracks appear to be more disruptive, although fewer in number, if the strain cycle is between 0 and 25 per cent. It has been frequently reported that a flexing cycle which includes zero strain accelerates failures of vulcanizates of crystallizable rubbers. But with vulcanizates of the noncrystallizing rubbers this effect is not so pronounced.[109]

The effect of ozone has been mentioned previously. Neal and Northam's[192] conclusion that ozone is not involved in flex cracking appears to have been based on insufficient evidence. Winn and Shelton[326] found that flexing in pure nitrogen greatly reduced the rate of crack growth of GR-S stocks and that as little as 0.4 per cent of oxygen had an appreciable accelerating effect.

The effect of temperature on flex-cracking tests of natural-rubber vulcanizates is not clear. Both increases and decreases of flex life have been reported with an increase in test temperature.[253, 215] With GR-S vulcanizates all investigators agree that the rate of crack growth increases with temperature.[326, 26, 46, 126, 114, 119]

The rate of crack growth correlates fairly well with several other properties. For a particular polymer, increasing states of cure (as judged by modulus[26, 11] or by temperature rise in a hysteresis test[126]) result in increasing rates of crack growth.

Evaluation of Test Results. The problem of determining when equal degrees of cracking have been obtained on two different materials or of rating differences in the degree of cracking is a most difficult one. The usual procedure involves preparing a graded set of standards representing increasing stages of cracking from 0 (representing no cracking) to 10 (representing complete failure). However, the character of the cracks developed during flexing differs with different materials, and this, combined with errors in judgment of what constitutes equal cracking, makes this procedure somewhat unsatisfactory. The difficulty is much more serious for tests in which the crack is not initiated beforehand and in which the test is carried through the initiation stage to complete failure or nearly to failure than for crack-growth tests in which a single crack is first initiated and measurements of its length are made periodically during the test.

Newton[196] has assessed the factors involved in the conventional DeMattia test, has suggested means of minimizing errors inherent in the procedure by improved testing techniques, and has developed a mathematical analysis which aids in determining a true measure of the flexing resistance of the material.

TESTS FOR SET, CREEP, AND STRESS RELAXATION

Set, creep, and stress relaxation are all manifestations of the flow properties of elastomers and their vulcanizates. Set is defined as the deformation remaining on removal of the deforming stress, creep as the increase of deformation with time under constant stress, and stress relaxation as the decrease of stress with time at a constant deformation. The theory of set, creep, and stress relaxation has been developed by several investigators.[84, 297, 5, 209, 4] A brief review of the findings of these investigators will be helpful in understanding the principles involved in these phenomena.

When a rubber or rubberlike material is deformed, there is a time- and temperature-dependent portion of elasticity which lags behind the deforming force on account of the stability of the secondary bonds. This time at a specific temperature is defined as the relaxation or orientation time of the material. At low temperatures the time is considerable, but at intermediate and high temperatures the orientation takes place rapidly. In creep or stress-relaxation tests initial readings which are made before the end of the orientation time will include the effect of this delayed elasticity; these fugitive effects are totally eradicable (i.e., are subpermanent), whereas permanent set is not. It is generally thought[297, 5, 169] that long-term set, creep, or stress relaxation is associated with a chemical reaction which breaks the primary valence bonds in the network. Since the presence of oxygen has been found[5] to be necessary for this reaction to proceed, oxidative scission and cross-linking are thought to be responsible for the changes.

The equivalent effects of time and temperature on the creep and recovery of elastomers has been determined experimentally.[62] It was also found that a low-temperature serviceability index could be calculated from the equation relating time and temperature for creep or recovery.

The effect of loading pigments has been found to be an important factor in the early deformation-stress relationship,[186] and the delayed elasticity exhibited by loaded stocks is due in part to this phenomenon.

Test Methods. Tests for measuring these effects are made on samples of various shapes, under different conditions of deformation, and with various methods of measurement. The ASTM methods D 412–51T and D 395–52T prescribe procedures for tension and compression set, respectively. These two methods are most widely used, particularly for specification purposes. The latter method covers procedures for compression set, either with constant load or with constant deflection.

Creep measurements have been made in tension[297, 5, 209, 4, 169, 96] by several techniques, both continuously and intermittently. Compression has been used in at least one investigation of creep.[169] Other investigators have employed specimens in shear[133, 112, 67] when this type of sample was thought to be advantageous.

Stress-relaxation measurements in tension have been made by Tobolsky and co-workers.[297, 5] In the apparatus first used a beam balance served to measure the stress; a later apparatus is essentially an inclined-plane type of instrument. Both methods utilize a ring specimen of thin cross section. Mooney and co-workers[169] also used specimens in tension, and later[168a] used specimens in torsion.

Blow and Fletcher[18] describe an apparatus for testing a specimen in compression with a simple beam system for measuring the load. The sample may be exposed to air or to any desired liquid during the test. Macdonald and Ushakoff[155] describe an apparatus for testing stress relaxation in compression in which the changes in deflection of a steel beam due to decrease in

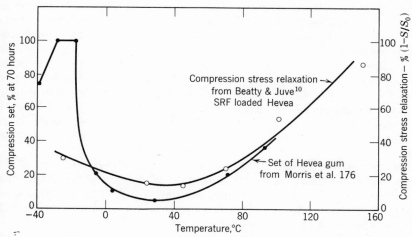

FIG. 20. Effect of Temperature on Compression Set and Compression Stress Relaxation

stresses of the specimen are followed by means of an electric strain-gage system. An automatic load-adjusting apparatus for measuring stress relaxation in compression has been developed.[322] Air pressure, which is used to load the test samples, is adjusted automatically to balance the stress at constant deformation of the samples.

Test Temperatures. There are three basic temperature regions in which work has been done on set, creep, and stress relaxation: (1) a low-temperature region in which the highly elastic component of elasticity due to the stability of the secondary bonds is time-dependent and in which crystallization phenomena may influence the results, (2) an intermediate-temperature range in which the secondary bonds are sufficiently unstable to allow practically complete orientation to occur before measurements are taken and in which oxidative scission is so slow that long time tests are necessary to note its effect, and (3) a high-temperature region in which the set, creep, or stress relaxation is associated with molecular changes. The total temperature range of these three regions is roughly from the second-order transition temperature to 150° C.

Data on compression set from a report of Morris and co-workers[176] and data on stress relaxation in compression from the work of Beatty and Juve,[10] which include data in the three temperature regions on natural-rubber compositions, are shown in Fig. 20. The higher set and stress-relaxation values at high temperatures represent permanent changes in the molecular structure of the material, but the corresponding values at low temperatures are practically all recoverable if the temperature is raised and sufficient time is allowed for orientation to occur.

Creep decreases with a lowering of the temperature and becomes a very slow process at low temperatures in the region of the large increase in modulus corresponding to the loss of rubberlike elasticity.[63, 108, 4] Primary creep, resulting from delayed elasticity rather than from chemical changes or viscous flow, approaches completeness very rapidly in the vicinity of room temperature, and hence the rate of primary creep if measured by the change in deformation in a given time interval shows a maximum at a moderately low temperature. When the selected time interval is long or is late in the creep process, the temperature of the maximum creep rate is lower than when the interval is short or early in the process. At the temperature of the observed maximum, the relaxation time of the rubber is of the same order as the effective rate of testing. The techniques of low-temperature creep measurements are described further in the section on low-temperature tests.

Interpretation of Data. Excluding research objectives, tests of set, creep, and stress relaxation are run principally for development and specification purposes, and, when applicable, the standard ASTM procedures are used. They have been prescribed, sometimes indiscriminately, for their value in predicting product performance.

Tension-set tests are but seldom used in present-day practice, probably because the conditions under which they are run do not resemble service conditions very closely and because they are normally run at room temperature, which for most applications is too low. Set tests in compression in the meantime have become more popular because they lack the above-mentioned limitations. The fact that in both cases either constant-load or constant-deformation conditions may be chosen somewhat complicates the interpretation of the results, particularly when stocks varying in stiffness are being compared. In general, harder stocks are penalized in constant-deformation tests while softer stocks are penalized in constant-load tests.

For those applications of rubber in compression in which either creep or stress relaxation is involved, the conventional compression-set tests either under constant load (creep condition) or constant deflection (stress relaxation) give results that are difficult to translate into useful engineering data. In both instances the time and temperature of exposure are arbitrarily chosen, as well as the recovery-time interval after removal of the load. The set is expressed as a percentage of the original height for constant-load tests and as a percentage of the deflection in the constant-deflection test. For control-test purposes the standard tests are probably quite satisfactory, but for supplying the design engineer with data that can be used directly they are unsatisfactory and will probably be replaced in the future with direct measurements of creep and stress relaxation at temperatures encountered in the service of the product.

LOW-TEMPERATURE TESTING

Many rubber products, such as boots, gaskets, electrical insulation, tubing, vibration mountings, and tires, are used at freezing and subfreezing temperatures, which may cause the rubber to harden and lose its elasticity to such an extent that the material will no longer function properly. In this respect the synthetics are generally found to be inferior to natural rubber, except for a few special compounds of the more cold-resistant synthetics. Measuring the low-temperature properties of rubberlike materials and determining the polymer structures and compounding variations beneficial to low-temperature behavior have consequently become of increasing interest in recent years.

Perrine and Scheirer[181] examined the shattering of rubber sheets at −20° C. under the impact of a bullet. McCortney and Hendrick[154] measured the vacuum pressure required to reverse various convex rubber automative diaphragms on exposure at −40° C. Conant, Dum, and C. M. Cox[61] determined the dynamic- and static-friction coefficients of tire-tread compounds on ice. R. J. Adams and co-workers[1] investigated the effect of composition, pressure, acceleration, plasticizer, and lubricant on the service life of Butyl inner tubes for winter use in Canada where failure was caused by buckling strains not being relieved as the wheel rotated. A similar investigation of these effects in the northern part of the United States has also been reported.[29] Morris and co-workers[174] and Greenleaf[109] described apparatus for measuring the sealing ability of gaskets for hatches on prolonged low-temperature exposure. These are examples of tests for specific applications, but, though desirable and in many cases necessary, they cannot be conducted in the ordinary laboratory and are not considered in this section. Laboratory studies at low temperatures are usually devoted to the more fundamental characteristics of brittleness, stiffness, hardness, creep, resilience, and compression set.

The loss of elasticity can be attributed mainly to a second-order transition of the rubber molecule, during which there is a decrease in its degrees of freedom. Thermal contraction is accompanied by a diminution and disappearance of certain vibrational and kinetic motions of segments of the molecule. There is a tendency for the transition to occur within a limited temperature range, but such properties of compounded stocks as are not very sensitive to the flexibility of the molecular structure show a progressive rather than a sharp change. In low-temperature testing the behavior of the rubber depends on its ability to respond to a deformation within the effective time allowed by the test, and, if the stress or strain exceeds the ultimate for the time interval involved, the material must fail. Boyer and Spencer[23] have described the various second-order transition phenomena and how they are influenced by rate of testing, molecular weight, copolymerization, vulcanization, plasticization, and fillers.

Crystallization of the rubber may superimpose additional effects which result in hardening at temperatures above those associated with second-order transitions. Wood[327] has reviewed in detail the crystallization phenomena in rubbers. On cooling, there is generally an optimum temperature for crystallization below which the crystallization decreases in rate and eventually becomes a very long-time process. Because of orientation effects, stressed rubber can crystallize much more readily than unstressed. However, short-term tests for brittleness, stiffness, and resilience can be properly designed without complications from crystallization, and, furthermore, some rubbers will not crystallize except under most extreme conditions. The investigator should be alert to detect the presence of crystallization effects in data, especially when studying new or unknown polymers.

Brittleness Tests. Brittleness tests determine the lowest temperature at which the rubber sample, usually as a simple beam, withstands a sudden impact without fracturing. Many tests[105, 159, 283, 241, 52] have been devised without regard to the importance of the deformation speed. These tests resorted to hand-operated and free-falling mechanisms which are adequate only for very thin or micro samples with which the energy absorbed by the deformation is a small fraction of the total energy of the striker. Morris, James, and Werkenthin[179] reported that a slow-bend hand brittle point was 25 to 45° F. lower than the corresponding rapid-bend brittle point.

In improved designs[104, 16] an electric solenoid is used to actuate the striker and a reproducible impact velocity obtained. Kemp, Malm, and Winspear[135] described a motor-driven tester developed at the Bell Telephone Laboratories which has been incorporated[284] into ASTM method D 746–52T. They doubled the speed of the striker from 75 to 150 r.p.m. and thus raised brittle temperatures as much as 6° C. By changing from a sharp right-angle bend to one of 3.8 in. radius, they lowered the brittle temperature more than 20° on a natural-rubber gum stock. In ASTM D 746–52T the striker's speed, edge radius, and clearance are specified, and nominal variations in sample thickness of about 0.075 in. do not change the brittle point of most materials more than 2° C. Because of the large number of samples to be tested, it is desirable to provide a fast heat-transfer medium, such as a liquid bath of acetone or alcohol with Dry Ice, but only when the liquid is known to have no swelling effect on the test sample.

A procedure complying with D 746–44T has been described by Smith and Dienes,[249] who used a five-place apparatus which allows a statistical study of per cent failures versus temperature to be made at the brittle point. They show that many elastomers do not possess a sharp brittle point but are characterized by an exponential distribution of failures over a temperature interval as broad as 35° C. The temperature at an arbitrary percentage of failures such as 1 or 50 per cent may be taken as the brittle temperature. A Neoprene and a Hycar stock were the only rubbers tested, and they had a transition interval of 6° C. which may be too narrow to be of concern in routine testing. A statistical calculation is incorporated in D 746–52T to determine the 50 per cent failure temperature of those polymers with a broad brittle point range.

Modulus Tests. Stiffness, hardness, and creep may be measured by adapting one of a variety of modulus tests to cold temperatures. The stiffening or freezing points obtained in such tests are usually arbitrarily defined as the temperatures at which the short-time modulus of a rubber stock increases by selected multiples of its room-temperature value. Measurements over a range of temperatures can give valuable additional information.

As Mullins[185] has shown in a review of tests at low temperatures, a stock having a lower freezing point than another is not necessarily more flexible at higher temperatures. The effect of low-temperature aging on the modulus may be used in crystallization and permanent-set studies. A simple test on these lines for resistance to crystallization at low temperatures is described by Beatty.[8a]

Green and co-workers[107] converted a du Noüy tensiometer to measure the force required to deflect a thin beam sample clamped at one end. A similar deformation is imposed in ASTM method D 747–50, but this test has not found wide application to rubber at low temperatures. Gregory and co-workers[110] determined the stress-deflection curves on loading and unloading a loop formed from a strip, 6 × 1 in., cut from a test slab. The behavior of various rubber compounds was studied during 1 to 3 months' exposures down to −22° C.

Koch[138] employed the principle of centrally loading a small rectangular beam, and Kish[137] used a cylindrical beam. Their methods have been refined by Liska,[151] who designed an apparatus that holds 30 samples on a turntable enclosed in a cold-air chamber and rotates each in turn under a loading foot at temperatures as low as −60° C. Young's modulus can be readily calculated from a beam-bending formula, after the observed deflection for indentation of the sample and compression of the loading foot has been corrected.

$$E = \frac{Wl^3}{4\,dbh^3}$$

where E is Young's modulus, W the load producing a deflection d, l the distance between sample supports, b the width of the sample, and h its thickness.

Using this apparatus with a counterbalance which minimizes the correction of the deflection, Conant and Liska[63] were also able to measure time-dependent primary-creep effects associated with stress relaxation. Figures 21 and 22 are taken from their article, to illustrate the technique employed with a butadiene-acrylonitrile copolymer. Figure 21 shows typical modulus-versus-temperature curves as measured at different experimental time intervals after the load is applied. At very low temperatures primary creep is negligible, so that for all practical purposes a measurement at 10 seconds gives the instantaneous modulus. In the vicinity of room temperature the 10-second modulus approaches the so-called total modulus. Both total and instantaneous moduli are relatively independent of temperature. At intermediate temperatures the 10-second modulus depends on both the total and the instantaneous. If the temperature is moderately low, the rate of creep is slow, and, if it is moderately high, the creep approaches completeness and the rate is also slow. A comparison of the 10-second modulus with that at some longer time is a measure of this rate of creep. For any practical application, the rate of creep as a function of temperature goes through a maximum, as shown in Fig. 22. The term $L/(L + 55)$ corrects the readings for the different loads at the various temperatures. R_n is the reading at time n, R_{10} the reading at 10 seconds, and R_0 the reading at "zero" load. The choice of different measuring times up to 6 minutes affects the creep constant but not the temperature of its maximum. This apparatus has been adopted in ASTM method D 797–46, with recommended conditions and procedure for operation.

Moll and LeFevre[164] modified the conventional laboratory analytical balance into a one-place flexural tester which with good control of time and temperature should be satisfactory for a smaller-sized sample.

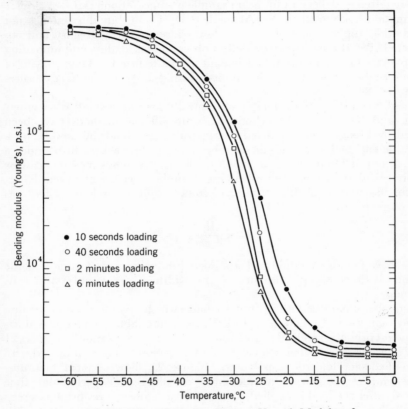

- ● 10 seconds loading
- ○ 40 seconds loading
- □ 2 minutes loading
- △ 6 minutes loading

FIG. 21. Effect of Loading Time on Young's Modulus of Butadiene-Acrylonitrile Stock Containing No Softener[63]

Torsion Tests. Torsion tests are finding increasing acceptance for determining the modulus at low temperatures by measuring the angular displacement of a strip of rubber when a torque is applied to one end. If Poisson's ratio is assumed to equal 0.5, the modulus of elasticity is three times the modulus of rigidity and can be calculated from the following beam-bending formula:

$$E = \frac{3TL}{ab^3\mu\phi}$$

where T is the torque producing a twist ϕ, L the length of the specimen, *a* its width, *b* its thickness, μ a function of b/a, and E Young's modulus. Since the results are generally reported as a relative modulus in comparison to the modulus at room temperature, μ need not be known. However, the twists should be small in order not to superimpose an appreciable tensile force that would result from the tendency of the sample to contract. The time of the reading must again be definitely specified. Yerzley and Fraser[330]

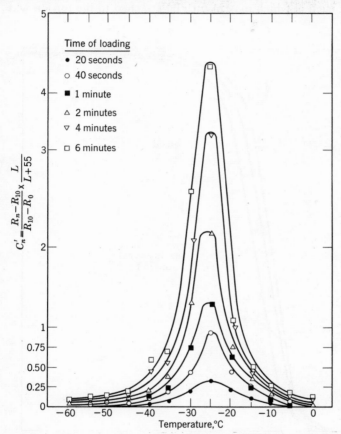

Time of loading
- • 20 seconds
- ○ 40 seconds
- ■ 1 minute
- △ 2 minutes
- ▽ 4 minutes
- □ 6 minutes

$$C'_n = \frac{R_n - R_{10}}{R_{10} - R_0} \times \frac{L}{L+55}$$

Temperature, °C

FIG. 22. Creep Constants for Butadiene-Acrylonitrile Stock Containing No Softener, at Various Loading Times, as a Function of Temperature

described an autographic torsional tester but did not make extensive application to evaluating low-temperature behavior. Bilmes,[15] Clash and Berg,[54-5] and Mullins[185] have been successful in interpreting torsional results, including time-effect studies.

Gehman Torsion Test. The procedure that is becoming most widely used for low-temperature modulus tests was originally described by Gehman, Woodford, and Wilkinson[100] and has been adopted as an ASTM test method.[291] In it a 1.625 × 0.125 × 0.079-in. test piece is connected in series with a torsion wire 2.38 in. long so that, when the torsion axis is turned

180°, the twist is distributed between the wire and sample in a manner depending on the modulus of the rubber. Nominal variations in the dimensions of the sample do not affect the results. Ten-second readings of the twist of the rubber are taken 5 minutes after temperature equilibrium has been established by adjusting the rate of air supply through tubing in a Dry Ice reservoir. In the compact design most commonly used, a fixture holds five samples, each of which may in turn be attached to the torsion wire. A

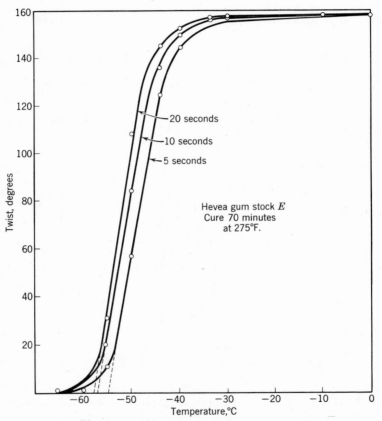

FIG. 23. Effect of Time Interval on Experimental Curves[100]

number of calibrated wires are available which allow comparable results to be made on different instruments. After the room-temperature reading is recorded, for which the wire in use gives a deflection between 170 and 120 degrees, the readings are taken first at the lowest temperature and then at 5 or 10° C. intervals as the temperature is raised. The apparatus can easily be adapted to long-time cold-hardening tests,[97, 132] but creep tests may be complicated on this design by the change in stress that accompanies any change in strain.

The modulus is calculated from the measurements by the relation

$$\text{Modulus} = \frac{\text{stress}}{\text{strain}} \, \alpha \, \frac{180° - \text{twist}}{\text{twist}}$$

The torsion wire constant and the geometry of the test piece may be omitted, since only relative values of the modulus at different temperatures are of

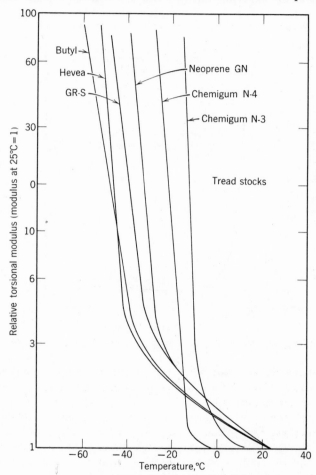

FIG. 24.   Cold Hardening of Various Elastomers[100]

interest.   The ASTM method[291] is provided with convenient tables for determining the twist corresponding to a relative modulus of 2, 5, 10, and 100 times the room-temperature modulus.   The temperatures at which the rubber attains these relative values may be picked from the graph of twist versus temperature and are designated T_2, T_5, T_{10}, and T_{100}, respectively.   From the same graph Gehman also defines a freezing point as the temperature of zero twist determined by extrapolation of the transition portion of the curve. Figures 23, 24, and 25 are from the author's article to illustrate the nature of the results. Figure 23 shows typical twist-temperature curves for a Hevea gum stock at various test times and illustrates the manner of extrapolating to this freezing point. Figure 24 shows a comparison of various synthetic tread stocks with one of Hevea. In Fig. 25 the relative total (long-time) modulus of a Hevea gum stock is observed to decrease slightly at low temperatures in an experimental procedure involving cooling the sample under stress.

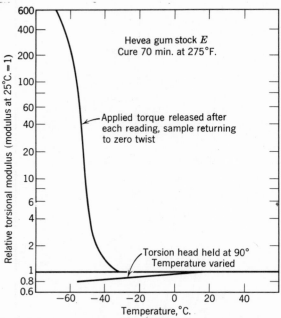

FIG. 25. Experimental Test of Magnitude of Modulus Changes[100]

Temperature-Retraction Test. The T–50 test,[280] ASTM D 599–40T, and modifications of it[28, 330] are not considered satisfactory as a means for describing the low-temperature behavior of rubber, because of extraneous effects of time, the state of cure, and the large stresses involved.[185, 154] Svetlik and Sperberg,[276] and Smith and co-workers[250] studied the temperature-retraction behavior of several rubbers and showed that limited, empirical correlations can be made with results from compression-set and torsional-modulus tests if the latter are conducted in a manner such that crystallization conditions are comparable. Proper interpretation of temperature-retraction curves requires data over a range of stresses and conditionings, in order for crystallization effects to be separated from changes associated with second-order transitions. The absence of a temperature equilibrium during the test also complicates the measurements.

Low-Temperature Tensile Tests. However, the usual techniques of tensile testing may be used at low temperatures with excellent results, and the principal interest in them has been for measuring the effect of temperature on the modulus. For proper analysis of the results, it is necessary to take the same precautions concerning time, temperature, and stress effects as in modulus tests at ordinary temperatures. Morris and co-workers[177] ran tensile modulus in a cold-air box on a T-50 dumbbell, taking readings 3 minutes after loading on samples conditioned 2 hours at the test temperature. They also measured the residual set at various time intervals after removing the load. Russell,[223] Graves and Davis,[105] Somerville,[254] and Morron and co-workers[181] adapted the conventional Scott tester to cold temperatures for both modulus and ultimate tensile. Graves and Davis[105] arbitrarily expressed the stiffening as the percentage increase of the modulus at 25 per cent elongation on reducing the temperature from +25 to −25° C. Morron and co-workers[181] also compared, on a tensiometer of their own design, the stress-strain curves on loading and unloading various synthetic- and natural-rubber stocks (in a liquid bath) over a stress range of 0 to 700 p.s.i. and over a temperature range down to −60° F. They claim good reproducibility, although the machine was hand-operated by turning a crank. Harris[113] has designed an autographic stress-strain tester which measures both stress and strain on a single-wire cable connected to the sample. The tester was used on elastomers at low temperatures, but neither the loading nor deformation rates were constant.

Greene and Loughborough[108] used a similar tester in a cold-air chamber to determine the secant modulus at 100 p.s.i. by loading granulated lead at a 1-lb.-per-minute rate after the sample was given a definitely prescribed stressing history of 4 cycles to 100 p.s.i. Hysteresis was defined from the ratio of the area of stress-strain loop to the area under the loading curve, and percentage set from the remaining elongation on recovering 100 seconds after 5 minutes at 100 per cent strain. Figure 26 shows the dependence of the modulus on the elapsed time of the reading, and Figs. 27 and 28 are plots of hysteresis and set as functions of temperature in a natural-rubber gum stock. Figure 29 compares the low-temperature modulus relative to that at room temperature for various synthetic-rubber gum stocks.

Low-Temperature Hardness Tests. Tests with the Shore durometer and other hardness testers at cold temperatures have been described by various workers, but they lack the precision and the possibilities of interpretation that can be achieved with other modulus tests. Yerzley and Fraser[330] designed a cylinder for holding 20 rubber samples $3/_4$ in. in diameter by $1/_2$ in. thick in a cold box and bringing them successively under the durometer, which was actuated by a 2-lb. weight. The deflections were observed to drift to a final position from an initial high reading. Both were used to calculate a "freezing factor," which was defined as the ratio of (a) the increase in hardness from that at room temperature to that of the final reading and (b) the maximum possible increase in hardness from that at room temperature to 100 Shore. Readings were taken after 1, 2, 4, 6, 24, and 48 hours' exposure to temperatures as low as −45° C., but all of the freeze factors reported were for 10 hours' exposure, derived by interpolation.

A hardness-decadence test in a cold-air box has also been described by Leeper.[145] In it the freeze resistance of vulcanized polymers was measured by primary creep. The creep was recorded as the decrease in hardness between readings taken at 5 and 30 seconds after the pressure foot contacted

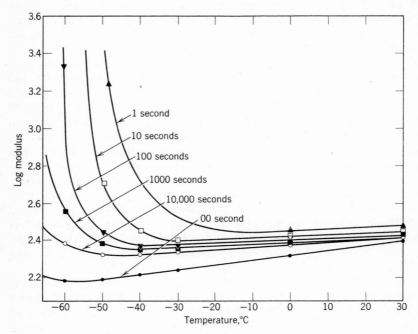

FIG. 26. Time Dependence of Modulus versus Temperature Curve for a Pure-Gum Natural-Rubber Stock. Modulus Calculated from Strain after Recorded Times Have Elapsed

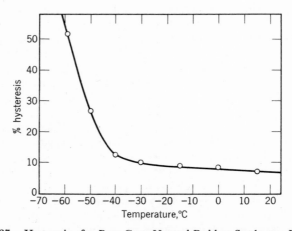

FIG. 27. Hysteresis of a Pure-Gum Natural-Rubber Stock at a Number of Temperatures[108]

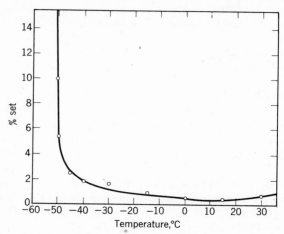

FIG. 28.   Set of a Pure-Gum Natural-Rubber Stock at a Number of Temperatures.[108]
Set equals final elongation divided by original elongation 5 minutes under
100 per cent strain, recovery after 100 seconds

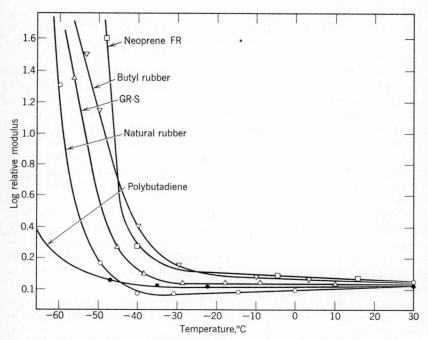

FIG. 29.   Modulus versus Temperature Curve for Six Pure-Gum Elastomers
as Determined by Single-Point Method.[108]
Ordinate is log of ratio of the modulus to modulus at 30° C.

the specimen.   Sample thickness was at least $1/_2$ in. and the surface large enough to permit 18 to 24 readings.   At successively lower temperatures, the creep was found to increase gradually, reach a maximum, and then decrease at a relatively high rate.   The temperature of rapid drop indicated the point at which the polymer ceased to function as a rubberlike material.

Laine and Roux[140] have photographically recorded the creep and recovery by hardness measurements up to 80 seconds after applying the load on seven rubber compounds to $-50°$ C.   They concluded that, although their results are of practical interest, it is hardly possible to deduce the true mechanical properties of the samples studied without resorting to other test methods.

Significance of Low-Temperature Tests.   Many new and improved tests are now available for more adequate measurement of the low-temperature behavior of rubbers.   In most instances no one test suffices.   When possible, the investigator should select and give emphasis to the tests for those properties that are most important for the service of the material in the particular problem at hand.

Attempts to relate the various tests in theory have been qualitatively successful when applied to rubbers in general.   Gehman,[100] Morron,[181] Liska,[151] and their co-workers compared torsional-modulus, stress-strain, and flexural-modulus data, respectively, with the brittleness temperatures for the stocks they tested, and they were unable to find an exact correlation. Larchar[143] has reported the results of a round-robin program in which several rubber laboratories tested a series of natural and synthetic stocks and from which rough comparisons were possible.   Similar values for the same stocks were obtained for the temperature at which the flexural, torsional, or tensile modulus increased to 10,000 p.s.i. or to ten times the room-temperature modulus, depending on how the particular laboratory defined the stiffening.   Another comparative group of tests included the brittle point, Gehman freeze point, durometer decadence point, and a torsional-creep temperature.   In general the results for each of these two groups were within a spread of less than $10°$ C., while the results from the second group were about 8 to $10°$ C. lower than those of the first.

The results of a useful study of nine different methods of testing the low-temperature behavior of vulcanizates, in which four laboratories cooperated, has been given by Helin and Labbe,[118a] who conclude that at present the following three methods are the most acceptable, viz., (1) the temperature-retraction test, (2) the Gehman torsion method, (3) compression set as measured by method B of ASTM D 395–40T.

Parallelism in Absorption of Electric and of Mechanical Energy.   Working in the more rapid range of testing, Mullins[187] showed that close similarities exist between the processes of electric and mechanical energy absorption and that both show a maximum at a temperature at which the period of the applied field or force is similar to the relaxation time of the molecular segments which take part in the process.   The temperatures in Table III are seen to increase with the effective rate of testing, and the differences in the results of the three tests are of the same order of magnitude, regardless of the rubber.

Table III. Effect of Speed of Test on Temperatures of Maximum Energy Absorption

Period of applied force	Torsional Freeze Point 20 sec.	Resilience Minimum 20 msec.	Power Factor Maximum 1 msec.
Perbunan	−34° C.	−10° C.	21° C.
Hycar OR-15	−23	3	30
Hycar EP	−16	8	45
Neoprene GN	−41	−15	
Thiokol RD	−11	22	52

Second-Order Transition Phenomena. The second-order transition is involved in a large number of tests, including many not covered in this discussion. Boyer and Spencer[23] have summarized its effects and depicted a second-order phase diagram for a high polymer. They point out that, although rotation of molecular segments is necessary in materials subjected to these physical tests if failure is not to occur, the tests differ in the demands they make on this rotation, that is, on how fast it must occur and on the segment length and displacement involved. Table IV is their comparison of the characteristic features of the more common tests. They have also plotted the transition temperatures T_m (from thermal-expansion curves) and T_b (from brittle-point measurements) as functions of the molecular weight for a typical polymer, as in Fig. 30, and have interpreted the behavior of the polymer from the diagram. In the region below T_m the polymer is a brittle solid; in the space between T_m and T_b the polymer behaves internally as a liquid characterized by a sluggish rotation of chain segments; above T_b the polymer is an internally tough liquid capable of developing rubberlike

Table IV. Comparison of Some Common Tests For Second-Order Transitions

Test	Linear Deformation Required	Speed of Test, cycles per second	Responsible Phenomena
Heat capacity	Negligible	Slow (0.01)	Rotation of segments
Thermal expansion	Negligible	Slow (0.01)	Viscous flow by rotation of segments
Young's modulus	Medium	Slow (0.1– 1.0)	Effect of short-range forces
Brittle point	Large	Rapid (10^2)	Uncoiling of polymer chains in highly elastic deformation
Dielectric loss	Nil	Very rapid (10^2–10^6)	Rotation of dipoles
Refractive index	Nil	Very rapid	Change in mobility of electrons

elasticity at high molecular weights. In the region below T_m time effects are negligible; at T_m time effects become important for measurements involving slight displacements; at T_b the time factor is quite important for large deformations; somewhere above the T_b line time again becomes a negligible variable (in the absence of viscous flow) with large, highly elastic deformations resulting almost instantly from an applied load. Below T_m Young's modulus is high and relatively independent of temperature; above T_b it is low and again relatively invariant with temperature. Finally, dielectric constant and loss factor are small and insensitive to temperature below T_m

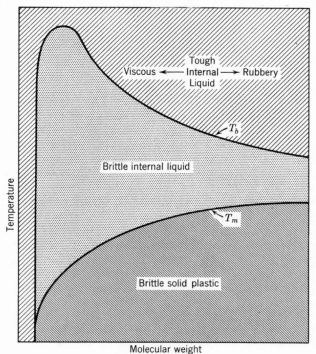

FIG. 30.   Proposed Second-Order Phase Diagram for High Polymers.
The Molecular-Weight Scale Is Logarithmic[23]

but rise abruptly at T_m and reach a maximum somewhere above T_b. The area between T_m and T_b is one in which a polymer shows the greatest variation of properties with temperature.

This plot of Boyer and Spencer[23] illustrates the type and extent of data needed to characterize the second-order transition of high polymers and suggests the contributions from the different molecular-weight species to the ultimate physical properties of a heterogeneous polymer. Their analysis also pertains to low-temperature tests for creep and resilience and to the other tests described in this chapter. Thus, the low-temperature characteristics of a rubber are a complex combination of distributions of several properties, including strength, moduli, relaxation times, and first- and second-order phase changes, as well as the variation of these with temperature.

TESTS FOR RESILIENCE, HYSTERESIS, AND DYNAMIC MODULUS

The use of rubber products in applications in which the rubber is subjected to deformations of varying frequency and amplitude has necessitated some knowledge of their dynamic properties. The difference in performance of the various synthetic rubbers and the general inferiority of the synthetics with respect to these properties have stimulated much work in this field.

Slow-Speed Hysteresis Loop. The earliest measurements of hysteresis were stress-strain tests in tension in which the ratio of the area of the loop between the loading and unloading curves to the area under the stress-strain curve was used. Hysteresis as determined by this method is defined as energy loss per cycle, and this is the usual definition, regardless of frequency of vibration. The slow-speed stress-strain hysteresis method has many drawbacks, the chief of which is that rubber products in service are seldom used under conditions resembling those used in the test.

Impact Tests for Resilience and Hysteresis. Impact tests have been used for many years for measuring resilience and hysteresis. In these tests resilience is defined as the ratio of energy returned to the energy required to produce the deformation. It is generally expressed as a percentage, and the hysteresis then is the difference between the resilience and 100. Typical of the impact tests are the falling-ball and the pendulum-impact types. The falling-ball type may either involve dropping the ball on a horizontally placed specimen and measuring the height to which it rebounds or dropping the ball on a specimen mounted at an angle to the horizontal and measuring the horizontal component of the rebound. The devices described by Breuil[160] and by Dillon, Prettyman, and Hall[77] are typical of the first, while the device of Hock[118] is typical of the second. The pendulum-impact testers have been developed by Schob,[231] Lupke,[153] and others; one of them, the Goodyear-Healy machine described by Fielding,[86] is the basis of an ASTM method[292] for measuring resilience.

Correlation of results of the various impact tests is good, and precise measurements are easily and quickly obtained. Unfortunately, however, there are several weaknesses in tests of this type. The chief of these is that resilience is a function not only of the internal friction or hysteresis but also of the dynamic modulus (the ratio of stress to strain under vibratory conditions). Explanations and equations showing the part of dynamic modulus and hysteresis in impact resilience tests have been given by Dillon and Gehman.[76] Several of the pendulum-impact devices, among which is the Goodyear-Healy, have facilities for measuring the depth of penetration of the striker during the test. This gives an indication of the dynamic modulus but does not permit its direct calculation.

Additional factors affecting test results are the size of ball or hammer, preworking, temperature, magnitude of stress, friction losses between striker and the surface of the specimen, air drag, movement of the specimen, and vibration.

The low-temperature rebound resilience of rubber has been investigated[187] with a modified Lupke pendulum by surrounding the test specimen with a temperature bath except for a small opening to admit the striker. From high values at room temperature, the resilience decreases steadily on cooling to a low value, characteristic of the material, at which practically all of the energy of the pendulum is absorbed by the rubber during the impact. At this value the resilience-temperature curve passes through a minimum, and further reduction of the temperature results in an increase in the resilience. The low resilience is due to the incompleteness of recovery of the delayed elastic component of the deformation during the period of impact. At very low temperatures the relaxation time of the rubber is so long that very little delayed elastic deformation can occur during impact, and most of the deformation is ideally elastic and reversible. Instruments with longer periods of deformation measure the minima at lower temperatures. Violent vibrations of the rubber at low temperatures, which appreciably reduce the resilience, were shown to be caused by the use of a compound pendulum.

Free-Vibration Methods. Free-vibration tests are widely used in measuring hysteresis, dynamic modulus, and resilience of rubber. The relative simplicity of the methods employed and the precise data that may be obtained make these tests of great practical importance and utility. The mathematics involved in the calculations are simple and have been reviewed by several authors.[136, 168, 51, 329] The instruments developed for testing of rubber by the free-vibration method deform the rubber in torsion,[168] torsion and extension,[51] compression, or shear.[329] The choice of method of strain depends on the type of information desired. Shear has the advantages that the shape factor need not be considered and that rubber is also often used in this manner.

One of the most widely used of these devices is the Yerzley oscillograph.[329] Static properties and dynamic modulus, resilience and hysteresis can be measured in compression or shear. The apparatus consists essentially of a balanced beam supported on knife edges, with weights which may be added at one end to strain the test specimen on the opposite side of the knife edges. A second knife edge and stabilizing arm are introduced to maintain the compressing surfaces parallel. On external excitation of the beam, a trace of the damping curve is recorded autographically. Figure 31 is an illustration of a typical curve. Hysteresis, resilience, and dynamic modulus may be calculated by well-known formulas from measurements obtained from these curves. An ASTM method D 945–52T has been published[289] for determining these properties.

Another widely used method employs the torsional apparatus of Mooney and Gerke.[168] This test does not permit a static load to be imposed on the sample other than that due to the weight of the pendulum system. The method is of interest because of its simplicity and its utility for control testing.

Moyal and Fletcher[183] have shown the agreement between the dynamic properties of samples in shear with free vibrations and with forced resonant vibrations. The free-vibration method of determining dynamic properties has the advantage over the impact method that the various quantities may be separated.

Since the frequency in a free-vibration test depends on the properties of the specimen being tested as well as on the characteristics of the method, comparisons of different materials are ordinarily made at different frequencies. Though this is probably undesirable from a theoretical standpoint, it has been found in practice that variations of frequency in devices of this type over a threefold range have no discernible effect on the dynamic properties. These properties also depend on the amplitude of vibration, but no experimental evidence is available in the literature to indicate that this has any appreciable effect at the low amplitudes ordinarily used in these methods.

The chief drawback to this class of tests is that with materials having high hysteresis (whether due to the rubber itself, to the materials with which it is compounded, or to the test temperature) critical damping is approached and erratic results are obtained.

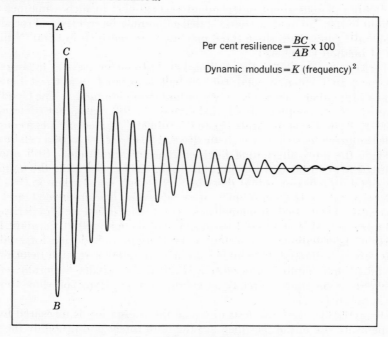

Per cent resilience $= \dfrac{BC}{AB} \times 100$

Dynamic modulus $= K$ (frequency)^2

Fig. 31. Typical Yerzley Damping Curve

Hysteresis has been measured at low temperatures by imposing free vibrations on rubber torsion springs.[66] A dynamic rate calculated from similar experiments has been used[9] to follow time and stress effects which promote crystallization. The dynamic rate is an indeterminate average of the dynamic modulus because of the stress distribution in the spring, but the test is particularly sensitive to the stiffening of rubbers which ordinarily show no evidence of crystallization.

Flexometers and Forced Nonresonance Vibration Methods. Testers in which the frequency of deformation carefully avoids the resonant frequency of the system and in which the hysteresis defect is measured by the temperature rise are sometimes called flexometers. Some tests are continued until blowout, but, because of the complexity of the forces causing the ultimate failure, the data obtained are not considered to be of particular value as an indication of hysteresis. Three widely known testers of this type are the Goodrich,[149] Firestone,[64] and St. Joe[115] flexometers. Results obtained on these three machines are comparable, but the tests differ in sample size, manner of operation, and adaptability of test conditions. The Goodrich machine operates under constant compressive amplitude, with either the load or the deflection initially constant. The Firestone flexometer is operated under constant static load in compression with a rotary shear motion of flexing, the magnitude of which is adjustable for each test. The St. Joe machine operates at either constant static load or constant deflection, and the horizontal shear motion of the sample is accomplished by rotating the top and bottom platens about independent vertical axes in such a manner that constant horizontal load or constant deflection may be secured. Procedures for operating tests with these three machines are specified in detail[281] in the ASTM method D 623–52T.

The Martens[157] compression rolling-ball test can be used to measure the heat generated when a solid rubber ball is rotated in a track between platens with either constant load or constant strain imposed. The Goodyear flexometer[98] is another device belonging to the class that determines hysteresis from the temperature rise in the rubber. It incorporates a constant static compression, and the magnetic drive at 60 cycles per second can be used to secure the same alternating force or constant amplitude. Still another device which can be used to determine hysteresis by measuring the temperature rise during flexing is that described by Gough and Parkinson.[102] This utilizes a rather large specimen at a relatively low frequency and has apparently been used principally to study the fatigue characteristics of vulcanizates. It is of interest because it can be operated at constant load, constant deformation, or constant energy input. A device for studying hysteresis as a function of temperature and frequency at high deformation and fairly high amplitudes in tension is called the "spider hysterometer."[166] Hysteresis is measured directly as energy loss per cycle of elongation in this apparatus.

Theoretically, hysteresis tests in which the energy loss is measured by the temperature rise of the specimen are the most direct and probably the best. Practically, this measurement is extremely difficult of accomplishment without complications, owing to variation in the heat capacity, conductivity, and radiation of the materials being tested.

The Roelig machine,[217–8] though in the same class as the above forced, non-resonance vibrators, does not depend on temperature measurements for determining hysteresis. The dynamic hysteresis loop is obtained by the use of a ring dynamometer and a mirror system. It is possible to secure data in compression, torsion, or shear, for hysteresis as well as for fatigue life. A study of this machine has been reported by Wilkinson and Gehman,[321] and a comparison by Waring[312] has been made with the I.C.I. electric compression vibrator with respect to the effect of the greater strain cycle of the Roelig machine.

Mullins[188] has described a machine that tests rubberlike materials at constant alternating deflection. He points out some of the difficulties associated with the measurement of dynamic properties and the fact that these tests are valid only if they use the rubber in a manner similar to service conditions. The after effects of previous deformations influence markedly subsequent measurements of static modulus.

The effect of temperature on rubber subjected to a sinusoidal loading has been studied in compression[4] over the frequency interval 0.1 to 1000 vibrations per minute. Increasing the frequency over this range increases by 30 to 35° C. the temperature of the rapid rise in modulus which occurs at low temperatures. By the same technique, rubber which is partially crystallized at room temperature has a more rapid rise on exposure to cold temperatures than an amorphous sample.

A shear resiliometer has been designed[181] which measures the transmission of mechanical energy through a rubber mounting at low temperatures by comparing the amplitudes of vibration at 1800 cycles per second. The percentage transmission increases with decrease in temperature to a characteristic temperature at which the rubber loses its ability to absorb vibration. The designers were able to determine the barrier temperature below which a mounting will not warm up, which is a consequence of the low hysteresis on the low-temperature side of the hysteresis maximum or resilience minimum. This temperature depends on the stock, the shape of the mounting, the initial deflection, the heat losses by conduction and radiation, and the frequency and amplitude of the vibrations. The results in Table V indicate the temperature range and differences that were observed.

Table V.   Barrier Temperatures

Natural rubber	−42° F.
GR-S	−24
GR-M	−18
GR-I	+2
Perbunan	+16

Forced Vibrations at Resonance. Several testers for measuring dynamic properties have been developed in which the rubber is made to vibrate at the resonant frequency of the system. Hysteresis as well as resilience and dynamic modulus may be deduced from the measured quantities of amplitude, frequency, mass, and driving force. Most of these testers operate at 60 cycles per second, and resonance is secured by changing the mass, but several may be operated at lower or higher frequencies. Kosten,[139] who was the pioneer in this work, used a mechanical system for securing vibration with the sample in compression. Fletcher and Schofield[89] have used mechanically actuated systems more recently. Naunton and Waring[190] developed the first electrically driven resonant-frequency tester. Their sample was stressed in compression. Other devices which are similar in principle to their equipment are the Firestone resonance vibrator,[77] the Goodyear machine,[99] and that of Moyal and Fletcher.[183] In these testers hysteresis serves to limit the amplitude of vibration to a finite value. The use of data from these tests is based on the assumptions that the product of hysteresis and frequency is constant and that dynamic modulus does not vary with frequency, neither of which is strictly correct.

The effect of low temperature on the dynamic modulus has been studied[100] on the Goodyear machine. The dynamic modulus increases more rapidly on cooling than the static modulus relative to their respective values at room temperature, showing that the rubber ceases to have the property of high elasticity for rapidly applied forces at a higher temperature than for slow deformations. However, it has been pointed out that the accuracy and sensitivity of the test are somewhat limited as the temperature decreases.

Enabnit and Gehman[81] studied the dynamic properties of raw and vulcanized polymers and found reasonable agreement. This makes it possible to evaluate a polymer without the variables of compounding and state of cure affecting the results.

Interpretation of Test Data. The many methods available for dynamic testing covering a wide variety of testing conditions makes possible the selection of a method for a particular problem that will meet the desired characteristics. There are many applications for rubber products in which subjection to vibratory conditions is involved, and the combinations of frequencies, amplitudes, conditions of constant force or constant deformation and temperature are most numerous. In the selection of a test for dynamic properties for the purpose of evaluating a variety of materials as to their suitability for a particular service, it is necessary that a careful analysis of the conditions of service should be made, to assure that a method capable of testing under the proper conditions is chosen.

An analysis of the conditions prevailing in tire-tread service has been made by Gehman, Jones, and Woodford[98] who concluded that for the center of the tread section approximately 40 per cent of the temperature rise was due to deformations of constant amplitude and the other 60 per cent to deformations of constant force. Other applications in which the rubber is obviously subjected to constant-deformation conditions are transmission belts and the rubber between the plies of tire carcasses. Constant-load applications include motor mountings, vibration insulators, and torsion springs.

As pointed out in a previous section on set, creep, and stress relaxation, comparisons of materials of different hardness will give very different results when made under constant-force or under constant-deformation conditions. This also applies to hysteresis measurements. Figure 32 is an illustration taken from the paper of Dillon and Gehman[76] showing the hysteresis determined under these two conditions for a series of stocks containing increasing quantities of carbon black.

Some difficulty is encountered when the data secured on simple shapes by any of the test methods are applied to the complicated shapes of products in which the rubber is usually subjected to two or more types of strain. In these cases the data secured from the simple shapes may only serve as a general guide to product or compound design.

In the range of low temperatures where time and temperature effects are pronounced, many of the test methods are of limited utility because of the increased modulus and relaxation times. Furthermore, in the case of repeated stressing it is necessary to distinguish between the actual temperature of the rubber and ambient temperatures which may be considerably different on account of hysteresis. Thus, for some applications the suitability of a rubber may be determined by behavior during the first few flexings rather than at equilibrium conditions.

An interesting application of dynamic testing is that reported by Liska[152] where two or more stocks, not only differently compounded but also perhaps employing different rubbers, are to be combined in the fabrication of a product, e.g., a tire, which is subject to dynamic conditions. Better results are secured by "matching" or "grading" adjoining compounds on the basis of dynamic modulus rather than on the basis of tensile modulus.

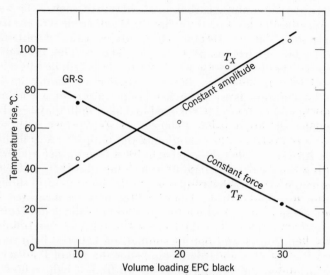

FIG. 32. Goodyear Flexometer Comparison at Constant Amplitude and Constant Force[76]

DISCOLORATION AND STAINING

The increased demand for white and light-colored rubber articles of good permanence and the requirement that certain products shall not impart a stain to lacquered or enameled surfaces with which they are in contact during use have led to the development of several test methods for evaluating these characteristics. Haworth and Pryer[116] have proposed definitions for the terms used to designate these phenomena. *Coloration* is defined as the color of the vulcanizate immediately after cure. *Discoloration* is the change in color that occurs on the surface of the rubber article when it is exposed to light. *Staining by migration* refers to the staining that becomes evident on the surface of some material, not necessarily rubbery in constitution, which has been in contact with a rubber article containing a staining ingredient.

Testing Techniques. Tests for discoloration are commonly made by exposing a sample to sunlight and noting the change in color. Because of variations in the intensity of natural sunlight both seasonally and geographically, these results are erratic and difficult to reproduce. For this reason it has become common practice to use accelerated tests under an artificial light source. These are quite satisfactory for comparative purposes but do not correlate well with tests in natural sunlight.[226, 116] Discoloration is probably the result of a wide range of photochemical reactions which differ in sensitivity to different wavelengths of light. Bowditch and co-workers[22] point out that the spectral energy distributions of several light sources are quite different from natural sunlight and different from one another. This probably accounts for the poor correlation of discoloration with natural and artificial light.

Weathering units containing a carbon-arc light source, such as the National Accelerated Weathering Unit or the twin-arc Weatherometer, are used for discoloration tests. Various spectral distributions can be obtained by using different types of carbons and filters.[175] The samples, mounted on a drum, are rotated about the light source and can, if desired, be subjected to a water spray at regular intervals. Several artificial sun lamps have been adapted for discoloration tests. The S-1 lamp consists of a combination tungsten filament and mercury arc enclosed in a glass bulb that absorbs most of the ultraviolet radiation below 2800 A. A transformer with suitable characteristics is required for the operation of this lamp.[80] A sunlight lamp known as the RS type, which does not require a transformer and which is equipped with a standard screw base to fit a standard socket, is reported to produce approximately the same ultraviolet radiation as natural sunlight.[288]

The amount of radiation from a light source may be measured by the uranyl oxalate actinometer, in which the amount of oxalic acid decomposed by the light is determined. The data are useful only for determining the constancy of a light source and not for comparison between light sources.[22] The National Bureau of Standards recommends the use of light-sensitized papers for determining the amount of radiation from a light source. The exposed papers are compared with a set of master papers of known exposure.[182]

The ASTM method D 925–51T describes two test procedures[288] for developing stain on metal panels painted with an organic finish, such as a lacquer or enamel. The first, for determining contact stain, i.e., the stain on an area directly in contact with the rubber, involves exposing the specimen, sandwiched between two panels, to a temperature of 175° F. in a circulating-air oven for a specified time. The second, for determining migration stain, i.e., the stain beyond the area directly in contact with the rubber, involves exposing the specimen placed on the test panel to light from an artificial sun lamp under controlled conditions.

In the procedure described by Moses and Rodde,[182] both discoloration and staining can be evaluated on a single strip of rubber. The strip is coated for half its length with the desired organic finish. The specimen is then mounted on a revolving disk in a suitable cabinet, where it is exposed to two RS sun lamps, but with half of the coated area and half of the uncoated area shielded from the light. Discoloration is observed on the uncoated, exposed area and staining on the coated, exposed area. The extent of the changes can be readily estimated by comparison with the corresponding unexposed areas.

Another method commonly used for detecting one variety of migration staining (probably not the same as that described above) is to vulcanize a nondiscoloring white stock in contact with a black backing stock.[306] The white side of the laminate is then exposed to light. Discoloration of the white layer greater than that of the control is then due to the migration of a component of the black stock through the white layer.

Estimation of Degree of Discoloration or Staining. The degree of stain or discoloration obtained by any of these methods is usually judged visually and generally compared with that of a control. It is well known that irregularities in the vision of different individuals makes it difficult to determine small differences in color.[325] For greater reliability and precision in measuring the extent of color changes, several photoelectric reflectance meters have been developed.[121, 101] These instruments give the reflectance of light from the sample as a percentage of that reflected from a standard white surface, such as magnesium oxide. By the use of appropriate filters these readings can be extended to give a crude spectrophotometric analysis. These tests can be made rapidly by an instrument operator. For more precise work spectrometers give the most fundamental colorimetric data, but measurements are time-consuming and require trained personnel.

IMMERSION TESTS

The influence of various liquids on rubber products has been of interest for many years because of its great practical importance. Although much work had been done on this problem in the years before the advent of the synthetics, the appearance of the early special-purpose synthetics and the later general-purpose synthetics provided a new stimulus to much work in this field. In general, immersion tests are concerned with measurements of the changes in volume (or weight) and in mechanical properties of a material after contact with a liquid under controlled conditions. The effect of immersion in oil on the stress relaxation of vulcanizates of natural rubber, GR-S, Neoprene, and nitrile rubber under compression has been studied by Beatty and Juve.[10a]

Test Methods. Excluding composition of the specimen and the nature of the fluid, the more important factors that influence the results obtained in an immersion test are the following:

1. *Time of Immersion.* The effect of time on the swelling of a natural-rubber stock in a typical organic solvent is shown in Fig. 33, taken from the work of Scott.[234] The values used by Scott to characterize these curves are the swelling maximum (defined as the intercept at zero time of the straight portion of the curve following the initial rapid rise) and the swelling increment (which is the slope of this same straight portion of the curve). The swelling maximum depends on the nature of the liquid and of the specimen being tested. The swelling increment is attributed by Scott[234] to oxidative breakdown of the rubber, although Whitby and co-workers[319] find it difficult to believe that it can be due entirely to this cause. The middle curve in Fig. 33 is the curve that Scott believes would result if the oxidative factor were absent.

The synthetic rubbers exhibit similar time-swelling curves, although the swelling increment is usually less, and some instances have been reported[129] of a decrease following the maximum. This decrease was attributed to stiffening of the vulcanizate with consequent syneresis, to loss of softeners by extraction or volatilization, or to both these factors.

The time required to reach the swelling maximum for a particular rubber is primarily a function of the viscosity of the fluid.

In the absorption of water the time-swelling curves do not exhibit maxima but continue to increase indefinitely. An apparent exception to this is a material containing a softener or plasticizer which may be leached out or evaporated at such a rate as to give a curve which is flat or which has a

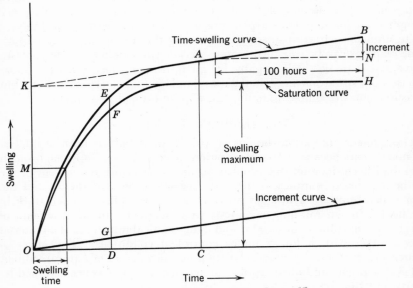

Fig. 33. Typical Time-Swelling Curve[234]

negative slope until the plasticizer is exhausted. The absorption of water is apparently an osmotic phenomenon, in which the rubber is the membrane, and the water-soluble components present in the raw rubber or added to the compound are the dilutable materials.

2. *Temperature.* The effect of increased temperatures is to reduce the time required to reach the swelling maximum and to increase the value of the latter. Increased temperature also accelerates the aging of the rubber, and this latter effect may either increase or decrease the capacity of the material to imbibe the fluid. In addition, the loss of volatile materials from the specimen is accelerated. In the case of immersion in water an increase in the temperature of immersion greatly accelerates the process of absorption.

3. *Size and Shape of the Specimen.* This factor affects the time required to reach the swelling maximum[234] but does not affect the value of it. The lower the ratio of surface area to volume, the longer is this time for a particular system. This generalization may not apply to cases such as those reported by Juve and Garvey,[129] in which the process is complicated by the extraction of plasticizers from the surface layers of the specimen, with consequent changes in permeability of the surface to the fluid and to the plasticizer.

In these cases the extent of either swelling or shrinking depends on the size and shape of the specimen.

For water-absorption tests, the ratio of the surface area to the volume of the specimen has a controlling effect, for the reasons described above. The results are therefore frequently reported on the basis of the weight of water absorbed per unit of surface area exposed.

4. *Volume Ratio of Liquid to Specimen.* The ratio between the quantity of liquid and the volume of the specimen would not be expected to have any great effect on the results as long as there is present an appreciable excess of the liquid over the quantity imbibed. However, in the case of mixed fluids in which one of the components is imbibed more readily than the others, this factor may be quite important. In the ASTM method[279] for determining swelling, the volume of liquid and the size of specimen are purposely specified to eliminate possible test variations from this source.

Measurement of Changes Produced by Immersion. The methods for measuring the changes in volume in an immersion test are classified by Proske[214] as of two types: (1) direct measurement of the volumes before and after immersion, and (2) indirect methods involving the measurement of the weight of liquid absorbed and from this a calculation of the volume change.

The methods of the first type include the following:

(*a*) Measurement of the dimensions of a specimen before and after immersion. This method is subject to the errors resulting from the unequal swelling in the three directions, irregular swelling, and the mechanical difficulties of measuring the swollen dimensions accurately. The procedure described as method B in the ASTM designation D 471–52T is of this type except that only one dimension, the length (of a 100 × 1.6 × 1.9-mm. specimen), is measured. The method prescribes the testing of two specimens cut at right angles to each other to permit an estimation and correction for grain effects.

(*b*) Measurement of the volume of the specimen before and after immersion by the quantity of liquid displaced.[19, 75, 304]

(*c*) The less direct methods of measuring volume by weighing the specimen in air and in distilled water before and after immersion. These are exemplified by method A of the ASTM designation D 471–52T and by the method described by Berens[13] for use with the Jolly balance.

The second or indirect type involves merely weighing the specimen before and after immersion and calculating the change in volume from the density of the absorbed fluid and that of the specimen. This procedure is subject to errors if appreciable extraction occurs from the specimen during immersion and also if the fluid is a mixture of materials, one or more of which may be absorbed to a greater extent than the others.

By any of the methods, errors may occur when volatile solvents are being used as the immersion media, on account of the loss of solvent between the time of removal of the specimen from the immersion medium and the time of measuring. This possibility of loss is minimized in the ASTM D 471–52T method A by requiring the use of a stoppered weighing bottle.

Of equal interest with the change in volume on immersion are the changes in physical properties that occur. Determinations of tensile strength,[279] hardness, and tear[180] are commonly made on rubber stocks before and immediately after immersion or after a period of recovery following immersion. For a particular rubber the reduction in tensile strength and tear is closely correlated with the change in volume, provided the immersion is conducted at a low temperature so that aging effects are not introduced. However, different rubbers vary markedly in the effects produced on these properties by an equal degree of swelling.

Standard Immersion Media. The greatest commercial interest in immersion tests has been to determine the effects of various petroleum fractions on rubber vulcanizates. The fractions may be automotive and aircraft fuels, Diesel fuels, hydraulic fluids, or lubricating oils. Since all these materials are mixtures of many individual chemical compounds, and since their nature varies with the source of the petroleum and the method of refining, it was necessary for specification, research, and development purposes to establish standard reference materials. This has been done in the ASTM method D 471–52T, which provides three reference oils and two reference fuels, representing the range of swelling potency normally encountered in service applications. The two fuels are made up from isoöctane of closely specified characteristics. One is simply isoöctane while the other is a mixture of isoöctane with toluene. These can be made up by any laboratory for its own uses.

The three reference oils are in a somewhat different category since pure hydrocarbons of the high molecular weights occurring in oils are not available commercially. For this reason commercial grades of oils representing high, medium, and low swelling effects are made available by their manufacturers and are controlled within narrow specification limits.

Interpretation of Immersion Data. The extent of swelling or shrinkage or the change in physical properties of a product exposed during service to an oil or solvent will not necessarily agree quantitatively with the results of laboratory tests because of differences in the conditions of exposure. If the product is confined by metal and is under compression, the pressure may be sufficient to prevent the absorption of any of the fluid, or in the more usual case the product can swell to a limited extent until the limits of its confinement are reached. The loss of material by extraction may be more or may be less efficient than in the laboratory test, depending on the conditions of the service. The ratio of surface area to volume of the product is usually different from that of the laboratory specimen, and this will affect both the swelling (or shrinkage) and the degree of extraction, particularly in services involving exposure to water. Other conditions which may result in effects different from what would be anticipated from laboratory data are the alternate wetting and drying of the product, flexing during exposure, and simultaneous exposure to the fluid and to light.

ELECTRICAL-CONDUCTIVITY MEASUREMENTS

The use of semiconducting compositions of the various rubbers for products such as belts, hose, tires, footwear, de-icing devices on airplanes, and rubber products used in hospital operating rooms where sufficient electrical conductivity is required to prevent the buildup of dangerous static potentials has stimulated interest in methods of measurement of conductivity or its reciprocal, resistivity. For applications such as these, resistivities of the order of 10^7 ohm-cm. or less are required.

Resistivity measurements[264] based on the ASTM method D 257–46 become subject to increasing error from contact resistance as the resistivity falls below 10^5 ohm-cm. Recently ASTM has adopted[290] a tentative standard D 991–48T which employs separate electrodes for current and voltage and eliminates contact resistance from the measurement. The circuit as described appears to cover a range from about 10^2 to 10^6 ohm-cm. This method is similar to the methods described by Newton[198] and Miller.[163] Miller[163] has described a considerably simplified circuit designed to measure resistivity over the range from 1 to 10^7 ohm-cm. In this circuit a 60 cycle alternating current is used to measure the resistivity, which is shown to be the same as the d-c resistivity.

Factors other than contact resistance influence the reproducibility of resistivity measurements. The effect of the measuring current on resistivity is somewhat controversial,[198, 163, 142] although there is agreement that the current should be limited so that the sample temperature does not rise appreciably because of the power dissipated in the specimen. Small changes in temperature have little effect. Flexing the specimen may raise the resistivity by several decades. Variations in relative humidity are reflected in the resistivity.

Recipes and methods of preparing semiconducting materials are discussed by Newton,[198] Bulgin,[42] and Wack and co-workers.[311]

METHODS FOR MEASURING PERMEABILITY

Interest in the phenomena of permeability, solubility, and diffusion of gases and vapors in elastomers has been stimulated by the advent of synthetic rubbers and other high-polymeric materials. The performance of GR-I as an air barrier in inner tubes, the use of materials as barriers to the diffusion of solvents in fuel cell construction, and the use of polymeric materials as food-packaging materials to minimize water-vapor transmission are illustrations of applications of these characteristics to the performance of the product.

The amount of gas that goes through a membrane when the pressure is higher on one side than on the other and when the flow has reached a steady rate is given by the expression

$$\frac{dQ}{dt} = \frac{AP(p_2 - p_1)}{d}$$

where Q is the quantity of gas, t the time, A the area of the membrane, P the permeability, d the thickness of the membrane, and p_1 and p_2 the pressures on the two sides of the membrane.

In the case of permanent gases, such as nitrogen, oxygen, and hydrogen, the permeability P depends only on the gas, the membrane material, and the temperature. However, in the case of water and probably of many organic gases and vapors, P depends also on p_1 and p_2, and therefore these values are usually reported along with the permeability.

The permeability can be described as the product of two factors, the solubility h and the diffusion constant k. Thus, when two of these quantities have been measured, the third can be calculated. Van Amerongen[301] has measured all three quantities for several gases and has shown that the above relationships hold.

Permeability to Gases. Gas permeability is determined by subjecting a membrane to a gas-pressure differential and measuring the amount of gas that goes through. Usually the flow is measured on the low-pressure side because of the greater sensitivity of measuring instruments at low pressures. The apparatus consists of a gas supply, a permeability cell, and a detector of some kind for measuring the quantity of gas. In addition, temperature control may be used, and means for obtaining a vacuum on one side of the sample at the start of the experiment may be provided. The permeability cell contains a porous supporting member for the membrane, which is held against it by the differential gas pressure. The edges of the sample are sealed in some way to prevent gas flow around them. The detector may be a gas burette, a McLeod gage, or a heat-conductivity cell.

In the hydrogen permeameter as manufactured by the Cambridge Instrument Company, pure hydrogen is used on one side of the permeability cell and air at atmospheric pressure on the other. Although the gage pressure of the hydrogen is only a few millimeters, the hydrogen-pressure differential is essentially 1 atmosphere. The detection depends on the effect of hydrogen content on the heat conductivity of the air. This instrument can be modified to measure gases other than hydrogen with reduced sensitivity. Although it is convenient and fast, it lacks temperature control and is subject to errors if not calibrated frequently. It is more suitable for measurement of high diffusion rates than of low ones.

The McLeod gage as a detector has been used by Van Amerongen[301] and by Barrer and Skirrow.[7] This type of detector can be used only when the low-pressure side of the sample is initially a fairly high vacuum. The great advantage of this detector is its high sensitivity to small amounts of gas, and the results obtained are precise to 5 per cent or better. The gas burette has been used as a detector by Belotserkovsky and Gorchakov.[12] In principle, it should be possible to obtain greater precision with a gas burette than with a McLeod gage, but in actual practice it is doubtful whether this has been done.

Solubility and Diffusion Constant. Henry's law states that $q = hp$, where q is the quantity of gas dissolved in a unit volume of liquid or solid, h is the solubility, and p the pressure. The solubility can thus be found by measuring q and p. The diffusion constant k is a measure of the rapidity with which solution takes place, or of the rapidity with which the flow rate in the permeability experiment becomes constant. The time required for either process is proportional to d^2/k, where d is the thickness of the sample.

A. S. Carpenter and Twiss[47] have devised a very simple method for measuring the solubility and the diffusion constant in a single experiment. A rubber rod of square cross section is put in a glass container connected to a graduated tube. The container is kept evacuated long enough to remove all gas from the sample; then gas is admitted to the container and the graduated tube, and the end of the tube is dipped in mercury. The rise of mercury in the tube measures the amount of solution of the gas in the rubber. Another method used by these experimenters with heavy gases like sulfur dioxide is to make a torsion pendulum of the sample by suspending it on a quartz fiber. Solution of gas increases the weight of the sample, and this is measured by observing the period of the pendulum.

Van Amerongen[301] and Barrer and Skirrow[7] have measured diffusion constants by observing the time required for the attainment of a steady rate of flow in their permeability experiments. Van Amerongen[301] also measures solubility, by the use of a shredded sample and a gas burette, thus obtaining a check on his diffusion-constant measurements. Table VI compares the permeability, diffusion constant, and solubility for hydrogen, nitrogen, and oxygen at two temperatures in natural rubber, Buna S, Perbunan, Neoprene G, and polyisobutylene (Oppanol B-200), as measured by Van Amerongen.[301]

Table VI. Permeabilities, Diffusion Constants, and Solubilities of Several Rubbers

Material	Gas	Permeability $P\left(\dfrac{cm^2}{Sec.\ Atm.}\right)$		Diffusion Constant $k\left(\dfrac{Cm^2}{Sec.}\right)$		Solubility h	
		25° C.	43° C.	25° C.	43° C.	25°	43° C.
Natural rubber	H_2	3.9×10^{-7}	7.7×10^{-7}	1.05×10^{-5}	1.85×10^{-5}	0.039	0.042
Buna S	H_2	3.05×10^{-7}	5.95×10^{-7}	1.00×10^{-5}	1.65×10^{-5}	0.036	0.039
Perbunan	H_2	1.15×10^{-7}	2.55×10^{-7}	4.2×10^{-6}	8.6×10^{-6}	0.028	0.031
Neoprene G	H_2	1.03×10^{-7}	2.3×10^{-7}	3.8×10^{-6}	7.4×10^{-6}	0.029	0.032
Polyisobutylene	H_2	4.9×10^{-8}	1.2×10^{-7}	1.4×10^{-6}	3.1×10^{-6}	0.035	0.040
Natural rubber	N_2	6.6×10^{-8}	1.6×10^{-7}	1.15×10^{-6}	2.8×10^{-6}	0.052	0.055
Buna S	N_2	4.8×10^{-8}	1.15×10^{-7}	1.0×10^{-6}	2.1×10^{-6}	0.048	0.050
Perbunan	N_2	8.9×10^{-9}	2.5×10^{-8}	2.3×10^{-7}	6.2×10^{-7}	0.035	0.040
Neoprene G	N_2	8.9×10^{-9}	2.55×10^{-8}	2.4×10^{-7}	7.2×10^{-7}	0.036	0.038
Polyisobutylene	N_2	2.2×10^{-9}	7.8×10^{-9}	4.3×10^{-8}	1.5×10^{-7}	0.052	0.054
Natural rubber	O_2	1.8×10^{-7}	3.9×10^{-7}	1.75×10^{-6}	3.6×10^{-6}	0.099	0.100
Buna S	O_2	1.3×10^{-7}	2.75×10^{-7}	1.4×10^{-6}	2.8×10^{-6}	0.093	0.093
Perbunan	O_2	3.2×10^{-8}	7.7×10^{-8}	3.6×10^{-7}	9.1×10^{-7}	0.079	0.080
Neoprene G	O_2	3.0×10^{-8}	7.7×10^{-8}	3.8×10^{-7}	1.00×10^{-6}	0.075	0.076
Polyisobutylene	O_2	9.0×10^{-9}	2.6×10^{-8}	7.8×10^{-8}	2.4×10^{-7}	0.107	0.102

The permeability given in the table is the value of P when the pressures are measured in atmospheres, the area is measured in square centimeters, the thickness in centimeters, and the quantity of gas as the number of cubic centimeters it would occupy at 0° C. and 1 atmosphere pressure. The diffusion constant is in square centimeters per second, and the solubility is the quantity of gas (in the above units) that dissolves in 1 cubic centimeter of elastomer at 1 atmosphere pressure.

It is seen that the permeability and the diffusion constant increase sharply with temperature and are very much dependent on the nature of the elastomer and of the gas. The solubility does not change much with temperature or with the type of elastomer but varies widely for different gases. The permeabilities of different elastomers are in the same order for different gases and for different temperatures, although the lower permeabilities change more (percentage-wise) with temperature and with type of gas. For example, polyisobutylene at 25° C. has a hydrogen permeability 1/8 of the value for natural rubber, while its nitrogen permeability is only 1/30 of the natural-rubber value. At 43° C. the factor 1/30 increases to about 1/20. There have been few published data on Butyl-rubber permeability; however, the values are probably close to those for polyisobutylene (cf. chapter 24).

Permeability to Water Vapor. Measurements of water-vapor permeability[302, 206] are made with simple apparatus in which the membrane to be measured is fastened to a cup and forms a closure for the cup. Either water is put in the cup and the cup is kept in a dry atmosphere or a desiccant is put in the cup and the cup is kept in a humid atmosphere. In the former case a salt may be added to the water to control the humidity in the cup.[302] The rate of transfer of water vapor in either case is measured by weighing. Because P depends on the vapor pressures, these are kept constant when different samples are compared.

PRODUCT PERFORMANCE TESTS

Tests run on different products such as V belts or tires which are either carried out under the same conditions as in normal use or under simulated (and usually accelerated) service conditions are nearly as numerous as the products themselves. Tests in this class should be considered as distinct from standard laboratory tests, such as those discussed in the previous pages, which might be conducted on specimens cut from the finished product.

An example of a highly specific service test is the testing of automobile radiator hose by flexing the hose through an amplitude of $^3/_8$ in., while it is full of a mixture of 33 parts water, 33 parts ethylene glycol, and 1 part petroleum oil at 220° F. and 20 p.s.i., and noting whether it will withstand such conditions for 80 hours without failure of any kind.[152]

Tests of this type are resorted to for several reasons, among which are the following: (*a*) The performance of the product is more dependent on its design than on the character of the rubber compositions used. (*b*) The performance is dependent on certain unknown qualities in the rubber.

In tires the performance properties of carcass bruise resistance, skid resistance, noise, cornering characteristics, and stability depend almost exclusively on design whereas treadwear performance depends partly on design but mostly on qualities of the tread composition, the nature of which are as yet unknown.

Because of the importance of tires in the rubber industry and because they provide an excellent exemplification of the two reasons for conducting product performance tests given above, a short discussion of tire testing is included here.

Tire performance tests in general can be classified as to purpose into the following two groups: (1) control tests on samples selected at random from current production, and (2) development tests designed to evaluate the effects of experimental features of design, compounding, or changes in processing procedures. In both cases observations of the performance of all elements of the tire are usually made, but quite frequently special emphasis is given to performance in some particular aspect. In either case the performance may be evaluated by any of the following tests: (1) an indoor test, (2) a road test on a commercial fleet, (3) a road test on a test fleet operating exclusively for tire testing.

Indoor Tire Tests. Indoor tests are run on a machine which consists essentially of a large motor-driven flywheel 60 to 120 in. in diameter. Against this wheel either single or multiple tires mounted on standard wheels are pressed at controlled loads. The machines are designed to permit running the tires under constant deflection or under constant load by the use of dead weights, hydraulic cylinders, screws, or pneumatic cylinders.

The purposes for which indoor tests are run include evaluations of carcass performance, tread cracking, tread and carcass adhesion, and bead performance. Speed, load, and inflation conditions are chosen to test the particular aspect of behavior concerned. Cleats are sometimes bolted to the flywheel to provide successive impacts and to test the adhesion of the various components. Tests for treadwear performance on these machines have not given reliable results. However, a unit designed specifically for this purpose has been built by the National Bureau of Standards, but no test data from it have been reported as of early 1954.

Sjothun and O. D. Cole[248] describe the details of such a test as used for the evaluation of various combinations of rubbers and carbon blacks in the treads. These were as follows: (1) test wheel—10 ft. diameter, (2) ambient temperature—100° F., (3) inflation and load—normal, (4) starting speed— 75 m.p.h., (5) speed regulation—increased 5 m.p.h. after each 6 hours' duration of test. Failures were by tread separation or crown breaks. The averages of a number of tests with different blacks in three different rubbers were reported by these authors as follows:

	Natural Rubber	GR-S-AC	Low-Temperature GR-S
Average miles to failure	2600	1800	2000
Average failing speed, m.p.h.	100	90	95

An interesting report has been made by Gardner and Worswick[91] of the occurrence of standing or stationary waves in the tire structure when tests are made on indoor machines at speeds of approximately 90 m.p.h. or higher. Conditions which, they report, will intensify such waves include a decrease in the cord angle, a decrease in inflation pressure, increasing loads, and increasing tread weights. Since vigorous stationary waves require a good deal of energy which must then be dissipated in the tire structure, their occurrence must contribute substantially to the equilibrium temperature of the tire.

The advantages of indoor tests include the following: (1) relative rapidity, (2) low cost, (3) no interference from adverse weather conditions, (4) equipment accurately controllable as to load, speed, tire inflation, etc., (5) operating temperatures in different parts of the tire easily measurable.

The disadvantages include the following: (1) Treadwear tests cannot be successfully conducted. (2) Conditions are not precisely comparable with road conditions. For example, the effects of road curves cannot be simulated. (3) The steel wheel surface, regardless of how large the diameter of the wheel may be, does not have the same effect as a road surface on carcass performance.

In addition to the rolling-wheel tests, other tests made on finished tires which may be classed as indoor tests include the following: (1) carcass burst strength, obtained by inflating a tire with hydraulic pressure until it bursts; (2) carcass bruise resistance, a comparative test made by slowly pushing a thick steel pin through the inflated tire until rupture of the fabric occurs (The energy necessary to break the fabric, calculated from the area of the pin, the distance penetrated, and the pressure, is a measure of the relative bruise resistance); (3) tread footprint, carcass molded size, crown thickness, carcass stretch after inflation, weight, etc.

Airplane Tires. The testing of airplane tires is in a somewhat different category from that of over-the-road tires in that service testing on field equipment is economically impractical and the hazards involved are extremely high. For this reason special indoor testing machines have been devised which simulate the conditions occurring during the landing of an airplane. The performance requirements of an airplane tire are quite different from those of land-vehicle tires. An airplane tire must withstand high shock loads and carry large loads for its size. Treadwear and prolonged carcass heating are not important considerations.

The machine generally used consists of a large motor-driven flywheel which can be rotated until the peripheral speed is the same as that of the landing plane's forward speed. This flywheel is variable in weight so that its inertia at the plane's landing speed can be made equal to that portion of the plane's inertia which would normally be absorbed by the tire under test. The test tire is mounted on a standard airplane wheel, which in turn is mounted on a movable arm or carriage so that the tire (free to rotate) can be rapidly pressed against the rotating flywheel when the proper landing speed is attained. The tire is forced against the flywheel with a load equivalent to that portion of the plane's take-off load which it would normally carry. This load is usually greater than that at landing, owing to gasoline consumption during flight. At the moment of forcing the tire against the flywheel, the driving motor is cut off and the flywheel allowed to coast (with the tire against its rim face) until a predetermined mileage has been run by the tire or a minimum speed is reached, at which time the carriage or arm holding the tire is retracted. The tire is then allowed to cool for a predetermined period and the cycle repeated. Airplane tire quality is rated by "landings," and specification requirements are based on the number of cycles repeated before tire failure.

Road Tests on Commercial Fleets. Most large bus and taxi fleets obtain their tires on a mileage rental basis. For this reason they offer an economical means of obtaining large-scale tests for control purposes and for new developments that have nearly reached the production stage. In addition to these advantages, the service conditions to which they are subjected are more nearly normal than those for accelerated tests. The disadvantages of commercial fleet tests are as follows: (1) Since the tests are not accelerated, they require a relatively long time for completion. (2) The service conditions relative to maintenance of inflation pressures, driving speeds, driver characteristics, road surfaces, etc., are not under the direct control of the tire manufacturer. (3) Tire sizes and types are limited. (4) Service records may not be so well kept as for special tire-test fleets.

Road Tests on Tire-Testing Fleets. Most development testing is done on special fleets of vehicles operated exclusively for tire testing and consisting of general-purpose-type trucks and standard passenger cars. Most of the large tire manufacturers maintain such fleets. In addition, a test fleet is operated for the Office of Synthetic Rubber for testing new polymers developed in the course of their investigations. During the early years of the synthetic-rubber program, Army Ordnance maintained test fleets at Camp Normoyle in Texas and at Camp Seely in California. At the former location some 200 vehicles were operated at one time with 700 employees.[298] This was later discontinued, but at the present time a test fleet is being operated at Camp Bullis.

Special fleets have at times been organized for purposes other than for the evaluation of tires, as for example in the work done by Moyer and Tesdall,[184] who were interested in determining the quality of various road surfaces from their influence on treadwear.

Since these fleets are operated exclusively for tire testing, the operating conditions can be chosen and controlled to provide the test conditions desired. Unfortunately, the testing cost is high. For this reason the tests are accelerated, usually by overloading, underinflation, and operation at speeds as near the legal limit as possible. Since high speeds demand sparse traffic as well as freedom from ice, snow, and other hazards, the southwestern part of the United States has become the location for all the major test fleets. Vehicles are usually operated 24 hours a day and 6 days a week, which gives approximately the equivalent of two and one-half years of normal driving in a single month. Under these conditions, variations in performance due to differences in tire construction and compounding are soon evident, and a fairly reliable estimate may then be made of probable performance under normal driving conditions.

The course or the character of the road surface chosen for testing depends on the information desired. For measurements of treadwear, tread cracking, sidewall performance, and carcass fatigue, a course consisting of 90 per cent pavement and 10 per cent gravel might be selected. For testing resistance to cutting, chipping, stone pickup, and carcass bruising, a gravel or unimproved road would be selected.

Observations and measurements made during the course of a test include determination of tread loss by measurements of the depth of the tread grooves (or less frequently by weighing the tires at suitable intervals) and observations of the number and length of cracks, tread radius, growth of cross section, and appearance of any abnormal developments, such as blisters, ply separation, or tread separation.

Records of the atmospheric conditions prevailing during the tests are important. The rate of wear during periods when the road surface is wet is very much slower than when it is dry. Temperature is one of the controlling factors and is usually carefully recorded. Since the temperature at the tread surface is the important consideration, it would be more desirable to have data on its value rather than on the air temperature. Data accumulated by a committee studying abrasion testing[224] show that the road-surface temperature may be as much as 30° F. higher than the air temperature when the sun is shining.

The effect of wheel position is important. More relative movement between the tread surface and the road takes place on the rear driving wheels than on the front wheels, with consequent greater wear. Misalignment may also be responsible for excessive wear and must be carefully checked. The usual procedure for minimizing these differences is to rotate the position of the test tires on the vehicle at regular intervals. Irregular wear across the face of the tread may occur as the result of under- or overinflation or because of design features of the tire as well as by misalignment.

Ratings of treadwear are usually based on a comparison of the loss of material from the experimental tread with that from a control run simultaneously. The latter may be in the form of half of the tread on the test tire with the experimental material forming the other half, or it may be a full tread on another tire. The rating of the experimental tread is usually expressed as an abrasion index, with the control rated as 100. Thus an experimental tread that lost 0.150 in. as compared with a control that lost 0.120 in. would have an index of 80. Buist and co-workers[38] point out that the ratio of the wear rate of the experimental to that of the control, which they refer to as the "wear index," is not related linearly to its reciprocal, the commonly used "tread rating" or "abrasion index." Hence in attempted correlations with other properties or tests it may be important which index is used.

The device of using a control stock, the performance of which is well known, is resorted to for the reasons that: (a) Seasonal differences in tests run at different times of the year will be canceled out if it is assumed that such differences have the same influence on each of the materials being tested. This assumption is probably not strictly true when different base polymers are compared. (b) A statistical picture can be built up whereby an estimate of the significance of a single test can be made. This may include both variations in testing and variations in successive lots of the control.

The comparison between two tread materials may be made at equal mileages or on the basis of the number of miles required to remove a specified fraction of the tread design. The differences between materials will not necessarily be the same when compared by the two methods.

Some of the factors influencing the wear of tires in road tests are discussed in a report by Mandel, Steel, and Stiehler[156] on tests with truck tires and a report by Stiehler, Steel, and Mandel[274a] on tests with both passenger-car and truck tires. Among other conclusions reached is that the influence of temperature on treadwear may be affected by the composition of the tread. Natural-rubber treads containing channel black wore faster as the ambient temperature rose, whereas GR-S treads containing furnace black appeared to wear faster as the temperature fell. These reports offer a comparison of the two methods of determining tread wear, namely: (a) weighing and (b) measuring groove depth.

Advantages and Disadvantages. The advantages of testing by these procedures are as follows: (1) closely controlled test conditions, (2) accurate and detailed operation records, (3) accelerated tests, (4) except for the accelerated features, normal operating conditions. The disadvantages are: (1) high cost, which limits the number of units that can be tested; (2) accelerated conditions, which may provoke failures that would not occur under normal conditions; (3) the limited area of operation, which restricts conditions to a narrow variety of road surfaces and climatic conditions.

Special Tests. Special tests not ordinarily run on test fleets for the measurement or estimation of certain performance characteristics include the following:

1. *Skid Resistance.* This is usually obtained by measuring the drawbar pull generated by the test tire, as compared to that of a standard tire, when mounted on a specially designed trailer and towed behind a heavy truck over the desired surface, which may be wet or dry pavement, ice or packed snow. Drawbar pull is measured as the trailer brakes are gradually applied to produce a full locked wheel slide.

2. *Noise.* This is tested by driving a car equipped with the tires under test over various road surfaces at various fixed speeds.

3. *Cornering Power and Stability.* This is the property of the tire that permits the car to negotiate corners at high speed.

4. *Ride.* This is a combination of cushioning power and stability. In general, an improvement in the ride is obtained at a sacrifice in cornering power and stability.

HAND TESTS AND VISUAL OBSERVATION

Although there have been no publications dealing with them, hand tests or visual observations are an important class of tests. Most experienced rubber technologists find them indispensable as supplements to or substitutes for formal physical tests. In some instances visual observations form a part of the test methods previously discussed. For example, the degree of cracking in a flexing test is judged visually, and so also are the extent of cracking in an ozone or weathering test and the degree of discoloration following exposure to sun or artificial illumination.

With a little experience it is quite easy to estimate with hand and eye the relative hardness, resilience, tear resistance, and tensile strength of a series of different compounds or cures, and this subjective appraisal is frequently very useful in the absence of objective physical tests or as confirmation of the reliability of the latter. The occasional mislabeling of stocks or the mis-arrangement of a series of cures can sometimes be unraveled by this means.

Hand Tear Tests. Perhaps the most common hand test is a tear test. There are two methods in common use. The first is to make two cuts in the sheet at right angles to each other, then pull each one in turn, and estimate the resistance to tearing and the effect of the direction of tearing. The other method is to stretch strips (cut both with and across the grain) to a fixed elongation, nick them with a pair of scissors, and observe whether or not the tear grows. The values obtained in machine tear tests for different compounds do not necessarily correlate with hand tear ratings. Both hand and machine tear results are complicated by the tensile strength, modulus, and elongation of the material as well as by its "pure tear resistance." These factors are present to different degrees in the two types of test, but it is probable that the hand test, particularly of the first type mentioned above, most closely approaches "pure tear." Also the machine test gives no indication of the presence or absence of the phenomenon of "knotty tear" (unless the broken specimens are examined), while the hand test shows this clearly. The tendency for a stock to exhibit lamination can also frequently be observed in a hand test more readily than in a machine test.

Hand tear tests are most frequently used for the purpose of estimating the best cure. The resistance to hand tear for most compounds falls off more rapidly on overcures than other properties, thus making this a more sensitive test for the selection of the best cure than the conventional machine tests.

Hot tear tests are also used and are perhaps more sensitive to overcures than room-temperature tests. They are run by heating specimens in an oven at the desired temperature or by taking specimens directly from the mold at the end of the cure, nicking them, pulling, and estimating the resistance to tear.

Hand Tests for State of Cure. Several hand tests for judging the state of cure are used extensively. One is the pencil test, which involves indenting the specimen with the point of a pencil or similar instrument and noting the rapidity and completeness of recovery. This is particularly useful in noting the degree of cure in scorched stocks. Since the ease of indentation depends on the hardness of the material, this test also permits an estimate of hardness. Another method, applicable only to materials in the form of sheets, involves making a tongue-shaped cut in the sheet. The "tongue" is pulled out several times and the state of cure estimated by the retractive force, the snap, and the amount of set remaining. If the width and length of the tongue are controlled, as well as the degree and number of stretches, a fairly quantitative estimate of the permanent set may be obtained.

Resiliency. An estimate of resiliency can be made by bending a corner of a sheet through 90 degrees, allowing it to flip back, and noting the degree of snap and the rate at which the material recovers, particularly in the final stages of recovery. This test is also used for products such as heels which fit easily in the palm of the hand.

Bending Test. When the material is bent through 180 degrees it becomes a bend test which is extremely useful and is required in many customers' specifications after heat aging or oil-immersion tests. It is used on relatively stiff materials such as soling, hard packing, and flooring and indicates whether or not a material has the necessary minimum elongation to permit this treatment. A sheet of the thickness suitable for tensile testing, when bent double will crack if the elongation is appreciably less than 100 per cent.

Adhesion. It is frequently necessary to estimate the adhesion between component parts of rubber products such as between the cover and carcass of a hose or between the plies of transmission belting or tires. Particular products may require some differences in technique, but usually the procedure applied involves preparing a ring or a strip about 1 in. wide, and starting the separation by means of a knife, then grasping the separated members by hand or with pliers, pulling them apart, and noting the ease of separation.

The "smear" test used on cord coatings and a similar test used on frictions in general are useful in estimating the state of cure in products being cut down for examination. In the former test, the separation of the cords from the rubber leaves a series of ridges of the rubber. These are "smeared" by wiping a knife blade, under some pressure, across the surface. If the ridges do not recover readily, a degree of undercure or surface tackiness is indicated. For frictions on woven duck, if the separated surfaces are pressed together and then separated, the degree of tack remaining gives an indication of the state of cure. In the latter test the noise made when the sample is held close to the ear during separation is sometimes noted.

Identification of Polymers by Burning. The identification of polymers by the odor of a burning sample is most useful when stocks have lost their identity or when a competitors' product is being examined. With a little practice, the odor produced by burning a natural-rubber stock can be readily distinguished from that produced by a GR-S stock or a nitrile-rubber stock. Neoprene stocks can be distinguished by the fact that they will not support combustion. The partially burned residue from GR-S, nitrile, or Neoprene stocks is hard and brittle, whereas that from the natural-rubber stocks is soft and gummy. Mixtures of polymers and certain compounding modifications will interfere with observations of this kind, which nevertheless are extremely useful.

Visual Observations during Testing. Visual observation of phenomena associated with the preparation and testing of specimens is extremely important and most helpful in the understanding of many of the problems of testing. Some of these are: (1) estimation of the degree of dispersion of the compounding ingredients, (2) character of defects occurring during vulcanization, such as porosity, blowholes, and flow cracks, (3) odor and color of the vulcanizates, (4) appearance of the broken ends of tested dumb-bells, (5) appearance and character of bloom, (6) extent and character of cracking during outdoor exposure, ozone exposure, flexing, etc., (7) discoloration of light-colored materials on exposure to sun or artificial light, (8) general appearance of aged samples (development of tackiness, resinification, etc.).

INTERLABORATORY COMPARISONS OF TEST METHODS

In the work of standardizing testing methods it is frequently necessary to conduct interlaboratory comparisons to determine the reproducibility of a particular method between laboratories. Such programs provide data on the effects of minor, unknown, or unsuspected variables in the procedures, all of which should be known in order to prescribe the details of a test method. For the most satisfactory comparisons of this type it is desirable to have a single source for the material to be tested so that all the laboratories involved will be testing a material that is as uniform as it is possible to make it. In response to this need the National Bureau of Standards in cooperation with committee D-11 of ASTM has embarked on a program to provide standard specimens from a series of standard compositions of the principal rubbers. A progress report on the project[59] states that 13 formulas have been tentatively agreed on (including 6 of natural rubber, 2 of GR-S, 2 of Neoprene, 2 of GR-I, and 1 of nitrile rubber) and also that 7 of the compounding ingredients necessary to prepare these materials are available from the Bureau of Standards.

The utility of the standard formulas, the standard compounding ingredients, and the standard vulcanizate samples will extend beyond their use for interlaboratory comparisons. For example, the standard vulcanizate samples may be used in a single laboratory for calibration purposes; the standard formulas and compounding ingredients for checking laboratory procedures of mixing, curing, and testing, and the standard formulas as controls or bases of comparison for compounding studies involving the evaluation of experimental rubbers, pigments, accelerators, and other chemicals.

POLYMER EVALUATION METHODS

Experimental polymers requiring evaluation are prepared in great numbers both by the production plants and by the laboratories associated with the synthetic-rubber industry. Some polymers are radically different in composition and properties from the standard synthetics; others represent minor variations in composition or processing technique. In all cases some estimate of the value of the experimental polymer is required in order to answer or to assist in answering the question that prompted its preparation.

A complete and comprehensive evaluation of an experimental rubber is virtually impossible to accomplish on a laboratory scale. This is due to the fact that rubbers are used in such a wide variety of products, each requiring for its successful performance a different combination of physical properties, that to test the variety of compositions and properties involved in a complete evaluation would be an endless task. Even on a full production scale a complete and final evaluation of a new rubber is probably never approached as long as new applications of rubber and new techniques of compounding and processing are being developed. Even after the consumption by the American rubber industry of over 3,000,000 tons of GR-S, there were still discussions and differences of opinion as to how GR-S compares with natural rubber both for processibility and for product performance.

The approach to the problem of evaluation employed in this section is that which involves the steps of compounding, mixing, vulcanization, and testing of the vulcanizates. The evaluation of experimental polymers by measurements of molecular weight, molecular-weight distribution, *cis-trans* isomerism, 1,2- and 1,4-addition, branching, etc., has been dealt with in earlier chapters and is not considered here.

The full-scale factory evaluation of experimental polymers will not be dealt with here in any detail, since the procedures to be used must be tailored to fit both the objectives for which the polymer was prepared and the procedures in use in the particular plant in which it is to be evaluated. In general, the basic procedure involves either a direct substitution of the experimental polymer for the standard polymer (with observations of differences in processing characteristics and product quality) or an adjustment of compounding formulas and processing techniques designed to develop the best characteristics in the experimental polymer, with particular emphasis on product quality. It might be mentioned that a large-scale factory evaluation is frequently necessary to bring out differences in behavior that cannot be observed in laboratory tests. This is particularly the case with respect to processibility but also applies to product quality when procedures of mixing, curing, etc., used in the laboratory differ materially from those used in the factory and when laboratory-measured properties are unreliable for predicting product performance.

Selection of Test Recipes. The first step in the evaluation process is the selection of a suitable test recipe or recipes. Some knowledge of the character of the polymer is necessary to determine the kind of vulcanization system to which it will respond, and this is usually available from those who prepared it. In the case of a completely unknown material some preliminary investgation may be required in order to find a suitable vulcanization system. In the more common case in which modifications of an established rubber are to be evaluated, the following considerations govern the selection of a test recipe:

1. For a rubber that is most likely to be used commercially for a specific product, e.g., tire treads, the recipe should resemble in its essential details a typical commercial recipe for that product. The essential characteristic of tire-tread compositions is the presence of a quantity of reinforcing carbon black, along with such adjustments in the vulcanization system as are made necessary by the presence of the black.

2. The loading of pigment and softeners should not be so excessive that differences between polymers are likely to be obscured.

3. The recipe should be suitable for as wide a variety of polymers as possible, only minor modifications being required to adjust for different polymers.

4. The use of materials which affect the cure so vigorously that normal errors in weighing affect the reproducibility of results should be avoided.

5. The curing ingredients and the curing times and temperature should be selected so that the changes in properties with time of cure permit the ready selection of comparable states of cure.

Pure-Gum Test Recipes. For the evaluation of some polymers, a "pure-gum" recipe (i.e., one that contains only the polymer and the ingredients necessary to affect a satisfactory cure) is quite satisfactory and, because of its greater simplicity, is preferable to one containing a reinforcing pigment. This is the case for certain variations of the Neoprenes, GR-I, and the condensation rubbers such as the Vulcollanes and Vulcaprene. For those polymers that give poor gum properties, such as GR-S and the nitrile rubbers, the pure-gum recipes are little used. Whitby and Budewitz[318] have reported data that indicate a fairly close parallel between stress-strain data in both pure-gum and black loaded recipes for a series of GR-S-type polymers. But in spite of the real advantage of not having to contend with the complication of using carbon black, little use has been made of pure-gum formulations in the evaluation of these polymers. There are indications that the advantages that have been found in carbon black stocks for GR-S-type polymers prepared at low temperatures would not have been recognized had pure-gum recipes been used in their evaluation. On the other hand, the exclusive use of a black loaded formulation may obscure the presence of good pure-gum properties.

Tread-Type Recipes. By far the greatest volume of evaluation work in the recent past has involved variations of GR-S polymers which have been evaluated in tread-type recipes. The essentials of such a recipe are as follows:

Polymer	100 parts
Carbon black	40–50
Sulfur	1–3
Accelerator and activator	Various kinds and amounts
Softener	0–10

It is customary for convenience to write compounding recipes on the basis of 100 parts of polymer. This does not necessarily mean that 100 parts of elastomer are present since the polymer usually contains small proportions of fatty acids and stabilizers. However, this practice makes it unnecessary to determine the elastomer content by analysis and conforms to commercial practice. Some laboratories, however, prefer to correct for the nonhydrocarbon materials present.

The type of carbon black most widely used for this recipe is EPC (easy-processing channel), although recently some technologists have expressed a preference for VFF (very fine furnace), HAF (high-abrasion furnace), or HMF (high-modulus furnace) blacks. The proportions used are in the same range as would be used in production tread recipes.

The proportion of sulfur that will give the optimum properties depends on the activity of the accelerator used and the viscosity of the polymer (and perhaps on its composition). The more active the accelerator and the tougher the polymer (i.e., the higher its Mooney viscosity), the less is the sulfur required. For a particular polymer in a typical tread-type recipe, increasing the sulfur ratio from 1 to 3 parts in small increments gives a tensile strength-versus-sulfur ratio curve having a sharp maximum, while modulus increases and elongation decreases steadily with increased sulfur ratios.

The most commonly used softeners are the coal-tar types, oil-refining residues, and wood products such as pine tar and wood rosin. The last materials are strong retarders of cure, and their use requires an upward adjustment of the accelerator ratio. For greatest ease in handling during weighing and during incorporation on the mill, a softener which is liquid at room temperature is preferable.

The original Rubber Reserve specification test recipes for the standard varieties of GR-S have been modified to omit the softener and decrease the carbon black ratio.[256] This was done in the interest of simplifying the recipe and improving the reproducibility between laboratories. Since the softener used was not a pure chemical, it was difficult to be certain that variations in the quality of successive shipments would not be a source of error.

The following considerations govern the selection of the accelerator-activator combination: (1) It should resemble in general the combinations that are commercially used. (2) It should function satisfactorily with a wide variety of polymers. (3) The rate of cure imparted to the composition should be neither too slow nor too fast at the desired curing temperature, and the rate of change of properties with time of cure should be such as to permit the selection of comparable states of cure. (4) Materials that are extremely active in small concentrations should be avoided because of possibilities of weighing errors and loss of material during mixing.

Zinc oxide is almost universally used as the inorganic activator, and in proportions of 2.5 to 5 parts per hundred of the polymer. The use of a mixture of litharge and zinc oxide without an organic accelerator has given satisfactory results with a wide variety of experimental polymers and has the advantages that normal errors in weighing have an insignificant effect and that variations in pH, which greatly influence organically accelerated mixes, have much less effect on this combination. The objection to using the litharge-zinc combination for evaluation work is that it is not used commercially.

The organic accelerators most commonly used for evaluation studies are, 2-mercaptobenzothiazole, 2,2'-benzothiazolyl disulfide, N-cyclohexyl-2-benzothiazolyl sulfenamide, B.J.F. (phenylamine methyl-2-benzothiazolyl disulfide), tetramethylthiuram monosulfide, and mixtures of these with secondary accelerators, such as the guanidines and the aldehyde–amines.

The reports of several laboratories describing the evaluations procedures used by them discuss the recipes chosen. Reynolds[216] states that for small samples of experimental polymers on which preliminary work cannot be done he has used the Rubber Reserve test recipe (for GR-S). This is as follows:

Polymer	100 parts
Zinc oxide	5
Sulfur	2
EPC black	40
2,2'-Benzothiazolyl disulfide	1.75

Recipe in effect since January 1, 1949. Previously the recipe contained 50 parts EPC black and 5 parts BRT no. 7 (a coal-tar softener) and used mercaptobenzothiazole 1.5 parts instead of 2,2'-benzothiazolyl disulfide.

For larger samples, the above recipe may be adjusted to provide the same rate of cure and approximately the same modulus as a reference bale of GR-S would give in the standard recipe. For still larger samples, recipes for a tread stock, a carcass stock, and a nonblack stock would be used, with the cure rate adjusted to match that of the standard GR-S in use.

Sperberg and co-workers[261] state that in their evaluation procedure the standard Rubber Reserve recipe is modified to use HMF black instead of EPC black because the former is easier to disperse than the latter; and a more active accelerator (N-cyclohexyl-2-benzothiazolyl sulfenamide) is used than that specified in the standard recipe.

Juve and Schroeder[131] report the following as a typical evaluation recipe:

Polymer	92.6 parts
EPC black	50
Accelerator masterbatch	6 (20% accelerator, 80% GR-S)
Zinc oxide masterbatch	6.25 (80% zinc oxide, 20% GR-S)
Sulfur masterbatch	3.33 (60% sulfur, 40% GR-S)
Softener	10
Stearic acid	1.5

Although not mentioned specifically in the above reference, the accelerator used was N-cyclohexyl-2-benzothiazolyl sulfenamide and the softener Paraflux. The use of fatty acid, as exemplified in the above recipe, is essential with all the accelerators mentioned previously. However, in many cases the polymer itself contains an adequate quantity, and additional quantities need not then be added.

Compounding and Mixing Techniques. The specified procedure for compounding the standard Rubber Reserve recipe[255] requires that all ingredients be weighed to within 1 per cent of the weight specified. Juve and Schroeder[131] report that for relatively small batches the rubbers and pigments weighing more than 5 grams are weighed to the nearest tenth of a gram on a beam balance or torsion balance and that smaller amounts are weighed to the nearest 5 mg. on an analytical balance or to the nearest 10 mg. on a torsion balance.

The standard batch size for the standard Rubber Reserve procedure[255] is based on 400 grams of polymer, and the mixing mill specified is a 6 × 12-in. mill having a front-roll speed of 24 r.p.m. and a ratio of the back-roll speed to the front-roll speed of 1.4. The specified mixing procedure is as follows:

Roll temperatures maintained at 120° F. $\pm$ 10°.

Operation	Time in Minutes
1. Pass the polymer through the rolls twice, without banding, at a mill setting of 0.008 plus or minus 0.002 in.	1
2. Band the polymer on the front roll with the mill set at 0.055 plus or minus 0.005 in. and make $3/4$ cuts every $1/2$ minute from alternate sides	10
3. Add the carbon black evenly across the mill and at a uniform rate. Open the mill at intervals so as to maintain an approximately constant bank. Make one $3/4$ cut from each side when half of the black has been added and another $3/4$ cut from each side when all of the black has been added except that which has fallen through the rolls. Then add the black in the pan. Do not cut the stock with free black on the bank. (Note: To add the black at a uniform rate, it is convenient to use an "hourglass" which will deliver 160 grams of standard channel black in 8 minutes. The "hourglass" may be made from a 500-ml. reagent bottle having a plastic screw cap through which a hole of suitable size has been drilled. A mark should be placed on the bottle to indicate when 80 grams of the black has been added.)	10

Operation	Time in Minutes
4. Add the zinc oxide, sulfur, and accelerator, taking care to avoid loss	4
5. Make three $3/4$ cuts each way	2
6. Cut the batch from the mill. Set the mill at 0.030 plus or minus 0.005 in., and pass the rolled stock endwise through the mill 6 times	2
7. Check the batch weight	$1/2$
8. Band the stock, and sheet it out to a minimum thickness of 0.25 in.	$1/2$
Total	30

9. Allow the stock to rest for at least 2 hours and for not more than 8 hours.

Mixing techniques different from that given above are followed by the different laboratories, depending on the quantity of polymer available, its behavior on the mill, the type of recipe used, and the mixing equipment available. Sperberg and co-workers[261] report that the carbon black (70 grams HMF black) and the polymer (140 grams) are mixed in a midget Banbury, and notations made of the power consumption, the rate of black incorporation, and the temperature developed during mixing. In addition, the performance of the rubber is noted on a roll mill at different temperatures. Reynolds[216] reports that during mixing (presumably on a 6-in. roll mill) the time required for the polymer to form a hole-free band is noted and also the time required for the incorporation of the black. The change in Mooney viscosity due to the addition of the black is measured, as well as the Mooney viscosity of the final mix. In addition, the rugosity[165] and shrinkage of a calendered specimen of the finished batch is measured. Juve and Schroeder[131] report that for batches containing up to 40 grams of polymer a 4 × 5-in. mill is used; for batches of 40 to 100 grams of polymer a 4 × 9-in. mill is used, and for batches of 200 to 500 grams a 6 × 12-in. mill. The mixing procedure involves breaking down the rubber on a cold mill, or milling 10 minutes if little breakdown occurs, followed by the addition of the zinc oxide and accelerator masterbatches. After blending, the softener is added and then the black. When no free black is visible on the bank the batch is cut back and forth for thorough blending. The sulfur masterbatch is then added and the batch rolled up and passed endwise through the mill six times. It is then allowed to stand for 24 hours and remilled for 3 minutes before samples are prepared for curing.

The operations of compounding and mixing are both potential sources of error due to mistakes in weighing, the use of wrong materials, the use of materials that vary in quality from shipment to shipment, and carelessness in mixing. It is essential, if reliable results are to be obtained, that the utmost care be exercised in carrying out these operations.

In dealing with a variety of experimental polymers it will be found that few of them behave on the mill as the reference rubber does. Some will be tough and dry, others soft and sticky. An experienced mill man can be permitted to use his judgment as to the most satisfactory procedure to be followed with unusual polymers. But, since the degree of breakdown of the rubber affects the physical properties of the vulcanizate, uniform results can be expected from lot to lot or batch to batch only if a uniform procedure is followed.

Curing and Testing. The tests to be run on cured specimens of the mixed batches depend on the quantity of material available and the tests that the evaluator considers to be most important.

Screening Tests. Reynolds[216] reports that on the first screening the following tests would be run: (1) cut growth on flexing, (2) hysteresis (by torsional method),[168] (3) resilience at 80 and 212° F., (4) hardness at 80 and 212° F., (5) modulus at 300 per cent tested at 80° F., (6) aged and unaged tensile strength at 80° F., (7) elongation (ultimate) at 80° F., (8) stiffening point by ASTM D 747–43T, (9) brittle point by ASTM D 736–43T, (10) hysteresis by ASTM D 623–41T method C.

Sperberg and co-workers[261] report that the following tests are conducted on routine evaluations: (1) stress-strain at 80 and 200° F. original and at 80° F. after aging, (2) hysteresis on the Goodrich flexometer or the Yerzley oscillograph or by speed of retraction, (3) crack growth at 210° F., (4) state of cure by compression set (2 hours at 212° F., 35 per cent compression), ASTM D 395–47T method B, (5) freezing test by T–50 method, (6) abrasion by Goodyear angle abrader.

Legge and co-workers[148] report the following list of tests: (1) rate of cure based on curves of modulus versus time of cure, (2) stress-strain at 82° F., (3) rebound at 82° F., (4) stiffening at low temperature by the Gehman method,[100] (5) crack growth (original and aged) at 212° F.

Micro-Scale Testing. The above procedures are based on the use of a quantity of the experimental polymer that is sufficient to permit mixing on a standard 6 × 12-in. laboratory mill and to permit the preparation of standard-size test specimens. When smaller samples of the experimental polymer are to be evaluated, the so-called micro techniques are used. Garvey[92] has described such a procedure, and later modifications are described by Juve and Schroeder.[131] By this procedure batches consisting of 9 to 100 grams of polymer are evaluated. By using micro specimens, the following tests can be run on a 9-gram sample of the experimental polymer: (1) stress-strain at three cures (one break on the lowest cure and two on the higher cures), (2) durometer hardness on pellets at two cures, (3) rebound and compression set on pellets at two cures, (4) on an extra strip given the longest cure, immersions run in hexane, benzene, acetone, and alcohol at room temperature and in Circo light processing oil at 180° F., (5) freezing characteristics measured on the tabs of the broken dumbbells.

The strips for the stress-strain and immersion tests were 4 × 0.5 × 0.025 in. From each of these strips a single dumbbell was cut having a restricted section $1/_8$-in. wide. Immersions were run according to the ASTM designation D 471–46T method B. Hardness, rebound, and compression set were run on pellets $1/_2$ in. in diameter by $3/_8$ in. high. When desired, hysteresis tests were run on pellets 0.7 in. in diameter by 1 in. high. Later modifications of this procedure involved the use of sheets $1^3/_8$ × 3 × 0.025 in. from which two dumbbells can be cut for stress-strain tests. In this case the dumbbells were shortened so that the distance between bench marks was $^3/_4$ in.

A procedure was also developed for testing the crack-growth character-istics of a standard DeMattia test strip 1 in. wide, of which only the center section containing the half-round groove is made from the experimental material, while the two ends are made from a standard tread-type stock.

Choice of Tests. The list of tests described by Reynolds,[216] Sperberg,[261] Legge,[148] and their co-workers are obviously designed to test the basic physical properties of the vulcanizates of polymers intended to be similar to the general-purpose rubber GR-S. The list described by Garvey[92] was similarly intended for oil-resisting polymers.

As stated earlier in this section, the purposes for which experimental polymers may be prepared are numerous. When the objective is the improve-ment of a specific property, it is necessary in planning the evaluation procedure to select the tests and test methods that will best demonstrate whether or not the desired improvement has been effected. Among the objectives that might be sought are the following: (1) low water absorption, (2) nondiscoloring when compounded into light-colored stocks and exposed to light, (3) nonstaining when the vulcanizate is held in contact with a white or light-colored finish, (4) superior performance at low temperatures, (5) low hysteresis, (6) better processing characteristics.

In each case procedures are available for the special tests. In some cases special test recipes must be developed, as for the nondiscoloring test mentioned above, which requires a white or light-colored stock.

Along with each experimental polymer or with each group of experimental polymers, it is necessary or at least desirable to run one or more controls using a polymer the general performance of which is well known.

Interpretation of Results. One of the complications of the problem of evaluating experimental polymers is not knowing with any precision the significance of physical test data in terms of product performance. For example, if one should test an experimental polymer and find that all its physical properties are in the same range as standard GR-S but that its hysteresis loss is appreciably less, the problem of how much benefit would be derived from the use of such a rubber in bus tires, for example, is one that can only be answered by performance tests in the tires themselves. Thus the final and conclusive test in many cases is a test on the product in actual service.

There are numerous other properties not mentioned in the preceding section which are important in the practical applications of a rubber but which cannot be satisfactorily measured or estimated on a laboratory scale. Among these properties are adhesion to other component parts in multi-component products, such as tires, belts and hose; blistering and porosity; tendency to crack under the particular conditions imposed by the service of the product; and various properties associated with processing behavior such as tack and flow in molds.

In considering the results of a laboratory evaluation such as has been described above, the data obtained should permit an estimate of the degree to which the objective of the originator has been accomplished and should also indicate whether the accomplishment of the desired objectives has resulted in the sacrifice of some other important property.

Related Properties.   It is useful in judging the results of a laboratory evaluation to consider together those properties that are closely related. This is helpful in checking possible errors in the testing procedures and in spotting real differences between polymers.

The properties of GR-S tread compounds exhibit the following close relationships: (1) High modulus is associated with low elongation, low hysteresis, high rebound, poor resistance to crack growth and poor hand tear. Conversely low modulus is associated with high elongation, high hysteresis, etc.   (2) High tensile strength is associated with high machine tear.   (3) High hot elongation is associated with high elongation at room temperature.

The relationship between hysteresis temperature rise, as measured in the Goodrich flexometer, and crack growth has been found[126] to be particularly useful in assessing test results on polymers intended to be superior to GR-S with respect to the balance between these two properties.   For a particular rubber and loading of a specific black, the logarithm of the flexures required for an initiated crack to grow to a specified length is linearly related to the temperature rise.   Thus, if time or state of cure is varied by changes in the curing or accelerating agents, the relationship between these properties remains unchanged.   When an experimental polymer is tested, the flex life found for a particular level of temperature rise may be higher or lower than that of the control, and this permits an estimate of its quality with respect to this pair of properties in which GR-S is deficient.   One of the most advantageous features of this relationship is that it is not necessary to adjust the recipe or cure of the experimental material with any great care in order to obtain valid results.   A further study of this relationship has been reported by Storey[275] who suggests refinements in the crack-growth and temperature-rise tests and in the interpretation of the test data to obtain a more reliable estimate of the balance between these properties.

Another useful generalization is that of Borders and R. D. Juve[21] with respect to the relation between on the one hand tensile strength and on the other hand the difference between the temperature at which the tensile strength is measured and the brittle point.   These workers report that for a wide variety of emulsion polymers and copolymers of butadiene (and also for Neoprene-GN and Butyl) a fairly definite relationship was found. Figure 34 is taken from this report.   The curve for natural rubber is appreciably higher than the average curve for the synthetics, while the values for synthetic polyisoprene 'and isoprene copolymers (not shown) are lower. This relationship is useful in evaluation, since an improvement in tensile strength when accompanied by poorer low-temperature properties should not be considered a legitimate improvement.

Buist and Davies,[33] in a study of natural rubber and Neoprene-GN stocks containing various carbon blacks, pointed out that certain properties are closely related.   For example, for natural rubber the following groups are highly correlated: (1) Shore hardness and abrasion resistance, (2) swelling in benzene, elongation, and modulus at 300 per cent, (3) all tests of hardness,

(4) plasticity compression, resilience, and high-frequency modulus, (5) the Shore hardness test with all other tests except tensile strength and detrition. Tensile strength is not highly correlated with any other properties.

For Neoprene-GN the following groups are highly correlated: B.S. hardness, Shore hardness, resilience and detrition. Tensile strength is not highly correlated with other properties except tear resistance and Dunlop abrasion resistance.

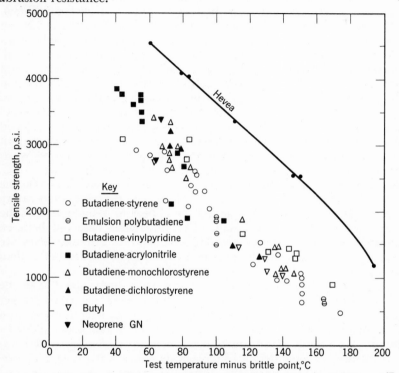

FIG. 34. Tensile–Brittle-Point Comparison of Tread Stocks of Hevea and Butadiene Polymers[21]

In discussing this point, the authors of *Rubber in Engineering*[221] suggest that the various pairs of properties can be grouped into three categories: (1) those that are directly compatible, in which an improvement in one property is accompanied by an improvement in the other; (2) those that are directly incompatible, an improvement in one property inducing a degradation in another; (3) those that have a haphazard or unpredictable relationship.

Laboratory Evaluation and Factory Behavior of Polymers. Though experimental polymers are often evaluated solely for research purposes, more frequently the ultimate goal of the evaluation is to provide information that will aid in predicting the performance of the polymer in the factory processes or in the product. For this reason, when the results of laboratory evaluation are considered, the differences in the conditions of mixing between the laboratory and the factory must be taken into account. Most laboratory mixing is done on laboratory roll mills at low temperatures, whereas factory mixing is done in internal mixers at relatively high temperatures. Those laboratories that are equipped with internal mixers are able to duplicate factory procedures quite satisfactorily, and in this case there is no problem. However, when insufficient material is available to permit mixing in an internal mixer or when the equipment is not available the disparity in the results obtained in the laboratory and in the factory present a serious problem. Braendle and co-workers[24] point out that the evolution of factory processing equipment in the recent past has tended to increase the divergence between the conditions under which laboratory batches and factory batches are mixed.

Although laboratory comparisons are usually based on a control stock made from the same type polymer being used in production, and a relation may be established between a roll-mill-mixed control and a factory-mixed control, one has no assurance that the differences observed with the control polymer will carry over to an experimental polymer. In the paper cited above it is shown that the relative ratings of two carbon blacks with respect to abrasion resistance in several different polymers were reversed when mixed cold on a roll mill and when mixed hot in an internal mixer.

This difficulty has also plagued those who have attempted to predict the processing characteristics of experimental polymers on the basis of laboratory tests which did not include subjection of the rubber to temperatures of the order attained in factory processing. The differences observed between low-temperature roll-mill mixes and high-temperature internal mixes are apparently much greater for the present butadiene-styrene synthetics than for natural rubber.

Evaluations in Different Laboratories. It will have been noted that the different laboratories whose testing procedures were described earlier selected different tests to be run and that certain of the tests that were run in all the laboratories were run by different methods and under different conditions. As a result of this situation it is not uncommon for two laboratories to disagree in some details of their evaluations of the same polymer. This is particularly true when the differences between the experimental polymer and the control polymer are small. However, when a polymer appears on the scene (such as the low-temperature modification of GR-S) which represents a substantial improvement over GR-S, all laboratories are in substantial agreement.

REFERENCES

1. Adams, R. J., Buckler, E. J., and Wanless, G. G., *Proc. Second Rubber Technol. Conf.*, 34–48 (1948).

1a. Adams, J. W., Reynolds, J. A., Messer, W. E., and Howland, L. H., *Rubber Chem. and Technol.*, **25**, 191–208 (1952).

2. Admiralty Rubber Meter, *J. Rubber Research*, **14**, 83–4 (1945).

3. Albert, H. E., *Ind. Eng. Chem.*, **40**, 1746–50 (1948).

4. Aleksandrov, A. P., and Lazurkin, J. S., *Acta Physiochim. U.S.S.R.*, **12**, 648–99 (1940); *Rubber Chem. and Technol.*, **13**, 886–98 (1940).

4a. Amerongen, G. J. van. See Van Amerongen, G. J.

5. Andrews, R. D., Tobolsky, A. V., and Hanson, E. E., *J. Applied Phys.*, **17**, 352–61 (1946).
6. *ASTM Bull. No. 155*, 2 (1948).
7. Barrer, R. M., and Skirrow, G., *J. Polymer Sci.*, **3**, 549–63 (1948).
7a. Bartel, A. W., and Temple, J. W., *Ind. Eng. Chem.*, **44**, 857–61 (1952).
8. Bashore, H. H., *India Rubber World*, **98**, 49–50 (1938).
8a. Beatty, J. R., *India Rubber World*, **125**, 438–9 (1952).
9. Beatty, J. R., and Davies, J. M., *J. Applied Phys.*, **20**, 533–9 (1949).
10. Beatty, J. R., and Juve, A. E., *India Rubber World*, **121**, 537–43 (1950).
10a. Beatty, J. R., and Juve, A. E., *India Rubber World*, **127**, 357–62, 423 (1952).
11. Behre, J., *Proc. Rubber Technol. Conf.*, 795–804 (1938).
12. Belotserkovsky, G. M., and Gorchakov, N. D., *Trudy LKKhTI*, No. 7, 80–4 (1939). See *Summary of Current Literature*, **18**, 487 (1940).
13. Berens, A. S., *ASTM Bull. No. 140*, 55–6 (1946).
14. Bierer, J. M., and Davis, C. C., *Ind. Eng. Chem.*, **16**, 711–7 (1924).
15. Bilmes, L., *J. Soc. Chem. Ind. Trans.*, **63**, 182–5 (1944).
16. Bimmerman, H. G., and Keen, W. N., *Ind. Eng. Chem., Anal. Ed.*, **16**, 588–90 (1944).
17. Blake, J. T., Symposium on Aging of Rubber, Spec. Tech. Pub. No. 89, Am. Soc. Testing Materials, Philadelphia, 1949, pp. 48–58.
18. Blow, C. M., and Fletcher, W. P., *India-Rubber J.*, **106**, 403–4 (1944).
19. Bobin, J., *Chemie et Industrie*, **32**, 270–5 (1934).
20. Boonstra, B. S. T. T., *India Rubber World*, **121**, 299–302 (1949).
21. Borders, A. M., and Juve, R. D., *Ind. Eng. Chem.*, **38**, 1066–70 (1946).
22. Bowditch, F. T., Greider, C. E., and Ollinger, C. G., *Am. Soc. Testing Materials Proc.*, **42**, 845–50 (1942).
23. Boyer, R. F., and Spencer, R. S., in *Scientific Progress in the Field of Rubber and Synthetic Elastometers* (*Advances in Colloid Science*, Vol. II), edited by H. Mark and G. S. Whitby, Interscience, New York and London, 1946, pp. 1–55.
24. Braendle, H. A., Steffen, H. C., and Dewender, J. G., *Rubber Age N.Y.*, **66**, 177–81 (1949).
25. Braendle, H. A., Valden, E., and Wiegand, W. B., *India Rubber World*, **110**, 645–6 (1944).
26. Breckley, J., *Rubber Age N.Y.*, **53**, 331–4 (1943).
27. British Standard Methods of Testing Vulcanized Rubber No. 903 (1950).
28. Buckley, D. J., and Chaney, A. L., *India Rubber World*, **109**, 60 (1943) (Abstract).
29. Buckley, D. J., Marshall, E. T., and Vickers, H. H., *Ind. Eng. Chem.*, **42**, 2407–13 (1950).
30. Buist, J. M., *Trans. Instn. Rubber Ind.*, **20**, 155–72 (1945).
31. Buist, J. M., *India-Rubber J.*, **120**, 451–8 (1951).
32. Buist, J. M., *India Rubber World*, **120**, 328–33 (1949).
33. Buist, J. M., and Davies, O. L., *Trans. Instn. Rubber Ind.*, **22**, 68–81 (1946).
34. Buist, J. M., and Davies, O. L., *India-Rubber J.*, **112**, 447–52, 454 (1947).
35. Buist, J. M., and Kennedy, R. L., *J. Sci. Instruments*, **23**, 242–3 (1946).
36. Buist, J. M., and Kennedy, R. L., *India-Rubber J.*, **110**, 809–12 (1946).
37. Buist, J. M., and Kennedy, R. L., Brit. Pat. 617,465, Feb. 16, 1949.
38. Buist, J. M., Newton, R. G., and Thornley, E. R., *Trans. Instn. Rubber Ind.*, **26**, 288–304 (1950).
39. Buist, J. M., and Powell, E. F., *Trans. Instn. Rubber Ind.*, **27**, 49–54 (1951).
40. Buist, J. M., and Welding, G. N., *Trans. Instn. Rubber Ind.*, **21**, 49–66 (1945).
41. Buist, J. M., and Williams, G. E., *India Rubber World*, **124**, 320–2, 447–9 (1951).
42. Bulgin, D., *Trans. Instn. Rubber Ind.*, **21**, 188–218 (1945).
43. Burr, G. S., *Electronics*, **22**, No. 5, 101–5 (1949); Gehman, S. D., and Clifford, R. P., Symposium on the Evaluation of Natural Rubber, Spec. Tech. Pub. No. 138, Am. Soc. Testing Materials, 97–111 (1935).
44. Bush, G. F., Associates, Princetown, N.J. and Scott Testers, Inc., Providence, R.I.
45. Busse, W. F., *Ind. Eng. Chem.*, **26**, 1194–9 (1934).

46. Carlton, C. A., and Reinbold, E. B., *India Rubber World*, **108**, 141–2 (1943).
47. Carpenter, A. S., and Twiss, D. F., *Ind. Eng. Chem.*, *Anal. Ed.*, **12**, 99–108 (1940).
48. Carpenter, A. W., in Davis and Blake, *Rubber Chemistry and Technology*, Reinhold, New York, 1937, pp. 777–847.
49. Carpenter, A. W., and Sargisson, Z. E., *Am. Soc. Testing Materials Proc.*, **31**, Pt. II, 897–907 (1931).
50. Carpenter, H. V., *Wash. State Coll. Bull.*, **7**, No. 8, Jan. 1925.
51. Cassie, A. B. D., Jones, M., and Naunton, W. J. S., *Trans. Instn. Rubber Ind.*, **12**, 49–84 (1936).
52. Chatten, C. K., Eller, S. A., and Werkenthin, T. A., *Rubber Age N.Y.*, **54**, 429–32 (1944).
53. *Chem. Eng. News*, **25**, 3315 (1947).
54. Clash, R. F., and Berg, R. M., *Ind. Eng. Chem.*, **34**, 1218–22 (1942).
55. Clash, R. F., and Berg, R. M., Symposium on Plastics, Am. Soc. Testing Materials, Philadelphia, 1944, pp. 54–65.
56. Cohan, L. H., and Steinberg, M., *Ind. Eng. Chem.*, *Anal. Ed.*, **16**, 15–20 (1944). Cf. Schade, J. W., *India Rubber World*, **126**, 67–72 (1952).
57. Columbian Carbon Company, *Columbian Colloidal Carbons*, New York, 1943, pp. 123–5.
58. Columbian Carbon Company, *Columbian Colloidal Carbons*, IV, 26 (1943).
59. Committee on Samples Tenders Report to ASTM, *Rubber Age N.Y.*, **65**, 437 (1949).
60. Compatibility and Incompatibility of Properties, *Rubber in Engineering*, Chemical Publishing Co., Brooklyn, 1946, pp. 166–74.
61. Conant, F. S., Dum, J. L., and Cox., C. M., *Ind. Eng. Chem.*, **41**, 120–6 (1949).
62. Conant, F. S., Hall, G. L., and Lyons, W. J., *J. Applied Phys.*, **21**, 499–504 (1950).
63. Conant, F. S., and Liska, J. W., *J. Applied Phys.*, **15**, 767–78 (1944).
64. Cooper, L. V., *Ind. Eng. Chem.*, *Anal. Ed.*, **5**, 350–1 (1933).
65. Cooper, L. V., Symposium on the Applications of Synthetic Rubbers, Am. Soc. Testing Materials, Philadelphia, 1944, pp. 17–26.
66. Cornell, D. H., and Beatty, J. R., *Rubber Age N.Y.*, **60**, 679–88 (1947).
67. Cornell, D. H., and Beatty, J. R., *Trans. Am. Soc. Mech. Engrs.*, **69**, 799–804 (1947).
68. Crabtree, J., and Kemp, A. R., *Ind. Eng. Chem.*, **38**, 278–96 (1946). Cf. Cuthbertson, G. R., and Dunnom, D. D., *Ind. Eng. Chem.*, **44**, 834–7 (1952).
69. Crabtree, J., and Kemp, A. R., *Ind. Eng. Chem.*, *Anal. Ed.*, **18**, 769–74 (1946). Cf. Crabtree, J., and Erikson, R. H., *India Rubber World.*, **125**, 719–20 (1952).
69a. Dalfsen, J. W. van. See Van Dalfsen, J. W.
70. Dart, S. L., Anthony, R. L., and Wack, P. E., *Rev. Sci. Instruments*, **17**, 106–8 (1946).
71. Dawson, T. R., and Porritt, B. D., *Rubber, Physical and Chemical Properties*, Croydon, Research Assoc. Brit. Rubber Manuf., 1935, pp. 331–49.
72. Dawson, T. R., and Porritt, B. D., *Rubber, Physical and Chemical Properties*, Croydon, Research Assoc. Brit. Rubber Manuf., 1935, p. 540.
73. Dawson, T. R., and Scott, J. R., *Trans. Instn. Rubber Ind.*, **16**, 198–210 (1940).
74. Daynes, H. A., Johnson, E. B., and Scott, J. R., *Trans. Instn. Rubber Ind.*, **6**, 63–81 (1930).
75. Determining the Volume Increase of Vulcanized Rubber in Solvents, *Vanderbilt Rubber Handbook*, R. T. Vanderbilt Co., New York, 1942, pp. 320–1.
76. Dillon, J. H., and Gehman, S. D., *India Rubber World*, **115**, 217–22 (1946).
77. Dillon, J. H., Prettyman, I. B., and Hall, G. L., *J. Applied Phys.*, **15**, 309–23 (1944).
78. Dock, E. H., and Scott, J. R., *J. Rubber Research*, **16**, 134–41 (1947).
79. Eccher, S., *Rubber Chem. and Technol.*, **13**, 566–75 (1940).
79a. Eccher, S., and Oberto, S., *Trans. Instn. Rubber Ind.*, **27**, 325–37 (1951). See also Juve, A. E., and Shearer, R., *India Rubber World*, **128**, 623–5 (1953).
80. Ellis, C., and Wells, A. A., *The Chemical Action of Ultra-violet Rays*, Reinhold, 1941, New York, p. 143.
81. Enabnit, R. S., and Gehman, S. D., *Ind. Eng. Chem.*, **43**, 346–51 (1951).

82. Evans, B. B., *India-Rubber J.*, **64**, 815–9 (1922).
83. Evans, R. D., *Proc. Twenty-Second Annual Meeting Highway Research Board*, Dec. 1942.
84. Eyring, H. L., *J. Chem. Phys.*, **4**, 283–91 (1936).
85. Fackler, M. B., and Rugg, J. S., *Anal. Chem.*, **23**, 1646–9 (1951).
86. Fielding, J. H., *Ind. Eng. Chem.*, **29**, 880–5 (1937).
87. Fielding, J. H., *Ind. Eng. Chem.*, **35**, 1259–61 (1943).
88. Fielding, J. H., *India Rubber World*, **115**, 802–5 (1947).
89. Fletcher, W. P., and Schofield, J. R., *Rubber Chem. and Technol.*, **18**, 306–17 (1945).
90. Flory, P. J., *Ind. Eng. Chem.*, **38**, 417–36 (1946).
91. Gardner, E. R., and Worswick, T., *Trans. Instn. Rubber Ind.*, **27**, 127–46 (1951).
92. Garvey, B. S., *Ind. Eng. Chem.*, **34**, 1320–3 (1942).
93. Garvey, B. S., and Emmett, R. A., *Ind. Eng. Chem.*, **36**, 209–11 (1944).
94. Gavan, F. M., Eby, S. W., and Schrader, C. C., *ASTM Bull. No. 143*, 23–9 (1946).
95. Geer, W. C. et. al., *India Rubber World*, **55**, 127–30 (1916).
96. Gehman, S. D., *J. Applied Phys.*, **19**, 456–63 (1948).
97. Gehman, S. D., Jones, P. J., Wilkinson, C. S., and Woodford, D. E., *Ind. Eng. Chem.*, **42**, 475–82 (1950).
98. Gehman, S. D., Jones, P. J., and Woodford, D. E., *Ind. Eng. Chem.*, **35**, 964–71 (1943).
99. Gehman, S. D., Woodford, D. E., and Stambaugh, R. B., *Ind. Eng. Chem.*, **33**, 1032–8 (1941); Gui, K. E., Wilkinson, C. S., and Gehman, S. D., *ibid.*, **44**, 720–3 (1952).
100. Gehman, S. D., Woodford, D. E., and Wilkinson, C. S., *Ind. Eng. Chem.*, **39**, 1108–15 (1947).
101. Gibson, K. S., *Instruments*, **9**, 309–11, 322, 335–8 (1936).
102. Gough, V. E., and Parkinson, D., *Trans. Instn. Rubber Ind.*, **17**, 168–242 (1941).
103. Graves, F. L., *India Rubber World*, **111**, 305–8, 317 (1944).
104. Graves, F. L., *India Rubber World*, **113**, 521 (1946).
105. Graves, F. L., and Davis, A. R., *India Rubber World*, **109**, 41–4 (1943).
106. Gray, H., Karch, H. S., and Hull, R. J., *Ind. Eng. Chem., Anal. Ed.*, **6**, 265–7 (1934).
107. Green, B. K., Chollar, R. G., and Wilson, G. J., *Rubber Age N.Y.*, **53**, 319–27 (1943).
108. Greene, H. E., and Loughborough, D. L., *J. Applied Phys.*, **16**, 3–7 (1945).
109. Greenleaf, E. F., *Rubber Age N.Y.*, **68**, 557–61 (1951).
110. Gregory, J. B., Pockel, I., and Stiff, J. F., *India Rubber World*, **117**, 611–6 (1948).
111. Griffith, T. R., Storey, E. B., Barkley, J. W. D., and McGilvray, F. M., *Anal. Chem.*, **20**, 837–47 (1948).
112. Hahn, S. H., and Gazdik, I., *India Rubber World*, **103**, 51–5 (1941).
113. Harris, C. W., *Rev. Sci. Instruments*, **16**, 5–6 (1945).
114. Harrison, S. R., and Cole, O. D., *Ind. Eng., Chem.* **36**, 702–7 (1944).
115. Havenhill, R. S., and MacBride, W. B., *Ind. Eng. Chem., Anal. Ed.*, **7**, 60–7 (1935).
116. Haworth, J., and Pryer, W. R., *Trans. Instn. Rubber Ind.*, **25**, 265–86 (1949).
117. Higuchi, T., Leeper, H. M., and Davis, D. S., *Anal. Chem.*, **20**, 1029–33 (1948).
118. Hock, L., *Z. Tech. Physik*, **6**, 50 (1925).
118a. Helin, A. F., and Labbe, B. G., *India Rubber World*, **126**, 227–31, 365–8 (1952).
119. Holt, W. L., and Knox, E. O., *Rubber Age N.Y.*, **60**, 689–92 (1947).
120. Holt, W. L., Knox, E. O., and Roth, F. L., *India Rubber World*, **118**, 513–7, 578 (1948).
121. Hunter, R. S., *J. Research Natl. Bur. Standards*, **25**, 581–618 (1940).
122. *India Rubber World*, **118**, 768 (1948).
123. Ingmanson, J. H., and Kemp, A. R., *Ind. Eng. Chem.*, **28**, 889–92 (1936).
124. Ingmanson, J. H., and Kemp, A. R., *Ind. Eng. Chem.*, **30**, 1168–73 (1938).
125. Juve, A. E., Private Communication to O.R.R., May 29, 1945.
126. Juve, A. E., *Ind. Eng. Chem.*, **39**, 1494–8 (1947).
127. Juve, A. E., Boxser, H., et al., *ASTM Bull. No. 147*, 51–3 (1947).
128. Juve, A. E., Fielding, J. H., and Graves, F. L., *ASTM Bull. No. 146*, 77–9 (1947).

129. Juve, A. E., and Garvey, B. S., *Ind. Eng. Chem.*, **34**, 1316–9 (1942).
130. Juve, A. E., and Garvey, B. S., *Ind. Eng. Chem.*, **36**, 212–8 (1944).
131. Juve, A. E., and Schroeder, C. H., *India Rubber World*, **115**, 515–7, 524 (1947).
132. Juve, R. D., and Marsh, J. W., *Ind. Eng. Chem.*, **41**, 2535–8 (1949).
133. Keen, W. N., *Trans. Am. Soc. Mech. Engrs.*, **68**, 237–40 (1946).
134. Kemp, A. R., Ingmanson, J. H., and Mueller, G. S., *Ind. Eng. Chem.*, **31**, 1472–8 (1939).
135. Kemp, A. R., Malm, F. S., and Winspear, G. G., *Ind. Eng. Chem.*, **35**, 488–92 (1943).
136. Kimball, A. L., *Vibration Prevention in Engineering*, Wiley, New York, 1932.
137. Kish, G. D., *Rubber Age N.Y.*, **53**, 131–5 (1943).
138. Koch, E. A., *Kautschuk*, **16**, 151–6 (1940); *Rubber Chem. and Technol.*, **14**, 799–810 (1941).
139. Kosten, C. W., *Rubber Chem. and Technol.*, **12**, 381–93 (1939).
140. Laine, P., and Roux, A., *Rev. gén. caoutchouc*, **21**, 189–91 (1944); *Rubber Chem. and Technol.*, **19**, 933–7 (1946).
141. Lambourn, L. J., *Trans. Instn. Rubber Ind.*, **4**, 210–34 (1928).
142. Lane, K. A., and Gardner, E. R., *Trans. Instn. Rubber Ind.*, **24**, 70–91 (1948).
143. Larchar, T. B., Sr., Private Communication to O.R.R., Oct. 3, 1947.
144. Larrick, L., *Am. Soc. Testing Materials Proc.*, **40**, 1239–50 (1940).
145. Leeper, H. M., *India Rubber World*, **115**, 215, 222 (1946).
146. Leeper, H. M., *Rubber Age N.Y.*, **59**, 73–4 (1946).
147. Lefcaditis, G., and Cotton, F. H., *Trans. Instn. Rubber Ind.*, **8**, 364–89 (1932).
148. Legge, N. R., Maher, E. D., and Watson, W. H., Private Communication to O.R.R., June 17, 1946.
149. Lessig, E. T., *Ind. Eng. Chem.*, *Anal. Ed.*, **9**, 582–8 (1937).
150. Lewis, T. R. G., *Trans. Instn. Rubber Ind.*, **21**, 375–90 (1946).
151. Liska, J. W., *Ind. Eng. Chem.*, **36**, 40–6 (1944).
152. Liska, J. W., S.A.E. Preprint No. 574 (1951), *SAE Journal*, **59**, No. 5, 22–5 (1951).
153. Lupke, P., *Vanderbilt News*, **3**, No. 6, 10–4 (1933); *Rubber Chem. and Technol.*, **7**, 591–8 (1934).
154. McCortney, W. J., and Hendrick, J. V., *Ind. Eng. Chem.*, **33**, 579–81 (1941).
155. Macdonald, W. S., and Ushakoff, A., *Anal. Chem.*, **20**, 713–7 (1948).
156. Mandel, J., Steel, M. N., and Stiehler, R. D., *Ind. Eng. Chem.*, **43**, 2901–8 (1951). Cf. I. J. Sjothun and P. S. Greer, *Rubber Age N.Y.*, **74**, 77–83 (1953).
157. Martens, A., *Sitzber. Königl. Preuss. Akad. Wiss.*, **14**, 346–66 (1911).
158. Martin, F. A., U.S. Pat. 2,340,401, Feb. 1, 1944.
159. Martin, S. M., *Rubber Age N.Y.*, **52**, 227–8 (1942).
160. Memmler, D. K., *Science of Rubber*, Reinhold, New York, 1934, pp. 566–77.
161. Mesrobian, R. B., and Tobolsky, A. V., *Ind. Eng. Chem.*, **41**, 1496–500 (1949).
162. Messenger, T. H., *J. Rubber Research*, **9**, 61–71 (1940).
163. Miller, R. F., *ASTM Bull. No. 151*, 91–5 (1948).
164. Moll, H. W., and LeFevre, W. J., *Ind. Eng. Chem.*, **40**, 2172–9 (1948).
165. Mooney, M., *Ind. Eng. Chem.*, *Anal. Ed.*, **17**, 514–7 (1945).
166. Mooney, M., and Black, S. A., *Can. J. Research*, **28**, F, 83–100 (1950); *Rubber Chem. and Technol.*, **23**, 744–59 (1950).
167. Mooney, M., and Gerke, R. H., *India Rubber World*, **103**, 29–32 (1941).
168. Mooney, M., and Gerke, R. H., *Rubber Chem. and Technol.*, **14**, 35–44 (1941).
168a. Mooney, M., and Wolstenholme, W. E., *Ind. Eng. Chem.*, **44**, 335–42 (1952).
169. Mooney, M., Wolstenholme, W. E., and Villars, D. S., *J. Applied Phys.*, **15**, 324–37 (1944).
170. Morley, J. F., *J. Rubber Research*, **17**, 61–3 (1948).
171. Morley, J. F., Porritt, B. D., and Scott, J. R., *J. Rubber Research*, **15**, 215–35 (1946).
172. Morley, J. F., and Scott, J. R., *J. Rubber Research*, **16**, 129–30 (1947).
173. Morris, R. E., and Bonnar, R. U., *Anal. Chem.*, **19**, 436–8 (1947).
174. Morris, R. E., Hollister, J. W., and Barrett, A. E., *Ind. Eng. Chem.*, **42**, 1581–7 (1950).

175. Morris, R. E., Hollister, J. W., Barrett, A. E., and Werkenthin, T. A., *Rubber Age N.Y.*, **55**, 45–52 (1944).
176. Morris, R. E., Hollister, J. W., and Mallard, P. A., *India Rubber World*, **112**, 455–8 (1945).
177. Morris, R. E., James, R. R., and Evans, E. R., *India Rubber World*, **110**, 529–32 (1944).
178. Morris, R. E., James, R. R., and Werkenthin, T. A., *Rubber Age N.Y.*, **51**, 205–8 (1942).
179. Morris, R. E., James, R. R., and Werkenthin, T. A., *Ind. Eng. Chem.*, **35**, 864–7 (1943).
180. Morris, R. E., Mitton, P., Montermoso, J. C., and Werkenthin, T. A., *Ind. Eng. Chem.*, **35**, 646–9 (1943).
181. Morron, J. D., Knapp, R. C., Linhorst, E. F., and Viohl, P., *India Rubber World*, **110**, 521–5 (1944).
182. Moses, F. L., and Rodde, A. L., *India Rubber World*, **119**, 201–4, 260 (1948).
183. Moyal, J. E., and Fletcher, W. P., *Rubber Chem. and Technol.*, **19**, 163–9 (1946).
184. Moyer, R. A., and Tesdall, G. L., Tire Wear and Cost on Selected Roadway Surfaces, *Iowa Eng. Exp. Sta. Bull.*, **161**, Iowa State College of Agriculture and Mechanic Arts, Ames, Aug. 1945.
185. Mullins, L., *Trans. Instn. Rubber Ind.*, **21**, 247–66 (1945).
186. Mullins, L., *J. Rubber Research*, **16**, 275–89 (1947).
187. Mullins, L., *Trans. Instn. Rubber Ind.*, **22**, 235–65 (1947).
188. Mullins, L., *Trans. Instn. Rubber Ind.*, **26**, 27–44 (1950).
189. Naugatuck Chemical Company, Distributors.
190. Naunton, W. J. S., and Waring, J. R. S., *Trans. Instn. Rubber Ind.*, **14**, 340–64 (1939).
191. Neal, A. M., Bimmerman, H. G., and Vincent, J. R., *Ind. Eng. Chem.*, **34**, 1352–7 (1942)
192. Neal, A. M., and Northam, A. J., *Ind. Eng. Chem.*, **23**, 1449–51 (1931).
193. Neal, A. M., and Ottenhoff, P., *Ind. Eng. Chem.*, **36**, 352–6 (1944).
194. Neal, A. M., and Vincent, J. R., Symposium on Aging of Rubber, Am. Soc. Testing Materials, Spec. Tech. Pub. 89, Mar. 1949, pp. 3–11.
195. Newman, S. B., and Taylor, R. H., *India Rubber World*, **119**, 345–6 (1948).
196. Newton, R. G., *Trans. Instn. Rubber Ind.*, **15**, 172–84 (1939).
197. Newton, R. G., *J. Rubber Research*, **14**, 27–62 (1945). Cf. Rugg, J. S., *Anal. Chem.*, **24**, 818–21 (1952).
198. Newton, R. G., *J. Rubber Research*, **15**, 35–60 (1946).
199. Newton, R. G., *J. Rubber Research*, **16**, 29 (1947).
200. Newton, R. G., *J. Rubber Research*, **17**, 178–94 (1948).
201. Newton, R. G., *Proc. Second Rubber Technol. Conf.*, 233–43 (1948).
202. Newton, R. G., and Scott, J. R., *J. Rubber Research*, **9**, 91–7 (1940).
203. Newton, R. G., Scott, J. R., and Willot, W. H., *J. Rubber Research*, **17**, 69–75 (1948).
204. Newton, R. G., and Wake, W. C., *J. Rubber Research*, **19**, No. 2, 9–16 (1950).
205. Nijveld, H. A. W., *Proc. Second Rubber Technol. Conf.*, 256–68 (1948).
206. Noll, A., *Papierfabr. Wochbl. Papierfabr.*, 151–3 (1944).
207. Patrikeev, G. A., and Melnikov, A. I., *India-Rubber J.*, **103**, 138–40, 157–60, 176–8, 181 (1942); *Kauchuk i Rezina*, **14**, 12–20 (1940); *Rubber Chem. and Technol.*, **14**, 863–76 (1941).
208. Perrine, V. H., and Scheirer, T. J., *India Rubber World*, **109**, 153–4 (1943).
209. Peterson, L. E., Anthony, R. L., and Guth, E., *Ind. Eng. Chem.*, **34**, 1349–52 (1942).
210. Pollack, L. R., McElwain, R. E., and Wagner, P. T., *Ind. Eng. Chem.*, **41**, 2280–6 (1949).
211. Poules, I. C., *India Rubber World*, **103**, 41–4 (1941).
212. Prettyman, I. B., *Ind. Eng. Chem.*, **36**, 29–33 (1944).
213. *Proc. Am. Soc. Testing Materials*, **48**, 421–2 (1948).
214. Proske, G., *Gummi-Ztg.*, **54**, 141–2, 167–8 (1940).
215. Rainier, E. T., and Gerke, R. H., *Ind. Eng. Chem., Anal. Ed.*, **7**, 368–73 (1935).
216. Reynolds, R. A., Private Communication to O.R.R., June 12, 1946.
217. Roelig, H., *Rubber Chem. and Technol.*, **12**, 394–400 (1939).

218. Roelig, H., *Rubber Chem. and Technol.*, **18,** 62–70 (1945).
218a. Rossem, A. van. See Van Rossem, A.
219. Roth, F. L., and Holt, W. L., *J. Research Natl. Bur. Standards*, **23,** 603–16 (1939).
220. Roth, F. L., and Stiehler, R. D., *India Rubber World*, **118,** 367–71 (1948).
221. *Rubber in Engineering,* Chemical Publishing Co., Brooklyn, 1946.
222. Rubber Reserve Rept. on Test 110.
223. Russell, J. J., *Ind. Eng. Chem.*, **32,** 509–12 (1940).
224. SAE-ASTM Tech. Comm. on Automotive Rubber, Sect. IV-A.
225. SAE-ASTM Tech. Comm. on Automotive Rubber, Sect. IV-F.
226. Sanders, P. A., *Rubber Chem. and Technol.*, **22,** 465–76 (1949).
227. Schade, J. W., *India Rubber World*, **123,** 311–4 (1950); **126,** 67–72 (1952).
228. Schade, J. W., and Roth, F. L., Symposium on Rubber Testing, Spec. Tech. Pub. 74, Am. Soc. Testing Materials, Philadelphia, 1947, pp. 27–35.
229. Scheu, D. R., and Schade, J. W., *India Rubber World*, **112,** 65–6 (1945).
230. Schiefer, H. F., *Textile Research J.*, **17,** 360–8 (1947).
231. Schob, A., *Mitt. kgl. Materialprüfungsamt*, **37,** 227 (1919).
232. Schoch, M. G., and Juve, A. E., Symposium on Aging of Rubber, Spec. Tech. Pub. 89, Am. Soc. Testing Materials, Mar. 1949, pp. 59–72.
233. Schwarz, H. F., *India Rubber World*, **110,** 412–5 (1944).
234. Scott, J. R., *Trans. Instn. Rubber Ind.*, **5,** 95–118 (1929).
235. Scott, J. R., *J. Rubber Research*, **17,** 7–9 (1948).
236. Scott, J. R., *J. Rubber Research*, **17,** 9–15 (1948).
237. Scott, J. R., *J. Rubber Research*, **17,** 75–82 (1948).
238. Scott, J. R., *J. Rubber Research*, **17,** 87–90 (1948).
239. Scott, J. R., *J. Rubber Research*, **17,** 145–9 (1948).
240. Scott, J. R., *J. Rubber Research*, **18,** 12 (1949).
241. Selker, M. L., Winspear, G. G., and Kemp, A. R., *Ind. Eng. Chem.*, **34,** 157–60 (1942).
242. Shearer, R., *India Rubber World*, **116,** 498, 560 (1947).
243. Shearer, R., Juve, A. E., and Musch, J. H., *India Rubber World*, **117,** 216–9 (1947); **117,** 491 (1948).
244. Shelton, J. R., Symposium on Aging of Rubber, Spec. Tech. Pub. 89, Am. Soc. Testing Materials, Mar. 1949, pp. 12–28.
245. Shelton, J. R., and Winn, H., *Ind. Eng. Chem.*, **36,** 728–30 (1944).
246. Shelton, J. R., and Winn, H., *Ind. Eng. Chem.*, **38,** 71–6 (1946).
247. Shelton, J. R., and Winn, H., *Ind. Eng. Chem.*, **39,** 1133–6 (1947).
248. Sjothun, I. J., and Cole, O. D., *Ind. Eng. Chem.*, **41,** 1564–7 (1949).
249. Smith, E. F., and Dienes, G. J., *ASTM Bull. No. 154*, 46–9 (1948).
250. Smith, O. H., Hermonat, W. A., Haxo, H. E., and Meyer, A. W., *Anal. Chem.*, **23,** 322–7 (1951).
251. Soden, A. L., *India-Rubber J.*, **115,** 555–6, 559 (1948).
252. Soden, A. L., *India-Rubber J.*, **119,** 1143 (1950); **120,** 13, 55, 92, 137, 173, 212, 254, 292, 332 (1951).
253. Somerville, A. A., *Trans. Instn. Rubber Ind.*, **6,** 130–69 (1930).
254. Somerville, A. A., *Proc. Rubber Technol. Conf.*, 773–86 (1938).
255. Specifications for Gov. Synthetic Rubbers, Sect. D-5a, R.F.C., O.R.R., Washington, D.C., 1949.
256. Specifications for Gov. Synthetic Rubbers, Sects. D-5a and D-5b, R.F.C., O.R.R., Washington, D.C., 1949.
257. Specifications for Gov. Synthetic Rubbers, Sect. D-7, R.F.C., O.R.R., Washington, D.C., 1949.
258. Specifications for Gov. Synthetic Rubbers, Fig. 2, R.F.C., O.R.R., Washington, D.C., 1949.

259. Specifications for Gov. Synthetic Rubbers, Tables I, II, R.F.C., O.R.R., Washington, D.C., 1949.
260. Specifications for Gov. Synthetic Rubbers, Table II, R.F.C., O.R.R., Washington, D.C., 1947.
261. Sperberg, L. R., Burleigh, J. E., and Svetlik, J. F., Private Communication to O.R.R., Sept. 5, 1946.
262. Sperberg, L. R., Svetlik, J. F., and Bliss, L. A., *Ind. Eng. Chem.*, **41**, 1641–6 (1949).
263. Stambaugh, R. B., Rohner, M., and Gehman, S. D., *J. Applied Phys.*, **15**, 740–8 (1944).
264. Standard Methods of Test for Insulation Resistance of Electrical Insulating Materials, ASTM Designation D 257–46, 1946 ASTM Book of Standards III-B, 135. Cf. Tentative Method D 257–52T, *ASTM Standards on Rubber Products*, 1952, 94–112.
265. Standard Methods of Test for Hardness of Rubber, ASTM Designation D 314–39, 1946 ASTM Book of Standards III-B, 293.
266. Standard Methods of Test for Abrasion Resistance of Rubber Compounds, ASTM Designation D 394–47, 1947 Suppl., ASTM Book of Standards III-B, 7.
267. Standard Methods of Tension Testing of Vulcanized Rubber, ASTM Designation D 412–41, 1946 ASTM Book of Standards III-B, 247; Tentative Method D 412–51T, *ASTM Standards on Rubber Products*, 1952, 201–9.
268. Standard Methods of Dynamic Testing for Ply Separation and Cracking of Rubber Products, ASTM Designation D 430–51T, *ASTM Standards on Rubber Products*, 1952, 227–37.
269. Standard Method of Air Pressure Heat Test of Vulcanized Rubber, ASTM Designation D 454–52, *ASTM Standards on Rubber Products*, 1952, 238–41.
270. Standard Method of Test for Indentation of Rubber by Means of the Pusey and Jones Plastometer, ASTM Designation D 531–49, *ASTM Standards on Rubber Products*, 1952, 286–9.
271. Standard Method of Test for Accelerated Aging of Vulcanized Rubber by the Oxygen Pressure Method, ASTM Designation D 572–52, *ASTM Standards on Rubber Products*, 1952, 301–4.
272. Standard Method of Test for Accelerated Aging of Vulcanized Rubber by the Oven Method, ASTM Designation D 573–52, *ASTM Standards on Rubber Products*, 1952, 305–8.
273. Standard Methods of Test for Tear Resistance of Vulcanized Rubber, ASTM Designation D 624–48, 1948 Suppl., ASTM Book of Standards III-B, 18.
274. Stevens, W. L., *India Rubber World*, **102**, 36–41 (1940).
274a. Stiehler, R. D., Steel, M. N., and Mandel, J., *Trans. Instn. Rubber Ind.*, **27**, 298–311 (1951).
275. Storey, E. B., *Rubber Age N.Y.*, **66**, 653–8 (1950).
276. Svetlik, J. F., and Sperberg, L. R., *India Rubber World*, **124**, 182–7 (1951).
277. Taylor, R. H., *ASTM Bull. No. 123*, 25–30 (1943).
278. Tentative Methods of Testing Rubber Insulated Wire and Cable, ASTM Designation D 470–52T, *ASTM Standards on Rubber Products*, 1952, 248–64.
279. Tentative Methods of Test for Changes in Properties of Rubber and Rubber-Like Materials in Liquids, ASTM Designation D 471–52T, *ASTM Standards on Rubber Products*, 1952, 265–73.
280. Tentative Method of Test for Physical State of Cure of Vulcanized Rubber (T–50 Test), ASTM Designation D 599–40T, 1946 ASTM Book of Standards III-B, 986.
281. Tentative Method of Test for Compression Fatigue of Vulcanized Rubber, ASTM Designation D 623–52T, *ASTM Standards on Rubber Products*, 1952, 333–40.
282. Tentative Method of Test for Indentation of Rubber by Means of a Durometer, ASTM Designation D 676–49T, *ASTM Standards on Rubber Products*, 1952, 353–6.
283. Tentative Method of Test for Low-Temperature Brittleness of Rubber and Rubber-Like Materials, ASTM Designation D 736–46T, 1946 ASTM Book of Standards III-B, 1026.

284. Tentative Method of Test for Brittle Temperature of Plastics and Elastomers by Impact, ASTM Designation D 746–52T, *ASTM Standards on Rubber Products*, 1952, 391–6.

285. Tentative Method of Test for Resistance to Accelerated Light Aging of Rubber Compounds, ASTM Designation D 750–43T, 1946 ASTM Book of Standards III–B, 980.

286. Tentative Method of Test for Resistance of Vulcanized Rubber or Synthetic Elastomers to Crack Growth, ASTM Designation D 813–44T, 1946 ASTM Book of Standards III–B, 1011. Cf. ASTM Designation D 813–52T, *ASTM Standards on Rubber Products*, 1952, 439–43.

287. Tentative Method of Heat Aging of Vulcanized Natural or Synthetic Rubber by Test Tube Method, ASTM Designation D 865–52T, *ASTM Standards on Rubber Products*, 1952, 475–8.

288. Tentative Methods of Test for Contact and Migration Stain of Vulcanized Rubber in Contact with Organic Finishes, ASTM Designation D 925–51T, *ASTM Standards on Rubber Products*, 1952, 481–4.

289. Tentative Methods of Test for Mechanical Properties of Elastomeric Vulcanizates under Compressive Shear Strain by the Mechanical Oscillograph, ASTM Designation D 945–52T, *ASTM Standards on Rubber Products*, 1952, 492–501.

290. Tentative Method of Test for Volume Resistivity of Electrically Conductive Rubber and Rubber-Like Materials, ASTM Designation D 991–48T, *ASTM Standards on Rubber Products*, 1952, 502–5.

291. Tentative Method of Measuring Low-Temperature Stiffening of Rubber and Rubber-Like Materials by the Gehman Torsional Apparatus, ASTM Designation D 1053–52T, *ASTM Standards on Rubber Products*, 1952, 547–53.

292. Tentative Method of Test for Impact Resilience and Penetration of Rubber by the Rebound Pendulum, ASTM Designation D 1054–49T, 1949, ASTM Book of Standards Part 6, 1205.

293. Tentative Method of Test for the Accelerated Ozone Cracking of Vulcanized Rubber, ASTM Designation D 1149–51T, ASTM Standards on Rubber Products D 590, May 1, 1951.

294. Thornley, E. R., *Trans. Instn. Rubber Ind.*, **24,** 241–55 (1949).

295. Throdahl, M. C., *Ind. Eng. Chem.*, **39,** 514–6 (1947).

296. Throdahl, M. C., *Ind. Eng. Chem.*, **40,** 2180–4 (1948).

297. Tobolsky, A. V., Prettyman, I. B., and Dillon, J. H., *J. Applied Phys.*, **15,** 380–95 (1944).

298. Torrance, P. M., Symposium on the Applications of Synthetic Rubbers, Spec. Tech. Pub. No. 61, Am. Soc. Testing Materials, Philadelphia, 1944, pp. 59–66.

299. Trepp, S. G., Ward, A. L., and Chaney, N. K., *India Rubber World*, **111,** 63–4, 112 (1944).

300. Tuley, W. F., *Ind. Eng. Chem.*, **31,** 714–6 (1939).

301. Van Amerongen, G. J., *J. Applied Phys.*, **17,** 972–85 (1946); *Rubber Chem. and Technol.*, **20,** 494–514 (1947).

302. Van Dalfsen, J. W., *Rubber Chem. and Technol.*, **16,** 388–99 (1943).

303. Van Rossem, A., and Talen, H. W., *Rubber Chem. and Technol.*, **4,** 490–504 (1931).

304. Van Wijk, D. J., *Rubber Chem. and Technol.*, **6,** 406–11 (1933).

305. *Vanderbilt News*, **10,** No. 2, 5–7 (1940).

306. *Vanderbilt News*, **11,** No. 6, 4–7 (1941).

307. *Vanderbilt News*, **14,** No. 5, 22–23 (1948).

308. *Vanderbilt Rubber Handbook*, R. T. Vanderbilt Co., New York, 1948, p. 453.

309. Vila, G., *Ind. Eng. Chem.*, **34,** 1269–76 (1942).

310. Villars, D. S., *J. Applied Phys.*, **21,** 565–73 (1950).

311. Wack, P. E., Anthony, R. L., and Guth, E., *J. Applied Phys.*, **18,** 456–69 (1947).

312. Waring, J. R. S., *Trans. Instn. Rubber Ind.*, **26,** 4–26 (1950).

313. Weaver, J. V., *Rubber Age N.Y.*, **48,** 89–95 (1940).

314. Werkenthin, T. A., *Rubber Age N.Y.*, **60**, 197–202 (1946).

315. Werkenthin, T. A., *Rubber Age N.Y.*, **59**, 697–702 (1946).

316. Werkenthin, T. A., Richardson, D., Thornley, R. F., and Morris, R. E., *Rubber Age N.Y.*, **50**, 103–8, 199–202 (1941).

317. Whitby, G. S., *Plantation Rubber and Testing of Rubber*, Longmans, Green, New York, 1920, pp. 340–3.

318. Whitby, G. S., and Budewitz, E. P., Private Communication to O.R.R., Apr. 23–May 4, 1945.

319. Whitby, G. S., Evans, A. B. A., and Pasternack, D. S., *Trans. Faraday Soc.*, **38**, 269–75 (1942).

320. Wiegand, W. B., *Ind. Eng. Chem.*, **18**, 1157–63 (1926).

320a. Wijk, D. J. van. See Van Wijk, D. J.

321. Wilkinson, C. S., and Gehman, S. D., *Anal. Chem.*, **22**, 283–9 (1950).

322. Wilkinson, C. S., and Gehman, S. D., *Anal. Chem.*, **22**, 1439–43 (1950).

323. Williams, I., *Ind. Eng. Chem.*, **18**, 367–9 (1926).

324. Williams, I., *Ind. Eng. Chem.*, **19**, 674–7 (1927).

325. Willott, W. H., *Proc. Rubber Technol. Conf.*, 169–76 (1938).

326. Winn, H., and Shelton, J. R., *Ind. Eng. Chem.*, **37**, 67–70 (1945).

327. Wood, L. A., *Advances in Colloid Science, Vol. II*, Interscience, New York, 1946, pp. 57–93.

328. Woodford, D. E., Wilkinson, C. S., and Gehman, S. D., *Ind. Eng. Chem.*, **39**, 1110, 1111, Figs. 5, 7, 12 (1947).

329. Yerzley, F. L., *Rubber Chem. and Technol.*, **13**, 149–58 (1940).

330. Yerzley, F. L., and Fraser, D. F., *Ind. Eng. Chem.*, **34**, 332–6 (1942).

331. Zimmerman, E. C., *Rubber Age N.Y.*, **12**, 130–2 (1922).

CHAPTER 13

AGING AND STABILIZATION OF GR-S

John O. Cole
Goodyear Tire and Rubber Company

Because of the presence of unsaturation in the GR-S hydrocarbon, a progressive change in its physical properties occurs on aging, as a result of the combined effects of heat, light, oxygen, and ozone. Oxygen and ozone are the agents primarily responsible for deterioration of the elastomer. Heat and light furnish the energy of activation required to initiate the chemical action of oxygen and ozone. Reactions involving curing agents or the sulfur groups in the vulcanizate or both these are probably associated with the aging process to a minor degree. This chapter is concerned primarily with thermal oxidation. Photooxidation and the action of ozone are discussed less extensively, since chemical literature on these aspects of aging is rather limited.

DETERIORATION OF GR-S BY OXYGEN

Effect of Oxidation on Physical Properties. The thermal oxidation of GR-S causes marked changes in its physical properties. It has been shown that the unvulcanized polymer first undergoes softening, but, as the extent of oxidation increases, it gradually becomes harder until a brittle, resinous material is finally formed.[11, 29, 32] The softening which occurs during early stages of oxidation is shown by decrease in the Mooney viscosity and intrinsic viscosity of the polymer. As oxidation proceeds, reactions causing hardening of the polymer predominate with the result that the Mooney viscosity rises, the gel content increases, and, correspondingly, the intrinsic viscosity of the soluble portion of the polymer falls. The effect of air oxidation on the properties of unvulcanized GR-S is illustrated in Fig. 1, which shows the change in Mooney viscosity, and in Table I, which shows change in solubility and in intrinsic viscosity. The decrease in intrinsic viscosity during early stages of oxidation, when the polymer remains completely soluble, indicates breakdown of the polymer.

Table I. Effect of Air Oxidation at 100° C. on Benzene Solubility and Intrinsic Viscosity of Unvulcanized GR-S Containing Antioxidant.[11]

Time, hr.	0	2	4	6	10	20	30	60	90
Solubility in benzene, %	100	100	100	100	100	100	98	75	60
Intrinsic viscosity	1.70	1.58	1.55	1.46	1.34	1.17	1.10	1.02	0.82

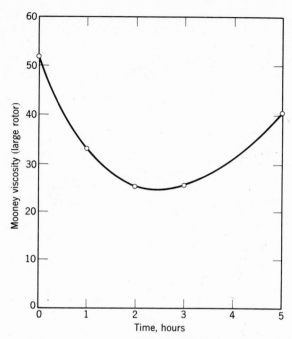

FIG. 1. Effect on Plasticity of Heating GR-S Containing 1.5 per cent Phenyl-β-naphthylamine in Air[29] at 149° C.

Table II. Comparison of Aging of GR-S and Natural-Rubber Tread Stocks[15]

				Aged (% of Unaged)					
Physical Property	Curing Time, Min.	Unaged Values Rubber	GR-S	7 Hr. Air Bomb, 236° F., 80 P.S.I. Rubber	GR-S	16 Hr. Air Bomb, 236° F., 80 P.S.I. Rubber	GR-S	7 Days Geer Oven Rubber	GR-S
Tensile strength, p.s.i.	20	2850	1100	70	200	40	220	131	215
	30	3900	2350	58	106	35	98	99	126
	40	4030	2750	58	92	24	89	98	100
	60	4050	3000	46	82	17	70	92	83
	80	4050	3000	48	77	15	72	79	77
Elongation, %	20	600	600	97	98	58	93	105	90
	30	620	680	86	75	47	56	94	75
	40	590	600	88	67	37	60	95	68
	60	565	530	71	64	32	51	92	69
	80	560	480	70	61	30	54	73	63
Modulus at 300%, p.s.i.	20	810	460	97	196	121	257	157	230
	30	1200	720	90	202	...	244	139	201
	40	1300	1020	92	181	...	200	179	182
	60	1520	1300	86	164	...	...	132	166
	80	1620	1580	90	158	...	...	133	146

	Rubber	GR-S
Rubber	100.0	
GR-S		100.0
EPC black	50.0	50.0
Stearic acid	3.0	
Paraflux	3.0	6.0
Zinc oxide	3.0	3.0
Phenyl-β-naphthylamine	1.0	
Sulfur	2.8	2.8
Mercaptobenzothiazole	0.9	0.9
Optimum cure, min.	40 at 275° F.	40 at 300° F.

The tendency toward hardening as aging proceeds, which is observed with the unvulcanized polymer, is to be seen during the aging of vulcanized GR-S in the progressive rise in modulus.[23, 37, 41]   The phenomenon of hardening in vulcanizates sometimes has been referred to in the literature as "heat embrittlement."   Typical data showing the change in physical

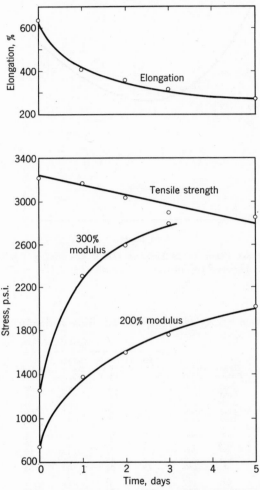

FIG. 2.   Change in Physical Properties of a GR-S Tread Stock during Air-Oven Aging[41] at 100° C.

properties of a GR-S tread stock during aging in a circulating-air oven are shown in Fig. 2.   The tensile strength decreases moderately at a rate which may fall off as oxidation proceeds.   The elongation at break decreases progressively.   The modulus rises and may sometimes reach rather high values; this rise is the most serious feature of GR-S aging.   Aging also results in a decrease in flex life.

The effect of oxidation on the physical properties of GR-S differs from that observed in natural rubber. Oxidation of natural-rubber vulcanizates causes a sharp drop in tensile strength and elongation. Modulus may either increase or decrease. The net result of oxidation of natural rubber is predominantly one of softening instead of hardening as observed with GR-S. A comparison of the aging of GR-S and of natural rubber is given in Table II. The rate of oxidation of GR-S is less than that of natural rubber.[12, 31, 44] It is generally considered that GR-S is more resistant to high-temperature oxidation than natural rubber.[31, 37] Scott[37] has suggested that the superiority of GR-S over natural rubber in this respect is due to the flattening of the tensile-strength–time-of-aging curve as aging proceeds.

Importance of Oxygen in the Aging of GR-S. The change in physical properties of GR-S vulcanizates can be attributed to the following factors or to combinations of these factors: (1) continued reaction of curing agents, (2) changes in molecular structure due to rearrangement of sulfur groups in the vulcanizates, (3) structural changes resulting from oxidation of the polymer hydrocarbon. Although conclusive evidence has not been obtained that oxygen is the only factor in the deterioration of GR-S at elevated temperatures, the evidence summarized below indicates that it is the most important factor.

Early investigations on the aging of GR-S indicated that reducing the sulfur content, employing flat-curing accelerator systems, and increasing the curing time result in improved age resistance. On this account, continued cure was considered to be the main cause for the increase in modulus that occurs during aging. Hunter[27] and Scott[38] have shown, however, that heating in air causes greater deterioration than continued cure in a mold under comparable conditions. This effect is illustrated by the data in Table III. Vulcanizates cured for a period equal to three times the optimum cure were compared with optimum vulcanizates heated in air at

Table III. Comparison of Continued Cure and of Air Aging at the Same Temperature[38]

Base Stock: GR-S 100, EPC Black 50, Zinc Oxide 3, Stearic Acid 1, Pine Tar 3

Mix	C	D	E	F
Sulfur	1.5	1.5	3.0	3.0
Mercaptobenzothiazole	1.5	...	1.5	...
Tetramethylthiuram disulfide	...	0.5	...	0.5
Optimum cure	60'/153° C.	45'/135° C.	35'/153° C.	30'/135° C.
Overcure	180'/153° C.	125'/135° C.	105'/153° C.	90'/135° C.

	Tensile Strength P.S.I.		Elongation, %		Modulus at 100%, P.S.I.	
	Optimum Cure Aged	Overcure Unaged	Optimum Cure Aged	Overcure Unaged	Optimum Cure Aged	Overcure Unaged
Mix C	1620	2260	280	400	320	250
D	1440	1780	300	350	275	240
E	1000	2240	140	280	545	440
F	1320	1760	200	240	435	420

vulcanizing temperature for a period equal to the difference in time between the optimum cure and the overcure.   It will be noted that in each case heating in air has resulted in a greater change in physical properties than continued cure at the same temperature.   Scott[37] had earlier presented evidence that increase in combined sulfur can account for only a small part of the modulus increase observed during aging in the presence of oxygen.   No evidence is available to prove or disprove the hypothesis of Hendricks[26] concerning the rearrangement of sulfur groups in the molecule.

Shelton and Winn[41] compared the effect of aging vulcanizates in nitrogen at 80 and 100° C. with aging in air at 100° C. and aging in the oxygen bomb at 80° C.   In general, the changes observed during aging in the absence of oxygen were less than in the presence of oxygen.   Typical data are given in Table IV to illustrate the effect of oxygen.   Results obtained by Shelton and Winn for oxygen-bomb aging, which might appear to indicate that oxygen is not a major factor in the modulus rise during aging, can probably be explained on the basis of the effect of high oxygen pressure on the nature of the reaction.   This effect will be discussed further in

Table IV.   Effect of Oven Aging in Air and in Nitrogen on Properties of GR-S Vulcanizates[41]

Composition of Stocks: GR-S 100, Coal-Tar Softener 5, Fatty Acid 1.5, Channel Black 50, Zinc Oxide 5, Cyclohexyl benzothiazolyl sulfenamide 1.2, Sulfur 2, Phenyl-β-naphthylamine 1 or 2.

Stocks B and B' were duplicates made up at different times.   They were cured for 55 min. at 298° F.   The GR-S in them was free from antioxidant.

Stock C was cured for 50 min. at 298° F.   The GR-S in it was regular GR-S, containing antioxidant.

The added phenyl-β-naphthylamine in stocks B and B' was 1 part; in stock C 2 parts.

	Tread Stock	Unaged	2 days at 100° C. Nitrogen	Air	5 days at 100° C. Nitrogen	Air
Tensile strength, p.s.i.	B	2720	93	92	93	73
	B'	2760	101	92	95	79
	C	3290	92	85	89	89
Elongation, %	B	510	70	54	63	41
	B'	555	68	52	61	52
	C	615	64	52	58	47
Modulus at 200%, p.s.i.	B	820	140	194	154	228
	B'	720	157	221	170	282
	C	790	156	201	172	244

another section.   Cole and Parks[12] observed that the change in physical properties during aging in oxygen at 100° C. was greater than during aging in vacuum at the same temperature.   Tobolsky and Andrews[48] have shown that the rate of stress relaxation for a sample held at constant elongation is reduced a thousandfold if aging is conducted in the absence of oxygen. Vulcanizates prepared without antioxidant were observed to deteriorate rapidly in the presence of oxygen.[41] Though Sturgis, Baum, and Vincent[45] observed no significant difference between aging in oxygen and in nitrogen, the balance of evidence indicates that oxygen plays a major role in the aging of GR-S. Reactions not involving oxygen are probably responsible for part of the observed changes in physical properties during aging.

The influence of the partial pressure of oxygen (up to 1 atm.) on the rate of oxygen absorption and the rate of change of physical properties during the heat aging of GR-S have been studied by Shelton and W. L. Cox.[39]

Winn and Shelton[50] have shown that the flex life of GR-S is increased approximately 50 per cent if oxygen is present only in traces. Thus oxygen is also an important factor in flex life.

Hydroperoxide Theory of Oxidation. Older theories concerning the mechanism of oxidation were based on the hypothesis originally proposed by Bach[3] and by Engler and co-workers[16] that absorption of molecular oxygen occurs at the double bond to give a product commonly formulated as a saturated peroxide. As the result of work during the last decade it

$$>C : C< \quad \xrightarrow{\ O_2\ } \quad >C \cdot C< \\ \qquad\qquad\qquad\qquad\quad O \cdot O$$

seems clear that the mechanism of oxidation cannot be adequately represented on the basis of the Engler-Bach hypothesis. Studies of Farmer and his co-workers[7, 19, 20] on the oxidation of simple olefins indicate that the course of oxidation of unconjugated olefins may be represented as follows. Peroxidation occurs exclusively or almost exclusively at the carbon atom adjacent to the double bond, with the formation of a hydroperoxide. In the oxidation of polyolefins, Bolland and Hughes[6] have shown, a more complex peroxide is formed.

$$-CH_2 \cdot CH : CH \cdot CH_2- \ + \ \cdot OO \cdot \ (or\ R\cdot) \ \longrightarrow$$
$$-CH \cdot CH : CH \cdot CH_2- \ + \ \cdot OOH \ (or\ RH) \qquad (1)$$

$$-CH \cdot CH : CH \cdot CH_2- \ + \ \cdot OO \cdot \ \longrightarrow \ -CH \cdot CH : CH \cdot CH_2- \qquad (2)$$
$$\qquad\qquad\qquad\qquad\qquad\qquad\qquad\qquad\qquad OO \cdot$$

$$-CH \cdot CH : CH \cdot CH_2- \ + \ -CH_2 \cdot CH : CH \cdot CH_2- \ \longrightarrow$$
$$OO \cdot$$
(I)
$$\qquad\qquad -CH \cdot CH : CH \cdot CH_2- \ + \ -CH \cdot CH : CH \cdot CH_2- \qquad (3)$$
$$\qquad\qquad OOH$$

Although the reaction probably has a free-radical chain mechanism, the nature of the initiating step is not clear.[4, 5, 17-8, 49] Termination of the reaction chain occurs by combination of the radicals involved in the propagation step. Thermal decomposition of the hydroperoxide accelerates the reaction by formation of a free radical capable of starting an oxidation chain. This complex decomposition is not well understood, but, in general, the hydroperoxide group reverts to hydroxyl. Under some conditions carbonyl and probably other oxygen-containing groups result from peroxide decomposition. When oxidation has reached an advanced stage, a complex mixture of oxidation products is formed, in which most of the common oxygen-containing groups are present.

Antioxidants have usually been considered to function as chain-terminating agents by reacting[8, 9, 11] with the peroxide radical, ROO·. If AH represents an anti-oxidant and ROO· a radical similar to (I) in reaction 3, chain termination can be formulated as follows:

$$ROO \cdot + AH \to ROOH + A \cdot$$

Shelton and Cox[40] have suggested that the above reaction should be considered not as a chain-termination reaction but as a chain-transfer process. Radical A· is considered capable of abstracting a hydrogen atom from the hydrocarbon to initiate a new reaction chain. Termination is postulated to occur by reaction of radical A· with another radical to give a stable product. In addition, the process of chain initiation is probably modified by reactions involving the antioxidant.

On the basis of the hydroperoxide theory, oxidation of GR-S probably occurs at points in the polymer chain where secondary or tertiary hydrogen atoms are adjacent to a double bond. Such points are indicated by an asterisk.

$$\overset{*}{-CH_2} \cdot CH : CH \cdot CH_2- \qquad \overset{*}{-CH} \cdot CH_2- \qquad \overset{*}{-CH} \cdot CH_2- \qquad \overset{*}{-CH} \cdot CH_2-$$

Mechanism of Deterioration of Physical Properties. Deterioration of physical properties of elastomers as the result of oxidation has commonly been attributed to reactions which cause the following structural changes: (1) scission or rupture of polymer chains, (2) cross-linking, i.e., the formation of primary valence links between polymer chains, (3) cyclization or the formation of rings, with a resultant shortening of polymer chains.

The fall in tensile strength during the oxidation of both GR-S and natural rubber has been generally attributed to chain scission. In GR-S the observed increase in modulus seems to be associated with a cross-linking reaction. Under most conditions the latter reaction predominates during the oxidation of GR-S. These effects have been clearly shown in measurements of continuous and intermittent stress relaxation carried out by Tobolsky and his associates.[48] If an elastomer at an elevated temperature is held at constant elongation, a gradual decrease in stress is observed. On the basis of the kinetic theory of elasticity, it can be shown that any change in stress is the result of a change in the number of network chains per unit volume which are maintaining the stress.   Stress relaxation at elevated temperatures is thus associated with a decrease in the number of network chains, resulting from a reaction in which primary valence bonds in the network structure are severed.   The rate of stress relaxation, therefore,

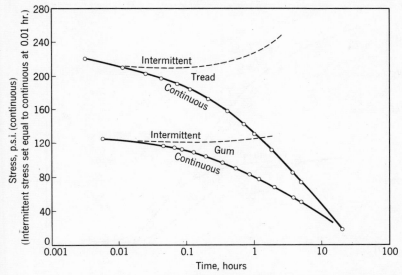

FIG. 3.   Continuous and Intermittent Stress Relaxation of GR-S Stocks[48] 50 per cent elongation at 130° C.

provides a measure of the rate of chain scission during oxidation.   Any cross-links which may be formed between polymer chains by a parallel reaction appear to be formed in a relaxed portion of the network chain and do not contribute to the stress.   When stress is measured intermittently, i.e., by merely making a periodic measurement of the modulus, any cross-links formed contribute to the stress.   The observed stress is the net result of both cross-linking and chain-scission reactions.   Stress-relaxation curves for GR-S and natural rubber are shown in Figs. 3 and 4.   For natural rubber, a decay of stress is observed by both continuous and intermittent measurements. A more rapid rate of relaxation is noted for the continuous measurement. This indicates that, although chain scission is the predominant reaction, cross-linking also occurs.   For GR-S, stress measured intermittently actually increases in contrast to the progressive decrease in stress measured continuously.   Chain scission hence occurs during oxidation of GR-S, although cross-linking is the predominant reaction.   These results indicate that chain-scission and cross-linking reactions occur during the oxidation of both GR-S and natural rubber.   Chain scission is the predominant reaction with natural rubber, whereas cross-linking is the major reaction with GR-S.   Shelton and Winn[41] concluded that differences in aging characteristics of natural rubber and GR-S reflect a variation in degree and relative rates rather than a fundamental difference in the reactions concerned.

There is no evidence to indicate that cyclization occurs during oxidation of GR-S. It is difficult to see how such a process could account for the changes in the physical properties of GR-S observed during its aging.

The mechanism of the reactions causing chain scission and cross-linking

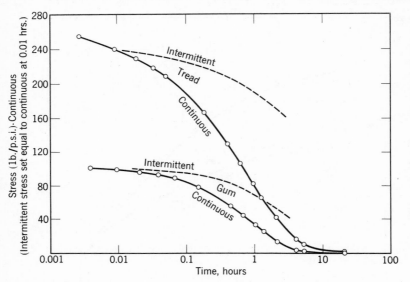

FIG. 4. Continuous and Intermittent Stress Relaxation of Hevea Stocks[48]
50 per cent elongation at 130° C.

are as yet unknown. The attempts which have been made by Farmer and his co-workers[17-8] and also by Taylor and Tobolsky[47] to explain these reactions follow two somewhat different lines of thought. The first theory assumes that chain scission and cross-linking occur through reactions of the peroxide or its decomposition products. The second theory supposes that the role of oxygen is to form free radicals which may disproportionate to cause chain scission or may attack the double bond to bring about cross linking.

FACTORS AFFECTING STABILITY OF GR-S

Antioxidants. GR-S must be stabilized by the addition of an antioxidant, to prevent oxidation of the polymer during manufacture and storage. In the absence of an antioxidant, unvulcanized GR-S undergoes oxidation, even at room temperature, to form a hard resinous material devoid of rubberlike properties. Even very small quantities of antioxidant afford considerable protection.[22] The problem of protecting GR-S by means of antioxidants is somewhat different from that encountered with natural rubber. Antioxidant is usually added to natural rubber only during compounding, since there is naturally present an antioxidant which provides sufficient protection to the rubber in its unvulcanized state.

The effect of an antioxidant on the stability of a GR-S tread stock is illustrated in Table V, from which it will be seen that a GR-S vulcanizate prepared without an antioxidant deteriorates rapidly on accelerated aging.

Table V. Oven Aging of GR-S Tread Stocks in Presence and Absence of Antioxidant[41]

			Aged (% of Unaged)			
	Anti-		2 Days at 100° C.		5 Days at 100° C.	
Physical Property	oxidant*	Unaged	Air	Nitrogen	Air	Nitrogen
Tensile strength, p.s.i.	None	2630	55	103	40	97
	1%	2720	92	95	73	93
	2%	3290	85	92	89	89
Elongation, %	None	490	35	67	25	64
	1%	510	54	70	41	63
	2%	615	52	64	47	58
Modulus at 200%,	None	730	...	172	...	184
p.s.i.	1%	820	194	140	228	154
	2%	790	201	156	244	172

* Phenyl-β-naphthylamine.

The stability of GR-S in the raw or vulcanized condition increases with increase in the antioxidant concentration up to a certain level. The optimum concentration of antioxidant is usually in the range from 1 to 2 per cent, depending on the particular antioxidant.[1, 51] Further increase in the concentration of antioxidant results in no additional stabilization or a decrease in stability. In general, the antioxidants used to stabilize GR-S are similar to those employed for natural rubber.

Temperature. Unvulcanized GR-S stabilized by an antioxidant undergoes a slight gradual change in properties on aging at room temperature. For periods of one to two years GR-S shows little or no change, but after five years' aging appreciable changes have been noted.[1] Natural aging of vulcanizates results in moderate changes in physical properties. The extent of the changes observed are indicated in Table VI.

Table VI. Effect of Shelf Aging on Properties of a GR-S Tread Stock[37]

	Tensile Strength, P.S.I.			Elongation, %			Modulus at 100%, P.S.I.			Modulus at 300%, P.S.I.		
		Aged (% Unaged)			Aged (% Unaged)			Aged (% Unaged)			Aged (% Unaged)	
Cure, Min.	Unaged	415 Days	753 Days	Unaged	415 Days	753 Days	Unaged	415 Days	753 Days	Unaged	415 Days	753 Days
15	1100	126	146	1100	78	78	60	217	150	180	178	188
30	2700	92	88	860	85	78	120	150	125	420	157	167
45	2920	84	75	750	83	75	150	160	130	670	134	131
90	2960	84	81	640	84	81	160	169	138	840	130	129
180	2760	88	84	550	87	86	190	153	132	1090	120	116

The aging process proceeds more rapidly as the temperature increases. Data on the temperature coefficient for the deterioration of the physical properties of GR-S vulcanizates have been summarized by Schoch and Juve.[36] Values ranging from 2.0 to 3.1 per 10° C. have been found, depending on the physical property measured and the conditions employed. With the exception of a few values derived from tensile-strength measurements, an average value of 2.1 is in good agreement with most of the literature data relating to aging in air at 70 to 125° C. Shelton, Wherley, and Cox[42] have shown that the same reaction appears to be rate controlling for the oxygen absorption of GR-S stocks over the temperature range 50 to 110° C. Thus it appears probable that the rate of oxygen absorption at room temperature could be predicted in short-term tests at higher temperatures, provided data were available at several temperatures. The change in physical properties that accompanies the absorption of a given amount of oxygen varies with temperature.[42] Aging at higher temperatures produces a softer stock with lower modulus and higher elongation than is obtained by the absorption of the same amount of oxygen at a lower temperature. The predominant reaction at higher temperatures appears to be chain scission, while cross linking becomes of greater importance at lower temperatures.

Oxygen Pressure. The rate of oxidation of GR-S rises with increasing oxygen pressure.[10, 39, 44] In addition, the change in physical properties observed during accelerated aging is somewhat dependent on the oxygen pressure employed. Shelton and Winn[43] have shown that oxygen-bomb aging results in a greater drop in tensile strength and a smaller increase in modulus than air-oven aging at the same temperature. The effect of oxygen pressure on the change in modulus and tensile strength during aging at 100° C. is shown in Fig. 5. The data here seem to indicate that the rate of chain scission, as measured by fall in tensile strength, increases with increase in oxygen pressure. The rate of cross-linking, determined from the increase in modulus, is approximately independent of oxygen pressure above that in air. The temperature coefficient, determined from fall in tensile strength and elongation, increases with increasing oxygen pressure. Since the nature of the changes in the physical properties of GR-S during aging at elevated temperatures varies with oxygen pressure, the use of the oxygen-bomb aging test for the practical evaluation of the aging of GR-S would seem inadvisable.

Metal Impurities. Neal and Ottenhoff[31] have shown that soluble copper and manganese accelerate the deterioration of GR-S vulcanizates, although the effect is appreciably less than with natural rubber. GR-S vulcanizates are virtually unaffected at a soluble metal concentration ten- to twentyfold greater than can be tolerated in natural rubber. Albert, Gottschalk, and Smith[2] observed that soluble iron has a more adverse effect on the stability of GR-S than soluble copper or manganese. Iron was found to have a more detrimental effect on the aging of the GR-S polymer than on GR-S vulcanizates. GR-S-type polymers prepared at low temperatures (cold GR-S) by means of iron-containing polymerization recipes have aging properties somewhat inferior to those of standard GR-S, particularly in the unvulcanized state.[22]

According to Rao, Winn, and Shelton,[35] the coagulant used to isolate GR-S from its latex may, by determining the nature of the metallic soap present, be a factor in the resistance of the rubber to the absorption of oxygen. Aluminum and zinc soaps, arising from the use of solutions of aluminum and zinc salts as coagulants, have a favorable effect on the stability of the raw polymer in storage and are without effect on the oxidation of vulcanizates derived from it.

Carbon Black. Winn, Shelton, and Turnbull[52] have shown that the rate of oxidation of GR-S vulcanizates rises with increasing carbon black loading. The increase in rate of oxidation depends on the surface area of the carbon

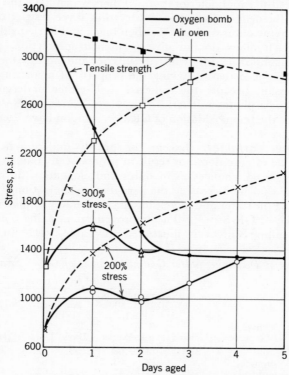

FIG. 5.   Effect of Oxygen Pressure on Aging[43] of GR-S at 100° C.

rather than on the weight of carbon in the vulcanizate. The relative activity per unit of surface area varies with the type of carbon black. On the basis of available data, furnace blacks are less active than channel blacks.

Light. Aging effects often attributed to light are now generally accepted as being due to the result of oxidation accelerated by light. There is no evidence to indicate that the basic mechanism of photooxidation differs from that of thermal oxidation. The effect of light on the oxidation rate of GR-S has been investigated by Mesrobian and Tobolsky.[30] The quantum yield for photooxidation of GR-S has been measured.[25] Crabtree and Kemp[13] have shown that the photooxidation rate of GR-S is less than that of natural rubber.

The investigators just named have shown that the type of filler employed in the compounding of elastomers has an effect on the photooxidation rate. Fillers having approximately the same refractive index as that of the polymer modify the reflectance characteristics to only a small degree. They increase the initial rate of deterioration both because of greater absorption of light by scattering and of greater permeability to air. The initial rate of oxidation may be decreased by fillers of high refractive index, such as zinc oxide and titanium oxide, which increase the total reflectance. As oxidation proceeds, however, the absorption of light may increase, as a result of discoloration, and ultimately lead to rapid deterioration. Fillers which have the property of absorbing all incident radiation in a very thin layer restrict oxidation largely to the extreme outer surface. Carbon black and ferric oxide belong to this class. Such fillers do not give permanent protection, probably because peroxidic substances diffuse into the interior and increase the thermal oxidation rate. Some materials having the property of increasing opacity may be unsuccessful, because the material itself either undergoes auto-oxidation or acts as a photosensitizer.

Some aspects of the photooxidation of GR-S in solution have been studied by Jones and Friedrich.[28]

Compounding Variables. Except for grossly undercured samples, increasing the time of cure does not result in pronounced changes in aging properties. In general, changes in modulus and elongation during aging are reduced somewhat by increasing the period of vulcanization.[37] Simultaneously reducing the sulfur and increasing the accelerator content improves the age resistance of GR-S.[31, 37] The improvement in aging obtained by decreasing the sulfur content is illustrated in Table VII. In general,

Table VII. Effect of Decreasing Sulfur Content on Aging of GR-S Tread Stocks[31]

				Aged (% of Unaged)					
				Oxygen Bomb, 70° C., 300 P.S.I.		Air Oven 121° C.		Air Bomb 127° C., 80 P.S.I.	
Sulfur Level	Accel-erator Level	Physical Property	Unaged	3 Weeks	5 Weeks	3 Days	6 Days	8 Hr.	24 Hr.
2.0	1.2	Tensile	3700	78	68	59	52	69	46
1.5	1.6	strength,	3600	72	63	73	63	74	54
1.0	3.0	p.s.i.	3300	67	57	72	62	84	63
2.0	1.2	Elongation, %	525	68	61	26	20	51	21
1.5	1.6		520	79	72	39	29	69	30
1.0	3.0		500	87	79	46	39	74	47
2.0	1.2	Modulus at	225			490	840	240	700
1.5	1.6	100%, p.s.i.	225			430	720	195	530
1.0	3.0		200			290	430	160	320
2.0	1.2	Modulus at	1450	240	250				
1.5	1.6	300%, p.s.i.	1550	205	215				
1.0	3.0		1425	145	155				

Base Stock: GR-S 100, Zinc Oxide 5, MPC Black 50, Stearic Acid 1, Blend of Mercaptobenzothiazole and Diphenylguanidine Various, Sulfur Various
Optimum Cure at 274° F.: 45 Min. with 3 Parts Accelerator and 1 Part Sulfur; 60 Min. in Other Cases

sulfur-accelerator combinations which have flat curing characteristics appear to give the best aging properties to the vulcanizate.[37] Vulcanizates prepared without free sulfur, by means of curing agents such as the thiuram disulfides, give superior aging properties.[12, 53] Nonsulfur vulcanizates have been observed to have better aging properties than the conventional vulcanizates cured with sulfur and accelerator.[45-6]

DETERIORATION OF GR-S BY OZONE

The multiple cracks which appear when stretched natural or synthetic rubber is exposed out of doors are now generally accepted to be due, not to the effects of light, but to atmospheric ozone.[13, 32] Ozone is the main factor in the deterioration of elastomers during weathering. Although there is no evidence that light has any direct effect on the reaction between ozone and double bonds in the polymer, indirect effects may be noted. Light causes a rapid oxidation of a very thin, outer layer, with a corresponding change in physical properties of the outer layer of the sample. This may modify to some extent the character of the changes observed in the surface of the elastomer as weathering proceeds.[13, 24]

Though the effect of ozone on GR-S is similar to that observed for natural rubber in most respects, some differences have been noted. During the early stages of deterioration GR-S is damaged to a smaller degree than natural rubber, but the reverse is true in the later stages. In general, GR-S is more susceptible to ozone damage than natural rubber. However, the degree of protection afforded GR-S by waxes is greater than what they confer on natural rubber. Hence properly compounded GR-S has greater weathering resistance than natural rubber.[13-4, 34] It should be pointed out that, under dynamic flexing, wax offers no protection to either GR-S or natural rubber.

Fisher and Scott[21] have compared the change in properties of GR-S during weathering with the deterioration resulting from natural and accelerated aging. Weathering differs from other forms of aging in that, for a given decrease in tensile strength, the increase in modulus is much less. Weathering is predominantly a deterioration of the surface.

REFERENCES

1. Albert, H. E., *Ind. Eng. Chem.*, **40,** 1746–50 (1948).
2. Albert, H. E., Gottschalk, G. W., and Smith, G. E. P., *Ind. Eng. Chem.*, **40,** 482–7 (1948).
3. Bach, A., *Compt. rend.*, **124,** 951 (1897).
4. Baxter, S., Morgan, W. McG., and Roebuck, D. S. P., *Ind. Eng. Chem.*, **43,** 446–52 (1951).
5. Bolland, J. L., and Gee, G., *Trans. Faraday Soc.*, **42,** 236–43 (1946).
6. Bolland, J. L., and Hughes, J., *J. Chem. Soc.*, **1949,** 492–7.
7. Bolland, J. L., and Koch, H. P., *J. Chem. Soc.*, **1945,** 445–7.
8. Bolland, J. L., and ten Have, P., *Trans. Faraday Soc.*, **43,** 201–10 (1947).
9. Bolland, J. L., and ten Have, P., *Faraday Soc. Discussions*, **2,** 255 (1947).
10. Carpenter, A. S., *Ind. Eng. Chem.*, **39,** 187–94 (1947).
11. Cole, J. O., and Field, J. E., *Ind. Eng. Chem.*, **39,** 174–9 (1947).
12. Cole, J. O., and Parks, C. R., Private Communication to O.R.R., Apr. 9, 1947.
13. Crabtree, J., and Kemp, A. R., *Ind. Eng. Chem.*, **38,** 278–96 (1946).
14. Cuthbertson, G. R., and Dunnom, D. D., *Ind. Eng. Chem.*, **44,** 834–7 (1952).
15. Dinsmore, R. P., and Fielding, J. H., *India Rubber World*, **119,** 457–61, 506 (1949).

16. Engler, C., and Wild, W., *Ber.*, **30,** 1669–81 (1897).
17. Farmer, E. H., *Trans. Faraday Soc.*, **42,** 228–36 (1946).
18. Farmer, E. H., Bloomfield, G. F., Sundralingam, A., and Sutton, D. A., *Trans. Faraday Soc.*, **38,** 348–56 (1942).
19. Farmer, E. H., and Sundralingam, A., *J. Chem. Soc.*, **1942,** 121–39.
20. Farmer, E. H., and Sutton, D. A., *J. Chem. Soc.*, **1942,** 139–48; **1943,** 119–22.
21. Fisher, D. G., and Scott, J. R., *J. Rubber Research*, **16,** 44–54 (1947).
22. Glazer, E. J., Parks, C. R., Cole, J. O., and D'Ianni, J. D., *Ind. Eng. Chem.*, **41,** 2270–2 (1949).
23. Harris, T. H., and Stiehler, R. D., *India Rubber World*, **118,** 365–6, 371 (1948).
24. Harrison, H. C., *Trans. Instn. Rubber Ind.*, **21,** 93—101 (1945).
25. Hart, E. J., and Matheson, M. S., *J. Am. Chem. Soc.*, **70,** 784–91 (1948).
26. Hendricks, J. G., *Rubber Age N.Y.*, **54,** 521–5 (1944).
27. Hunter, J. S., *India-Rubber J.*, **107,** 429–33 (1944).
28. Jones, G. D., and Friedrich, R. E., *Ind. Eng. Chem.*, **43,** 1600–4 (1951).
29. Juve, R. D., *India Rubber World*, **115,** 657–8 (1947).
30. Mesrobian, R. E., and Tobolsky, A. V., *J. Polymer Sci.*, **2,** 463–87 (1947).
31. Neal, A. M., and Ottenhoff, P., *Ind. Eng. Chem.*, **36,** 352–6 (1944).
32. Newton, R. G., *Trans. Instn. Rubber Ind.*, **21,** 113–8 (1945).
33. Parks, C. R., Cole, J. O., and D'Ianni, J. D., *Ind. Eng. Chem.*, **42,** 2553–7 (1950).
34. Popp, G. E., and Harbison, L., *Ind. Eng. Chem.*, **44,** 837–40 (1952).
35. Rao, N. V. C., Winn, H., and Shelton, J. R., *Ind. Eng. Chem.*, **44,** 576–80 (1952).
36. Schoch, M. G., and Juve, A. E., Symposium on Aging of Rubbers, Spec. Tech. Pub. No. 89, Am. Soc. Testing Materials, Philadelphia, 1949, p. 60.
37. Scott, J. R., *Trans. Instn. Rubber Ind.*, **21,** 78–93 (1945).
38. Scott, J. R., *J. Rubber Research*, 18, 117–30 (1949).
39. Shelton, J. R., and Cox, W. L., *Ind. Eng. Chem.*, **45,** 392–6, 397–401 (1953).
40. Shelton, J. R., and Cox, W. L., *Rubber Age N.Y.*, **73,** 215 (1953).
41. Shelton, J. R., and Winn, H., *Ind. Eng. Chem.*, **36,** 728–30 (1944).
42. Shelton, J. R., Wherley, F. J., and Cox, W. L., *Ind. Eng. Chem.*, **45,** 2080–6 (1953).
43. Shelton, J. R., and Winn, H., *Ind. Eng. Chem.*, **39,** 1133–6 (1947).
44. Shelton, J. R., Symposium on Aging of Rubbers, Spec. Tech. Pub. No. 89, Am. Soc. Testing Materials, Philadelphia, 1949, p. 21.
45. Sturgis, B. M., Baum, A. A., and Vincent, J. R., *Ind. Eng. Chem.*, **36,** 348–51 (1944).
46. Sturgis, B. M., Baum, A. A., and Trepagnier, J. H., *Ind. Eng. Chem.*, **39,** 64–8 (1947).
47. Taylor, H. S., and Tobolsky, A. V., *J. Am. Chem. Soc.*, **67,** 2063–7 (1945).
48. Tobolsky, A. V., and Andrews, R. D., *J. Chem. Phys.*, **13,** 12–23 (1945).
49. Warner, W. C., and Shelton, J. R., *Ind. Eng. Chem.*, **43,** 1160–4 (1951).
50. Winn, H., and Shelton, J. R., *Ind. Eng. Chem.*, **37,** 67–70 (1945).
51. Winn, H., and Shelton, J. R., *Ind. Eng. Chem.*, **40,** 2081–5 (1948).
52. Winn, H., Shelton, J. R., and Turnbull, D., *Ind. Eng. Chem.*, **38,** 1052–6 (1946).
53. Wolf, G. M., and de Hilster, C. C., *India Rubber World*, **120,** 196–7 (1949).

CHAPTER 14

USE OF GR-S IN RUBBER MANUFACTURING

R. A. Crawford

B. F. Goodrich Company

Present-day American-made general-purpose rubbers, including GR-S and its postwar variations, seem to have won a definite and permanent place in rubber manufacturing.   In 1948, 243,400 tons of GR-S was used in the United States in transportation products and 101,913 tons in nontransportation products, making a total of 345,313 tons.   During the same period 627,332 tons of natural rubber was used in these items.   In this period the price of GR-S was lower than that of natural rubber.   But in 1949 GR-S was higher priced than competitive grades of natural rubber.   Even no. 1 ribbed smoked sheet was cheaper[6] than GR-S in the last 8 months of 1949. In spite of this, the ratio of GR-S to natural-rubber consumption remained almost constant.   And when, during 1951, the price of natural rubber was at a much higher level than that of GR-S, the ratio in which GR-S was consumed went much higher.   Data on the consumption of natural rubber and GR-S in the United States are given in Table I.

Table I.   Consumption of Natural Rubber and GR-S in the United States

Year	1947	1948	1949	1950	1951	1952
Transportation Uses						
Natural, tons	442,742	443,006	382,591	477,466	308,573	303,208
GR-S, tons	319,817	243,400	225,047	275,141	434,595	470,247
GR-S as percentage of total	40.6	35.5	37.0	36.5	58.5	60.8
Nontransportation Uses						
Natural, tons	119,919	184,326	191,931	242,802	145,442	150,638
GR-S, tons	128,772	101,913	96,090	141,089	191,849	196,173
GR-S as percentage of total	51.8	35.6	33.4	36.7	56.9	53.7
Grand Total						
Natural, tons	562,661	627,332	574,522	720,268	454,015	453,846
GR-S, tons	448,589	345,313	321,137	416,230	626,444	666,420
GR-S as percentage of grand total	44.4	35.5	35.8	36.6	58.0	59.6
Price per Lb. (Cents)						
Natural*	20.97	22.01	17.56	41.10	59.07	38.57
GR-S	18.5	18.5	18.5	19	25	23.5

* Ribbed sheet at New York.

The use of synthetic rubber in nontransportation items by type and by product division was sampled for the last 2 months of 1947 and 1948 as percentages of the total rubber used in such items. The figures relating to GR-S and natural rubber are reproduced in Table II by courtesy of the Rubber Manufacturers Association, Inc. The results of a later and more detailed survey made in the third quarter of 1951 are reviewed in Chapter 1. They show that the proportion of GR-S to natural rubber used in almost all classes of rubber goods had risen above the proportions given in Table II.

Table II. Consumption of Natural Rubber and GR-S as Percentage of Total Consumption in Various Classes of Nontransportation Products

| | GR-S | | Natural Rubber | |
Division	1947	1948	1947	1948
Coated materials	48.2	42.7	42.6	48.8
Sundries	8.8	5.4	90.2	92.2
Belt and hose	19.8	18.7	61.3	68.4
Hard rubber	69.3	68.1	30.3	31.4
Heel and sole	32.7	40.5	63.3	55.9
Molded and extruded	37.0	32.5	55.9	59.6
Footwear	12.0	10.7	88.0	89.2
Wire and cable	38.0	35.2	20.4	18.0
Sponge rubber	42.7	38.4	53.6	59.2
Sporting goods	82.2	72.1	15.6	25.0
Flooring	92.0	89.7	9.8	10.3
Total nontransportation	35.1	34.2	50.1	51.4

Percentage of nontransportation volume represented by sample companies for 1948:

GR-S   69%
Natural rubber   57.5%

GR-S and its postwar variations have their own characteristics, which differ in many respects from those of natural rubber, and it is these differences that make the general-purpose synthetic rubber more desirable than natural rubber for a large number of end products.

GR-S excels most grades of natural rubber in uniformity and cleanness. Therefore it is preferred in manufacturing operations where it is as cheap to use as natural rubber and yields end products of equal quality. It is less thermoplastic than natural rubber and therefore tends to resist collapse better in hollow extrusions. Other properties in which GR-S is superior to natural rubber include aging at elevated temperatures; resistance to wear in many end products, especially when cold rubber is used in their manufacture; and resistance to premature vulcanization during processing operations. It is inferior to natural rubber in pure-gum strength, tensile strength and elongation when hot, tear resistance, hysteresis, cut growth in treads (but not in soling), and tackiness in the uncured state. Sometimes characteristics that render a material unfit for certain uses make it desirable for others, and this has been found to be true of GR-S. Thus its relatively poor resilience makes it unfit for thread but is an advantage in some other products, as will appear later. Similarly, its poor tackiness makes it more difficult to use in plied constructions but is an advantage where nontacky surfaces are needed.

The choice of the material used in any given article of manufacture depends on the particular properties required in the article, on the manufacturing operations involved, and on economic factors. In 1948, when the use of GR-S in certain transportation products was required by Government regulation, transportation items alone actually consumed more than the minimum specified by law; all other uses of GR-S were voluntary and were made because of either processing or quality considerations. It is hardly feasible to discuss in detail the many kinds of rubber goods manufactured from GR-S. But a review of examples chosen from the major classes of goods is offered, in order to illustrate the ways in which the characteristic properties of GR-S serve to advantage. This review has been prepared with the cooperation of specialists of the B. F. Goodrich Company technical staff.

Tires. Originally GR-S was used in tires only of necessity and with great difficulty. Compounding, processing, building techniques and tire constructions in common use for natural rubber resulted in expensive, unsatisfactory handling and in tires that would wear at most about 80 per cent as well as natural-rubber tires.

Continuous and large-scale research and the development of polymerization recipes and other aspects of polymer manufacture resulted in gradual improvement of the polymer. Other long studies in compounding, processing, tire building, and curing techniques have taught the tire industry how and where to use GR-S and have solved the various manufacturing problems. So that today many manufacturers produce passenger tires with GR-S treads that equal natural-rubber tires or even outlast them by as much as 10 per cent in treadwear. The use of "cold rubber" and new blacks has brought still further improvement, and synthetic-rubber treads can now be made that outwear natural-rubber treads by 15 to 25 per cent.

Because of improvements in the manufacture and use of GR-S, only recent treadwear comparisons are of any value. Published reports are rather meager and do not always agree. Thus Dinsmore and Fielding (January 1949) rated GR-S 95 per cent as good as rubber,[4] but Fielding[5] (August 1949) rated it only 90 per cent as good. In an analysis of 17 tests on the Government test fleet participated in by 11 companies, where comparisons of GR-S and natural rubber are possible, a composite rating shows GR-S the equal of natural rubber. Other Government tests indicate that GR-S is somewhat better than natural rubber in summer and also wears better on rear wheels. Other unpublished data confirm these Government tests, and it is considered that Fielding's rating for GR-S is too low and that GR-S must be rated equal to natural rubber.

An analysis made by the Bureau of Standards[3, 5] of a group of road tests led to the conclusion that the relative rate of treadwear of Hevea and GR-S is influenced by the service conditions under which the rubber is used and by the type of black in the rubber. Thus, for example, the influence of temperature is such that, in road tests made in Texas, treads of natural-rubber compounded with channel black wore faster in the summer than in the winter, whereas the reverse was true for treads made of GR-S compounded with furnace black. The analysis led to the following conclusion. "Treads made from GR-S polymerized at 41° F. wear about 20 per cent longer under all conditions than those made from GR-S polymerized at 122° F. Hevea treads may wear better than treads made from GR-S polymerized at 41° F. or poorer than treads made from GR-S polymerized at 122° F. depending on the service conditions."

"Cold rubber" is rated variably by various authors[1-5] but an average of all reports and unpublished data would be about 15 per cent improvement in treadwear as compared to GR-S or natural rubber.

HAF black produces better treadwear than EPC black in all three rubbers, but is relatively somewhat less effective in natural rubber than in the other two. Here again published information is not voluminous, and agreement is not perfect.

The following table gives the approximate relative values of the various rubbers using EPC and HAF blacks:

Polymer	Treadwear Rating
Natural rubber (EPC)	100
Natural rubber (HAF)	105–110
GR-S (EPC)	100
GR-S (HAF)	110–115
Cold rubber (EPC)	115
Cold rubber (HAF)	115–125

The greatest usage of GR-S by the tire industry is in passenger tires and recapping material, but it is also used to a high percentage in farm service tires.

Passenger-Car Tires. GR-S is used in both treads and carcasses. In treads it is usually the sole source of rubber hydrocarbon, but in carcasses it is customary to use it in mixtures with natural rubber. Handling properties are satisfactory—in most cases equal to those of corresponding natural-rubber compounds. In quality GR-S compounds possess both advantages and disadvantages compared to natural rubber, but, where used, the advantages are the most important, as shown below.

Treads. As mentioned above, GR-S passenger-car tire treads are equal or superior to natural-rubber treads for treadwear, and "cold rubber" is superior to standard GR-S in this property. GR-S is superior to natural rubber for skid resistance on wet or dry pavement. On wet ice it is somewhat inferior, and its relative skid resistance decreases as temperature decreases. However, in most instances the difference is too slight to be noticed by the consumer. Heat buildup is considerably higher for GR-S than for natural rubber, but this is unimportant in passenger-car tires because service temperatures are not too high for the strength of the material. Improved constructions have minimized the well-known crack-growth tendency of GR-S tread compounds, and no tread-cracking troubles are encountered in today's passenger-car tires.

Carcasses. GR-S, when used alone, has never been completely satisfactory in carcass compounds because they require cement coatings for adequate building properties and because their strength decreases with increased temperature and becomes too low to stand stresses imposed by service conditions. GR-S carcasses tend to become brittle and bruise easily. Therefore it is customary to mix the GR-S with natural rubber to obtain a more flexible carcass. The introduction of some natural rubber also improves building tack. This type of compound has proved extremely satisfactory; used in the proper tire construction, it produces a better tire than tires that were made entirely from natural rubber before World War II.

Farm Service Tires. Farm service tires in which 75 per cent or more of the rubber is GR-S have performed in an extraordinarily satisfactory manner, perhaps because the demands made on them by service are not so great as for those made on other large tires, used for road service. They operate at low speed under light load.

Industrial Tires. Where low rolling resistance is not a requirement, GR-S can render satisfactory service in some types of industrial tires, but it has no inherent advantages for this use.

Truck Tires. Truck tires frequently operate at high temperature under heavy loads, and GR-S with its relatively high hysteresis and power loss and its relatively low strength at high temperatures is not so suitable for this service as natural rubber.[4a]

The development of the use of GR-S in tires has resulted not only from improved quality of the polymer and from improved and specialized compounding, but also from processing refinements, and long studies of tire constructions and, above all, the right type of cure. The improved curing technique developed for GR-S has been a large factor in making the tire the satisfactory product it is today. Further modifications involving lower hysteresis, greater freeze resistance, and better elevated-temperature properties may be expected to advance still further the use of GR-S by the tire industry.

Shoes. GR-S seems to have won at least a limited place in rubber footwear, even when it costs appreciably more than natural rubber and in spite of the somewhat costlier operations which its use entails. It maintains embossed design, eliminates fabric impressions, improves handling, and reduces tendency to blow when used in conjunction with natural rubber. In black stocks the rubber hydrocarbon of which is entirely GR-S, it produces better abrasion resistance than natural rubber, has less tendency to crack (under both static and dynamic conditions), and is at least as good in oil resistance. There is almost complete elimination of scorching with black GR-S stocks. In combinations with natural rubber, it does not increase scorching. It is cleaner and is more uniform in cure than natural rubber.

Mechanical Goods. GR-S is used in a considerable variety of mechanical goods, including belting, hose, molded goods, lathe cut goods, platens, roll covering, unvulcanized gum, sheet rubber, packing, and flooring. In these articles it possesses advantages over natural rubber for processing, for quality of the end product, or because it has gained better general customer acceptance. In its application to the manufacture of mechanical goods GR-S is mostly used in black compounds, since the advantages indicated below for the synthetic product are not present in colored articles except in isolated cases, because of the fact that there is a much greater difference between black and colored compounds with GR-S than with natural rubber. The advantages of GR-S in typical mechanical goods products are shown in the following discussion.

In transmission belting GR-S possesses no advantage over natural rubber, but its superior wearing qualities and heat-resistance have dictated its continued use in conveyor belting.

For hose uses GR-S possesses both processing and quality advantages over natural rubber. It has less tendency to scorch during processing operations and thus less tendency than natural rubber to produce defective stock. It is less thermoplastic than rubber and hence extruded hose tubes of GR-S hold their shape better than natural-rubber tubing. Finished GR-S hose displays superior heat and aging resistance.

In various types of molded goods the resistance of GR-S stocks to premature cure effects, scorching, and back rinding means fewer defective molded parts and hence lower costs. Advantages in quality have also been found in a number of instances. This is illustrated by the following examples:

1. *Refrigerator Parts.* GR-S will "soak up" more pigment than natural rubber and hence will enable harder cured stock to be produced without resorting to a hard-rubber cure. These hard stocks perform better in lids, collars, etc.

2. *Steam Valve Disks.* GR-S is more steam-resistant than natural rubber.

3. *Wringer Rolls.* GR-S discolors less in high-temperature cures than natural rubber. The finished roll has better service life; it is less affected by mild grease conditions and is superior in resistance to detergents and to abrasion.

In lathe cut goods in general, GR-S compounds are used because of superior cleanness and batch-to-batch uniformity, and because of better extruded surface smoothness and dimensional stability than is displayed by natural-rubber compounds. In food gaskets, GR-S is superior to most natural rubbers for odor and taste. In business-machine rolls, GR-S has better abrasion resistance than natural rubber. GR-S typewriter platens are superior in cleanness and general all-round performance in service. Unvulcanized gum users, in many cases, have preferred to continue use of GR-S because of a long record of satisfactory service.

In sheet rubber and packing the lack of tack displayed by GR-S is an advantage. Other advantages as compared to natural rubber are superior compression set and heat resistance, less crystallization tendency, and greater ease in meeting customers' specifications. GR-S is used to some extent for roll coverings and is said to have specific advantages in this use.

During the period of rubber shortage flooring manufacturers were forced to turn to GR-S. Now most of the rubber floor tiling sold is based on this material. Its ability to absorb high ratios of filler and thus yield firm products, its uniformity, its resistance to aging, discoloration, indentation, abrasion, and cleaning materials combine to make it superior to natural rubber.

Soles and Heels. Because of its excellent service qualities, present-day rubber shoe soling is based almost universally on GR-S. Owing partly to the fact that GR-S is less resilient than natural rubber, and partly to the fact that it is stiffened more efficiently with the high-styrene resins than natural rubber, stiffer soling with a more "leathery feel" results than can be made from natural rubber. It is also interesting to note that, contrary to general tire experience, GR-S withstands flexing longer than natural rubber in soling.

There is no particular advantage in the use of GR-S in heel compounds; in fact, the higher resilience of natural rubber makes it somewhat the more desirable material, and little GR-S is used.

Coated Fabrics. The lack of tack which brought difficulties in the use of GR-S in the fabrication of plied-up constructions is a distinct and considerable advantage in the production of coated fabrics and unsupported sheeting. Little if any dust is required to keep the uncured surfaces from sticking together, and it is possible to secure a cleaner cured surface than with natural rubber, which requires a heavy coating of a lubricating material.

The same lack of tack also allows more latitude in the compounding of unsupported opaque sheeting. A more liberal use of plasticizer can be tolerated, and strong calendered sheets with no fabric insert or backing are made and used extensively for hospital and nursery sheeting. A further economic advantage, in the case of spreading materials, is found in the fact that GR-S can be used in more concentrated solution than natural rubber.

Advantages of GR-S over natural rubber in quality are: (a) improved aging (GR-S is over twice as durable as natural rubber), (b) lower gas permeability, (c) equal or better waterproofness, (d) nontacky surfaces. These advantages have established a permanent place for GR-S in coated fabrics and sheeting.

Cements and Adhesives. Although GR-S is lacking in tack and strength compared to natural rubber, proper compounding makes it a fair substitute for rubber adhesives, and, where aging properties are of paramount importance, it is an established material in the cement field. Thus for can sealing cements, coatings, and sealants, GR-S is a permanent fixture.

Sundries. In most sundries items, GR-S has no inherent advantages over natural rubber and so is not used.

Sponge. GR-S can be produced in almost any desired consistency, and advantage of this fact has been taken to produce a rubber that does not require costly premastication for use in blown sponge manufacture. GR-S sponge compounds also require less critical curing conditions than natural-rubber stocks and yield cured sponge with somewhat superior resistance to aging and sunlight. If GR-S is used in a solid rubber covering for sponge, it can be made to approach the sunlight cracking resistance of Neoprene.

Extruded Goods. GR-S will undoubtedly continue to be used in extruded goods because of several inherent advantages. It is less subject to overcuring than natural rubber, it is more heat-resistant, and, with equivalent compounding, especially with the addition of waxes and other ozone resisters, it is more resistant to sun cracking than natural rubber. It is less prone to change in shape during vulcanization than natural rubber. This is an outstanding advantage where complicated extruded cross sections are involved.

The relatively high resistance of GR-S compounds to softening by mastication and to scorching permits reworking of extruded or calender scrap with a minimum production of stock that cannot be used in end products. GR-S is often used in combination with nitrile rubbers to modify the properties of the latter. For this purpose, lack of compatibility precludes the use of natural rubber.

Examples of GR-S Compounds. Some typical GR-S recipes for a variety of end products are shown in Table III.

Table III.   Typical GR-S Compounds.

Tan Soling—High Grade

GR-S 50	100
Hard extending resin	10
High-styrene resin	30
Zinc oxide	5
Calcium silicate (precipitated)	60
Red iron oxide	3
MPC black	0.1
Age resister	1
Benzothiazolyl disulfide	2
Diphenylguanidine	1
Sulfur	2.5

Tan Soling—Medium Grade

GR-S 50	100
Hard extending resin	30
Zinc oxide	5
Calcium silicate (precipitated)	50
Clay	75
Red iron oxide	3
MPC black	0.1
Age resister	1
Benzothiazolyl disulfide	2
Diphenylguanidine	1
Sulfur	4

Passenger Tread

GR-S 10	100
EPC black	50
Zinc oxide	5
Stearic acid	1
Softener	10
Age resister	1
Activated mercaptobenzothiazole	1.2
Sulfur	2

Table III—continued

Chemically Blown Sponge

GR-S	100
Clay	40
Coarse thermal black	60
SRF black	10
Zinc oxide	3
Petrolatum	10
Cumar resin	10
Light mineral oil	30
Stearic acid	10
Sodium bicarbonate	20
Benzothiazolyl disulfide	1.5
Tetramethylthiuram monosulfide	0.25
Sulfur	3

Tiling

GR-S (nondiscoloring)	100
White clay	200
Asbestine	200
Zinc oxide	10
Titanium dioxide	10
Nonstaining resinous extender	15
Stearic acid	3
Paraffin	1
Age resister (nonstaining)	1
Benzothiazolyl disulfide	2.5
Diphenylguanidine	0.5
Sulfur	9

Platen

GR-S	100
Zinc oxide	5
Clay	50
Whiting	150
SRF black	100
Softeners	20
Sulfur	20
Benzothiazolyl disulfide	2.5
Diphenylguanidine	1

Wringer Roll—White

GR-S	100
Zinc oxide	5
Clay	50
Titanium dioxide	40
Whiting	50
Softeners or resins	10
Sulfur	1.5
Benzothiazolyl disulfide	1.5

Table III—continued

Belt Coat—Skim

GR-S	100
Zinc oxide	5
EPC black	33.5
Rosin oil	7.5
Other softeners	15
Age resister	1
Sulfur	2
Benzothiazolyl disulfide	2

Water-Hose Tube

GR-S	100
Zinc oxide	5
SRF black	50
EPC black	35
Rosin oil	10
Other softeners	6
Age resister	2
Sulfur	2
Cyclohexyl benzothiazolyl sulfenamide	1.5

Conclusion. In view of the fact that GR-S is a factory-made product differing widely in composition from natural rubber, it is astonishing, not that it does not duplicate the latter, but rather that it is amazingly like the natural product in many of its major properties. At our present state of knowledge, GR-S may still be regarded as deficient in certain properties we have come to associate with rubber, but it is these very deficiencies that make it more advantageous, as shown above, in a host of miscellaneous uses; and it appears to be here to stay, even when it carries a moderate premium in cost over natural rubber. As research and development continue, it is to be expected that more and better polymers will result and that general-purpose synthetic rubber will assume an increasingly important place in rubber-goods manufacture.

REFERENCES

1. Dinsmore, R. P., and Fielding, J. H., *India Rubber World*, **119,** 457–61, 506 (1949).
2. Fielding, J. H., *Ind. Eng. Chem.*, **41,** 1560–3 (1949).
3. Mandel, J., Steel, M. N., and Stiehler, R. D., *Ind. Eng. Chem.*, **43,** 2901–5 (1951).
4. Sjothun, I. J., and Cole, O. D., *Ind. Eng. Chem.*, **41,** 1564–7 (1949).
4a. Sjothun, I. J., and Greer, P. S., *Rubber Age N.Y.*, **74,** 77–83 (1953).
5. Stiehler, R. D., Steel, M. N., and Mandel, J., *Trans. Instn. Rubber Ind.*, **27,** 298–311 (1951).
6. Synthetic Rubber—Recommendations of the President, Jan. 1950, p. 40.

CHAPTER 15

SYNTHETIC RUBBER
IN WIRES AND CABLES

John T. Blake
Simplex Wire & Cable Company

Synthetic rubbers are today used extensively in the construction of insulated wires and cables. At the present time approximately 75 per cent of the rubber used in the wire and cable industry is synthetic. Before World War II the employment of synthetic rubber by the industry was represented only by a minor use of Neoprene as a jacket for portable cords and cables where resistance to oil was important. The large increase in the use of Neoprene is due not so much to newly developed properties as to the substantial reduction in its price brought about by large-scale production. The large present use of GR-S and Butyl rubber in wire insulation has been on the basis of their technical virtues. Fluctuations in the price of natural rubber have influenced to a slight extent the amounts of synthetic rubber used, but even when the synthetics were at a substantial premium over natural rubber their consumption was not reduced greatly.

REQUIREMENTS OF AN INSULATION

In discussing the performance of synthetic rubber in wires and cables, it is necessary to consider the qualities that are essential to the serviceability of the product. As a practical matter, the purchaser desires a wire or cable that can be installed without damage and will withstand the various hazards of its use. Suitable electrical and physical properties are necessary, and these should be stable over a long period of time under the particular conditions of service.

The usual article consists of a copper conductor (although sometimes aluminum, silver, or bronze is used), an insulation to provide the electrical properties necessary for the service, and one or more coverings for mechanical protection. Frequently a number of insulated conductors are protected by a single rubber jacket or other covering. The assembly should be sufficiently flexible so as not to be damaged by the bending necessary for installation or for use in service.

Electrical Properties. Four electrical properties are of importance in a rubber insulation.

1. *Dielectric strength* should be high enough to ensure that electric breakdown due to voltage will not occur either during normal use or under the surges that frequently develop on power systems. The dielectric strength

553

may be measured with direct current, alternating current (usually 60 cycles), or under impulse.

2. The *dielectric resistance* of an insulation should be sufficient to prevent unduly great leakage losses. Actually, many specifications demand as a measure of quality resistance values far in excess of service needs.

3. The *power factor*, or loss angle, should be low in power cables for alternating currents, so that losses in the insulation will not be excessive.

4. The *dielectric constant* of an insulation is the principal factor in determining the attenuation of signals in communication cables. The lower the dielectric constant, the greater the distance over which signals may be transmitted satisfactorily.

For power cables, the dielectric constant of the rubber insulation may be as high as 5 and the power factor as high as 3.5 per cent. The dielectric strength should be high. The a.c. dielectric strength should be approximately 500 volts per mil; the d.c. dielectric strength about 1500 volts per mil, and the impulse strength 1200 volts per mil. In a communication cable the rubber insulation should have a dielectric constant of not over 3.5 and a power factor of less than 1.5 per cent. The dielectric strength is relatively unimportant.

Physical Properties. The physical properties necessary in rubber insulations vary greatly, depending on the particular service conditions. Tensile strength and elongation are commonly used criteria of quality, although as a measure of practical serviceability tensile strength in itself is of little significance. Elongation values of 100 per cent would be adequate for practically all uses, but high values are demanded ordinarily as a measure of quality.

Electrical insulation should be tough enough to withstand installation and service hazards. It is difficult to define the property of toughness. Modulus is a rough measure of it. Also, resistance to tear, to abrasion, and to compression are useful in evaluating it.

Age Resistance. There are a number of other hazards to which rubber compounds used in wires and cables may be exposed. Efficient operation of a power cable demands high conductor-current loading, which results in the evolution of heat, so that the insulation may be exposed to high temperatures for long periods of time. In this connection rubber insulations must have excellent resistance to aging at high temperatures. Further, many rubber insulations and jackets may undergo long exposure to sunlight, and they should not deteriorate physically under these conditions. Since cables are frequently exposed to various chemicals, they should resist the action of such materials.

Ozone Resistance. High-voltage cables generate ozone at the surface of the insulation. Rubbers suitable for such cables should not be cracked by ozone when under physical stress, and it is almost impossible to be certain that this condition of exposure does not exist. Even low-voltage insulation near any other conductor carrying high voltage is subjected to this hazard.

Water Resistance. Many cables are laid in water—in rivers, harbors, and the ocean, for various submarine services. Cables laid in the ground are also usually exposed to water at some portion of their length. Years ago it was essential for these reasons to cover the rubber insulation with an impervious lead sheath, in order to prevent deterioration of its electrical properties through the absorption of water. Today the lead sheaths have been largely eliminated, and it is customary now to expose rubber insulations directly to water or to bury them directly in the ground. In this connection they must have low water absorption.

The essential requirement for rubber insulation continuously exposed to water is that its electrical properties shall not deteriorate materially through absorption of water. In any type of rubber insulation, the smaller the amount of absorbed water, the less will the electrical properties be affected. In different rubber compounds, however, the extent of the deterioration caused by the absorption of a given amount of water may vary widely. This deterioration can be determined with the best reproducibility by capacitance measurements. Many specifications now set maximum values both for actual water absorbed and for change in capacitance during an accelerated water-exposure test.

Requirements for Rubber Jackets. Rubber jackets are essential for cables subject to mechanical hazards either during installation or in service. The jackets need many of the qualities of a tire tread, such as tear, abrasion, and bruise resistance. Such cables may be exposed to the weather for long periods of time, and hence sunlight resistance in the jacket is important. Many cables come in contact with oil and chemicals. For such services the jackets should maintain their physical properties under these conditions.

Low-temperature flexibility is of importance. Rubber compounds tend to stiffen at low temperatures. In a portable cord or cable such stiffening should be minimized. As the temperature is decreased further, rubber insulations and jackets become brittle. For many services the brittle point should be as low as possible.

Great Importance of Age Resistance. Too much emphasis cannot be placed on the resistance to aging of both insulation and jackets. A tire usually wears out from abrasion long before it has deteriorated excessively through heat and oxidation. On the other hand, rubber insulation and rubber jackets are expected to be unusually long-lived. The wire industry is probably more conscious than any other branch of rubber manufacturing of the need for heat-resistant and long-lived rubber compounds. Accelerated aging tests on many of the rubber compounds used for insulation predict a useful life of thirty to fifty years. The need for good age resistance in rubber compounds to be used on wire and cable has dominated much of the thinking in the design of them.

GR-S FOR ELECTRICAL PURPOSES

Influence of Mode of Preparation on Water Absorption. During the latter part of World War II most of the rubber insulation was made from the general-purpose synthetic rubber, GR-S, because of the scarcity of natural rubber. GR-S was developed primarily for use in tires, since transportation consumes by far the largest part of the rubber used in industry. Standard GR-S, suitable for such use, was not particularly suitable for rubber insulation. The salt-acid method of coagulation with sodium chloride and sulfuric acid is commonly used in the preparation of GR-S. The sulfuric acid converts the sodium soaps used in the polymerization to sodium sulfate, which, together with some of the sodium chloride, is retained in the rubber. The presence of these electrolytes in GR-S results in low dielectric resistance. Exposure to water for only a short period of time reduces the latter even further. Some improvement in the dielectric resistance of GR-S compounds has been effected by reducing the water-soluble salts to a minimum, either by careful washing of the finished rubber, by precipitation of the polymer as a fine granular material, or, as at the Naugatuck polymer plant, by the use of special dewatering extruders. Some of the original GR-S polymers had a mechanical water absorption as high as 120 mg. per sq. in. after exposure to water for 7 days at 70° C. Improvement in the processing reduces this to as low as 25 to 30 mg. per sq. in.

Kemp and his associates called attention[9] to this basic problem by showing that the original electrical properties as well as the resistance to water absorption are related to the water-soluble ash content of the GR-S.

The alum process for coagulating GR-S gives a better product from an electrical viewpoint than even the best of the standard salt-acid-coagulated polymers. The absence of sodium chloride is advantageous, and the alum completely converts the sodium soaps used in polymerization to insoluble aluminum soaps. The water absorption of alum-coagulated GR-S averages 18 to 25 mg. per sq. in.

Madigan and his co-workers evolved[10] a system of coagulation which produces a GR-S of better electrical properties and lower water absorption than that obtained through the previous procedures. In order to remove water-soluble materials, extensive contact between the latex and the coagulating medium must be obtained. Such contact is secured by diluting the GR-S latex and atomizing it beneath the surface of 0.25 per cent sulfuric acid. When so prepared, the coagulum tends to be tacky, and the serum may be somewhat milky. The addition, however, of 0.05 per cent glue to the sulfuric acid assists flocculation and gives a clear serum and a nontacky coagulum. Much of the water-soluble salts is discarded in the serum, and passage of the coagulum through a dewatering extruder eliminates still more of this material. GR-S 65-SP (GR-S 1016), which is prepared in this way, has a water-soluble ash content of only about 0.05 to 0.08 per cent, contrasted with about 0.75 per cent in regular salt-acid-coagulated GR-S. Similarly the soap content is only about 0.04 per cent, whereas in regular salt-acid-coagulated GR-S it is about 0.4 per cent. The mechanical water absorption of this grade of GR-S is 6 to 14 mg. per sq. in.

"*Cold*" *GR-S.* Special GR-S polymers X-511, X-512, and X-530 were prepared at a low temperature and with low water-soluble ash contents for testing in insulating compounds. Compounds made with them average approximately 25 per cent higher in tensile strength and elongation than those obtained with GR-S 65 (GR-S 1007). Their electrical properties are somewhat inferior, probably owing to the use of a formulation containing sugar. To overcome the tendency of the polymer latex to ferment, alkali is added to maintain a proper pH. This also harms the electrical properties.

A new polymer[8] (GR-S X-565, now GR-S 1503) was prepared by an iron-free, sugar-free formulation. Tests on this rubber indicate that it has as good electrical properties as GR-S 65-SP (GR-S 1016) and that the improved physical characteristics due to the low temperature of polymerization are maintained.

The actual practical experience of the wire and cable industry with the grades of GR-S best adapted to its requirements has been reviewed by Schatzel.[11]

Rosin Acids in GR-S. Rosin soaps have been used extensively as the dispersing agent for the polymerization of standard grades of GR-S because they produce a more tacky material, which helps adhesion in tire building. Rosin soap acids tend to affect the aging and the dielectric resistance adversely. They are not suitable therefore for use in insulating rubbers.

GR-S of Higher-Than-Normal Styrene Content. The standard types of GR-S used for rubber insulation contain about 25 per cent styrene and 75 per cent butadiene. When the styrene content is increased, GR-S compounds not only extrude more satisfactorily but also have better physical properties. The cold brittle temperatures are not so low, however, as those of stocks prepared from the 75/25 polymers. For special purposes a GR-S with a styrene content of about 35 per cent may be advantageous, provided the low-temperature characteristics are not critical.

Cleanliness. An important requirement in the rubber used for insulation is physical cleanness. Although in the case of natural rubber the top grades have always been selected, for many insulations even these are not sufficiently free of mechanical impurities since natural rubber is produced on plantations operated by natives. In the past considerable effort has been expended in additional washing and the necessary vacuum drying. On the other hand, GR-S, which is made in clean factories, is substantially free of mechanical impurities, and no advantage has been demonstrated for further washing and vacuum drying. This results in a real saving in processing costs.

COMPOUNDING OF GR-S INSULATION

The greater part of the output of GR-S is used in the construction of tires. To impart sufficient tear and abrasion resistance and tensile strength, it is compounded with carbon black. Compounds of standard GR-S and carbon black show tensile strengths of only 2700 to 3300 p.s.i., as contrasted with values of 4000 to 4500 p.s.i. obtainable with natural rubber. The lower tensile strength of the GR-S–carbon black tire compounds is not serious. But compounds for wire insulation usually contain little or no carbon black, since it has an adverse effect on the electrical properties.

Pure-gum compounds of natural rubber have tensile strengths of approximately 3000 p.s.i. The mineral fillers normally used in rubber insulation are whiting, zinc oxide, clay, calcium silicate, etc., and their total proportion may be as high as 70 per cent by weight of the insulating compound. In these proportions they can be considered to have only minor reinforcing value. The addition of the above quantities of such mineral fillers dilutes natural rubber sufficiently to lower the tensile strength to 1200 to 1500 p.s.i.

Pure-gum GR-S compounds, on the other hand, have tensile strengths of only 200 to 300 p.s.i., and the addition of inorganic fillers to the above extent increases the tensile strength only slightly. It is evident that compounding GR-S in a manner similar to that employed with natural rubber would result in products far inferior physically to corresponding compounds of natural rubber and probably of unsatisfactory quality for extensive use.

The Organic-Reinforcement Effect. Fortunately, certain organic materials produce a reinforcement phenomenon in GR-S which is absent in

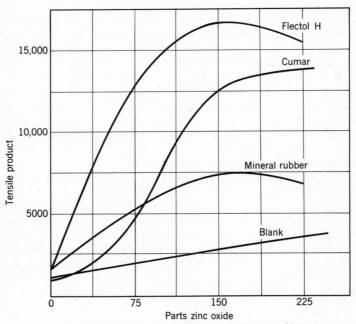

FIG. 1.   Effect of Organic Reinforcing Agents in GR-S Loaded with Zinc Oxide

natural rubber. This is sometimes known as the "organic-reinforcement effect." Resins made by the polymerization of coumarone or coumarone-indene mixtures are examples of materials that exhibit this effect. In a pure-gum GR-S compound they do not enhance the tensile strength, but when they are added to a GR-S compound containing mineral fillers they increase the tensile strength markedly. In some manner the combined effect of coumarone resin and mineral fillers in GR-S is to give appreciably greater tensile strength than can be obtained by the use of either alone. There is no satisfactory explanation for this action.

If the tensile product (the product of tensile strength and elongation) is used as a criterion of reinforcement, the effect of three of the organic-reinforcing agents can be demonstrated. The agents selected were Cumar MH $2^1/_2$, mineral rubber, and Flectol H (a condensation product of acetone and aniline). In Figs. 1 and 2 the tensile product obtained in the presence of 10 per cent of each of these materials on the GR-S illustrates the effect for various loadings of zinc oxide and of Kalvan (a fatty acid-coated whiting). It is evident that the use of either the mineral filler alone or the organic reinforcing agent alone results in a comparatively insignificant improvement in physical properties. The combination of the two materials, however, brings about an increase in the tensile product of from two to seven times. It is this effect that has allowed the wire and cable industry to produce satisfactory non-carbon black insulations from GR-S. The newer cold rubbers

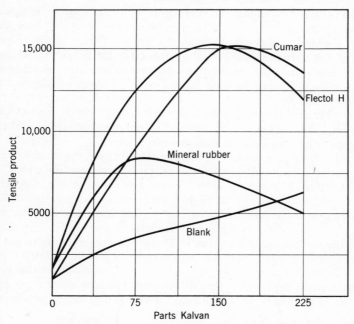

FIG. 2. Effect of Organic Reinforcing Agents in GR-S Loaded with Whiting

give higher tensile and elongation values with non-carbon-black compounding ingredients than are obtained with GR-S made at the standard temperature.

Because of its reasonable cost and substantial reinforcing effect, coumarone resin is universally used in GR-S insulating compounds containing mineral fillers. The proportion varies from 10 to 20 per cent on the GR-S. The addition of more than 20 per cent does not increase the effect. In some cases asphaltic materials can act as organic reinforcing agents in the presence of moderate—not too large—quantities of inorganic fillers.

Inorganic Fillers. Inorganic fillers are used in GR-S insulating compounds to increase tensile strength, elongation, modulus, and resistance to compression; to improve extrudability; and to reduce costs. Their selection must be made with care so that the electrical properties also will be satisfactory.

Zinc oxide is used universally in insulating compounds as an activator for acceleration and to some extent as an inorganic filler. It gives good electrical properties, but its high volume cost limits its use.

Whiting is used extensively, since its cost is moderate. Water-ground whitings are to be preferred. Water grinding, in addition to giving small particle size, which produces somewhat better reinforcement, also leaches out water-soluble materials harmful both to the original electrical properties and to the electrical stability in water. Fatty acid-coated whitings are somewhat better as reinforcing agents than whitings not so coated, but the allowable quantity may be limited because of a tendency to produce stickiness in the compound, which may interfere with processing.

Clays have a greater reinforcing effect than whitings, but their quality must be examined carefully. Clays tend to be somewhat hygroscopic, and this is to be avoided. Thoroughly washed clays are better in this respect than many of the varieties which might be suitable for nonelectrical uses. The recently developed clays prepared by acid treatment and calcining are even more satisfactory, and so their higher cost is frequently justified for special applications.

Bitumen. Mineral rubber is excellent electrically, since it is essentially a hydrocarbon. Because of its low cost and its organic reinforcing effect, it is used frequently in combination with mineral fillers.

Fatty Acids. In the compounding of natural rubber it is customary to add 1 or 2 per cent of a fatty acid to assist the vulcanization reaction, since the quantity of fatty acid occurring naturally (of the order of 2 per cent) is variable. In the acid-salt or acid-glue methods of coagulating GR-S latex the fatty acid soap, present during the polymerization to the extent of approximately 5 per cent, is converted largely to fatty acid. In alum coagulation the sodium soap is converted to aluminum soap. It might be expected that the 4 or 5 per cent fatty acid present in acid-coagulated GR-S would be sufficient for accelerator activation. For some reason this fatty acid is comparatively inactive, so that it is desirable to add about 2 per cent stearic, lauric, or similar fatty acid during the compounding for accelerator activation.

Antioxidants. Approximately 1 per cent of a standard antioxidant, such as phenyl-β-naphthylamine, is added during the preparation of GR-S to prevent oxidation during the drying operation and in subsequent storage. Since, in the compounding of GR-S insulation, much emphasis is placed on the possession of excellent resistance to aging, it is customary to add up to 2 per cent additional antioxidant to improve this property.

Waxes. Waxes, notably paraffin, are ordinarily added to insulating compounds to the extent of 2 to 3 per cent on the rubber hydrocarbon in the compounding of natural rubber. The wax assists processing. The solubility of paraffin in natural rubber is approximately 1 per cent. The excess gradually blooms to the surface to form a thin film which acts to protect natural-rubber insulation from sunlight.

GR-S insulation without wax is very susceptible to so-called sun cracking or atmospheric cracking. The addition of waxes in the amounts customary in the compounding of natural-rubber insulation furnishes little or no protection in GR-S compounds, so that such compounds are far more susceptible to atmospheric exposure than the corresponding natural-rubber compounds. If wax is added to the extent of 5 to 7 per cent on the GR-S, however, excellent protection is obtained, the resulting compounds being far more resistant than those prepared from natural rubber. Mixtures of amorphous or microcrystalline waxes are particularly effective. Paraffin alone is of much less value, but a portion of the 5 to 7 per cent of amorphous wax may be replaced with paraffin without loss of quality.

Accelerators. Accelerators and accelerator combinations used in the vulcanization of GR-S insulation resemble closely those used with natural rubber. The presence of the saturated styrene groupings in GR-S decreases by about 25 per cent the number of double bonds available, on a statistical basis, for vulcanization. It is desirable, therefore, to increase the quantity of accelerator 20 to 50 per cent over that normally used in natural rubber. Accelerators that give satisfactory electrical properties in natural rubber are suitable for use in GR-S. These include metal dithiocarbamates, thiuram mono- and disulfides, and mercaptobenzothiazole derivatives.

Compounding for Age Resistance. Since resistance to heat and oxidation is of much more importance in the wire industry than in other branches of the rubber industry, the accelerator compounding principles involved in the production of so-called "superaging" compounds are used extensively. Basically the reduction in the sulfur content used in such compounds for vulcanization results in better resistance to heat and oxidation. To obtain full vulcanization in a reasonable period of time, the amount of accelerator must be increased from about 1 per cent to 2 or 3 per cent. Whereas in the compounding of many rubber goods 1.5 to 3 per cent sulfur is common, GR-S insulating compounds of the better grades contain much less than this amount. Many compounds are formulated with 0.1 per cent or less elementary sulfur along with thiuram disulfides to furnish most of the sulfur necessary for vulcanization. Selenium thiuram disulfides also are vulcanizing agents; they probably furnish selenium sulfide as the active material. A comparison of the oven aging of "superaging" natural-rubber and GR-S insulations is shown in Figs. 3 and 4.

Polyethylene. Polyethylene has been used to some extent as a compounding ingredient for both GR-S and Butyl rubber. It aids extrudibility and is itself ozone-resistant. On the other hand, it is not very compatible with either material, and therefore two hydrocarbon phases may exist in the insulation. According to the conditions of processing, either one or the other ingredient may become the external phase. Since it is the external phase in such a mixture that controls the physical properties, widely differing characteristics may be obtained in a given compound, so that it is difficult to produce a uniform and satisfactory material.

Extrudability. The ease of processing of rubber insulating compounds is of extreme importance. It is essential that they should be capable of being extruded at high speeds, which, in the continuous vulcanization process, may reach 700 ft. per min. The compound should flow easily, form a smooth surface, and have a minimum nerve under such conditions. In the compounding of rubber insulation, particular attention must be paid to its plasticity. In the wire industry it is the practice to use compounds of higher plasticity than in many other branches of the rubber industry. The Firestone extrusion plastometer and the Garvey die are of great value in predicting the factory performance of an insulation.

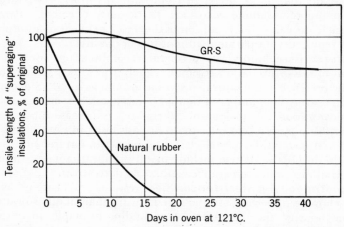

Fig. 3. Accelerated Aging of Natural-Rubber and GR-S "Superaging" Insulation Compounds—Tensile Strength

In the polymerization of GR-S, the presence of a small quantity of divinyl-benzene, which has two active groups, results in a polymer containing a

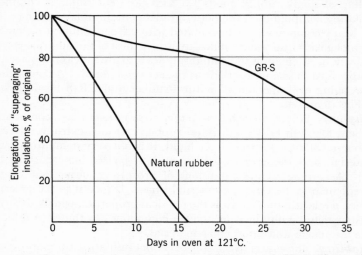

Fig. 4. Accelerated Aging of Natural-Rubber and GR-S "Superaging" Insulation Compounds—Elongation

proportion of cross-links. The use of 10 to 20 per cent of such rubber (GR-S 1017 and 1018, formerly GR-S 60 and 61) in a conventional GR-S compound promotes smoother extrusion, which is especially advantageous in the manufacture of thin-walled insulation and is beneficial in insulations of greater thickness

"High-styrene" resins (see chapter 18), containing approximately 85 per cent styrene and 15 per cent butadiene are also used as compounding ingredients to obtain smoothness of extrusion. Since these materials have low water absorption and good electrical properties, they are advantageous where their greater cost is justified. They also exhibit the organic-reinforcing effect.

Contact of Insulation with Copper. It is common practice to separate natural-rubber insulation and the copper conductor which it covers by coating the latter with tin or tin-lead alloys. Such coatings are not always mechanically perfect. They have a twofold purpose in the case of natural rubber. The contamination of natural rubber by copper decreases sharply its resistance to aging. Residual sulfur in the rubber may cause corrosion of the copper conductor, particularly where moisture is present. GR-S and Butyl rubber are not themselves harmed appreciably by copper contamination, but, if not suitably compounded, they may produce corrosive effects on the copper with which they are in contact. Corrosiveness of the rubber insulation may be reduced sharply by compounding and vulcanizing in such a manner that free sulfur is at a minimum. A standard test for the corrosiveness of rubber insulation has been developed[13] by the U.S. Coast Guard, and is in universal use.

Influence of Elevated Temperatures on Physical Properties. Natural-rubber insulating compounds may have their physical properties impaired by increasing temperature. Low-sulfur vulcanizations minimize this effect. GR-S insulating compounds are somewhat inferior to natural rubber in this respect, although polymers made at low temperatures show appreciable improvement.

BUTYL-RUBBER INSULATION

Butyl rubber is prepared by the copolymerization in solution, under the influence of an ionic catalyst at a very low temperature, of isobutylene with a small proportion of a diolefin such as butadiene or isoprene. The amount of unsaturation introduced is normally small, ranging between 1.5 and 3.5 per cent of that occurring in natural rubber or GR-S, and hence the residual unsaturation after vulcanization is minimal. The advantages of such a material are evident. The aging of rubber is due mainly to oxidation, whether at room temperature or at higher temperatures. Attack by light also involves oxidation. The tendency to oxidize is associated with the unsaturation in the molecule. Of more importance, the extreme susceptibility of natural rubber and GR-S insulation to ozone can be avoided by the use of Butyl rubber. Butyl rubber should, therefore, provide useful insulation for wires and cables. The very low rate of diffusion of gases through Butyl rubber, which makes the latter so valuable for use in inner tubes, is of no advantage in electrical insulation.

Since Butyl rubber is not prepared by an emulsion polymerization process, it does not contain the dispersing agents employed in the production of GR-S. Although there is some residue from the polymerization catalyst, the amount is small. Being almost pure hydrocarbon, Butyl rubber has inherently good electrical properties and low water absorption.

Vulcanization. Butyl rubber has a low rate of vulcanization because of the small mass-action effect of the few double bonds available. It can be vulcanized with many of the conventional rubber accelerators and sulfur but requires appreciably longer time and higher temperature than are normally used for natural rubber or GR-S. Since the development of good ozone resistance depends on the presence of only a minimum of residual double bonds, the vulcanization of Butyl insulation must be more complete than is necessary in other Butyl-rubber articles.

p-Quinonedioxime and *p,p'*-dibenzoylquinonedioxime are active vulcanizing agents for Butyl rubber. They tend to reduce the residual unsaturation more than when vulcanization is brought about by sulfur and conventional accelerators. This may be due to their ability to cross-link at a greater distance than sulfur. The use of these materials is therefore beneficial in the production of ozone-resistant Butyl-rubber insulation.

It is probably impossible in the vulcanization of Butyl-rubber compounds to eliminate the unsaturation completely, since, as Flory has pointed out,[4] the relatively few double bonds originally present may not be sufficiently close to each other to allow complete cross-linking.

Compounding Ingredients. The choice of compounding ingredients is much more restricted for Butyl rubber than for natural rubber or GR-S. Those containing even small amounts of unsaturation (such as mineral rubber) cannot be used because they absorb the vulcanizing agents, and consequently the Butyl rubber remains unvulcanized. Similarly, the presence of even a small amount of natural rubber or GR-S inhibits the vulcanization of Butyl rubber. Because of the low rate of diffusion of gases in Butyl rubber, any combination of ingredients that generates a gas may result in porosity, even though such ingredients would not harm GR-S or natural rubber.

Extrusion. Unlike natural rubber and, to a lesser extent, GR-S, Butyl rubber is not made appreciably more plastic by mastication. The addition of a substantial proportion of fillers is necessary in order to produce a compound that extrudes smoothly. Inorganic fillers differ greatly in this respect. The diatomaceous earths and calcium silicates are the most effective. Clays and whiting, both plain and surface-coated, are somewhat effective in this direction, but zinc oxide is practically ineffective. Waxes are beneficial in promoting extrudability. Asphaltic materials are of some benefit, but they have a tendency to produce porosity during vulcanization.

Aging. Since Butyl rubber is nearly saturated, it is highly resistant to oxidation. Antioxidants seem to be without effect in increasing the resistance to oxidation in the conventional accelerated aging tests. Various combinations of accelerators and vulcanizing agents differ in their effect on the age resistance of Butyl compounds. With some combinations tensile strength is practically unaffected during aging, while elongation decreases appreciably, whereas with other combinations the reverse is true. It is desirable to adjust the vulcanizing ingredients so that the tensile strength and elongation decrease proportionally during accelerated aging.

Corrosion. The use of relatively large percentages of sulfur and accelerators in order to reduce the unsaturation as much as possible and thereby increase the ozone resistance may result in an insulation that has a tendency to corrode copper conductors or braids when it is exposed to water or a humid atmosphere. It is necessary, therefore, to adjust carefully the vulcanizing ingredients and the time and temperature of vulcanization, to obtain both satisfactory ozone resistance and low corrosiveness.

Ozone Resistance. A properly vulcanized Butyl-rubber insulation has excellent ozone resistance. Even when bent sharply it will withstand a concentration of over 0.04 per cent ozone for an indefinite period of time. Its resistance to sunlight is excellent, and it withstands severe accelerated aging conditions. Even after 30 to 40 weeks in an air oven at 121° C., it remains flexible and rubbery, whereas an exposure of only 10 days to such conditions embrittles the type of ozone-resisting insulation containing vulcanized oils.

Water Absorption. Such Butyl insulation has a mechanical water absorption in 70° C. water about one-sixth that of the so-called oil-base compounds and therefore is particularly adapted for underwater use or direct burial in the ground. Its dielectric constant and power factor are reasonably low, and its dielectric resistance is high. It has good resistance to distortion at elevated temperatures. Thus Butyl rubber promises to furnish an insulation which will solve well the long-standing problem of high-voltage rubber insulation.

DETERIORATION OF RUBBER INSULATION

Effect of Ozone. Insulated wires operating at high voltage, unless perfectly shielded electrically, generate ozone by the electrical stress on the air at the surface of the insulation. When rubber is under physical stress, ozone can attack it vigorously and produce deep cuts, which destroy its usefulness. Even low-voltage insulated wires may be destroyed in the vicinity of high-voltage insulation by the ozone generated. Much attention has been given to the development of rubber insulations that resist the action of ozone.

If there is no physical stress in the rubber, ozone produces no obvious physical effect, although there may be a very slight increase in weight owing to the formation of a thin surface film of rubber ozonide. If, however, the rubber is under stress, exposure to ozone produces cracks which gradually deepen and lead to an appearance similar in some respects to that caused by atmospheric cracking. The latter is in fact generally attributed to dilute ozone. The action may be so rapid that in a few minutes a $1/4$-in. layer of rubber may be cut completely through to the conductor, with a resulting failure of the wire.

The mechanism by which the cracks develop and grow is not entirely clear. Probably no piece of rubber is completely homogeneous, and, therefore, when it is stretched there are portions that are under more stress than others. Ozone probably acts on the whole surface to form the rather brittle ozonide. At the points of greatest stress the surface ruptures most easily, thereby exposing a new surface which has high stress and which is attacked immediately and breaks. The process is repeated, and the crack deepens.

There are two general methods of making rubber insulation insensitive to ozone. The first is to compound it in such a manner that its physical properties do not allow physical stresses to be maintained. This leads to an insulation with high permanent set. In the past this method has been used almost exclusively in the formulation of rubber insulation on cables designed for voltages high enough to produce ozone. Compounds of this type contain various vulcanized vegetable oils, vulcanized bitumens, and reclaimed rubbers. They do not possess inherent chemical resistance to ozone, but their ability to relax physically allows them to pass the various specification tests. These compounds tend to be soft and weak, and their ozone resistance may decrease on aging. The other method is to make the insulation chemically resistant to ozone. Such resistance has been partially achieved by coating ozone-susceptible insulation with a thin layer of a flexible, chemically saturated material. Of course, if this layer is abraded away during installation or in service the protection fails.

An ideal rubber insulation from the point of view of resistance to ozone would be one having no unsaturation at all. Because its unsaturation is low initially and can be reduced by vulcanization to extremely low values, Butyl rubber offers great promise of solving this long-standing problem of attack by ozone.

Effect of Sunlight. Many artificial and accelerated tests have been proposed for use in the development of insulations and jackets adapted to resist sunlight. Exposure to sunlight itself, especially of highly resistant compounds, requires long periods of time before failure can be produced. In addition, such exposures are not capable of good control, and the results are not easily reproducible. Various sun lamps, mercury-vapor lamps, and the Weatherometer have been used to some extent in the appraisal of the sunlight or weather resistance of rubber compounds. Although these may give clues to quality, a consistently good correlation with natural exposure is not obtained.

Crabtree and Kemp propose[3] exposure to dilute ozone as a tool for use in the development of sunlight-resisting compounds, on the basis that atmospheric cracking is caused by dilute ozone. A concentration of about 25 parts of ozone per hundred million of air is used. This, although several times the concentration occurring naturally, is still of the same order of magnitude as the latter. It gives good acceleration to the process without appearing to change the essential nature of the action. The correlation between natural exposure and this artificial test seems to be good, and the method has served well in the development of wire insulations and jackets adapted to withstand outdoor exposure.

Rubber compounds differ greatly in their resistance to sun cracking. Although light-colored rubbers are more susceptible than dark ones, black asphaltic compounding ingredients do not improve the resistance. Carbon black is beneficial. Certain antioxidants which may retard the oxidation of unstretched rubber and other antioxidants which retard cracking during flexing are not necessarily effective in continuously stretched samples.

A practical method of retarding the latter type of atmospheric cracking is to produce a protective surface layer. This may be accomplished by painting with a flexible lacquerlike material or by developing thin protective films of wax on the surface. The solubility of most petroleum waxes in rubber is about 1 per cent at room temperature and increases rapidly as the temperature is raised. By the use of 2 to 8 per cent wax on the rubber, enough diffuses to the surface in a short time to form a film. As long as the protective film is unbroken, the rubber resists the action of sunlight, weather, and dilute ozone. Even if the film is broken by abrasion or occasional flexing, more wax diffuses to restore the film as long as there is any reserve left in the mass of rubber. Mixtures of various waxes are more effective than a single type, since they give a more continuous and adherent film. Similarly amorphous waxes are better than crystalline waxes.

Effect of Corona. Specifications for high-voltage rubber insulation occasionally demand a test for corona resistance in addition to that required for resistance to ozonized air. In this test, called the vertical U test, the insulated wire is bent to a definite curvature and placed with the bottom of the U resting on a grounded metal plate. An alternating voltage is applied between the conductor and plate to produce the desired stress, usually 100 volts per mil. Failure in this test is indicated by the dielectric breakdown which follows cracking of the rubber insulation. Since the appearance of the rubber in such a failure resembles that produced on exposure to ozone, it has been assumed that the failure is caused by ozone generated by the action of corona on air. Erosion of the surface of the rubber may sometimes occur in the test. The mechanism of the action is not entirely understood; some compounds which successfully pass the standard ozone tests fail rapidly under the influence of corona.

Effect of Oxidation. The aging of natural rubber in service usually involves oxidation. Oxidized rubber has a higher rate of water absorption than unoxidized rubber. Even a rubber insulation formulated to have low initial water absorption may on aging become less resistant to water absorption and may thus show poor electrical stability in water. It is important, therefore, that such insulation shall have maximum resistance to oxidation, to insure electrical stability. A natural-rubber compound, such as one made with deproteinized rubber, should be of the superaging type if it is to preserve its initial properties to the greatest extent.

Since GR-S is much more resistant to oxidation than natural rubber, it shows much greater stability in water. The mechanical water absorption of the best low-water-absorption GR-S compounds is not initially quite so low as that of the best deproteinized natural-rubber compounds, but after only a moderate accelerated aging test the GR-S compounds are superior both in mechanical water absorption and in stability of capacitance during soaking. Butyl-rubber compounds resist oxidation even better than GR-S compounds and hence their stability toward water absorption is still less affected by aging.

Resistance to Low Temperatures. If insulated wires and cables are to be used or installed at low temperatures, the brittle point of the rubber and its stiffening at low temperatures are both of importance. The stiffening may be of two varieties, first, the increase in the viscosity which occurs immediately on attaining a low temperature, and second, a progressive hardening which takes place with time. The latter is probably due to the partial or complete crystallization of the rubber. GR-S and Butyl exhibit little of the gradual stiffening. On the other hand, natural rubber, Neoprene, and Thiokol crystallize slowly with time. In GR-S, as the proportion of styrene is reduced and as the temperature of polymerization is lowered, a greater tendency to crystallize is exhibited.

In addition to the brittle point there is a temperature at which a rubberlike material will shatter under severe impact. In most of the rubbers this is 3 to 5° C. higher than the brittle point. This is in sharp contrast to plasticized polyvinyl chlorides, where the corresponding temperature difference may be as high as 45° C.

The brittle point of rubber insulations may be lowered somewhat by compounding with ester plasticizers and certain petroleum oils. Normal rubber insulating compounds have approximately the following brittle points: natural rubber about −55° C., Neoprene −40° C., GR-S −50° C., Butyl rubber −50° C., and Buna N −20° C.

Effect of Soil Microorganisms on Rubber Insulation. It has been shown[1, 2] that rubber insulation buried directly in the ground occasionally fails electrically and that such failures are due not to water absorption but to the action of soil microorganisms. Only in rare cases are conditions favorable to such attack. By the use of soils with high microbial activity, such deterioration has been duplicated in the laboratory in relatively short periods of time. It has been shown that bacteria, Actinomycetes, and especially fungi can participate in such electrical failures. Natural rubber may be consumed completely by soil microorganisms, and attack on rubber insulation of this type results in a deep pitting of the surface. GR-S and Butyl rubber hydrocarbons are not consumed by soil microorganisms, but microbial action on some compounding ingredients may cause failures. It is possible by suitable compounding and techniques of design to minimize or eliminate failures in rubber insulation from this cause.

Neoprene cannot be consumed by soil microorganisms. But the use with it of susceptible compounding ingredients may allow the electrical failure of some Neoprene-jacketed cables by the action of soil microorganisms. Since failure of Butyl rubber or GR-S insulation is due not to the hydrocarbon but to food streaks present, and since Neoprene can fail only from the same cause, Neoprene-jacketed insulations of Butyl or GR-S are relatively safe. The chance of susceptible points coinciding in the jacket and the insulation is slight. Since, however, natural-rubber hydrocarbon can actually be consumed by soil microorganisms, Neoprene-jacketed natural-rubber insulation is relatively unsafe unless protected by fungicidal or bactericidal agents.

OTHER INSULATIONS

The electrical properties of Neoprene, nitrile rubbers, and Thiokol are poor because these rubbers are not hydrocarbons. The latter two have not been used as electrical insulation to any extent. On the other hand, some compounds of Neoprene extrude smoothly in thin walls and have, therefore, been used for some low-voltage installations where good electrical properties are not necessary.

The silicone rubbers are finding use as insulation for wires in cases where their special properties warrant the increased cost. Their tensile strengths and elongations are low, and frequently the insulation must be protected mechanically. They will withstand an operating temperature of 150° C. successfully, and their brittle temperature may be lower than the brittle temperatures of natural rubber and of the other synthetic rubbers. They are moderately susceptible to the absorption of water, but the newer silicone polymers are appreciably better than the earlier ones in this respect.

JACKETS

Insulated wires and cables for portable use must frequently withstand much physical abuse. This applies also to cables intended for direct burial in the ground, where mechanical damage may be produced during the installation. It is necessary to provide mechanical protection through the use of a tough rubber jacket. Before World War II such a jacket was composed of a tire-tread type of compound of natural rubber, which furnishes high abrasion, compression, and tear resistance. For such use the materials of stocks of this type can be compounded to produce appreciably greater toughness than that ordinarily possessed by tire treads, because no attention need be paid to heat buildup during flexing, crack growth, etc. Neoprene jackets were used to some extent where oil resistance was a factor, but the high price of Neoprene prevented at that time any large usage.

A brief comparison of the essential properties of synthetic-rubber jackets along with those of natural rubber is given in the accompanying table.

	Tear	Abrasion	Aging	Sunlight	Low Temperature	Flame Resistance
Natural rubber	Good	Good	Fair	Fair	Excellent	Poor
GR-S	Good	Good	Excellent	Good	Good	Poor
Neoprene	Excellent	Excellent	Good	Excellent	Fair	Excellent
Butyl rubber	Excellent	Good	Excellent	Excellent	Excellent	Poor

The effects of the oven aging at 70° C. of GR-S, Neoprene, and natural-rubber jackets on tensile strength and elongation are shown in Figs. 5 and 6.

Flame Resistance of Neoprene Jackets. A sharp drop in the price of Neoprene as a result of quantity production has made practicable the extensive use of Neoprene jackets, and the advantages of such a material have become more available. Neoprene is inherently flame-resistant because of its high chlorine content. Jackets for cables used in coal mines are now universally made from Neoprene, since regulations of the Pennsylvania Bureau of Mines and the U.S. Bureau of Mines compel such cables to pass a rigid flame test.

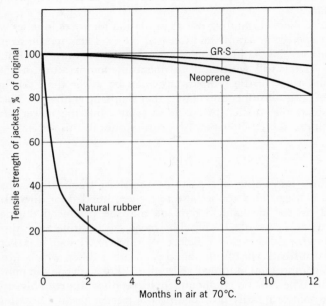

FIG. 5.   Oven Aging of Cable Jackets at 70° C.—Tensile Strength

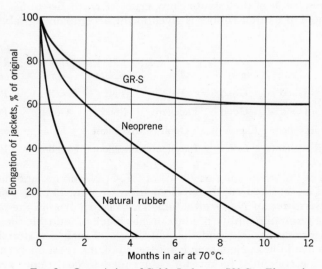

FIG. 6.   Oven Aging of Cable Jackets at 70° C.—Elongation

When Neoprene jackets are compounded with inflammable plasticizers, such as petroleum oils, paraffin, and petrolatum, the flame resistance is decreased. It can be improved by the use of chlorinated hydrocarbons as plasticizers. The addition of antimony oxide improves the flame resistance, presumably by a catalytic effect on the chlorinated materials present.

Use of Neoprene Jackets. Neoprene jackets are now almost universally used as the protective jacket on heavy portable cords and cables, underground cables, and aerial cables. They can be compounded to have higher tear resistance and abrasion resistance than are obtainable with either natural rubber or GR-S. They are also more resistant to sunlight than natural-rubber jackets, and in certain applications their oil resistance is advantageous. Neoprene jackets over an oil-susceptible insulation cannot impart complete oil resistance to the assembly, since oil can diffuse through the jacket and swell the insulation.

"Structure" carbon blacks should not normally be used in a cable jacket except for low-voltage operation. Such carbon blacks give high conductivity to the jacket, and in the event of an insulation failure sufficient voltage might appear on the jacket to be dangerous.

Resistance to Water. Jackets on underground or submarine cables should have both low water absorption and a low rate of water diffusion. The principles followed in compounding natural-rubber and GR-S insulation for low water absorption apply equally to compounding jackets. Butyl-rubber jackets, when properly compounded, have both low water absorption and a low rate of diffusion of water. Thiokol jackets may have rather high water absorption, but their water diffusion coefficient is low. There has been some use of Thiokol as a jacketing material for this purpose in the past, but this practice has been abandoned.

Neoprene jackets as ordinarily compounded have rather high water absorption. Starkweather and Walker found[12] that this is due primarily to the almost universal use of magnesium oxide in Neoprene compounds. By eliminating the magnesium oxide and carefully adjusting the other vulcanizing agents, the water absorption of Neoprene jackets can be lowered substantially.

For many applications ordinary Neoprene compounds possess sufficient ozone resistance. For use in the jackets of high-voltage cables, however, Neoprene compounds must be specially formulated to make them capable of resisting both ozone and corona. The addition of high percentages of the antioxidant Acroflex C (a 65/35 mixture of phenyl-α-naphthylamine and diphenyl-p-phenylenediamine) has been recommended for increasing the ozone resistance of Neoprene.

Influence of Thiuram Disulfides in Neoprene. The use of Neoprene in cable jackets has introduced complications because of the fact that Neoprene GN contains an appreciable amount of a thiuram disulfide as a stabilizer. Such a jacket in contact with conventional rubber insulating compounds containing moderate amounts of sulfur produces overvulcanization of the insulation surface during the vulcanization of the jacket. This results in a sharp deterioration of the physical properties of the insulation. It has been necessary, therefore, either to make the insulation compatible with such a jacket by vulcanizing it with thiuram disulfide in the absence of sulfur or to provide mechanical separation of the insulation from the jacket.

The newer Neoprene type W[5] offers promise as a material for cord and cable jackets. It has the advantage of a much lower compression set than that obtainable with Neoprene GN. In addition it does not contain a thiuram disulfide as a stabilizer, and so the complications occurring with Neoprene GN are eliminated. Compounds of type W are vulcanized with agents different from those used with type GN. Instead of metallic oxides, sulfur and various accelerators are preferred.

Colored Jackets. It is frequently necessary or desirable to have colored jackets on certain types of cables. By the use of reinforcing clays, satisfactory jackets may be made with either natural rubber or Neoprene, although their tear resistance and abrasion resistance are inferior to those of jackets containing carbon black. Because of the poor physical properties of pure-gum GR-S compounds, colored jackets from GR-S are not sufficiently tough, even when organic-reinforcing agents are used. There is some indication that, by the use of GR-S polymerized at low temperatures, satisfactory colored jackets may be obtained.

Nitrile-Rubber-PVC Blends. Polyvinyl chloride is compatible with the nitrile rubbers. With the ordinary types of the latter, an ester plasticizer is necessary to make a compound that can be processed satisfactorily. The use of polymers of lower molecular weight makes it possible to omit the plasticizer, whereby the disadvantage of a diffusible compounding ingredient is avoided. When such materials are compounded with carbon black, they are suitable for the jacketing of cables. They may be used, depending on the purpose, in either the vulcanized or the unvulcanized condition.

REFERENCES

1. Blake, J. T., and Kitchin, D. W., *Ind. Eng. Chem.*, **41**, 1633–41 (1949).
2. Blake, J. T., Kitchin, D. W., and Pratt, O. S., *Elec. Eng.*, **69**, 782–7 (1950); *Trans. Am. Inst. Elec. Engrs.*, **69**, Pt. II, 748–54 (1950); **72**, 321–8 (1953).
3. Crabtree, J., and Kemp, A. R., *Ind. Eng. Chem.*, **38**, 278–96 (1946).
4. Flory, P. J., *Ind. Eng. Chem.*, **38**, 417–36 (1946).
5. Forman, D. B., Radcliff, R. R., and Mayo, L. R., *Ind. Eng. Chem.*, **42**, 686–91 (1950).
6. *India Rubber World*, **120**, 350 (1949).
7. *India Rubber World*, **120**, 476 (1949).
8. *India Rubber World*, **122**, 553 (1950).
9. Kemp, A. R., Ingmanson, J. H., Howard, J. B., and Wallder, V. T., *Ind. Eng. Chem.*, **36**, 361–9 (1944).
10. Madigan, J. C., Borg, E. L., Provost, R. L., Mueller, W. J., and Glasgow, G. U., *Ind. Eng. Chem.*, **40**, 307–11 (1948).
11. Schatzel, R. A., *India Rubber World*, **126**, 369–70 (1952).
12. Starkweather, H. W., and Walker, H. W., *Ind. Eng. Chem.*, **29**, 1380–4 (1937).
13. U.S. Coast Guard Specification 15C–1, Corrosion Test, Par. F9A, Sept. 1947.

CHAPTER 16

SYNTHETIC HARD RUBBER

Frank S. Malm and Henry Peters

Bell Telephone Laboratories

Synthetic hard rubber had its early beginning during the last year of World War I, when the Germans used methyl rubber H, prepared by the polymerization of dimethylbutadiene at room temperature, for the manufacture of submarine and other types of storage-battery jars, containers and covers, as well as for the manufacture of sheet, rod, and tubing for dielectric use. According to Memmler[15] they did not experience any difficulty in applying methyl rubber H to the manufacture of these products. It was further asserted that the vulcanizates could be easily processed, machined, and polished, and possessed sufficient strength for the uses to which they were applied. At the end of the war German use of methyl rubber H was terminated because of its high price in comparison with natural rubber.

During 1938–39 the Germans were manufacturing hard-rubber products from Buna 85, a butadiene-sodium polymer. According to Stöcklin[23] hard rubber from this synthetic polymer had excellent heat and chemical resistance.

With the attack on Pearl Harbor and the fall of Singapore in World War II, the American chemical industry produced man-made rubber suitable for the numerous needs of the hard-rubber industry.

ELASTOMERS FOR HARD-RUBBER MANUFACTURE

Butadiene-Styrene Rubbers. GR-S containing approximately 75 per cent butadiene and 25 per cent styrene, which is the equivalent of the German Buna-S synthetic rubber, was developed in the United States during World War II and proved to be a satisfactory substitute for plantation rubber in the manufacture of many hard-rubber products. Chemical, physical, and electrical properties of GR-S hard rubbers are approximately of the same order of magnitude as those of hard natural rubber, and hence GR-S was used for making most of the hard rubber during the war and to a considerable extent thereafter. However, Winspear, Herrmann, Malm, and Kemp[26] have shown that fully cured GR-S hard rubber is brittle on bending through a sharp angle and will fracture more easily than natural hard rubber. Hard rubbers can also be made from copolymers having a higher or a lower styrene content than that of GR-S. Winspear, Herrmann, Malm, and Kemp[26] made hard rubbers from butadiene copolymers containing as much as 50 per cent styrene, which are comparable to the German Buna-SS synthetic rubbers.

573

Butadiene-Acrylonitrile Rubbers. GR-A, containing approximately 75 per cent butadiene and 25 per cent acrylonitrile, the equivalent of the German Buna-N or Perbunan synthetic rubber, was developed in the United States and was used during World War II for the manufacture of hard rubber. Although its electrical properties are inferior to those of either GR-S or natural-hard-rubber vulcanizates, its softening point is slightly higher and its resistance to chemical solvents is excellent. Hard rubbers have been made from copolymers having as much as 40 per cent acrylonitrile, which are comparable to the German Perbunan Extra.

Polybutadiene Rubbers. Existing information on hard rubbers made from polybutadiene polymers concerns those prepared from German synthetic rubbers such as Buna 85 and Buna 115. (Number denotes the viscosity value.) In many respects these hard rubbers have properties similar to those of hard rubbers obtained from GR-S and natural rubber.

2-Chlorobutadiene Rubber. According to Fisher, Mullins, and Scott[6] hard rubber can be made from GR-M, which is prepared from 2-chlorobutadiene. Information on the physical properties of the product is not available, but the dielectric properties are stated to be very poor.

2,3-Dimethylbutadiene Rubber. This rubber, which was known during the latter part of World War I as methyl rubber H, was used exclusively by the Germans in the manufacture of hard-rubber products. As far as is known, it is not in current use.

Polysulfide Rubbers. According to Martin[14] polysulfide rubbers when compounded with high percentages of sulfur do not yield vulcanizates comparable to hard rubber. This conclusion has been substantiated by Fisher, Mullins, and Scott.[6]

Isobutylene-Isoprene Copolymers. GR-I, containing approximately 98 per cent isobutylene and 2 per cent isoprene, cannot be made into a hard rubber because of the lack of sufficient unsaturation (double bonds).

NOMENCLATURE

Hard rubber, whether natural or synthetic, may be defined as highly vulcanized rubber containing large proportions of combined sulfur. Hard rubber prepared from simple mixtures of sulfur and rubber is often referred to as ebonite or vulcanite. Hard rubbers prepared from natural rubber are vulcanizates that have vulcanization coefficients[12] between 25 and 47. The theoretical vulcanization coefficient value for natural hard rubber is 47, while the values for the synthetics are more or less than this amount, as shown in Table I. The coefficient of vulcanization is usually defined as the number of units of weight of sulfur combined with 100 units by weight of unsaturated hydrocarbon. As indicated in Table I, the theoretical coefficient values are all corrected for the impurities existing in the various raw rubbers.

Certain semihard-rubber compositions made from natural rubber may be vulcanized to coefficients lower than 25. The vulcanizates have poor chemical stability and physical strength as shown by Gibbons[9] and by Gibbons and Cotton.[10] Similar studies were conducted by Rostler[19] with

Table I. Coefficient of Vulcanization of Natural and Synthetic Rubbers at Saturation.

Types of Rubber	Theoretical Coeff. of Vulcanization to Ebonite Stage	Approx. % of Nonrubber Constituents	Coeff. after Correction for Nonrubber Content
Buna 85 (butadiene)	59.3	0.5	59.0
Buna 115 (butadiene)	59.3	0.5	59.0
Natural rubber (isoprene)	47.0	6	44.2
GR-S (75% butadiene, 25% styrene)	44.5	8	41.0
Buna S (75% butadiene, 25% styrene)	44.5	6	41.8
Copolymer (50% butadiene, 50% styrene)	29.7	4	28.5
GR-A (75% butadiene, 25% acrylonitrile)	44.5	3	43.2
Nitrile rubber (82% butadiene, 18% acrylonitrile)	48.6	3	47.2
Nitrile rubber (65% butadiene, 35% acrylonitrile)	38.6	3	37.4
Nitrile rubber (60% butadiene, 40% acrylonitrile)	35.6	4	34.2

GR-S and a butadiene-acrylonitrile (60/40) copolymer. The vulcanization coefficients varied between 10 and 25, and the physical strength as well as the chemical stability were far superior to those obtained with natural semihard rubbers. Winspear, Herrmann, Malm, and Kemp[26] obtained similar results with semihard rubber prepared from GR-S.

VULCANIZATION

When a mixture of rubber, either natural or synthetic, and sulfur is heated, the sulfur present combines with the rubber in such a manner as to render it unextractable with acetone. With natural rubber, this reaction results in one sulfur atom adding chemically, presumably stoichiometrically but not necessarily structurally, at the double bond of each isoprene unit (C_5H_8) and at saturation forms $(C_5H_8S)_x$, which corresponds to 32 per cent combined sulfur, i.e., to a vulcanization coefficient of 47. When a hard-rubber mixture contains an excess of sulfur and is heated for a long period of time at high temperatures, it is possible to exceed the theoretical combined sulfur content, owing to substitution of hydrogen atoms in the rubber molecule.

Present-day synthetic rubbers have butadiene (C_4H_6) as one of their constituents, and, when they are heated for a prolonged time with sulfur, one atom of sulfur adds chemically at the double bond and at saturation forms $(C_4H_6S)_x$, which corresponds to 37.3 per cent combined sulfur, i.e., to a vulcanization coefficient of 59.3. Hard rubbers from polybutadiene as well as from butadiene copolymers containing either styrene or acrylonitrile have some properties, such as chemical, physical, and electrical, similar to those of hard natural rubber.

HEAT OF EXOTHERMIC REACTION

The exothermic heat of vulcanization may be defined as the heat evolved by the chemical combination of sulfur with an elastomer during the process of vulcanization. The heat is liberated at an early stage in the reaction, and this indicates that substitution or dehydrogenation is taking place in the rubber molecule. The temperature rise due to the exothermic heat depends mostly on the temperature of vulcanization and the thickness of the sample. When excessive substitution takes place in the rubber molecule, a considerable amount of hydrogen sulfide is liberated, and this in turn usually destroys the vulcanized specimen, by making it porous.

Winspear, Herrmann, Malm, and Kemp[26] have shown that the time required for the beginning of the exothermic reaction is considerably longer for GR-S than for natural rubber, but that the reaction, when it does occur, produces a much higher rise in temperature in GR-S containing 45 and 50 parts of sulfur and vulcanized at 150° C.

COMPOUNDING AND PROCESSING

Mixing. The compounding of synthetic or natural hard rubber is similar to that of soft rubber, the chief difference being the sulfur ratio and the use in hard-rubber stocks of hard-rubber dust as a filler. The same equipment is used in fabricating either type of hard or of soft rubber. Mixing mills, Banbury mixers, extruders, and calenders are all standard equipment.

Breakdown of Rubbers. In the fabrication of synthetic-hard-rubber compounds, it is important that the raw elastomer should be sufficiently broken down before it passes the mixing operation. Since the synthetic elastomers have a higher degree of "nerve," a considerable amount of shrinkage results if the breakdown of the elastomer is not adequate. Natural rubber in this respect is more easily broken down and has much less "nerve." Breakdown of the elastomers is generally accomplished in a Banbury mixer or on a mixing mill. The Banbury is preferred because of the higher temperatures and greater capacity. A hot breakdown is claimed to be more efficient than a cold breakdown.

Compounding of Ingredients. The amount and the order of adding sulfur, fillers, plasticizers, activators, and accelerators to synthetic-hard-rubber mixtures are practically the same as those for natural rubber. Accelerated synthetic-hard-rubber mixtures are compounded against liability to scorching in the same manner as in the preparation of hard rubber from natural rubber.

Calendering. The calendering of synthetic-hard-rubber compounds is essentially the same as that for natural-hard-rubber stocks. Although there may be variations in temperatures, the general procedures are practically the same. However, to avoid excessive shrinkage of the synthetic-hard-rubber compounds during and after the calendering operation, it is necessary to use highly loaded stocks of controlled plasticity.

Molding. Molding of hard-rubber compounds may be relatively simple or difficult, depending on the thickness of the specimen to be cured. The thicker the specimen, the more precautions must be taken to prevent porosity or blowing. The same precautions must be taken with synthetic hard rubber as with natural hard rubber. Hard-rubber dust is generally used in molded articles to prevent excessive shrinkage, to impart satisfactory physical properties, and to keep the exothermic reaction to a minimum. Unloaded stocks, such as ebonite, have shrinkage values as high as 8 per cent and are vulnerable to blowing. Hard-rubber dust prepared from natural or synthetic rubber may be used in synthetic-hard-rubber compounds.

Extrusion. Synthetic-hard-rubber stocks require the same type of loading that is used in natural-rubber compounds. To obtain products such as extruded rods and tubes, it is necessary to use compounds that show the minimum amount of shrinkage. Hard-rubber dust is generally used to obtain this property in addition to producing goods having high dimensional stability.

METHODS OF VULCANIZATION

Whether the hard rubber is made from synthetic or natural rubber, factory methods of vulcanization with open steam, water, and mold are practically the same.

Open Steam. This method is to be used for products that do not require a lustrous surface such as tank linings, drums, battery cases, tubes, and rods.

Water Cure. Hot water, heated by steam is used in industry for curing hard-rubber products such as combs, trays, and high-grade sheet material, pressed between highly polished tinfoil. After vulcanization, the tinfoil is stripped from the products, leaving on the surface a satisfactory finish which can be further buffed to a high polish.

Mold. The number of piece parts is usually the deciding factor as to the method of vulcanization. If the required number of products is large, it is necessary to resort to a molding operation. Since this molding cycle must be kept to a minimum, an accelerated rubber compound must be used. Typical products cured by this procedure are magneto parts, pipe stems, syringes, tool handles, etc.

PHYSICAL PROPERTIES OF GR-S HARD RUBBERS

Effect of Sulfur. Cheyney and Robinson[2] pointed out that GR-S vulcanizates increase continuously in tensile strength and decrease continuously in elongation as the sulfur content is increased from 2.5 through 35 per cent. Later data by Rostler[19] are in agreement with this conclusion. Over the sulfur range mentioned by Cheyney and Robinson, the tensile strength rose from 200 to 6200 p.s.i., the Shore hardness rose from 34 to 100, and the elongation fell from 400 to 8 per cent. The samples prepared with 25 per cent sulfur were in these respects practically the same as those prepared with 35 per cent sulfur.

This behavior of GR-S with increase in sulfur combination is considerably different from the behavior of natural rubber. As vulcanization of the latter proceeds, the rubber, after passing through the soft-rubber stage, in which the tensile strength and elongation are both high, enters a lengthy stage in which the tensile strength and the elongation are both very low (in this stage the products are in effect highly overcured, very "short," soft rubber); it then passes into a stage in which the elongation rises appreciably and the material has good resistance to flexing (this is the true semiebonite or leathery stage); it then enters the actual hard-rubber stage, in which the strength increases greatly and the elongation falls to a very low figure. Some authors have failed to distinguish these stages properly and have designated the whole range of vulcanizates between soft rubber and hard rubber as "leathery."[11] In order to clarify the situation, data from Gibbons and Cotton,[10] which show that the leathery stage, in which the elongation rises and the rubber has marked resistance to flexing, is relatively narrow, are quoted in Table II. The unaccelerated stocks A and A' represent badly overcured soft rubber. When sulfur combination is carried further, as in stocks B, B', C and C', the rubber enters the leathery stage, in which the extensibility and the ability to withstand flexing increase. The accelerated stocks E and E' are already in the leathery stage.

Table II. *Range of Natural-Rubber Vulcanizates Including the Leathery Stage*

Pale crepe	100	100	100	100	100	100	100	100
Sulfur	17.5	17.5	20	20	22.5	22.5	25	25
Nonox*	...	1	...	1	...	1	...	1

Unaccelerated Stocks (Cured 5 Hr., 148° C.)

	A	A'	B	B'	C	C'	D	D'
Tensile strength p.s.i.	170	235	205	210	480	470	510	500
Elongation, %	8.3	4.2	50.0	54.2	104.2	104.2	83.3	91.7
Flexures to rupture	12,742	11,631	230,274	275,909	112,680	129,481	37,728	47,926

Accelerated Stocks (Cured 2 Hr., 148° C.)
(with Vulcafor Resin, 2 Parts†)

	E	E'	F	F'	G	G'	H	H'
Tensile strength, p.s.i.	405	408	516	517	1145	1228	1264	1364
Elongation, %	128	144	128	144	184	200	184	200
Flexures to rupture	140,643	138,696	48,097	48,289	15,724	14,084	13,808	13,707

* An antioxidant consisting of a mixture of aldol-α-naphthylamine and aldol-β-naphthylamine.

† Apparently an aldehyde-amine accelerator.

Winspear, Herrmann, Malm, and Kemp[26] investigated the physical properties of GR-S ebonites over a sulfur range of 20 to 33 per cent. Their results are recorded in Table III, along with a natural rubber ebonite for

Table III. Physical Properties of GR-S Ebonites

Compd.	Sulfur as Compounded, %	Cure Hr.	Cure °C.	% Cold Flow 49°C.	% Cold Flow 71°C.	Impact Strength (Izod), Ft.-Lb. per In.	Flexural Properties, P.S.I. Strength	Flexural Properties, P.S.I. E* × 10³	Tensile Strength, P.S.I.	Elongation, %	Rockwell Hardness, HR
100 : 25	20.0	16	153	39.3	73.3	1.119	9,130	242	5,980	5.8	83
		20		35.4	48.0	0.963	10,100	263	6,525	4.0	89
		24		26.2	39.4	0.759	10,230	295	7,380	3.9	94
100 : 30	23.07	8	153	35.8	43.7	0.871	9,300	255	6,670	7.2	83
		10		34.6	45.7	0.932	9,830	265	7,560	6.0	88
		24		2.8	21.6	0.684	12,770	301	8,280	5.4	104
100 : 35	25.92	6	153	18.0	33.2	0.732	10,570	319	8,380	5.3	92
		8		17.4	35.5	0.661	11,800	296	8,280	5.8	95
		10		5.0	27.1	0.583	11,800	305	7,920	3.9	101
100 : 40	28.57	6	153	3.0	20.5	0.552	13,070	323	8,740	7.0	103
		8		1.4	9.9	0.523	12,470	317	7,870	5.2	107
		10		1.4	9.5	0.515	14,000	325	9,100	8.5	107
100 : 45	31.03	8	148	4.7	26.6	0.517	13,300	318	7,570	7.2	103
		10		2.2	17.2	0.501	13,700	323	8,420	4.8	108
		12		2.2	17.8	0.496	13,870	325	8,100	3.2	109
100 : 50	33.33	8	148	12.0	27.5	0.547	13,030	332	8,325	5.7	101
		10		1.5	11.2	0.562	14,270	337	8,130	3.1	109
		12		1.8	8.4	0.563	14,700	333	7,410	2.4	111
Natural-rubber ebonite 100 : 47		10	148	2.5	18.0	0.53	15,000	330	10,000	4.0	99

*E = modulus.

comparison. They conclude that there is a striking similarity between the over-all properties of the GR-S ebonites compounded with 40 (28 per cent) or more parts of sulfur per 100 parts of whole polymer and the corresponding data for a fully cured natural ebonite. Physical tests were made in accordance with ASTM[1] methods.

Morris, Mitton, Seegman, and Werkenthin[16] investigated the tensile strength, elongation, Rockwell hardness, and flexural stiffness of GR-S ebonites over a sulfur range of 16.6 to 37.1 per cent. They used micronized sulfur, and obtained their best results with 50 (33 per cent) parts of sulfur per 100 parts of polymer. They obtained unusually high tensile values of approximately 10,000 p.s.i., stiffness values of approximately 60 lb., elongations of 5.5 per cent, and hardness values slightly greater than a hundred.

Effect of Pigments. The effect of nine different pigments at 60 volume loadings on the properties of a GR-S stock containing 50 parts of sulfur was investigated by the last-mentioned authors. The pigments generally shortened the time of cure needed to reach optimum tensile strength and had a tendency to smooth out the raw stocks. None of the loaded stocks quite equaled the base stock in tensile strength, elongation, or flexibility. Stock loaded, respectively, with channel black, SRF black, and calcium silicate gave the highest tensile strength (8000 to 9500 p.s.i.) while stocks loaded with whiting and barytes gave the lowest tensile strength (about 4500 p.s.i.) and highest elongation (4.5 to 4.75 per cent). Stocks loaded with hard clay, magnesium silicate, iron oxide, and ground silica gave intermediate values for tensile strength. None of the loaded stocks was much harder than the base stock and several were softer.

Effect of Softeners. The effects of seven different softeners at 15 parts loading were investigated. The base stock consisted of 100 parts of polymer, 50 parts of sulfur, and 115 parts of semireinforcing black. All raw stocks containing softeners had more tack than the base stock. A sulfur-reactive unsaturated hydrocarbon oil decreased the time to reach optimum tensile strength and gave good tensile strength and increased elongation. Rosin oil was an outstanding softener from the standpoint of tackiness, and the vulcanizate containing it had good tensile strength, but it did slow down the curing rate. A study of the effects of various plasticizers in GR-S ebonite has been published by Norman, Westbrook, and Scott.[18]

Effect of Accelerators. The effects of 6 inorganic and 12 organic accelerators on the base stock containing 50 parts of micronized sulfur, 115 parts of semireinforcing black, 15 parts of a sulfur-reactive unsaturated hydrocarbon oil, and 100 parts of polymer have been examined by Morris and his co-workers.[16] Zinc dimethyldithiocarbamate, although the fastest, was undesirable because it set the stock up in storage. The aldehyde–amines were satisfactory accelerators and produced vulcanizates that had good tensile properties. The unaccelerated stock required 2.5 hours at 307° F. to reach its optimum tensile strength; 2 parts of zinc dimethyldithiocarbamate reduced the time to 0.75 hour; 2 parts of butyraldehyde–aniline or of butyraldehyde–butylamine reduced the time to 1 hour. Tetramethylthiuram disulfide did not speed up the cure very much but gave stocks of good tensile strength and better elongation than all the other stocks except that containing zinc dimethyldithiocarbamate.

Cold GR-S and Oil-Extended GR-S. The GR-S used in investigations reviewed in this chapter was so-called "hot" GR-S, i.e., GR-S prepared at a polymerization temperature of 122° F. Hard rubber can be made from cold GR-S but the available physical data concerning it are not sufficient to justify a review of it.

Cold (43° F.) oil-extended GR-S containing 40 parts of naphthenic processing oil can be vulcanized to hard rubber. The product has a tendency to exude oil shortly after cure.

PHYSICAL PROPERTIES OF BUTADIENE-ACRYLONITRILE HARD RUBBERS

The American nitrile rubbers used to make hard rubber samples mentioned in this chapter were prepared by polymerization at elevated temperatures, such as 122° F. Hard rubber can be made from butadiene-acrylonitrile polymers prepared at low temperatures, but sufficient physical data concerning it are lacking.

A study was undertaken by Garvey and Sarbach[8] to determine some of the important properties of ebonites made from a copolymer containing approximately 40 per cent acrylonitrile and 60 per cent butadiene. The investigation, which is reviewed herewith, deals with the effect of various proportions of sulfur, accelerators, and a variety of softeners and pigments on the tensile strength, impact strength, elongation, hardness, and softening temperature.

Effect of Sulfur. Ebonites containing 20 to 50 parts of sulfur on 100 parts of rubber were investigated. Strips $0.25 \times 1 \times 6$ in. were cured at 320° F. for 60, 90, and 120 minutes. As shown in Fig. 1 the best combination of physical properties was found in the compound with 35 parts of sulfur at the 120-minute cure.

Effect of Accelerators. A base stock consisting of 100 parts of rubber and 35 parts of sulfur was used for the evaluation of accelerators. Inorganic accelerators (2 parts), such as zinc oxide and magnesium oxide, accelerated the cure mildly, but litharge failed to do so. Selenium and tellurium (1, 3, and 5 parts) accelerated the cure slightly but gave inferior physical properties along with a high degree of softening. Twelve organic accelerators (2 parts) were investigated, and all of these shortened the time of cure from 120 minutes to approximately 20 and 30 minutes. The aldehyde-amine and the thiocarbamate types have a strong accelerating action and at the same time maintain the good physical properties evident in the pure-gum control containing no accelerator.

Effect of Pigments. The base stock consisting of 100 parts of rubber and 35 parts of sulfur was used to evaluate 11 different compounding ingredients, all stocks being cured for 90 minutes at 320° F. except those containing channel black and zinc oxide, which, because of a retarding effect, were cured 120 minutes at 320° F. Weight loadings of pigments varied between 10 and 125 parts. Whiting, iron oxide, SRF carbon black, and butadiene-acrylonitrile dust are the most promising from the standpoint of the physical

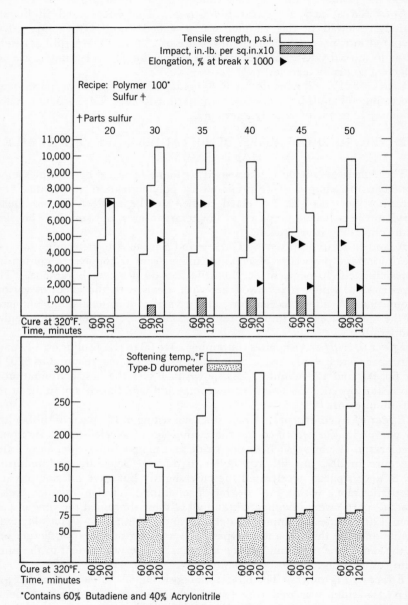

FIG. 1. Effect of Sulfur Ratio on Properties of Nitrile-Rubber Vulcanizates

properties of the product. Hard-rubber dust from natural rubber and asbestos can also be used but produce lower physical properties. Whiting produces an unusually good processing stock coupled with good physical properties; clay produces unusually poor physical properties, including a very low softening point. However, because of its acid-resisting properties, clay is one of the most widely used pigments.

Effect of Softeners. Synthetic rubbers are not so tacky as natural rubber. For this reason and because of their low plasticity, a considerable amount of softener is necessary to impart good processing qualities to the compound. Twenty different softeners were investigated in compounds consisting of 100 parts of rubber, 35 parts of sulfur, and 15 parts of softener, cured for 90 minutes at 320° F. Dipolymer oil (a coumarone resin oil) and dibutyl-*m*-cresol are the best softeners for maximum tensile strength; for heat resistance coal tar and rosins are best. For increased tack it is advisable to use a mixture of coal tar, coumarone resin, or dibutyl-*m*-cresol with a light ester-type softener such as dioctyl phthalate, or even a blend of these three. A minimum amount of softener should be used to prevent porosity.

Colors. These synthetic ebonites also take a high polish and can be made in various colors by the use of colored inorganic pigments, such as iron and chromic oxides, antimony sulfides, and cadmium selenide.

The authors conclude with a statement that, based on the above evidence, ebonites made from a copolymer of 40 per cent acrylonitrile and 60 per cent butadiene have physical properties equal to those of natural-rubber ebonites and in addition have considerably higher softening temperatures.

PHYSICAL PROPERTIES OF BUNA HARD RUBBERS

Various Buna compounds, shown in Table IV, were investigated by Scott[22] with regard to plastic yield, cross-breaking strength, elongation, and impact strength. Some results of the investigation are given.

Table IV. Hard-Rubber Formulations

	A	B	C	D	E	F
Buna 85	100	...	...	...	...	...
Buna 115	...	100	100	100	...	...
Buna S	...	...	...	...	100	...
Buna N	...	...	...	...	...	100
Sulfur	47	35	47	59	47	47

Plastic Yield. The plastic yield (determined by the torsion method[3] over a temperature range of 140 to 150° C.) of the Buna ebonites is, at least at the higher temperatures, much smaller than, and it does not increase so rapidly with rise of temperature as, the plastic yield of natural-rubber ebonites. As in the natural-rubber ebonites, increasing the proportion of sulfur (mixes B, C, D) reduces the yield at any given temperature and raises the yield temperature. At a temperature of 150° C., C, F, and E cured 11 hours and containing 47 parts of sulfur gave yield values of 0.7°, 1.4°, and 3.6°.

In a later study Scott[23] found that the resistance to plastic deformation at elevated temperatures of unloaded butadiene-styrene ebonites is greater, the higher the styrene content of the polymer, at least up to 46 per cent styrene. In the most favorable case butadiene-styrene ebonite had a yield temperature as much as 30° C. higher than the yield temperature of natural-rubb ebonite. The resistance to plastic yield in butadiene-styrene ebonites was best when the sulfur was as high as 1.2 or even 1.4 atoms per butadiene unit.

Cross-Breaking Strength. Determinations were made over a curing range of 3 to 11 hours at 155° C. The results are considerably lower than those for natural-rubber ebonites, which range from 18,000 to 21,000 p.s.i. The values for the Buna ebonite varied between 4,700 and 11,000 p.s.i.

Elongation. Elongations for the Buna rubbers were also lower than those, around 10 to 15 per cent, obtained with natural rubber.

Impact Strength. The impact strength of unnotched specimens of Buna ebonites is much less than that of natural-rubber ebonite, for which values of 19 to 33 ft.-lb. per sq. in. have been found. Notched specimens, however, generally give values of the same order as those for natural-rubber ebonite, namely, 0.8 to 1.8 ft.-lb. per sq. in.

WATER ABSORPTION

Scott[22] carried out some water-absorption experiments on ebonites prepared from German synthetic rubbers in accord with the formulations shown in Table IV. The results obtained are recorded in Table V. The weight increase was used to calculate the amount (cubic centimeters) of water absorbed by 100 cc. of ebonite, this being termed the "apparent water absorption" (apparent because any weight increase due to oxidation would be reckoned as water). All of the ebonites used were vulcanized for 11 hours at 155° C.

Table V. Apparent Water Absorption (Percentage) of Synthetic-Rubber Ebonites

	97.2% Relative Humidity				0% Relative Humidity	
Mix	1 Day	4 Days	5 Days	6 Days	5 Days	7 Days
A	1.4	1.35	1.25	1.3	0.18	0.18
B	1.5	1.45	1.5	1.5	0.22	0.23
C	1.05	1.05	1.15	1.15	0.17	0.17
D	0.82	0.86	0.88	0.93	0.14	0.16
E	0.52	0.59	0.62	0.64	0.13	0.13
F	3.35	4.05	4.25	4.5	1.0	1.0

The results for all except the Buna-S ebonite (E) are higher than results previously obtained for ebonites made from washed smoked sheet, namely, 0.45 to 0.50 per cent. This greater tendency to absorb water may offset the superiority of the Buna ebonites in electrical properties in situations where the ebonite is exposed to damp conditions. The high absorption value for Buna-N ebonite (F) had not been anticipated since a soft-rubber vulcanizate made from a similar copolymer absorbs less water than a natural-rubber vulcanizate.[17] It is conceivable that the emulsifying agents used in making the emulsion polymers may be responsible for the anomalous water-absorption effects.

SWELLING IN ORGANIC SOLVENTS

Scott[22] also investigated the effect of organic solvents on the ebonites made from German synthetic rubbers shown in Table IV. Slabs of ebonite 20 × 20 × 5 mm. were placed in the liquids and weighed periodically. The swelling, i.e., volume of liquid absorbed by 100 volumes of ebonite, was calculated from the weight increase and the densities of the liquid and ebonite. All tests were carried out at 25° C. except those with transformer oil, for which 70° C. was used.

Table VI. Swelling in Organic Liquids of Ebonites from Synthetic and Natural Rubbers

Liquid Period of Immersion, Days	Benzene	Petroleum Ether	Gasoline	Transformer Oil*	Carbon Disulfide	Nitrobenzene
	150	56	64	28	3	50
Mix A	2.3	0.4	0.2	−0.2	37	0.4
Mix B	56	3.2	16.6	0.1	54	9.8
Mix C	12.5	0.2	0.9	0.0	46	0.7
Mix D	1.5	0.3	0.1	0.0	32	0.2
Mix E	10.0	0.3	0.2	0.0	52	0.6
Mix F	0.5	0.0	0.0	−0.5	0.6	0.2
65/35 rubber–sulfur†	62	−0.1			83	27

* At 70° C.
† Vulcanized for 5 hours at 155° C.

The results, in Table VI, clearly indicate that the Buna ebonites are generally much more resistant to the swelling action of benzene and nitrobenzene than natural-rubber ebonite, but there is less marked superiority of the Buna ebonites in resistance to carbon bisulfide. As might be expected, in series B, C, D, the swelling nearly always decreases with increasing sulfur content of the mix, just as in the case of natural-rubber ebonites. Furthermore, it should be noted that the acrylonitrile-butadiene ebonites (mix F) surpass all of the other synthetic ebonites in resisting the swelling action of the solvents.

Gartner[7] has shown that Buna 85, Buna S, Buna SS, and natural rubber have satisfactory stability and resist swelling when immersed in water for 8 weeks at 95° C. Under these same conditions of immersion Perbunan and Perbunan Extra proved to be unsatisfactory. With the exception of Perbunan and Perbunan Extra, the Buna types just mentioned fail to resist swelling when immersed in benzene for 8 weeks at 20° C.

ACID-RESISTANT HARD RUBBERS

Since hard rubbers are widely used because of their resistance to corrosion, Gartner[7] further studied the behavior in contact with various acids and chemicals of hard rubbers made from German elastomers. The results obtained are shown in Table VII. It is interesting to note that the results show Perbunan or acrylonitrile hard rubber to be unsatisfactory in acids as well as in chlorine gas.

Table VII. Effect of Acids on Hard Rubbers

	Satisfactory	Unsatisfactory
Hydrochloric acid (12 weeks at 20° C.)	Natural rubber Buna 85 Buna S Buna SS (All equally good)	Perbunan Perbunan Extra
Sulfuric acid, dilute (12 weeks at 20° C.)	Natural rubber Buna 85 Buna S Buna SS (All equally good)	Perbunan Perbunan Extra
Hydrofluoric acid (2 weeks at 20° C.)	Heavily loaded blends of natural rubber with Buna 85 Neoprene Buna SS	Perbunan Perbunan Extra
Nitric acid (8 weeks in 32% at 20° C.)	Neoprene Buna SS (Both vulcanized in hot air)	Perbunan Perbunan Extra
Chlorine gas (4 weeks at 70° C.)	Buna 85 vulcanized in steam Natural rubber	Perbunan Perbunan Extra (Both vulcanized in steam)

DIELECTRIC PROPERTIES

Winspear, Herrmann, Malm, and Kemp[26] studied the variation of dielectric constant and power factor with frequency and sulfur content in ebonites made from GR-S and from a copolymer of 50 per cent butadiene and styrene. The results are shown in Table VIII. Data on a 30 per cent acrylonitrile hard rubber, as well as data of A. H. Scott, McPherson, and Curtis[22] on natural-rubber ebonite, are included for comparison. With the exception of the acrylonitrile hard rubber, the sulfur content appears to have only a small effect on the dielectric properties of the products under the given conditions.

Winspear and co-workers[26] have shown in another series of experiments that the change in dielectric constant and power factor with time of cure of the GR-S and the 50 per cent butadiene-styrene-copolymer ebonites is also slight.

The dielectric constant and power factor of a GR-S hard rubber was measured over a frequency of 1 kc. to 14 Mc. The data are plotted in

Table VIII. Variation of Dielectric Properties of Natural and Synthetic Hard Rubbers with Sulfur Content and Frequency at 25° C.

	Sulfur Content %	Dielectric Constant					Power Factor, %				
		1 Kc.	3 Kc.	10 Kc.	30 Kc.	100 Kc.	1 Kc.	3 Kc.	10 Kc.	30 Kc.	100 Kc.
GR-S ebonites (cured 8 hr. at 153° C.)	20	2.86	...	2.84	...	2.82	0.48	...	0.43	...	0.53
	23	2.85	...	2.84	...	2.82	0.39	...	0.37	...	0.47
	26	2.94	...	2.91	...	2.89	0.68	...	0.55	...	0.63
	29	2.98	...	2.95	...	2.93	0.66	...	0.56	...	0.61
	31	2.97	...	2.94	...	2.92	0.67	...	0.56	...	0.64
Ebonites* (cured 8 hr. at 153° C.)	23	2.94	...	2.92	...	2.92	0.31	...	0.34	...	0.44
	26	2.87	...	2.86	...	2.85	0.31	...	0.33	...	0.48
	29	2.86	...	2.84	...	2.83	0.31	...	0.33	...	0.40
	31	2.91	...	2.89	...	2.88	0.36	...	0.35	...	0.39
Nitrile hard rubbers (cured 10 hr. at 142° C.)	30	3.81	3.77	3.71	3.65	3.58	1.62	1.79	2.10	2.44	2.79
	26	3.94	3.89	3.80	3.72	3.61	3.03	2.83	3.09	3.37	3.53
Natural rubber ebonites[22] (cured 25 to 40 hr. at 140° C.)	20	2.73	2.72	...	...	2.70	0.30	0.35	...	...	0.58
	23	2.73	2.72	...	...	2.70	0.25	0.33	...	...	0.63
	26	2.76	2.75	...	...	2.74	0.32	0.39	...	...	0.74
	29	2.79	2.78	...	...	2.75	0.42	0.48	...	...	0.75
	32	2.82	2.81	...	...	2.78	0.43	0.47	...	...	0.72

* 50% butadiene and 50% styrene.

Fig. 2, with corresponding data for a natural hard rubber. It is evident from these data that the dielectric behavior of natural and of GR-S rubber is similar. Both show a region of dielectric absorption spanning the greater part of the given range of frequency. The dielectric constant decreases slowly but steadily with increasing frequency, and the power factor rises to a maximum in the neighborhood of 3×10^6 cycles for natural rubber and 10^7 cycles for GR-S. This dielectric absorption is small compared to that found

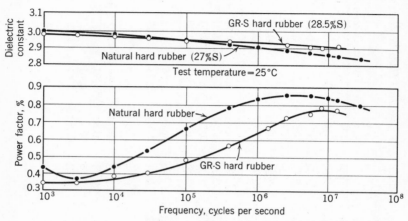

FIG. 2. Frequency Variation of Dielectric Constant and Power Factor of GR-S and Natural-Rubber Ebonites

at higher temperatures by Kitchin[13] and A. H. Scott, McPherson, and Curtis.[24] The data show that, at 25° C., GR-S hard rubber has over a wide frequency range insulating characteristics at least as good as natural hard rubber.

Ebonites A, B, C, D, E, made from the German elastomers shown in Table IV, were investigated by Scott[22] with regard to dielectric constant and power factor. Measurements were made at 1 million cycles per second, using alternating voltage of sine-wave form. The temperature of test was 20° C. In most cases the dielectric constant was only slightly below that of ordinary rubber-sulfur ebonite, i.e., 2.80 to 2.88, but the power factor was considerably lower, ordinary ebonites giving values between 0.77 and 0.95 per cent.

Fisher, Mullins, and Scott[6] have examined natural-rubber ebonites and those made from American and German synthetic elastomers with regard to dielectric constant and power factor over a wide range of temperature (20 to 120° C.) and frequency (1 to 1000 kc. per sec.). They have concluded with a statement that a rise in temperature produces a slight increase in dielectric constant and power factor, although the power factor may reach a maximum and then show signs of an eventual decrease. The course of the change in power factor with temperature divides the ebonites into three groups as follows: (1) Natural-rubber ebonite shows a rapid rise beginning between 60 and 70° C. (2) Ebonites from polybutadiene and from co-polymers of butadiene containing 25 and 40 per cent styrene show little change up to about 90° C., and consequently GR-S ebonite, although inferior to natural-rubber ebonite at room temperature, is superior at temperatures above 50° C., while ebonites from polybutadiene and the copolymer of butadiene and 40 per cent styrene become better than natural-rubber ebonite as the temperature is raised. (3) Ebonites from butadiene-acrylonitrile copolymers show more or less continuous increase.

SURFACE DETERIORATION OF HARD RUBBER BY LIGHT

Stöcklin[25] in 1939 observed that, like natural-rubber ebonite, the ebonites made from German synthetic elastomers suffer deterioration of electrical surface resistivity when exposed to light.

Hard rubbers made from the synthetic rubbers GR-S (butadiene–styrene) and GR-A (butadiene–acrylonitrile) have been compared by Crabtree[5] with natural-rubber ebonite from the standpoint of their rate of light-activated oxidation. The effect of calcium and barium carbonate additions to offset the effect of oxidation have been compared also. The only improvement practicable seems to be such as results from neutralizing the sulfuric acid arising from oxidation of the combined sulfur, by means of additives which will precipitate the acid as insoluble sulfates. The limit of oxidation of GR-S hard rubber is at a wavelength of 4600 A., and for natural rubber it is about 5200 A. The oxidation rate rises rapidly with decrease in wavelength. Hard rubber made with selenium in place of sulfur showed a decided improvement in resistivity since selenic acid is much less soluble than sulfuric. However, the selenium hard rubber was very poor in its physical properties.

In discussions on hard rubber the question of the contribution of free sulfur to the rate of oxidation usually arises. Accordingly, blocks of elemental sulfur were exposed outdoors under Corning glass filters. In comparison with an ebonite made with natural rubber, it was found that, while some oxidation takes place at short wavelengths, the rate of oxidation of sulfur is much slower. Behind window glass the rate should be zero.

Winspear, Herrmann, Malm, and Kemp[26] found that after 800 hours' exposure to an ultraviolet light, whiting-filled natural and whiting-filled GR-S hard rubbers exhibited surface resistivity values of 3.4×10^8 and 3.8×10^7 ohms, respectively, as compared to 4.8×10^5 ohms for natural-rubber ebonite after 800 hours and 3.3×10^5 ohms for GR-S ebonite after 110 hours. The importance of carbonates for the neutralization of the sulfuric acid arising from the oxidation of the free and combined sulfur on the surface of hard rubber is clearly seen.

According to Scott[22] all German Buna ebonites (Table IV) give good values of surface resistivity before exposure to light and are similar to natural-rubber ebonite in respect to this property. They all, however, deteriorate more quickly when exposed to light, and, judged by the number of hours required to bring the resistivity down to 10^{14} ohms, the deterioration of some of the Buna ebonites is almost twice as rapid as that of natural-rubber ebonite, as is shown in Table IX. After 60 hours' exposure, the Buna ebonites as well as the natural-rubber ebonite have a surface resistivity of approximately 10^{10} ohms.

Table IX. Fall in Surface Resistivity of Synthetic- and Natural-Rubber Ebonites (Table IV) on Exposure to Light

Hours to 10^{14} Ohms

A	B	C	D	NR*	E	F
$12\frac{1}{2}$	$9\frac{1}{2}$	9	$10\frac{1}{2}$	$17\frac{1}{2}$	$12\frac{1}{2}$	8

* Natural-rubber ebonite, 68/32 rubber–sulfur.

CELLULAR HARD RUBBER

Blowing agents may be incorporated into a synthetic- or natural-hard-rubber stock in order to produce on heating a cellular or sponge hard rubber. The cellular structure results from the fact that these agents, such as sodium bicarbonate and diazoaminobenzene, decompose with heat and liberate carbon dioxide, nitrogen, ammonia, etc. The cells may be separate or interconnecting, depending on the type of blowing agent and the technique used. Production of cellular hard rubber presents a greater problem than that of cellular soft rubber because the stock must be held inflated for a longer period of time before cure is completed.

According to Schwarz[20] cellular hard rubber having individual closed cells and being made from synthetic or natural rubber can be floated on water almost indefinitely. Such low-density hard rubbers are also considered to be good acoustical and thermal insulators. When used as a core between thin sheets of aluminum, the cellular hard synthetic rubbers produce sandwich constructions of good strength, although not so strong as end-grain balsa wood.[21] It is reported by Clark, McCuistian, and Cheyney[4] that the thermal conductivity for cellular ebonite is approximately 0.20 B.t.u./hr./sq. ft./F.°/in., which is superior to that of most plastics. In Table X, Sebrell[21] has compiled some physical data for synthetic cellular hard rubbers.

Table X. Properties of Cellular Synthetic Hard Rubbers

	Density, Lb. per Cu. Ft.	Compressive Strength, P.S.I.	Compressive Modulus, P.S.I.	Tensile Strength, P.S.I.
Blown GR-A	6.8	120	5680	69
Blown GR-A	13.1	170	5680	157
Blown GR-S	9.0	200	...	110
Foamed GR-S latex hard rubber	9.0	45	1830	7

REFERENCES

1. Am. Soc. Testing Materials, D 530 (1944).
2. Cheyney, L. E., and Robinson, A. L., *Ind. Eng. Chem.*, **35**, 976–9 (1943).
3. Church, H. F., and Daynes, H. A., *J. Rubber Research*, **8**, 41–51 (1939).
4. Clark, R. A., McCuistion, T. J., and Cheyney, L. E., *India Rubber World*, **117**, 361–2 (1947).

5. Crabtree, J., Bell Telephone Lab., Private Communication, 1949.
6. Fisher, D. G., Mullins, L., and Scott, J. R., *J. Rubber Research*, **18**, 37–43 (1949).
7. Gartner, E., *Kautschuk*, **16**, 109–16 (1940).
8. Garvey, B. S., and Sarbach, D. V., *Ind. Eng. Chem.*, **34**, 1312–5 (1942).
9. Gibbons, P. A., *Trans. Instn. Rubber Ind.*, **10**, 494–516 (1935).
10. Gibbons, P. A., and Cotton, F. H., *Trans. Instn. Rubber Ind.*, **11**, 354–76 (1935–36).
11. Houwink, R., *Elasticity, Plasticity and Structure of Matter*, Cambridge Univ. Press, London, 1937, p. 183.
12. Kemp, A. R., and Malm, F. S., Hard Rubber, in *The Chemistry and Technology of Rubber*, edited by C. C. Davis and J. T. Blake, Reinhold, New York, 1937, pp. 635–55.
13. Kitchin, D. W., *Ind. Eng. Chem.*, **24**, 549–55 (1932).
14. Martin, S. M., Thiokol Corp., Private Communication.
15. Memmler, K., *Science of Rubber*, Reinhold, New York, 1936.
16. Morris, R. E., Mitton, P., Seegman, I., and Werkenthin, T. A., *Rubber Age N.Y.*, **54**, 129–33 (1943).
17. Newton, R. G., and Scott, J. R., *J. Rubber Research*, **13**, 1–19 (1944).
18. Norman, R. H., Westbrook, M. K., and Scott, J. R., *J. Rubber Research*, **19**, 89–100 (1950).
19. Rostler, F. S., *India Rubber World*, **117**, 492–7 (1948).
20. Schwarz, H. F., *India Rubber World*, **114**, 211–2 (1946).
21. Sebrell, L. B., *India Rubber World*, **114**, 388–90 (1946).
22. Scott, J. R., *J. Rubber Research*, **13**. 23–6, 27–31 (1944).
23. Scott, J. R., *J. Rubber Research*, **19**, 128–30 (1950).
24. Scott, A. H., McPherson, A. T., and Curtis, H. L., *J. Research Natl. Bur. Standards*, **11**, 173–209 (1933).
25. Stöcklin, P., *Trans. Instn. Rubber Ind.*, **15**, 51–66 (1939).
26. Winspear, G. G., Herrmann, D. B., Malm, F. S., and Kemp, A. R., *Ind. Eng. Chem.*, **38**, 687–94 (1946).

CHAPTER 17

RECLAIMED SYNTHETIC RUBBER

F. L. Kilbourne, Jr.

Xylos Rubber Company

Reclaimed synthetic rubber may be described in exactly the same terms as those used in early patents to describe reclaimed natural rubber, viz., as "the product resulting from restoring or utilizing waste vulcanized rubber." The title of this chapter, if construed literally, is not an exact one, because little unadulterated reclaimed synthetic rubber is being made in the United States. Butyl inner-tube reclaim and Neoprene reclaim are reclaimed synthetic rubber, strictly so-called. But most reclaim is made from scrap tires, and, since relatively few all-GR-S tires were made except during the early days of the synthetic-rubber program, little all-GR-S synthetic reclaimed rubber has been made. True, during the period 1945 through 1947 reclaims were available, made exclusively from S-3 passenger-car tires (in which a maximum of 2.5 per cent of new natural rubber was permitted[58]) or GR-S no. 1 peelings, but even these scraps were diluted by the original manufacturer with natural-rubber hydrocarbon, added either as rubber or in the form of reclaimed natural rubber. Thus reclaims derived entirely from synthetic rubber are relatively rare and will be considered only briefly later. Most of what is said in this chapter refers to the processing of mixed natural and GR-S rubber scraps and to the reclaims made therefrom.

The importance of the reclaimed-rubber industry is well illustrated by reference to the war period, when, during 1942, 1943, and 1944, reclaimed rubber was a strategic war material. Reclaimed rubber was used extensively in conjunction with natural rubber before the large-scale synthetic-rubber industry was born and later was used extensively with GR-S, as the stockpile of natural rubber became smaller and smaller. Reclaim even helped to alleviate a carbon black shortage in 1945. Table I shows the annual consumption of reclaimed, natural, and synthetic rubber in the period 1940 to 1952. Almost one out of every four pounds of rubber used in this period was reclaimed rubber. Obviously, the industry is an important part of the whole rubber industry. These figures, even when they are discounted to allow for the fact that reclaim does not consist entirely of rubber hydrocarbon, are still convincing evidence of the importance to rubber manufacturing of the reclaimed product.

As reclaimers foresaw the rapid change in type of hydrocarbon which was to enter their scrap piles during the war years, they began intensive research into methods for reclaiming the new type of rubber. It was found possible to use existing equipment to a large extent; as a matter of fact, the industry did its research with the desirability of making use of such equipment in mind,

592

Table I. Reclaimed Rubber Consumption in the United States, 1940–52

Long Tons

Year	Reclaim	Natural Rubber	Synthetic	Total	Ratio Reclaim to Total
1940	190,200	648,500	2,904	841,604	22.6
1941	251,231	775,000	6,259	1,032,490	24.3
1942	254,820	376,791	17,651	649,262	39.2
1943	291,082	317,634	170,891	779,607	37.4
1944	251,083	144,113	566,670	961,866	26.2
1945	241,036	104,429	693,580	1,040,045	23.2
1946	275,410	277,597	761,699	1,314,706	20.9
1947	288,395	562,661	559,666	1,410,722	20.4
1948	261,113	627,332	442,072	1,330,517	19.6
1949	222,679	574,522	414,381	1,211,582	18.4
1950	303,733	720,268	538,289	1,562,290	19.4
1951	346,121	454,015	758,897	1,559,033	22.2
1952	280,002	453,846	807,037	1,540,885	18.1
13-year totals	3,456,905	6,037,708	5,739,996	15,234,609	22.7

because of the wartime shortage of steel and facilities for building new equipment.

METHODS OF MANUFACTURE

As mentioned above, the introduction of GR-S into the scrap rubber available did not make necessary any radical changes in the processes of reclaiming or the equipment used. The resistance of synthetic cured rubber to reversion by heat was overcome by using larger percentages of softening oils and by using more effective oils and hence the existing reclaiming plants were fully utilized. The reader is, therefore, referred to older descriptions of reclaiming processes given by Miller[43] and Palmer,[45] since they still adequately describe the several unit processes by which most vulcanized rubber is reclaimed. A more recent description of reclaiming processes is that of Hader and le Beau.[26] A brief outline of the operations commonly used for reclaiming rubber is given here only to emphasize certain operations which are more important today than formerly because of synthetic scrap.

Purchase of Scrap. Scrap is bought to definite specifications, or, in the case of miscellaneous grades, on the basis of samples. Tires and tire parts which are obtained by splitting tires with sharp knives may be separated into types, e.g., natural and synthetic, or they may be purchased as "mixed." Considerable care is necessary in checking purchases against the samples.

Debeading. This operation serves to remove the heavy wire bead of tires. Some reclaimers today salvage the rubber around the bead by grinding the whole tire, bead and all, the bead wire being discarded later on in the grinding process. Other reclaimers have more effectively utilized the hand labor at the debeaders by simultaneously inspecting and sorting the tires at this point.

Sorting. Some manufacturers sort tires as they are received and are placed in storage yards. Others pay the scrap dealers for performing this operation. Shortly after the war, the presence of substantial percentages of both natural-rubber passenger-car tires and GR-S or "red dot" passenger-car tires made it advisable for most reclaimers to sort such scrap, often during the debeading operation. This made it possible for them to continue to supply the prewar type of natural-rubber reclaim, while segregating the synthetic tires in piles for special treatment. In some plants, for sorting purposes, visual examination was supplemented by an odor test, in which smoke from a burned portion of the tire was identified by smell, so that the tire could be marked either synthetic or natural. In other plants the reactivity, with a drop of a 1:1 mixture of sulfuric and fuming nitric acids, of the tread or sidewall of tires, the nature of the rubber in which was unknown, was used to separate the natural (fast-reacting) from the synthetic (slow-reacting) tires. Methods for identifying synthetic rubber have been discussed by Burchfield[11] and Kilbourne.[33] By 1949 most manufacturers of reclaimed rubber had abandoned chemical methods of sorting and had developed reclaiming recipes capable of handling mixtures of natural and GR-S hydrocarbons. By 1949, however, substantial quantities were available of large-size truck tires almost free from synthetic rubber, and such were adapted to sorting by size alone. A premium was paid at this time for them. The reason for sorting tires, when possible, during the period 1945 to 1948 was not that synthetic reclaim is inferior to the natural variety but that natural scrap tends to become oversoftened when exposed to the high concentrations of oils needed for reclaiming the synthetic scrap. Therefore, as long as many natural and many high GR-S tires were being received, better reclaims·could be made more economically when the scraps were separated. Less sorting is done today, because markings have been removed, and because most passenger-car and small-size truck tires are mixtures anyway. Steps are always taken to charge to the devulcanizers a mixture of synthetic and natural scrap as uniform as possible from day to day, so that the most effective concentrations of devulcanizing agents may be found and held constant and a uniform product secured. Skill in handling mixed scraps has now been acquired, and chemical sorting of individual tires is less necessary now than it was earlier. Whenever sorting can be done, it is still found to "pay off" in greater efficiency in the finishing process.

Grinding. Heat alone will sometimes produce sufficient devulcanization by breaking molecular bonds in vulcanized natural rubber. Synthetic rubber, however, invariably requires swelling or the action of chemicals in addition to heat, to become replasticized. For this reason, reclaimers have paid greater attention in recent years to the problem of grinding their scrap fine before heating. Since the rate of penetration of oils is proportional to the exposed surface, most reclaimers now subdivide their scrap to a maximum size of at most $1/_4$ in. diameter. Some scraps are ground to pass through a 30-mesh screen (approximately 0.016-in. opening). The finer the grinding, the more uniform is the distribution of the chemicals or swelling oils introduced. Furthermore, the finer the grinding of a mixture of natural and synthetic scrap, the more closely is the final objective of the reclaiming process approached, viz., the production of a homogeneous blend of well plasticized natural and synthetic rubber, with minimum loss of strength and elongation.

These principles are important features of several recently patented processes for reclaiming, which also depend on mechanical grinding or the generation of heat to accomplish the devulcanization of scrap rubber. For example, Evans[20-1] claims that vulcanized rubber may be plasticized by working it on a tight mill under oxidizing conditions with admixed acid material (fat and rosin acids) and a mercaptan at a temperature below 180° F.   Warren and the U.S. Rubber Reclaiming Company[2, 57] describe a process in which ground inner-tube scraps are mixed with oils and dioctyl disulfide and in which the scrap is further comminuted, to promote penetration of heat and reclaiming agents, by passing the mixture through a tubing machine provided with a nozzle having an annular opening a few millimeters in radius, several inches in length, and, in order to bring the composition quickly to a temperature of about 400° F., steam-jacketed. The time of treatment in the heating zone is approximately 3.1 minutes at approximately 407° F.

Sverdrup[2, 50] discloses the use of an extrusion plasticator for periods of 2 to 4 minutes at temperatures of 340 to 385° F., to reclaim ground (24-mesh) scrap rubber in the presence of thiophenethiol. Banbury, Comes, and Schnuck[8] treat vulcanized or semivulcanized scrap rubber in a Banbury mixer with oils, tackifiers, and pigments under close confinement, so as to require at least 1.5 h.p. per lb. of charge, for the purpose of creating a high shearing action on the charge, continuing such action until the temperature of the charge reaches 425 to 550° F.   The period of treatment at this temperature is 3 to 12 minutes. The process has been described by Comes.[15] Kelly[28] masticates ground scrap rubber, either natural or synthetic, at temperatures of 250 to 400° F. in the presence of a catalyst consisting of a mixture of a water-insoluble aliphatic amine and a phosphoric acid.   The process is intended to be less strenuous than Banbury reclaiming at higher temperatures and pressures.   Navone[44] has patented a continuous process of reclaiming which consists in conveying ground rubber, by means of a screw, against progressively increasing pressures with a means for limited counterflow, in order to produce a kneading action.   The entire process requires 13 minutes at temperatures approaching 400° F.

It remains to be seen whether these mechanical processes will supplant the older digestion or open-steam processes. The intimate blending of all parts of the charge before completing the plasticization is certainly an important contribution to the technique of reclaiming vulcanized scrap rubber, and this is especially true of mixed types of rubber.

A biological process for the defibering of scrap has been described by Stewart, Crawford, and Miller.[49a]   The scrap, ground to 0.25 in., is treated as a compost and the cellulose is decomposed in a period of 6 weeks by the action of suitable aerobic and sulfur-tolerent fungi in the presence of nutrient solution.

Devulcanisation. *Alkali Process.* This process has become somewhat less popular than it formerly was because synthetic scrap requires more plasticizers when alkali solutions are used than when neutral solutions are used. The reason for this is not the action of the alkaline solution on the rubber but that alkaline solutions inhibit the action of most chemical peptizers. Actually, in the absence of chemical peptizers, alkaline solutions have a greater softening action than neutral solutions (e.g., solutions of calcium chloride).

Acid Process. This process can still be used to remove fabric, usually followed by digestion or open-steam heating at high temperatures, to accomplish the devulcanization.

Neutral Process. Zinc chloride or calcium chloride solutions (1 to 3 per cent), like alkali solutions, result in destruction of the cotton fibers, so that the latter may be washed from the rubber. Neutral reclaiming solutions have no effect on most chemical peptizers such as xylyl mercaptan, naphthyl mercaptan, or alkylphenol sulfides, which are used whenever possible, to reduce the total amount of softening oil needed. Generally, such peptizers are expensive compared with ordinary softening oils.

Pan Process. This process continues to be used on fabric-free or low-fabric scraps. Here also chemical peptizers and oxygen, trapped in the heater or added as air under pressure, are effective chemical softening agents.

Mechanical Devulcanization. Mention of the processes of Kelly,[29] Evans,[21] U.S. Rubber Reclaiming,[59] Navone,[44] and Banbury and associates[8] has already been made under the subject of "Grinding," since all of these processes start out with a grinding action on the scrap. When this action generates sufficient heat, the particles tend to coalesce, and a milling action occurs which in some cases at least develops sufficient heat to cause devulcanization. The essential requirement of all these processes is the development of machinery that will stand the severe wear and strain involved in developing the required temperatures or pressures. It is a little early to know whether the use of mechanically developed energy will prove to be as practical as heating with steam. The Banbury mechanical process of reclaiming has come into limited use for producing reclaim from fabric-free scrap arising in factory operations.[5]

Washing and Drying. Synthetic scrap has caused no notable changes in these operations. Neutral digestion appears to have reduced the amount of suspended solids in waste liquor from devulcanizers and washing systems.

Refining. In general, refining has been more difficult since the advent of synthetic scrap. One reason for this is the necessity for care to prevent "streaks" of tailings, caused by synthetic scrap insufficiently ground or improperly softened, from getting into the finished slab. Such particles are recooked or reworked until they are soft and fine enough to disperse in the matrix of more normally plasticized reclaim. Furthermore, excessively tight refining increases the percentage of tailings and causes excessive wear of the refiner. For these reasons, a greater investment in refining (and grinding) equipment and in its maintenance seems to be necessary.

Limited Heating. In addition to the new emphasis on fine grinding, conventional reclaiming methods have been modified by limiting the amount of heating. Restricting the amount of heating to which scrap is subjected has the effect of limiting the hardening or cross-linking of the synthetic parts of it and also of limiting the degradation of tensile strength during reclaiming. Several investigators (see later) have proved that prolonged heating is unnecessary and even detrimental, especially to the plasticity of the reclaimed product. In some instances devulcanization periods have been reduced as much as 50 per cent.

Mention has already been made of the Evans and Warren patents. Another patent[30] features limited heating when synthetic rubber is reclaimed with softening agents, swelling agents, and tackifiers at temperatures over 212° F. for periods of less than 2 hours. Sverdrup and Elgin[51, 57] limit the heating period, when reclaiming is being conducted in the presence of available oxygen, to that time which elapses when the "first point of maximum plasticity" is reached. According to Sverdrup, Plumb, and Elgin,[52] when reclaiming is carried out for a "normal" period of time, there is first a rapid increase in plasticity until a maximum plasticity is reached; then the material hardens and its plasticity falls, and only later and more slowly does the plasticity again increase. Campbell[13] has patented an improvement in the refining operation which comprises cooling the bank of rubber in the refiner with partially dehumidified air cooled to below 40° F. The cooling of the rubber strengthens the refined sheet, especially when mixtures of natural and synthetic reclaim are being processed, and makes it easier to wind it up to form a dense plastic slab of reclaimed rubber.

Market Grades. A descriptive listing of 185 brands of reclaimed rubber, marketed in the United States in 1953 and made by various processes, is given by Drogin,[19a] who also offers a review of current reclaiming practice.

NEW RECLAIMING OILS OR CHEMICAL PLASTICIZERS

As has been indicated above, synthetic scraps are softened only with difficulty by heat in conjunction with low concentrations of prewar types of swelling agents or softeners. Hence attention has been directed to (1) the development of more effective oils, and (2) chemicals that (apparently) attack the molecules of vulcanized rubber under the influence of heat, oxygen, or mechanical working, with a resulting softening effect. The chief reason for thinking that a chemical action is involved is the large degree of softening action produced in comparison with the small amounts of chemical used.

When, during World War II, research was undertaken on the problem of reclaiming synthetic scraps, trials of aromatic mercaptans as catalysts or peptizers in the reclaiming process were promptly undertaken, since such chemicals were already known to accelerate the plasticization of raw natural rubber.[24, 49] A parallel development in England[9, 10] and in Germany[10, 25, 26] also occurred. The English materials were known as RRA's and the German materials as Renacits. Renacit I, II, and III were, respectively, naphthyl, trichlorophenyl, and 9-anthryl mercaptans, and were claimed to be decreasingly less irritating to the skin. As reclaiming agents for Buna S, they were said to be similar in potency. Renacit II, sold as a mixture with paraffin, displaced Renacit I and was in turn displaced by Renacit III in reclaiming. Renacit II was also used for peptizing new Buna S and Perbunan.

In Germany, much use in reclaiming was made of oxygen (air), usually in conjunction with Renacits. A typical process consisted in using about 15 parts of a mixture of softening oils with 4 parts of Renacit III, mixing these in an internal mixer with the ground rubber (100 parts), spreading the mixture on perforated iron sheets to a depth of 3 cm., and then heating with an air-steam mixture (60 lb. of air and 75 lb. of steam) for about 90 minutes. The devulcanized mixture was then refined in the usual manner. In the reclaiming of mixed natural and synthetic scraps the concentration of Renacit was reduced to 0.3–0.6 part, and this was well distributed in the scraps by grinding the latter extremely fine.

The Allied Commission investigating the German rubber industry reported[25] that experimental work had been carried out in Germany, seeking to find an organic chemical containing an active thiol group and also an acid group such as would cause hydrolysis of the cellulose in tire scrap. It is not clear whether lack of suitable digesters operating at 190 to 200° C. or lack of zinc or calcium chlorides, dilute solutions of which are used in the United States for recipes involving chemical reclaiming agents, was responsible for the German interest in organic hydrolyzing agents for cellulose.

Active interest in development of chemical plasticizers for reclaiming in the United States has been considerable. Since all of the published information is readily accessible to those especially interested in this field, only brief references to each separate investigation will be made. The research on this subject, however, has been intense, commensurate with the importance of the reclaiming industry and the difficulty of the problem of softening polymers (synthetic) which seem highly prone to become tougher when heated.

Kirby and Steinle[35] disclose the effectiveness of di(hydroxaryl) sulfides for plasticizing synthetic rubbers. The same investigators[35] find that vulcanized polychloroprene may be plasticized more easily in the presence of water and lecithin. They also find[55] that cotton or other cellulose fibers assist in the softening of vulcanized Neoprene which is being treated with pine oil and rosin oil in the presence of water at ordinary reclaiming temperatures (300 to 420° F.).

Sverdrup and Elgin[51, 57, 59] have disclosed the effective action of organic sulfides, disulfides, and mercaptans for reclaiming rubber, either natural or artificial, when the heating is limited. Sverdrup has found thiophenol and the lower alkylthiophenes to be satisfactory agents for reclaiming synthetic scrap.[50] Tewksbury and Howland disclose a process for reclaiming vulcanized rubber, including both natural and synthetic, in the presence of small amounts of bisalkoxyaryl disulfides in which the alkoxy groups contain 1 to 4 carbon atoms.[53]

Albert discloses the reaction products of sulfur chloride and dibutyl-*m*-cresol as reclaiming agents for synthetic rubber.[1] This and similar chemicals are discussed in detail as reclaiming agents by Cook, Albert, Kilbourne, and Smith, who draw the conclusion that oxidative scission of polymer vulcanizates plays a predominant role in most of the reclaiming processes.[17] Kilbourne has shown how Albert's reaction product may be used in reclaiming both natural and GR-S passenger-car tires and has compared the effect of alkali and neutral solutions and of long and short devulcanization periods on natural and synthetic scrap.[33]

D. S. le Beau[40] has tested as reclaiming oils some of the pure hydrocarbons which are possible components of coal-tar naphtha and found that in general their effectiveness in reclaiming seems to parallel their susceptibility to autoxidation—to peroxide formation. Petroleum solvent naphtha, which lacks such compounds, is not an effective reclaiming agent. In these experiments GR-S scrap ground to pass through a opening of 10 mm. was heated with 15 parts of the hydrocarbon per 100 parts scrap for periods of $^1/_2$ to 5 hours in open steam at 185 p.s.i. (194° C.). In most cases a maximum plasticity was reached in about one hour. Among the most effective hydrocarbons were dicyclopentadiene and indene.

Campbell finds that vulcanized synthetic-rubber–natural-rubber blends may be reclaimed more easily by using a blended oil containing 10–50 parts of a resin of the coumarone-indene type, partially above 700 and partially below 500 in molecular weight, and 90–50 parts of an oil in which the resin is soluble, which is high boiling, penetrates the scrap, and contains less than 4 per cent of naphthalene.[14]

Corkery has discovered a novel reclaiming oil blend for reclaiming synthetic rubber.[18] The blend may be a mixture of 20 to 80 per cent of a crude solvent naphtha boiling within a range of 145 to 210° C. and 20 to 80 per cent of a higher-boiling aromatic oil containing compounds similar in structure to biphenyl or bibenzyl.

Randall has patented the process of reclaiming mixed vulcanized rubbers by heating with 1 to 15 per cent of abietic acid and 3 to 30 per cent of Gray-tower resin oil, a residual by-product obtained in the refining of cracked gasoline.[48] This oil comprises compounds containing bi- and tricyclic fused-ring structures with short aliphatic side chains and is highly unsaturated.

R. V. le Beau finds alkyl- and arylamines to assist in the softening of vulcanized chloroprene and vulcanized butadiene copolymers in the presence of water[41] between 366 and 388° F. Le Beau also reports the same polymers to be softened by solvents carrying fat acids which contain 4 to 18 carbon atoms in the chain and which do not contain hydroxyl groups in the chain other than in the carboxyl groups.[42] Kelly disclosed that devulcanization of natural and synthetic rubber may be catalyzed by heating with a mixture of a water-insoluble aliphatic amine and a phosphoric acid.[29]

Dasher has patented the reclaiming action of acrylic esters on vulcanized nitrile rubber.[19]

Baldwin[3] has patented a process for reclaiming cured Butyl rubber by heating it in the presence of hydrogen sulfide in the temperature range of 250 to 350° F. for 1 to 5 hours or until the polymer becomes plastic and millable. (Cf. Chapter 24, p. 874.) Cured Butyl rubber reverts easily in the absence of added hydrogen sulfide at temperatures above 350° F.

DIGESTION CONDITIONS

The following data, taken from recent literature, are quoted in order to illustrate some of the statements already made regarding (a) the use of active reclaiming chemicals in pan reclaiming, (b) the use of such chemicals in digester reclaiming with neutral solutions, (c) the application of alkali solutions in the reclaiming of synthetic scrap, and (d) the need for larger proportions of softening oils and tackifiers in the reclaiming of synthetic- as compared with natural-rubber scrap.

Cook, Albert, Kilbourne, and Smith[17] report the following (among other) properties for reclaims made from GR-S scrap by heating the latter for 4 hours in open steam at 175 p.s.i. (378° F., 192° C.) with a phenol sulfide or an aromatic mercaptan as a reclaiming agent together with a swelling oil and a tackifier. Judging from the Mooney plasticity and the chloroform extract, it appears that the order in which the three reclaiming agents are listed is the order of their diminishing effectiveness. In parallel tests, aliphatic mercaptans were less effective than aromatic mercaptans.

GR-S tire scrap (5-mesh)	100 parts
Dipentene fraction, b.p. 173–201° C.	6 parts
Coumarone-indene resin, m.p. 115–125° C.	6 parts
Reclaiming agent	1.5

Reclaiming Agent	Mooney	Uncured chloroform extract, %	Modulus, 300%	Cured* T	E
4,6-Di-*tert*-butyl-3-methylphenol sulfide	41	14.85	375	800	400
α-Naphthyl mercaptan	50	12.68	400	925	480
Xylyl mercaptan	55	10.90	425	1000	460

* 30′/287° F. in the test recipe given on p. 602.

A neutral formula applicable to the reclamation of GR-S and mixed tires by the digester process is given by Kilbourne[33] as follows:

Scrap (4-mesh)	100
Water	203
Zinc chloride	2.73
Oil	19
Resin tackifier	6
Digestion conditions	10 hr. at 375° F.

The following data given by the same author illustrate the facts that, judging from the values of Mooney viscosity and chloroform extract, (1) the use of an active chemical reclaiming agent enables the devulcanization of GR-S scrap to be effected in neutral digestion with a much smaller proportion of oils than is needed without such an agent, (2) alkaline digestion reclaiming (where reclaiming chemicals are not effective) is for GR-S scrap somewhat more efficient than neutral digestion without reclaiming chemicals, (3) in both neutral and alkaline digestion, natural rubber needs less oil than GR-S. The data also illustrate the fact that the acetone extract of GR-S reclaims is higher than that of natural-rubber reclaims (because of the higher oil requirement for the former).

Reclaiming Recipe

Natural-rubber passenger-car tires (5-mesh)	100	100			
GR-S passenger-car tires (5 mesh)	...	...	100	100	100
Solvent naphtha (D-4 oil)	12	12	30	30	...
Paraflux	5	5	15	15	10
Alkylphenol sulfide	...	...	...	...	1.5
Calcium chloride solution (pH 7.5)	213	...	213	...	213
Sodium hydroxide solution (7%)	...	213	...	213	

Properties of Reclaim

After 5 hr. at 375° F.					
Mooney viscosity	63	55	78	63	65
Chloroform extract	13.15	20.85	10.53	12.38	11.45
Acetone extract	13.38	13.95	20.90	22.45	16.95
After 15 hr. at 375° F.					
Mooney viscosity	49	48	70	62	60
Chloroform extract	14.90	22.35	12.30	12.05	11.00
Acetone extract	16.45	15.10	19.83	23.88	22.80

PROPERTIES OF SYNTHETIC RECLAIMS

Chemical Analysis. The chemical analysis of prewar and present-day whole-tire reclaims shows some changes because of the higher synthetic content of the scrap tires of today. Typical analyses of reclaim in these periods are shown in Table II. The comparison shows modern, as compared

Table II. Typical Whole-Tire Reclaim Analyses

	Prewar		Postwar	
	1942	1942	1949	1949
Acetone extract	9.4	10.5	16.0	18.7
Cured chloroform extract	1.1	0.7	1.2	1.0
Ash	16.9	13.9	13.0	11.4
Total sulfur	1.7	1.7	1.5	1.5
Carbon black	13.7	17.2	18.4	17.1
Cellulose*	4.0	0.9	2.8	0.7
Rubber content (by difference)	53.2	55.1	47.1	49.6
Alkalinity	0.11	0.28	0.07	0.05
Specific gravity	1.17	1.15	1.19	1.18

* This determination was not generally run on reclaimed rubber before the war but is usually reported in complete analyses today.

with prewar reclaims, to have a higher content of softening or tackifying oils (acetone extract), to reflect the higher carbon black loading characteristic of synthetic tire compounds, and to have lower alkalinity because alkaline solutions are avoided, on account of the fact that chemical softeners are not generally compatible with them.

Le Beau[39] has studied the direct or chromic oxide method of determining the natural hydrocarbon content of synthetic reclaims. Data in Table III show how the synthetic content of a typical whole tire reclaim varied over the period 1940–50. The ratio of synthetic consumption to natural-rubber consumption by years in tires is shown in the same table.

Table III. Synthetic Content of Typical Whole-Tire Reclaim

Year	Ratio of Synthetic- to Natural-Rubber Hydrocarbon*	Ratio of Synthetic to Natural in Tires Made in U.S.A. (New Rubber Only)†	
		Pass. Tires	Truck Tires
1940–42	...	...	...
1943	...	40 : 1	1 : 2.65
1944	...	50 : 1	2 : 1
1945	...	50 : 1	3 : 1
1946	1 : 4	7 : 1	1 : 1
1947	1 : 3	1 : 1	1 : 7
1948	1 : 2	1 : 1.2	1 : 12
1949	1 : 2	1 : 1.5	1 : 12
1950 (est.)	...	1 : 2.0	1 : 12

* Data in this column supplied by D. S. le Beau of the Midwest Rubber Reclaiming Co.
† Data in these columns supplied by D. A. MacDougall of the economics department of the Firestone Tire & Rubber Co.

Vulcanization Tests. Since 1942 reclaimed rubber has generally been tested in a standard formula[31] as follows:

Reclaimed rubber hydrocarbon	100
Zinc oxide	5
Stearic acid	2
Sulfur	3
Mercaptobenzothiazole	0.5
Diphenylguanidine	0.2

Cure 10, 15, 20, 25, etc., minutes at 40 p.s.i. (287° F.) until optimum cure is reached.

Data given in Table IV show physical tests on typical prewar and postwar reclaims in this test formula.

Table IV. Typical Reclaim Physical Tests

Description	Date	Specific Gravity	Cure, Min. at 287° F.	Elongation, %	Tensile Strength, P.S.I.
Whole tire (alkali)	1942	1.16	25	340	1045
Whole tire (neutral)	1942	1.17	20	313	810
Modified tire	1942	1.25	20	310	965
Black tube	1942	1.14	20	525	1525
Black blend	1942	1.35	20	330	740
Brown blend	1942	1.30	20	455	1100
Whole tire (alkali)	1949	1.19	25	385	830
Whole tire (neutral)	1949	1.19	20	350	900
Modified tire	1949	1.24	25	295	940
Black tube	1949	1.18	20	585	1465
Black blend	1949	1.49	20	285	615
Brown blend	1948	1.31	20	485	935
S-3 tire reclaim	1948	1.19	20*	375	855

* At 312° F.

It will be seen in Tables II and IV that the chemical and physical tests do not show any remarkable differences between prewar and postwar (synthetic) reclaims, partly because the postwar reclaim is only partially synthetic. The last reclaim listed in Table IV was made out of S-3 tires and was relatively high in GR-S content, viz., about 1 part to 0.6 part of natural hydrocarbon. This and the fact that it was a neutral reclaim account for a slower rate of curing so the optimum cure was reached only after curing 35 minutes at 312° F. Such slow curing was characteristic only of high GR-S reclaims. Most whole-tire reclaims, whether of the alkali or the neutral type, now cure at about the same rate as prewar reclaims.

Quality of Synthetic Reclaim. Typical recipes containing natural and synthetic reclaims were reported in 1944 and 1945 by Randall,[47] Busenburg,[12] and Kilbourne.[31] These chemists concluded as follows:

1. GR-S reclaim possesses physical properties equal to those of natural-rubber reclaim and is less tacky and more inert than the natural-rubber product; its milling, calendering, and extruding properties are excellent; its lack of tack is usually an advantage.

2. The absence of alkali in reclaims of high GR-S content results in slower curing, which, however, can be offset by lengthening the time of cure, raising the temperature of cure, or increasing the percentage of sulfur.

3. Satisfactory soft- and hard-rubber formulations can be made with GR-S reclaim in place of natural-rubber reclaims.

4. GR-S reclaim contains more carbon black than natural reclaim. This was a real advantage during World War II when carbon black was scarce.

The above authors were discussing reclaim made from the early (S-3) synthetic tire, which contained relatively more GR-S than postwar tires.

Conover[16] and Plumb[46] also studied the early reclaiming of GR-S. They pleaded for cooperation of tire manufacturers in marking of tires to permit easy sorting, and pointed out the inferior quality of reclaims made before that time (1945) from mixed tires. In more recent years, as pointed out in the introduction to this chapter, reclaims from mixed tires are now produced with as high quality as prewar natural-rubber reclaims. Such a reclaim is discussed by Kilbourne,[33] who also gives typical mechanical goods recipes containing this reclaim to meet various specifications.

Ball and Randall,[6] by reclaiming natural and synthetic tread vulcanizates and mixtures of the two in the laboratory and then using the resulting reclaims in new compounds made with natural, synthetic, or mixed rubbers, reached the conclusion that the changing composition of whole-tire reclaim during the postwar years would not greatly affect the final properties in typical tread compounds, whether these were made out of natural, synthetic, or mixed rubbers. Fuhrmann and Randall[22] have compared postwar and prewar whole-tire reclaims in tire carcass, undertread, black-heel, automotive mat and hard-rubber storage-battery compounds. All of these recipes contained substantial percentages of reclaim, and the synthetic or postwar reclaim was equivalent to the prewar natural reclaim in all respects.

Ball and Randall[7] published a very interesting study which proved that reclaimed rubbers, especially postwar synthetic reclaims, have a beneficial effect on the aging of compounds in which they are used. According to Torrence and Schwartz, whole-tire reclaim blended with Neoprene gives especially advantageous results in resistance to ozone cracking, weathering, cut growth, and heat aging, compared with blends of reclaim with natural rubber and with GR-S.[54]

Plasticity of Reclaim. The measurement and control of reclaim plasticity has been of interest since World War II, probably because of the adoption of the Mooney viscometer for measuring the plasticity of GR-S during the war. Kilbourne, Misner, and Fairchild[34] compared the Williams, Mooney, extrusion, and milling-test methods for measuring plasticity of reclaim and reached the conclusion that the Mooney (or the extrusion) method and the milling test should both be used. The Mooney method and the extrusion method appear to measure plasticity, but the milling test seems to measure residual elasticity, which appears to increase on aging in all reclaims to a varying degree. Little increase in power consumption occurs in reclaim aged for 2 months. Plasticity samples should be cut from the full width of the slab because the center of the slab is usually less plastic than the edges. The writer feels that synthetic reclaims are generally more plastic than prewar natural reclaims, possibly because they contain more plasticizer.

ADVANTAGES OF RECLAIMED RUBBER

Because there are many young technologists in the rubber industry and because the experience of many of these has been mostly with synthetic rubber, it seems advisable to emphasize the reasons why reclaimed rubber is used extensively. These reasons, as important today with synthetic reclaim as they were before 1945, are as follows.

1. Reclaimed rubber used alone or with new rubbers speeds up the mixing operation. It breaks down quickly, wets the pigments, and blends the

mixture together, thereby saving several minutes on the mixing rolls or in the Banbury.

2. Reclaimed rubber, since it contains pigments already dispersed and because it has been plasticized, saves power. An experiment showed that whole-tire reclaim consumed only 55 per cent as much power as a crude-rubber compound of approximately the same material cost and analysis.[34] Palmer reported that reclaimed rubber consumed only 60 to 65 per cent as much power as crude rubber.[45]

3. Reclaimed rubber permits faster extrusion and calendering.[44] This is an observation made also over many years by factory men, who attribute the result to the stabilizing influence of reclaim on the gage of the compound during processing.

4. Reclaimed rubber imparts firmness to compounds, making them easier to handle during processing. For example, on repeated milling, reclaim stocks tend to break down less than stocks prepared from new rubber; they thus exhibit a more consistent plasticity. Another way in which the improved handling characteristics of reclaimed rubber compounds show up is in decreased sagging and loss of shape. In other words, reclaim decreases the thermoplasticity of compounds in which it is used.

5. Reclaim decreases shrinkage of uncured compounds during their life in the uncured state and also reduces the shrinkage that takes place during cure.

6. Reclaim speeds up the curing of compounds into which it is introduced. This saves press or mold time.

7. Reclaim has excellent aging properties, when used in either the cured or the uncured condition.[7] The severe treatment to which it has already been subjected includes oxidation, heating, digestion, washing, and mastication in air, and these appear to have stabilized the hydrocarbon against further changes. Reclaim is used extensively in friction tape, where retention of tackiness is important. Cured reclaim compounds were used for many years in auto topping, where resistance to sunlight was important.

8. Reclaimed rubber is plastic and may be worked with solvents (usually in the presence of resins) to form useful, high-solids cements. It may also be easily dispersed in water, yielding important adhesives with or without the addition of natural or synthetic latexes. Smooth dispersions in water may be made from synthetic reclaims, but natural reclaims are still preferred for solvent cements.

9. Reclaimed rubber is a manufactured product and subject to rigid control and inspection to a degree that natural rubber was formerly not subjected. The present trend is toward closer control and inspection of all types of rubber, whether natural, synthetic, or reclaimed.

10. Reclaimed rubber can be modified to suit individual customer requirements, with respect to physical or chemical properties, color, cost, size of package or slab and can often be purchased in regularly spaced shipments, thus reducing warehouse space requirements.

11. Reclaimed rubber is low and stable in price because it is a domestic manufactured product made from a plentiful scrap product. GR-S should also be stable in price. And in a freely competitive market these two raw

materials together could exert a more effective stabilizing influence on the price of natural rubber than reclaimed rubber alone was able to accomplish before 1940.

THEORY OF DEVULCANIZATION

Synthetic-rubber scrap breaks down to a lesser degree than natural-rubber scrap under identical reclaiming conditions.[33] The breakdown is probably similar in the two kinds of scrap and is measurable by the proportion of chloroform extract in the final product. The additional oils required in the reclaiming of high-synthetic scraps are needed in part to increase the degree of breakdown of the synthetic vulcanizate and in part because their presence in the final product supplements the low content of chloroform-soluble rubber-decomposition products which make reclaim plastic and workable. The accurate relevant data available in the literature is insufficient to show the rate at which maximum plastic properties are developed in synthetic as compared with natural scrap under identical conditions of time and temperature, although le Beau[36] indicates that GR-S softens faster than natural rubber but quickly becomes harder with prolonged heating. In general, it may safely be said that, in contrast to natural reclaiming processes, prolongation of the heat treatment seems to confer little benefit in synthetic reclaiming recipes.[25]

Several investigators[26, 30, 33–4, 37–8, 51] have shown that both natural and synthetic scrap rubber tends, either in process or after reclaiming, to become tougher rather than more plastic. This change is especially noticeable in the milling properties of the reclaim[34] and is accelerated by elevated temperatures of storage.[26, 33] Reclaims that have been more thoroughly plasticized, whether by longer heating, the use of alkali, or the use of swelling oils, seem to be more stable than less thoroughly plasticized reclaims with respect to decrease in plasticity[33–4] or change in chloroform extract[33, 38] on aging. Inadequate devulcanization results in very unstable reclaim, which apparently oxidizes rapidly. The presence of swelling agents during devulcanization results in softer more stable reclaim.[38] The instability of short-cooked natural-rubber reclaims lends support to the view that oxygen acts during preliminary stages of reclaiming by forming hydroperoxides, which engage in chain reactions, with resultant scission of the molecules.

Le Beau has suggested[37] that the high rate of molecular breakdown during the early stages of the reclaiming process is caused by decomposition of hydroperoxides formed possibly in the grinding process or during the initial heating. The faster rate of production of reclaim by open-steam heating as compared with water digestion seems to accord with the possibility that greater concentration of oxygen is present in the heater process. Further, it is known that the decomposition of olefin hydroperoxides occurs faster in alkaline than in neutral media, and this seems to agree with the greater rate of molecular breakdown (as measured by chloroform extract or plasticity) in alkaline as compared with neutral reclaiming solutions. The atom in the isoprene chain most likely to undergo hydroperoxidation is the α-methylenic carbon atom. Since GR-S does not have substitutional groups in the carbon atoms at the double bond, hydroperoxidic reactions may be expected to proceed at a different rate from those in natural rubber.[37] Hence, while alkali solutions greatly increase the degree of devulcanization of natural-rubber vulcanizates, they cause only a slight increase in degree of devulcanization of synthetic vulcanizates.[33]

The role played by mercaptans and disulfides in reclaiming has been assumed to be catalytic in nature, in view of the fact that oxidative scission is almost surely of great importance in reclaiming and because mercaptans and disulfides catalyze oxidation reactions.[17] It is noteworthy that most chemical plasticizers used in reclaiming also have a pronounced softening effect on natural rubber or GR-S when these are masticated at temperatures from 100 to 150° C. Thus it appears that the action of chemical softeners is on the hydrocarbon chain rather than on the sulfur linkages of the vulcanizates. As mentioned above, several authors[17, 33, 36-8] have attempted to explain (1) the role of oxygen in promoting devulcanization, (2) the decrease in plasticity during the aging of reclaim, (3) the greater instability of reclaim in the production of which only a short devulcanization period has been employed, (4) the decrease in unsaturation of reclaimed rubber during reclaiming, (5) the role of peptizers as catalysts for oxidative scission of the rubber molecules during reclaiming, and (6) the effect of the pH of the medium in which devulcanization takes place. The development and application of improved analytical methods for determining oxygen and peroxides in vulcanized rubber at various stages of the reclaiming process is essential in order that we may learn the true mechanism by which rubber is replasticized. With such knowledge we may even learn how to speed up molecular scission of vulcanized GR-S and slow down that of vulcanized natural rubber so that the two scraps can be more easily reclaimed together.

OTHER SYNTHETIC RECLAIMS

Reclaimed Neoprene and Nitrile Rubber. Neoprene was one of the first synthetic rubbers to be reclaimed. The tendency of Neoprene to harden when heated and its resistance to softening by oils make it necessary to use large quantities of oils in reclaiming Neoprene scrap. In experiments on the reclaiming of Neoprene scrap with the aid of an efficient reclaiming chemical, as much as 35 parts of oils and tackifiers per 100 parts of scrap has been specified.[30] Kirby and Steinle[35] have found that the hydrolytic decomposition products of cellulose assist in plasticizing Neoprene vulcanizates. Fraser[23] has compared Neoprene reclaimed rubber from several different sources and found Neoprene reclaim useful in reducing cost and conserving new Neoprene. However, Neoprene scrap is not plentiful and is not so important commercially as other synthetic reclaims. The same comment may be made concerning reclaims made from nitrile-rubber scrap, which is also scarce and resistant to softening by reclaiming oils and heating. A typical high-grade Neoprene reclaim has the following properties:

Acetone extract	35.4	per cent
Ash	11.5	per cent
Polymer	50.0	per cent
Specific gravity	1.24	per cent

Reclaimed Butyl Rubber. Butyl rubber has been used widely in inner tubes and therefore constitutes a fairly large source of scrap. Little information has appeared in the literature concerning Butyl reclaim. However, this new type of synthetic reclaim has been made and used in large quantity since 1950, not only in inner tubes but in mechanical goods also. More publications on the properties and uses of Butyl reclaim may be expected in the next few years.

Butyl reclaim is usually more nervy (elastic) than tire reclaims. The unsaturation of Butyl rubber itself is low; consequently, that of reclaimed Butyl may be even lower. Difficulty in curing Butyl reclaim is sometimes experienced, and hence substantial quantities of curing ingredients are usually used. A satisfactory test formula is given below:

Test Formula for Butyl Reclaim

Butyl reclaim	182
Zinc oxide	5.5
Sulfur	2.3
Mercaptobenzothiazole	1.1
Tetramethylthiuram disulfide	1.1
Selenium diethyldithiocarbonate	1.1

When cured 10 to 30 minutes at 312° F., this recipe gives vulcanizates with a tensile strength of 900 to 1500 p.s.i. and an elongation of 500 to 600 per cent, depending on the type of Butyl in the original scrap, the degree of devulcanization, and the extent of contamination with natural or GR-S reclaim. It is well known that unsaturated rubbers or oils are harmful to the curing properties of Butyl compounds because they remove the free sulfur from the compound more rapidly than it can combine with the Butyl hydrocarbon. Data given by Busenberg[12a] show that the presence of 5 per cent of natural rubber tube reclaim in a sample of Butyl tube reclaim reduced the maximum tensile strength in the above test formula from 1630 to 1100 p.s.i.

The Technical Committee of the Rubber Reclaimers' Association has adopted the following test formula for Butyl reclaim. It is not as rapid as and does not produce vulcanizates as snappy as those produced by the test recipe already given, but it is closer to trade practice.

Reclaimers' Standard Test Recipe for Butyl Reclaims[12a]

Butyl reclaim hydrocarbon	100.0
Zinc oxide	5.0
Sulfur	2.0
Tetramethylthiuram disulfide	1.0
Mercaptobenzothiazole	0.5

A typical Butyl reclaim has the following chemical analysis:

Typical Analysis of Butyl Reclaim

Acetone extract	2–4	per cent
Ash	4–7	per cent
Carbon black	29–30	per cent
Total sulfur	1.5–1.9	per cent
Hydrocarbon	60–65	per cent
Specific gravity	1.15	

Butyl reclaim possesses the same properties, low air permeability and good aging, that make the parent rubber desirable for some rubber products. Because of the low unsaturation, Butyl reclaims possess outstanding resistance to oxidative aging and ozone. The low content of oils in the early Butyl tube reclaims resulted in good resistance to migration or contact staining of enamels or light-colored rubber veneers. This may not be true of future Butyl tube reclaim because of the trend toward the use of tougher Butyl with larger proportions of softeners. Like natural-rubber tube reclaims, Butyl reclaim forms smooth cements, which may have some application in the future.

SUMMARY

The complete "devulcanization" of the vulcanized synthetic-rubber molecule still eludes research investigators, like that of vulcanized natural rubber. New reclaiming processes have appeared which are applicable to natural or synthetic scrap rubber, and some of them have been put into commercial operation. Synthetic scrap rubber is more heat-resistant than natural scrap rubber and sometimes becomes harder on heating, but both scraps behave alike under the influence of oils and mechanical working, so that only minor changes in process are necessary to handle synthetic scraps. The industry successfully reclaims mixed scraps, but it is still looking for a means of softening synthetic scraps more rapidly than natural scraps. Fine grinding and extensive refining are even more important today than formerly.

The reclaimed product today, although facing stiff competition, still offers low cost, stability of processing and curing properties, faster processing, and other valuable qualities which give promise of a healthy future for the nearly 100-year-old reclaiming industry.

With the rapid development of the large-scale synthetic-rubber industry, the reclaiming industry was faced with the problem of coping with the transition from raw material consisting of natural-rubber scrap to raw material comprising large quantities of synthetic-rubber scrap. This problem it has successfully solved. Reclaimed rubber is still of great value in rubber manufacturing and represents one of the three important sources of rubber hydrocarbon.

REFERENCES

1. Albert, H. E., Brit. Pat. 635,129 (1950).
2. Augenstein, J. G., and Sverdrup, E. F., *India Rubber World*, **120,** 92 (1949); *Rubber Age N.Y.*, **65,** 63 (1949).
3. Baldwin, F. P., U.S. Pat. 2,493,518 (1950).
4. Ball, J. M., *Reclaimed Rubber*, Rubber Reclaimers' Assoc., New York, 1947, p. 25.
5. Ball, J. M., *India Rubber World*, **126,** 77–8 (1952).
6. Ball, J. M., and Randall, R. L., *India Rubber World*, **114,** 368–73 (1946).
7. Ball, J. M., and Randall, R. L., *Rubber Age N.Y.*, **64,** 718–22 (1949).
8. Banbury, F. H., Comes, D. A., and Schnuck, C. F., U.S. Pats. 2,461,192–3 (1949); T. Robinson, U.S. Pat. 2,221,490, Nov. 12, 1940; F. H. Cotton and P. A. Gibbons, U.S. Pat. 2,408,296, Sept. 24, 1946.
9. British Ministry of Supply, Private Communication, June 1943.
10. British Ministry of Supply, Private Communication, May 1946.

11. Burchfield, H. P., *Ind. Eng. Chem., Anal. Ed.*, **16**, 424–6 (1944); **17**, 806–10 (1945).
12. Busenberg, E. B., *Rubber Age N.Y.*, **57**, 181–5 (1945).
12a. Busenberg, E. B., *Rubber Age N.Y.*, **70**, 608 (1952).
13. Campbell, C. H., U.S. Pat. 2,468,482 (1949).
14. Campbell, C. H., U.S. Pat. 2,471,392 (1949).
15. Comes, D. A., *India Rubber World*, **124**, 175–7 (1951).
16. Conover, F. S., *Rubber Age N.Y.*, **57**, 308–10 (1945).
17. Cook, W. S., Albert, H. E., Kilbourne, F. L., and Smith, G. E. P., *Ind. Eng. Chem.*, **40**, 1194–1202 (1948).
18. Corkery, F. W., U.S. Pat. 2,449,879 (1948).
19. Dasher, P. J., U.S. Pat. 2,465,205 (1949).
19a. Drogin, I., *India Rubber World*, **128**, 772–4; **129**, 63–7 (1953).
20. Evans, W. W., Brit. Pat. 571,784 (1945).
21. Evans, W. W., U.S. Pat. 2,414,145 (1947).
22. Fuhrmann, A., and Randall, R. L., *Rubber Age N.Y.*, **64**, 201–4 (1948).
23. Fraser, D. F., *India Rubber World*, **105**, 150–2 (1941).
24. Garvey, B. S., U.S. Pat. 2,193,624, Mar. 12, 1940.
25. German Reclaiming Industry as reported by the Allied Commission sent to investigate the German rubber industry.
26. Gillman, H. H., and Haines, W., Private Communication.
27. Hader, R. N., and le Beau, D. S., *Ind. Eng. Chem.*, **43**, 250–63 (1951). See also Dorris. T. B., *Rubber Age N.Y.*, **71**, 773–80, 821 (1952).
28. I.G. Farbenindustrie A.-G., Kautschuk Zentrallaboratorium, Rept. on Reclaiming of Buna Vulcanizates with the aid of Renacit, Aug. 28, 1945.
29. Kelly, J. H., U.S. Pat. 2,477,809 (1949).
30. Kilbourne, F. L., Jr., U.S. Pat. 2,324,980 (1943).
31. Kilbourne, F. L., Jr., *India Rubber World*, **111**, 687–99 (1945).
32. Kilbourne, F. L., Jr., *Rubber Age N.Y.*, **62**, 541–2 (1948).
33. Kilbourne, F. L., Jr., *Rubber Age N.Y.*, **64**, 581–9 (1949).
34. Kilbourne, F. L., Jr., Misner, J. E., and Fairchild, K. W., *Rubber Age N.Y.*, **66**, 423–8 (1950).
35. Kirby, W. G., and Stienle, L. E., U.S. Pats. 2,359,122 (1944); 2,363,873 (1944); 2,372,584 (1945); 2,414,428 (1947).
36. Le Beau, D. S., *Rubber Age N.Y.*, **62**, 51–5 (1947).
37. Le Beau, D. S., *India Rubber World*, **118**, 59–65 (1948).
38. Le Beau, D. S., *India Rubber World*, **119**, 69–74 (1948).
39. Le Beau, D. S., *Anal. Chem.*, **20**, 355–8 (1948).
40. Le Beau, D. S., *Rubber Age N.Y.*, **68**, 49–56 (1950).
41. Le Beau, R. V., U.S. Pat. 2,423,032 (1947).
42. Le Beau, R. V., U.S. Pat. 2,423,033 (1947).
43. Miller, G. W., Chap. 22 in *Chemistry and Technology of Rubber*, edited by C. C. Davis and J. T. Blake, Reinhold, New York, 1937.
44. Navone, J. O., U.S. Pat. 2,487,666 (1949).
45. Palmer, H. F., *Rubber Age N.Y.*, **41**, 25–29, 93–8 (1937).
46. Plumb, J. S., *India Rubber World*, **112**, 307–8 (1945).
47. Randall, R. L., *Rubber Age N.Y.*, **56**, 65–6 (1944).
48. Randall, R. L., U.S. Pat. 2,471,496 (1949).
49. Simmons, H. E., War Production Board, Office of Rubber Director, Private Communications, Feb. 23 and June 3, 1943.
49a. Stewart, W. D., Crawford, R. A., and Miller, H. A., *India Rubber World*, **127**, 794–6, 801 (1953).
50. Sverdrup, E. F., U.S. Pat. 2,494,593 (1950).
51. Sverdrup, E. F., and Elgin, J. C., U.S. Pat. 2,415,449 (1947).

52. Sverdrup, E. F., Plumb, J. S., and Elgin, J. C., *India Rubber World*, **119**, 223 (1948); *Rubber Age N.Y.*, **65**, 64 (1948).
53. Tewksbury, L. B., and Howland, L. H., U.S. Pat. 2,469,529 (1949).
54. Torrence, M. F., and Schwartz, H. G., *Rubber Age N.Y.*, **71**, 357–60 (1952).
55. U.S. Rubber Co., Brit. Pat. 557,803 (1943).
56. U.S. Rubber Co., Brit. Pats. 575,545–7 (1946).
57. U.S. Rubber Reclaiming Co., Brit. Pat. 610,812 (1948).
58. War Production Board, R–1 Orders of Dec. 4, 1943; July 1, 1944; Feb. 1, 1946.
59. Warren, S. F., and U.S. Reclaiming Co., Brit. Pat. 610,901 (1948).

CHAPTER 18

SYNTHETIC-RUBBER RESINS

J. D. D'Ianni

Goodyear Tire & Rubber Company

Before World War II natural rubber was used extensively for the production of chemical derivatives by chlorination (Parlon,[75] Tornesit,[41] Pergut[85]), hydrochlorination (Pliofilm[66]), isomerization (Pliolite,[66] Marbon B,[109] Thermoprene[52]), and oxidation (Rubbone[37]). The critically short supply of natural rubber during the war eliminated the production of almost all these products, except for essential needs which could not be fulfilled by substitute materials. As GR-S and other synthetic rubbers became available during the war, it was logical for the rubber chemist to turn to them as starting materials for chemical derivatives which could be used in place of their natural-rubber counterparts. This change was made with reasonable success for cyclized and chlorinated derivatives, but since the synthetic products showed no real advantage over the natural-rubber derivatives, production of them was discontinued at the end of the war.

Of more permanent interest are the resinous products which were concurrently developed as a result of the availability in large quantities and at low cost of the monomers employed in synthetic-rubber production—butadiene, styrene, and acrylonitrile. Attention will be particularly directed to the resins of low-butadiene and high-styrene content (hereafter referred to as high-styrene resins) which were developed during the war and have since enjoyed an ever-expanding market. In addition to these two major topics, brief mention will be made of other chemical derivatives of synthetic rubbers which are reported in the literature but have not achieved commercial stature. Reference to natural-rubber derivatives will be made only for comparative purposes; more detailed information regarding them is available in reviews already published by Fisher,[53] Jones,[88] Sibley,[143] Memmler,[114] Farmer,[50] Dawson,[37-39] Schidrowitz,[39, 136] and, most recently, Le Bras and Delalande.[101] Comparisons of natural- and synthetic-rubber derivatives were made by Thies[158] and by Endres.[48]

ISOMERIZATION OF SYNTHETIC RUBBERS

The literature describes several attempts to isomerize synthetic rubbers to resinous materials of commercial utility. Buizov and Kusov[24] heated sodium–polybutadiene to 230 to 270° C. and obtained 33 to 67 per cent yields of isomerized product. Natural rubber underwent the same reaction at a much lower temperature (160° C.). Addition of sulfuric acid, sulfonic acids, etc. lowered the isomerization temperature to 100 to 140° C. for natural rubber, but not for polybutadiene. This difference was attributed to the absence of methyl substituents in the synthetic rubber. The effect of chemical structure on the relative ease of reaction with isomerizing, i.e., cyclizing, reagents was further emphasized in work on GR-S reported by Endres[48] and on isoprene polymers by D'Ianni, Naples, Marsh, and Zarney.[46] The ready tendency of isoprene polymers to cyclize was found to require a correction factor, in a careful study by Lee, Kolthoff, and Mairs[103] of the determination of unsaturation with iodine monochloride. Fromandi[59] obtained cyclorubbers by subjecting rubber and synthetic polyisoprene to a silent electric discharge.

Isomerized GR-S.[48] Practically no reaction occurred when GR-S was subjected to the conditions that readily isomerized natural rubber, viz., treatment of a benzene solution at reflux temperature with chlorostannic acid or boron fluoride. If the temperature of reaction was raised to 160 to 180° C. by the use of higher-boiling solvents (phenol, cresol, neutral coal-tar oils, or naphthalene), the addition of chlorostannic acid readily effected the desired reaction; within a short time the temperature began to rise, and the solution formed a gel which was difficult to agitate. Product isolated during the gel stage was insoluble in the common organic solvents. On further reaction, the temperature decreased and the viscosity fell until the reaction mixture became a thin, brownish solution from which the isomerized product was isolated by steam distillation or by precipitation with a nonsolvent. The final product was a brown resin, soluble in aromatic and chlorinated hydrocarbons and insoluble in aliphatic hydrocarbons and alcohols.

The properties of isomerized GR-S depend on the experimental conditions used in its preparation. The hardness and softening point depend on reaction temperature, reaction time, and proportion of catalyst. With increasing severity in the conditions of preparation, the products can be varied from soft, rubberlike materials to hard, brittle resins, with softening points as high as 105° C.

In addition to chlorostannic acid, stannic chloride and boron fluoride could be used under the conditions described above. GR-S was converted into a hard rubbery product when treated with concentrated sulfuric acid at 180° C. for 3.5 hours. Hydrogen chloride did not, as it does with natural rubber, form a resinous product at 180° C. Aluminum chloride in a refluxing toluene solution converted GR-S to a resinous material which had a high softening point and was insoluble in ordinary solvents.

Table I lists the characteristics of GR-S isomerized by one of the procedures outlined above.

Paper coatings containing cyclized GR-S are transparent, tasteless, and odorless and possess a low rate of moisture vapor transmission. Other suggested uses are in moisture-, alkali-, and acid-resistant paints and lacquers, and as a reinforcing agent in GR-S wire coating stocks and other rubber stocks.

The production of cyclized GR-S did not proceed beyond the pilot-plant stage. Laboratory work was done with other synthetic rubbers, such as butadiene-acrylonitrile copolymers and polychloroprene, but cyclization either did not occur or required extremely large proportions of catalyst.

Table I.   Properties of Cyclized GR-S[48]

Specific gravity	0.99–1.03
Softening point	55–105° C.
Solubility	Aromatic hydrocarbons, chlorinated hydrocarbons,
Iodine number	
Original GR-S	350
Soft cyclo GR-S	200
Hard cyclo GR-S	180–60
Compatibility	Paraffin, GR-S

Water-Vapor Transmission Rate on Glassine
(Compounded with 10 per cent paraffin)

Coating Weight, Lb. per Ream	W.V.T.R. G./Sq. M./24 Hr.
0.7	4.08
2.4	1.79
3.5	1.09

Isomerized Isoprene Polymers[46]

Preparation.   In contrast to GR-S and other butadiene polymers, synthetic isoprene polymers are isomerized under almost the same conditions as those applicable to natural rubber.   During the war stannic chloride and (as the ether complex) boron fluoride were used to isomerize polyisoprene to resins of considerable commercial utility (Pliolite S-1 and S-2, respectively).[66] It was desirable, however, to start with polymer of a lower solution viscosity and to react to a lower viscosity in the product than is the practice with natural rubber.   As in the case of natural rubber,[160] isomerization causes an enormous decrease in solution viscosity.   To obtain the resin in the form of discrete particles, the cyclized polymer solution was acidified, and glue or gelatin was added before removal of the solvent by steam distillation.[43]

Polyisoprene is also isomerized to a resin by hydrogen fluoride, a catalyst employed by Lawson,[100] Garvey,[60] and Söll[147] with natural rubber. Addition of hydrogen fluoride to the double bond apparently does not take place.   Under the same conditions GR-S remains unaffected.

Properties.   Table II lists a number of properties of Pliolite (from natural rubber) and Pliolite S-1 (from synthetic polyisoprene) and compares unmilled resin with resin hot-milled for 5 minutes.   The chief effect of milling was to make the resin more soluble and to give lower viscosity solutions.

The two products are similar in specific gravity, index of refraction, ash content, iodine number, chlorine content, and percentage of acetic acid formed by chromic acid oxidation.[25]   Pliolite S-1 possesses a lower molecular weight than Pliolite, as indicated by viscosity and Olsen stiffness data.

The solubility characteristics of Pliolite S-1 and Pliolite are very similar. Pliolite S-1 is soluble in benzene, toluene, gasoline, mineral spirits, and chloroform; slightly soluble in ethyl acetate (4 per cent), methyl ethyl ketone (1.7 per cent), and acetic acid (0.7 per cent), and insoluble in dioxane,

Table II. Properties of Isomerized Natural Rubber (Pliolite) and Isomerized Synthetic Polyisoprene (Pliolite S-1)[46]

| | Pliolite S-1 | | Pliolite | |
	Unmilled	Milled	Unmilled	Milled
Color	Cream	Brown	Cream	Amber
Specific gravity	1.09	1.04	1.12	1.06
Refractive index	...	1.535	...	1.545
Ash, %	2.12	1.89	1.70	1.69
Iodine no.	157	163	180	180
Chlorine, %	1.52	1.28	1.01	1.63
Intrinsic viscosity	0.36	0.25	0.49	0.41
Gel, %	0	0	0	0
Heat distortion point, °C.	...	62	...	54
Acetic acid by chromic acid oxidation, %	28.3	24.9	29.7	24.1
Solution viscosity (Ford cup no. 4) min.				
Toluene, 10%	0.21	0.18	0.26	0.21
20%	0.62	0.25	2.13	0.64
30%	3.46	0.58	15.75	3.44
Textile spirits, 10%	0.18	0.17	0.20	0.19
20%	0.36	0.21	0.91	0.37
30%	1.87	0.35	12.97	3.31
Resin/Paraffin, 30/70				
Brookfield viscosity, centipoises				
180° F.	2520	2250	6775	2370
200° F.	2015	870	5050	2240
235° F.	965	760	1810	775
W.V.T.R. (5 lb./ream coating), g./100 sq. m./24 hr.	0.40	0.47	0.78	0.70
Glassine Coating				
W.V.T.R. (3 lb./ream coating)	0.32	0.21	1.33	0.19
Tinius-Olsen stiffness test				
Bending moment, in.-lb.		1.32		5.68
Angle at break, degrees		5		17
Water absorption (20 hr. at 70° C.), mg./sq. in. (Thickness, 0.08 in.)		8–10		3–5

alcohol, and acetone. It is completely compatible with mineral waxes such as paraffin and ceresin.

A Pliolite S-1/paraffin 30/70 mixture exhibits considerably lower Brookfield viscosities at 180 to 235° F. than a corresponding Pliolite mixture and is, therefore, more advantageous to use in hot-melt dips. The water-vapor transmission rate (W.V.T.R.) of a 5-lb.-per-ream coating on glassine is considerably lower for a Pliolite S-1 than for a Pliolite coating.

The electrical properties of Pliolite S-1 are excellent and compare favorably with those reported for Pliolite by Thies.[157] The electrical properties of a Pliolite S-1/low-water-absorption GR-S 50/50 masterbatch are as follows:[46]

	1 Kilocycle	1 Megacycle
Dry Sample		
Dielectric constant	2.60	2.63
Power factor, %	0.167	0.274
Loss factor	0.0043	0.0072
D.c. specific resistivity at		
540 volts ohm.-cm.	3.6×10^{14}	3.6×10^{14}
After 24-Hr. Immersion in		
Water at 50° C.		
Dielectric constant	2.68	2.73
Power factor, %	0.195	0.493
Loss factor	0.0052	0.0135
D.c. specific resistivity at		
540 volts, ohm.-cm.	10.1×10^{14}	10.1×10^{14}

Applications. Isomerized polyisoprene was very useful for a number of applications during World War II. It is an excellent reinforcing agent for natural rubber, GR-S, Butyl rubber, and other synthetic rubbers. It can be used as a dispersing medium for pigments in paints, printing inks, etc. Its compatibility with waxes allows its use in hot paraffin dips. It has found use as coatings for kraft paper, labels, etc. When it is used in rubber-coated fabrics, improvements in feel, embossing characteristics, and abrasion resistance are obtained.

When used with many synthetic rubbers, this resin improves processing, decreases tack, reinforces while maintaining elongation and flexibility, maintains good electrical properties, and results in stocks of low gravity. Extrusion operations are improved, since the addition of the resin results in smoother extrusion surfaces or higher rate of extrusion. In high-gum compounds for thin-wall tubing, it imparts stiffness in the uncured state and makes possible perfect centering during extrusion.

Compounding. Rubber-resin masterbatches are prepared by banding the resin on a hot mill (200 to 215° F.) and adding the rubber in small amounts. During mixing, the batch is cut back and forth in the usual way until a homogeneous stock is obtained. A smooth, flexible stock is obtained by a few final passes through a cold mill. The reverse procedure of adding the resin to the rubber on the mill is usually not successful because the resin is merely dispersed in the rubber as a pigment, and there results a grainy, heterogeneous masterbatch. Incorporation of the resin in a GR-S gum stock increases the hardness, tensile strength, and ultimate elongation. In a GR-S black stock, if 10 volumes of Wyex black are replaced by 10 volumes of resin, the elongation is increased and the modulus decreased, while the tensile strength and hardness remain unchanged. Stress-strain values for molded, uncompounded stocks show that Pliolite reinforces natural rubber more efficiently than does Pliolite S-1 or S-2, but that in GR-S the resins are equivalent. Butyl rubber and isomerized polyisoprene are compatible when mixed on the mill. A tough, rubberlike compound is obtained with

75 parts Butyl rubber and 25 parts resin. Larger proportions of resin give hard, leatherlike stocks.

Mechanism of Isomerization. The chemical change taking place when polyisoprene or natural rubber is treated with stannic chloride, boron fluoride, phenolsulfonic acid, etc. is probably best represented by the following reaction:

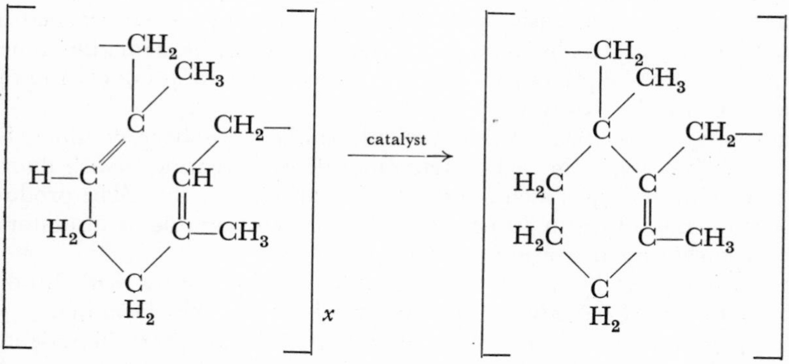

This reaction mechanism explains most of the following changes in properties when polyisoprene or natural rubber is isomerized with chlorostannic acid: (1) increase in specific gravity, (2) increase in refractive index, (3) decrease in unsaturation to approximately one-half the original value, as measured by infrared absorption, iodine number, and percentage of acetic acid formed by chromic acid oxidation, (4) disappearance of the $RR_1C = CHR_2$ type of double bond, by infrared absorption data, (5) no substantial change in C_5H_8 empirical formula, (6) lower molecular weight, from intrinsic viscosity measurements, (7) increase in heat distortion point. Gordon, in a study of the kinetics of cyclization in Hevea latex, offered additional support for the reaction mechanism just cited.[68]

Cyclized natural and synthetic polyisoprene solutions are much less viscous than the corresponding rubber solutions; the final viscosity values may be as low as 0.05 to 0.001 of the original. Intrinsic viscosity measurements in dilute benzene solutions also show a marked drop; thus:

| | Intrinsic Viscosity | |
	Before Reaction	After Reaction
Natural rubber	4.45	0.49
Polyisoprene	1.51	0.36

The decrease in chain length due to the cyclization reaction proposed is not sufficient to account for the drastic reduction in viscosity obtained, and presumably some chain scission accompanies the cyclization reaction.

HALOGENATION OF SYNTHETIC RUBBERS

The halogenation of synthetic rubbers has been the subject of intensive investigation, as is evidenced by the number of literature references on the subject. However, industrial application of the products has been quite limited, since they were manufactured only during the war period when natural rubber was not available. In the United States chlorinated polyisoprene (Pliochlor,[66] Parlon-X[75]) was produced in limited quantities during the war. In Germany Buna S was chlorinated to a product called Bunalit at the production rate of 3 tons per month. No information is available on production of similar products in other countries. The chlorination of GR-S was studied extensively during the war but never reached the production stage. In this section a brief review of the available general literature will first be presented, and then the commercial products and their applications will be described.

Fluorination[146] of natural and synthetic rubbers has been conducted by dissolving the polymer in carbon tetrachloride and treating it with fluorine diluted with an inert gas such as nitrogen or carbon dioxide. The products, which may contain up to 30 per cent fluorine, are suitable as insulators in electrolytic cells for the production of fluorine.

The bromination[110] of natural- or synthetic-rubber latex with bromine water produces a derivative containing up to 70 per cent bromine.

Special synthetic rubbers were chlorinated by various investigators. Polymethylpentadiene[29] formed resinous chlorine-containing derivatives by reaction with chlorine in carbon tetrachloride solution, with tert-butyl hypochlorite in benzene solution, and with hydrogen chloride in toluene solution. Polychloroprene[14, 30, 94] readily chlorinated in chloroform solution at 45° C. in diffuse daylight to form a derivative containing 68 per cent chlorine, which corresponds closely to the theoretical figure for the addition product $(C_4H_5Cl_3)_x$. The derivative was suggested for use in coating, impregnating, adhesive, and molding applications. A butadiene-acrylonitrile copolymer[23, 34] when lightly chlorinated to 1 to 3 per cent chlorine content yielded a tough, elastic product suitable for cable sheathing, molding compounds, gaskets, tires, and hose.

The chlorination of polyisobutylene has been the subject of patent investigations. Frolich[58] found that chlorination in the presence of an organic acid (acetic or aminoacetic acid) improved the thermal stability of the material. Morrell, Frolich, and Bannon[118] obtained enhanced stability by chlorinating in the presence of 1 to 10 per cent natural rubber. Gleason and Rosen[65] chlorinated isobutylene polymers in carbon tetrachloride solution by mixing with chlorine, exposing a thin film to light, and quickly removing it from the reaction zone to prevent undesirable side reactions. These fire-resistant resinous products are suitable for impregnating cloth, paper, and wood. Murray[119] cautioned that chlorination of synthetic rubber (in this case Butyl rubber) in liquid chlorine at elevated temperatures (ca. 100° C.) proceeded with explosive violence. The chlorination of synthetic rubbers with liquid chlorine at atmospheric pressure was described in other patents.[63, 127]

Russian investigators have studied the chlorination of butadiene rubbers, but no information is available regarding the commercial exploitation of the products. Koshelev[98] reported that chlorination of Soviet natural and synthetic rubbers proceeded best in dichloroethane solution. Poor results were obtained in the absence of solvent. The presence of lead decreased solubility but had no effect on the stability of the product. The presence of iron reduced the solubility and lessened the stability of the product. Neither metal affected the percentage of chlorine combined. The products were stable to heating for 12 hours at 60 to 80° C. Zaionchkovskii[169] obtained insoluble products containing as much as 40 per cent chlorine when a butadiene-type rubber was chlorinated in benzene solution.

The stability of chlorinated rubbers can be improved by the addition of ethylene oxide derivatives[115] (epichlorohydrin, phenoxypropene oxide, etc.) or organic isonitroso compounds,[102] or by carrying out the chlorination and subsequent operations in the presence of nickel surfaces.[120]

Chlorinated GR-S. The chlorination of GR-S was extensively studied during the years of World War II, but an economical process suitable for production was not developed. Endres[48] described the properties of chlorinated GR-S, as listed in Table III, but gave no details on its preparation. As compared with chlorinated natural rubber, the synthetic material exhibited higher tensile strength and ultimate elongation than the natural-rubber derivative and had similar chemical resistance. The superior flexibility of the synthetic material was considered a significant advantage which would allow its use where chlorinated natural rubber was not applicable.

Table III. Properties of Chlorinated GR-S[48]

Chlorine content	53%
Viscosity (20% solution in toluene)	0.50–20.0 poises
Solubility	Aromatic hydrocarbons, chlorinated hydrocarbons, ketones, esters of organic acids.
Compatibility (5% or more)	Paraffin, chlorinated paraffins, tricresyl phosphate, dibutyl phthalate, isobutyl linoleate.
Stability	Equal to chlorinated natural rubber.
Softening point	75°–85° C.
Water-vapor transmission on glassine	
Coating weight, lb. per ream	W.V.T.R.g./sq. m./24 hr.
3.3	114.7
5.6	70.4
10.5	49.9
Flexibility	Greater than chlorinated natural rubber.
Specific gravity	1.36–1.39
Adhesion	Equal to chlorinated natural rubber.
Tensile strength	5100 p.s.i.
Elongation	18.7%

Remy[110] stated that synthetic rubbers such as GR-S could be chlorinated in latex or solution form in the presence of a metallic naphthenate of group 1 of the periodic table of elements, e.g., sodium and copper. The solution viscosity was reduced by the treatment, and the need for oxidation processes was obviated. Lichty claimed the chlorination of GR-S-type polymers dissolved in ethylene dichloride and in the presence of iodine.[105] Other patents concerning the chlorination of butadiene polymers have been issued to Reid[128] and to Briant.[22]

Chlorinated Buna Polymers. The chlorination of Buna polymers was studied extensively by German workers because, with the war demands, natural rubber was no longer available for the production of Pergut, Tornesit, and similar materials. These studies culminated in the successful, though expensive, production of substantial quantities of chlorinated Buna S, known as Bunalit, during World War II. Hebermehl[73] has discussed the preparation, properties, and uses of chlorinated Buna.

A survey of the literature reveals a number of patents, mostly assigned to I. G. Farbenindustrie A.-G., relating to the chlorination of Buna-type polymers. Schweitzer[137] preferred the simultaneous addition of chlorine and oxygen to the rubber solution, stating that less initial milling was required and that the final product was superior to the product obtained in the absence of oxygen. Polybutadiene[84] could be chlorinated with chloroform, ethylene dichloride, or chlorobenzene as the solvent, but carbon tetrachloride was not suitable. The products are soluble in acetone and aromatic hydrocarbons and compatible with softeners like dibutyl phthalate, tritolyl phosphate, chlorinated diphenyl, and drying and nondrying oils. Some patents[16] stressed that the chlorination of butadiene polymers should be conducted at temperatures above 60° C., to obtain products with high softening point and chlorine content, and soluble in drying oils. In another patent[17] the chlorination was begun at a temperature below 60° C. and finally at least as high as 60° C., to produce a derivative suitable for use in paints and lacquers. Heat treatment[18] to depolymerize the butadiene polymer before solution was beneficial, as was the preparation[15] of the polymer in the presence of a modifier (diisopropylxanthogen disulfide) to increase its plasticity and solubility. The stability[117] toward concentrated sulfuric acid of rubber chloride from synthetic rubber improved with increasing chlorine content (60 per cent minimum value for good stability).

Manufacture of Chlorinated Buna S. Specific information on the chlorination of Buna rubbers and the properties and applications of the commercial products was made available as the result of the investigations of British[21] and American[40] teams of scientific workers sent to Germany shortly after the close of World War II. Pertinent details are briefly summarized in the following paragraphs.

Large-scale production of Bunalit (chlorinated Buna S) by I. G. Farbenindustrie A.-G. started in 1939 when natural rubber ceased to be available for chlorination. The process and equipment were generally similar to those used for natural rubber but were modified in view of the much more serious corrosion problems encountered. Buna S was heat-softened to a Defo number 250 and cut into strips 1 cm. wide before its solution. The rubber was dissolved in chloroform at 50° C., and the quantities per charge were 200 kg. of Buna S and 6000 kg. of solvent. The batch temperature was kept at 20 to 25° C. by passing brine through the jacket, until 265 kg. of chlorine had been added. Addition of chlorine was stopped and the batch heated to the boiling point under total reflux, to remove hydrogen chloride. The second chlorination stage was carried out at a pH of 3.5, and 335 kg. of sodium acetate was added after cooling to 20 to 30° C. This step was necessary, since it was found that chlorination could not be completed in the presence of hydrogen chloride. A further amount of 265 kg. of chlorine was then added at a temperature of 30 to 40° C.

The total time in the chlorinator was about 1 week, of which 36 to 48 hours and 24 hours were required for the first and second chlorination steps, respectively. When chlorination was complete, the solution was fed to the precipitator containing distilled water maintained at 80 to 90° C. Vigorous agitation caused the product to separate as granules. The mixture of water and precipitated product was cooled to 30° C., treated with sodium hypochlorite solution for about 30 minutes, and then neutralized with acetic acid. Sodium thiosulfate was introduced to remove free chlorine, and finally the stabilizer, phenoxypropene oxide, was added. After being stirred for an hour, the batch was filtered and washed with distilled water. The product was centrifuged to remove surface moisture and fed to the drier. Drying was accomplished at 60 to 70° C. under a pressure of 40 mm. of mercury in a horizontal stainless-steel cylinder vacuum drier heated by circulating hot water.

The product was variously stated to contain between 55 and 62 per cent chlorine, but the solvent content was not known. The output of the Leverkusen plant was 3 to 4 tons per month and could be raised to 5 tons per month maximum. It was reported that some 1000 tons per year was produced at Schkopau. The yield of chlorinated Buna was 2.4 to 2.5 tons per ton of Buna S. No data were found on mechanical losses, but with natural rubber the chlorine consumption was estimated at 10 per cent more than the theoretical amount, and the carbon tetrachloride loss was 1 ton per ton of product.

Bunalit was made in two viscosities, grades N and H (viscosity values not given). Plasticator 32E, a low-molecular-weight Buna made at Leverkusen by emulsion polymerization with Diproxid as modifier, was probably admixed with Buna S in the manufacture of the lower-viscosity product. Bunalit was found to be a close match to chlorinated natural rubber (Tornesit and Pergut) for most purposes, except that a little alcohol had to be added to its solutions in xylene, toluene, butyl acetate, etc., to obtain a clear lacquer. Solid samples of Bunalit were of a pale buff color.

Chemistry of the Chlorination. Some information is available on the chemical changes occurring during the chlorination of Buna rubbers. The I.G. staff used their well-known K value from viscosity and osmotic-pressure measurements to control the process of chlorination and to determine its effect on molecular weight. The K value increased in the case of natural rubber but considerably more so in the case of the Buna rubbers. A chlorine content of 30 to 35 per cent was considered the danger point from the viewpoint of gelling for both types of rubbers. The solvent was considered to influence the degree of cyclization and of cross-linking or of both during chlorination. Solvents containing less hydrogen were considered to cause more cyclization. Though carbon tetrachloride was satisfactory for natural rubber, chloroform was used in the manufacture of Bunalit, since Buna was considered to be more easily cyclized than natural rubber. This viewpoint does not agree with the facts noted above on the relative ease of cyclization of natural rubber and polyisoprene, as compared with a butadiene rubber, GR-S. Substitution reactions with the formation of hydrogen chloride readily occur when natural rubber is chlorinated. However, much less hydrogen chloride is formed when Buna rubbers are chlorinated, although quantitative figures for this comparison are not available.

Uses. The main uses of Bunalit were in protective coatings of a wide variety and in rubber-to-metal adhesives. For many of these applications, Bunalit was used as a straight substitute for chlorinated natural rubber, although certain differences were noted. The thermal stability of Bunalit was considered by some to be inferior to that of Pergut or Tornesit. Lacquers made from Bunalit were darker and of poorer clarity than those made from Pergut. Bunalit was not considered to be so resistant to acids and alkalies as Pergut, and it was less flexible than the latter.

Use in Protective Coatings. Bunalit found application in the production of paint for anticorrosive purposes and in this form was used to a large extent by I.G. within its own organization. Chlorinated rubber paints were used for acid-resistant and alkali-resistant coatings and also as coatings for new concrete. In preparing the paints, Bunalit was dissolved in normal-type paint mixers. Pigments were mixed with one-half of the plasticizer and a small quantity of the chlorinated rubber solution. This was milled on a three-roll mill and mixed with the remaining solution to form the finished paint. Bunalit paints were considered suitable for under-water use, for the protection of steel in and near chemical plants, and for gas holders. Other recommended applications were in paints and lacquers for superstructures and underground structures, in paints for the chemical and allied industries and for vehicles, for the impregnation of textiles, paper, etc., and in paints for the wood industry. Bunalit products were also used for special cable lacquers and printing colors. Chlorinated rubber paints were proved outstanding for weather resistance as well as for chemical and mechanical requirements. In some cases they also showed to advantage in resistance to the action of seawater and to damage by termites.

Use in Metal-to-Rubber Bonding. A typical formula for the use of chlorinated rubber as a primer for metal, especially iron, is given in the accompanying table. Bunalit was also used as an adhesive for coated abrasives, such as sand paper and emery paper, and as an extender for polyvinyl ethers, to improve their adhesive quality.

Bunalit (or Pergut H)	100
Chlorinated biphenyl	
(Clophen A.60)	50
Dimethylthianthrene (Sintol)	30
Red lead	520
Xylol	270–300
Butanol or ethylene glycol	30–0

The use of chlorinated rubber in rubber-to-metal adhesives is referred to as the Dartex process and is also associated with the trade names, BX Bonding and BDZ Bonding. Bunalit, according to some reports, found use in such adhesives, either alone or with Desmodur R (triphenylmethane triisocyanate). In the manufacture of molded goods, the pickled or sand-blasted components were coated with a 15 per cent solution of Bunalit in xylene, followed by two more coatings with Bunalit dissolved in xylene and methylene chloride. After aging for 3 to 24 hours, the metal and Perbunan blank were cured at a temperature not exceeding 145° C. and removed from the mold after cooling to 80° C. Adhesion strengths as high as 35 kg. per sq. cm. were obtained in mass production with Perbunan stocks. Use with other rubbers was not recommended.

Another rubber-to-metal adhesive application was as a solution prepared from equal parts of a 15 per cent solution of Bunalit in xylene and a 20 per cent Desmodur solution in methylene chloride. One coating of the adhesive was sprayed on the cleaned metal surfaces. The rubber compound was applied to the metal quickly, but not before 15 minutes. Protection from moisture was essential. The adhesive hardened in the heated mold within a very short time, and articles could be removed from the mold hot without affecting the adhesion. Various metals and rubbers were bonded by this technique.

Chlorinated Isoprene Polymers.[46, 99] The chlorination of isoprene polymers, particularly emulsion polyisoprene, was extensively studied in the United States during World War II, in the attempt to obtain a satisfactory substitute for chlorinated natural rubber. The most noteworthy point of interest about this work was that polyisoprene behaved in the chlorination process very similarly to natural rubber, whereas other commercially available polymers, such as GR-S, failed to give the desired results. This work culminated in the limited production of chlorinated polyisoprene (Parlon-X, Pliochlor) during the war period.

Polyisoprene was prepared in the GR-S type of polymerization recipe with the modifier content adjusted to give a product of 40 to 45 Mooney viscosity. When a 5 per cent solution of this polymer in carbon tetrachloride was prepared, an extremely viscous, gelled solution was obtained. Several methods were studied for reducing the viscosity, so that the chlorination could be readily carried out and would produce a chlorinated rubber in the desired viscosity range (less than 150 centipoises). The most practical way found to "break down" the polymer was to hot-mill or Banbury it to a Mooney viscosity of 20 or below.

As with natural rubber, the presence of oxygen exerted a profound influence on the chlorination of polyisoprene. The action of oxygen appeared to be twofold: depolymerization, which reduced the molecular weight and viscosity; and chlorination catalysis, which was of particular value in obtaining highly chlorinated products. It was concluded that "the addition of controlled amounts of oxygen during the chlorination is unquestionably the most practical way of controlling product viscosity as well as of shortening chlorination time."[46] Other techniques studied were the prolonged milling of the rubber prior to solution, treatment of the carbon tetrachloride solution with benzoyl peroxide before chlorination, and the action of visible light during chlorination. Light had a beneficial effect but was not necessary if traces of oxygen were admitted during chlorination. Patents concerning the chlorination of isoprene polymers have been issued to Bartovics[13] and to D'Ianni.[44]

After the reaction had progressed to the desired viscosity and chlorine content in the product, further change was stopped by heating the solution with agitation at the reflux point to remove unreacted chlorine and hydrogen chloride. Addition of 1 to 2 per cent soda ash with subsequent heating removed the last traces of free acid and possibly some loosely combined chlorine. Several methods were studied for isolating the chlorinated product from solution, such as precipitation in a nonsolvent (alcohol, gasoline), vacuum distillation, and steam distillation. The last method was preferred because it eliminated the solvent with no loss of product and yielded a more stable product. The washed and centrifuged product was dried at 70 to 75° C. in a circulating air oven.

Comparison of Chlorinated Polyisoprene and Chlorinated Natural Rubber. For as direct a comparison as possible, polyisoprene and natural rubber of the same plasticity were chlorinated under identical conditions and samples removed at intervals. In each case a solution containing 125 grams of rubber and 1500 cc. carbon tetrachloride was chlorinated at 35° C. with agitation by admission of chlorine at the rate of 1.2 grams per minute. Table IV summarizes the results of the comparative chlorinations and emphasizes the great similarity in the two systems. With increasing chlorine content, the reaction proceeded through an insoluble stage and then formed soluble products of diminishing viscosity. With continuing reaction, the color of the product became lighter, the iodine number rapidly decreased and finally approached zero, the softening point increased, and the molecular weight (as measured by intrinsic viscosity) decreased regularly beyond the gel stage.

Table IV. Comparative Chlorinations of Polyisoprene and Rubber[46]

Chlorination Time, Hr.	Gross Chlorine Input, Grams	Chlorine, %	Product Viscosity, Centipoises	Stability, Min.	Color	Iodine No.	Soft Point, °C.	Intrinsic Viscosity
Polyisoprene								
0	...	...	...	...	...	364	...	1.13
1.85	133	37.2	Insol.	0	Brown	118.9	30.0	(73% gel)
3.05	220	53.2	Insol.	2	Tan	35.2	58.5	(29% gel)
4.48	323	63.4	146	2	Cream	2.4	87.0	0.39
5.22	376	64.1	137	2	Cream	1.9	92.5	0.36
6.22	448	65.8	105	4	Cream	0.6	94.0	0.31
Natural Rubber								
0	...	...	...	...	...	357	...	1.23
1.85	133	38.3	Insol.	1.5	Red-brown	135.8	35.5	(87% gel)
3.17	228	57.9	Insol.	4	Light brown	14.2	44.0	(21% gel)
4.27	307	70.4	670	4	White	1.25	104.5	0.49
5.00	360	69.5	300	4	White	1.20	97.5	0.41
6.00	432	70.8	77	2.5	White	1.25	99.0	0.19

The properties of chlorinated polyisoprene are similar to those of chlorinated natural rubber. Proper control of the factors mentioned above allows the consistent preparation of products with 60 to 70 per cent chlorine content and centipoise viscosity of 10 to 150. Chlorinated polyisoprene is readily soluble in benzene, toluene, xylene, chloroform, carbon tetrachloride, ethylene dichloride, ethyl acetate, and methyl ethyl ketone, but insoluble in methyl alcohol, ethyl alcohol, butyl alcohol, gasoline, kerosene, and acetic acid. The natural and synthetic products are similar in thermal stability, which is greatly enhanced by the addition of magnesia, styrene oxide, or epichlorohydrin.

Chlorinated polyisoprene is apparently saturated, as judged by iodine number and infrared absorption data, although the first method of analysis is probably affected by the large amount of chlorine present, and the second method is not sensitive enough to detect a small amount of unsaturation.

The chlorinated products of polyisoprene and natural rubber are compatible with a wide variety of plasticizers and resins, such as dibutyl phthalate, dioctyl phthalate, tricresyl phosphate, perilla oil, linseed oil, coumarone resins, Bakelite resins, Chlorowax 70, Aroclors, and some Glyptal resins. They are incompatible with oleic acid, chinawood oil, bodied dehydrated castor oil, blown soybean oil, Vistac, Vistanex, and Butyl rubber. Both products are similar in resistance to corrosive agents, such as concentrated hydrochloric acid, concentrated sulfuric acid, concentrated and fuming nitric acid, and chromic acid solution. Both are darkened by fuming sulfuric acid.

Uses of Chlorinated Polyisoprene. Chlorinated polyisoprene can be used for the same applications as the chlorinated natural product, such as protective coatings, printing inks, and rubber-to-metal adhesives. Satisfactory rubber-to-metal adhesions were obtained with various rubber stocks to soft steel, brass, cast iron, aluminum, stainless steel, and bronze. The table[46] summarizes results on the bonding of rubber to steel (as determined by a modification of ASTM test D 429) with adhesive cements in which the natural and synthetic products were compared directly on four different rubber stocks. The synthetic product gave better results with the synthetic stocks; chlorinated rubber was better on a natural rubber stock.

| | | Adhesion, P.S.I. | | |
	Neoprene	Nitrile Rubber	GR-S	Natural Rubber
Chlorinated polyisoprene.	1175*	837*	432	313
Chlorinated rubber	440	913	410	430

*Substantial failure in the stock.

Chemistry of the Chlorination. Since natural rubber and polyisoprene possess substantially the same chemical structure as far as the repeating unit in the chain is concerned, they probably undergo chlorination by the same reaction mechanism. Research by Bloomfield[19] has thrown considerable light on this reaction in natural rubber. It is likely that the course of reaction involves a completely substitutive initial reaction, followed by simultaneous substitution and addition, and then, finally, by substitution; thus:

$$C_{10}H_{16} + 2Cl_2 \longrightarrow C_{10}H_{14}Cl_2 + 2HCl \qquad (34.6\% \text{ Cl})$$

$$C_{10}H_{14}Cl_2 + 2Cl_2 \longrightarrow C_{10}H_{13}Cl_5 + \quad HCl \qquad (57.2\% \text{ Cl})$$

$$C_{10}H_{13}Cl_5 + 2Cl_2 \longrightarrow C_{10}H_{11}Cl_7 + 2HCl \qquad (65.4\% \text{ Cl})$$

The lower than theoretical unsaturation during the first part of the reaction, as measured by iodine number, is attributed partially to steric hindrance of chlorine in the product and to a simultaneous cyclization reaction resembling that suggested above for the cyclization of polyisoprene. Another mechanism

proposed by Bloomfield involves the initial formation of an activated additive dihalide, which then simultaneously loses hydrogen chloride and forms the cyclic monohalide given as the last product in the above series of reactions.

Staudinger[153] visualized chlorinated rubber as a highly chlorinated cyclorubber, or more specifically, as a highly chlorinated, branched polycyclic polyterpene. Two types of ring structures were proposed: a six-membered ring formed through the side methyl groups by loss of hydrogen chloride, and cyclization of long side chains with the main chain, which markedly shortens the chain length. Thus the viscosity of the product is relatively low because of the presence of short, compact molecules, although osmotic molecular-weight determinations indicate that chlorinated rubber is of the same order of molecular magnitude as the original rubber. Staudinger's hypothesis that chlorination involves cyclization of long side chains with the main chain is probably even more applicable to synthetic polyisoprene than to rubber, in view of the likelihood that the synthetic polymer is more highly branched than natural rubber. However, the enormous effect of traces of oxygen, or of subsequently formed peroxides, during the chlorination of both rubber and polyisoprene, in solubilizing the gel stage and reducing the viscosity of the products seems ample evidence that oxidative degradation plays an important role in the reaction.

HYDROCHLORINATION OF SYNTHETIC RUBBERS

Rubber hydrochloride has been marketed for a number of years as Pliofilm[66] and has found widespread use, particularly in the food-packaging field,[4] but similar reaction products from synthetic rubbers have not passed the laboratory stage.

As early as 1913 Harries and Fonrobert[55, 72] described the successful hydrohalogenation of synthetic polyisoprene and polydimethylbutadiene. Several patents[62] were issued on the hydrochlorination of synthetic polymers, particularly with liquid hydrogen chloride. More recently[42, 46] the hydrochlorination of isoprene polymers and other synthetic polymers was investigated. It was shown that butadiene polymers, such as polybutadiene, GR-S, and butadiene-acrylonitrile copolymers, do not add hydrogen chloride, whereas the isoprene polymers markedly resemble natural rubber in the rapid addition of this agent. The presence of a methyl group attached to one of the unsaturated carbon atoms apparently is necessary for the rapid addition of hydrogen chloride. The same situation exists in simple molecules, since isopropyl ethylene does not react with hydrochloric acid, whereas trimethylethylene and 2-methyl-1-butene readily react. (Lately it has been found that under more drastic conditions polybutadiene and GR-S will add hydrogen chloride.[135] The conditions required are use of a dioxane-toluene solution saturated with hydrogen chloride at $-10°$ C. and then heated under pressure to 70 to 100° C. The chlorine in the products is more stable than that in the hydrochloride of Hevea or polyisoprene. Bases such as aniline readily remove tertiary chlorine from the latter products, whereas the secondary chlorine in the products from the butadiene polymers is much more resistant to removal.)

Isoprene polymers and rubber satisfactorily add hydrogen chloride in benzene or chloroform solution, but in carbon tetrachloride the reaction is much slower and incomplete. The theoretical chlorine content of polyisoprene hydrochloride $(C_{10}H_{18}Cl_2)_x$ is 33.97 per cent. When emulsion

Polymer		Solvent	Chlorine in Product, %
Polyisoprene		Benzene	26.87
Isoprene–styrene, 50/50		Benzene	16.91
	75/25	Chloroform	23.15
	75/25	Carbon tetra-chloride	5.10

polyisoprene was completely reacted, its hydrochloride attained a chlorine content of 28.14 per cent, which is 83 per cent of the theoretical value. The difference of 17 per cent is undoubtedly due to the presence of nonreactive vinyl groups formed by 1,2-addition polymerization of isoprene. Isopropylidene groups formed by 3,4-addition probably add hydrogen chloride because of the methyl group attached at the double bond. Isoprene polymers

$$—CH_2 \cdot C(CH_3)— \qquad\qquad —CH_2 \cdot CH—$$
$$| \qquad\qquad\qquad\qquad\qquad |$$
$$CH : CH_2 \qquad\qquad\qquad C(CH_3) : CH_2$$

1,2-Addition   3,4-Addition

prepared by various polymerization techniques add varying proportions of hydrogen chloride, and the results have been of interest in connection with a study of their chemical structures.[42] For instance, sodium polyisoprene adds only 63 per cent of the theoretical amount of hydrogen chloride, and polyisoprene prepared with an organometallic (Alfin) catalyst, 77 per cent.

In the preparation of rubber hydrochloride, after saturation with hydrogen chloride the product passes through a "ripening" period, in which it is converted from an amorphous, rubberlike material to a solid, crystallized product suitable for films. If the reaction is carried too far, the product becomes insoluble, probably as the result of condensation or cross-linking reactions. On the other hand, synthetic polyisoprene hydrochloride did not exhibit this ripening phenomenon and remained as a soluble rubberlike product with poor film characteristics. The product, however, has a low moisture-vapor transmission rate similar to amorphous rubber hydrochloride and, when it is suitably compounded, can be used in moistureproof coatings.

MISCELLANEOUS REACTIONS OF SYNTHETIC RUBBERS

Hydrogenation. Hydrogenation of synthetic and natural rubbers has frequently been proposed in the patent literature,[51, 54, 69, 83, 87, 126] but no commercial applications are known. Flint[54] suggested, for example, the hydrogenation of a rubberlike polymer (in the presence of supported nickel catalyst) in two stages at 100 to 400° C. and 500 to 5000 p.s.i. pressure in the presence of an aromatic hydrocarbon, the second stage being conducted at higher temperatures and pressures than the first. The products are useful for insulations and coatings. Graves[69] suggested decahydronaphthalene as a solvent for hydrogenation with a nickel catalyst at 240 to 350° C. and 750 p.s.i. pressure. Kirchhof[92] in a review article described hydrorubbers obtained from natural and synthetic polymers. An interesting study of the hydrogenation of emulsion polybutadiene has been made by Jones, Moberly, and Reynolds,[89a] who obtained products having the character of thermoplastic resins, resembling polyethylene but generally more flexible than the latter, especially at very low temperatures.

Miscellaneous Reactions. Butyl rubber[167] treated in petroleum naphtha with sulfur chloride yielded a derivative that is stated to make an excellent tie gum between butyl rubber and natural rubber. Polyisobutylene[47] treated with sulfur dioxide and chlorine in carbon tetrachloride solution formed a sulfonyl chloride, which on further reaction with alkali or ammonia was converted into a product suitable as a softener for rubbers. Reactions of synthetic rubbers with perbenzoic acid,[95, 134] iodine, iodine chloride, and thiocyanogen[31, 112] have been of interest primarily for the analysis of polymers. Treatment of Neoprene, Perbunan, and Vistanex (polyisobutylene) with aromatic phosphine halides has been stated[111] to produce tough rubberlike materials possessing increased modulus and hardness. Trichloroalkylsilanes[12] reacted with GR-S to form products useful as hydrophobing agents and as film-forming materials with strong adherence to glass and ceramic coatings. The patent literature also describes the reaction of natural and synthetic polymers with sulfur dioxide,[76, 121] nitrous oxides,[121] sulfuric acid,[82] benzyl chloride,[91] acetic anhydride, and acetylsulfuric acid.[81]

Oxidation. The oxidation of natural rubber to form thermoplastics known as Rubbone has achieved some prominence in England especially for use in paint formulations, but similar attention has not been devoted to synthetic rubbers. Palmer[123] claimed the use of oxidized isoprene polymers as tackifiers for synthetic rubber. Rust and Pfeifer[131] asserted that latexes of the GR-S type after oxidation with hydrogen peroxide are useful for treating wool to secure shrinkproofness with normal handle. White powders or fibrous products[80] suitable for molding were obtained by oxidizing natural or synthetic rubbers with an aliphatic percarboxylic acid, such as peracetic acid.

Reactions with Mercaptans. The reaction of mercaptans with synthetic rubbers has attracted considerable interest. Burke[26] described the reaction of rubbers with ethanethiolic acid in the presence of oxygen or peroxides. By further reaction of the product with a monomer, such as styrene, acrylonitrile, or the acrylates, materials suitable as films, fibers, or coating compositions were obtained. Serniuk and co-workers[141, 142] found that thioglycolic acid added exothermally to butadiene polymers in benzene solution to give apparent double-bond unsaturation values of 38 to 47 per cent. A similar extent of reaction in mass or latex was obtained with aliphatic mercaptans of C_2 to C_{16} chain length. The reactive double bonds are predominantly those in the vinyl side chains. However, natural rubber and polyisoprene reacted only to a slight extent with ethyl mercaptan, and natural rubber showed very little reaction with thioglycolic acid. Cunneen[33] confirmed the nonreactivity of natural rubber toward thioglycolic acid and aliphatic mercaptans, even in the presence of organic peroxides, but Holmberg[77] claimed the addition of thioglycolic acid to rubber under conditions such that atmospheric oxidation also occurred. The experimental evidence indicates that butadiene polymers are surprisingly more reactive to mercaptans than is natural rubber.

Cunneen[33] reports that, although thioglycolic acid fails to react with natural rubber, thiolacetic acid and bisthiol acids react with it readily.

HIGH-STYRENE RESINS

World War II drastically reduced the conversion of natural rubber to useful chemical derivatives, such as isomerized rubber, which had become established before the war period as a valuable reinforcing and processing aid for rubber stocks, as a rubber-to-metal adhesive, and as the pigment vehicle for corrosion-resistant coatings. As the GR-S production program reached its full growth, butadiene and styrene monomers became available in large quantities and at low prices. Consideration was then given to the commercial production and utilization of resins containing these monomers, since it was well recognized that from them one could prepare at the one extreme a rubbery polymer (polybutadiene) or at the other extreme a relatively hard brittle plastic (polystyrene) and that by varying the proportions of the two monomers copolymers of intermediate properties could be prepared. The first description of the properties and applications of a commercially available styrene-butadiene resin (Pliolite S-3, a 15/85 butadiene-styrene copolymer) was published in 1946 as the outgrowth of work by Borders, R. D. Juve, and Hess.[20] Descriptions of other high-styrene resins were subsequently made by Jones and Pratt,[89] and Fox.[57] General information on high-styrene polymers was summarized by Winkelmann,[166] Fordyce,[56] and others.[5]

A review of the patent literature indicates that interest in butadiene-styrene resinous polymers paralleled that in butadiene-styrene rubbery polymers of the Buna and GR-S types, although commercial application was not made until the war period. For example, Tschunkur and Bock[162] described the emulsion copolymerization of butadiene and styrene to give rubbery polymers and in a companion patent[161] stated that, if large amounts of styrene were copolymerized with small amounts of butadiene, there was obtained a resinous, insoluble polymer, less brittle and more elastic than polystyrene prepared in a similar manner. Konrad and Ludwig[97] prepared emulsion copolymers of butadiene–styrene containing 47.5 to 70 per cent styrene but were primarily interested in the rubberlike properties of the products. A more recent patent[165] claimed copolymers of butadiene and styrene containing 80 to 95 per cent styrene. Waterman[163] coagulated with brine a latex of a butadiene-styrene copolymer containing 65 to 90 per cent styrene to obtain a slurry having a particle size optimal for filtering and washing. Other patents concerning early applications of high-styrene resins will be referred to later.

Polymerization Studies. The emulsion polymerization recipe usually employed in the manufacture of high-styrene resins is patterned after the one used for the production of GR-S. Tschunkur and Bock indicated, however, that one monomer could first be polymerized and then the other one added to complete the polymerization. Guss and Amidon[71] prepared a new type of thermoplastic resin flexible over a wide temperature range by first polymerizing styrene and then adding a mixture of butadiene and styrene to the polymerizing system. Other patents referring to increment addition of monomers were granted to LeFevre and Harding[104] and to Stanton and Lowry.[152]

Comparatively little information on emulsion polymerization processes leading to high-styrene resins can be found in the literature. MacLean, M. Morton, and Nicholls[108] studied butadiene-styrene copolymers over the entire range of monomer composition. By the use of *tert*-hexadecyl mercaptan as modifier and the isolation of polymer at low conversion, products of relatively narrow range of molecular weight and relatively homogeneous in comonomer composition of the chains were obtained. The effect of increasing styrene content was to increase solubility in benzene-methanol systems, and to decrease the intrinsic viscosity of benzene solutions of the copolymer. Meehan[113] and Koningsberger and Salomon[96] also studied butadiene-styrene copolymers of various ratios.

Mitchell and H. L. Williams[116] have conducted a laboratory study of high-styrene, low-butadiene copolymers and observed that the polymerization characteristics were similar to those found with GR-S, although by no means identical. The modifier, dodecyl mercaptan, was less rapidly consumed than with GR-S. The regulating index (ratio of the logarithm of residual mercaptan over conversion) was definitely lower than for GR-S, and there did not appear to be a waste factor. From bound-styrene and increment-styrene curves, monomer reactivity ratios were calculated which, when corrected for the bifunctional nature of butadiene, gave constants[3] for butadiene of Q equal to 0.9 and e equal to -1. The growth of the copolymer chain was considered to be quite random with respect to the order of succession of the monomer units in it.

The gel-viscosity data were similar to those obtained with GR-S except that, with increasing styrene in the charge, the pre-gel rise in viscosity, the formation of gel and the slope of the viscosity-conversion curves diminished. The regulating effect of the mercaptan was apparent in all the systems studied, but with increasing styrene content the chain-transfer action of this monomer became more evident. The dodecyl mercaptan requirement of three butadiene-styrene systems is shown in Table V. As the proportion of styrene was increased, the rate of reaction increased, and the amount of mercaptan required for the same dilute solution viscosity and solubility in the polymeric product fell.

Table V. Dodecyl Mercaptan Requirements for Copolymers of Similar Viscosity[116]

	Butadiene-Styrene Ratio		
	70/30	50/50	30/70
Dodecyl mercaptan, parts	0.425	0.320	0.214
Reaction time, hr.	13	8.67	5.5
Conversion, %	74.3	74.0	70.3
Soluble in benzene, %	96	95	89
Gel, %	4	5	11
Dilute-solution viscosity	1.92	2.16	2.20

A few references are to be found to high-styrene-type resins containing other monomers. D'Alelio[35] prepared vulcanizable polymers from large proportions of styrene along with smaller proportions of butadiene and ethyl acrylate or maleate. Copolymers of butadiene and 2-vinylfluorene in various ratios to give rubbery or plastic products were suggested by Kern and Abbott.[90]

Solution and mass polymerization techniques have also been employed to prepare high-styrene resins. Perry[125] obtained resinous polymers of low intrinsic viscosity by the solvent polymerization of isoprene-styrene mixtures with tertiary-butyl peroxides as catalysts. Soday[145] obtained molding and coating resins by partially polymerizing a diene (butadiene, isoprene, piperylene, etc.) with the aid of a metal acid catalyst, then adding styrene or a homolog of it, and finally completing the polymerization.

Properties of High-Styrene Resins. The properties of high-styrene resins depend in large measure on the ratio of styrene to butadiene employed in their manufacture. A polymer containing equal amounts by weight of the two monomers appears to be a rubber at room temperature, although certain characteristics in its cured stocks, such as resilience, heat buildup, and low-temperature flexibility, are much poorer than in more truly rubberlike polymers, such as GR-S. As the proportion of styrene increases and that of butadiene decreases, the polymer becomes stiffer, less extensible, and harder, thus approaching the properties of polystyrene. If films are cast from latexes of various butadiene-styrene copolymers, it is found that the polymer must contain at least 25 to 30 per cent butadiene, to obtain a continuous film on drying at room temperature. Copolymers containing small proportions of butadiene (10 to 15 per cent) have the physical appearance of polystyrene but are less soluble and less brittle and have lower softening points. Other differences will be noted in a discussion of applications of these new resins. D'Ianni, Hess, and Mast[45] have reported the latex and solid polymer properties of a series of resins containing more than 50 per cent styrene.

Storey and H. L. Williams[154] studied, not only a series of butadiene-styrene copolymers having various monomer ratios up to ratios representing high-styrene resins, but also blends of these copolymers themselves and blends of them with GR-S. They found that beyond a styrene content of 50 per cent there occurred on vulcanization a fall in tensile strength and ultimate elongation and an increase in modulus. A copolymer from a 30/70 butadiene-styrene charge could not be flexed, and from a 20/80 charge the product resembled ebony. Latex blends of such polymers exhibited properties at least as good as those of blends made on a mill. The properties of blends were determined primarily by the total styrene content of the blend rather than by the composition of the specific copolymers blended.

The first of the commercial high-styrene resins[20] to be identified was a 15/85 butadiene-styrene copolymer (Pliolite S-3). It is described as a white thermoplastic crumb or powder with a heat distortion point of 50° C. It is practically odorless and nontoxic in normal handling and processing. A small amount of nonhydrocarbon, including 0.5 per cent antioxidant, is present in the resin, which has a specific gravity of approximately 1.05. The resin is only moderately affected by aliphatic solvents but is much less resistant to aromatic and chlorinated solvents. Hot milling causes the resin to be readily soluble in benzene, toluene, ethylene dichloride, and methyl ethyl ketone. Solutions of 20 per cent solids content and of low viscosity can be made in these solvents.

Pliolite S-3 exhibits excellent resistance to acids and alkalies. Being less unsaturated, it is more resistant to oxidation than cyclized rubbers, natural or synthetic. No appreciable change in it is noted on prolonged exposure to air, even at elevated temperatures. Severe conditions reduce its solubility but have no effect on the stiffening action of the resin in rubber compounds. Moisture-absorption tests on the molded resin (thickness 0.08 in.) show absorption of only 3 to 6 mg. of water per sq. in. of surface. The electrical properties are typical of those of hydrocarbon polymers, as shown by the following values obtained at 1 kc. at 35° C.:

Dielectric constant	2.5–2.6	Loss factor	0–0.004
Power factor, %	0–0.2	Specific resistivity, ohm-cm.	0.8–1.6 × 10^{15}

Another type of resin, Pliolite S-5, has found widespread use in concrete floor enamels and other protective coatings and has been described[49] as possessing the following properties:

Specific gravity	1.03	Acid value	Neutral
Softening point	50° C.	Sward hardness	90
Tensile strength	1450 p.s.i.	Color of film	Clear and colorless
Elongation		Index of refraction	1.585
at 20° C.	0.0%	Solution viscosity	Can be varied to meet
Iodine number	57.8		requirements

The resin is stated to possess remarkable resistance to acid, alkalies, and other chemicals. Paints made from this resin dry by solvent evaporation rather than by oxidation; are highly resistant to water, acids, alkalies, alcohol, and vegetable, animal, and mineral oils and greases; and show good adhesion, resistance to abrasion, and durability with age. More recently a modified type of the resin, Pliolite S-5B, has been placed on the market; it possesses a lower solution viscosity than the original product.

Table VI. Properties of Commercial High-Styrene Copolymers

Resin	Volatile* %	Heat distortion,† °C.	Ash	Rupture strength,‡ Lb., Mullen	Dielectric Properties at 1 Megacycle D.C.	% P.F.	Iodine No.§
A	0.96	51	0.47	135	2.66	0.22	46.0
B	0.62	61	0.10	70	2.59	0.07	54.0
C	0.66	40.5	2.38	186	2.81	0.20	66.0
D	0.73	48	0.35	190	2.68	0.20	62.0
E	0.46	51	1.20	221	2.57	0.10	68.0
F	0.0	83	0.58	345	2.63	0.20	85.0
G	0.85	65.5	0.44	80	2.73	0.07	53.0
H	0.50	51	0.13	249	2.55	0.24	46.0
I	0.35	50	0.24	170	2.50	0.07	45.0

* 30 minutes at 200° C.

† Temperature of collapse of a strip 1 × 4 × 0.04 in. under load of 30 grams midway between supporting edges.

‡ On a sheet of molded resin 0.04 in. thick.

§ Values for unextracted resins.

A distinction is drawn by Fox[57] between plastic and elastic high-styrene resins. One type is brittle at room temperature but becomes soft at elevated temperatures, and these characteristics are reflected in rubber compounds containing them. Most of the resins produced by the isomerization of natural rubber also have these characteristics. A second type is considered to be elastic, since a thin sheet of the resin, unlike ordinary styrene resin, can be bent double without breaking. Rubber compounds containing the elastic type of resin are not brittle at temperatures as low as −5° F. and soften little when heated as high as 140° F.

Data, summarized by Susie and Wald[155] for nine typical commercial high-styrene resins, are given in Table VI.

RUBBER COMPOUNDING WITH HIGH-STYRENE RESINS

The use of the various commercial high-styrene resins in rubber compounding has been described by a number of authors.[2, 20, 57, 89, 139, 155, 159] The usual procedure in mixing a compound is to band the resin on a hot mill and then slowly add the rubber. Natural rubber and a wide variety of synthetic polymers (GR-S, Neoprene, nitrile rubbers, polyisoprene, Butyl, Vistanex, polyethylene, etc.) show various degrees of mechanical compatibility with high-styrene resins, particularly after the hot-milled masterbatch is cooled and remilled cold. The smoothness of a rubber stock during processing is markedly improved by the presence of these resins. Milled sheets of a 50–50 mixture of GR-S and the resin show little shrinkage after removal from the mill. The rest of the compounding ingredients (sulfur, accelerator, etc.) can be added on a cool mill. The rubber and resin can readily be blended by Banbury mixing, in which case everything except the sulfur may be added initially, and the sulfur added later by mill mixing.

Some of the high-styrene resins have been introduced as so-called easy-processing resins. Some of them may be added directly to the rubber banded on a mill; in the Banbury they allow satisfactory dispersion to be obtained in shorter mixing cycles. In their preparation, fusion of the resin is avoided, and hence the granules are readily friable. Also important is control of the softening point and molecular weight. One resin of this type is described by Holt, Susie, and Jones.[78] Sell and McCutcheon[140] studied the factors, such as their physical form, particle size, softening point, and flow characteristics, which influence the Banbury dispersion of high-styrene resins in rubber.

The rate of vulcanization of the rubber in the compound is not appreciably affected by the presence of the resin. Because of the unsaturated nature of the resin, however, it probably enters into the vulcanization reaction. Tests[155] on the resin alone show that it is capable of undergoing curing (Table VII). It was also shown[20] that curing the resin alone had little effect on the heat-distortion point but greatly decreased its solubility in toluene and at a full cure rendered it completely insoluble in this liquid.

Compounding with GR-S and Other Synthetic Rubbers. The high-styrene resins have been shown to be of particular benefit when compounded with synthetic rubbers, although valuable results are also obtained with natural rubber, as will be described later. Several investigators have concerned themselves particularly with synthetic rubbers.[20, 155, 159]

Table VII. Curing Characteristics of a High-Styrene Resin*

Parts by Weight	A	B	C	D
Marbon S-1	100	100	100	100
Zinc oxide	...	5	5	5
Sulfur	...	...	2	2
Cyclohexyl benzothiazolyl sulfenamide	...	...	...	1.5
Properties†				
Tensile strength, p.s.i.	3770	5240	5470	5690
Rupture strength, lb.‡	105	150	165	180
Heat distortion, °C.	54	57.5	64.5	60
Shore "D" hardness	81	82	83	84
Surface shrinkage, %	25	8	6	3.5
Solubility in toluene	Soluble-gel	Gels	Swells	Swells

* All stocks milled same length of time.

† All stocks cured 15 minutes at 320° F., then cooled in press for all determinations but shrinkage.

‡ On sheet 0.04 in. thick with Mullen tester.

High-styrene resins blended with GR-S greatly increase stiffness and hardness, even though the stock is not cured, as the data in the accompanying table show.[20] Other properties studied were flow characteristics, impact

Resin-Rubber Ratio	Tensile Strength, P.S.I.	Elongation, %	Relative Torsion Modulus	Olsen Stiffness, In.-Lb.
10 Pliolite S-3/90 GR-S	60	925	0.023	0.016
30 Pliolite S-3/70 GR-S	60	385	0.023	0.021
50 Pliolite S-3/50 GR-S	550	70	0.50	0.215
70 Pliolite S-3/30 GR-S	1900	20	22	0.90
90 Pliolite S-3/10 GR-S	4350	5	44	2.41

resistance, brittle point, and torsional modulus over a range of temperature. As much as 50 parts, or as little as 10 parts, of resin in 100 parts of rubber had no appreciable effect on the brittle point or the rate at which the torsional modulus increased with lowered temperature. The effect on impact resistance is discussed later.

Uncured blends of a high-styrene resin and Neoprene (GR-M) were also studied.[155]   General reinforcement is obtained and in addition the dielectric properties of the blend are significantly improved with increasing amounts of resin, as shown in Table VIII.

Table VIII.   Properties of Uncured Marbon S-1/GR-M Blends[155]

Percentage GR-M Replaced by Marbon S-1	Tensile Strength, P.S.I.	Ultimate Elongation, %	Tear, Lb. per In.	Hardness (Shore A)	Dielectric Properties (at 1 Megacycle)	
					D.C.	% P.F.
0	840	1070	30	44	...	...
10	820	910	20	60	...	...
20	1210	535	82	73	4.85	1.085
30	1295	450	86	85	4.53	0.614
40	1514	500	114	95	3.99	0.538
50	2070	425	201	98	3.79	0.364
60	2110	130	222	99	3.50	0.304
70	3330	35	337	100	3.44	0.256
80	4210	0	348	100	2.97	0.164

The high-styrene resins are more commonly compounded in rubber stocks containing curing ingredients.   In general, the effects of adding increasing proportions of the resin are to improve the tensile strength, elongation at rupture, tear resistance, hardness, stiffness, and flex life.   As little as ten parts resin per hundred of rubber has a measurable effect.   Even in carbon black stocks the resin gives additional reinforcement.   Other advantages noted are improved processing, primarily due to less nerve in the stock, and improved electrical properties of the finished article.   Calender shrinkage[155] is reduced from a comparative value of 100 for a resin-free GR-S stock, to 65 for one containing 20 parts of a styrene resin, and to as low as 30 for one containing 60 parts of resin.

By way of example, reference may be made to a study of the reinforcing effect of Pliolite S-6 in the following gum stock.[159]

GR-S	100	Sulfur	2.0
Zinc oxide	5	Tetramethylthiuram	
Phenyl-β-naphthylamine	1	disulfide	0.1
Stearic acid	2	Benzothiazolyl disulfide	1.5

The addition of 30 parts of the high-styrene resin to this stock greatly increased tensile strength, modulus, and elongation; it also improved the flex life tenfold.   Ten parts of the resin was sufficient to double the tear strength.   A stock containing 25 parts of the resin had about the same hardness and modulus as a corresponding stock containing 25 parts of carbon black.   It has been found advantageous to employ cold rubber in place of GR-S in rubber-resin blends.

High-styrene resins appear to be compatible in all proportions with GR-S, nitrile rubbers, and Neoprene but to have only limited compatibility (about 10 per cent) with Butyl rubber.

Compounding with Natural Rubber. Before World War II, natural-rubber stocks were reinforced where necessary with cyclized natural rubber, such as Pliolite and Marbon B. The development of high-styrene resins has led to their use[20, 57, 155, 159] in natural-rubber compounds, as well as in synthetic rubbers.

Uncured blends of a high-styrene resin and natural rubber were studied[20] and found to give results similar to those with GR-S, as the data in the accompanying table show. As noted previously for GR-S, blends of resin

Resin-Rubber Ratio	Tensile Strength, P.S.I.	Elongation, %	Relative Torsion Modulus	Olsen Stiffness, In.-Lb.
10 Pliolite S-3/90 Pale Crepe	75	780	0.023	0.023
30 Pliolite S-3/70 Pale Crepe	300	420	0.233	0.058
50 Pliolite S-3/50 Pale Crepe	530	180	0.73	0.205
70 Pliolite S-3/30 Pale Crepe	1590	50	8.0	0.61
90 Pliolite S-3/10 Pale Crepe	3700	3	89.0	2.61

and natural rubber containing 10 to 50 parts of resin per hundred of rubber had the same impact brittle point as well as the same rate of relative increase in modulus with fall of temperature.

The effects of high-styrene resins on the properties of natural-rubber vulcanizates are different in some respects from those obtained with GR-S or nitrile rubbers. With increasing proportions of resin, the natural-rubber stocks show increasing stiffness, hardness, and abrasion resistance, but the tensile strength, ultimate elongation, tear resistance, and flex life tend to decrease, although the deterioration is not severe if only 10 to 20 parts of resin per 100 parts of rubber is used. These differences are not unexpected, if one recalls that the gum-stock properties of natural rubber are excellent, whereas those of the synthetic rubbers mentioned are relatively poor. The resins can be used in gum stocks or in stocks loaded with carbon black or other fillers. In highly loaded stocks the resin appears to act as plasticizer and aids in the incorporation of the heavy filler content, without sacrifice in hardness, compression set, or other desirable properties. Aiken[1] summarized the properties imparted to rubber stocks by the use of high-styrene resins. Details of compounding recipes, physical test data, etc., of specific resin-rubber blends are readily available from publications referred to previously.

Impact-Resistant Resin-Rubber Blends. Up to this point the discussion of the compounding of high-styrene resins has been concerned primarily with the use of relatively small amounts of resin with larger amounts of rubber. However, resin plasticized with relatively small proportions of rubber has received considerable attention[2, 139] for the formulation of solid plastics of high-impact resistance.

A typical high-styrene resin itself possessed a notched-Izod impact resistance of 2.5 in.-lb., whereas an 80–20 blend of the resin with natural rubber gave a value of 32.5 in.-lb. Curing the resin-rubber blend produced a still greater improvement in impact resistance; it also produced other desirable changes, such as improvement in tensile strength, in resistance to tearing or breaking on bending or flexing, and in increased hardness (to a value approaching that of the resin itself). Excellent results in these respects were obtained in blends with natural rubber, GR-S, nitrile rubbers, and Neoprene, but blends with Butyl rubber gave relatively poor impact resistance, presumably because of incomplete compatibility of the resin with this rubber.

For purpose of illustration, Table IX lists data on vulcanized blends of Pliolite S-6 with GR-S. Good results have also been obtained in stocks containing reinforcing pigments (SRF black, Silene EF) and loading pigments (natural whiting).

Table IX.   Vulcanized Blends of High-Styrene Resin with GR-S[139]

Pliolite S-6	90	85	80	75
GR-S	10	15	20	25
Tensile strength, p.s.i.	6360	5680	4870	4140
Ultimate elongation, %	0	0	0	35
Durometer hardness (Shore D)	80	77	76	72
Unnotched impact resistance, in.-lb.	16	20	52	100+
Stiffness in flexure, in.-lb.	5.68	5.52	4.00	3.36
Taber abrasion, cc. per 500 rev., 1000 g. wt., H 22 wheels	0.299	0.249	0.237	0.213

APPLICATIONS OF HIGH-STYRENE RESINS

Before discussing in detail some of the more recent applications of high-styrene resins, it is of interest to consider the patent literature on the subject. As early as 1930, for example, it was stated[149] that a high-styrene low-diene copolymer or polystyrene could be mill-mixed with natural rubber, gutta-percha, or balata for use as a cable dielectric. More recent patents[124, 151] have been issued for compositions based on polystyrene with natural rubber or synthetic polymers. Smith[144] claimed a waterproof coating consisting of a high-styrene low-butadiene copolymer admixed with polyisobutylene and a wax or waxlike material. Sparks, Gleason, and Frolich[148] asserted that electrical insulating compositions of high-styrene resins and polyisobutylene were of particular value in the construction of ultrahigh-frequency radio-transmitting and receiving systems. Daly[36] claimed a composition of a styrene-acrylonitrile resin admixed with a butadiene-acrylonitrile rubbery polymer as a hard, resilient, tough, thermoplastic molding compound which does not deform or soften in boiling water and is an excellent leather substitute.

According to Te Grotenhuis,[156] a relatively tough butadiene-styrene copolymer is mixed in latex form with a well-modified butadiene-styrene copolymer and the product obtained by co-coagulation. Hetro TT[64] is a commercial rubber-resin blend. Polysar SS 250 is a 60/40 blend of GR-S and high-styrene resin, the two components being mixed while in the form of latexes. Bacon, Farmer, and Schidrowitz[11] had previously shown that polystyrene latex when blended with natural rubber in small proportions increased the modulus of the resulting compound, with little effect on tensile strength. Others[138, 150] have made resin-rubber blends by polymerizing styrene with natural rubber or synthetic polymers such as polyisobutylene. A blend suitable for shoe soles, wire covering, etc. was claimed[61] from mixing GR-S with 25 to 100 parts of a high-styrene–low-butadiene resin (5 to 30 per cent butadiene content) and 10 to 25 parts of a cellulosic flock.

During recent years the use of high-styrene resins has shown a continuous growth because of their enthusiastic reception by compounders in the rubber industry and related fields. Although complete figures are difficult to obtain, it is conservatively estimated[6] that 25 million lb. of these resins was manufactured in 1949—a growth of more than 100 per cent over the estimated figure of 12 million lb. for 1947. Production increased to 35,000,000 lb. in 1950, and in 1953 was estimated as 33 million pounds. As new and better resins specifically tailored for certain applications become available, it is certain that the total production of these resins will show a substantial increase in the near future.

Rubber Reinforcement. By the use of high-styrene resins, rubber stocks can be adequately reinforced to give products of low specific gravity, light color, good dielectric properties, excellent flex life, abrasion resistance, and tear resistance, and improved aging properties. Processing is improved because of reduced mill shrinkage, improved extrusion behavior, and more efficient reworking of scrap in the millroom. The resins have, therefore, found particular application in the manufacture of shoe soles and heels, rubber flooring, hard board stocks, gaskets, caster wheels, electrical insulation, hard-rubber stocks, luggage, and many other mechanical goods.

Rubber shoe soles containing high-styrene resins as reinforcing agents have acquired much popularity and have been found to have at least twice the wear resistance of leather soles and to be better than the latter in resisting water and slipping. Hoover[79] stated that 35 per cent of "leather" soles made in 1949 probably consisted of synthetic rubber compounded with special butadiene-styrene resins. More recently it has been estimated that 58.4 per cent of the shoes and slippers manufactured in November 1951 had nonleather soles.[10] The figure for 1953 is estimated as 55 per cent. Not all these soles, however, were of the rubber-resin blends under discussion. The application of rubber-resin blends in the footwear industry has been discussed in recent articles. The formulation and properties of a typical rubber-resin shoe-sole stock are given[67] in Table X.

Moakes[116a] has studied soling stocks made from natural rubber reinforced with high-styrene resin, the latter being introduced as such or as Polysar SS 250, a blend of high-styrene resin and GR-S. Formulations for soling stocks

Table X. Light-Colored High-Grade Shoe-Sole Stock

Cold rubber (GR-S 1505, formerly X-625)	100
High-styrene resin (Pliolite S-6B)	50
Calcium silicate, pptd. (Silene EF)	75
Coumarone resin (Cumar MH 2½)	5
Light process oil	1
Paraffin wax	1.5
Stearic acid	1
Zinc oxide	5
Benzothiazolyl disulfide	2
Tetramethylthiuram disulfide	0.2
Sulfur	3
	247.7

	Cure at 315° F.	
	12 Min.	24 Min.
Hardness (Shore A/B)	93/70	94/71
Ross flex (12 iron sole aged 24 hr. at 212° F., 250,000 flexes) (Rating)	9	7
Abrasion (Bureau of Standards)	39.5	36.5
Tensile strength, p.s.i.	1640	1650
Elongation, %	495	480
Modulus at 200%, p.s.i.	800	840
Tear resistance, Crescent B, lb. per in.	160	183

Olsen stiffness (12 iron sole cured 12 min., 8 in. lb. wt., 4 in. span)

10°	20°	30°	40°	50°	60°	70°	80°	90°
25	36	43	48	51	53	54	54	54

based on natural rubber reinforced with Polysar SS 250 and flex-cracking and cut-growth tests on such stocks are given by Buist.[23a]

Pliolite S-6B is a valuable compounding agent in rubber floor tiling and in jackets for electric cables.

Impact-Resistant Plastics. In rubber reinforcement, the proportion of resin to rubber is rather small. With impact-resistant blends, the ratio is reversed, and the rubber can be considered as a plasticizer for the resin. Many interesting applications[2] of such blends have been made, such as for golf ball covers as a replacement for scarce and expensive balata, for cutting blocks or beam punch pads, football helmets, bowling balls, luggage, special shipping cases, printing plates, golf club heads, resurfacing worn bowling pins, and protective football equipment (shoulder pads, shin guards, hip pads, etc.). The compositions are of such hardness and toughness that they can be readily machined, buffed, sawn, drilled, and threaded with wood- or metal-working tools. Even after being cured, the stock is thermoplastic, and hence "postforming" can be done at elevated temperatures.

The addition of Pliolite S-6 to hard rubber improves its impact resistance.[140a] Uncured blends of Pliotuf and Neoprene have shown unusually interesting properties.[140b]

Solution Applications. In addition to their use in rubber and plastics compounding, high-styrene resins have found considerable utility in protective coatings and other applications in solution. Before World War II cyclized natural rubber (Thermoprene and Pliolite) found use in paints, particularly for concrete floors, because of the unsaponifiable nature of the resinous vehicle. During the war, as development work on high-styrene resins was intensified, it was found that soluble resins could be made with exceptional properties. These developments were described by Thies and Aiken,[159] and paint applications of the new resin (Pliolite S-5) were described by Endres,[49] Burr,[27] Workman,[168] and Kirschner.[93] Properties of the resin have been described earlier. One respect in which this resin differs from the high-styrene resins used for rubber reinforcement is that it has been modified during preparation in order to insure ready solubility in organic liquids.

The incorporation of pigments into Pliolite S-5 by mill or Banbury mixing was found to be exceptionally advantageous, since excellent dispersion of the pigment is obtained. Electron micrographs of masterbatches of carbon blacks and other ultrafine pigments prepared by this technique have exhibited uniform and complete dispersion, which results in a film of high gloss and quality. However, good results can also be obtained with the conventional wet-grinding method.

One of the largest uses of Pliolite S-5 is in concrete floor finishes, where its characteristics of alkali resistance, waterproofness, abrasion resistance, and good adhesion show up to the best advantage. Because of its extreme chemical inertness, it is used also in finishes for corrosion resistance and for special applications where ordinary paints are inadequate. Other promising applications are in stucco paints, traffic paints, and plaster sealers.

Typical formulations contain the styrene resin (100), regular paint pigments (65), chlorinated paraffin (3), raw tung oil (1.5), high-flash aromatic naphtha (100) as solvent, and mineral spirits (100) as thinner, the figures in parentheses representing approximately the quantities by weight of the various components.

Another application of solutions of high-styrene resins of considerable interest is in the formulation of heat-sealing, creaseproof, moisture-vaporproof coatings for paper. For this use there has appeared on the market an entirely new resin (Pliolite S-7), which is available only in solution form or in compounded forms. This also is a high-styrene resin of undisclosed composition. It is readily soluble in ordinary solvents, such as toluene, and paper coatings of any desired drying rate can be made. To obtain the best resistance to the passage of water vapor, a small amount (10 to 15 per cent) of wax, such as a microcrystalline wax with melting point of about 170° F., is usually added. Modifications can be made by adding other resins, such as coumarone-indene resins, modified phenolic resins, modified styrene resins, and rosin esters.

Magnesium castings can be treated with a solution of a new synthetic resin-styrene copolymer to eliminate porosity. The nature of the copolymer is not disclosed.[32]

Latex Applications. Since high-styrene resins are usually prepared by emulsion polymerization methods, the unique opportunity exists of applying them to latex compounding. Little has been published in this field as yet, but large-scale applications are rapidly developing.

Irvin[86] described a water dispersion (Marmix) of a high-styrene resin and its application to latex compounding with GR-S and nitrile-rubber latexes. Finished products of superior properties were obtained. Owen[122] states that fibrous sheets formed from an aqueous suspension of fibers and a butadiene-styrene copolymer, preferably in the range 70/30 to 50/50, exhibit greatly improved initial tear resistance, as judged by hand tear tests.

Weatherford and Knapp[164] discuss the properties and applications of a latex (Pliolite latex 190) containing a resinous copolymer of 10/90 butadiene–styrene. When compounded with various other latexes (natural rubber, Neoprene, GR-S, and nitrile rubber), this new latex acts as an effective reinforcing agent. The hardness, rigidity, and tear strength are increased significantly; the tensile strength is increased slightly or moderately, and other properties remain substantially unchanged. As little as 15 to 20 per cent of the resin has an appreciable effect. Applications mentioned are molded toys and sport equipment, dipped goods, latex threads, paper impregnants, fabric coatings, latex foam, and wire and cable insulation. Pliolite 170 latex is also of interest for paper coatings.

The following are examples of the latex compounding of a high-styrene copolymer latex (Pliolite latex 190) with several rubber latexes.[164]

1. Blends with the following natural-rubber latex compound were cured in air at 200° F. for 15 minutes:

Natural rubber	100 (dry weight)
Potassium hydroxide	0.5
Sulfur	1
Zinc oxide	5
Setsit-5	3
Heptylated diphenylamine	1

Addition of up to 15 parts (dry weight) of the resin as latex had only a slight effect on the tensile strength and ultimate elongation but greatly increased the stiffness, as shown by a doubling of the modulus. Twenty parts of the resin raised the tear strength by about 50 per cent and the Shore hardness by about 25 points.

2. Blends with GR-S type-III latex (GR-S 2000) compounded as follows were cured in air at 220° F. for 15 minutes:

GR-S type III	100 (dry weight)
Casein	2
Sulfur	2
Zinc oxide	5
Setsit-5	3
Heptylated diphenylamine	1

The addition of up to 15 parts of resin as latex had little effect on the tensile strength and ultimate elongation. With 20 parts of resin, the modulus was doubled, the tear resistance increased about 55 per cent, and the Shore hardness increased about 15 points.

3. Blends with a nitrile-rubber latex compounded as follows were cured in air at 250° F. for 5 minutes:

Nitrile rubber (Chemigum latex 200)	100 (dry weight)
Casein	1
Sulfur	2
Zinc oxide	5
Setsit-5	3
sym-Di-β-naphthyl-p-phenylenediamine	1

The addition of 15 parts of resin doubled the tensile strength, had little effect on the elongation, and greatly stiffened the stock, as shown by a doubling of the modulus. The addition of 20 parts doubled the tear resistance and increased the hardness about 25 points.

The addition of an alkaline casein solution to the synthetic-rubber latexes is desirable, in order to stabilize and thicken them.

Latex Paints. Finally, the use of high-styrene latexes in emulsion paints has rapidly assumed importance. The styrene content of the styrene-butadiene copolymers in these latexes, although higher than that of synthetic rubbers, is not so high as that of the high-styrene resins, strictly so-called. Ludwig,[106] and also Ryden, Britt, and Visger[133] have described the use in paints of Dow latex 512-K (containing a 40/60 butadiene-styrene copolymer), with and without the use of high-styrene resin latex. Such paints show promise for use on stucco, plaster, and cement surfaces, where alkaline materials destroy saponifiable vehicles. Many emulsion paints based on similar latexes, such as paints based on Chemigum Latex 101, described by Burr and Matvey,[28] are on the market for inside applications. Articles describing synthetic latexes and their use in paints have been written by Henson,[74] Rinse,[130] Zwicker,[170] and others.[7, 8] Ryden[132] claims an emulsion paint containing a butadiene-styrene copolymer with a definite range of composition. McIntyre, Taber, and Young[107] discuss styrene-butadiene copolymer latexes as raw materials for the coatings industry.

The paints are typically prepared by mixing the latex (which has a high solids content) with a dispersion of regular paint pigments, such as titanium dioxide, lithopone, and mica, in water containing a wetting agent and sodium pyrophosphate. The paints are easy to apply, dry quickly, and are washable.

The paints have very rapidly become popular—so much so that, according to estimates, in 1952 polymer-emulsion paints (in most of which the polymer was butadiene–styrene) were used in the United States to the extent of 30 to 40 million gallons and represented[8] considerably more than half the paint used for interior work. The volume increased to 45 million gallons in 1953.

Systematic study of the scientific and technical problems that emulsion paints present is being actively undertaken. A Symposium on Emulsion Paints[155a] held by the Division of Paint, Varnish and Plastics Chemistry of the American Chemical Society included papers on such aspects of them as pigmentation, mechanism of film formation, resistance to water penetration, stability of viscosity, and resistance to freezing followed by thawing.

COMMERCIAL BRANDS

In Table XI are listed high-styrene resins and related materials being produced commercially.

Table XI. High-Styrene Resins

Trade Name	Composition	Manufacturer
Butaprene SLF, SD	High-styrene copolymer	Firestone
Darex X34, X43, 43G	85% styrene	Dewey and Almy
Darex Copolymer 3	70% styrene	Dewey and Almy
Darex Copolymer 9L	Latex of Copolymer 3	Dewey and Almy
Dow Latex 512	40/60 Butadiene–styrene	Dow
Dow Latex 529	20/80 Butadiene–styrene	Dow
Goodrite Resin 50	Butadiene-styrene copolymer	Goodrich
Hetro TT	Rubber-resin blend	General
Isopol	25/75 Isoprene–styrene	Union Bay State Chem.
Kralac A	15/85 Butadiene–styrene	U.S. Rubber
Kralite*	Kralac A–rubber blend	U.S. Rubber
Marbon S, S-1, "8000"	High-styrene copolymers	Marbon
Marmix	Latex of Marbon S	Marbon
Pliolite S-3	15/85 Butadiene–styrene	Goodyear
Pliolite S-5, S-5A, S-5B, S-5C, S-5D, S-6, S-6B, S-7	High-styrene copolymers	Goodyear
Pliolite Latex 170	Latex of styrene copolymer	Goodyear
Pliolite-Latex 160, 160-FR	Latex of styrene copolymer	Goodyear
Pliolite Latex 150	Latex of styrene copolymer	Goodyear
Pliotuf*	Styrene copolymer	Goodyear
Polyco 350	High-styrene copolymer latex	American Polymer Corp.
Polysar SS250	High-styrene resin—GR-S	Polymer Corp. Ltd.
Styraloy 22, 22-A	Butadiene-styrene copolymer	Dow

* Impact-resistant plastics.

REFERENCES

1. Aiken, W. H., *Modern Plastics*, **24,** No. 6, 100–2 (1947); **27,** No. 6, 72–4 (1950).
2. Aiken, W. H., *Modern Plastics*, **26,** No. 2, 99–103 (1948).
3. Alfrey, T., and Price, C. C., *J. Polymer Sci.*, **2,** 401–6 (1947).
4. Anon., *Modern Packaging*, **17,** 103 (1943).
5. Anon., *Modern Plastics*, **25,** No. 11, 93–6 (1948).
6. Anon., *Modern Plastics*, **27,** No. 6, 62 (1950).
7. Anon., *Chem. Week*, **69,** No. 8, 27 (1951).
8. Anon., *Chem. Week*, **70,** No. 1, 22–23; **71,** No. 6,51 (1952).
9. Anon., *Rubber Developments*, **4,** No. 2, 40–53 (1951).
10. Anon., *India Rubber World*, **126,** 76 (1952).
11. Bacon, R. G. R., Farmer, E. H., and Schidrowitz, P., in *Proc. Rubber Technol. Conf.*, edited by T. R. Dawson and J. R. Scott, Heffer, 525–37 (1938).
12. Barry, A. J., Hook, D. E., and DePree, L. (Dow), U.S. Pat. 2,475,122, July 5, 1949.
13. Bartovics, A. (Firestone), U.S. Pat. 2,537,641 (1951).
14. Becker, W. (I.G. Farbenindustrie), Ger. Pat. 641,801, Feb. 13, 1947.
15. Blömer, A., and Becker, W. (I.G. Farbenindustrie), U.S. Pat. 2,222,345, Nov. 19, 1940.

16. Blömer, A., and Becker, W. (I.G. Farbenindustrie), Ger. Pat. 728,701, Oct. 29, 1942; Brit. Pat. 519,175, Mar. 19, 1940; Fr. Pat. 843,262, June 28, 1939.

17. Blömer, A., Becker, W., and Hebermehl, R. (I.G. Farbenindustrie), U.S. Pat. 2,292,737, Aug. 11, 1947.

18. Blömer, A., and Konrad, E. (vested in APC), U.S. Pat. 2,301,926, Nov. 17, 1942.

19. Bloomfield, G. F., *J. Chem. Soc.*, **1943**, 289–96; **1944**, 114–20.

20. Borders, A. M., Juve, R. D., and Hess, L. D., *Ind. Eng. Chem.*, **38**, 955–8 (1946).

21. Brazier, S. A. (team leader), BIOS Final Rept. No. 1626, Item No. 22, German Chlorinated Rubber. See also Dawson, T. R., BIOS Overall Rept. No. 7, The Rubber Industry in Germany during the Period 1939–1945, pp. 29, 61; James, A. A. (team leader), BIOS Final Rept. No. 1530, Item No. 21, Investigation of German Methods of Rubber/Metal Bonding, pp. 51, 56–7; German Plastics Industry, BIOS Misc. Rept. No. 1.

22. Briant, P. C. (Firestone), U.S. Pat. 2,578,653 (1951).

23. British Thomson-Houston Co., Brit. Pat. 556,684, Oct. 15, 1943.

23a. Buist, J. M., *Trans. Instn. Rubber Ind.*, **29**, 72–91 (1953).

24. Buizov, B. V., and Kusov, A. B., *J. Rubber Ind. U.S.S.R.*, **12**, 46–51 (1953).

25. Burger, V. L., Donaldson, W. E., and Baty, J. H., *ASTM Bull. No. 120*, 23–6 (1943); *Rubber Chem. and Technol.*, **16**, 660–7 (1943).

26. Burke, W. J. (du Pont), U.S. Pat. 2,478,038, Aug. 2, 1949.

27. Burr, W., *Offic. Dig. Federation Paint & Varnish Production Clubs*, **277**, 198–207 (1948).

28. Burr, W. W., and Matvey, P. R., *Offic. Dig. Federation Paint & Varnish Production Clubs*, **304**, 347–58 (1950).

29. Calfee, J. D., and Thomas, R. M. (Standard Oil Devel. Co.), U.S. Pat. 2,447,610, Aug. 24, 1948.

30. Carothers, W. H. (du Pont), U.S. Pat. 2,067,172, Jan. 12, 1937.

31. Clayton, J. H., and Bann, B. (Manchester Oxide), Brit. Pat. 527,935, Oct. 18, 1940.

32. Crawford, R. W., and Glick, S. E., *Product Eng.*, **16**, 53–5 (1945).

33. Cunneen, J. I., *India-Rubber J.*, **114**, 543–6 (1948); *J. Chem. Soc.*, **1947**, 134–41; *J. Applied Chem.*, **2**, 353–7 (1952).

34. D'Alelio, G. F. (General Electric), U.S. Pat. 2,380,726, July 31, 1945.

35. D'Alelio, G. F., U.S. Pats. 2,414,803, Jan. 28, 1947; 2,419,202, Apr. 22, 1947; 2,457,872, Jan. 4, 1949.

36. Daly, L. E. (U.S. Rubber), U.S. Pat. 2,439,202, Apr. 6, 1948.

37. Dawson, T. R., *Rubber Developments*, **1**, No. 2, 15–21 (1947).

38. Dawson, T. R., in *Elastomers and Plastomers*, Vol. II, edited by R. Houwink, Elsevier, New York, 1949, Chap. 11.

39. Dawson, T. R., and Schidrowitz, P., in *Chemistry and Technology of Rubber*, edited by C. C. Davis and J. T. Blake, Reinhold, New York, 1937, pp. 656–77.

40. De Bell, J. M., Goggin, W. C., and Gloor, W. E., *German Plastics Practice*, PB Rept. 12,467, Murray Printing Co., Cambridge, Mass., 1946, pp. 445–6. See also Schonol (I.G. Farbenindustrie), Method of preparation of Suitable Solutions for Buna Chlorination, PB Rept. 50,494; Blömer, A., and Becker, W. (I.G. Farbenindustrie), Research on the Chlorination of Synthetic Rubber, PB Rept. 33,443; Chlorination of Synthetic Butadiene Polymer, PB Rept. L–67,935.

41. Deutsche Tornesit G. m.b.H., trademark.

42. D'Ianni, J. D., *Ind. Eng. Chem.*, **40**, 253–6 (1948).

43. D'Ianni, J. D. (Wingfoot), U.S. Pat. 2,484,614, Oct. 11, 1949.

44. D'Ianni, J. D. (Wingfoot), U.S. Pat. 2,571,346 (1951).

45. D'Ianni, J. D., Hess, L. D., and Mast, W. C., *Ind. Eng. Chem.*, **45**, 319–24 (1951).

46. D'Ianni, J. D., Naples, F. J., Marsh, J. W., and Zarney, J. L., *Ind. Eng. Chem.*, **38**, 1171–81 (1946).

47. Elmore, N. H., and Cessler, A. M., U.S. Pat. 2,458,841, Jan. 11, 1949.
48. Endres, H. A., *Rubber Age N.Y.*, **55**, 361–6 (1944).
49. Endres, H. A., *Am. Paint J.*, **32**, 86, 88–92 (1947).
50. Farmer, E. H., *Endeavour*, **3**, 72–9 (1944).
51. Felton and Guilleaume Carlswerk A.-G., Brit. Pat. 369,647, Nov. 17, 1930; Brit. Pat. 362,049, Sept. 20, 1930.
52. Fisher, H. L., *Ind. Eng. Chem.*, **19**, 1325–33 (1927).
53. Fisher, H. L., *Chem. Revs.*, **7**, 51–138 (1930).
54. Flint, R. B. (du Pont), U.S. Pat. 2,046,257, June 30, 1936.
55. Fonrobert, E., Inaugural dissertation for doctor's degree, Konigl. Christian-Albrechts-Universität, Kiel, 1913.
56. Fordyce, R. G., *India Rubber World*, **118**, 377–8 (1948).
57. Fox, K. M., *India Rubber World*, **117**, 487–91 (1948).
58. Frolich, P. K. (Jasco, Inc.), U.S. Pat. 2,327,517, Aug. 24, 1943.
59. Fromandi, G., *Kolloidchem. Beih.*, **27**, 189–222 (1928); *Rubber Chem. and Technol.*, **2**, 161–5 (1929).
60. Garvey, B. S. (Goodrich), U.S. Pat. 2,168,279, Aug. 1, 1939.
61. Gates, G. H. (Wingfoot), Can. Pat. 459,736, Sept. 13, 1949.
62. Gebauer-Fuelnegg, E., and Moffett, E. W. (Marsene), U.S. Pat. 1,980,396, Nov. 13, 1934. See also Fr. Pat. 779,826, and Brit. Pats. 446,818 and 447,110.
63. Gebauer-Fuelnegg, E., and Moffett, E. W. (Marbo Patents, Inc.), U.S. Pat. 2,072,355, Mar. 2, 1937.
64. General Tire & Rubber Co., trademark.
65. Gleason, A. H., and Rosen, R. (Jasco, Inc.), U.S. Pat. 2,291,574, July 28, 1942.
66. Goodyear Tire & Rubber Co., trademark.
67. Goodyear Tire & Rubber Co., Chem. Div., Techni-Guide PR–600–6.
68. Gordon, M., *Ind. Eng. Chem.*, **43**, 386–93 (1951).
69. Graves, C. D. (du Pont), U.S. Pat. 2,046,160, June 30, 1936.
70. Grimwade, D., *Rubber Age and Synthetics Lond.*, **32**, 51–6 (1951).
71. Guss, C. O., and Amidon, R. W. (Dow), U.S. Pat. 2,388,685, Nov. 13, 1945.
72. Harries, C. D., and Fonrobert, E., *Ber.*, **46**, 733–43 (1913).
73. Hebermehl, R., *Farben, Lacke, Anstrichstoffe*, **3**, 105–9 (1949).
74. Henson, W. A., *Offic. Dig. Federation Paint & Varnish Production Clubs*, **316**, 298–300 (1951).
75. Hercules Powder Co., trademark.
76. Hilton, F. (Brit. Rubber Producers' Research Assoc.), U.S. Pat. 2,379,354, June 26, 1945.
77. Holmberg, B., *Rubber Chem. and Technol.*, **20**, 978–81 (1947).
78. Holt, C. R., Susie, A. G., and Jones, M. E., *India Rubber World*, **121**, 416–8, 423 (1950).
79. Hoover, J. R., as quoted in *India-Rubber J.*, **116**, 595 (1949).
80. Hopff, H., Ebel, F., and Valko, E. (I.G. Farbenindustrie), U.S. Pat. 1,988,448, Jan. 22, 1935.
81. I.G. Farbenindustrie, Fr. Pat. 723,838, Oct. 3, 1931.
82. I.G. Farbenindustrie (Konrad, E., and Kleiner, H.), Ger. Pats. 585,622–3, Oct. 6, 1933.
83. I.G. Farbenindustrie, Ger. Pat. 597,086, May 16, 1934.
84. I.G. Farbenindustrie, Brit. Pat. 495,085, Nov. 7, 1938; Fr. Pat. 836,869, Jan. 27, 1939.
85. I.G. Farbenindustrie A.-G., trademark.
86. Irvin, H. H., *India Rubber World*, **114**, 660–2 (1946).
87. Isolonium A.-G., Fr. Pat. 728,760, Nov. 9, 1931.
88. Jones, F. A., *Trans. Instn. Rubber Ind.*, **17**, 133–8 (1941).
89. Jones, M. E., and Pratt, D. M., *India Rubber World*, **117**, 609–10 (1948).
89a. Jones, R. V., Moberly, C. W., and Reynolds, W. B., *Ind. Eng. Chem.*, **45**, 1117–22 (1953).

90. Kern, E. A., and Abbott, R. K. (General Electric), U.S. Pat. 2,476,737, July 19, 1949.
91. Kirchhof, F., Ger. Pat. 557,270, Sept. 14, 1930.
92. Kirchhof, F., *Kautschuk*, **12**, 80–5 (1936).
93. Kirschner, W. C., *Am. Paint J.*, **35**, No. 2, 62–70 (1950).
94. Klebanskii, A. L., and Rakhlina, M., *Org. Chem. Ind. U.S.S.R.*, **2**, 392–4 (1936); *Rubber Chem and Technol.*, **10**, 467–70 (1937).
95. Kolthoff, I. M., Lee, T. S., and Mairs, M. A., *J. Polymer Sci.*, **2**, 199–228 (1947).
96. Koningsberger, C., and Salomon, G., *J. Polymer Sci.*, **1**, 353–63, 364–79 (1946).
97. Konrad, E., and Ludwig, R., U.S. Pat. 2,335,124, Nov. 23, 1943.
98. Koshelev, F. F., Provorov, V. N., and Solov'eva, A. S., *Kauchuk i Rezina*, **1939**, No. 8, 21–4, 88; see also *India Rubber World*, **108**, No. 1, 74–5 (1943).
99. Kraus, G., and Reynolds, W. B., *J. Am. Chem. Soc.*, **72**, 5621–6 (1950).
100. Lawson, W., U.S. Pat. 2,018,678, Oct. 29, 1935.
101. Le Bras, J., and Delalande, A., *Les Dérivés chimiques du Caoutchouc naturel*, Paris, 1950, 485 pp.
102. Le Claire, C. D. (Firestone), U.S. Pat. 2,476,829, July 19, 1949.
103. Lee, T. S., Kolthoff, I. M., and Mairs, M. A., *J. Polymer Sci.*, **3**, 66–84 (1946).
104. LeFevre, W. J., and Harding, K. G. (Dow), U.S. Pat. 2,460,300, Feb. 1, 1949.
105. Lichty, J. G. (Wingfoot), U.S. Pat. 2,560,869 (1951).
106. Ludwig, L. E., *Offic. Dig. Federation Paint & Varnish Production Clubs*, **276**, 122–4 (1948).
107. McIntyre, O. R., Taber, D. A., and Young, A. E., Programme, Chem. Inst. Canada, Toronto, June 19–22, 1950, p. 29.
108. MacLean, D. B., Morton, M., and Nicholls, R. V. V., *Ind. Eng. Chem.*, **41**, 1622–5 (1949).
109. Marbon Corp. trademark.
110. Martin, G., Davey, W. S., and Baker, H. C. (Brit. Rubber Producers' Research Assoc.), Brit. Pat. 476,269, Dec. 6, 1937.
111. Martin, G. D. (Monsanto), U.S. Pat. 2,375,572, May 8, 1945.
112. Medvedchuk, P. I., Aldoshin, F. D., Marovich, V. P., and Repman, A. V., *J. Gen. Chem. U.S.S.R.*, **12**, 220–6 (1942); *Rubber Chem. and Technol.*, **18**, 24–31 (1945).
113. Meehan, E. J., *J. Polymer Sci.*, **1**, 318–28 (1946).
114. Memmler, K., *The Science of Rubber*, trans. by R. F. Dunbrook and V. N. Morris, Reinhold, New York, 1934, pp. 159–230.
115. Meyer, G. (I.G. Farbenindustrie), U.S. Pat. 2,166,604, July 18, 1939; Brit. Pat. 418,230, Oct. 22, 1934; Fr. Pat. 755,486, Nov. 25, 1933.
116. Mitchell, J. M., and Williams, H. L., *Can. J. Research*, **27F**, 35–46 (1949).
116a. Moakes, R. C. W., in *Rubber Technology*, edited by R. C. W. Moakes and W. C. Wake, Butterworth, London, 1952, pp. 47–55.
117. Mol, E. A. J., *Ingenieur* (s'Gravenhage), **55**, No. 33, 55–7; No. 38, 59–62 (1940); *Chem. Zentr.*, **1941**, I, 132–3.
118. Morrell, C. E., Frolich, P. K., and Bannon, L. A. (Jasco, Inc.), U.S. Pat. 2,334,277, Nov. 16, 1943.
119. Murray, R. L., *Chem. Eng. News*, **26**, 3369, 3646 (1948); also Private Communication from A. H. Maude, Hooker Electrochemical Co., Jan. 10, 1950.
120. Newton, E. P. (Hercules Powder), Brit. Pat. 576,744, Apr. 17, 1946.
121. N. V. deBataafsche Petroleum Maatschapij, Brit. Pat. 611,919, Nov. 5, 1948.
122. Owen, A. F., U.S. Pat. 2,474,801, June 28, 1949.
123. Palmer, H. F. (Firestone), U.S. Pat. 2,459,761, Jan. 18, 1949.
124. Patentverwertungs G.m.b.H. Hermes, Belg. Pat. 444,548, Mar. 31, 1942.
125. Perry, L. H., *Ind. Eng. Chem.*, **41**, 1438–41 (1949).
126. Plauson, H., Brit. Pat. 397,136, Aug. 9, 1933.
127. Pratt, D. D., and Handley, R., Brit. Pat. 474,979, Nov. 8, 1937.
128. Reid, R. J. (Firestone), U.S. Pat. 2,537,630 (1951).
129. Remy, T. P., U.S. Pat. 2,470,952, May 24, 1949.
130. Rinse, J., *Paint Varnish Production*, **42**, No. 5, 25–8, 62 (1952).

131. Rust, J. B., and Pfeifer, C. W. (Montclair Research), U.S. Pat. 2,447,772, Aug. 24, 1948.
132. Ryden, L. L. (Dow), U.S. Pat. 2,498,712, Feb. 28, 1950.
133. Ryden, L. L., Britt, N. G., and Visger, R. D., *Offic. Dig. Federation Paint & Varnish Production Clubs*, **303**, 292–301 (1950).
134. Saffer, A., and Johnson, B. L., *Ind. Eng. Chem.*, **40**, 538–41 (1948).
135. Salomon, G., and Koningsberger, C., *Rec. trav. chim.*, **69**, 711–23 (1950).
136. Schidrowitz, P., *Trans. Instn. Rubber Ind.*, **11**, 458–77 (1936).
137. Schweitzer, O. (Metallgesellschaft A.-G.), U.S. Pat. 2,037,599, Apr. 14, 1936.
138. Scott, T. R., and Field, M. C. (Internatl. Standard Elec. Corp.), U.S. Pat. 2,282,002, May 5, 1942.
139. Sell, H. S., and McCutcheon, R. J., *India Rubber World*, **119**, 66–8, 116 (1948).
140. Sell, H. S., and McCutcheon, R. J., *Ind. Eng. Chem.*, **43**, 1234–43 (1951).
140a. Sell, H. S., and McCutcheon, R. J., *Rubber Age N.Y.*, **72**, 494 (1953).
140b. Sell, H. S., and McCutcheon, R. J., *Rubber Age N.Y.*, **73**, 503 (1953).
141. Serniuk, G. (Standard Oil), U.S. Pat. 2,589,151 (1952).
142. Serniuk, G. E., Banes, F. W., and Swaney, M. W., *J. Am. Chem. Soc.*, **70**, 1804–8 (1948).
143. Sibley, R. L., *India Rubber World*, **106**, 244–7, 347–9 (1942).
144. Smith, W. C. (Standard Oil), U.S. Pat. 2,396,293, Mar. 12, 1946.
145. Soday, F. J. (United Gas Improvement), U.S. Pats. 2,317,857–9, Apr. 27, 1943; 2,338,741–2, Jan. 11, 1944.
146. Söll, J. (I.G. Farbenindustrie), U.S. Pat. 2,129,286, Sept. 6, 1938; Brit. Pat. 456,536, Nov. 11, 1936; Fr. Pat. 788,840, Oct. 18, 1935.
147. Söll, J. (I.G. Farbenindustrie), U.S. Pat. 2,187,185, Jan. 16, 1940; J. Söll and A. Koch, Ger. Pat. 619,211, Sept. 25, 1935.
148. Sparks, W. J., Gleason, A. H., and Frolich, P. K. (Standard Oil), U.S. Pat. 2,477,316, July 26, 1949; Brit. Pat. 577,860, June 4, 1946.
149. Standard Telephones & Cables, Ltd., Brit. Pat. 345,939, Dec. 16, 1929; Brit. Pat. 357,624, June 20, 1930.
150. Standard Telephones & Cables, Ltd., Brit. Pat. 526,959, Sept. 30, 1940.
151. Standard Telephones & Cables, Ltd., Brit. Pat. 529,649, Nov. 26, 1940.
152. Stanton, G. W., and Lowry, C. E. (Dow), U.S. Pat. 2,454,486, Nov. 23, 1948.
153. Staudinger, Hermann, and Staudinger, Hansjürgen, *J. prakt. Chem.*, **162**, 148–80 (1943); *Rubber Chem. and Technol.*, **17**, 15–37 (1944).
154. Storey, E. B., and Williams, H. L., *Rubber Age N.Y.*, **68**, 571–7 (1951).
155. Susie, A. G., and Wald, W. J., *Rubber Age N.Y.*, **65**, 537–40 (1949).
155a. Symposium on Emulsion Paints, *Ind. Eng. Chem.*, **45**, 709–54 (1953).
156. Te Grotenhuis, T. A., U.S. Pat. 2,457,097, Dec. 21, 1948.
157. Thies, H. R., *Ind. Eng. Chem.*, **33**, 393 (1941).
158. Thies, H. R., *Vanderbilt Rubber Handbook, 1948*, pp. 595–603.
159. Thies, H. R., and Aiken, W. H., *Rubber Age N.Y.*, **61**, 51–8 (1947).
160. Thies, H. R., and Clifford, A. M., *Ind. Eng. Chem.*, **26**, 123–9 (1934).
161. Tschunker, E., and Bock, W., Ger. Pat. 588,785, Nov. 27, 1933.
162. Tschunker, E., and Bock, W., Ger. Pat. 570,980, Feb. 27, 1933; U.S. Pats. 1,938,730–1, Dec. 12, 1933.
163. Waterman, W. W. (Standard Oil), U.S. Pat. 2,393,208, Jan. 15, 1946.
164. Weatherford, J. A., and Knapp, F. J., *India Rubber World*, **117**, 743–4, 748 (1948).
165. Wingfoot Corp., Brit. Pat. 606,980, Aug. 24, 1948.
166. Winkelmann, H. A., *India Rubber World*, **113**, 799–804 (1946).
167. Wolf, R. F., and Sparks, W. J. (Standard Oil), U.S. Pat. 2,442,068, May 25, 1948.
168. Workman, R. E., *Offic. Dig. Federation Paint & Varnish Production Clubs*, **291**, 177–87 (1949).
169. Zaionchkovskii, A. D., *Trudy Moskov. Teknol. Inst. Leghoi Prom. im L.M. Kaganovicha*, **1941**, No. 2, 201–17; *C.A.*, **40**, 2677 (1946).
170. Zwicker, B. M. G., *Ind. Eng. Chem.*, **44**, 774–86 (1952).

CHAPTER 19

GR-S LATEX

L. H. Howland

United States Rubber Company, Synthetic Rubber Division

HISTORY

Before World War II, the use of natural rubber in latex form was well established in industry. The product was used in many applications such as tire-cord solutioning, foam sponge, impregnation, adhesives, and the forming of many articles by dipping. The manufacture of foam sponge was a relatively new industry and was growing faster than almost any other branch of the rubber business. Latex was obtained mainly from large plantations operated by such companies as U.S. Rubber, Firestone, Goodyear, and Dunlop. It was furnished both at normal solids content and in concentrated form, the latter being made by creaming, centrifuging, or evaporating the former on the plantations. The recent trend has been strongly toward the use of concentrated latex, and today little Hevea latex of normal solids content is shipped.

The plantations are mainly in the East Indies, Ceylon, and Liberia. During the war the supply of latex as well as of dry natural rubber was largely cut off, and as long as German submarines were in the Atlantic, it was difficult to get latex even from the remaining available sources. Although numerous civilian products made from latex could be and were dispensed with, latex was necessary for many essential products—tires, military wire, etc. The shortage of natural latex made it necessary to turn to a synthetic product. Some synthetic latexes had become available in small volume before the war. These were mainly Neoprene and Buna-N latexes. But the potential volume was inadequate, and they were not satisfactory for all applications. As soon as large-scale manufacture of GR-S was undertaken, the fact that it was made in latex form offered the opportunity to produce in quantity usable synthetic latexes of this polymer. Also a wide variety of latexes, tailored to many specific uses, could be made by moderate variations in the process.

The development of GR-S latexes was mainly pioneered by the companies who were operating Government-owned GR-S plants for Rubber Reserve. Those varieties that reached the plant stage were manufactured in these Government-owned plants. They were sold through established latex distributors—drum lots from the distributors' plants and carload quantities directly from the synthetic-rubber plants. Originally the price established by Rubber Reserve on tank-car quantities of latexes under 50 per cent

649

solids was $18^1/_2$ cents per pound dry weight and of those over 50 per cent solids $20^1/_2$ cents per pound, plus $^3/_4$ cent per pound wet weight for transportation. These prices have since been revised upward. During the war and the early postwar period, GR-S latexes were made at the Government plants at Naugatuck, Conn., and Los Angeles, Calif., operated by U.S. Rubber; at Akron, Ohio, operated by Firestone; at Louisville, Ky., operated by B. F. Goodrich; and at Akron and Los Angeles operated by Goodyear. These plants were able to make as much latex as was required, in addition to a variety of specialty dry polymers. As of January 1954 the synthetic-rubber plants operated by the Goodyear and Firestone companies at Akron, Ohio, and by the U.S. Rubber Company at Naugatuck, Conn., were devoted largely to the manufacture of GR-S-type latexes, while some latex manufacture was carried out at the plant at Baton Rouge operated by the Copolymer Corporation and at the plant at Los Angeles operated by the Midland Rubber Corporation.

LOW- AND MEDIUM-SOLIDS TYPES

Types I and II. The GR-S synthetic latexes first made on a production scale, known as type I and type II, had a low content of solids—around 25 per cent. The polymer in type I was identical with standard GR-S, and type II differed only in that it contained no antioxidant. Articles from type-I latex discolored in sunlight and stained lacquer and fabric owing to the antioxidant present; articles from type II were much better in this respect. Type I was not in fact used extensively, and production of it has now been discontinued. Type-II latex[1, 26, 43] was made with fatty acid soap as the emulsifier and a butadiene-styrene charge ratio of 71/29, the conversion being carried to 72 per cent. The Mooney viscosity (ML-4) of the polymer was around 50, and its content of combined styrene 23.5 per cent. Type II has now been replaced by latex X-695, which differs from it only in containing a little (40 p.p.m.) of formalin, to prevent bacterial action during storage.

Type III. The next latexes made were of medium-solids content—from 38 to 50 per cent. The first of these, known as type III,[26, 43] had about 38 per cent solids. It was made to this solids content directly in the polymerization reactor. The butadiene-styrene charge was 50/50 and the combined styrene in the polymer 45 per cent. The reason for this ratio is that it gives cured films of better tensile strength than it is possible to obtain from type-II latex. It is a general characteristic of butadiene-styrene copolymers that tensile strength improves as the bound styrene of the polymer increases. The low-temperature behavior of the polymer, however, generally becomes poorer as the styrene ratio is increased, thus limiting the amount of bound styrene that can be used advantageously. As the styrene increases above 50 per cent, the properties of the product move away from those of a rubber toward those of a plastic.

Type-III latex is made with wood rosin soap as the emulsifier, and the polymerization is carried to about 80 to 90 per cent conversion. It should be pointed out that all latexes of low- and medium-solids type are stripped by steam distillation of unreacted butadiene and styrene. Usually some antifoam material is needed to aid in the efficiency of this operation.

Several modifications of type-III latex have been made in order to give certain users a more suitable product. The most important of these is X-381 GR-S latex (now latex 2001). This has a lower Mooney viscosity, 25 to 40 instead of 90, and is made with *tert*-dodecyl mercaptan in place of *n*-dodecyl mercaptan as the modifier.

Type IV. Demand for somewhat higher-solids latex led to the production of type-IV latex.[43] This was identical with type-III latex except that the solids were 39.1 to 42 per cent instead of 37.0 to 39.0 per cent. A modification of type IV, X-395 GR-S latex, had a maximum content of volatile unsaturated hydrocarbons of only 0.05 per cent. These "free unsaturates" appear to be mainly styrene and tetrahydrostyrene (butadiene dimer) and are believed to be responsible for a good deal of the odor in most latexes. They are removed by exhaustive stripping.

Still further efforts to increase the solids content of this class of latex (types III and IV) resulted first in X-409 latex[2, 58] made to approximately 46 per cent solids. The higher concentration was reached by using a suitable emulsifier combination, employing less water than in type IV, and carrying the polymerization to as near 100 per cent as possible. The production of X-409 has been discontinued in favor of X-446 GR-S latex[3, 27] which is easier to make and has 47 to 49.9 per cent solids. This is made by polymerizing to approximately 95 per cent conversion in a recipe containing a reduced amount of water and employing the potassium salt of disproportionated rosin acid (Dresinate 214) as the emulsifier and *tert*-dodecyl mercaptan as the modifier. Extra soap is added after polymerization to improve stability. The polymer in the latex usually has a Mooney viscosity between 75 and 100. This latex is being used in rather large volume. It is now known as latex 2002.

Latex of types III and IV has now been replaced by latex 2000 (see Table I), which is essentially similar to type IV.

Control of Fluidity. The time of polymerization of low- and medium-solids latexes such as type III and X-466 varies from 20 to 48 hours. The time for individual latexes is adjusted mainly through the temperature of polymerization and the selection of emulsifiers. In GR-S latex manufacture, a very viscous stage occurs at about 30 per cent solids, and, the lower the water in the system, the more trouble is caused by such thickening, which interferes with proper agitation and heat transmission. If polymerization is too rapid, the temperature cannot be controlled at this stage. A factor that accentuates the thickening stage is the inherently fine particle size of GR-S latex. The thickening is favored by even small amounts of fatty acid soap, which result in the formation of small polymer particles, and is lessened by the addition of electrolytes, which tend to increase particle size. In order to get through the thickening stage without loss of temperature control, the reaction velocity must be adjusted to a proper rate by selecting the right polymerization recipe and conditions. For latexes of types III and IV (latex 2000) polymerization is carried out at 150° F. The polymerization of X-446 (2002) is an example of the way temperature may be used to control polymerization rate and latex viscosity. The temperature is maintained at 110° F. until the latex solids reach 30 per cent; it is then raised to 130° F., and, when the solids content reaches 40 per cent, it is raised to 140° F.

Shortstops and Antioxidants. As is well known, GR-S type-polymerizations are usually shortstopped at the desired conversion by addition of a suitable substance, such as hydroquinone or sodium sulfide. Type-I and type-II latexes and their modifications are shortstopped with hydroquinone. The other latexes referred to above are not shortstopped. This is possible because of the high conversion. In the past shortstops were avoided in the production of GR-S latexes because the available substances imparted to the latexes or the finished products an undesirable color or odor. More recently,[52] however, it has been found that certain dithiocarbamates are satisfactory shortstops for low-temperature polymerizations and that they do not produce any odor or any great amount of discoloration. Some of the newer types of latex made at 120° F. and above (type VIII, X-534, X-621) have been shortstopped with a substance of this sort, and many of the cold latexes contain a water-soluble dialkyldithiocarbamate as a shortstop.

Also, with the exception of type-I latex, stabilizers or antioxidants for the polymer are not added to GR-S latexes intended for use as such. The polymer while in the form of latex appears to resist aging well. The main reason, however, that stabilizers are omitted is that a universally suitable material has not been found. This leaves the user in a position to add whatever stabilizer is desired for his application.

Some of the characteristics of the most important of the latexes mentioned above are given in Table I.

Table I. GR-S Latexes of Low- and Medium-Solids Content Prepared at Elevated Temperatures

			Type III	
Original designation	Type II	X-381	Type IV	X-446
Present designation	X-695	2001	2000	2002
Solids, min.–max., %	26–28	37–39	39–42	47–50
pH	8.5–9.28	8.5–12.0	10.0–11.0	9.5–11.5
Free unsaturates, %, max.	0.50	0.50	0.50	0.10
Bound styrene, %	22.5–24.5	44–48	44–48	44–48
Average particle size, A.*	700	1000	1100	1450
Mooney viscosity of polymer	46–54	20–40	60–90	50–80

* Representative value or range; not a specification.

HIGH-SOLIDS TYPES

Preparation by Creaming. Many applications of latex demand higher than 50 per cent solids. The first synthetic latexes to meet this requirement were made by creaming. One way in which synthetic latexes differ from natural latex is in the fact that the average particle size (about 700 A.) is much smaller than in natural latex (in the neighborhood of 2500 A.). These fine-particle-size latexes, unlike natural latex, do not cream readily with alginate. The latex must be agglomerated to a larger average particle size before the creaming agent is added. This is accomplished by decreasing the stability of the system by adding electrolytes or by reducing the pH. After agglomeration, the stability of the system may be restored again and creaming agents added. Commercial creams have been made from type-III latex by creaming with a combination of sodium silicate and ammonium alginate, and from type-II latex by first agglomerating with ammonium chloride–formaldehyde, then increasing the pH in order to stabilize the system, and then adding ammonium alginate.[5, 18, 45, 50a, 53, 63] In the latter procedure the solids content of the product is increased by adding about 30 per cent of unagglomerated latex to the agglomerated latex, in order to favor particle packing, before adding creaming agent. In this way a solids content of 58 to 62 per cent can be reached. Another commercial cream has been made from GR-S-85-type latex. These creams have been made by latex distributors and not by the Government synthetic-rubber plants. Agglomerated latexes can be concentrated by centrifuging, but there have appeared to be no advantages in this process over that of creaming.

Direct Preparation. Although the creaming process gave very desirable latexes, the products were more expensive than latex made to the same solids content directly, without creaming, would be expected to be. This fact stimulated efforts to make high-solids latexes (50 to 60 per cent or above) directly in the synthetic-rubber-plant reactors by using the minimum amount of water in the polymerization recipe. An intensive research and development program throughout the industry culminated in satisfactory high-solids latexes so made.[12, 19, 39] The first trouble encountered in developing these latexes was loss of temperature control because the emulsions became very viscous, especially at intermediate conversions as mentioned earlier, and proper agitation could not be obtained in the standard plant reactor. A modification of the agitator improved conditions considerably,[9] but, before satisfactory results could be obtained, more fundamental changes in the recipes employed in the production of low-solids latex had to be made. One procedure investigated was seeding.[16, 20, 46, 56] It consists of adding to a fresh polymerization charge a certain proportion of latex from a previous polymerization as "seed." The average particle size of the polymer in the latex is increased by this procedure, and this improves the fluidity of the latex. However, there is a practical limit to the amount of seed that can be employed because it reduces plant productivity.

Studies of the GR-S emulsion polymerization system led Harkins to develop a theory of its mechanism[24] according to which most of the polymerization is initiated in the soap micelles. This means that the number and character of the soap micelles present at the beginning is an extremely important factor in determining the number, average size, and size distribution of the polymer particles at later stages. The ultimate aim in most work with high-solids latex has been to change or control the number and character of the soap micelles and to adjust the initial polymerization rates so that there is obtained the proper distribution of polymer particle size to give a fluid latex at the desired solids content. At the same time the latex stability must be preserved. The usual practice has been to use a total amount of water approximating that desired in the finished latex along with a minimum amount of an emulsifier, such as will permit a suitable rate of polymerization but will not give too small an average particle size. In many cases further amounts of emulsifier are added later (at 40 to 50 per cent solids) in a "booster solution." Such a solution in some cases consists of a small amount of water, emulsifier, catalyst, and modifier or initiator. Synthetic emulsifiers and inorganic salts are often used in the charge formula to reduce the viscosity, and the choice of these and their amounts are important factors in securing the greatest possible fluidity and stability in the latex. Vigorous agitation, as mentioned earlier, is usually needed to help prevent gelling at intermediate conversions and to insure adequate heat exchange. In some recipes, developed more recently, the total emulsifier present during polymerization is comparatively low, and so more emulsifier is added after the reaction is over in order to improve the stability of the latex during storage and use. Brief descriptions of some of the commercial high-solids latexes are given below.

Types of High-Solids Latex. Type V or latex 2003 is an emulsion of a copolymer of butadiene and styrene with soaps of fatty acids and disproportionated rosin acids as emulsifiers.[8, 43] Of the hydrocarbon present, approximately 30 per cent by weight is styrene. Neither shortstop nor antioxidant is used. The latex is made to a minimum of 59 per cent solids, and the polymer in it has a Mooney viscosity (small rotor) of 65 to 95. Incremental addition of soap and seeding are employed in its manufacture. This latex finds its largest use in foam sponge, but it has numerous other applications.

Type VI (latex 2005) is a general-purpose latex[43] made to 60 to 63 per cent solids directly in reactors without resort to a seeding technique. It is stabilized with synthetic emulsifiers (e.g., a formaldehyde condensation product of a sodium alkylnaphthalene sulfonate) and disproportionated rosin soaps. Polymerization is started at 140° F. and finished at 150° F. The hydrocarbon consists approximately of 45 per cent by weight of styrene. Like type V it has neither shortstop nor antioxidant.

Type VII is also an emulsion[43] of a copolymer of butadiene and styrene stabilized with synthetic emulsifiers and soaps of fatty and disproportionated rosin acids. It contains about 21 per cent by weight of styrene and has no

Table II. High-Solids GR-S Latexes made at Elevated Temperatures

	Type V	Type VI	Type VII	Type VIII
Original designation				
Current code	2003	2005		
Bound styrene, %	28–30	41–45	21 approx.	45 approx.
Total solids, %	59 min.	60–63	60–63	50 min.
Total soap, %	1.5–2.0	...	...	2.3–2.7
pH	9.5–11.0	10.0–11.0	9.0–10.5	9.0–10.5
80-mesh coagulum, max., %	0.14	0.15	0.15	...
Filterability (felt)	...	300 g.*	100 g., min.	...
Rubbing stability	Poor	Good	Good	...
Surface tension, dynes per cm.	...	40*	50–60*	35, min.
Mooney viscosity of polymer	58–88 (MS)	45–75	35–46*	...
Particle size, A.*	2500	2300	2300	...

* Not a specification.

shortstop or antioxidant. The solids content is 60 to 63 per cent, and free unsaturates are less than 0.10 per cent. The copolymer of this latex was designed for good low-temperature characteristics. Type VII is made by a procedure similar to that followed for type VI.

Type VIII is a 50 per cent solids (minimum) emulsion[43] of a copolymer of butadiene and styrene stabilized with synthetic emulsifiers and dispro-portionated rosin soaps. About 48 per cent of its hydrocarbon content is styrene. It contains a nonstaining shortstop but no antioxidant.

In general high-solids latexes have larger average particle size than low-solids latexes. The specifications for these latexes and other features of them are listed in Table II.

COLD GR-S LATEXES

After it had been discovered that GR-S (dry polymer) made at 41° F. was superior to GR-S made at 122° F., it was natural that attention should be given to the production of GR-S latex at low temperatures.[25, 51–2] The manufacture of GR-S latex at 40 to 50° F. was undertaken and is being expanded considerably. And the development of cold latexes is being actively pursued.

Whereas most of the latexes mentioned previously were polymerized, at elevated temperatures, to a conversion of 72 per cent or higher, low-tempera-ture latexes are polymerized to as low as 60 per cent conversion. The reason for this is that at higher conversion branching and cross-lining tend to affect the quality adversely. However, work is in progress on latexes of higher conversion (90 to 100 per cent), particularly with polymerization recipes designed to produce high solids (60 per cent or above) directly in the reactor. The development at present is mainly concerned with butadiene-styrene ratios in the neighborhood of 70/30, although other ratios from 100/0 up to 50/50 have been used. Polymers with high Mooney viscosities have been favored.

Good film tensile strength in 122° F. polymers is obtained only with high styrene ratios such as 50 per cent. Such strength is at the expense of good performance at low temperatures. The films of low-temperature polymers (41° F.) from latex prepared from a 70/30 butadiene-styrene charge ratio have much better tensile strength than films from corresponding GR-S latex made at 122° F. and are not appreciably poorer in low-temperature performance. Also low-temperature latexes usually give films of better wet strength than higher-temperature latexes. In addition, latexes pro-duced at low temperatures make foam sponge with greatly improved resistance to static cracking. Other advantages are hoped for as the development proceeds. It should be pointed out that tear resistance is not improved so much as tensile strength.

Films of good tensile strength from low-temperature latexes are best obtained by the cast-film method. A fundamental study of film strength is needed, however, since frequently latexes that should theoretically give good film tensile strength unexplainably do not. Tensile strengths of around 3000 p.s.i. have been obtained from 41° F. latexes, and of around 4000 p.s.i.

from 0° F. latexes.[51] These results were obtained with creams prepared from latex of 20 to 25 per cent solids content made to 60 per cent conversion and a Mooney viscosity of 60, with either Dresinate or fatty acid soap as the emulsifier. It has been shown that the quality of films from low-temperature latex is not dependent on creaming. A more extensive study[10b] has shown that commercial cold high-solids latexes will give film tensile strengths of 2000 to 3000 p.s.i.

The first commercial latexes made at low temperature were X-544 and X-547 GR-S latexes.[52] The latter soon superseded the former because of its much greater ease of manufacture. The polymerization of latex X-547 (now designated as 2100) is carried out in an organic hydroperoxide-ferrous sulfide recipe[51] to a conversion of 60 per cent and a solids content of 38.5 per cent which later is raised to 47 to 50 per cent by heat concentration. This latex can be concentrated satisfactorily to approximately 60 per cent solids by creaming according to the alginate-silicate method,[44] which is fairly cheap because the quantity of the creaming aids required for concentrating from approximately 50 per cent solids to around 60 per cent is small and also because serum loss is very small. In commercial practice, concentration by heat rather than by creaming is being used at the present time,[52] and by its means a concentrated (60 to 63 per cent solids) type of X-547 containing polymer of high Mooney viscosity (about 140) is being produced[25] as X-619.

Polyamine-activated recipes suitable for the production of latexes at low temperatures of polymerization have been developed, the initiator combination consisting of diisopropylbenzene hydroperoxide and diethylene triamine.[25] The first two latexes made by such recipes to be carried to the production scale[4] were X-633 and X-635. A monomer charge of 50/50 butadiene–styrene is used in the production of X-633, and the latex has a solids content of 49 per cent maximum. X-635 was prepared from a 70/30 butadiene-styrene charge carried to 80 per cent conversion and 70 to 100 Mooney viscosity. The solids content of 47.6 per cent was raised by stripping and concentration with the aid of a heat exchanger to 60 to 63 per cent. It has been reported that such latex yielded good foamed-rubber sponge without admixture with Hevea latex. The production of latex X-635 has now been discontinued and other high-solids cold latexes suitable for the manufacture of foamed sponge have been introduced. A systematic study of GR-S latexes in foamed sponge rubber shows that the superiority of cold latex to regular (hot) latex in sponge is relatively much greater than the superiority, generally, of cold to regular dry GR-S.[55a]

Other latexes made by low-temperature polymerization in peroxamine recipes are X-667, X-678, X-684, X-710, X-711, and X-749, all of high-solids content and used in foamed-sponge manufacture; and X-617, X-683, and X-701, of low- to medium-solids content. X-617 and X-701 are employed for dipping tire cord. The rubber in all these latexes contains about 25 per cent styrene in combination, except that in X-667, where the bound styrene is only 14 per cent, and X-711, in which the rubber is polybutadiene. The Mooney viscosity of the polymer in most of these latexes is high. The emulsifier is a fatty acid soap or a mixture of fatty acid and rosin acid soap and in addition several of the latexes contain a synthetic emulsion-stabilizer.

The application of peroxamine recipes to the preparation of latexes at 14° F. and 0° F. is under development.[25]

Another polymerization recipe that shows considerable promise in the preparation of cold high-solids latex with improved color and viscosity characteristics depends on a redox system employing sodium formaldehyde sulfoxylate as the reducing agent.[10a] Latexes using this recipe that have reached the production stage are X-753 (sulfoxylate activated counterpart of X-667) and X-758 (similar to X-619).

SPECIAL LATEXES

Volatile-Base Soaps. There are a number of applications where a GR-S latex is required that will deposit polymer free from soap and other water-soluble materials in order to obtain low water absorption in the final product. To prepare such latex it is necessary that the soap and electrolytes used in polymerization shall decompose during the drying and curing of the deposited film. This is accomplished by using for emulsification soaps of volatile bases such as ammonia, diethylamine or, better, morpholine, and, for control of latex viscosity during polymerization, volatile electrolytes such as ammonium carbonate or acetate. The first volatile-base latexes were polymerized in the high-temperature range (up to 150° F.) using as catalyst a persulfate, or a free radical type of initiator such as a diazothioether or bisazoisobutyronitrile[47, 61] More recently[25] systems using organic hydroperoxides have been developed for both high and low (41° F.) polymerization temperatures.

A difficulty in manufacturing the early latexes of this type containing a highly volatile base such as ammonia or diethylamine was in the steam stripping of unreacted styrene. Soap hydrolysis and consequent volatilization of amine led to instability of the latex and the formation of coagulum. This could be compensated for by continually adding base during stripping, but such an operation was costly and none too satisfactory. The use of such materials as morpholine as the volatile base has now been found to give excellent stability during stripping without loss of low water absorption properties in the final product. Latexes of this type have been prepared with solids contents as high as 50 per cent. It is expected that in the future development of such latexes emphasis will be placed on the use of low temperatures of polymerization and moderately volatile bases such as morpholine (b.p. 128° C.).

Acid Latexes. Acid latexes, in the preparation of which cationic agents such as laurylamine acetate, rosin amine hydrochloride, or quaternary ammonium soaps serve as emulsifiers, are in the process of development.[21, 25, 29, 37, 42] For their production at 122° F. a moderately active peroxide such as *tert*-butyl hydroperoxide has been used as the catalyst, and for their production at 41° F. a more active hydroperoxide, such as diisopropylbenzene hydroperoxide, activated by a ferrous salt. Such latexes are of particular interest in impregnating and coating processes because of the fact that many organic fibers are electronegative.

Latex from Other Monomers. In the preparation of the latexes described hitherto, the monomers used have been mixtures of butadiene and styrene in various proportions. However, latexes have been prepared from other monomers, mainly butadiene alone or isoprene-styrene mixtures. The manufacture of these latexes is similar to that of butadiene-styrene latexes and has been conducted at 122° F. Polybutadiene latexes X-534, X-621, and X-653 reached plant-scale production, and the last mentioned has been continued in full-scale production under the designation latex-2004. This latex and also the low-styrene cold latex, X-667, and the polybutadiene cold latex, X-711, mentioned previously, are, because of considerations of odor, used in blends with natural-rubber latex in the manufacture of foamed sponge.

An isoprene-styrene latex has been manufactured under the designation X-388 GR-S latex.[48] It was made to 50 per cent solids with Dresinate emulsification and was intended for tire-cord treatment. Although an improvement in this application, it has not proved better than standard GR-S latexes to an extent sufficient to warrant the higher cost of substituting isoprene for butadiene.

Emulsifier-Free Latex. A novel latex has been made without emulsifier by agitating a monomer mixture containing a water-soluble monomer, for example, butadiene and styrene with a small proportion of acrylonitrile, in water with persulfate catalyst and sodium bisulfite.[59, 60, 62] This type of latex is still in the development stage.

CODE DESIGNATIONS FOR SYNTHETIC LATEXES

There have been changes in the code employed to identify synthetic latexes produced in Government plants. Those varieties of latex that are considered to be established most firmly are given numbers, so-called hot latexes, i.e., latexes prepared by polymerization at elevated temperatures, being put in a 2000 series, and cold latexes, prepared by polymerization at reduced temperatures, being put in a 2100 series. Other latexes are given X numbers. If and when an X-numbered latex meets with wide acceptance, it is given a number in the 2000 or 2100 series. New X-latexes are announced from time to time by the Reconstruction Finance Corporation, and lists of them are presented in the *India Rubber World* and the *Rubber Age*. Convenient summary lists of the main latexes are to be found in various articles.[14a, 24a, 66]

CONCENTRATION

As inferred previously, it is often of advantage to concentrate low-solids latex to a higher-solids content, because the resulting product is more suited for some uses than the high-solids latex produced directly in the reactor. The three general methods, creaming, centrifuging, and heat concentration, that have been used on Hevea latex have been tried with GR-S latex. The first two of these processes have already been described (p. 652). The heat-concentration method involves simply the removal of some of the water by a suitable evaporation process. This increases not only the rubber but also the nonrubber content of the latex solids, which is a disadvantage in some cases. The application of these concentration methods to latexes made at 122° F. is of relatively less importance than it is to latexes made at 41° F. because most of the high-quality low-temperature latexes have been made to 60 to 80 per cent conversion. In general, the heat concentration involves passing the latex under reduced pressure from the stripper through a heat exchanger and spraying it back into the stripper so as to strip and to concentrate the latex simultaneously. Both internal and external steam may be employed for the evaporation. The method has been successfully applied[25, 52] at the Naugatuck, Conn., plant by the use of the Walker-Wallace plate type of heat exchanger.

The investigation of the German rubber industry at the end of the European war disclosed a novel process for concentrating synthetic-rubber latexes,[22] the so-called *Stockpunkt* method. Buna-S latex was gelled by chilling, and whipped into a foam until the volume had increased some sevenfold. This caused a granulation to take place, and the latex on standing at a low temperature separated into a concentrate and a clear serum. The serum was drained off, and the semisolid concentrate reverted to a liquid latex when it was warmed to room temperature. Maron and co-workers[33] have made a careful study of this method and have introduced a number of fundamental improvements which make possible its application to GR-S latexes. The soap content and the pH of the GR-S latex are carefully adjusted and a solution of an electrolyte such as sodium chloride is slowly added with stirring at room temperature until the salt content reaches a certain value. The gelled mix is then cooled with moderate agitation, the formation of foam being carefully avoided. The mass gradually thickens until, at 5 to 10° C. below room temperature, it suddenly becomes fluid again. The temperature is lowered to a point between 5° and 10° C., and the mass is then filtered through paper or a similar medium. The clear serum separates rapidly until the filter cake contains 60 per cent solids; thereafter separation is slower and, when the cake contains 70 per cent solids, is very slow. The filter cake when warmed to room temperature gives a fluid latex of 60 to 70 per cent solids and good stability. The process has been successful on a pilot-plant scale.

EVALUATION

A number of tests are employed to characterize GR-S latexes. Those used for specifications are described in "Specifications for Government Synthetic Rubbers" issued by the Office of Rubber Reserve, Reconstruction Finance Corporation, 811 Vermont Avenue, Washington 25, D.C. Some tests of special interest are reviewed in the following paragraphs.

Particle Size. The particle size of latexes has been determined by a method suggested by Ewart and White and described by Harkins.[24] The latex is titrated with a soap solution and the surface tension plotted against the soap addition. At first the soap is adsorbed on the particles and the surface tension drops, but finally, on saturation, the curve breaks, marking the end of the titration. If the surface area of a soap molecule and the relationship between surface area and particle diameter for spheres are known, it is a simple matter to calculate the average particle diameter. Maron[31] developed this method so as to get more precise results by differentiating between adsorbed and unadsorbed soap. The method has been the subject of several published studies.[28, 35, 64]

A method involving light scattering[6, 7, 14, 38] for determining particle size has been developed by Debye. It is much less time-consuming than the soap-titration method and involves the use of photoelectric cells to measure the direct and scattered intensities of a beam of light impinging on a cell containing the latex. The turbidity of the latex is measured and the average particle diameter calculated from this value, the refractive indexes of the diluted latex and the medium, the wavelength of the incident light, and the concentration of polymer in the diluted latex. This method is reasonably accurate up to an average particle diameter of about 1000 A. It will give a useful approximation up to about 2000 A., provided it is calibrated with latexes of known particle size or correction is made for the change in scattered intensity with angle of observation. Above 2000 A. it is highly erratic. The results obtained have been compared with those calculated from electron-microscope measurements.[15]

Particle size and its distribution can also be determined by the electron microscope.[23, 30] This requires in the case of latex a somewhat involved procedure as follows: (1) bromination of latex to stiffen the particles, (2) evaporation of diluted, stabilized, stiffened latex on a collodion film, (3) shadow-casting the mounted specimen with a moderately heavy deposit of gold, (4) preparation of electron micrographs for statistical measurement and counting, (5) calculation of average particle size.

A simple, fairly rapid method for determining latex particle size and its distribution is by centrifugal fractionation employing a high-speed tube centrifuge[36] running at about 20,000 r.p.m. The average particle size of the fractions removed from each sixth of the tube by a long hypodermic needle are determined by light scattering. A more precise method for the determination of particle-size distribution by centrifugal means has been developed.[41a] It depends on the quantitative application of Stokes's law to the migration (as determined by change in solids content at a given point) in the gravitational field generated by an ordinary laboratory centrifuge running at about 2000 r.p.m.

Mechanical Stability. A test for the mechanical stability of latex has been developed by Maron.[34] It is based on the coagulation of latex between two surfaces nearly in contact, one stationary and the other revolving. The stationary surface, in a stainless-steel cup which holds the latex, is made of rubber while the rotor is a chrome-plated tool steel disk. Temperature control is provided and stability is measured by the weight of coagulum formed under defined standard conditions—1000 r.p.m. for 5 minutes at 25° C. using 75 cc. of latex.

Filterability. This test is a modification of the strainability test of Brass and Slovin.[10] The test measures the weight of latex that will pass through a 0.5-in. disk of standard felt before the felt is plugged. The filtration is carried out under a constant head of 12 in. of latex. A measure of the amount of microscopic coagulum present in the latex is obtained in this way. Another test of a similar nature is the filtration of the latex through an 80-mesh screen. The result ("screens value") is expressed as the weight per cent of the latex sample retained on the screen.

Viscosity. A large amount of work has been done on methods of determining the viscosity of latex.[32, 40, 65] The classic method of Ostwald was used in most early work. It is not entirely satisfactory, as it assumes Newtonian flow. Since this may not always prevail the method appears often to give different viscosities with different rates of flow: the viscosity varies with the rate of shear. A study of these non-Newtonian latexes using the Mooney-Ewart viscometer or modifications of it has shown that their flow characteristics can be described in terms of a "viscosity" and a "yield point." The theoretical significance of this viscosity value is open to some question, and the "yield point" probably has no fundamental meaning, but nevertheless both these values have proved extremely useful to producers and users of latex. Up to a concentration of 25 per cent, latex is usually approximately Newtonian and may be so up to considerably higher-solids content. Inasmuch as the viscosities of many of the higher-solids latexes are not independent of the rate of shear, it has been proposed that a very simple method such as that of Brookfield be used to get quick reproducible results in arbitrary units.[17]

Film Tensile Strength. For practical evaulation, latexes are tested in the type of product in the production of which they are to be employed. This is usually done by the user, not at the synthetic-rubber point. Tests frequently employed comprise film laying of the compounded latex, vulcanizing, and running various physical tests such as tensile strength, tear, and aging.[44] Some workers have found the cast-film method to give the best results. Poor tensile strengths have often been obtained in acid-dipped films, but this difficulty may be overcome by proper compounding. Latexes polymerized at 122° F. with a 71/29 butadiene-styrene charge ratio should give film tensile strengths of about 500 p.s.i., whereas similar products from a 50/50 butadiene-styrene charge ratio should give 2000 p.s.i. or more. As previously stated, latexes from 41° F. polymerization[10b, 51] with a 72/28 butadiene-styrene ratio should have film tensile strengths of about 3000 p.s.i., while corresponding 0° F. latexes may have about 4000 p.s.i.

Foamed-Latex Sponge Properties. Conant and Wohler[13] have presented a study of the mechanical properties of foamed-latex sponge under three aspects: (1) the tensile strength calculated on the actual rubber present in the cross section of a test piece, (2) compression modulus, and (3) fatigue resistance. They find a considerable difference in tensile strengths between a foam sponge made from Hevea latex rubber and one made from a 50/50 mixture of Hevea rubber and GR-S. The compression modulus showed considerably less difference while both had great resistance to flexing in compression, in shear, and in combined compression and shear. Only when the flexing conditions included tension did failures occur. It should be pointed out that this work was done before the low-temperature latexes were available and that Conant and Wohler do not indicate the type of GR-S latex they used.

A study of the accelerated aging of foamed sponge showed that the change of compression resistance with aging gave the best correlation of the methods evaluated.[49a]

BACTERIAL ACTION IN LATEX

There has been some complaint of hydrogen sulfide odor in certain GR-S latexes. A bacteriological study has been made, and it has been found that bacteria are present that convert sulfur compounds to hydrogen sulfide.[49] The sulfur probably comes from both mercaptan used as the modifier and sulfate resulting from decomposition of the persulfate catalyst. These bacteria may be controlled by use of chlorinated water in the system or by pasteurizing the latex.

LATEX CONSUMPTION

As indicated earlier, the application of rubber in latex form was a rapidly expanding prewar business. This is illustrated by Table III which covers both natural and synthetic latexes over a number of years.

Table III. Use of Latex in the United States by Years

Long Tons—Dry Weight

Year	Natural Latex Imports*	Con-sumption*	GR-S Con-sumption‡	Neoprene Con-sumption‡	Nitrile-Rubber Con-sumption*	Total Con-sumption
1936	19,852†					
1937	23,185†					
1938	11,878†	16,161				16,161
1939	27,438†	21,952				21,952
1940	33,789†	25,210				25,210
1941	34,789†	Not available				
1942	9,595	9,392				9,392
1943	1,890	9,578	15§			9,593
1944	3,090	6,085	6,598§			12,683
1945	4,768	3,886	15,197§	7,077		26,160
1946	8,012	5,724	24,810	13,603		44,137
1947	17,675	13,909	22,474	6,087		42,470
1948	32,630	28,489	23,441	4,500		56,430
1949	29,974	36,117	21,500	3,750		61,360
1950	54,410	56,138	23,680	2,530		82,350
1951	54,963	46,750	31,031*	6,279*	2,628	86,688
1952	48,228	53,567	40,562*	7,368*	3,093	104,680
1953	*75,000*	*68,000*	*46,000**	*8,100*	*3,800*	*125,900*

* Rubber Division, U.S. Department of Commerce, except items†.
† *Rubber Age N.Y.*, **64**, 764 (1949).
‡ G. R. Vila, *India Rubber World*, **120**, 592 (1949); **124**, 446 (1951), except items*.
§ Production figures, probably similar to consumption.
Figures in italics are estimated.

We are indebted to Everett G. Holt and Saul Grimmer of the rubber division, O.D.C., U.S. Department of Commerce for the production figures of GR-S latex for 1943 to 1945, inclusive, and for the figures for consumption of natural rubber latex during 1938 to 1940, inclusive. These latter are estimates based on consumption figures obtained from 147, 181, and 209 firms for each of the three years. Many users of natural-rubber latex did not report separate figures for consumption of latex and of dry rubber, and so the figures only indicate the trend in usage. The synthetic-latex figures for most years do not include butadiene-nitrile copolymer latexes and others made in small volume. No attempt was made to estimate the small amount of synthetic latex available before 1943.

This table shows the growth in volume of the total latex business. In 1948 the use of natural latex had approximately reached its prewar volume, and the total consumption of latex (natural and synthetic) was much larger than before the war. The ratio of GR-S to natural-rubber usage in the form of latex dropped off much less at this time than it did in the form of dry rubber. The indications are that synthetic GR-S latex is here to stay because it is lower in cost and because for many applications it is more economical than natural latex or provides qualities not obtained with the latter. This is particularly true since the advent of "cold"-rubber latex.

LATEX APPLICATIONS

Table IV will serve to give an idea of the uses to which synthetic latexes have been put, and the types most suitable.

Table IV. Applications of GR-S Latexes

General Uses	Specific Applications	Types
Paper	Saturating to improve strength, coating, as a binder for pigments, replaces casein in some places. Beater applications: results similar to saturation.	2000, 2002
Containers	Sealing.	Misc. specialties
Carpets	Sizing. To mix with starch to give more permanent sizing, or to use as a binder for pigments in sizing. Nonskid backings. Binder. To hold pile into backing of rug, thereby permitting less costly construction. To adhere pile to backing in certain nonwoven run construction.	2000, 2002, type VII
Pile fabrics	To coat back of pile fabrics to give a strong but less costly construction.	2000, 2002
Combining	To build up strength by combining costly to less costly fabrics. May be mixed with starch for adhesive and stiffening.	2000, 2002, 2005

Table IV—continued

General Uses	Specific Applications	Types
Tires	Cord dipping.	X-695, 2000, 2002, X-523, 2100, X-617
Footwear	Combining fabrics.	2000, 2002
Asbestos fiber	Brake linings, gaskets, etc.	Special, 2100
Barrels	Liners.	2000
Jute	Shoe insoles, etc.	2000
Jute and sisal	Upholstery pads, rug underlays	2000, 2002, 2100
Foam sponge	Cushions, etc.	Type VII, 2003, 2004, X-684, X-710, X-711, X-749, X-753, X-758
Batteries	Separator sheets	2000
Chewing gum		X-695, 2000
Linoleum base-coat		2000
Nonwoven cotton fabrics	Binder, wallpaper, drapes, etc. To impart stretch and strength. Cotton pads, etc.	2000
Textiles	Shrinkproofing woolens, improving wear, etc.	Misc. specialties

In comparison with natural-rubber latex, GR-S latexes have a smaller average particle size and possess greater mechanical stability. There is considerable evidence to show that GR-S latex penetrates much better into a fibrous structure such as paper than natural-rubber latex. It possesses the further advantage of much better aging properties especially in the presence of trace amounts of copper compounds. This latter property is of distinct value in many textile uses. However, the polymer in the latexes polymerized at elevated temperatures gives cured films that have either a low gum tensile strength or a poor low-temperature flexibility, depending on the amount of styrene in the polymer. (As mentioned previously, high-styrene content results in improved tensile strength but poor low-temperature properties.) For these reasons, such GR-S latexes are most often used where the latex film is supported by some other material or where low-temperature stiffness is not an important factor. This includes such uses as tire-cord dipping, paper saturation, and treatment of textiles and various other fibrous materials. The cold latexes represent a considerable improvement in film tensile strength, with little or no loss in low-temperature properties, and for this reason give better unsupported products, such as foam sponge, than does hot latex. Compared with the latter, they also appear to have better adhesion to tire cord, less odor, and less tendency to form prefloc when mixed with starch solutions.

Foamed-Latex Sponge Rubber. One of the most interesting and important applications of latexes, both natural and synthetic, is in the manufacture of foamed latex sponge. The process in most general use consists of beating the thickened latex into a foam. The proper coagulation of the latex so as to give a stable foam (commonly referred to as gelation) is the key to the successful application.[11, 41, 57] An alternative process calls for the use of a vacuum to cause the generation of gas in the latex. The resulting foam is stabilized by the presence of certain chemicals or by other means.[54] The rubber used in foam sponge must have good wet strength to allow removal of the articles from the mold and in many cases good low-temperature flexibility.

Until the advent of cold latex no foam sponge was made from GR-S latex alone; Hevea-GR-S blends containing 30 to 50 per cent Hevea were used. However, with cold latex it has become possible to use blends with considerably more GR-S (up to 100 per cent), in products such as foam sponge. The GR-S latex confers considerable improvement in aging and makes an appreciable reduction in the material cost. It also allows the manufacture of a foam sponge having a higher modulus (as measured by compression under a constant load) with little or no increase in density. Those working in this field have had varying degrees of success, some reporting very excellent results in comparison to Hevea latex. However, most of these reports have not been published, and to date the only comprehensive studies available are those presented by Talalay and Talalay.[54a, 55]

General. It should be pointed out in any general discussion of the utilization of latexes that much of the success in their application depends on the techniques and apparatus used. These often vary widely from user to user, and this fact, coupled with the fact that many techniques are not disclosed, makes any detailed description of limited value.

CONCLUSION

The development during and since the war of GR-S latexes as complete or partial replacements for natural-rubber latexes has been very interesting and an important phase in the history of the rubber industry. Beginning as wartime makeshifts some of these products have now become well established. It seems highly probable that GR-S latexes are to have a permanent place in our economy for several reasons: (1) the recent discovery of high-quality synthetic latexes, (2) the frequent tendency of natural latex to go to a high price, and (3) the seriousness of wartime shortages that have existed and may recur. Furthermore, the interest is sufficient to continue to stimulate research, which should result in improvements destined to increase the value of such products to a still greater degree.

The author wishes gratefully to acknowledge the assistance in the preparation of this chapter given him by Morris G. Shepard, Donald E. Fowler, William E. Messer, and Victor S. Chambers of the U.S. Rubber Company.

REFERENCES

1. Anon., *India Rubber World*, **111**, 700–1 (1945).
2. Anon., *India Rubber World*, **117**, 226 (1948).
3. Anon., *India Rubber World*, **118**, 254 (1948).
4. Anon., *India Rubber World*, **124**, 587 (1951).
5. Arundale, E., U.S. Pat. 2,444,801 (1948).
6. Bardwell, J., and Sivertz, C., *Can. J. Research*, **25B**, 255–65 (1947).
7. Billmeyer, F. W., Jr., Private Communication to O.R.R., Mar. 1, 1945.
8. Borders, A. M., and Pierson, R. M., *Ind. Eng. Chem.*, **40**, 1473–7 (1948).
9. Borders, A. M., and Roberts, H. P., Private Communication to O.R.R., July 13, 1945.
10. Brass, P. D., and Slovin, D. G., *Anal. Chem.*, **20**, 172–4 (1948).
10a. Brown, R. W., Bawn, C. V., Hansen, E. B., and Howland, L. H., Paper presented before Rubber Div., Am. Chem. Soc., Chicago, Sept. 11, 1953; *India Rubber World*, **128**, 638 (1953) (abstract).
10b. Brown, R. W., Messer, W. E., and Howland, L. H., *Ind. Eng. Chem.*, **45**, 1322–9 (1953).
11. Chapman, W. H., Pounder, D. W., and Murphy, E. A., Brit. Pat. 332,525 (1929); U.S. Pat. 1,852,447 (1932).
12. Chittenden, F. D., McCleary, C. D., and Smith, H. S., *Ind. Eng. Chem.*, **40**, 337–9 (1949).
13. Conant, F. S., and Wohler, L. A., *India Rubber World*, **121**, 179–84, 192 (1949).
14. Debye, P., *J. Applied Phys.*, **15**, 338–42 (1944); Private Communication to O.R.R., Sept. 13, 1943.
14a. Drogin, I., *India Rubber World*, **127**, 505–10 (1953).
15. Elder, M. E., and Maron, S. H., Private Communication to O.R.R., Oct. 5, 1950.
16. Ellslager, W. M., Private Communication to O.R.R., Feb. 18, 1947.
17. Foley, H. K., and Woltz, F. E., Private Communication to O.R.R., Jan. 7, 1949.
18. Freeman, R. D., U.S. Pat. 2,423,766 (1947).
19. Gay, C. W., Private Communication to O.R.R., Oct. 28, 1943.
20. Glasgow, G. U., Private Communication to O.R.R., Oct. 5, 1945.
21. Goodrich Chemical Co., B. F., Private Communication to O.R.R., June 7, 1945.
22. Handley, E. T., Rowzee, E. R., Fennebresque, J. D., Garvey, B. S., Juve, R. D., Monrad, C. C., and Troyan, J. E., *P.B. Rept.* 189, pp. 79–82.
23. Hanson, E. E., and Daniel, J. H., *J. Applied Phys.*, **18**, 439–43 (1947).
24. Harkins, W. D., *J. Am. Chem. Soc.*, **69**, 1428–44 (1947).
24a. Howland, L. H., and Neklutin, V. C., *India Rubber World*, **126**, 371–2 (1952).
25. Howland, L. H., Neklutin, V. C., Brown, R. W., and Werner, H. G., *Ind. Eng. Chem.*, **44**, 762–9 (1952).
26. Howland, L. H., Peaker, C. R., and Holmberg, A. W., *India Rubber World*, **109**, 579–81 (1944).
27. Jackson, D. L., Private Communication to O.R.R., Apr. 9, 1948.
28. Klevens, H. B., *J. Colloid Sci.*, **2**, 365–74 (1947).
29. Laundrie, R. W., Private Communication to O.R.R., Mar. 19, 1947.
30. Maron, S. H., Private Communication to O.R.R., Aug. 5, 1948.
31. Maron, S. H. et al., Private Communications to O.R.R., Aug. 13, 1945, Sept. 23, 1946, May 4, 1949.
32. Maron, S. H., and Krieger, I. M., Private Communication to O.R.R., Oct. 10, 1947.
33. Maron, S. H., Moore, C., Kingston, J. C., Ulevitch, I. N., Trinastic, J. C., and Borneman, E. H., *Ind. Eng. Chem.*, **41**, 156–61 (1949).
34. Maron, S. H., and Ulevitch, I. N., Private Communication to O.R.R., Mar. 27, 1944; *Anal. Chem.*, **25**, 1087–91 (1953).
35. Mattoon, R. W., Stearns, R. S., and Harkins, W. D., *J. Chem. Phys.*, **15**, 209–10 (1947).
36. Meadors, V. G., and Messer, W. E., Private Communication to O.R.R., Oct. 11, 1946.
37. Messer, W. E., and Howland, L. H., Private Communication to O.R.R., Dec. 28, 1942.
38. Meyer, G. E., Private Communication to O.R.R., Jan. 2, 1945.

39. Mitchell, G. R., Private Communication to O.R.R., Nov. 14, 1947.

40. Mooney, M., and Ewart, R. H., *Physics*, **5**, 350–4 (1934).

41. Murphy, E. A., and Owen, E. W. B., Brit. Pat. 332,526 (1929).

41a. Nisonoff, A., Messer, W. E., and Howland, L. H., Paper presented before Rubber Div., Am. Chem. Soc., Boston, May 29, 1953; *India Rubber World*, **128**, 217 (1953) (abstract).

42. Ober, H. E., Private Communication to O.R.R., June 1, 1945.

43. Office of Rubber Reserve, Specifications for Govt. Synthetic Rubbers, GR-S Latex, Types II–VIII, Jan. 1, 1949.

44. Peaker, C. R., *Rubber Age N.Y.*, **57**, 423–4 (1945).

45. Peaker, C. R., U.S. Pats. 2,393,261 (1946), and 2,446,101 (1948).

46. Pierson, R. M., Coleman, R. J., and D'Ianni, J. D., Private Communications to O.R.R., May 7, Aug. 6, Oct. 12, Dec. 8, 1948.

47. Reynolds, R. A., Private Communication to O.R.R., Dec. 9, 1946.

48. Reynolds, J. A., and Mauger, F. A., Private Communication to O.R.R., Oct. 28, 1947.

49. Rodde, A. L., Private Communication to O.R.R., Feb. 21, 1949.

49a. Rogers, T. H., and Heineman, H. H., Paper presented before Rubber Div., Am. Chem. Soc., Chicago, Sept. 10, 1953; *India Rubber World*, **128**, 638 (1953).

50. Rumbold, J. S., U.S. Pat. 2,446,107 (1948).

50a. Schmidt, E., and Kelsey, R. H., *Ind. Eng. Chem.*, **43**, 406–12 (1951).

51. Smith, H. S., Werner, H. G., Madigan, J. C., and Howland, L. H., *Ind. Eng. Chem.*, **41**, 1584–7 (1949).

52. Smith, H. S., Werner, H. G., Westerhoff, C. B., and Howland, L. H., *Ind. Eng. Chem.*, **43**, 212–6 (1951).

53. Svendsen, E. C., U.S. Pat. 2,446,115 (1948).

54. Talalay, J. A., U.S. Pat. 2,140,062 (1938).

54a. Talalay, J. A., Paper presented before Rubber Div., Am. Chem. Soc., Chicago, Sept. 10, 1953; *India Rubber World*, **128**, 638 (1953) (abstract).

55. Talalay, L., *Rubber Age N.Y.*, **69**, 331 (1951).

55a. Talalay, L., and Talalay, A., *Ind. Eng. Chem.*, **44**, 791–5 (1952).

56. Thompson, R. D., Smith, H. F., and Costanza, A. J., Private Communication to O.R.R., May 3, 1946.

57. Untiedt, F. H., U.S. Pat. 1,777,945 (1930).

58. Werner, H. G., and Smith, H. S., Private Communication to O.R.R., Oct. 22, 1947.

59. Whitby, G. S., and Gross, M. D., Private Communication to O.R.R., Dec. 2, 1946.

60. Whitby, G. S., Gross, M. D., Miller, J. R., and Costanza, A. J., Internat. Cong. Chemistry, New York, Sept. 1951.

61. White, L. M., U.S. Pat. 2,393,133 (1946).

62. Willis, J. M., *Ind. Eng. Chem.*, **41**, 2273–6 (1949).

63. Willson, E. A., U.S. Pat. 2,357,861 (1944).

64. Willson, E. A., Miller, J. R., and Rowe, E. H., *J. Phys. & Colloid Chem.*, **53**, 357–74 (1949).

65. Worrell, W., Smith, H. S., and McCleary, C. D., Private Communication to O.R.R., Feb. 22, 1946.

66. Zwicker, B. M. G., *Ind. Eng. Chem.*, **44**, 774–86 (1952).

CHAPTER 20

LATEX MASTERBATCHING

J. W. Adams and L. H. Howland

United States Rubber Company, Synthetic Rubber Division

HISTORY

In the course of manufacturing useful articles from rubber or rubberlike substances, various combinations of vulcanizing, protective, processing, loading, coloring, and reinforcing agents must be mixed with the elastomers. Conventional procedures for incorporating these compounding ingredients have generally depended entirely on the use of powerful masticating equipment such as Banbury mixers or rubber mills. In this type of compounding either the ingredients are blended directly with the elastomer or portions of previously prepared, concentrated ingredient-elastomer mixtures, known as masterbatches, are added during mixing. The deficiencies of such methods requiring large amounts of power to work the very tough rubbery compounds were recognized at an early date, and procedures have been investigated for incorporating many of the materials with rubber latexes before flocculating them. Since reinforcing agents constitute the largest volume of ingredients used, improved methods for adding these materials have received considerable attention.

From the patents that have been issued on processes for incorporating various reinforcing agents in natural rubber by coprecipitating them with latex[1, 1a, 5, 6, 8, 10–1, 12a, 13, 18, 20, 20a, 23, 28] the amount of work that has been done in this field is evident. Actual production of such compounds has not, however, been realized to any great extent because of the distances separating the latex-producing plantations and the pigment-manufacturing plants.

With synthetic rubbers the situation has been different, since latexes representing various classes of synthetic rubber (butadiene-styrene, butadiene-acrylonitrile, and chloroprene polymers) are produced near sources of reinforcing agents; hence there is presented an ideal opportunity for exploiting the advantage of mixing these agents in the latex stage and thus minimizing the use of heavy, power-consuming equipment. The necessity for milling the mixtures at a later stage still remains, but mixing at the latex stage saves time and power in the later milling.

Colloidal carbons, generally referred to as carbon blacks, are the best reinforcing agents. Improvements imparted by them to natural-rubber vulcanizates are increased modulus, tensile strength, and resistance to abrasion. A comparison of the properties of unpigmented vulcanizates of general-purpose GR-S and of natural rubber shows the latter to be vastly superior to the former. However, when both of these rubbers are compounded with carbon black, GR-S is reinforced to such an extent that the differences between the two are greatly diminished. For this reason carbon black plays an even more important role in the utilization of GR-S than in that of natural rubber. With many other reinforcing agents the situation is similar. Further, nitrile rubbers are analogous to GR-S in their response to reinforcing pigments. Reinforcing agents are not so necessary in the use of Neoprene, but they are still required to obtain desired wearing properties.

During World War II one of the factors limiting production of rubber goods for military and civilian use was the amount of equipment available for dry-mixing reinforcing agents with rubberlike material. The hope that latex masterbatching might alleviate the condition stimulated interest in research and development work along this line. Incorporation of carbon blacks in synthetic-rubber latexes had been investigated in Europe during the period 1924–29, and in 1935 a U.S. patent covering a basic method was issued to Beck and Mueller-Conradi.[3] This method, which specifies the use of aqueous dispersions of carbon blacks containing surface-active agents to render the slurries compatible with latex, was found to be well suited to GR-S types of latex. The same procedure was also suited to elastomer emulsions of other types. Methods for incorporating carbon black in GR-S latex were studied by practically all collaborators in the Government Synthetic Rubber Program. Over 200 reports on miscellaneous latex-masterbatching methods have been submitted to the Office of Rubber Reserve. In addition there are several papers[14, 16, 19, 22] and patents[25, 26] dealing with the coprecipitation of elastomers and carbon black from synthetic-rubber latexes.

THE LATEX-MASTERBATCHING PROCESS

Latex masterbatching of carbon blacks in GR-S is a convenient method for reinforcing elastomers of this type because of the size of the elastomer and carbon black particles and the relative number of each present in the carbon black slurry-latex mixtures required for properly pigmented compounds. Typical diameters for easy-processing-channel (EPC), high-modulus-furnace (HMF), and semireinforcing-furnace (SRF) grades of rubber-reinforcing blacks and GR-S particles are given in Table I. The approximate number of particles of each present in an ultimately dispersed black slurry-latex mixture for preparing masterbatches containing 50 parts by weight of carbon black per 100 parts GR-S are also listed.

Table I.   Size of Particles in Blacks and in Latex

Particle	Diameter, Millimicrons	Number of Particles per Gram 50 Parts Masterbatch $\times 10^{-13}$
EPC black	30	1290
HMF black	40	545
SRF black	80	69
GR-S latex	70	393

The carbon black particles actually exist as small agglomerates in normal slurries. For this reason the schematic drawing (Fig. 1) serves to illustrate theoretical microscopic conditions in a black slurry-latex mixture for producing a pigmented compound containing 50 parts (33.3 per cent by weight) of EPC carbon black. Reasonably good mixtures of carbon black in GR-S are produced when dispersions such as these are properly coprecipitated.

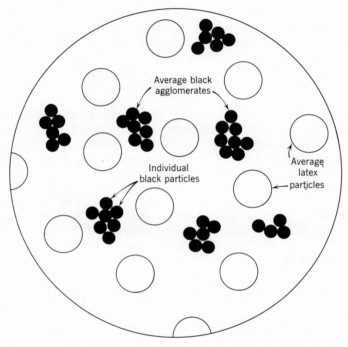

Fig. 1.   Theoretical Particle-Size Relation—GR-S Latex and EPC Black

The process for making "latex masterbatches" of GR-S and carbon black consists of blending aqueous dispersions of the black with GR-S latex in the desired proportions and subsequently flocculating, washing, filtering, and drying the mixture. Since GR-S latexes contain anionic surface-active agents such as sodium salts of fatty and rosin acids, anionic or non-ionic surface-active agents must be used for preparing the necessary slurries of black, if the latter are to be compatible with the latexes. The agents most frequently used are salts of sulfonated alkylnaphthalene-formaldehyde condensation products and sulfolignins. Since these materials and their acidic derivatives impart no improvement, or very little, to the physical properties of compounds and vulcanizates from masterbatches, it is customary to use no more of these surface-active agents than is necessary to produce slurries that are fluid, nonsettling, and latex-compatible. Their use is further restricted to maximum quantities of approximately 2.5 per cent on the carbon black in order to avoid excessive losses of carbon black and copolymer as fines during flocculating operations.

The manufacture of GR-S latex has been described in Chapter 19 and the preparation of black masterbatch has been described in detail in previous publications.[14, 22] The following is a condensed conventional description of these processes, which are diagrammed in Fig. 2.

Emulsions of 71 parts butadiene and 29 parts styrene are copolymerized to 72 per cent conversion at a temperature of 117 to 121° F. in 3750-gal.

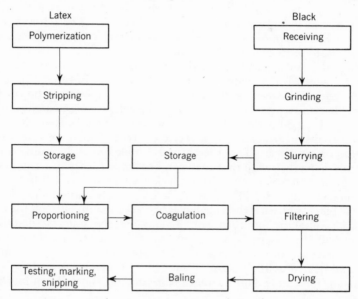

FIG. 2. Latex Masterbatching Process—Condensed Over-All Flow Diagram

glass-lined reactors. The reaction is stopped by adding a solution of hydroquinone. The raw latex is stripped by venting and by steam distillation to remove unreacted butadiene and styrene, respectively. It is then blended and stored in 30,000-gal. concrete tanks in which gentle agitation is applied. At this point antioxidant is added to the latex which contains about 28 per cent solids.

The black in pellet form comes in 80,000-lb. carload lots and is discharged into a screw conveyor which carries it to a bucket elevator. The material is elevated to a grinder (Micro-Pulverizer) where it is ground directly into a slurry make-up tank of 9000 gal. capacity having an impeller and draft-tube type of agitator. Fourteen thousand (14,000) pounds of black are slurried in 6300 gal. of water in which 300 lb. of dispersing agent (sodium alkylnaphthalene sulfonate type) and 54 lb. of flake caustic soda have been dissolved. The solids content of this slurry is finally adjusted to 20.5 ± 0.2 per cent. It is convenient to store black in the form of a slurry; such slurries are easy to handle, and, because of the low apparent density of dry carbon black, a 20 per cent slurry does not occupy much more space in storage than dry black would.

The mixing, flocculating, and finishing operations are continuous.

GR-S latex and black slurries are measured in open weirs or gravimetric measuring towers and are mixed in such proportions as to give a ratio of 1 lb. of black solids to 2 lb. of latex solids. Flocculation is accomplished by adding solutions of sodium chloride and sulfuric acid to the carbon black-latex mixture. A more uniform distribution of black in the coagulum is obtained by rapid or "shock" flocculation, such as can be obtained by the addition of a mixture of salt and acid, than by the two-step flocculating procedure used in the usual manufacture of GR-S. The latter procedure consists in first creaming the latex with sodium chloride and then adding sulfuric acid to form crumbs of coagulum.

During the subsequent masterbatch washing operation the friable coagulum, which is denser than water, must be kept in suspension by mild agitation to avoid excessive disintegration. It is dewatered on an Oliver vacuum filter and washed with cold water to produce a pigmented compound containing about 30 per cent water. The filter cake is broken up by means of a Jeffrey disintegrator and conveyed to the top flight of a continuous, 3-pass forced-air drier. The dried crumb is baled in 90-lb. units which are sufficiently nonadhesive so that they can be shipped without wrapping.

The masterbatch is analyzed for black content on an ash-free basis by pyrolyzing samples at 550° C. under a blanket of carbon dioxide. The residue of black and ash is then heated at 550° C. in air to burn off the carbon. Methods of analysis for other components are the same as described in Office of Rubber Reserve specifications for GR-S except for the determinations of moisture and antioxidant. To expel moisture, samples are hot-milled and heated at 105° C. for 30 minutes. Provision is made to analyze latex blends for antioxidant content before the incorporation of black.

TYPES OF GR-S-BLACK MASTERBATCHES PRODUCED

The general procedure outlined above has been used by various GR-S-producing plants to formulate about 90 different types of carbon black masterbatches. The early major producers were General Tire & Rubber Company and U.S. Rubber Company in plants operated for the Government at Baytown, Texas; at Institute, W. Va., and at Borger, Texas. Later the copolymer plant at Houston, Texas, operated by the Goodyear Tire & Rubber Company, undertook the production of these compounds, the Phillips Chemical Company assumed the operation of the Borger, Texas, plant, and production at the Institute, W. Va., plant was discontinued. More recently the copolymer plant at Los Angeles, Calif., operated by the Midland Rubber Company, has gone into the production of these compounds.

Six types of rubber-reinforcing carbon blacks as follows have been used in making commercial masterbatches: (1) easy-processing channel black (EPC), (2) medium-processing channel black (MPC), (3) high-modulus furnace black (HMF), (4) semireinforcing furnace black (SRF), (5) reinforcing furnace black (RF), (6) high-abrasion furnace black (HAF). Originally the product manufactured in largest amount by the process of latex master-batching was a mixture of 50 parts (33.3 per cent) EPC black in 100 parts of a type of general-purpose GR-S. It was designated GR-S Black 1. More lately there has been a rapidly growing interest in latex masterbatches composed of RF and HAF types of black with "cold" GR-S. These now constitute more than three fourths the output of latex-masterbatched rubber. The number of possible variations that may be introduced in GR-S-black masterbatches is of course considerable. Not only the type of black but also its concentration in the final pigmented compound is important. Loadings in GR-S may vary from 10 to 60 per cent of the masterbatch, depending on the final use for which the material is intended. Another factor that may be varied is the type and quantity of surface-active agent used in preparing the black slurries. Again, the type of GR-S used in latex masterbatching may be varied widely. Although hitherto GR-S latexes representing copolymers of butadiene and styrene in the ratio 71/29 carried to 60 to 75 per cent conversion have usually been employed, almost any variety of GR-S latex is susceptible to the treatment. The most important variations here are in regard to the emulsifier, the temperature of polymerization, and the Mooney viscosity of the polymer.

It will thus appear that the number of latex-masterbatch types is limited only by the practical consideration of keeping the variety of products made in a single plant to a minimum. Table II shows test results for four of the important earlier types of black masterbatches. It also gives a list of the black masterbatches that had been adopted as standard products in 1953. X-Numbered masterbatches are omitted from this list.

Table II. GR-S-Black Masterbatches

	GR-S-Black 1*	GR-S-Black 20	GR-S-Black 3*	X-537 Black GR-S
Carbon black type	EPC	HMF	EPC	HAF
Carbon black loading (parts per 100 parts GR-S)	50	55	50	55
Dispersing agent in black slurry	Sodium alkylnaphthalene sulfonate (Daxad 11 or Triton R-100)		Sodium lignin sulfonate (Marasperse CB)	
GR-S polymerization temp., °F.	122	122	122	41
Compounded Mooney viscosity	50	43	54	83
Vulcanizate properties at optimum cure:				
Tensile strength, p.s.i.	2920	2020	3140	3910
Modulus at 300%, p.s.i.	850	700	850	2600
Elongation at break, %	690	740	600	450

Standard Grades of GR-S Black Masterbatches, 1953

No.	1100	1101	1102	1103 and 1104	1600 and 1601	1801
Polymerization temp., °F.	122	122	122	122	41	41
Carbon black type	EPC	HMF	SRF	MAF†	HAF	HAF OIL

* Now replaced by grade No. 1100.

† Medium-abrasion furnace.

FACTORS IN THE PREPARATION OF CARBON BLACK SLURRIES

A procedure for comparing the efficiency of various dispersing agents in the preparation of stable, fluid dispersions of black in water has been described by Adams, Messer, and Howland.[2] It involves agitation of the slurry in a high-speed mixer—a Waring Blendor.

Recent studies[2, 7] of factors influencing the slurrying of carbon blacks have led to conclusions as follows:

1. The lower the bulk density of a black after micropulverization, i.e., the more completely the aggregates of black particles which occur in pelletized black have been broken up by dry grinding, the more slowly is the black wetted by an aqueous solution of a dispersing agent.

2. The time required for wetting a black is notably reduced by employing water at 40° C. as against water at 25° C.

3. The occurrence on the surface of fine furnace blacks of benzene-soluble material (probably derived from polymerization of the oil which largely forms the raw material for the black) is a factor in the ease of wetting. In comparing two blacks micropulverized to the same apparent density but differing in their content of benzene-soluble material, the black containing the smaller quantity of such material will wet more rapidly.

4. The quantity of dispersing agent required to produce a stable slurry is markedly influenced by the ash content and the pH of the black. The fine furnace blacks, such as HAF black, have a relatively high content of ash and produce a high (alkaline) pH when they are slurried in water. Such blacks require relatively large amounts of dispersing agent. For example, a sample of HAF black having the high ash content of 1.25 per cent required 4.1 per cent of dispersing agent, whereas the same black after it had been extracted with water and its ash content thus reduced to 0.11 per cent required only 2.9 per cent of the same agent. The ash of furnace blacks is derived largely from the water used to cool the carbon-laden gases emerging from the cracking furnace, and variation in the degree and type of hardness in the water used at different plants is probably a factor in determining the ash content of black produced by the plant. Analysis of the above-mentioned HAF black sample indicated that the ash was largely sodium sulfate.

5. With blacks made by the same process, the amount of dispersing agent required to yield a stable, fluid slurry follows closely the size of the ultimate particles, provided that soluble ash is first extracted from the black. Thus, for example, considering thermal blacks, a sample of MT thermal black having a surface area of 6.3 square meters per gram required only 0.3 per cent of dispersing agent, whereas a sample of HAF black having a surface area of 164 square meters per gram required 3.9 per cent. Considering channel blacks, a sample of EPC black having a surface area of 106 square meters per gram had a requirement of 0.7 per cent, whereas a sample of CC black having a surface area of 172 square meters per gram had a requirement of 2.9 per cent.

Over the whole particle-size range of channel blacks the requirement of dispersing agent is linearly related to the surface area of the black (as determined by the adsorption of iodine), but in the furnace blacks a linear relationship holds only up to a surface area of about 70 square meters per gram, owing, it is suggested, to internal porosity in the very fine blacks—porosity which, while available to iodine, is not penetrated by the dispersing agents used in the preparation of black slurries.

6. As is indicated by the figures quoted above, channel blacks in general require less dispersing agent than furnace blacks of similar particle size. This has been attributed to the fact that in the channel process the surface of the black is partly oxidized and that the consequent polarity facilitates dispersion in aqueous media. In accord with this concept, it is found that, when channel black is subjected to calcination at 1700° F. in nitrogen and the chemically combined oxygen thus removed, its requirement of dispersing agent rises to that of HAF black of the same particle size, and conversely that the subjection of HAF black to oxidation reduces its requirement of dispersing agent to about the same level as that of channel black of the same particle size.

7. In the preparation of slurries from both channel blacks, which are acidic, and furnace blacks, which are alkaline, it is advantageous, from the viewpoint of the fluidity of the slurry and the completeness of dispersion of the black, to add sodium hydroxide to the solution of the dispersing agent. The optimum proportion of sodium hydroxide varies with different types of black and even with shipments of the same type of black.

8. When rubber compounds are made by the regular, dry-mixing process, HAF blacks, which are alkaline, in general lead to quicker curing than channel blacks, which are acidic; and in fact there is danger of scorching with HAF blacks. However, when HAF black is mixed with GR-S by the latex-masterbatching process, it becomes subjected to a pH of 2.3 to 3.5 at the stage of coagulation, and accordingly the scorching tendency and higher rate of cure normally associated with such black largely disappears.

9. Even after micropulverization of dry black, the aggregates originally present are not entirely broken down, and aggregates survive in slurries made by mild agitation such as is employed in practical latex masterbatching. Such aggregates, however, can be broken down to a large extent (although probably not to the ultimate black particles) in 30 seconds or so by vigorous agitation, such as can be applied by a Waring Blendor, as shown by examining the light transmission of highly diluted samples of the slurries. In view of this, it seems reasonable to suppose that the aggregates occurring in technical slurries are largely broken down by the shearing action applied when masterbatched rubber–black is worked on a mill or in an internal mixer.

10. The manner in which the coagulation of a mixture of latex and black is brought about influences the thoroughness with which the black remains dispersed. It has been found that by rapid or "shock" coagulation, in which salt and acid are added simultaneously, better dispersion, as determined by microscopic examination and as reflected in the properties of the vulcanized rubber, is obtained than by the procedure normally used in the coagulation of GR-S latex alone, viz., by the addition of sodium chloride first to produce creaming and the addition afterwards of acid to produce coherent crumbs. Experiments indicate that, in the last-mentioned coagulation procedure, precipitate formed during creaming may take up more than its share of black and form highly loaded agglomerates which persist to a greater or less extent during subsequent dry milling, with the result that the final distribution of black in the stock is uneven.

11. In practical carbon black slurry recipes used in latex masterbatching, over 90 per cent of the dispersing agent is adsorbed on the surface of the black.

12. The occurrence of filter-passing fines of black which have failed to precipitate along with the GR-S but have remained in the serum is more likely to happen when the dispersing agent is a sodium alkylnaphthalene sulfonate than when it is of the lignin or sulfolignin class, and also when an excess of dispersing agent (beyond the quantity that is adsorbed on the black) is used in the preparation of the slurry. When trouble with fines is encountered, it has been found that the addition of certain amines, especially the higher polyethylene polyamines, is effective in deactivating the unadsorbed dispersing agent and eliminating or minimizing the occurrence of fines. It is also advantageous, in avoiding fines, to have the latex-black mixture at 50° C. instead of at room temperature when coagulation is brought about.

NON-BLACK FILLERS AND REINFORCING AGENTS

Experiments with GR-S latex incorporation of nonblack reinforcing agents[16] have indicated that certain benefits may be derived by this method of compounding. Many pigments such as inorganic oxides, carbonates, silicates, and clays are difficult to disperse in elastomers by dry-mixing methods, and latex incorporation appears to yield compounds in which the pigments are better dispersed. Research and development projects carried out by Firestone Tire & Rubber Company and U.S. Rubber Company in the Government Synthetic Program have resulted in actual plant production of 13 different nonblack pigmented masterbatches. The results of much of this work have been published,[4] and a patent has been issued.[15]

Methods for incorporating zinc oxide, calcium carbonate, calcium silicate, clay, and fine-particle silica in GR-S latexes are similar to those described above for black masterbatches, the major differences being in the slurry-preparation and product-flocculation steps. Because surface-active agents used for preparing aqueous dispersions are frequently specific for the type of material dispersed, it has often been necessary to develop special slurrying formulas. Special methods for flocculating each kind of pigment-latex mixture have also been developed, to minimize or eliminate losses as fines

Table III. Latex-Masterbatching of Some Nonblack Reinforcing Agents

Reinforcing Agent	Dispersion Stabilizers	Flocculant
Zinc oxide	Daxad 23	Magnesium or aluminum sulfate
Calcium carbonate	Rosin soaps, Aquarex ME or Darvan 2	Aluminum sulfate
Buca A or Dixie clay	NaOH, sodium silicate or fatty acid soaps	Hydrochloric acid or calcium chloride
Calcium silicate (Silene)	Gum arabic	Zinc sulfate
Hydrated silica (Hi-Sil)	Tetraethylenepentamine	Acetic acid

in the filtrates. Some of the slurry make-up and flocculating materials actually used in producing the various nonblack compounds are listed in Table III.

In a study of latex masterbatching with Buca-A clay it has been reported that slurries of low viscosity with as high as 65 to 70 per cent solids content can be prepared.[4] Since improved pigment retention in the coagulum was obtained by increasing the clay concentration in the latex, a slurry of 51 per cent solids content was used in plant operations. The use of fatty acid soaps, rather than other materials, for dispersing the clay in water also decreased pigment and copolymer losses during filtering operations.

FORMATION OF REINFORCING AGENTS IN SITU

A mode of latex masterbatching different from the procedures hitherto discussed involves forming reinforcing agents in situ. This has received some attention.

When dry, pulverized lignins from wood-pulping operations are dry-mixed with elastomers, the amount of reinforcement obtained is not great. However, the lignins can be brought into solution as salts by treating them with alkalies; such solutions can be mixed with latexes and, on coprecipitating lignin and GR-S from the mixtures, highly reinforced compounds are obtained. Various synthetic-rubber and paper manufacturers in the United States and Canada have carried on research and development projects along these lines. Several papers have been published.[9, 12, 17, 21, 24, 27] One obstacle to the use of lignin coprecipitation processes has been the problem of flocculating alkali lignin-latex mixtures. Alkali lignins are surface-active agents for which reason mixtures containing these materials when treated with solutions of inorganic acids or salts yield fine precipitates (more like sludges than the crumbs obtained in other latex-masterbatching operations) which are very difficult to dewater and wash. The advantages of lignin-reinforced elastomers over those containing carbon black reside in their lighter color, lower specific gravity, high tensile strength (as compared with that of other nonblack reinforced stocks), high tear resistance, as well as the retention of high tensile strength at high-lignin loadings. In addition to these properties, Murray and Watson[17] have found that some types of lignin are effective GR-S antioxidants.

It has also been said that white pigmented stocks reinforced to a high degree can be prepared by incorporating water-soluble inorganic salts with elastomer latexes and coprecipitating the materials. A specific example of this is the use of sodium silicate-latex mixtures that produce reinforced compounds when coprecipitated with calcium chloride or acids. The coagulants have the effect of forming calcium silicate and silicic acid, respectively, in situ. Most compounds of this type have not advanced beyond the laboratory stage.

GENERAL

From this brief description of the progress in latex incorporation of fillers and reinforcing agents, particularly carbon black, in synthetic rubbers, it will be evident that this method of rubber compounding has made important strides and has now become an established procedure in the rubber industry. Not only has the use of heavy, power-consuming equipment been substantially reduced, but also better dispersions of the pigments in elastomers have been obtained in many instances, and this has resulted in imparting better physical (stress-strain) and wearing properties to articles made from these products.

Although in the production of latex-masterbatched mixtures of GR-S and carbon black a relatively intimate blend of the two materials is obtained, nevertheless, in order to secure a proper dispersion of the black in the rubber, to incorporate curing and other ingredients, and to render the rubber processible for the fabrication of rubber goods, it is still necessary to mill the mixture. However, compared with regular milling and compounding procedures, the use of latex masterbatch gives a saving in the time and power required, enables lower mixing temperatures to be used if desired, avoids loss of black to the dust-collecting system, and generally provides cleaner and simpler compounding in the rubber factory. Further, according to Adams, Messer, and Howland,[2] experience has indicated that tire treads prepared from latex masterbatch are about 7 per cent better in road wear than similar treads made by normal dry mixing. Also latex masterbatches have certain advantages in shipping over long distances: in contrast to bulky, dirty carbon black, GR-S-black masterbatches are compact and clean.

As a practical enterprise, latex masterbatching has been definitely associated with and made feasible by the development of large-scale synthetic-rubber production. Obviously the end is not yet reached. Experimenters are still enthusiastically active, and further improvements may be expected as time goes on.

Statistics showing the output of latex-masterbatched GR-S are given in Table IV.

Table IV. Production of Latex-Masterbatched GR-S (Long Tons)

	1949	1950	1951	1952	1953
Gross production of black masterbatch	96,611	97,817	125,004*	132,617†	127,298‖
Net content of GR-S	62,737	64,244	81,357§	84,544‡	82,542**
Total production of GR-S	288,881	350,802	694,583	621,867	631,289
GR-S in black masterbatches as percentage of total production of GR-S	21.7	18.3	11.7	14.0	13.1

* Includes 13,431 oil-black masterbatch.
† Includes 16,189 oil-black masterbatch.
‡ Includes 9,300 in oil-black masterbatch.
§ Includes 7,746 in oil-black masterbatch.
‖ Includes 27,012 oil-black masterbatch.
** Includes 15,406 in oil-black masterbatch.

Of the GR-S contained in black masterbatches 56.8 per cent was cold rubber in 1951. The corresponding figure rose to 67.7 per cent in 1952 and to 72 per cent in 1953.

OIL MASTERBATCHES

Considerable quantities of GR-S of high Mooney viscosity are being latex-masterbatched with oil or with oil and black (see also pages 49, 219, 387). The production figures given in Table IV for oil-black master-batches and in Table V for oil masterbatches show the importance that these types of masterbatches have assumed.

Table V. Production of Oil-Extended GR-S (Long Tons)

	1951	1952	1953
Oil masterbatch (gross)	19,212	59,315	154,261
GR-S content (net)	15,378	46,539	116,689
Total GR-S (net)	694,583	621,867	631,289
Per cent of total GR-S in oil masterbatch (net)	2.2	7.5	18.5

Extenders [11a, 22a] other than petroleum oils have been investigated for this application, and plant-scale quantities of masterbatches containing a disproportionated resin acid extender and containing this extender plus HAF black have been produced. Such masterbatches have properties significantly superior in some respects to the oil-extended masterbatches.[11a]

Table VI. Oil and Oil-Black GR-S Masterbatches (1953)

All grades: Polymerization temperature, 43° F.
Sugar-free iron-pyrophosphate recipe (except 1708)
Dithiocarbamate shortstop
1.25 per cent antioxidant
Salt-acid coagulation
Styrene content of polymer, 22.5 to 24.5 per cent

GR-S	Pts. oil per 100 polymer	Oil Character	Emulsifier (as Soap)	Mooney Viscosity	Antioxidant
1703	25	Naphthenic	Rosin and fatty acids	50–70	Nonstaining*
1704	25	Aromatic	Rosin acids	50–70	Slightly staining
1705	25	Aromatic	Rosin and fatty acids	50–70	Staining
1706	25	Highly aromatic	Rosin and fatty acids	50–70	Staining
1707	37.5	Naphthenic	Rosin acids	45–65	Nonstaining†
1708‡	37.5	Naphthenic	Fatty acid	50–70	Nonstaining†
1709	37.5	Aromatic	Rosin acids	45–65	Staining
1710	37.5	Aromatic	Rosin and fatty acids	45–65	Staining
1711	37.5	Highly aromatic	Rosin acids	45–65	Staining
1712	37.5	Highly aromatic	Rosin and fatty acids	45–65	Staining
1801	25 (plus 50 pts. HAF black)	Naphthenic	Rosin and fatty acids	55–75	Staining§

* Mixed alkylphenol type.
† Tris-nonyl phenyl phosphite type.
‡ Peroxamine polymerization recipe.
§ 1.5 per cent.

In the latex masterbatching of oil-type extenders, an emulsion of the oil (prepared by dissolving an emulsifier acid, usually oleic acid, in the oil and adding this solution to a dilute aqueous caustic solution to form the emulsifier in situ) is mixed with the latex, and this mixture is then flocculated and dried in the same manner as used for unextended GR-S.

In the preparation of oil-black masterbatches the oil emulsions and carbon-black dispersions are sometimes prepared separately and then added to the latex before coprecipitation and processing by the procedures described above for black masterbatching. A preferable method, which was specified in the production of GR-S X-691, is to add to the latex a single slurry of oil and black. This is accomplished by emulsifying the oil containing dissolved emulsifier acid in an alkaline carbon-black dispersion after the black slurry has been prepared in the manner described above for black masterbatching. Tread-type vulcanizates of oil-black masterbatches prepared with an aqueous dispersion of an HAF black that had been coated with the oil in either the dry state or in an aqueous medium have shown in laboratory tests abrasion resistance superior to that of a corresponding vulcanizate of a masterbatch prepared in the usual manner.

The currently available oil and oil-black masterbatches are shown in Table VI (X polymers are excluded).

The authors wish to acknowledge gratefully the indispensable assistance in the preparation of this chapter given them by associates.

REFERENCES

1. Acheson, G. S., U.S. Pat. 1,623,517.
1a. Adams, J. W., U.S. Pat. 2,658,040.
2. Adams, J. W., Messer, W. E., and Howland, L. H., *Ind. Eng. Chem.*, **43**, 754–65 (1951).
3. Beck, A., and Mueller-Conradi, M., U.S. Pat. 1,991,367.
4. Borg, E. L., Madigan, J. C., Provost, R. L., and Meeker, R. E., *Ind. Eng. Chem.*, **38**, 1013–6 (1946).
5. Cohen, E. S. A., U.S. Pat. 1,610,226.
6. Cohen, E. S. A., Brit. Pat. 214,210.
7. Dannenberg, E. M., and Seltzer, K. P., *Ind. Eng. Chem.*, **43**, 1389–96 (1951).
8. Darling, J. F., and Powers, D. H., U.S. Pat. 1,846,820.
9. Dawson, T. R., *Trans. Instn. Rubber Ind.*, **24**, 227–40 (1949).
10. Gibbons, W. A., U.S. Pat. 1,802,761.
11. Hopkinson, E., U.S. Pat. 1,567,506.
11a. Howland, L. H., Reynolds, J. A., and Provost, R. L., *Ind. Eng. Chem.*, **45**, 1053–9 (1953).
12. Keilen, J. J., and Pollak, A., *Ind. Eng. Chem.*, **39**, 480–3 (1947).
12a. Leukhardt, W. H., and Adams, J. W., U.S. Pat. 2,616,860, Nov. 4, 1952.
13. Loomis, C. C., and Stump, H. E., U.S. Pat. 1,558,688.
14. Madigan, J. C., and Adams, J. W., *Chem. Eng. Prog.*, **44**, 815–20 (1948).
15. McMahon, W., Can. Pat. 457,146.
16. McMahon, W., and Kemp, A. R., *Ind. Eng. Chem.*, **36**, 735–8 (1944).
17. Murray, G. S., and Watson, W. H., *India Rubber World*, **118**, 667–9 (1948).
18. Novotny, C. K., and Cox, J. T., U.S. Pat. 2,354,424.
19. O'Connor, H. F., and Sweitzer, C. W., *Rubber Age N.Y.*, **54**, 423–7 (1944).
20. Peterson, A. H., U.S. Pat. 1,611,278.

20a. Pollak, A., U.S. Pat. 2,608,537, Aug. 26, 1952.

21. Raff, R. A. V., Tomlinson, G. H., II, Davies, T. L., and Watson, W. H., *Rubber Age N.Y.*, **64**, 197–200 (1948).

22. Rongone, R. L., Frost, C. B., and Swart, G. H., *Rubber Age N.Y.*, **55**, 577–82 (1944).

22a. Rostler, F. S., *Rubber Age N.Y.*, **69**, 559–78 (1951).

23. Russell, R., and Broomfield, H., Brit. Pat. 212,597.

24. Sohn, A. W., *Gummi u. Asbest*, **4**, 42–4 (1951).

25. Te Grotenhuis, T. A., and Frost, C. B., U.S. Pat. 2,441,090.

26. Vesce, V. C., U.S. Pat. 2,419,512.

27. Waeser, B., *Gummi u. Asbest*, **4**, 197–9 (1951).

28. Wiegand, W. B., Brit. Pat. 250,279.

CHAPTER 21

DIENE POLYMERS AND COPOLYMERS OTHER THAN GR-S AND THE SPECIALTY RUBBERS

W. K. Taft and G. J. Tiger

University of Akron, Government Laboratories

The early years of World War II were devoted almost exclusively to obtaining the raw materials necessary for the production of GR-S and to improving the manufacturing techniques and the quality of the polymer. Later, extensive work was conducted[10, 54, 296] on the preparation of substitutes for the butadiene and styrene monomers and for other raw materials and on their evaluation from the standpoint of both polymer quality and considerations of supply and cost. The effects of substitution of either or both of the monomers currently used in the GR-S system are reviewed herein. The voluminous data from bottle polymerizations are considered first; then pilot-plant and plant polymerizations are discussed, and, finally, work on the nonemulsion polymerization and copolymerization of dienes under the influence of sodium, sodium hydride, and the Alfin catalysts is reviewed. Early investigations on the preparation of synthetic rubber by the polymerization of dienes are here omitted from review. The earliest investigations centered mainly about isoprene and dimethylbutadiene; later the study was extended to halo-, alkyl-, and aryl-substituted 1,3-butadienes. Bulk polymerization was mostly employed in the early work, which has been described elsewhere.[4, 22, 33–9, 55, 83, 94, 109, 247–8, 293, 336–8, 343]

EMULSION POLYMERIZATION

Du Pont Survey. The du Pont organization has reported the results of a survey in which 214 different unsaturated compounds, alone or in admixture with butadiene, were subjected to the conditions of emulsion polymerization.[297] Many of these polymerizations were carried out before the recent modifiers, catalysts, and activators were developed, and hence the results are not definitive as regards the relative value of the various monomers. The results, however, are of interest and value as showing that a very wide variety of unsaturated compounds are capable of copolymerizing with butadiene to yield vulcanizable elastomers. Among the comonomers which stood out as giving polymers of potential usefulness were halostyrenes, alkylstyrenes, esters of methacrylic acid, methyl vinyl ketone, vinylpyridine, vinyl alkylpyridines, dimethyl fumarate, and dimethylvinyl ethynyl carbinol. Other classes of monomers found to copolymerize with butadiene included (1) vinylidene chloride and certain other haloalkenes, (2) alkyl-, halo-, cyano-, and acetoxybutadienes, (3) unsaturated aliphatic nitriles, (4) esters of

682

fumaric and maleic acids, (5) unsaturated aliphatic amides, (6) unsaturated ketones containing —CH : CH · CO—, (7) certain derivatives of cinnamic acid, (8) certain acrylates.

LABORATORY-SCALE STUDIES

Data relating to laboratory studies made in recent years and covering a wide range of monomers and monomer mixtures are presented in Tables I through IV. Laboratory-scale polymerization techniques have been

Table I. *Effect of Impurities in Butadiene on Rate of Polymerization of 75/25 Butadiene–Styrene at 122° F.*

Contaminant	Concentration, %	Decrease in Conversion, %	Concentration, %	Decrease in Conversion, %
Isobutane	1	0.4	2	0.8
n-Butane	1	0.9	2	1.5
Isobutylene	1	0.8	5	5.0
	2	1.9	10	10.0
Butadiene-1,2	1	0.9	2	3.8
High-boiling butene-2	1	2.0	2	4.7
Butene-1	1	1.5	2	5.6
Low-boiling butene-2	1	2.2	2	5.8
1-Pentene	1	5.0		
2-Pentene	1	3.0		
Allene	1	8.0		
1,4-Pentadiene	0.01	Nil	0.1	1.6*
			1.0	13.0*
Vinylacetylene	0.01	Nil	1.0	Nil†
	0.1	Nil		
Butadiene dimer	0.003	Nil	1.0	3.1
	0.03	Nil	2.0	6.3
	0.3	2-3	4.0	12.0
Propylene	1	Nil		
Methylallene	1	Nil		
Piperylene	1	Nil		
Isoprene	1	Nil		
Ethylacetylene	1	Nil		
Dimethylacetylene	1	Nil		
Acetaldehyde	0.1	Nil	1.0	Nil*
Ammonia	0.10	Nil	No effect on rate or properties	
Methyl mercaptan	0.001	Nil	0.01-0.1 Serious retarder	
Ethyl mercaptan	0.001	Nil	0.01-0.1 Serious retarder	
Phenyl-β-naphthylamine	0.2	24.3		

* Reduces intrinsic viscosity of polymer.
† 50 per cent gel in polymer.

adequately described elsewhere.[54, 74, 78, 215]   In general, the charge formula in most of the studies here discussed was as follows:

Ingredient	Parts by Weight
Monomers	100
Soap flakes	5.0
n-Dodecyl mercaptan, pure	0.15–0.50
Potassium persulfate	0.30
Water	180

Butadiene of special purity, supplied by the Phillips Petroleum Company, was customarily used in the laboratory polymerizations, which were conducted at 122° F. in bottles of 4- to 32-oz. capacity.

Influence of Purity of Butadiene and Isoprene. The effects of impurities in butadiene and isoprene have been determined[54, 74, 78] and are shown in Tables I and II. Hydrocarbon impurities containing an allylic methylene group ($>C : \overset{|}{C} \cdot CH_2-$) have some retarding effect on polymerization, and the influence of the allylic methinyl group ($>C : \overset{|}{C} \cdot CH<$) is somewhat greater. The effects on the copolymerization of butadiene and styrene of ethylbenzene, o-xylene, methyl phenyl carbinol, and acetophenone in the styrene were found to be negligible. Small amounts of phenylacetylene (0.01 per cent of the styrene) accelerated the copolymerization but 1 per cent had no effect. Divinylbenzene had little effect on reaction rate, but produced a cross-linked insoluble polymer.[54] (This reagent was later used in GR-S 60, to give a highly gelled, easily processed polymer.)

Table II. Effect of Impurities (1%) in Isoprene on Polymerization of 75/25 Isoprene–Styrene at 122° F.

No Effect on Conversion

n-Pentane	Dicyclopentadiene	n-Propylacetylene
Isopentane	1,2-Butadiene	Isopropylacetylene
Cyclopentane	Dimethylacetylene	

Slight Retarding Effect

1-Butene	1-Pentene	Cyclopentane
cis-2-Butene	trans-2-pentene	3-Methyl-1-butene
trans-2-Butene	2-Methyl-1-butene	
Isobutene	2-Methyl-2-butene	

Substantial Retardation

Isoprene dimer	Piperylene dimer	Carbon disulfide
1,3-Pentadiene	Piperylene dimer	
(piperylene)		

Strong Retardation

1,4-Pentadiene	Cyclopentadiene	Ethyl mercaptan

Physical Testing of Polymers. The following explanations may be given of the testing conditions and the definitions used in connection with the evaluation of polymers prepared either in the laboratory or in the pilot plant.

The coagulated polymers, containing about 2 per cent of antioxidant, were compounded according to the following test recipe. When the available samples were not large enough for the regular laboratory-scale testing procedure,[7] tests were made by a microtechnique.[76, 104, 119, 284]

Ingredient	Parts by Weight
Polymer	100.0
EPC carbon black	50.0
Zinc oxide	5.0
Softener	5.0
Stearic acid	0–1.5
Accelerator	1.0–1.5
Sulfur	1.0–3.0

The *dilute-solution viscosity* (DSV) or the inherent viscosity η, determined by means of an Ostwald viscometer on a 0.25 per cent solution of polymer in benzene, is calculated as the natural logarithm of the ratio of the flow time of the solution to the flow time of the solvent divided by the concentration in grams per 100 ml. The intrinsic viscosity $[\eta]$ is obtained by extrapolating the viscosity-versus-concentration curve to zero concentration.

The *mill-processing index* of a polymer is the total of the ratings assigned to it in respect of (a) carbon black stiffening expressed in Mooney viscosity units, (b) mill shrinkage, (c) roughness, (d) banding time, and (e) time to incorporate the carbon black, each rating being on the basis of 0.1 as best and 4 as poorest.[285] The *extrusion index* is determined on the uncured, uncompounded stock, using a no. $^1/_2$ Royle extruder and a Garvey die. The total of ratings for surface, edge, corners, and porosity of the specimen is designated as the index, and is 4 for the poorest and 16 for the best stock.[85]

Stress-strain properties (300 per cent modulus, tensile strength, and elongation) were determined on a Scott tester at room temperature (77 or 82° F.) or at 212° F., as noted in the text. The tests were conducted as described in the Rubber Reserve Specifications.[7]

Goodrich hysteresis temperature-rise tests were conducted at 212° F. by means of the cyclic compression of a test pellet. The hysteresis value was taken as the heat rise of the specimen above 212° F. either after 30 minutes or, if necessary, after additional 10-minute periods until the specimen had attained equilibrium.[5, 165]

The *DeMattia cut-growth test* was made in a DeMattia flexometer at 82° F. for the samples from polymers prepared in the laboratory and at 212° F. for the specimens from polymers made in the pilot plant. The number of flexures, in thousands, required to produce a cut growth of 0.8 in. in the pierced sample was taken as the cut-growth rating. A *quality index*, based on the relationship between the Goodrich hysteresis–heat-rise and the DeMattia cut-growth values, is defined as the ratio of the flexures of the experimental polymer to the flexures of the GR-S control at cures showing equal heat-rise values.[119]

The *Shore A hardness* is determined with the Shore durometer in which a pointer is forced into the test specimen. A scale from 0 to 100 units reflects the hardness, the higher readings indicating harder compounds.[6]

The *Goodyear rebound* determination[284] is made on a wall-mounted instrument in which the test specimen is held securely 40 in. below the axis on which the pendulum swings. The pendulum is dropped through 15 degrees, and the angle of rebound is read on a scale at the end of this pendulum. Results are expressed as percentage rebound according to the following formula: Rebound (%) = (1 − cos of angle of rebound)/(1 − cos of starting angle).

The *Gehman low-temperature test* is conducted by fastening a thin specimen to a standard wire and twisting the specimen and wire through an angle of 180 degrees. The angle of twist of the polymer specimen 10 seconds after release is plotted against the test temperature, the latter being reduced slowly during the test. The intercept of the extrapolated straight portion of the resultant curve with the temperature axis is considered as the "freeze point." The T_2, T_5, T_{10}, and T_{100} values also obtained represent the temperatures at which the relative modulus (stiffness) of the test specimen is 2, 5, 10, and 100 times, respectively, the modulus of the specimen[87] at 25° C.

DIENE POLYMERS AND COPOLYMERS

In Table III are summarized results on the polymerization and copolymerization of certain dienes and on the properties of the polymeric products. The monomers studied are the hydrocarbons butadiene and its more readily available homologs, also myrcene, and 2-cyanobutadiene, various mixtures of these monomers themselves, and also various mixtures of them with styrene, halostyrenes, and vinylpyridines. Some comments as follows are offered.

Diene Hydrocarbons. In the GR-S type of emulsion polymerization formula, considerable differences in reaction rate were observed between, on the one hand, purified monomers, prepared usually from the cyclic sulfones, and, on the other hand, monomers purified merely by distillation (see butadiene, isoprene, and 2,3-dimethylbutadiene). The dienes may be arranged in the following order in regard to their rate of polymerization at 122° F., but it should be remarked that the presence of impurities is a complicating factor in some cases.

Monomer	Time to 75% Conversion, Hr.	Monomer	Time to 75% Conversion, Hr.
2-Cyano-1,3-butadiene	1	2-Isopropyl-1,3-butadiene	27
1-Cyano-1,3-butadiene	9 max.	2-n-Amyl-1,3-butadiene	44
2-Methyl-1,3-butadiene (isoprene)	12	1-Phenyl-1,3-butadiene	62
Myrcene	14	2, 4-Dimethyl-1,3-butadiene	83
1,3-Butadiene	15	1-Methyl-1,3-butadiene (cis)	146
2,3-Dimethyl-1,3-butadiene	16.5	1-Methyl-1,3-butadiene (trans)	540
2-Ethyl-1,3-butadiene	20		

The above relationship is in general agreement with early investigations, in which bulk polymerization was mostly employed.   Further, it is seen that increase in the chain length of an alkyl group (CH_3, C_2H_5, etc.) substituted in the butadiene molecule results in progressively slower reaction.

The *trans* isomer of piperylene (1-methylbutadiene) was much less reactive than the *cis* isomer.   Isomerization of *cis* to *trans* and the subsequent formation of dimer probably accounts for the large differences in rates shown by these monomers in the GR-S system on the one hand, where virtual cessation of the reaction occurs at a low conversion, and in activated formulas on the other hand.   Poly-*cis*-piperylene was superior to polybutadiene with respect to stress-strain properties, hysteresis–heat rise, and quality index (flex-hysteresis balance).   The styrene copolymer was superior to GR-S in flexing and quality index but exhibited poorer stress-strain properties.   Both polymers appeared to be inferior to GR-S in low-temperature properties.

Polyisoprene was similar to polybutadiene in stress-strain but inferior in low-temperature properties.   With the exception of its copolymers with 2-methyl-5-vinylpyridine, the co- and tripolymers of isoprene were equal to or poorer than GR-S in stress-strain and low-temperature properties. Although a decrease (improvement) in hysteresis–heat rise was shown by them, little improvement in quality index was evident.   Other work has indicated that isoprene polymers should be useful where elastomers of low-hysteresis characteristics are required without undue sacrifice in flexing properties compared to GR-S.[84, 256, 342]

The polymers of 2-ethyl-, 2-isopropyl-, and 2-*n*-amylbutadiene were equal to or slightly poorer than polybutadiene in stress-strain properties.   The low-temperature properties of all were similar to those of polyisoprene. Poly-1-phenylbutadiene of 89 per cent conversion was obtained in 78 hrs. At all conversions the product was a white powder, reminiscent of polystyrene.

The use of commercially available 2,3-dimethylbutadiene alone or with butadiene resulted in slow reactions and much softer polymers than were obtained with cyclic sulfone-purified material.   The rate of reaction of the latter was similar to that of butadiene and of isoprene and faster than the rates of 2-ethyl-, isopropyl-, or amylbutadiene.   Dimethylbutadiene showed the strongest self-modifying tendency of all the alkylbutadienes studies; i.e., it required the least amount of modifier.   Other studies have shown dimethyl-butadiene polymers to be superior to polybutadiene in strength but deficient in elastic and low-temperature properties.[297, 338]   Homologs of 2,3-di-methylbutadiene, viz.; 3-ethyl-, 3-isopropyl-, and 3-*n*-butylisoprene, have been reported to polymerize more slowly than 2,3-dimethylbutadiene and to give polymers of lower inherent viscosity.[212b]

· The methylpentadiene monomer, as received, was reported to contain 85 per cent of 2,4- and 15 per cent of 4,4-dimethylbutadiene.   As expected, the rates of reaction found in polymerization experiments with it were very low.   No reaction was obtained with 1,2,3,4-tetramethylbutadiene.[189]

Conversions of only 7 and 23 per cent were obtained in 11 hours with 75/25 charges of, respectively, 2,3-di-*tert*-butylbutadiene–styrene and 1,3-butadiene–2,3-di-*tert*-butylbutadiene.[195]   Substitution of the sodium

Table III. Polymerization, Chemical, and Physical Test Data for Diene Polymers and Copolymers Polymerization Temperature 122° F.

Monomers Type	Ratio	React. Time, hr.	Final Conv., %	DDM, Part	Mooney Visc., ML-4	Combined Comonomer, %	Gel, %	DSV	T/300[a] P.S.I.	T[a] P.S.I.	E[a] %	Dur. Hardness	Hysteresis, °F.[b]	DeMattia Flexures,[c] $\times 10^{-3}$	Quality Index	Low-Temp. Index minus °C.	Ref.
1,3-Butadiene (commercial)	100	16	73	0.6	...	...	25	...	1150	1550	350	64	78	2	0.38	76	170, 283
Butadiene (pure)	100	13, 15	60, 78	0.5	...	...	0, 0	1.9, 2.4	...	...	...	...	...	...	...	...	227
Butadiene-styrene (BD/S)	72/28	15[d]	72[d]	0.5[d]	50[d]	23.5(S)	...	...	960	3360	620	67	52	230	2.7	43	170, 221
cis-Piperylene (1-methyl-1,3-BD)	100	144	74	0.15	Too soft	...	...	...	840	2240	600	70	53	190	2.1	19/30 (T_b)[g]	79, 80
trans-Piperylene	100	288	40[d]	0.15	...	...	...	...	...	...	...	...	...	...	...	...	79, 80
cis/trans-Piperylene/S	50/50	288	65[d]	0.15	...	...	...	...	...	...	...	...	...	...	...	...	79, 80
cis-Piperylene/S	75/25	161	100	0.15	11	19.6(S)	...	...	1370	2780	660	70	55	780	6.2	3/30 (T_b)[g]	79, 192
trans-Piperylene/S	75/25	23	63	0.15	...	25.4(S)	0	1.145	...	...	...	...	...	...	...	...	79, 192
cis-Piperylene[e]	100	24	73	...	...	...	...	...	...	...	...	...	...	...	...	...	196
trans-Piperylene[e]	100	20	51	...	...	...	0	...	...	...	...	...	...	...	...	...	196
cis-Piperylene/S[e]	75/25	23	77	...	...	...	0	1.38	...	...	...	...	...	...	...	...	196
Isoprene (commercial 2-Methyl-1,3-BD)	100	15[h]	73	0.2	...	...	0	2.44	...	1500	280	...	...	i	...	...	170, 43
Isoprene (pure)	100	23	75	0.30	...	...	0	1.2	1025	2325	520	...	...	...	...	57	227
I/S	70/30	25	76	0.27	56	...	1	2.3	1110	1870	430	...	...	...	...	...	170
BD/I	95/5	24	72	0.34[f]	42	...	...	...	...	...	...	...	...	...	...	...	43
BD/I	90/10	24	72	0.33[f]	75	...	...	...	...	...	...	...	...	...	...	74	43
BD/I/S	90/5/5	145	72	0.38[f]	62	...	2	...	1000	2740	580	...	...	...	...	...	43, 325
I/m-Methylstyrene (pure)	75/25	12	79	0.15	42	15.6(MS)	5	2.80	1060	2440	490	62	44	230	3.4	34	221, 325
I/m-fluorostyrene (pure)	75/25	23	71	0.15	...	26.3(FS)	45	2.03	...	2025	280	61	41	170	2.6	28	189, 221
I/Monochlorostyrene	85/15	14	79	0.25	...	...	0	1.6	1080	2720	570	...	...	...	...	...	170
I/o-MCS (pure)	75/25	14	74	0.15	53	29.5(MCS)	7	2.16	1180	3460	610	...	...	...	...	...	189, 221
I/3,4-dichlorostyrene (pure)	75/25	23	80	0.125	...	27.9(DCS)	14	1.87	1250	3125	540	...	...	...	...	...	187, 189
BD/I/DCS	40/35/25	13	84	0.18	51	24.1(DCS)	3	2.6	1575	2875	460	63	46	190	2.6	32	170
BD/I/DCS	60/25/15	13	80	0.30	...	16.1(DCS)	1	2.4	900	2200	550	65	54	270	2.8	45	170
I/2-vinyl-5-ethylpyridine	75/25	23	70	0.50	...	...	36	0.52	1080	1080	360	...	...	...	...	...	170
I/5-vinyl-2-methylpyridine	75/25	11	70	0.15	6	18-24(VMP)	3	3.20	1220	1550	360	...	...	...	...	...	170
2-Ethyl-1,3-butadiene[j]	100	97	77	0.10	...	...	0	1.90	440	1690	800	57	108	1150	5.8	57	76, 183
EBD/S (pure)	75/25	20	85	0.20	...	...	0	1.42	1500	2280	390	...	...	...	...	...	76
BD/EBD[j]	95/5	19	75	0.75	...	...	0	1.40	...	...	...	...	...	...	...	...	228
2-Isopropyl-1,3-butadiene	100	26	70	0.10	...	...	0	1.42	490	1520	740	...	...	...	...	33	228
IBD/S	75/25	20	86	0.20	...	...	0	1.8	...	...	...	...	...	...	...	70	228
BD/IBD	95/5	19	73	0.50	...	...	1	1.95	...	...	...	...	...	...	...	...	228
2-n-Amyl-1,3-butadiene	100	48	81	0.10	...	...	0	1.69	...	...	...	...	...	...	...	55	228
ABD/S	75/25	39	74	0.20	...	...	0	1.1	...	...	...	...	...	...	...	...	228
BD/ABD	95/5	19	71	0.50	...	...	1	2.13	...	1860	240	...	...	...	...	71	228
BD/ABD	90/10	19	75	0.75	...	...	0	1.5	1470	2150	370	...	...	...	...	69	302

1-Cyano-1,3-butadiene	100	12	95	0.35	Dry, viscous		100	...									185, 186
BD/CBD	75/25	16	73	0.35	...	App. 36 (CBD)	2	0.9	...	2520	...	...	>52	App. 230	0.65	...	185, 186 93, 280
CBD/S	75/25	12	100	0.35	...	...	92	...									187
BD/CBD/S	75/15/10	12	55	0.35	...	...	4	...									187
BD/CBD/S	75/12.5/12.5	12	45	0.35	...	...	7	...									187
1-Phenyl-1,3-butadiene	100	53, 78	68, 89	0.5	White powder at all conversions												194
2,3-dimethyl-1,3-butadiene (commercial)	100	18	68	0.01	...	...	0	2.5									198, 227
DMBD (pure)	100	19	87	0.05	...	...	...	2.0									227
DMBD/S	75/25	19	87	0.06	...	...	0	0.59									227
BD/DMBD	25/75	16	60	...	...	...	0	0.73									198, 227
BD/DMBD	50/50	15	61	...	...	...	0	1.06									198, 227
BD/DMBD	75/25	14	62	...	...	...	0	1.95									198, 227
BD/methylpentadiene[m] (2,4-dimethyl-1,3-BD)	90/10	20/12[k]	53/73[k]	0.5	50	...	...	...									42
BD/MPD[m]	80/20	20/12[k]	39/69[k]	0.4	50[d]	...	...	...									42
BD/MPD[m]	50/50	20/20[k]	18/54[k]	...	...	...	...	...									42
1-Cyano-2-methyl-1,3-butadiene	100	12	100	...	Hard, brittle	...	...	...									188
BD/CMBD	75/25	12	72	...	Resembled GR-S	...	0	...									188
Myrcene[l] (2-methyl-6-methylene-2,7-octadiene)	100	24	79[l]	...	...	...	...	...	...	600	300	...	...	...	...	...	115
Myrcene/S[l]	75/25	24	86–90[l]	0.15	...	...	...	...	...	1030	475	54	...	...	...	...	115
Myrcene/S	95/5	22	80	0.15	...	...	...	...	...	965	175	62	...	...	...	...	115
Myrcene/S	75/25	22	83	0.15	...	...	...	...	...	1000	475	54	...	...	...	...	115
Myrcene/S	60/40	22	85	0.15	...	...	...	...	...	925	575	58	...	...	...	...	115
BD/S/myrcene	74/25/1	16	90	0.5	...	...	...	...	...	2370	400	69	...	...	...	...	115
BD/S/myrcene	50/25/25	18	86	0.5	...	...	...	...	...	2270	500	61	...	...	...	...	115

a Stress-strain tests conducted at room temperature (77° to 82° F.). *b* Goodrich hysteresis–heat-rise test conducted at 212° F.

c DeMattia flexure data shown for polybutadiene were obtained at 212° F. rather than 82° F. as for the other data shown.

d Approximate value.

e Piperylene polymerized in redox formula which contained 0.5 part of benzoyl peroxide, 0.5 part of ferrous sulfate heptahydrate, 1.0 part of sorbose, 0.7 part of sodium pyrophosphate decahydrate, 5.0 parts of fatty acid soap, and 200 parts of distilled water.

f *tert*-C-12 Mercaptan used in place of DDM.

g T_5 is the temperature at which the modulus of the test stock is 5 times the 25° C. modulus value.[7] Values shown are for the experimental polymer/GR-S control.

h Other work with isoprene purified through its cyclic sulfone resulted in 67 per cent conversion in 11 hours.[84]

i The running temperature for polyisoprene was 245° F. as compared with 274° F. for polybutadiene and 264 to 288° F. for GR-S; cut growth was poorer than for either polybutadiene or GR-S.

j 2-Ethyl-1,3-butadiene obtained from Carbide & Carbon Chemicals Co. and reported to contain small amounts of 3-methyl-1,3-pentadiene. Additional pure material prepared in laboratory resulted in a conversion of 85 to 90% of polyethyl butadiene in 24 hours, and a dilute solution viscosity of 0.9 when 0.1 part of DDM was used.[226]

k Reaction time and conversion in GR-S-type formula/that in redsol-persulfate catalyzed formula.

l Polymerized at 140° F.

m Methylpentadiene obtained from the Shell Development Co. and reported to contain about 85 per cent of 2- and 15 per cent of 4-methyl-1,3-pentadiene.

salt or the methyl ester of sorbic acid (1-methyl-4-carboxylic acid-1,3-butadiene) for styrene in the GR-S formula resulted in conversions of 0 and 39 per cent in 11 and 40 hours, respectively.[198, 200]

Myrcene [2-methyl-6-methylene-2,7-octadiene, $CH_2 : CH \cdot C(: CH_2)$ $\cdot CH_2CH_2CH : C(CH_3)_2$] reacted to 79 per cent conversion in 24 hours at 140° F. The copolymerization of myrcene and styrene took place only slightly faster. Even at low concentrations of modifier, the polymers were relatively soft. In tensile strength all the myrcene polymers were weaker than polybutadiene. Experiments in which part of the butadiene in the GR-S formula at 122° F. was replaced by myrcene indicated that probably little of the myrcene entered the copolymers; tripolymers from charges of 74/1/25 and of 50/25/25 butadiene–myrcene–styrene were similar. The use of small quantities of myrcene to increase the tack of films made from high-conversion GR-S polymers has been reported.[329]

1-Cyanobutadiene. Poly-1-cyanobutadiene produced tough, insoluble, and resinous stocks, which when heated could be drawn into filaments but which reset on cooling. Its use with styrene in place of butadiene gave a tough, leathery polymer, which, however, appeared rubberlike when warmed. When cyanobutadiene was substituted for styrene in copolymerization with butadiene, a copolymer was obtained equal to GR-S in flex life but inferior in stress-strain properties and hysteresis. It is to be expected that polymers containing this material will show improved oil resistance but poorer brittle point. Poly-2-cyanobutadiene has been reported[122] to be obtainable in 80 per cent yield on polymerization for 1 hour at 59° F., but the polymer was insoluble and not rubberlike. A light-catalyzed mass polymerization of this monomer yielded a tough rubberlike product.

2-Fluorobutadiene. Polymers of 2-fluorobutadiene have been reported.[248] Like polybutadiene, but unlike polychloroprene (Neoprene), polyfluorobutadiene did not crystallize when stretched at low temperatures. It was similar to GR-S in low-temperature properties and superior to Neoprene GN (polychloroprene) and Neoprene FR (isoprene–chloroprene). Its tensile strength was equal to that of GR-S but inferior to that of Hevea and of Neoprene GN. The oil resistance was better than that of GR-S and Neoprene FR and equal to that of Neoprene GN. Copolymers of 95/5 fluoroprene–dimethyl vinylethynyl carbinol exhibited tensile properties better than those of polyfluoroprene with no change in low-temperature or oil-resistance characteristics. Somewhat poorer properties were found in styrene copolymers. The introduction of small proportions of acrylonitrile resulted in marked improvement in oil resistance. Only little further improvement was noted with more than 20 per cent acrylonitrile, while the freeze resistance continued to decrease.

BUTADIENE COPOLYMERS WITH SUBSTITUTES FOR STYRENE

In Table IV is summarized data on the copolymerization of butadiene with a large number of monomers which were substituted for the styrene comonomer of GR-S. Some comments on the data are as follows.

Alkyl- and Aryl-Substituted Styrenes; Also Vinylnaphthalene. In copolymerization with butadiene, the mono- and disubstituted alkyl- and arylstyrenes show reaction rates about the same as those for GR-S when 25 parts are substituted for styrene. *m-sec*-Butylstyrene polymerized appreciably less rapidly than the other monomers and entered the copolymer more slowly. Monomers containing a phenyl or benzyl group and also β-vinylnaphthalene seemed to copolymerize slightly faster than the others. All the comonomers yielded softer polymers than those obtained from butadiene–styrene (GR-S) at a similar modifier level. The copolymers of the *p*-isopropyl-, *m-sec*-butyl-, *p*-phenyl-, and *p*-benzylstyrenes, and of β-vinylnaphthalene were inferior in stress-strain properties. The other copolymers in this group appeared to be similar to GR-S in stress-strain properties, hysteresis, flexing, and (when corrected for combined styrene content) low-temperature properties.

Halostyrenes. Of the halogenated styrenes, the dichlorostyrenes (DCS) reacted faster and entered the copolymer to a slightly greater extent than the monosubstituted fluoro-, chloro-, or bromostyrenes, or than styrene itself. No appreciable differences among the various isomers were apparent. 2,6-Dichlorostyrene and the pentachlorostyrene entered the copolymers more slowly than any of the other comonomers tested. The comonomer content of the monochlorostyrene (MCS) copolymers was nearly constant over the conversion range, whereas in the styrene copolymers it increased and in the dichlorostyrene copolymers decreased as conversion proceeded. With α-chlorostyrene, the emulsion broke and a mixture acid to litmus was obtained.[24, 59, 215] *p*-Chloro-α-ethoxystyrene gave a tough polymer only after an extended reaction period.

Only the 2,3-, 2,5-, and 3,4-dichlorostyrene copolymers appeared to be significantly better than GR-S; the 2,5- and 3,4-copolymers showed both higher quality indexes and better stress-strain properties.[118, 246] Copolymers of the monosubstituted styrenes seemed to be essentially similar to GR-S; those of 2,6-dichlorostyrene and the mixed di- and the tetrachlorostyrenes appeared inferior. In general, the copolymers of all the halogenated styrenes showed higher hysteresis temperature-rise values but longer flex life than GR-S.

Cyanostyrene. *p*-Cyanostyrene gave a reaction rate and stress-strain properties similar to those for GR-S, but a superior quality index. However, its copolymers would be expected to show inferior freeze resistance, and, further, the monomer was reported to polymerize very readily during preparation in the presence of hydroquinone, indicating poor stability in storage.[208, 215, 218, 220]

Nitrostyrene. In confirmation of earlier work with nitro compounds,[297] *m*-nitrostyrene, when substituted for styrene,[215] completely inhibited polymerization. In work designed to determine the cause of this inhibition, it was found that nitromethane and nitromesitylene inhibit the emulsion polymerization of butadiene and of styrene, in combination or singly, but do not affect the solution polymerization of styrene.[179] A mixture of butadiene or styrene with 2-nitro-1-butene failed to polymerize in an acid-emulsion formula or by photoactivation.[181, 207] Nitrobenzene or nitromesitylene inhibited the emulsion polymerization of styrene in alkaline

Table IV. Polymerization, Chemical, and Physical Test Da

Polymerizatic

Comonomer[a] Type	Ratio BD/X	React. Time, Hr.	Conv., %	DDM, Part	Mooney Visc., ML-4	Combined Comonomer,[b] %	Gel, %	DSV	T/300[c] P.S.I.
Styrene	75/25	8/12	49/78	0.35[e]	48	23[e]	0	2.0[e]	890
m-Methylstyrene	75/25	13	78	0.35	...	27.0	4	1.91	750/800
m-Methylstyrene	85/15	14	73	0.35	...	11.1	0	1.90	700/650
p-Methylstyrene	75/25	12	76	0.35	31	...	0	1.57	1070/960
p-Ethylstyrene	75/25	11	74	0.35	42	24.8	0	1.84	1000/101
p-Isopropylstyrene	75/25	11	65	0.35	28	...	0	1.38	1100/105
m-sec-Butylstyrene	75/25	25	79	0.35	36	9.16	3	1.73	880/960
m-tert-Butylstyrene	75/25	14	77	0.35	33[f]	19.6	4	1.90	860/890
p-tert-Butylstyrene	75/25	12	73	0.35	20	...	0	1.37	1000/960
m-Phenylstyrene	75/25	11	73	0.35	g	...	0	0.95[g]	1390/119
p-Phenylstyrene	75/25	11	75–81	0.35	18	...	0	2.21	780/108
p-Benzylstyrene	75/25	11	81	0.375	20	22.2	0	2.13	870/960
β-Vinylnaphthalene	75/25	12	75	0.35	...	32.4	6	1.58	525/600
2,4-Dimethylstyrene	75/25	14	76	0.35	38	19.5	2	2.05	1020/940
2,5-Dimethylstyrene	75/25	13	76	0.35	35	26.4	1	1.81	900/940
3,4-Dimethylstyrene	75/25	13	79	0.35	35	28.6	3	1.91	700/810
3,5-Dimethylstyrene	75/25	15	75	0.35	...	19.9	3	1.75	630/110
Ethyl vinyl ether	75/25	45	64	0.7	...	max. 6.5	22–30	1.50	
Isopropyl vinyl ether	75/25	30	65	0.7	...	8.5	7–32	1.8– 1.5	
n-Propyl vinyl ether	75/25	24	55	0.7	...	7.5	8	2.2	
o-Methoxystyrene	75/25	10	76	0.35	17	24.8	0	1.97	1300/138
m-Methoxystyrene	75/25	12	76	0.35	38	22.6	0	1.94	1000/102
p-Methoxystyrene	75/32.5	12	73	0.35	...	...	4	2.18	1250/950
p-Phenoxystyrene	75/25	10	78	0.35	12	19.3	2	1.32	750/960
4-Methoxy-3-methylstyrene	75/25	12	67	0.35	...	19.5	6	1.45	600/600
6-Methoxy-3-methylstyrene	75/25	12	76	0.35	...	21.9	9	2.20	1000/600
3,4-Dimethoxystyrene	75/25	13	70	0.35	...	...	0	1.71	600/850
2-Methyl-4-methoxy-5-isopropylstyrene	75/25	21	89	0.35	...	...	20	1.80	
Anethole	75/25	30	70	0.35	...	4.1	0	2.30	
4-Propenylveratrole	75/25	24	0	0.35	...	...	...	...	
o-Fluorostyrene	75/25	10	84	0.35	25	27.2	0	1.90	1275/925
m-Fluorostyrene	75/25	12	70	0.35	...	24.3	4	1.44	1050/950
p-Fluorostyrene	75/25	14	71	0.35	...	...	4	2.01	825/950
m-Trifluoro methylstyrene	75/25	15	75	0.35	...	26.8	0	1.83	810/650
o-Chlorostyrene	75/25	11	66	0.35	...	23.3	4	1.09	600/600
o-Chlorostyrene	85/15	15	72	0.35	59	14.2	8	1.89	890/950
m-Chlorostyrene	75/25	11	78	0.35	...	25.0	6	1.82	900/110
p-Chlorostyrene	75/33.7	12	67	0.35	...	26.7	0	1.64	1125/950
Mixed MCS	95/5	22	72	0.40[i]	62	...	...	...	1500
Mixed MCS	75/25	12	77	0.35	...	25.8	5	1.55	1000/900
2,3-Dichlorostyrene	75/25	12	75	0.5	25	27.5	6	1.16	920/940
2,4-Dichlorostyrene	75/25	12	76	0.35	...	23.0	5	1.65	1050/950
2,5-Dichlorostyrene	75/25	11	80	0.35	...	28.2	1	1.52	1250/950
2,5-Dichlorostyrene	75/41.6	12	89	0.35	...	38.1	1	2.05	1200/950
2,6-Dichlorostyrene	75/25	16	72	0.35	...	14.3	10	1.92	1200/950
3,4-Dichlorostyrene	75/41.6	12	81	0.35	...	36.2	0	1.89	1175/950
3,4-Dichlorostyrene	85/15	12	77	0.35	59	16.4	6	2.06	1000/950
3,5-Dichlorostyrene	75/25	10	75	0.35	23	27.1	1	1.24	940/940
Mixed DCS	75/25	12	75	0.35	...	26.2	4	1.52	1200/110
Mixed DCS	95/5	21	72	0.40[i]	49	...	...	...	1040
Mixed trichlorostyrene	75/25	15	71	0.35	...	20.0	7	1.60	675/600
Mixed tetrachlorostyrene	75/25	37	80	0.35	48	24.1	54	1.13	1330/950
Pentachlorostyrene	75/25	12	65	0.35	...	3.9	0	Low	

T^c P.S.I.	E^e %	Dur. Hardness	Hysteresis, °F.^{d}	DeMattia Flexures ×10^{-3}	Quality Index	Low Temp. Index, minus °C.	Comonomer Prep. Ref.	Ref.
3380	620	60	65	310	2.5	48	...	
550/3450	660/600	...	52/52	500/280	5.8/3.4	46	218	214
140/3050	730/755	68/61	44/50	145/275	2.2/2.9	60	218	214
310/3750	600/650	64/65	47/43	140/170	1.9/2.6	...	252	214
580/3550	620/625	55/60	59/60	390/300	3.7/2.7	52	224	214
810/3640	560/670	60/61	47/52	170/210	2.3/2.4	53	135	214
680/3360	600/610	61/67	58/52	165/230	1.6/2.7	55	209	214
100/3380	610/620	58/52	58/52	110/310	1.1/3.8	49	209	214
470/3760	620/650	65/65	49/43	105/170	1.3/2.6	46	252	214
760/3790	565/665	66/62	56/47	75/300	0.8/4.0	44	106	214
890/3820	655/625	58/58	56/70	200/270	2.1/1.8	48	106	214
770/3820	605/695	57/60	48/60	80/300	1.0/2.7	61^h	213	214
895/3150	760/750	...	83/45	1200/350	3.4/...			212, 288
620/3170	620/575	65/65	53/47	240/200	2.6/2.6	52	223	214
400/3740	650/675	66/57	59/66	280/420	2.6/3.2	52	223	214
620/3510	710/695	59/61	54/61	310/290	3.4/2.6	51	223	214
200/3200	755/680	...	63/55	270/290	2.3/3.0	...	223	214
							i	181
							i	181
							i	181
190/4170	535/585	61/57	64/63	430/430	3.5/3.7	51	213, 331	214
160/3380	660/650	64/61	68/58	680/440	4.9/4.2	53	73	214
300/3050	575/630	...	54/58	310/210	3.4/2.1	...	73	214
310/2900	595/550	67/69	38/37	165/105	2.8/1.9	57	73	214
900/3150	720/750	...	70/...	460/...	3.2/...	36^j	73	214
450/3150	595/750	...	49/...	360/...	4.6/...	46	73	214
200/3100	805/670	...	70/57	250/210	1.7/2.1	...	73	214
								180
							k	214
							...	179
575/3850	615/730	65/60	67/61	120/200	0.9/1.8	51	213	214
700/3050	700/630	...	110/58	2000/210	4.2/2.0	...	29, 176	215
800/3050	550/630	...	57/58	270/210	2.7/2.0	51	29	215
180/3050	710/755	66/61	50/50	160/275	2.0/2.9	51	218	214
750/3150	685/750	...	70/...	730/...	5.1/...	...	29, 224	215
320/3140	665/630	64/60	76/62	160/190	0.9/1.6	...	29, 224	215
800/3550	720/620	...	122/53	2000/430	3.0/4.6	46	29, 224	215
000/3050	550/630	...	81/58	570/210	2.8/3.0	...	29, 224	215
1500	300	...	...	...	...	74	...	41
500/3000	670/640	...	44/51	230/260	3.3/3.0	...	m	215
790/3170	655/575	62/65	51/47	250/200	3.0/2.6	...	219	215
850/3050	530/630	...	58/58	180/210	1.7/2.0	...	219	215
400/3050	635/630	...	99/78	2000/330	4.0/2.0	45	n	215, 287
400/3050	635/630	...	99/58	1450/210	4.2/2.0	30	n	215
450/2800	505/575	...	56/57	100/410	1.0/4.1	...	219	215
750/3050	610/630	...	83/58	1950/210	9.2/2.0	...	219	215
470/3140	610/630	63/60	74/62	270/190	1.7/1.6	...	219	215
300/3050	630/670	67/57	83/66	750/420	3.5/3.2	...	219	215
800/3550	480/620	...	73/53	320/430	2.0/4.6	...	m	215
2050	450	...	...	...	...	71	...	41
000/3150	700/750	...	64/...	240/...	1.9/...	53	m	215
150/3020	510/650	61/62	56/58	70/640	0.7/6.1	62	m	215
							166	215

Table IV—

Comonomer[a] Type	Ratio BD/X	React. Time, Hr.	Conv., %	DDM, Part	Mooney Visc., ML-4	Combined Comonomer,[b] %	Gel, %	DSV	T/300[c] P.S.I.
o-Bromostyrene	75/25	12	79	0.35	...	26.8	1	1.63	1200/950
m-Bromostyrene	75/25	13	75	0.35	...	26.6	8	1.55	1400/600
p-Bromostyrene	75/25	10	74	0.35	...	23.4	8	1.19	925/900
p-Cyanostyrene	75/25	11	72	0.35	26	26.1	3	1.34	1120/940
Acrylonitrile	95/5	9	72	0.49[l]	53	...	...	...	...
Methacrylonitrile	75/25	5	75	0.35	...	26.8	0	1.61	1500/860
Vinylidene chloride	75/25	15	71	0.35	...	17.1	0	1.74	1100
Methyl vinyl ketone	90/10	6	50	0.44[l]	41	...	...	...	1000
Methyl isopropenyl ketone[q]	75/25	17	75	0.60	...	...	...	...	
Methyl acrylate	75/25	11	67	0.35	...	...	1	2.26	
Ethyl acrylate	70/30	10	76	0.5	49	...	2	2.35	1430
Ethyl acrylate	85/15	11	74	0.5	52	...	1	2.24	1420
Isopropyl acrylate	75/25	11	75	0.35	...	...	1	2.5	
Methyl methacrylate	75/25	10	86	0.35	...	...	1	2.2	
Methyl methacrylate	95/5	24	72	0.40	57	...	...	...	1400
Butyl acrylate	90/10	12	69	0.5	62	...	...	...	1420
Isobutyl dichloroacrylate	75/25	18[s]	72	0.5	54	23	...	...	830/975
Isobutyl dichloroacrylate	60/40	21	78	0.35	61	36	...	...	1480/975
Styrene-vinyl linseed ester	65/20/15	19	76	0.7	...	...	12	1.56	...
Styrene-vinyl linseed ester	65/25/10	16	78	0.7	...	...	10	1.82	...
p-Carboxystyrene	75/25	22	41	0.35	...	25[e]	87	0.44	
Sodium-p-carboxystyrene	75/25	20	100	0.35	...	...	88	0.47	
p-Carbomethoxystyrene	81/19	15	59	0.35	...	25[e]	0	1.33	
2-Vinylfuran	75/25	20	74	0.35	...	...	50	...	1600
2-Vinylthiophene	75/25	15	77	0.35	...	27–29	6	1.98	1100/1020
2-Vinylpyridine	75/25	8	82	...	...	24	...	...	
3-Vinylpyridine	75/25	8	79	0.5	88	20–29	36	1.15	1360
4-Vinylpyridine	75/25	8	91	...	...	24	...	...	
2-VP/4-VP	75/12.5/12.5	8	79	...	...	...	...	...	
2-Me-6-VP	75/25	7	71	0.5	...	27	21	1.91	1050
2-VP/3-VP	75/12.5/12.5	8	78	...	...	...	...	...	
2-VP/5-Et-2-VP	75/12.5/12.5	8	78	...	...	...	...	...	
2-Me-5-VP	75/25	9	76	0.35	72	21–28	9	2.05	1920
5Et-2-VP	75/25	10	79	0.4	...	21–22	4	2.05	1350
2,4-DiMe-6-VP	75/25	8	84	0.5	88	18	28	...	1190
2-VP/DCS	66/21/13	9	74	0.21[l]	59	23/10[e]	...	...	1930/1420
2-VP/DCS	66/14/20	8	72	0.21[l]	53	...	...	...	1920/1420

[a] Comonomers with butadiene in the ratios shown.
[b] Combined monomer other than butadiene.
[c] Stress-strain tests conducted at room temperature (77 to 82° F.). Values presented for stress-strain, hardness, hysteresis, flexures, and quality index are those for the experimental polymer/aGR-S control.
[d] Goodrich hysteresis temperature-rise test conducted at 212° F.
[e] Approximate value.
[f] Mooney viscosity value after 1 minute at 212° F. (ML-1) as compared with a value for GR-S of 48 ML-1.
[g] m-Phenylstyrene copolymer appeared to be highly overmodified by a trace of impurity in the monomer.
[h] Value appears to be low (absolute scale).
[i] From the General Aniline & Film Corp.
[j] Value appears high (absolute scale) as compared with other data.
[k] Anethole, described as 100 per cent pure, from the Newport Industries, Inc.
[l] tert-C-12 Mercaptan substituted for DDM.

T^e P.S.I.	E^e %	Dur. Hard-ness	Hyster-esis °F.[d]	DeMattia Flexures $\times 10^{-3}$	Quality Index	Low Temp. Index, minus °C.	Comon-omer Prep. Ref.	Ref.
850/2800	525/575	...	60/57	320/410	2.9/4.1	...	217	215
750/3150	450/750	...	51/...	170/ ...	2.0/–	...	217	215
000/3050	600/640	...	66/51	180/260	1.3/3.0	48	217	215
150/3170	580/575	67/65	83[t]/47	2000/200	9.2/2.6	...	218	215
1830	290	...	...	...	...	67	...	41
500/3200	515/680	...	62/52	550/140	4.7/1.6	...	o	75
3175	570	...	64	90	0.7	...	p	75
2240	320	...	...	...	...	67	...	98
								70
								178
2940	480	...	...	...	...	51	...	46
2060	390	...	...	...	...	58	...	46
							...	178
								178
2100	390	...	...	...	...	75	...	41
2410	400	...	...	...	...	73	...	249
550/2700	595/615	...	...	...	...	eq. to GR-S	...	25, 26
810/2700	465/615	...	...	...	...	eq. to GR-S	...	25, 26
500/3380	730/620	65/60	83/52	160/310	0.55/3.5	48	r	184, 216, 258
150/3380	610/620	68/60	55/52	165/310	1.6/3.5	55	r	184, 216, 258
...	...		...	...	...		218	187, 215
...	...		...	...	...		218	187, 215
...	...		...	...	...		218	198, 215
3100	485	...	71[t]	90	1.0	...	u	75
500/3380	635/650	65/61	58/58	230/...	2.2	40	17	75
							x	76
3780	590	59	64	480	5.5	52[v]	76, 107,	76
							299	76
							76	76
								76
3100	675	...	54	300	2.5	...	x	76
							...	76
							...	76
3610	475	66	57	280	5.4	...	76	76
3050	550	...	55	1400	6.5	47	77	76
2060	490	66	69	200	1.9	42	76	76
550/3660	540/560	72/69	99/65	65/7[w]	6.0/2.1	38	...	64, 295
470/3660	550/560	72/69	97/65	75/7[w]	7.4/2.1	37	...	64

From the E. I. du Pont de' Nemours & Co.
From the Monsanto Chemical Co.
From the Rohm & Haas Co. and E. I. du Pont de Nemours & Co.
From the Dow Chemical Co.
Methyl isopropenyl ketone copolymerized at 95° F.
From the Armstrong Cork Co.
Some hydrolysis encountered in the isobutyl dichloroacrylate polymerization; final pH of 7 as compared to 8 to 8.5 for GR-S.
Hysteresis sample blew out before completion of the test, actual hysteresis value is probably higher.
From the Northern Regional Research Laboratory, U.S. Department of Agriculture.
Low-temperature index shown is for 2-VP copolymer.[64]
DeMattia flexures determined at 212° F. rather than at 82° F.
Vinylpyridines and 2-(2-pyridyl)-allyl alcohol were obtained from Reilly Tar & Chemical Co.

mixtures but not in acid-emulsion or in solution polymerizations; hence the nitro group per se apparently does not cause inhibition.[180]

Hydroxy-, Acetoxy-, and Carboxystyrenes. It has been noted that only those resins of the Koresin type which contain free hydroxyl groups show good tackifying properties.[222] Therefore, 2-hydroxystyrene and 2-acetoxy-styrene seemed to be of interest as comonomers, the copolymers of which with butadiene might possibly possess a useful degree of tackiness. Neither monomer could be copolymerized with butadiene in an alkaline redox-activated or in the GR-S formula at 122° F. or in an acid formula at 86° F. The introduction of 1 to 5 parts of the hydroxystyrene into the GR-S system completely inhibited polymerization or stopped polymerization which had been started before it was injected. A 75/15/10 butadiene-styrene-acetoxy-styrene mixture, in an acid formula, reacted to 75 per cent conversion in 48 hours. Infrared analysis indicated the presence of *o*-acetoxystyrene in the polymer. Since free phenol arising from hydrolysis was thought to be at least partially the cause of inhibition with *o*-hydroxystyrene and acetoxy-styrene, the effect of phenol was checked in the alkaline GR-S system and in an acid formula. However, the polymerization of styrene or of a 75/25 butadiene-styrene mixture was not inhibited by the reagent in the acid system.[197, 202–4]

p-Carboxystyrene reacted slowly to an intermediate conversion and yielded a tough brittle polymer; its sodium salt led to 100 per cent conversion in 20 hours. High gel was encountered in both cases. The methyl ester gave an elastic polymer, but attempts to obtain a conversion higher than 60 per cent were unsuccessful.[201, 205]

Ethers. Ethyl, isopropyl, and *n*-propyl vinyl ethers reacted slowly in copolymerization with butadiene, and the rates tended to decrease in the order listed. The reactions appeared to approach 75 per cent conversion, but little of the comonomer had entered into reaction, as the products contained a maximum ether content of about 7.5 per cent only. With the exception of 2-methyl-4-methoxy-5-isopropylstyrene, anethole (1-methoxy-4-propenylbenzene), and 4-propenylveratrole (1,2-dimethoxy-4-propenyl-benzene), the unsaturated aromatic ethers examined copolymerized at about the same rate as styrene. The rate of entry into the copolymer was slightly lower than that of styrene. Anethole entered the copolymer only to a small extent, and the propenylveratrole completely inhibited the polymerization, possibly because of the presence of traces of free phenol. With the exception of the phenoxystyrene copolymer, which was poorer in stress-strain properties, the alkoxy and phenoxy copolymers were in the same range as GR-S with respect to stress-strain, hysteresis, flex-life, and low-temperature properties.

Chlorine-Substituted Isopropenylbenzenes. Copolymers of buta-diene with *o*-, *m*-, and *p*-chloro- and with 2,3-, 2,4-, 3,4-, and 3,5-dichloro-isopropenylbenzenes have been described.[17, 243] Unlike the chlorinated styrenes and similar to α-methyl- and α-methyl-*p*-methylstyrenes, all the chloroisopropenylbenzene monomers were resistant to homopolymerization, but all, except such as contained *o*-alkyl substituents, copolymerized with butadiene at a rate similar to that of styrene and yielded products superior to GR-S in stress-strain properties and flex life but inferior in hysteresis and cold resistance. A reduction in the substituted styrene charged to about 15 parts results in copolymers similar to GR-S in quality.

Halo- and Cyanoölefins. Methacrylonitrile appeared to copolymerize with butadiene as rapidly as acrylonitrile and at about three times the rate of styrene.[321] Vinylidene chloride as the comonomer gave a rate similar to styrene but entered the copolymer to a lesser amount than styrene. Reaction was slow with trichloroethylene as the comonomer, and in fact the limiting conversion in its case indicated that probably little of the chloroölefin had entered into copolymerization. A mixture of butadiene and 1,1-diphenyl-ethylene failed completely to undergo polymerization.[174] Conversions of 47 and 32 per cent were obtained in 36 hours with 75/25 blends of butadiene and 1,2-dichloro-3-butene and 1,4-dichloro-2-butene, respectively. Chlorine analyses indicated that neither monoölefin entered into the product.[173-4] Of this group of comonomers only methacrylonitrile showed evidence of superiority over styrene, its copolymer with butadiene exhibiting a higher quality index and slightly better tensile properties than GR-S. The vinylidene chloride copolymer appeared to be inferior, possibly because of its extreme sensitivity to sulfur variations during compounding.

Carbonyl Compounds. Methyl isopropenyl ketone at 95° F. and methyl vinyl ketone at 122° F. copolymerized with butadiene much more rapidly than did styrene. Acrolein in a similar charge at 122° F. yielded only a small amount of very soft polymer.[179] $\alpha\beta$-Unsaturated ketones in which the carbonyl group is also conjugated with an aromatic ring have been found to copolymerize readily with butadiene. Vulcanizates prepared from copolymers of butadiene and benzalacetophenone showed good tensile strength and were somewhat better than GR-S in hysteresis behavior.[221a]

Esters. Of the unsaturated esters, methyl, ethyl, isopropyl, and Butyl acrylates were in the same range of activity as styrene and slower than methyl methacrylate. Reactions with butyl methacrylate were slightly faster than corresponding reactions with styrene.[341] With ethyl acrylate little increase in reaction rate was obtained when the ester in the charge at 122° F. was raised to 70 or to 100 parts.[46] Slightly better copolymers were produced with butyl acrylate than with ethyl acrylate. The ethyl acrylate copolymers had slightly better stress-strain properties than isobutyl dichloro-acrylate copolymers. All the acrylate copolymers had low-temperature properties similar to those of styrene copolymers of the same combined comonomer content.

The properties of copolymers of the 2-ethoxyethyl esters of acrylic and methacrylic acids and of tetrahydrofurfuryl acrylate with butadiene have been reported on.[212a]

Dibutyl itaconate (dibutyl methylenesuccinate) reacted at 104° F. more slowly than the above monoesters; 77 per cent conversion was reached in 17 hours.[70] *p*-Vinylphenyl acetate gave only a small yield when it was co-polymerized with butadiene at 122° F.[175, 206] Butadiene-styrene-vinyl linseed ester reacted slowly and gave polymers similar to GR-S in stress-strain properties but inferior in quality index.

Vinylpyridines. The rates of copolymerization of the vinylpyridines (VP) with butadiene were higher than the rate of copolymerization of styrene with butadiene and fell in the following decreasing order: 4-VP, 2-VP, 50/50 2-VP/4-VP, 2-methyl-6-VP, 50/50 2-VP/3-VP, 3-VP, 50/50 2-VP/5-ethyl-2-VP, 2-methyl-5-VP, 5-ethyl-2-VP. Attempts to copolymerize 2-(2-pyridyl)-allyl alcohol failed.[76] All the polymers were relatively tough in the crude state but broke down readily on a cold mill. The polymers generally cured faster and showed higher moduli and tensile values than GR-S. Reductions in sulfur and accelerator levels resulted in moduli and curing rates similar to those of GR-S. Exceptionally high tensile values were obtained with tripolymers of butadiene–2-vinylpyridine–dichlorostyrene. The copolymers, except those of 2-methyl-6-vinylpyridine and 2,4-dimethyl-6-vinylpyridine, all showed exceptionally high quality indexes, although the hysteresis–heat rise was generally higher than that of GR-S.[76, 273]

Other Heterocyclic Comonomers. Vinylquinoline,[18] related to pyridine as naphthalene is to benzene, reacted to about 35 per cent conversion in 20 hours to give a copolymer with a dilute solution viscosity of 2.3 and apparently of good quality.[178, 181] N-Vinylphthalimide and vinylsuccinimide, when substituted for styrene in the GR-S formula, caused the emulsion to break. When 4 parts only of the styrene charge was replaced by vinylphthalimide, rapid polymerization occurred; the resulting polymer was of poor quality. A similar substitution by vinylsuccinimide resulted in a polymer of low conversion.[184]

The substitution of styrene in the GR-S formula by dihydropyran resulted in only 8 per cent conversion in 11 hours.[198] Severe retardation of the polymerization of butadiene–styrene, both in the GR-S and redox systems, by the introduction of 2.5 to 10 parts of dihydropyran has been reported.[199] The use of 2-vinylfuran as the comonomer with butadiene gave a highly gelled polymer in 20 hours at 74 per cent conversion. Its sulfur analog, 2-vinylthiophene, however, was only slightly less reactive than styrene and yielded a soluble polymer. Both copolymers were similar to GR-S in stress-strain properties. In hysteresis and quality index the copolymer from vinylfuran was inferior and that from vinylthiophene similar to GR-S.

Carboxylic Acids. It has been found that acrylic acid and methacrylic acid will undergo copolymerization with butadiene in acid redox-activated emulsions and that the copolymer with methacrylic acid exhibits good resistance to oils. If the acid comonomer exceeds 10 per cent of the monomer charge, the emulsion system is unstable.[212a] Copolymers of butadiene with 10 per cent of cinnamic acid, its esters or nitrile, and with cinnamic aldehyde have been reported to give vulcanizates similar in quality to vulcanizates of GR-S.[216a]

PILOT-PLANT-SCALE STUDIES

The ensuing discussion of experimental polymers made on a pilot-plant or production scale refers mainly to work conducted at the Government Laboratories. This work is selected for presentation, not because much similar work has not been done elsewhere, and in some cases earlier, in connection with the American synthetic-rubber program, but because it is thought that results obtained by a single organization are most likely to be comparable among themselves. The road tests on tires, for which average values are presented, were done elsewhere. Small differences in the polymerization formulas employed in different cases and compounding differences in testing the polymers are not considered in the discussion.

POLYBUTADIENE—INFLUENCE OF POLYMERIZATION TEMPERATURE ON PROPERTIES

A typical polybutadiene made at 122° F. in accordance with the GR-S formula (soap flakes as the emulsifier) to the usual specification of 51 Mooney viscosity at 72 per cent conversion exhibited poor mill-processing behavior. Generally the dispersion of carbon black into the polymer was not so good as that obtained with standard GR-S. The polymer would not adhere to the mill rolls when black was added, and the milled sheets were full of holes. Treatment of this type of polymer in a laboratory "B" Banbury mixer did not cause the polymer-black mixture to adhere to the rolls during subsequent milling.[238]

Reduction in the degree of conversion at 122° F. from 72 per cent to 50 to 60 per cent resulted in an improvement in the properties of the polymer. Tests of polymer prepared in 5-gal. reactors to 60 per cent conversion, with fatty acid soap as the emulsifier, indicated that the viscosity of the polymer must be reduced to about 20 ML-4 to insure sufficient adherence to the mill rolls. With a rosin acid type of soap as the emulsifier, somewhat higher viscosity levels were satisfactory; at viscosities of 35 to 50 ML-4, pretreatment of the raw polymers in a laboratory Banbury gave rosin acid stocks that would stick satisfactorily to the mill rolls. Breakdown of the polybutadienes of 50 to 60 per cent conversion was more rapid than that of GR-S of similar original viscosity. Differences in the mill-processing characteristics of the untreated polybutadiene stocks were not significant, since the values were all so poor. When broken down or when polymerized to a Mooney viscosity of 20 ML-4 or slightly higher, the mixed stocks extruded about as well at standard GR-S.[60, 237]

In comparison with polybutadiene prepared by the Mutual formula as 122° F., polymer made at 41° F. according to a redox-type formula,[163] with 3 parts of dextrose, exhibited higher tensile-strength values, slightly higher Goodyear rebound, marked improvement in hysteresis values, and a quality index considerably closer to that of GR-S 10 (Table V). It is uncertain whether a still lower polymerization temperature (14° F.) gave continued improvement of tensile values, but undoubtedly higher temperatures of polymerization worsened the polymers. The processibility (data not shown) was reduced by lowering the polymerization temperature to 41° F., in that the polymer tended to come free from the mill rolls when it was Banbury-treated and mixed with carbon black,[305] whereas polybutadiene made at 122° F. to the same conversion and viscosity levels adhered well. Also, the extrudability, as judged by the Garvey scale, was poorer for the 41° F. polymers.

Lowering the reaction temperature below 86° F. raised the freeze point, as determined by the Gehman method. Over the entire range of polymerization temperatures studied, the T_{10} values also increased as the temperature of

Table V.   Effect of Polymerization Temperature on Properties of
Polybutadiene[21, 233, 239, 240, 266, 268]

O.R.R. Test Recipe; 2 Parts Sulfur

Code	XP-171	XP-148	AU-454	XP-170	XP-150	XP-169
Polymerization temperature, °F.	145	122	122	86	41	14
Conversion, %	48.5	50	51.5	51	50	51
Raw ML-4 (212° F.)	38	29	51	39	31	45
Modulus* at 300%, p.s.i.	660	610	1690	1140	740	1400
Tensile strength,* p.s.i.	1300	1580	1830	2000	2420	2400
Elongation,* %	450	570	300	450	570	450
Shore A hardness†	62	58	72	108	62	66
Goodyear rebound,‡ % (RT/212° F.)	48/60	46/57	53/66	49/60	59/67	58/69
Goodrich hysteresis						
Temp. rise,† °F.	94	104	74	81	60	43
Initial compression,† %	25.6	29.3	16.1	17.7	17.4	14.2
Set,† %	25.6	...	10.5	5.7	6.8	4.7
Flexures† × 10⁻³	5	4	1	5	3	3
Quality index	0.5	0.6	0.2	0.8	1.0	1.8
Freeze point, −°C.	71	72	71	72	66	62
T_{10} values, −°C.	63	60	...	63	44	42

* Stress-strain tests conducted at 77° F. on samples cured for the optimum time, as determined from 300 per cent modulus versus time-of-cure curve at the point where a tangent to the curve is parallel to the line connecting the values for 0 and 150 minutes of cure.

† Stocks cured at 292° F. for 30 minutes more than the optimum.

‡ Stocks cured at 292° F. for 10 minutes more than the optimum.

polymerization was lowered (Fig. 1). At the three highest polymerization temperatures, up to 145° F., the freeze point is practically constant, but it is definitely greater for 41 and 14° F. polymers.

As mentioned later, Meyer has shown that, in the polymerization of butadiene, reduction in the temperature of polymerization has only a relatively small effect on 1,2-addition, the slight effect observed being a decrease. Change in the temperature of polymerization has, however, a marked effect on the sterical configuration of the 1,4-butadiene units in the polymer, the *trans* 1,4-content being 50 per cent at 100° C. and 80 per cent at −20° C. The ratio of *trans* to *cis* was 4 to 2 at 122° F. and 10 to 3 at 41° F. The tendency of polybutadiene to crystallize was found to begin with polymers made at about 68° F. and to increase with reduction of the polymerization temperature.[23, 244] This tendency, which X-ray photographs of polybutadiene samples prepared at different temperatures confirm,[117] indicates increased linearity in these polymers with decreasing polymerization temperature to −40° F.

Some tire tests[348] have been conducted with polybutadiene made to relatively high viscosities and conversions. In one test, the resistance to wear of polybutadiene made at 122° F. was only 81 per cent that of standard GR-S; in another, the resistance of polybutadiene made at 14° F. was rated at 115 per cent. The trend toward improvement with reduction in the temperature of polymerization which is indicated by these meager data for polybutadiene is in accord with that shown by other more extensive tests on copolymers made at various polymerization temperatures.

Later work by Schulze and co-workers[289-90] has substantiated the existence and the importance of the effect of reduction in the viscosity and conversion of emulsion polybutadiene on the ability of compounds of this

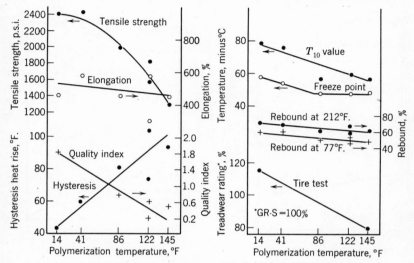

FIG. 1. Effect of Polymerization Temperature on Physical Properties of Polybutadiene—Approximately 50% Conversion

type of polymer to be mixed properly and on the resultant quality shown by actual treadwear. Polymers made to the approximate viscosity and conversion levels necessary for good processing and at temperatures such as to take advantage of all the improvement in physical properties obtainable without affecting unfavorably the low-temperature behavior have shown road wear about 130 per cent that of GR-S 10.[348]

BUTADIENE-STYRENE COPOLYMERS

GR-S Formula: Influence of Polymerization Temperature. This section has been divided into two parts. The first covers large-scale pilot-plant polymerizations over a temperature range of 86 to 190° F. The polymers prepared were subjected to laboratory evaluation only. The second part covers large-scale pilot-plant as well as plant-scale polymerizations over the temperature range of 122 to 0° F. The polymers, after evaluation in the laboratory, were also manufactured into tires which were subjected to road tests. Data are given in Table VI and Fig. 2.

Temperatures 86 to 190° F. A series of polymers was made in 500-gal. reactors[308] according to the usual GR-S formulation; temperature control at the two higher temperatures was maintained by means of a reflux condenser. Over the range from 86 to 190° F., only a negligible increase in the amount of dodecyl mercaptan was required as the temperature was raised, in order to produce polymers of a desired viscosity at the same degree of conversion. Analyses conducted throughout the reactions showed that, as the temperature was increased, respectively, from 86 to 122, 150, 170, and finally to 190° F., the combined styrene in the polymers at 10 per cent conversion increased from 18 to 20, 21, and 22 per cent. Calculations (not given here)

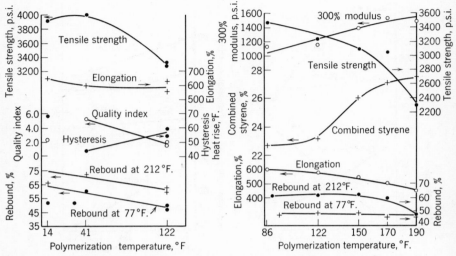

FIG. 2. Effect of Polymerization Temperature on Physical Properties of GR-S Polymers

showed that the instantaneous rates of combination of butadiene (dBD/dt) and of styrene (dS/dt) are the same at approximately 140° F. Above this temperature the rate for butadiene was relatively lower, below 140° F. the rate for styrene was smaller. In these calculations the relationship of refractive index to combined styrene in polymers made at 122° F. was used in determining the composition of the products. These results are the opposite of those reported by Rabjohn,[259] who used the amount of unreacted monomers as the basis of his data for polymer composition.

The effect of temperature on reaction rate is pronounced, and plots of the data indicate a linear relation between the temperature and the logarithm of the reaction time. The gel content of the polymers, at about 72 per cent conversion, remained virtually zero for polymerization temperatures up to 150° F. but increased markedly at 170 and 190° F.

The stress-strain properties deteriorated slightly with each successive increase in polymerization temperature from 86 to 170° F., and at 190° F. the deterioration was more marked. An effect of polymerization temperature on the rebound of the compounded stocks was noticeable only at a test temperature of 212° F. Within the range under discussion, the effect of the temperature of polymerization on hysteresis–heat rise, cut growth and quality index is relatively small, although possibly the flex life of polymers made at 190° F. is poorer than that of polymers made at lower temperatures.[308]

Table VI. Effect of Polymerization Temperature on Properties of GR-S-Type Polymers[159, 261, 263, 308]

O.R.R. Test Recipe; 2 Parts Sulfur

Code	4HH12J	4HH6B	4HH10D	2HH72N	4HH8M	AU-422	HF55	XP-181	XP-137
Polymerization temperature, °F.	190	170	150	122	86	122	122	41	14
Reaction time, hr.	0.9	2	3.5	11	63	⋯	⋯	⋯	⋯
Conversion, %	73	77	70	73	82	72 ± 3	61	61.2	61.3
Raw ML-4 (212° F.)	62	48	41	49	49	50	50	47	56
Combined styrene,* %	27.5	27.1	26	23.2	22.7	22.5	21.9	22.2	23.4
Gel, %	64	57	5	0	0	0	⋯	1	3
Dilute-solution viscosity	0.63	0.95	2.27	2.16	2.18	2.18	⋯	1.99	2.01
Compounded ML-4 (212° F.)	⋯	⋯	⋯	⋯	⋯	65	61	65	79
Mill shrinkage, %	⋯	⋯	⋯	⋯	⋯	39	40.6	41	25
Rugosity	⋯	⋯	⋯	⋯	⋯	38	⋯	34	27
Mill-processing index†	14.5	15.5	11.7	10.1	8.4	10.0	9.7	5.4	5.8
Extrusion index‡	13.5	13	11	13	12	12.0	13.5	9.5	10.0
Modulus at 300%, p.s.i.	1490	1520	1390	1170	1130	1250	1055	1250	1040
Tensile strength, p.s.i.	2300	3050	3090	3240	3470	3300	3320	3990	3920
Elongation, %	450	500	540	580	600	560	630	600	650
Shore A hardness	70	69	69	68	69	67	71	66	72ᵐ
Goodyear rebound, % (RT/212° F.)	47/48	47/59	49/62	49/62	48/61	48/60	50/63	60/72	52/66
Goodrich hysteresis									
Temp. rise, °F.	65	66	65	62	80§	59	55	44	69‖
Initial compression, %	18.1	18.7	17.2	17.5	18.6	19	16.9	14.4	15.1
Set, %	11.5	11.7	10.8	9.9	15.0	13.3	9.4	5.9	6.9
Flexures	2000	4000	4000	3000	4000§	4800	5000	6000	9000‖
Quality index	0.5	1.0	1.0	0.9	0.5§	1.8	2.0	5.3	2.3‖

* Values based on refractive index-styrene relationship for BD/S polymers made at 122° F.

† Ratings: 1 (excellent) to 20 (poor).

‡ Ratings: 1 (poor) to 16 (excellent).

§ Values appear to be out of line.

‖ Replacement of EPC with fine-surface black and use of other softeners increased the flexlife and quality index of XP-137 from two- to fivefold with little effect on hysteresis.

The mill processibility of the polymers became worse with increase in reaction temperature, but the temperature of polymerization had little or no effect on the extrudability of the mill-mixed compounds. When the variations in original viscosity were taken into account, the amount of Mooney viscosity breakdown was similar for all the polymers.

Temperatures 0 to 122° F. There are considerable data to show that the properties of GR-S made at 41° F. represent an improvement over those of GR-S made at the standard temperature of 122° F. Further, there are indications that the trend toward improved properties will continue at polymerization temperatures lower than 41° F., although at present the evidence is limited and the quantitative values relating to polymerizations below 41° F. are at present somewhat doubtful, since changes in formula, emulsifier, and antifreeze in the lower-temperature polymerizations, as well as variations in compounding of the polymers for tire testing, introduce variables that may affect the data available. White[339] in his analysis of the effect of temperature in the polymerization of butadiene–styrene has shown the results for average experimental polymers compared to GR-S given in Table VII.

Table VII.   Influence of Temperature of Polymerization on Properties of Butadiene-Styrene Polymers

Average Treadwear, Hysteresis, and Crack-Growth Ratings
GR-S rated as 100

Polymerization temperature, °F.	122	104	86	68	41	14	0
Treadwear resistance	100	108	103	126	119	126	...
Hysteresis temp.-rise	100	96	108	78	93	89	72
Crack growth	100	...	72	85	50	14	33

Average Stress-Strain Properties of Unaged Stocks at 77° F.

Tensile strength, p.s.i.	2870	3310	3140	3720	3570	3660	3320
Modulus at 300%, p.s.i.	1170	1350	1110	1430	1320	1270	1470
Elongation, %	560	540	640	590	600	600	540

Average Stress-Strain Properties of Unaged Stocks at 200 to 212° F.

Tensile strength, p.s.i.	1190	...	1380*	1670*	1500*	1650	1590
Elongation, %	345	...	375	370	370	445	385

Average Stress-Strain Properties (at 77° F.) of Stocks Aged for 24 Hours at 212° F.

Tensile strength, p.s.i.	2580	2810	3140*	3500*	3190	3230	3190*
Elongation, %	360	350	410	400	380*	440	410
Rebound, %	49	50	47	54	53	54	53

* Values calculated from stress-strains data for unaged stocks tested at 77° F. and percentage retention of property shown in original tables.

These results indicate that reduction in the polymerization temperature from 122 to 41° F. has definitely improved road-wear resistance and tread cracking and that reduction to 0° F. has possibly carried the improvement further. Moreover, similar improvement is indicated in the stress-strain properties of unaged stocks tested at room temperature or at 212° F. and of compounds tested at room temperature after aging for 24 hours at 212° F. The hysteresis temperature-rise and flex-cracking properties are improved markedly by reduction of the polymerization temperature, while the rebound and elongation are improved somewhat.

The polymers made at 41° F. in general break down readily at the low temperatures encountered in milling. The extrudability of their compounded stocks, as measured by the Garvey die, is poorer than that of standard GR-S, and the stocks tend to run hotter and to be scorchy.[3] Banbury mixing of the low-temperature polymers under conditions that develop relatively high temperatures produces poor stocks.[303, 304] This behavior can be controlled by chemical means or by lower Banbury loadings. Under conditions that will break down GR-S to produce a smooth running stock, the low-temperature polymers do not break down but stiffen.[303] No attempt has been made here to review the excellent comparisons of 41° F. polymers with GR-S and natural rubber reported by Fielding[66] and Sjothun.[294] (Cf. Chapter 11.)

Rabjohn[259] has reported the results of polymerizations in bombs at temperatures up to 150° C. of butadiene alone and of butadiene and styrene mixtures. Using the type of polymerization formula and modification used for the GR-S polymerizations, there were obtained polymers, which, when examined by the tumbling method, were found to be soluble. The copolymers were made to the same conversion range as GR-S, but the butadiene polymers reached only low conversions (under 45 per cent). Surprisingly good results were obtained with n-butyl mercaptan as modifier for butadiene-styrene polymerizations at temperatures of 195 to 250° F. (At 122° F. this mercaptan has an inhibiting action.) The physical properties of a blend of polymers prepared separately at 230 to 266° F. (110 to 130° C.) were essentially those to be expected from the results shown in Table VI. Road tests[348] of this copolymer indicated the surprising result of 116 per cent of the treadwear rating of standard GR-S.

Effect of Variation in Styrene Content—122° F. Polymerizations. As discussed previously, if polybutadiene produced at 122° F. is to stick to mill rolls and extrude similarly to GR-S, it must be made with a rosin acid emulsifier to 50 to 60 per cent conversion, and either it must be polymerized to a low viscosity, of 20 to 25 ML-4, or after polymerization the viscosity must be reduced to this level by mechanical treatment. Replacement of 5 parts of the butadiene by styrene in the charge markedly improved the processing characteristics as compared with those of polymer from butadiene alone in the charge.[272] The copolymer had a Mooney viscosity of 30 at 60 per cent conversion.

When 10 parts of styrene was used to replace an equal weight of butadiene in the charge, a further improvement in processibility was secured. With this charge ratio, 60 per cent conversion polymer with a viscosity of 40 ML-4 was as satisfactory as 72 per cent conversion GR-S made to 50 ML-4. Replacement of 15 parts of the butadiene by styrene resulted in processing characteristics such that a 60 per cent conversion 42 ML-4 polymer was, in laboratory tests, easier processing than and equal in extrudability to GR-S. When 40 parts of styrene replaced an equal quantity of butadiene, processing and extrusion characteristics somewhat better than those of GR-S were attained.

Table VIII. *Effect of Styrene Content on Butadiene-Styrene Copolymers Prepared at 122° F.*[60-63, 156, 265]

O.R.R. Test Recipe; 2 Parts Sulfur

Code	XP-147	XP-142	XP-143	XP-140	XP-138	AU-422	XP-98
Butadiene-styrene charge	100/0	95/5	95/5	90/10	85/15	71.5/28.5	60/40
Reaction time, hr.	19	17.3	18.3	15.8	14.5	11.5	10.4
Conversion, %	58	60	51	61	59	72 $\pm$ 3	72
Raw ML-4 (212° F.)	43	40	30	43	42	50	37
Combined styrene, %	...	4.9	4.8	8.1	11.4	22.5	31.5
Gel, %	0	...	...	...	...	0	24
Dilute-solution viscosity	2.13	...	...	...	...	2.18	1.51
Compounded ML-4 (212° F.)	65	50	45	51	48	65	50
Mill shrinkage, %	44	43	41	39	39	39	32
Rugosity	300+	24	18	24	20	38	17
Mill-processing index	12.6	7.8	6.3	8.2	6.2	10.0	8.9
Extrusion index	8.0	10.5	11.0	11.0	12.5	12.0	14.5
Modulus at 300%, p.s.i.	1000	860	790	760	750	1250	1360
Tensile strength, p.s.i.	1640	1820	1920	2190	2370	3300	3690
Elongation, %	440	540	540	600	610	560	610
Shore A hardness	61	62	62	60	61	67	71
Goodyear rebound, % (RT/212° F.)	48/62	44/54	42/53	44/54	44/56	48/60	53/64
Goodrich hysteresis							
Temp. rise, °F.	88	109	92	88	77	59	53
Initial compression, %	24.1	27.2	29.3	28.7	26.6	19	14.7
Set, %	19.3	...	22.4	23.7	21.4	13.3	5.9
Flexures $\times$ 10^{-3}	4	7	7	6	5	4.8	5
Quality index	0.5	0.5	0.8	0.8	1.0	1.8	2.2
Freeze point, minus °C.	72	69	69	66	64	50	38
T_{10} values, minus °C.	60	59	59	53	53	45	32

As shown in Table VIII and Fig. 3, the reaction time was reduced almost 50 per cent by replacing 40 parts of butadiene with styrene; the reduction in reaction time was linear with the amount of butadiene replaced. When the styrene in the charge was between 0 and 29 parts, the polymers up to 72 per cent conversion and 50 ML-4 were gel-free, whereas, at the same conversion, with 40 parts of styrene the viscosity was somewhat lower and the gel was 24 per cent.[272]

Increasing the styrene content improved the reaction rate, tensile strength, rebound, hysteresis temperature-rise, quality index, and processing qualities of the resultant polymers but involved a sacrifice in their resistance to low temperatures. These results are in general agreement with published data.[169]

Limited tire tests[348] conducted on polymers of varying monomer ratios made at 122° F. from various monomer charge ratios indicate the following:

Ratio BD/styrene	100/0	85/15	71/29	63/37
Treadwear resistance (% GR-S)	81	86	100	107

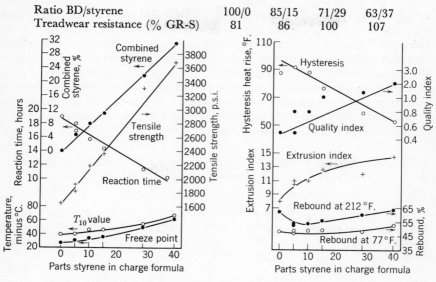

FIG. 3.   Effect of Variation in Styrene Content—122° F. Polymerization

Effect of Variation in Styrene Content—41° F. Polymerizations.

Lowering the temperature of polymerization of the various butadiene-styrene charge ratios increased the tensile strength of the polymers markedly

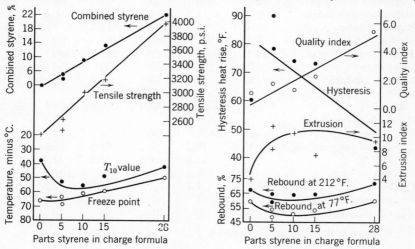

FIG. 4.   Effect of Variation in Styrene Content—41° F. Polymerization

and the rebound values somewhat; it reduced the hysteresis temperature-rise and improved the quality index significantly but affected unfavorably both the processing and the extrusion properties.   Within the experimental error, lower polymerization temperatures had very little effect on the freeze point and the T_{10} values, except with polybutadiene, which, as the temperature of polymerization was reduced, stiffened and became brittle at higher temperatures (shown by comparison of Tables VIII and IX and Figs. 3 and 4). X-ray studies of various butadiene-styrene copolymers made at $-4°$ F. indicated[23] that 6.8 per cent of combined styrene does not prevent crystallization and preferred orientation in the copolymer, such as occur in polybutadiene made below 104° F., but that 16 per cent of combined styrene does prevent these phenomena. More recent work has indicated that the presence of styrene may not prevent crystallization so completely as is implied by the results just noted, and that, given enough time at the optimum conditions, all polymers probably crystallize somewhat.[86]

Table IX. *Effect of Variation in Styrene Content of Butadiene-Styrene Copolymers Prepared at 41° F. Redox Formula*[65, 159, 237, 264, 266]

O.R.R. Test Recipe; 2 Parts Sulfur

Code	XP-150	XP-144	XP-145	XP-141	XP-139	XP-181
Butadiene-styrene charge	100/0	95/5	95/5	90/10	85/15	71.5/28.5
Conversion, %	50	62	52	57	63	61.2
Raw ML-4 (212° F.)	31	42	25	34	46	47
Combined styrene, %	...	5.3	4.1	9.0	13.4	22.2
Gel, %	...	4	0	2	...	1
Dilute-solution viscosity	...	1.94	1.19	1.70	...	1.99
Compounded ML-4 (212° F.)	50	83	50	56	67	65
Mill shrinkage, %	38	27	39	33	30	41
Rugosity	300+	38	12	15	53	34
Mill-processing index	15.9	7.3	5.7	6.2	6.6	5.4
Extrusion index	4.0	8.3	11.5	10.5	7.5*	9.5
Modulus at 300%, p.s.i.	740	810	690	1150	760	1250
Tensile strength, p.s.i.	2420	2630	2580	2970	3180	3990
Elongation, %	570	570	660	650	650	600
Shore A hardness	62	66	63	67	66	66
Goodyear rebound, % (RT/212° F.)	59/67	53/64	48/59	50/63	53/64	60/72
Goodrich hysteresis						
Temp. rise, °F.	60	78	90	74	73	44
Initial compression, %	17.4	19.1	20.3	17.0	16.6	14.4
Set, %	6.8	13.0	20.4	12.5	71.3	5.9
Flexures $\times$ 10^{-3}	3	9.0	14	6	10	6
Quality index	1.0	1.7	1.7	1.3	2.2	5.3
Freeze point, minus °C.	66	64	68	62	60	50
T_{10} values, minus °C.	38	53	62	56	49	42

* Value lower than that usually obtained.

A 95/5 butadiene-styrene copolymer made at 41° F. to 72 per cent conversion with soap emulsification processed badly at Mooney viscosities that would be practical for preparation in a plant. It was necessary to reduce the conversion level to between 50 and 60 per cent at viscosities below

35 ML-4, to obtain stock which extruded properly. A 90/10 butadiene-styrene copolymer at a conversion level of 60 per cent and Mooney viscosity of 41 ML-4 and an 85/15 copolymer of 55 per cent conversion and 49 ML-4 viscosity were similar to GR-S in extrudability.[265] These relationships applied to mill-mixed polymers and not to those mixed in a Banbury or heat-plasticized.

The effect of the type of emulsifier is shown by the following examples. Polybutadiene made at 122° F. to 74 per cent conversion and 51 ML-4 viscosity with a fatty acid emulsifier was broken down in a laboratory Banbury to below 30 ML-4, but the stock from it containing the carbon black and softener would not band when subsequently milled.[238] Corresponding polymers made with a rosin acid emulsifier, however, adhered to the mill after similar treatment. Polybutadiene made at 41° F. with a rosin acid emulsifier to 50 per cent conversion and a Mooney viscosity of 40 to 45 ML-4 could be broken down to a lower viscosity, but would not adhere to the mill.[305] However a 90/10 butadiene-styrene copolymer made at 41° F. to the same conversion and viscosity levels adhered to the mill.

90/10 Butadiene-Styrene Polymers—Effect of Temperature of Polymerization. Fewer data are at hand regarding the influence of the

Table X.   Effect of Polymerization Temperature on Properties of 90/10 Butadiene-Styrene Polymers[20, 63, 236, 264, 268]

O.R.R. Test Recipe; 2 Parts Sulfur

	XP-175	XP-140	XP-173	XP-141	XP-172
Polymerization temperature, °F.	145	122	86	41	14
Conversion, %	55	61	52.3	57	53
Raw ML-4 (212° F.)	41	43	38	34	37
Combined styrene, %	8.7	8.1	8.9	9.0	9.3
Gel, %	0	...	0	2	0
Dilute-solution viscosity	1.77	...	1.70	1.70	1.66
Compounded ML-4 (212° F.)	48	51	56	56	57
Mill shrinkage, %	52	39	39	33	29
Rugosity	24	24	84	15	40
Mill-processing index	8.9	8.2	6.9	6.2	5.9
Extrusion index	8.0	11.0	11.0	10.5	10.5
Modulus at 300%, p.s.i.	800	760	975	1150	1170
Tensile strength, p.s.i.	1780	2190	2870	2970	2940
Elongation, %	490	600	570	650	560
Shore A hardness	63	60	74	67	70
Goodyear rebound, % (RT/212° F.)	45/58	44/54	47/60	50/63	49/62
Goodrich hysteresis					
Temp. rise, °F.	98	88	76	74	71
Initial compression, %	25.8	28.7	15.8	17.0	15.3
Set, %	26.1	23.7	15.6	12.5	9.8
Flexures	7000	6000	4000	6000	6000
Quality index	0.7	0.8	0.8	1.3	1.4
Freeze point, minus °C.	63	66	68	62	61
T_{10} value, minus °C.	57	53	58	56	50

temperature of polymerization on copolymers from a 90/10 butadiene-styrene charge than on copolymers of the normal GR-S ratio of butadiene-styrene, but such data as are available indicate the same general trends in properties as found for the GR-S ratio of monomers, but at a somewhat lower level. (See Table X and Fig. 5.) The 90/10 copolymer made at 14° F. appears to be about as good as GR-S in all respects, as judged by

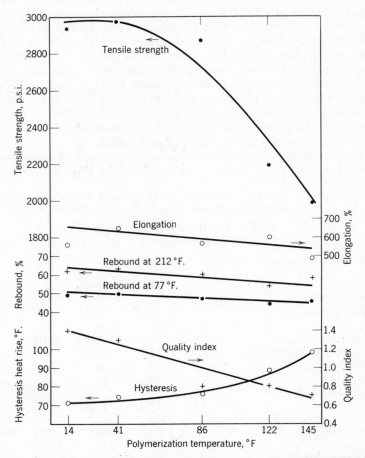

FIG. 5. Effect of Polymerization Temperature on Properties of 90/10 Butadiene-Styrene Polymers

laboratory tests, and to have a freeze point and T_{10} value about 10° C. lower than GR-S. Very limited data indicate the treadwear of the 14° F., 90/10 butadiene-styrene polymer to be about 33 per cent better than that of GR-S.[348] White,[339] as noted previously, from the average of many more tests, concluded that the treadwear of GR-S made at 41° F. was 119 per cent that of standard GR-S. The 90/10 copolymer made at 14° F. is probably similar in treadwear to the 71/29 butadiene-styrene polymer made at 41° F.

Oil-Extended Polymers. Reports by D'Ianni, Hoesly, and Greer[50] and by Swart, Pfau, and Weinstock,[300] as well as work conducted at the Government Laboratories,[160] have shown that the blending of processing oils with GR-S-type polymer made at 41° f. to a high viscosity, not only extends the GR-S and thus makes a larger volume of rubber available, but also seems to improve the quality with respect to treadwear, hysteresis, and other dynamic properties. The oil can be introduced by masterbatching it as an emulsion with latex.[153] Tire tests conducted from the summer of 1947 through the next several years have indicated that the use of up to about 25 per cent of oil, based on the rubber, yields stock that is satisfactory in processing and, after vulcanization, in treadwear.

DICHLOROSTYRENES COPOLYMERS

GR-S Ratio of Monomers. Styrene substituted with 2 atoms of chlorine has been made by various processes, which yield either a relatively pure isomer or a mixture of isomers. The polymerizations and laboratory tests, as well as tire tests, of the polymers made from the chlorostyrenes obtained from various sources[311, 313, 316, 319, 326] indicated little or no difference in results attributable to the source of the monomer or its isomeric composition.

Butadiene-dichlorostyrene polymers made from about the GR-S charge ratio (71/29) at 95 and 122° F. are definitely higher in tensile properties (Table XI) than standard GR-S and possibly possess better elongation at equal modulus. These polymers were made to a Mooney viscosity substantially higher than that of standard GR-S. The retentions of tensile strength and elongation at 212° F. (data not shown) were considerably better than in GR-S in absolute value as well as percentage-wise. The stress-strain values of the dichlorostyrene copolymers tested at room temperature are similar to those of 41° F.-polymerized GR-S. The hysteresis–heat-rise rating at 30-minute overcure is substantially poorer than that of GR-S, but the resistance to flex cracking is significantly better, and the over-all quality index somewhat better. These polymers broke down on the mill easier than GR-S of similar viscosity and even at higher viscosities yielded processibility values within the range of GR-S (Table XI). They required somewhat more power to break them down in a laboratory "B" Banbury to about the same extent, but not to the same value as GR-S (data not shown). In extrudability through the Garvey die, the mill-mixed stocks were equal to or better than GR-S. Reports indicated that at high loadings of zinc oxide, some of the dichlorostyrene copolymers have merit.[1]

Dichlorostyrene combined with butadiene during polymerization much faster than did styrene at the same charge ratio. When the charge contained 29 parts of dichlorostyrene, the polymer at 10 per cent conversion contained about 40 per cent, whereas for styrene the corresponding amount is about 17 per cent. At 72 per cent conversion about 28 per cent of the dichlorostyrene charged was unreacted, whereas in a standard GR-S polymerization this percentage is about 42 per cent. While the recovery of styrene is efficient, the amount of the dichlorostyrene recovered on stripping the latex was negligible, and the material obtained formed an oil-in-water emulsion that was difficult to break.

Storage of the dichlorostyrene presented a serious problem,[311, 315] since this monomer tends to react with itself readily, and, at the time most of the chlorostyrene polymerizations were made, a satisfactory inhibitor was not available.

Table XI. Dichlorostyrene Copolymers of about 71/29 Butadiene-Dichlorostyrene Ratio[142, 148, 261, 314, 316, 332]

O.R.R. Test Recipe; 2 Parts Sulfur

Code	XP-17	XP-8	XP-39	XP-41	XP-42	AU-422
Polymerization temperature, °F.	122	95*	122	95†	122	122
BD/DCS charge ratio	73/27	70/30	71/29	70/30	71/29	71.5/28.5‡
Reaction time, hr.	11.4	...	10.6	10.1	9.4	12
Conversion, %	69.7	70	72.9	70.9	73	72 ± 3
Raw ML-4 (212° F.)	70	69	62	61	55	50
Dichlorostyrene, %	...	32.0	28.6	29.8	31.1	22.5‡
Gel, %	43	...	1	2	6	0
Dilute-solution viscosity	1.78	...	2.03	1.87	2.01	2.18
Compounded ML-4	86	79	72	53	63	65
Mill shrinkage, %	45	29	38	25	33	39
Rugosity	97	60	53	124	19	38
Mill-processing index	11.5	9.0	10.9	7.0	6.8	10.0
Extrusion index	10.0	15.5	12	15	14.0	12.0
Modulus at 300%, p.s.i.	1270	1130	1230	1130	1320	1250
Tensile strength, p.s.i.	3680	3990	4060	4150	3800	3300
Elongation, %	600	640	640	645	620	560
Shore A hardness	70	69	74	73	70	67
Goodyear rebound, % (RT/212° F.)	...	47/57	...	...	43/54	48/60
Goodrich hysteresis Temp. rise, °F.	69	91	73	74	80	59
Initial compression, %	16.6	16.8	20.3	18.4	18.7	19
Set, %	11.0	21.1	16.1	15.7	17.9	13.3
Flexures × 10⁻³	9	32	14	20	18	4.8
Quality index	2.2	3.9	2.6	3.6	3.2	1.8
Freeze point, minus °C.	48	42	48	42	43	50
T_{10} value, minus °C.	42	36	39	37	34	45

* Special activated formula (all others Mutual GR-S).
† Ferricyanide-activated formula.
‡ Styrene.

The relatively poor low-temperature properties of the polymers with about 30 parts of the chlorostyrene in the charge ratio should be noted in comparison with the corresponding properties of GR-S.   The low-temperature properties,[322] as determined by the Gehman method, are more nearly those of a copolymer from a 65/35 butadiene-styrene charge.

Table XII. Effect of Charge Ratio on Butadiene-Dichlorostyrene Copolymers Prepared at 122° F.[1, 145, 315, 322]

O.R.R. Test Recipe; 2 Parts Sulfur

	XP-63	44G4	19G8B	19G5B	19G1B	XP-21	19G15B	19G18B
Reference Code								
Polymerization temperature, °F.	122	122	122	122	122	122	122	122
BD/DCS ratio	100/0	95/5	90/10	80/20	70/30	63/37	60/40	50/50
Reaction time to 72% Conversion, hr.	15.5	13.0	11.5	12.5	8.0	6.8	7.8	6.9
Conversion, %	72.9	72	71.9	74.2	73.9	74	74.4	73.2
DDM, parts*	0.80	0.75(0.55)	0.44	0.36	0.32	0.23	0.25	0.16
Raw ML-4 (212° F.)	51	55(77)	67	70	58	68	72	76
Dichlorostyrene, %	0	3	9.8	20.6	31.8		42.0	51.6
Gel, %	0		59	11	2	20	50	1
Dilute-solution viscosity		2.21	1.60	2.39	1.97	⋯	1.58	2.06
Compounded ML-4		76	83	88	63	83	80	73
Mill shrinkage, %		44	32	20	30	28	25	19
Rugosity		102	56	56	16	17	13	7
Mill-processing index	>16	16.1	15.2	14.6	13.0	7.2	8.2	4.5
Extrusion index		8.5	⋯	⋯	⋯	16	⋯	⋯
Optimum cure, min.		57	60	60	60	45	60	63
Modulus at 300%, p.s.i.	1400	1520	2250	1750	1410	1350	1890	1660
Tensile strength, p.s.i.	1460	2050	2490	3560	3440	3510	3590	3230
Elongation, %	310	380	330	505	550	570	490	480
Shore A hardness†		⋯	70	72	73	75	74	77
Goodrich hysteresis† Temp. rise, °F.			64	64	83	104	77	77
Initial compression, %			15.4	14.1	19.5	22.5	20.0	23.0
Set, %			9.8	7.7	21.0	27.7	17.0	24.9
Flexures† × 10⁻³			1.0	5.0	21.5	55	42.5	93
Quality index			0.9	0.3	3.4	4.3	8.0	12.5
Freeze point, minus °C.	77	71	65	57	45	34	30	16
T_{10} value, minus °C.	70	62	56	40	32	23	23	9

* Per 100 parts of monomers.

† These values are for polymers at the optimum cure.

Effect of Variation of Dichlorostyrene in the Charge Ratio.

The addition of dichlorostyrene to butadiene in the charge increased the reaction rate markedly and to a greater degree than did the same quantity of styrene. At a reaction temperature of 122° F., a decreasing amount of dodecyl mercaptan was needed for modification, as the chlorostyrene content was increased. The same kind of relationship was true for styrene, but, with other conditions the same, less mercaptan was required for the dichlorostyrene than for the styrene copolymer.

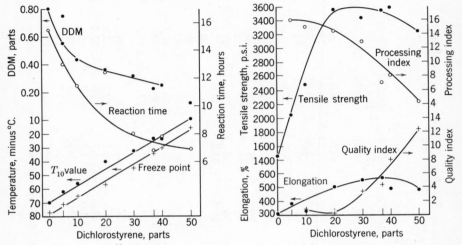

Fig. 6. Effect of Variation of Dichlorostyrene Charge in Polymerizations at 122° F.

It will be noted that for the 70/30 butadiene-dichlorostyrene copolymer the properties shown in Table XII are somewhat inferior to those shown in Table XI. However, it may be mentioned that all the results shown in Table XII and Fig. 6 were obtained at one time under comparable conditions with the same materials and that the general conclusions they indicate should be valid. The optimum balance of tensile strength and elongation is at about the 70/30 charge ratio, which led to a copolymer containing about 32 per cent of combined dichlorostyrene. This peak was at a lower comonomer ratio than when styrene itself was used; in fact, the values discussed previously for styrene do not indicate a maximum up to the highest ratio tested.

The ease of breakdown of the dichlorostyrene copolymers, compared to the styrene copolymers, is indicated by the mill-processing index. In comparing these data with the styrene data (Tables VIII and IX), it should be remembered that, at the same charge-weight ratio of the respective monomers, the combined dichlorostyrene content in the copolymers is higher than the combined styrene. This difference indicates better processing for the dichlorostyrene copolymers when the charge ratio is used as the basis. Moreover, the dichlorostyrene copolymers have been made to considerably higher Mooney viscosities than the styrene copolymers. With about the same charge ratios, a dichlorostyrene copolymer of about 70 Mooney viscosity will exhibit mill breakdown equivalent to that of 50 Mooney GR-S. When broken down in laboratory "B" Banbury, the dichlorostyrene copolymers of higher Mooney viscosities, as compared to GR-S, required more power initially, broke down further (but not to so low a value), ran 40 to 50° F. hotter, and increased in gel content to about 50 per cent of gel compared to a negligible gel increase in GR-S.

The marked improvement in the resistance of the copolymers to flex-crack growth with increase in dichlorostyrene content is significant and is more pronounced than the corresponding improvement in styrene copolymer. This improvement is accompanied by apparently greater hysteresis temperature-rise, although the quality index (balance of the two properties) is markedly improved by use of the dichlorostyrene as the comonomer.

The low-temperature properties of the dichlorostyrene copolymers are considerably poorer than those of similar copolymers made with styrene. It will be noted that about 18 to 20 parts of dichlorostyrene in the charge ratio is the maximum at which are obtained low-temperature properties approximating those of GR-S.

Average results of tire tests[348] conducted on stocks made from dichlorostyrene copolymers of various monomer ratios are:

BD/DCS ratio	63/37	70/30	73/27	80/20
Resistance to treadwear (% GR-S)	108	126	118	105
Cracking comparison (Experimental polymer/ GR-S)	Equal	Better	Better	Better

In judging the results for the copolymers of 70/30 and 73/27 ratio, it should be borne in mind that there are two inherent factors that make these results favorable when compared to those for GR-S: (1) The higher comonomer content in the dichlorostyrene copolymers is relatively the same as that from a 63/37 butadiene-styrene charge. Limited tire testing[348] indicated this copolymer to wear about 107 per cent as well as GR-S. (2) The viscosities of these polymers (approximately 70 Mooney) were selected to insure processing equal to that of GR-S. It has been shown[116] that polymers of higher viscosity, i.e., of higher molecular weight, display better tire wear. GR-S of 80 to 100 Mooney viscosity would be expected to show resistance to wear approximately 108 per cent that of similar polymer made to the usual 50 Mooney level.[348]

Effect of Polymerization Temperature on Dichlorostyrene Copolymers. A series of butadiene copolymers[57] was prepared with dichlorostyrene in 5-gal. reactors at 41 to 68° F. to 72 per cent conversion according to an activated formula over a range of viscosities, for comparison with similar copolymers made with styrene. With dichlorostyrene, the charge ratio was 73/27; with styrene, 71.5/28.5. The results (Table XIII) bring out again the better mill-processing and extrusion properties of copolymers of dichlorostyrene. The improvement in the copolymer made with dichlorostyrene at low temperatures, compared to the styrene copolymer,

Table XIII. 73/27 Butadiene-Dichlorostyrene Polymerized at Low Temperatures to Various Mooney Viscosities[57]

O.R.R. Test Recipe; 2 Parts Sulfur

Code	GR-S	Special	20G28	20G11	20G8	20G10	20G31
Type Monomer	BD/Styrene	BD/Styrene	BD/DCS	BD/DCS	BD/DCS	BD/DCS	BD/DCS
Formula	Mutual	Activated	Activated	Activated	Activated	Activated	Activated
Polymerization temperature, °F.	122	41–68	41–68	41–68	41–68	41–68	41–68
Reaction time, hr.	11.0	10.2	7.5	10.2	9.5	8.8	10.7
Conversion, %	72.0	69.0	72.6	71.0	71.9	73.3	75.0
DDM, part	0.47	0.40	0.40	0.31	0.32	0.30	0.10
Raw ML-4 (212° F.)	50	85	39	60	82	111	170
Other monomer, %	23	...	32.7	32.2	31.9	31.9	31.6
Gel, %	0	2.8	...	27	8.29	39	65
Dilute-solution viscosity	2.0–2.2	1.6	...	1.3	2.4, 1.8	1.6	1.2
Compounded ML-4	65	96	60	72	82	101	175+
Mill shrinkage, %	40	28	26	29	27	25	21
Rugosity	35–40	50	23	14	36	...	37
Mill-processing index	11.5	8.6	6.8	6.7	7.5	8.9	12.7
Extrusion index	13–14	10.5	...	15	13.5	12	6.5
Optimum cure, min.	60	65	57	57	58	58	49
Modulus at 300%, p.s.i.	1200	1260	1260	1425	1510	1615	2125
Tensile strength, p.s.i.	3400	4150	3585	3800	4240	4240	4930
Elongation, %	600	620	600	600	595	550	500
Shore A hardness			...	75	75	75	76
Goodyear rebound, % (RT/212° F.)	49/59	50/67	...	43/56	44/59	45/59	47/64
Goodrich hysteresis Temp. rise, °F.	58	48	...	73	68	68	55
Initial compression, %			...	17.8	16.2	16.0	13.8
Set, %			...	12.9	10.3	9.1	5.4
Flexures × 10^{-3}	6	9	...	25	25	20	9
Quality index	1.0	3.5	...	5.6	6.6	5.3	3.7

occurs almost entirely in the resistance to flex cracking and, concurrently, in quality index. In this respect, the improvement over GR-S resulting from lower temperature of polymerization and the replacement of styrene by dichlorostyrene are additive in effect. With respect to other physical properties, use of either variable alone seems to cause improvement, but the results are not additive.

Compared to the controls made with styrene, polymers of 82/18 butadiene-dichlorostyrene ratio made[270] according to the same activated formula exhibited lower tensile strengths, similar quality indexes (better crack-growth resistance but greater hysteresis temperature-rise), and generally no improvement. The reduction in chlorostyrene content did, however, improve the Gehman low-temperature properties of the product to the point of equality with the styrene copolymer.

Complete Conversion of Butadiene–Dichlorostyrene. As discussed previously, the recovery of unreacted dichlorostyrene by the regular stripping procedure applied to the latex at conversion levels around 72 per cent was negligible. In view of the added cost that would thus be imposed by the use of a monomer basically more expensive than styrene, a series of polymers was made at 122° F. to as high a conversion as possible. At conversions of 92 ± 2 per cent, the dichlorostyrene in the copolymer was 95 per cent or more of the dichlorostyrene charged.[155] The use of *n*-dodecyl mercaptan was preferred to that of either mixed tertiary mercaptans or a tertiary C_{16} marcaptan, because the reaction times to the high conversions were significantly shorter with the first-named. When the finished polymers contained 70 per cent or more of gel below 50 in swelling volune, the polymers were reasonably satisfactory with respect to mill processing. However, the modulus was higher, the tensile strength and elongation lower, the hysteresis higher, the flex life poorer, and the resilience poorer than for GR-S; the Gehman low-temperature properties were similar to those of GR-S. A copolymer[157] with 19.5 per cent of combined dichloro-styrene made to a viscosity of 76 Mooney at 92.6 per cent conversion in 25.5 hours exhibited treadwear[348] equivalent to that of GR-S but poorer cut growth than the latter. Similar copolymers were made to 31 and 36 Mooney—the highest viscosities at which gel-free polymers were obtained at these high conversions. Generally, the physical properties of these polymers were unattractive.

Conclusions. Thus the dichlorostyrene copolymers, when first made according to the usual GR-S charge ratio, showed, as compared to GR-S, favorable stress-strain properties, flex-crack resistance, treadwear, and reten-tion of properties when heated. The unfavorable factors included higher hysteresis–heat rise, poorer low-temperature properties, high cost of the monomer, difficulty of storing the raw monomers, and difficulty of recovering unreacted monomer after copolymerization to conventional levels of conver-sion. When the dichlorostyrene in the finished copolymer was lowered to the level of the styrene in GR-S, the favorable results were less marked. If polymerization conditions were such as to obviate the necessity of recovery of dichlorostyrene, the tire results were similar to those of GR-S.

Dichlorostyrene Tripolymers. Tripolymers were made[340] in 5-gal.

Table XIV.   *66/21/13 Butadiene-Dichlorostyrene-2-Vinylpyridine Tripolymers Made at 122° F.*[340]

Compounding Recipe

Ingredient	Parts
Polymer	100
EPC black	50
BRT no. 7 (coal-tar softener)	3.75
Stearic acid	1.25
Zinc oxide	5.0
Mercaptobenzothiazole	0.2
Sulfur	1.6

Code	10N4–12	10N8–11	10N3–6	GR-S
Reaction time to 72% conversion, hr.	6.2	7.0	7.0	...
Conversion, %	75.0	71.0	72.5	72 ± 3
$tert$-C_{12}-Mercaptan,* part†	0.235	0.26	0.31	...
Raw ML-4 (212° F.)	73	47	28	50
Gel, %	0	2	2	0
Dilute-solution viscosity	2.01	1.47	1.35	2.20
Compounded ML-4	83	62	47	65
Mill shrinkage, %	35	32	18	40
Rugosity	24	21	7	35–40
Mill-processing index	12.1	6.8	4.5	11.5
Extrusion index	10.5	12.5	14.5	11–13
Mooney viscosity ML-4, after 5 min. in Banbury	68	40	27	30
Power consumption, kw. (0 min.)	4.2	4.4	5.5	4.7
(1 min.)	3.8	2.8	2.6	3.1
(5 min.)	2.4	1.8	1.7	1.7
Pyrometer temperature, °F.	365	330	270	360
Optimum cure, min.	30	30	30	60
Modulus at 300%, p.s.i.	1770	1495	1290	1210
Tensile strength, p.s.i.	3790	3720	3280	3390
Retention at 212° F., %	45	43	44	36
Elongation, %	520	590	605	600
Shore A hardness	75	76	72	67
Goodyear rebound, % (RT/212° F.)	52/61	52/60	49/59	48/61
Goodrich hysteresis Temp. rise, °F.	91	90	89	50
Initial compression, %	16.8	17.1	19.5	19.5
Set, %	22.1	20.4	...	10.7
Flexures × 10^{-3}	18	30	40	5
Quality index	2.2	3.8	5.2	1.0

* Mixed tertiary mercaptans.

† Per 100 parts of monomers.

reactors, with *tert*-C_{12}-mercaptan as modifier, from a charge of 66/21/13 butadiene–dichlorostyrene–2-vinylpyridine in the Mutual formula at 122° F. The reaction times were approximately 50 per cent of those for GR-S to the same conversion level. The modifier requirements were similar to those for GR-S of equal viscosity. The tripolymers, as usual for any polymer containing a monomer of the pyridine class, were scorchy and rapid curing, as shown in Table XIV by the optimum cure data and by the viscosity data for the Banbury breakdown tests. Mill-processibility and Garvey die-extrusion values were similar or superior to corresponding values for GR-S, depending on the Mooney viscosity of the tripolymer. The breakdown in the "B" Banbury was poor. At equal or higher polymer viscosities the tensile strength of the tripolymers at room temperature was superior to that of GR-S, and the retention of tensile strength at 212° F. was definitely superior. The Goodyear rebound values were in the same range as those for GR-S. The hysteresis temperature-rise was poorer, but the resistance to crack growth was superior to that of GR-S, with the net result of markedly better quality index. No information is available on building tack or adhesion in tires after cure, but presumably the results would be similar in this respect to other polymers made with the vinylpyridines (cf. pp. 725-6).

A tripolymer[64] made in bottles from a 66/14/20 ratio of the same constituents was shown to have mill-processing characteristics and stress-strain values similar to those of the 66/21/13 tripolymer and to be equally fast curing. A 75/15/10 charge of the same monomers in 5-gal. reactors at 122° F. yielded a tripolymer showing no marked improvement[311] compared with GR-S, except in hysteresis–flex-crack growth balance; it had the disadvantage of being very rapid curing.

Several charges of butadiene-styrene-dichlorostyrene mixtures (70/20/10, 70/16/14, 70/10/20) were polymerized in 5-gal. reactors. Tests on the resulting tripolymers indicated that the stress-strain properties improved somewhat with increase in the amount of dichlorostyrene charged.[316]

BUTADIENE-MONOCHLOROSTYRENE COPOLYMERS

Preliminary tests[312, 316, 318] on the copolymerization of butadiene with various samples of monochlorostyrene (ortho and para mixtures, pure meta, pure para, and mixtures of meta and para, all obtained from the Dow Chemical Company) indicated little significant difference in the polymerization characteristics of the different samples, except that the pure para sample required about 18 per cent more dodecyl mercaptan than any other sample; in the stress-strain properties of the copolymers after vulcanization, and in their processing behavior as determined by the mill-processing index, the Garvey die-extrusion index, and the 5-minute "B" Banbury breakdown test. No definite trends attributable to individual isomers could be recognized in regard to hysteresis temperature-rise, flex-cut growth rate, or quality index. Other samples of mixed isomers containing 0.01 to 0.02 per cent of the cross-linking agent divinylbenzene produced copolymers containing possibly more loose gel than the previously mentioned samples but not sufficient to affect control of viscosity or processibility.

Table XV. Butadiene-Monochlorostyrene Copolymers[143, 144, 145, 147, 267]

O.R.R. Test Recipe; 2 Parts Sulfur

Code*	36T	36C	37C	38C	49GA(C)	49GB(C)
Polymerization temperature, °F.	122	122	95	95	41	41
Polymerization formula	GR-S	GR-S	Special	activated	TDN	Redox
Charge ratio (BD/MCS)	71/29	71/29	70/30	70/30	80/20	80/20
Reaction time to 72% conversion, hr.	10.5	9.2	10.2	13.1	7.2†	14.4†
Conversion, %	71.6	71.0	73.0	70.4	60.8	60.4
DDM, part‡	0.36	0.34	...	0.32	0.329§	0.229§
Raw ML-4 (212° F.)	63	64	70	64	56	47
Combined MCS, %	25.0	27.1	28.4	28.4	24.5	23.9
Gel %	14	0	0	2	...	...
Dilute-solution viscosity	2.06	2.20	2.09	2.10	...	...
Compounded ML-4	71	54	74	72	66	59
Mill shrinkage, %	41	44	29	31	44	33
Rugosity	31	28	35	26	17	20
Mill-processing index	10.2	10.1	7.1	6.8	5.3	3.9
Extrusion index	10.5	14.0	11.5	11.5	10.0	11.0
Optimum cure, min.	58	60	59	63	65	70
Modulus at 300%, p.s.i.	1390	1370	1120	1240	1130	1210
Tensile strength, p.s.i.	3500	3980	4170	4170	4310	4390
Elongation, %	580	600	680	640	660	650
Shore A hardness	70	72	70	71	66	67
Goodyear rebound, % (RT/212° F.)	49/62	...	48/57	58/83	58/70	59/72
Goodrich hysteresis Temp. rise, °F.	66	63	81	59	54	56
Initial compression, %	17.1	19.2	17.9	17.7	16.6	16.3
Set, %	9.9	11.9	18.4	9.3	5.4	6.1
Flexures × 10⁻³	7	12	23	14	6	8
Quality index	2.0	3.8	3.9	5.0	2.6	3.2
Freeze point, minus °C.	46	46	49	48	41.5‖	41‖
T_{10} value, minus °C.	40	39	38	40	33‖	35‖

* "T" in code means the monochlorostyrene was prepared by the thermal process; "C" designates the chemical process. The purity of chemical-process monochlorostyrene was 99.0 to 99.5 per cent MCS. The material contained no divinylbenzene and varied in composition from 32.5 to 49.4 per cent para, the remainder being ortho. The thermal process material contained 92.1 per cent MCS, 54.7 per cent of which was ortho, 40.7 per cent para, and 4.6 per cent saturates. These analyses were by the Dow Chemical Co. It contained also 175 p.p.m. of a cross-linking agent similar to divinylbenzene.

† To conversion shown, not 72 per cent.

‡ Per 100 parts of monomers.

§ Mixed tertiary mercaptans.

‖ The freeze point of GR-S tested at the same time was minus 45° C. and the T_{10} value was minus 35° C. These values appear high by 5 and 10° C., respectively.

Polymerizations in 500-gal. reactors[316] indicated that at low conversions monochlorostyrene and butadiene had combined in about the same ratio as that in which they occurred in the monomer charge. For comparison, it may be recalled that at low conversions dichlorostyrene combines in a higher and styrene in a lower ratio than they occur in the charge. During the course of copolymerization the ratio of monochlorostyrene to butadiene in the copolymer diminished slightly; the ratio of dichlorostyrene to butadiene decreased markedly, and that of styrene to butadiene increased. The polymerization rates, using the GR-S recipe, were somewhat less with mono- than with dichlorostyrene as the comonomer. Unreacted monochlorostyrene could not be recovered satisfactorily by stripping.

The stress-strain values for the monochlorostyrene copolymers were similar to those obtained with the dichlorostyrene polymers and substantially better than those of GR-S. The retention of tensile strength at 212° F. of the monochlorostyrene copolymers was about the same as for GR-S, while the dichlorostyrene copolymers showed better retention. The hysteresis and rebound values seemed to be quite irregular, but the quality indexes were definitely superior to those of GR-S. The copolymers of higher Mooney viscosities had mill-processing characteristics similar to those of the dichlorostyrene copolymers. The effect on the physical properties of the vulcanizates produced by the combined influence of a lower temperature of polymerization, less monochlorostyrene in the charge, and lower Mooney viscosity in the polymer is shown in Table XV.[58, 267]

The average of several tire tests[348] carried out with 72/28 to 70/30 butadiene-monochlorostyrene copolymers made at 95 to 122° F. indicates resistance to treadwear 116 per cent that of GR-S. No apparent trend in the properties of the polymers resulted from changes in charge ratio, in polymerization temperature within these limits, or in the composition of the mixture of isomers. Cracking and cut growth were less than with standard GR-S. Tires made from an 80/20 butadiene-monochlorostyrene mixture in a diazo thioether formula at 41° F. (49GA) showed treadwear resistance 122 per cent that of GR-S, and a redox polymer (49GB) prepared under similar conditions 127 per cent. In both cases cut growth was similar to that of GR-S.

ALKYL- AND ALKOXY-SUBSTITUTED STYRENES AS COMONOMERS

The most interesting of the substituted styrenes other than the mono- and dichlorostyrenes is α-methylstyrene, $C_6H_5C(CH_3) : CH_2$. This monomer was considered because of its relative inactivity with respect to self-polymerization. In copolymerization with butadiene at 122° F. the reaction rate[141, 317] of α-methylstyrene of approximately 95 per cent purity, furnished by the Dow Chemical Company, was somewhat lower than that of styrene (see Table XVI). The stress-strain properties of the copolymer were within the range of those of GR-S, the hysteresis temperature-rise was somewhat greater than that of GR-S, the resistance to flex cracking somewhat better, and the quality index in the same range as GR-S. The processing characteristics of the copolymer also were in the same range as those of GR-S. Preliminary tire tests[348] indicated a little better treadwear and less cracking than for GR-S. A 75/25 butadiene-α-methylstyrene copolymer[327] made at 41° F. had physical properties related to those of 41° F. butadiene-styrene copolymer in about the same manner as those of the corresponding 122° F. copolymers are related, except for the quality index, which was not so good as that of the styrene copolymer. Preliminary tire tests[348] did not indicate in the α-methylstyrene copolymer the improvement with reduction in temperature that has been obtained in styrene copolymers.

Table XVI.
Copolymers of Butadiene with Substituted Styrenes[44, 112, 141, 232, 261, 317, 324, 327, 348]

O.R.R. Test Recipe

Code	AU–422	29G–A518	33G–A461	30G–A504	17G–54–55
Monomer	Styrene:	p-Methoxy-styrene: Univ. of Ill.	2,4-Di-methyl-styrene:	2,4-Di-methoxy-styrene:	α-Methyl-styrene:
Source	Com-mercial	Gov't. Lab.	Gov't. Lab.	Univ. of Ill.	Dow Chem. Co.
Purity, %	99.0	90.0	92.8, 92.5	89.2	96.0–99.2
Specific gravity (25°/25° C.)	...	0.9993	0.9009, 0.9008	1.072	0.903–0.908
n_D^{25}	...	1.5580	1.5428 1.5425	1.5665	
Polymerization temperature, °F.	122	122	122	122	122
Charge ratio (BD/special monomer)	71.5/28.5	70/30	75/25	75/25	72/28
Reaction time, hr.	11.5	11.9	14.3	25.6	15
Conversion, %	72 ± 3	73	74	72[a]	71
DDM, part	0.47	0.54	0.54	0.34	0.50
Raw ML-4 (212° F.)	50	46	51	61	51
Special monomer, %	22.5	...	...	...	24.4
Gel, %	0	3.6[b]	0	...	0
Dilute-solution viscosity	2.18	2.32[b]	2.29	...	1.87
Compounded ML-4 (212° F.)	65	59	60	74	76
Mill shrinkage, %	39	42	37	50	37
Rugosity	38	11	19	35	23
Mill-processing index	10.0	9.7	8.1	15.5[c]	9.8
Extrusion index	12.0	...	11.5	...	
Optimum cure, min.	60	57	50	51	63
Modulus at 300%, p.s.i.	1250	1270	1330	940	1160
Tensile strength, p.s.i.	3300	3300	3250	3120	3120
Elongation, %	560	580	545	640	580
Shore A hardness	67	67	...	65	68
Goodyear rebound, % (RT/212° F.)	48/60	53/66[d]	51/68	52/66	...
Goodrich hysteresis Temp. rise, °F.	59	73	...	113	70
Initial compression, %	19	18.8	...	21.5	21.1
Set, %	13.3	12.4	...	24.0	17.3
Flexures × 10⁻³	4.8	4	...	6	4
Quality index	1.8	0.9	...	0.3	0.8
Freeze point, minus °C.	50	54	54	56	...
T_{10} value, minus °C.	45	48	47	46	...
Resistance to treadwear (% GR-S)	...	99	91	97	...
Cut growth, experimental/ standard	...	5.5/0.5	2.25/2.5	1.0/1.25	...

Table XVI—continued

Code	17G-56-63	17G-50-51	XP-44	47G-A625	48G-A643
Monomer	α-Methyl-styrene:	α-Methyl-styrene:	α-Methyl-styrene:	α-Methyl-styrene:	α-Methyl-p-Methyl-styrene:
Source	Dow Chem. Co.	Dow Chem. Co.	Dow Chem. Co.	Dow Chem. Co.	Newport Industries
Purity, %	Same as 17G–54–55			94.5–9.6	...
Specific gravity (25°/25° C.)	Same as 17G–54–55			0.9072–0.9076	0.8986, 0.8985
n_D^{25}	...	...	...	1.5329–1.5358	1.5313, 1.5322
Polymerization temperature, °F.	122	104	122	41	122
Charge ratio (BD/special monomer)	72/28	72/28	72/28	75/25	75/25
Reaction time, hr.	11[e]	11.5[e]	19.25[f]	poor[g]	14.2
Conversion, %	73.5	73.5	71.6	59.9	72.7
DDM, part	0.45	0.44	0.78	0.24[h,i]	0.475
Raw ML-4 (212° F.)	48	48	49	57	51
Special monomer, %	23.4	24.1	22.0	19.2	22[j]
Gel, %	0	0	0	0	17
Dilute-solution viscosity	...	2.00	2.07	1.94	1.80
Compounded ML-4 (212° F.)	50	48	54	71	56
Mill shrinkage, %	32	34	34	27	50
Rugosity	40	32	30	28	21
Mill-processing index	7.8	7.2	6.8	5.6	8.0
Extrusion index	...	...	13.0	7.0	8.0
Optimum cure, min.	66	64	55	72	66
Modulus at 300%, p.s.i.	1110	1110	1090	960	1130
Tensile strength, p.s.i.	3290	3680	2980	3800	2930
Elongation, %	610	670	590	620	540
Shore A hardness	68	69	62	64	63
Goodyear rebound, % (RT/212° F.)	...	...	...	64/75	...
Goodrich hysteresis Temp. rise, °F.	70	75	68	55	79
Initial compression, %	20.3	20.1	23.6	18.0	21.0
Set, %	17.2	18.4	18.2	5.3	14.0
Flexures × 10⁻³	7	9	9	7	5
Quality index	1.5	1.5	2.0	2.9	0.9
Freeze point, minus °C.	...	...	55	51	51
T_{10} value, minus °C.	...	...	48	44	44
Resistance to treadwear (% GR-S)	...	...	110	113	...
Cut growth, experimental/standard	...	...	6/2	3/2	...

[a] Made in 8-oz. bottles; other special polymers made in 5-gal. reactors.

[b] Average of individual batches; 48 per cent gel and 0.31 DSV for final blend.

[c] Difficult to break down on mill and band.

[d] 30-minute overcure.

[e] Formula same as that for 17G-54-55, except activated. Other formulas, unless noted otherwise, same as GR-S with fatty acid soap as emulsifier.

[f] Rosin-type soap as emulsifier.

[g] High-sugar redox formula.

[h] Mixed tertiary mercaptan.

[i] Viscosity control poor.

[j] Refractive-index method, using curve established for 122° F. polymer.

p-Methoxystyrene (made by Marvel at the University of Illinois and made at the Government Laboratories), dimethylstyrene (made at the Government Laboratories), and dimethoxystyrene (prepared at the University of Illinois) were polymerized satisfactorily with butadiene, even though these substituted styrenes[44, 112, 234] were of a lower purity than that of the styrene. The dimethoxy monomer required less modifier, but its rate of copolymerization was a little less than one-half that of styrene. The stress-strain values obtained with the three copolymers were similar to those of GR-S. The processing characteristics of the copolymers made with the first two monomers seemed to be normal, but those of the dimethoxystyrene copolymer were definitely poor. Their low-temperature properties were slightly better than those of GR-S. The results of tire tests,[348] given in Table XVI, were not outstanding.

The physical properties and processing characteristics of the co-polymer[310, 324] of α-methyl-p-methylstyrene, $CH_3C_6H_4C(CH_3) : CH_2$, furnished by the Dow Chemical Company, were similar in general to those of GR-S. In tire tests the copolymers of butadiene and α-methyl-p-methylstyrene were similar (96 per cent) to butadiene-styrene copolymers. Tire tests[348] on a butadiene-p-phenylstyrene copolymer indicated treadwear 96 per cent as good as that of GR-S.

VINYLPYRIDINE COPOLYMERS

Investigations in the laboratory and on a small pilot-plant scale of 75/25 copolymers of butadiene with 2-, 3-, or 4-vinylpyridine, methyl- or ethyl-substituted vinylpyridines or 2-vinylquinoline showed none of these comonomers to exhibit any superiority over 2-vinylpyridine.[54, 81-2, 100] Since the latter appeared to be the most readily available of these monomers, larger quantities of a copolymer of it with butadiene were prepared in a pilot plant for full-scale testing. Data for such a copolymer and for a tripolymer of 72/18/10 butadiene–styrene–2-vinylpyridine are summarized in Table XVII.

Polymerization rates were high[51] (about twice the rate of GR-S) and varied somewhat more than is common in GR-S production. Little difference was noted between distilled and undistilled monomer; however, some evidence was obtained that polyvinylpyridine formed very rapidly after distillation if the distilled monomer was not charged immediately or was not stabilized against spontaneous polymerization.[82] In some cases, where these precautions were not taken, slightly longer polymerization cycles were encountered along with heavy preflocculation; and the resultant polymer contained hard particles high in vinylpyridine.[82, 134]

Dodecyl mercaptan, which was used in making the polymers shown in Table XVII, was found to react with the vinylpyridine, and this led to poor control of Mooney viscosity.[133-4] The use of tertiary mercaptans apparently overcame this difficulty. The latexes prepared in accordance with a substituted GR-S formula at 40 to 50° C. were generally less stable than GR-S latex and required the inclusion of electrolytes or stabilizing agents.[49, 82, 273]

Table XVII. Copolymers of 2-Vinylpyridine with Butadiene and Butadiene–Styrene[133, 134]

Code	AU-43	AU-42
Polymerization temperature, °F.	113	104*
Monomer	BD/VP†	BD/S/VP†
Monomer ratio	75/25	72/18/10
Reaction time, hr.	11	15
Conversion, %	72	67
DDM, part	0.68	0.78
Raw ML-4	62	50
Gel, %	19	12
Dilute-solution viscosity	1.8	1.8
Compounded ML-4	App. 91	App. 76
Mill-processing index	7.3	9.8
Extrusion index	12	9.5
Cure at 292° F., min.	60	60
Modulus at 300%, p.s.i.	1480	1070
Tensile strength, p.s.i.	2880	3080
Elongation, %	500	630
Shore A hardness	...	61
Goodyear rebound, % (RT/212° F.)	51/57‡	55/61‡
Goodrich hysteresis		
Temp. rise, °F.	88	107
Initial compression, %	19.3	24.3
Set, %	18.9	...
DeMattia flexures $\times$ 10^{-3}	25	22
Quality index	3.3	1.9
Gehman freeze point, minus °C.	52	...
Gehman T_{10}, minus °C.	45	...

  * Ferricyanide activation employed.

  † Polymers made in 80- and 500-gal. reactors according to the substituted GR-S or GR-S 10 formula.

  ‡ Cured for 70 minutes.

Because of the difficulties encountered in the copolymerization, later work centered about butadiene-styrene-vinylpyridine tripolymers. The quality of the tripolymer improved with increase in the vinylpyridine content and reached an optimum at about 15 parts, i.e., at a charge 75/10/15. Even use of 5 parts of vinylpyridine produced a substantial improvement in quality with a minimum of polymerization difficulties and cost.[273]

The main difficulties encountered with vinylpyridine copolymers were an extremely high rate of cure and incompatibility with other rubbers.[54, 82, 273] Substitution of as little as 5 parts of vinylpyridine for some of the styrene in GR-S resulted in polymers with curing characteristics similar to those of the 75/25 butadiene-vinylpyridine copolymers. Except for scorching tendencies, the polymers containing vinylpyridine processed as well as or better than GR-S. The considerable tackiness exhibited by the uncompounded stocks was lost after compounding, and the mixed stock had too little tack for tire-building operations.

In general, the vinylpyridine stocks were superior to GR-S in tensile strength, flex cracking, tear resistance, and quality index (hysteresis–flex-life balance); they were equal to or slightly poorer than GR-S in rebound, hysteresis–heat buildup, and brittle point. While the quality of the polymer improved with decreased conversion and increased Mooney viscosity, the advantage of higher Mooney viscosities was counterbalanced by greater sensitivity to scorching and by rapid curing.[54, 82, 273] A tendency to become hard and brittle during continued exposure to heat was noted in the vinyl-pyridine copolymers and tripolymers. Isoprene-vinylpyridine copolymers were even faster curing and more scorchy than butadiene-vinylpyridine copolymers and showed reversion, i.e., fall in modulus on prolonged curing or aging.[49, 82]

Tires made with butadiene-vinylpyridine polymers have been unsatisfactory because of the poor compatibility of those polymers with GR-S, with natural rubber, and even with isoprene–vinylpyridine. The resistance to treadwear[348] of tires made with vinylpyridine polymers throughout was rated at 112 as compared to 100 for GR-S. However, tire failure was encountered at very low mileage because of body breaks and tread separation. Other work showed that the polymers became excessively hard and brittle during running.[49] No results for isoprene-vinylpyridine polymers in tires have been reported.

KETONES, ESTERS, AND OTHER COMONOMERS

Ketones. As mentioned previously, the copolymer of methyl vinyl ketone ($CH_2 : CHCOCH_3$) with butadiene appeared to be interesting in the survey made by Starkweather and co-authors.[297] A sample (supplied by E. I. du Pont de Nemours Company) distilled to 98.2 per cent purity required an excessive amount of $tert$-C_{12}-mercaptan in copolymerization at 104° F. (Table XVIII).[262] The viscosity control was poor and the average reaction time was under 14 hours. The polymer exhibited rather poor stress-strain properties and was very poor in processing compared to GR-S. The low-temperature properties, as determined by the Gehman method, were similar to those of GR-S. Tire tests[348] indicated the copolymer to be inferior to GR-S in treadwear and similar in cut growth.

In Table XIX the swelling volumes, determined by ASTM methods D 88, D 611, and D 92, in oils 1, 2, and 3 as well as in a 60/40 isooctane-toluene mixture, are compared to those of GR-S 10 and to those of commercial nitrile rubbers containing 18 and 26 per cent of acrylonitrile.

The Hercules Powder Company[95] reported that the copolymer made from butadiene and methyl isopropenyl ketone [$CH_2 : C(CH_3)COCH_3$] was superior to GR-S in stress-strain properties, cut-growth resistance, and resistance to swelling. A 75/25 mixture of butadiene and this ketone (furnished by the Hercules Powder Company) was polymerized in an unactivated GR-S type formula; it reacted in a normal time[234] at 104° F. About 50 per cent more of a tertiary C_{12} mercaptan was required than in the preparation of GR-S. The copolymer, as compared with GR-S, had higher tensile strength (by 400 to 700 p.s.i.), similar moduli, elongation, and set, and a higher rate of cure. The hysteresis temperature-rise was higher, but the resistance to cut growth was much better. This observation was borne out by tire tests. The processibility was similar to that of GR-S. The low temperature properties were poorer. Tire tests[348] indicated the treadwear to be better than that of standard GR-S.

A 75/25 butadiene-methyl isopropenyl ketone mixture reacted[335] rapidly at 41° F. in a low-sugar redox polymerization formula. The polymer thus made did not exhibit better stress-strain properties than polymer made at 104° F., despite its possession of a higher viscosity. (In the copolymerization of butadiene–styrene, improved stress-strain properties would be expected in polymer made at the lower temperature to the higher viscosity.) The processing behavior, as determined by a plastograph (an internal mixer), was found to be poorer than that of a butadiene-styrene copolymer made at 41° F. and considerably poorer than that of standard GR-S 10.

Esters. Diethyl chloromaleate, $C_2H_5OOC \cdot CH : CCl \cdot COOC_2H_5$, supplied by the research division of the Goodyear Tire & Rubber Company, was copolymerized[231] with butadiene in the ratio of 30/70 at 100° F. with 0.29 part of dodecyl mercaptan (considerably less than that used for comparable GR-S) to 50 Mooney viscosity and at 72 per cent conversion in 16 hours (Table XVIII). The polymer was high in modulus and low in tensile strength and elongation compared to GR-S; it was a little harder to mill than GR-S, and its vulcanizate had poorer low-temperature properties than GR-S. Tires[348] made with the copolymer gave treadwear similar to that of polybutadiene prepared at 122° F. Neither GR-S nor this copolymer showed evidence of cut growth under the test conditions. Although the monomer is a skin irritant, both the monomer and the polymer were handled without any ill effects.

Diethyl fumarate, from Charles Pfizer Company, was copolymerized[271] with butadiene at 104° F. About one-half of the amount of dodecyl mercaptan required for GR-S at 122° F. was used. The rate of cure of the copolymer was higher than that of GR-S, and the vulcanizate produced had a higher modulus, lower tensile strength, and lower elongation than GR-S. The hysteresis temperature-rise was greater, and the resistance to cut growth less, resulting in a lower quality index than in the case of GR-S. The mill-processing, extrusion, and low-temperature properties were fairly good. In road tests,[348] treadwear resistance was 85 per cent and cut growth about one-third that of GR-S when tested in half tires—results contrary to those found in laboratory tests.

Other Monomers. Methacrylonitrile, furnished by Rohm & Haas Company, was copolymerized with butadiene at 104° F.[321] This monomer, like acrylonitrile, reacted much more rapidly than styrene in a 75/25 butadiene-comonomer mixture. The modifier (dodecyl mercaptan) usage at 104° F. was in the range of that required for GR-S, and control of viscosity was good. The stress-strain properties, the quality index, and the rebound values of the copolymer were similar to those of GR-S. The low-temperature properties and the hysteresis were poorer than those of GR-S. The milling properties and Garvey die-extrusion rating equaled or excelled those of GR-S.

Table XVIII. Copolymers of Butadiene with Esters, Ketones, and Other Monomers[110, 111, 113, 161, 231, 234, 262, 271, 320, 328, 335, 348]

O.R.R. Test Recipe

Type	Ketones		Esters		
Code	385-A-457	42G-A-560	39G-A-419	37G-A-399	26G-A-358
Monomer	Methyl Vinyl Ketone	Methyl Isopropenyl Ketone	Diethyl Chloromaleate	Diethyl Fumarate	β-Vinylnaphthalene
Purity, %	98.2ᵃ	97.5	99.0ᵇ	98.25	94.5ᶜ
Specific gravity (25°/25° C.)	0.859ᵈ	0.8526	1.177	1.0526ᵈ	...
n_D^{25}	...	1.4200	...	1.4389	...
Polymerization temperature, °F	104	41	100	104	122
Charge ratio (BD/special monomer)	75.7/24.3	75/25	70/30	70/30	75/25
Reaction time, hr.	13.6	6.8	16.2	14.4	14.2
Conversion, %	70.3	62.0	71.8	72.1	73.0
DDM, parts	2.4ᵉ	0.16	0.29	0.26	0.58
Raw ML-4 (212° F.)	49	74	50	46	47
Special monomer, %	...	...	...	22	31.9
Gel, %	0	<5	2	8	3
Dilute-solution viscosity	1.59	1.66	2.30	1.66	2.31
Compounded ML-4 (212° F.)	76	64	90	78	58
Mill shrinkage, %	46	40	35	39	30
Rugosity	143	18	82	49	12
Mill-processing index	18.3ᵍ	7.4	13.6	10.0	4.1
Extrusion index	7.0	...	...	8.5	11.0
Optimum cure, min.	50	39	55	45	55
Modulus at 300%, p.s.i.	1110	1610	2290	2010	870
Tensile strength, p.s.i.	2220	3780	2290	2500	3310
Elongation, %	470	550	315	360	680
Shore A hardness	...	69	...	72	68
Goodyear rebound, % (RT/212° F.)	...	52/70ʰ	...	49/64	44/53
Goodrich hysteresis Temp. rise, °F.	...	68	...	70	110
Initial compression, %	...	16.3	...	17.1	24.7
Set, %	...	10.8	...	9.4	29.1
Flexures × 10⁻³	...	14	...	2	20
Quality index	...	3.7	...	0.5	2.0
Freeze point, minus °C.	48	35	40	53	48
T_{10} value, minus °C.	42	29	33	47	40
Resistance to treadwear (% GR-S)	85	114	80	85	107
Cut growth (experimental/standard)	0.75/0.75	0.25/0	0/0	1.5/5	5/5
Remarks			Skin irritant	Orig. work by Std. Oil Div. gave poor results.	

	25G-A-262 Methacrylonitrile	36G-A-460 Vinylidene Chloride	27G-A-354 2-Vinyl Thiophene	24G-A-349 Dimethyl Vinylethynylcarbinol	24-G-2A318 Vinylethynylcarbinol	7Q-A433 Methylpentadiene	7Q-A433 Methylpentadiene
Code / Monomer							
Purity, %	91.5	99.4	…	…	…	…	…
Specific gravity (25°/25° C.)	0.7990	1.2182[d]	1.036[d]	0.8846	0.8846	0.715	0.715
n_D^{25}	1.3993	1.4230	1.5717[f]	1.4733	1.4733	1.4430	1.4430
Polymerization temperature, °F.	104	122	122	122	122	122[k]	122[k]
Charge ratio (BD)/special monomer	75/25	75/25	75/25	75/25	75/20/5	90/10	80/20
Reaction time, hr.	13.8	15.7	12.8	6.9	9.2	10.5	20.4
Conversion, %	74.8	73.4	73.5	72.2	73.0	72.0	73.2
DDM, parts	0.47	0.77	0.42	0.55	0.50	0.50	0.49
Raw ML-4 (212° F.)	47	56	53	47	53	50	47
Special monomer, %	24.6	17.0	25.4	24.0	…	…	…
Gel, %	0	55	53	2	3	60	26, 32
Dilute-solution viscosity	1.68	0.70	1.14	1.94	2.25	1.38	2.04, 1.78
Compounded ML-4 (212° F.)	67	64	79	56	50	78	64
Mill shrinkage, %	44	40	43	37	42	…	34
Rugosity	22	28	65	10	22	Too rough	91
Mill-processing index	8.2	10.4	16.1	5.2	9.7	…	14.2
Extrusion index	12.0	10.5	8	12.0	…	…	…
Optimum cure, min.	30	…	60	45	60	57	57
Modulus at 300%, p.s.i.	1150	2000	1520	1200	1400	1490	1240
Tensile strength, p.s.i.	3250	…	3420	3600	3500	1920	2260
Elongation, %	600	110	530	600	560	370	470
Shore A hardness	73	71	71	68	67	…	…
Goodyear rebound, % (RT/212° F.)	43/54	…	…	…	53/69	…	…
Goodrich hysteresis — Temp. rise, °F.	117	59	62	60	57	…	…
Initial compression, %	19.3	14.7	17.6	22.8	19.4	…	…
Set, %	…	9.3	10.5	10.4	8.9	…	…
Flexures × 10⁻³	14	…[l]	2	7	5	…	…
Quality index	1.2	…	0.65	2.3	1.9	…	…
Freeze point, minus °C.	30	60	44	56	57	…	…
T_{10} value, minus °C.	24	48	36	48	47	…	…
Tire test (% GR-S)	95	Excess cracking	106	115	…	…	…
Cut growth (experimental/standard)	1.5/6	4/116	3.5/8.5	4/51	…	…	…

[a] As used after distillation; purity of original sample 83 per cent.
[b] Calculated from chlorine analysis.
[c] Calculated from bromination.
[d] Measured at 20°/20° C.
[e] tert-C-12-Mercaptan. Viscosities ranged from 11 to 175 + ML-4 for same modifier charge.
[f] Mixed tertiary mercaptans.
[g] After addition of black on the mill, material would not band.
[h] Determined on sample overcured for 30 minutes.
[i] Measured at 20° C.
[j] Styrene.
[k] Activated GR-S formula.
[l] Failed at start of test.

The oil resistance of this copolymer approached that of commercial nitrile rubber containing 18 per cent of acrylonitrile. The relative resistances[2] to the ASTM oils and to a 60/40 isooctane-toluene mixture, as well as corresponding data for the controls, appear in Table XIX.

β-Vinylnaphthalene, supplied by Koppers Company, was copolymerized in a 70/30 charge ratio with butadiene at 122° F. About 17 per cent more time and about 20 per cent more dodecyl mercaptan than necessary for the usual GR-S formula were required (Table XVIII). The resultant polymer[113] contained about 32 per cent of the comonomer at 73 per cent conversion. The tensile strength and elongation of the compounded polymer and the quality index were in the upper part of the range found for samples of GR-S; the modulus was low; the processing properties were good. Tire tests[348] indicated the polymer to be a little better than GR-S in treadwear and equal to it in cut growth.

Vinylidine chloride, $CH_2 : CCl_2$, obtained from the Dow Chemical Company, yielded with butadiene a copolymer[111] having very poor physical properties and quality index, and tire tests[348] confirmed the poor results obtained in the laboratory tests.

2-Vinylthiophene, supplied by the Texas Company, copolymerized[161] with butadiene in a manner similar to styrene, except that about 12 per cent less dodecyl mercaptan was required. This monomer reacted faster than styrene. The copolymer gave good stress-strain results, but processibility, cut growth, and quality index were rather poor. In tires treadwear[348] was slightly better than that of GR-S, but resistance to cracking was only about 40 per cent as good.

Dimethyl vinylethynylcarbinol, $CH_2 : CH . C : C . C(OH)(CH_3)_2$, supplied by the E. I. du Pont de Nemours & Company, reacted with butadiene in about 60 per cent of the time required for the GR-S mixture of monomers and produced a copolymer[320] that exhibited slightly better stress-strain properties than those of GR-S and good processing properties. However, the satisfactory cut growth and quality index indicated in the laboratory were not duplicated by tire tests.[348] The treadwear was better than that of GR-S.

A 75/20/5 butadiene-styrene-dimethyl vinylethynylcarbinol tripolymer[110] was made in a shorter reaction time than that required for standard GR-S. The results of physical tests conducted on the tripolymer in the laboratory were slightly better than those for GR-S in regard to most of the properties examined. No tire tests were made.

The dimethyl vinylethynylcarbinol polymers were tested for oil resistance. The results[2] (Table XIX) indicate the 75/25 copolymer to be slightly better in oil resistance than GR-S 10 but not comparable to commercial nitrile polymers. The tripolymer was slightly poorer than GR-S 10.

McMillan[241] of the Shell Development Company prepared at 90 to 110° C. polymers from methylpentadiene and copolymers from high proportions of this monomer in admixture with butadiene. The products exhibited very flat curing characteristics, retention of properties after aging much superior to that of GR-S, and processing properties and tackiness that were of interest. The low-temperature properties were poor.

Mixtures of butadiene and methylpentadiene in the ratios 90/10 and 80/20 were polymerized at the Government Laboratories in an activated GR-S recipe.[328]   Even with activation, the 80/20 ratio required over 20 hours at 122° F. to attain 72 per cent conversion.   The Shore-hardness decadence freeze method indicated the freeze points, i.e., the temperatures at which a definite decrease between the 5- and 30-second Shore hardness values sets

Table XIX.   Oil and Solvent Resistance of Special Copolymers[2]

Polymer	ASTM Oil or Solvent	Change in Volume, % 1 Day	2 Days	7 Days	30 Days
	Controls				
GR-S 10	No. 1	3	3	9	19
	2	8	13	21	50
	3	17	55	57	137
	Iso./tol.*	173	178	180	180
Commercial polymer	No. 1	0	1	0	0
with 18% acrylonitrile	2	3	5	3	10
	3	7	13	19	31
	Iso./tol.	87	89	90	93
Commercial polymer	No. 1	0	0	0	0
with 26% acrylonitrile	2	2	2	0	3
	3	2	3	3	9
	Iso./tol.	46	51	52	52
Experimental Polymers					
BD/methylvinyl ketone	No. 1	5	7	10	17
	2	9	13	21	35
	3	25	37	57	78
	Iso./tol.	166	166	167	177
BD/methacrylonitrile	No. 1	5	5	5	5
	2	4	5	8	11
	3	6	6	15	29
	Iso./tol.	92	99	100	102
75/25 butadiene–dimethyl-	No. 1	5	5	5	10
vinylethynylcarbinol	2	7	8	15	26
	3	17	25	36	62
	Iso./tol.	127	128	136	139
75/20/5 butadiene–styrene–	No. 1	7	9	16	24
dimethyl vinylethynyl-	2	13	20	33	60
carbinol	3	42	58	108	137
	Iso./tol.	190	193	194	196

* 60/40 Isooctane–toluene.

in, to be —67 and —62° F., respectively, for the copolymers made at the two ratios. By the same test the freeze point of polybutadiene prepared at 122° F. is —90° F. and that of GR-S —45° F. The stress-strain characteristics of the copolymers were somewhat better than those of polybutadiene made at 122° F., the higher pentadiene ratio giving better results than the lower. At 72 per cent conversion the polymers processed poorly. As the viscosity was increased from 30 to 50 Mooney, the gel content increased from zero to 60 per cent or more. No information is available in respect to polymers at lower conversion levels or made in redox formulations at lower polymerization temperatures.

ISOPRENE POLYMERS AND COPOLYMERS

Since isoprene was readily available and early work had indicated isoprene-styrene polymers to show improvement over the butadiene polymers in heat-aging, flex-resistance, and hysteresis properties,[84, 114, 256, 342] considerable quantities of polyisoprene, isoprene-butadiene, and isoprene-styrene polymers were prepared on pilot-plant and plant scales.[19, 30–1, 40, 240, 245, 274] Data for several of the pilot-plant polymers are summarized in Table XX.

The isoprene used in the polymerizations had generally been washed with caustic soda to remove inhibitor or distilled to remove inhibitor and dimer. Usually reaction rates were low, because of the presence of dimer, the relatively low purity of the monomer used, or both. Higher temperatures of polymerization and activation were often necessary. As noted earlier in regard to other alkyl-substituted butadienes, considerably less mercaptan was required to modify the isoprene polymers than is necessary with butadiene. The polymerization rate varied with the pH of the emulsion, and boosters of caustic soda were often required to obtain the desired conversion.[40, 149, 152] Modified techniques for the coagulation of the latex and special precautions during the drying of the isoprene polymers were applied in order to overcome the excessive tackiness and tendency of the isoprene polymers to heat-plasticize at the temperatures usually employed for drying GR-S-type polymers.[19, 30, 32, 40, 240, 274, 330]

Polyisoprene and isoprene-butadiene copolymers cured faster than GR-S or polybutadiene and were essentially similar to polybutadiene in processing and in stress-strain and flex properties, slightly superior in rebound and possibly in hysteresis–heat rise, and inferior in Gehman low-temperature properties. Isoprene-styrene copolymers of 75/25 ratio also cured faster than GR-S and were slightly poorer in processing, stress-strain properties, flex life, and Gehman low-temperature properties and slightly better in rebound, hysteresis–heat rise, and aging behavior.[67–69, 256] When tested in passenger-car tires[348] (either complete or on half treads), the polyisoprene polymers showed an average treadwear rating of 72, as compared to 99 for the 75/25 isoprene-styrene copolymer, 100 for GR-S, and 81 for polybutadiene.

Polymers from several combinations of isoprene with substitutes for styrene have been made on a pilot-plant[82, 100] scale. With the possible exception of isoprene–vinylpyridine,[49] none showed appreciable improvement over corresponding copolymers of butadiene and the special monomers.

Table XX. Isoprene Polymers and Copolymers[60, 150, 151, 156, 158, 233, 235, 281]

O.R.R. Test Recipe

Monomer	Isoprene[a]	I/BD[a]	I/BD[b,c]	I/BD[b,c]	BD[a]	I/DCS[a]	I/S[b]	I/S[d,e]	BD/S (GR-S 10)[f]
Polymerization temperature, °F.	131	122	122	122	122	122	122	122	122ᵍ
Monomer ratio	100	50/50	40/60	20/80	100	90/10	85/15	75/25	75/25ᵍ
Reaction time, hr.	16	19	18	16	19	30	28	9	14ᵍ
Conversion, %	58	58	72	72	58	69	73	74	72ᵍ
DDM, part	0.52	0.75	0.64	0.81	1.0	0.32	0.2ʰ	0.2	0.50ᵍ
Raw ML-4	48	58	52	53	43	64	49	65	57
Gel, %	0	1	1	0	0	26	0	…	0
DSV	2.2	2.4	2.4	2.3	2.1	1.8	2.3	…	2.0ᵍ
Compounded ML-4	59	68	79	90	65	72	62	87ʰ	63
Mill-processing index	8.5	13.8	13.8	17.6	12.6	15.4	13.0	10.4	6.3–8.6
Extrusion index		7.0	7.5	7.5	8.0	6.5	…	9.5	13.5
Cure at 292° F, min.	60	90	60	60	90	45	45	60	60
T/300, p.s.i.	1380	1160	1410	1400	1160	1270	1750	1520	1180
Tensile strength, p.s.i.	1890	2150	1930	1750	1850	2480	2280	2930	3370
Elongation, %	380	450	360	350	410	520	370	500	600
Shore A hardness	63	61	65	63	61	64	…	65ⁱ	62
Goodyear rebound, % (RT/212° F.)	53/67	53/66	54/70ⁱ	52/64ⁱ	49/62	47/65ʲ	…	48/70ⁱ	54/66
Goodrich hysteresis Temp. rise, °F.	83	91	71	84	88	61	…	47ⁱ	71
Initial compression, %	22.7	23.5	19.4	22.2	24.1	23.0	…	17.2ⁱ	24.3
Set, %	21.3	25.0	22.2	20.0	19.3	19.5	…	9.0ⁱ	19.2
DeMattia flexures × 10⁻³	4	2	2	1	4	4	…	3ⁱ	5
Quality index	0.6	0.2	0.3	0.2	0.5	1.3	…	1.6	1.2
Gehman freeze point, minus °C.	57	62	71	70	72	43	45	35	48
Gehman T_{10}, minus °C.	46	51	59	59	60	35	39	30	38

[a] Prepared in 500-gal. reactors.
[b] Prepared in 5-gal. reactors.
[c] Fatty acid soap in place of rosin soap for emulsification.
[d] Prepared in 80-gal. reactors.
[e] Ferricyanide-activated formula.
[f] Prepared in plant (3750-gal.) reactors.
[g] Approximate values.
[h] tert-C_{12}-mercaptan used in place of n-dodecy mercaptan.
[i] Cured for 70 to 80 minutes at 292° F.
[j] Cured for 60 minutes at 292° F.

BULK AND SOLUTION POLYMERIZATION

ALKALI METAL-CATALYZED POLYMERIZATION

The observation in 1910 by Matthews[230] and by Harries[102] of the fact that sodium will catalyze the polymerization of isoprene and its application in early studies of the polymerization of isoprene, butadiene, and dimethyl-butadiene have been adequately described in detail elsewhere.[309, 343, 346] Briefly, well-purified isoprene and clean sodium were needed to produce a transparent, water-white rubber in a period of several weeks. Agitation increased the polymerization rate. A comparison of the effectiveness of various alkali metals presented by Midgley[247] showed that, sodium required several weeks to effect the polymerization of isoprene, potassium and rubidium a few days, and cesium or the eutectic sodium-potassium alloy (both liquids) only a few hours. Midgley found lithium to produce no polymerization, but Ziegler[347] has indicated contrary findings. Calcium is very slow in effecting the polymerization.[253]

According to Midgley, the presence of oxygen has an unfavorable influence on the quality of polymer formed by alkali metal catalysis, but its removal from the system reduces the rate of polymerization. Nitrogenous compounds, particularly aniline, in infinitesimal amounts, almost completely inhibited reaction at room temperature.

At an early stage in the polymerization of isoprene under the influence of sodium it is possible to isolate, by the addition of alcohol, three isomeric octadienes in which two isoprene units are combined in the 1,1-, 1,4-, and 4,4-sense.[247] In polybutadiene prepared by means of sodium some of the diene units enter into the polymer chains in the 1,4- and some in the 1,2-sense, and (Table XXXIII) the proportion of the latter increases with decrease in the temperature of polymerization, from about 15 per cent at 110° C. to approximately 100 per cent at −70° C.[309, 347] A similar change with temperature in the ratio of 1,2- to 1,4-addition has not been reported for polyisoprene or for butadiene-styrene copolymers. In polybutadiene the ratio was not affected by the use of diethyl ether, benzene, or hexahydrotoluene as solvents, by variation in the rate of addition of butadiene, or by the use of lithium or potassium as the polymerization catalyst in place of sodium.[309]

The use of phenylisopropyl potassium, triphenylmethyl sodium, lithium butyl, amyl sodium, and like compounds as catalysts for the polymerization of dienes has been reported,[309] but no superiority on their part over the alkali metals has been indicated. Acceleration of the organosodium-catalyzed polymerization of butadiene has been accomplished, without appreciable change in the mode of addition. As opposed to the stepwise process involved in the use of metallic sodium or amyl sodium, phenyl sodium has been shown to exercise a catalytic effect in the polymerization. Amyl potassium is reported to involve a stepwise polymerization at 0° C. but a catalytic process at −30° C. The presence of sodium chloride (arising from the preparation of the organosodium catalyst) had an accelerating effect on the reaction, whereas the presence of sodium bromide did not.

German Practice. The practical application of alkali metal catalysis in Germany to the manufacture of polybutadiene is described in Chapter 26.

A few references to the subject may be noted here.[4, 8, 9, 11a–p, 12–3, 45–6, 83, 94, 99, 136, 140, 167–8, 239, 254, 276–7, 333–4, 338, 343, 345]

Russian Practice. In Russia, at all events up to the end of World War II, practically the only kind of synthetic rubber manufactured was sodium-catalyzed polybutadiene.

Considerable information,[105a–z, 309] based on Russian literature before 1944–45, concerning monomer preparation and polymerization techniques has been collected and published.   In 1926–28 the "rod" process for the sodium-catalyzed mass polymerization of butadiene was selected as the basis for synthetic-rubber production in the U.S.S.R.   The reactors are described as being about 1.35 meters (about 4 ft.) in internal diameter and height and to have a divided upper and lower jacket for steam, water, or brine.   An iron-plate liner was used to facilitate unloading of the reactor.   About 650 iron rods, 5 to 6 mm. in diameter, mounted in radially arranged "combs," were plated with 0.5 to 0.8 mm. of sodium for about 100 cm. of their length by dipping into a bath of the molten metal at 120° C.   The coated surface corresponded to an area of about 45 sq. cm. per liter of butadiene charged and a ratio of sodium to butadiene of about 0.5 per cent.   The sodium-plated rods were dipped into paraffin at 70° C. to prevent reaction of the sodium with moisture before the reactors were charged.

After insertion of the "combs" into the reactor, the butadiene was charged, and the temperature was increased to about 30° C. for a maximum of 6 hours in order to shorten the induction period.   As the reaction started, the temperature and pressure increased (about 65° C. and 8 atm. were regarded as optimum).   Early polymerizations required 10 to 12 days and yielded polymers with Karrer plasticities[55] of 0.6 or higher.   Later, reaction times of 4 to 5 days and a plasticity of 0.4 became a common standard for the "rod" polymerization.   Difficulties were encountered with control of temperature and with the formation of nonuniform insoluble polymers as the result of overheating.[309]   The polymerization was considered complete when the pressure of the head gas became constant; thereafter the reactor was heated to about 45° C., and the excess butadiene and other volatile impurities were distilled off.

After the liner had been removed from the reactor, the polymer was dissected into segments and charged into a vacuum kneader to remove the volatile matter, to blend the polymer, and for the purpose of adding about 0.5 per cent of an antioxidant.   During subsequent mill refining, over-polymerized strands (tendons) were either removed or blended.

"Rodless" polymerization, developed later, initially involved the use of dispersions of 1 part of molten sodium in 2 parts of kerosene made by means of colloid mills at 125 to 130° C.   The dispersion under pressure was charged intermittently into the reactor.   Settling of the sodium particles and insufficient surface area resulted in unpredictable reactions and products of poor quality.   Soaps and other materials were then used as protective coatings, and the catalyst dispersions were spread over dried wool or cotton gauze.   Apparently the textile carrier, amounting to 0.1 per cent of the amount of polymer, could be satisfactorily milled into the polymer.

The "rodless" polymerization technique, as it apparently has been used since about 1935, involved the use of a dispersion of sodium thinly spread over removable trays placed in a chamber of several tons capacity. Butadiene vapor was continuously recirculated over the trays and through a cooling unit. Reaction time was a matter of hours instead of days, and polymers with plasticities as low as 0.16 to 0.25 (Karrer) were obtained. Temperature control was more efficient, and presumably the sodium requirement was smaller and the quality of the product more uniform than in earlier methods of operation. Production of polybutadiene in Russia was reported to have reached 90,000 tons[331] per year by 1939.

Information available concerning products made from either the "rod" or the "rodless" SK or Sovpol (sodium polybutadiene) polymers indicate maximum tensile strengths of 2100 to 2400 p.s.i. and elongations of 200 to 800 per cent in stocks loaded with 60 parts per cent of their weight of gas black.[105a,d,e,q, 286] Natural rubber in similar stocks exhibited tensile strengths of 5000 to 6000 p.s.i. Satisfactory oil resistance and improved cold resistance were gained by blending natural rubber with either SK or Sovprene (poly-2-chlorobutadiene). In gum stocks the best blends were brittle at −60° C. The SK polymers apparently stiffen badly on heating, whereas natural rubber softens under similar treatment. In contrast to hysteresis data for other sodium-polymerized polymers, SK was reported to show a heat rise about 36° F. higher than that for natural rubber when flexed in a DeMattia machine and also to run hotter than natural rubber in tires.[105z,p]

Tests on Russian Rubber. Tests carried out in the United States[286] on part of a shipment of several thousand pounds of SK polymer from Russia revealed a refractive index of 1.5090, gel of 0 per cent, and dilute solution viscosity of 1.5. When compounded according to a tread test recipe similar to that used in testing GR-S, except for the addition of stearic acid, the polymer, of 22 to 32 Mooney viscosity, was soft and sticky on the mill and was difficult to remove even in the presence of carbon black. Tensile-strength values up to about 2000 p.s.i., 200 per cent moduli of 1000 to 1500 p.s.i., and elongations of 350 to 500 per cent were obtained. Goodrich hysteresis temperature-rise values of about 50° F and DeMattia flex tests indicated the polymers to be essentially equivalent to GR-S in these properties. But, when the SK polymer was more tightly cured, hysteresis temperature-rise values better than those for GR-S were obtained. The polymers were described as "liquid-phase refined sodium polybutadiene, types SOR, SOSR, SSR, and SSSR" of 0.47 to 0.55 (Karrer) plasticity. Extraction of the polymer with an ethanol-toluene azeotropic mixture (ETA) gave strongly alkaline solutions. Mooney viscosities ranged from 22 to 32 ML-4. Because of the tacky, soft nature of the polymers, storage for even 6 months led to severe difficulties with respect to adhesion of bales and cold flow. Noteworthy is the fact that considerable variation, both chemical and visual, existed within every bale investigated.

Blends of "rodless" SK and low-molecular-weight polydienes are apparently widely used in the Russian tire industry, the best results being obtained with viscous diene oils of approximately 30- to 40-centipoise viscosity. The dispersion of carbon black improved with increased plasticity, i.e., with softer stocks. Automobile tires made with SK were reported as satisfactory for 22,000 miles on paved roads and 10,500 miles on cobblestone roads.[105p] Apparently, 100 per cent synthetic stocks are unsatisfactory for truck tires; for them blends with natural rubber must be used.[14, 105l]

Recent Studies. In the United States in recent years some study has been devoted to the sodium polymerization of not only butadiene but also of 75/25 butadiene-styrene mixtures.[88, 96, 211, 291-2] Succeeding sections of this chapter are largely concerned with this work. The vapor-phase polymerization of butadiene by means of sodium has been the subject of a single recent study.[56]

RECENT STUDIES OF SODIUM POLYMERIZATION AND COPOLYMERIZATION

Purity of Monomers. In the polymerization and copolymerization of butadiene by means of sodium the purity of the butadiene appears to be highly important. Butadiene of research grade (99.7 per cent minimum purity) and that made by the alcohol process gave satisfactory reaction rates; butadiene from the petroleum processes apparently gave variable results, depending on its source.[211, 292] Small quantities of butanes and butadiene dimer had little effect on the reaction. Commercial styrene (99.5 per cent or somewhat lower in purity), whether it was or was not washed with caustic in order to remove the small amount of inhibitor present, appeared to be satisfactory.[211]

Conduct of Polymerization in the Laboratory. Sodium sand was prepared by high-speed agitation or vigorous shaking of clean molten sodium (1 part) with dry toluene or xylene (4 parts) at 110 to 140° C., followed by cooling under agitation. The resultant sand particles had diameters ranging from 10 to 50 microns.[211, 291] Glass bottles or steel bombs were charged with varying amounts of catalyst, diluent, styrene, and a slight excess of butadiene, in that order. The excess butadiene was allowed to vent off and thus remove the air before the vessel was sealed. The charge was then placed in a water bath to be rotated or rocked in order to maintain temperature control.

Description of the Progress of the Polymerization. The polymerization has been described by Marvel[211] as follows:

There are two distinct phases of polymerization. During the induction period no increase in viscosity of the monomer is observed; this period varies in length, but is ordinarily longer than the one which follows. The period of actual polymerization first makes itself known by the cohesion between catalyst particles. The particles become surrounded by a clear, solid gel, so that the mass bears a strong resemblance to frog eggs. Meanwhile the liquid phase becomes more viscous and when reaction is complete the contents are entirely converted to a transparent solid. These two periods for butadiene alone at 30° C. are in the neighborhood of ten and four hours, and those for a 75/25 mixture of butadiene and styrene at the same temperature, are roughly six and four hours, respectively. Since the conversion is quantitative, it is evident that the rate even at this low temperature is extremely rapid.

This high rate causes considerable generation of heat in a short time and, if this is not dissipated, it may be detrimental to the polymer and may also produce uncontrollable pressures. Polybutadienes prepared without agitation or in large batches

(150 grams in 500-ml. bottles), so that heat transfer was reduced, were found to be partially insoluble; moreover, the product from the large-scale preparation could be separated into a highly insoluble, darkened core and a partially soluble, outer shell.

A typical reaction time versus conversion relationship for sodium-catalyzed polymerizations of undiluted butadiene[309] is shown in Fig. 7.

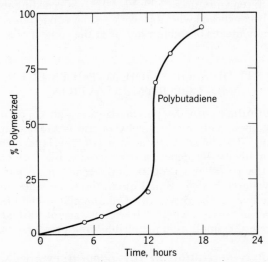

FIG. 7.   Sodium-Catalyzed Polymerization at 35° C. as a Function of Time

Effect of Temperature and Catalyst Concentration. With sodium sand as the catalyst, an increase in the concentration of the catalyst or in its surface area, i.e., in its fineness of subdivision, produced only a slight increase in the rate of polymerization but, as shown in Table XXI, a large decrease in the molecular weight of the polymer.

Table XXI.   Sodium-Catalyzed Polymerization of 75/25 Butadiene–Styrene at 40° C.[292]

Sodium, Parts	Inherent Viscosity	Polymer Appearance
0.10	6.13	Solid
0.20	4.27	Solid
0.30	4.26	Solid
0.40	3.92	Solid
0.60	3.33	Solid
1.0	2.25	Solid
2.0	1.32	Soft, sticky
5.0	...	Viscous oil

The use of sodium wire in place of sodium sand leads to localized concentrations of catalyst, which become the centers of masses of nonuniform polymer.[56, 211, 292]

Increase in the polymerization temperature from 10 to 50° C. was found to result in decreased reaction times and softer polymers.[56, 211]   Later work

in a continuous unit on a much larger scale at reaction temperatures up to
215° F. has led to polymers of substantially the same viscosity as those for the
production of which in earlier work the use of lower temperatures was
necessary.[269]

Influence of Diluents. Toluene, xylene, methylcyclohexane, isobutane,
n-pentane, n-hexane, and n-heptane as diluents facilitated temperature
control and generally resulted in decreased reaction rates and softer
products (Table XXII). When diluent sufficient for a fluid dispersion of

*Table XXII. Effect of Diluents on Sodium-Catalyzed Polymerization
at 30° C.*[88, 96, 211]

0.3 Part Sodium per 100 Parts Monomers

Compound	Approx. Ratio, Diluent to Monomer	Polymerization Time, Hr.	Intrinsic Viscosity*	Raw Mooney Viscosity, ML-4
Polybutadiene				
None	...	10	5.1†	...
n-Hexane	1 : 1	App. 14	4–5	...
n-Heptane	1.5 : 1	16	2.1–5.1	...
n-Heptane	4 : 1	144	0.68	...
Xylene	0.25 : 1	6	2.6	54
Toluene	0.25 : 1	8	2.4	51
Methylcyclohexane	0.50 : 1	...	3.4	91
Methylcyclohexane	1 : 1	App. 14	4–5	88
90/10 Butadiene–Styrene				
Xylene	1 : 1	6–8	1.1–1.4	...
Methylcyclohexane	1 : 1	10–14	4.2	75
75/25 Butadiene–Styrene				
None	...	10	4.5‡	...
Isobutane	1 : 1	24	3.4	98
n-Pentane	1 : 1	6–9	4.1	92
n-Pentane	2 : 1	30	0.62	...
n-Pentane	4 : 1	54	0.28	...
n-Heptane	2 : 1	34	0.73	...
n-Heptane	4 : 1	74	0.30	...
Xylene	0.05 : 1	9	1.9	38
Xylene	0.25 : 1	4	1.6	...
Xylene	0.50 : 1	9	1.7	Very soft
Xylene	1 : 1	12	Fluid	...
Toluene	1 : 1	15	2.1	34
Toluene	1.5 : 1	32	1.7	...
Toluene	2 : 1	20–30	Fluid	...
Methylcyclohexane	1 : 1	6–15	3.7	69
Methylcyclohexane	2 : 1	12	2.9	...

* Polymers were essentially gel-free.
† 5 to 15 per cent of gel.
‡ 10 to 20 per cent of gel.

the final product was used, the polymerization time was multiplied several-fold and the resultant polymer was soft and tacky. Nitrogen, hydrogen, butene, ethyl ether, and lower aliphatic ethers have been reported to produce similar effects.[309]

Influence of Other Components. Other materials, such as aldehydes, ketones, alcohols, acetylenes, and possibly certain aromatic hydrocarbons react with the sodium catalyst faster than does butadiene, delaying the beginning or sometimes completely inhibiting the reaction. Ketones, anthracene, phenanthracene, and some of the acetylenes also acted as modifiers, yielding liquid products. Compounds that do not react with metallic sodium but do react with sodium hydrocarbons are mainly primary amines and halogen and cyano derivatives.[309] These materials would be expected to slow down the polymerization without causing an induction period and to give softer polymers. Olefins such as isobutylene, 2-butene, 2-pentene, stilbene, styrene, and asymmetrical diphenylethylene, by entering the polymer chain to a certain extent,[211, 309] modified the polymer and thus lowered the viscosity. The effects of some of these substances is indicated in Table XXIII.

The addition of steel nails, stainless-steel screens, and copper wire that had been washed with benzene, immersed in dilute hydrochloric acid, and rewashed with water resulted in no change in rate; copper that had not been treated with acid reduced the rate of reaction. A reduction in the viscosity of the polymeric product was noted in some cases but seemed to be due to the increased agitation caused by the loose particles in the bottles.[211]

Effect of Oxides of Carbon. The action of carbon dioxide[309] and carbon monoxide[275] on the sodium polymerization of butadiene is somewhat different from that of other impurities or additives. Concentrations of 0.04 per cent of carbon dioxide caused no noticeable delay, but 0.09 to 0.01 per cent of carbon dioxide or 0.01 to 0.06 per cent of carbon monoxide caused delays of up to 1 month. In both cases there are formed, on the surface of the sodium, reaction products of the type $(C_4H_6)_n$—$(COONa)_2$[309] or —$(CONa)_2$[275] and a spongy sodium-carbon (di)oxide "autopolymer" which disintegrate to produce a sodium powder. If the concentration of carbon dioxide is not too high, the normal reaction will start after a prolonged induction period and proceed rapidly because of the large surface of the sodium powder. The "autopolymer" formed is reminiscent of the "popcorn" polymer encountered in butadiene and GR-S plants.[54, 97, 132] It is described as being insoluble, with little or no swelling, easily powdered, not vulcaniz-able, and oxidized rapidly in air to a yellow, brittle material. Even after being washed to remove the sodium, the autopolymer can be used to induce further polymerization to a similar type of material.

Inhibition by carbon monoxide can reportedly[275] be successfully counter-acted by the use of either water or potassium hydroxide, the latter of which, it is thought, fixes the carbon monoxide as potassium formate. Presumably the same materials could be used to counteract the inhibiting effect of carbon dioxide. Normal polymer is obtained after the removal of the oxide. The over-all effect of a given quantity of carbon dioxide apparently depends on the equilibrium conditions existing at the time it is introduced. Thus the addition of 3 per cent of the gas at the start of a butadiene polymerization at

Table XXIII. *Effect of Various Compounds on the Polymerization of Butadiene by Sodium at 30° C.*[88, 96, 211, 309]

0.2 to 0.3 Part Sodium

Impurity	Concentration of Impurity %	Polymerization Time, Hr.	Solubility, %	Intrinsic Viscosity
None	...	13	96	5.1
1-Butene	15	18	100	3.0–4.0
2-Butene	15	14	100	2.9–3.7
Acetylene	0.01–1.6	Delayed up to 1 month	...	...
Vinylacetylene	1	>500; trace of conversion		
Dimethylacetylene	0.4–1.6*	Slight delay	Liquid polymer*	
Phenylacetylene	0.4–1.6	Slight delay	...	...
Phenanthrene	0.4–1.6*	Slight delay	Liquid polymer*	
Anthracene	20	240	74	0.60
Acetaldehyde	0.2–0.3	>400; 12% conv.	98	2.3–5.6
Furfural	0.12–0.58	Delayed 1.5–2.5 times	...	...
Methyl ethyl ketone	0.08–0.33*	...	Liquid polymer*	
Acetophenone	0.50	Delayed 10 times	...	...
Menthane	0.09–0.36*	Delayed 2 times	Liquid polymer*	
Methanol	0.10–0.25	Accelerated	High	High
Ethanol	0.24–0.40	Delayed 1.5–4 times	...	...
sec-Octanol	0.1	Delayed 2 times	...	...
Pyrrole	...	Strong delay	...	...
unsym.-Dimethyl allene	0.4–1.6	No delay	Liquid polymer	
Aniline or xylidine	5	Slow, stopped in some cases		
Piperidine	0.5–1.0	Delayed 2–4 times	Viscous liquid	
Water	0.09–0.15	18	98	5.3–6.1
Water	0.50–1.0	Delayed 2.5–5 times	Soft to liquid	
Dioxane†	0.7–1.0	No delay†	97	2.6
Koroseal‡		Delayed up to 144 hours	...	Soft

* Highest concentration gave liquid polymers.

† Dioxane used in 75/25 BD/S polymerizations with 100 parts of hexane to give polymer of 45 ML-4 viscosity.

‡ Gaskets used for bottles.

23° C. resulted in zero conversion in 51 hours, whereas the addition of a similar amount of carbon dioxide after 8 hours of normal reaction showed little or no effect on rate.[131]

Styrene Content of Butadiene-Styrene Copolymers. In contrast to emulsion polymerization where the styrene content of a 75/25 butadiene-styrene polymer at low conversions is also low (17 per cent)[242] and increases with conversion, the styrene content of similar sodium-polymerized polymer has been shown[211, 291] to start at a high value (about 32 per cent) and decrease with conversion; the final polymers contained about 23 per cent of combined styrene in both cases.

Aging. In the presence of uniformly dispersed antioxidant, sodium-catalyzed butadiene-styrene copolymer (30° C.) appeared to be superior to GR-S with respect to age resistance in air at 70° C. The addition of 5 per cent of fatty acid was accompanied by poorer aging properties.[211]

Molecular Weight of Sodium Polymers. Number-average molecular weights, determined by means of osmotic pressures, are shown in Table XXIV. The molecular-weight distribution of sodium-catalyzed butadiene-styrene copolymer prepared at 50° C., as reflected by heterogeneity indexes,[211] is similar to that of natural rubber; that of the corresponding copolymer made at 30° C. is nearly like that of GR-S.

Table XXIV. *Number-Average Molecular Weights and Heterogeneity Indexes*

Polymer	Poly. Temp., °C.	Intrinsic Viscosity	Number-Average Molecular Weight	Heterogeneity Index
Natural rubber	...	6.18*	60,800*	1.5–1.9
Na polyBD	30	...	...	2.5–2.6
Na BD/S (75/25)	30	6.34	76,900	1.1–2.0
Na BD/S (75/25)	50	2.07	48,500	4.4
GR-S	50	2.44	42,200	4.4–12.4

* Determined for the benzene-soluble portion of smoked sheet (39.5 per cent).

Processing. The sodium-catalyzed butadiene-styrene polymers processed on the mill more readily than GR-S and exhibited considerably more tack. Results indicated that these compounds could be satisfactorily calendered and extruded and that they could be fabricated with less filler than is required for GR-S.[120] Sodium-catalyzed polybutadiene was tough on the mill.

Vulcanization. When compounded according to the specification test recipe of the Office of Rubber Reserve for GR-S (except for addition of 3 parts of stearic acid per 100 parts of polymer), sodium-catalyzed 75/25 butadiene-styrene polymer showed maximum tensile strengths of 3600 to 3900 p.s.i. at a 30-minute cure and a severe overcure characterized by very high moduli and a sharp drop in tensile strengths at longer cures—of 45 to 150 minutes. Reduction to 1.5 parts of the stearic acid added or its complete omission resulted in slightly higher tensile strengths at the same curing time with little or no reduction in the rate of cure or overcure (data not shown). Reduction in the accelerator level (HG9, 14, and 15 in Table XXV) lowered the curing rate and broadened the peak in the tensile strength–time curve. Reduction of the amount of sulfur from 2 to 1 part (HG9, 10, 11, and 12 in Table XXV) resulted in progressively improved moduli and also broadened the plateau of the tensile strength–time curve.

Table XXV. Variation of Sulfur and Accelerator in the Vulcanization of Sodium 75/25 Butadiene–Styrene * 96*

Polymerized at 30° C.

Code	Min. Cured at 292° F.	HG9	HG10	HG11	HG12	HG14	HG15	HG16	GR-S
Sulfur, parts†		2	1.5	1.25	1.0	2.0	2.0	1.5	2.0
Benzothiazolyl disulfide, parts		1.75	1.75	1.75	1.75	1.25	0.75	1.25	1.75
Modulus at 300%, p.s.i.	15	440	390	330	270	380	210	350	...
	30	1020	930	830	720	800	460	680	860
	45	1450	1210	1100	990	1140	1020	990	1020
	60	1750	1490	1320	1180	1510	1500	1230	1100
	90	2170	1840	1580	1450	1950	2000	1640	1170
	150	...	2270	1920	1710	2530		2110	...
Tensile strength, p.s.i.	15	3320	3130	2620	2400	2780	1550	2590	...
	30	3720	3410	3420	3430	3900	3160	3530	4070
	45	3090	3510	3020	3520	3930	3760	3960	3930
	60	2870	3100	3560	3400	3260	3950	3400	3620
	90	3230	2830	2640	2900	3400	3220	2750	3600
	150	2430	2390	2580	2460	2680	2620	2600	...
Elongation, %	15	810	880	900	990	880	1110	930	...
	30	610	620	640	680	690	790	720	710
	45	460	550	520	610	590	690	640	630
	60	400	460	530	490	470	620	530	590
	90	370	400	400	450	420	420	440	570
	150	280	310	350	340	320	360	350	...
Set, %	15	19	22	24	31	27	52	29	...
	30	14	12	14	14	15	21	14	24
	45	10	12	10	11	13	16	14	19
	60	8	10	11	8	10	14	12	15
	90	8	7	6	8	9	9	7	13
	150	5	6	6	5	6	7	6	...
Cure at 292° F., min.		30	60	60	60	60	60	60	45
Shore A hardness		60	63	62	63	62	62	61	62
Goodrich hysteresis‡ Heat rise, F.		32	25	27	29	24	35	28	65
Initial compression, %		25.3	20.6	21.7	25.1	23.5	27.5	23.6	22.5
Set, %		12.8	5.1	5.9	6.5	8.0	12.6	7.5	19.9

* All polymers tested by microtechnique at room temperature.
† Per 100 parts of polymer.
‡ Goodrich hysteresis test run at 212° F.

While reduction of the sulfur and/or accelerator levels led to more reasonable cure rates and a broader tensile-strength plateau for the sodium-catalyzed butadiene-styrene copolymers, similar variations with sodium-catalyzed polybutadiene (Table XXVI) reduced the rate of cure but did not alter the peak tensile-strength values in magnitude or the period of cure at which they occurred. Emulsion GR-S and polybutadiene are included in Tables XXV and XXVI for comparison.

The sodium-catalyzed butadiene-styrene copolymer was shown to be similar in tensile strength to emulsion GR-S, as reported earlier.[120, 292] At normal periods of cure the sodium-catalyzed polybutadiene showed little or no improvement on emulsion polybutadiene made at 122° F. but at its (much shorter) optimum period of cure its tensile strength was 600 to 1000 p.s.i. higher than that of emulsion polybutadiene. Such improvement had been indicated earlier in evaluations of sodium-catalyzed polybutadiene made by vapor-phase polymerizations.[56] Some variations in stress-strain properties have been reported for polymers prepared with different diluents.[88, 96]

The Goodrich hysteresis–heat-rise values for the cures nearest to those yielding optimum tensile strength are shown in Tables XXV and XXVI. (These cures in general yielded similar modulus and elongation values.) Typical variations in the Goodrich hysteresis value with curing time are listed in Table XXVII for sodium-catalyzed butadiene–styrene (HG16) and for polybutadiene (HG20). The effect of overcuring, which is shown by the stress-strain data (Tables XXV and XXVI), is reflected in the hysteresis and set values (Table XXVII), which both decrease rapidly with increased cure time. At the optimum cures for which hysteresis data are shown in Tables XXV and XXVI, the sodium butadiene-styrene polymer gives temperature-rise values of about 30° F., and the sodium polybutadiene values of 20 to 23° F. At overcures, the two polymers give low values of about 20 and 16° F., respectively. Thus, both sodium-catalyzed polymers are potentially superior in heat buildup to emulsion polybutadiene, GR-S, or natural rubber. The DeMattia flexures and quality indexes (flex-hysteresis balance) shown in Table XXVII are much better than corresponding values for GR-S, in confirmation of other work, which showed sodium-catalyzed butadiene-styrene copolymer to be similar to natural rubber in this respect.[120] More recent work has indicated that sodium polybutadiene made by continuous polymerization at 168° F. has a quality index in the neighborhood of 1, and sodium butadiene-styrene (charge ratios 90/10 and 75/25) has quality indexes somewhat higher, in the range found for GR-S.[307]

The brittle points of both the sodium-catalyzed butadiene–styrene and polybutadiene were poorer than those of corresponding emulsion polymers, presumably because of the higher proportion of 1,2-addition in the sodium polymers (see Table XXXIII). Later investigations have shown that the application of higher temperatures in the conduct of sodium polymerization lowers the freeze point of the product and lowers the proportion of 1,2-addition in it. The replacement of part of the sodium by potassium increases this effect.[162] Whereas the Gehman freeze point of polymer prepared at 30° C. by the action of sodium on a 75/25 butadiene-styrene charge was −13° C., that of polymer prepared at 87° C. was −29° C. An extensive study[269a] of

Table XXVI. Variation in Sulfur and Accelerator in the Vulcanization of Sodium Polybutadiene [88]*

Polymerized at 30° C.

Code		HG17	HG18	HG19	HG20	Emul. PolyBD	NR
Sulfur, parts[†]		2.0	1.25	2.0	1.25	2.0	3.0
Benzothiazolyl disulfide, parts[†]		1.75	1.75	1.25	1.25	1.75	0.75
Stearic acid, parts[†]		3	3	3	3	...	3
	Min. Cured at 292° F.						
Modulus at 300%, p.s.i.	15	840	650	610	480	...	...
	30	1330	1120	1042	900	520	1230
	45	1610	1290	1310	1100	710	1400
	60	...	1440	1610	1280	840	1485
	90	...	1730	...	...	940	1540
	150	...	...	...		1090	1510
Tensile strength, p.s.i.	15	2530	2530	2500	2810	...	...
	30	1970	1960	2050	2270	1330	4650
	45	1840	1760	1570	1970	1790	4840
	60	1430	1730	1730	1900	1810	4600
	90	1560	1730	1480	1670	1940	4120
	150	1470	1500	1470	1370	1970	3850
Elongation, %	15	530	570	590	670	...	...
	30	360	400	410	470	540	610
	45	320	340	330	400	540	590
	60	260	320	310	360	500	560
	90	250	300	260	320	480	520
	150	220	260	240	270	460	510
Set, %	15	6	7	8	10	...	...
	30	2	3	5	5	8	31
	45	3	2	3	3	9	36
	60	2	3	2	3	7	34
	90	1	2	1	1	6	30
	150	1	1	1	1	5	25
Cured at 292° F., min.		30	30	30	30	90	60
Shore A hardness		61	58	59	60	58	62
Goodrich hysteresis[‡]							
Temp. rise, °F.		23	36	20	23	67	30
Initial compression, %		18.3	26.1	20.5	19.0	25.0	18.6
Set, %		4.3	5.2	5.8	5.1	16.4	12.3

* All polymers tested by microtechnique at room temperature.

† Parts per 100 parts of polymer.

‡ Goodrich hysteresis test run at 212° F.

Table XXVII.   Variation in Goodrich Hysteresis Temperature-Rise with Curing Time[88, 96]

Polymers prepared at 30° C.

	Min. Cured at 292° F.	Na-BD/S HG16	Na-PolyBD HG20	Natural Rubber	X-539 GR-S	Emulsion Poly-BD
Shore A Hardness	30	60	60	56	62	52
	60	61	62	62	63	57
	90	64	62	61	65	58
	150	65	63	61	66	59
	300	...	...	61	...	...
Initial compression, %	30	31.5	19.0	25.1	22.5*	33.6
	60	23.6	18.5	18.6	21.1	26.5
	90	21.7	17.5	18.0	17.8	25.0
	150	19.4	16.7	20.2	16.7	24.1
	300	...	...	21.7	...	...
Temperature rise, °F.	30	46	23	48	65*	101
	60	28	18	30	59	87
	90	23	17	29	51	67
	150	18	16	33	47	56
	300	...	...	37	...	...
Set, %	30	23.5	5.1	30.3	19.9*	37.0
	60	7.5	3.9	12.3	15.2	29.2
	90	3.7	2.0	8.1	10.0	16.4
	150	1.5	1.5	6.4	5.1	10.9
	300	...	...	3.4	...	...

* Specimen cured for 45 minutes.

the compounding of butadiene-styrene polymer prepared at 75.5° C. by sodium showed the properties of its vulcanizates to be essentially similar to those outlined above of polymers prepared at 30° C.

Passenger-car tires carrying half-and-half treads of GR-S and 75/25 (charge ratio) sodium butadiene-styrene polymer showed the latter to run cooler than the GR-S control but to have a poorer rating for treadwear, viz., 91 compared with 100 for GR-S.[348]   In later tests, in which sodium-catalyzed copolymer was compared with cold GR-S in 9.00 × 20-in., 10-ply truck tires, the rubber of which was 68 per cent synthetic and the balance natural, the sodium polymer ran about twice as far as the GR-S.

Sodium Polymerization and Copolymerization of Other Monomers. Data for sodium-catalyzed polymerizations of isoprene and of butadiene–α-methylstyrene are summarized in Table XXVIII along with data for sodium butadiene–styrene and polybutadiene.   The use of pure isoprene resulted in reaction cycles similar to those of butadiene, while the relatively impure commercial material reacted only in a matter of days and has been reported to inhibit the polymerization completely.[88]   The sodium polyisoprene seems to be similar to sodium polybutadiene in stress–strain properties but inferior in the flex-hysteresis relationship and in low-temperature properties.   A 75/25 mixture of butadiene–α-methylstyrene showed an induction period of about 3 days followed by a reaction of 12 hours.   The polymer, samples of which showed intrinsic viscosities of 1.5 to 2.4, was soft and tacky and showed considerable cold flow; its low-temperature index was −26° C.[186, 211, 325]

Table XXVIII.   Sodium-Catalyzed Polymerizations at 30° C. of Monomers Other than Butadiene–Styrene[47, 88, 96, 120, 190, 192, 211, 280, 287]

	Poly-isoprene*	I/Styrene*	PolyBD	BD/S	BD/AMS†
Ratio (diene/X)	100	75/25	100	75/25	75/25
Time to 100% conversion, hr.	14, 30–60	15	10–20	App. 10	9
Na, %	0.2, 0.35	0.35	0.3–0.4	0.3–0.4	0.3
Gel, %	0, 4–19	0	0.10	3–14	3
DSV	1.0, 1.4–2.1	0.5	4.4–5.7	4.6–5.9	2.25
Modulus at 300%, p.s.i.	1000	...	...	1015	1140
Tensile strength, p.s.i.	1900–2100	...	2600	3060	2270
Elongation, %	540–895	...	520	580	450
Shore A hardness	60	...	...	...	61
Hysteresis					
Temp. rise, °F.	86‡	...	35	49 (25)§	29
Flexures × 10^{-3}	240	...	270	770 (7)§‖	1§
Quality index	>1.3	...	6.1	9.6 (7.8)§	1.0
Low-temp. index, minus °C.	1	...	55	24–30	...

* Pure (99.8 mole per cent) isoprene used.
† Butadiene–α-methylstyrene polymer of 37 ML-4 viscosity.
‡ Hysteresis sample ruptured.
§ Tests run on polymers made in presence of pentane.
‖ DeMattia test at 212° F. rather than at 82° F.

Hard, granular, insoluble polymers were obtained from 1-cyanobutadiene and from a 75/25 mixture of the monomer and butadiene, and also from mixtures of butadiene and either 2-vinyl-5-ethylpyridine or 2,5-dichlorostyrene.   The latter monomer required an excess (2 per cent) of sodium. Both *p*-N,N-dimethylaminostyrene and *p*-methoxystyrene gave very soft, colored copolymers with butadiene after prolonged reaction.[210-1]   *p*-Acetoxystyrene, *m*-nitrostyrene, and *p*-vinylphenyl acetate, in admixture with butadiene, inhibited reaction nearly completely.   The nitrostyrene resulted in a black powder.   In the presence of the acetate, the sodium developed a bright green color, and there was no formation of polymer.[183, 210-1]

ALFIN-CATALYZED POLYMERIZATIONS

Alfin Catalysts. Alfin catalysis was discovered and has been extensively described by A. A. Morton.[250, 251] The catalysts are derived from the salts of a secondary *al*cohol paired with that from an ole*fin*; hence the name "alfin." To prepare a typical catalyst, amyl chloride and sodium are reacted to give amylsodium. Isopropyl alcohol is added to destroy at least half of the amylsodium, thereby producing sodium isopropoxide in an extremely fine state. Propylene is then passed into the mixture to complete the preparation:

$$C_5H_{11}Cl + 2Na \qquad \rightarrow C_5H_{11}Na + NaCl$$

$$C_5H_{11}Na + (CH_3)_2CHOH \rightarrow (CH_3)_2CHONa + C_5H_{12}$$

$$C_5H_{11}Na + CH_2 : CHCH_3 \rightarrow CH_2 : CHCH_2Na + C_5H_{12}$$

The operations have been carried out in a high-speed stirring apparatus under an atmosphere of dry nitrogen. The reagents are insoluble aggregates of ions which are associated with sodium chloride; half of each aggregate is composed of that salt. Coordination valences probably bind the aggregate together.

In general, the catalysts must be made from secondary alcohols, at least one branch of which is a methyl group. The olefin contains as an essential the system $\cdot$ CH : CH $\cdot$ CH$_2$ $\cdot$, which may even be part of a ring system, as in toluene. Substitution of potassium for sodium in the preparation of the catalyst proved unsuccessful. In the absence of sodium chloride, the catalysts do not promote polymerization. Certain other alkali-metal halides may, however, be used in place of sodium chloride, provided that their cations and anions are of suitable sizes.[251a] The alfin catalyst is a three-component system, variation in any component of which affects both the rate of polymerization and the intrinsic viscosity of the polymer formed. Morton pictures the catalyst complex as an insoluble cluster of ions onto which the monomer is adsorbed. He has offered a working hypothesis designed to explain in some detail the factors controlling the mode of action of effective alfin catalysts.[251b]

Alfin polymerization reactions are normally completed in a matter of minutes, as compared with hours required for polymerization by sodium at the same temperature. Those catalysts producing the fastest reactions generally yield polymers with the highest intrinsic viscosities.

Characteristics of Alfin Polymerization. Polymerizations by means of alfin catalysts have mostly been carried out at 30° C. Approximately 500 to 1000 parts of a solvent, such as pentane, is charged to the reaction vessel, which may be a bottle in laboratory-scale work, and this is followed by 100 parts of monomer. A slight excess of the volatile monomer (butadiene) is added and allowed to vent off, in order to remove air before addition of 50 to 100 parts of the catalyst suspension. Agitation is accomplished either with a high-speed stirrer or by rotation of the bottle. Very careful preparation of the bottle is necessary, to remove the small amounts of moisture occluded to the glass.[52]

A complicating factor in alfin polymerization is a secondary reaction in which the polymer reacts with the sodium of the catalyst. Metalation of various polymers with amylsodium or phenylsodium resulted in larger amounts of insoluble product, as the concentration of metalating agent or the severity of the reaction conditions was increased. Since the gel-forming tendency has also been shown to be increased by the alfin catalyst, the concentration of catalyst is kept as low as possible. The use of increased proportions of alkoxide in the catalyst mixture has been found to give progressively lower gel contents. Generally, the concentration of alkoxide giving the lowest gel content has been found to correspond to that which is optimal as regards reaction rate.

In contrast to polymerizations catalyzed by an alkali metal or an organo-alkali metal compound, which are generally considered to proceed by a stepwise process,[346] and in which progressively lower intrinsic viscosities and faster reactions occur as the amount of catalyst is increased, the alfin poly-merization is catalytic, and probably takes place on the catalyst surface.[250] An increase in the concentration of the alfin catalyst resulted in a higher reaction rate but produced little change in the intrinsic viscosity of the polymeric product.

According to Stewart and H. L. Williams,[298] in the polymerization of butadiene by means of an alfin catalyst the intrinsic viscosity of the polymer falls as conversion proceeds, and the converse is true in the alfin polymeriza-tion of butadiene–styrene. In experiments on the alfin polymerization of butadiene–styrene over the temperature range 40 to −60° C., the same authors found the intrinsic viscosity of the polymer to increase as the tempera-ture of polymerization was reduced.

A comparison of a series of alfin polymers of butadiene, isoprene, and butadiene–styrene (90/10, 80/20, and 70/30) with corresponding emulsion and sodium polymers has been reported.[47, 52] Sodium isopropoxide-allyl sodium catalyst (7/1 molar ratio) acting at 30° on butadiene or butadiene-styrene gave 65 to 90 per cent conversion in 30 minutes, and acting on iso-prene 79 per cent conversion in 1 hour. Presumably because of metalation and other side reactions, the butadiene–styrene and the isoprene polymers required more catalyst than polybutadiene for similar conversions in equal times. The intrinsic viscosities of all the alfin polymers ranged from 10 to 15—values not obtainable in corresponding polymers made by the emulsion or the sodium process. The content of gel was high in the polybutadiene; it fell as the styrene content of the copolymers was increased. The poly-isoprene was gel-free.[52] Other work has indicated dilute-solution viscosities of 10 to 12 for polybutadiene and 8 to 10 for butadiene-styrene copolymers prepared in a similar fashion. The use of hot benzene in determining gel in the polybutadiene samples reduced the value for the apparent gel content to a negligible quantity.

A moderate amount of hot milling was found to reduce substantially the intrinsic viscosity and gel content of the 70/30 butadiene-styrene and the isoprene polymers.[48, 52] On milling, the gel in the polybutadiene rose; it showed little change in the 90/10 and 80/20 butadiene-styrene copolymers.

Structure of Alfin Polymers. Molecular-weight-distribution deter-minations showed the lowest fractions of the alfin polymer to be about

1.5 $\times$ 10^6 and to extend to 2.0 $\times$ 10^6 or higher, whereas the highest fraction of GR-S is about 750,000 by the same test.[71]   Other work has shown the highest fraction of GR-S to be 1.5 $\times$ 10^6 and the average about 100,000.[344]

All the alfin polymers (butadiene, isoprene, and butadiene–styrene) contained higher proportions of 1,2- (and 3,4-) addition (see Table XXXIII) and a greater proportion of *trans* 1,4-configuration than corresponding emulsion polymers made at 43° C.[52]   The proportion of 1,2-addition was, however, much less than in sodium polymers, indicating the alfin polymers to be more like emulsion than sodium polymers in this respect.[47]   The occurrence of crystallinity in stretched alfin polybutadiene is shown by the fact that it gives an X-ray diffraction pattern.   The amount of crystallinity decreases as increasing proportions of styrene are copolymerized with butadiene.   In contrast, all the emulsion polymers (made at 43° C.) and also alfin polyisoprene showed amorphous patterns.   The increased disorder in the polyisoprene caused by the presence in it of 1,2-, 1,4-, and 3,4-addition as well as of *cis* and *trans* 1,4-configurations presumably would explain its amorphous character.   It is of interest to note that, as described later, alfin polymers prepared at 30° C. showed more crystallinity than similar emulsion polymers prepared at $-20°$ C.[23]

Processibility of Alfin Polymers and Properties of Their Vulcanizates.   The difficulties in the practical use of Alfin polymers arise from the very high molecular weight and toughness of the latter.   Attempts to regulate the polymerization by any of the conventional means have thus far been unsuccessful.   Softer products have been reported, as the temperature of polymerization was increased[53] from 30 to 50° C.   Some improvement in the millability of the polymers has also been reported with decreased polymerization temperature (to 10° C.) or by the use of other organosodium reagents.

The butadiene-styrene copolymers showed some improvement in mill breakdown as the styrene content was increased.[52]   Heat softening of the polymer prior to milling did not greatly improve milling properties, although the intrinsic viscosity was reduced.   On milling the gel content first dropped but on continued milling rose rapidly.   The use of liquid plasticizers such as tributyl phosphate greatly improved milling properties but gave inferior tensile strength and rebound values, as compared to those for unplasticized polymers;[52] it also markedly increased the elongation and the tensile product (tensile strength $\times$ percentage elongation).[96]

When compounded according to either a gum- or a tread-stock recipe, alfin polybutadiene appeared to be superior to emulsion polybutadiene in stress-strain properties.   Little difference could be found between polymers prepared by the two processes from either butadiene–styrene or isoprene, probably because of overcure and poor dispersion of the carbon black in the alfin polymers.[52]   Other work of a preliminary nature[96] has indicated that improved mixing and carbon black dispersion may be obtained by blending alfin rubbers with liquid polymers or softeners, or by prolonged periods of mill breakdown and mixing.   The stress-strain and gel data shown in Table XXIX give an indication of results obtained with polymers prepared by the use of sodium isopropoxide-propenylsodium catalyst.

Table XXIX.   Alfin Polymers from Various Butadiene-Styrene Charge Ratios[96]

O.R.R. Test Recipe: 2 Parts Sulfur; Microtechnique

Polymer	BD*	BD†	BD/S	BD/S	BD/S
Monomer ratio	100	100	90/10	80/20	75/25
Gel, %	20	...	14	...	3
DSV	12.6	...	8.0	...	8.2
Modulus at 300%, p.s.i.	...	1270	1290	1170	1350
Tensile strength, p.s.i.	2460	3350	4210	4450	5030
Elongation, %	160	490	530	600	590
Set, %	8	12	15	23	21
Freeze point (Gehman, −°C.)[25,26,52]	65	...	61	55	40

* 0.5 parts sulfur in test recipe.

† 100 parts 15 ML-4 polybutadiene.

Special Banbury treatment has been used to soften the raw alfin copolymers. Such treatment is accompanied by deterioration of the stress-strain properties of the compounded stock but not of its resistance to flex cracking or abrasion.[306] The same treatment with breakdown agents has reduced the viscosity to practical factory limits, but the physical properties were badly degraded. Different compounding ingredients have been useful in producing from alfin copolymers stocks with tensile strengths of 4600 p.s.i. and 300 per cent moduli of 1600 p.s.i. at room temperature.[255]

The use in the alfin catalyst of benzylsodium in place of allylsodium gave polymers, of lower intrinsic viscosity, which were more readily milled and compounded. Apparently, though, a marked decrease in stress-strain level also results from such modification of the catalyst.[96]

As would be expected from their structure, the low-temperature properties of the alfin polymers are about midway between those of corresponding emulsion polymers and sodium polymers. Alfin polybutadiene has been reported to be similar to GR-S in rebound and flex life and to be poorer in dynamic properties. Polyisoprene was similar to GR-S in flex life and superior in rebound and dynamic properties.[25, 26, 47, 52] A Goodrich hysteresis temperature-rise value of 64° F. and a Shore hardness value of 65 have been reported for a butadiene-styrene copolymer containing about 25 per cent combined styrene.[96] But, when the polymer was compounded in such a way as to cure more slowly, the hysteresis of this alfin polymer was similar to that of the corresponding sodium polymer.[255] Alfin poly-butadiene samples tested were stiff, and hysteresis specimens ruptured.

Passenger-car-size tires containing half-and-half treads[103] of GR-S and an alfin 80/20 butadiene-styrene copolymer have shown considerably better treadwear than the GR-S control—a rating of 117 compared to 100 for GR-S.[52]

Alfin polybutadiene when properly compounded or when merely heated in a press or when mixed with black and then pressed with heating has shown good oil resistance and excellent low-temperature flexibility.[255]

Oil-Extended Alfin Polymers. Since the preceding was written, it has been shown that polymerization by the alfin catalyst can be conducted in the presence of oils.[15, 255] This offers a promising approach to the problem of meeting the difficulties that the high molecular weight of the alfin polymers and their toughness, as previously produced, places in the way of their practical use. For the oil-extended polymers possess reasonably good processibility and show promising results on vulcanization. Experiments on the extension of alfin polymers by means of oil introduced before or immediately after polymerization have also been reported by Stewart and H. L. Williams.[298] These authors show further that, in the copolymerization of butadiene–styrene by an alfin catalyst, the presence of oil markedly decreases the proportion of styrene that enters into the copolymer, compared with that present in copolymer formed by the alfin catalyst under prior techniques of operation.

ALKALI METAL HYDRIDE-CATALYZED POLYMERIZATION

The replacement of sodium by sodium hydride as a polymerization catalyst was proposed in a French patent[309] in 1930. Early laboratory work in the United States showed sodium and lithium hydrides, in small proportions, to be excellent catalysts for the bulk polymerization of butadiene. The polymer obtained was colorless. Sodium hydride was effective for both liquid and vapor-phase polymerization; lithium hydride was effective only in the liquid phase.[130] Sodium methoxyborohydride, $NaHB(OCH_3)_3$, was ineffective for the polymerization of butadiene–styrene[130] at 50° C. In general, the polymerizations with sodium hydride have been found to proceed more gradually than those catalyzed with sodium.[89] (Sodium hydride, in powdered form, was handled in the complete absence of air at all times. Whereas small quantities of the hydride and large amounts of water were safe, large amounts of hydride and a small amount of water proved to be dangerous.[89]) Air was a powerful inhibitor, particularly for polymerizations conducted in the vapor phase. Data are summarized in Table XXX.

Increased amounts of sodium hydride overcame to some extent the inhibitory effects of oxygen.[123] The effect of carbon dioxide was similar to that encountered with sodium catalysis. The addition of carbon dioxide at the start of a butadiene polymerization resulted in inhibition; addition during the polymerization resulted in temporary inhibition, or no effect at all. Without agitation or at higher temperatures, the effect was much less pronounced.[131]

The use of 0.77 part each of dianisyl disulfide (DADS) and sodium hydride resulted in a clearer polymer, highly elastic, soluble in benzene or toluene, and considerably softer than unmodified material. When cooled in liquid nitrogen, it could be struck without the occurrence of shattering or cracking.[130] Various other materials have been tried as modifiers or diluents. Of these, 10 to 20 per cent of toluene, heptane, styrene, or carbon black retarded the polymerization when the agitation was relatively inefficient. With efficient agitation, little or no retardation was noted with 50 to 65 per cent of heptane as a diluent; a polymer of 70 to 80 per

Table XXX. Polymerization of Butadiene by Alkali Metal Hydrides at 50° C.[130]

Hydride	Parts	Air	Reaction Time, Hr.	Conversion %	Product
Liquid-Phase Polymerization					
NaH	0.4	Present	3 mo.	0	
			1 yr.	100	Popcorn-type polymer
LiH	2.0	Present	72 hr.	0	
NaH	0.4	Degassed BD	22 hr.	100	Insol., transparent, elastic
LiH	2.0	Degassed BD	22 hr.	100	Insol., white, opaque
Vapor-Phase Polymerization*					
NaH	0.67	Present	72 hr.	5	
LiH	2.0	Present	288 hr.	Approx. 60	Partially clear (72-hr. induction)
NaH	0.8	Twice-degassed butadiene	3 hr.	50	Faint yellow, slightly sol.
			5 hr.	66	Opaque, curdy
			14 hr.	100	

* Reactor kept at 50° C. and butadiene reservoir at 25 to 30° C.

cent conversion (gel content, 40 per cent) was obtained.[123] About 12 per cent by weight of dianisyl disulfide (based on the amount of sodium hydride) or about 10 per cent of heptane (based on the amount of butadiene) gave polymers, of 10 to 25 per cent solubility in benzene, that could be milled without crumbling.[124] The use of 50 to 60 per cent of either material (similarly based) gave gel-free polymers.[126]

Although ligroin as solvent resulted in little change in reaction time, the low-gel polymers that were formed contained hard nodules and were difficult to remove from the reaction vessels. The use of 0.5 per cent of allyl chloride completely inhibited polymerization. Dilauryl disulfide and phenylhydrazine, applied as modifiers, caused lower reaction rates and yielded polymers with higher gel content than those obtained with dianisyl disulfide.[123] In general, symmetrical disulfides were found to be effective in reducing the gel content, whereas the unsymmetrical disulfides were not.[128] In the absence of modifiers, faster reactions and more soluble polymers were obtained as the temperature or the degree of agitation was increased or as both were increased (see Table XXXI).[123, 129]

Table XXXI. Effect of Temperature and Agitation on Polymerization of Butadiene by Sodium Hydride (1 Part)

Temperature, °C.	Reaction Time, Hr.	Conversion, %	Remarks
23	24	85–90	
50	7	85–90	
100	0.5	90–100	
23	21	90	Stationary tub
23	0.5	90	Round-bottom flask; beads shaken through small arc

Physical test data for sodium hydride-catalyzed polybutadiene and butadiene-styrene polymers are summarized in Table XXXII. Briefly, both types of polymers seemed to be similar to emulsion polymers in stress-strain properties and superior in hysteresis. The quality indexes (flex-hysteresis balance) for the butadiene-styrene polymers were outstanding; the indexes for the sodium hydride polybutadiene samples were similar to those for emulsion polybutadiene. The values for 1,2-addition listed in

Table XXXII. Physical Test Data for Polymers Prepared by Means of Sodium Hydride (1 Part) [91, 92, 124, 125, 127, 129]*

	Polybutadiene			75/25 BD/S	
Polymerization Temperature, °C.	50	50	23	50	50
Reaction time, hr.	...	...	...	6–7	6–11
Conversion, %	65–90	65–90	65–90	70–95	60–100
Modifier, type/parts	DADS/0.12	None	None	DADS/0.1	None
Diluent, type/parts	Heptane/50	Heptane/50	Heptane/50	Ligroin/100	Ligroin/100
1,2-addition,† %	...	21–23	34–42‡	...	...
Raw ML-4 (212° F.)	54	100+	100+	87	74
Gel, %	6	46	...	2	22
Intrinsic viscosity	4.5	3.7	...	3.4	3.5
$T/300$, p.s.i.	...	...	...	1090	1150
Tensile strength, p.s.i.	1315	1090	1410	2820	3020
Elongation, p.s.i.	340	230	300	530	540
Goodrich hysteresis, °F.	<GR-S (64)	<GR-S (64)	...	54	49
DeMattia flexures × 10^{-3}	<GR-S (290)	<GR-S (290)	...	1950	1040
Quality index	0.75	0.69	...	14 (avg.)	12 (avg.)

* Compounding according to the O.R.R. tread test recipe plus 5 per cent of stearic acid.
† Determined by ozonolysis.
‡ 18 to 19 per cent of 1,2-addition found on similar sample modified with about 0.1 part of DADS.[125] Recent incomplete work on sodium polymerizations of butadiene and butadiene–styrene has also shown appreciably lower refractive indexes for modified polymers.[96]

Table XXXII indicate that sodium hydride-catalyzed polybutadiene samples are more nearly similar in this regard to emulsion or alfin polymers than to sodium-catalyzed polybutadiene.

Similar polymerizations with isoprene proceeded at lower rates than those with butadiene and yielded softer products. Isoprene-styrene (75/25) mixtures reacted somewhat faster and gave partially gelled products, which appeared to be hard and plastic and to be rubbery only at elevated temperatures.[129]

POLYMER STRUCTURE AND BRITTLE POINTS

Factors that arise for consideration in connection with the readiness of the molecular chains of a polymer to fall into a crystallite lattice on cooling are (a) the regularity or irregularity of the structure of the chains, (b) the nature, size, and number of substituent side groups attached to the main chains. In diene polymerization random 1,2-, cis 1,4-, and trans 1,4-addition may be considered as a major structural irregularity leading to noncrystallizable polymers. The proportion of 1,2-units in such polymers has been determined by ozonolysis and by titration with perbenzoic acid. And more recently there has been developed an infrared technique suitable for determining all the three types of addition mentioned.[101, 244] In Table XXXIII are assembled data on the brittle point and the proportion of 1,2-addition for certain diene polymers and copolymers. Included in the table are data on the proportion of 1,2-addition in a number of polymers to which reference has been made earlier in the chapter but for which information on the brittle point is lacking. Data on the content of 1,2-units in still other polymers is to be found in Chapter 9.

The temperature of polymerization (−2 to 206° F.) has been found to have relatively little effect on the proportion of 1,2-addition. A maximum of 23 per cent of such addition was found experimentally in both polybutadiene and 71/29 butadiene-styrene polymer prepared in emulsion

Table XXXIII. 1,2-Addition and Brittle-Point Data for Various Polymers

Polymer	Description	Ozonization Number* or % 1,2-Addition	Method†/Reference	Brittle Point, Minus °C.
Natural rubber	Crepe and smoked sheet	10, 3–6	O,PB/244, 278	47, 52, 279, 282
GR-S	80/20 to 65/35 BD/S; 50 to 80% conv., 50 to 110° C.	18–22	O,PB,IR/47, 244, 278	57
GR-S	Sodium, 75/25 BD/S, 50° C.	57, 80	PB,IR/47, 244	27
GR-S	90/10 to 75/25 BD/S, Alfin	23–26, 32	PB/47, 52,	46
Polybutadiene	Emulsion, 15 or 50° C.	19–23	O,PB,IR/47, 164, 227, 244	80
Polybutadiene	Sodium, 50° C.	59, 20	PB,O/181, 244	51
Polybutadiene	Sodium, minus 50° C.	95	PB,IR/171	...
Polybutadiene	Sodium, minus 70° C.	100	O/309	...
Buna 85	German K polybutadiene	40, 40–50, 60	O,PB,IR/47,164	63
Buna 115	German Na polybutadiene	50	O/225	40
Russian polybutadiene	Sodium	54, 41	PB,O/164, 286	44
Polybutadiene	Alfin	35, 32, 30	O,PB,IR/26, 47	54
Polybutadiene	Sodium hydride, 23° C.	37, 18	O/180, 182	...
Polybutadiene	NaH, 23° C., DADS‡	18	O/180, 182	...
Polybutadiene	NaH, 50° C.	22, 19	O/180, 182	...
Polybutadiene	NaH, 50° C., DADS‡	24	O/182	...
Polyisoprene	Alfin	20, 10–18	PB,IR/27, 47, 52	46–51
Polyisoprene	Emulsion	16, 10–14, 12	PB,IR/47, 52, 137, 278–9	57–69
Polyisoprene	Sodium, 30 to 50° C.	41–55, 64§	PB/47, 137, 139, 278	1
Poly-2,3-dimethylbutadiene		12–15, 9	PB/227, 278, 279	30
Butadiene/2,3-dimethylbutadiene (25/75)		10BD/23 DMBD	PB/227	...
Butadiene/2,3-dimethylbutadiene (50/50)		11BD/18 DMBD	PB/227	...
Butadiene/2,3-dimethylbutadiene (75/25)		15BD/9 DMBD	PB/227	...
Poly-cis- or poly-trans-piperylene		Nearly all 1,4-add.	IR/193	...
BD/2-chloroBD/styrene (69/11/20 BD/Cl BD/S)		14	O/164	...
Neoprene GN§ (poly 2-chlorobutadiene)		4	O/164	40
Poly-2-ethyl 2-ispropyl or 2-amylBD		15 ± 2	PB/228	...
2-ethyl, or 2-n-amylbutadiene–styrene (75/25)		15 ± 2	PB/228	...
2-isopropylbutadiene–styrene (75/25)		18 ± 5	PB/228	...
Butadiene–2-vinylpyridine (70/30 BD/VP)		14	O/172	...
Gutta-percha or balata (trans form of NR)		0	PB/139	54
Isoprene–styrene (75/25 I/S)		20, 5	O,PB/139, 164	...
Buna N (74/26 BD/AN)		25, 23	O,PB/139, 164	30
GR-N		13	O/177	...
BD/o- or m-chlorostyrene, or m-bromostyrene		9–12, 20	O,PB/139, 181	...
BD/B-vinylnaphthalene		18	O/181	...
Isoprene–isobutylene (Various isoprene levels)		0	O/260	40

* Considered as indicating only the relative amounts of 1,2-groups of the various polymers. The value for natural rubber is not construed to indicate the presence of terminal vinyl groups.
† O designates ozonization; PB, perbenzoic acid; IR, infrared.
‡ DADS—sodium hydride-catalyzed polybutadiene was modified in some cases with dianisyl disulfide.
§ 40 per cent 3,4-addition; 24 per cent 1,2-addition.
‖ Neoprene predicted to polymerize by 1,4-addition; X rays indicate all the double bonds to be in *trans* configuration.[247]

at 122° F. The 1,2-addition decreased with the temperature of polymerization to a value of 19.6 per cent at −2° F. On the basis of a linear relationship, it would be no higher than 25 per cent at 206° F. Ozonolysis yielded values in the same range for 80/20 to 65/35 butadiene-styrene copolymers of 50 to 80 per cent conversion prepared[164] at from 50 to 110° C.

The *trans* 1,4-content of 71/29 butadiene-styrene polymer and of polybutadiene increased continuously from a value of about 51 per cent at 206° F. to about 80 per cent for polymer prepared at −2° F.[101] The slightly greater content of *trans*-1,4-butadiene units found in GR-S than in poly-butadiene possibly explains the reported[23, 117] crystallization in polybuta-diene prepared at 68° F. or lower.

The ratio of *trans* to *cis* unions in polybutadiene was found to be 4/2 in samples prepared at 122° F., and 10/3 in those made at 41° F. The introduction of 10 parts of styrene in copolymerization with butadiene at —4° F. did not inhibit crystallization, whereas, under the same conditions of test, 20 parts of styrene did.

The data obtained by oxidative breakdown,[47] by titration with perbenzoic acid,[138-9] and from infrared spectra reveal one of the basic differences believed to exist between some of the polymers listed in Table XXXIII. Substitution of methyl, ethyl, isopropyl, or amyl groups in the 2-position of 1,3-butadiene apparently had a diverting influence on 1,2-addition, reducing the content from about 22 to 15 per cent. Substitution of methyl groups in both the 2- and 3-positions further reduced the 1,2-addition to about 11 per cent, as predicted earlier.[279] Polymers of both *cis*- and *trans*-piperylene (1-methyl-1,3-butadiene) appear to contain substantially all 1,4-addition. Some evidence of similar reductions in 1,2-addition caused by the substitution of a chlorine atom in butadiene or by the use of acrylonitrile, chlorostyrene, or bromostyrene in a butadiene copolymer are also indicated.

With sodium polybutadiene the effect of polymerization temperature on a 1,2-addition is opposite that noted for emulsion polybutadiene, 1,2-addition being favored by low-polymerization temperatures. No such effect of temperature was apparent for the sodium-catalyzed butadiene-styrene or isoprene polymers. It has been predicted that polybutadiene of 100 per cent 1,4-addition, in the *cis* configuration, would, if it could be prepared, have a brittle point, or freeze point, lower than —100°C. Over the temperature range from —2° F. to about 200° F., in which a maximum decrease of 6 per cent in 1,2-addition and a maximum increase of 29 per cent in *trans* 1,4-addition was found, the freezing points of polybutadiene and GR-S, determined by the Gehman technique, were found to increase 5 to 10° C. at the lower as compared with the higher polymerization temperatures.[117, 301, 323] The brittle points of smoked sheet and crepe (*cis* 1,4-) rubbers were reported to be —53 to —58° C., whereas that of gutta-percha,[279] (*trans* 1,4-) was —54° C. Apparently the brittle point of polymers is influenced much more by the type and number of substituents in the diene molecule or by the comonomer than by the structure of the polymer itself. This view supports the opinion of German investigators.

CONCLUSION

Semon[293] reported some of the early work in the United States, by Kyriakides and Earle, in which isoprene, butadiene, and dimethylbutadiene were prepared and were polymerized, apparently by means of sodium. Polydimethylbutadiene was made into overshoes. Early German polymers also were made from dimethylbutadiene. Since that time polymers and copolymers from a wide variety of monomers have been prepared and studied. Both Weidlein's[333] report on the German development work and also other opinions are in accord with the results reported in the present chapter that polymers made from materials other than butadiene and styrene do not offer marked advantages over the butadiene-styrene polymers as general-purpose rubbers and that quality may depend more on polymerization conditions than on the types of monomers used. For instance, the copolymers of chlorosubstituted styrenes with butadiene were exceptional as compared with copolymers prepared from butadiene and the same amount of styrene in the charge formula, when flex life, stress-strain properties, and processing characteristics were considered. The same could be said for the vinylpyridine copolymers, except for their processing characteristics. However, when costs, low-temperature behavior, hysteresis temperature-rise, monomer recovery, etc. were considered, the chlorostyrenes were not favorable as comonomers in comparison with styrene. If adjustments were made to overcome some of the weaknesses of the copolymers, the favorable properties were reduced, in general, to the same range as GR-S, and the net result was mainly an increase in costs. In the case of the vinylpyridine copolymers, processing difficulties, because of scorchiness and incompatibility, render these materials of little practical value as general-purpose elastomers. Of course, the advantages of "tailor-made" rubbers, prepared from special monomers to meet specific uses, as, for instance, for oil resistance, resilience at low temperatures, and so on, should not be overlooked.

REFERENCES

1. Aisenberg, I. M., Private Communication, Govt. Lab. to O.R.R., Sept. 15, 1948.
2. Alden, G. E., Private Communication, Gov. Lab. to O.R.R., May 19, 1949.
3. Alden, G. E., Private Communication, Gov. Lab. to O.R.R., May 19, 1949.
4. Anon., *Ind. Eng. Chem.*, **32**, 291-2 (1940).
5. Anon., *ASTM Standards on Rubber Products*, D 623-41T (1944).
6. Anon., *ASTM Standards on Rubber Products*, D 314-39 (1944).
7. Anon., *India Rubber World*, **109**, 375-9 (1944); **111**, 446 (1945).
8. Anon., PB Rept. 1896, June 13, 1945.
9. Anon., PB Rept. 13328, microfilm reel No. 8, frame 822.
10. Anon, *Chem. Eng. News*, **26**, 1696-7 (1948).
11. Anon., *India Rubber World* (*a—c*), **114**, 373, 522, 812-3 (1946); (*d—e*), **115**, 75-6, 410 (1946); (*f*), **115**, 710 (1947); (*g—h*), **116**, 659,854 (1947); (*i—j*), **117**, 116, 274 (1947); (*k—m*), **118**, 298, 406-10, 564 (1948); (*n—o*), **119**, 244, 359 (1948); (*p*), **119**, 762 (1949).
12. Anon., *Chem. Eng. News*, **27**, 1575 (1949).
13. Anon., *Chem. Eng. News*, **27**, 1636 (1949).
14. Anon., *Chem. Eng. News*, **27**, 2129 (1949).
15. Anon., *Chem. Eng. News*, **29**, 5058 (1951).
16. Bachman, G. B., and Heisey, L. V., *J. Am. Chem. Soc.*, **70**, 2378-80 (1948).
17. Bachman, G. B., and Hellman, H. M., *J. Am. Chem. Soc.*, **70**, 1772-4 (1948).
18. Bachman, G. B., and Micucci, D. D., *J. Am. Chem. Soc.*, **70**, 2381-4 (1948).
19. Bandre, F. A., Jackson, D. L., and Lawson, C. W., Private Communication, U.S. Rubber Co. to O.R.R., Nov. 5, 1945.
20. Bauchwitz, P. S., Private Communication, Govt. Lab. to O.R.R., June 9, 1949.
21. Bauchwitz, P. S., Private Communication, Govt. Lab. to O.R.R., June 10, 1949.
22. Berchet, G. J., and Carothers, W. H., *J. Am. Chem. Soc.*, **55**, 2004-8 (1933).
23. Beu, K. E., Reynolds, W. B., Fryling, C. F., and McMurry, H. L., *J. Polymer Sci.*, **3**, 465-80 (1948).

24. Blitz, H., *Ann.*, **296,** 259 (1897).

25. Borders, A. M., et al., Private Communication, Goodyear Tire & Rubber Co. to O.R.R., July 1–31, 1945.

26. Borders, A. M., et al., Private Communication, Goodyear Tire & Rubber Co. to O.R.R., Aug. 1–31, 1945.

27. Borders, A. M., et al., Private Communication, Goodyear Tire & Rubber Co. to O.R.R., Jan. 1–31, 1946.

28. Brock, L. W., Swart, G. H., and Osberg, E. V., *India Rubber World*, **119,** 464–5, 599–603, 725–6 (1949); **120,** 70–2 (1949).

29. Brooks, L. A., *J. Am. Chem. Soc.*, **66,** 1295–7 (1944).

30. Caldwell, I. P., Private Communication, Goodyear Synthetic Rubber Corp. to O.R.R., Sept. 2, 1944.

31. Caldwell, I. P., Private Communication, Goodyear Synthetic Rubber Corp. to O.R.R., Oct. 14, 1944.

32. Caldwell, I. P., Mitchelson, J. B., and Reed, R. C., Private Communication, Goodyear Synthetic Rubber Corp. to O.R.R., July 14, 1944.

33. Carothers, W. H., *Ind. Eng. Chem.*, **26,** 30–3 (1934).

34. Carothers, W. H., and Berchet, G. J., *J. Am. Chem. Soc.*, **55,** 2807–13 (1933).

35. Carothers, W. H., and Berchet, G. J., *J. Am. Chem. Soc.*, **55,** 2813–7 (1933).

36. Carothers, W. H., and Coffman, D. D., *J. Am. Chem. Soc.*, **54,** 4071–6 (1932).

37. Carothers, W. H., Kirby, J. E., and Collins, A. M., *J. Am. Chem. Soc.*, **55,** 789–95 (1933).

38. Carothers, W. H., Williams, I., Collins, A. M., and Kirby, J. E., *J. Am. Chem. Soc.*, **53,** 4203–25 (1931).

39. Coffman, D. D., *J. Am. Chem. Soc.*, **57,** 1981–4 (1935).

40. Coincon, J., Private Communication, Firestone Tire & Rubber Co. to O.R.R., Nov. 2, 1946.

41. Costanza, A. J., Private Communication, Govt. Lab. to O.R.R., May 13, 1948.

42. Costanza, A. J., and Gyenge, J. M., Private Communication, Govt. Lab. to O.R.R., Feb. 18, 1947.

43. Costanza, A. J., Gyenge, J. M., and Mooney, H. R., Private Communication, Govt. Lab. to O.R.R., Nov. 23, 1948.

44. Costanza, A. J., and McCann, R. F., Private Communication, Govt. Lab. to O.R.R., Jan. 21, 1948.

45. Costler, V. A., *India Rubber World*, **95,** No. 3, 43–5 (1936).

46. Dean, D. J., Private Communication, Govt. Lab. to O.R.R., Jan. 6, 1948.

47. D'Ianni, J. D., *Ind. Eng. Chem.*, **40,** 253–6 (1948).

48. D'Ianni, J. D., et al., Private Communication, Goodyear Tire & Rubber Co. to O.R.R., July 6, 1949.

49. D'Ianni, J. D., and Cousins, E., Private Communication, Goodyear Tire & Rubber Co. to O.R.R., May 6, 1946.

50. D'Ianni, J. D., Hoesly, J. J., and Greer, P. S., *Rubber Age N.Y.*, **69,** 317–21 (1951).

51. D'Ianni, J. D., Marsh, J., and Findt, W., Private Communication, Goodyear Tire & Rubber Co. to O.R.R., June 24, 1943.

52. D'Ianni, J. D., Naples, F. J., and Field, J. E., *Ind. Eng. Chem.*, **42,** 95–102 (1950).

53. D'Ianni, J. D., Zarney, J., Naples, F. J., and Marsh, J. W., Private Communication, Goodyear Tire & Rubber Co. to O.R.R., Apr. 20, 1945.

54. Dunbrook, R. F., *India Rubber World*, **117,** 203–7, 355–9, 486, 552, 617–9, 745–8 (1947–48).

55. Dykstra, H. B., *J. Am. Chem. Soc.*, **57,** 2255–9 (1935).

56. Eberly, K. C., and Johnson, B. L., *J. Polymer Sci.*, **3,** 283–96 (1948).

57. Ellslager, W. M., Private Communication, Govt. Lab. to O.R.R., June 2, 1947.

58. Ellslager, W. M., Private Communication, Govt. Lab. to O.R.R., June, 2, 1947.

59. Emerson, W. S., and Agnew, E. P., *J. Am. Chem. Soc.*, **67,** 518–20 (1945).

60. Feldon, M., Private Communication, Govt. Lab. to O.R.R., May 25, 1948.

61. Feldon, M., Private Communication, Govt. Lab. to O.R.R., June 2, 1948.

62. Feldon, M., Private Communication, Govt. Lab. to O.R.R., July 22, 1948.

63. Feldon, M., Private Communication, Govt. Lab. to O.R.R., July 22, 1948.

64. Feldon, M., and Gyenge, J. M., Private Communication, Govt. Lab. to O.R.R., May 27, 1948.

65. Feldon, M., and Reich, M. H., Private Communication, Govt. Lab. to O.R.R., Sept. 21, 1948.

66. Fielding, J. H., *Ind. Eng. Chem.*, **41**, 1560–3 (1949).

67. Firestone Tire & Rubber Co., Private Communication to O.R.R., May 18, 1944.

68. Firestone Tire & Rubber Co., Private Communication to O.R.R., Jan. 11, 1945.

69. Firestone Tire & Rubber Co., Private Communication to O.R.R., July 13, 1945.

70. Firestone Tire & Rubber Co., Private Communication to O.R.R., Apr. 3, 1946.

71. Firestone Tire & Rubber Co., Private Communication to O.R.R,. Sept. 10, 1948.

72. Fisher, H. L., *Ind. Eng. Chem.*, **39**, 1210–2 (1947).

73. Frank, R. L., Adams, C. E., Allen, R. E., Gander, R., and Smith, P. V., *J. Am. Chem., Soc.*, **68**, 1365–8 (1946).

74. Frank, R. L., Adams, C. E., Blegen, J. R. Deanin, R., and Smith, P. V., *Ind. Eng. Chem.*, **39**, 887–93 (1947).

75. Frank, R. L., Adams, C. E., Blegen, J. R., Smith, P. V., Juve, A. E., Schroeder, C. H., and Goff, M. M., *Ind. Eng. Chem.*, **40**, 420–1 (1948).

76. Frank, R. L., Adams, C. E., Blegen, J. R., Smith, P. V., Juve, A. E., Schrodeer, C. H., and Goff, M. M., *Ind. Eng. Chem.*, **40**, 879–82 (1948).

77. Frank, R. L., Blegen, J. R., Dearborn, R. J., Myers, R. L., and Woodward, F. E., *J. Am. Chem. Soc.*, **68**, 1368–9 (1946).

78. Frank, R. L., Blegen, J. R., Inskeep, G. E., and Smith, P. V., *Ind. Eng. Chem.*, **39**, 893–5 (1947).

79. Frank, R. L., and Emmick, R. D., Private Communication to O.R.R., Sept. 30, 1946.

80. Frank, R. L., Emmick, R. D., and Johnson, R. S., *J. Am. Chem. Soc.*, **69**, 2313–7 (1947).

81. Frank, R. L., and Smith, P. V., *Organic Syntheses*, **27**, 38 (1947).

82. Friedman, L. A., Private Communication to O.R.R., Aug. 1, 1945.

83. Frolich, P. K., *Ind. Eng. Chem., News Ed.*, **18**, 285 (1940).

84. Fryling, C. F., *Ind. Eng. Chem.*, **39**, 882–6 (1947).

85. Garvey, B. S., Whitlock, M. H., and Freese, J. A., *Ind. Eng. Chem.*, **34**, 1309–15 (1942).

86. Gehman, S. D., Jones, P. J., Wilkinson, C. S., and Woodford, D. E., *Ind. Eng. Chem.*, **42**, 475–82 (1950).

87. Gehman, S. D., Woodford, D. E., and Wilkinson, C. S., *Ind. Eng. Chem.*, **39**, 1108–15 (1947).

88. Goldsmith, H., Private Communication, Govt. Lab. to O.R.R., June 29, 1949.

89. Goldsmith, H., Private Communication to Govt. Lab., July 7, 1949.

90. Goldsmith, H., Private Communication, Govt. Lab. to O.R.R., Aug. 4, 1949.

91. Goodrich Co., B. F., Private Communication to O.R.R., May 1945.

92. Goodrich Co., B. F., Private Communication to O.R.R., June 1945.

93. Goodrich Co., B. F., Private Communication to O.R.R., Dec. 1945.

94. Gottlob, K., *India-Rubber J.*, **58**, 305–8, 348–50, 391–5, 433–6 (1919).

95. Gould, G. W., and Hulse, G. E., *Ind. Eng. Chem.*, **41**, 1021–4, 1025–7 (1949).

96. Government Laboratories, Investigations in Progress.

97. Graham, W., and Winkler, C. A., *Can. J. Research*, **26B**, 564–80 (1948).

98. Gyenge, J. M., Private Communication, Govt. Lab. to O.R.R., June 20, 1949.

99. Habgood, B. J., *Annual Report on the Progress of Rubber Technology*, Heffer, Cambridge, England, 1947, pp. 32–50.

100. Hale, N. H., Private Communication to O.R.R., Mar. 31, 1947.

101. Hampton, R. R., *Anal. Chem.*, **21**, 923–6 (1949).

102. Harries, C. D., U.S. Pat. 1,058,056 (1913).

103. Harrison, S. R., Am. Chem. Soc. Meeting, New York, Sept. 1947.

104. Higuchi, T., Leeper, H., and Davis, D. S., *Anal. Chem.*, **20**, 1029–33 (1948).

105. Hoseh, M., *India Rubber World* (*a—c*), **108**, 253–4, 461–3, 559–60 (1943); (*d—e*), **109**, 45–8, 155–7 (1943); (*f—g*), **109**, 368–70, 478–9 (1944); (*h—k*), **110**, 59–60, 176–7, 533–4, 647–8 (1944); (*l*), **111**, 178–9 (1944); (*m—n*), **111**, 564–5, 691–2 (1945); (*o—p*), **112**, 185, 450 (1945); (*q—r*), **113**, 74, 390–1 (1945); (*s*), **113**, 508 (1946); (*t—v*), **114**, 63–6, 374–5, 521 (1946); (*w—x*), **115**, 73–4, 216 (1946); (*y*), **115**, 518 (1947); (*z*), **116**, 62 (1947).

106. Huber, W. F., Renoll, M., Rossow, A. G., and Mowry, D. T., *J. Am. Chem. Soc.*, **68**, 1109–12 (1946).

107. Iddles, H. A., Lang, E. H., and Gregg, D. C., *J. Am. Chem. Soc.*, **59**, 1945, 1946 (1937).

108. Inskeep, G. E., and Deanin, R., *J. Am. Chem. Soc.*, **69**, 2237, 2238 (1947).

109. Jacobson, R. A., and Carothers, W. H., *J. Am. Chem. Soc.*, **55**, 1624 (1933).

110. Jensen, R. A., Private Communication, Govt. Lab. to O.R.R., Mar. 21, 1947.

111. Jensen, R. A., and Alden, G. E., Private Communication, Govt. Lab. to O.R.R., Oct. 20, 1947.

112. Jensen, R. A., and Alden, G. E., Private Communication, Govt. Lab. to O.R.R., Oct. 20, 1947.

113. Jensen, R. A., and Tiger, G. J., Private Communication, Govt. Lab. to O.R.R., May 7, 1947.

114. Johanson, A. J., and Goldblatt, L. A., *Ind. Eng. Chem.*, **40**, 2086–90 (1948).

115. Johanson, A. J., McKennon, F. L., and Goldblatt, L. A., *Ind. Eng. Chem.*, **40**, 500–2 (1948).

116. Johnson, B. L., *Ind. Eng. Chem.*, **40**, 351–6 (1948).

117. Johnson, P. H., and Bebb, R. L., *Ind. Eng. Chem.*, **41**, 1577–8 (1949).

118. Jones, H. C., *Rubber Age N.Y.*, **63**, 735–40 (1948).

119. Juve, A. E., *Ind. Eng. Chem.*, **39**, 1494–8 (1947).

120. Juve, A. E., Goff, M. M., Schroeder, C. H., Meyer, A. W., and Brooks, M. C., *Ind. Eng. Chem.*, **39**, 1490–3 (1947).

121. Juve, A. E., and Schroeder, C. H., *India Rubber World*, **115**, 515–7 (1947).

122. Juve, R. D., et al., Private Communication, Goodyear Tire & Rubber Co. to O.R.R., June 30, 1946.

123. Kharasch, M. S., Private Communication to O.R.R., Dec. 12, 1944.

124. Kharasch, M. S., Private Communication to O.R.R., Jan. 10, 1945.

125. Kharasch, M. S., Private Communication to O.R.R., Feb. 7, 1945.

126. Kharasch, M. S., Private Communication to O.R.R., Apr. 12, 1945.

127. Kharasch, M. S., Private Communication to O.R.R., May 10, 1945.

128. Kharasch, M. S., Private Communication to O.R.R., June 22, 1945.

129. Kharasch, M. S., Private Communication to O.R.R., June 24–Aug. 3, 1945.

130. Kharasch, M. S., and Nudenberg, W., Private Communication to O.R.R., Sept. 14–15, 1944.

131. Kharasch, M. S., and Nudenberg, W., Private Communication to O.R.R., May 10–12, 1945.

132. Kharasch, M. S., Nudenberg, W., Jensen, E. V., Fischer, P. V., and Mayfield, D. L., *Ind. Eng. Chem.*, **39**, 830–7 (1947).

133. Kirschner, W. C., Private Communication, Govt. Lab. to O.R.R., June 1, 1945.

134. Kirschner, W. C., Private Communication, Govt. Lab. to O.R.R., June 1, 1945.

135. Klages, A., and Keil, R., *Ber.*, **36**, 1632 (1903).

136. Koch, A., *Ind. Eng. Chem.*, **32**, 464–7 (1940).

137. Kolthoff, I. M., and Meehan, E. J., Private Communication to O.R.R., Mar. 31, 1946.

138. Kolthoff, I. M., and Lee, T. S., *J. Polymer Sci.*, **2**, 206–19 (1947).

139. Kolthoff, I. M., Lee, T. S., and Mairs, M. A., *J. Polymer Sci.*, **2**, 220–9 (1947).

140. Konrad, E., PB Rept. 13355, Aug. 18, 1945.
141. Laundrie, R. W., Private Communication, Govt. Lab. to O.R.R., Jan. 14, 1947.
142. Laundrie, R. W., Private Communication, Govt. Lab. to O.R.R., Feb. 21, 1947.
143. Laundrie, R. W., Private Communication, Govt. Lab. to O.R.R., Mar. 10, 1947.
144. Laundrie, R. W., Private Communication, Govt. Lab. to O.R.R., Mar. 17, 1947.
145. Laundrie, R. W., Private Communication, Govt. Lab. to O.R.R., Mar. 25, 1947.
146. Laundrie, R. W., Private Communication, Govt. Lab. to O.R.R., May 5, 1947.
147. Laundrie, R. W., Private Communication, Govt. Lab. to O.R.R., May 13, 1947.
148. Laundrie, R. W., Private Communication, Govt. Lab. to O.R.R., May 20, 1947.
149. Laundrie, R. W., Private Communication, Govt. Lab. to O.R.R., June 18, 1947.
150. Laundrie, R. W., Private Communication, Govt. Lab. to O.R.R., July 29, 1947
151. Laundrie, R. W., Private Communication, Govt. Lab. to O.R.R., July 29, 1947.
152. Laundrie, R. W., Private Communication, Govt. Lab. to O.R.R., Aug. 27, 1947.
153. Laundrie, R. W., Private Communication, Govt. Lab. to O.R.R., Oct. 21, 1947;
    Taft, W. K., Duke, J., Snyder, A. D., Feldon, M., and Laundrie, R. W., *Ind. Eng. Chem.*,
    45, 1043–53 (1953).
154. Laundrie, R. W., Private Communication, Govt. Lab. to O.R.R., Oct. 27, 1947.
155. Laundrie, R. W., Private Communication, Govt. Lab. to O.R.R., Jan. 15, 1948.
156. Laundrie, R. W., Private Communication, Govt. Lab. to O.R.R., June 25, 1948.
157. Laundrie, R. W., Private Communication, Govt. Lab. to O.R.R., Oct. 6, 1948.
158. Laundrie, R. W., Private Communication, Govt. Lab. to O.R.R., Nov. 29, 1948.
159. Laundrie, R. W., Private Communication, Govt. Lab. to O.R.R., June 10, 1949.
160. Laundrie, R. W., Private Communication, Govt. Lab. to O.R.R., Apr. 2, 1951.
161. Laundrie, R. W., and Jensen, R. A., Private Communication, Govt. Lab. to O.R.R.,
    Apr. 30, 1947.
162. Laundrie, R. W., and Murray, D., Private Communication, Govt. Lab. to O.R.R.
    Sept. 27, 1950.
163. Laundrie, R. W., and McCann, R. F., *Ind. Eng. Chem.*, **41**, 1568–70 (1949).
164. Lawson, J. K., Rabjohn, N., and Marvel, C. S., Private Communication to O.R.R.,
    Dec. 23, 1943.
165. Lessig, E. T., *Ind. Eng. Chem., Anal. Ed.*, **9**, 582–8 (1937).
166. Levine, A. A., and Cass, O. W., U.S. Pats. 2,290,759 (1943) and 2,193,823 (1940).
167. Livingston, J. W., *Chem. Eng. News*, **23**, 1627 (1945).
168. Livingston, J. W., *Chem. Eng. News*, **27**, 2444 (1949).
169. Maher, E. D., and Davies, T. L., *Rubber Age N.Y.*, **59**, 557–62 (1946).
170. Marquardt, D. N., Poirier, R. H., and Wakefield, L. B., *Ind. Eng. Chem.*, **41**, 1475–8
    (1949).
171. Marvel, C. S., Private Communication to O.R.R., Feb. 26, 1944.
172. Marvel, C. S., Private Communication to O.R.R., Mar. 30, 1944.
173. Marvel, C. S., Private Communication to O.R.R., May 2, 1944.
174. Marvel, C. S., Private Communication to O.R.R., May 23, 1944.
175. Marvel, C. S., Private Communication to O.R.R., June 23, 1944.
176. Marvel, C. S., Private Communication to O.R.R., July 22, 1944.
177. Marvel, C. S., Private Communication to O.R.R., Oct. 9, 1944.
178. Marvel, C. S., Private Communication to O.R.R., Nov. 1, 1944.
179. Marvel, C. S., Private Communication to O.R.R., Jan. 10, 1945.
180. Marvel, C. S., Private Communication to O.R.R., Feb. 10, 1945.
181. Marvel, C. S., Private Communication to O.R.R., Mar. 10, 1945.
182. Marvel, C. S., Private Communication to O.R.R., Apr. 10, 1945.
183. Marvel, C. S., Private Communication to O.R.R., May 10, 1945.
184. Marvel, C. S., Private Communication to O.R.R., June 10, 1945.
185. Marvel, C. S., Private Communication to O.R.R., July 10, 1945.
186. Marvel, C. S., Private Communication to O.R.R., Aug. 10, 1945.

187. Marvel, C. S., Private Communication to O.R.R., Sept. 10, 1945.
188. Marvel, C. S., Private Communication to O.R.R., Oct. 10, 1945.
189. Marvel, C. S., Private Communication to O.R.R., Nov. 10, 1945.
190. Marvel, C. S., Private Communication to O.R.R., Jan. 10, 1946.
191. Marvel, C. S., Private Communication to O.R.R., Feb. 10, 1946.
192. Marvel, C. S., Private Communication to O.R.R., Mar. 10, 1946.
193. Marvel, C. S., Private Communication to O.R.R., Apr. 10, 1945.
194. Marvel, C. S., Private Communication to O.R.R., May 10, 1946.
195. Marvel, C. S., Private Communication to O.R.R., June 10, 1946.
196. Marvel, C. S., Private Communication to O.R.R., July 10, 1946.
197. Marvel, C. S., Private Communication to O.R.R., Dec. 10, 1946.
198. Marvel, C. S., Private Communication to O.R.R., Feb. 10, 1947.
199. Marvel, C. S., Private Communication to O.R.R., Mar. 10, 1947.
200. Marvel, C. S., Private Communication to O.R.R., May 10, 1947.
201. Marvel, C. S., Private Communication to O.R.R., June 10, 1947.
202. Marvel, C. S., Private Communication to O.R.R., Aug. 10, 1947.
203. Marvel, C. S., Private Communication to O.R.R., Sept. 10, 1947.
204. Marvel, C. S., Private Communication to O.R.R., Nov. 10, 1947.
205. Marvel, C. S., Private Communication to O.R.R., Sept. 10, 1948.
206. Marvel, C. S., et al., Private Communication to O.R.R., Sept. 14–15, 1944.
207. Marvel, C. S., et al., Private Communication to O.R.R., Feb. 5, 1945.
208. Marvel, C. S., et al., Private Communication to O.R.R., July 31, 1945.
209. Marvel, C. S., Allen, R. E., and Overberger, C. G., *J. Am. Chem. Soc.*, **68,** 1088–91 (1946).
210. Marvel, C. S., and Bailey, W. J., Private Communication to O.R.R., Jan. 11–12, 1945.
211. Marvel, C. S., Bailey, W. J., and Inskeep, G. E., *J. Polymer Sci.*, **1,** 275–88 (1946).
212. Marvel, C. S., Deanin, R., and Inskeep, G. E., Private Communication to O.R.R., Jan. 30, 1945.
212a. Marvel, C. S., Fukuto, T. R., Berry, J. W., Taft, W. K., and Labbe, B. G., *J. Polymer Sci.*, **8,** 599–605 (1952).
212b. Marvel, C. S., and Fuller, J. A., *J. Am. Chem. Soc.*, **74,** 1506–9 (1952).
213. Marvel, C. S., and Hein, D. W., *J. Am. Chem. Soc.*, **70,** 1895–8 (1948).
214. Marvel, C. S., Inskeep, G. E., Deanin, R., Heim, D. W., Smith, P. V., Young, J. D., Juve, A. E., Schroeder, C. H., and Goff, M. M., *Ind. Eng. Chem.*, **40,** 2371–3 (1948).
215. Marvel, C. S., Inskeep, G. E., Deanin, R., Juve, A. E., Schroeder, C. H., and Goff, M. M., *Ind. Eng. Chem.*, **39,** 1486–90 (1947).
216. Marvel, C. S., Inskeep, G. E., Overberger, C. G., and Deanin, R., Private Communication to O.R.R., Mar. 26, 1946.
216a. Marvel, C. S., McCain, G. H., Passer, M., Taft, W. K., and Labbe, B. G., *Ind. Eng. Chem.*, **45,** 2311–7 (1953).
217. Marvel, C. S., and Moon, N. S., *J. Am. Chem. Soc.*, **62,** 45–9 (1940).
218. Marvel, C. S., and Overberger, C. G., *J. Am. Chem. Soc.*, **67,** 2250–2 (1945).
219. Marvel, C. S., Overberger, C. G., Allen, R. E., Johnston, H. W., Saunders, J. H., and Young, J. D., *J. Am. Chem. Soc.*, **68,** 861–4 (1946).
220. Marvel, C. S., Overberger, C. G., and Carlin, R. B., Private Communication to O.R.R., Dec. 29, 1943.
221. Marvel, C. S., Overberger, C. G., Saunders, J. H., and Saunders, R., Private Communication to O.R.R., Apr. 4, 1946.
221a. Marvel, C. S., Petersen, W. R., Inskip, H. K., McCorkle, J. E., Taft, W. K., and Labbe, B. G., *Ind. Eng. Chem.*, **45,** 1532–8 (1953).
222. Marvel, C. S., and Rao, N. S., Private Communication to O.R.R., Feb. 26, 1948; *J. Polymer Sci.*, **4,** 703–7 (1949).
223. Marvel, C. S., Saunders, J. H., and Overberger, C. G., *J. Am. Chem. Soc.*, **68,** 1085–8 (1946).

224. Marvel, C. S., and Schertz, G. L., *J. Am. Chem. Soc.*, **65**, 2054–8 (1943).
225. Marvel, C. S., Smith, P. V., and Frank, R. L., Private Communication to O.R.R., July 21, 1943.
226. Marvel, C. S., and Williams, J. L. R., Private Communication to O.R.R., June 8, 1948.
227. Marvel, C. S., and Williams, J. L. R., *J. Polymer Sci.*, **4**, 265–72 (1949).
228. Marvel, C. S., Williams, J. L. R., and Baumgarten, H. E., *J. Polymer Sci.*, **4**, 583–95 (1949).
229. Matsui, E., *Chem. Zentr.*, **114** (I), 1111 (1943).
230. Matthews, F. E., and Strange, E. H., Brit. Pat. 24,790 (1910).
231. McCann, R. F., Private Communication, Govt. Lab. to O.R.R., July, 30, 1947.
232. McCann, R. F., Private Communication, Govt. Lab. to O.R.R., Feb. 16, 1948.
233. McCann, R. F., Private Communication, Govt. Lab. to O.R.R., Apr. 29, 1948.
234. McCann, R. F., Private Communication, Govt. Lab. to O.R.R., May 11, 1948.
235. McCann, R. F., Private Communication, Govt. Lab. to O.R.R., Sept. 3, 1948.
236. McCann, R. F., Private Communication, Govt. Lab. to O.R.R., Nov. 1, 1949.
237. McCann, R. F., Reich, M. H., and Watson, C. A., Private Communication, Govt. Lab. to O.R.R., Oct. 6, 1948.
238. McCann, R. F., and Taft, W. K., Private Communication, Govt. Lab. to O.R.R., Oct. 13, 1947.
239. McKenzie, J. P., and Samuels, M. E., *Ind. Eng. Chem.*, **40**, 769–77 (1948).
240. McKeown, C. B., et al., Private Communication, B. F. Goodrich Chemical Co. to O.R.R., Feb. 26, 1946.
241. McMillan, F. W., Bishop, E. T., Marple, K. E., and Evans, T. W., *India Rubber World*, **113**, 663–9, 714 (1947).
242. Meehan, E. J., *J. Polymer Sci.*, **1**, 318–28 (1946).
243. Metzger, R. E., et al, and Cross, R. B., et al., Am. Chem. Soc. Meeting, New York, Sept. 1947.
244. Meyer, A. W., *Ind. Eng. Chem.*, **41**, 1570–7 (1949).
245. Meyer, G. E., and McCleary, C. D., Private Communication, U.S. Rubber Co. to O.R.R., Feb. 18, 1946.
246. Michalek, J. C., and Clark, C. C., *Chem. Eng. News*, **22**, 1559–63 (1944).
247. Midgley, T., Synthetic and Substitute Rubbers, in Davis and Blake, *The Chemistry and Technology of Rubber*, Reinhold, New York, 1937, pp. 677–704.
248. Mochel, W. E., Salisbury, L. F., Barney, A. L., Coffman, D. D., and Mighton, C. J., *Ind. Eng. Chem.*, **40**, 2285–9 (1948).
249. Mooney, H. R., Private Communication, Govt. Lab. to O.R.R., July 29, 1949.
250. Morton, A. A., Private Communication to Govt. Lab., Sept. 1949.
251. Morton, A. A., et al., *J. Am. Chem. Soc.*, **68**, 68, 93–6 (1946); **69**, 160–1, 161–7, 167–72, 172–6, 950–61, 1675–81 (1947); **70**, 3132–5 (1948); **71**, 481–6, 487–9 (1949); **72**, 3785–92 (1950); *Ind. Eng. Chem.*, **42**, 1488–96 (1950).
251a. Morton, A. A., Bolton, F. H., Collins, F. W., and Cluff, E. F., *Ind. Eng. Chem.*, **44**, 2876–82 (1952).
251b. Morton, A. A., *Rubber Age N.Y.*, **72**, 473–6 (1953).
252. Mowry, D. T., Renoll, M., and Huber, W. F., *J. Am. Chem. Soc.*, **68**, 1105–9 (1946).
253. Nagai, H., *J. Soc. Chem. Ind. Japan*, **45**, 1147–57 (1942).
254. Ort, Dr., PB Rept. 13323, microreel No. 3, frame 465.
255. Parrish, C. I., Private Communication, Govt. Lab. to O.R.R., July 20, 1951.
256. Poirier, R. H., Wakefield, L. B., and Willis, J. M., Private Communication, Firestone Tire & Rubber Co. to O.R.R., Oct. 16, 1946.
257. Powers, P. O., *Chem. Eng. News*, **22**, 1992–3 (1944).
258. Powers, P. O., *Ind. Eng. Chem.*, **38**, 837–9 (1946).
259. Rabjohn, N., Dearborn, R. J., Inskeep, G. E., Synder, H. R., and Marvel, C. S., *J. Polymer Sci.*, **2**, 488–502 (1947).

260. Rehner, J., *Ind. Eng. Chem.*, **36**, 46–51 (1944).
261. Reich, M. H., Private Communication, Govt. Lab. to O.R.R., July 24, 1947.
262. Reich, M. H., Private Communication, Govt. Lab. to O.R.R., Oct. 20, 1947.
263. Reich, M. H., Private Communication, Govt. Lab. to O.R.R., June 1, 1948.
264. Reich, M. H., Private Communication, Govt. Lab. to O.R.R., Aug. 27, 1948.
265. Reich, M. H., Private Communication, Govt. Lab. to O.R.R., Sept. 14, 1948.
266. Reich, M. H., Private Communication, Govt. Lab. to O.R.R., Sept. 21, 1948.
267. Reich, M. H., Private Communication, Govt. Lab. to O.R.R., Feb. 7, 1949.
268. Reich, M. H., Private Communication, Govt. Lab. to O.R.R., July 27, 1949.
269. Reich, M. H., Schneider, R. E., and Mills, E. O., Private Communication, Govt. Lab. to O.R.R., Sept. 15, 1950.
269a. Reich, M. H., Schneider, R. E., and Taft, W. K., *Ind. Eng. Chem.*, **44**, 2914–22 (1952).
270. Reich, M. H., and Watson, C. A., Private Communication, Govt. Lab. to O.R.R., Nov. 25, 1947.
271. Repar, J., Private Communication, Govt. Lab. to O.R.R., July 1, 1947.
272. Repar, J., Private Communication, Govt. Lab. to O.R.R., May 13, 1948.
273. Rinne, W. W., and Rose, J. E., *Ind. Eng. Chem.*, **40**, 1437–40 (1948).
274. Robinson, H. H., et al., Private Communication, Firestone Tire & Rubber Co. to O.R.R., Nov. 14, 1944.
275. Rokityanskii, I. V., *J. Applied Chem.*, **21**, 139–45 (1948).
276. Rubber Bureau, War Production Board, and O.R.R., PB Rept. 13336, Aug. 1945.
277. Rubber Subcommittee, Tech. Ind. Intelligence Comm., PB Rept. 13342, July 11, 1945.
278. Saffer, A., and Johnson, B. L., *Ind. Eng. Chem.*, **40**, 538–41 (1948).
279. Salomon, G., and Koningsberger, C., *J. Polymer Sci.*, **2**, 522–41 (1947).
280. Saunders, J. H., Private Communication to O.R.R., Oct. 15, 1946.
281. Schade, J. W., Private Communication, Govt. Lab. to O.R.R., Mar. 12, 1948.
282. Schade, J. W., Private Communication, Govt. Lab. to O.R.R., Apr. 16, 1948.
283. Schade, J. W., Private Communication to C. S. Marvel, Sept. 28, 1948.
284. Schade, J. W., and Labbe, B. G., Private Communication, Govt. Lab. to O.R.R., May 3, 1946.
285. Schade, J. W., and Labbe, B. G., Private Communication, Govt. Lab. to O.R.R., June 20, 1946.
286. Schade, J. W., Labbe, B. G., and Hill, N. C., Private Communication, Govt. Lab. to O.R.R., Aug. 13 and Nov. 25, 1946.
287. Schroeder, C. H., Private Communication, B. F. Goodrich Co. to O.R.R., Sept. 14–15, 1944.
288. Schroeder, C. H., Private Communication, B. F. Goodrich Co. to O.R.R., Jan. 11–13, 1945.
289. Schulze, W. A., Private Communication, Phillips Petroleum Co. to O.R.R., June 10, 1945.
290. Schulze, W. A., Private Communication, Phillips Petroleum Co. to O.R.R., Dec. 10, 1949.
291. Schulze, W. A., and Crouch, W. W., *J. Am. Chem. Soc.*, **70**, 3891–3 (1948).
292. Schulze, W. A., Crouch, W. W., and Lynch, C. S., *Ind. Eng. Chem.*, **41**, 414–6 (1949).
293. Semon, W. L., *Chem. Eng. News*, **21**, 1613–9 (1943).
294. Sjothun, I. J., and Cole, O. D., *Ind. Eng. Chem.*, **41**, 1564–7 (1949).
295. Smith, H. F., and Thompson, R. D., Private Communication, Govt. Lab. to O.R.R., June 14, 1946.
296. Soday, F. J., *Trans. Am. Inst. Chem. Engs.*, **42**, 647–64 (1946).
297. Starkweather, H. W., Bare, P. O., Carter, A. S., Hill, F. B., Hurka, V. R., Mighton, C. J., Sanders, P. A., Walker, H. W., and Youker, M. A., *Ind. Eng. Chem.*, **39**, 210–22 (1947).
298. Stewart, R. A., and Williams, H. L., *Ind. Eng. Chem.*, **45**, 173–82 (1953).

299. Strong, F. M., and McElvain, S. M., *J. Am. Chem. Soc.*, **55**, 816–22 (1933).
300. Swart, G. H., Pfau, E. S., and Weinstock, K. V., *India Rubber World*, **124**, 309–19 (1951). See also Weinstock, K. V., Storey, E. B., and Sweeley, J. S., *Ind. Eng. Chem.*, **45**, 1035–43 (1953).
301. Taft, W. K., Private Communication, Gov. Lab. to O.R.R., Jan. 29, 1948.
302. Taft, W. K., Private Communication to C. S. Marvel, Oct. 27, 1948.
303. Taft, W. K., Private Communication, Govt. Lab. to O.R.R., June 10, 1949.
304. Taft, W. K., and Alden, G. E., Private Communication, Govt. Lab. to O.R.R., Apr. 8, 1949.
305. Taft, W. K., and Goldin, J., Private Communication, Govt. Lab. to O.R.R., July 13, 1948.
306. Taft, W. K., and Goodsmith, H., *Ind. Eng. Chem.*, **42**, 2542–6 (1950).
307. Taft, W. K., Reich, M. H., Schneider, R. E., and Goldsmith, H., Private Communication, Govt. Lab. to O.R.R., Oct. 5, 1951.
308. Taft, W. K., Snyder, A. D., and Oates, W. E., Private Communication, Govt. Lab. to O.R.R., May 31, 1946.
309. Talalay, A., and Magat, M., *Synthetic Rubber from Alcohol: Survey Based on the Russian Literature*, Interscience, New York, 1945.
310. Thorn, J. P., Private Communication, Govt. Lab. to O.R.R., May 28, 1945.
311. Tiger, G. J., Private Communication, Govt. Lab. to O.R.R., Aug. 1, 1945.
312. Tiger, G. J., Private Communication, Govt. Lab. to O.R.R., Aug. 29, 1945.
313. Tiger, G. J., Private Communication, Govt. Lab. to O.R.R., Sept. 2, 1945.
314. Tiger, G. J., Private Communication, Govt. Lab. to O.R.R., Sept. 4, 1945.
315. Tiger, G. J., Private Communication, Govt. Lab. to O.R.R., Sept. 16, 1945.
316. Tiger, G. J., Private Communication, Govt. Lab. to O.R.R., Dec. 10, 1945.
317. Tiger, G. J., Private Communication, Govt. Lab. to O.R.R., Oct. 24, 1946.
318. Tiger, G. J., Private Communication, Govt. Lab. to O.R.R., Nov. 18, 1946.
319. Tiger, G. J., Private Communication, Govt. Lab. to O.R.R., Nov. 19, 1946.
320. Tiger, G. J., Private Communication, Govt. Lab. to O.R.R., Apr. 30, 1947.
321. Tiger, G. J., Private Communication, Govt. Lab. to O.R.R., May 13, 1947.
322. Tiger, G. J., Private Communication, Govt. Lab. to O.R.R., July 25, 1947.
323. Tiger, G. J., Private Communication, Govt. Lab. to O.R.R., May 25, 1948.
324. Tiger, G. J., Private Communication, Govt. Lab. to O.R.R., Dec. 8, 1948.
325. Tiger, G. J., Private Communication, Govt. Lab. to O.R.R., Dec. 21, 1948.
326. Tiger, G. J., and KixMiller, R. W., Private Communication, Govt. Lab. to O.R.R., Feb. 2, 1945.
327. Tiger, G. J., and Watson, C. A., Private Communication, Govt. Lab. to O.R.R., Oct. 12, 1948.
328. Tiger, G. J., and Whitney, R. H., Private Communication, Govt. Lab. to O.R.R., Aug. 27, 1947.
329. Trumbull, H. L., Private Communication, B. F. Goodrich Co. to O.R.R., Oct. 1946.
330. Trumbull, H. L., Private Communication, B. F. Goodrich Co. to O.R.R., June 10, 1947.
331. Walling, C., and Wolfstirn, K. B., *J. Am. Chem. Soc.*, **69**, 852–4 (1947).
332. Watson, C. A., Private Communication, Govt. Lab. to O.R.R., Apr. 6, 1948.
333. Weidlein, E. R., Jr., *Chem. Eng. News*, **24**, 771–3 (1946).
334. Weidlein, E. R., Jr., and KixMiller, R. W., PB Rept. 13340, Apr. 21, 1945.
335. Wheelock, G. L., Private Communication, B. F. Goodrich Co. to O.R.R., Apr. 7, 1949.
336. Whitby, G. S., and Crozier, R. N., *Can. J. Research*, **6**, 203–25 (1932).
337. Whitby, G. S., and Gallay, W., *Can. J. Research*, **6**, 280–91 (1932).
338. Whitby, G. S., and Katz, M., *Can. J. Research*, **6**, 398–408 (1932).
339. White, L. M., *Ind. Eng. Chem.*, **41**, 1554–9 (1949).
340. Whitney, R. H., and Tiger, G. J., Private Communication, Govt. Lab. to O.R.R., July 7, 1947.

341. Wilks, W. W., Chaudoir, C. C., and McKenzie, J. P., Private Communication, Copolymer Corp. to O.R.R., Apr. 28, 1945.
342. Willis, J. M., Wakefield, L. B., Poirier, R. H., and Glymph, E. M., *Ind. Eng. Chem.*, **40,** 2210–5 (1948).
343. Wolf, R. F., *Scientific Monthly*, **66,** 221–31 (1948).
344. Yanko, J. A., *J. Polymer Sci.*, **3,** 576–601 (1948).
345. Youker, M. A., and Copeland, N. A., PB Rept. 16029, Mar. 14, 1946.
346. Ziegler, K., and Bahr, K., *Ber.*, **61B,** 253–63 (1928).
347. Ziegler, K., Grimm, H., and Willer, R., *Ann.*, **542,** 90–122 (1940).
348. Private Communications by various rubber companies and other organizations cooperating in the Government Synthetic Rubber Program.

CHAPTER 22

NEOPRENE

Arthur M. Neal and Leland R. Mayo

E. I. du Pont de Nemours & Company

The term Neoprene is a generic one which denotes a synthetic rubberlike polymer made by polymerizing chloroprene (2-chloro-1,3-butadiene) or by polymerizing a mixture of polymerizable monomers, the major component of which is chloroprene. By common usage the term has been broadened to include commercial rubberlike compounds, the major elastomeric constituent of which is Neoprene as above defined. Thus it is proper to speak of a Neoprene product such as a Neoprene wire jacket or a Neoprene gasket. With the introduction of Neoprene in 1931, it became clear that the properties of rubber that make it so important to our industrial development and social well-being do not depend on its specific chemical composition and that a successful synthetic rubber, unlike synthetic indigo, would not necessarily be an exact duplication of the natural product but rather would duplicate those properties of natural rubber that are responsible for its utility. Furthermore, the earliest tests on Neoprene showed that in many properties, notably age resistance, sunlight resistance, resistance to attack by most oils and fats, ozone resistance, and resistance to chemicals, it is far superior to natural rubber.

STRUCTURE OF NEOPRENE

In spite of the fact that the elastic properties of Neoprene demonstrated that it is not necessary to duplicate the chemical composition of natural rubber in order to produce a practical synthetic rubber, nevertheless it is understandable that the question of the molecular structure of this polymer has been the subject of considerable investigation. In their original paper, published in 1931, Carothers, Williams, Collins, and Kirby[11] reported the results of experiments on the molecular structure of polychloroprene. Their work was carried out on so-called mu polymer, which is formed when chloroprene is allowed to polymerize spontaneously at room temperature in the absence of a modifier, and which more closely resembles vulcanized than unvulcanized Neoprene. On the basis of oxidation experiments with hot nitric acid (which yielded succinic acid) they concluded that the polymer was formed by 1,4-addition. On the basis of the fact that boiling the polymer for 6 hours with alcoholic potash or with pyridine removed only slight traces of chlorine, they concluded that the chlorine atoms of the polymer are still attached to carbon atoms bearing double bonds. They likewise reported that mu polychloroprene, when stretched about 500 per cent, exhibited a fiber diagram showing a number of definite layer lines. Their measurements gave a value of 4.8 A. for the identity period along the fiber axis, which corresponds rather closely with the calculated length of one chloroprene unit. They concluded that the X-ray diffraction pattern indicates a *trans* rather than a *cis* configuration, which corresponds in many respects with the structure assigned to β-gutta-percha.

The validity of these early conclusions regarding the structure of Neoprene has been fully substantiated by more recent investigators. Bunn[9] found that all the diffraction spots in an X-ray photograph of a stretched specimen of Neoprene could be accounted for by a 4-molecule orthorhombic cell having dimensions 8.84, 10.24 and 4.79 A. and showed that the relative intensities could all be accounted for by a structure analogous to that of β-gutta-percha. However, the orientation of the molecules is considerably different from that in β-gutta-percha, presumably because of the difference in size and electrical characteristics between the chlorine and the methyl substituents. Salomon and Koningsberger,[52] working with Neoprene types GN-A and FR, showed that with both of these polymers at least 95 per cent of the chlorine has the expected stability of vinyl chlorine, the chlorine atom being attached to a carbon atom bearing a double bond, and that not over 5 per cent is of the allylic type, such as could result from 1,2-addition during polymerization.

The most extensive work on the structure of Neoprene is that carried out by Mochel and his co-workers. Thus Mochel, Nichols, and Mighton[40] studied the molecular-weight distribution of Neoprene type GN and reported that the range of molecular weight was 20,000 to 959,000, with the most abundant species at 100,000. Their results were obtained by fractionating a sample of Neoprene type GN by partial precipitation from solution in benzene and examining the fractions both osmotically and viscometrically. The extension of the distribution curve at the high-molecular-weight end suggests the presence of soluble branched or cross-linked material. Mochel and Nichols[39] investigated the structure of Neoprene type W. They show that this Neoprene has a higher average molecular weight than Neoprene type GN or CG and furthermore that the molecular-weight distribution in it is more uniform that in other types of Neoprene and that Neoprene type W approximates natural rubber in this respect. They believe that this more uniform distribution and higher molecular weight are responsible for the improved processing characteristics of Neoprene type W. The results of fractionating Neoprene type W by precipitation from dilute solution in benzene at constant temperature with methanol are given in Table I.

Neoprene type W also has a greater uniformity of molecular structure than other types of Neoprene. Ozonolysis yields a somewhat higher proportion of succinic acid, and X-ray diffraction studies show that type W crystallizes more readily than Neoprene type GN, indicating greater uniformity of structure. Both these measurements, however, indicate that there is no difference in the crystalline structure of the various Neoprenes, i.e., in the sense that only *trans* 1,4-addition is involved.

It has been established that variation in the temperature at which chloroprene is polymerized provides a range of polymers that differ markedly in

Table I. Fractionation of Neoprene Type W

Fraction	Per Cent of Whole	Intrinsic Viscosity	Molecular Weight
Unfractionated	100	1.35	206,000
Fraction 1	3.5	2.86	1,050,000
2	8.6	1.94	...
3	16.0	1.89	...
4	14.2	1.55	415,000
5	16.0	1.20	275,000
6	11.8	0.98	200,000
7	8.5	0.80	151,000
8	7.9	0.68	120,000
9	4.7	0.54	113,000
10	4.8	0.35	47,000
11	2.9	0.24	...
12	1.1	...	...
	100.0		

the raw state in their rate of crystallization and dispersion in solvents and in the physical properties of their vulcanizates. Walker and Mochel[64] investigated the effect of the temperature of polymerization on the structure of the resulting polymers. Ozonolysis disclosed that at least 94 per cent of the polychloroprene made at both 10 and 40° C. is a 1,4-addition product. Even the most careful investigation was unsuccessful in establishing the presence of side vinyl groups, longer side branches, or cross-linkages. It is interesting to note that omega or "popcorn" polychloroprene[11] gives on ozonolysis approximately the same high amount of succinic acid as the soluble plastic emulsion polymers, indicating that even in this highly insoluble hard polymer the number of branches and cross-links is extremely small. Infrared spectroscopic results were consistent with the ozonolysis experiments and showed that the 10 and 40° C. polymers have no significant structural differences. Careful examination in the 3-micron region failed to reveal the existence of vinyl groups in either polymer. X-ray examination showed that the identity period along the fiber axis of stretched films of both 10 and 40° C. polychloroprenes is 4.75 ± 0.08 A., corresponding to a *trans*-2-chloro-2-butenylene unit structure. No evidence of head-to-head or tail-to-tail structure or of the presence of *cis* isomers was obtained, but the authors consider that their work does not completely exclude the possibility of the existence of such configurations.

TYPES OF NEOPRENE AND METHOD OF MANUFACTURE

Chloroprene is a very versatile monomer, from which a wide variety of polymers can be produced, depending on the manner in which the polymerization process is carried out. Since 1931, when the first Neoprene was introduced under the trade name of DuPrene, the number of products that have been made commercially is rather large. Currently there are available ten Neoprenes of the dry-polymer type and nine different latexes.

The most familiar of the dry-polymer types is the general-purpose *Neoprene, type GN* (identical with GR-M, i.e., Government Rubber—Monovinylacetylene, under which designation it was known during World War II), and the following brief description of its method of manufacture can well serve as a basis of discussion for all of the dry polymers.

The process for the manufacture of monovinylacetylene from acetylene described by Nieuwland, Calcott, Downing, and Carter[46] is still followed as the first step in the manufacture of Neoprene. This process consists in polymerizing acetylene in the presence of an aqueous catalyst solution of cuprous chloride, acidified with hydrogen chloride and containing additional chlorides as solubilizing agents. Monovinylacetylene is then made to react with hydrogen chloride in the presence of a cuprous chloride solution to form chloroprene, a water-white liquid, according to the process described by Carothers and co-workers.[11]

After suitable purification by careful fractionation, chloroprene is polymerized according to a method described by Walker and Mochel.[64] In this process the chloroprene is emulsified in water by means of sodium rosinate soap and polymerized at 40° C. with the aid of potassium persulfate as a catalyst and in the presence of elemental sulfur as a modifier. The basic polymerization recipe is as follows.

Chloroprene	100 parts
N Wood rosin	4 ⎱ Dissolved in
Sulfur	0.6 ⎰ the monomer
Water	150
Sodium hydroxide	0.8
Sodium salt of naphthalene sulfonic acid-formaldehyde condensation product	0.7
Potassium persulfate	0.2 to 1.0

The progress of the polymerization is followed by means of specific-gravity changes, according to the method disclosed by Barrows and Scott.[4] After completion of the polymerization, an emulsion of tetraethylthiuram disulfide is introduced, and the alkaline latex is allowed to age for a controlled period of time, during which the plasticity of the polymer is increased to a desirable level by the action of the thiuram disulfide in cleaving sulfur linkages.[41] (In this connection it may be mentioned that, as a result of experiments involving the use of radiosulfur, Mochel and Peterson[41] concluded that sulfur used as a modifier in the polymerization of chloroprene actually enters into the polymer chains to the extent of about one atom of sulfur per 100 units of chloroprene.) The thiuram disulfide also acts as a stabilizer in the finished dry polymer.

The finishing operation in the production of Neoprenes is unique and has been described in some detail by Youker.[72] The alkaline latex is acidified with acetic acid to a point just short of coagulation, and ultimate coagulation is accomplished by freezing on the surface of a large rotating brine-cooled drum partially immersed in the latex. The resulting film is stripped from the freeze roll by means of a doctor blade, thoroughly washed with water, passed through squeeze rolls, and finally dried in air at 120° C.

The dried film is gathered into rope form and cut into short lengths for bagging. The resulting sticks are flexible, free from odor, and light amber in color and have a specific gravity of 1.23.

Neoprene type GN-A (GR-M 10) is produced by a process substantially identical with that by which type GN is made with the exception that a small amount of phenyl-α-naphthylamine is present as an additional stabilizer. The discoloration caused by this stabilizer on exposure to light or its fluorescence in ultraviolet light provides a convenient means of distinguishing between these two types of Neoprene.

Most Neoprenes, including the general-purpose types already described, result from the polymerization of the single monomer, chloroprene. Such polymerization yields a fairly regular linear polymer which, like natural rubber, is subject to the reversible process of crystallization. *Neoprene type RT* is a third general-purpose Neoprene in which crystallization is minimized by copolymerizing a small amount of styrene with the chloroprene. In this polymer the basic regularity of the polychloroprene chain is broken up to a degree sufficient to reduce significantly the rate and extent of crystallization.[34]   Otherwise, Neoprene type RT differs from type GN only in that it contains a stabilizer of the nondiscoloring antioxidant type, in addition to a thiuram disulfide. A slightly different version of type RT is Neoprene type GRT, which has now replaced the former.

Neoprene type FR, a copolymer of chloroprene and isoprene, is a specialty Neoprene developed primarily for low-temperature service. Its vulcanizates have brittle points 10 to 15° C. lower than those of corresponding vulcanizates from Neoprene type GN. This Neoprene is stabilized with both a thiuram disulfide and phenyl-α-naphthylamine. It is dark in color and, as a result of its isoprene content, has a lower specific gravity (1.14) than the other Neoprenes. Its odor is characteristic. *Neoprene type Q*, introduced in 1952, is a copolymer of chloroprene and acrylonitrile having good processibility and oil resistance.

All of the other dry polymers are produced from the single monomer, chloroprene. Although, in general, their method of manufacture is similar to that of Neoprene type GN, the polymerization environment is modified sufficiently to produce special properties in the individual finished products. It has been found[64] that, the lower the temperature at which the polymerization is conducted, the greater is the tendency of the polymer to crystallize and the higher is its tensile strength. *Neoprene types CG and AC* are representative of Neoprenes that crystallize rapidly. They find their widest use in adhesive cements because of their rapid development of high, uncured bond strength as the result of the rapid setup of the polymer. *Neoprene type KNR* is a soft (chemically plasticizible) type suitable for the production of cements and spreading compounds. The use as cements of the three last-mentioned types of Neoprene is discussed by Mitchell.[37]

Neoprene type S is a tough unmodified cross-linked polymer suitably stabilized to permit its use in the uncured state for crepe soles or other wearing surfaces. It serves in addition as an effective stiffener for compounds of other Neoprenes.

The oldest of the Neoprenes to remain, until recently, in commercial production was *Neoprene type E*, a stable, slow-curing polymer containing both phenyl-α-naphthylamine and a thiuram disulfide as stabilizers, the latter in a concentration considerably lower than that in Neoprene type GN.

Neoprene type W differs from type GN both in the method of modification employed in its preparation and in its stability. It contains no sulfur, thiuram disulfide, or other compound capable of decomposing to yield either free sulfur or a vulcanization accelerator. Neoprene type W is a general-purpose Neoprene which has markedly improved stability, improved processing characteristics, and greater adaptability to variations in vulcanizing systems than the earlier general-purpose types. It is silver-gray in color but is otherwise similar in appearance to Neoprene type GN. A crystallization-resistant counterpart of type W is Neoprene type WRT.[50]

Neoprene Latexes. The Neoprene latexes, like the dry polymers, are designed to function in processes normally employed with natural rubber. Thus polymer may be isolated from all of the Neoprene latexes by conventional spreading, coating, spraying, gelling, and coagulant dipping techniques. The Neoprene latexes are more concentrated than the emulsions used in the polymerization of the dry types. They do not contain a thiuram disulfide, are somewhat more alkaline (*p*H when fresh: 12.2 to 12.4), and contain a buffer to aid in maintaining proper alkalinity during storage and processing. Latexes of 50 per cent solids are made directly by polymerization. Less sulfur modification is used in their preparation than in the preparation of the dry Neoprenes, so that the rubber in them is nonplastic and will not disperse in solvents.

Neoprene latex type 571 is an alkaline 50 per cent dispersion of polymerized chloroprene. As in the Neoprene type-GN process, the emulsifier is sodium rosinate and the modifier is elemental sulfur. The polymerization environment, however, is adjusted sufficiently to produce a polymer which is tougher and less plastic, being somewhat akin to partially prevulcanized natural rubber.

Neoprene latex type 572 bears the same relation to type 571 as Neoprene type CG does to type GN. Thus it is made to the same solids content and in the same catalyst-modifier-emulsifier system as type 571. Like CG, type 572 is particularly suited to adhesive applications.

Neoprene latex types 842 and 842A are similar to type 571 except that they contain a modifier which produces a more easily vulcanized polymer with considerably less tendency toward crystallization. Type-842A polymer is noticeably superior to type-842 polymer (the manufacture of which has now been discontinued) and markedly superior to type-571 polymer in both these respects; however, it has somewhat lower tensile strength than either of these.

Neoprene latex types 60 and 601A are the 60 per cent creamed counterparts of types 571 and 842A, respectively. Creaming is accomplished through the use of ammonium alginate solutions. These latexes are used where, as in foam manufacture, high solids contents are required.

Neoprene latex type 700 is a modification of type 842A especially designed for saturating paper and other fibrous materials. It is emulsified with a larger proportion of sodium rosinate than type 842A. Further, the type-700 polymer contains some ester gum added to the chloroprene monomer before emulsification. This results in unusually efficient plasticization of the polymer and decreases its tendency to crystallize.

Neoprene latex type 735 is a 36 per cent sodium rosinate latex containing a sol type of chloroprene polymer. This type polymer is especially suitable for addition to paper pulp according to procedures described by Walsh, Abernathy, Pockman, Galloway, and Hartsfield.[68]

ELASTOMERS FROM OTHER HALOPRENES

A natural question arises concerning the possibility of preparing elastomers from haloprenes other than chloroprene. Elastomers have been prepared from all of the other haloprenes analogous to chloroprene. The most interesting of this group are the fluoroprene polymers described by Mochel and co-workers.[42] None of these polymers, however, has shown sufficient promise to justify commercial development.

UNVULCANIZED NEOPRENE

Stability. Carothers and his co-workers[11] showed that α-polychloroprene, the plastic, soluble polymer which can be isolated after chloroprene in bulk has polymerized to the extent of only about 30 per cent, is capable of spontaneously changing to the mu (nonplastic, "vulcanized") type. It is not surprising, therefore, to find that the commercial types of Neoprene undergo changes on storage. Catton[12] investigated this phenomenon and showed that, from a practical standpoint, these changes result in a reduction of processibility, an increase in rate of cure, a reduction in processing safety, and changes in plasticity. The effects vary considerably from one Neoprene type to another, and among the general-purpose types, Neoprene type W has by far the least tendency to change on storage. These changes are apparently the result of chemical reactions and, therefore, are influenced markedly by the temperature of storage, elevated temperatures accelerating the phenomenon and reduced temperatures retarding it.

The physical properties of vulcanizates, e.g., their tensile strength, also change as a result of extended aging of raw polymer, and these effects are most noticeable in gum-type stocks. Figure 1 shows the effect of a 10 minutes' at 307° F. cure on the tensile strength, the stress at 600 per cent elongation, and the ultimate elongation of gum-type vulcanizates prepared from Neoprene type GN which had been stored[51] for periods up to one year at 82° F.

Oxidation undoubtedly plays a major role in the changes that are observed to accompany the extended aging of raw uncompounded Neoprene. It is perhaps simplest to think in terms of two types of oxidation phenomena. The first is a chain scission, which reduces the average polymer chain length and results in a softer polymer having a low potentiality for the development of tensile strength. The second oxidation reaction may be considered as one that creates additional centers of reactivity. In the raw polymer this induces the formation of additional cross-links, with a consequent increase in nerve and toughness. In compounded stocks such increased number of reactive centers accelerates curing at both processing and curing temperatures. As would be expected, the presence of an antioxidant type of stabilizer, such as the phenyl-α-naphthylamine in Neoprene type GN-A, effects a general improvement in stability over that of the corresponding Neoprene type GN which is not so stabilized.

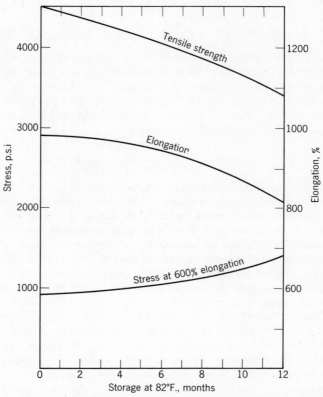

FIG. 1. Influence of Storage of Raw Neoprene, Type GN, on the Properties of Vulcanizates Obtained by Curing 10 minutes at 307° F. in a Gum Stock

Processing. All the commercial varieties of Neoprene can be processed on conventional rubber equipment and by methods similar to those used in preparing corresponding products from natural rubber. For most applications, Neoprene as supplied can be compounded directly without preliminary breakdown. If breakdown is required, however, it may be accomplished by mechanical working at low temperatures or by means of chemical peptizing agents. The latter are normally more effective at high temperatures. The most effective Neoprene peptizers are guanidines, such as di-o-tolylguanidine, and dithiocarbamates, such as piperidinium pentamethylenedithiocarbamate. Aromatic mercaptans, widely used in peptizing natural rubber, will also peptize Neoprene but exert such a strong retarding action on the rate of cure of stocks containing them that they find little practical application. For reasons undoubtedly associated with its greater stability, Neoprene type W resists both mechanical breakdown and chemical peptization.

Catton and co-workers[12, 15] have made an extensive study of processing problems with Neoprene and have concluded that the most important single rule to follow is to carry out the processing operations as rapidly as possible. They recognize that Neoprene is capable of existing in three distinct phases which depend on the temperature. Below 150 to 160° F. it exists in what is called the *elastic phase*. In this phase Neoprene coheres tightly so that strong running bands are obtainable on a mixing mill. Between approximately 160 and 200° F. it exists in a so-called *granular phase*. In this condition Neoprene loses cohesion to itself but tends to stick tightly to the mill rolls. Above 200° F. the Neoprene enters a *plastic phase*, in which it is very soft, has only weak cohesion to itself, shows very little tendency to stick to the mill rolls, and has almost no nerve.

Similar transitions have been noted with natural rubber, although the temperatures at which they occur are less clearly defined and, in general, are somewhat higher. A clear apprehension of this three-phase cycle in Neoprene gives a good clue to proper processing of all types of Neoprene stocks. Operations in a range from 160 to 200° F., where Neoprene exists in the granular phase, should be avoided as far as possible. The best dispersions are obtained when fillers are incorporated in the elastic phase, i.e., below 160° F. The fine calendering of high-quality stocks requiring minimum shrinkage and the most careful control of gage is normally carried out in the plastic phase above 200° F., but most sheet calendering, where careful control of gage is not required, is best carried out at temperatures where the elastic phase prevails.

Crystallization, which takes place in Neoprene at a temperature considerably higher than in natural rubber, sometimes influences the processing of Neoprene stocks. This phenomenon may result in a marked stiffening and loss of tack in an uncured compound as a consequence of exposure at room temperatures for several days. This effect, which is readily reversible with heat, may be a factor in building and laminating operations.

VULCANIZATION AND BASIC COMPOUNDING

The vulcaniztion of Neoprene differs fundamentally from that of other elastomers. This is illustrated by the fact that Neoprene, without any added ingredients, may be vulcanized by heat alone. Bridgwater and Krismann[7] studied the effects of metallic oxides on the vulcanization of Neoprene and recommended the use of combinations of zinc oxide and magnesium oxide for the production of practical Neoprene vulcanizates. This recommendation has never been supplanted. In mixing Neoprene stocks, the magnesia should be added at the first and the zinc oxide last.

Any of the grades of zinc oxide suitable for use in natural rubber is satisfactory for use in Neoprene compounds. Neoprene vulcanization is, however, markedly influenced by variations in the type of magnesia used.

The best results are obtained with the so-called extra-light calcined types. Heavy magnesias, frequently called "rubber grade," do not give either the safety during processing or the stability in vulcanizates afforded by the light calcined types. Since all magnesias are rapidly converted to the carbonate or hydroxide by the action of carbon dioxide and water, respectively, and since neither of these latter compounds is effective in Neoprene vulcanization, due care must be exercized to protect from long exposure to moisture and air any magnesia destined for use in Neoprene.

The mechanism by which these oxides benefit the vulcanization process is not clearly understood, but the influence they exert may be briefly described as follows. The use of zinc oxide as the sole curing agent produces both a fast cure and a uniform state of cure on extended vulcanization; i.e., such stocks are "flat curing." The limitations of such compounds include a pronounced tendency to be "scorchy," and the leveling off of cure at a relatively low point. On the other hand, the use of magnesia as the sole curing agent produces Neoprene compounds which are very slow curing and are safe processing, but in which curing activity tends to persist. In addition, magnesia improves vulcanizate aging properties, presumably by acting as a scavenger for hydrogen chloride liberated by Neoprene under oxidizing conditions. In combination, zinc oxide and magnesia supplement each other to produce well-balanced stocks, the properties of which can be varied, frequently to advantage, by adjustment in the ratio of one oxide to the other.

Many other metallic oxides affect the cure of Neoprene.[54] With the exception of litharge[54] and red lead, [33] however, which are used to produce vulcanizates having low water absorption, none of the other oxides alone or in combinations approach the desirable over-all balance of the zinc oxide-magnesia curing system. Red lead is much less scorchy than litharge in Neoprene stocks.[30a]

The vulcanization of Neoprene with sulfur is of comparatively minor importance. The reaction of this agent with Neoprene is slow, even in the presence of accelerators. It is commonly used only in the presence of metallic oxide curing agents for the purpose of producing special properties in the vulcanizate. The metallic oxide curing system is sufficiently active for many and perhaps most Neoprene compounds. Both speed of vulcanization and eventual state of cure can be favorably influenced, however, by the use of additional special accelerators.

The unique character of Neoprene vulcanization is again illustrated by the fact that rubber accelerators as a class are not effective. As a matter of fact, several have measurable retarding effects and one, mercaptobenzothiazole,[62] is an extremely potent retarder of Neoprene vulcanization. In general, the most effective accelerators for dry Neoprene stocks are special organic compounds. Neal,[44] however, was unsuccessful in an attempt to discover a relationship between chemical structure and accelerating activity. Torrence[60] has shown that antimony sulfide is a strong accelerator. The most widely used accelerators are the di-o-tolylguanidine salt of dicatechol borate (Permalux), 2-mercaptoimidazoline (NA-22), and p,p'-diaminodiphenylmethane. Catechol itself is a powerful accelerator but is too scorchy for most applications.   Sodium acetate is an efficient retarder for Neoprene compounds.[48]   It is ordinarily preferred to mercaptobenzothiazole or benzothiazolyl disulfide since its retarding effect does not persist at normal curing temperatures.

The vulcanization of Neoprene type W varies in some details from that of the earlier types of Neoprene.   In general, this Neoprene has a slower rate of cure than the latter, and it is almost always necessary to use organic acceleration in order to produce satisfactory vulcanizates in a reasonable length of time.   Two basic systems have been advocated: first, the use of sulfur in conjunction with typical rubber accelerators, such as the guanidines, tetramethylthiuram monosulfide, and 2-mercaptothiazoline,[22, 24] and, second, the use of accelerators such as Permalux or NA-22 without sulfur.[44]

A noteworthy feature of Neoprene type W is the character of the stress-strain curve exhibited by its vulcanizates.   Whereas Neoprene GN, for example, gives a stress-strain curve such that at high elongations the stress increases with the strain at a markedly slower rate than it does with natural rubber, the ratio of stress to strain at high elongations of Neoprene type W is close to that found with natural rubber.[22]

PROPERTIES OF NEOPRENE VULCANIZATES

In the ultimate analysis the utility of any elastomer depends on the properties that can be developed in its vulcanizates.   Although the art of compounding has advanced to such a state that rubber technologists can influence many of the specific properties through the proper choice of compounding ingredients, nevertheless there remain many fundamental properties which appear to be inherent in the elastomer itself.   In the paragraphs that follow a brief description of a few of these fundamental properties of Neoprene vulcanizates is presented.

Tensile Strength.   Neoprene, unlike GR-S and most of the other synthetic rubbers based on butadiene as the predominant monomer, is capable of producing vulcanizates having high tensile strength in the absence of reinforcing fillers.   Gum tensile strengths as high as 5200 p.s.i. have been reported[10] on a freshly prepared sample of Neoprene type GR-M (type GN).   The average value for the tensile strength of gum-type Neoprene vulcanizates lies between 3500 and 4000 p.s.i.

Resistance to Oxidation.   One of the serious drawbacks of the early German synthetic rubbers (based on 2,3-dimethyl-1,3-butadiene) which preceded Neoprene was the rapidity with which they deteriorated on aging. In contrast to this, Neoprene vulcanizates suitably protected by an antioxidant are extremely resistant to deterioration.   Unlike natural rubber, which tends to become soft and sticky on oxidation, Neoprene stocks after prolonged periods of aging tend to increase in modulus, decrease in ultimate elongation, and become dry and hard.   Neal, Bimmerman, and Vincent[45] reported that it took approximately 40 days' exposure in the oxygen bomb at 70° C. and 300 p.s.i. oxygen pressure to reduce the tensile strength of a lightly loaded Neoprene stock to 50 per cent of its original value.   Corresponding rubber stocks required about 20 days for the same amount of degradation.   Kowalski[28] has shown that it takes about a year in the oxygen bomb at 70° C. and 300 p.s.i. oxygen pressure to destroy completely a Neoprene gum vulcanizate containing 1.0 per cent of phenyl-β-naphthylamine.

Mayo, Griffin, and Keen[36] studied the catalytic effect of copper on the deterioration of Neoprene vulcanizates and found that, although this metal exerts a catalytic effect on the oxidation of Neoprene, its effect is not so great as it is on natural-rubber stocks. Its catalytic action can be overcome through the use of the same type of metal deactivators that have proved effective in natural-rubber vulcanizates, viz., copper-complexing agents, such as disalicylal ethylenediamine.

Resistance to Ozone Cracking. Ever since Williams[70] showed that the cracking of natural rubber on exposure to sunlight in a strained condition is due to the attack of ozone, rubber technologists have been interested in the effect of this chemical on elastomeric vulcanizates. Carothers and his co-workers,[11] in their first paper on Neoprene, reported that its vulcanizates were not detectably affected during an exposure of 3 hours to a concentration of ozone which would rupture a natural-rubber vulcanizate after 3 minutes' exposure. More recent work in our laboratory has confirmed the correctness of this early observation but has shown that it is possible to produce cracking in Neoprene vulcanizates by prolonged exposure to high concentrations of ozone. Progressively increasing the strain in the vulcanizate being tested increases the tendency for cracking to take place.

Since resistance to weathering represents a combination of ozone resistance, oxidation resistance, and sunlight resistance, it naturally follows that Neoprene vulcanizates have proved to be extremely resistant to weathering.[47, 59] Tests reported by Morris, James, and Werkenthin[43] show Neoprene to be much superior to nitrile rubber in resistance to deterioration on exposure.

Flame Resistance. Although all organic materials burn if subjected to a high enough temperature for a sufficient length of time, nevertheless, within practical limits, there is a wide divergence in the flame resistance of rubber-like materials. As would be expected from its high chlorine content, Neoprene vulcanizates possess a high degree of flame resistance. C. S. Williams[71] showed that Neoprene vulcanizates, in the absence of highly flammable compounding ingredients, do not propagate a flame, though they do burn if held in the flame. Skinner and McNeal[53] showed that no phosgene was produced when Neoprene vulcanizates were decomposed at high temperatures.

Resistance to Oils and Chemicals. In recent years the effect of oils and chemicals on vulcanizates has become an important factor in evaluating the utility of the latter. Deterioration on exposure is generally characterized by swelling and loss in physical properties. Malcolmson,[32] who has made the most extensive study hitherto published of the resistance of Neoprene vulcanizates to the action of oils and chemicals, concludes that service tests are the best criteria for determining the suitability of any given vulcanizate to a particular application. This is especially true of Neoprene vulcanizates, since they do not show the same linear-type relationship between swelling index and loss in physical properties commonly displayed by elastomers.

In general, Neoprene vulcanizates are much more resistant to swelling in oils than vulcanizates of natural rubber or of the butadiene-styrene polymers and are somewhat less resistant than vulcanizates of butadiene-acrylonitrile polymers. The presence of aromatic hydrocarbons in oils tends to increase their swelling effect on all types of rubber. Neoprene is more resistant to chemicals such as alcohols, ketones, nitro compounds, aromatic amines, and sulfurized lubricants than butadiene-acrylonitrile rubbers.

Dynamic Properties. The rapid expansion in the use of elastomeric compositions to dampen or distribute vibration, absorb shock, and carry loads in dynamic applications has led to an increased interest in the dynamic properties of such materials. Such properties as heat buildup, resilience, dynamic modulus, and compression set are ordinarily included under this heading. Catton, Krismann, and Keen[16] investigated the dynamic properties of Neoprene vulcanizates extensively. They showed that in these properties Neoprene vulcanizates are superior to those of any other synthetic rubber and are only slightly inferior to those of natural rubber. Furthermore, it has been established that the dynamic properties of Neoprene vulcanizates are affected less than those of natural-rubber vulcanizates by increasing the temperature of service.

COMPOUNDING FOR SPECIFIC PROPERTIES AND APPLICATIONS

The development of any rubberlike product almost always involves a special compounding study either to produce the initial properties required or to meet the expected service conditions. Such service conditions often require that superior behavior shall be displayed in more than one property. Thus, for example, wedge-type V belts for automotive fans, for which Neoprene has given the best results, require not only long flex life but also resistance to heat aging. In practice, the rubber technologist usually finds that his final compositions represent a compromise designed to meet the over-all demands on the product under consideration. As is evident from the practical formulations at the end of this section, such problems face the technologist in the development of Neoprene products. However, many problems are common to a variety of products and a brief discussion of the compounding techniques applicable to each is of interest. A more complete study of the many possible variations will be found in Catton's brochure "The Neoprenes" and in his later book with the same title.[13] A survey of the principles of Neoprene compounding is also presented by Lanning.[29]

Resistance to Oxidative Degradation. Under the general term oxidative degradation are grouped not only simple attack by oxygen but also the related effects due to heat, light, ozone, and weathering. To obtain adequate service from products designed to meet exposure to such degradative agencies, it is necessary to use a good antioxidant in the stock. Most of the commonly used rubber antioxidants are even more effective in protecting Neoprene stocks than they are in protecting rubber stocks. For most applications 1.0 to 2.0 per cent of antioxidant is sufficient, and the most widely used antioxidants are phenyl-α- and phenyl-β-naphthylamines.

Although, on heat aging, Neoprene vulcanizates maintain good breaking strength, they are prone to suffer marked increase in modulus, decrease in extensibility, and increase in hardness. Forman,[20] working with Neoprene GN, has studied the compounding problems associated with the development of good heat resistance in a Neoprene vulcanizate. His work shows that the tendency of such vulcanizates to stiffen, lose elongation, and finally become brittle on exposure to air at elevated temperatures can be minimized by using relatively high concentrations of zinc oxide, curing to a high state of vulcanization, holding filler loading to a minimum, avoiding volatile softeners, and using adequate concentrations of antioxidants. This work indicates that compounds such as $p(p'$-tolylsulfonylamido)diphenylamine (Aranox) are the best antioxidants for this type of service. Although high proportions of magnesia tend to increase the rate of stiffening during the initial stages of heat aging, they delay the point of final embrittlement.

McCormack[30] has shown that nonreinforcing types of fillers, particularly whiting, contribute to superior retention of elongation during heat aging. He also concludes that vulcanizates of Neoprene type W accelerated with 2-mercaptoimidazoline (NA-22) are to be preferred for this type of service.

Exposure to light catalyzes surface oxidation in Neoprene vulcanizates. Thompson and Catton[59] made an extensive study of the factors influencing the changes that take place as a result of outdoor exposure. They found that the most effective means of protecting Neoprene stocks is through the use of fine-particle carbon blacks as fillers. Semireinforcing blacks and hard clay are the next most effective fillers, though they permit a gradual surface deterioration which is manifested as a form of shallow crazing. The inclusion of 10 parts of fine-particle black in clay-loaded Neoprene gives a degree of weather resistance approaching that of Neoprene in which black is the sole loading. The use of fillers such as whiting, calcium silicate, barytes, and magnesium carbonate results in a fairly rapid erosion of the exposed surface and should be avoided in vulcanizates designed for this type of service. According to Thompson and Catton, the weather resistance of blends of Neoprene and GR-S is better than that of Neoprene-natural-rubber blends.

Thompson[56] has also studied methods of enhancing resistance to attack by ozone. He concludes that stocks of low modulus and high set are desirable where permissible, that antioxidants, particularly those containing diphenyl-p-phenylenediamine, are highly desirable, and that high concentrations of fillers and softeners should be avoided. Ester plasticizers of the type generally used to produce vulcanizates having low brittle points are the most deleterious in services involving exposure to ozone. For static exposures the use of a microcrystalline wax which will bloom to the surface of the vulcanizate is effective although this method of compounding cannot be used for stocks that will be exposed under dynamic conditions.

Effect of Compounding on Physical Properties. McCune, Catton, and Keen[31] studied the effect of adding various fillers to a Neoprene base formula and concluded that, in general, the effects are similar to those obtained when the same fillers are added to natural rubber. Table II,

Table II

Neoprene type GN	100.0
Stearic acid	0.5
Phenyl-β-naphthylamine	2.0
Magnesium oxide	4.0
Zinc oxide	5.0

Cure: 35 minutes at 153° C.

Filler	Volumes Filler /100 Volumes Neoprene	Stress at 400% Strain	Tensile Strength, P.S.I.	Elongation, %	Hardness (Shore A)
MT black	20	900	2850	920	50
	40	1475	2200	575	58
	80	...	1900	320	73
FT black	20	875	3100	950	50
	40	1175	2200	680	59
	80	...	1450	240	76
SRF black	10	1050	3600	920	48
	20	1675	3450	800	55
	40	2750	3000	440	68
HMF black	10	1050	3550	890	50
	20	1900	3350	720	58
	40	...	2900	340	73
EPC black	10	1125	4300	880	51
	20	2075	4000	700	61
	40	...	3500	320	78
Clay	20	900	3200	870	56
	40	1200	2600	860	66
	80	...	1550	300	83
Whiting	20	350	2300	880	51
	40	325	1550	820	59
	80	300	900	750	73
Calcium silicate	10	500	3250	880	50
	20	800	2850	800	59
	40	1125	2150	700	77
Blanc Fixe	20	450	2500	825	52
	40	650	1950	780	60
	80	825	1000	600	77
Zinc oxide	5	350	3400	930	46
	10	575	3350	920	48
	20	725	3000	875	55
Titanium dioxide	5	325	3550	925	45
	10	450	3500	920	49
	20	600	2725	860	54
Light calcined magnesia	5	750	3900	840	51
	10	925	3575	785	54
	20	1225	3250	775	62

taken from their work, shows the effect of adding a variety of fillers to a base stock compounded as shown in the table.

A detailed study of the influence of a variety of blacks on the properties of Neoprene has also been presented by Buist and Mottram.[8]

The effect of compounding ingredients on tear resistance and abrasion resistance is even more striking than their effect on tensile properties. Resistance to tearing approximately seven times that of the gum vulcanizate is obtained by the addition of 30 volumes of channel black to a Neoprene type-GN stock. Optimum wear resistance is obtained in a tread type of vulcanizate at a loading of 20 volumes of channel black.[61]

It was first mentioned by Whitby and Katz[69] that Neoprene in a gum stock suffered a marked loss in tensile strength when the temperature was raised. Boonstra[6] has emphasized the importance of measuring the physical properties of rubber vulcanizates at elevated temperatures. He shows that the tensile strength of a gum type of Neoprene vulcanizate decreases rapidly as the temperature at which it is measured increases (Fig. 5, Chapter 12). This change becomes particularly pronounced as the temperature is raised from 25 to 70° C. The use of EPC black in the stocks markedly decreases their temperature sensitivity. Other work in this laboratory has indicated that similar effects will be obtained with other types of fillers and that even the so-called inert fillers have a marked effect on this property. Fillers also exert the same kind of an effect on the relationship between tear resistance and temperature.

The generally accepted explanation for the sharp lowering in tensile strength of Neoprene gum vulcanizates with increasing temperature involves considerations of crystallization. High tensile strength in such stocks is believed to be caused by the formation of reinforcing crystallites due to forced orientation as a result of stretching. At elevated temperatures these crystallites are not formed or are so unstable that they do not contribute any reinforcing action.

Effects of Compounding on Solvent and Water Resistance. It was recognized early that compounding variations had a marked effect on the solvent resistance of Neoprene vulcanizates.[27] Most of these effects are similar to those that had previously been established for natural-rubber vulcanizates. Catton and Fraser[14] studied the effect of compounding variations on the swelling of Neoprene vulcanizates in petroleum products. They concluded that the volume of Neoprene in a vulcanizate has a greater effect on swelling than the modulus of the stock or the type of filler used. More recent work[58] shows that, when a reinforcing type of filler, such as channel black, is used, the amount of swelling is less than for an equal volume loading of a nonreinforcing type of filler. Fraser[23] studied the effect of variations in the type of softener used in vulcanizates being exposed to petroleum products. He points out that, by the proper selection of the type and proportion of softener, it is possible to produce a Neoprene vulcanizate having no measurable swelling in any given petroleum product.

A practical aspect of the effect of compounding on solvent resistance is found in the development of sole stocks which will be subjected to flexing in the presence of oil. For this service it has been established that the use of

Carbonex S (a polymerized hydrocarbon material of coal-tar origin) as the softener results in an outstanding increase in the serviceability of the product.

Neoprene vulcanizates are, in general, somewhat less water-resistant than natural-rubber vulcanizates. However, by the proper selection of curing systems and compounding ingredients, this property can be modified substantially.[30a] Starkweather and Walker[54-5] have pointed out the importance of avoiding water-sensitive compounding ingredients and have shown that the use of litharge or red lead in place of magnesium oxide and zinc oxide as the curatives markedly improves the water resistance of the resulting vulcanizates. Bake[3] has shown that the use of calcium silicate not only reduces the extent of swelling but also produces vulcanizates in which the amount of water absorbed is not a logarithmic function of the time of immersion.

Compounding for Low-Temperature Exposure. Since most Neoprenes are readily crystallizable, it is necessary to take account of three types of changes in their vulcanizates as a result of exposure to low temperatures. These are classified by ASTM in method D 832–46T as simple temperature effects, second-order transition or vitrification, and first-order transition or crystallization. All three of these phenomena are markedly affected by compounding. Forman[21] made an extensive study of the influence of various plasticizers on these changes in Neoprene vulcanizates and showed that, in so far as simple temperature effects are concerned, there is no apparent method of classifying plasticizers. Furthermore, on extended exposure, some plasticizers become less effective in reducing stiffening, a phenomenon that Forman classifies as a plasticizer time effect.

As a class, the ester type of plasticizers are most effective in reducing the temperature of embrittlement (second-order transition). The lowest brittle-point temperatures are obtained through the use of Neoprene type FR containing substantial percentages of an ester plasticizer. It has been shown by Mayo and Avery[35] that the extreme softening effect of the very large proportions of ester plasticizer which may be required to retain flexibility at very low temperatures may be offset by replacing a substantial proportion of a general-purpose Neoprene with the inherently firm Neoprene type S. Plasticizers of the ester type are, however, completely ineffective in inhibiting the effects of crystallization in Neoprene compounds. Mayo[34] has shown that, as in Neoprene types RT and FR, crystallization may best be retarded or prevented by copolymerizing a second monomer with the chloroprene. He has shown also that certain polymerized hydrocarbons, when used as softeners, will inhibit crystallization. Further, he confirms the early observation of Forman[21] that combination with sulfur during vulcanization markedly increases resistance to crystallization.

Effect of Compounding on Dynamic Properties. Regardless of whether one is interested in high resilience, low heat buildup, or low compression set, the first requirement in a Neoprene vulcanizate is to obtain a very full cure. This generally means that it is necessary to use accelerators, such as Permalux or 2-mercaptoimidazoline, in order to develop the maximum properties with a reasonable length of cure. Catton and Fraser[14] made the first detailed study of the effect of fillers on these properties. Their work was done on Neoprene type G, a forerunner of the present Neoprene type GN, but subsequent work has shown their conclusions to be applicable to other types of Neoprene. They found that loading with a reinforcing type of carbon black reduces resilience to a greater extent than loading with mineral fillers, whereas the reverse is true of the effect of loading on compression set.

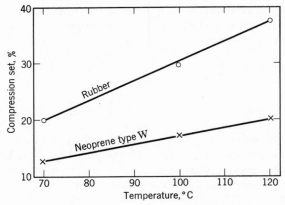

FIG. 2. Effect of Temperature on Compression Set of Neoprene Type W and Hevea

Catton, Krismann, and Keen's[16] findings regarding the effect of increased amounts of carbon black loading on the dynamic properties of Neoprene are particularly noteworthy, since this type of loading is required in most practical applications. Their work shows that increased proportions of carbon black have a less deleterious effect on the dynamic properties of Neoprene than on those of rubber. As a result, at loadings above 15 to 20 volumes, which represents the amount present in most practical vulcanizates, the dynamic properties of the Neoprene stocks frequently excel those of corresponding natural-rubber vulcanizates.

Neoprene type-W vulcanizates are significantly superior to those of other Neoprenes in resistance to compression set. Neal[44] pointed out the effect of the temperature at which the stress is applied and the importance of proper acceleration in developing type-W vulcanizates having superior resistance to compression set. Figure 2, taken from his work, reveals that a Neoprene type-W stock accelerated with 2-mercaptoimidazoline is not only superior to the best natural-rubber vulcanizate in tests carried out at 70° C. but also shows less change than the natural-rubber vulcanizate when the temperature at which the stress is applied is increased over the range of 70 to 120° C. Both rubbers were loaded with SRF black.

PRACTICAL COMPOUNDS

Tables III to X give a brief summary of pertinent test data on eight stocks which illustrate Neoprene compounds designed for specific end uses. The practicability of these or closely related formulations has been demonstrated by the successful factory production of finished articles.

Table III. Black Wire Jacket Compound

Neoprene type GN-A	100.0
Phenyl-β-naphthylamine	2.0
Magnesium oxide	4.0
SRF black	25.0
MAF black	25.0
Hard clay	35.0
Light process oil	12.0
Heliozone (a mixture of waxy materials)	3.0
Zinc oxide	5.0
NA-22 (2-mercaptoimidazoline)	0.5

High pressure steam cure: 40 seconds at 203° C.

Original Properties

Stress at 200%, p.s.i.	1575
Tensile strength at break, p.s.i.	2300
Elongation at break, %	340

Oil-Immersion Test—18 Hr. at 121° C. in ASTM No. 2 Oil

Original tensile strength, %	100
Original elongation, %	77

Water Absorption—7 Days at 70° C.

Mg. picked up per sq. in.	54

Electrical Properties

D.c. resistivity, ohm.-cm.	9.6×10^7
Dielectric strength, volts per mil	113

This compound, suitable for continuous curing in high-pressure vulcanizers, is of good quality and meets most of the requirements for jacket applications with respect to resistance to weathering and physical abuse.

Table IV. Neoprene 90° C. Appliance Wire Insulation Compound

Neoprene type W	100.0
Magnesium oxide	4.0
Aranox (p-(p'-tolylsulfonylamido)diphenylamine)	1.5
Finely ground natural whiting	90.0
MAF black	20.0
Process oil (low volatility)	8.0
Heliozone (a mixture of waxy materials)	3.0
Zinc oxide	10.0
NA-22 (2-mercaptoimidazoline)	1.0

Press cure: 10 minutes at 153° C.

Table IV—continued

Original Properties

D.c. resistivity, ohm.-cm.	3.0×10^{12}
Dielectric strength, volts per mil	450
Modulus at 200%, p.s.i.	700
Tensile strength at break, p.s.i.	2000
Elongation at break, %	500
Hardness (Shore A)	67

After Aging 10 Days in 121° C. Air Oven

Tensile strength retained, %	91
Elongation retained, %	60
Hardness (Shore A)	82

After Aging 60 Days in 100° C. Air Oven

Tensile strength retained, %	80
Elongation retained, %	65
Hardness (Shore A)	80

This compound is basically designed to resist stiffening and loss of elongation during high-temperature service. The presence of MAF black does not contribute to this end but produces a better processing stock with generally superior physical properties.

Table V.   Water-Resistant Black Wire Jacket

Neoprene type GN-A	100.0
Phenyl-β-naphthylamine	2.0
Heliozone (a mixture of waxy materials)	3.0
Stearic acid	0.75
MPC black	40.0
Light process oil	12.0
Red lead	20.0

Press cure: 15 minutes at 153° C.

Original Properties

Stress at 200%, p.s.i.	560
Tensile strength at break, p.s.i.	3200
Elongation at break, %	700

After Air-Oven Aging 7 Days at 70° C.

Tensile strength retained, %	100
Elongation retained, %	77

After Oil Immersion 18 Hr. at 121° C. in ASTM No. 2 Oil

Tensile strength retained, %	81
Elongation retained, %	68

Water Absorption after 7 Days at 70° C.

Water (wt. increase) mg. per sq. in.	19
D.c. resistivity, ohm.-cm.	9×10^9
Dielectric strength, volts per mil	140

This high-quality black-jacket stock illustrates the use of a red lead type of cure to minimize water absorption.

Table VI. Neoprene White Sidewall Compound

Neoprene type GN	50.0
Pale crepe	50.0
Stearic acid	0.5
Magnesium oxide	2.0
Zinc oxide	70.0
Titanium dioxide (rutile)	50.0
Sulfur	0.5
Du Pont rubber blue GD	0.02
Benzothiazolyl disulfide	1.0

Press cure: 45 minutes at 138° C.

Stress at 400%, p.s.i.	750
Tensile strength at break, p.s.i.	2175
Elongation at break, %	620
Hardness (Shore A)	61

	Reflectance (%)* after Exposure to	
Exposure Period	Direct Sunlight	Indirect Sunlight
Original	75.0	75.0
1 week	73.0	74.5
2 weeks	73.0	71.5
4 weeks	73.0	71.5

* Reflectance measured using monochromatic blue light having a wavelength of 4570 A.

This compound has adequate resistance to exposure cracking because of the Neoprene present. Discoloration is minimized by dilution of the Neoprene with rubber, the use of fillers of high covering power, and the presence of a trace of blue color.

Table VII. Colored Garden Hose

Neoprene type GN	100.0
Stearic acid	0.5
Flectol H (polymerized trimethyldi- hydroquinoline)	1.5
Magnesium oxide	4.0
Hard clay	130.0
Process oil	18.0
Heliozone (a mixture of waxy materials)	2.0
Petrolatum	2.0
Zinc oxide	5.0
Color	As desired

Press cure: 20 minutes at 153° C.

Stress at 300%, p.s.i.	1100
Tensile strength at break, p.s.i.	1800
Elongation at break, %	700
Hardness (Shore A)	67

This stock illustrates a colored compound with good weather-aging properties. In this competitive field the replacement of up to 25 per cent of the Neoprene with GR-S is frequently made in order to reduce cost. As would be expected, the weather resistance of such stocks is impaired by this substitution.

Table VIII.   Extreme Cold-Resistant Tubing

Neoprene type FR	100.0
Phenyl-α-naphthylamine	2.0
Stearic acid	1.0
Petrolatum	1.0
Magnesium oxide	5.0
MT black	90.0
Ester plasticizer	12.0
Zinc oxide	5.0
NA-22 (2-mercaptoimidazoline)	0.5

Press cure: 20 minutes at 153° C.

Stress at 300%, p.s.i.	700
Tensile strength at break, p.s.i.	1500
Elongation at break, %	670
Hardness (Shore A)	48
Solenoid brittle point	−60° C.

The use of Neoprene type FR eliminates crystallization problems and gives a low brittle point without using ester plasticizer in excessive concentration.

Table IX.   Resilient Mounting Stock

Neoprene type W	100.0
Stearic acid	0.5
Phenyl-α-naphthylamine	2.0
Magnesium oxide	2.0
FF black	20.0
Process oil	5.0
Zinc oxide	5.0
NA-22 (2-mercaptoimidazoline)	0.5

Press cure: 20 minutes at 153° C.

Stress at 400%, p.s.i.	1600
Tensile strength at break, p.s.i.	3500
Elongation at break, %	620
Hardness (Shore A)	48
Resilience (Yerzley)	81
Static modulus (Yerzley, 20% deformation), p.s.i.	775
Dynamic modulus (Yerzley, 20% deformation), p.s.i.	950
Heat buildup (Goodrich flexometer, 20 min. at 3/16 in. stroke, °C.)	25
Compression set (method B, 70 hr. at 100° C.), %	23

This stock exhibits a combination of properties desirable in many dynamic applications requiring the transmission of vibrational energy.

Table X.   Automotive Windshield Channel and Stripping

Neoprene type GN	90.0
Nonstaining GR-S	10.0
Petrolatum	1.0
Heliozone (a mixture of waxy materials)	2.0
Zinc salt of mercaptobenzothiazole	3.0
Magnesium oxide	4.0
MT black	100.0
Neophax A (vulcanized vegetable oil)	30.0
Process oil	30.0
Zinc oxide	5.0
NA-22 (2-mercaptoimidazoline)	0.5

Press cure: 20 minutes at 153° C.

Stress at 300%, p.s.i.	1160
Tensile strength at break, p.s.i.	1450
Elongation at break, %	370
Hardness (Shore A)	50

In this relatively low-cost stock, resistance to atmospheric cracking is enhanced by the use of a high set compound.   The zinc mercaptide acts as a nonstaining antioxidant and, because of its retarding action, favors stress relaxation of the strip when placed in service.   Other practical Neoprene compounds of low elastomer content are described by Thompson.[57]   The compounding of Neoprene with reclaimed rubber is discussed by Torrence and Schwartz.[63]

COMPOUNDING AND USE OF NEOPRENE LATEXES

Dales, Abernathy, and Walsh[18] have given a general description of the compounding and use of Neoprene latex type 571.   Additional descriptions of this and other Neoprene latexes have been given by Walsh,[66] Hartsfield and Galloway,[26] Pockman and co-workers,[49] Dales,[17] and Abernathy and associates.[2]   An examination of these reports indicates that, in general, compounding ingredients have the same effect in polymer from Neoprene latex that they do in dry Neoprene.   These investigators, however, indicated a few differences which are worthy of discussion.

There are some limitations to the use of compounding ingredients in Neoprene latex which arise from the fact that the latexes are highly alkaline colloidal dispersions and that ingredients to be used in them must be capable of being stable and nonreactive when dispersed in such a system.   For example, light calcined magnesia cannot be used, because alkaline dispersions of it contain enough soluble, positively charged ions to flocculate or coagulate the latex severely.   Fortunately, this limitation is not serious, since zinc oxide alone serves very well as the major vulcanizing agent in Neoprene latex compounding.

Vulcanizing Ingredients. Since scorching is not a problem in curing systems for use in latex, it is possible to develop vulcanizing systems for Neoprene latex that are much faster than those employed with dry Neoprene compounds. Further, Neoprene latex does not precure in latex form as natural-rubber latex does. Hence, as described by Walsh[65] and by Abernathy and co-workers,[2] very fast accelerators such as Polyac (active ingredient, dinitrosobenzene) and thiocarbanilide may be used to obtain cures in relatively short times at low temperatures. Sulfur combines with the Neoprene polymer from latex faster than with the dry polymer.[67] This is particularly true of polymer from types 842, 842A, 601, and 601A latexes. Thus, an accelerating system comprised of at least 5 parts of zinc oxide with 1 or 2 parts of Polyac or similar fast accelerator, and 1 part of sulfur will produce a very satisfactory cure in type-842A latex films in 30 minutes at 100° C., and the vulcanizate will have superior low-temperature properties by virtue of its combined sulfur.[34]

Dithiocarbamate accelerators are much more effective in Neoprene latex than in dry Neoprene, although not so effective as in rubber latex. The accelerating effect of piperidinium pentamethylenedithiocarbamate, a peptizer for dry Neoprene, is described by Pockman and associates.[49]

Fillers. As is usual in latexes, carbon black and other reinforcing fillers have very little reinforcing effect. However, in contrast to natural-rubber latex, when up to about 10 parts of reinforcing agent per 100 parts of elastomer is used in Neoprene latex, there is some improvement in tensile strength. Also, in contrast to rubber latex, the tearing strength of Neoprene-latex films can be increased through the use of blacks and clays.[18] Above 10 parts of filler the expected results—increased modulus, set, and relatively rapid loss of tensile strength—are obtained with increasing quantities of fillers. Latex compounds, of either natural rubber or Neoprene, cannot be loaded nearly so highly with fillers without severe loss of physical properties as can dry polymers.

In compounding for specific properties such as improved heat, oil, sunlight, and water resistance, the same devices can be used in Neoprene latex as are used in dry Neoprene compounding.

Latex pH and Stability. Considering Neoprene latexes as colloidal systems, they are more alkaline than rubber latex and must remain so during storage and use if good stability is to be obtained. Neoprene latexes are made in a fixed-alkali rosin soap system in contrast to the protein system of natural-rubber latex. The Neoprene-latex soap system begins to break down when its *p*H drops below about 9.7. Since during storage small amounts of hydrogen chloride are evolved in Neoprene latexes, it is necessary to buffer the latex with diethanolamine or a similar agent. On long storage it may be necessary to add small amounts of fixed alkali as a dilute solution to maintain the *p*H above 10.

Stabilizing Agents. As indicated by Hartsfield and Fitch,[25] the same sorts and quantities of stabilizing, wetting, and colloidal protective agents as are normally used in rubber latex are effective in Neoprene latexes. An exception is Aquarex SMO (33 per cent water solution of the monosodium salt of sulfated methyl oleate), which is much more effective in Neoprene than in rubber latex. It is specifically useful in Neoprene latexes in preventing small surface imperfections in films formed by coagulating dip procedures. Further, vegetable gums, such as karaya and locust bean, used to thicken natural-rubber latex, cause creaming and flocculation in Neoprene latexes. Neoprene latexes will not tolerate nearly so much salt as rubber latex. This is particularly true of salts of monovalent metals such as sodium chloride. Compounding ingredients to be used in Neoprene latex must be dispersed in water in an alkaline system, just as materials used in rubber latex are.

Uses of Neoprene Latexes. Walsh,[66] Dales,[18] Pockman,[49] and Abernathy[2] have discussed a wide variety of uses for Neoprene latexes. Such latexes have been used with a high degree of success in the production of dipped goods, such as gloves, balloons, and industrial parts; in foam; in making saturated paper; and in the treatment of paper pulp prior to the formation of the finished paper. Neoprene latex type 572 has been widely used in the preparation of adhesives for bonding leather, fabric, and paper.[19] Neoprene latex can be used as a carrier for resorcinol–formaldehyde resins in tire-cord treatment.[1]

REFERENCES

1. Abernathy, H. H., and Radcliff, R. R., E. I. du Pont de Nemours & Co., Rubber Chem. Div. Rept. No. 47–4 (1947).
2. Abernathy, H. H., Walsh, R. H., Galloway, J. R., and Pockman, W. W., E. I. du Pont de Nemours & Co., Rubber Chem. Div. Rept. No. 51–1 (1951).
3. Bake, L. S., E. I. du Pont de Nemours & Co., Rubber Chem. Div. Rept. No. 45–1 (1945).
4. Barrows, R. S., and Scott, G. W., *Ind. Eng. Chem.*, **40**, 2193–6 (1948).
5. Best, L. L., and Moakes, R. C. W., *Trans. Instn. Rubber Ind.*, **27**, p. 124 (1951).
6. Boonstra, B. S. T. T., *India Rubber World*, **121**, 299–302, 313 (1949).
7. Bridgwater, E. R., and Krismann, E. H., *Ind. Eng. Chem.*, **25**, 280–3 (1933).
8. Buist, J. M., and Mottram, S., *Trans. Instn. Rubber Ind.*, **22**, 82–110 (1946).
9. Bunn, C. W., *Proc. Roy. Soc. London*, **A180**, 40–66 (1942).
10. Carbon Reinforcement of Neoprene GN, *Columbian Colloidal Carbons*, **6**, No. 1 (1945).
11. Carothers, W. H., Williams, I., Collins, A. M., and Kirby, J. E., *J. Am. Chem. Soc.*, **53**, 4203–25 (1931).
12. Catton, N. L., *Rubber Age N.Y.*, **65**, 39–42 (1949).
13. Catton, N. L., E. I. du Pont de Nemours & Co., Rubber Chem. Div. Rept. No. 51–5 (1951); *The Neoprenes—Principles of Compounding and Processing*, E. I. du Pont de Nemours & Co., Wilmington, Del., 1953, 245 pages.
14. Catton, N. L., and Fraser, D. F., *Ind. Eng. Chem.*, **31**, 956–60 (1939).
15. Catton, N. L., Fraser, D. F., and Forman, D. B., E. I. du Pont de Nemours & Co., Rubber Chem. Div. Rept. No. 42–2 (1942).
16. Catton, N. L., Krismann, E. H., and Keen, W. N., *Rubber Chem. and Technol.*, **22**, 450–64 (1949).
17. Dales, B., E. I. du Pont de Nemours & Co., Rubber Chem. Div. Rept. BL-201 (1945).
18. Dales, B., Abernathy, H. H., and Walsh, R. H., E. I. du Pont de Nemours & Co., Rubber Chem. Div. Rept. No. 43–2 (1943).
19. Dales, B., Abernathy, H. H., and Walsh, R. H., E. I. du Pont de Nemours & Co., Rubber Chem. Div. Rept. BL-184 (1944).
20. Forman, D. B., *Ind. Eng. Chem.*, **35**, 952–7 (1943).
21. Forman, D. B., *Ind. Eng. Chem.*, **36**, 738–41 (1944).
22. Forman, D. B., Radcliff, R. R., and Mayo, L. R., *Ind. Eng. Chem.*, **42**, 686–91 (1950).

23. Fraser, D. F., *Ind. Eng. Chem.*, **32**, 320–3 (1940).
24. Fritz, F. H., and Mayo, L. R., *Ind. Eng. Chem.*, **44**, 831–3 (1952).
25. Hartsfield, E. P., and Fitch, J. C., E. I. du Pont de Nemours & Co., Rubber Chem. Div. Rept. No. 51–2 (1951).
26. Hartsfield, E. P., and Galloway, J. R., E. I. du Pont de Nemours & Co., Rubber Chem. Div. Rept. No. 48–1 (1948).
27. Hayden, O. M., and Krismann, E. H., *Ind. Eng. Chem.*, **25**, 1219–23 (1933).
28. Kowalski, L. J., Unpublished Results from this laboratory.
29. Lanning, H. J., *Trans. Instn. Rubber Ind.*, **26**, 151–74 (1950).
30. McCormack, C. E., Unpublished Results from this laboratory.
30a. McCormack, C. E., Baker, R. H., and Graff, R. S., *Rubber Age N.Y.*, **74**, 72–6, 84 (1953).
31. McCune, S. W., Catton, N. L., and Keen, W. N., E. I. du Pont de Nemours & Co., Rubber Chem. Div. Rept. No. 48–7 (1948).
32. Malcolmson, R. W., E. I. du Pont de Nemours & Co., Rubber Chem. Div. Rept. BL–223 (1948).
33. Mayo, L. R., E. I. du Pont de Nemours & Co., Rubber Chem. Div. Rept. BL–227 (1948).
34. Mayo, L. R., *Ind. Eng. Chem.*, **42**, 696–700 (1950).
35. Mayo, L. R., and Avery, E. C., E. I. du Pont de Nemours & Co., Rubber Chem. Div. Rept. BL–232 (1949).
36. Mayo, L. R., Griffin, R. S., and Keen, W. N., *Ind. Eng. Chem.*, **40**, 1977–80 (1948).
37. Mitchell, A., *Rubber Age N.Y.*, **71**, 67–70 (1952).
38. Mochel, W. E., *J. Polymer Sci.*, **8**, 583–92 (1952).
39. Mochel, W. E., and Nichols, J. B., *Ind. Eng. Chem.*, **43**, 154–7 (1951).
40. Mochel, W. E., Nichols, J. B., and Mighton, C. J., *J. Am. Chem. Soc.*, **70**, 2185–90 (1948).
41. Mochel, W. E., and Peterson, J. H., *J. Am. Chem. Soc.*, **71**, 1426–32 (1949). Cf. Maynard, J. T., and Mochel, W. E., *Rubber Age N.Y.*, **72**, 495 (1953).
42. Mochel, W. E., Salisbury, L. F., Barney, A. L., Coffman, D. D., and Mighton, C. J., *Ind. Eng. Chem.*, **40**, 2285–9 (1948).
43. Morris, R. E., James, R. R., and Werkenthin, T. A., *Rubber Age N.Y.*, **51**, 205–8 (1942).
44. Neal, A. M., *Rubber Age N.Y.*, **67**, 569–72 (1950).
45. Neal, A. M., Bimmerman, H. G., and Vincent, J. R., *Ind. Eng. Chem.*, **34**, 1352–7 (1942).
46. Nieuwland, J. A., Calcott, W. S., Downing, F. B., and Carter, A. S., *J. Am. Chem. Soc.*, **53**, 4197–202 (1931).
47. Northam, A. J., E. I. du Pont de Nemours & Co., Rubber Chem. Div. Rept. No. 38–2 (1938).
48. Nowlen, J. P., E. I. du Pont de Nemours & Co., Rubber Chem. Div. Rept. BL–63 (1942).
49. Pockman, W. W., Walsh, R. H., and Abernathy, H. H., E. I. du Pont de Nemours & Co., Rubber Chem. Div. Rept. No. 48–8 (1948).
50. Radcliff, R. R., *India Rubber World*, **125**, 311–4 (1951).
51. Radcliff, R. R., Unpublished Results from this laboratory.
52. Salomon, G., and Koningsberger, C., *Rec. trav. chim.*, **69**, 711–23 (1950).
53. Skinner, G. S., and McNeal, J. H., *Ind. Eng. Chem.*, **40**, 2303–8 (1948).
54. Starkweather, H. W., and Walker, H. W., *Ind. Eng. Chem.*, **29**, 872–80 (1937).
55. Starkweather, H. W., and Walker, H. W., *Ind. Eng. Chem.*, **29**, 1380–4 (1937).
56. Thompson, D. C., E. I. du Pont de Nemours & Co., Rubber Chem. Div. Rept. BL–238 (1950).
57. Thompson, D. C., *Rubber Age N.Y.*, **72**, 638–42 (1953).
58. Thompson, D. C., and Catton, N. L., *Ind. Eng. Chem.*, **40**, 1523–6 (1948).
59. Thompson, D. C., and Catton, N. L., *Ind. Eng. Chem.*, **42**, 892–5 (1950).
60. Torrence, M. F., *Ind. Eng. Chem.*, **41**, 641–3 (1949).
61. Torrence, M. F., Unpublished Results from this laboratory.
62. Torrence, M. F., and Fraser, D. F., *Ind. Eng. Chem.*, **31**, 939–41 (1939).
63. Torrence, M. F., and Schwartz, H. G., *Rubber Age N.Y.*, **71**, 357–60 (1952).

64. Walker, H. W., and Mochel, W. E., *Proc. Second Rubber Tech. Conf.*, 69–78 (1948).
65. Walsh, R. H., E. I. du Pont de Nemours & Co., Rubber Chem. Div. Rept. BL–194 (1945).
66. Walsh, R. H., *Rubber Age N.Y.*, **61,** 187–93 (1947).
67. Walsh, R. H., Unpublished Results from this laboratory.
68. Walsh, R. H., Abernathy, H. H., Pockman, W. W., Galloway, J. R., and Hartsfield, E. P., *Tech. Assoc. Pulp Paper Ind.*, **33,** 232–8 (1950).
69. Whitby, G. S., and Katz, M., *Ind. Eng. Chem.*, **25,** 1204–11, 1338–48 (1933).
70. Williams, I., *Ind. Eng. Chem.*, **18,** 367–9 (1926).
71. Williams, C. S., E. I. du Pont de Nemours & Co., Rubber Chem. Div. Rept. No. 38–5 (1938).
72. Youker, M. A., *Chem. Eng. Prog.*, **43,** 391–8 (1947).

CHAPTER 23

NITRILE RUBBER

Waldo L. Semon

B. F. Goodrich Company

The introduction of the chemical rubber polychloroprene in 1931 marked the beginning of a philosophy that has had a profound influence on the rubber industry, namely, that specialty rubbers could be manufactured and used to perform functions that could not be matched with natural rubber. Products from Neoprene with excellent resistance to sunlight and corona discharge and low swelling in oil educated rubber technologists to some of the advantages that could be obtained from a specialty rubber. It soon became apparent that a special rubber of higher cost might be cheaper on a performance basis than a lower-priced but less appropriate rubber. Hence, when the German oil-resistant rubber, Buna N, was announced, the reports of its use were followed with considerable interest. ·

COMMERCIAL NAMES AND COMPOSITION OF NITRILE RUBBERS

The properties of Buna N were described in 1936 by Konrad,[126-8] Koch,[118-20] and others[9-13, 103] before its chemical composition was disclosed. Twiss[225, 226] on the basis of the high nitrogen content deduced that Buna N was probably an acrylonitrile copolymer made by a method disclosed broadly[124] in British patent 360,821 (May 30, 1930). In 1937 the name Perbunan was assigned[175] to a grade of Buna N being commercialized in Germany. However, its composition as a copolymer of butadiene and acrylonitrile was not acknowledged by the Germans until 1938,[97, 212-4] by which time the German patent 658,172 (March 25, 1938) had issued.[125]

After Perbunan was introduced into the United States and a number of related polymers, such as Ameripol, Butaprene, Chemigum, and Hycar, made their appearance there, it became desirable to have a generic name to describe these rubbers. The class name *nitrile rubber* was assigned by the nomenclature committee of the American Chemical Society to rubberlike copolymers of unsaturated nitriles with dienes. The most common nitrile rubbers are copolymers of acrylonitrile with butadiene. The following are some that have been mentioned in the technical literature:

Ameripol. A trademarked name applied to vulcanized articles made with synthetic rubber. Articles for oil-resisting service were made from stocks compounded with crude nitrile rubber.

Buna N. The early name assigned by I.G. Farbenindustrie to a copolymer of butadiene with about 25 per cent of acrylonitrile. It was made by emulsion polymerization. When it was commercialized, the name was changed to Perbunan.

Buna NN. An emulsion copolymer of butadiene with, in this case, about 35 per cent of acrylonitrile. It was later sold as Perbunan Extra.

Butaprenes. Copolymers developed by the Firestone Tire & Rubber Company and sold through the Xylos Rubber Company. A number of types have been offered, among which may be mentioned the following:[236, 250]

Type	NF	NL(ML-4, 60)*	NL(ML-4, 100)	NAA	NXM
Acrylonitrile combined (nominal), %	24	30	30	35	40

* ML-4 refers to Mooney viscosity of the polymer at 212° F. with the large rotor after 4 minutes in the machine.

Chemigums. Copolymers developed and sold by the Goodyear Tire & Rubber Company.[85, 47, 198-200, 236]

Type, Original	N4NS	N4NS	N1NS	N3NS
	MS30*	MS50	Rapid Mooney	...
As of 1954	N-6	N-7	breakdown on milling	
Acrylonitrile combined (nominal), %	30	30	30	40

* MS refers to Mooney viscosity of the polymer at 212° F. with the small rotor.

GR-A. A copolymer of acrylonitrile with butadiene as made in the Government Synthetic Rubber Program.

Hycar OR. Emulsion copolymers of acrylonitrile and butadiene manufactured and sold by the B. F. Goodrich Chemical Company. It should be noted that not all Hycars are nitrile rubbers. Thus, Hycar OS is an oil soluble variety, a copolymer of butadiene with styrene, and Hycar PA is a polyacrylate heat-resisting variety. The composition of the oil-resistant Hycar rubbers has been variously reported.[236, 104, 67a]

Type of Hycar	OR-24	OR-25EP	OR-25NS	OR-15
		Easy processing	Nonstaining	
Acrylonitrile combined (nominal), %	33	33	33	40
Type as of 1954	1043	1042	1432	1041	1012X41
Acrylonitrile combined	Medium low	Medium	Medium	High	?
	Easy processing	Easy processing	Crumb Directly soluble	Easy processing	Liquid

Paracril (Perbunan). The manufacture of Perbunan in the United States was taken over at an early date by the Enjay Company. Several grades have been offered at various times.[236, 67a] The trade name was later changed to Paracril, and in 1951 the plant and business were bought by the U.S. Rubber Company.

Type	Paracril A	Paracril B	Paracril C
	Perbunan 18	Perbunan 26NS60	Perbunan 35NS90
Acrylonitrile com-bined (nominal), %	18	26	35
Type as of 1954	18-80 AJ	B BJ BV	C CJ CV CS
Oil resistance	Medium	Good	Maximum

J—Easy processing.   V—Crumb with soluble coating.

S—Crumb with insoluble coating.

Paracril D is reported to have an even higher nitrile content. The NS grades contain a nonstaining antioxidant.

Perbunan (German) was the emulsion copolymer of 25 per cent acrylonitrile with butadiene as made originally by the I.G. Farbenindustrie.

Polysar N. Nitrile rubbers are produced by the Polymer Corporation, Sarnia, Ontario, under the general trade name of Polysar N.[252] Polysar Krynac is described as a "cold nitrile rubber" of improved physical characteristics.

Type	N-301	NP-350	NP-450	Krynac
Combined acrylonitrile	Medium	Medium	High	
Mooney viscosity	85	Low	Low	76

Part or all of the butadiene in these oil-resistant copolymers might be replaced by isoprene, the acrylonitrile might be wholly or partially replaced by methacrylonitrile, and various other dienes or vinyl compounds might be included in the composition to impart specific processing properties or added value to the finished vulcanizates. All such varieties would be termed nitrile rubbers. However, if the product were resinous or not capable of vulcanization such as the copolymer of styrene and acrylonitrile, the class name would not apply.

HISTORY IN GERMANY

The issuance of the French patent[125] in 1931 was the first published disclosure of rubbery copolymers of acrylonitrile and butadiene. Both the British[124] and United States patents[125] would indicate that nitrile rubber was developed by Konrad and Tschunkur early in 1930.

The story of German Buna is interestingly told by Fritz Hofmann,[95-6] by Konrad[126-8] who headed up the work done by I.G. Farbenindustrie, by Stöcklin,[214] and others.[9-13] Reports of the properties that had been realized in vulcanized stocks created worldwide interest. Buna-N stocks were stated to be better than stocks from natural rubber for aging, resistance to light and ozone, heat resistance, resistance to abrasion, low permeability to gases, and resistance to oils and other organic solvents.[9, 103, 118-9, 14, 174-5] Hard rubber from Buna N was reported to be more chemically resistant[9] and to have better high-temperature properties than ebonite from natural rubber. Tests on various products such as packings, rolls, cables, belts, and even tires were outstanding.[12, 10, 90] The chief obstacle mentioned at this early date was the high cost[13, 103, 136, 225] of Buna N, which Twiss estimated to be 3s. 7d. per lb.[227] To encourage German manufacture of synthetic rubber in 1937, a Buna company was organized for which the German Government advanced 30 million marks. A duty of about $0.22 per pound was put on natural rubber, to be removed when the Buna plants reached full operation. It was believed that, when production reached a large enough scale, no subsidies would be required to enable Buna to compete with natural rubber.[15]

Articles on the compounding, processing, and uses of Perbunan appeared in rapid succession.[175, 77, 162, 216–8, 5, 32] Production capacity in Germany was expanded rapidly. Importation into the United States began in March 1937 and by 1939 had reached an annual volume of 150 tons,[157] distribution being made through the Advance Solvents Corporation. The supply of Perbunan was shut off by the European war in 1939. Development and expansion continued in Germany all through the war[233] (for statistics, see p. 53), and, on October 30, 1941, the Germans converted their manufacture of Perbunan to a continuous process.[55] Special grades, such as the nonstaining[216] Perbunan M1038 and the odorless product for the food industry,[55] Perbunan GF, were introduced.

HISTORY IN THE UNITED STATES

Negotiations between I.G. Farbenindustrie, Standard Oil Company, and various American rubber companies for licensing rights to the manufacture of I.G. synthetic rubber in United States were initiated in 1932. In 1939 Standard Oil Development Company obtained control of Jasco, Inc. (Joint Account Stock Company) to which the American rights had been assigned. Both Firestone Tire & Rubber Company and U.S. Rubber Company then obtained sublicenses (1940). The Goodyear Tire & Rubber Company and the B. F. Goodrich Company decided to proceed on their own.[228]

To supply the American demand, Standard Oil Company of New Jersey built a plant to manufacture 5 tons of Perbunan per day. This plant, at Baton Rouge, La., began operation in March 1941 and was shortly expanded to 15 tons per day. The B. F. Goodrich Company built a small plant for producing nitrile rubber for factory evaluation. This started operation in January 1939 and by March 1940 was producing 250 lb. a day. In June 1940 this was increased to 1 ton a day, and this capacity was maintained until Hydrocarbon Chemical & Rubber Company took over the facilities. The latter, a company jointly held by the B. F. Goodrich Company and Phillips Petroleum Company, started producing Hycar OR in December 1940 at a rate of 5 tons a day and by September 1941 had increased production capacity to 25 tons a day. This plant was subsequently sold to the B. F. Goodrich Chemical Company, which since December 1945 has been manufacturing and selling the various Hycars. In 1940 the Goodyear Tire & Rubber Company built their first plant for the production of Chemigum and at about the same time Firestone Tire & Rubber Company started production of Butaprene.

ACRYLONITRILE

The emulsion polymerization process used in the production of nitrile rubber is similar to that which gives rise to other emulsion polymers such as

GR-S or acrylate copolymers. The raw materials other than the nitrile have been discussed elsewhere. The nitrile that has found widest use in the production of nitrile rubber is acrylonitrile or vinyl cyanide, $CH_2 : CH \cdot CN$. Most of this is prepared from ethylene oxide and hydrogen cyanide.[141a, 251]

$$CH_2\!\!-\!\!CH_2 + HCN \rightarrow HO \cdot CH_2 \cdot CH_2 \cdot CN \rightarrow H_2O + CH_2 : CH \cdot CN$$
$$\diagdown\!\!O\!\!\diagup$$

The ethylene oxide used in the process is available on a large scale since it has other extensive commercial uses such as in the manufacture of ethylene glycol, ethanolamines, and numerous polyethylene oxide derivatives. Most of the hydrogen cyanide required is made from black cyanide, in turn prepared from calcium cyanamide by reaction with salt and coke. Another process for making hydrogen cyanide which is in large-scale commercial operation in Texas is the simultaneous oxidation of ammonia and methane over a platinum catalyst.[132]

$$2CH_4 + 2NH_3 + 3O_2 \rightarrow 2HCN + 6H_2O$$

If ethylene oxide and hydrogen cyanide are added to ethylene cyanohydrin containing a trace of an alkaline catalyst such as sodium cyanide or alkyl amine, there is formed smoothly more ethylene cyanohydrin. This cyanohydrin after neutralization can be readily dehydrated catalytically to yield acrylonitrile of high quality.

Another process for the preparation of acrylonitrile, namely, the direct reaction of hydrogen cyanide and acetylene, has been used extensively in Germany and is assuming increased importance in the United States.[141a] This reaction is realized by passing a mixture containing a large excess of acetylene through a concentrated solution of cuprous chloride in sodium, ammonium, or calcium chloride. The acrylonitrile is scrubbed from the excess acetylene, and the latter is recycled. The crude reaction product contains numerous impurities of which divinylacetylene is most troublesome, for it is difficult to remove and if present even to the extent of a fraction of a per cent has a profoundly adverse effect on the processing properties of nitrile rubber made from such acrylonitrile.

PREPARATION OF NITRILE RUBBERS

Butadiene and acrylonitrile will copolymerize when heated together in bulk with a trace of benzoyl peroxide. Under these conditions the rate may be rather slow, and considerable cyclic codimer is formed,[123] as would be expected from simultaneous thermal, nonradical initiated codimerization. However, if butadiene and acrylonitrile are stirred with an aqueous solution of an emulsifying agent in the presence of a peroxide, polymerization occurs at a much lower temperature and proceeds at a more rapid rate.[98] The amount of cyclic codimer is also greatly decreased. This briefly is the basis for the emulsion process for preparing nitrile rubber. However, many refinements are necessary, and many accessory conditions must be met if a uniform rubber of definite properties is to be produced.

In place of butadiene, other dienes, such as isoprene, piperylene, or 2,3-dimethylbutadiene, can be used. Commercially, butadiene is used almost exclusively, since substituted dienes tend to yield a more leathery product. In any case the diene must be quite pure, since certain trace impurities inhibit polymerization, and others lead to a tough, gelled rubber.

The nitrile used for comonomer is usually acrylonitrile, although, when part of this is replaced by methacrylonitrile or ethacrylonitrile, the plasticity and cement-making properties of the rubber are improved. Large proportions of such homologs give rubbers with higher hysteresis, poorer low-temperature properties, and more oil resistance.

Small proportions of other monomers can be used along with the two main components. Thus, styrene, methyl methacrylate, ethyl acrylate, or vinylidene chloride in minor amounts can lead to certain desirable properties in the copolymer. The specific improvements are, however, usually attained at the expense of some other property. Thus, though incorporation of, say, 5 per cent styrene along with the acrylonitrile improves plasticity, it detracts from low-temperature flexibility.

The ratio of butadiene to acrylonitrile has a profound effect on the properties of the vulcanized polymer. This is shown in Table I in which an attempt has been made to classify nitrile rubbers broadly on the basis of combined nitrile content.

Table I. Relation of Oil Resistance of Vulcanizate to Combined Acrylonitrile in Copolymer

Fraction of Acrylonitrile in Copolymer, %	Oil Resistance of Vulcanizate
50 to 60	Leathery plastic with high resistance to aromatics
35 to 40	High oil resistance
25	Medium oil resistance
15	Fair oil resistance
2 to 5	High swelling in oil

The ratio of butadiene to acrylonitrile combined in the polymer will usually differ from the ratio charged to the polymerizer. Thus, in a polymerizer charged batchwise, the molecular structure of the copolymer formed tends to vary from the beginning to the end of the polymerization cycle. If a continuous polymerization is being run, both of the monomers can be introduced at a definite rate in order to maintain a uniform monomer ratio in the reactor even though the product formed may have a composition of some ratio other than that in the charge. Only under special so-called azeotropic conditions will the polymer formed have a composition similar to that of the monomer mixture charged. This mixture varies with the recipe used and the temperature of polymerization but is usually about butadiene 60 ± 5 per cent, acrylonitrile 40 ± 5 per cent.

Water. In emulsion polymerization, water is used as the dispersing medium in which to carry out the reaction. It also serves to keep the rubber in a fluid form so that the heat of reaction can be controlled by repeated passage of the reacting mixture over the cooling surface. It is usual to use 180 to 200 parts of water per 100 parts of monomer. Smaller proportions can be used if the recipe is carefully adjusted and precautions are taken to avoid precoagulation or gelling.

The requirements for purity of the water are very stringent. Distilled water free from metallic contamination is satisfactory; however, it has been found that deionized or zeolite-softened water can be used in most cases with equivalent results. The water is often vacuum deaerated to remove traces of dissolved oxygen and thereby speed up the polymerization.

Emulsifying Agents. The fatty acid soap or other emulsifying agent used in the emulsion polymerization process serves to solubilize the reactants so that the reaction starts in monomer oriented in the micelles. After a sufficient number of particles have been formed, such that adsorption of soap on their surface reduces the residual concentration of soap to a point where essentially no more micelles are present, further polymerization occurs in the particles already formed. It is possible to employ a mixture of emulsifying agents of two types, one that gives the micelles in which the monomer dissolves, the other that coats the surface of the rubber particles and prevents sticking.

Soaps of fatty acids are quite satisfactory for use in making nitrile rubber.[204] The sodium salt of oleic, myristic, or palmitic acid is excellent. A salt of linoleic acid can be used to provide a more plastic polymer, although at the expense of a slower rate of polymerization. An excess of fatty acid in the soap in many cases gives more rapid polymerization than a neutral soap. Nonconjugated polyunsaturation in the emulsifying agent slows up the polymerization reaction. Consequently, many fats are hydrogenated or isomerized before being employed for making the soaps to be used in polymerization. Rosin itself can be used in making nitrile rubber if the nonconjugated unsaturation is removed by hydrogenation or disproportionation.

A large number of anionic emulsifying agents other than soap can be used by themselves or in addition to soap to alter the course of the polymerization. Higher alkylbenzene sulfonates, dialkylnaphthalene sulfonates, or higher alkyl sulfates have been suggested and used in special cases.

Various cationic soaps can also be used in making nitrile rubber. Thus, the hydrochloride of diethylaminoethyl oleamide was one of the first emulsifying agents used. Dodecylamine hydrochloride, lauryl pyridinium chloride, and other cationic soaps can be used and are reported to impart special benefits to the nitrile rubber.

So far it has not been possible to use nonionic emulsifying agents alone in performing the polymerization. They are, however, useful to add after the reaction in order to stabilize the latex or to impart special properties to it.

Modifiers and Conversion. Certain substances which serve as chain terminators or chain-transfer agents have a profound effect on the speed of polymerization and the properties of the rubber formed. These are called modifiers or regulators and, as in the preparation of GR-S, are necessary in order to insure that the nitrile copolymer shall be plastic and soluble. If xanthogen disulfides are used for modifiers at a relatively low temperature, the polymerization can be carried practically to completion and still yield a plastic, processible nitrile rubber. When higher mercaptans are used, it is customary to stop the polymerization when it has proceeded 60 to 80 per cent; otherwise there is a tendency for high gel content in the rubber. It is often advantageous to add the mercaptan stepwise as the polymerization proceeds.[16] Tertiary mercaptans are often used instead of primary mercaptans because of their slower rate of consumption during polymerization. Carbon tetrachloride has been suggested as a modifier; however, it has been found by several workers to be relatively inactive. Whereas, xanthogen disulfides and mercaptans are effective at 0.1 to 1 per cent on the polymer, carbon tetrachloride, to show appreciable activity, must be used at 10 to 20 per cent.

As in the preparation of GR-S, a suitable mercaptan modifier is n-dodecyl mercaptan. This is illustrated by the data[157] in Table II, which also show that, as with GR-S, the lower and higher homologs of n-dodecyl mercaptan are less suitable than n-dodecyl mercaptan itself but that lauryl mercaptan (prepared from "lauryl" alcohol derived from coconut oil and containing, in addition to n-dodecyl mercaptan, certain proportions of n-tetradecyl and n-hexadecyl mercaptans) is still better than pure n-dodecyl mercaptan.

Table II. Influence of Various Mercaptans on the Plasticity and Stress-Strain Properties of Butadiene-Acrylonitrile Copolymers

Mercaptan	Conversion (%) in 16 Hr. at 28° C.	Williams Plasticity at 80° C. with 10 Kg. Weight	Tensile Strength P.S.I. (in Stock with 50 Pts. EPC Black; Dibenzothiazolyl Disulfide Acceleration)	Elongation, %	Modulus at 300%, P.S.I.
n-Decyl	66	86–0	2650	775	495
n-Dodecyl	62	74–0	3150	735	500
n-Tetradecyl	60	160–67	4200	620	980
n-Hexadecyl	63	214–116	3900	420	2200
"Lorol"	65	64–0	3940	705	755

Polymerization Initiators. Polymerization may be initiated by a chemical reaction which forms free radicals. Thus, a peroxide, such as benzoyl peroxide, hydrogen peroxide, potassium persulfate, or cumene hydroperoxide, may be used.[141] Such reagents are especially effective if there is present some reductant which can react at the proper rate to maintain the necessary level of radicals. Mercaptans may serve as both reductants and modifiers. Other reductants which may be used are aldehyde, sulfite, polyamines, or sugars. This action is intensified by the presence of traces of ions which can exist in more than one valence. Hence ferrous,[26] cupric,[52] and cobalt salts serve as activators when used in sufficiently small amounts. Salts and organic compounds which form complexes with metal ions can be used to buffer the concentration of the ion; thus sodium pyrophosphate, sodium cyanide, and amino acids are often used in polymerization recipes. In certain recipes secondary aliphatic amines or their condensation products with acrylonitrile in amounts as low as 0.02 per cent have an accelerating effect on the speed of polymerization.

A Typical Polymerization Recipe. A typical polymerization recipe suitable for producing nitrile rubber with a nominal 25 per cent acrylonitrile content is shown in Table III.   The reasons for the various ingredients have already been discussed.   In this recipe all factors are combined and balanced.

Table III.   A Typical Recipe for Producing Nitrile Rubber

	Parts by Weight
Butadiene (freshly distilled)	75
Acrylonitrile	25
Soap flakes	4.5
"Stearic acid"	0.6
tert-Dodecyl mercaptan	0.5
Potassium chloride	0.3
Sodium pyrophosphate (anhyd. basis)	0.1
Ferric sulfate (anhyd. basis)	0.02
Hydrogen peroxide (20% soln.) (anhyd. basis)	0.35
Water (deionized or distilled)	180

If the parts mentioned in the recipe are taken in grams, the polymerization can be run in a 16-oz. crown-cap bottle.   The solutions are made up and charged in the following manner: The soap and stearic acid are dissolved in about 50 parts of water at 50° C.   The sodium pyrophosphate, potassium chloride, and ferric sulfate are dissolved in about 5 parts of water.   Charging is done in the following order: (1) The excess water above what was used in making the solutions is placed in the reactor.   (2) The soap solution is added, followed by (3) the ferric pyrophosphate solution.   (4) The acrylonitrile is added, followed by (5) the mercaptan and (6) the butadiene. (7) The hydrogen peroxide is added and the reactor sealed and run under pressure with mild agitation and cooling to maintain an internal temperature of 30° C.   About 90 per cent conversion of the monomers will be attained in 24 hours.   The latex is then discharged and stabilized by adding immediately a dispersion containing 2 parts of phenyl-β-naphthylamine.

In order to coagulate the latex in the form of small crumbs suitable for washing, there can be added 0.5 part of sodium alkylbenzene sulfonate, followed by 40 parts of sodium chloride as saturated brine.   Dilute sulfuric acid (0.25 per cent) is now added with vigorous stirring until a pH of about 3 is reached.   The crumbs and mother liquor are then made alkaline to a pH of 11 using sodium hydroxide.   The crumbs are filtered off, washed with hot soft water until free from soap, dried at about 60° C., and sheeted by passing through a tight mill.

Temperature of Polymerization. As can be seen by the example in Table III, a polymerization recipe for making nitrile rubber is likely to be rather complex.   Many ingredients are used, each to perform a specific function.   The polymerization is then carried out at as low a temperature as practical. In general, the lower the temperature at which the polymerization is performed, the more regular the structure of the rubber and the better the over-all quality that can be realized. The limiting factors are the speed of polymerization, freezing of the charge, or gelling in the reactor. Nitrile rubber has been made commercially at as high a temperature as 50° C., although the trend is toward polymerization at 25° C. or even as low as 5° C. The higher the temperature at which polymerization is performed, the higher the vapor pressure and the stronger the reactor that will be required.

Time of Reaction. The highest speed of polymerization that can be tolerated in a reactor is limited by the rate at which the cooling medium can remove the heat of polymerization. The other limit often is an economic one and represents the time that the charge can be left in the reactor and still turn out a competitive product. Polymerization recipes are usually adjusted so that the rate will not exceed 12 per cent per hour. A polymerization requiring more than 48 hours is usually considered uneconomical.

Treating the Latex. When the polymerization has proceeded to completion or as far as desired, a "shortstop" such as hydroquinone is added, and the excess reactants are removed by releasing the pressure and distilling in a vacuum under conditions to recover the unreacted butadiene and acrylonitrile. Dispersed antioxidant is then added. Phenyl-β-naphthylamine or an alkyldiphenylamine are often used if staining is of no consequence. If a nonstaining rubber is required, a hindered alkylphenol or other phenol derivative may be used.

Coagulation of the latex is accomplished by first flocculating the particles with brine. Conditions can be set so the flocculation is complete and the particles are large enough to be filtered off on a rotary filter, then reslurried in water, and refiltered to remove the soap. The consumption of salt can be decreased if the latex is lightly flocculated and then treated with a small amount of acid to yield crumbs containing fatty acid. Adding an excess of alkali causes resaponification of the fatty acid. The crumbs can then be filtered off, wet-sheeted, and wash-milled to remove the salt. Or the alkaline crumbs can be sheeted on a Foudrinier, washed, and pressed to a porous sheet.

The crumbs from the rotary filter process are dried in a continuous tunnel drier and then sheeted or pressed. The sheets from the wash-mill process are vacuum-dried to yield a marketable rubber. The sheet from the Foudrinier is dried in a tunnel drier and then cut into sheets which are plied up into bales.

Nitrile rubber is rather sensitive to the coagulation and drying processes, and these are closely controlled in order to assure proper solubility and processibility of the rubber.

Crude Nitrile Rubber. Crude nitrile rubber appears on the market in a number of forms, the particular form varying with the supplier. It can be in crumbs, thin sheets, pellets, rolls, or blocks. These can all be used easily for weighing up batches and are designed for ease in mixing.

Nitrile rubber varies from an almost colorless product to a dark-brown one, depending on the grade and the use in mind. It usually has a typical odor imparted by traces of codimer, modifier, and fatty acid. Where odor is a factor in use, special odorless grades can be made. Nitrile rubber is tasteless, and the polymer itself does not impart taste to objects that contact it.

The specific gravity of nitrile rubber varies with the content of acrylonitrile, in general being close to 1. A product with an acrylonitrile content of 25 per cent has a specific gravity of about 0.97, and one with 40 per cent about 1.00.[104, 244]

THEORY OF THE COPOLYMERIZATION

As is so often true in a new and commercially important field, the art of manufacturing nitrile rubber (like that of other high-molecular compounds) progressed more rapidly than the scientific explanation of the phenomenon. It was known that peroxides or diazo compounds would initiate polymerization and that the polymerization itself is a chain reaction. Theories to explain chain initiation, growth, transfer, and termination developed more slowly. These are discussed in Chapter 8. In mass polymerization where the reactants and catalyst are all in the same phase, chain initiation and transfer to acrylonitrile is more rapid than to butadiene, for Gindin, Abkin, and Medvedev[87] have shown, as illustrated in Table IV, that in a system initiated with benzoyl peroxide the rate of polymerization is vastly accelerated as the ratio of acrylonitrile to butadiene is increased.

Table IV. Effect of Monomer Ratio on Rate of Polymerization to Nitrile Rubber

Composition	Rate of Mass Polymerization
10% acrylonitrile, 90% butadiene	0.26% per hr.
80% acrylonitrile, 20% butadiene	8.2% per hr.

Alekseeva[2] polymerized an equimolecular mixture of acrylonitrile and butadiene with 1 per cent benzoyl peroxide in homogeneous phase for 116 hours at 60° C. After lower-weight impurities were removed, the rubber was ozonized, and the ozonide was hydrolyzed and oxidized with hydrogen peroxide. There were thus obtained succinic acid, butanetricarboxylic acid, hexanetetracarboxylic acid, and dodecanepentacarboxylic acid in ratios which indicated that in about 50 per cent of the product the acrylonitrile and butadiene alternated in the chains, while in about 33 per cent of the product two to three acrylonitrile units were between butadiene units. This shows that, in a homogeneous system, terminal acrylonitrile on a growing chain has a slightly greater tendency to react with acrylonitrile than with butadiene and that a terminal butadiene has a much greater tendency to react with acrylonitrile than with another butadiene.[123]

In emulsion polymerization systems many other factors enter to complicate the process.[181] The point of initiation, the locus of the reaction, and the distribution to this locus are only a few of the added variables.[98] However, Wall, Powers, Sands, and Stent[232] show that, even in a case where one of the comonomers is as soluble in water as is acrylonitrile, nevertheless, a copolymerization equation such as holds for mass polymerization applies in the modified form.

$$\frac{dM_1}{dM_2} = \frac{M_1(r_1'M_1 + M_2)}{M_2(M_1 + r_2'M_2)}$$

where M_1 and M_2 represent the total amounts of the respective monomers and r_1' and r_2' are copolymerization parameters but not necessarily ratios of specific reaction-rate constants. r_1' was found to be 0.4 for butadiene and r_2' to be -0.1 for acrylonitrile. The negative value is probably due to the water solubility of the nitrile. In a polymerization system run at 50° C., using soap flakes for the emulsifying agent, n-dodecyl mercaptan as the activating modifier, and potassium persulfate as the initiator, polymers of the average composition shown in Table V were obtained.[232]

Table V. Influence of Composition of Charge on Nitrile Content of Emulsion Copolymer from Butadiene and Acrylonitrile

	Time, Hr.	Fraction Converted	Fraction Acrylonitrile	Time, Hr.	Fraction Converted	Fraction Acrylonitrile	Time, Hr.	Fraction Converted	Fraction Acrylonitrile
Reactants charged			0.150			0.250			0.350
Product from initial reaction	1.0	0.21	0.214	0.5	0.16	0.301	0.5	0.18	0.357
Product at almost complete reaction	13.7	0.99	0.144	18.3	1.00	0.233	9.0	1.00	0.341

These data show how the average composition of the butadiene-acrylonitrile copolymer is affected by the composition charged and the degree of conversion.[231] The effect is even more pronounced if the composition of the polymer formed at any instant is plotted against the degree of conversion.[203] Thus, for a specific recipe run at 25° C., Fig. 1 shows the heterogeneity of nitrile rubber formed if the charging composition departs from 37 per cent acrylonitrile. This latter composition might be called the azeotropic copolymerization composition (termed in this case by Morgan[144] a pseudo-azeotrope), i.e., the composition at which the comonomers enter into copolymerization in the same ratio as that in which they are charged.

As might be expected, the details of the polymerizing system affect the composition and uniformity of the polymer formed. This is noted especially with changes in the emulsifying agent, as in going from an anionic to a cationic soap.

Embree, Mitchell, and Williams[69] have investigated incremental composition of polymer formed at any instant as a function of the residual monomer ratio in a redox recipe at 5° C. Their data suggest an azeotropic polymer in this case of approximately 44 per cent acrylonitrile. From this and the results obtained by others when different recipes and temperatures were used, it is indicated that the apparent relative reactivities of acrylonitrile and butadiene toward the growing chain[3] are the resultant of a number of factors, there being a strong tendency toward alternation, modified, however, by the relative solubility in the water phase and rate of transfer between the two phases.

When butadiene is copolymerized with increasing amounts of acrylonitrile, there is a very marked improvement in tensile strength as the proportion of nitrile is increased. This has been ascribed to a 1,4-directive effect exercised by acrylonitrile on the polymerization of butadiene.[183] Even small additions of acrylonitrile accelerate the polymerization of butadiene and also the copolymerization of systems such as butadiene–styrene.[76] The inclusion of a low percentage of styrene (about 5 per cent)[75] or of an acrylate ester such as methyl methacrylate[51] gives nitrile rubbers with improved processing properties, as evidenced by higher plasticity and better solubility characteristics.

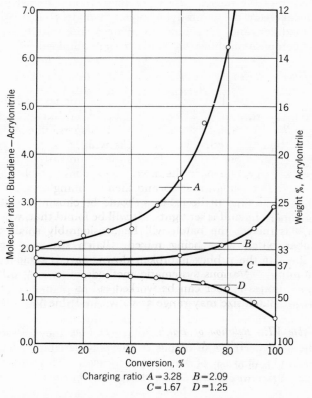

Charging ratio $A = 3.28$ $B = 2.09$
$C = 1.67$ $D = 1.25$

Fig. 1. Showing Composition of Butadiene-Acrylonitrile Copolymer Formed at Any Instant as Affected by the Composition Charged and the Degree of Conversion at that Instant

MILLING AND PROCESSING

In general nitrile-rubber stocks can be mixed and handled on the same rubber machinery that is used for natural-rubber stocks. Those nitrile rubbers that are less plastic than natural rubber will of course develop more heat when mixed under similar conditions. Preliminary breakdown of nitrile rubber is even more important than that of natural rubber. Chemical softening is not so effective with nitrile rubber as with natural rubber, although Campbell has recently shown that esters of dialkylcarbamates and dithiocarbamates increase the plasticity manyfold.[43] Milling on cold rolls set tight is perhaps the most efficient method of breaking down the crude rubber. When first put on the rolls, nitrile rubber appears knotty and rough. As mastication proceeds, the stock smoothes out and under ideal conditions gives a glossy rolling bank. Usually the stock runs more easily on the faster or back roll. A well masticated nitrile rubber is characterized by strings or webs that can be noticed when a piece is pulled out between the fingers.

Good dispersion of the pigments is extremely important if optimum physical properties are to be attained. It is more essential to add sulfur and zinc oxide first rather than the accelerator. Dispersion of the sulfur is much more important than in natural rubber since its solubility in nitrile rubber is much lower.[114] Poor sulfur dispersion leads to a grainy or knotty vulcanized stock, in which undispersed particles of sulfur have caused local overcured areas. Carbon black tends to disperse in agglomerates if it is added too fast or if the rubber stock is too hot. If the first portions of black are added slowly and dispersed well, it will be found that the later portions can be added more rapidly. Carbon black dispersion can be checked rapidly by cutting the freshly milled stocks with scissors. A dull surface indicates poor dispersion, while a glossy surface indicates good dispersion. If the dispersion appears poor in the mixed stock, considerable improvement can be obtained by cooling the stock and then remilling on a cold tight roll.

Batch sizes for mixing nitrile rubbers should be chosen so that there is a rolling bank when the mill is set tight. It will be found that, when a nitrile-rubber stock is mixed, the batch will be considerably smaller than if a natural-rubber batch were being mixed. Batch size can be increased somewhat if, after the rubber is broken down, part is removed until the pigment is mixed. Various systematic methods of cutting off stock and blending back excess rubber can be worked out to give most efficient and rapid mixing. Batch size may range as shown in Table VI.

Table VI. Relation of Batch Weight to Length of Mill Roll When Milling a Nitrile-Rubber Stock

Length of roll, in.		48	60	84
Batch weight (lb.)	Min.	30	45	110
	Max.	50	75	160

Nitrile-rubber stocks can be mixed in the Banbury if proper precautions are observed. In any case, the Banbury should be run as cool as possible. Addition of suitable softener assists in making the batch knit together. The black can well be added in at least two portions. In Banbury-mixed stocks cooling and remilling for improving dispersion is even more important than in mill-mixed stocks. Nitrile rubber of Mooney viscosity (ML-4) of 85 to 120 often works better in the Banbury than such rubber in the 60 to 80 Mooney range.

For use in making molded articles, the mixed nitrile-rubber stocks can be batched off, calendered, or extruded. In some cases the blank should be cut so that minimum flow occurs during the molding operation; however, it is surprising how often a superior molded article will be produced if the blank is cut so that *maximum* flow occurs during the molding and curing operation. In this respect nitrile rubber behaves quite differently from natural rubber.

CALENDERING AND EXTRUSION

Calendering operations vary considerably according to the type of nitrile-rubber stock being run. The calender rolls should usually be run at a lower temperature than in the case of natural rubber. It will be found difficult to obtain good ply adhesion in built-up stocks, although operation at a proper temperature, use of a minimum of blooming oils, high buildup roll pressure, and wrapping in liners in a tight roll all aid in reducing lamination. In nitrile rubber it is especially true that good calendering depends on proper compounding. Structure black, ample zinc oxide, and compatible softener help the stock to run smooth. If fabric is to be frictioned or skim-coated, the rolls should be run rather hot, and the fabric should be hot and dry.

Mill-plasticized nitrile rubber is in general more difficult to extrude than natural rubber.[112] However, if properly compounded and mixed, nitrile-rubber stocks extrude smoothly and rapidly. Use of structure black, thermal black, and certain waxes or oils improve the extrusion properties. For calendering or extrusion, improvement can often be obtained by including a small amount of reclaim, natural rubber, or high-styrene copolymer in the recipe. Certain phenolic resins also are helpful when they can be included.

Where unvulcanized nitrile-rubber constructions have to be built up before vulcanization, it will be found helpful to coat the surfaces to be adhered with a cement made from a nitrile-rubber stock. The cement may be made up in ketones, esters, aromatics, or mixtures. Toluene, methyl ethyl ketone, or mixtures of them are often helpful in freshening up the surface of nitrile stocks to be adhered if cement is not used. The solvent functions by removing dust, waxes, and oils and probably forming a thin layer of cement on the surface. In any case, building tack will not be so good as when natural-rubber stocks are used.

COMPOUNDING OF NITRILE RUBBER

Detailed compounding data obtained with the grades of nitrile rubber on the market are included in compounding manuals prepared by the suppliers[48, 166, 104] or in numerous articles in the technical literature.[59, 117, 122, 142, 78, 133, 193, 60, 198-201] Certain general principles, however, are common to the entire class.

Sulfur. Nitrile rubbers are usually vulcanized with sulfur. Because of their lower unsaturation,[113] the proportion of sulfur required is generally less than would be used for natural rubber. From 1 to 2 p.h.r. (parts per 100 parts of rubber) is usually sufficient unless stiffer stocks or hard rubbers are desired. In general, those nitrile rubbers with higher combined acrylonitrile require relatively smaller amounts of sulfur for proper cure than nitrile rubbers with lower nitrile content, and nitrile rubbers of lower Mooney viscosity require greater amounts of sulfur than those of higher viscosity. Excessive amounts of sulfur decrease heat and tear resistance as well as elongation. The general effect of increasing sulfur content is shown in Table VII.

Table VII. Influence of Sulfur Content on the Cure of a Nitrile-Rubber Stock

Nitrile rubber (33% combined acrylonitrile)	100
Carbon black (SRF)	65
Zinc oxide	5
Dibutyl phthalate	10
Coumarone-indene resin	10
Stearic acid	1
Benzothiazolyl disulfide	1.5
Sulfur	Variable

Vulcanized at 310° F.

Sulfur Content, P.H.R.	1.25	1.50	1.75	2.00
Modulus at 300%, p.s.i.				
15 at 310° F.	950	1000	1100	1250
30	1100	1300	1450	1650
45	1425	1500	1625	1850
60	1450	1550	1675	1950
Ultimate tensile strength, p.s.i.				
15 at 310° F.	2200	2350	2350	2600
30	2600	2650	2500	2750
45	2650	2700	2600	2850
60	2700	2800	2700	2900
Elongation at break, %				
15 at 310° F.	680	640	600	560
30	610	580	530	500
45	580	540	500	470
60	550	530	500	460
Shore hardness				
15 at 310° F.	52	53	56	56
30	54	55	58	58
45	55	56	59	60
60	56	56	59	60

Sulfur donors[28, 222, 243] such as tetramethylthiuram disulfide (TMTD)[104] may be used in place of free sulfur as the vulcanizing agent. In this specific case about 3 p.h.r. of TMTD is required. Stocks so vulcanized have exceptionally good physical properties and are characterized by low set and excellent heat resistance. Nitrile rubbers may, like natural rubber, be vulcanized at room temperature by alternate exposure to sulfur dioxide and to hydrogen sulfide, in which case the combined sulfur for optimum cure of a 40 per cent acrylonitrile rubber is 1.37 per cent and for a 33 per cent acrylonitrile rubber 2.55 per cent.[30] This method of vulcanization has not, however, found practical application.

Accelerators. Almost any accelerator used with natural rubber can be used with nitrile rubber. Benzothiazolyl disulfide, mercaptothiazoles, thiuram disulfides, sulfenamides, aldehyde–amines, and guanidines may be used by themselves or in combination. 1 to 1.5 p.h.r. of a thiazole or 0.25 to 0.50 p.h.r. of a thiuram accelerator will usually be found sufficient. Good stocks can be obtained with litharge acceleration, 10 p.h.r. of litharge giving a flat-curing high-tensile stock.

Zinc Oxide. About 5 p.h.r. of zinc oxide is required for proper activation of a nitrile-rubber stock. Large amounts can be used to impart particular properties; thus, 100 p.h.r. in the absence of carbon black gives a stock of interesting high plasticity and flow characteristics.

Fatty Acid. A fatty acid or soap is required for proper activation of nitrile rubbers. Since many grades of nitrile rubber on the market contain sufficient fatty acid for proper cure, the recommendations of the supplier should be followed.

Pigments. Pigmentation is of utmost importance for developing the best properties in nitrile rubber. Of all pigments, carbon black is most outstanding because of the remarkable increase in tensile strength that it imparts. The effect of various proportions of typical pigments is shown in Table VIII.[166]

Table VIII. Influence of Various Loadings of Typical Pigments on the Properties of Vulcanized Nitrile Rubber

Nitrile rubber (26% combined acrylonitrile), parts by wt.	100
Zinc oxide, parts by wt.	5
Stearic acid, parts by wt.	1
Sulfur, parts by wt.	1.5
Benzothiazolyl disulfide, parts by wt.	0.25 to 1.25
Pigment, parts by volume	10 to 60

Accelerator adjusted to produce approximately optimum tensile strength
Cures: 15 to 120 min. at 287° F. (optimum chosen for test).
Properties selected from smoothed curves.

Pigment Volumes	Channel Black EPC	Furnace Black SRF	Thermal Black FT	Zinc Oxide	Clay	Silene	Fine Particle Whiting (Kalvan)	Whiting (Gilder's)
			Modulus at 300%, P.S.I.					
10	400	450	300	260	280	220	300	170
20	900	1000	580	250	400	500	220	190
30	2100	2000	850	250	750	580	300	150
40	2700	2400	900	300	1000	900	500	180
60	...	...	1600	700	...	1500	1100	210
			Ultimate Tensile Strength, P.S.I.					
10	2000	1900	1200	1100	1600	1100	700	600
20	3000	2500	1800	1300	1800	1700	1600	650
30	3600	2500	2000	1500	2000	1600	1900	500
40	3400	2500	1800	1800	1900	2000	2000	750
60	3200	2400	1800	1800	1800	1800	1850	500
			Elongation at Break, %					
10	550	620	600	570	720	700	580	560
20	520	580	600	650	770	600	800	600
30	450	400	560	600	700	550	750	550
40	380	320	450	650	600	550	700	680
60	200	200	350	600	250	400	500	600
			Shore Hardness					
10	54	54	50	54	57	53	49	54
20	64	56	54	59	62	58	54	56
30	70	69	54	60	69	65	55	58
40	78	70	64	62	71	70	62	64
60	90	84	70	69	75	86	74	68

Softeners and Plasticizers. Proper choice and use of softener or plasticizer is essential with nitrile-rubber stocks. Since the polymer itself is oil-resisting, special types of softeners are required. Aromatic oils, esters, and polar-type derivatives are preferable to paraffinic oils or petrolatum.[49]

Table IX. Effect of a Typical Softener (Dioctyl Phthalate) on the Properties of Nitrile Rubber

	No Softener	30 P.H.R. of DOP
Nitrile rubber (33% combined acrylonitrile)	100	100
Zinc oxide	5	5
Carbon black (SRF)	60	60
Stearic acid	1	1
Sulfur	1.5	1.5
Benzothiazolyl disulfide	1.5	1.5
Dioctyl phthalate	...	30
Modulus at 300%, p.s.i.		
20 min. at 310° F.	1700	1175
30	1830	1250
45	2050	1350
Ultimate tensile strength, p.s.i.		
20 min. at 310° F.	1890	2050
30	2050	2275
45	2080	2275
Elongation at break, %		
20 min. at 310° F.	375	510
30	325	500
45	310	450
Shore hardness	68	46
Rebound (Lupke), %	38	55
Compression set (%) ASTM method B	12	10

	\multicolumn 4 Passed −65° F.				\multicolumn 4 Passed −65° F.			
ASTM low-temperature test	T	E	H	ΔV	T	E	H	ΔV
Aged in paraffin lub. oil 70 hr. at 212° F.	1900	275	62	5	2600	340	60	−14
Aged in distilled water 70 hr. at 212° F.	2160	275	65	10	1700	290	46	17
Aged in 100-octane gasoline 70 hr. at room temperature	1300	260	55	5	1575	385	43	3
Aged in air oven 48 hr. at 300° F.	600	10	90	...	1700	160	63	...
Aged in air oven 70 hr. at 212° F.	2110	175	75	...	2275	300	56	...
Heat loss 48 hr. at 300° F, %	2				1			

T—Ultimate tensile strength, p.s.i.   *H*—Durometer hardness
E—Elongation at break, % ΔV—Increase in volume, %

Detailed tables showing the effect of various softeners are given in the suppliers' manuals. In general, rather surprisingly large amounts of softeners (up to 50 p.h.r.) can be used in a nitrile-rubber stock without adversely affecting the physical properties. In general softeners are relied on for controlling the hardness and low-temperature flexibility of nitrile-rubber stocks. In choosing a softener, special attention should be paid to volatility, chemical stability, rate of hydrolysis, and extraction. Table IX shows the effect of an ester softener such as dioctyl phthalate when used in a typical nitrile-rubber stock. Table X is a brief guide to the selection of softeners adapted to secure specified properties in nitrile-rubber stocks.

Table X.   Selection of Softeners for Nitrile-Rubber Stocks

	For Ease of Incorporation	For High Tensile Strength	For High Elongation	For Low Durometer Hardness	For High Durometer Hardness	For Resilience	For Low-Temp. Flexibility	For Heat Service	For Low Compression Set	For Low Extraction
Coumarone-indene resin	x	x	x			x				
Dibutyl phthalate				x			x		x	
Dibutyl sebacate				x			x	x	x	
Dioctyl phthalate		x					x	x	x	x
Paraplex G-25*	x					x			x	x
Octadecene nitrile			x	x		x	x			
Piccocizer 30†	x	x	x			x			x	
Plasticizer SC‡			x	x		x	x	x		
Tricresyl phosphate		x						x	x	
Tributoxyethyl phosphate			x			x	x		x	
Plasticizer A-118§								x		x

* Polyester resin.
† Polymerized aromatic distillate (high boiling).
‡ A triglycol ester of a vegetable-oil fatty acid (high-boiling liquid).
§ California Flaxseed Products Co.

TYPICAL COMPOUNDS

The application of the principles discussed in the previous paragraphs can be illustrated by a series of oil-resisting stocks compounded from nitrile rubber containing 40 per cent combined acrylonitrile[78] (Table XI). For high resistance to abrasion, medium-processing channel black (MPC) is used in a rather high proportion. For a soft stock, fine thermal black (FT) is used, along with a high proportion of softener. A pure-gum stock

Table XI. Typical Nitrile-Rubber Compounds Showing Physical Properties and Swelling in Oils and Solvents

	Tread Stock	Low Set	Gasoline Hose Tube	Heat-resistant	Soft Stock	Clay-Loaded	Soling	Pure Gum	Fast Curing
Nitrile rubber (40% combined acrylonitrile)	100.0	100.0	100.0	100.0	100.0	100.0	100.0	100.0	100.0
Zinc oxide	5.0	5.0	5.0	5.0	5.0	5.0	5.0	5.0	5.0
Antioxidant*				5.0					
Mercaptobenzothiazole				3.0					
Benzothiazolyl disulfide	1.25	1.0	1.0		1.5	1.5	1.25	1.0	1.5
Tetramethylthiuram disulfide				3.0					
Di-o-tolylguanidine									0.25
Lauric acid	1.5	1.5	1.5				1.5	1.0	
Sulfur	1.25	2.0	1.0	1.5	1.5	1.5	1.5	1.0	1.25
Carbon black (MPC)	50.0						75.0		50.0
Carbon black (SRF)			100.0	50.0			50.0		
Carbon black (FT)		50.0							
Clay						75.0			
Soft coal tar	3.5				25.0		25.0		
Dibutyl phthalate		20.0	20.0	10.0	50.0	20.0			20.0
Cure = min. at 275° F.	162.50	179.5	228.5	177.5	183.0	203.0	257.75	108.0	178.00
min. at 310° F.	60	60	60	60	60	60	60	60	45
Modulus at 300%, p.s.i.	2600	1700	1900	1000	200	1300		200	1000
Tensile strength, p.s.i.	4200	2400	2200	3800	500	2500	3000	600	3500
Elongation, %	445	405	400	650	450	560	300	600	600
Shore hardness	77	58	69	71	36	58	91	50	71
Crescent tear (lb. per 0.1 in.)	30	18	35	39	5	18	40	7	37
Schopper rebound, %	14	34	27	13	54	37	12	24	13
Compression set (40% constant deflection), %	9	7	7	10	12	18	12	9	23
Hysteresis (Goodrich flexometer, ΔT° F. at 212° F.)	60	26	60	65	25	45	75		75
Abrasion (Williams) (natural rubber tread = 1000)	2100	1300	1200	1500	1100	900	1000	6000	1200
Low-temperature brittleness, °C.	−55	−55	−50	−25	−55	−55	−25	−55	−40
Flex. life (deMattia) $\times 10^{-3}$	43.6	83.3	221	112	490	83.5	2.3	130	65.8

Volume Change after 48 Hr. Immersion at Room Temperature, %

	Tread Stock	Low Set	Gasoline Hose Tube	Heat-resistant	Soft Stock	Clay-Loaded	Soling	Pure Gum	Fast Curing
Hexane	1.5	0.0	0.0	3.0	−6.1	0.0	0.0	4.5	0.0
Benzene	127	105	86	102	115	91	76	210	135
Carbon tetrachloride	37	26	23	37	12	21	25	64	37
Acetone	176	151	145	235	192	147	105	345	225
Ethanol, 95%	9	3	3	8	−10	3	3	15	8
Distilled water	0	−1.5	0	1.5	−0.8	−0.8	0.8	0	1.5
S.A.E. 20-W oil	4.5	1.5	3.0	0.8	−6	3.0	0		0
Gasoline (X-70)	0.8	0	0	4.5	−9.2	0	2.2	9.2	3.7
Kerosene 95% + benzene 5%				1.5	−4.5		0	2.2	0

Volume Change after 48 Hr. Immersion at 212° F., %

	Tread Stock	Low Set	Gasoline Hose Tube	Heat-resistant	Soft Stock	Clay-Loaded	Soling	Pure Gum	Fast Curing
Light oil (Circo X)	9.2	0	0	5.3	−19	0	4.5	15	4.5
Kerosene 95% + benzene 5%	12	0	0	6	−17	18	6	0	4.5

* Polymer of 2,2,4-trimethyl dihydroquinoline.

without carbon black has a low tensile strength. Note that the heat-resistant stock is vulcanized with tetramethylthiuram disulfide (TMTD). It should be observed that a polar softener such as dibutyl phthalate increases the resilience of a stock. All of the stocks have good resistance to typical solvents except to acetone, benzene, and carbon tetrachloride.

BLENDS CONTAINING NITRILE RUBBER

By using nitrile rubber as a blending agent with other synthetic rubbers or plastics, it is possible to obtain products with many interesting and useful properties.

Blends with Natural Rubber. While natural rubber is essentially incompatible with nitrile rubber, homogeneous two-phase blends can easily be made, especially if both rubbers are mixed with carbon black before blending. Thus, if a plasticized nitrile-rubber stock shrinks too much when operated in hot oil, it is possible to add a small proportion of natural rubber or of GR-S and obtain a stock with essentially no shrinkage or a definite prescribed swelling under service conditions. At the other end of the range, if a small amount of nitrile rubber is added to a natural or GR-S rubber tread, there is formed a stock which, although it quickly micro-checks, is quite resistant to the development of large cracks on flexing.[189]

Compounding with High-Styrene Copolymers. As the proportion of combined styrene in a butadiene copolymer is increased, the resultant rubber has improved blending characteristics with nitrile rubber. Thus, a copolymer made from equal parts of butadiene and styrene apparently gives a one-phase mixture with nitrile rubber. Nitrile rubber may be replaced by up to 50 per cent of such a polymer with comparatively minor effect on the rate of cure, modulus, tensile strength, elongation, compression set, hardness, and low-temperature properties. Such blending gives superior molding behavior, but swelling in solvents or in oil is increased.[104]

Copolymers of acrylonitrile and styrene containing 50 to 85 per cent styrene blend well with nitrile rubbers to yield very tough leathery stocks that have good tensile properties and resistance to ozone and are suitable for exterior covers of wires or cables.[62]

Compatability with Other Polymers. Polysulfide rubbers such as polyethylene polysulfide blend with nitrile rubber to yield stocks in which most of the properties are additive. The tensile strength is better than that of straight polysulfide stocks, and resistance to some solvents is superior to nitrile-rubber stocks alone.[211, 140]

Stocks having improved heat resistance can be obtained by blending polyacrylate with nitrile rubber.[162] Polyesters serve as nonextractable plasticizers; however, they tend to yield rather leathery stocks. Polyiso-butylene has extremely limited compatibility when blended into nitrile rubber; however, it does give some improvement in ozone resistance at the expense of increased dielectric loss and decreased tensile strength.[162]

Nelson and Vanderbilt[157] have shown that, in spite of limited miscibility with nitrile rubber, 10 to 20 p.h.r. of polyisobutylene of molecular weight 10,000 to 20,000 improves the processing and extrusion characteristics of nitrile-rubber stocks even more than a corresponding amount of dibutyl phthalate. A heavily loaded nonblack nitrile-rubber stock can be plasticized with a mixture of Butyl rubber (28 to 40 p.h.r.) and mineral oil (20 to 30 p.h.r.) to give a compound suitable for floor tiling.

Blends with Polyvinyl Chloride. Polyvinyl chloride plasticized with tricresyl phosphate when added in the proportion of 33 to 50 p.h.r. improves the sunlight and ozone resistance of nitrile-rubber stocks.[70] Such blends, however, tend to vulcanize more slowly than the straight nitrile-rubber stock, and hence adjustment of sulfur and accelerator is usually necessary. Mixing such stocks is rather difficult to perform on a mill in the factory; accordingly there has been introduced a product in which the polyvinyl chloride and nitrile rubber have been premixed in the latex form before coagulation. This, of course, simplifies processing on manufacturing equipment.[154] Minor proportions of polyvinyl chloride modify the properties of the nitrile-rubber stock, increasing sunlight resistance, ozone resistance, and resistance to fire, abrasion, and tearing. On the other hand, specially plasticized nitrile rubbers serve as especially useful nonextractible plasticizers for polyvinyl chloride.[115, 153-4] Such mixtures are quite useful for forming clear films deposited from cement. The uses of blends of polyvinyl chloride and nitrile rubber are numerous:[71] as insulation, bookbinding, unsupported upholstery fabric, and leather substitute.[240, 168, 171, 245-6]

Blends with Phenolics. Phenol-aldehyde resins are often mixed with nitrile rubber to improve the plasticity of the stock during processing and to stiffen it after vulcanization.[158] Such resins can be used to vulcanize the stock.[234] To give a single example to illustrate these statements, when 30 parts of a reactive phenol-formaldehyde resin (Durez 12687) was mixed with 100 parts of a nitrile rubber containing 35 per cent acrylonitrile and the blend heated at 325° F. for 15 minutes, the product had a tensile strength of approximately 1600 p.s.i., an elongation of approximately 305 per cent, and a tear strength of 270 lb. per in. A similar blend containing the sulfur, accelerator, and zinc oxide necessary to vulcanize the rubber had a tensile strength of approximately 2800 p.s.i. and an elongation of 340 per cent. For comparison, it may be noted that the tensile strength of the nitrile rubber alone when cured in a gum stock[158] is only of the order of 400 p.s.i.

When high proportions of thermosetting phenolic resins are used, it is possible to prepare a thermosetting adhesive that can be applied in cement or sheet form for bonding metal to metal, plastic to glass, or asbestos brake linings to steel brake shoes. When a minor proportion of nitrile rubber is mixed into a thermosetting phenolic molding powder, it greatly improves the impact strength. Strong hard blends can be made which are oil-resistant and suitable for machine parts.[25]

SYNTHETIC EBONITES FROM NITRILE RUBBER

Nitrile rubbers can be compounded and cured to yield synthetic ebonites having quite interesting properties. The amount of sulfur that combines chemically in nitrile-rubber ebonite corresponds to one sulfur atom per double bond.[195] Considerable heat is evolved during the reaction.[38]

About 35 parts of sulfur per 100 of rubber and a curing time of 120 minutes at 320° F. are required.[79] Various extenders and pigments can be used to obtain products with a wide range of properties. Colored pigments can be incorporated to get colored ebonites,[79] or organic diluents such as unsaturated petroleum products may be added to obtain tough semi-ebonites.[177]

Details of compounding for nitrile-rubber ebonite can be found in the manuals of various suppliers of nitrile rubber.[104] Properties of the unpigmented ebonites have been carefully studied. Tensile strength, elongation, bending modulus, impact strength, hardness, electrical properties, and resistance to water and chemicals have been measured and tabulated.[82, 241] In general, these nitrile ebonites maintain their physical properties better at high temperatures and have poorer impact strength and higher electric power loss than ebonites from natural rubber.[195, 241, 73] (See Chapter 16.) Table XII, compiled from data by Fisher, Mullins, and Scott,[73] shows the power loss of a number of ebonites.

Table XII. Power Loss at 1000 Cycles in Various Ebonites

| | Loss Factor at 1000 Cycles | |
Ebonite made from:	At 20° C.	At 90° C.
Natural rubber	0.030	0.134
Nitrile rubber (40% acrylonitrile)	0.068	0.213
Nitrile rubber (30% acrylonitrile)	0.076	0.198
Styrene-butadiene copolymer (40% styrene)	0.017	0.032

Because of their exceptional properties, nitrile ebonites are used in steam valve disks, bonding abrasive grinding and cutting wheels,[4] and elsewhere where the products must withstand high temperatures.[110]

ADHESION

Viscosity of Cements. Concentrated cements of nitrile rubber can be made with solids content as high as 50 per cent if nitroparaffins are used as solvents.[81] A peptizing effect of nitro compounds is also apparent when nitro compounds are added to nitrile-rubber cements in the usual aromatic or ketone solvents. 1-Chloro-1-nitropropane is especially effective and will stabilize nitrile-rubber cements against gelling.[42] Both milling and the use of polar solvents reduce the viscosity of nitrile-rubber cement, and this lowering of viscosity can be brought about by what is usually considered rather mild treatment. Chemical softeners reduce the viscosity of nitrile-rubber cements as does also the action of light. A benzene solution of nitrile rubber undergoes as high as a thousandfold decrease in viscosity if the cement is exposed to sunlight.[111]

Cements and Adhesives. Nitrile-rubber cements have been developed for a wide range of uses. In general, nitrile rubbers having high plasticity are most convenient to use for making cements. On the other hand, a nitrile rubber can be designed especially to give rapid breakdown when milled (Chemigum N1NS). In a Banbury mixer at 300° F. the Mooney viscosity

(ML-212° F.) of such a rubber will decrease from 70 to 12 in a period of 10 minutes. The commercial polymer of this type, however, has higher water absorption than one that does not break down so rapidly.

For cementing unvulcanized nitrile-rubber stocks in preparation for cure, a cement can be made from a properly compounded nitrile-rubber stock using solvents such as toluene, methyl ethyl ketone, chloroform, isopropyl acetate, or mixtures of them. The bonding layer can thus be made as oil-resistant as the stock. Various softeners or tackifiers can be used to increase the tack of the uncured stock. Thus, wood rosin,[170] coal tar,[93] chlorinated dioctyl carbonate,[188] butylphthalylbutyl glycollate[187] and allyl 3-sulfolanyl ether[149] have been suggested to improve the building tack. Although the tack thus obtainable is sufficient for factory operations, the nature of the tack is quite different from that obtained with natural rubber and is more like a resin tack.

Nitrile-rubber cements have the added advantage that they can be used for cementing together or bonding a wide range of products. Thus, leather can be bonded to itself with a nitrile cement. Various materials can be attached to metal, wood, porcelain, or even glass if the hard surface is first coated with a pigmented chlorinated rubber cement. The nitrile-rubber cement is then applied and serves as the step-off bond. Addition of a phenolic resin gives cements having high bond strength and resistance to heat.[207, 134] Cements containing heat-hardening phenol-formaldehyde resins can be used to bond metal to (1) nitrile-rubber stocks, (2) plasticized vinyl chloride compositions, (3) mixtures of these two, (4) leather, (5) upholstery, or (6) plywood. A cement of this type has been used for bonding brake linings to brake shoes and even to take the place of riveting for joining together aluminum sheet. The bond after heat treatment is not thermoplastic, is not brittle at high or low temperature, and is resistant to oil and solvents.

Adhesive Bonds. For many adhesive uses it is not necessary to use a cement. A calendered tape made from a stock composed of nitrile rubber, thermosetting phenolic resin, pigment, and vulcanizing agents is quite satisfactory and avoids the mess, time loss, and hazard encountered in using a cement. Such tape is used in making brake shoes for General Motors cars and for adhering plastic knobs to glass.

If a nitrile-rubber stock is to be bonded to metal, methods other than those utilizing a nitrile-rubber cement can be used. For instance, the metal surface can be plated with brass of about 70 per cent copper content. If a nitrile-rubber stock containing sufficient sulfur, proper pigments, and activators, but free from ingredients specifically inhibiting adhesion, is vulcanized in contact with such a brass surface, an excellent heat-resisting bond is obtained.[39] If the nitrile-rubber stock has been vulcanized ahead of time, the surface of the stock may be crazed by treating with sulfuric acid and then some suitable cement used for obtaining the bond.[192] Chlorinated rubber cement, thermosetting phenolic resin, or a special chlorine containing rubber may be used in such a case and also for obtaining a vulcanized bond of a nitrile-rubber stock to steel.[151] Of the three, the thermoset phenolic bond was the best in aromatic solvents and at a high temperature, while the chlorinated rubber and special chlorine containing rubber bonds were most satisfactory of the group for use in boiling water. Under other conditions, polyvinyl butyral[94] or even a specially compounded Butyl-rubber stock is suggested.[91]

NITRILE-RUBBER LATEX

One of the largest uses for nitrile rubber is in the form of latex, in which form it finds application as an adhesive, as a modifier for other water-dispersed resins, and as an impregnant for paper, textiles, and leather. Of the total quantity of nitrile rubber (13,866 long tons) consumed in 1952 in the United States, 22 per cent (3,093 long tons) was in the form of latex.

Nitrile-rubber latex as usually made in the course of preparing the solid rubber may have a concentration of 20 to 25 per cent of rubber. It often has a rather high ratio of soap to rubber and an extremely small particle diameter of from 400 to 600 A. Because of these properties it was long thought that the latex would have only a few and special uses.[19] As the material was studied and processes were developed for growing the particles to larger size[238] (up to 2500 A.), as methods were found for producing more concentrated latex (up to 50 to 54 per cent total solids), and as more stable dispersed systems were prepared,[242] specific uses were found for the various types of latex.[57]

The dried latex films do not give high physical properties unless compounded and vulcanized. The tensile strength of unvulcanized, unloaded films of nitrile rubber is, however, appreciably higher than that of corresponding films of GR-S. In this connection, it may be noted not only that nitrile rubber comprises polar groups which GR-S lacks, but also that the molar proportion of the comonomer to the diene is markedly higher in nitrile rubber than it is in GR-S. Thus, whereas GR-S containing 23.5 per cent of combined styrene contains only 1 styrene unit to every 6.4 butadiene units, a butadiene-acrylonitrile copolymer containing 33 per cent of combined nitrile contains 1 acrylonitrile unit to every 2 butadiene units.

The compounding ingredients used with latex are quite similar to those used in a mill-mixed stock. Dry pigments must be dispersed with proper protective colloids in water and ground before being added, in order to prevent incipient coagulation. Oils or waxes must be emulsified before being added to the latex. From 1 to 2 per cent of sulfur and 1 to 2 per cent of accelerator are required for vulcanization. The accelerator may be of the ultrarapid type such as zinc dimethyldithiocarbamate, since there is no danger of scorching as a result of heat evolution during mixing. Zinc oxide is required for activation. Channel black is not ordinarily used, although fine thermal black adds considerably to physical properties. Details of compounding are described in the suppliers' manuals.[104]

Maron[139] has reported an unpigmented uncured nitrile-rubber latex film having a tensile strength of 760 p.s.i. at 2000 per cent elongation. Special compounding can be used to develop higher strength and to yield a water-dispersed rubber suitable for cementing fabrics and other cellular materials. The bond is of special value because of its oil and heat resistance.

Spectacular improvements in the properties of paper have been realized by incorporating nitrile-rubber latex. The addition can be made in the beater (1) if special cationic latex is used or (2) if the paper pulp is treated so that anionic latex particles will be adsorbed on the fibers. Perhaps the simplest procedure is to impregnate porous paper sheet with latex. When 4.5 parts of a 33 per cent nitrile rubber are added to 100 parts of paper by running a porous paper through a latex of 10 per cent solids content, the dried sheet has its elongation improved to 1118 per cent, its fold resistance to 780 per cent, its tear to 147 per cent, and its bursting strength to 239 per cent of the values obtained from the basis paper. This paper can be printed with ordinary ink or made oilproof and waterproof by printing with a polyvinyl chloride organosol.

Compounded nitrile-rubber latex can be used for improving the appearance and wearing properties of leather and fabrics. Mixtures with phenolics can be used as adhesives for bonding fibrous materials which if plied up and heated under pressure can be cut into heat-, oil-, and wear-resistant products such as gears.

Lists of commercially available nitrile rubber latexes are given by Zwicker[249] and Drogin.[67a]

OIL AND SOLVENT RESISTANCE OF NITRILE RUBBER

Their possession of oil resistance is the prime factor responsible for the extensive use of nitrile rubbers. Oil resistance implies low solubility, low swelling, and good tensile strength and resistance to abrasion after immersion in gasoline or oils. The effect of various hydrocarbons, oils, and solvents on nitrile rubbers has, therefore, been widely studied.[191, 102] Theoretical studies have correlated swelling of the rubber in solvents with the entropy and interaction of various groups,[182] and methods have been worked out for predicting volume increase.[101] The swelling characteristics of a nitrile rubber depend considerably on its compounding, the pigments incorporated, softeners used, and type of cure,[210, 160, 191] although the content of combined acrylonitrile in the rubber is by far the most important factor[68a] in its oil resistance (see Table XI). High pigmentation and tight cure tend to reduce swelling. Softener or plasticizer in the original vulcanized stock often reduces swelling and may even cause shrinkage when the stock is immersed.[109] Such shrinkage or reduced swelling is due to extraction of plasticizer and its replacement by solvent.[220] Naturally this replacement affects the stock in other ways than in its dimension. It may cause hardening or softening or may affect the hardening temperature.[53] The swelling of nitrile rubber in various lubricating oils correlates with the aniline point of the oil.[44, 68a] In general, there is no straight-line relationship between the logarithm of the increase in volume and the aniline point,[92] and the curve if traced for different stocks prepared from a given rubber does not change with change in temperature, volume loading, or degree of cure.[169]

When extractable softeners are used, the swelling can be reduced to a negligible value.[220] Aromatic solvents cause much more swelling than paraffinic hydrocarbons. Swelling in various mixtures of the two has been studied.[102] This is of importance when nitrile rubber is used for gasoline tanks in airplanes. The inclusion of plasticizer in the stock may also be utilized to reduce the swelling in gasoline, but the material extracted may cause difficulty.[239] Most plasticizers do not in fact affect the value as airplane fuel of gasoline stored in contact with nitrile-rubber stocks containing them, but phosphate plasticizers tend to lower the octane number of the gasoline.[105]

Resistance to Polar Solvents. Nitrile rubbers in general swell more in polar solvents than in nonpolar ones. However, the swelling in water and in antifreeze solutions is low enough to allow nitrile-rubber stocks to be used in contact with these liquids. The oil resistance of the stocks makes them particularly valuable for such service, since oil is often deliberately added to antifreeze solutions to reduce corrosion. Resistance to ethylene glycol is also good, and nitrile stocks are preferable to natural rubber for this service[196] up to 130° C.

GAS DIFFUSION

The diffusion of gases through nitrile rubber is quite different from that through hydrocarbon rubbers. Numerous measurements have been made of the solubility of gases in and their rate of diffusion through nitrile rubbers at different temperatures.[24, 180, 6, 8, 100] Van Amerongen[8] has tabulated the relative ratio of permeability at 25° C. as given in Table XIII. As a result of these measurements it has been shown that the polar groups in nitrile rubber reduce the solubility of nonpolar gases and, hence, retard their

Table XIII.   Permeability of Vulcanized Rubbers at 25° C.

| | Compared to Natural Rubber = 100 | | | | | |
	H_2	O_2	N_2	CO_2	CH_4	He
Natural rubber[a]	100	100	100	100	100	100
Nitrile rubber[b]	29	18	14	23	11	38
Butadiene-styrene rubber[c]	78	72	73	92	73	76
Polychloroprene[d]	26	17	14	19	11	...

	Compared to Hydrogen = 100					
Natural rubber	100	46	17	260	56	59
Nitrile rubber	100	28	7.7	200	21	76
Butadiene-styrene rubber	100	43	16	310	52	57
Polychloroprene	100	29	8.6	190	24	...

	a	b	c	d
Natural rubber	100	...	...	...
Perbunan	...	100	...	...
Buna S	...	...	100	...
Neoprene G	...	...	...	100
Zinc oxide	5	5	5	...
Stearic acid	2	2	2	...
Sulfur	1.5	1.5	1.5	...
Benzothiazolyldiethyl sulfenamide	1.8	1.8	1.8	...
Phenyl-β-naphthylamine	1	1	1	...
Vulcanized at 142° C. in min.	30	30	30	15

rate of diffusion.[8] On the other hand, highly polar gases are more soluble in and diffuse more rapidly through nitrile rubbers than through hydrocarbon rubbers.[8, 100] Permeability is independent of the length of the primary valence chains in the rubber; however, it increases with increasing curvature of the primary valence chain and with the size of the side chains but decreases with increased energy of internal bonding.[173]

Combinations of nitrile rubber and polyvinyl chloride can be formed into films which are specially useful for food packaging because of their oil resistance and toughness. At one time not long ago they were widely used in the United States for packaging oleomargarine. Microcrystalline wax is often added to decrease still further the water-vapor transmission.[116]

LOW-TEMPERATURE PROPERTIES

Brittle-Point and Second-Order-Transition Temperature. Next to oil resistance, the low-temperature properties of nitrile rubber have probably been studied more than any others. Wiley and Brauer's data on brittle temperature[236] is shown in Table XIV. Zhurkov[248] has explained the general effects on a molecular basis. The specific heat for nitrile rubber was determined by Bekkedahl and Scott.[29] Using a commercial "40 per cent" nitrile rubber which analyzed carbon 80.1, hydrogen 9.3, and nitrogen 10.2 per cent, the specific heat in the range of -23 to $67°$ C. is given by $C_p = 0.00283\,T + 1.126$, in which C_p is joules gram^{-1} degree^{-1} and T is absolute temperature. The second-order transition as determined from the specific-heat–temperature curve is $-23°$ C. Heat capacity measurements were made to $-183°$ C. and entropy calculations to absolute zero. In comparable stocks from various nitrile rubbers the temperature of hardening is a function of the combined nitrile.[121, 89, 47, 147, 155, 209, 135] For instance,

Table XIV. *Second-Order-Transition and Brittle Temperatures of Various Commercial Nitrile Rubbers*

	Nitrogen, %	Acrylonitrile, % Calc.	Second-Order Transition, °C.	Brittle Temperature, °C.
Perbunan 18	5.31	20.1	−56	−55
Perbunan 26 NS 60	7.55	28.6	−46	−46
Perbunan 35 NS 90	9.66	36.6	−34	−26.5
Hycar OR-25	8.66	32.8	−37	−33
Hycar OR-15	10.3	39.0	−26	−23
Butaprene NF-NSP75	5.91	22.4	−52	−49.5
Butaprene NL	7.97	30.2	−41	−38
Butaprene NL-NSP84	6.91	26.2	−50	−47
Butaprene NAA	6.81	33.4	−39	−29
Butaprene NXM	10.3	39.1	−33	−26
Butaprene 20–90	13.7	52.0	−16	−16.5
Chemigum N4NS	8.13	30.8	−43	−35.5
Chemigum N3NS	10.6	39.7	−27	−24.5

in a stock made with a 40 per cent acrylonitrile rubber the temperature of hardening was $-5°$ C., whereas with a 30 per cent acrylonitrile rubber it was $-15°$ C.[85]

The second-order-transition temperatures of a number of nitrile rubbers were determined by Wiley and Brauer[236] using the refractive-index method. The reference refractive index for Hycar OR-15 was n_D^{26} 1.5187.

When plotted on rectilinear coordinates there is to be seen a linear relation between the second-order transition and the acrylonitrile content. This same linear relation carries over to vulcanized products from rubbers with various acrylonitrile content,[36] there being an increase of 6 to 13°.C. in this transition point as a result of vulcanization.

Effect of Softeners and Compounding Ingredients. Carbon black tends to lower the brittle point.[202, 36] Softeners, especially of the ester type, are quite effective in lowering the hardening temperature.[54, 36, 50a] Incorporating other polymers such as butadiene–low-styrene copolymer[21] or vinyl acetate[86] causes a slight improvement in low-temperature flexibility. When nitrile-rubber stocks are used in contact with solvents, their resistance to cold is often decidedly improved, because of solvent imbibed by or exchanged into the sample.[146, 247] The effect may decrease the brittle point of a gasoline tank stock as much as from $+8$ to $-65°$ F. when used in aromatic gasoline.[146]

Table XV shows how various softeners affect the brittle point of a typical

Table XV. The Effect of Various Softeners on the Low-Temperature Properties and Air-Oven-Aged Properties of Typical Nitrile-Rubber Stocks

Base Recipe

Nitrile rubber (33% combined acrylonitrile)	100
Zinc oxide	5
Stearic acid	1
Benzothiazolyl disulfide	1.5
Sulfur	1.5
Carbon black (SRF)	60
Softener (as specified)	20
Cure 30 minutes at 310° F.	

Softener	Shore Hardness		Original Properties	After Air-Oven Aging		Heat Loss (%) in 48 Hr. at 300° F.	Low-Temp. Test,* °F. (ASTM)
				70 Hr. at 212° F.	48 Hr. at 300° F.		
Control	69	300% modulus, (p.s.i.)	2550			0.6	P–60
No softener		Tensile strength (p.s.i.)	2740	2900	...		F–70
		Elongation (%)	315	240			...
Dibutyl sebacate	51	300% mod.	1425			15	P–60
		Tensile	2150	2175	1290		F–70
		Elongation	400	265	125		
Tributoxyethyl	53	300% mod.	1500			9	
phosphate		Tensile	2425	2200	425		P–60
		Elongation	455	300	25		F–70
Octadecene	49	300% mod.	875			9	P–60
nitrile		Tensile	2025	2275	1525		F–70
		Elongation	600	345	170		
Plasticizer SC	52	300% mod.	1325			7	P–60
		Tensile	2200	2150	900		F–70
		Elongation	450	245	60		
Piccocizer 30	52	300% mod.	1100			10	P–50
		Tensile	2290	2500	1610		F–60
		Elongation	515	300	160		
Cumar P–25	56	300% mod.	1300			5	P–30
		Tensile	2625	2775	1790		F–40
		Elongation	530	325	175		

* P = Passed, F = Failed

nitrile-rubber stock.[104]   Esters, nitriles, and many other polar compounds tend to lower the brittle point or at any rate do not cause it to be higher than a control sample.   Coal-tar resins, petroleum oils, greases, and tars, on the other hand, tend to give a stock that more readily becomes brittle in the cold. Different softeners affect the Shore hardness and also the modulus, tensile strength, and elongation of the stock.   The effect of these typical certain softeners on the air-oven aging of the stock and the heat loss of the plasticizer are shown for comparison.

RESILIENCE

Nitrile rubbers, whether tested by rebound elasticity or by other methods, have been found to be inherently less resilient than natural rubber by a factor of 33 to 60 per cent.[206]   The resilience is of course a function of temperature,[64, 106-7] becoming less as the temperature is lowered.   It has been measured for most nitrile rubbers in gum stocks and in black compounds.[33, 155]

The ability of nitrile-rubber stocks to absorb vibration is decidely diminished by lowering the temperature.   Thus, in a nitrile-rubber stock of 60 Durometer hardness, at 5° F., 25 per cent of applied vibration energy was absorbed, whereas for a natural-rubber stock a corresponding point was not reached until —10° F.[152]

MOLECULAR STRUCTURE AND PHYSICAL PROPERTIES

Tensile Strength.   The tensile strength of rubbers that crystallize is considerably augmented by the formation of crystallites.[2a]   Rubbers in general fall into two classes: (1) those that give vulcanized stocks of high strength without added pigments and (2) those that give vulcanized stocks of low strength unless compounded with carbon black.   Aleksandrov and Lazurkin[1] have shown that rubbers of the first class crystallize on stretching and are thus self-reinforcing.   However, if they are heated above their "melting point" there is a sharp drop in the tensile strength of these stocks. Nitrile rubbers fall in the second class and respond in a spectacular manner to reinforcement by carbon black.[66-68, 133]   Nitrile rubber does not possess flexing properties so good as those rubbers that crystallize on stretching.[72]

J. C. Williams[237] has pointed out that theoretically a 1,4-copolymer of butadiene and acrylonitrile when compounded and vulcanized to optimum strength should have a higher tensile strength than even natural rubber. That it does not indicates, in his opinion, the presence in it of a certain amount of 1,2-addition.   The ratio of 1,4- to 1,2-addition has been measured by several means, and it appears to be true that a completely 1,4-polymer has not yet been prepared.

Further, geometric isomerism appears to affect the temperature coefficient of the tensile strength of rubbers.   Infrared spectra indicate that the molecules of nitrile rubber are largely *trans*, and experience shows that, as is common among *trans* polymers, high temperature has an inordinate effect on tensile properties.   Natural rubber, a *cis* polymer, shows less effect of temperature on tensile strength.

Other Properties.   Molecular structure affects physical properties in general.[183]   In a series of polymers made from various dienes and acrylonitrile, the brittle point was found to bear a definite relation to the acrylonitrile content, decreasing as the nitrile content decreases and reaching a value of nearly —80° C. for polybutadiene.   The elastic recovery of vulcanized nitrile rubber stocks, i.e., the rate of recovery from tensile set at a given temperature, is the property most sensitive to composition.[186]   The introduction of acrylonitrile has a softening effect on polymers made from various methyl-substituted butadienes, and copolymers with as high as 65 per cent nitrile are easily extensible[186] at 100° C.

Molecular Linearity.   By measuring the streaming anisotropy of nitrile rubber in benzene cement, Tsvetkov and Petrova[224] conclude that nitrile rubber has the most linear structure and polybutadiene the most branched of the synthetics studied.

Barnes, Liddel, and V. Z. Williams[22–3] investigated the infrared spectra of various rubbers.   The spectra in general were found to be characteristic and can be used to identify the type of rubber in an unknown composition. Nitrile rubbers give spectra indicating (1) their preparation from acrylonitrile and (2) their content of combined acrylonitrile.   Thompson and Torkington[221] determined the spectrum of acrylonitrile and also of nitrile rubber.   The band at 970 cm.$^{-1}$ was attributed to 1,4-addition in the polymer while that at 915 cm.$^{-1}$ implies some 1,2-addition.

Sudzuki[219] spread a film of Buna N on water and studied its properties in the range 3 to 32° C.   The compressibility of the film was ten times as great as that of a protein film under corresponding conditions.   Wall and Zelikoff[230] spread films of nitrile rubber on water from a benzene cement.   Low-conversion, soluble polymers gave a film 6 A. thick, whereas high-conversion rubbers gave a thickness of 30 A.   This indicates cross-linkage in the high-conversion polymers and a "bottle-brush" structure.

MOLECULAR WEIGHT AND GEL CONTENT

Considerable variation occurs in the molecular weight and gel content of various grades of nitrile rubber.   Products with high gel require more milling to give cements and in general do not dissolve in solvents so readily as products with low gel.

The molecular weight of various closely fractionated cuts of nitrile rubber has been determined by Scott, Carter, and Magat,[197] using osmotic pressure as a measure.   In the range of molecular weight 20,000 to 900,000 there is a linear relation between the logarithm of the intrinsic viscosity in various solvents and the logarithm of the molecular weight.   The constants for the viscosity equation $\eta = KM^a$, where η = intrinsic viscosity and M = osmotic molecular weight, are given in Table XVI.

Landler[130] has made a chromatographic fractionation of nitrile rubber in a mixed toluene and methanol cement, using gas black 75 per cent and granular activated carbon 25 per cent in the column.   From a low-concentration cement the low-molecular fraction was adsorbed first; however, from more concentrated cements much more rubber was adsorbed and the low-molecular-weight portion was least adsorbed.   Difficulty was

Table XVI.   Constants for Viscosity Equation of Nitrile Rubber
in Various Solvents

Solvent	$K \times 10^4$	a
Toluene	4.9	0.64
Acetone	5.0	0.64
Chloroform	5.4	0.68
Benzene	1.3	0.55

encountered with molecular degradation of unsaturated copolymers in the adsorption process.

BEHAVIOR IN SOLVENTS

Salomon[185] has investigated the effect of the nitrile group in nitrile rubber on swelling in solvents.   He concludes that there is a specific interaction with the solvent.   Thus, a rubber with a 20 per cent nitrile content swells a maximum in benzene, whereas one with 50 per cent nitrile swells a minimum.   This is interpreted in terms of Langmuir's theory of independent surface action.   A profound difference is noted when chloroform is compared with carbon tetrachloride, the chloroform showing a much greater swelling effect, due possibly to a bonded hydrogen in the molecule.   Copolymers of dimethylbutadiene and acrylonitrile show a quite similar effect.

Gee[83] has investigated the interaction between nitrile rubber and liquids. In mixtures of solvents, Buna N shows a critical solubility.   This is interpreted mathematically.   Gee and Treloar[84] have investigated the cohesive energy density of various rubbers.   Natural rubber shows 63.7 cal. per cc., whereas Buna N is 88 cal. per cc.   This difference is reflected in the fact that, in a mixture of petroleum ether 40 parts and benzene 60 parts, a vulcanized nitrile-rubber compound shows only low swelling, whereas a natural-rubber one swells highly.

Sarbach and Garvey[191] compared the swelling of natural rubber, nitrile rubbers, and other synthetics in 80 typical solvents.   Their extensive tabulation is of great practical importance.   While it does show nitrile rubbers to be soluble in aromatic hydrocarbons, chlorinated hydrocarbons, ketones, esters, and nitro compounds and to be insoluble in aliphatic hydrocarbons, hydroxyl compounds, and acids, nevertheless there are vast differences among solvents in rate and degree of swelling.   A high acrylonitrile content in the rubber has the effect of reducing swelling in petroleum hydrocarbons.   The tests were run on the crude unvulcanized rubbers, but, if certain allowances are made for the presence of plasticizers, pigments, and degree of vulcanization, the tests are quite helpful in predicting the swelling of vulcanized stocks.   Solvents that swell nitrile rubber and dissolve it rapidly are of use in making cements.   The table also helps to predict classes of compounds that would be of value for use as plasticizers or softeners. Salomon and van Amerongen[182] have made a similar study, and explain the variable solubility results in terms of entropy and the interaction of groups in the rubber with groups in the solvent.

SOME ASPECTS OF STRESS-STRAIN PROPERTIES

Molecular Basis for Stress-Strain Behavior. The molecular explanation for the forcible recovery of strained nitrile rubber is the same as for natural and other rubbers. However, in the case of nitrile rubber there are different limits. Thus, Peterson, Anthony, and Guth[167] found the equation of state for nitrile rubber to be the same as for natural rubber at low elongations but at high elongations found nitrile rubber to show greater plastic flow. Stress could be resolved into predominant thermal motion of the molecules augmented by minor molecular forces. Wildschut[235] found that the stress-elongation curves for natural rubber and for nitrile rubber are nonrectangular hyperbolas practically up to the point of rupture.

Set. It has been found that nitrile rubbers compounded and vulcanized with customary pigment, sulfur, and accelerator ratios flow under constant load more than natural rubber.[17] However, cold set in samples compressed 40 per cent was retained more by natural-rubber stocks than by nitrile-rubber stocks, which latter gradually recovered when the stress was removed.[150]

In view of the fact that in some important applications nitrile rubbers are used under compression, considerable study has been devoted to the influence of various curing schemes on the resistance of nitrile rubbers to compression set at elevated temperatures. These studies indicate that a tight cure is best for achieving low compression set. Morris, James, and Seegman[148] found the compression set to be much lower when nitrile-rubber stocks were cured by a high dosage of tetramethylthiuram monosulfide and low sulfur than when cured with a conventional combination of dibenzothiazolyl disulfide and sulfur. Cashion[45] has shown that nitrile rubber vulcanized with thiuram disulfide in the absence of sulfur has extremely low compression set and that compositions can be made having satisfactory properties even at 250 to 300° F. The proportion of combined acrylonitrile has a negligible effect on the set in such stocks.

Stress Relaxation. Stress relaxation of nitrile rubber has been studied by Tobolsky, Prettyman, and Dillon[223] who find that relaxation under tension of both primary and secondary bonds occurs in 100 hours at 100° C. Oxidative scission is suggested as the cause. Beatty and Juve[27] studied stress relaxation of pellets of nitrile rubber compressed 40 per cent. In a nitrile-rubber stock (40 per cent combined acrylonitrile) containing 45 p.h.r. of SRF black and vulcanized with benzothiazolyl disulfide 1 part and sulfur 1.5 parts, the stress retention after 100 hours at 70° C. was 55 per cent. However, in a similar stock vulcanized with tetramethylthiuram disulfide (4 parts), it was found that 88 per cent of the stress was maintained after a similar test period. These findings are in accord with the results discussed earlier on compression set and indicate that nitrile rubber vulcanized by the latter method would be exceptionally suitable for gaskets and seals which must maintain their stress under compression.[50] (Cf. 27a.)

Modulus. The modulus of nitrile rubbers increases with decreasing loading time.[37] The Nutting constants for strain and relaxation of nitrile and other synthetic rubbers have been tabulated[31] and the effect of temperature on the dynamic modulus and resilience of several nitrile rubbers discussed.[74]

Compressibility and Sound Velocity. Copeland[56] has measured the adiabatic compressibility coefficients for several vulcanized nitrile rubbers with and without carbon black up to 5000 p.s.i. Whereas the internal pressure of natural rubber is 83, that for nitrile rubber (26 to 40 per cent combined acrylonitrile) is 105 to 108. The velocity of sound in nitrile rubber increases as the temperature is lowered.[161] The acoustic propagation constants for the range of 10 to 30 megacycles have been measured in the range of −60 to +100° C. for nitrile-rubber stocks with and without carbon black. The ultrasonic velocity and attenuation in swollen samples of nitrile rubber has been measured by Nolle and Mifsud[160a] and related to the mechanism of rubberlike elasticity.

THE EFFECT OF HEAT ON NITRILE RUBBERS

On mild heating nitrile rubbers are only slightly deteriorated, but if heating is continued to 380° C. exothermal decomposition is initiated, and at 430° C. it becomes sufficient to permit ignition in the flash-cup test. Small amounts of hydrogen cyanide are evolved in this thermal decomposition.[208]

Many compounds have been devised for stability at high temperature. In general, low-sulfur stocks or stocks vulcanized entirely with tetramethylthiuram disulfide are most resistant to heat deterioration.[215] Combination of nitrile rubber with polychloroprene gives some improvement in heat stability;[190] however, combination with polyethylene polysulfide is not so satisfactory since mixed compounds appear to partake of the bad properties of both of the types of rubber included.[63]

The effect of temperature on the tensile properties of a nitrile-rubber stock (26 per cent combined acrylonitrile) is given in Table XVII.[34–5]

Table XVII. The Effect of High Temperature on Tensile Properties of a Nitrile-Rubber Stock

	Temperature, °F.	Ultimate Tensile Strength, P.S.I.	Elongation at Break, %
Nitrile-rubber stock reinforced with channel black	0	2850	450
	200	725	200

Heat Conductivity and Thermal Expansion. The thermal diffusivity of nitrile rubbers is independent of their nitrile content.[172] The thermal expansion for nitrile rubber is approximately equivalent to that of natural rubber.[229] The average linear expansion in the range 20 to 120° C. is for natural rubber 140 to 195 $\times$ 10^{-3} and for nitrile rubber (26 per cent combined acrylonitrile) 130 $\times$ 10^{-3}.

AGING OF NITRILE RUBBER

Nitrile rubber if made and used without antioxidant deteriorates rapidly, especially in the presence of certain metallic impurities or peroxides. To overcome this difficulty, (1) a "shortstop" is usually added to the latex in the normal course of manufacture to destroy radicals and traces of peroxides, and (2) an antioxidant is incorporated to protect against other deteriorating influences. Fortunately, nitrile rubber is extremely responsive to protection by antioxidants. Thus, 1 to 3 per cent of phenyl-β-naphthylamine added to the polymer yields a rubber that is even more stable and resistant to oxidation than natural rubber. (Cf. 128a.)

The deterioration of unvulcanized nitrile rubber is evidenced by a hardening or cross-linking which makes processing more difficult. Cements made from such deteriorated rubber are more than normally viscous and have a tendency to gel.

Vulcanized nitrile-rubber stocks are quite resistant to oxidative deterioration, as shown by their excellent retention of physical properties. This is shown in comparison to natural rubber in Table XVIII from the report by Schoch and Juve.[194] Stocks were aged in air at elevated temperatures, after being given approximately optimum cures, with the results shown in the table.

Volumetric oxygen absorption can be used as a measure to compare the resistance of various nitrile rubbers to oxidative deterioration. The protective effect of different antioxidants and the effect of various conditions of storage or use can thus be compared.

Exposing nitrile rubber to sunlight increases its rate of oxidation, whereas subsequent heating counteracts this effect.[7] The effect of sunlight on the aging of nitrile-rubber stocks has been quite thoroughly investigated.[58] In themselves such stocks are not particularly resistant unless microcrystalline wax is present.[145, 80] The cracking caused by ozone is quite similar to that caused by sunlight,[159] and in general there is a correlation between the two effects.[80] Improved resistance to sunlight and ozone cracking can be accomplished by the incorporation of 50 to 100 p.h.r. of polyvinyl chloride in the compound. Such stocks are suitable for cable coverings.[18]

In high-temperature aging tests in air, the behavior of nitrile-rubber stocks is in general quite good, as shown in Table XV; however, such tests sometimes lead to results that are anomalous and do not parallel behavior in service.[194] A hardening of the surface often occurs under these test conditions, whereas, if the compound is aged in oil,[137] the deterioration is slower and more uniform throughout the sample. Graves[88] has investigated the effect of oven aging on the tear resistance of nitrile-rubber stocks and finds the results qualitatively the same as for natural rubber.

Ozonization of uncured nitrile rubber as followed by infrared spectroscopy[3a] shows a progressive increase in hydroxyl and carbonyl groups.

ELECTRICAL PROPERTIES

In comparison with hydrocarbon types of rubber, nitrile rubber has a low specific resistance—of the order[176, 108] of 3 to 50×10^{10} ohms per sq. cm.

Table XVIII.   *Comparative Aging of Natural-Rubber and Nitrile-Rubber Stocks*

	No. 1, Natural Rubber Without Added Antioxidant	No. 2, Natural Rubber with Added Synthetic Antioxidant	No. 3, Nitrile Rubber Containing Only Antioxidant Added during Manufacture
Smoked sheet	100	100	...
Nitrile rubber (40% combined acrylonitrile)	...	...	100
Zinc oxide	5	5	5
Stearic acid	...	3	1
Sulfur	3.5	3.0	1.5
Mercaptobenzothiazole	...	0.8	
Accelerator 808	1	...	...
Benzothiazolyl disulfide	...	...	1.5
EPC black	50	...	...
MPC black	...	50	...
SRF black	...	...	75
Pine tar	5	4	...
Phenyl-β-naphthylamine	...	0.5	...
Age-Rite HP	...	1	...
Plasticizer*	...	...	30
Cure, minutes	45 at 292° F.	45 at 275° F.	45 at 275° F.

	No. 1			No. 2			No. 3		
	Mod. 300%, P.S.I.	Tensile, P.S.I.	Elong., %	Mod. 300%, P.S.I.	Tensile, P.S.I.	Elong., %	Mod. 300%, P.S.I.	Tensile, P.S.I.	Elong., %
Original values	1887	3485	490	1457	4070	615	1547	2450	510
After aging in air									
512 hr. at 70° C.	...	781	127						
672 hr. at 70° C.				1740	2380	390	...	2470	200
32 hr. at 100° C.	...	572	176						
168 hr. at 100° C.				...	920	180	...	2750	237
24 hr. at 121° C.	...	379†	20	...	960	290	...	2650	265
48 hr. at 121° C.				...	570	70	...	2935	230

* Equal parts of tributoxyethyl phosphate and Plasticizer SC.
† At 125° C.

Whereas the conductivity of the former class is due to the presence of conducting fillers or of migrating ions, nitrile rubber conducts because of its molecular polarity. If the various types of conduction are combined in a nitrile-rubber stock[108, 99] it is possible to secure stocks having fairly high electrical conductivity. Soft stocks containing acetylene black may have a resistance as low as 100 to 500 ohms per cc., while hard stocks containing a higher proportion may have a resistance of 1 to 10 ohms per cc. These stocks are useful for dissipating static charges or for electric heating units. For a discussion of the specific electrical uses of synthetic rubber and of compounding for such uses the literature should be consulted.[46, 163, 179, 138]

RECLAIMING

Methods of reclaiming specifically designed for nitrile rubber have not been used extensively, although various methods have been suggested.[178, 143, 20]

Slightly scorched nitrile-rubber stocks can generally be utilized by grinding them into fresh nitrile rubber. If 100 parts of cured or scorched nitrile-rubber stock is remilled with about 15 parts of coumarone-indene resin on a tight mill, this reclaimed batch may be used up to 25 per cent in a fresh stock with almost inappreciable loss in physical properties. Furthermore, vulcanized nitrile rubber from used articles can be ground on a roll mill and used as filler in oil-resisting stocks where high tensile strength is not required.

CHEMICAL DERIVATIVES

Apparently nitrile rubber has not been studied extensively as a chemical raw material. It does react readily with chlorine; 1 to 3 per cent causes a toughening and 22 to 25 per cent leads to a material that has been suggested as a cable sheathing.[61]

Certain double bonds in nitrile rubber are reactive toward mercaptans,[205] and the evidence indicates that only those bonds resulting from 1,2-addition are reactive toward thiogycolic acid. Such a reaction carried out to a slight degree softens the rubber and improves processing. No specific use is known for a rubber in which all the double bonds have been saturated by mercaptan addition.

IDENTIFICATION OF NITRILE RUBBER

A highly distinctive feature of nitrile rubber is its relatively large content of nitrogen. As a matter of fact, in crude uncompounded rubber the nitrogen analysis can be used as an index of combined acrylonitrile. However, if the rubber has been compounded or vulcanized, it becomes more difficult to distinguish. If a mixture of rubbers including nitrile rubber is used, the task of determining the composition becomes more difficult. Burchfield[40] found that, if a sample of the stock was dry-distilled from a hot tube, tests run on the droplets of condensate gave a quick and positive identification. Later, as he perfected the procedure,[41] he found that, if the vapors from a nitrile-rubber stock were passed into a hydrochloric acid solution of p-dimethylaminobenzaldehyde, a red color was obtained. Other rubbers tried gave blue, green, and yellow colors. The color is quite characteristic for nitrile rubbers and distinguishes them from natural rubber, GR-S, polychloroprene, Butyl, and vinyls. As a simplified test, all that is necessary is to brand the stock with a hot iron and hold a piece of suitable test paper in the fumes. With the fumes from nitrile rubber, test paper, impregnated by a solution of cupric acetate, Metanil Yellow, and benzidine hydrochloride, shows a green color.

Other methods can be used, such as the reaction toward a mixture of equal volumes of concentrated nitric and sulfuric acids at 70° C. Parker[164] reported the time in seconds to disintegrate a sliver of stock to be as follows for various rubbers: natural rubber, 0 to 5 seconds; GR-S, 10 to 25 seconds; various nitrile rubbers, 30 to 1200 seconds, those with a higher nitrile content disintegrating most rapidly. Softeners (which can be removed by extraction with acetone) must be absent when the test is made. Finally Parker[165] made a study of the swelling in solvents and found that by using benzene (B), petroleum ether (P) (b.p. 40 to 60 °C.), and aniline (A) different degrees of swelling were obtained with different types of rubber. If $\log \dfrac{\text{swelling in } B}{\text{swelling in } A}$ is plotted against $\log \dfrac{\text{swelling in } B}{\text{swelling in } P}$, there is obtained a pattern where each type of rubber has its distinct position. If two types of rubber are known to be used in combination, this procedure can be used to determine the proportion of each.

NITRILE-MODIFIED NATURAL RUBBER

The study of acrylonitrile copolymers has demonstrated that their oil resistance is a function of the combined nitrile. If this is the essential feature, perhaps acrylonitrile could be caused to react with natural rubber, so as to incorporate nitrile radicals into the molecules and thus convert natural rubber to an oil-resistant rubber. This problem was studied both in England and France. Le Bras and Compagnon[131] report that acrylonitrile can be combined with rubber by heating a rubber cement containing acrylonitrile and benzoyl peroxide or by mixing acrylonitrile with rubber latex and polymerizing. When this is done, there is an appreciable decrease in the swelling characteristics of stocks made from the altered rubber. This is shown in Table XIX.

Table XIX. Swelling in Hydrocarbon Solvents of Vulcanized Stocks from Natural Rubber Containing Combined Acrylonitrile

Acrylonitrile Combined in the Rubber	Swelling of a Vulcanized Stock In Benzene	In Kerosene
0	346%	190%
20%	221%	134%
56.5%	110%	50%

REFERENCES

1. Aleksandrov, A. P., and Lazurkin, Ya S., *Compt. rend. acad. sci. U.R.S.S.*, **45,** 291–4 (1944); *Rubber Chem. and Technol.*, **19,** 42–5 (1946).

2. Alekseeva, E. N., *J. Gen. Chem. U.S.S.R.*, **15,** 1426–30 (1909); *Rubber Chem. and Technol.*, **20,** 927–32 (1947).

2a. Alfrey, T., and Mark, H., *J. Phys. Chem.*, **46,** 112–8 (1942).

3. Alfrey, T., and Price, C. C., *J. Polymer. Sci.*, **2,** 101–6 (1947).

3a. Allison, A. R., and Stanley, I. J., *Anal. Chem.*, **24,** 630–5 (1952); *Rubber Chem. and Technol.*, **25,** 908–19 (1952).

4. Allison, H. V., U.S. Pat. 2,229,880, Jan. 28, 1941.

5. Ambros, O., *Rev. gén. caoutchouc*, **16,** 213–8 (1939).

6. Amerongen, G. J., van, *Rev. gén. caoutchouc*, **21,** 50–6 (1944); *Rubber Chem. and Technol.*, **20,** 479–93 (1947).

7. Amerongen, G. J., van, *Rubber-Stichting*, No. 39, 8 pp. (1944); *Rubber Chem. and Technol.*, **19,** 170–5 (1946).

8. Amerongen, G. J., van, *J. Applied Phys.*, **17,** 972–85 (1946); *Rubber Chem. and Technol.*, **20,** 494–514 (1947).

9. Anon., *Gummi-Ztg.*, **50**, 218–9 (1936).
10. Anon., *Gummi-Ztg.*, **50**, 723 (1936).
11. Anon., *Gummi-Ztg.*, **50**, 1051 (1936).
12. Anon., *Kautschuk*, **12**, 44–5 (1936).
13. Anon., *India-Rubber J.*, **92**, 803 (1936).
14. Anon., *Rubber Age London*, **18**, 206 (1937).
15. Anon., *India Rubber World*, **96**, No. 3, 74 (1937).
16. Arundale, E., U.S. Pat. 2,434,536, Jan. 13, 1948.
17. Badum, E., and Leilich, K., *Kautschuk*, **17**, 145–7 (1941).
18. Badum, E., U.S. Pat. 2,297,194, Sept. 29, 1943.
19. Bächle, O., *Kautschuk*, **13**, 174–6 (1937).
20. Bächle, O., U.S. Pat. 2,273,506, Feb. 17, 1942.
21. Baldwin, F. P., and Howlett, R. M., *Rubber Age N.Y.*, **54**, 433–5 (1944).
22. Barnes, R. B., Liddel, U., and Williams, V. Z., *Ind. Eng. Chem., Anal. Ed.*, **15**, 83–90 (1943).
23. Barnes, R. B., Williams, V. Z., Davis, A. R., and Giesecke, P., *Ind. Eng. Chem., Anal. Ed.*, **16**, 9–14 (1944); *Rubber Chem. and Technol.*, **17**, 253–66 (1944).
24. Barrer, R. M., *Trans. Faraday Soc.*, **35**, 628–43 (1939).
25. Bascom, R. C., *Modern Plastics*, **27**, 84–6 (1949).
26. Baxendale, J. H., Evans, M. G., and Park, G. S., *Trans. Faraday Soc.*, **42**, 155–69 (1946).
27. Beatty, J. R., and Juve, A. E., *India Rubber World*, **121**, 537–43 (1950).
27a. Beatty, J. R., and Juve, A. E., *India Rubber World*, **127**, 357–62 (1952); *Rubber Chem. and Technol.*, **26**, 336–49 (1953).
28. Beaver, D. J., and Throdahl, M. C., *Rubber Chem. and Technol.*, **17**, 896–902 (1944).
29. Bekkedahl, N., and Scott, R. B., *J. Research Natl. Bur. Standards*, **29**, 87–95 (1942); *Rubber Chem. and Technol.*, **16**, 310–7 (1943).
30. Bekkedahl, N., Quinn, F. A., and Zimmerman, E. W., *J. Research Natl. Bur. Standards*, **40**, 1–7 (1948); *Rubber Chem. and Technol.*, **21**, 701–10 (1948).
31. Blair, G. W. S., Veinoglou, B. C., and Caffyn, J. E., *Proc. Roy. Soc. London*, **189A**, 69–87 (1947).
32. Blank, J. H. E., *Automotive Ind.*, **80**, 696–702, 719–21 (1939).
33. Blow, C. M., Fletcher, W. P., and Schofield, J. R., *Rubber Chem. and Technol.*, **18**, 471–85 (1945).
34. Boonstra, B., *India Rubber World*, **121**, 299–302 (1949).
35. Boonstra, B., *India-Rubber J.*, **117**, 460, 463 (1949).
36. Borders, A. M., and Juve, R. D., *Ind. Eng. Chem.*, **38**, 1066–70 (1946); *Rubber Chem. and Technol.*, **20**, 515–24 (1947).
37. Brenschede, W., *Kolloid-Z.*, **101**, 64–72 (1942).
38. Bruce, P. L., Lyle, R., and Blake, J. T., *Ind. Eng. Chem.*, **36**, 37–9 (1944); *Rubber Chem. and Technol.*, **17**, 404–11 (1944).
39. Buchan, S., *Trans. Instn. Rubber Ind.*, **19**, 25–38 (1943).
40. Burchfield, H. P., *Ind. Eng. Chem., Anal. Ed.*, **16**, 424–6 (1944).
41. Burchfield, H. P., *Ind. Eng. Chem., Anal. Ed.*, **17**, 806–10 (1945).
42. Campbell, A. W., and Burns, J. W., *India Rubber World*, **107**, 169–70 (1942).
43. Campbell, A. W., and Tryon, P. F., *Ind. Eng. Chem.*, **45**, 125–30 (1953).
44. Carman, F. H., Powers, P. O., and Robinson, H. A., *Ind. Eng. Chem.*, **32**, 1069–72 (1940).
45. Cashion, C. G., *Rubber Age N.Y.*, **65**, 307–12 (1949).
46. Chapman, H. B., Brit. Pat. 530,512, Dec. 13, 1940.
47. Chatten, C. K., Eller, S. A., and Werkenthin, T. A., *Rubber Age N.Y.*, **54**, 429–32 (1944).
48. Goodyear Tire & Rubber Co., Chemigum Compounding Manual, 1949.
49. Cheyney, L. E., *Ind. Eng. Chem.*, **41**, 670–5 (1949); *Rubber Chem. and Technol.*, **23**, 217–28 (1950).
50. Clark, R. A., and Cheyney, L. E., *Rubber Age N.Y.*, **65**, 531–6 (1949).

50a. Clark, R. A., and Dennis, J. B., *Ind. Eng. Chem.*, **43,** 771–8 (1951).
51. Clifford, A. M., and Wolfe, W. D., U.S. Pat. 2,386,661, Oct. 9, 1945.
52. Clifford, A. M., and Wolfe, W. D., U.S. Pat. 2,370,010, Feb. 20, 1945.
53. Cole, O. D., *Trans. Am. Soc. Mech. Engrs.*, **65,** 15–20 (1943).
54. Conant, F. S., and Liska, J. W., *J. Applied Phys.*, **15,** 767–78 (1944).
55. Conference for Tech. Applic. Com. for Rubber, 7th, PBL 42815.
56. Copeland, L. E., *J. Applied Phys.*, **19,** 445–9 (1948); *Rubber Chem. and Technol.*, **22,** 79–85 (1949).
57. Cornic, Y., *Rev. gén caoutchouc*, **24,** 362–7 (1947).
58. Crabtree, J., and Kemp, A. R., *Ind. Eng. Chem.*, **38,** 278–96 (1946).
59. Crossley, R. H., and Cashion, C. G., *Ind. Eng. Chem.*, **36,** 55–9 (1944).
60. Crossley, R. H., and Cashion, C. G., *Rubber Age N.Y.*, **58,** 197–203 (1945).
61. D'Alelio, G. F., U.S. Pat. 2,380,726, July 31, 1945.
62. Daly, L. E., U.S. Pat. 2,439,202, Apr. 6, 1948.
63. Diehl, K., *Kunststoffe*, **37,** 49–53 (1947); *Chimie et industrie*, **58,** 577 (1947).
64. Dillon, J. H., Prettyman, I. B., and Hall, G. L., *J. Applied Phys.*, **15,** 309–23 (1944); *Rubber Chem. and Technol.*, **17,** 597–616 (1944).
65. Dinsmore, H. L., and Smith, D. C., *Anal. Chem.*, **20,** 11–24 (1948).
66. Drogin, I., *India Rubber World*, **106,** 561–9 (1942); **107,** 42–9 (1942); *Rubber Age N.Y.*, **51,** 483–8 (1942).
67. Drogin, I., *India Rubber World*, **107,** 272–7 (1942).
67a. Drogin, I., *India Rubber World*, **127,** 797–801 (1953).
68. Drogin, I., Grote, H. W., and Dillingham, F. W., *Ind. Eng. Chem.*, **36,** 124–8 (1944).
68a. Duke, N. G., and Mitchell, W. A., *India Rubber World*, **128,** 485–91 (1953).
69. Embree, W. H., Mitchell, J. M., and Williams, H. L., *Can. J. Chem.*, **29,** 253–69 (1951).
70. Emmett, R. A., *Ind. Eng. Chem.*, **36,** 730–4 (1944).
71. Field, G. E., *Modern Packaging*, **22,** 149–50, 190, 193 (1948).
72. Fielding, J. H., *Ind. Eng. Chem.*, **35,** 1259–61 (1943).
73. Fisher, D. G., Mullins, L., and Scott, J. R., *J. Rubber Research*, **18,** 37–43, 13–37 (1949); *Rubber Chem. and Technol.*, **22,** 1084–91 (1949).
74. Fletcher, W. P., and Schofield, J. R., *J. Sci. Instruments*, **21,** 193–8 (1944); *Rubber Chem. and Technol.*, **18,** 306–17 (1945).
75. Fryling, C. F., U.S. Pat. 2,384,543, Sept. 11, 1945.
76. Fryling, C. F., U.S. Pat. 2,384,544, Sept. 11, 1945.
77. Garner, T. L., and Westhead, J., *Proc. Rubber Technol. Conf.*, 423–33 (1938); *Kautschuk*, **14,** 173 (1938).
78. Garvey, B. S., Juve, A. E., and Sauser, D. E., *Ind. Eng. Chem.*, **33,** 602–6 (1941).
79. Garvey, B. S., and Sarbach, D. V., *Ind. Eng. Chem.*, **34,** 1312–5 (1942).
80. Garvey, B. S., and Emmett, R. A., *Ind. Eng. Chem.*, **36,** 209–11 (1944).
81. Garvey, B. S., U.S. Pat. 2,360,867, Oct. 24, 1944.
82. Gartner, E., *Kautschuk*, **16,** 109–16 (1940).
83. Gee, G., *Trans. Faraday Soc.*, **40,** 463–80 (1944); *Rubber Chem. and Technol.*, **18,** 236–55 (1945).
84. Gee, G., and Treloar, L. R. G., *India-Rubber J.*, **108,** 289–93, 319–22, 349–52, 359–61, 375–9 (1945); *Rubber Chem. and Technol.*, **18,** 707–30 (1945).
85. Gehman, S. D., Woodford, D. E., and Wilkinson, C. S., *Ind. Eng. Chem.*, **39,** 1108–15 (1947); *Rubber Chem. and Technol.*, **21,** 94–111 (1948).
86. Gidley, P. T., U.S. Pat. 2,474,309, June 28, 1949.
87. Gindin, L., Abkin, A., and Medvedev, S., *J. Phys. Chem. U.S.S.R.*, **21,** 1269–87 (1947).
88. Graves, F. L., *India Rubber World*, **111,** 305–8, 317 (1944); *Rubber Chem. and Technol.*, **18,** 414–23 (1945).
89. Green, B. K., Chollar, R. G., and Wilson, G. J., *Rubber Age N.Y.*, **53,** 319–27 (1943).
90. Grodzinski, P., *India-Rubber J.*, **91,** 279–80 (1936).

91. Hall, F. M., and Griffith, R. W., U.S. Pat. 2,442,083, May 25, 1948.

92. Hanson, A. C., *Ind. Eng. Chem.*, **34,** 1326–7 (1942).

93. Henderson, D. E., U.S. Pat. 2,331,979, Oct. 19, 1943.

94. Herschberger, A., U.S. Pat. 2,430,053, Nov. 4, 1947.

95. Hofmann, F., *Chem. Ztg.*, **60,** 693–6 (1936).

96. Hofmann, F., *Naturwissenschaften*, **24,** 423–6 (1936).

97. Hopff, H., *Angew. Chem.*, **51,** 432 (1938).

98. Hohenstein, W. P., and Mark, H., *J. Polymer Sci.*, **1,** 549–80 (1946).

99. Houwink, R., *Kunststoffe*, **34,** 25–6 (1944).

100. Houwink, R., *Ind. plastiques*, **3,** 409–14 (1947).

101. Howlett, R. M., *Ind. Eng. Chem.*, **37,** 223–5 (1945).

102. Howlett, R. M., and Cunningham, E. N., *Rubber Age N.Y.*, **60,** 562–4 (1947).

103. Huth, F., *Kunststoffe*, **26,** 166 (1936).

104. Hycar Blue Book, B. F. Goodrich Chemical Co., 1943–50.

105. James, R. R., and Morris, R. E., *Ind. Eng. Chem.*, **40,** 405–11 (1948).

106. Jones, H. C., and Snyder, E. G., *India Rubber World*, **108,** 137–40 (1943); *Rubber Chem. and Technol.*, **16,** 881–7 (1943).

107. Jones, H. C., and Snyder, E. G., *India Rubber World*, **110,** 405 (1944).

108. Juve, A. E., *India Rubber World*, **103,** No. 5, 47, 50 (1941).

109. Juve, A. E., and Garvey, B. S., *Ind. Eng. Chem.*, **34,** 1316–9 (1942).

110. Juve, W. H., Symposium on Application of Synthetic Rubbers, Am. Soc. Testing Materials, 104–8 (1944).

111. Kambara, S., *J. Soc. Chem. Ind. Japan*, **43,** 359–61 (1940).

112. Kelsey, R. H., and Dillon, J. H., *J. Applied Phys.*, **15,** 352–9 (1944); *Rubber Chem. and Technol.*, **17,** 621–31 (1944).

113. Kemp, A. R., and Peters, H., *Ind. Eng. Chem., Anal. Ed.*, **15,** 453–9 (1943); *Rubber Chem. and Technol.*, **17,** 61–75 (1944).

114. Kemp, A. R., Malm, F. S., and Stiratelli, B., *Ind. Eng. Chem.*, **36,** 109–13 (1944); *Rubber Chem. and Technol.*, **17,** 693–703 (1944).

115. Kenney, R. P., *Modern Plastics*, **24,** No. 1, 106–7 (1946).

116. Kinzinger, S. M., U.S. Pat. 2,445,727, July 20, 1948.

117. Klebsattel, C. A., *India Rubber World*, **102,** No. 6, 37–9; **103,** No. 1, 45–8 (1940).

118. Koch, A., *Gummi-Ztg.*, **50,** 573 (1936).

119. Koch, A., *Z. Ver. Deut. Ing.*, **80,** 963 (1936).

120. Koch, A., *Caoutchouc gutta-percha*, **34,** 78–9 (1937).

121. Koch, A., *Kautschuk*, **16,** 151–6 (1940).

122. Koch, A., *Ind. Eng. Chem.*, **32,** 464–7 (1940).

123. Koningsberger, C., and Salomon, G., *J. Polymer Sci.*, **1,** 353–63 (1946); *Rubber Chem. and Technol.*, **20,** 380–91 (1947).

124. Konrad, E., and Tschunkur, E., Brit. Pat. 360,821, May 30, 1930.

125. Konrad, E., and Tschunkur, E., U.S. Pat. 1,973,000, Sept. 11, 1934; Fr. Pat. 710,901, Feb. 4, 1931; Ger. Pat. 658,172, Mar. 25, 1938.

126. Konrad, E., *Gummi Ztg. Jubiläumsheft*, 13–15, 62, 64 (1936).

127. Konrad, E., *Angew. Chem.*, **49,** 799 (1936).

128. Konrad, E., *Kautschuk*, **13,** 1 (1937).

128a. Kuzminskii, A. S., and Popova, E. B., *Doklady Academii Nauk S.S.S.R.*, **58,** 1077–9 (1952); *Rubber Chem. and Technol.*, **26,** 840–2 (1953).

130. Landler, I., *Compt. rend.*, **225,** 629–31 (1947); *Rubber Chem. and Technol.*, **21,** 682–3 (1948).

131. Le Bras, J., and Compagnon, P., *Bull. soc. chim.* (5), **2,** 553–61 (1944); *Rubber Chem. and Technol.*, **20,** 938–48 (1947).

132. Lee, J. A., *Chem. Eng.*, **56,** 134–6 (1949).

133. Lightbown, I. E., *Rubber Age N.Y.*, **47,** 19–21 (1940).

134. Lindner, G. F., Schmelzle, A. F., and Wehmer, F., *Rubber Age N.Y.*, **65,** 424–6 (1949).

135. Liska, J. W., *Ind. Eng. Chem.*, **36,** 40–6 (1944).
136. Luttringer, A., *Caoutchouc gutta-percha*, **33,** 17,493 (1936).
137. McCarthy, G. D., Juve, A. E., Boxser, H., Sanger, M., Doner, S. R., Cunningham, E. N., McWhorter, J. F., and Crossley, R. H., *ASTM Bull. No. 132*, 33–7 (1945).
138. McKinley, R. B., *Elec. World*, **124,** No. 25, 52–5 (1945).
139. Maron, S. H., and Madow, B., *Anal. Chem.*, **20,** 545–7 (1948); *Rubber Chem. and Technol.*, **22,** 224–8 (1949).
140. Martin, S. M., and Laurence, A. E., *Ind. Eng. Chem.*, **35,** 986–91 (1943).
141. Marvel, C. S., Deanin, R., Claus, C. J., Wyld, M. B., and Seitz, R. L., *J. Polymer Sci.*, **3,** 350–3 (1948).
141a. Messing, R. F., and James, R. L., *Chem. Ind. Week*, **68,** No. 2, 19–24 (1951).
142. Moll, R. A., Howlett, R. M., and Buckley, D. J., *Ind. Eng. Chem.*, **34,** 1284–91 (1942).
143. Moore, D. V., and Thompson, H. H., Can. Pat. 429,743, Aug. 28, 1945.
144. Morgan, L. B., *Trans. Instn. Rubber Ind.*, **23,** 219–30 (1948); *Rubber Chem. and Technol.*, **22,** 392–401 (1949).
145. Morris, R. E., James, R. R., and Werkenthin, T. A., *Rubber Age N.Y.*, **51,** 205–8 (1942).
146. Morris, R. E., James, R. R., Berger, E. H., and Werkenthin, T. A., *India Rubber World*, **108,** 553–5, 558 (1943); *Rubber Chem. and Technol.*, **17,** 116–23 (1944).
147. Morris, R. E., James, R. R., and Evans, E. E., *India Rubber World*, **110,** 529–32, 534 (1944); *Rubber Chem. and Technol.*, **18,** 192–203 (1945).
148. Morris, R. E., James, R. R., and Seegman, I. P., *India Rubber World*, **119,** 466–72 (1949).
149. Morris, R. C., and Shokal, E. C., U.S. Pat. 2,391,330, Dec. 18, 1945.
150. Morris, R. E., Hollister, J. W., and Mallard, P. A., *India Rubber World*, **112,** 455–8 (1945).
151. Morris, R. E., Hollister, J. W., and Mallard, P. A., *Rubber Age N.Y.*, **64,** 53–6, 96 (1948).
152. Morron, J. D., Knapp, R. C., Linhorst, E. F., and Viohl, P., *India Rubber World*, **110,** 521–5 (1944); *Rubber Chem. and Technol.*, **18,** 83–96 (1945).
153. Moulton, M. S., *Modern Plastics*, **24,** No. 2, 117–20 (1946).
154. Moulton, M. S., *India Rubber World*, **116,** 371–3 (1947).
155. Mullins, L., *Trans. Instn. Rubber Ind.*, **21,** 247–66 (1945).
156. Mullins, L., *Trans. Instn. Rubber Ind.*, **22,** 235–58 (1947); *Rubber Chem. and Technol.*, **20,** 998–1019 (1947).
157. Nelson, J. F., and Vanderbilt, B. M., *Proc. Second Rubber Tech. Conf.*, 49–59 (1948).
158. Newberg, R. G., Young, D. W., and Fairclough, W. A., *Rubber Age N.Y.*, **62,** 533–9 (1948).
159. Newton, R. G., *J. Rubber Research*, **14,** 27–39, 41–62 (1945); *Rubber Chem. and Technol.*, **18,** 504–56 (1945).
160. Newton, R. G., *J. Rubber Research*, **14,** 63–80 (1945).
160a. Nolle, A. W., and Mifsud, J. F., *J. Applied Phys.*, **24,** 5–14 (1953); *Rubber Chem. and Technol.*, **26,** 884–901 (1953).
161. Nolle, A. W., and Mowry, S. C., *J. Acoust. Soc. Am.*, **20,** 432–9 (1948).
162. Nowak, P., and Hofmeier, H., *Kautschuk*, **14,** 193–7 (1938).
163. Nowak, P., U.S. Pat. 2,281,375, Apr. 28, 1942.
164. Parker, L. F. C., *J. Soc. Chem. Ind.*, **63,** 378–9 (1944); *Rubber Chem. and Technol.*, **18,** 659–62 (1945).
165. Parker, L. F. C., *J. Soc. Chem. Ind.*, **64,** 65–7 (1945); *Rubber Chem. and Technol.*, **18,** 896–901 (1945).
166. Perbunan-Compounding and Processing, Stanco Distributors, Inc., 1942.
167. Peterson, L. E., Anthony, R. L., and Guth, E., *Ind. Eng. Chem.*, **34,** 1349–52 (1942).
168. Pittenger, J. E., and Cohan, G. F., *Modern Plastics*, **25,** No. 1, 81–6 (1947); *Rubber Age N.Y.*, **61,** 563–6 (1947).
169. Powers, P. O., and Billmeyer, B. R., *Ind. Eng. Chem.*, **37,** 64–7 (1945).
170. Puddefoot, L. E., and Swire, W. H., Brit. Pat. 569,666, June 4, 1945.
171. Reed, M. C., *Modern Plastics*, **27,** No. 4, 117, 162–4 (1949).

172. Rehner, J., *J. Polymer Sci.*, **2**, 263–74 (1947); *Rubber Chem. and Technol.*, **21**, 82–93 (1948).

173. Reitlinger, S. A., *J. Gen. Chem. U.S.S.R.*, **14**, 420–7 (1944).

174. Roelig, H., *Kautschuk*, **13**, 154–9 (1937).

175. Roelig, H., *Kautschuk*, **13**, 179–80 (1937).

176. Roelig, H., *Kautschuk*, **16**, 26–33 (1940); *Rubber Chem. and Technol.*, **13**, 948–61 (1940).

177. Rostler, F. S., *India Rubber World*, **117**, 492–7 (1948).

178. Rudy, *Gummi-Ztg.*, **51**, 335–6 (1937).

179. Särnö, B. O., *Tek. Tid.*, **75**, 613–8 (1945).

180. Sager, T. P., *J. Research Natl. Bur. Standards*, **25**, 309–13 (1940).

181. Salomon, G., and Koningsberger, C., *J. Polymer Sci.*, **1**, 364–79 (1946); *Rubber Chem. and Technol.*, **20**, 392–408 (1947).

182. Salomon, G., and Amerongen, G. J., van, *J. Polymer Sci.*, **2**, 355–70 (1947); *Rubber Chem. and Technol.*, **21**, 66–81 (1948).

183. Salomon, G., and Koningsberger, C., *J. Polymer Sci.*, **2**, 522–41 (1947); *Rubber Chem. and Technol.*, **21**, 377–97 (1948).

185. Salomon, G., *J. Polymer Sci.*, **3**, 173–80 (1948); *Rubber Chem. and Technol.*, **21**, 805–13 (1948).

186. Salomon, G., *J. Polymer Sci.*, **3**, 32–8 (1948); *Rubber Chem. and Technol.*, **21**, 814–20 (1948).

187. Sarbach, D. V., U.S. Pat. 2,332,263, Oct. 19, 1943.

188. Sarbach, D. V., U.S. Pat. 2,395,070, Feb. 19, 1946.

189. Sarbach, D. V., U.S. Pat. 2,397,050, Mar. 19, 1946.

190. Sarbach, D. V., U.S. Pat. 2,482,600, Sept. 20, 1949.

191. Sarbach, D. V., and Garvey, B. S., *India Rubber World*, **115**, 798–801 (1947); *Rubber Chem. and Technol.*, **20**, 990–7 (1947).

192. Saunders, S. G., and Morrison, H., U.S. Pat. 2,429,897, Oct. 28, 1947.

193. Sauser, D. E., *Rubber Age N.Y.*, **53**, 42–4 (1943).

194. Schoch, M. G., and Juve, A. E., Am. Soc. Testing Materials Tech. Pub. No. 89, 59–72 (1949).

195. Scott, J. R., *J. Rubber Research*, **13**, 23–31 (1944); *Rubber Chem. and Technol.*, **17**, 719–30 (1944).

196. Scott, J. R., *J. Rubber Research*, **16**, 216–8 (1947).

197. Scott, R. L., Carter, W. C., and Magat, M., *J. Am. Chem. Soc.*, **71**, 220–3 (1949); *Rubber Chem. and Technol.*, **22**, 660–6 (1949).

198. Sebrell, L. B., and Dinsmore, R. P., *Automobile Engr.*, **31**, 135–6 (1941).

199. Sebrell, L. B., and Dinsmore, R. P., *India Rubber World*, **104**, No. 1, 45–50 (1941).

200. Sebrell, L. B., and Dinsmore, R. P., *India Rubber World*, **103**, No. 6, 37–40 (1941).

201. Sebrell, L. B., *Ind. Eng. Chem.*, **35**, 736–50 (1943).

202. Selker, M. L., Winspear, G. G., and Kemp, A. R., *Ind. Eng. Chem.*, **34**, 157–60 (1942); *Rubber Chem. and Technol.*, **15**, 243–50 (1942).

203. Semon, W. L., *Chem. Eng. News*, **24**, 2900–5 (1946); *India Rubber World*, **115**, 364–9, 373 (1946).

204. Semon, W. L., *J. Am. Oil Chemists' Soc.*, **24**, No. 2, 33–6 (1947); *India Rubber World*, **116**, 63–5, 132 (1947).

205. Serniuk, G. E., Banes, F. W., and Swaney, M. W., *J. Am. Chem. Soc.*, **70**, 1804–8 (1948); *Rubber Chem. and Technol.*, **22**, 148–54 (1949).

206. Shaw, R. F., *India Rubber World*, **118**, 796–801, 868 (1948); *Rubber Chem. and Technol.*, **22**, 1045–59 (1949).

207. Shepard, A. F., and Boiney, J. F., *Modern Plastics*, **24**, No. 2, 154–6, 210, 212 (1946).

208. Skinner, G. S., and McNeal, J. H., *Ind. Eng. Chem.*, **40**, 2303–8 (1948); *Rubber Chem. and Technol.*, **22**, 667–79 (1949).

209. Smith, E. F., and Dienes, G. J., *ASTM Bull. No. 154*, 46–9 (1948); *Rubber Chem. and Technol.*, **22**, 820–7 (1949).

210. Springer, A., *Österr. Chem. Ztg.*, **43**, 96–8 (1940).
211. Stöcklin, P., and Konrad, E., U.S. Pat. 2,080,363, May 11, 1937.
212. Stöcklin, P., *Kautschuk*, **14**, 172 (1938).
213. Stöcklin, P., *Kunststoffe*, **28**, 292 (1938).
214. Stöcklin, P., *Ind. Chem.*, **14**, 248 (1938).
215. Stöcklin, P., *Kautschuk*, **15**, 1–7 (1939).
216. Stöcklin, P., *Kautschuk*, **15**, 117–9 (1939).
217. Stöcklin, P., *Trans. Instn. Rubber Ind.*, **15**, 51–75 (1939).
218. Stöcklin, P., *India-Rubber J.*, **98**, 344–9 (1939).
219. Sudzuki, K., *J. Chem. Soc. Japan*, **63**, 1058–60 (1942).
220. Tallant, J. A., *Petroleum Engr.*, **16**, No. 8, 210–4 (1945).
221. Thompson, H. W., and Torkington, P., *J. Chem. Soc.*, **1944**, 597–600.
222. Throdahl, M. C., and Beaver, D. J., *Rubber Chem. and Technol.*, **18**, 110–5 (1945).
223. Tobolsky, A. V., Prettyman, I. B., and Dillon, J. H., *J. Applied Phys.*, **15**, 380–95 (1944).
224. Tsvetkov, V. N., and Petrova, A., *J. Tech. Phys. U.R.S.S.* **12**, 423–63 (1942).
225. Twiss, D. F., *India-Rubber J.*, **92**, 157–8 (1936).
226. Twiss, D. F., *Trans. Instn. Rubber Ind.*, **11**, 491–504 (1936).
227. Twiss, D. F., *India-Rubber J.*, **93**, 593 (1937).
228. U.S. Senate, Hearings Before the Committee on Patents, 77th Cong., 2d Sess. on S.2303 and S.2491, p. 2675.
229. Vieweg, R., and Schneider, W., *Kunststoffe*, **31**, 215–9 (1941).
230. Wall, F. T., and Zelikoff, M., *J. Am. Chem. Soc.*, **68**, 726 (1946).
231. Wall, F. T., Powers, R. W., Sands, G. D., and Stent, G. S., *J. Am. Chem. Soc.*, **69**, 904–7 (1947).
232. Wall, F. T., Powers, R. W., Sands, G. D., and Stent, G. S., *J. Am. Chem. Soc.*, **70**, 1031–7 (1948).
233. Weidlein, Jr., E. R., *Chem. Eng. News*, **24**, 771–4 (1946).
234. Wildschut, A. J., *Rec. trav. chim.*, **61**, 898–909 (1942).
235. Wildschut, A. J., *Physica*, **10**, 65–78 (1943); *Rubber Chem. and Technol.*, **17**, 826–36 (1944).
236. Wiley, R. H., and Brauer, G. M., *J. Polymer Sci.*, **3**, 704–7 (1948); *Rubber Chem. and Technol.*, **22**, 402–4 (1949).
237. Williams, J. C., *India Rubber World*, **113**, 805–7 (1946).
238. Willson, E. A., U.S. Pat. 2,444,689, July 6, 1948.
239. Wilson, G. J., Chollar, R. G., and Green, B. K., *Ind. Eng. Chem.*, **36**, 357–61 (1944); *Rubber Chem. and Technol.*, **18**, 182–91 (1945).
240. Winkelmann, H. A., *India Rubber World*, **113**, 799–804 (1946).
241. Winspear, G. G., Herrmann, D. B., Malm, F. S., and Kemp, A. R., *Ind. Eng. Chem.*, **38**, 687–94 (1946).
242. Wohler, L. A., *India Rubber World*, **116**, 66–8, 120 (1947).
243. Wolf, G. M., Deger, T. E., Cramer, H. I., and DeHilster, C. C., *Ind. Eng. Chem.*, **38**, 1157–66 (1946).
244. Wood, L. A., Bekkedahl, N., and Roth, F. L., *Ind. Eng. Chem.*, **34**, 1291–3 (1942).
245. Young, D. W., Newberg, R. G., and Howlett, R. M., *Ind. Eng. Chem.*, **39**, 1446–52 (1947).
246. Young, D. W., Buckley, D. J., Newberg, R. G., and Turner, L. B., *Ind. Eng. Chem.*, **41**, 401–8 (1949); *Rubber Chem. and Technol.*, **22**, 735–55 (1949).
247. Zhurkov, S. N., and Lerman, R. I., *Compt. rend. acad. sci. U.R.S.S.*, **47**, 106–9 (1945).
248. Zhurkov, S. N., *Compt. rend. acad. sci. U.R.S.S.*, **47**, 475–7 (1945).
249. Zwicker, B. M. G., *Ind. Eng. Chem.*, **44**, 774–86 (1952).
250. Personal Communication from Firestone Tire & Rubber Co.
251. *The Chemistry of Acrylonitrile*, American Cyanamid Co., Beacon Press, New York, 1951.
252. *Survey of Synthetics*, Rubber Age and Synthetics, London, Oct. 1952, 20–5.

CHAPTER 24

BUTYL RUBBER

Robert M. Thomas and William J. Sparks

Standard Oil Development Company

The name "Butyl" is used throughout this chapter as a convenient brief designation for Butyl rubber. It refers to vulcanizable hydrocarbon polymers of low unsaturation used to make inner tubes and other products that must be chemically stable and resistant to gas permeation. These polymers, originated and perfected by the Standard Oil Development Company, are made by the copolymerization of a large proportion of olefin with a small proportion of diolefin.[113, 114] The reaction is brought about by catalysts of the Friedel-Crafts type[112] and is unique in that it takes place with great speed at low temperatures. At the present time all commercial grades of Butyl are produced by the copolymerization of isobutylene with small proportions of isoprene at a temperature of about 150° F. below zero. Products similar to Butyl can be made by employing butadiene, dimethylbutadiene, or piperylene instead of isoprene in the copolymerization, but on balance the isobutylene-isoprene copolymers have the most advantageous properties.[126]

COMPOSITION AND STRUCTURE

Since both the physical and chemical behavior of a copolymer depend on the ratio in which the monomer units enter the structure, and on their geometrical configurations in the polymer chain, early work was undertaken to clarify these factors. Thomas and co-workers[115] pyrolyzed a rubbery sample of polyisobutylene at 350° C., and obtained about 50 per cent of the monomer and 20 per cent of octenes. Although these degradation products are consistent with the head-to-tail structure, the results are not conclusive,

$$-CH_2 \cdot \underset{\underset{CH_3}{|}}{\overset{\overset{CH_3}{|}}{C}} \cdot CH_2 \cdot \underset{\underset{CH_3}{|}}{\overset{\overset{CH_3}{|}}{C}} \cdot CH_2 \cdot \underset{\underset{CH_3}{|}}{\overset{\overset{CH_3}{|}}{C}}-$$

because (a) the pyrolysis products were not all accounted for, (b) the occurrence of certain higher-boiling octenes in the products may have resulted from isomerization during pyrolysis, and (c) some random structure may have existed in the polymer.

Clearer evidence of structure was obtained from X-ray diffraction studies. Brill and Halle[20] examined stretched samples of polyisobutylene and detected a fiber pattern similar to that of natural rubber, with a fiber period of 18.5 A. They concluded that helical rather than zigzag chains are present, and their X-ray spacings indicated a head-to-tail structure, as shown above. Fuller and co-workers[56] explain their own observed repeat distance of 18.6 A. by starting with a plane zigzag chain with pairs of methyl groups on alternate chain atoms. This allows rotation about the single bonds such that the pairs of methyl groups form a helix around the chain axis. The helix repeats after eight monomer units. There still appears to be some question[27] about the correctness of this spiral picture, since certain observed intensities do not support the suggested configuration. Also the unique character of the diffraction pattern (by far the sharpest and most detailed fiber pattern known to date) indicates remarkably perfect crystallization on stretching. This suggests not only the geometrical regularity of the head-to-tail structure but also unique chain flexibility. However, the idea of unrestricted chain flexibility appears to conflict with the fact that the methyl side groups in the polymer are grossly overcrowded[21, 43, 96, 103] It is clear that more work will need to be done to clarify the finer details of chain structure, even though the present evidence for head-to-tail configuration is satisfactory.

The current evidence for head-to-tail configuration in polyisobutylene can readily be extended to include Butyl. It has been found that Butyl and polyisobutylene give indistinguishable X-ray patterns on stretching.[8, 92] While it is plausible to assume that the head-to-tail structure exists in that part of the Butyl molecule adjacent to the isoprene units, three possible structures can be written to represent the mode of isoprene addition. Rehner[93]

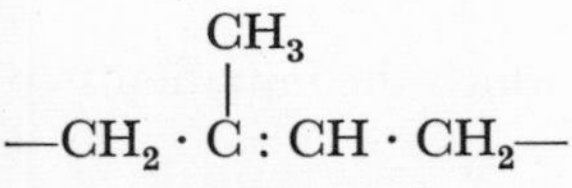

$$
\begin{array}{cc}
& CH_3 \\
& | \\
-CH_2 \cdot C : CH \cdot CH_2- \\
& \text{Structure 1}
\end{array}
$$

$$
\begin{array}{c}
| \\
CH_2 \\
| \\
CH_3 \cdot C \cdot CH : CH_2 \\
| \\
\text{Structure 3}
\end{array}
$$

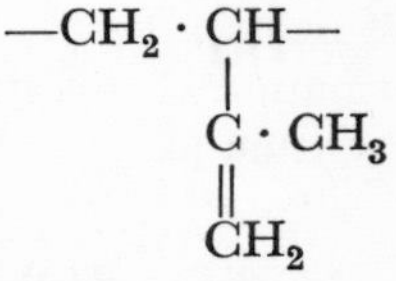

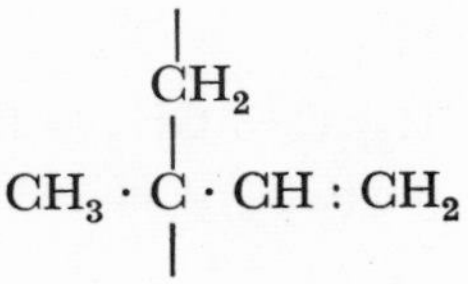

$$
\begin{array}{c}
-CH_2 \cdot CH- \\
| \\
C \cdot CH_3 \\
\| \\
CH_2 \\
\text{Structure 2}
\end{array}
$$

investigated this structural problem by the method of ozone degradation, and concluded that the isoprene units in the copolymer are almost exclusively, if not entirely, 1,4 (structure 1), the proportion of any 1,2- and 3,4-addition units being much less than 1 per cent of the diolefin present. He also demonstrated that there is no tendency for the isoprene units to occur in uninterrupted sequences. The diolefin therefore enters the growing polymer chain in a random manner during polymerization.

The random entry of isoprene units suggests that the proportion of these units in various molecular-weight fractions of a given polymer should be very nearly the same. Measurements on such fractions by the ozone-degradation method[94, 98] have in fact confirmed this expectation.[54] However, the polymerization may not be quite so simple, because analysis of the small percentage of nonrubbery fractions of low molecular weight that are always present in the polymers has shown that their isoprene content exceeds that of the higher fractions by a significant amount.[92]

Less certainty exists concerning the nature of the end groups in the Butyl molecule. The small isoprene content of the latter appears to make it virtually certain that the end groups are derived from isobutylene units. The difficulties in making direct analyses of end groups in polymers of high molecular weight are too well known to require comment. However, some indications are to be found in the results of Dainton and Sutherland.[37] They made an infrared examination of a very low-molecular-weight polyisobutylene prepared by polymerization of isobutylene vapor at room temperature with BF_3 and D_2O as cocatalysts. The end groups in this polymer gave intensity bands comparable to those arising from the main polymer chain. These bands were easily identified as those not occurring in the generally accepted spectrum for polyisobutylene[116] of high molecular weight. Analysis of the infrared results proved that in the product examined one terminus of the chain was a primary olefinic unit, while the other consisted of a trimethyl-substituted carbon atom. The chain structure is therefore as given in the formula. Dainton and Sutherland explain the

$$(CH_3)_3C \cdot \left[CH_2 \cdot \overset{\overset{\displaystyle CH_3}{|}}{\underset{\underset{\displaystyle CH_3}{|}}{C}} \cdot \right]_n \cdot CH_2 \cdot \overset{\overset{\displaystyle CH_3}{|}}{C} : CH_2$$

formation of this structure by a reaction scheme in which the initiation step is written as

$$BF_3 \cdot ROH + (CH_3)_2C = CH_2 \rightarrow BF_3 \cdot RO^- + (CH_3)_3C^+$$
$$[\text{or } (CH_3)_2CH - CH_2^+]$$

where $BF_3 \cdot ROH$ is the generalized catalyst complex which usually contains a hydroxyl group. The propagation step is shown in the formulas. The

$$(CH_3)_3C \cdot \left[CH_2 \cdot \overset{\overset{\displaystyle CH_3}{|}}{\underset{\underset{\displaystyle CH_3}{|}}{C}} \cdot \right]_n^+ + (CH_3)_2C : CH_2 \longrightarrow (CH_3)_3C \cdot \left[CH_2 \cdot \overset{\overset{\displaystyle CH_3}{|}}{\underset{\underset{\displaystyle CH_3}{|}}{C}} \cdot \right]_{n+1}^+$$

termination reaction requires charge neutralization. This involves the anionic fragment $BF_3 \cdot RO^-$ formed originally from the catalyst complex. Charge neutralization occurs by proton expulsion from the growing chain.

$$(CH_3)_3C \cdot \left[CH_2 \cdot \overset{\overset{\displaystyle CH_3}{|}}{\underset{\underset{\displaystyle CH_3}{|}}{C}} \cdot \right]_{n+1} + BF_3 \cdot RO^-$$

$$\longrightarrow BF_3 \cdot ROH + (CH_3)_3C \cdot \left[CH_2 \cdot \overset{\overset{\displaystyle CH_3}{|}}{\underset{\underset{\displaystyle CH_3}{|}}{C}} \cdot \right]_{n} \cdot CH_2 \cdot \overset{\overset{\displaystyle CH_3}{|}}{C} : CH_2$$

Although measurements of chemical unsaturation in polyisobutylenes invariably show evidence for the presence of residual olefinic groups,[94] in keeping with the end-group structures discussed above, the small values observed approach or exceed the reproducibility of the analytical methods. It is therefore uncertain whether olefinic terminal groups are present in polymers such as Butyl or higher-moleculer-weight polyisobutylene, which are usually synthesized by a heterogeneous liquid-phase polymerization at low temperatures. Also, liquid-phase synthesis always results in a net consumption of catalyst. This disagrees with the reaction scheme shown above, in which regeneration of active catalyst is indicated to be complete; and analysis of exhaustively purified polymer samples of high molecular weight always shows the presence of very small but detectable amounts of catalyst residues, which may be present as component parts of the chain structure.

It is usually assumed that polyisobutylene and Butyl molecules are devoid of cross-linking or branching, although direct proof is lacking. Indirect evidence for this assumption is that (a) all samples of these polymers are completely soluble in the usual hydrocarbon solvents, and (b) the highly developed state of crystallization observed on stretching demands a high degree of geometrical regularity.

Though, according to the best modern polymer theories, complete solubility can be accepted as a good argument for the absence of more than a trivial amount of cross-linking, it tells little or nothing about branching. Also, as has been shown in the case of polyethylene, a moderate amount of branching is not necessarily inconsistent with the development of a high order of crystallization. Since branching is quite elusive, both as to definition and measurement, conclusive evidence on the structures of these chain molecules will have to await the development of more refined experimental techniques.

PHYSICAL PROPERTIES

Molecular Weight. Experience has shown that both the average molecular weight and the molecular-weight distribution of Butyl bear an important relationship to processing characteristics such as plasticity and elasticity in unvulcanized compositions, as well as to vulcanizate properties such as modulus, tensile strength, resilience, and hysteresis. Some of these relationships are discussed elsewhere in this chapter.

Flory[54] has found that the molecular weight M of Butyl or its fractions can be calculated from the intrinsic viscosity η in diisobutylene at 20° C., by the equation

$$\log M = 5.378 + 1.56 \log \eta \qquad (1)$$

which he previously established for polyisobutylene.[52] Equally applicable to Butyl is his fractionation procedure, in which the polymer is precipitated from a dilute benzene solution by increasing proportions of acetone. This procedure yields data from which the differential distribution curve can be derived directly. It has the disadvantage, common to all such fractionation procedures, that several weeks are required for a single complete determination.

To expedite the examination of many Butyl samples, prepared under various conditions of synthesis and treatment, Rehner has extended to Butyl a procedure first developed for other polymers by Wall.[123] Aliquot samples of a benzene solution of the Butyl are treated with increasing proportions of acetone. In each case the amount of polymer precipitated is determined by analysis of the supernatant solution. The equilibrium percentage f of polymer remaining in solution at 25° C. is plotted against the precipitant ratio (cubic centimeters of acetone per 100 cc. of benzene). Values of Δf between successive pairs of this ratio are taken from the curve. The corresponding values of ΔM are obtained from the Schultz relationship

$$\text{Precipitant ratio} = 1.14 + 5.28/\eta_i \qquad (2)$$

where η_i is the threshold intrinsic viscosity of the smallest molecular species in the precipitated polymer phase. These parameters for Butyl were established by Flory.[49] Distribution curves obtained by this method of analysis are satisfactorily reproducible under rigidly standardized conditions. The complete fractionation and analysis can be made in a day or two.

The molecular-weight distribution for a Butyl of about 300,000 viscosity-average molecular weight can be represented, at least roughly, by the familiar Poisson distribution. However, extensive work has subsequently shown that both the nature and shape of the distribution curve depend on the composition and temperature of the polymerizing system, the extent of conversion, the nature of the catalyst, the presence of impurities, the agitation, and other factors. It is interesting that under certain conditions the distribution curve has more than one maximum. Explanations for this behavior can be based on the possible existence of multiple termination reactions or phase heterogeneity in the reaction system.

A method of estimating the molecular weight of Butyl from measurements of elastic compression rate under constant load was developed by Rehner.[95] The objective is simplicity and speed suitable for routine control, rather than extreme accuracy. Also Andrews and co-workers[5] found stress relaxation in polyisobutylene to be sensitive to molecular weight and recognized that such measurements can be used as a practical method for determining the molecular weight of the polymer directly in the solid state, with an accuracy comparable to that obtained from measurements of the viscosities of dilute solutions.

Physical Constants. Various fundamental and derived physical constants for Butyl have been compiled in Table I.   In some cases the values apply to the raw polymer, in others to vulcanized gum or carbon black stocks, and in some instances to polyisobutylene.   Though Butyl and polyisobutylene are outstandingly different products, it is considered that the small amount of diolefin in Butyl modifies certain physical measurements so slightly that the difference in values obtained from the two polymers would probably be less

Table I.   Some Physical Constants of Butyl and Polyisobutylene

Constant	Value	Composition	Reference
Unit rectangular cell, A.			
a	6.94		
b	11.96	Polyisobutylene	56
c	18.63		
Density, g. per cc.			
(26–27° C.)	0.910	Raw Butyl	61
Coefficient of expansion (per °C.)			
Linear	1.74×10^{-4}	Gum vulcanizate	38
Volumetric	5.67×10^{-4}	Gum vulcanizate	17
Linear	1.17×10^{-4}	50 parts black vulcanizate	38
Thermal diffusivity, sq. cm. per sec.			
60° C.	0.89×10^{-3}	Gum vulcanizate	
140° C.	0.51×10^{-3}		
60° C.	1.34×10^{-3}	50 parts black + 20 parts oil, vulc.	96
140° C.	0.65×10^{-3}		
Specific heat, cal. per g. per °C.			
26–27° C.	0.464	Raw Butyl	61
Heats, kg.-cal. per mole			
Formation (1 atm., 25° C.)	21.25	Polyisobutylene	102
Combustion (1 atm., 25° C.)	628.14	Polyisobutylene	102
Polymerization (in hexane)	12.8	Polyisobutylene	43
Isothermal compressibility, sq. cm. per dyne	53×10^{-12}	46 parts black	35
Adiabatic compressibility, sq. cm. per dyne	49×10^{-12}	Vulcanizate	
Refractive index (n_D)			
at 27.1° C.	1.5045	Polyisobutylene	45
Dielectric constant			
At 1 kc.	2.3–2.35		
At 50 Mc.	2.2–2.3	Raw Butyl	107
At 1300 Mc.	2.12		
Power factor			
1 kc.	0.0005–0.0009		
50 Mc.	0.0003–0.0009	Raw Butyl	107
1300 Mc.	0.0004		

than that attributable to experimental error. Reference should be made to the original sources for further details such as the variation of the specific heat of polyisobutylene with temperature,[45] the velocity and attenuation of sound waves in Butyl and their dependence on temperature and frequency,[70, 127] and the pressure dependence of adiabatic compressibility,[35] for both pure-gum and carbon black Butyl vulcanizates.

ANALYSIS

Identification. Butyl and polyisobutylene can be detected by the pyrolysis of a sample containing either of these polymers, and passing the resulting isobutylene vapor into a methanol solution of mercuric acetate.[86] The derivative, probably methoxyisobutylmercuric acetate, $CH_3O \cdot C(CH_3)_2 \cdot CH_2 \cdot Hg \cdot OOCCH_3$, can be confirmed by determining the mixed melting point with the derivative prepared from isobutylene. Burchfield[28] described a similar test in which he used paper strips treated with mercuric sulphate. The isobutylene from the pyrolysis reacts with the mercuric salt to form the characteristic[48] compound, $C_4H_8(HgSO_4 \cdot HgO)_3$. Workers at the National Bureau of Standards[79, 124] have described a related technique which is applicable to polyisobutylene and presumably to Butyl. Microsamples are decomposed by pyrolysis under a high vacuum. A volatile fraction from the decomposition products is analyzed in a mass spectrometer by observing the spectrum characteristic of a given polymer. A nitric acid digestion method[86] has also been described which gives good quantitative estimates of polyisobutylene in uncompounded mixtures with natural rubber and is moderately successful with vulcanized and black compounded mixtures. The method is based on the fact that polyisobutylene and Butyl are far more resistant to nitric acid than are other synthetic rubbers and natural rubber.

Unsaturation. The determination of chemical unsaturation in Butyl is complicated because the polymer contains only about 1 mole per cent of diolefin units and because halogen reagents applied to it usually cause substitution as well as addition reactions. Rehner[94] has shown that unsaturation as determined by the use of nitrosobenzene is high, because of accompanying oxidation, and that unsaturation based on the thiocyanogen reaction is complicated by the simultaneous formation of polymerized thiocyanogen. He also found that iodine chloride tends to give high values, because of the occurrence of substitution reactions. Later, however, a procedure by which reliable results for unsaturation can be obtained by means of iodine chloride despite the occurrence of substitution was described by Lee, Kolthoff, and E. Johnson.[75]

Values based on ozone degradation appear to be reliable. The ozone method depends on the fact that practically all of the isoprene units occur in the 1,4-configuration within the main chain. When the polymer is dissolved in carbon tetrachloride and ozonized, these units split. The degradation proceeds to a limit which corresponds to complete reaction of the diolefin units and beyond which further ozonization produces little change. The percentage of diolefin units can be calculated readily from viscosity data for the original and the degraded solution.[54, 94, 98] Though this method appears to give the closest approach to the true unsaturation, it is not easy to apply in routine analysis. However, a correlation was established[98] between unsaturation values obtained by the use of ozone and of iodine chloride. Thus, the latter reagent, which is more practical, can be used. Vasil'ev[121] has utilized iodine bromide dissolved in carbon tetrachloride (Hanus solution) for determining unsaturation in some Russian copolymers stated to be of the Butyl type. His isoprene copolymer gave a value of 1.2 per cent, or, after correction for substitution, 0.2 per cent. The value is the ratio expressed as percentage of the iodine number of the polymer to that of natural rubber. In view of the large corrections involved, his method is of questionable value.

Gallo, Wiese, and Nelson[57] developed a rapid, reproducible unsaturation method based on a reaction with iodine in the presence of mercuric acetate and trichloroacetic acid. Unpublished data by these authors also show good agreement between values obtained by their method and those by the ozone-degradation test. Currie[36a] has found that the degree of unsaturation of samples of Butyl as determined by essentially this method shows a close correlation with the modulus of the samples after vulcanization under standard conditions and that the method is well suited to controlling the quality of Butyl in plant operation.

Carbon Black. Several procedures have been described for determining carbon black in vulcanized Butyl. Louth, who developed a suitable method,[78] pointed out that the earlier ASTM test[3] cannot be used for determining carbon black in vulcanized Butyl and that the Galloway and Wake[58] method based on nitric acid digestion is not desirable when only carbon black values are wanted. Louth's method is superior to the earlier ones because the nitric acid decomposition of the vulcanized product proceeds rapidly and completely if the sample is first digested properly with tetrachlorethane.

Sulfur. The rapid determination of sulfur in vulcanized Butyl is hampered by the low rate of diffusion of free sulfur. Cohan, Sohn, and Steinberg[32] calculated combined sulfur by difference between the sulfur originally present and the free sulfur as determined by a sodium sulfite-extraction method. They pointed out that this procedure is subject to error if an accelerator such as tetramethylthiuram disulfide is present. Also, at least 8 hours is required for complete extraction by the sulfite. Rehner and Holowchak[99] have independently developed a method in which the free sulfur is extracted by methyl ethyl ketone. Both total and combined sulfur remaining after extraction are determined by combustion and conversion to barium sulfate. The extractable sulfur (which may also be derived in part from an accelerator containing sulfur) is determined by difference. The ketone extraction step takes about 8 hours. Except for this time requirement, the procedure is not difficult and gives reliable sulfur analyses.

MANUFACTURE OF BUTYL

Commercial grades of Butyl are made by copolymerizing isobutylene with a small amount of isoprene (1.5 to 4.5 per cent). The reaction is conducted at a low temperature, approximately 150° F. below zero, using as catalyst a dilute solution of aluminum chloride in methyl chloride.[112] Methyl chloride is used also in the feed as an inert diluent to control the reaction and to serve as a nonsolvent medium for the polymer product. The copolymerization is conducted on a continuous basis. More detailed information is provided in the discussion that follows.

Raw Materials. The primary raw materials required for the manufacture of Butyl, viz., isobutylene and isoprene, are both derived from cracked refinery gases. Information on the recovery and purification of isoprene is

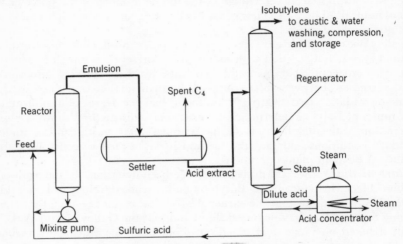

Fig. 1. Simplified Flow Plan of an Isobutylene Extraction Unit

given in Chapter 3. The C_4 cut containing isobutylene can come from various refinery cracking operations, such as (1) catalytic or thermal cracking of gas oil for high-octane gasoline production, (2) thermal reforming of virgin gasoline for increasing octane number, and (3) low-pressure cracking of various streams for the purpose of producing olefins. In the last case the fraction that remains after the removal of ethylene and propylene is sent to an extraction plant for the removal of butadiene. The C_4 cut obtained by these methods contains about 10 to 35 per cent by weight of isobutylene, mixed with normal butenes and varying amounts of saturates. The isobutylene is extracted by absorption in sulfuric acid of such a strength that it does not absorb other components of the mixture to a significant degree. Absorbed isobutylene is then liberated from the acid extract by treatment with steam. Product from this operation is caustic and water-washed, compressed, condensed, and delivered to storage. At this stage purity of the isobutylene is approximately 96 per cent by weight the impurities being chiefly normal butenes. A simplified flow plan of a typical unit for the extraction of isobutylene is shown in Fig. 1. A final distillation step removes butene-2 and results in isobutylene of over 99 per cent purity.

Process. Feed stock for the manufacture of Butyl is prepared by mixing isobutylene and isoprene in the desired proportions with an inert diluent, methyl chloride, which serves as an aid in controlling the violence of the polymerization reaction. This blended feed, cooled to a temperature of about −140° F., is supplied continuously to individual reactors, in which excellent agitation with indirect cooling by liquid ethylene is applied. Various sizes and types of reactors have been used in the commercial plants. The reactors are supplied also with a chilled solution of catalyst, which is mixed rapidly and thoroughly with the cold blended feed. Catalyst solution is prepared by dissolving anhydrous aluminum chloride in methyl chloride of high purity.

On contacting the feed with dilute catalyst solution, the polymer forms almost instantaneously. The reaction is exothermic. The temperature of the reactants is maintained at about −130° F. or lower by vaporizing liquid ethylene in a jacket surrounding the reactor compartment. The polymer, which is generated in the form of extremely fine particles suspended in the reaction medium, emerges from the reactor as a slurry and passes through an overflow line at the top into a considerable volume of hot, vigorously agitated water in the bottom of a large vessel called the flash tank. Unreacted feed components are flashed off and sent to a compression, drying, and refractionation system. A small quantity of zinc stearate (under 1 per cent on the polymer) is injected into the flash tank as a suspension, to prevent agglomeration of the small particles of Butyl in the resulting hot-water–polymer slurry. A very small amount of an antioxidant is generally introduced in the emulsion to stabilize the polymer and prevent its deterioration during finishing and subsequent storage. The aqueous slurry of Butyl particles is then put through a vacuum stripping section, for more complete removal of unreacted components. After stripping, the slurry is pumped to a finishing section, where a wet polymer crumb is isolated by rotary vacuum filters or by vibrating screens. This crumb is passed through a tunnel drier at 200 to 350° F. to remove most of the remaining water. It is fed ultimately through an extruder to large rubber mills. The product at this stage of the operation is removed in a continuous strip from the mills and is passed by a cooling conveyor to a cutting and stacking machine, which cuts the strip into squares and then stacks the squares for insertion into cardboard containers.

The unreacted hydrocarbons and inert diluent which are vaporized in the flash drum are compressed, dried by passing through alumina driers, further compressed, and delivered to fractionating towers. The major portion of the recycle gas is fractionated and returned to the original feed drum. A small stream of highly purified methyl chloride is taken overhead from the fractionation system for subsequent use as a catalyst solvent. Streams containing accumulated impurities are purged from the fractionation system and can be returned to extraction plants for recovery of isobutylene and isoprene.

Control Problems. To produce polymer that conforms to stringent specifications, control of a number of variables is required. Among these is maintenance of the composition of the feed. Automatic analyzers provide a continuous record of the concentrations of isobutylene and isoprene. In general, the feed blending and the polymerizing system are regulated through extensive use of instruments rather than by manual control. Continuous gas analyzers have also been installed on the flash drum as an aid in control of the reactor. Thus, a continuous record of conversion level in the reactors is available. This is important in controlling quality of the polymer.

The two most significant quality variables of the polymer product are molecular weight and unsaturation. Mooney viscosity is employed as an approximate index of molecular weight, and rate of vulcanization as a measure of degree of unsaturation. The molecular weight of the polymer depends on the temperature maintained during synthesis; it increases as the temperature is lowered. Molecular weight is influenced also by the presence of depressants or "poisons," such as normal butylenes, numerous oxygenated compounds, and many compounds containing sulfur. In general, it is important to keep the poison content of the reactor feed at a low and uniform concentration. The degree of unsaturation in the polymer depends on the ratio of isobutylene to isoprene in the feed. Because of a difference in the reaction rate of the two monomers, isoprene being the slower of the two, the conversion level obtained in the reactor affects both the unsaturation and the molecular weight of the polymer produced from a given feed.

Preparation and handling of catalyst is another significant variable in the control of product quality and reactor operation. The solvent must be of extremely high purity, particularly as regards content of both moisture and unsaturated hydrocarbons. In general, the presence of impurities results in a lowering of the molecular weight of the polymer product. Variable moisture in the catalyst solvent causes fluctuations in catalyst activity which make it difficult to control conversion of the reactants at a desired level.

Available Grades. Various grades of Butyl polymers representing different degrees of unsaturation and different ranges of molecular weight are commercially available.[40a] They are all copolymers of isobutylene and isoprene. The degree of unsaturation depends, of course, on the content of isoprene, and controls the rate of cure. GR-I (now GR-I 50) was formerly the "regular" or "standard" grade; it contains the unsaturation derived from a feedstock containing 2 per cent of isoprene and has a Mooney viscosity (8 minutes at 212° F.) of 41 to 49. But at present GR-I 15, with a somewhat higher isoprene content and a faster rate of cure, ranks as the "standard" grade. GR-I 25 cures still more rapidly.

Normal grades, shown in Table II, are derived as such from the reactor. It is, however, possible to reduce the viscosity of a high-viscosity polymer to a specified level by hot-working with agents such as xylyl mercaptan; and no significant difference has ever been established between such plasticized polymer and polymer made directly to the same viscosity.

Still another grade of Butyl known as B-1.45 or GR-I R2, and now called GR-I 35, has been manufactured from time to time for special applications such as tank linings and electrical insulation. It possesses superior resistance to ozone and is suitable for the production of insulation for high-voltage conductors. This grade is made from feedstock that contains less isoprene than that used for other grades, i.e., feedstock having an isoprene content of about 1.45 per cent based on isobutylene.

Table II. Classification of Grades of Butyl*

Grade	Rate of Cure	Mooney Range (8 Min. at 212° F.)
GR-I 35	Slow	38–47
GR-I 40	Intermediate	30–40
GR-I 50	Intermediate	41–49
GR-I 15	Regular	41–49
GR-I 17	Regular	61–70
GR-I 18	Regular	71+
GR-I 25	Fast	41–49

* Excluding experimental products and discontinued grades.

Grades of Butyl containing higher proportions of isoprene vulcanize more quickly and yield tighter cures than grades containing lower proportions. There has been a marked tendency for the more highly unsaturated GR-I 15 to replace the less highly unsaturated GR-I. (The still more highly unsaturated grade GR-I 25 has hitherto found only limited use, either because of a proneness to scorch during processing or because, all things considered, the physical properties of processible compounds prepared from it are unsatisfactory.) These statements are illustrated by the data given in Table III.

Table III. Trend in Butyl Production to Polymers of Higher Unsaturation[126]

Recipe

Polymer	100	Tetramethylthiuram disulfide	1
EPC black	50	Mercaptobenzothiazole	0.5
Zinc oxide	5	Sulfur	3
Stearic acid	3	Cured at 307° F.	

Grade	GR-I,	GR-I 15,	GR-I 25,
Avg. Mole %, Unsaturation	1.1	1.6	2.2
Tensile strength, p.s.i., minimum, 40-min. cure	2500	2400	2300
Elongation, %, minimum, 40-min. cure	650	550	500
Modulus at 300%, p.s.i., min.-max.			
20-min. cure	575–775	750–950	900–1100
40-min. cure	875–1125	1125–1375	1325–1750
60-min. cure	1200–1500	1475–1775	1750–2050
Fraction of total U.S. production, %			
1946	98.4	1.6	0.0
1947	66.4	32.0	1.6
1948	22.2	75.0*	2.2

* Includes GR-I 17, which has a Mooney viscosity of 60 to 70 and a rate of cure similar to that of GR-I 15.

COMPOUNDING

Butyl is easy to handle on standard rubber machinery by conventional methods. Special equipment for mixing, extruding, calendering, molding, and vulcanizing is not required. To obtain optimum performance in some instances, slight modifications of compounding and processing procedures have been recommended.[29]

By virtue of production controls, synthetic rubber offers advantages over natural rubber in uniformity, cleanness, and stability. These advantages have been fully realized with Butyl. This product is sorted into uniform grades and packaged in specially treated containers which prevent contamination and facilitate ultimate removal from the package. Packaging of Butyl with plastic film wrapping (polyethylene) has shown that extremely high standards of cleanness and ease of processing are possible. Stability during storage and processing is afforded by antioxidant incorporated in the Butyl at the time of its manufacture. Additional advantages include the elimination of mastication before compounding and the reduction of scorching difficulties during processing.

The plasticity of raw Butyl can be altered by different methods. Petroleum oils can be used as softeners or the polymer itself can be chemically plasticized by hot working in the presence of agents such as xylyl mercaptan. On the other hand, the stiffening of Butyl in either loaded or unloaded form can be brought about by heating it with small amounts of p-dinitrosobenzene, which is available as a commercial preparation, called Polyac, said to contain 30 per cent of p-dinitrosobenzene and 70 per cent of inert mineral filler.[34] In general, the stiffening treatment is carried to a level such that the solubility characteristics of the polymer are preserved. The use of more than 0.1 per cent of p-dinitrosobenzene for this purpose is seldom required. The chemical is of a rather insoluble nature, and possibly for this reason it reacts with Butyl most rapidly and most efficiently when heated under dynamic rather than static conditions. The reaction is best conducted at temperatures in the range of 120 to 170° C. Below 120° C. reaction proceeds quite slowly. At considerably higher temperatures, e.g., 170 to 200° C., a decomposition of p-dinitrosobenzene presumed to be of thermal origin has been reported to occur.[106]

Accelerators. As in the case of natural rubber, the rate at which Butyl is vulcanized by cross-linkage with sulfur is governed by using accelerator systems. Of course, Butyl vulcanizes more slowly than natural rubber because of its limited unsaturation. This is compensated for by using ultra-accelerators, by resorting to higher vulcanization temperatures (up to 350° F.), and by curing for longer times than are usual. Good accelerators for Butyl are almost exclusively of the thiuram sulfide and dithiocarbamate types. Accelerators selected from these two classes and used at a concentration of 1.0 to 1.5 parts per 100 parts of polymer result in Butyl vulcanizates of excellent quality. Normally a concentration of 1.0 to 2.5 parts of sulfur per 100 parts of polymer is used. If desired, the vulcanization can be speeded somewhat by the use of 0.5 to 1.0 part of an auxiliary accelerator, selected from the aromatic thiazole, guanidine, or aldehyde-amine types. The optimum rate of vulcanization with sulfur is obtained by combining a primary accelerator such as tetramethylthiuram disulfide with a selenium, tellurium, copper, or bismuth dialkyldithiocarbamate. Such combinations are useful in applications that require curing temperatures in the range 260 to 290° F.

If stability at high temperatures is an objective, the primary accelerator may be increased to 3 parts. A minimum of 2 to 3 parts of zinc oxide is required in all Butyl compounds to suppress reversion.[13] Maximum modulus is obtained with approximately 5 parts of zinc oxide, but up to 20 parts is used advantageously to obtain a desired degree of heat resistance. As discussed later in the vulcanization section, oxidizing agents such as calcium peroxide are effective in suppressing reversion.

Fillers and Reinforcing Agents. Reinforcement of Butyl by carbon blacks, fillers, or pigments does not result in an observable increase in tensile strength. The highest tensile values are obtainable by vulcanizing an essentially pure-gum compound.[65] However, reinforcement does increase modulus and tear resistance. Channel and furnace blacks are most effective. Similar but less pronounced reinforcement is obtained by using inorganic fillers such as calcium metasilicate or finely divided forms of silica, whiting, and clay. The general effect of filler loading on the properties of GR-I vulcanizates cured for 60 minutes at 307° F. is illustrated in Fig. 2. The ability of carbon blacks, especially channel blacks, to render uncured natural rubber insoluble is well known.[47, 125] Channel black has a similar effect on Butyl. This phenomenon has been attributed to the presence of oxygen on the surface of the carbon black[129] and to the inherently large ratio of surface area to volume for the fine particles.

Drogin[40] and also Turner and co-workers[120] have shown the effect of different types of carbon black on vulcanized Butyl. This work was done with a polymer of relatively low unsaturation. Reinforcing effects became more pronounced as polymer unsaturation increased. Cohan and Mackey[31] compared various blacks in Butyl with a fine particle whiting, adjusting the acceleration to compensate for adsorption. Despite large variations in crystalline and surface structures, a fairly distinct trend was observed, relating particle size and such properties as rebound, plasticity, and tear. In attempting to develop a conductive type of Butyl vulcanizate, Cohan and Steinberg[33] evaluated a series of channel blacks ranging in particle size from 10 to 35 mμ. They found that extremely fine particle size did not increase either abrasion resistance or modulus. This may have been due to dispersion difficulties. As expected, rebound decreased with decrease in particle size.

In some significant studies on the dispersion of black (SRF) in Butyl, Ford and co-workers[55a] have shown that, in milling, the size of the roll opening has a profound influence on the extent to which agglomerates of black particles are broken up and to which therefore dispersion of the black as discrete particles is approached. The more complete the dispersion, the greater is the plasticity and the electrical resistivity of the mixture. The extent of the difference between a stock in which good dispersion has been attained by the use of a small roll opening and a stock in which dispersion is poor because of the use of a wide roll opening is reduced by vulcanization. But nevertheless vulcanizates from the former stock show appreciably higher tensile strength, lower modulus (especially at small extensions), and better resistance to abrasion than vulcanizates from the former stock.

As previously indicated, the reinforcement of Butyl produced by inorganic

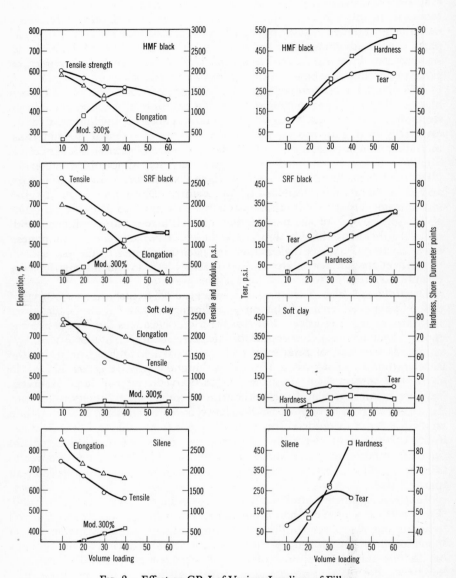

FIG. 2.   Effect on GR-I of Various Loadings of Fillers

Butyl 100, zinc oxide 5, stearic acid 3, tetramethylthiuram disulfide 1, sulfur 1.5 parts by weight.   Filler, various parts by volume per 100 parts by volume of Butyl.   Cured 60 minutes at 307° F.

fillers is less than that produced by channel and furnace blacks. The most effective inorganic fillers are the magnesium carbonates, ultrafine silicas, and calcium metasilicate. All of these are difficult to disperse adequately in the Butyl matrix. These fillers also tend to cause severe sticking of the unvulcanized Butyl compound to milling equipment. Wolf and Gage, however, state, in regard to fine silica, that sticking can be prevented by inclusion in the stock of one part of butoxyethyl diglycol carbonate.[128] The silicas have adsorptive characteristics similar to those exhibited by channel blacks of high volatile content. Stocks containing silicas, therefore, require adjusted curatives.[128] Other inorganic fillers, including various clays, are attractive because they deaden the elasticity of Butyl and thus contribute to smooth processing, and because vulcanized products of reasonably high modulus can be produced by their use. One mineral filler in particular, Whitetex (presumably a calcined aluminum silicate), is exceptionally good for facilitating Butyl processing. Diatomaceous earths are also good deadeners but are not attractive in regard to ultimate physical properties. Ultrafine whitings produce good tear resistance, resilience, and high tensile properties. Calcium carbonates of large particle size, lithopones, barytes, and the hydrated aluminas, are essentially bulking agents. All of these fillers provide a cohesiveness in the raw stock state which surpasses that generally obtainable with clays.

Plasticizers. Many substances have been used to plasticize rubber, but some of them are unsuitable for Butyl. Plasticizers having a high degree of unsaturation tend to deprive Butyl of the use of curatives during vulcanization. The most widely accepted plasticizers for Butyl include essentially saturated materials of relatively low polarity, such as hydrocarbon oils, waxes, asphalts, metal salts of fatty acids, higher aliphatic esters, aromatic esters, and certain resins, including those derived from alkylated phenols, terpenes, and hydrogenated ester gum.

In general, the use of plasticizers in Butyl follows a pattern similar to that which has been established for natural rubber. From 2 to 5 per cent of hydrocarbon oils or esters in ordinary grades of Butyl (such as GR-I) is generally adequate for facilitating calendering and extrusion or for creating greater resiliency or softness in the final product. With such polymers, from 10 to 15 parts of plasticizer may be used for frictioning work. For Butyl grades of higher viscosity (GR-I 17 and 18, for example), a substantially increased concentration of plasticizer is required. Petroleum oils of paraffinic nature are effective for improvement of the elasticity characteristics of Butyl at low temperatures and for protecting against embrittlement at extremely low temperatures, e.g., $-60°$ C. or lower. Such oils are inexpensive and are used extensively. On a quality basis they are inferior to aliphatic esters, of which butyl cellosolve pelargonate, dioctyl sebacate, and trioctyl phosphate are outstanding examples. In general, plasticizers are considered to improve the low-temperature performance characteristics of Butyl by permitting greater freedom of molecular rotation through the separation of polymer chains.

Metallic soaps such as zinc or aluminum stearate are useful agents in Butyl to produce smooth compounding by reducing nerve and tack. In cement and adhesive work the inherent tack of Butyl is increased by the addition of oil-soluble resins of low unsaturation. The latter form the basis of many adhesive formulations when used with fillers and subjected to dinitrosobenzene prevulcanization to reduce cold flow.

Blends with other Elastomers. Because of the low unsaturation of Butyl, satisfactory curing of blends with natural rubber (or GR-S) has not been accomplished. Natural rubber deprives Butyl of curatives in the same manner as do highly unsaturated plasticizers. Neoprene is an exception to the rule. Blends of this product with Butyl can be made that cure reasonably well. The unsaturation of Neoprene appears to be hindered and is relatively inactive.

The low unsaturation of Butyl also restricts the degree of cross-linkage attained during vulcanization and thus prevents the production of a hard-rubber product similar to ebonite. Where hardness and stiffness are required, a high loading of filler is sometimes supplemented by polymers of ethylene, or copolymers of isobutylene and styrene, which are compatible with Butyl and therefore useful. Such polymeric products have excellent electrical characteristics, resembling those of Butyl itself. Polyisobutylene, despite its general similarity to Butyl in these and certain other respects, is sometimes used as a compounding material for Butyl. Grades of low molecular weight can be used in cement work to improve adhesiveness. Grades of higher molecular weight enhance resistance to ozone and general heat aging.

VULCANIZATION

The vulcanization of Butyl is unique, although it is conducted by procedures that are commonplace to both natural and certain varieties of synthetic rubber. The limited unsaturation of Butyl polymer offers fewer active locations for potential cross-links. For this reason, more active accelerations and higher temperatures are required than with other rubbers, to obtain useful vulcanizates in reasonable periods of time.

Rate of Vulcanization. The term rate of vulcanization usually means the time required to vulcanize a rubber compound to a specified state of cure. There are many criteria for state of cure, the most common being extension modulus, tensile strength, functions of tensile and elongation, and energy of rupture. Criteria of vulcanization which have a theoretical relationship to the actual concentration of cross-links include equilibrium or elastic modulus[15, 55, 71, 73, 118, 122] at low extensions and swelling capacity in "good solvents."[53-4]

Unsaturation. The effect of unsaturation on the time of vulcanization of Butyl has been studied using polymers of similar molecular weight[130] obtained by fractionation. The chemical unsaturation of these polymers was determined by ozonolysis[98] and expressed as molar unsaturation per 1000 monomer units. In this study, time required for the vulcanization reaction to reach a concentration of cross-links defined by a swelling capacity of 1000 per cent volume increase in cyclohexane was determined. To polymers of varying unsaturation, sulfur was added in excess of the quantity required to satisfy, on the basis of a disulfide bridge, all potential points of cross-linkage. A highly active accelerator such as tetramethylthiuram disulfide was used in conjunction with 2 per cent of sulfur and 5 per cent of zinc oxide. Under these conditions, the following relationship between time of reaction and molar unsaturation was obtained:

$$t = C/n^{1.8} \tag{3}$$

where t = time in minutes, n = molar unsaturation per 1000 monomer units, C = a constant depending on the concentration of cross links or state of cure. This relationship is plotted in Fig. 3. On a theoretical basis, assuming

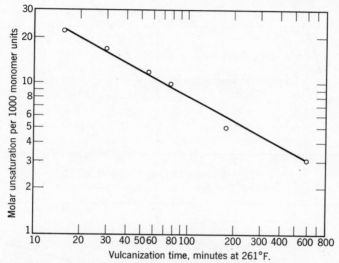

FIG. 3.   Effect of Molar Unsaturation in Butyl on the Time of Vulcanization Required to Produce 18.2 Relative Cross-Links

Stock: Butyl fraction 100, zinc oxide 5, sulfur 2, tetramethylthiuram disulfide 1, antioxidant 0.5

random motion and thus equal reaction potentialities for all active centers, the time of reaction required to reach a given state of vulcanization is related to the molar unsaturation by the expression

$$t = C/n^2 \tag{4}$$

When a less active accelerator was used, the relationship between time and unsaturation departed still further from the theoretical expression given in equation 4. Similar Butyl polymers, for example, were accelerated with benzothiazolyl monocyclohexylsulfenamide, and the relation between time and unsaturation was reduced to approximately

$$t = C/n \tag{5}$$

In other words, the type of accelerator has a marked effect on the time required to vulcanize Butyl polymers of varying unsaturation. This phenomenon appears to be connected with reversion as discussed later on.

Accelerations and Sulfur Content. Another relationship of interest is that of the time of vulcanization to a given state of cure, keeping the unsaturation constant. Such a comparison has been made[13] in a compound using 54 parts of SRF carbon black per 100 parts of Butyl polymer at a sulfur content of 2 per cent. The criterion for state of cure was simply extension modulus

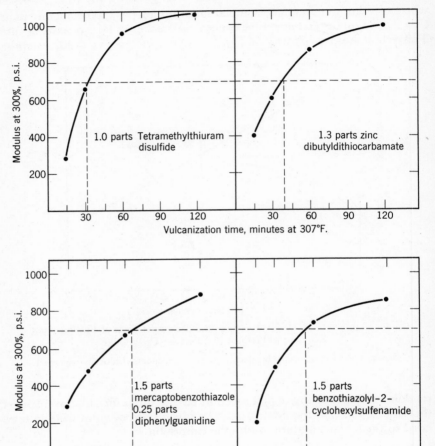

FIG. 4. Effect of Various Accelerators on the Vulcanization of GR-I

Stock: GR-I 100, zinc oxide 5, stearic acid 3, sulfur 2, SRF black 54, accelerator various

at 300 per cent elongation. Compounds derived from dithiocarbamic acid were observed to be more active accelerators than the thiazole types (Fig. 4).

With a rubber polymer of limited unsaturation it is possible to obtain a maximum state of cure such that the presence of more sulfur fails to produce a greater concentration of cross-links. The network is still in a soft elastic

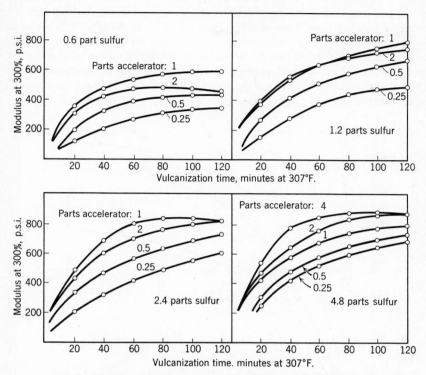

FIG. 5. Relation between Vulcanization Time and Modulus for Butyl
Containing Various Proportions of Sulfur and Tetramethylthiuram Disulfide

Stock: GR-I 100, zinc oxide 5, stearic acid 3, sulfur various, tetramethylthiuram disulfide various

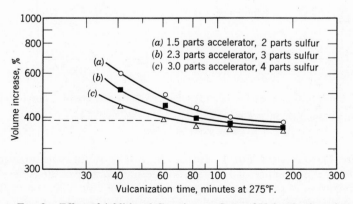

FIG. 6. Effect of Additional Curatives on State of Vulcanization of Butyl

state. In Fig. 5 it is evident[13] that further addition of sulfur beyond about 2 parts per hundred of polymer fails to increase the extension modulus. A similar illustration based on swelling capacity in cyclohexane[132] is given in Fig. 6. It is concluded that cross-linking does not proceed beyond a limiting point (as defined by 400 per cent volume increase), despite an increase in concentration of sulfur and accelerator. With the higher concentrations of sulfur and accelerator, of course, there is an increase in the rate at which this maximum state of cure is developed.

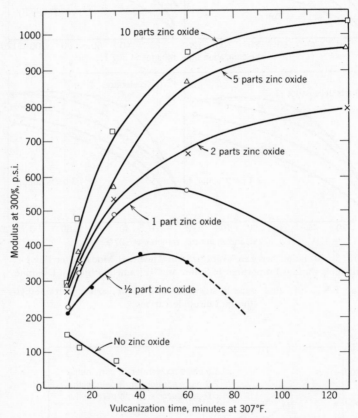

FIG. 7.   Effect of Concentration of Zinc Oxide on State of Cure of a Butyl Stock

Stock: Butyl 100, SRF black 54, sulfur 2, tetramethylthiuram disulfide 1, zinc oxide as shown

Metallic Oxides and Fatty Acids. As in the sulfur vulcanization of other rubbers, the presence of a metallic oxide such as zinc oxide is necessary for the realization of good physical properties in the vulcanization of Butyl. In Fig. 7 cure curves are presented wherein zinc oxide content varies from 0 to 10 parts per hundred of polymer.[13] The effect of increasing amounts of zinc oxide on the modulus characteristics is marked up to about 5 parts of zinc oxide. Only small advantages are noted with zinc oxide contents in excess of 5 parts.

Unlike the state of affairs with natural rubber and with GR-S, the presence of fatty acid is not necessary for vulcanizing Butyl. This is true even for specially prepared samples of polymer which do not contain the small quantities of zinc stearate (1 per cent or less) ordinarily introduced as slurry additive during Butyl-manufacturing operations. In Fig. 8 cure curves show that the effect of increasing quantities of stearic acid is a negative one.[13]

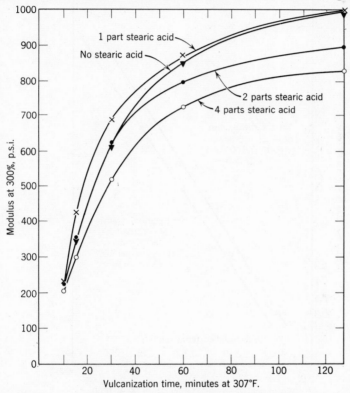

FIG. 8. Effect of Various Concentrations of Stearic Acid on the Vulcanization of a Butyl Stock

Stock: Butyl 100, zinc oxide 5, sulfur 2, tetramethylthiuram disulfide 1, stearic acid as shown

The slight reductions in modulus and the increases in elongation are indicative of a dilution or softening of the rubber matrix.

Temperature Dependence. As of any chemical process, the rate of vulcanization is a function of temperature. The vulcanization reaction, although complicated, follows a regular Arrhenius type of relationship when the time to obtain a certain degree of cross-linkage is related to temperature. Reliable measurements of cross-linkage by volume swelling technique were obtained for Butyl polymers which were vulcanized at 250 to 400° F. for different periods of time.   The logarithm of time required to reach 1000 per cent volume increase in cyclohexane at 25° C. was plotted against the reciprocal of the absolute temperature.   Figure 9 shows the linear relationships obtained for two accelerators.   Temperatures in degrees Fahrenheit are imposed on the abscissa.   The graph is so constructed that, if a certain time of cure at one temperature is known, a straight line through that point parallel to the existing lines will yield equivalent times of cure at other

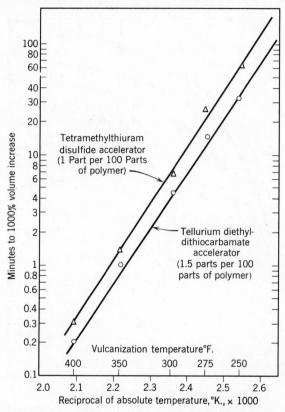

FIG. 9.   Effect of Temperature on the Rate of Vulcanization of Butyl as Measured by Swelling Capacity of the Vulcanizate

Stock: Butyl 100, zinc oxide 5, sulfur 2, accelerator as shown

temperatures.   Caution must be observed in extrapolating data to very high temperatures where reversion might occur if cure is prolonged.

Stoichiometry of Vulcanization.   In early work on Butyl vulcanization[130] it was observed that the more efficient accelerators gave vulcanized products possessing relatively little combined sulfur for a given state of cure. It was noted also that, when no accelerator was used, sulfur addition occurred in an inefficient manner.   These observations are shown in Fig. 10, where state of cure is given in terms of a relative cross-linkage index derived from measurements of volume swelling. For a given combined sulfur value, the thiuram type of acceleration results in a greater concentration of cross-links than the thiazole type. When no accelerator at all is used, only a marginal state of vulcanization for a corresponding amount of combined sulfur is obtained.

Molecular Characterization of Vulcanization. As previously mentioned, it is possible to obtain a maximum state of cure with a Butyl polymer

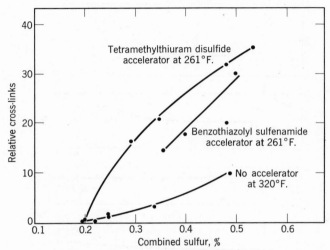

FIG. 10. Influence of Accelerators on the Formation of Cross-Links in the Combination of Sulfur with Butyl of 1.7 Mole Per Cent Unsaturation

Stock: Butyl 100, zinc oxide 5, sulfur 2, accelerator nil or 1

of low unsaturation. This makes it possible to characterize vulcanization in a molecular manner. The average distance between points of unsaturation becomes the chain molecular weight between cross-links when complete vulcanization has occurred, and, if the true unsaturation of the polymer is known through ozonolysis, this distance can be expressed in terms of molecular weight. The network so calibrated may then be used in the Flory-Rehner relationship[55] of swelling capacity, chain molecular weight, and interaction coefficient of a solvent-polymer system. Once obtained, the interaction coefficient may be used to characterize lower states of vulcanization on a molecular basis. By comparing states of vulcanization expressed in this manner with the quantity of sulfur combined in the polymer chain, the amount of sulfur in a cross-link can be calculated.[132]

Combined Sulfur Per Cross-Link. Vulcanization reactions have been conducted over a range of temperatures from 250 to 400° F. with several of the dithiocarbamate type of ultra-accelerators.[132] The vulcanization was followed in the molecular manner described previously. The results were compared with the amount of sulfur combined in the polymer chain. This was accomplished by differentiating between total combined sulfur and organically combined sulfur. A swollen gum vulcanizate, from which free sulfur had been extracted, was heated with oleic acid to form zinc oleate from residual zinc oxide in the compound. Upon extraction, only zinc in the form of sulfide remained. This zinc sulfide was determined by polarographic means.[72] The difference between total combined sulfur and sulfur in the form of zinc sulfide was considered to be organically combined sulfur. Volume swelling at various times of cure was converted to cross-links per average molecule, the number-average molecular weight of the polymer

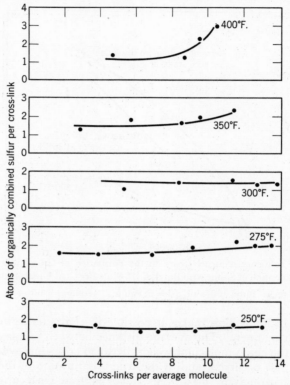

Fig. 11. Relation between Cross-Links and Combined Sulfur in Butyl Vulcanized at Various Temperatures Using Tellurium Diethyldithiocarbamate Accelerator

Stock: Butyl 100, zinc oxide 5, sulfur 2, accelerator 1.55

used being known. Data are given in Table IV. A comparison of cross-links with content of organically combined sulfur revealed that the amount of sulfur per cross-link was close to 2 atoms, regardless of the state of cure. This ratio persisted with other dithiocarbamate accelerators and at other temperatures, as shown in Fig. 11. Reversion occurred at high temperatures.

Relationship of Combined Sulfur to Physical Properties. The orderly character of the vulcanization of Butyl is contrary to observations made from using other rubberlike polymers.[7, 22] In Butyl, at least, a direct relation between organically combined sulfur and physical properties is indicated. This is demonstrated in Fig. 12, where cross-links are plotted against organically combined sulfur as a linear relationship.[132] The direct relationship of certain physical properties, such as modulus at low extensions, to the concentration of cross-links has been established by other investigators.[15, 54]

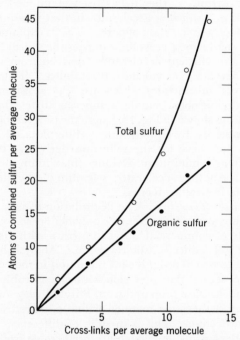

FIG. 12. Relation between Cross-Links and Combined Sulfur in Butyl Vulcanized at 250° F. Using Tellurium Diethyldithiocarbamate as Accelerator

Stock: As in Fig. 11

Table IV. Data Establishing Amount of Sulfur per Cross-Link in Vulcanized Butyl

Time, Min. at 250° F.	Volume Increase, %	v_2 Volume Fraction	M_c Chain Length	$\frac{1}{2}(M_n/M_c)$ Cross-Links per Avg. Molecule	Combined Sulfur %	Organic Sulfur %	Atoms Combined Sulfur per Avg. Molecule	Atoms Organic Sulfur per Avg. Molecule	Atoms Organic Sulfur per Cross-Link
20	1460	0.064	56,400	1.6	0.09	0.05	4.9	2.7	1.7
40	840	0.106	22,200	3.9	0.18	0.13	9.8	7.1	1.8
60	640	0.135	13,700	6.4	0.25	0.19	13.6	10.4	1.6
80	570	0.149	11,640	7.5	0.31	0.22	16.9	12.0	1.6
120	490	0.170	9,100	9.6	0.45	0.28	24.5	15.3	1.6
180	450	0.182	7,560	11.6	0.68	0.40	37.2	21.8	1.9
240	410	0.196	6,600	13.3	0.82	0.42	44.8	22.9	1.7
300	400	0.200	6,300	13.9	0.44		-		

STABILITY OF VULCANIZATES

As the vulcanization of rubber progresses, a reversion process takes place, whereby the molecular network is actually degraded to yield a softer matrix with impaired physical properties. This reversion process is particularly evident at higher temperatures and can be accelerated or hindered by compounding techniques.   It is very important in Butyl-rubber technology. With Butyl there is no hardening or compensating reaction during exposure to high temperatures such as there is with most other synthetic or natural rubber materials.

Effect of Acceleration.   Earlier it has been mentioned that a thiazole accelerator differs from a thiuram accelerator in that it requires more combined sulfur to produce a given state of cure.   This is explained on the basis of an accelerated breakdown of cross-links during the vulcanization process. Among even the more efficient accelerators, derived from dithiocarbamic acid, there are differences in vulcanizate stability.   Tellurium diethyldithiocarbamate gives vulcanizates which are very resistant to reversion. That tellurium diethyldithiocarbamate is superior to tetramethyl thiuram disulfide in this regard is shown in Fig. 13, where the progress of the vulcanization reaction is represented by equilibrium swelling data (cf. ref. 132).   At high temperatures, the curve for the tellurium derivative does not sweep upward sharply after a minimum swelling (maximum degree of cross-linkage) is reached.   Another accelerator, selenium diethyldithiocarbamate, appears intermediate in its effectiveness.

Unsaturation and Reversion.   The degradation of vulcanizate quality does not depend on any properties of the raw polymer per se.   This was shown by heating a sample of commercial Butyl in a press with zinc oxide under conditions expected to produce drastic reversion.   Analysis before and after the treatment failed to reveal a significant reduction in molecular weight.[132]

It has also been observed that reversion is not measurably dependent on unsaturation.   A vulcanizate prepared from Butyl of very low unsaturation exhibits degradation in the same manner as one prepared from a Butyl of relatively high unsaturation.   This is in spite of differences in ultimate or maximum degree of cross-linkage, which are predictable from differences in unsaturation.

Chemistry of Reversion.   When a vulcanization process is carried through to the point of reversion at high temperatures, it has been observed that the combined sulfur continues to increase.[132]   The curves of Fig. 11 show that the effective amount of combined sulfur per cross-link definitely increases during reversion.   This has been interpreted as a severance of disulfide linkages rather than as a formation of polysulfide bridges.   Experimental evidence supports this interpretation.   It has been determined[131] analytically that, as reversion occurs, the amount of mercaptan sulfur increases.   Such evidence points to a breakdown of disulfide linkages into mercaptan groups.

For the analysis methods were used similar to those employed for the determination of mercaptans in petroleum products[18, 80] by argentometric titration.   A vulcanizate in a finely divided and highly swollen state was suspended in naphtha or other aliphatic hydrocarbon and shaken overnight with an excess of standard silver nitrate.   The liquid layer was separated and the naphtha suspension washed repeatedly with distilled water.   The amount of silver nitrate consumed was determined by back titration using standard ammonium thiocyanate solution with ferric alum indicator.   Some results are given in Table V. It is shown that degraded samples of high swelling capacity are rich in mercaptan sulfur. At optimum cure the organically combined sulfur content of this particular Butyl is estimated to be about 0.8 per cent. This calculation is based on disulfide linkages. It is assumed

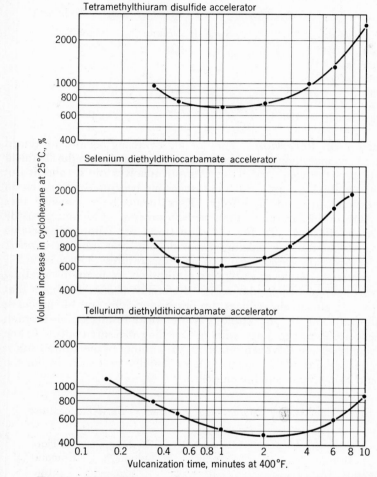

FIG. 13. Effect of Various Accelerators on the Stability of Butyl Vulcanizates at High Temperature

that there was no reversion. In the degraded sample, which exhibited 1560 per cent volume increase, most of the disulfide linkages have probably been severed to thiol groups as indicated by a mercaptan sulfur content of 0.7 per cent.

Literature[74] shows that organic disulfides can be broken down by the action of hydrogen sulfide or mercaptans in accordance with the following equation:

$$R \cdot S \cdot S \cdot R + H_2S \longrightarrow 2RSH + S \qquad (6)$$

Furthermore, it is well known that hydrogen sulfide is often a by-product of the vulcanization process. Alkali metal sulfides have been shown to decompose dialkyl sulfides in a manner analogous to the action of hydrogen sulfide.[74]

Table V. Effect of Reversion on Thiol Content of Butyl Vulcanizates

Cure	Swelling, Volume %	Thiol, %
80 min. at 300° F.	360	0.23
80 min. at 300° F.	...	0.17
80 min. at 300° F. plus 30 min. at 400° F.	1560	0.83
80 min. at 300° F. plus 30 min. at 400° F.	...	0.70

Retardation of Reversion. The use of oxidizing agents to retard reversion has been investigated, since analytical results as well as the chemical literature support the view that its mechanism involves the breakdown of disulfide links to thiol groups. Such use is feasible because the Butyl chain proper is resistant to oxidative effects. Experiments have shown[131] that oxidizing agents do in fact retard the reversion process. Calcium peroxide in particular is effective. In Fig. 14 cure states are compared for two compounds, one containing calcium peroxide and the other a control. These two compounds were vulcanized at temperatures (350 to 400° F.) critical to reversion. State of cure was established in terms of swelling capacity expressed on a molecular basis. Reference to Fig. 14 shows that calcium peroxide effectively reduces the degradation portion of the curve at both 400 and 350° F. In Table VI analytical results on mercaptan sulfur content are shown to agree with the swelling measurements. In the stock in which reversion has been retarded, the content of mercaptan sulfur has been reduced by about one-half. Other metallic peroxides (if not too stable) should behave in a similar manner.

Table VI. Data Showing Retardation of Reversion in Butyl Vulcanizates

Cure	Calcium Peroxide Content*	Swelling, Volume %	Thiol Content, %
10 min. at 350° F. (optimum cure)	0	520	0.16 0.12
10 min. at 350° F. (optimum cure)	8	490	0.16 0.10
80 min. at 350° F. (reversion evident)	0	1440	0.54
80 min. at 350° F. (reversion retarded by CaO₂)	8	140	0.26

* Parts by weight per 100 parts of polymer.

As mentioned earlier, zinc oxide is necessary for good vulcanization. It has been proposed[13] that at least part of the utility of zinc oxide is due to a reaction with hydrogen sulfide to form zinc sulfide and thus reduce reversion tendencies during the vulcanization process. The metallic peroxides and other oxidizing agents serve to oxidize the mercaptan groups once they are formed. This is of practical value when the vulcanization reaction is prolonged or when the vulcanizate is subjected to high-temperature service for extended periods.

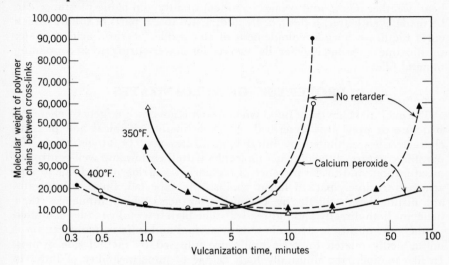

FIG. 14.   Retarding Effect of Calcium Peroxide on Reversion during
Vulcanization of Butyl

VULCANIZATION WITHOUT SULFUR

When desirable, Butyl can be vulcanized without the use of sulfur. Haworth[63-4] found that nitrogen compounds such as quinone dioxime will cure Butyl in the absence of sulfur if an oxidizing agent is present. Rehner and Flory[97] introduced the concept that the active agent is p-dinitrosobenzene, into which, it was postulated, quinone dioxime would be converted in the presence of an oxidising medium. Later work showed that this assumption could not be verified experimentally. Quinone dioxime is normally used[29] at a concentration of 2 parts per 100 parts of polymer, with lead oxides as the oxidizing agents. With a concentration of 4 parts of lead dioxide activity is such that the vulcanization of Butyl at room temperature is possible. Cures of this kind are desirable in cement work and are made possible through the use of a split batch technique.

The use of red lead oxide as an oxidizing agent at a concentration of 8 parts is permissible. The resulting curing is very fast. To adapt it to normal processing one part of a retarder is added. Suitable retarders include octadecylamine, dibenzylamine, and thiocarbanilide. A more controllable system can be obtained by esterifying quinone dioxime with benzoic acid. About 6 parts of the resulting product, dibenzoyl quinone dioxime is used, with about 10 parts of red lead oxide as the oxidizing agent. Another valuable system, which avoids the use of lead oxides, is a combination of 2 parts of quinone dioxime with 4 parts of benzothiazolyl disulfide.

All of the nonsulfur curing agents here described have a tendency to cause scorching in compounds containing carbon blacks and especially channel blacks. The various dioxime types of cure are of interest, however, because the vulcanized products are remarkably resistant to the effects of heat, weather aging, and ozone. Product quality and aging resistance can be made even better, if desired, by adding 1.0 to 2.5 parts of sulfur. Such cures require a normal complement of zinc oxide. Stearic acid activates the dioxime cures and is generally reserved for slower-curing stocks containing mineral fillers.

PROPERTIES OF VULCANIZATES

The over-all behavior of Butyl vulcanizates is governed largely by chemical structure as previously described. The distinctive chemical and physical characteristics exhibited by Butyl vulcanizates may be attributed to the paraffinic nature of the polymer molecules with their low and well-controlled unsaturation, and to the presence of regularly occurring methyl side groups attached to closely packed linear chains. As expected, Butyl vulcanizates are distinguished chiefly by chemical inertness, which influences their behavior both directly and indirectly. The high internal viscosity exhibited by the vulcanizates in dynamic tests at normal and subzero temperatures is undoubtedly related to steric hindrance imposed by methyl side groups. In like manner, the unusually high degree of impermeability of Butyl to gases is an important physical property which may be traceable to dense molecular packing.

CHEMICAL PROPERTIES

Resistance to Acids, Alkalis, and Salt Solutions. Haworth and Baldwin[65] have published data on the behavior of Butyl vulcanizates exposed to a variety of acids and bases at various temperatures. Results are shown in Table VII. A natural-rubber vulcanizate was included for comparison. Only sulfuric, nitric, and hydrochloric acids had pronounced effects on either of the polymers. Hevea was transformed rapidly in hydrochloric acid to a transparent hydrochloride, and it was decomposed completely in sulfuric and nitric acids. In essentially all instances Butyl retained a high degree of its original strength. Resistance to hydrochloric acid was especially good. It was concluded that Butyl is far superior to natural rubber for acid stability.

Pigmented Butyl vulcanizates have been shown by Baldwin[9] to be less resistant than pure-gum mixtures to sulfuric acid. This was attributed to the sorptive effects of pigments. The more strongly adsorptive channel blacks brought about the most rapid deterioration. In this work it was further noted that resistance to sulfuric acid was much better at a concentration level below that corresponding to the acid monohydrate (84.4 per cent

Table VI. Resistance of Butyl Gum Vulcanizates to Acids and Bases at 78° F.

| | Original | 2 and 4 Weeks in: | | | | | | |
		37% HCl	95.5% H_2SO_4	70% HNO_3	85% H_3PO_4	28% NH_4OH	30% NaOH	85% Lactic Acid
Butyl A*								
Tensile, p.s.i.	3460	2970–2850	3300–3640	2830–2770	3630–3720	2810–2810	3710–3650	3280–3340
Elongation, %	1050	1020–950	1025–1050	1050–1050	1050–1050	1030–1050	1030–1050	1030–1030
Butyl B-1.45								
Tensile, p.s.i.	2500	2860–2340	†	1670–1110	2480–2910	2350–2440	3160–2920	2910–2680
Elongation, %	880	850–870	†	910–850	850–850	875–855	850–900	900–900
Natural rubber								
Tensile, p.s.i.	4240	2790–1690	Decomposed	Decomposed	…	4050–3500	4020–4270	4440–3860
Elongation, %	800	740–510	Decomposed	Decomposed	…	850–830	840–850	840–810

* Polymer prepared using butadiene rather than isoprene as the diolefin in the feed, but otherwise comparable to Butyl B-1.45.

† Complete cleavage at points along the sample.

H_2SO_4 by weight). The effect of pH of the Butyl compound was likewise determined by Baldwin. Evidence was obtained to prove that, for the least deterioration, the mixtures should be kept nonalkaline.

Butyl vulcanizates displayed excellent resistance to a wide variety of inorganic salts, including metal salts which are known to accelerate the oxidative breakdown of Hevea even when present in extremely low concentrations. Copper and manganese, for example, have no apparent effect on the aging of Butyl vulcanizates. Unpublished work by Hubbard,[68] however, has shown that crude Butyl degrades more rapidly on a hot mill when manganese salts are present.

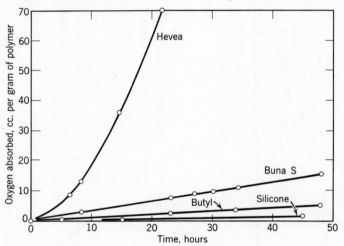

FIG. 15. Rate of Absorption of Oxygen at 130° C. by Gum Vulcanizates of Butyl and Other Elastomers

Other sources of information on the chemical resistance of Butyl are the work of Thomas and co-workers,[111] Lightbown,[76] Turner and Cunningham,[119] Parker,[90] and Ball and Maassen.[14]

Resistance to Oxygen and Air. The great practical importance of the aging of polymeric substances in contact with the oxygen-containing atmosphere has led many investigators to its study. Not all of the references on the subject are cited, but only those that have a direct bearing on Butyl. One interesting study was published in 1947 by Mesrobian and Tobolsky.[85] The results of this work and their interpretation help greatly to clarify the behavior of Butyl vulcanizates under oxidative conditions. The oxygen-absorption rates of various pure-gum elastomers were found to increase in the following order: polysilicones, polyesters, Butyl, Thiokol, GR-S, Neoprene, Butaprene-NM, and Hevea. Fig. 15 shows the wide contrast between Hevea on the one hand and Butyl or silicone on the other.

The contrasting oxygen-absorption rates were accounted for as follows: "The conclusion that can be drawn, then, is that the presence of a double bond in the skeletal structure of a polymer and of a methyl side group both enhance the rate of absorption of oxygen, the double bond being more important." Further substantiation of this conclusion was evident in the comparison by Mesrobian and Tobolsky of Hevea, polyisobutylene, and polyethylene. In the order in which they are given, these substances are representative of (1) a polymer chain containing a high concentration of both double bonds and methyl side groups, (2) a chain containing methyl side groups without unsaturation, and (3) a chain with neither double bonds nor methyl side groups. As expected, oxygen absorption was at a maximum for Hevea and a minimum for polyethylene. This is shown in Fig. 16.

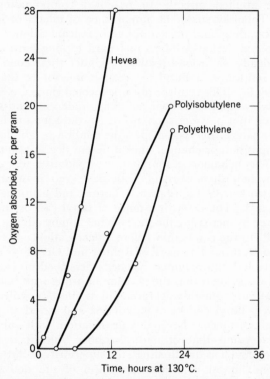

FIG. 16. Rates of Oxygen Absorption by Polymers of Various Chain Structure

Such information on unvulcanized polymer serves to emphasise the basis of the behavior of Butyl-type polymers in contact with oxygen. It is important to note the chemical and physical changes that accompany the oxidation reaction in Butyl and to compare these changes with those occurring in other elastomers. Whereas Butyl invariably softens when heated in an oxidative atmosphere, polymers such as GR-S and Paracril (Buna N) become brittle. Hevea, it has been observed, initially softens and then hardens gradually to brittleness. Mesrobian and Tobolsky as well as others have attributed these contrasting reactions to variations in the ratio between cross-linking and chain scission. Cross-linking tends to harden the polymer, whereas chain scission softens and tackifies the mass. Since the two reactions occur simultaneously, their net effect determines the ultimate change in physical characteristics. As pointed out by the authors, polymer structure is not the only factor affecting the type of chemical and physical change. Methyl side groups favor scission, and carbon-to-carbon double bonds favor cross-linking. The behavior of Butyl conforms to this conception of structural influence. In general, numerous mechanisms have been proposed in the literature for the oxidative degradation of Butyl and other elastomers.[2, 16, 19, 44, 60, 81, 83–4, 109–10]

Obviously, the theory of oxygen reaction with crude polymers cannot in all instances be applied directly to practical compounding problems. Mesrobian and Tobolsky stress the importance of effects of antioxidants, the type of compounding, and the conditions of vulcanization. Rehner and Robison[100] compared various rubbers in regard to long-term creep rate at 25° C., using a dead-weight-load method. Under these conditions it was shown that the resistance of Butyl to creep is inferior to that of natural rubber and of GR-S. This result is not unexpected since the characteristic softening of Butyl would be shown as creep under a dead-weight load, whereas hardening in a polymer such as GR-S would tend to inhibit creep. It is difficult to analyze this type of data in relation to absolute degree of structural decomposition or change, since different degradation phenomena are involved. The behavior of Butyl under oxidative test conditions is shown further in published work of Tobolsky, Prettyman and Dillon,[117] Thomas and co-workers,[111] Lightbown,[76] Turner and Cunningham,[119] and Ball and Maassen.[14] The creep resistance of Butyl can be improved to a pronounced degree by increasing the state of cure and by adding antioxidants (e.g., Agerite HP) to the compound before vulcanization.

Resistance to Ozone. The work of Rehner and Gray[94, 98] on the ozone degradation of crude Butyl polymers has been described previously. From that description it is evident that the reaction of ozone with uncompounded and unvulcanized Butyl is highly degrading in its effect on molecular weight. Actually, however, Butyl can be compounded and cured to yield a high degree of ozone resistance. Normal vulcanization with sulfur produces resistance far exceeding that of natural-rubber vulcanizates. By using special accelerators, such as the dioximes, the superiority of Butyl is increased still further. The almost complete saturation of the double bonds on vulcanizing Butyl is presumably responsible for the high resistance of vulcanized products. Regardless, however, of the mode of compounding and vulcanization, Butyl is not completely inert to attack by ozone, as has been shown in unpublished experiments[42] conducted over a period of years with a wide variety of Butyl types. It is, of course, easily shown that the degree of unsaturation of the Butyl is of great importance. Highly unsaturated Butyls (2.0 to 2.5 mole per cent) are inferior to the lower unsaturated types (0.7 mole per cent), but both are far better than natural rubber. Compounding and the mode of vulcanization have various effects which cannot be discussed in detail.

A rough description of the relative resistances of Butyl and Hevea to ozone is as follows. Specimens of simple pure-gum vulcanizates of Butyl and Hevea were folded in a 180 degree loop and exposed simultaneously to ozone. Cracking and crazing were noted in the Hevea samples after an exposure of

5 to 15 minutes. The Butyl specimens showed no weakening for several hours. Ultimately, however, the Butyl specimens developed sporadic cracks.

Resistance to Halogens. In the presence of light, halogens attack both vulcanized and unvulcanized Butyl. (It is well known that polyisobutylene can be made to react quite rapidly with halogen gases.) Whether vulcanized or unvulcanized, the effect of the reactions is to produce rapid deterioration in molecular weight. A strip of a Butyl vulcanizate suspended in bromine gas, for example, degrades rapidly. In a matter of minutes, the specimen becomes fluid enough to drip to the bottom of the vessel.

At first glance it appears surprising that Butyl polymer chains should be attacked by halogens so much more easily than simple unbranched hydrocarbons, such as paraffin wax. The explanation lies in the numerous activating methyl side groups in the Butyl molecules. Also, the small concentration of double bonds remaining in the vulcanizates may increase the reactivity toward halogens.

Hydrogen Sulfide and Other Gases. Among the numerous other gases, hydrogen sulfide is of appreciable importance in its chemical effect on Butyl. The role of hydrogen sulfide in the vulcanization of Butyl and in the aging of the vulcanizates has been discussed by Baldwin, Turner, and Zapp,[13] who concluded that the hydrogen sulfide formed during vulcanization was at least partly responsible for reversion observed during high-temperature or long-time vulcanization cycles. A possible mechanism for the decomposition of disulfides is as follows:

$$R \cdot S \cdot S \cdot R + H_2S \longrightarrow 2RSH + S \tag{6}$$

Baldwin et al. pointed out that since monosulfides undergo a similar reaction

$$R \cdot S \cdot R + H_2S \longrightarrow 2RSH \tag{7}$$

it is possible that hydrogen sulfide may destroy either sulfide or disulfide cross-links which form during vulcanization. In support of such a mechanism, the authors described results of an experiment in which specimens of crude and vulcanized Butyl were compared in resistance to hydrogen sulfide gas. Samples were heated for 2 hours at 255° F. in the presence of hydrogen sulfide at 50 lb. pressure. The reagent had no appreciable effect on the crude polymer as judged by molecular-weight measurements, but did cause the vulcanizate to undergo a high degree of reversion as evidenced by the modulus falling off and the swelling in solvents increasing. Baldwin and co-workers have shown also that zinc oxide in quantities exceeding that needed for vulcanization serves to minimize the tendency to revert. The zinc oxide is believed to combine with hydrogen sulfide and thereby decrease its effective concentration.

$$H_2S + ZnO \longrightarrow ZnS + H_2O \tag{8}$$

But even a large excess of zinc does not prevent the eventual degradation of the vulcanizates. This is attributed to oxidation of the chain proper, or to some free-radical mechanism.

In other work Baldwin[10] has demonstrated that by using sufficiently severe conditions Butyl vulcanizates can actually be reclaimed with hydrogen sulfide.

As to the effect of other types of gases, reference is made to unpublished work of Baldwin and Thomas.[12] These investigators determined the effect of carbon dioxide, nitrogen, sulfur dioxide, ammonia, and hydrogen sulfide on the creep and tensile properties of Butyl inner-tube vulcanizates at 120° C. Carbon dioxide and nitrogen showed effects definitely less harmful than air. Exposure to ammonia increased the modulus of the vulcanizates; exposure to sulfur dioxide produced an initial degradation, followed by an equilibrium state during which no further breakdown occurred.

Resistance to Organic Liquids. The behavior of Butyl in contact with organic chemicals or solvents is typical of that of hydrocarbon elastomers. Butyl vulcanizates are swollen by hydrocarbon solvents, the paraffinic type being the most effective. Polar liquids, such as alcohols, ketones, esters, and glycols, have little or no solvent action. Some alcohols may actually cause a decrease in volume because of extraction. Fatty acids and vegetable and animal oils are intermediate in their effect, exhibiting slight swelling power.

Flory[51] and Huggins[69] have utilized an "interaction coefficient" μ to describe the interaction between a solvent and a polymer. Osmotic-pressure methods were employed to measure μ for dilute solutions of soluble polymers. Flory and Rehner[55] later devised a more convenient method for determining interaction coefficients by extending the Flory-Huggins theory to the swelling of vulcanized or cross-linked networks. This latter method has been used by McGovney and Zapp,[82] in unpublished work, to determine the equilibrium volume swelling and interaction coefficients for a pure-gum Butyl vulcanizate in a variety of organic liquids at 25° C. These results are shown in Table VIII. Good solvents are characterized by a μ value of 0.5 or less, whereas poor solvents have higher values. A value of about 0.7 represents marginal solubility. The results show clearly the relation of swelling and interaction coefficient to solvent properties, such as degree of paraffinicity, polarity, and aromaticity. Cyclohexane gives high swelling or low-interaction coefficient, whereas acetone, which contains the polar carbonyl group, has practically no effect.

An unexpected characteristic of Butyl vulcanizates is their relatively high resistance to swelling by animal and vegetable oils. Results published by Haworth and Baldwin[65] show the volume increase of Butyl in a variety of oils compared with corresponding results for Hevea and Perbunan. "After three months' immersion at room temperature, the swelling of Butyl in linseed oil and in soybean oil was only 10 per cent and 6 per cent (by volume), respectively, whereas the swelling of natural rubber was 124 and 106 per cent, respectively."

Rostler and White[104] have published a highly informative and comprehensive comparison of natural and synthetic elastomers in regard to their resistance to swelling by organic liquids. Other publications are available which have either a theoretical or a practical bearing on the subject of the behavior of Butyl vulcanizates in organic liquids.[59, 62, 66, 87, 91] Data on the influence of various fillers on the swelling of Butyl vulcanizates has been given by

Zapp and Guth, who find that carbon blacks restrict volume swelling to the extent of their surface area and that mineral fillers possess little or no ability to restrict swelling on a comparable basis.[133]

Table VIII. Equilibrium Volume Increase and Interaction Coefficients for Butyl in Various Solvents

Standard GR-I at 25° C.

	Volume increase, %	μ
n-Hexane	367	0.394
n-Heptane	313	0.422
n-Octane	324	0.422
Isoöctane	277	0.463
Cyclohexane	540	0.350
Cyclohexanone	45	0.995
Benzene	188	0.606
Toluene	291	0.505
Acetone	7	2.010
Methyl ethyl ketone	17	1.440
n-Butyl acetate	58	0.902
Ethyl formate	14	1.600
Ethyl acetate	21	1.337
Carbon tetrachloride	470	0.395
n-Decane	326	0.376
50-50 MEK-acetone	8	1.968
50-50 Benzene-butyl acetate	84	0.782

PHYSICAL PROPERTIES

In many respects the physical properties of Butyl vulcanizates are much like those of Hevea. There are, however, some outstanding differences in behavior. Ordinary vulcanized sheets of Butyl and Hevea are soft, flexible, extensible, and elastic. When stretched, both rubbers display considerable strength and toughness, and both reach an elastic limit characteristic of the vulcanized product. Although both materials retract rapidly when released from stretching, Hevea recovers more quickly and with more "snap." Even though very superficial, such comparisons offer simple but important demonstrations of fundamental similarities and dissimilarities which can be measured quantitatively by other means. On the other hand, casual examination of Butyl gives no clue to the phenomenal degree of impermeability to the passage of gases or to other characteristics such as excellent electrical properties, good tear resistance after aging and at elevated temperatures, or relative chemical inertness. These are the properties that distinguish Butyl among elastomers.

Stress-Strain Curves. In accordance with publications by Flory[54] on Butyl, Anthony, Caston, and Guth[6] on Hevea, and Roth and Wood[105] on GR-S, it is possible to characterize the pure-gum stress-strain curves for these elastomers qualitatively as is shown diagrammatically in Fig. 17. Although not identical in shape, the curves for both Butyl and Hevea show a sharp rise in slope at high extension due to crystallization. The occurrence of crystallization (proved by means of X rays) and its effect on the properties of Butyl and Hevea have been demonstrated and explained by Flory,[54, 50] Field,[46] and others. The importance of crystallization to the development of high tensile strength is illustrated by the low stress values given by pure-gum GR-S, which is known not to crystallize. Butyl and Hevea do not depend on

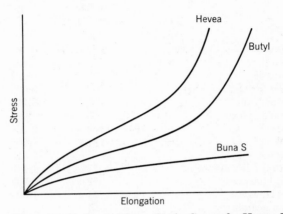

Fig. 17. Diagrammatic Typical Stress-Strain Curves for Hevea, Butyl, and Buna-S Gum Stocks

reinforcing pigments for high tensile strength as GR-S does. Flory[54] showed that the stress developed in an unpigmented vulcanizate is as high as 4500 p.s.i. for Butyl of high molecular weight. Unpigmented GR-S usually fails at stresses under 500 p.s.i. Since both molecular weight and unsaturation can be controlled in Butyl polymers, it is important to know the effect of these variables on stress-strain characteristics. In this regard, reference is again made to Flory,[54] who investigated rather fully the effects of these factors on stress-strain and other physical properties.

Baldwin and co-workers[13] demonstrated the effect of reversion and overcure on the form of the stress-strain curves for Butyl.

As mentioned previously, the tensile strength of Butyl is not increased by the addition of pigments. Filler reinforcement is displayed in other ways. For example, reinforcing agents increase the modulus, tear strength, and abrasion resistance of Butyl. Haworth and Baldwin[65] and also Zapp[129] have published data on the effects of pigments in Butyl. Other authors who have contributed to the literature on the stress-strain characteristics of Butyl are Thomas and co-workers,[111] and Turner and Cunningham.[119]

Tear Resistance. The tear resistance of properly compounded and vulcanized Butyl is excellent. And—what is important—Butyl has the ability to retain its good tear strength at elevated temperatures and after extended aging periods. A room-temperature test of comparable unaged samples shows a slight advantage in tear strength for Hevea over Butyl, but tests run after a suitable aging period show a distinct superiority for Butyl.

The good tear resistance of Butyl at elevated temperatures is attributed by Buist[25] to the fact that the tearing is of the knotty type. Its retention of tear strength after aging is undoubtedly related to the high over-all resistance of Butyl to degradation of air or oxygen. Lightbown et al.[77] have depicted graphically the relative effects of air-oven and air-bomb aging on the tear strength of Butyl and Hevea inner-tube vulcanizates. These effects are shown in Fig. 18.

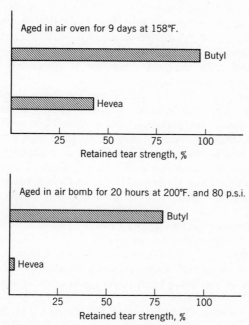

FIG. 18. Effect of Aging on the Tear Resistance of Butyl and Natural-Rubber Inner-Tube Stocks

Electrical Properties. As would be expected of a nonpolar essentially saturated hydrocarbon, Butyl possesses excellent electrical properties. Relevant data, of which an example is given in Table IX, have been published from a number of sources.[14, 26, 33, 65, 76, 111, 119] Besides having inherently

Table IX. *Electrical Property Comparison of Butyl with Natural Rubber*[119]

	Pure-Gum Butyl	Pure-Gum Natural Rubber
Dielectric strength, volts per mil thickness	600	500
Dielectric constant (10 volts, 1,000 cycles)		
Dry	2.11	2.46
After 88 hr. in distilled water	2.10	2.76
Power factor (10 volts, 1000 cycles)		
Dry	0.04	0.04
After 88 hr. in distilled water	0.05	0.16

good original electrical properties, Butyl resists water, ozone, weathering, bacteria, and fungi, all of which facts combine to make the polymer attractive in the field of electrical insulation.

Influence of Temperature on Properties. Butyl, like other elastomers, is sensitive to changes in temperature. Although the rebound at room temperature of an ordinary unplasticised Butyl, for example, is poor, at elevated temperatures Butyl becomes highly resilient and is equivalent to Hevea.[119] As temperatures are decreased, Butyl vulcanizates stiffen rapidly, and at subzero levels they become hard and brittle. The unusual property of Butyl, however, is not its brittleness at, say, $-70°$ F. but rather its high degree of sluggishness in the temperature range of $+100$ to $-30°$ F. It has been conceded generally that this behavior is associated with molecular structure, as discussed earlier in this chapter.

Much published and unpublished work has been done in measuring internal friction in Butyl vulcanizates. Most of these studies have covered a wide range of temperatures, and many have offered methods and techniques whereby the high internal viscosity may be suppressed. Eby and Buckley[41] separated the extension modulus of Butyl at low temperatures into elastic and viscous components and compared these values with results for other elastomers. Independently determined values for the internal viscosities of various rubbers have agreed at least qualitatively with respect to the properties of Butyl.

The low-temperature performance of Butyl vulcanizates is improved by a number of means, employed either individually or in combination. Among valuable techniques are (1) the use of freeze-resistant plasticizers, (2) the use of coarse-particle pigments, and (3) vulcanization to a higher degree. These and other methods have been proved efficacious both experimentally and commercially.[1, 11, 23–4, 70]

Dynamic Properties. The characteristics displayed by Butyl in dynamic tests have been indicated in the preceding discussions of rebound and retraction rate. Physical constants of Butyl have been studied by many investigators at various temperatures, frequencies, and amplitudes. Baldwin[11] measured the elastic constant, the coefficient of internal viscosity, and absolute and relative damping under conditions of free vibration. In this work the helpful effects of plasticizers in reducing modulus and internal viscosity at low temperatures were demonstrated. With plasticizer present, an increase in relative damping was observed as temperature was lowered, but the actual numerical values for relative damping were reduced all along the curve.

Guth and co-workers[70, 127] determined the dynamic viscoelastic properties of Butyl, GR-S, and Hevea by measuring velocity transmission and attenuation of sound in the polymers. Both audiofrequency and ultrasonic pulsations were employed. Dart and Guth[38] measured the rise of temperature on stretching of Butyl at a high rate. By such means, they were able to indicate the onset of crystallization and to obtain an estimate of internal friction. Stambaugh[108] demonstrated the low internal friction of Hevea compared with Butyl by studying the wave motion of stretched samples when released. Hevea retracted at the released end first and progressed in a wave, whereas Butyl retracted along its entire length without wave motion.

In tests of the forced vibration type, Morron and co-workers[88] have defined a "barrier point" for various elastomers as a point on the temperature scale at which a given specimen would not absorb vibrational energy and hence would not heat up by flexing under a given set of conditions. The barrier point depends on the size and shape of the specimen, as well as on the composition and other factors. Mrowca, Dart and Guth[89] compared the retraction rates of natural and synthetic rubbers, and Chilton[30] showed how the dynamic strains compare in descending order as follows: Hevea, Buna S, Neoprene, Buna N, and Butyl.

Regardless of the kind of measurement employed, the mechanical behavior of Butyl is mainly that of a polymer having high internal friction and high damping power. However, results have been consistent in showing that the internal viscosity of Butyl may be drastically reduced, if desired, by compounding with appropriate plasticisers and pigments and by vulcanization to a high degree.

Permeability to Gases. Any discussion of the properties of Butyl must necessarily call attention to its high impermeability to gases. Numerous references have been made to this phenomenon in the literature.
14, 36, 39, 65, 67, 76, 101, 111, 119

Reitlinger[101] and van Amerongen[4] have both obtained experimental evidence showing interesting relationships between gas permeability and microstructure in polymers. From such work it is concluded that the factors responsible for the gas impermeability of polyisobutylene, and presumably of Butyl also, are (1) the essentially linear nature of the principal valence chains, and (2) the moderate dimensions of the side groups on these principal chains. Butyl has close, unstrained molecular packing which makes possible maximum intermolecular bonding and consequent high interference with the diffusion of gases. Data on the permeability of Butyl and Hevea to different gases including helium, hydrogen, nitrogen, and carbon dioxide are given in Fig. 19.

APPLICATIONS

Although Butyl has been called a special-purpose rubber to distinguish it from GR-S in the Government program, its commercial utility is actually quite wide. Among present-day industrial products, inner tubes are by far the most firmly established. Other products include electrical wire insulation, tire-curing bags, tank lining, sports goods, cements, and proofed fabric. Of particular note is the increasing popularity of Butyl for use in mechanical rubber goods, including steam hose, belting, and automobile parts. In each of these cases, Butyl has one or more special properties that make it more valuable than natural rubber or other synthetics.

Among the uses that have not yet reached the commercial state but are at an advanced experimental stage, tires for farm equipment are significant. Though tractor tires are not subjected to high driving speeds, they are exposed to all sorts of rigorous weather conditions year in and year out. Consequently, cracking and deterioration caused by sunlight or high-temperature aging and weathering in general are more often the cause of failure than, as is the case with automobile tires, abrasion or wearing out. Further, tractor tires must stand up under off-the-road conditions, since the tractor's "highway" is the open field and rocky terrain. Full-scale experimental field tests on Butyl tractor tires are very promising.

Inner Tubes. Whether the inner tube or the casing of a pneumatic tire is the principal part of the assembly is open to discussion. On the one hand, it has been said that the inner tube is nothing more than an air container for the more important tire casing. On the other hand, the opposite point of view has been taken that the tire casing is no more than a protective shield for the more significant inner tube. In any event, whichever view is taken, it is apparent that the pneumatic characteristic of the tire assembly is its essential

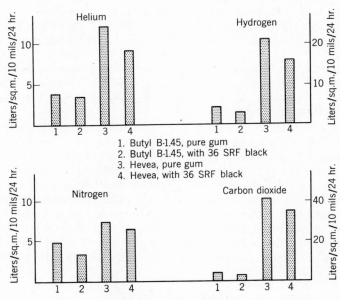

1. Butyl B-1.45, pure gum
2. Butyl B-1.45, with 36 SRF black
3. Hevea, pure gum
4. Hevea, with 36 SRF black

Fig. 19.   Permeability of Butyl and Hevea to Various Gases

property. No one who has had a flat tire at midnight on a lonely road far from a service station will question this statement. No matter how good the casing, a tire assembly that has lost its air is worthless. But there are also gradations in the matter of air loss. A tire that has lost part of its air may not be worthless, but it is less valuable than a tire that has maintained its correct inflation pressure. Retention of air pressure is, of course, a function of the inner tube and one on which the tire casing has no real influence. That natural-rubber tubes lose air fairly rapidly by direct permeation through the rubber is well known. It is also well known from the experience of millions of motorists driving billions of tire miles that Butyl is a much better air barrier than natural rubber. For this reason and because of superior physical and chemical properties, inner tubes of better quality can be produced from Butyl than from other rubbers.

The difference in air retention between natural rubber and Butyl inner tubes is demonstrated by the following data taken from actual road tests under carefully controlled conditions.

Type of Tube	Original Pressure, P.S.I.	Air-Pressure Loss, p.s.i.		
		1 Week	2 Weeks	1 Month
Natural rubber	28.0	4.0	8.0	16.5
Butyl	28.0	0.5	1.0	2.0

These data show that Butyl is about eight times better than natural rubber. They were obtained under severe conditions with cars driven 60 m.p.h. for 1000 miles per day. Other tests under driving and standing conditions that more nearly approach normal practice have indicated prevention of air leakage even more favorable to Butyl. These latter tests indicate that the average motorist with Butyl inner tubes need reinflate his tires only two or three times a year.

Automobile manufacturers specify not only tire size but also the optimum inflation pressure at which the tire should be maintained for safe and economical operation. Overinflation is undesirable; it reduces resistance to punctures and blowouts, increases tire strain, causes rapid and uneven treadwear, and decreases riding quality. Similarly, underinflation is undesirable. It reduces puncture and blowout resistance. It also increases tire strain and operating temperatures because of increased flexing. Tread-wear, driving safety, and ease of steering are consequently impaired. With an air container such as natural rubber, which permits air to permeate its walls at a fairly rapid rate, it is virtually impossible to maintain optimum tire inflation. Motorists often overinflate their tires to compensate for air leakage, but this is about as undesirable as underinflation. Even so, air leakage is rapid enough to result in underinflation after a short time. Unless the motorist is careful to inflate his tires every two or three days when using natural-rubber tubes, he is bound to be operating at pressures that are either too high or too low. In either case, treadwear and tire life are adversely affected.

Since the superior air holding characteristics of Butyl make it a relatively simple matter to maintain proper inflation pressure, it was anticipated that an over-all improvement in treadwear would result from the use of Butyl tubes. To establish the validity of this expectation, treadwear was measured during the course of the following special air-holding test.[77] Two natural-rubber and two Butyl tubes were mounted in tires on the same automobile. The tires were especially made to be identical. During the test there was a greater loss of air through the natural-rubber than the Butyl tubes. To make the test correspond to the driving practice of the ordinary motorist, as well as to maintain safety, all tubes were reinflated to the original pressure of 28 lb. when any tire showed 22 lb. pressure or less. It was, of course, always one of the tires containing natural-rubber tubes that fell to this lower figure. The tires were rotated in the normal manner every 1000 miles. Treadwear was measured every 2000 miles by the accepted method. Briefly, this method consists of gaging the change in depth of the nonskid grooves. The tests were made at both 40 and 60 m.p.h. As may be seen from a photograph (Fig. 20), the results confirm predictions that the maintenance of proper inflation pressure by Butyl tubes results in an increased tread life.　This increase varies between 10 and 18 per cent, depending on test conditions.

Aside from its ability to hold air, an inner tube must also be strong and durable.　One important characteristic of this durability is high tear resistance.　In this property, Butyl is outstandingly superior to natural rubber. Since an inner tube is expected to last at least as long as the tire casing and perhaps as long as several tire casings, it must maintain its original good

Natural rubber tube

Butyl tube

17,095 miles

Fig. 20.　GR-S Tires with Inner Tubes of Natural Rubber
and of Butyl Rubber

properties throughout many miles or years of service on the road.　This is an important matter.　The average car owner seldom sees his inner tubes. Any evidence of deterioration after long periods of driving is not noticed, since it cannot be seen with convenience.　A natural-rubber inner tube, after having been driven for, say, 20,000 to 30,000 miles, loses a great deal of its original strength and durability, particularly its tear resistance.　If, therefore, a natural-rubber inner tube that has seen typical service on the road is punctured by a nail, there is a good possibility that the small puncture will be the focus of a tear that will proceed rapidly and cause a serious blowout.　On the contrary, Butyl tubes maintain most or all of their original physical properties, notably tear resistance and tensile strength.　Even after long driving periods such as 50,000 miles, a small puncture of a Butyl tube will not result normally in the type of tear that causes a blowout.

A typical formula used initially in the manufacture of Butyl tubes consists of the following:

Butyl	100
Zinc oxide	5
EPC black	20
SRF black	30
Paraffin wax	1
Petrolatum	2
Sulfur	2
Tetramethylthiuram disulfide	1
Mercaptobenzothiazole	0.5

A Butyl tube made according to early formulations performs well in most parts of the United States, but, in Canada and the north central tier of the United States, a modified type of Butyl and changes in compounding are necessary to improve retractability at low temperatures and thus obviate the buckling of tubes and consequent chafing and failure which had previously been experienced sometimes in subzero weather. The improvement desired is achieved through the use of higher-molecular-weight Butyl polymers, plasticizers, and an increased state of vulcanization. A typical formula follows.

Butyl (GR-I 18)	100
Zinc oxide	5
HMF black	25
SRF black	35
Tetramethylthiuram disulfide	1.25
Mercaptobenzothiazole	0.5
Sulfur	2
Polyac	0.2–0.7
Petroleum oil plasticizer*	25

* Necton 45 or Forum 40.

These changes result in a softer, more elastic end product. They are of value also in leading to significant economies. Raw-material costs are lowered through the use of inexpensive petroleum oil plasticizers, while fabrication costs are reduced because of the facility of processing the softer compound.

Practical factory problems in the fabrication of Butyl inner tubes are discussed in a helpful manner by Iknayan.[69a]

Curing Bags. Natural rubber exhibits surface hardening, cracking, checking and forms deterioration products of a resinous nature when exposed to elevated temperatures in the presence of high-sulfur or accelerator combinations. Because of this deterioration, natural-rubber tire-curing bags have a relatively short life. It is common practice in the tire industry to buff off the surface of the bag after a few curing cycles. This is to remove the hardened surface resulting from high-temperature exposure in the presence of sulfur which migrates from the curing tire. Sometimes the bag is "retreaded" with a new rubber surface to prolong its life; but in the end, in spite of these expedients, the bag must be replaced.

Butyl compounds, on the other hand, do not harden on heat or oxygen aging, even in the presence of high-sulfur or accelerator concentrations. Because of this, Butyl is used widely as a tire-curing-bag material. The introduction of the Bag-O-Matic press is said to have been made possible by the availability of Butyl for making the diaphragm which takes the place of a conventional curing bag. Instead of becoming harder under the conditions to which a curing bag is subjected, Butyl usually becomes progressively softer. Consequently, in developing stocks to meet severe aging

FIG. 21.   Resistance of Butyl Rubber to Ozone Cracking

Semireinforcing black compounds after exposure to 0.015 per cent ozone for 17 hr.   From left to right, samples are Butyl, natural rubber, and GR-S

requirements, emphasis is placed on modulus loss after aging. Butyl stocks should be compounded to the highest initial stiffness compatible with the service for which they are intended. Some compounders prefer GR-I 25 for making curing bags because of the tighter cure that can be obtained.

Typical Butyl curing-bag stocks are illustrated by the following:

	1	2
Butyl (GR-I 15 or 25)	100	100
Zinc oxide	25	25
SRF black	25	25
HMF black	25	25
Sulfur	1.5	1
Selenium diethyldithiocarbamate	3	...
p-Quinonedioxime (GMF)	...	2
Benzothiazolyl disulfide	...	4
Softener	5	5

Electrical Insulation. Inherently good electrical properties and exceptional resistance to deterioration by heat, ozone (Fig. 21), and outdoor weathering (Fig. 22) have made Butyl attractive for widespread use in the electrical-insulation field. The largest outlet for Butyl in nontransport items, in fact, has been in the production of insulated wire and power cable. A recent and important development in the electrical field, that of using Butyl to make injection-molded transformers, offers promise of extending further the utility of the polymer.

Difficulties from two sources have so far retarded the use of Butyl in insulation commodities. These difficulties are those of getting good processing and accomplishing vulcanization at high speeds. On both problems progress has already been made. New types of fillers have been introduced on the market which aid greatly in the solution of extrusion difficulties, and modifications of compounding and processing techniques are making it more feasible to use Butyl in continuous vulcanization machines.

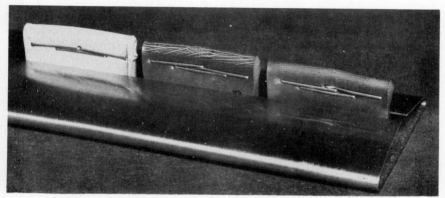

FIG. 22. Resistance of Butyl Rubber to Ultraviolet Light Aging

White mineral pigment compounds after exposure to ultraviolet light in fadeometer for 7 days. From left to right, samples are Butyl, natural rubber, and GR-S

Mechanical Goods. In the mechanical goods field Butyl is used for numerous purposes, either because its properties are unique or because it permits a definite advantage over competing materials such as natural rubber, GR-S, or other synthetics. Its heat resistance makes it particularly adaptable to the manufacture of steam hose and conveyor belts for carrying hot materials. In the hose and belt applications, the otherwise desirable property of impermeability to gases sometimes presents a problem in the development of building operations. Trapped air between the plies of the hose or belt must be removed before cure to prevent blister formation. Because Butyl holds air and other gases so well, it is used for making bladders for inflatable sports goods, carbon dioxide hose, and laboratory tubing.

Extruded and molded goods are also made from Butyl. Butyl is extremely flex-resistant and is therefore used in a wide variety of automobile parts that are subject to flexure. Also, Butyl is used in automobile parts that are exposed to continuous weathering. When compounded with carbon black, Butyl does not crack or otherwise deteriorate in sunlight (see Fig. 22).

Like all other rubber and plastic products, and indeed like any other material of commerce, Butyl has found and is still finding uses where, for either economical or utility reasons, it is specially advantageous. These uses extend virtually across the whole field of rubber products.

Acknowledgement. We are indebted to I. E. Lightbown of the Enjay Company, and to our Standard Oil Development Company associates, D. J. Buckley, A. W. Hubbard, J. Rehner, Jr., W. C. Smith, and R. L. Zapp for highly valued assistance in the preparation of this chapter.

POSTSCRIPT

Heat Treatment in Compounding. Since the above was written, a development of an unusual nature in the compounding of Butyl has been made public [A. M. Gessler, *Rubber Age N.Y.*, **74**, 59–71 (1953)]. It has been disclosed that vulcanizates of improved properties can be prepared from Butyl rubber by heating and mixing under prescribed conditions (before compounding for curing) masterbatches of the rubber and carbon black. Such treatment, it has been demonstrated, increases the reinforcing effect of carbon black in Butyl.

In laboratory work, for example, it has been possible to prepare vulcanizates showing increases in moduli (at extensions above 100 to 150 per cent) of approximately 100 per cent, tensile strength increases up to 20 per cent, and abrasion resistance enhancement as much as 40 per cent above the values normally obtained. These changes are accompanied by substantial improvement in resilience (reflected in a reduction of at least 50 per cent in absolute damping under specific conditions of measurement) and by a very large increase in electrical resistivity, e.g., from 10^7 to 10^{14} ohm centimeters.

The effects are attributed to carbon-polymer bonding, brought about by the heat treatment, and to improved dispersion resulting from the milling of agglomerates of black anchored to polymer by the bonding process. The idea that improved dispersion is involved is supported by an observation of increased resistance of the heat-treated vulcanizate to chemical attack. Three samples of a conventional Butyl vulcanizate containing 50 parts of MPC black and three corresponding samples that had been heat treated before curing were immersed in 96 per cent sulfuric acid for 7 days at room temperature. On removal from the acid, the control samples were greatly swollen and split into layers, whereas the heat-treated samples showed little evidence of attack. Indications are that the black in the control samples was less adequately dispersed than in the heat-treated samples and served as a wick to imbibe acid.

These favorable effects are developed with Butyl only because of its low degree of unsaturation and inherent stability, which permit it to withstand without apparent deterioration the high temperatures (up to 450° F.) of mixing involved.

Blacks other than channel black do not respond to heat treatment with Butyl unless modifications are made either in the black or in the nature of the compound. The enhancement of physical properties by heat treatment appears to depend on the presence of chemi-sorbed oxygen on the surface of the black [R. L. Zapp and A. M. Gessler, *Rubber Age N.Y.*, **74**, 243–51 (1953)]. Furnace blacks to which oxygen has been added respond readily to heat treatment, yielding Butyl vulcanizates of high modulus, improved tensile strength, and improved resilience. X-Ray patterns given by stretched vulcanized Butyl show that the heat treated compositions containing oxygenated carbon blacks are capable of superior molecular alignment. It is possible that more orientation at a given elongation, attributed to better pigment-polymer bonding, accounts for the superior tensile and dynamic properties.

Furnace and thermal blacks, even though their surfaces are practically devoid of oxygen, can be effectively heat-treated in Butyl with the aid of certain chemical promoting agents. [A. M. Gessler and F. P. Ford, *Rubber Age N.Y.*, **74**, 397–408 (1953).] These include sulfur, *p*-dinitrosobenzene (Polyac), and *p*-quinone dioxime (GMF). Conditions required for getting the maximum effect depend on both the agent used and the black. It is thought to be desirable in all cases to have the concentration of promoting agent at a level conducive to maximum polymer-black bonding and minimum polymer-polymer cross linkage.

Technique of Heat Treatment. On a laboratory scale, heat treatment may be conducted by alternately heating and milling a Butyl-carbon black mix in repeated cycles. A typical cycle consists of a 30-minute heat at 320° F. in an open-steam vulcanizer, followed by 5 minutes of milling on a cool laboratory mill with the rolls set to allow a free rolling bank (0.030 to 0.040-in. roll clearance). As shown in Fig. 23, moduli (above about 100 per cent elongation) increase sharply in the resulting vulcanizates with the first few cycles of treatment; the moduli continue to rise with additional cycles but on a basis of diminishing returns.

For larger scale work, a more practical although less effective method of heat treatment, viz., treatment in a Banbury mixer, has been employed. Conditions required have been found to vary from plant to plant, but the initial ingredients used are typically as follows.

Butyl (preferably GR-I 15 or GR-I 17)	100 parts
Carbon black (preferably MPC)	50 parts
Stearic acid	0.5 part
Polyac (*p*-dinitrosobenzene on clay)	0.5 part

Generally speaking, these ingredients are mixed together in a Banbury, with adjustment of the charge and the cooling water such that the temperature builds up steadily to about 380° F. after 10 minutes' elapsed time. Mixing is continued for another 10 minutes in the range of 380 to 450° F., and the batch is dropped at 450° F. If possible, it is desirable to let the batch "heat soak" for an additional 10 to 20 minutes before sheeting out on a mill. As experience is gained in any given plant, conditions can be set to shorten the schedule. Batches of more than 1000 lb have, for example, been made in a 15-minute cycle in a No. 27 Banbury. At the end of each run the masterbatch was allowed to stand hot for 15 minutes before milling. This technique gave a change in modulus similar to that of approximately 6 laboratory heat cycles.

Influence of Polymer Unsaturation. The responsiveness of Butyl rubber to heat treatment with blacks is highly dependent on the amount and nature of the polymer unsaturation [J. Rehner and A. M. Gessler, *Rubber Age N.Y.*, **74**, 561–6 (1954)]. Heat treatment appears to increase the number of carbon black-polymer bonds through some mechanism involving the double bonds in the polymer chain rather than through an enhancement of van der Waal's forces.   In a conventionally mixed compound, the modulus of a given Butyl vulcanizate system increases with the molecular weight of the polymer used.   Heat treatment, in all instances observed, raises modulus to a common level regardless of the molecular weight of the original polymer.

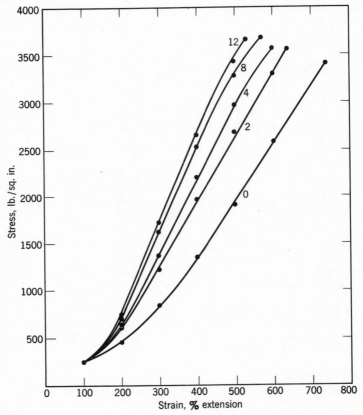

FIG. 23.   Effect of Cyclic Heat Treatment on Stress-Strain Properties of Butyl Loaded with MPC Black (50 parts)

The numerals show the number of cyclic treatments

It thus appears that heat treatment becomes less efficient with increase in polymer molecular weight. This is explained on the basis that longer chains are subject to a greater amount of entanglement or other restraint, which reduces the frequency with which reactive sites on the chain can collide with carbon black surfaces.

Compounding with Non-Black Fillers. In a recent study by W. C. Smith [*India Rubber World*, **129,** 55–60 (1953)] of Butyl compounded with various non-black fillers, a comparison is given of various curative schemes adapted to curing in steam, in air under pressure, and in hot air at ordinary pressure, and also adapted to continuous vulcanization at 400° F. The study also gives a comparison of the effect of various non-black fillers on the processibility, gas permeability, and ozone resistance of Butyl.

REFERENCES

1. Adams, R. J., Buckler, E. J., and Wanless, G. G., *Proc. Second Rubber Technol. Conf.*, 34–46 (1948); *Rubber Chem. and Technol.*, **23**, 670–82 (1950).
2. Alfrey, T., and Goldfinger, G., *J. Chem. Phys.*, **12**, 205–9 (1944).
3. Am. Soc. Testing Materials, 1946 Standards, Part III-B, D 833-46T, Sect. 17, 969–76.
4. Amerongen, G. J. van, *J. Applied Phys.*, **17**, 972–85 (1946).
5. Andrews, R. D., Hofman-Bang, N., and Tobolsky, A. V., Private Communication, 1947.
6. Anthony, R. L., Caston, R. H., and Guth, E., *J. Phys. Chem.*, **46**, 826–40 (1942).
7. Armstrong, R. T., Little, J. R., and Doak, K. W., *Ind. Eng. Chem.*, **36**, 628–33 (1944).
8. Baker, W. O., Private Communication, 1942.
9. Baldwin, F. P., Esso Laboratories, Unpublished Data.
10. Baldwin, F. P., U.S. Pat. 2,493,518 (1950).
11. Baldwin, F. P., Presented before Meeting-in-Miniature, New Jersey Sect. Am. Chem. Soc., Jan. 1949.
12. Baldwin, F. P., and Thomas, J. A., Esso Laboratories, Unpublished Data.
13. Baldwin, F. P., Turner, L. B., and Zapp, R. L., *Ind. Eng. Chem.*, **36**, 791–5 (1944).
14. Ball, J. M., and Maassen, G. C., Symposium on Application of Synthetic Rubbers, Am. Soc. Testing Materials, 27–38 (1944).
15. Bardwell, J., and Winkler, C. A., *India Rubber World*, **118**, 509–12 (1948).
16. Barnes, C. E., *J. Am. Chem. Soc.*, **67**, 217–20 (1945).
17. Bekkedahl, N., *J. Research Natl. Bur. Standards*, **43**, 145–56 (1949).
18. Bergstrom, P., and Reid, E. E., *Ind. Eng. Chem., Anal. Ed.*, **1**, 186–7 (1929).
19. Bolland, J. L., and Gee, G., *Trans. Faraday Soc.*, **42**, 236–43 (1946).
20. Brill, R., and Halle, F., *Naturwissenschaften*, **26**, 12 (1938).
21. Brown, H. C., and Barbaras, G. K., *J. Chem. Phys.*, **14**, 114 (1946).
22. Brown, J. R., and Hauser, E. A., *Ind. Eng. Chem.*, **30**, 1291–6 (1938).
23. Buckler, E. J., *Trans. Instn. Rubber Ind.*, **24**, 52–8 (1948).
24. Buckley, D. J., Marshall, E. T., and Vickers, H. H., *Ind. Eng. Chem.*, **42**, 2407–13 (1950).
25. Buist, J. M., *Trans. Instn. Rubber Ind.*, **20**, 155–72 (1945).
26. Bulgin, D., *Trans. Instn. Rubber Ind.*, **21**, 188–218 (1945).
27. Bunn, C. W., in H. Mark and G. S. Whitby, *Advances in Colloid Science*, Vol. II, Interscience, New York, 1946, pp. 125–8.
28. Burchfield, H. P., *Ind. Eng. Chem., Anal. Ed.*, **17**, 806–10 (1945).
29. Butyl Manual, Enjay Co., New York, 1948.
30. Chilton, E. G., *J. Applied Phys.*, **17**, 492–5 (1946).
31. Cohan, L. H., and Mackey, J. F., *Rubber Age N.Y.*, **55**, 583–5 (1944).
32. Cohan, L. H., Sohn, M., and Steinberg, M., *Ind. Eng. Chem., Anal. Ed.*, **16**, 562 (1944).
33. Cohan, L. H., and Steinberg, M., *Ind. Eng. Chem.*, **36**, 7–15 (1944).
34. *Compounding Ingredients for Rubber*, compiled by the editors of *India Rubber World*, 2d. ed., 1947.
35. Copeland, L. E., *J. Applied Phys.*, **19**, 445–9 (1948).
36. Corwin, A. H., and Karr, C., *Anal. Chem.*, **20**, 1116 (1948).
36a. Currie, L. L., *Anal. Chem.*, **24**, 1327–30 (1952).
37. Dainton, F. S., and Sutherland, G. B. B. M., *J. Polymer Sci.*, **4**, 37–43 (1949).
38. Dart, S. L., and Guth, E., *J. Chem. Phys.*, **13**, 28–36 (1945).
39. Davis, D. W., *Paper Trade J.*, **123**, No. 4, 33–40 (1946).
40. Drogin, I., *India Rubber World*, **106**, 561–9 (1942).
40a. Drogin, I., *India Rubber World*, **127**, 646–50 (1953).

41. Eby, L. T., and Buckley, D. J., Presented before Div. Rubber Chem., Am. Chem. Soc. Meeting, Chicago, Apr. 1948.
42. Esso Laboratories, Unpublished Data.
43. Evans, A. G., and Polanyi, M., *Nature*, **152**, 738–40 (1943).
44. Farmer, E. H., and Sundralingam, A., *J. Chem. Soc.*, **1943**, 125–33.
45. Ferry, J. D., and Parks, G. S., *J. Chem. Phys.*, **4**, 70–5 (1936).
46. Field, J. E., *J. Applied Phys.*, **12**, 23–34 (1941).
47. Fielding, J. H., *Ind. Eng. Chem.*, **29**, 880–5 (1937).
48. Fischer, H., and Orth, H., *Die Chemie des Pyrrols*, Vol. I, Akad. Verlag, Leipzig, 1934, p. 66.
49. Flory, P. J., Unpublished Results.
50. Flory, P. J., Esso Laboratories, Unpublished Results, 1942.
51. Flory, P. J., *J. Chem. Phys.*, **10**, 51–61 (1942).
52. Flory, P. J., *J. Am. Chem. Soc.*, **65**, 372–82 (1943).
53. Flory, P. J., *Chem. Rev.*, **35**, 51–75 (1944).
54. Flory, P. J., *Ind. Eng. Chem.*, **38**, 417–36 (1946).
55. Flory, P. J., and Rehner, J., *J. Chem. Phys.*, **11**, 521–6 (1943).
55a. Ford, F. P., and Mottlau, A. Y., *Rubber Age N.Y.*, **70**, 457–63 (1952); Ford, F. P., and Gessler, A. M., *Ind. Eng. Chem.*, **44**, 819–24 (1952).
56. Fuller, C. S., Frosch, C. J., and Pape, N. R., *J. Am. Chem. Soc.*, **62**, 1905–13 (1940).
57. Gallo, S. G., Wiese, H. K., and Nelson, J. F., *Ind. Eng. Chem.*, **40**, 1277–80 (1948).
58. Galloway, P. D., and Wake, W. C., *Analyst*, **71**, 505–10 (1946).
59. Gee, G., *Trans. Instn. Rubber Ind.*, **18**, 266–81 (1943).
60. George, P., and Walsh, A. D., *Trans. Faraday Soc.*, **42**, 94–7 (1946).
61. Hamill, W. H., Mrowca, B. A., and Anthony, R. L., *Ind. Eng. Chem.*, **38**, 106–10 (1946).
62. Hanson, A. C., *Ind. Eng. Chem.*, **34**, 1326–7 (1942).
63. Haworth, J. P., U.S. Pat. 2,393,321, Jan. 22, 1946.
64. Haworth, J. P., *Ind. Eng. Chem.*, **42**, 2314–9 (1948).
65. Haworth, J. P., and Baldwin, F. P., *Ind. Eng. Chem.*, **34**, 1301–8 (1942).
66. Hildebrand, J. H., *Solubility*, Reinhold, New York, 1936, p. 73.
67. Houwink, R., *Ind. plastiques*, **3**, 409–14 (1947).
68. Hubbard, A. W., Esso Laboratories, Unpublished Data.
69. Huggins, M. L., *Ann. N.Y. Acad. Sci.*, **43**, 1–32 (1942).
69a. Iknayan, A. N., *India Rubber World*, **126**, 505–7 (1952).
70. Ivey, D. G., Mrowca, B. A., and Guth, E., *J. Applied Phys.*, **20**, 486–92 (1949).
71. James, H. M., and Guth, E., *J. Chem. Phys.*, **11**, 455–81 (1943).
72. Kolthoff, I. M., and Ligone, J. J., *Polarography*, Interscience, New York, 1941, pp. 271–3.
73. Kuhn, W. J., *J. Polymer Sci.*, **1**, 380–8 (1946).
74. Lecher, H., *Ber.*, **53**, 591–3 (1920).
75. Lee, T. S., Kolthoff, I. M., and Johnson, E., *Anal. Chem.*, **22**, 995–1001 (1950).
76. Lightbown, I. E., *Rubber Age N.Y.*, **51**, 377–80 (1942).
77. Lightbown, I. E., Verde, L. S., and Brown, J. R., *Ind. Eng. Chem.*, **39**, 141–6 (1947).
78. Louth, G. D., *Anal. Chem.*, **20**, 717–9 (1948).
79. Madorsky, S. L., Straus, S., Thompson, D., and Williamson, L., *J. Research Natl. Bur. Standards*, **42**, 499–514 (1949).
80. Malesoff, W. M., and Andring, C. E., *Ind. Eng. Chem., Anal. Ed.*, **7**, 86–8 (1935).
81. Mayo, F. R., and Lewis, F. M., *J. Am. Chem. Soc.*, **66**, 1594–1601 (1944).
82. McGovney, C. L., and Zapp, R. L., Esso Laboratories, Unpublished Data.
83. Medvedev, S., and Zeitlin, P., *Acta Physicochim. U.R.S.S.*, **20**, 3–30 (1945).
84. Mesrobian, R., and Tobolsky, A., *J. Am. Chem. Soc.*, **67**, 785–7 (1945).
85. Mesrobian, R., and Tobolsky, A., *J. Polymer Sci.*, **2**, 463–87 (1947).
86. Ministry of Supply, Users' Memorandum No. U9, Identification and Estimation of Natural and Synthetic Rubbers, London, 1944.

87. Morris, R. E., Mitton, P., Montermoso, J. C., and Werkenthin, T. A., *Ind. Eng. Chem.*, **35**, 646–9 (1943).
88. Morron, J. D., Knapp, R. C., Linhorst, E. F., and Viohl, P., *India Rubber World*, **110**, 521–5 (1944).
89. Mrowca, B. A., Dart, S. L., and Guth, E., *J. Applied Phys.*, **16**, 8–19 (1945).
90. Parker, L. F. C., *J. Soc. Chem. Ind.*, **63**, 378–9 (1944).
91. Parker, L. F. C., *J. Soc. Chem. Ind.*, **64**, 65–7 (1945).
92. Rehner, J., Esso Laboratories, Unpublished Results, 1942–3.
93. Rehner, J., *Ind. Eng. Chem.*, **36**, 46–51 (1944).
94. Rehner, J., *Ind. Eng. Chem.*, **36**, 118–24 (1944).
95. Rehner, J., *J. Polymer Sci.*, **1**, 225–8 (1946).
96. Rehner, J., *J. Polymer Sci.*, **2**, 263–74 (1947).
97. Rehner, J., and Flory, P. J., *Ind. Eng. Chem.*, **38**, 500–6 (1946).
98. Rehner, J., and Gray, P., *Ind. Eng. Chem., Anal. Ed.*, **17**, 367–70 (1945).
99. Rehner, J., and Holowchak, J., *Ind. Eng. Chem., Anal. Ed.*, **16**, 98 (1944).
100. Rehner, J., and Robison, S. B., Esso Laboratories, Unpublished Data.
101. Reitlinger, S. A., *J. Gen. Chem. U.S.S.R.*, **14**, 420–7 (1944); *Rubber Chem. and Technol.*, **19**, 385–91 (1946).
102. Richardson, J. W., and Parks, G. S., *J. Am. Chem. Soc.*, **61**, 3543–6 (1939).
103. Rossini, F. D., quoted by Flory, P. J., *J. Am. Chem. Soc.*, **65**, 372–82 (1943).
104. Rostler, F. S., and White, R. M., *Rubber Age N.Y.*, **61**, 313–21 (1947).
105. Roth, F. L., and Wood, L. A., *J. Applied Phys.*, **15**, 749–57 (1944).
106. Ruggli, P., and Petitjean, C., *Helv. Chim. Acta*, **21**, 711–32 (1938).
107. Smith, W. C., Esso Laboratories, Unpublished Data.
108. Stambaugh, R. B., *Phys. Rev.*, **65**, 250 (1944).
109. Staudinger, H., *Ber.*, **58**, 1075–9 (1925).
110. Stevens, H. P., *India-Rubber J.*, **108**, 9–10, 12–3, 35–8, 65–6, 68–9, 91–3, 122, 124–5 (1945).
111. Thomas, R. M., Lightbown, I. E., Sparks, W. J., Frolich, P. K., and Murphree, E. V., *Ind. Eng. Chem.*, **32**, 1283–92 (1940).
112. Thomas, R. M., and Slotterbeck, O. C., U.S. Pat. 2,243,658, May 27, 1941.
113. Thomas, R. M., and Sparks, W. J., U.S. Pat. 2,356,128, Aug. 22, 1944.
114. Thomas, R. M., and Sparks, W. J., U.S. Pat. 2,356,130, Aug. 22, 1944.
115. Thomas, R. M., Sparks, W. J., Frolich, P. K., Otto, M., and Mueller-Cunradi, M., *J. Am. Chem. Soc.*, **62**, 276–80 (1940).
116. Thompson, H. W., and Torkington, P., *Trans. Faraday Soc.*, **41**, 246 (1945).
117. Tobolsky, A. V., Prettyman, I. B., and Dillon, J. H., *J. Applied Phys.*, **15**, 380–95 (1944).
118. Terloar, L. R. G., *Trans. Faraday Soc.*, **39**, 36–41 (1943).
119. Turner, L. B., and Cunningham, E. N., *Product Eng.*, **16**, 764–9 (1945).
120. Turner, L. B., Haworth, J. P., Smith, W. C., and Zapp, R. L., *Ind. Eng. Chem.*, **35**, 958–63 (1943).
121. Vasil'ev, A. A., *J. Gen. Chem. U.S.S.R.*, **17**, 923–8 (1947); *Rubber Chem. and Technol.*, **22**, 287–98 (1949).
122. Wall, F. T., *J. Chem. Phys.*, **10**, 485–8 (1942).
123. Wall, F. T., 1945, Private Communication.
124. Wall, L. A., *J. Research Natl. Bur. Standards*, **41**, 315–22 (1948).
125. Wiegand, W. B., *Can. Chem. Process Ind.*, **25**, 579–81 (1941).
126. Welch, L. M., Nelson, J. F., and Wilson, H. L., *Ind. Eng. Chem.*, **41**, 2834–40 (1949).
127. Witte, R. S., Mrowca, B. A., and Guth, E., *J. Applied Phys.*, **20**, 481–5 (1949).
128. Wolf, R. F., and Gage, F. W., *India Rubber World*, **123**, 565–9 (1951).
129. Zapp, R. L., *Ind. Eng. Chem.*, **36**, 128–33 (1944).
130. Zapp, R. L., *Ind. Eng. Chem.*, **40**, 1508–17 (1948).
131. Zapp, R. L., and Ford, F. P., *J. Polymer Sci.*, **9**, 97–113 (1952).
132. Zapp, R. L., Decker, R. A., Dyroff, M. S., and Rayner, H. A., *J. Polymer Sci.*, **6**, 331–48 (1951).
133. Zapp, R. L., and Guth, E., *Ind. Eng. Chem.*, **43**, 430–8 (1951).

CHAPTER 25

MISCELLANEOUS SYNTHETIC ELASTOMERS

C. H. Fisher, G. S. Whitby, and E. M. Beavers

Eastern Regional Research Laboratory, University of Akron, and Rohm & Haas Company

Described in what follows are several specialty elastomers not given consideration elsewhere in this volume. They include elastothiomers, polyacrylates, polyesters, silicones, and polymers in which a variety of ester, amide, and urethane linkages occur. The last group has been identified as "diisocyanate-linked condensation elastomers." These various specialty elastomers differ from the more conventional rubbers in several respects: They are not based on conjugated dienes; their preparation mostly depends on condensation polymerization, although one group, the polyacrylates, depends on the addition polymerization of vinyl monomers; they are mostly saturated and hence require special methods for their vulcanization or cross-linking.

POLYSULFIDE ELASTOMERS

Although reactions between ethylene dihalides and alkali sulfides were studied as early[33, 86, 105] as 1839, the possibility of obtaining elastomers through them was not appreciated until the 1920's. In 1920, Patrick allowed ethylene dichloride to react with sodium polysulfide, in an attempt to prepare ethylene glycol.[9, 43, 145] The product was a more or less rubbery solid, insoluble in most organic liquids. The Thiokol Corporation, organized by Bevis Longstreth, offered the new polysulfide rubber to industry in 1930 under the name Thiokol. Despite certain drawbacks, especially odor, the production of Thiokols increased rapidly[43, 145]—largely because of their outstanding oil resistance—from 2 tons in 1930 to 700 tons in 1940 and about 2400 tons in 1949. Since the first production of Thiokol, Patrick and his associates in the Thiokol Corporation[37, 44, 78, 87, 90–3, 113–5, 141–4] have investigated extensively and systematically the reaction of alkali polysulfides with a variety of organic dihalides, with the result that a number of different types of polysulfide rubbers have been introduced commercially.

In 1938 the Dow Chemical Corporation undertook the manufacture of Thiokols A and FA for the Thiokol Corporation; this activity was terminated in 1947.[43] During the war, the Dunlop Rubber Company in England and Naugatuck Chemical Limited in Canada manufactured Thiokol under license. In 1942 the Baruch Committee recommended, as an emergency measure, that up to 60,000 long tons of GR-P, a polysulfide rubber, should be manufactured annually in the United States, to provide material for recapping civilian tires (cf. p. 45). For various reasons, including success in collecting adequate supplies of scrap for reclaiming, the original plan was never fully executed and in fact did not proceed very far.

At about the time of Patrick's early work,[113] Baer in Switzerland developed similar elastothiomers[7] that were later commercialized in Germany by I.G. Farbenindustrie. Polysulfide rubbers are being or have been manufactured in several other countries.

PREPARATION

The basic reaction in the preparation of polysulfide elastomers is the condensation of an aliphatic dihalide and sodium polysulfide. Both reactants being bifunctional, they give rise to long-chain polycondensation products, thus:

$$x\mathrm{HalRHal} + x\mathrm{Na_2S}_n \longrightarrow (-\mathrm{RS}_n-)_x + 2x\mathrm{NaHal}$$

The dihalide is added slowly, with vigorous agitation, to an aqueous solution of sodium polysulfide, which may be made from caustic soda and sulfur or otherwise. The specific gravity of the polysulfide solution is adjusted to approximately the density of the halide. It is preferable to employ a dispersing agent (magnesium hydroxide) to facilitate reaction and, if latex is desired as the final product, to assist dispersion. Excess sodium polysulfide is used for economy and because a viscous liquid instead of rubber is obtained if the halide is in excess. The reaction between aliphatic dihalides and sodium polysulfides, which is exothermic, requires 2 to 6 hours and is best carried out at about 70° C. At the end of the reaction, the suspension is allowed to settle, and the product is washed free from sodium chloride and excess sodium polysulfide.

The "rank" of the polysulfide used, i.e., the value of n in $\mathrm{Na_2S}_n$, varies in different types of polysulfide elastomers, as indicated later. A considerable variety of dihalides may be used. (Even mustard gas, $\mathrm{Cl(CH_2)_2S(CH_2)_2Cl}$, has been shown to yield a polysulfide elastomer when reacted with sodium polysulfide, although more useful products are obtained by replacing part of it with the corresponding sulfoxide.[1]) Dihalides which are or have been used commercially in the production of polysulfide rubbers are (1) ethylene dichloride, $\mathrm{Cl(CH_2)_2Cl}$, (2) propylene dichloride, $\mathrm{ClCH_2CH(CH_3)Cl}$, (3) bis(2-chloroethyl) ether, $\mathrm{Cl(CH_2)_2O(CH_2)_2Cl}$, (4) di-2-chloroethyl formal, $\mathrm{Cl(CH_2)_2OCH_2O(CH_2)_2Cl}$, (5) 1,3-glycerol dichlorhydrin, $\mathrm{ClCH_2CH(OH)CH_2Cl}$.

Certain variations on the basic reaction, viz. (*a*) the inclusion of a proportion of a trifunctional halies in the reaction mixture, and (*b*) controlled scission of sulfide linkages in the polymer, will be discussed later.

STRUCTURE

The regular linearity of their structure is such that, on being stretched, some of the elastothiomers give X-ray diffraction patterns indicative of a more or less crystalline fiber structure. The pattern shows that in polyethylene polysulfide the identity period along the fiber axis is 4.32 A. and that there is one $\mathrm{C_2H_4S_4}$ unit per unit cell.[104]

Treatment of the elastic polysulfide derived from ethylene dichloride and sodium tetrasulfide with caustic alkali removes 2 atoms of sulfur per polymer unit and yields a product, $(C_2H_4S_2)_x$, which, although of high molecular weight, lacks rubbery properties. Patrick[93, 113] has suggested the following formulas for the two materials:

$$HS(CH_2)_2S \cdot S(CH_2)_2S \cdot S— \text{ (rubbery)} \quad HS(CH_2)_2S \cdot S(CH_2)_2S \cdot S—$$
$$\underset{S\ \ S}{\|\ \|} \quad \underset{S\ \ S}{\|\ \|} \qquad \text{(not rubbery)}$$

If the latter material is treated with sulfur by passage between hot rolls, it can be reconverted to the original rubbery material. For such reconversion it is not necessary to bring the sulfur content up to its original level; half an atom of sulfur per $(C_2H_4S_2)$ unit is sufficient. The added sulfur is insoluble in carbon disulfide.

By treating the above polydisulfide with warm sodium sulfide, the disulfide linkages can be completely cleaved and reduced, with the formation of the simple dimercaptan, $HS(CH_2)_2SH$. The latter, in turn, can, by means of alkaline oxidizing agents, be reconverted to a product closely similar to the nonrubbery polydisulfide, which, as already stated, can, by treatment with sulfur, be transformed back again to the original, rubbery polypolysulfide.

The state of affairs with some of the other polysulfide polymers is not in all respects parallel to that just outlined. Not only the polytetrasulfides, $—RS_4—$, but also the polydisulfides, $—RS_2—$, are rubbers when R, instead of being $(CH_2)_2$, is longer and is represented by $(CH_2)_5$, $(CH_2)_2O(CH_2)_2$, $(CH_2)_2OCH_2O(CH_2)_2$, or $(CH_2)_2S(CH_2)_2$.

Patrick supposed that, as shown in the formulas given above for the polyethylene polysulfides, the molecular chains of the polysulfides terminate in thiol groups. Fettes and Jorczak, however, think it more likely that the chains terminate in hydroxyl groups, arising in the course of the preparation of the polysulfides from slight hydrolysis of the aliphatic dihalides by the highly alkaline solution of sodium polysulfide.[44]

TYPES OF POLYSULFIDE ELASTOMERS[44, 78]

The following is a review of the various types of polysulfide elastomers which are now or have been manufactured commercially. Variables that may be involved in producing the different types are (1) the nature of the dihalide, (2) the use of a mixture of dihalides, (3) the "rank" of the sodium polysulfide, (4) the inclusion of a small proportion of a trihalide, in order to cross-link the polymer chains, (5) controlled scission of the disulfide bonds in the initially produced polymer, in order to adjust its molecular weight.

The simplest type is represented by Thiokol A, derived from ethylene dichloride and sodium tetrasulfide, and composed of $—C_2H_4S_4—$ units. This, the original polysulfide rubber, is still made in the United States. Thiokol B, no longer made in the United States, was derived from di-2-chloroethyl ether and sodium tetrasulfide. The British Novaplas A and the German Perduren G are derived from the same dihalide. Thiokol D, no longer manufactured, was the polydisulfide made by treating the poly-tetrasulfide Thiokol B with sodium hydroxide. The German Perduren H is made from di-(2-chloroethyl) formal and the British Vulcaplas from 1,3-glycerol dichlorhydrin.

It may be advantageous to use a mixture of dihalides in the manufacture of polysulfide rubber. Thiokol FA, manufactured in the United States, is derived from a mixture of ethylene dichloride and di-2-chloroethyl formal by reaction with sodium polysulfide of the composition $Na_2S_{1.8}$. As compared with Thiokol A, it has markedly improved resistance to low temperatures and a less objectionable odor. Whereas polymers prepared mainly from ethylene dichloride harden at about $-20°$ F. and soften at about $+180°$ F., polymers from mixtures of ethylene dichloride and the formal are serviceable over a range approximately from -35 to $+300°$ F., the degree of resistance to temperature change being influenced by the ratio of the two dihalides. Polymers derived from the formal alone, such as Thiokol ST (*infra*) have a still wider range of serviceability—from, say, -65 to $+350°$ F.

Thiokol F, which presumably represented an earlier version of FA, is no longer manufactured. Its composition had not been announced, but the suggestion has been made that this type, like Thiokol FA, was derived from a mixture of ethylene dichloride and di-2-chloroethyl formal.

GR-P (Government Rubber—Polysulfide), made on a limited scale during the war, was a form of Thiokol N (no longer manufactured) and was based on the use of a mixture of ethylene dichloride and propylene dichloride. It is now considered to be a relatively inferior elastothiomer, and has not been commercialized. However, the polymers in Thiokol latex MX and Thiokol latex WD-6 are derived from mixtures of ethylene and propylene dichlorides. The proportion of propylene dichloride used in the preparation of the latter is greater than in the preparation of the former, with the result that the latter latex, unlike the former, deposits a continuous film on drying. Thiokol WD-6 is a 67/33 mixture of polyethylene disulfide and polypropylene disulfide.[93a]

In the preparation of several types of Thiokol a trihalide is included in the reaction mixture, for the purpose of producing a limited degree of cross-linking of the polymer chains and thus of reducing markedly the tendency of vulcanizates to undergo cold flow and suffer compression set. Thiokol ST, made from di-2-chloroethyl formal by reaction with sodium polysulfide of rank 2.25, and Thiokol PR-1, made[18] from a mixture of ethylene dichloride and the formal by reaction with sodium polysulfide of rank 2.0, are thus cross-linked. The nature of the cross-linking agent here used has not been announced, but it is presumably 1,2,3-trichloropropane. The liquid poly-mers, Thiokol LP-2 and LP-3, also are cross-linked, a mixture of di-2-chloroethyl formal and 1,2,3-trichloropropane in the molar ratio 98/2 being used in their preparation.

In addition to the use of a cross-linking agent in the preparation of the four types of Thiokol just mentioned, treatment causing chain scission is involved in their preparation. It has been known for a considerable time that the linear polysulfide elastomers, such as Thiokols A and FA, which cannot be broken down to a satisfactory degree of plasticity on a rubber mill, can be plasticized by the addition of small proportions of substances such as benzothiazolyl disulfide and tetramethylthiuram disulfide. And it is now recognized that in such plasticization scission of a proportion of the disulfide links occurs, thus reducing the molecular weight and converting the original polymer chains into smaller ones terminating in thiol groups. (On vulcanization, the disulfide linkages become regenerated.) Another and more recent method of bringing about such scission is by reductive cleavage of disulfide links by treating the polysulfide in latex form with sodium hydrosulfide and sodium sulfite.[114]

In preparing the solid polymers Thiokols ST and PR-1, treatment to effect chain scission is carried only to a point sufficient to confer millability. In the preparation of the newer, liquid polymers, treatment is carried much further—to a point where the molecular weight is so low that the products are fluid. The molecular weight of Thiokol LP-2 is 4000; that of LP-3 is 1000. The liquid polymers have special uses described later.

Controlled scission of the chains is also practised in connection with Thiokol latexes. It has already been indicated that latex MX fails to form a continuous film on drying. The latexes MF (containing polymer derived from a mixture of ethylene dichloride and dichloroethyl formal) and WD-2 (containing polymer derived from a mixture of the formal and trichloropropane) similarly fail. But, if these latexes are given a preliminary treatment with ammonium hydrosulfide or tetramethylthiuram disulfide, in order to bring about a certain amount of chain scission, they will deposit a continuous film. The softening effect diappears on aging at room temperature, presumably because of regeneration of disulfide links from the terminal thiol groups formed during the softening treatment. (It is of interest to note in passing a certain parallelism in the Thiokols and Neoprene in regard to chemical plasticization. In both classes of material thiuram disulfides function as plasticizers by acting on sulfide linkages in the polymer chains, these linkages in Neoprene being derived from the sulfur used as a modifier in its preparation (see Chapter 22).

Polysulfide rubbers of unstated composition which are or have been made abroad are: in Japan Thionite, Hydrite,[145] and Hikatol[102–3]; in Belgium Ethanite; and in Russia Resinit. It should be noted in passing that the designation Thiokol has, confusingly, been attached to a commercial butadiene-acrylonitrile rubber, viz., Thiokol RD.[6]

CURING

It has already been mentioned that polysulfides such as Thiokols A and FA, the nature of the terminal groups in which is unknown, are, as a preliminary step to compounding, first chemically plasticized on the mill, usually by the incorporation of a little benzothiazolyl disulfide, often along with a little diphenylguanidine, which promotes the softening action. The curing of such polymers is brought about by heating them with zinc oxide. The mode of action of the zinc oxide is uncertain but may be primarily that of an oxidizing agent serving to re-form disulfide linkages which during plasticization were reduced to thiol groups. Quantities of zinc oxide up to 10 parts per hundred of the polysulfide can be used. Amounts greater than this act merely as loading in the stock. Although a small amount of sulfur has an accelerating effect, it is seldom used, because it restricts the curing range. Cupric oxide and lead peroxide also may be used as curing agents. A trace of moisture is necessary for curing by metal oxides, which are supposed to function as mild oxidizing agents.

It has been stated that organic oxidizing agents, such as polynitro-benzenes and benzoyl peroxide, are effective as curatives[145] and that vulcanization is inhibited by reducing agents, such as pyrogallol and zinc dust. Further, the elastomer can be vulcanized by merely heating at about 140° C. In practice, zinc oxide is mostly relied on for curing polysulfides of the types mentioned above.

The types of Thiokol the molecules of which terminate in thiol groups can be vulcanized by a wide variety of oxidizing agents. For the solid Thiokols, ST and PR-1, zinc peroxide (say, 4 per cent) is an effective nondiscoloring vulcanizing agent, but a combination of zinc oxide (say, 0.5 per cent) and p-quinone dioxime (say, 1.5 per cent) is most generally used, because of the wide range of temperatures over which it is effective. For the liquid Thiokols, LP-2 and LP-3, the oxidant chosen for the purpose of curing depends on the kind of application to which the polymer is to be put. For many purposes lead peroxide (say, 5 to 10 per cent) is chosen. For mixing with the liquid polymer, it is conveniently used as a paste in dibutyl phthalate containing a little stearic acid. Cumene hydroperoxide along with an activating agent (diphenylguanidine or triphenylguanidine) is effective, and, for curing films and coatings of liquid polymer, there can be used a cobalt drier or a mixture of zinc peroxide and diphenylguanidine. The liquid polymers can also be cured by certain agents the action of which does not depend on oxidation of the terminal thiol groups. Thus, furfural will produce curing by linking polymer chains through thioacetal linkages.

COMPOUNDING

Regular rubber mills are suitably used to incorporate reinforcing and curing agents in the solid polysulfide polymers. Stearic acid (0.5 to 1.0 per cent) is used, to prevent sticking to mill rolls and molds and to facilitate the dispersion of pigments. Stocks with smaller quantities of stearic acid have better tack and ply adhesion than stocks with larger quantities. The compounding of the liquid polymers can be done on a three-roll paint mill, or in a colloid mill, ball mill, or internal mixer.

Semireinforcing and soft carbon blacks are used almost universally in the compounding of polysulfide polymers; they greatly enhance the physical properties of the vulcanizates. Channel blacks used alone have less rein-forcing effect, apparently because their acidity is unfavorable to the occur-rence of curing. Laurence and Perrins[84] observed that suitable mixtures of channel and semireinforcing furnace blacks produce better vulcanizates than either type of black alone. When the proportion of furnace black (alkaline) is sufficient to bring the pH to the alkaline side, a good cure is obtained, and the reinforcement is greater than that secured with furnace black alone

(Table I). Channel blacks will give satisfactory results if buffered by diphenylguanidine (Table I).

Table I. Influence of pH on the Curing of Thiokol FA Stocks Loaded with Blacks[84]

Thiokol FA	100	Benzothiazolyl disulfide	0.4
Zinc oxide	10	Diphenylguanidine	0.4
Stearic acid	0.5	Black	70

Black, Parts			pH of Black	Tensile Strength at
HPC	SRF	EPC	Slurry	Optimum Cure, P.S.I.
70	0	0	4.0	100
0	70	0	9.0	1385
35	35	0	7.0	1550
70*	0	0	...	1560
0	0	70	4.8	1350
0	35	35	9.3	1725

* 1 part diphenylguanidine.

Aside from carbon black, zinc oxide and certain other pigments can be used, but the choice of pigments is not so wide as with natural rubber. Curing times for the solid polysulfide polymers are in general the same as for natural-rubber stocks. The values for GR-P, the Government wartime grade of Thiokol, given in Table II, will serve to illustrate the composition, rate of cure, and tensile properties of a typical polysulfide rubber stock.

Table II. Vulcanization of GR-P[137]

GR-P	100	Stearic acid	0.5
Zinc oxide	10	Benzothiazolyl disulfide	0.4
Channel black	35	Diphenylguanidine	0.1

Cure at 287° F., min.	Modulus at 300%, P.S.I.	Modulus at 500%, P.S.I.	Tensile Strength, P.S.I.	Elongation, %
10	185	295	550	1160
20	380	595	920	840
30	555	850	1110	680
40	700	1070	1240	590
50	780	1210	1250	520
60	815	...	1250	480

The liquid polymers when compounded and cured become converted, without shrinking, to elastic solids, the tensile strength of which is not, however, so high as that of the solid polymers.[44] A stock consisting of Thiokol LP-2 (100), stearic acid (1), SRF black (50), and lead peroxide (8), after standing for 48 hours at room temperature and then heating in a press for 10 minutes at 310° F. gave a vulcanizate with a tensile strength of 870 p.s.i., an elongation of 470 per cent, a tear index of 169, and a hardness (Shore A) of 61.

PROPERTIES

The specific gravity of the polysulfide rubbers ranges from 1.33 to 1.38. The rubbers can be obtained in a wide range of hardness by proper compounding. From the standpoint of their usefulness, the polysulfide rubbers are in some properties excellent; in others quite poor. The rubbers have unexcelled solvent and oil resistance and gas impermeability. They possess good aging characteristics and exceptional resistance to ozone. Some types stiffen on aging, although this behavior is less marked in vulcanizates. Newer types, such as Thiokol FA, are an improvement in this respect over the older types, although some stiffening is still encountered in GR-P. Still newer types, such as Thiokol ST, retain rubbery properties down to −60° F. Stocks comparable in resilience to natural rubber can be made, and suitable flexibility also can be attained. Tensile strengths up to 1500 p.s.i. and elongations of 500 to 600 per cent can be obtained. These values are largely retained upon immersion in oil. The resistance of Thiokol FA to oils and solvents is illustrated by the data in Table III.

Table III.   Swelling of Cured Thiokol FA after 6 Months' Immersion at Room Temperature[91]

Liquid	Increase in Volume, %
Motor gasoline	1.5
Gasoline–benzol (50/50)	8
Linseed oil	0
Ethyl acetate	6.3
Acetone	6.8
Carbon tetrachloride	12.3

On the debit side, most of the elastothiomers have a disagreeable odor, and one of them, Thiokol A, liberates a lachrymatory substance on heating or milling. Some improvement in odor has been achieved in later types. Relatively poor heat resistance, lack of abrasion resistance in tire treads, and low tensile strength are other shortcomings of the materials. For some applications, of course, some of these characteristics are relatively unimportant or even advantageous, e.g., cold flow in gaskets and cements.

APPLICATIONS[44, 78, 140]

The most important uses of the elastothiomers depend largely on their excellent resistance to oils, solvents, and water, and their impermeability to gases. They are or have been used in gasoline hose in automobiles and service stations, in oil-loading hose; in printers' rolls and newspaper blankets; as the hydrocarbon-resistant lining of bulletproof tanks for airplanes, military portable fuel depots, and boxcars adapted for the transportation of fuel oil; in protective coatings; for caulking; and as adhesives and binders. Elastothiomers are used also in paint-spray hose, electric-cable covering, and miscellaneous diaphragms and gaskets. For the last application, paper coated with elastothiomers serves admirably under unfavorable conditions of solvent and temperature, plastic flow being in this case a valuable characteristic, which insures satisfactory sealing. Asbestos impregnated with Thiokol is used as a flexible seal for floating roofs of gasoline storage tanks. Thiokol putties, which may be prepared from Thiokol latex or from the liquid polymers, are largely employed to seal cracks and seams in aluminum airplane wings used as integral fuel cells, and to seal pressurized aircraft cabins. The low permeability of elastothiomers is of advantage in balloon fabrics.[145]

The polysulfides are available, not only as solids for rubber processing, but also as latexes, putties, molding powders, solutions, and as fluid polymers of low molecular weight. Thiokol latex has found several applications, one of the most important being in lining underground gasoline-storage tanks and concrete tankers. For such use, the concrete is sprayed first with water and then two coats of latex. The surface is lined with light cotton sheeting, and then additional coats of latex are applied. The latex has also been applied to lining steel tanks used for the storage of sour crude oil. The application to the protection of steel of latex of Thiokol WD-6, either alone or in conjunction with a synthetic-resin latex, is discussed by Massa, Colon, and Schurig,[93a] who also report the results of corrosion tests, in water and in sea-water, carried out on steel thus protected.

The liquid polymers have been used to impregnate leather[19] in the manufacture of oil seals. They can be used for casting, in which application they are compounded with ester-type plasticizers (in addition to particulate fillers) in order to increase their fluidity. They are used in the preparation of flexible molds. The liquid polymers can be used as cold-setting adhesives, and, for the purpose of protecting metals including aluminum and magnesium, as air-drying coating preparations.

The polysulfides are available also in powdered form for flame spraying with the Schori and Freeman equipment. The most important application of flame-sprayed Thiokol is in coating rudders, struts, and steel shafts fitted with bronze propellors for use in mine sweepers, destroyer escorts, and other vessels. Corrosion caused by cavitation and electrolytic action between dissimilar metals in sea water is thus eliminated. Other uses of the flame-sprayed material include linings of salt-water pipe-line valves, bearings, and low-pressure condenser plates.

ACRYLIC ELASTOMERS

INTRODUCTION

A new type of specialty elastomer based on polymers of methyl or ethyl acrylate was announced in 1944 by the Department of Agriculture's Eastern Regional Research Laboratory.[101] The preferred method of making the elastomer comprised copolymerizing ethyl acrylate with about 5 per cent of a chlorine-containing monomer, such as 2-chloroethyl acrylate or 2-chloroethyl vinyl ether, and vulcanizing the resulting copolymer with sulfur and suitable accelerators. The chlorine was introduced into the polymer molecule to facilitate vulcanization.

Somewhat similar elastomers were described[5, 6, 119, 133-4] in 1946. The product introduced by the B. F. Goodrich Company[52] was a halogen-free polyacrylic ester capable of being vulcanized or cross-linked with either sodium metasilicate pentahydrate or litharge used in conjunction with mixed ethyl and dimethyl mercaptothiazoles.

Being available in semicommercial quantities, 2-chloroethyl vinyl ether[121] was adopted for the preparation of the chlorine-containing type of acrylic elastomer. The 95/5 ethyl acrylate-chloroethyl vinyl ether copolymer, called Lactoprene EV, was found to be much more amenable to vulcanization than ethyl polyacrylate. Suitable curing agents reported for the new elastomer included amines and sulfur in conjunction with many of the conventional accelerators,[40, 94-7] sulfur, tetramethylthiuram monosulfide, and Trimene Base being used most frequently.[53, 97, 99, 100]   (Trimene Base is a reaction product, of indefinite structure, from ethyl chloride, formaldehyde, and ammonia.)

Lactoprene EV has been made experimentally in pilot plants at one time by the B. F. Goodrich Company, The University of Akron's Government Laboratories,[2] and the Eastern Regional Research Laboratory.[71] The B. F. Goodrich Chemical Company began the manufacture of a product similar to Lactoprene EV in 1948, renaming it Hycar PA21.[46, 53]

The acrylic elastomers are of particular interest because of their outstanding ability to withstand the deteriorating effects of sustained high temperatures (up to 400° F.) in air and various nonaqueous immersion media. The acrylic elastomers are much superior in heat resistance to all the commercial rubbers with the exception of the silicones. They are of interest also because of their resistance to flexural breakdown, compression set, ultraviolet light, ozone, mineral oils, and gas diffusion; their compatibility with plasticizers[99] and other elastomers,[95] and the electrical properties of specially compounded stocks prepared from them.[95] Because of their usefulness for certain specialty applications, both the solid and the latex forms of the acrylic elastomers have been well received by industry.[46, 54]

PREPARATION

The monomeric acrylic esters required for the preparation of the acrylic elastomers are made commercially from ethylene cyanohydrin.[109] The dehydration, hydrolysis, and esterification involved in the preparation are effected in a single operation by heating the cyanohydrin with alcohol and sulfuric acid.

Methyl acrylate can be made from lactic acid,[45] which is obtainable by fermentation from materials such as molasses, whey, starch hydrolyzates, and sulfite waste liquor. Pyrolysis of the acetyl derivative of methyl lactate at about 500° C. yields methyl acrylate and acetic acid.[20, 120]

The new Reppe synthesis[13, 109, 121a] has been modified and used for the manufacture of acrylic esters directly from acetylene. In this process acetylene and an alcohol are treated with a metal carbonyl or with carbon monoxide in the presence of metal compounds capable of forming carbonyls:

$(C_2H_2 + CO + ROH \longrightarrow CH_2 : CHCOOR)$

It is of considerable interest that the three efficient methods outlined above for making acrylic esters involve the use of petroleum, coal, and by-product carbohydrates as raw materials. The world seems assured of a permanent supply of acrylic esters and their polymers.

Acrylic esters have been manufactured in the United States by the Rohm & Haas Company since about 1933. In 1949 the price of the ethyl ester (colorless liquid boiling at 100° C.) dropped several cents to approximately 50 cents per pound. The year 1949 was marked also by the entry into the field of a second manufacturer, Carbide & Carbon Chemicals Corporation. Because of the great promise of acrylic esters as monomers and chemical intermediates[127] and on the basis of statements of the manufacturers, additional price reductions were to be anticipated. In 1953 the price was 42 cents a pound.

The acrylic esters polymerize readily under the influence of heat, light, and peroxide-type catalysts, such as hydrogen peroxide, benzoyl peroxide, cumene hydroperoxide, ammonium persulfate, and sodium perborate. In general, the preparation of acrylic elastomers is much simpler than that of the diene rubbers. The emulsion polymerization of acrylic esters[100] is complete in about 3 hours, auxiliary agents such as regulators and shortstops are not used, the operation is conducted at atmospheric pressure, and over 90 per cent of the monomer mixture is converted into polymer.

A simplified method, called granulation polymerization,[97] has been recommended for the preparation of ethyl acrylate polymers. The method comprises polymerizing in a rugged mixer about 8 parts of monomer in the presence of 1 part water and, as a polymerization initiator, a little potassium persulfate. The use of emulsifiers and other agents is avoided, and the polymer is obtained directly in granular, easily handled form.

VULCANIZATION METHODS

Although polyacrylic esters are substantially saturated, nevertheless they contain reactive groups (ester and active α-hydrogen), which can be used for vulcanization with special recipes. It has been reported, for example, that ethyl polyacrylate can be vulcanized with the following: amines,[59, 97] benzoyl peroxide,[101] oxides and hydroxides of certain bivalent metals,[6, 97] glycols,[5] quinone dioxime and red lead,[101] litharge and alkyl mercaptothiazoles,[52] sodium metasilicate,[52] sodium metastannate or orthovanadate,[52] litharge,[135] litharge and an aniline-butyraldehyde product,[97] and 2,6-dichloroquinone chloroimide and lead chromate.[97]

Although a variety of special agents are thus capable of vulcanizing ethyl polyacrylate, both the number of suitable recipes and the opportunity for preparing different types of vulcanizate are limited. For this reason much attention has been directed to the preparation of polyacrylates containing reactive functional groups that facilitate vulcanization. In most instances these functional groups have been introduced by copolymerizing ethyl acrylate with about 5 per cent of a monomer containing the desired group. For example, olefinic linkages,[98] halogen,[95-6, 101] methyl ketone,[97] carboxyl,[58] and cyano groups[101] have been introduced into the polymer by using about

5 per cent of butadiene (or isoprene), 2-chloroethyl vinyl ether, methyl vinyl ketone, acrylic acid, or acrylonitrile, respectively, in the original monomer mixture. Ethyl polyacrylate has been halogenated to increase its amenability to vulcanization.[97, 119]

The mechanism of the vulcanization of polyethyl acrylate by means of alkaline reagents such as sodium metasilicate has been discussed by Semegen and Wakelin.[136] One possible mechanism is a Claisen type of condensation between carbethoxy groups in one molecule and α-hydrogen atoms in another, with the elimination of ethyl alcohol, to form cross links, as shown in the formula. In accord with this view is the fact that polyethyl methacrylate,

$$
\begin{array}{ccc}
\begin{array}{c}
-CH_2 \cdot CH \cdot CH_2 \cdot CH- \\
| \quad\quad | \\
CO \cdot OEt \quad CO \cdot OEt \\[6pt]
-CH_2 \cdot CH \cdot CH_2 \cdot CH- \\
| \quad\quad | \\
CO \cdot OEt \quad CO \cdot OEt
\end{array}
& \longrightarrow &
\begin{array}{c}
-CH_2 \cdot CH \cdot CH_2 \cdot CH- \\
| \quad\quad | \\
CO \quad\quad CO \cdot OEt \\[6pt]
-CH_2 \cdot CH \cdot CH_2 \cdot CH- \\
| \quad\quad | \\
CO \cdot OEt \quad CO \cdot OEt
\end{array}
\end{array}
$$

which lacks α-hydrogen, is not vulcanizable by sodium metasilicate. The mechanism just quoted is not, however, definitely proved.

EXPERIMENTAL STUDIES

Some of the experimental work on acrylic elastomers is described below because it provides information on the effects of the polymerization method, the proportion of chloroethyl vinyl ether used in the copolymerization, the purity of the vinyl ether,[121] or some other variable.

Both emulsion and granulation polymerization[97] were employed to prepare the copolymers shown in Table IV. Pure chloroethyl vinyl ether and a technical grade containing about 66 per cent of the ether were used. The proportion of vinyl ether was either 2 or 5 parts with 98 or 95 parts of ethyl acrylate. The reinforcing agents and curatives listed below were incorporated into the copolymers with a Banbury mixer and conventional rubber mill. The properties of the vulcanized stocks are shown in Table V.

| | Vulcanization Recipe | | |
	A	B	C
Copolymer	100	100	100
Stearic acid	...	1	1
Tetramethylthiuram disulfide	1	...	...
Sulfur	2	1	2
Trimene Base	2	4	2
SRF black	50	50	50

From the results (Tables IV and V) it is evident that the impure chloroethyl vinyl ether is satisfactory for making readily vulcanizable acrylic elastomers. Moreover, as little as 2 parts ether to 98 of ethyl acrylate is sufficient in spite of the fact that only 73 per cent of the chloroethyl vinyl ether entered the copolymer chain. The products made both by emulsion and granulation polymerization[97] had excellent heat resistance at 350° F. (Table V).

The results shown in Table VI were obtained with XP polymers prepared at the Government Laboratories in 80-gal. reactors with 95 parts ethyl acrylate and 5 parts chloroethyl vinyl ether, emulsifiers, water, and persulfate.[2]

Table IV. Preparation of Ethyl Acrylate-2-Chloroethyl Vinyl Ether Copolymers*

Copolymer No.	R74	T102	T235	T236
Polymerization method	Granula-tion	Granula-tion	Emulsion	Emulsion
Polymerization recipe:				
Ethyl acrylate, g.	1900	1960	380	392
Ethyl acrylate, %	95	98	95	98
Chloroethyl vinyl ether, g.	100	40	30.3	12.1
Chloroethyl vinyl ether, %	5	2	5	2
Purity of the ether, %	100	100	66†	66†
Calcium stearate, g.	20	20	...	...
Water, cc.	250	250	800	800
Potassium persulfate, g.	0.65	0.75	0.02	0.02
Triton 720,‡ g.	...	...	18	18
Tergitol 4,§ g.	...	...	12	12
Copolymer yield, %	Good	Good	90.4	90.1
Intrinsic viscosity‖	...	Insol.	4.63	5.72
Raw Mooney viscosity (ML-4 at 212° F.)	57.8	50	55	55

* The emulsion and granulation methods previously described were used. The chlorine contents of copolymers T235 and T236 were 1.11 and 0.49 per cent, indicating that 66.5 and 73.1 per cent, respectively, of the chloroethyl vinyl ether had entered into polymerization.

† The refractive index (n_D^{20}) of this sample was 1.4345; the principal impurity was dioxane.

‡ The Triton 720 had a solids content of 28 per cent.

§ The Tergitol 4 had a solids content of 50 per cent.

‖ In toluene.

The emulsifying system and conditions (refluxing at approximately atmospheric pressure) were those previously recommended.[71] Experimental batches of the copolymer were also made in 5-gal. reactors at 140 and 104° F., in a polymerization recipe similar to that used for XP-101.

The following recipe was used in compounding the polymers:

Copolymer	100	Tetramethylthiuram monosulfide	1
EPC black (Wyex)	50	Trimene Base	2,* 1
Stearic acid	2	Sulfur	2

* Used for the XP polymers.

When compounded in an Office of Rubber Reserve test recipe and cured, standard GR-S broke at 3100 p.s.i. and had an elongation at break of 540 per cent. After being heat-aged for 72 hours at 300° F., the acrylic elastomers showed increases in tensile strength, whereas the GR-S had only about 25 per cent of its original strength.

It is of interest that the copolymers made at 140 and 104° F. had higher dilute-solution viscosities but lower Mooney values than those made under refluxing conditions, i.e., at 180 to 200° F. The Mooney values (54 and 58)

Table V. Vulcanization of Chloroethyl Vinyl Ether-Ethyl Acrylate Copolymers

Copolymer No.	R74	T102	T102	T102	M-16-1*	T235	T236
Recipe	A	A	B	C	B	B	B
Scorch (MS at 300° F.), min.	14.3	30	3.3	14	5.4	4.0	10.3
Vulcanizate no.	2496	2637	2751	2638	2704	2827	2828
Tensile strength, p.s.i.							
30 min.	1580	...	1290	1630	1570	1430	1410
60 min.	1810	1450	1440	1700	1580	1360	1360
Ultimate elongation, %							
30 min.	530	...	300	480	210	200	400
60 min.	440	630	270	490	190	180	370
Modulus at 100%, p.s.i.							
30 min.	...	...	390	...	650	620	260
60 min.	...	...	390	...	750	750	230
Modulus at 200%, p.s.i.							
30 min.	490	...	850	650	1490	1430	690
60 min.	750	460	1090	690	...	...	690
Durometer hardness (30 sec.)							
30 min.	45	...	58	48	62	64	53
60 min.	46	45	57	50	62	64	53
Swelling in water (48 hr. at 212° F.),%	...	78.4	...	60.1	...	48.5	37.0
Properties of Vulcanizates after Aging for 72 Hours at 350° F.†							
Tensile strength, p.s.i.	1690	240	1490	250	1560	1590	1250
Ultimate elongation, %	250	670	240	720	130	160	400
Modulus at 100%, p.s.i.	...	...	420	...	1000	840	210
Modulus at 200%, p.s.i.	1400	...	1300	110	...	...	580
Durometer hardness (30 sec.)	55	31	62	33	69	69	50

* Copolymer M16-1 was prepared by emulsion polymerization in a 10-gal. glass-lined autoclave; the raw Mooney was 56.

† Specimens prepared by vulcanization at 298° F. for 120 minutes were used in the aging tests.

of the XP copolymers (Table VI) are rather typical of those observed with most experimental samples of Lactoprene EV, whereas the Mooney values of the GA copolymers (Table VI) approximate those of the commercial product, Hycar PA21. The resilience was relatively low and the heat buildup relatively high, a fact that limits to some extent the practical applications of the acrylic vulcanizates.

Careful studies on the relaxation of Lactoprene EV have been made by Andrews,[3] and information on this subject and on birefringence have been published by Stein, Krimm, and Tobolsky.[139] It was concluded that the stress and birefringence properties of the elastomer undergo changes with variations in temperature near the second-order transition point. These phenomena can be interpreted in terms of a change in the mechanism of deformation accompanying reversible changes in chain configuration.

Dietz and Hansen[38] describe copolymers, which they designated as

Lactoprene BN, of *n*-butyl acrylate with small proportions of acrylonitrile, as possessing a combination of resistance to hot oils, dry heat, water, and low temperature which should make them useful for the manufacture of oil seals, coolant seals, O rings, gaskets, and similar mechanical specialties. Poly-*n*-butyl acrylate has a much lower brittle point but is less resistant to oils than polyethyl acrylate. If a small proportion of acrylonitrile is copolymerized

Table VI. *Properties of Pilot-Plant Batches of Lactoprene EV (95/5 Ethyl Acrylate-Chloroethyl Vinyl Ether Copolymer)* *

Code	Made under Reflux†		Made at Constant Temperature‡	
	XP-101	XP-103	41GAZ	41GA6
Polymerization temperature, °F.	180–200	195–210	140 (+20)	104
Reaction time, hr.	1.8	3.0	4	7
Conversion, %	85.5	90	100	96
Raw Mooney (ML-4 at 212° F.)	54	58	45	42
Tensile strength,§ p.s.i.	1720	1630	1630	1160
Elongation,§ %	390	260	630	710
Goodyear rebound‖ (RT/212° F.)	25/54	25/49	27/71	29/70
Goodrich hysteresis:				
Heat rise, °F.	>138	>135	>105	>110
Initial compression, %	...	...	32.6	34.7
Set, %	...	...	43.9	...
Flexures × 10^{-3}	Very high	Very high	Over 200; no growth	Very high
Hardness (Shore A)	...	...	55	56
Compounded Mooney	77	92	62	58
Mill shrinkage, %	25	30	42	29
Rugosity	Too rough	17	116	Too rough
Extrusion index	11.5	9.5	11.5	8.0
Gel, %	0,1	44	7	6.5
Dilute-solution viscosity	4.55, 4.80	2.97	6.0	7.3
Gehman freeze point, °C.	−19	−21	−19	−19
Gehman T_{10} value, °C.	−13	−11	−13	−13

* Made in the University of Akron Government Laboratories.
† In 80-gal. vessel.
‡ In 5-gal. vessel.
§ Reported for 120-minute cure at 292° F.
‖ Cure for 240 minutes.

with *n*-butyl acrylate, oil resistance is improved, but the brittle point is affected unfavorably. A suitable balance of these effects can be secured from, e.g., a mixture of 87.5 parts *n*-butyl acrylate and 12.5 parts acrylonitrile, which yields a copolymer having a brittle point of −27° C. compared with 18° F. for Lactoprene EV and −60° F. for a 75/25 butadiene-acrylonitrile copolymer. The butyl ester copolymer is superior to the nitrile rubber in oil resistance, but inferior to Lactoprene EV. Butyl acrylate-acrylonitrile copolymers can suitably be vulcanized by sulfur (1 part) and triethylene-tetramine (1 part). Filachione and co-workers[44a] have described the preparation and vulcanizates of products made by copolymerizing small proportions of acrylonitrile with members of the homologous series of esters of acrylic acid from ethyl to octyl.

Lactoprene BN and a copolymer of ethyl acrylate and acrylonitrile have lately become commercially available from Monomer-Polymer, Inc., under the names of Acrylon BA and Acrylon EA, respectively.

COMMERCIAL PRODUCTS

Hycar PA, Hycar PA21, and Hycar PA31 are supplied by the B. F. Goodrich Chemical Company.[54] Hycar PA, an elastomeric polymer of an acrylic acid ester, was made available in 1947.[52] The production of Hycar PA21 (now known as Hycar 4021), similar in composition to Lactoprene EV, was begun in 1948.[53] Hycar PA31, a modification of Hycar PA21, was developed for easier processing. The easy-processing form permits much faster mixing cycles and the attainment of products having a better finish. Hycar PA31 is somewhat inferior to PA21 with respect to physical properties, low-temperature flexibility, and resilience. Being cross-linked, the easy-processing form cannot be applied as solutions. It is, however, roughly equal to PA21 in heat resistance, and it has better water resistance than the latter. Hycar PA and PA21 are readily soluble in toluene, methyl ethyl ketone, ethyl acetate, and similar solvents.[46, 111] Latexes of Hycar PA and PA21 are available commercially.[54] Emulsions of polymeric acrylic esters are supplied also by Rohm & Haas Company and American Polymer Corporation.

The chlorine-containing elastomers, Hycar PA21 and PA31, are more versatile with respect to vulcanization than Hycar PA. Other noteworthy features of the chlorine-containing elastomers as compared with the acrylic ester polymer include (1) better milling behavior, (2) less tendency to scorch on the mill, (3) better moldability because of improved flow and reduced likelihood of delamination, and (4) vulcanizates several times more water-resistant.[54]

Many chemicals exercise some degree of curative action on Hycar PA, but thus far sodium metasilicate pentahydrate and litharge have given the most promising results.[54, 111] When heat resistance is the prime service factor, sodium metasilicate pentahydrate is preferred to litharge. The use of the metasilicate is inconvenient because it must be added on the mixing mill as a hot melt in its water of hydration.[54] Its curative action is augmented by hydrated lime. Normally, 10 parts of silicate plus 4 parts of hydrated lime per 100 of Hycar PA (by weight) are used. This ratio is varied slightly according to the state of cure desired.

Hycar PA21 and PA31, in addition to responding to the curatives for PA, can be vulcanized with numerous accelerators and curing agents used with conventional rubbers. For maximum heat resistance, combinations of sulfur and Trimene Base are used most frequently. The use of the proper proportions of these two agents results in the maximum stability of cure on prolonged exposure to high temperatures. If a given stock tends to revert or soften at high temperatures, the amine should be increased and sulfur decreased, whereas, if the stock hardens, adjustment in the reverse sense is necessary.

To attain maximum effectiveness in the performance of acrylic elastomers, selective pigmentation is essential. Properties of polyacrylic-rubber vulcanizates are much more sensitive to pigmentation changes than those of most conventional rubbers, principally because their vulcanization depends largely on maintenance of a basic pH in the uncured stocks. Fillers, reinforcing pigments, plasticizers, lubricants, etc., should be chosen with this in mind. Clays, acidic plasticizers, or other pigments that absorb or react with bases should be avoided. Inert or slightly basic pigments are to be preferred. Although maximum heat resistance is attained with loadings of 25 to 30 volumes of dry pigments per 100 of rubber, many variations in the base recipes can be made to produce vulcanizates having optimum properties for specialty applications.[111]

Polyacrylic rubbers are inherently soft when vulcanized, and, therefore, the chief advantage of using plasticizers is to lower the brittle point. Plasticizers known to impart a marked reduction in brittle point (Thiokol TP-90B, butoxyethyl sebacate, Flexol 4GO, and butoxyethyl diglycol carbonate) are volatile at service temperatures of 300° F. and above.[96] The nonvolatile plasticizers thus far tested are relatively inefficient in lowering the brittle point. Blends of polyacrylic rubbers with other polymeric materials have been investigated only briefly. Neoprene (GR-M) seems best with respect to compatibility and noninterference of curative systems.[111]

Processing acrylic elastomers requires a reasonable amount of care if ease in manufacture is to be obtained. The standard equipment of the rubber industry is used for processing and in much the same manner as for other rubbers, with these exceptions: (1) No breakdown period is required on a mill or in a Banbury; (2) during the addition of curatives to a batch, the temperature should be between 150 and 175° F.; (3) temperatures for calendering and extrusions must be maintained within a narrower range than usual to prevent sticking and scorching; (4) curing cycles are usually longer than for other rubbers. Vulcanization is accomplished in 30 to 60 minutes at 300 to 310° F. in pressure molds or an autoclave. Air cures are impractical if advanced states of cure are desired.

Typical recipes[111] for Hycar PA and PA21 are given below. Table VII shows properties of the vulcanizates and the effects of aging the latter in a circulating air oven for periods up to 28 days at the high temperature of 300° F.[111] Under the test conditions of aging, the most heat-resistant of the conventional rubbers become either hard and brittle or soft and tacky in less than 1 week. Vulcanizates of PA21 retain their original elongation and hardness longer than vulcanizates of PA but lose a greater percentage of tensile strength on exposure to elevated temperatures. This may be desirable where a softer stock is needed over long periods of heat service.[111]

Recipe 1 (black molding): Hycar PA, 100; EPC black (Wyex), 45; wool grease, 4; hydrated lime, 4; sodium metasilicate pentahydrate, 10.

Recipe 2 (white molding): Hycar PA, 100; precipitated whiting (Whitcarb R), 30; powdered asbestos, 30; titanium dioxide, 10; wool grease, 5; hydrated lime, 4; sodium metasilicate pentahydrate, 10.

Recipe 3 (black molding): Hycar PA21, 100; sulfur, 1; HMF black (Philblack A), 50; stearic acid, 1; Trimene Base, 4.

Table VII. Original and Aged Properties of Hycar PA and Hycar PA21

| | Days at 300° F. | | | |
	0	7	14	28
Hycar PA, recipe 1*				
Tensile strength, p.s.i.	2200	2260	2160	1900
Elongation, %	500	360	255	110
Hardness	72	80	85	90
Compression set,† %	85	38	30	32
Compression set,‡ %	51	27	23	18
Hycar PA, recipe 2*				
Tensile strength, p.s.i.	1600	1860	1600	1600
Elongation, %	380	170	130	100
Hardness	70	75	78	82
Hycar PA21, recipe 3§				
Tensile strength, p.s.i.	2030	1990	1900	1510
Elongation,‖ %	130	100	100	130
Hardness	68	72	78	77

* Press cured 45 minutes at 310° F.
† 70 hr. at 212° F.
‡ Method B, 22 hr. at 158° F.
§ Press-cured 45 minutes at 298° F.
‖ Low original elongation is due to reinforcement with HMF black.

It is evident from Table VII that aging at 300° F. is effective in improving compression set, which originally was relatively high. Indications are that this improved compression set can be attained in a few hours at 350° F. instead of the 2 to 3 days required at 300° F. In view of the foregoing, an air-oven aging period for finished acrylic-rubber articles requiring low set would be worth while.[111]

Inertness of polyacrylic-rubber vulcanizates to oils at temperatures of 300° F. and above is unexcelled in rubbery materials. Table VIII shows the volume changes that occurred when Hycar PA (compounded in recipes 1 and 2) was immersed in ASTM oils 1 and 3 at 300° F. and Standard Reference fuels SR-6 and SR-10 at room temperature.[111] Only the aromatic-containing SR-6 had a significant swelling effect. In other tests it was

Table VIII. Resistance of Hycar PA Vulcanizates to Hydrocarbon Oils

| | Per Cent Volume Change | | | |
| | 300° F. | | Room Temperature | |
	ASTM No. 1 Oil	ASTM No. 3 Oil	SR-10	SR-6
Recipe 2*				
70-hr. immersions	−2	8	17	78
200-hr. immersions	−1	9	21	83
Recipe 1*				
70-hr. immersions	−1	11	15	82
200-hr. immersions	−1	8	27	84

* Hycar PA (polyacrylic ester) as shown in Table VII.

found that the service life of these rubbers in oils is even better than in air at a given temperature.[111]

Excellent heat and ozone resistance, coupled with fair electrical properties, render the acrylic rubbers suitable for special-purpose wire insulation. The effect on the dielectric strength of Hycar PA31 of aging at 300° F. has been determined. The PA elastomers and a vinyl insulation compound designed for high-temperature service were extruded on wire for testing. Short lengths of the wire were aged in an air oven 1, 2, and 4 weeks and tested for dielectric strength. The PA insulations gained in dielectric strength as they were aged, whereas the vinyl lost. After 2 days' aging the PA materials were as good dielectrically as the vinyl, and after several days they were much better. A further advantage is that after 4 weeks the PA31 stock lost only 5 to 6 per cent of its original weight and showed no indications of shrinkage or brittleness, whereas the plasticized vinyl lost 42 per cent of its weight, shrank considerably, and became very brittle.[111]

White vulcanizates of acrylic rubber have unsurpassed resistance to discoloration in sunlight. Exposure to ultraviolet light in an Atlas Fade-O-Meter for 100 hours has been observed to have little or no effect on acrylic vulcanizates. Conventional rubbers were discolored under these conditions.[111]

Because of their unique and interesting properties, the acrylic elastomers should be useful, and possibly preferable to other elastomers, for the following products: O rings, gaskets, grommets, hose, transmission and conveyor belts, actuating diaphragms, valve seats, packings, oil seals and other mechanical goods that come into contact with oils and/or high-operating temperatures, printing rolls, hot-melt applicator rolls, mats, pads, air bags, protective coatings on heat-resistant fabrics such as glass and asbestos, sunlight resistant temporary coatings on rubber goods, such as de-icers and tires; transformer leads, insulation on wire and bus bars, and white or pastel articles in which permanence of color is required.[111]

Like all other materials, the acrylic elastomers have limitations and disadvantages. Most prominent are slow acceleration, relatively high brittle point, and affinity for water. Many improvements in the raw elastomers as well as in methods of vulcanization and utilization have already been made,[54] and, with the steadily growing interest in these materials, more can be expected. It is already known that the curatives are primarily responsible for water affinity. It seems likely that curatives having high acceleration and yielding both water- and heat-resistant vulcanizates will be found. Moreover, it seems probable that feasible methods of lowering the brittle point—by plasticization[99] or by copolymerization with a higher alkyl acrylate[99]—will be developed. As the volume of production increases, the disadvantage of relatively high price, $1.35 a lb., also should be partially overcome.

SILICONE RUBBERS

No other elastomer surpasses the silicone rubbers in the range of temperature over which the useful, rubbery character is retained. This property of low sensitivity to temperature change, coupled with good electrical properties, has been a great factor in the commercial development these products have experienced since their introduction in 1944. The manufacture of the products is still undergoing expansion.

Two general types of silicone rubbers are available, one for uses where flexibility is required between -60 and $550°$ F. and a more recent variety useful between -130 and $550°$ F. These wide ranges of usefulness are achieved without plasticizers. The absence of such foreign, and often volatile, materials contributes to the long service life of the extremely inert silicone polymers.

There are two suppliers of silicone rubbers at present in the United States, the Dow-Corning Corporation and the General Electric Company.

CHEMISTRY AND PREPARATION

The silicones are polymers formed by the polycondensation of silicols (with loss of the elements of water), which in turn are obtained by hydrolysis of the chlorosilanes. The reactions may be illustrated in the following way (in most present-day cases, R is methyl):

$$R_2SiCl_2 + 2H_2O \longrightarrow R_2Si(OH)_2 + 2HCl$$
$$2R_2Si(OH)_2 \longrightarrow HO \cdot SiR_2 \cdot O \cdot SiR_2 \cdot OH$$
$$3R_2Si(OH)_2 \longrightarrow HO \cdot SiR_2 \cdot O \cdot SiR_2 \cdot O \cdot SiR_2 \cdot OH, \text{ etc.}$$
$$nR_2Si(OH)_2 \longrightarrow HO(SiR_2 \cdot O)_nH$$

The hydrolysis of the chlorosilane occurs very rapidly and easily. While it may be considered that the first products of the hydrolysis are the corresponding silicols (or silanols), the latter usually begin to condense immediately in the manner shown. The lower alkyl silanediols are so reactive that they have never been isolated (but diphenylsilanediol is a stable, crystalline powder).

The siloxanes resulting from condensation of the silanediols may also be cyclic in form, as shown in the formula. Such rings have been observed with a distribution of sizes. The reversible conversion of open-chain methylpolysiloxanes to cyclic compounds has been described by Patnode and Wilcock.[112]

The halosilane may be prepared via other organometallic compounds. Friedel and Crafts,[48] and Ladenburg,[83] reacted zinc and mercury alkyls with silicon tetrachloride or silicon orthoesters. A number of investigators[80, 118, 130-1] have applied the Wurtz-Fittig synthesis—the action of sodium on a mixture of silicon chloride and an alkyl chloride—to the preparation of the organosilicon compound. Kipping and co-workers[79] employed the Grignard reagent with silicon tetrachloride, and this is one of the methods that have been adopted by the Dow-Corning Corporation for the commercial production of chlorosilanes.[29]

The preparation by means of the Grignard reaction occurs stepwise as follows:

$$RMgCl + SiCl_4 \longrightarrow RSiCl_3 + MgCl_2$$
$$RMgCl + RSiCl_3 \longrightarrow R_2SiCl_2 + MgCl_2$$
$$RMgCl + R_2SiCl_2 \longrightarrow R_3SiCl + MgCl_2$$
$$RMgCl + R_3SiCl \longrightarrow R_4Si + MgCl_2$$

The mixture of substituted silanes may be separated by fractional distillation.

The monomeric halosilanes may also be derived from elemental silicon by reaction of alkyl or aryl halides with copper-silicon or silver-silicon alloys.[122-3, 125] This is a method developed and used commercially by the General Electric Company. It has been suggested that the mechanism of the reaction is as follows.[74]

$$2Cu + RCl \xrightarrow{250\text{-}350°C} CuCl + CuR$$
$$Si + CuCl \longrightarrow Cu + (SiCl) \text{ active intermediate}$$
$$(SiCl) + R \longrightarrow (RSiCl)$$
or
$$(SiCl) + CuR \longrightarrow (RSiCl) + Cu$$
or
$$(SiCl) + CuCl \longrightarrow (SiCl_2) + Cu$$

Such a series of reactions can continue until the silicon atom is tetrasubstituted, resulting in a mixture of the various possible alkylchlorosilanes, R_nSiCl_{4-n}. Under favorable conditions, the dialkyldichlorosilane can be made the predominating product, with lesser amounts of $RSiCl_3$ and R_3SiCl, and some R_4Si and $SiCl_4$. In general, the lowest temperature that will suffice for the reaction is most satisfactory. At higher than the minimum practical temperatures, pyrolysis results in organosilicon compounds richer in halogen and poorer in alkyl groups. Copper has been found the best catalyst for the preparation of alkyl silicon halides, and silver the best for that of aryl silicon halides.[122]

The direct synthesis of the organosilicon halides has been reported also to be possible by reacting silicon tetrachloride with appropriate hydrocarbons[106] at temperatures of 450 to 1000° C.

Organosilicon halides may be produced by the addition of silicochloroform to olefins.[10, 82, 116, 138] The reaction may be carried out in the liquid phase at around 40 to 50° C. under about 5 to 10 lb. pressure with the help of an acyl peroxide or ultraviolet light, or the reaction may be promoted by heat alone at temperatures of 160 to 400° C. under autogenous pressure. A variety of products may be prepared, but in general only one group is added in place of the initial hydrogen.

The formation of elastomeric polymers, such as "methyl silicone rubber," requires a dialkyldichlorosilane of exceptional purity. The presence of methyltrichlorosilane or of silicon tetrachloride leads to cross-linking in the condensation reaction, with loss of plasticity and extensibility in the final rubber. Trimethylchlorosilane, $(CH_3)_3SiCl$, if present, would act as a chain stopper and would reduce the molecular weight of the polymer.

There is little published information on the actual technique used in this purification. Fractionation of the methylchlorosilanes requires a very efficient column because of the proximity of their boiling points.[60] The separation may also be effected by fractionating some derivative and then regenerating the alkylchlorosilane.[61] The pure cyclic compound, $(Me_2SiO)_4$, b.p. 175° C., may be distilled from the initial hydrolysis mixture and can serve as the purest source of bifunctional Me_2SiO groups for making silicone rubber.[124]

Polymers comprising two thousand or more of the units, $\cdot (CH_3)_2Si \cdot O \cdot$, may be prepared. Scott[132] determined by osmotic-pressure measurements that the elastic fraction of a sample of polydimethylsilicone had a number-average molecular weight of 2,800,000. Fractions of polymer with molecular weights down to 610,000 were progressively softer and more plastic. A method of determining the molecular weight of silicone polymers by the examination of fractured specimens under the electron microscope has been described by T. G. and E. G. Rochow.[126]

COMPOUNDING AND VULCANIZING

Fillers and curing agents may be incorporated on a conventional rubber mill. In selecting suitable fillers, one must take into consideration the thermal stability of the filler and the temperatures to which the final stock will be exposed. Carbon blacks, for example, gas badly at about 200° C. Titanium dioxide exerts less reinforcing action than carbon but will withstand the high temperatures at which the silicone rubbers can be used.[42] Zinc oxide and silica are also used as fillers.[81] It has been stated that unusually high tensile strength is conferred on silicone rubber by a form of silica that has been treated in such a way as to make it hydrophobic.[78a] Some fillers, like carbon blacks, interfere with the action of the curing agent employed.[124]

The silicones are completely saturated and do not cure by mechanisms typical of the butadiene or isoprene copolymers. The curing agents incorporated in the stock are usually organic peroxides. These were considered at one time to cause oxidation of some of the methyl groups, and thus enable siloxane bridges to be formed.[60, 123] The more recent concept involves the action of free radicals from the peroxide catalyst.

$$-R_2SiO- + 2 \boxed{O} \longrightarrow -\left[\begin{array}{c} RSiO \\ | \\ O \\ | \end{array}\right] - + R'CHO$$

$$R\cdot \text{ (from peroxide)} + [-(CH_3)_2Si \cdot O-] \longrightarrow RH + [-CH_3Si \cdot O-]$$
$$\underset{CH_2}{\overset{|}{\underset{\cdot}{}}}$$

$$2[-CH_3Si \cdot O-] \longrightarrow [-CH_3Si \cdot O-]$$
$$\underset{CH_2}{\overset{|}{}} \qquad \underset{CH_2}{\overset{|}{}}$$
$$\underset{CH_2}{\overset{|}{}}$$
$$[-CH_3Si \cdot O-]$$

Benzoyl peroxide is said to be generally the most effective of the organic peroxides.[77] The amount of peroxide is fairly critical for best results, and 2 parts per hundred of gum is generally used when an appropriately inert pigment has been chosen. Curing is usually carried out in two steps. In the first step, the cure is advanced only to the stage where the molded or extruded piece has dimensional stability under its own weight. This step may be carried out in a press, for example, at 500 p.s.i. and 260° F. for 5 minutes.[81] The sample is then transferred to an oven held at about 500° F., where the cure is finished in several hours.

It has been stated[144a] that alkyl titanates will accelerate the curing of silicone resins, enabling it to be carried out at temperatures as low as 50 to 80° C. Information is lacking as to whether the same reagents will accelerate the curing of silicone rubbers.

Different varieties of silicone rubber are available which are specifically suited for different methods of fabrication, such as extrusion, molding, or coating on glass cloth. Both the gum base and compounded stocks are available from suppliers. For example, General Electric 9979 G silicone rubber is a highly elastic brown solid containing no filler or vulcanizing agent. Silastic 160, available from the Dow-Corning Corporation, contains 60 per cent of filler consisting of equal weights of zinc oxide and titanium dioxide.

PROPERTIES OF VULCANIZATES

As with other elastomers, the properties of cured stocks are dependent on a variety of factors, e.g., the choice of filler, the proportion of filler used, conditions of cure, etc. In general, the most outstanding properties of silicone rubber are (1) stability at high temperatures, (2) weather resistance, (3) low-temperature flexibility, and (4) resistance to lubricating oils.

Table IX describes some of the properties of Silastic 125. These data illustrate the excellent retention of properties over a wide temperature range. A study of the properties of some of the early Silastic products (silicone

Table IX.   Properties of a Silastic 125* Stock

	Temperature		
	−65°C.	25°C.	200°C.
Tensile strength, p.s.i.	600	575	500
Shore A hardness	69	49	45
Bashore resilience	17	34	41
Compression set, %	2	28	61
Modulus, p.s.i.	...	360	320

* Designation of the Dow-Corning Corporation.   The data in Table IX are interpolated from a figure shown in reference 42.

polymers supplied in an already-compounded condition) has been published by Moakes and Pyne.[107]

An improved material, Silastic 250, is described in Table X.

Table X.   Properties of Silastic 250

Tensile strength, p.s.i.	650	Dielectric strength	800 volts per mil
Elongation, %	300	Dielectric constant (1 Mc.)	3.0
Shore A hardness	40 to 55	Power factor (1 Mc.)	0.009
Compression set, %	40 to 60	Brittle point	−130° F.

In Table XI are shown some of the physical properties of several G-E silicone stocks.[51]

Table XI.   Properties of G-E Silicone Stocks*

Grade No.	12600	12601	12602	12603
Specific gravity	2.00	2.00	1.40	1.50
Shore A hardness	45 to 55	55 to 65	65 to 75	75 to 85
Compression set,† %	30	40	25	25
Tensile strength, p.s.i.	200	450	650	500
Elongation, %	225	150	110	100
Tear resistance, lb. per in.	35	45	45	35
Dielectric strength step by step,	300	250	510	380
volts per mil. (short time)	350	330	560	500
Power factor (1 Mc. dry)	0.0007	0.005	0.003	0.004
Dielectric constant (1 Mc. dry)	3.8	7.4	3.1	3.1
Volume resistivity, megohms per cc.	1.5×10^5	1.5×10^5	1.5×10^5	1.5×10^5

* The composition of these stocks, presumably compounded, is undisclosed.

† After 6 hours at 302° F. and 30 per cent compression.

One of the most outstanding characteristics of a silicone rubber is its retention of properties after exposure to heat.   Konkle, Selfridge, and Servais[81] have exposed silicone stocks to various elevated temperatures in a circulating-air oven and compared their behavior with that of a commercial Neoprene (GR-M) stock specially compounded for heat and oil resistance. After 50 days at 302° F., all the silicone samples were still flexible enough to be bent 180 degrees without cracking or breaking.   The average weight loss of three representative formulations was 2.5 per cent, the average shrinkage 1.7 per cent.   The increase in hardness, as measured by the Shore A durometer, averaged 15 points. After one day at 302° F., the Neoprene stock cracked badly on a mild bend, and suffered a loss in weight of nearly 9 per cent. Even on exposure to a temperature of 482° F. the silicone stocks remained flexible and rubbery for more than 42 days.

Work at the National Bureau of Standards[149] showed that with respect to low-temperature properties silicone rubbers are of three types: first, those showing crystallization beginning at −75 to −89° F.; second, those showing crystallization beginning at −103° F.; and third, those showing no clear-cut setting in of crystallization down to the second-order transition. Oddly enough, all types have the same second-order transition of −189° F., the lowest yet recorded for any polymer. The first type of silicone rubber is essentially polydimethylsiloxane. The latter two types represent slight changes in composition which are in the nature of the insertion of bulky side groups, e.g., phenyl groups, to interrupt the otherwise regular structure and so impede crystallization.[73, 147−8] Silicone rubbers with brittle points of −130° F. are of the second type above, showing crystallization beginning at −103° F.

Polmanteer, Servais, and Konkle[117] have published actual stress-strain curves for silicone rubbers at temperatures down to −130° F. Silastic 160, a silicone of the first type mentioned above, is stated to crystallize at about −76° F. At 75° C. this rubber possessed a tensile strength of 550 p.s.i. and an ultimate elongation of 195 per cent; at −67° F. its tensile strength was 830 p.s.i. and its elongation 180 per cent. Silastic 250, a silicone of the second type mentioned above, crystallizing at about −112° F., had at 75° F. a tensile strength of 880 p.s.i. and elongation of 350 per cent, and at −112° F. a tensile strength of 1490 p.s.i. and elongation of 370 per cent. The corresponding figures for Silastic 6-160, which belongs to the same type in resistance to hardening, were: tensile strength: 550 p.s.i. at 75° F., 960 p.s.i. at −112° F.; elongation: 225 per cent at 75° F., 215 per cent at −112° F. Another silicone rubber[4a] that resists very low temperatures is G-E's SE-550, which remains flexible at −120° F. These data show that the silicone rubbers retain extensibility and resist stiffening (increase in modulus) at low temperatures much better than other rubbers. Parallel experiments with natural rubber and GR-S reported by the same authors show both these rubbers to stiffen much more severely with fall in temperature than the silicones and to be substantially inextensible at −76° F.

The silicone rubbers have excellent weathering characteristics and resistance to oil, oxidation, ultraviolet, and ozone. They are waterproof and water-repellent. The adhesion of ice to a silicone surface is remarkably poor.

High-temperature silicone adhesives have been developed for bonding silicone rubber to glass, steel, aluminum, brass, and other materials, as well as to itself.

The electrical properties of stocks recommended for electrical applications are good.[41] The insulating and dielectric properties are well retained at low and at high temperatures. In preparing silicone-glass fabric compositions, care should be taken to fill voids as well as possible, since these detract substantially from dielectric strength. Carbon tracking cannot occur after a dielectric-strength breakdown because the product of the decomposition of the silicone polymer is merely silica.[42]

Most silicones will burn and support combustion. It is said, however, that polymers have recently been devised that are self-extinguishing under drastic conditions and do not support combustion when subjected to flame or very high temperatures.

Properties of silicone rubbers that may be improved in time are their at present comparatively low tensile strength, tear strength, and abrasion resistance, and their relatively high compression set at very high temperature. With respect to the latter property, it should be noted that the temperatures referred to are higher than the service temperatures of any other elastomer. Consequently, compression set at high temperature is not a disadvantage in silicone rubbers by comparison with any other rubbery material, but rather a property in which improvement would be desirable. Reinforcement of the rubber with glass or asbestos cloth increases its resistance to compression set and its resistance to tearing.

APPLICATIONS OF SILICONE RUBBER

The silicone rubbers are obviously not recommended for use in the temperature range where the less expensive, conventional rubbers are capable of serving (generally at about −25 to +125° C.), unless some special property of the silicone rubbers other than heat resistance is required. They are economical and outstandingly useful at temperatures where ordinary rubbers cannot serve.

The silicones are ideally suited for the construction of stationary seals for high-temperature oil lines, because of their characteristic of "cold" flow at high temperatures. The packing is supported within a mechanical framework underneath a spring loading device so that the rubber is at all times under mechanical stress. Under these conditions, it flows into the physical irregularities of the metal and effects an intimate seal.[42] An interesting application which takes good advantage of the unusual properties of extruded silicone rubber is the use of extruded silicone gaskets in sealing the steam chamber of a steam iron.

Caulking and sealing compounds have found use as elastic sealers of high dielectric strength capable of withstanding high temperatures. The compound may be forced into a void and partially vulcanized by a short heat treatment. Such compounds have been recommended as seals for small, liquid-filled capacitors and transformers, refrigeration equipment, and valve seats.

Gaskets of silicone rubber can withstand temperatures and, in many cases, solvents (such as Pyranol and chlorinated hydrocarbons) to which conventional rubbers are not resistant. At temperatures where cold flow is objectionable, reinforcement with glass cloth is recommended. Because of their combination of heat and oil resistance, the silicone rubbers are used in aircraft supercharger gaskets and in integral parts of aircraft hydraulic and engine systems, jet engines, and Diesel motors.

Electrical insulating tapes made from glass cloth and silicone rubber can function at temperatures much higher than the conventional organic materials. If due provision is made for lubrication, etc., electric equipment can be designed to operate at higher temperatures than usual and to withstand without danger to the insulation greater temporary overloads than usual.

Since the silicone rubbers are odorless, tasteless, and nontoxic, they have found use in wire and glass-cloth-reinforced conveyor belts for food processing. For handling hot glassware, the pads, gloves, or steel tools used may be coated with silicone rubber, in order to reduce thermal shock to the glass article and to eliminate the danger of slippage.

This is not an exhaustive list of current proposed applications of silicone rubbers, but it serves to illustrate the possibilities that this new class of elastomers has opened to the design engineer.

POLYESTER RUBBERS

Linear polyesters of high molecular weight prepared by the polycondensation of hydroxycarboxylic acids or of mixtures of dibasic acids and glycols were described in 1929 by Lycan and Adams[88] and by Carothers and co-workers.[21, 23-4] It was discovered in the Bell Telephone Laboratories,[15-6] in 1942, that polyesters of low melting point could be compounded with pigments and peroxides and cured, or "vulcanized," to form strong, rubbery masses.[14, 49, 50, 89] Such vulcanizable polyesters, originally designated Paracon and later known as Paraplex rubbers, were produced on a pilot-plant scale for several years during World War II. Chiefly because of their unfavorable raw-material cost, they are not now commercially available.

Outstanding features of the polyester rubbers are their stability on exposure to high temperatures, oxidation, and weathering, their resistance to hydrocarbon solvents, and their excellent low-temperature flexibility without plasticizers. Their serviceable range of temperature is much wider than that of the butadiene copolymers but not so wide as that of the silicones.

The polyesters offer an ideal opportunity for fundamental studies of the property of rubberiness, because their formation, depolymerization, and molecular-weight distribution can be treated in exact, quantitative terms.[47]

PREPARATION

To realize the most satisfactory properties in the final vulcanizate, the polyester must have a minimum weight-average molecular weight of about 10,000, as measured by its intrinsic viscosity. Molecular weights of two or three times this magnitude will result in improved tensile strength and low-temperature flexibility. One semicommercial product, Paraplex X-100, had a weight-average molecular weight of about 20,000.

The choice of glycols and dibasic acids for the preparation of the polyesters is determined by their thermal stability and by their individual contributions to the properties of the final polymer. Unsubstituted, straight-chain aliphatic glycols and acids, taken together, usually result in crystalline polymers which are hard and plastic, for example, at ordinary temperatures.

When side-chain substituents are present in one or both of the reactants, the freezing point of the polyester obtained is lower than that of the polyester from corresponding unsubstituted reactants.   A similar effect is achieved by using a mixture of glycols or acids of different lengths.

For example, the polyester from ethylene glycol and sebacic acid is a hard, crystalline solid, which shows a well-defined X-ray pattern and melts at 72° C.   The polyester from propylene glycol and sebacic acid, however, is a gum which crystallizes at about 12° C.   The lesser tendency of the latter polyester to crystallize is due to the fact that the side-chain methyl groups

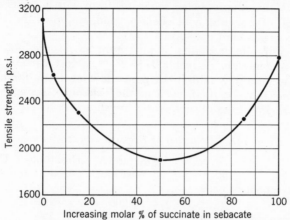

FIG. 1.   Maximum Tensile Strength at Optimum Cure of Polyesters Prepared by Condensing Various Ratios of Sebacic and Succinic Acids with Ethylene-Propylene Glycols

Stocks loaded with 75 parts Kalvan

derived from the propylene glycol expand the interchain spacings from 4.18 to 4.31 A. and so weaken the interchain forces.   On stretching cured polypropylene sebacate, however, it exhibits an X-ray pattern, indicating that considerable polar coordination has occurred, and this is desirable from the viewpoint of its rubbery qualities.   Too great a degree of irregularity in the structure of the polymer chains reduces the ability to "crystallize" on stretching and correspondingly reduces the tensile strength.   Thus, for example, polyesters prepared by condensing a mixture of ethylene and propylene glycols with mixtures of succinic and sebacic acids show, when cured, regularly decreasing tensile strengths as the composition approaches equimolar amounts of the two acids[15] (Fig. 1).

The polyesterification is carried out[72] at temperatures up to 250° C., under conditions to exclude air, the presence of which might mutilate the terminal, reactive groups and so restrict the polymer size.   The reactor is designed to facilitate the efficient removal of the esterification by-product (water, or, if the method of transesterification is used, alcohol or glycol). The progress of the reaction is followed by periodic viscosity determinations on samples taken from the batch.   When the optimum molecular weight has been reached, the reaction is stopped by cooling.   The product at this stage is a tacky gum. The ease with which it may be compounded on the mill is much improved by "precuring" in a power mixer with a very small amount of an organic peroxide.[12] The crumb which results may be sheeted on the mill. The sheet obtained has an appearance very similar to that of crepe rubber.

COMPOUNDING AND VULCANIZING

The polyesters may be vulcanized by incorporating in them a peroxide and heating the stock in a mold to a temperature at which the peroxide decomposes rapidly.[8, 146] With polypropylene sebacate, from 4 to 6 parts of benzoyl peroxide for 100 of polyester will effect a good, snappy cure in about 10 minutes at 20 p.s.i.g. steam pressure.

The tear and tensile strengths of pure-gum stocks are not high, and reinforcement is generally desirable.[14, 49, 50, 89] Most carbon blacks are unsatisfactory because they inhibit the peroxide cure, but a large variety of mineral pigments may be used, such as the calcium carbonates and iron oxides. Typical properties of such stocks are shown in Table XII.

Table XII. Properties of Polyester-Rubber Stocks

	Saturated Sebacate*	Unsaturated Sebacate†	Unsaturated Sebacate‡
Composition of compounds, parts			
Gum	200	200	200
Calcium carbonate (Kalvan)	150	150	150
Benzoyl peroxide	4	1	...
Sulfur	...	...	3
Tetramethylthiuram monosulfide	...	...	2
Cure, min. per p.s.i.g. steam	10/20	10/20	45/80
Physical properties			
Tensile strength, p.s.i.	2400	2000	940
Elongation, %	510	440	307
Set, %	14	15	ca. 10
Hardness (Shore A)	65	66	55
Compression set	Low	Low	Low
Resilience	High	High	Very high
Tear strength	Low	Low	Low
Low-temperature flex,§ °C.	−40	−40	−45
Properties after 72 hr. aging at 150° C.:			
Tensile strength, p.s.i.	Too weak	1200	740
Elongation, %	Too weak	490	200
Hardness (Shore A)		55	70

* From sebacic acid and an 80/20 mole ratio of propylene and ethylene glycols. Molecular weight approximately 20,000.

† 97/3 mole ratio of sebacic and maleic acids; otherwise same as above.

‡ Same as above, but contains 20 mole per cent maleic on total dibasic acids.

§ Strips 5 × 0.5 × 0.075 in. flexed by hand over half-inch mandrel after conditioning 24 hours at −3° C. The temperatures reported are the lowest (±5° C.) at which the strips would flex without breaking.

By including small amounts of maleic anhydride as one of the raw materials for the polyester, more reactive polymers are obtained which can be cured with smaller amounts of peroxide. In a typical gum of this kind, about 3 mole per cent of the dibasic acid units are maleate or fumarate. Only 0.5 to 1 per cent of benzoyl peroxide is required in such a polyester, to produce a good cure, and the resulting vulcanizate has better stability than that from a saturated polyester, because it contains less benzoic acid derived from the peroxide. Properties of compounded vulcanizates of this kind are shown in Table XII. If the ratio of unsaturated acid be increased to 10 to 20 mole per cent, the resulting polymer can be cured with sulfur and conventional rubber accelerators (see Table XII).

The peroxide cure has advantages over sulfur curing in speed and the absence of sulfur contamination in the product. The latter feature might be important, for example, in electrical applications. Sulfur-curable stocks, however, are less sensitive to scorching on the mill or to bin curing than peroxide-containing stocks.

Their resistance to water is good up to about 140° F., but in steam the cured polyesters disintegrate quickly. The stocks are not significantly affected in any way by exposure to oxygen at 3000 lb. pressure and 70° C. for 14 days. Their electrical properties are similar to those of Neoprene or of plasticized vinyl chloride.

POLYESTERS FROM VEGETABLE OILS

The dimeric acids and esters which can be prepared by polymerization from vegetable oils may, since they contain two carboxyl groups, serve for the production of polyesters by condensation with glycols, and, since such polyesters contain unsaturated centers, they can be vulcanized by heating them with sulfur and accelerators, and indeed can be polymerized and cross-linked (although less effectively) by merely heating them. Working along these lines, there can be obtained from vegetable oils products much superior in strength and extensibility to factice (prepared by heating the oils themselves with sulfur) and in fact possessing rubberlike properties sufficient to render them useful as rubber substitutes.

During 1942 and 1943 considerable effort was devoted to the development of such products. From the polyester prepared by the condensation with ethylene glycol of the dimeric acids or methyl esters of soybean oil there were obtained, by further polymerization followed by sulfur vulcanization, elastic materials having tensile strengths of 400 to 900 p.s.i. and elongations of from 100 to 175 per cent.[34] The work was centered at the Northern Regional Research Laboratory of the U.S. Department of Agriculture, and the product was termed Norepol (*N*orthern *R*egional *pol*ymer). It was demonstrated that various rubber articles having a reasonable degree of usefulness could be made from the polymer, but no extensive commercial development followed, because soybean oil was allocated to food uses. A limited quantity (apparently 500 to 1000 tons) of Norepol was actually manufactured and was used in the manufacture of jar rings (in which the Norepol was mixed with a minor proportion of reclaimed rubber), floor tiles, washers, and rubberized cloth.

In further experimental work,[35] it was shown that the dimeric fatty acids could be condensed, to yield vulcanizable polymers, with (a) other glycols, (b) aminoalcohols, and (c) diamines, and, further, that the polyesters could be reacted with diisocyanates, to yield improved products. From polyesters of especially high molecular weight ("superpolyesters") derived from the dimeric acids and ethylene glycol there were reported vulcanized products having tensile strengths up to 2200 p.s.i. and elongations up to 480 per cent.

DIISOCYANATE-LINKED CONDENSATION ELASTOMERS

The formation of polyesters by the condensation, by means of heat, of simple reactants such as dicarboxylic acids and dihydric glycols involves the elimination of water. Although, as Carothers showed,[22] it is possible, by applying very low pressure during the heating (by the use of a molecular still), to push the reaction to the point where the condensation products attain relatively high molecular weights, e.g., 25,000, yet, under ordinary conditions of condensation, such as those employed in the production of alkyds, the products have relatively low molecular weights—not high enough for the development of good rubbery properties. The treatment of such relatively short, linear polyester molecules with diisocyanates provides a device for uniting a number of such molecules and thus, under mild conditions of reaction, of producing linear, condensed molecules long enough to make possible the development of good rubbery properties by cross-linking. (Examples of earlier applications of similar treatment to polyesters and polyesteramides for purposes other than the development of rubbery materials are to be found in references 25 and 128.)

The production of elastomers from diisocyanate-linked condensation polymers has been developed in England and in Germany on somewhat different lines. In the preparation of the British materials, called Vulcaprenes, a "polyester" is prepared from reactants which introduce, in addition to ester linkages, a proportion of amide linkages; such linear molecules are then united—to form larger, linear molecules—by treatment with less than an excess of a diisocyanate, and finally curing (cross-linking) is brought about by the action of formaldehyde on the amide groups.[19, 27, 30, 56–7, 64, 67–8, 70, 76] (The application of formaldehyde to polyamides, such as nylon 66, for purposes other than the production of rubbery properties has been described earlier in reference 28; cf. also 129). In the preparation of the German materials, known as Vulcollanes, an excess of diisocyanate is used in bringing about the union of the initial polyester chains, and curing is effected by treatment with water, which reacts with the free, terminal isocyanate groups.

VULCAPRENES

Preparation. Although alternative reactants adapted to condensation through the formation of ester and amide linkages may be employed, the components actually chosen for the production of the polyesteramide used in the manufacture of commercial Vulcaprene are ethylene glycol, ethanolamine, and adipic acid. It appears that the ethanolamine is applied in a considerably smaller molar ratio than the ethylene glycol. Whereas in the preparation of the Vulcollane polyesters a deficit of the dicarboxylic acid component is used, in order to insure that the polyester molecules shall terminate in hydroxyl groups, in the preparation of the Vulcaprene poly-esteramides such an excess of hydroxyl groups is not stipulated. Hence the terminal groups of the polyesteramides may be, at random, OH, NH_2, or COOH. All these groups, however, are capable of reacting with diiso-cyanates, to form, respectively, urethane, carbamide, and, with the libera-tion of carbon dioxide, amide links. Such linkages unite the initial linear molecules into longer linear ones. After the initial condensation has been brought about by heating the polyesteramide-forming reactants under nitrogen, water is removed by, it is specified, azeotropic distillation with xylene.[57] The product at this stage is a waxy material with a molecular weight of, say, 5000 to 7000.

The polyesteramide is then heated in an internal mixer with about 5 per cent of its weight of hexamethylene diisocyanate, the temperature being gradually raised from 130 to 170° C. The product at this stage is a tough, rubbery mass. The reaction of the diisocyanate with molecules of poly-esteramide containing the various possible terminal groups may be shown schematically as follows. (R represents the nonterminal part of the poly-esteramide chains.) The specific gravity of the uncured, unloaded polymer is 1.15.

$$NH_2ROH + OCNR'NCO + NH_2RCOOH + OCNR'NCO + OHRCOOH$$
$$\rightarrow -NHRO \cdot CO \cdot NHR'NH \cdot CO \cdot NHRCO \cdot NHR'NH \cdot CO \cdot ORCO-$$

Curing. Curing of the polymer thus built up is brought about by incor-porating a formaldehyde-generating substance and, preferably, a small quantity of an acid or acid-generating substance, and then heating.[56] The formaldehyde-generating substance may be paraformaldehyde, but in technical practice a complex amine-formaldehyde condensation product, Vulcafor VHM, is recommended. Among acidic substances specified as cure accelerators are phthalic anhydride, sodium dihydrogen phosphate, and[56] styrene dibromide but, in technical practice, the use is recommended of a halogenated naphthol, Vulcafor VDC, which decomposes at curing temperatures to generate a minute trace of acid. (1-Bromo-2-naphthol[27], 2,4-dibromo-1-naphthol,[66] and 2,4-dichloro-1-naphthol[69] are specified elsewhere.)

The cross-linking produced during curing by the reaction of formaldehyde with amido groups in the diisocyanate-reacted polyesteramide can be shown as similar to that usually considered to be involved in the setting of urea-formaldehyde resins, namely, first the introduction of a methylol group, and then condensation of the latter with the amido hydrogen of another chain.

$$
\cdots\cdot CO \cdot NH \cdots\cdot \qquad \cdots\cdot CO \cdot N \cdots\cdot \qquad \cdots\cdot CO \cdot N \cdots\cdot
$$
$$
\underset{+}{\overset{+}{CH_2O}} \longrightarrow \underset{+}{CH_2OH} \longrightarrow H_2O + \qquad CH_2
$$
$$
\cdots\cdot CO \cdot NH \cdots\cdot \qquad \cdots\cdot CO \cdot NH \cdots\cdot \qquad \cdots\cdot CO \cdot N
$$

Cross-linking agents that have been specified in addition to those already mentioned include chromates and dichromates,[26] dimethylolurea dimethyl ether,[55] hexamethylolmelamine hexamethyl ether,[66] trinitrobenzene, benzoyl peroxide, and quinone dioxime.[32]

Physical Properties. The physical properties of mold-cured Vulcaprene stocks and suitable conditions of cure are shown in Table XIII.

Table XIII. Properties of Various Vulcaprene Stocks

Base Stock: Vulcaprene A,100; Dark Factice, 10; Stearic Acid, 0.75; Vulcafor VHM, 5; Vulcafor VCD, 0.75

Cure: 30 minutes at 141°C. in press, followed by 2 hr. at 125°C. in air

	No Filler	SRF Black		China Clay		Blanc Fixe	
Parts Filler by Weight	0	45	75	67.5	112.5	120	200
Parts Filler by Volume	0	30	50	30	50	30	50
Tensile strength, p.s.i.	1066	2019	2103	1849	1450	1408	1166
Elongation, %	385	300	195	295	190	415	385
Modulus at 200%, p.s.i.	213	1209	...	1422	...	455	498
Hardness (Shore)	40	75	87	64	90	60	71
Resilience at 50° C., % (Tripsimeter)	60	52	44	44	29	50	40

Set (10 min. stretch at 300%, 60 min. recovery), %	4.3
Tear (cold, hot), p.s.i.	512, 213
Brittle temperature (R.A.E. test)	−45 to −50° C.

Channel blacks are not suitable as reinforcing agents for Vulcaprene, since their acidity causes scorching. An SRF black, Kosmos 20, is recommended. It has not been stated how the newer fine furnace blacks behave in Vulcaprene. For white and colored stocks china clay (which has marked reinforcing power and also enables hard stocks to be produced) and blanc fixe (which has little reinforcing power but does not reduce the extensibility) are recommended, along with some titanium dioxide as a white pigment. Basic fillers must be avoided. A small proportion of brown factice is useful as a softener.

Cured Vulcaprene shows good resistance to dry heat aging but suffers deterioration when exposed to steam.

Applications. The chief technical applications of the Vulcaprenes[76] at present seem to be as coatings, in which connection they are used alone (together with pigments, dyes, and plasticizers if desired) or in combination with cellulose acetate,[70] cellulose nitrate,[70] or degraded leather.[69] The coatings, if properly formulated, can be cured by heating for 1 to 2 hours at 125° C., or they will cure at room temperature if a polyisocyanate, Vulcafor VCC, is incorporated in the stock in rather high proportion. Vulcaprene can also be applied as an adhesive,[63] which has high tack and which when cured gives, in the case of its application to leather, a bond stated to be stronger usually than the leather itself. Cements of Vulcaprene admixed with cellulose nitrate or cellulose acetate find application in the shoe trade for such purposes as sole laying. For use in coatings and cements several grades of polymer (Vulcaprene AC) are offered, all soluble in acetone but differing in the viscosity of their solutions. The viscosity of the uncured polymer can be reduced by milling.[62]

Bright-colored varnishes for rubber goods can be compounded from Vulcaprene with a relatively small proportion of cellulose acetate, titanium dioxide, and color. For articles such as rubber footwear the varnishes may be heat-cured. For rubber proofings cured by sulfur chloride or if desired for other rubber articles, cold-curing Vulcaprene varnishes may be used. The varnishes are applicable to druggists' sundries, rubber flooring, colored side walls, and so on. The varnishes are stated to withstand deformation without cracking or peeling, because they are themselves elastic and because they are, thanks to the curing process used, chemically bonded to the rubber. The inclusion of cellulose acetate in Vulcaprene varnishes involves sacrifice of some extensibility but much improves abrasion resistance. For a glossier finish but one less resistant to abrasion, cellulose nitrate may be used in place of cellulose acetate.

Similar varnishes (designated as lacquers) can be applied to rubberized fabrics and are stated to have excellent resistance to cracking by crumpling and flexing. In this application it is recommended that there shall first be applied a tie coat consisting of a weak solution of Vulcaprene and a very high concentration of polyisocyanate.

There is sufficient similarity in structure between Vulcaprene and proteins to make Vulcaprene compatible with hydrolyzed leather. This last can be suitably produced by boiling vegetable tanned leather in water for 30 to 45 minutes. Such treatment destroys the fibrous character of the leather and renders the latter thermoplastic and capable of being milled into Vulcaprene. A 50–50 mixture of Vulcaprene and leather, marketed as Vulcaprene PL, can, after being suitably compounded and colored, be applied to fabrics by spreading or calendering. The coatings are cured by heating for 2 hours in air at 125° C. It is stated that the resistance of such coatings to flexing, abrasion, and scrubbing is outstanding. The coatings are said to be one hundred times as good in these respects as nitrocellulose-coated imitation leather, and also to be superior to the latter in being entirely free from tack and any tendency to collect dust. The coatings may be embossed satisfactorily. In addition to the application of Vulcaprene PL-coated fabrics as imitation leather, they also find application for gasoline and oil pump diaphragms, printers' blankets and liners.

Unsaturated Vulcaprenes. Polymers analogous to the Vulcaprenes but vulcanizable by means of sulfur and regular organic accelerators have been described in the patent literature[31] but have not so far been marketed. In the formation of such polymers, the reactants used in making the first condensation product, which may be either a polyester or a polyesteramide, comprise an unsaturated component, such as maleic anhydride or fumaric acid. The products of the first stage are, as with the Vulcaprenes, built up by reaction with hexamethylene diisocyanate, after which curing is brought about by heating with sulfur and an organic accelerator. The cured products appear to show lower tensile strength than cured Vulcaprenes.

VULCOLLANES

Vulcollanes have not yet reached the stage of large-scale commercial production. But the properties of Vulcollane N—the representative of the group on which most development work has been done—are in at least certain respects so outstanding that the Vulcollanes deserve attention. In gum tensile strength, tear resistance, and abrasion resistance, Vulcollane N probably surpasses any other synthetic rubber hitherto developed. (Unless otherwise indicated, all the information here given on the Vulcollanes is based on reference 75.)

Preparation. The preparation of cured Vulcollane N can be looked on as involving four reaction steps, as follows:

1. A polyester is prepared by condensing adipic acid and ethylene glycol in a jacketed internal mixer. The acid is melted at 170° C., and the glycol is then added gradually, the temperature being raised until it finally reaches 220° C. The polyester is then dried (to about 0.02 per cent moisture[17]) by heating with agitation in an autoclave under vacuo for 1 hour at 130° C. The product is a waxlike material melting at 50 to 70° C. and having a low viscosity above 80° C. In order to insure the presence of terminal hydroxyl groups, an excess of the glycol is used in the preparation of the polyester, so that the product has a "hydroxyl number" of 55 to 65, corresponding to 1.5 to 1.7 hydroxyls per molecule. (The "hydroxyl number" is defined as the milligrams of potassium hydroxide per gram of polyester, and is determined by adding pyridine and acetic anhydride and titrating the acetic acid formed by potassium hydroxide solution.) The molecular weight of the product is "at least 1000." (In reference 17 the molecular weight is stated to be 3000 to 5000.) The polyester can be written as

$$HO(CH_2)_2O[CO(CH_2)_4COO(CH_2)_2O]_nH.$$

2. The molten polyester is now transferred to an internal mixer, where it is treated for, say, 30 minutes at 85° C. with 15.5 parts naphthalene-1,5-diisocyanate per 100 parts polyester. The reaction produces a marked increase in viscosity. The product is a highly viscous liquid above 70° C.; it hardens on standing at room temperature and thereafter can be worked on a rubber mill. In this step the diisocyanate reacts with the terminal hydroxyls of the polyester, forming urethane linkages and lengthening the chains. As the diisocyanate is in excess, the product formed carries terminal isocyanate groups. It has been indicated[11] that the molecular weight at this stage is 4000, although another reference[17] states it to be about 15,000. If the polyester of the first step is represented as HOXOH, the product of the second step may be written as

$$OCNC_{10}H_6NH \cdot CO \cdot OXO[CO \cdot NHC_{10}H_6NH \cdot CO \cdot OXO]_n$$
$$CO \cdot NHC_{10}H_6NCO.$$

3 and 4. These steps are brought about by incorporating a small amount of water into the product of the second step and then heating. The reactions that take place involve the terminal isocyanate groups and, although not yet fully elucidated, are thought to be as follows.

3. Terminal isocyanate groups react with water, uniting the chains through carbamide groups and thus producing longer chains. At the same time carbon dioxide is evolved. (The mechanism may involve first the conversion of one isocyanate group to an amino group and then reaction of the latter with the isocyanate group of another chain.) If the product of the second step is written as $OC : N \cdot Z \cdot N : CO$, the reaction of the third step may be shown as

$$2OCNZNCO + H_2O \longrightarrow OCNZNH \cdot CO \cdot NHZNCO + CO_2$$

4. Carbamide groups thus formed in one chain react with the terminal isocyanate groups in other chains, thus:

$$\cdots\cdots NH \cdot CO \cdot NH \cdots\cdots \qquad \cdots\cdots N \cdot CO \cdot NH \cdots\cdots$$
$$+ \qquad \longrightarrow \qquad |$$
$$OC : N \cdots\cdots\cdots \qquad\qquad CO \cdot NH \cdots\cdots$$

This is the actual cross-linking reaction, yielding the final, cured product.

The reaction with water is brought about by adding about 1 per cent of water to the product of the second step in an internal mixer and kneading the mass at, say, 85° C. for, say, 45 minutes (including a period of 10 minutes during which the water is being added gradually). The material becomes progressively stiffer. Reaction is carried to the point at which the material clings to the blades of the mixer. Reaction 4 proceeds more or less along with reaction 3, and the product from the treatment with water will set up if kept for long at room temperature. However, when freshly prepared it can be milled, sheeted, and molded, although it cannot be extruded satisfactorily. It is desirable to bring about curing, i.e., the completion of reaction 4, within 6 hours. Curing can be effected by a short period of heating at 130 to 170° C. The extent to which the isocyanate groups have reacted at any stage can be followed by adding an excess of dibutylamine to a weighed sample and then titrating unreacted amine with hydrochloric acid.

Physical Properties of Vulcollane N. The physical properties of unloaded Vulcollane N cured for 50 minutes at 150° C. under 3300 p.s.i. have been reported as given in Table XIV. Comparative data for a tread stock made from natural rubber are also reported.

Table XIV

	Unloaded Vulcollane N	Hevea Tire-Tread Stock
Tensile strength, p.s.i.	3500–4200	4000
Elongation, %	760	500
Set, %	22	8–10
Modulus at 300%, p.s.i.	925	850
Rebound (Schopper, 20° C., 70° C.)	60, 68	42, 50
Hardness (Shore)	69	62
Tear resistance, lb.	88–110	88
Stitch tear resistance, p.s.i.	728	

The above tensile data were obtained on ring test pieces. A strip from a small sample of Vulcollane N available to us gave the following (approximate) tensile results: T, 5000 p.s.i.; 300 per cent modulus, 740; E, 900; set, almost nil.[150]

The tear resistance is claimed to surpass that of any other synthetic rubber. The abrasion resistance is outstanding, being five to ten times that of natural rubber. In tests against an emery paper surface, the losses by volume were as follows: natural-rubber tread stock, 200; Buna-S tread stock, 150; Vulcollane N, 20 to 30. The resistance of Vulcollane N to aging is also outstanding. After 8 weeks in a Geer oven at 70° C., a sample showed no change in tensile strength or set, only a very slight (6 per cent) increase in modulus, and a moderate fall (23 per cent) in elongation. The high-temperature tensile properties are not very good. A sample which had a tensile strength of 4050 p.s.i. at 20° C. broke under 1560 p.s.i. at 70° C.

Vulcollane N resists petroleum oils better than any of the recognized oil-resistant elastomers except the polysulfides, as the data in Table XV show. The oil and solvent resistance of Vulcaprene is equal to that of Vulcollane N.

Table XV. Swelling of Vulcollane N after 7 Days Immersion at Room Temperature

Grams Solvent per Gram Elastomer

	Gasoline	Kerosene	Trans-former Oil	Benzene	CH_2Cl_2	CS_2
Vulcollane N*	0.0068	0.0089	0.0000	0.70	3.17	0.15
Perbunan Extra	0.06	0.05	0.003	0.99	2.52	0.55
Neoprene	0.16	0.29	0.027	1.00	1.51	1.45
Perduren G†	0.002	0.000	—0.004	0.17	0.37	

* With 10 parts CK-3 black. Composition of stocks for the other rubbers is not stated.
† A polysulfide elastomer.

We found a sample of Vulcollane N to show high swelling in tetrachlorethane (9.60 grams per gram elastomer in 8 days at room temperature) and dioxane (3.18) and fairly high swelling in methyl ethyl ketone (0.92).[150] The resistance of Vulcollane N to water is moderately good at ordinary temperature, but boiling water or steam brings about rapid deterioration.

Applications. Not much information is yet available concerning the technical applications of Vulcollane N. Its high resistance to abrasion suggests its use for such applications as conveyor belts and as a thin tread on tire carcasses. Vulcollane N lacks building tack but can be bonded to other materials by means of triisocyanates, such as Desmodur R, $CH(C_6H_4NCO)_3$. Solid tires built by winding newly calendered Vulcollane N sheet under tension onto a steel rim, when subjected to a wheel test at 22° C. at a speed of 35 km. per hour and a load of 100 kg., showed no signs of deterioration after 5000 km., and the temperature at different points in the tire was 45 to 60° C.

When, however, the load was doubled, considerable heat developed, and the tire went to pieces after 30 to 35 km. In other words, when the mechanical stress was below a certain level, Vulcollane N performed very well but above this level broke down rapidly, because, it is thought, it was not sufficiently cross-linked.

Fillers. Fillers (which should be well dried, e.g., for 5 hours at 120° C.) are best incorporated, in an internal kneader, after the polyester and diisocyanate have been condensed but before treatment with water. Preliminary studies of the use of particulate fillers in Vulcollane N indicated the following results:

1. As in natural rubber, the addition of fillers raises the modulus and (in contradistinction to those synthetic rubbers that have poor gum strength) reduces the elongation. The balance between the increased stiffness and the reduced extensibility, however, is such that, unlike the case of natural rubber, even active fillers do not raise the ultimate tensile strength, but, instead, reduce it somewhat.

2. Active fillers, such as good grades of carbon black, raise the tear resistance. They, however, apparently reduce abrasion resistance somewhat.

3. At present the maximum loading of active fillers which can be properly dispersed, at all events without a plasticizer, seems to be about 25 volumes. Even relatively small loadings of black tend to raise the stiffness to an unduly high level.

4. By means of inactive fillers, e.g., barytes, the hardness can be raised to a point suitable for shaft packings without much sacrifice of rebound. The data in Table XVI will serve to illustrate some of the points just mentioned. The data are for optimum cures.

Table XVI. Influence of Fillers on Properties of Vulcollane N

	Unloaded	With CK-3 Black 5 Vol.	With CK-3 Black 10 Vol.	With P-33 Black 5 Vol.	With P-33 Black 10 Vol.
Modulus at 300%, p.s.i.	768	1408	1934	1209	1280
Tensile strength, p.s.i.	3700	3635	3240	2845	2985
Elongation, %	765	575	570	590	680
Set, %	20	24	23	18	33
Rebound, % (20°, 70° C.)	59, 63	57, 59	55, 55	63, 67	57, 59
Hardness (Shore)	68	43	73	70	71
Tear resistance, lb.	73	66	121	86	84

Variations on Vulcollane N. Vulcollane N gives a distinct X-ray pattern when stretched, in accord with the linear and regular character of the main polymer chains. The facility of crystallization is sufficient to give the material a rather high brittle point[17] (about −25° C.) and also leads to a tendency, at least in some samples, to harden on storage. In order to explore the possibility of overcoming this, Vulcollanes were made from a number of other components in addition to those used in the preparation of Vulcollane N. These components were introduced as replacements for part of the adipic acid or part of the ethylene glycol in the preparation of the polyester. Because of the diminished degree of regularity in their structure, the polyesters thus prepared have lower melting points than polyethylene adipate, and, correspondingly, the Vulcollanes derived from them may be expected to have a diminished tendency to fall into a crystallite lattice. An advantage of such mixed polyesters is that satisfactory reaction with naphthalene-1,5-diisocyanate can generally be secured at a lower hydroxyl number (30 to 40) than is required with polyethylene adipate, and accordingly a small proportion of the diisocyanate (a relatively expensive reagent) can be used in making the Vulcollane. Some selected results from the experiments made in this connection are given in Table XVII.

Table XVII. Variations on Vulcollane N

Third Component	Moles per 100 Moles Adipic Acid	Properties of the Vulcollane		
		T	E	Tear Resistance
1,2-Propanediol	15	3140	810	84 lb.
	20	3860	745	64
1,4-Butanediol	15	2070	650	44
2,2'-Dihydroxyethyl ether	15	2555	720	42
1,4-Butenediol	14	4345	680	46
	33	3610	650	29
2-Hydroxyethylaniline	3	3680	700	79
	6	3780	630	57
	10	3150	660	46
Piperazine	1.5	5110	820	70
	2.2	4890	720	59
	3	2140	645	55

It will be noticed that, of the saturated glycols used, 1,2-propanediol, which is nearest in structure to ethylene glycol, gives a product quite close to Vulcollane N in mechanical properties, whereas the next two glycols on the list, which introduce larger carbon chains between the ester links, give markedly weaker products. However, 1,4-butenediol gives strong products, perhaps because of an influence of the rigidity of the double bond on the flexibility of chain links adjoining it. The last two components listed have the effect of introducing a proportion of amide linkages into the polymer chains, but these amide groups lack amide–hydrogen which would be open to attack by diisocyanate and which might lead to an excessive amount of cross-linking during cure. Finally, it will be noticed that the chain irregularity introduced by the third component reduces the tear resistance of the products.

The Vulcollanes have aroused considerable interest in the United States. Early in 1953 the Goodyear Tire and Rubber Company announced a development involving (1) modification of step 2 in the preparation designed to obviate the tendency of the product of this step to set up during processing and storage, and (2) a concomitant modification of the curing procedure. Experimental tires have been built from diisocyanate-linked polymer prepared according to these modifications, but, it appears, no commercial manufacture of the polymer is in immediate prospect.

The modifications involved are as follows.[101a]   (1) In step 2, i.e., in the diisocyanate-linking of the initial polymer of step 1, there is applied, not an excess of diisocyanate, which would leave terminal isocyanate groups in the product, but rather a proportion limited to a maximum of 0.99 mole of diisocyanate per mole of polyester.   The products thus formed in step 2 can be stored for 1 year or even longer and still retain good processibility. (2) Curing of such products is brought about by mixing them with a further quantity of a suitable diisocyanate (or polyisocyanate) and subjecting the mixture to heat.   The quantity of diisocyanate used in step 2 and in curing is in total 2.8 to 3.2 equivalents —NCO per mole of initial polyester.   The curing or cross-linking is considered to depend on reaction of the di- or polyisocyanate last introduced with the hydrogen atoms of the urethane links formed in the modified step 2.

To illustrate the development just outlined, the following example may be quoted. In step 1 a 1-to-1.19 molar ratio of (a) adipic acid and (b) a 55/45 mixture of ethylene and prolylene glycols are condensed by heating to a temperature of 200° C., the pressure being gradually reduced and nitrogen being bubbled through the melt.   In step 2 the product of step 1 is mixed for 10 minutes, in an internal mixer steam-heated to 200° C., with 12.35 per cent of its weight of 4,4'-diphenyldiisocyanate of 95.7 per cent purity, equivalent to 0.96 mole diisocyanate per mole of polyester.   The mixture is baked in a tray for 8 hours at 130° C.   For the purpose of curing it, the product (100 parts) is mixed on a rubber mill with 5.44 parts of 4,4'-diphenyldiisocyanate and cured for 70 minutes at 300° F.   The cured product, when tested, gave the results shown in Table XVIII.

Table XVIII.   Properties of Diisocyanate-Linked Polymer

Tensile strength, p.s.i.	5150
Elongation, %	705
Hardness (Shore A)	64
Rebound, % (room temp. and 200° F.)	72.4, 84.0
Abrasion (Bur. Standards, D394–47, method B)	200
Compression set (ASTM, D395–47T, method B)	5.8

In comparing these results with the results in Table XIV of tests on cured samples of the German product, Vulcollane N, the following should be borne in mind: (1) The fact that the tensile-strength data in Table XIV were obtained on ring test pieces whereas the strength data in Table XVIII were presumably obtained on dumbbell test pieces is probably sufficient to account[132a] for the difference in magnitude of the strength data in the two tables.   (2) The rebound data in Table XIV were obtained by the Schopper tester, whereas those in Table XVIII were presumably obtained by the Goodyear-Healey pendulum.

In the preparation of the initial "polyester," a limited proportion of the glycol or glycols may be replaced by an amino alcohol or a diamine, to yield a polyesteramide in place of a polyester strictly so-called, but, if then the result of step 2 is to be processible and stable, the proportion in which such replacement is used must not exceed in the case of an amino alcohol 15 molar per cent of the glycol and in the case of a diamine 7.5 molar per cent.

Polymer Structure and Rubberlikeness. Bayer, who has been closely associated with the use of di- and polyisocyanates and with the development of the Vulcollanes,[11] has expressed some interesting views on the structural features which contribute to rubberlikeness in polymers.[17] He considers that good tear resistance requires that the cross-links shall be regularly spaced and shall be separated by chain sections of considerable length (represented by a molecular weight of 20,000 to 30,000). The steps involved in the production of Vulcollane N have the effect of insuring considerable uniformity in the length of the primary chains and considerable regularity in the distribution of cross-links in the final network. Contrariwise, an I.G. product, I Gummi, developed before the Vulcollanes, had poor tear resistance, because, it is supposed, the cross-links were too numerous and too irregularly spaced. This product was made by cross-linking with a diisocyanate a polyester prepared from adipic acid, ethylene glycol and trimethylol propane. The random distribution of the trifunctional hydroxy component would be expected to cause randomness in the distribution of the cross-links.

According to Bayer, a factor favorable to good elastic properties in a linear polymer such as a polyester that otherwise would be prone to form a crystallite lattice is the presence of "spacious" groups at regular intervals—*in* the chain, not merely pendant to the chain. Such groups confer some rigidity on the structure and also separate adjacent chains. In Vulcollane N the naphthalene ring is such a group. The same result cannot be secured with a benzidine residue (derived from the use of benzidine diisocyanate in the preparation of the Vulcollane) because of the ease of rotation of the benzene rings in such a case. Again, the result is not achieved by using a proportion of a naphthalene dicarboxylic acid in place of some of the adipic acid, because the naphthalene residues are then less regularly spaced in the chains; at all events, the products have poor tear resistance.

REFERENCES

1. Abrams, J. T., Barker, R. L., Jones, W. E., and Woodward, F. N., *J. Soc. Chem. Ind.*, **68**, 237–44 (1949).
2. Alden, G. E., Private Communication, Govt. Lab. to O.R.R., Feb. 20, 1948; Howerton, W. W., Dietz, T. J., Snyder, A. D., and Alden, G. E., *Rubber Age N.Y.*, **72**, 353–63 (1952).
3. Andrews, R. D., Dissertation, Princeton University, 1948.
4. Anon., *Chem. Week.*, **70**, No. 7, 11–2 (1952); *Chem. Eng. News*, **31**, 2074–6 (1953).
4a. Anon., *Chem. Eng. News*, **31**, 370 (1953).
5. Atwood, F. C. (Natl. Dairy Products Corp.), U.S. Pat. 2,400,477, May 21, 1946.
6. Atwood, F. C., and Hill, H. A., U.S. Pat. 2,398,350, Apr. 16, 1946.
7. Baer, J., Brit. Pat. 279,406 (1926); Fr. Pat. 640,967, Sept. 14, 1927.
8. Baker, W. O., *J. Am. Chem. Soc.*, **69**, 1125–30 (1947).
9. Barron, H., *Modern Synthetic Rubbers*, Chapman & Hall, London, 1949, p. 495.
10. Barry, A. J., DePree, L., Gilkey, J. W., and Hook, D. E., *J. Am. Chem. Soc.*, **69**, 2916 (1947).

11. Bayer, O., Mueller, E., Peterson, S., Piepenbrink, H. F., and Windemuth, E., *Angew. Chem.*, **62**, 57–66 (1950).

12. Beavers, E. M., U.S. Pat. 2,454,539, Nov. 23, 1948.

13. Bigelow, M. H., *Chem. Eng. News*, **25**, 1038–42 (1947).

14. Biggs, B. S., U.S. Pat. 2,448,572, Sept. 7, 1948.

15. Biggs, B. S., Erickson, R. H., and Fuller, C. S., *Ind. Eng. Chem.*, **39**, 1090–7 (1947).

16. Biggs, B. S., and Fuller, C. S., *Chem. Eng. News*, **21**, 962–3 (1943).

17. BIOS Final Rept. No. 1166.

18. Boswell, W. E., and Jorczak, J. S., *India Rubber World*, **120**, 334–6 (1949).

19. Buist, J. M., Harper, D. A., Smith, W. F., and Welding, G. N., U.S. Pat. 2,424,885, July 29, 1947.

20. Burns, R., Jones, D. T., and Ritchie, P. D., *J. Chem. Soc.*, **1935**, 714–7.

21. Carothers, W. H., and Arvin, J. A., *J. Am. Chem. Soc.*, **51**, 2560–70 (1929).

22. Carothers, W. H., and Hill, J. W., *J. Am. Chem. Soc.*, **54**, 1557–9 (1932); Carothers, W. H., and van Natta, F. J., *J. Am. Chem. Soc.*, **55**, 4714–9 (1935).

23. Carothers, W. H., and Hill, J. W., *J. Am. Chem. Soc.*, **54**, 1559–66 (1932).

24. Carothers, W. H., and van Natta, F. J., *J. Am. Chem. Soc.*, **55**, 4714–9 (1933).

25. Christ, R. E., and Hanford, W. E., U.S. Pats. 2,333,639, Nov. 9, 1943, and 2,333,917, Nov. 9, 1943.

26. Coffey, D. H., Smith, W. F., and White, H. G., Brit. Pat. 581,146, Oct. 2, 1946.

27. Coffey, D. H., Smith, W. F., and White, H. G., Brit. Pat. 614,992, Dec. 30, 1948.

28. Coffman, D. D., U.S. Pat. 2,177,637, Oct. 31, 1939.

29. Collings, W. R., *Chem. Eng. News*, **23**, 1616–9 (1945).

30. Cook, J. G., Harper, D. A., Reynolds, R. J. W., and Smith, W. F., Brit. Pat. 573,811, Dec. 7, 1945.

31. Cook, J. G., Harper, D. A., and Smith, W. F., U.S. Pat. 2,424,884, July 29, 1947.

32. Cook, J. G., and Seymour, R. C., Brit. Pat. 579,857, Aug. 19, 1946.

33. Crafts, J. M., *Ann*, **124**, 110–4 (1862).

34. Cowan, J. C., Ault, W. C., and Teeter, H. M., *Ind. Eng. Chem.*, **38**, 1138–44 (1946).

35. Cowan, J. C., et al., *Ind. Eng. Chem.*, **41**, 1647–52 (1949).

36. Crosby, J. W., *India Rubber World*, **106**, 133–5 (1942).

37. Crosby, J. W., *Petroleum Engineer*, **14**, No. 9, 96, 98, 100 (1943).

38. Dietz, T. J., and Hansen, J. E., *Rubber Age N.Y.*, **68**, 699–700 (1951).

39. Dietz, T. J., Hansen, J. E., and Meiss, P. E., Water Resistance of Acrylic Vulcanizates, Presented before Third Meeting-in-Miniature, Philadelphia Sect. Am. Chem. Soc., Jan. 20, 1949.

40. Dietz, T. J., Mast, W. C., Dean, R. L., and Fisher, C. H., *Ind. Eng. Chem.*, **38**, 960–7 (1946).

41. Doede, C. M., DiNorscia, G., and Panagrossi, A., Some Properties of Silastic at Elevated Temperatures, Presented before Div. Rubber Chem., Am. Chem. Soc. Meeting, Atlantic City, Apr. 10, 1946.

42. Doede, C. M., and Panagrossi, A., *Ind. Eng. Chem.*, **39**, 1372–5 (1947).

43. Fettes, E. M., Thiokol Corp., Private Communication.

44. Fettes, E. M., and Jorczak, J. S., *Ind. Eng. Chem.*, **42**, 2217–23 (1950).

44a. Filachione, E. M., Fitzpatrick, T. J., Rehberg, C. E., Woodward, C. F., Palm, W. E., and Hansen, J. E., *Rubber Age N.Y.*, **72**, 631–7 (1953).

45. Fisher, C. H., and Filachione, E. M., Lactic Acid—Versatile Intermediate for the Chemical Industry, U.S. Bur. Agr. Ind. Chem., AIC–178, 1948, 22 pp.

46. Flanagan, G. W., *India Rubber World*, **120**, 702–5 (1949).

47. Flory, P. J., *Chem. Revs.*, **39**, 137–97 (1946).

48. Friedel, C., and Crafts, J. M., *Ann.*, **127**, 28–32 (1863); *Ann.*, **136**, 203–11 (1865); Friedel, C., and Ladenburg, A., *Ann.*, **159**, 259–74 (1871); 203, 241–55 (1880).

49. Frosch, C. J., U.S. Pat. 2,448,584, Sept. 7, 1948.

50. Fuller, C. S., U.S. Pat. 2,448,585, Sept. 7, 1948.

51. General Electric Bull. No. If/1 (CDC–121, 4–49–5M).

52. Goodrich Chemical Co., B. F., Polyacrylic Ester—Experimental Product, Service Bull. 47–SD3, July 1, 1947, 10 pp.

53. Goodrich Chemical Co., B. F., Hycar PA21, Service Bull. H–3, Dec. 1948, 12 pp.; Polyacrylic Rubber, Service Bull. H–11, Mar. 1953, 41 pp.

54. Goodrich Chemical Co., B. F., Properties and Utilization of Hycar Polyacrylic Rubbers, Bull., 1950.

55. Habgood, B. J., and Harper, D. A., Brit. Pat. 580,525, Sept. 11, 1946. Cf. Coffman, D. D., U.S. Pat. 2,275,008, Mar. 3, 1942.

56. Habgood, B. J., Harper, D. A., and Reynolds, R. J. W., Brit. Pat. 580,524, Sept. 11, 1946.

57. Habgood, B. J., Harper, D. A., and Reynolds, R. J. W., U.S. Pat. 2,424,883, July 29, 1947.

58. Hansen, J. E., Meiss, P. E., and Dietz, T. J., A Mechanism for the Amine Vulcanization of Acrylic Rubbers, Presented before Third Meeting-in-Miniature, Philadelphia Sect., Am. Chem. Soc., Jan. 20, 1949.

59. Hansen, J. E., Palm, W. E., and Dietz, T. J., Amine Vulcanization of Ethyl Polyacrylate, U.S. Bur. Agr. Ind. Chem., AIC–205, Sept. 1948, 6 pp.; Hansen, J. E., and Dietz, T. J., U.S. Pat. 2,579,492, Dec. 25, 1951.

60. Hardy, D. V. N., and Megson, N. J. L., *Quart. Rev.*, **2**, 29 (1948).

61. Hardy, D. V. N., and Megson, N. J. L., *Quart. Rev.*, **2**, 30 (1948).

62. Harper, D. A., Brit. Pat. 572,738, Oct. 22, 1945.

63. Harper, D. A., Brit. Pat. 581,134, Oct. 2, 1946.

64. Harper, D. A., *Trans. Instn. Rubber Ind.*, **24**, 181–95 (1948).

65. Harper, D. A., and Huggill, H. P. W., Brit. Pat. 580,526, Sept. 11, 1946.

66. Harper, D. A., and Huggill, H. P. W., Brit. Pat. 614,994, Dec. 30, 1948.

67. Harper, D. A., Naunton, W. J. S., Reynolds, R. J. W., and Walker, E. E., 11th Internatl. Cong. Pure Applied Chem., 1947.

68. Harper, D. A., and Reynolds, R. J. W., Brit. Pat. 574,739, Jan. 18, 1946.

69. Harper, D. A., and Smith, W. F., Brit. Pat. 581,410, Oct. 11, 1946.

70. Harper, D. A., and Smith, W. F., Brit. Pat. 585,205, Jan. 31, 1947.

71. Howerton, W. W., General Operating Conditions for Producing Lactoprene EV, Eastern Regional Lab., Mar. 1947; Howerton, W. W., Dietz, T. J., Snyder, A. D., and Alden, G. E., *Rubber Age N.Y.*, **72**, 353–62 (1952).

72. Howard, J. B., U.S. Pat. 2,410,073, Oct. 29, 1946.

73. Hunter, M. J., Barry, A. J., and Warrick, E. L., Paper in the Silicone Polymer Symposium, Am. Chem. Soc. Meeting, Sept. 7, 1951 (see Ref. 148).

74. Hurd, D. T., and Rochow, E. G., *J. Am. Chem. Soc.*, **67**, 1057–9 (1945).

75. I.G. Farbenindustrie, Repts. from Central Scientific Lab. and Central Rubber Lab., Mar. 23, May 1, and June 1, 1946; Rept. from Central Rubber Lab., July 1, 1946; Rept. from Central Scientific Lab., July 1, 1946. (Most of the material in these reports has been published in Ref. 11.)

76. Imperial Chemical Industries, *Vulcaprene Manual*.

77. Irby, G. S., Goss, W., and Pyle, J. J., *India Rubber World*, **117**, 605–8, 616 (1948); Irby, G. S., *Rubber Age N.Y.*, **65**, 63 (1949).

78. Jorczak, J. S., and Fettes, E. M., *Ind. Eng. Chem.*, **43**, 324–34 (1951).

78a. Kilbourne, F. L., *Chem. Eng. News.*, **30**, 4720 (1952); Spencer, W. B., Davis, W. B., Kilbourne, F. L., and Montermoso, J. C., *Ind. Chem. Eng.*, **45**, 1297–1304 (1953). Cf. Glime, A. C., Duke, N. A., and Doede, C. M., *India Rubber World*, **128**, 766–70, 774 (1953).

79. Kipping, F. S., *Proc. Chem. Soc.*, **20,** 15 (1904). See also Dilthey, W., *Ber.*, **37,** 319, footnote 2 (1904), and bibliography on pp. 290–1 of Krause and von Grosse, *Die Chemie der metallorganischen Verbindungen*, Berlin, 1937, and bibliography on p. 76 of Rochow, E.G., *Chemistry of the Silicones*, 2d ed., Wiley, New York, 1951.

80. Kipping, F. S., and Lloyd, L. L., *J. Chem. Soc.*, **79,** 449–59 (1901).

81. Konkle, G. M., Selfridge, R. R., and Servais, P. C., *Ind. Eng. Chem.*, **39,** 1410–3 (1947).

82. Krieble, R. H., and Burkhard, C. A., *J. Am. Chem. Soc.*, **69,** 2689–92 (1947).

83. Ladenburg, A., *Ann.*, **164,** 300–32 (1872); **173,** 143–66 (1874).

84. Laurence, A. E., and Perrine, V. H., *Rubber Age N.Y.*, **54,** 139–41 (1943).

85. Leeper, H. M., Private Communication, Govt. Lab. to O.R.R., Feb. 6, 1946.

86. Loewig, C., and Weidmann, S., *Pogg. Ann.*, **46,** 45–92 (1839).

87. Longstreth, B., *Chem. Eng. News*, **20,** 1362 (1942).

88. Lycan, W. H., and Adams, R., *J. Am. Chem. Soc.*, **51,** 3450–64 (1929).

89. Malm, F. S., U.S. Pat. 2,448,609, Sept. 7, 1948.

90. Martin, S. M., *India Rubber World*, No. 6, 30–4 (1941).

91. Martin, S. M., *Vanderbilt Rubber Handbook*, 1942, p. 161.

92. Martin, S. M., and Laurence, A. E., *Ind. Eng. Chem.*, **35,** 986–91 (1943).

93. Martin, S. M., and Patrick, J. C., *Ind. Eng. Chem.*, **28,** 1144–76 (1936).

93a. Massa, A. P., Colon, H., and Schurig, W. F., *Ind. Eng. Chem.*, **45,** 775–82 (1953). Cf. Allen, F. H., and Fore, D., *Ind. Eng. Chem.*, **45,** 374–7 (1953).

94. Mast, W. C., Dietz, T. J., Dean, R. L., and Fisher, C. H., *India Rubber World*, **116,** 355–60 (1947).

95. Mast, W. C., Dietz, T. J., and Fisher, C. H., *India Rubber World*, **113,** 223–9 (1945).

96. Mast, W. C., and Fisher, C. H., *Ind. Eng. Chem.*, **40,** 107–12 (1948).

97. Mast, W. C., and Fisher, C. H., Improved Preparation of Acrylic Rubber. Curing Methods and Properties of the Vulcanizates, U.S. Bur. Agr. Ind. Chem., AIC–206, Dec. 1948, 17 pp.; U.S. Pat. 2,588,398, Mar. 11, 1952.

98. Mast, W. C., and Fisher, C. H., *India Rubber World*, **119,** 596–8, 727–30 (1949).

99. Mast, W. C., and Fisher, C. H., *Ind. Eng. Chem.*, **41,** 703–8 (1949).

100. Mast, W. C., and Fisher, C. H., *Ind. Eng. Chem.*, **41,** 790–7 (1949).

101. Mast, W. C., Rehberg, C. E., Dietz, T. J., and Fisher, C. H., *Ind. Eng. Chem.*, **36,** 1022–7 (1944); Mast, W. C., and Fisher, C. H., U.S. Pat. 2,509,513, May 30, 1950.

101a. Mastin, T. G., and Seeger, N. V., U.S. Pat. 2,625,535, Jan. 13, 1953. See also Seeger, N. V., U.S. Pat. 2,525,531, Jan. 13, 1953; Seeger, N. V., Mastin, T. G., Fauser, E. E., Farson, F. S., Finelli, A. F., and Sinclair, E. A., *Ind. Eng. Chem.*, **45,** 2538–42 (1953).

102. Matsuyama, Y., *Hitachi-Hyoron*, **23,** 331–4 (1940).

103. Matsuyama, Y., and Matsushima, T., *J. Soc. Rubber Ind. Japan*, **15,** 170–6 (1942).

104. Meyer, K. H., *Natural and Synthetic High Polymers*, 2d ed., Interscience, New York, 1950, p. 260; J. R. Katz, *Trans. Faraday Soc.*, **32,** 77–96 (1926).

105. Meyer, V., *Ber.*, **19,** 3259–66 (1886).

106. Miller, H. C., and Schreiber, R. S., U.S. Pat. 2,379,821, July 3, 1945.

107. Moakes, R. C. W., and Pyne, J. R., *J. Rubber Research*, **19,** 77–86 (1950).

108. Mochulsky, M., and Tobolsky, A. V., *Ind. Eng. Chem.*, **40,** 2155—63 (1948).

109. Neher, H. T., in *Elastomers and Plastomers*, edited by R. Houwink, Elsevier, New York, 1949, Vol. II, Chap. 4.

110. Oehler, R., and Kilduff, T. J., *J. Research Natl. Bur. Standards*, **42,** 63–73 (1949).

111. Owen, H. P., Paper presented before Div. Rubber Chem., Am. Chem. Soc., Sept. 1949; *Rubber Age N.Y.*, **66,** 544–8 (1950).

112. Patnode, W., and Wilcock, D. F., *J. Am. Chem. Soc.*, **68,** 358–63 (1946).

113. Patrick, J. C., *Trans. Faraday Soc.*, **32,** 347–58 (1936).

114. Patrick, J. C., and Ferguson, H. R., U.S. Pat. 2,466,963, Apr. 12, 1949.

115. Patrick, J. C., and Mnookin, N. M., Brit. Pat. 302,270, Dec. 13, 1927.

116. Pietrusza, E. W., Sommer, L. H., and Whitmore, F. C., *J. Am. Chem. Soc.*, **70**, 484–6 (1948).

117. Polmanteer, K. E., Servais, P. C., and Konkle, M. G., *Ind. Eng. Chem.*, **44**, 1576–81 (1952).

118. Polis, A., *Ber.*, **18**, 1540–4 (1885).

119. Rainard, L. W. (Natl. Dairy Products Corp.), Methods of Curing Acrylate Polymers, U.S. Pat. 2,410,103, Oct. 29, 1946.

120. Ratchford, W. P., and Fisher, C. H., *Ind. Eng. Chem.*, **37**, 382–7 (1945).

121. Rehberg, C. E., and Fisher, C. H., Preparation and Purification of 2-Chloroethyl Vinyl Ether. Copolymers of 2-Chloroethyl Vinyl Ether and Ethyl Acrylate, U.S. Bur. Agr. Ind. Chem., AIC–255, 1950.

121a. Riddle, E. H., *Chem. Eng. News*, **31**, 2854–7 (1953).

122. Rochow, E. G., *J. Am. Chem. Soc.*, **67**, 963–5 (1945).

123. Rochow, E. G., *Chem. Eng. News*, **23**, 612–6 (1945).

124. Rochow, E. G., Private Communication.

125. Rochow, E. G., and Gilliam, W. F., *J. Am. Chem. Soc.*, **67**, 1772–4 (1945).

126. Rochow, T. G., and Rochow, E. G., *Science*, **111**, 271–5 (1950).

127. Rohm & Haas Co., *The Monomeric Acrylic Esters*, 2d ed., 1949, 24 pp.

128. Rothrock, H. S., U.S. Pat. 2,282,827, May 12, 1942.

129. Schneider, A. K., U.S. Pat. 2,441,085, May 4, 1948.

130. Schumb, W. C., Ackerman, J., and Saffer, C. M., *J. Am. Chem. Soc.*, **60**, 2486–8 (1938).

131. Schumb, W. C., and Saffer, C. M., *J. Am. Chem. Soc.*, **63**, 93–5 (1941).

132. Scott, D. W., *J. Am. Chem. Soc.*, **68**, 1877–9 (1946).

132a. Scott, J. R., *J. Rubber Research*, **18**, 30–2 (1949); Morley, J. F., Porritt, B.D., and Scott, J. R., *J. Rubber Research*, **15**, 215–35 (1946).

133. Semegen, S. T. (Goodrich), U.S. Pat. 2,411,899, Dec. 3, 1946.

134. Semegen, S. T. (Goodrich), U.S. Pats. 2,412,475–6, Dec. 10, 1946.

135. Semegen, S. T. (Goodrich), U.S. Pat. 2,451,177, Oct. 12, 1948.

136. Semegen, S. T., and Wakelin, J. H., *Rubber Age N.Y.*, **71**, 57–63 (1952).

137. Simmons, H. E., *India Rubber World*, **108**, 173–4 (1943).

138. Sommer, L. H., Pietrusza, E. W., and Whitmore, F. C., *J. Am. Chem. Soc.*, **69**, 199 (1947).

139. Stein, R. S., Krimm, S., and Tobolsky, A. V., *Textile Research J.*, **19**, No. 1, 8–12 (1949).

140. Stevens, W. H., *Trans. Instn. Rubber Ind.*, **18**, 17–31 (1942).

141. Thiokol Corporation, High Tensile Thiokol FA Compounds, Bull.

142. Thiokol Corporation, Importance of Plasticity in Processing Thiokol Compounds, Bull.

143. Thiokol Corp., Mixing Schedules for Thiokol FA Compounds, Bull.

144. Thiokol Corp., Thiokol Liquid Polymer LP–2, Bull.

144a. Union Chimique Belge, Belge Pat. 500,963 (1952). See R. Sidlow, *Chem. Products*, June 1953.

145. Wakeman, R. L., *The Chemistry of Commercial Plastics*, Reinhold, New York, 1947, 836 pp.

146. Warden, W. B., *India Rubber World*, **111**, 309–11, 317 (1944).

147. Warrick, E. L., U.S. Pat. 2,560,498, July 10, 1951.

148. Warrick, E. L., Hunter, M. J., and Barry, A. J., *Ind. Eng. Chem.*, **44**, 2196–202 (1952).

149. Weir, C. E., Leser, W. H., and Wood, L. A., *J. Research Natl. Bur. Standards*, **44**, 367–72 (1950).

150. Whitby, G. S., Private Communication to O.R.R., July 14, 1947.

CHAPTER 26

GERMAN SYNTHETIC-RUBBER DEVELOPMENTS

R. L. Bebb and L. B. Wakefield

The Firestone Tire & Rubber Company

INTRODUCTION

The development of the German synthetic-rubber industry up to 1945 is based on a long history of experiments extending forward from such early workers as Williams, Harries, and Ostromislensky. Much of the early laboratory work is discussed in detail in a number of books and reviews.[25, 29, 30, 60, 62–3, 69, 87, 95, 107, 133] Most of the laboratory work on which the actual manufacturing processes were based was carried out after the formation of the I.G. Farbenindustrie subsequent to World War I. The industrial development entered its most intensive phase with the German program directed toward self-sufficiency.

The evolution of this industry could be traced only through the patent literature up until the end of World War II. Since then a number of investigating committees have visited the German plants, and their reports have made available details concerning the preparation of the standard commercial varieties of synthetic rubber, the history of their development, and the research tools used in their study. Almost equal in interest to the details of the polymerization recipes used is information concerning unsuccessful leads which had been embarked upon and concerning the directions in which the Germans looked for future progress.

The German industry has seen the commercial application of both mass and emulsion polymerization techniques to the production of a number of different types of rubber. However, since emulsion polymerization was adopted for the large-scale development of tire rubber, this process is considered more fully than mass polymerization in the present chapter. Ionic catalysis of polymerization was used in the production of polyisobutylene (Oppanol), to which the American Butyl rubbers bear some relationship.

The information contained in the reports of the teams sent into Germany at the close of the Second World War reflects not only the judgment of the German workers who were interviewed, some of whom may not have been immediately connected with the work under discussion, but also the opinions of the various team members. Future publications from the German laboratories may possibly present data more complete than those in the teams' reports and differing from the latter in some points.[37] The review of the German work given in this chapter does not in general cover such aspects of the work as are discussed in other chapters.

GERMAN COMMERCIAL PRODUCTS AND PROCESSES

EMULSION BUTADIENE COPOLYMERS

The most important phase of the German work was the emulsion copolymerization of butadiene with styrene. The first commercial type of emulsion synthetic rubber was Buna S, made by similar recipes at Schkopau and Hüls. Buna S was next replaced by Buna S3 because of the latter's[138] improved adhesion and hysteresis characteristics.

Buna S. The original butadiene-styrene copolymer was made at Schkopau, the charge ratio varying, according to several reports, from 70/30 to 68/32, and the recipe being as follows.[58, 105]

Butadiene	70 or 68
Styrene	30 or 32
Nekal BX* (100% basis)	2.85
Linoleic acid	2.0
Sodium hydroxide	0.50
Potassium persulfate	0.45
Water	105

* See p. 965.

Polymerization was carried to 60 per cent conversion; the polymer had a Defo plasticity value of 4300 and contained 4.4 per cent of fatty acid calculated as stearic acid.[31, 34] The Buna S made at Hüls (designated Buna S2) differed from that made at Schkopau (Buna S1) in that the polymerization was carried to only 55 per cent conversion. As compared with natural rubber, the Buna-S types of synthetic rubber were stated to display improved heat and abrasion resistance in tires and to have slightly better resistance to swelling in gasoline than that normally observed for natural rubber.[71] The favorable heat resistance and good aging characteristics of Buna S led to its use in electric-cable applications. Water absorption was 65 per cent that of natural rubber—about in line with the usual value for deproteinized rubber.

The information on record concerning the German polymerization recipes is marked by considerable ambiguity as regards the ratio of monomers charged and the ratio in the final polymer. The situation is complicated by the practice of various plants of making slight changes in the charge ratio in order to control the properties of the product, and further by the German contention that polymerization variables produce polymers with different final ratios. An example may be found in the following production data for Hüls.[57]

Buna Type	S	SR	SS	S3
Charging ratio (Butadiene–styrene)	68/32	68/32	47/53	68/32
Consumption ratio per 100 pt. polymer	73.9/26.1	74.3/25.6	52.6/47.4	72.8/27.2
Polymer analysis	80/20 to 77.5/22.5		60/40 to 57.5/42.5	

Defo Plastometer. The entire German industry accepted the Defo test as a measure of the plasticity of synthetic rubbers.[20, 28, 34, 43, 55, 84, 92, 122] The name is a contraction of *Deformierung,* the test being designed to measure the deformation of polymers at elevated temperatures. The instrument employed resembles the Williams plastometer,[66, 134] widely used in the United States, with one major difference. The Defo value is obtained by measuring the weight in grams necessary to compress a cylinder of rubber 10 mm. in diameter and 10 mm. high to 40 per cent of its original height in 30 seconds. In the Williams plastometer, which is less flexible, a fixed weight (5 kg.) is applied, and the height of the test block is measured after a specified time. As in the Williams test, recovery is determined in the Defo test 1 minute after the load is removed and is expressed as a percentage of the original height. In Germany considerable emphasis is placed on the recovery value as a measure of polymer elasticity or nerve.[58] The Defo plastometer requires considerable skill on the part of the operator, and several tests are necessary before the actual plasticity reading can be established. A competent worker, however, can make the determination with speed and accuracy.

Heat Softening. The American butadiene-styrene copolymer, GR-S, was, in the polymerization process itself, made to a specified plasticity and supplied to users ready for compounding. The German approach, however, involved making a rubber of low plasticity—a tough rubber of high Defo number—and allowing the user, before compounding it, to soften it in his own preferred manner to the plasticity value most nearly suited to his processing equipment and product requirements. The Buna rubbers had to be used within 24 hours after they were softened; otherwise, they increased in nerve, became stiffer than they were originally, and could not be softened by a second heat treatment.[130] In general, lower temperatures and somewhat longer softening times produced a more uniform product.[9,18–20,35,43,50,94]

The heat-softening step involved an exothermic process, and care had to be taken to avoid an uncontrolled decomposition, the process being only partially reversible; if it was carried too far, the rubber resinified to a worthless material. The softening process was dependent on the presence of oxygen[55]; it would not proceed in the absence of air or in an inert gas, although the amount of oxygen absorbed during it was too small to be measured. Within certain limits, the quantity of air had no influence on the extent of softening. While a certain minimum amount of air was necessary to effect degradation, laboratory experiments showed that the solution viscosity of polymers softened by large volumes of air was only slightly lower than that of polymers degraded by relatively small volumes. In practice, at least 40 cubic meters of air was necessary for softening 100 kg. of Buna S.[101] The use of an oven was preferred to that of an autoclave for the softening process because it was applicable to continuous softening.[9, 34, 43, 59, 61, 99]

The Schildeschrank process started with finely divided Buna prepared by a Werner-Pfleiderer rotary cutter. A 2-in. bed of rubber on a metal screen belt was passed through an oven at 130° C. in 30 minutes. The Büttner process started with a sheet of rubber prepared on a Fourdrinier-type drier. The sheet was conveyed through an oven at 135° C. and rolled up at the desired Defo value of 700. The Kissell process at first resembled the original oven treatment of the polymer; in a later version, the rubber was prepared in strips or noodles, spread on trays, and heated at 130 to 150° C. for 30 to 40 minutes in an autoclave at 3 to 5 atm. of air pressure.

A study of the effect of polyvalent ions on the heat softening of polymers showed that magnesium and lead, introduced as linoleates, had a marked retarding action, zinc a rather weak retarding action, and calcium an accelerating action.[119] Consequently, the use of calcium chloride or a mixture of sodium and calcium chlorides was favored for coagulating latex in the course of preparing the rubber. Iron was also active as a heat-softening agent and was frequently introduced into the latex, either as a contaminant during polymerization or as an additive just before coagulation. The most satisfactory method of adding iron to the rubber appears to have been through the use of iron salts of saturated fatty acids.[126] The preferred method was to include 0.5 part of a paraffinic soap in the original polymerization charge and add an iron salt just before coagulation.[126]

Buna S3. During the summer of 1942 the production of Buna S was restricted by a shortage of linseed oil which served both as a polymerization regulator and as an auxiliary emulsifying agent.[104, 126] This shortage led to the development of the Buna-S3 recipe which required no linoleic acid. The new procedure made use, for auxiliary emulsification, of a mixture of synthetic saturated fatty acids prepared by the Fischer-Tropsch process[19] along with caustic alkali and, for regulation to improve the processibility of the product, of diisopropylxanthogen disulfide (Diproxid).

Buna S3 was produced in several of the plants by a continuous process with a butadiene-styrene ratio set at 68/32 although one report[40, 102] indicated a 70/30 charge ratio. Recipes are given in Table I.

While the ideal operation of the plants was based on pure monomers, actual production conditions necessitated the use of recycle monomers mixed with a constant amount of fresh material. The original purity of the butadiene was 98 ± 0.5 per cent; the recovered monomer analyzed about 1 per cent lower. The recycle butadiene was discarded as its purity approached 60 per cent. The fresh styrene was 99.5 per cent pure; the one-pass recycle styrene approached 95 per cent. According to one report, in normal practice 2 parts of fresh styrene for 1 part of recycle monomer was used.[40]

Agitation in all of the plants was extremely gentle in comparison with that available in the American plants; simple wooden paddles were attached to slowly rotating shafts. The process was a continuous one involving seven reactors (20 cubic meters, 5280 gal.) of chrome-vanadium steel alloy at Hüls and nine reactors at Schkopau. The Schkopau system of nine reactors involved the increment addition of modifier to the second, fifth, and seventh autoclaves. This corresponded to approximately 14, 25, and 43 per cent conversion. If all of the Diproxid were added initially, the polymerization rate was retarded, and the polymer viscosity increased throughout the run; a fairly constant viscosity material could be made by using increment additions. The Diproxid concentration was adjusted to give a product that softened to a Defo value of 320 after 50 minutes of heating. A modifier concentration of 0.1 part per 100 parts of monomer was regarded as the

Table I. Buna S3

	Ludwigshafen	Hüls	Schkopau
Butadiene	70	68	68
Styrene	30	32	32
Nekal BX	3.0	3.1	2.85
Paraffin acids	0.5	0.5	0.5
Sodium hydroxide	0.4	0.32	0.5
Potassium persulfate	0.4	0.45	0.45
Diproxid (total)	0.06–0.10	0.09	0.08
Water (treated)	106	105	105
Reaction temperature, °C.	50	45–50	47–50
Phenyl-β-naphthylamine	3.0	3.0	2.8
Coagulation	$CaCl_2$-NaCl-acetic acid	$CaCl_2$-NaCl $NaHSO_3$-$FeSO_4$	$CaCl_2$-acetic acid-$FeSO_4$

Per Cent Conversion by Reactor

Reactor No.			
1	7–11	20	9 out of 10
2	15–20	35	reactors in
3	25–30	44	series, to
4	32–37	52	58–60% conv.
5	40–45	57	
6	47–52	Standby	
7	52–60		
8	Standby		
Diproxid added in increments at Reactor no.	2, 4, 6	1, 3, 5	2, 5, 7
Total reaction time, hr.	24–30	25–30	37

maximum that could be used without sacrificing over-all quality for processibility.

Polymerization was effected at about 45° C. at Hüls and at 50° C. at Schkopau, to a final conversion of 58 to 60 per cent. A polymerization temperature of 50° C. was considered to be the maximum permissible in this system. If higher temperatures were used, dimerization increased with an undesirable effect on heat treatment and odor.[104]

Shortstopping agents were not mentioned as being used in Germany as they were in American practice; instead, a higher concentration of the antioxidant phenyl-β-naphthylamine was added at the end of the polymerization. At Hüls 3 per cent was added as a 20 per cent dispersion in Nekal BX; at Schkopau, 2.8 per cent was used. There is a reference to sodium hydrosulfite as a polymerization inhibitor.[86] And undoubtedly stopping agents became more important in the German program as activated polymerization formulas were used for low temperature operation or when latex was to be stored at a definite conversion for later use.

Butadiene was recovered from the finished latex to the extent of 90 to 95 per cent of the available monomer; styrene was removed by a vacuum steam distillation in which all but 10 to 25 per cent was recovered.

In the subsequent coagulation of the latex, crumb size was emphasized as being very important, since it was impossible to wash out all of the emulsifier if the crumb was too large. The final analysis of the crumb was set at 0.15 per cent sulfur and 1 per cent ash and moisture.

The change from Buna S to Buna S3 was generally favored throughout the industry although Dr. Konrad believed[103] that skid resistance had been sacrificed by the move. Buna S3 was preferred for tire bodies, having better tack than S1 but similar tensile strength, cold resistance, and abrasion resistance.[31, 34–5, 58, 83, 85, 105, 138] The German tires made from Buna S3 during World War II displayed poor ply adhesion, but the failure of heavy tires was attributed more to poor rayon than to inferior polymer. Although the rayon was specified as 3.0 grams per denier, during the war it often tested as low as 2.75 to 2.50 grams per denier. Stöcklin asserted that Buna-S3 tires were 35 per cent higher in road wear than the best natural-rubber tires.[129]

Buna S4. The several teams disagreed about the nature of the polymer bearing the designation Buna S4. It was stated[138] that Klein and Weinbrenner developed Buna S4 at Schkopau as a modification of Buna S3 by introducing a larger amount (0.16 part) of the modifier (Diproxid) in equal increments of 0.04 part, initially and at 20, 40, and 50 per cent conversion. The polymer was softer than Buna S3 and more nearly resembled GR-S in Defo and K values (see p. 967), although it was somewhat higher in osmotic molecular-weight value. The introduction of still more modifier (total of 0.24 part) all at the beginning of a Buna-S4 charge lowered the molecular weight as shown in Table II.

Table II

	Defo	K Value	Osmotic Molecular Weight
GR-S	625	104	79,000 $\pm$ 7,000
Buna S3	600*	99	700,000 $\pm$ 320,000
Buna S4 (0.16 increment-wise)	725	103	300,000 $\pm$ 55,000
Buna S4 (0.24 at outset)	700	114	110,000
American Neoprene	1500	73	450,000
Unmilled natural rubber	...	171	400,000
Milled natural rubber	...	75	80,000

* Heat-softened.

It was considered at Schkopau that the Klein and Weinbrenner Buna S4 gave lower running temperatures, but the Leverkusen group considered it to be no real advance over Buna S3, since it had lower rebound characteristics. One of the principal changes involved in Buna S4 was the elimination of the heat-softening step; this change, however, was not immediately approved, since considerable data indicated that a heat-softened polymer of high initial Defo value was superior to a polymer modified sufficiently to permit immediate processing.[89]

One of the other polymers which was designated Buna S4 was introduced by Jost[138] after a study of Fikentscher's investigation of polymer structure. An unmodified 69/31 butadiene-styrene charge was sampled at 10 per cent intervals of conversion, each sample being fractionated for examination by viscometric, osmotic-pressure, ultracentrifuge, and perbenzoic acid titration techniques. At 40 to 50 per cent conversion, the average molecular weight of the polymer was 160,000, whereas new chains being formed at this stage had a value of 400,000. Beyond 45 per cent conversion the polymers became insoluble. Hence 45 per cent conversion was selected for the initial introduction of Diproxid to reduce insolubility. Such a technique permitted carrying the conversion to about 80 per cent without loss of quality. The rubber needed no heat softening before being processed on a mill and showed better building tack than the other Buna-S types.[22, 138]

Jost concluded that (1) the amount and nature of the emulsifying agent (considering fatty acid soap, Nekal, Mepasin, and Esteramin types) had little effect on polymer structure and (2) to judge from Defo values as a molecular-weight indicator, there was little change in going from acid to neutral conditions; although a further increase in pH increased the Defo value. Neither the choice of catalyst nor the ratio of monomers to aqueous phase altered the relative amount of 1,2-units in the rubber. Jost considered that the polymerization temperature had a questionable effect.[138]

A third polymer assigned the designation Buna S4 [13, 17, 22, 103] consisted of a fatty acid-free version of Buna S3, but little work was done on it.

Buna SW (S10). Another polymer which was prepared in a fully modified condition and which did not require heat softening was a low-styrene butadiene-styrene copolymer known as Buna SW (W: Weich). It was originally prepared for low-temperature applications as a substitute for Buna S3 in cable insulations requiring low-temperature flexibility and in airplane packings and gaskets:[7, 58, 94]

Butadiene	90.0
Styrene	10.0
Water	140.0
Nekal BX	4.0
Sodium hydroxide	0.2
Potassium persulfate	0.2
Diproxid	0.33*

* 0.11 at 0, 20, 45 per cent conversion.

The charge was run for 60 to 70 hours to 60 per cent conversion. Phenyl-β-naphthylamine was used as the antioxidant, and coagulation was effected with calcium chloride. The Defo value was adjusted[53, 103] to 600 to 800. The tensile strength was 190 kg. per sq. cm. (2700 p.s.i.), the elongation 600 per cent, and the room-temperature rebound 42 per cent. Buna SW surpassed all other Buna types in laboratory abrasion tests, although in tire tests it was judged less satisfactory than Buna S3.[35]

Buna SW was outstanding in its resistance to stiffening at low temperatures. The properties at $-40°$ C. compared well with the properties of other rubbers:[103]

	Buna S	Natural Rubber	Buna SW
Rebound elasticity	8%	10–12%	20–25%
Tensile strength	350 kg. per sq. cm.	...	350 kg. per sq. cm.
Elongation	100–150%	...	300–400%

The development of improved polymers for tire applications was expected to continue in the direction of better tack, improved adhesion to decrease ply separation, and better processing characteristics. It was believed[138] that for tire applications a decreased styrene charge should favor wear resistance and resilience. It was hoped that this goal could be reached through low-temperature polymerizations and that low-temperature polymerization would also lead to better structural regularity in the polymer. Further, it was expected[75] that a decreased styrene ratio would increase the "netting" of the Buna-SW molecules. Therefore, the charge was varied to include enough modifying agent (regulator) to give a processible rubber, without having to resort to the heat-softening step. In a 1944 report, this move was declared not to be entirely satisfactory since the product lost some wear resistance with the change in monomer ratio. In spite of this deficiency, tire-test values were nearer to those of Buna S3 than S1. This special Buna was not generally used as a tire rubber but was used rather in carcass stocks and for airplane applications where low-temperature service was needed.

Buna SR. Early in the development of synthetic rubbers, an effort was made to improve the processing characteristics of Buna S by blending it with Buna R. The latter was a very high-conversion variation of Buna S polymerized at 50° C. up to 50 per cent conversion and then at 90° C. to 96 per cent conversion.[58] A blend of equal parts of Buna S and Buna R had a Defo value near 3500.[35, 58, 105] In line with Jost's experiments, this technique gave a more heterogeneous rubber with better milling characteristics.

Buna SR had[31, 34, 130] cold resistance superior to that of Buna S or Buna S3 and was therefore used in tire tubes. The importance of this polymer in World War II was emphasized by the insistence of the German Naval Testing Station on its use[80] because it most nearly duplicated natural rubber in a torpedo depth-control membrane.

Acid Buna. Acid Buna was developed as a Buna-S3-type polymer made under acid conditions, with Esteramin (p. 965) as the emulsifier, and coagulated with alkali.[33, 83] At a final Defo value of 2500 to 3000, it was considered better than Buna S3, giving stocks with lower hysteresis and higher elasticity. Indoor tests in one laboratory showed it to be better than Buna S1, Buna S3, or natural rubber. However, the Ludwigshafen laboratories did not favor the acid Buna types since they believed that their properties were not sufficiently outstanding to justify commercial application.[89]

Buna SS.[15] This synthetic rubber contained a higher proportion of styrene than any of the previously mentioned Bunas. It had better electrical properties than any other Buna type. Rubber goods made with it were inferior to natural rubber in electrical properties when first made, but after a short period of storage the properties changed in favor of the Buna, making it particularly favorable for the manufacture of cables. The higher styrene content of Buna SS reduced abrasion resistance, rebound, and low-temperature resistance but increased, even in the presence of carbon black, processibility and ease of calendering.[2, 130] The polymer had only a secondary interest in the tire field, where its main use was for inner tubes.

Polymerization was conducted in a formula similar to that used for Buna S but with a butadiene-styrene ratio[34, 57-8, 130] of 55–47/45–53. The Defo value was near 3700.

Buna SSE. The polymer was intended for pharmacological use or for uses where its nonstaining and nonconducting properties would be of value. Buna SSE (E: Eisenfrei) was made in a recipe similar to that for Buna SS but with iron-free water. The conversion reached 60 to 62 per cent in about 30 hours at 40 to 45° C. Oxycresyl camphane was added as the stopping agent and stabilizer; coagulation was effected with solutions made up with iron-free water. The final Defo value was approximately 2400 to 3500. The polymer had a high tack in solutions but was relatively unstable. A benzene solution underwent "cyclization" on standing with the precipitation of a benzene-insoluble product.

Buna SSGF. This polymer was intended for uses requiring as complete freedom from odor and taste as possible. The primary use of Buna SSGF (GF: Geruch- und Geschmackfrei)[58, 94, 105] was in bottle-cap liners and jar rings. It was made in a polymerization recipe similar to that for Buna S3, but the butadiene-styrene ratio was 50/50, and condensate water was used. The reaction temperature for this charge was 40 to 45° C. at which the conversion reached 62 per cent in approximately 30 hours. Phenyl-β-naphthylamine (3 per cent) was added before coagulation. Particular care was taken to remove all unreacted monomers during the stripping operation and to wash the polymer thoroughly after coagulation with calcium chloride (5 kg. calcium chloride per 100 kg. of polymer). The final plasticity was adjusted to a Defo of 3000.

In 1943, when their production was substantially at its maximum, the output of the four last-mentioned special grades of butadeine-styrene copolymers was as follows: Buna SS, 334; Buna SSE, 118; Buna SSGF, 337; Buna SW, 193 tons.

GERMAN COMMERCIAL PRODUCTION OF
BUTADIENE-STYRENE COPOLYMERS

The German synthetic-rubber industry consisted of four large producing units and two smaller experimental ones. The largest unit was at Schkopau; it was put into operation in 1939 and had a rated capacity of 6000 metric tons per month.[45] The second largest plant was at Hüls, with a capacity of 4000 tons; its construction was started in 1938 and operation began[57] in August 1940 and continued until March 29, 1945. The plant at Auschwitz was rated at 3000 tons a month, but operation had only reached the state of producing acetaldehyde from acetylene before the plant was seized by the Russian army. The smallest of the plants was at Ludwigshafen and had a capacity of 2000 tons; this plant contained the research and development departments of the German industry. Full production was never reached in the plants of the German industry. As shown by the data on p. 53, the maximum production in any one year (1943) was about 110,000 tons.

The operation of the Hüls plant is described in the literature more completely than that of any other German plant,[5, 6, 21, 39, 57, 84] and unless otherwise indicated the following description of the manufacture of Buna emulsion polymers refers to the procedure at that plant. Reference is made to the other plants only at points where the procedure therein differed significantly from that at Hüls. Hüls was a self-contained plant producing Buna S2, S3, SR, and SS.

Monomers. Hydrocarbon gases were cracked by the electric-arc process to acetylene, which was then converted to butadiene by the aldol process; at Schkopau[45, 57] acetylene was prepared in the conventional manner from calcium carbide. Styrene was made at Hüls from benzene obtained from the Ruhr coal-tar plants and ethylene prepared largely by the hydrogenation of acetylene.

A typical and a specification analysis for butadiene made at Hüls follows:

	Per Cent by Weight	
	Typical	Specification
Butadiene	98.5–99.0%	98.0% min.
Active oxygen	0.0005–0.001%	0.001 max.
Acetaldehyde	0.002–0.004%	0.005 max.
Water	0.01–0.03%	0.05 max.
Nesslers reagent	Colorless to yellow	Yellow
	Clear to faint turbidity	Slight turbidity
Jlosvay reagent	Colorless, clear	Faint rose
	Turbidity in methanol	No turbidity

Butadiene was stored in underground tanks without the addition of inhibitor. A sodium hydroxide solution was added to the storage tanks at Hüls to destroy peroxides of butadiene and of polybutadiene, which were observed to settle out as oily droplets. The diene was pumped into 50-cubic-meter tanks (13,200 gal.) for preblending with styrene before being metered into the reactors; four such blending tanks were supplied to allow the production of several different Buna types simultaneously.

Styrene was likewise stored without an inhibitor unless it was being shipped or stored for an unusually long period, in which case 0.01 per cent hydroquinone was introduced. Typical analyses for fresh styrene at Hüls were as follows.

	Typical	Specification
Styrene (bromine method)	99.5–100%	99.0% min.
(refractometer)	99.4–99.8%	
p-Divinylbenzene	0.001%	0.005% max.

Recycle hydrocarbons were recovered from the latex and redistilled. Normal practice involved mixing 5 to 30 per cent by weight of 95 per cent recovered butadiene with fresh monomer to an average purity of 98 per cent. Fresh styrene was also mixed with recycle styrene of 94 to 96 per cent purity.

Emulsifiers. Emulsifier solution for the polymerization process consisted of Nekal BX (p. 965) dissolved in water. The solution was known as Emulgator 1000 and was prepared in 50-cubic-meter (13,200 gal.) tile-lined tanks. At Schkopau, ion-exchange water was used in the Emulgator make-up and throughout the process, because of the natural hardness of the available supply. To the Emulgator 1000 were added linoleic acid (for Buna S, SS, or SR) or paraffinic fatty acids (for Buna S3) along with sufficient caustic for saponification. The pH was adjusted to 10 to 12, and the solution was discharged into the reactors at 40° C. The paraffin fatty acids consisted of a mixture of acids ranging from C_{10} to C_{15} prepared by the Witten Fettsäurewerke through oxidation of paraffin oils from the Fischer-Tropsch plants.

Catalyst and Modifier. The catalyst was potassium persulfate, added continuously as a 3 to 4 per cent aqueous solution to the first reactor. The solution was prepared in a rubber-lined tank and pumped through stainless-steel lines to the reactors.

Modifier, viz., diisopropylxanthogen disulfide ("Diproxid"), was made up as a 5 to 10 per cent solution in styrene, which was blown by nitrogen to calibrated feed tanks.

Polymerization. The temperature for all of the polymerizations was 45 to 50° C., except, as already explained, in the case of Buna SR. The polymerization equipment consisted of eight lines of six 20-cubic-meter reactors (5280 gal.), five of which were in use at one time. Three of the lines consisted of lead-lined steel reactors, two were chromium-plated, and three were made of stainless steel. The reactors were pressure-tested at 25 atm., being designed for an operating pressure of 15 atm. and usually operating at 8 atm. They were used in the continuous process full of liquid with the pressure being controlled by hand at the outlet valve. Agitation was supplied by V_2A metal stirrers with three horizontal paddles spaced along the shaft, operating at 30 r.p.m.

The Emulgator solution and the hydrocarbon blend were pumped continuously in the desired ratio into 200-gal. turbomixed emulsifying vessels, located at the head of each line of reactors. The catalyst was pumped into the first reactor to initiate polymerization. In Buna S3 preparation, Diproxid was also charged into three of the vessels (first, third, and fifth) by means of small piston pumps.

Approximately 1500 kg. of hydrocarbons and 1800 kg. of aqueous phase were charged each hour in the production of Buna S. The polymerization charge entered each reactor through a line extending to the bottom; the effluent left from the top, all reactors being operated full. Based on this feed rate and the volume of five vessels (100 cubic meters, 26,400 gal.), a holdup or reaction time of 30 hours was indicated. Because more water was included in the Buna SS recipe, the hourly feed rate of hydrocarbon phase had to be reduced somewhat. Whereas the polymer yield rate of Buna S at 60 per cent conversion was 9 kg. per cubic meter per hour, for Buna SS it was only 8 kg. per cubic meter per hour. On the basis of the latter production rate, the reaction time for Buna SS was calculated to be about 27 hours. Reaction times reported by the Hüls personnel as 25 to 30 hours are in good agreement with the values estimated above for Buna S and SS.

The conversion was followed in each reactor every 4 hours by coagulating a 50-gram sample of latex with 300 ml. of a mixture of methanol (900 ml.), calcium chloride (50 ml. of concentrated solution), and acetic acid (3 ml. glacial), basing the conversion on the weight of the washed and dried rubber.

Stripping. The latex from the last reactor was filtered and stabilized by the addition of 3 per cent phenyl-β-naphthylamine on the rubber. The stabilizer was originally introduced as a 20 per cent dispersion in Nekal, but this practice was supplanted by the use of a 17 per cent solution in styrene.

Unreacted monomer was recovered from the finished latex by a vacuum steam distillation. The latex was fed into the top of a column and mixed with steam to vaporize the monomers. The foam which developed in this step was broken down by discharging the latex into a large flash chamber at a high velocity. The latex was fed into a smaller tower and mixed with low-pressure steam to reduce the unreacted styrene to 0.1 per cent on the rubber. This process was conducted at 100 to 120 mm. pressure. At Schkopau stripping was done in a single-stage flashing process followed by a second-stage steam distillation, the recovery of unreacted styrene being 90 per cent and of butadiene 95 per cent.

Coagulation. For the coagulation of Buna S latex, solutions were made up, consisting of 11 per cent brine prepared from rock salt, 40 per cent sodium bisulfite, and 0.034 per cent ferrous sulfate. The consumption of coagulants per 100 parts of polymer was 35.4 parts sodium chloride and 6.9 parts sodium bisulfite. The latex was coagulated in a $2^1/_2$ in. glass pipe containing two special tees for the introduction of latex and coagulant. The latex entered the first tee at 30 to 40° C., receiving a spiral motion by the shape of the nozzle. Brine was introduced into the side of the same tee directly opposite to the latex stream. The creamed latex moved about 24 in. to the second tee where bisulfite solution was added to agglomerate the creamed particles. For the coagulation of the Buna-S3 latex, calcium chloride was added to the brine solution; 10.1 kg. of calcium chloride was added per 100 kg. of Buna S3, along with 38.4 kg. of sodium chloride and 11.4 kg. of sodium bisulfite. A ferrous sulfate solution was also added to the Buna-S3 coagulant to accelerate the subsequent heat-softening step. For this purpose it was customary to add 0.07 kg. $FeSO_4 \cdot 7H_2O$ per 100 kg. of Buna S3.

At Schkopau coagulation was effected in the same manner as at Hüls but with different reagents, the composition of chemicals being as follows:

Chemical Consumption for Coagulation at Schkopau

Parts per 100 Parts Rubber

	Buna S	Buna S3
Sodium chloride	60	...
Calcium chloride	...	6.0
Acetic acid	0.75	0.8
Ferrous sulfate	...	0.12
Dilution water	5000	5000

The coagulation step at Ludwigshafen was conducted in a continuous operation.[102] Latex was allowed to flow by gravity through a 3-in. polyvinyl chloride pipe, and a stream of coagulant was pumped into the center of the stream through a special fitting which imparted a spiral flow to the latex.

The flow rate of the various solutions was specified as latex, 2400 liters per minute; calcium chloride (36 per cent solution), 250 liters per minute; sodium chloride (saturated), 1000 liters per minute; and water, 3000 liters per minute. A second nozzle was located about 15 ft. further along the pipe, and through this was introduced a stream of 16 liters acetic acid in 12,000 to 16,000 liters of water. A third jet was used to add further water, if necessary, and to permit adjusting the final pH to 5 to 6.

The quantities of chemicals used in the coagulation process, on the basis of 100 kg. rubber, were:

	I.	Water	1000 liters
		Sodium chloride	5 kg.
		Calcium chloride	10–20 kg.
	II.	Water	4000 liters
		Acetic acid	1.0–1.5 kg.

Washing and Drying. The fine crumb produced by coagulation was carried as a slurry through a specially designed weir box overflowing onto a Fourdrinier machine equipped with a fine screen 2 meters wide and approximately 19 meters long, traveling at a rate of 8 to 11 meters per minute. Care was taken at this stage to wash all of the Nekal out of the rubber. The filtrate was passed through a settling basin and discarded. The rubber passed under a roller, and vacuum was applied from the under side for a distance of about 24 in.; the blanket was picked off the screen and passed through a 19-pass drier divided into four horizontal compartments. The temperature of the first, or top, compartment was maintained at 125° C. by means of steam coils. The second was held at 115° C., the third at 90 to 100° C., and the fourth at room temperature to cool the polymer. The finished blanket was slit down the middle and rolled up to a desired weight of about 50 kg. At Schkopau, a more usual type of drier was used; it was divided into three zones of decreasing temperature, viz., 120, 105, and 98 to 100° C.

PROPERTIES OF VULCANIZED BUNA-S-TYPE POLYMERS

There are several different sets of physical testing data for Buna S3 in the German literature. One report[89] gives the following data for a stock consisting of Buna S3 100, CK3 black 40, zinc oxide 5, Kautschol (softener) 5, stearic acid 2, sulfur 1.8, Vulkazit AZ (accelerator) 1.0:

Raw Defo Plasticity after Heat Softening	550, Recovery 26%
Compounded Defo Plasticity	1400, Recovery 28%

Cure at 2.1 atm.	Tensile Strength, Kg. per Sq. Cm.	Modulus at 300%, Kg. per Sq. Cm.	Elongation, %
40 min.	212	78	555
60 min.	231	84	565
80 min.	208	86	510

Shore hardness (40 min. cure)	67
Rebound, % (20° C.)	47
(70° C.)	52
Branching index	18

In Table III are given the results for several types of Buna rubber in a stock consisting of Buna rubber 100, Kautschol 5, stearic acid 2, zinc oxide 5, CK3 black 40, sulfur 1, and Vulkazit AZ 0.8, cured for 60 minutes[127] at 134° C.

Table III

	Buna S1	Buna S4	Buna S3	Buna SS
Defo plasticity of raw rubber	5000–7000	4000–5500	2700–3500	3400–5000
Defo plasticity after softening 50 min. in air at 3 atm. and 130° C.	500–1200	800–1800	400–2000	300–1100
Defo plasticity of compounded stock	5000–6000	6000–7000	3000–4000	3000–4000
Tensile strength, kg. per sq. cm.	250	210	240	230
Elongation, %	500	450	700	550
Modulus at 300%	90–110	90–110	60–90	50–70
Abrasion	100	90	140	105
Tack	100	90	...	120
Rebound (room temperature), %	51–53	47–48	49–51	22–23

COMPARISON OF BUNA S3 AND GR-S

At the end of the war samples of Bunas from current production were made available to American laboratories for compounding experiments and comparison with GR-S. In a comparison of Buna S3 with GR-S, the laboratories of the Copolymer Corporation reached the following conclusions:[103]

	S3 without Heat Treatment	S3 Heat-treated 1 hr. at 135° C.
Rate of cure	Slightly faster	Similar
Modulus	Slightly higher	Slightly lower
Tensile strength	Much higher	Much higher
Elongation	Similar	Higher
Tear resistance	Similar	Superior
Hysteresis	Slightly inferior	Superior
Crack-growth resistance	Superior	Greatly superior
Gel content	Much higher	Slightly higher

In view of the several advantages shown by Buna S3, a program of work was undertaken to prepare a Buna-S3 polymer from American starting materials and along with it a GR-S polymer carried to only 60 per cent conversion and having a high Mooney value (obtained by appropriate use of *tert*-dodecyl mercaptan as modifier). Advantages found for the Buna-S3 analog included better flex-crack resistance, better aged hysteresis, and better processibility at a plasticity of 55 Mooney. The Buna-S3 type was, however, more difficult to heat-soften; its heat buildup was high, and its polymerization slower, requiring a 20 per cent longer reaction time.

Another American laboratory[27] established the effect of varying the Diproxid in a polymerization charge of the Buna-S3 type as follows:

Diproxid	Conversion, % (18.5 Hr., 50° C.)	Plasticity (ML)	Gel, %	Swelling Index	Intrinsic Viscosity
0.10	57	134	35.4	87	3.28
0.15	58	90	0.77	...	3.02
0.20	57	63	0.40	...	2.42
0.25	55	33	0.53	...	1.90

A pilot-plant preparation was then made, using American starting materials in the Buna-S3 system. Twelve thousand pounds were made in

Charging Formula: X-343

Water	106
SA 178*	5.0
Potassium persulfate	0.4
Sodium hydroxide	0.4
Stearic acid	0.5
Diproxid†	0.05
Styrene	28
Butadiene	70

Add at 20% and 40%:

Styrene	1
Diproxid	0.05

* *sec*-Butylnaphthalene sodium sulfonate prepared by the General Aniline & Film Corp., as a substitute for Nekal BX.

† Diisopropylxanthogen disulfide.

24 to 29 hours at 133° F. to 60 per cent conversion. Hydroquinone (0.1) and phenyl-β-naphthylamine (1.2 part) were added as the shortstop and antioxidant, respectively. The principal difficulty encountered during the polymerization was the excessive foaming during stripping. The latex was coagulated by a salt and acetic acid solution, and the rubber was dried for 35 minutes at 225° F. The heat-softening characteristics at 240° F. were similar to those of German Buna S3.

The American preparation X-343 was built into experimental tires for comparison with a Buna-S3 control but did not give such good results as the control.

	Wear Rating,* 18,000–20,000 Miles
E-120D X-343, heat-softened	92
E-120E German Buna S3, heat-softened	135
E-120F Half X-343 heat-softened	92
Half Buna S3 heat-softened	132

* Against GR-S-AC as 100.

In 1946 the Goodyear laboratories compared a sample of Buna S3 with GR-S in a tire-tread formula, with the following conclusions:

	Buna S3
Processing	Very much poorer
Rate of cure	Equal
Modulus at 300% (room temperature)	Equal
Tensile strength (room temperature)	Slightly better
(hot)	Much better
Brittle point	Equal
Rebound, room temperature	Better
hot	Equal
Heat rise (Goodyear method)	Very slightly better
Pierced-groove flex-crack resistance	Very much better
Abrasion (ring)	Poorer (?)
Set, % (Goodrich flexometer)	Poorer (?)
Speed of retraction	Better

A preparation of Buna S3 was made by the Goodyear pilot plant, and the following conclusions were reached in a comparison[51] of its properties with those of German Buna S3 and GR-S.

1. In the pilot-plant polymerizations a domestic emulsifier SA-178 (sec-butylnaphthalene sodium sulfonate) was used satisfactorily in place of Nekal BX in the production of polymer of the Buna-S3 type. Myristic acid was used in place of the oxidized Fischer-Tropsch hydrocarbon employed in Germany and was found to contribute markedly to increasing the reaction rate, the stability of the latex, and the ease of heat softening.

2. The physical properties of Buna S3 thus made and of its vulcanizates were similar to those of German-made Buna S3.

3. No significant difference was found between GR-S and Buna S3 in regard to iodine number, infrared absorption, or content of side vinyl groups and of styrene.

4. Buna S3 has apparently a narrower molecular-weight distribution than GR-S and contains less than the latter both of low-molecular-weight and of high-molecular-weight polymer. This difference is thought to be responsible for the fact that increase in resilience with increase in temperature is relatively greater for vulcanizates of GR-S than for those of Buna S3. At room temperature Buna S3 is markedly superior to GR-S in resilience.

5. When Buna S3 is heat-softened to the same processibility as GR-S and vulcanized to the same state of cure as judged by set, it shows no advantages in laboratory flex-life tests over GR-S. However, there are definite indications that satisfactory properties can be obtained with Buna S3 at a lower state of cure than can be used for GR-S, with correspondingly improved flex properties.

The Leverkusen laboratories reported the following results in a preliminary comparison of GR-S with Buna S3 heat-softened to the same plasticity.[73]

	GR-S	Buna S3
Osmotic molecular weight	80,000	700,000
Viscosity (in terms of K)	104	99

It was concluded that GR-S was more highly branched in structure than Buna S3.

	Defo Plasticity	Defo Elasticity	Rebound at 20° C.	Rebound at 70° C.
GR-S	625	36	37%	45%
Buna S3 (heat-softened)	550	26	47	52

Both rubbers were reported to be similar in tensile strength and elongation. It was considered that the higher rebound of Buna S3 was an important favorable characteristic and was independent of the compounding recipe.

A more complete laboratory comparison subsequently reported by Ecker[42] covered GR-S, Buna S3, and low-temperature GR-S (X-485).

1. Chemical analysis of the polymers showed Buna S3 to have more antioxidant and higher ash than either GR-S or X-485. The latter contained less styrene and fewer 1,4-butadiene units than either GR-S or Buna S3; it contained significant amounts of rosin acid.

2. When mixed in a tread-type stock, cured, and tested, X-485 was found to have a somewhat higher tensile strength than either GR-S or Buna S3. In a carcass stock recipe, however, Buna S3 showed a definite superiority in tensile strength.

3. The hysteresis of X-485 at room temperature was between that of GR-S and of Buna S3, the latter being the best; at temperatures above 70° C. there was little or no difference between the last two polymers.

4. In abrasion resistance X-485 was inferior to Buna S3, but in cut-growth resistance (DeMattia) it was slightly better. The addition of rosin acid to Buna S3 improved its cut-growth resistance until it was superior to X-485.

5. No difference in age resistance was found between X-485 and Buna S3.

6. Philblack O was found to stiffen Buna S3 somewhat more than CK3 black, but there was little difference between the blacks in respect of other properties.

7. The reinforcing white pigment "Teg" alumina reinforced Buna S3 to a greater extent than X-485.

A laboratory and pilot-plant study by H. L. Williams and co-workers of systems of the Buna-S3 type indicates some of the special features of the system and the difficulty of duplicating the German Buna-S3 polymer exactly.[36, 47] Polymerization in this system, comprising the sodium salts of alkylnaphthalene sulfonic acids and of fatty acids along with persulfate, in contrast to the GR-S system, comprising sodium fatty acid soaps and persulfate, was independent of the presence of a mercaptan or Diproxid. Further, in the Buna-S3 type of emulsion, the rate of mercaptan disappearance was much faster than in the GR-S system; consequently, the regulating index of mercaptans was much higher in the Buna-S3 than in the GR-S charge. The rate of disappearance of Diproxid in the Buna-S3 emulsion was also very rapid; when as much as 0.5 part was added at the start of polymerization at 45° C., only 6 per cent remained when the conversion reached 30 per cent. Hence, increment addition of Diproxid is essential if a Buna-S3 polymer of satisfactory quality is to be obtained.

The study indicates the possibility that careful control of the Diproxid addition throughout the course of the polymerization will yield a polymer of substantially uniform viscosity and a final product of high viscosity, containing gel sufficiently "loose" (i.e., having a high swelling index) to insure easy thermal breakdown and yielding vulcanizates of good quality. To achieve such a result, however, would seem to call for very nice adjustment throughout the polymerization. One factor of significance, aside from increment addition of the modifier at proper times, is the rate and type of agitation applied during polymerization. In the manufacture of Buna S3 in Germany, the agitation is milder than that in the manufacture of GR-S. Also, the nature of the coagulants used to isolate the rubber from the latex significantly affects the ease of thermal breakdown of the Buna-S3-type polymer, the best results apparently being obtained with sodium chloride and sulfuric acid. In agreement with other workers and with German practice, the addition of ferrous sulfate in the coagulant was found to facilitate heat softening.

Considered broadly, the study just reviewed leaves the impression that the Buna-S3 system is more sensitive and requires a nicer degree of control than the GR-S system. Further, polymerization in the former system is markedly slower (with Diproxid as the modifier) than in the GR-S system, reaction times for 60 per cent conversion at 45° C. being of the order of 30 to 48 hours. With the development of cold GR-S, interest in the further study of the Buna S3 system has waned.

NITRILE RUBBERS

The butadiene-acrylonitrile copolymers developed in Germany were known first as Buna N and Buna NN, later as Perbunan and Perbunan Extra. They were reported to possess excellent resistance to numerous solvents and therefore found wide use in German industry.[96, 121] Because of this high solvent resistance, it was possible to manufacture from them many special articles for which rubber had been previously unsatisfactory, such as gaskets, rubber springs, bumpers, and clutches. Also, Perbunan had considerable value in marine cables, offering a heat- and oil-stable covering which weighed much less than lead. In electrical applications, Perbunan was classified as a half-conductor and used to make conducting rubber goods (conveying belts in mines and special tires such as tail wheels for airplanes). Perbunan Extra (60 butadiene/40 acrylonitrile) showed a gum tensile strength better than that of any other butadiene copolymer but not high enough to be of practical importance.

The plant at Leverkusen was responsible for the preparation not only of the specialty Buna-S types but also of the Perbunans, which are listed in Table IV and for which the polymerization recipes are given.

In the production of the nitrile rubbers the line of reactors was made up of six stainless-steel autoclaves, each of 10 cubic meters capacity, five being used for the continuous polymerization. They were completely filled with latex during the polymerization and the contents agitated at 35 r.p.m. Diproxid dissolved in acrylonitrile was pumped into the charge in three equal increments at 0, 20, and 45 per cent conversion. The latex was discharged from

Table IV

Perbunan[31, 58, 94, 105]	A solvent-resistant rubber prepared by copolymerizing butadiene (74) and acrylonitrile (26)
Perbunan Special[58]	Perbunan latex for impregnation purposes
Perbunan Extra[58, 94, 105]	A high-acrylonitrile polymer for increased solvent resistance
Igetex NN[58]	A high solids (45%) latex of the Perbunan-Extra type used for dipped goods of high film strength

Polymerization Charges for Perbunan[58]

	Perbunan (Buna N)	Perbunan Extra (Buna NN)	Igetex NN Latex
Butadiene	74	60	60–65
Acrylonitrile	26	40	40–35
Water	150	200	57
Potassium persulfate	0.20	0.20	0.5
Diproxid	0.30	0.27	0.27–0.39
Nekal BX (100% basis)	3.60	3.60	...
Mepasin sulfonate	...	...	3.75
Tetrasodium pyrophosphate	0.3	0.3	0.3
Sodium hydroxide	0.05–0.10	0.10	0.1
Antioxidant (on the rubber)			
Phenyl-β-naphthylamine	3	3	
Hydroquinone			1
Oxycresylcamphane			2
Reaction time, hr.	25–30	60	30
Temperature, °C.	30	24	25
Type of reaction	Continuous	Batch	Batch
Conversion, %	75	75	62–65
Bound nitrile, %	26	36	...
Defo plasticity	2400–2800	...	...

the last autoclave at 75 per cent conversion, filtered, and treated with a dispersion of phenyl-β-naphthylamine. The unreacted monomers were recovered by a steam distillation; it was stated that 90 per cent of the unreacted butadiene and 50 per cent of the acrylonitrile was recovered. The latex was coagulated by the addition of a 25 per cent solution of sodium chloride (150 kg. sodium chloride solution per 100 kg. polymer). The rubber was dried at 125° C., being cooled to 90° C. in the final stages of the drier.

The processing of Perbunan-type polymers was made difficult by their failure to soften during milling. One solution of this difficulty involved handling the polymers on a cold mill where there was little possibility of overmilling; low-friction and low-speed mills were considered desirable.[1] A second method of handling the rubber was through the introduction of softening agents during the compounding step.[70, 71, 106, 121] A detailed review of the various methods of handling polymers of the Perbunan type may be found in the review of Koch.[71]

The production at Leverkusen is shown in Table V.

Table V. Production of Perbunan-Types, Tons

	1939	1940	1941	1942	1943	1944 (to Oct. 26)
Perbunan	960	1692	2433	2535	3341	2802
Perbunan Sp. (dry basis)	97	114	72	49	93	93
Perbunan Extra	69	92	136	230	222	234
Igetex	Made irregularly at rate of 1–2 tons per month					

SYNTHETIC LATEX[28, 94]

German synthetic latexes, known as "Igetex," were prepared by normal emulsion polymerization processes as in the manufacture of dry Buna synthetic rubbers. Their type letters, therefore, refer to the particular Buna copolymers present in the dispersed phase. Some of the latexes were prepared directly at the concentration at which they were marketed, whereas others were concentrated, either by a creaming process or more often after treating by the "Stockpunkt" method described later.

Type of Igetex	Rubber Content as Marketed, %	Type of Buna in Dispersed Phase
Igetex S (normal)	35	Buna S
Igetex S (concentrated)	45	Buna S
Igetex SS	45	Buna SS
Igetex N	30	Perbunan
Igetex NN	45	Perbunan Extra

The Igetex S referred to in the above table passed through the same historical steps as did the dry polymers; that is, early samples were made with the use of sodium linoleate with Nekal, whereas later production was made with paraffinic fatty acid soaps, in the manner of Buna S3. There appear to have been few cases, however, in which the lettering code was changed to follow this development. No further description of the polymerization is necessary since this has already been covered in the discussion of the Bunas themselves.

After polymerization, the latexes were steam-stripped of unreacted monomers, so that the monomer ratios differ somewhat from the proportions charged. Table VI gives the constitution of the latexes as marketed.

Igetex S and SS were definitely more viscous than natural latex of the same concentration at room temperature, but the N and NN types were nearer

Table VI.   Styrene Copolymer Latexes

	Igetex S (35%)	Igetex S (45%)	Igetex SS (45%)
Butadiene	75	75	57
Styrene	25	25	43
Nekal BX	3.0	2.0	1.8
Sodium linoleate	2.0	1.3	1.6
Phenyl-β-naphthylamine	2.0	2.0	3.0
Styrene monomer	0.8	0.8	1.3
Sodium hydroxide	0.2	0.13	0.38
Salts (sodium chloride, sodium sulfate, potassium sulfate)	0.7	0.46	1.3
Water	177	115.4	112.6
	285.6	222.0	222.0
Total solids, %	38	48	49.3
Dry rubber content, %	35	45	45
pH	11	10	10
Rel. viscosity, cps.	6	30	15
Strength of vulcanized film kg. per sq. cm.	40–50	40–50	150–200

natural latex both at room temperature and when cooled. The film strength of NN was fairly good, but, if the latex were made alkaline, the film strength was reduced, and discoloration occurred. Both types could be thickened, if desired, by the addition of polyacrylate-type thickeners.

Recommended Uses of Igetex Latexes. *Igetex S* gave vulcanizates with the best cold resistance, elasticity, and lowest permanent set, but with the lowest tensile strength of any of the synthetic latexes. It was suggested for adhesives, textile impregnation, artificial leather, and sponge. *Igetex SS* was recommended for general dipping and spreading, for thread, and for sponge where the tensile strength of Igetex S was too low. *Igetex N* was suggested for impregnation, adhesives, artificial leather, hair coating, and general spreading where resistance to solvents was important. *Igetex NN* found application primarily in dipping uses, especially for thin-walled articles, since it gave even coatings, had the highest tensile strength in the series, and exhibited a desirable oil resistance. The foam from it was less stable than that from S-type latexes, and hence it was not recommended for sponge.

Heat Sensitization. Two variations of a heat-sensitization method were developed so that Igetex latexes could be used for dipping. Both depended on the addition of a water-soluble material which became insoluble at an elevated temperature and caused complete coagulation.

In one method two parts of Emulphor O (p. 965) was added to the latex first, followed by four parts of calcium chloride. The usual compounding ingredients could then be added and the mixture held for 3 or 4 days. Coagulation took place when the mixture was heated to 50° C. or higher. The coagulation temperature could be lowered by increasing the calcium chloride content and lowering the amount of Emulphor O, or raised by reversing this procedure. The process could not be used with Igetex NN.

The second process made use of Igevin M 50, a 50 per cent water solution of a polyvinyl methyl ether, with a K value of 50, which became insoluble in water at temperatures above 35° C. When a mixture of Igevin M 50 and natural or Igetex latex was heated, coagulation occurred at the temperature at which the Igevin precipitated. The latex had to be nearly neutral before the addition of the Igevin, and some Emulphor O was usually added, partly for protection of the latexes in compounding with fillers, and partly as a means of controlling the coagulation temperature and firmness of the coagulum. This process had the advantage of a much greater storage life than the one involving calcium chloride; if kept at room temperature the treated latex could be safely stored for months.

Latex-Concentration Techniques.[28] On cooling, Igetex S types were found to behave differently from N types and from natural latex, in that they suddenly thickened to a pasty form at a temperature in the neighborhood of 10° C., known as the "Stockpunkt." This behavior was the foundation of a unique method of latex concentration used by the I.G. Farbenindustrie in the production of high-solids S-type latex.

Centrifuging had been given up as unsatisfactory because of the small particle size of synthetic latexes; evaporation suffered from the objection of concentrating the nonrubber constituents; creaming was objectionable because of the high final viscosity of the latex, and electrophoresis, which was considered, was given up because of the difficulties in plant operation. Altogether, the Stockpunkt method was considered to be by far the most satisfactory technique for concentrating synthetic latex.

The actual Stockpunkt for any given latex (of a suitable type) was found to depend on the amount of Nekal and other salts present, and it could be raised by the addition of salts such as sodium sulfate or sodium carbonate. Also, it was higher, the greater the concentration of the latex. Thus, Igetex S at 30 per cent concentration might have a Stockpunkt of 15° C.; at 25 per cent concentration, 8 to 10° C. Igetex which in the original state had too high a Stockpunkt was not used alone, since the pasty concentrate might have a Stockpunkt at or above room temperature and would be permanently too thick. Such latex could be blended with other samples of lower Stockpunkt, however, for normal use. If held at very low temperatures (0° C.) too long, slow coagulation occurred, but recovery on warming from just below the Stockpunkt was quite reversible.

In normal operation, the Igetex was tested for its Stockpunkt, and if this was below 10 to 12° C. it was raised to this figure by the addition of sodium carbonate solution. The Igetex was then cooled to the Stockpunkt or a little below until it became a paste. The paste was stirred by large paddle-shaped blades for several hours until it was broken up into small lumps or granules; the stirring was then stopped and the mass allowed to stand until a creamy layer of concentrate had collected above a practically clear serum. The serum was drawn off and the concentrate warmed to room temperature and stirred well to insure uniformity. The result was a stable latex of 45 to 50 per cent concentration.

American experience with this technique has been reported by Maron and co-workers,[88] who found that, with some modifications necessitated by the different nature of the emulsifiers used in the United States, the process could be satisfactorily applied to the concentration of many GR-S-type latexes. It is believed that the process has received no commercial application in the United States, largely because of subsequent development of high-solids latexes produced by direct polymerization.

POLYBUTADIENES PREPARED BY ALKALI METALS

Three polybutadienes, often designated as "numbered Bunas,"[12, 45, 76, 82] were prepared in commercial quantities in Germany by mass polymerization, viz: Bunas 32, 85, and 115, the number representing a viscosity index that increases with the degree of polymerization.[76] Buna 32 (Plastikator 32) is a honeylike, viscous polymer with a molecular weight of about 30,000.[45] It was used as a softener for the mass or emulsion polymers. Buna 85 was a more rubberlike polymer with a molecular weight of about 80,000.[45] It was used at first chiefly in the preparation of hard-rubber compositions, where it had the advantage of giving products of high softening point and excellent chemical resistance. It also served as a polymer extender. Buna 85 comprised the largest part of the sodium polymer production. Buna 115, of higher molecular weight than Buna 85, was the stiffest of the group; it was considered to lie on the upper limit of processibility.[76] Buna 115 was not being produced in Germany at the time of the war's end.

Bunas 32 and 115 were made by sodium catalysis in a batch process at 80° and 20° C., respectively; Buna 85 was made by a continuous process with potassium as the polymerization catalyst. As might be expected, impurities had an important bearing on the process. Butadiene was obtained by distilling the plant-grade monomer to remove small amounts of aldehydes and water. Aldehydes widened the molecular-weight distribution of the polymer, and water poisoned the catalyst.[45, 82]

Batch Process.[16, 31, 41, 45, 53, 58, 82, 105] The first equipment employed for making the numbered Bunas was stainless-steel cylindrical autoclaves of 8 cubic meters capacity, equipped with two longitudinal mixing baffles and rotated in a water bath. Each reactor was tested to 80 atm., was provided with a pressure relief valve, and was enclosed in a concrete bay, with the control apparatus behind a concrete wall. This equipment was used at the end of World War II only for the manufacture of Buna 32, although Buna 115 could be made in the same equipment.

Fresh sodium was melted under xylene and stirred vigorously to form a sand of about 0.5 mm. diameter; the mixture was cooled; the xylene decanted, and the catalyst dried under nitrogen. Butadiene (3000 liters) containing 0.1 per cent vinyl chloride was charged into the autoclave, and about 0.5 per cent of the dried sodium sand was forced in by nitrogen pressure. The autoclave was started rotating and heated to 80° C. for Buna 32 or 20° C. for Buna 115. In the former case the pressure remained at 7 to 10 atm. for the reaction period, which was about 8 hours but varied with the purity of the butadiene. At the end of the polymerization, the pressure within the autoclave had dropped to 1 to 2 atm., about 5 per cent of the butadiene remaining unpolymerized. The butadiene was recovered, and the charge was cooled and blown through a dip leg into another similar vessel. Here it was washed with water to remove the sodium, and the rubber was discharged into an iron vessel equipped with a stirrer; phenyl-β-naphthylamine was added for normal usage, while for use in white rubber products, aldol-α-naphthylamine was used. Material consumption for 100 kg. of Buna 32 was 106 kg. butadiene, 0.83 kg. vinyl chloride, and 0.7 kg. sodium.

Continuous Process. The Schkopau plant had two continuous units[82, 89] for the preparation of Buna 85, the original one and an improved model being used at the end of the war for all Buna 85 production. These reactors could produce 120 tons per month on a 24-hour cycle.

The continuous polymerizer consisted of a horizontal jacketed cylindrical steel autoclave, which was approximately 15 ft. long and within which a close-fitting hollow screw revolved. The pitch of the screw as well as the distance from the reactor shell to the core or shaft of the screw increased gradually along the length of the reactor. The tapered shaft and wider pitch provided an increasingly greater volume for the polymer being formed in the annular space. Cooling water for controlling polymerization temperature passed through the shell jacket and hollow screw. Packing flanges on both ends of the screw shaft permitted rotation under 25 atm. operating pressure. A small amount of Plastikator 32 was forced around a lantern ring at the inlet end to seal in the butadiene; thick polymer present at the outlet end provided a satisfactory seal, so that no special packing gland design was required.

The catalyst for the reaction was potassium. It was dispersed (10 to 15 microns) by being melted with Buna 32 and mixed rapidly to give a 2 per cent emulsion of potassium in this polymer. One-hundred-liter batches of catalyst were made up at a time in an electrically heated 150-liter vessel.

Butadiene of 99.8 per cent purity was charged to the top of the polymerizer by means of one of three available high-pressure piston feed pumps. A brine precooler upstream from the pumps prevented vapor lock; a filter removed the dirt. Dioxane (0.5 per cent based on the butadiene) was simultaneously pumped from a calibrated feed tank into the butadiene inlet line, to serve as a protective agent by preventing local overheating during the polymerization. Catalyst paste was fed into the bottom front end of the reactor from a machine cylinder feed tank. Application of oil pressure on a piston moving in this cylinder forced the catalyst out. The feed rate (0.5 per cent potassium per 100 kg. of butadiene) was regulated by varying the amount of oil being pumped to the cylinder from a calibrated feed tank. The reaction was carried out at 70° C. The polymer began to form as soon as butadiene and catalyst came into contact at the front end of the reactor; it gradually increased in viscosity (molecular weight) as the reactants were forced through the unit by the rotating screw. After about 1.5 to 2 hours holdup time, the finished polymer was extruded from the reactor through a 3-in. outlet valve. The reactor was reported to operate at a rate of 600 kg. of Buna 85 per cubic meter per hour.

The crude Buna 85 coming out of the autoclave was practically gas-free and dark in color, the color being stated to come from the organometallic compound formed by reaction of potassium and butadiene. The crude product was worked up in a mixer and 2 per cent phenyl-β-naphthylamine added to serve as the antioxidant along with fatty acid (1 per cent) to neutralize the free potassium or potassium hydroxide in the finished rubber. The material from the mixer was sheeted off on a mill, cut into sections, and packed into paper cartons.

Material consumption for 100 kg. of Buna 85 has been given as: butadiene 106 kg., potassium 0.23 kg., dioxane 1.2 kg., phenyl-β-naphthylamine 2.0 kg., and paraffin fatty acid 1.4 kg.

Properties. The physical properties of polybutadiene from the alkali metal-catalyzed process were reported superior to emulsion polymers. Polybutadiene from the alkali process gave a tensile strength of 1960 p.s.i. as compared with a strength of 1120 p.s.i. for an emulsion product.

The physical properties and aging characteristics of Buna 85 and Buna 115 in carbon black stocks are shown by the data in Table VII. From analyses of the numbered Bunas it was concluded that the sodium polymers had 35 per cent of 1,4-units and the potassium polymers 40 to 45 per cent.

Table VII

	Natural Rubber	Buna 85	Buna 115
Tensile strength, kg. per sq. cm.	260	175	200
Elongation, %	600	600	700
Shore hardness	65	65	60
Aged in Geer oven 32 days at 90° C.			
Tensile strength	160	155	170
Elongation	350	600	625
Aged in oxygen bomb 32 days at 60° C.			
Tensile strength	50	150	180
Elongation	200	620	650

Butol. Butol, a liquid used as a plasticizer, was made by the sodium vapor-phase polymerization of recovered butadiene.[89] The process consisted in admitting a vapor of recycle butadiene at room temperature into a reactor containing sodium sand dispersed in polyisobutylene paste.

RESEARCH IN EMULSION POLYMERIZATION

The Central Research Laboratory for Rubber of I.G. Farbenindustrie, located at Leverkusen,[58, 94] was built in 1939 at a cost of 10 million marks, and impressed the investigating teams as being the largest, best equipped, and most modern plant of its kind in the world. It contained research facilities for making tires and other rubber goods. The total personnel toward the end of the war was about 500, of whom 35 were doctors engaged in directing major research projects, 150 technical workers, and 300 workmen. The director was E. Konrad; the director of the chemical research department, W. Becker, and the director of the technical development and physical testing department, P. Stöcklin.

MONOMERS

The primary source of information on the study of monomers other than butadiene and styrene in the German program is the records of the various field intelligence teams sent to Germany at the close of the war.[58, 89, 94, 103] In most instances the conclusions necessarily reflected the personal opinion of the particular men interviewed, and, consequently, some disagreement is found among them. Further, the conclusions are in general terms only and are not accompanied by numerical data on the various polymers and copolymers studied. Hence the results of the German program will be reviewed only briefly, especially as American studies comprising substantially the same list of monomers are reviewed in Chapter 21.

Dienes. In the first phase, considerable work was done with isoprene as a butadiene replacement, but no technical advantage was observed to offset its considerably greater cost.[23, 58] 2,3-Dimethylbutadiene, also more expensive than butadiene, gave polymers with inferior elastic properties.[129] Butadiene and isoprene polymers were found to be insoluble whereas dimethylbutadiene rubbers were soluble; this was attributed to a higher degree of cross-linking in the former. 1,3-Dimethylbutadiene, 1,1-dimethylbutadiene, 1,4-dimethylbutadiene, and 1,1,4,4-tetramethylbutadiene polymerized very slowly.[103] 1-Phenylbutadiene was prepared in the belief that the phenyl group, like the chlorine in chlorobutadiene, would impart interesting properties; the results did not support this conclusion. 2-Cyanobutadiene was reported to dimerize rapidly but the 1-cyano isomer was less troublesome in this respect. Both could be copolymerized with butadiene under controlled conditions to yield products resembling acrylonitrilebutadiene copolymers in oil resistance. Aminobutadiene and vinylamine did not form copolymers with butadiene. Polymers of 1- and 2-ethoxybutadiene, 1- and 2-acetoxybutadiene, and 2-phenoxybutadiene showed no special promise.[103]

A large number of chlorinated butadienes were prepared and evaluated at Leverkusen.[103] 2-Chlorobutadiene polymerized to more interesting products than 1-chlorobutadiene, 2-chloro-3-methylbutadiene, 1,2-dichlorobutadiene, 2,3-dichlorobutadiene, or 1-phenyl-2-chlorobutadiene. 2-Fluorobutadiene polymerized alone but failed to copolymerize with butadiene.[138]

Vinyl Compounds. The German investigators believed that the ratio of any vinyl monomer in a butadiene copolymer could be changed by varying the polymerization conditions.[89] Thus, it was thought that an increase in the polymerization temperature or the addition of an activator system would increase the ratio of vinyl to diene units in the copolymer. This is contradictory to observations made in the United States in which a decrease in polymerization temperature to 5° C. in a redox charge caused no material change in the monomer ratio.

Only a few vinyl monomers formed sufficiently promising copolymers with butadiene to justify tire tests. These included methyl isopropenyl ketone, methyl acrylate, methyl methacrylate, N-diisobutylacrylamide, β-vinylnaphthalene, and 2-chlorobutadiene.[89] Methyl isopropenyl ketone[89, 103] like methyl vinyl ketone[103] yielded oil-resistant copolymers which showed better processing characteristics than the butadiene-styrene control but could not be mixed with natural rubber. Methyl vinyl ketone copolymers (Buna K), did not match the oil resistance of high acrylonitrile types like Perbunan Extra. Buna K aged poorly and was not successfully built into tires.

Butadiene-methyl methacrylate copolymers were designated Buna M,[12, 33, 58, 89, 103, 105] Buna M25 containing 25 parts of methyl methacrylate, and Buna M35, 35 parts. A 1941 Kuko report characterized these as better in processibility than Buna S of the same Defo plasticity; later reports showed them inferior to Buna S3. Pilot-plant preparations were much inferior to small samples; the larger samples were low in tack and storage stability.

A number of miscellaneous monomers were evaluated as comonomers with butadiene. No more than 3 to 5 per cent nitrostyrene could be introduced into a butadiene copolymer.[103] Trichloroethylene could be copolymerized with butadiene only in a redox-activated polymerization system. Considerable interest was therefore displayed in this activation technique for the possible preparation of unusual polymers which could not be made in normal systems. Cyclopentadiene did not copolymerize but served as a regulator for the Buna-S polymerization, 3 to 4 per cent having an effect approximately equal to 0.1 per cent Diproxid. Acrolein, methacrolein, and their homologs were capable of being copolymerized with butadiene. The alkyl acrolein copolymers were stated to be interesting in being vulcanizable without sulfur, by diamines. It was also thought that a long side chain in the monomer would be particularly interesting; when, however, ethyl-, isopropyl-, and hexylacrolein were tested, polymerization was found to become more difficult as the chain length increased.[103] The use of N-diisobutylacrylamide as a comonomer appeared to be a promising development over Buna S, but it showed no interesting improvement[89] over Buna S3. Methacrylonitrile polymerized more rapidly than acrylonitrile; its copolymers were less compatible with natural rubber and less oil-resistant than those of the Perbunan type.[89] 2-Vinylpyridine polymers and its copolymers with butadiene were generally soluble in dilute acid. Polymers containing over 50 per cent vinylpyridine were water soluble.[103] Vinylidene chloride copolymerized readily, but the copolymer showed no unique properties.[89]

The work with dichlorostyrene was concentrated on a monomer containing one lateral and one ring chlorine[89, 103] and, consequently, different from the dichlorostyrenes tested in the American program where both chlorines were on the ring. This monomer was prepared by chlorinating styrene in a glass reactor in daylight or ultraviolet light and then continuing chlorination, using a ring-chlorination catalyst, until a chlorine atom entered the ring. Hydrochloric acid was split out in the presence of an organic base (dimethylaniline), an inorganic alkali such as caustic soda solution, or by passing the compound over solid barium chloride, in the vapor phase.[103] The butadiene copolymer was considered at first to be more processible and tacky than a styrene counterpart; later work showed that this improvement was not outstanding.

Copolymers of butadiene with α-chloroacrylic acid esters were particularly promising in the unvulcanized form. The vulcanizates, however, were hard and had poor mechanical properties. Cyanoethyl- and methylacrylic acid and diethylmuconic acid copolymers with butadiene were not of particular interest. No copolymers could be prepared from butadiene and ethyl crotonate, ethyl cinnamate, acrylamide, or cinnamic acid. The last acid was, however, found to be both a polymerization activator and a modifier.[103] Copolymers with the dichloroethyl ester of fumaric acid were sufficiently oil-resistant to be of interest in airplane fuel cells. The butadiene-alkyl cyanosorbate copolymers had about the same properties as the styrene counterparts.[103] Crotonaldehyde, vinyl acetate, vinyl chloride, vinyl ethyl ether, and vinyl isobutyl ether were not successfully copolymerized with butadiene.[89]

Benzalacetone copolymerized readily with butadiene yielding a colored copolymer, which however displayed no unique properties. p-Methyl- and ethylbenzalacetone polymerized equally readily; p-nitrobenzalacetone did not.[103]

Acrylic acid was included in a terpolymer with butadiene and styrene which was capable of being vulcanized without the use of sulfur. It was said that the polymer was not so interesting as copolymers of methacrylic esters.

An important observation on the thinking of the Leverkusen group is recorded in the report of the Marvel committee.[102] In this report it was stated that the Leverkusen workers believed that the best synthetic rubber would ultimately be made by the polymerization of a diene alone. This is somewhat in contradiction to the statements of other laboratories which favored copolymers. Both agree that, except for those cases where a group adds some desirable characteristic, the particular monomer selected is of somewhat less importance than the conditions under which the polymerization is effected.[129] This is reflected in the interest shown by the Germans toward the close of the war in acid polymerization systems and highly activated charges for low-temperature polymerization. These are discussed elsewhere in this chapter.

The bulk of the work on polydienes was done early in the program[138] and was abandoned in favor of copolymers, which in general had better processibility and properties. Becker was stated as believing that work on the polymerization of dienes alone should be initiated again using the "low-temperature redox systems and studying very carefully the structure of the resulting polymer." Only then did he believe that the properties of a polymer would surpass those of a copolymer.

Emulsifying Agents

Nekal. The production of synthetic rubbers in Germany differed from the production of GR-S in always involving the use of synthetic emulsifying agents,[129] the chief of which was diisobutylnaphthalene sodium sulfonate.

The commercial product, known as Nekal BX (X meaning that the position of the alkyl groups was not established) contained 65 per cent of the actual emulsifier (soluble in ether), the remainder consisting of a sodium sulfate, water, and a trace of sodium chloride. It was manufactured at Schkopau at the rate of almost 300 tons a month by alkylating naphthalene with butanol and sulfonating the product. Several lists of raw materials exist for the preparation of Nekal BX. One[129] includes the following for 100 parts of Nekal BX: naphthalene 42.6, oleum (100 per cent) 100.6, sulfur trioxide 69.4, sodium hydroxide 42.3, butanol 47.8, calcium hypochlorite 5.6 parts. Another [56] lists naphthalene 24.7, oleum (24 per cent) 92.9, 98 per cent sulfuric acid 51.7, sulfonic acid D 0.1, 34.5 per cent sodium hydroxide 74.3, sodium hypochlorite solution 10, butanol 28.6, powdered limestone 0.7, and sodium chloride 2.9. Neither reference indicates the nature of the butanol used. The structure of the butyl groups is uncertain; it is given as isobutyl,[4, 38, 49, 57, 128] as sec-butyl,[37, 89] and in many cases simply as butyl.[56, 104, 129] A purified grade known as Nekal BXG (G: Gereinigt) contained only 10 to 15 per cent of inert material.

Esteramin. As an emulsifier in acid polymerization systems, "Esteramin," $RCOOC_2H_4N(CH_3)_2$, made from dimethylamine, ethylene oxide, and a paraffinic acid containing approximately 12 carbon atoms, was preferred. The base was, of course, dissolved in a solution of hydrochloric or sulfuric acid when it was to be used for emulsification.

Emulphor O was a condensation product of oleyl alcohol with ethylene oxide, $C_{18}H_{35}(OCH_2CH_2)_nOH$. Its effective range was limited to temperatures below 60° C. since its solutions gelled above that point. The primary use of Emulphor was not in polymerization but for addition to finished latexes where heat sensitization was needed. Some variation exists in the literature on the proportion of ethylene oxide in Emulphor O; some references indicate 20 to 25 moles of ethylene oxide;[37, 135] others show 4 to 6 moles.[89, 129]

Mersolat. The redox polymerization formula studied in Germany was based on the use of a unique emulsifying agent. This agent was prepared from a C_{14}–C_{17} cut of Fischer-Tropsch hydrocarbons known as Kogasin which was first reduced to a mixture of saturated hydrocarbons (Mepasin); these latter were subsequently sulfonated by treatment with sulfur dioxide and chlorine in the Reed process, the product being an acid chloride mixture (Mersol). The sulfonation process was stated to produce all of the possible isomeric compounds,[89] although it is indicated elsewhere that the reaction at 30° C. produced a maximum addition of SO_2Cl groups to the end of the hydrocarbon chains and a minimum substitution of chlorine elsewhere. Terminal reaction is further favored by allowing the reaction to proceed only to approximately 50 per cent completion.[97] By treatment with alkali, the sulfuryl chloride yielded the corresponding sodium sulfonate, known as Mersolat or Mepasin sulfonate.

Modifying Agents.[54] The first effective move toward improving the processibility of emulsion butadiene polymers over the sodium analogs was the use of vinyl comonomers.[79, 123, 125] The copolymers had lower gel contents, fewer vinyl side chains, and fewer cross-links than polymers of the dienes alone. Accordingly, styrene might be classed as one of the first modifying or regulating agents. An almost ideal chemical modifier to control molecular weight was believed by Graulich and Becker[54] to have been first discovered when less than 1 part of diisopropylxanthogen disulfide (Diproxid) was introduced into a butadiene-acrylonitrile charge. The use of this reagent was later extended to butadiene-styrene polymerizations,[72] becoming the preferred reagent for German production. Diisopropyl-xanthogen disulfide was first prepared by Whitby and Greenberg[32, 132] in 1929 and is one of the few xanthogen disulfides, if not the only one, that is a solid of reasonably high melting point. It is, incidentally, a powerful accelerator of vulcanization.[131] In Germany, it (100 pt.) was manufactured[45] by the oxidation of potassium isopropylxanthate (162.6 pt.) with potassium persulfate (113.6 pt.). The German preference for Diproxid instead of the mercaptan modifiers employed in the American process is attributed[89, 103] to the higher concentrations of the latter class of chemicals which were necessary to produce a comparably modified polymer. It was the German belief that an effective modifying agent should be one that acts at very low concentrations. According to Becker, no other materials had been found that were as effective as diisopropylxanthogen disulfide,[89] but it was stated that diisobutyl- and diisoamylxanthogen disulfide are approximately equivalent to the standard diisopropyl analog.[103] Laboratory investigations of the possible mechanism for Diproxid modification were reported[54] to have established the formation of mercaptan and sulfinic acid during its decomposition; Wollthan showed that the rate of this decomposition was faster under acid and alkaline conditions than in neutral media. As a consequence of this lead, aliphatic mercaptans were considered and were shown to be modifying agents if the carbon chain contained more than eight carbons; shorter aliphatic mercaptans were poisons to the butadiene copolymerization.[135]

It was shown that mercaptans give rise to disulfides which, in turn, decompose into mercaptans and therefore act as modifying agents, and that mercaptans add to double bonds in both the monomer and the polymers.[54]

In addition to mercaptans and xanthogen disulfides, the German laboratories investigated numerous other chemicals as potential modifying agents. The first sulfur-free modifier was ω-nitrostyrene and its substitution products. Although the polymerization rate was satisfactory with this modifier in acid systems, nitrostyrene effectively poisoned alkaline charges.[54, 74] Polymers made with it showed the desired drop in Defo on heat softening and had none of the potential difficulties associated with the presence of the accelerator, Diproxid.

Sodium phenyl and tolyl diazotates were found to serve as modifying agents in a polymerization system catalyzed by potassium persulfate; in the absence of such a catalyst, polymerization still occurred, the decomposition of the diazotates initiating the polymerization.[89] An objection to this class of materials was the dye formation which occurred when the polymerization was stopped by phenyl-β-naphthylamine.

An additional class of potential modifying agents was the group of alkylphenols. Triisobutylphenol and dodecylphenol were effective in charges catalyzed by hydrogen peroxide or sodium hypochlorite.[103]

Work was reported[54] on the use of phenyl-substituted chlorobutanes as modifiers. The K-value curve with such materials was desirably flat throughout the polymerization, but high concentrations of the modifier were needed and the rubber scorched badly. Azodicarboxylic esters were stated to be useful as modifiers over a wide range of pH values, although observation of the K curve suggested that a stepwise addition, using as pure a chemical as possible to eliminate an induction period, would be preferable to initial addition. Some regulating action was shown by such hydrocarbons as cyclopentadiene, 1,2-dihydronaphthalene, anthracene, fluorene, and indene. No explanation was proposed for the action of substances of this class beyond the suggestion[54] that the high concentrations required made more for plasticizing than modifying action.

Another class of modifying agents recorded in the German literature are the chlorinated hydrocarbons[124] in connection with the polymerization of both butadiene and chlorobutadiene.[64] A special instance is cited where carbon tetrachloride served as a modifier in the polymerization of styrene:

CCl_4 Concentration	K Value	
	Styrene	Butadiene
0	114	53
10	102	105
20	75	156
30	38	...

In the polymerization of butadiene, contrary to the state of affairs in the polymerization of styrene, the K value rises as the concentration of carbon tetrachloride increases; this indicates no modifying action. It is known that chlorine remained in the polystyrene chains.[108]

The relative effectiveness of potential modifiers was judged in the German laboratories by a study of the K value and the branching index, both on the finished polymer and at various intervals throughout the course of polymerization.[54]

K Value. The concept of the K value was based on early work of Fikentscher[46, 78] on the relative viscosity η_r of colloid solutions. In the equation

$$\log \eta_r = \left(\frac{75K^2}{1 + 1.5K_c} + K \right) c$$

K is the parameter of individual curves $\times 10^3$ and is related to the specific viscosity at infinite dilution by the expression below, in which c is the concentration in grams per 100 ml. of solution. When the equation above is differentiated with respect to concentration and c is allowed to approach zero, it becomes:[49]

$$\lim_{c \to 0} \frac{\ln \eta_r}{c} = 2.3026(75K^2 + K) = Z_0$$

The expression $\lim\limits_{c \to 0} \dfrac{\ln \eta_r}{c}$ is equal to the intrinsic viscosity when concentration is expressed in grams per 100 cc. of solution and is equal to the Mark and Schulz specific viscosity η_{sp}/c at infinite dilution Z_0. Other references to the K value may be found,[78, 81, 89, 129] using different symbols and different forms of the above equation.

The preferred method[11, 15, 49, 78] for determining the K value involved shaking the stabilized latex sample (4 to 12 cc.) of known rubber content for 5 minutes with a solvent mixture consisting of 156 cc. chlorobenzene and 44 cc. pyridine. The solution was concentrated to 150 to 199 cc. at 30° C. under a vacuum. A 100-cc. portion of the concentrate was precipitated with glacial acetic acid; the resultant precipitate was separated from the supernatant solution after 24 hours' standing, dried, and weighed. The relative viscosity of a 0.4 to 0.7 per cent solution was used to calculate the K value. In order to shorten the calculation, curves were used from which both viscosity coefficients (K value) and limiting viscosities $\left(\lim\limits_{c \to 0} \dfrac{\eta_{sp}}{c} \right)$ could be obtained by interpolation. One set of curves for this purpose is given by K. H. Meyer.[91]

It was asserted that synthetic rubbers of equal K values were identical for most purposes and that the K values, being additive, permitted a prediction of the K values of polymer mixtures. Correlation with the usual physical tests showed that the tensile strength and elongation were proportional to the K value; the resilience, hardness, and Defo values were inversely proportional to it.[54] It was thought that the curve for the K value during the course of polymerization offered a better way of defining new polymers than the determination of the usual physical properties. The best rubber was believed to be one having the highest and most nearly horizontal curve. It was recognized that a compromise existed between processability and good vulcanized properties; a high K-value curve would indicate high properties but difficult processability. Consequently, a desirable polymer might show a low value early in the polymerization and rise slowly toward the end.

Branching Index. In early studies,[111] the average degree of polymerization was calculated for polybutadienes from the intrinsic-viscosity determination, using the constant 3.8×10^{-4} obtained from low-molecular-weight substances with fiber molecules. For high-molecular-weight substances the values thus calculated were only one-third to one-fifth of the true degree of polymerization obtained from osmotic-pressure determinations. The concept branching index (*Verzweigungsgrad*) was, therefore, introduced as the ratio of the Degree of Polymerization by osmotic methods to the DP as obtained from viscosity measurements. The procedure involved purifying the sample and determining the K_m and K_{equiv} values by viscosity determinations on dilute solutions.

$$\frac{\eta_{sp}}{C_g} = K_m M$$

$$\frac{\eta_{sp}}{c} = K_m P$$

$$\frac{\eta_{sp}}{c} = K_{equiv}\, n$$

Table VIII. Determination of K_m and K_{equiv} Constants for Toluene Solutions and Degree of Branching

	Degree of Polymerization (Osmotic)	No. of Chain Members	$\frac{\eta_{sp}}{c}$	$K_m \times 10^4$	$K_{equiv} \times 10^4$	Apparent DP $K_m = 3.8 \times 10^{-4}$ for Benzene	$\frac{\text{DP*}}{\text{DP}\dagger}$
Sodium polybutadiene							
Easily soluble fraction	1450	5800	0.144	1.0	0.25	400	3.5
Difficultly soluble fraction	2400	9600	0.330	1.4	0.34	850	2.8
Emulsion polybutadiene, highly modified charge							
Easily soluble fraction	1000	4000	0.161	1.6	0.40	400	2.5
Middle fraction	1400	5600	0.220	1.6	0.39	600	2.3
Difficultly soluble fraction	2100	8400	0.287	1.4	0.34	750	2.8
Emulsion polybutadiene, low modifier charge							
Easily soluble fraction	4300	17,200	0.360	0.8	0.21	950	4.5
Difficultly soluble fraction	8200	32,800	0.571	0.7	0.17	1500	5.5
Gutta-percha	1550	6200	0.210	1.4	0.34	550	2.8
Hevea (ether-soluble fraction)	1600	6400	0.126	0.8	0.20	350	4.6

* Osmotic.

† Viscometric.

Here η_{sp} is the specific viscosity; K_m a 'proportionality constant at infinite dilution,' C_g the "base mole" concentration as the molecular weight of the polymer in grams per liter of solution, c the concentration in grams per 100 ml. of solution, n the number of chain members per base molecule, and K_{equiv} the chain-equivalent molecular-weight constant.[110]

An example of the study of branching index may be seen in results of Staudinger given [109] in Table VIII. It was noted that K_m for polybutadiene and for rubber were about the same. Staudinger considered that this indicated that macromolecules of natural rubber were therefore branched and not fibrous in form. He also called attention to the polybutadiene made with small amounts of modifier in the charge. The low K values and high branching-index values were assumed to indicate the occurrence of more branching in this polymer than in the polybutadiene sample prepared in the presence of a large amount of modifier.

Garten and Becker studied the K value and branching index for a number of synthetic rubbers. In general, polymers with lower branching indexes showed a more rubberlike character. It was cautioned by Garten and Becker that the branching index could not be used alone as an indication of size and form and that other supporting evidence must be examined. One such piece of evidence comes from the decomposition of the polymer by oxidizing agents and the examination of products for 1,2/1,4 ratio. However, no quantitative relationship seems to exist between the branching index and the content of 1,2-diene units which apparently remains constant throughout the polymerization.

Table IX

Conversion, %	Mol. Wt.	K Value	Osmotic DP / Viscometric DP
6	150,000	136×10^{-3}	3.01
16	310,000	151	4.83
21.5	400,000	136	8.02
39.5	610,000	109.5	19.95
63.5	1,060,000	74.0	66.70
75.0	1,400,000	62.0	122.30

A K value-conversion relationship is given[54] in Table IX. It may be observed that the peak in the K value is at about 16 per cent conversion and that the molecular weight and branching index increased with increasing conversion.

Further examples of the use of the K value and branching index may be seen in the work of Garten and Becker.[49] For a polybutadiene made at a range of temperatures it was concluded that the weaker the activation, the longer the polymerization time and the larger the final molecule:

Persulfate %	Reaction Time, Hr.	Molecular Weight $\times 10^{-3}$	Final K Value	Branching Index
0.1	51.5	1100	89	34.7
0.3	32.5	875	91	26.5
0.6	22.25	610	88	19.6
0.9	20	540	85	18.5

Lower polymerization temperatures were also shown to increase the molecular weight of a polybutadiene:

Polymerization Temperature, °C.	Reaction Time, Hr.	Molecular Weight $\times 10^{-3}$	Final K Value	Branching Index
10	12–16	2350 ± 800	115–125	39.0
20	5–9	1600 ± 400	95–110	39.1
30	2.5–5	1150 ± 200	85–100	36.8
40	30–70 min.	870 ± 80	80– 90	30.0

The introduction of styrene in the preparation of Buna S1 or S3 made a substantial change in the branching index in the direction of natural rubber, as indicated by the following data.[49]

	K Value	Molecular Weight $\times 10^{-3}$	Branching Index
Natural rubber	171	404	3.37
Buna S1	113	650	13.1
Buna S3	107	350	7.8

Antioxidants. A number of antioxidants were evaluated in synthetic rubber, but none was found to surpass phenyl-β-naphthylamine,[58] which, however, did not permit the preparation of white stocks, since it discolored badly. The best antioxidant found for such stocks was oxycresylcamphane, prepared by the condensation of cresol with camphene under the influence of boron trifluoride; it was, however, considered inferior to phenyl-β-naphthylamine in aging tests.

Tack. The constant search for improved tack in synthetic rubber is frequently mentioned in the German literature. It was especially important because many of the tire failures in Germany resulted from low tack or "weldability." In the American synthetic-rubber development lack of tack was also of concern, but its importance was significantly less because enough natural rubber was available for blending throughout the entire war period.

The best tackifier developed in Germany was Koresin, an almost odorless, yellow-to-brown resin, with a softening range of 105 to 125° C. and a specific gravity of 1.03 to 1.04. It was produced at Ludwigshafen by the condensation of acetylene (6 moles) with *p-tert*-butylphenol (5 moles) in the presence of zinc naphthenate at temperatures[68] from 180 to 230° C. It produced a good degree of tack in Buna compounds without affecting the cure, and it also facilitated extrusion. Vulcanizates containing Koresin had good tensile strength, higher elongation, improved tear resistance, and excellent aging properties. Increase in the proportion of Koresin used caused the resilience to drop, but this could be counterbalanced to a large extent by the addition of suitable softeners.[94]

REDOX POLYMERIZATION

The research program at Leverkusen had indicated that no diene appeared likely to be better than butadiene and that no substantial improvement would be made by considering as a comonomer one of the alternatives for styrene. Instead it was expected that the major inprovement would result through a change in polymerization technique.[83, 103] Consequently, attention was directed toward polymerization conditions to improve the quality of synthetic rubbers.

At Leverkusen polymerization activators were believed to be an important part of the synthetic-rubber program, since it was held that each gave a different product, especially if the polymerization temperature could be reduced substantially. The temperature was necessarily limited in the early part of the program to rather narrow ranges; polymerization recipes developed for 50° C. required an unreasonable time at any lower temperature. "Redox" activation systems which made possible polymerization at low temperatures in reasonable periods of time were disclosed to a Technical committee in June 1945. They had been used in Germany for some time previously in laboratory and semiplant runs at temperatures as low as 0° C. The history back of their development has been described by Konrad and Becker.[77]

Discovery of Redox Polymerization in Germany. The "redox" systems were so named because they were developed along the line of Warburg's biological redox systems containing both a reducing and an oxidizing agent along with a small amount of a metal which could exist in several valence states.[89]

In the development of redox systems applicable to polymerization, Kern attributed considerable importance to work in 1941 on the polymerization of chloroprene[67, 98] where the presence of oxygen caused strong inhibition, but where polymerization was rapid in an inert atmosphere. This particular system was so sensitive to molecular oxygen that the variable amounts present in the aqueous phase and in the monomer made polymerization times very erratic. These last traces could be removed by the addition of less than 1 per cent of sulfites or hydrosulfites based on the aqueous phase. These facts were extended by Logemann into a generalization that molecular oxygen retarded the emulsion polymerization of unsaturated monomers initiated by peroxy compounds and that the inhibitory tendency could be avoided by the careful exclusion of molecular oxygen, using an inert gas. Logemann developed this further to include the addition of reducing agents such as hydroxylamine and hydrazine and divalent metallic ions, such as iron or chromium. Monheim and Sohnke at Höchst[67] added formaldehyde sodium sulfoxylate (Rongalit) and formamidine sulfinic acid to the list of reducing agents for use along with water-soluble peroxidic compounds.

The next development in this field was attributed by Kern[67] to independent work at both Leverkusen and Höchst where it was recognized that the reducing agent did more than simply remove molecular oxygen; the addition of a reducing agent led to faster polymerization than the mere exclusion of molecular oxygen. Hence it was necessary to postulate the reaction between the oxidizing and reducing agents (the redox reaction) as being the initiating step in the polymerization mechanism.

It was found that, besides the persulfate catalysts, other nonperoxidic compounds such as potassium premanganate, manganese dioxide, chlorates, and hypochlorites could function as normal polymerization initiators in the presence of reducing agents. Such work was presented[67] as further evidence of the role of the "redox" reaction in catalyzing the polymerization. Hypochlorites alone were capable of initiating polymerization without the addition of a reducing agent.

The list of reducing agents was expanded to include organic compounds containing thiol, sulfinic acid, and α-ketocarboxylic acid groups, the α-ketols or enediols $[—CO \cdot CH(OH)— \rightleftharpoons —C(OH) : C(OH)—]$ being especially promising. Comparative data for various reducing agents in an alkaline emulsion polymerization of butadiene-styrene mixture catalyzed by benzoyl peroxide at 10° C. are given in Table X.

Table X. Evaluation of Reducing Agents in Butadiene-Styrene Polymerization[103]

Nekal-Oleate Emulsifier, Buffered with Pyrophosphate,
pH 9, 0.6% Benzoyl Peroxide, 0.3% Reducing Agent

Reducing Agent	Yield (14 Hr. at 10° C.)	Reducing Agent	Yield (14 Hr. at 10° C.)
None	15	Sorbitol	21
Glucose	70	Mannitol	17
Mannal	55		
Galactose	48	Formaldehyde	7
Xylose	80	Benzaldehyde	11
Ribose	78	Aldol	18
Arabinose	56	Acetylacetone	2
Fructose	95		
Maltose	32	Formic acid	20
Ascorbic acid	42	Oxalic acid	9
Acetoin	40	Glycolic acid	17
Dihydroxyacetone	96	Lactic acid	23
Glyceraldehyde	90	Tartaric acid	27
Benzoin	91	Mucic acid	31

The aldehyde or ketone group alone was apparently not effective, nor were polyalcohols effective without carbonyl groups. Pentoses were extremely effective probably through cleavage under alkaline conditions to form reductone by way of dihydroxyacetone or glyceraldehyde.[44] Attention was called to the desirability in emulsion redox polymerization of separating the oxidizing and the reducing agents by using as the members of the redox pair one reagent that was oil-soluble and the other water-soluble. It was found, for example, that, whereas, as shown above, ketols greatly accelerated polymerization when the oxidant was the oil-soluble benzoyl peroxide, they caused only slight acceleration when used in conjunction with persulfates or hydrogen peroxide.[67] The effectiveness of redox catalysis is influenced not only by the choice of the redox pair but also by (a) the monomer to which it is applied and (b) the emulsifying agent employed. Thus, with persulfate–Rongalit chloroprene polymerizes vigorously and styrene well, but butadiene polymerizes only slowly.

Influence of Iron. In a further development of the redox principle of polymerization, it was found that the efficiency of almost all redox systems could be improved by the addition of small amounts of iron. This was in accord with the long-known fact that small quantities of iron often have a profound effect on oxidation reactions under the influence of hydrogen peroxide and on biological redox systems. As an illustration of the influence of iron in redox polymerization, the following, quoted by Kern, may be noted. In a Nekal-oleate emulsion of 75/25 butadiene–styrene buffered at pH 9 by sodium pyrophosphate, the yields of polymer after 4 hours at 40° C. were as follows: (a) with 0.2 part benzoyl peroxide 4 per cent, (b) with in addition 0.3 part dihydroxyacetone 54 per cent, (c) with further addition 0.05 part ferrous ammonium sulfate 85 per cent. The last-mentioned catalyst system, viz., redox pair plus iron, was capable of bringing about polymerization at a reasonable rate at 5° C.

The pyrophosphate, it appeared, served as more than a mere buffer. When orthophosphate or borate was used in its place, no enhancement of the rate of polymerization was to be observed. The pyrophosphate forms an iron complex, whereas the other two buffers do not.

Mechanism of Redox Polymerization. The mode of action of a redox pair is considered to involve reaction between the oxidant and the reductant with the donation of one hydrogen or the acceptance of one hydroxyl by the latter and the formation of a free radical capable of initiating polymerization of the monomer.

$$ROOR + XH \longrightarrow RO^{\cdot} + ROH + X$$

Conceivably, as a result of termolecular collisions, such as are not unlikely in view of the high concentration of monomer, the catalyst radical may add immediately to a monomer molecule, with the initiation of polymerization. Thus:

$$ROOR + XH + CH_2 : CHR \longrightarrow RO \cdot CH_2 \cdot CHR{-} + ROH + X$$

When ferrous iron is the reductant member of the redox pair, as, for example, when the catalyst system is benzoyl peroxide–ferrous ammonium sulfate (complexed with pyrophosphate), a similar direct reaction between the oxidant and the reductant is involved.

$$ROOR + Fe^{++} \longrightarrow 2RO^{\cdot} + Fe^{+++}$$

When, however, a small amount of iron is added to a regular redox pair i.e., when the catalyst system is oxidant-reductant-iron, then it is considered that the ferrous iron acts largely as a "carrier." The iron breaks up the peroxy compound to form free radicals and is itself oxidized to ferric iron; the latter is reduced to the ferrous condition by the reductant, and so the decomposition of the peroxide is maintained by a small amount only of iron.

$$ROOR + Fe^{++} \longrightarrow RO^{\cdot} + Fe^{+++} + RO^{-}$$

$$Fe^{+++} + XH \longrightarrow Fe^{++} + X \qquad + H^{+}$$

Hence, in the case of the ternary catalyst system, as contrasted with a simple redox pair, a *reversible* redox reaction is involved.

Mersolat in Redox Polymerization. The use of Mersolat (Mepasin sulfonate) as an emulsifier presents a special case in respect to redox polymerization, since most samples of the emulsifier were contaminated with Mepasin sulfinate. As mentioned earlier, sulfinates are effective reducing agents in redox polymerization. The rate of polymerization in an ordinary Buna-S charge in which Mersolat was used as the emulsifier was very sensitive to changes in pH, as the following results show:[45a]

pH	12	10	8	7.7	6
Conversion in 6 hr.	0	21	75	88	47

Redox recipes involving oxidant-reductant-iron catalysis which were recommended by the German workers are typified by the data in Table XI.

Table XI. Redox Polymerization Recipes

	I	II	III
Butadiene	75	75	75
Styrene	25	25	25
Water	?	?	200
Mersolat	Quantity not stated		...
Nekal BX	...	...	2–3
Sodium oleate	...	...	0.5–1.0
Benzoyl peroxide	0.5	0.5	0.1
Iron as salt of paraffin acids	0.05	0.05	...
Sorbose	0.6	0.6	...
Dihydroxyacetone	...	...	0.1
Sodium pyrophosphate	Small amount	...	2–3
Ferrous ammonium sulfate	Trace	Trace (?)	0.01
Sulfuric acid	...	to $pH4$	...
pH	12	4	
60% conversion at 40° C.	4 hr.	30–70 min.	1 hr.
60% conversion at 70° C.	30–40 min.	Less than 10 min.	
60% conversion at 5–10° C.			30–40 hr.

It will be noted that the emulsifier Mersolat is adapted to conducting redox polymerization in an acid medium.

Quality of Low-Temperature Polymer. Physical testing data on a 10° C., 12-hour polymer compared with a 50° C., 30-to-40-hour Buna S are given in Table XII.

Reagents for Redox Polymerization. It was concluded that the oxidizing agent must be capable of oxidizing ferrous to ferric iron; the reducing agent must be capable of reducing ferric iron. Oil-soluble iron salts such as the fatty acid soaps were the best media for transferring iron from one phase to the other. Benzoyl peroxide, hydrogen peroxide, and air, served as activators, although an excess of oxygen poisoned the polymerization; ascorbic acid, glucose, sorbose, dihydroxyacetone and benzoin could be used as reducing agents.

If polymerization were effected at subzero temperatures, glycerine was added as an antifreeze.[103] The speed of these charges at 5° C. was adjusted

Table XII. Comparison of Buna S Prepared at 50 and at 10° C.

	Buna S1	Redox Polymer[89] Unmodified, pH 4
Raw Defo	5000	5500
Defo after 10 mill passes	5000	2200
Compounded Defo	7000	5200
Modulus at 300%, kg. per sq. cm.	95	63
Tensile strength	250	338
Elongation, %	500	670
Shore hardness	70	68
Rebound, % (at 20° C.)	48	56
(at 70° C.)	50	60
K value	76	120
Molecular weight	920,000	2,350,000
Degree of branching	54	39

to match the normal Buna-S3 control; at the normal temperature (48 to 50° C.) 60 per cent conversion was reached in 1 hour.

Most of the German work on polydienes was done early in the program at 50° C. It was reported by Youker and Copeland[138] that Dr. Becker felt that this field needed additional study "in low temperature redox systems . . . studying very carefully the structure of the resulting polymer." Only then did he believe that a polymer would be made with better properties than those of existing copolymers. It was reported that a number of the workers at Leverkusen considered that the presence of an olefin in a copolymer was harmful to the general-purpose rubbers where high resilience is sought.

Oxygen as the Oxidant in Redox Polymerization. Additional information on the laboratory development of the redox charges was obtained by Dr. Kolthoff on a trip to Germany.[73] Great importance was attached to the presence of oxygen in the charge, especially for the development of a recipe wherein oxygen served as the sole oxidizing agent. As an example, a benzoyl peroxide-free charge containing Mersolat (the emulsifier), sodium pyrophosphate, sodium hydrosulfite ($Na_2S_2O_4$, 0.5 per cent), and ferrous ammonium sulfate, and having pH 9 was treated with different amounts of oxygen, with the following results. The polymerization rate showed an

Conversion after 4 Hours with Varying Oxygen Pressures

Initial Pressure, mm.	Conversion, %
0	5
50	5
80	8
85	50
100	45
150	12

optimum at a definite oxygen content for the system. No information existed on the fate of oxygen in this system, although it was observed that the optimum rate corresponded to the amount of oxygen equivalent to the hydrosulfite. In the presence of benzoyl peroxide, the addition of small amounts of oxygen caused no inhibition; larger amounts had an additive effect on the rate.

Attention was given to acid systems using Mersolat, Nekal, or Esteramin hydrochloride. Mersolat was preferred by the Leverkusen group as the emulsifier in acid systems because its sulfinate impurity accelerated the rate markedly. The recommended procedure involved loading the reactor 60 per cent full with the aqueous solutions and the styrene, leaving the upper portion filled with air. Butadiene was pumped in to 90 per cent of capacity, leaving the autoclave under a 3- to 4-atm. pressure. The conversion reached 60 per cent in 30 to 60 minutes at 40° C. at a pH of 3 (sulfuric or hydrochloric acid). The latex, after shortstopping, was coagulated with aluminum sulfate and sodium chloride and the polymer was washed with dilute caustic and water.

The advantages of the peroxide-free recipe were lower cost, safety through avoiding the need to store and handle benzoyl peroxide, a faster rate than in peroxide recipes, and the absence of benzoyl peroxide in the finished rubber.[67]

At 40 per cent conversion, the average molecular weight determined osmotically was about 800,000. The viscosity molecular weight was much smaller, indicating considerable branching in the polymer.[73] The osmotic molecular weight was found to increase appreciably with decreasing polymerization temperature, thus:

Temperature, °C.	Molecular Weight
2°	2.5×10^6
30°	1×10^6
40°	5×10^5
	(estimated; gel present)

Dr. Kern of Leverkusen felt that the acid polymer was slightly better than its counterpart made in an alkaline charge.[73] He preferred monohydroxyacetone as the reductant in acid systems and dihydroxyacetone (0.5 per cent) in alkaline charges, along with a trace of ferrous iron.

Discovery of Redox Polymerization Elsewhere. The work of Bacon,[24] of Morgan,[93] and of Baxendale, Evans, and Park[26] was published just after the German information became available at the end of the war.[89] It was largely limited to acrylonitrile in solution but used a range of redox techniques. Also, a series of patents became available[48, 100, 112-8] covering the work of the B. F. Goodrich Company on the redox activation of hydrogen peroxide-catalyzed charges. Some mention was made of a benzoyl peroxide-ferrous pyrophosphate system but no outstanding value was claimed.

Application of Redox Polymerization in the United States. Interest in redox polymerization as part of the United States Synthetic Rubber Program was greatly stimulated by the reports of the German enthusiasm for redox activation,[83, 103] and an additional commission was sent to Europe under the supervision of C. S. Marvel.[89] These reports led to a substantial change in the Rubber Reserve program. Part of the effort went in the direction of duplicating German polymerization recipes with chemicals available in the United States; part was directed toward modifying the existing GR-S practice in the direction of the more favorable leads of the German system.[65]

It was recognized[90] that in the GR-S charge the potassium persulfate, as the water-soluble catalyst, and the mercaptan, as the oil-soluble reducing agent, constituted in effect a "redox" system and that the addition of iron would complete Dr. Becker's requirements for a redox activating system.[89] However, the addition of iron did not, in practice, accelerate the GR-S charge enough to compare with the other redox recipes, and consequently consideration was given to the incorporation of some of the other refinements of the German system.

The development of a redox-type GR-S formulation was a cooperative project of a number of laboratories engaged in the Rubber Reserve program; its outcome is described in a review of the operation on a plant scale of polymerization at 5° C. written by the Phillips Petroleum Company and the Copolymer Corporation.[3]

The first move was to duplicate as far as possible a German recipe with American chemicals. Mersolat was replaced by a sample of mixed alkane-sulfonates made by the E. I. du Pont de Nemours & Company from a C_{16} petroleum fraction by chlorosulfonation followed by hydrolysis and bearing the trade name MP189S. Using as a basis the recipe given above, experiments with this emulsifier were carried out in the following recipe:

Butadiene	75.0	Sorbose	0.6
Styrene	25.0	Iron (as ferric laurate)	0.05
Water	200.0	Tetrasodium pyrophosphate	
Mixed alkanesulfonates	7.5	(10 H_2O)	0.5
Benzoyl peroxide	0.5	Ferrous ammonium sulfate	
		(6 H_2O)	0.5

At pH of 7.7 the conversion reached completion in 3 hours at 50° C. Variations in the iron and the pyrophosphate showed that both were necessary in the charge. Without ferrous ammonium sulfate the conversion was 44 per cent; without ferric laurate the conversion was 69 per cent; in the total absence of iron no polymerization occurred.

Further efforts made in the United States to duplicate the German work resulted in the following recipes:[65]

	Optimum BD/S Recipe	Optimum Isoprene Recipe
Butadiene	70	...
Styrene	30	...
Isoprene	...	100
Water	200	200
Sodium alkanesulfonate (MP189S)	3.5	4.0
Benzoyl peroxide	0.5	0.5
Sodium oleate	1.5	1.0
Sodium pyrophosphate	0.2	0.2
Sorbose	0.6	0.6
Ferrous ammonium sulfate (6 H_2O)	0.35	0.25
Conversion in 18 hr. at 10° C.	49	62

The polymers, when tested in a standard tread-stock recipe, showed modulus and tensile-strength values higher than those of a GR-S control, elongations lower, and rebounds better.

Another section of the American redox program involved the substitution of soap flakes for the synthetic detergent to bring the system more nearly into line with the GR-S practices. This change brought the pH to 9.0 and gave a 100 per cent conversion in 3 hours at 50° C. Considerable importance was placed on the order in which the ingredients were added.[90] Any variation from the optimum order retarded the rate. Ferric laurate was omitted from the charge when it was shown that the presence of a fatty acid soap and ferrous ammonium sulfate supplied sufficient oil-soluble iron.

An effort was directed toward lowering the concentration of iron in the charge to avoid a high amount in the finished polymer. A new basic formula was developed:

Low-Iron Redox Charge

Butadiene	75.0
Styrene	25.0
Water	200.0
Rubber Reserve Std. soap flakes	5.0
Benzoyl peroxide	0.5
Ferrous ammonium sulfate (6 H_2O)	0.05
Sodium pyrophosphate (10 H_2O)	0.5
Reducing agent	0.5

Testing a large group of reducing agents gave the results listed in Table XIII.

Table XIII.   Reducing Agents in Redox Charge

Conversion (%) in 5 Hr. at 30° C.

Sorbose	65	n-$C_{12}H_{25}SO_2Na$	33	Sucrose	23
Fructose	65	Maltose	32	Dextrin	16
Inositol	46	Sodium hydrosulfite	30	Dodecyl mercaptan	14
Glucose	33	Lactose	30	Inulin	8
Glycerol	33	Ethylene glycol	25	None	10

It was found at this point that ferrous sulfate could replace ferrous ammonium sulfate, and better recipes were developed (Table XIV).

More soluble polymers were prepared in this system by introducing mercaptan modifiers. Very little effect on the rate was noted with mercaptans of six or more carbon atoms.

The further development in the United States of redox polymerization formulas adapted to the low-temperature manufacture of synthetic rubber took place with the introduction of hydroperoxides, such as cumene hydroperoxide, in place of benzoyl peroxide and is described in Chapter 8.

Table XIV. *Optimum Recipes*

	I 20–50° C.	II 0° C.
Butadiene	75	70
Styrene	25	30
Water	200	200
SF flakes	5	...
K oleate	...	5
Benzoyl peroxide	0.25	0.5
Ferrous sulfate (7 H_2O)	0.5	...
Sodium pyrophosphate (10 H_2O)	3.0	1.25
Ferrous ammonium sulfate (6 H_2O)	...	1.25

Temperature, °C.	Recipe	Time, Hr.	Conversion, %
50	I	1.0	81
40	I	1.5	80
30	I	2.0	81
20	I	8.0	80
0	II	21.5	43

Polymerization in Continuous Tubular Reactors. Redox activation, being very rapid, was readily adapted to continuous polymerization and was used in an experimental tube reactor located in the Leverkusen pilot plant.[58]

The equipment consisted of two 150-liter mixing tanks, one for hydrocarbons and the other for the aqueous phase. Lines from each ran at rates of 0 to 5 liters per hour through micrometering pumps to a premixer equipped with rapid stirrers and into 16 stainless-steel jacketed tubes 2.5 meters in length and 25 mm. in inside diameter. These tubes were inclined, to allow the latex to fall continuously. The total length was 44 meters; the total volume was 22 liters. The setup was arranged to allow water at four different temperatures to be circulated around the various lengths of tubing, if desired.

The entire equipment was swept with nitrogen before being used and kept under nitrogen throughout the run. No agitation was applied during the course of the tubes, as it was unnecessary if the rate of flow was fast enough. The latex was discharged through control valves, an antioxidant dispersion was added through a metering pump, and the latex was admitted to stripping equipment for the recovery of the monomer.

Two formulas which were charged in this equipment follow:

Butadiene	75	Butadiene	75
Styrene	20	Styrene	25
Acrylonitrile	5	Mersolat	4.5
Nekal	6	Water	144
Fatty acid	2	Sodium pyrophosphate	0.5
Water	180	sec-Potassium phosphate	0.25
Ammonium persulfate	0.5	Potassium ferrocyanide	0.25
Sodium hydroxide	0.66	Ammonium persulfate	0.05
Time to 60% conversion at 70° C. 60 min.			30 min.
at 50° C. 4–5 hr.			

Rubber when made by this technique at 60° C. was comparable to a Buna-S control; when prepared at 70° C., its properties were inferior.

The continuous-tube polymerizer was also used for reactions as fast as 15 minutes at 50° C., using the redox activation system and very carefully adjusting the amount of oxygen to the optimum concentration.

POLYMER EVALUATION[14, 94]

A definite procedure for the evaluation of polymers had been established, requiring 2 to 3 kg. of polymer. It was considered that this quantity was absolutely necessary. The procedure is summarized below.

Raw-Material Properties. The polymer was examined for condition, color, odor, and volatile content. The Defo value (Rohfelldefo) was determined[15] at 80° C. Solubility (in benzene) and solution viscosity were determined before and after milling. Storage stability, as measured by Defo and solubility, was determined after 4, 8, and 12 weeks at room temperature and after 2 and 4 weeks at 50° C.

Ease of breakdown was determined by milling and, using crumbs 2.5 mm. in size, by thermal breakdown at 130° C. and 3 atm. air pressure. The rate of thermal breakdown was taken as the time required—normally about 50 minutes—to reach a Defo value of 750. The change in Defo on storage after breakdown was also measured.

Preparation and Properties of Mixtures. Three types of compound were used, viz., tread, carcass, and pure gum. Millability was determined on the tread stock using index numbers to indicate behavior at different stages of the milling and compounding. *Compound Plasticity* (Mischungsdefo) was determined at 80° C. *Tubing Quality* (Spritzbarkeit) was determined by running the tread stock through a small tube machine having a small tread profile and head temperatures of 50, 75, and 100° C. Appearance, length and weight extruded per minute were observed. *Calendering* (Kalandrierbarkeit) was done with the carcass-type stock on a small calender under carefully controlled conditions at different temperatures. Calender temperature, shrinkage after 5 and after 24 hours, and surface appearance after 24 hours were reported. *Tack* (Klebrigkeit) was measured by pressing calendered sheets together under constant load and checking the adhesion by a hand pull. An index of tack was used. *Solubility* of the carcass compound was determined in isooctane.

Vulcanization and Physical Properties. Cure. Test samples were cured 30, 60, and 90 minutes at 110° C. as a scorch test. As an additional test for scorching tendency (Anvulkanisation) the stock might be run through a Marzetti plastometer at increasing temperatures. For tensile properties, cures were run at 20, 40, 60, 80, and 100 minutes at 133° C.

Mechanical Properties. Tensile (Festigkeit), modulus at 300 per cent (Belastung), and elongation (Dehnung) were determined on Schopper machines. The Schopper tear (Struktur: Einreiszfestigkeit) test was also used. Rebound (Elastizität) was measured at 20 and 70° C. Hardness (Härte) was determined on a Shore durometer with a dead weight.

Static elastic-plastic properties were determined by applying a load to a cylinder at different temperatures.   The height of the cylinder was measured 1 second and 1 hour after the load was applied and 1 hour after it was released, and from these were calculated the mean elastic modulus, the creep, and the permanent set.   Dynamic hysteresis and modulus were also determined.

Aging tests were run in the Geer oven at 100° C. for 1, 3, and 7 days. Bomb aging was used only in special cases.   *Swelling* tests were run in benzene, isooctane and water.   The change in mechanical properties after swelling was also measured.

Electrical properties were measured on a special compound pigmented with talc and clay.   Specific resistivity, dielectric properties, and breakdown voltage were measured after drying over P_2O_5 and after 24 hours in air saturated with water at 20° C.

Special tests such as low-temperature behavior, ozone and weather resistance, gas permeability, and heat conductivity were run in special cases only. For tire rubbers it was necessary to include Schopper detrition, flex life (Ermüdungsbeständigkeit), and low-temperature resistance.   *Evaluations* had to be made by experienced men on the basis of the various tests.

Full details of the above and other physical tests employed at Leverkusen are given in reference 94, where also a comparison of the Defo and Williams plasticity values is to be found.

POLYMER STRUCTURE

The team led by Dr. Marvel interviewed a number of prominent German chemists on their methods of studying the chemical structure of synthetic rubbers and ascertained that, although some attention had been given to chemical methods of determining polymer structure, little had been devoted to physicochemical methods, such as infrared absorption, and little to questions of the ratio of *cis* to *trans* addition and to the degree of crystallinity. Little satisfaction had been obtained from ozonolysis.   Using the older ozonolysis procedures of Harries and Pummerer, only 20 per cent of the total carbon had been accounted for.   By applying oxidation with permanganate after decomposition of the ozonide with steam and by then preparing esters of the carboxylic acids thus formed, still no more than 50 per cent of the carbon of Buna S was accounted for as succinic acid, β-phenyl-adipic acid, and 1,2,4-butanetricarboxylic acid.   The better results from ozonolysis obtained by Marvel and his co-workers are reviewed in Chapter 9.

A more complete oxidative scission was obtained in Germany by treating polymers in nitrobenzene solution with an aqueous solution of potassium permanganate for 15 hours at 20° C.   By this method, it was stated, as much as 93 per cent of the carbon in emulsion polybutadiene and 80 per cent of that in sodium polybutadiene could be accounted for.

More conclusive and direct results were obtained by titration with perbenzoic acid as a method of distinguishing between 1,4- and 1,2-diene addition.   This method has since been extensively studied in the United States and has been found to be of marked value, being relatively simple experimentally and giving consistent and apparently reliable results (cf. Chapter 9).   It depends on the fact that perbenzoic acid adds much more rapidly to double bonds in the chain of diene polymers and copolymers than to side vinyl groups. Some German results obtained by the two methods last mentioned are as follows.

	Per Cent, 1,4-Units	
	By Perbenzoic Acid Titration	By Permanganate Oxidation
Buna S	73	52
S3	75	...
SS	85	...
85	50	24
32	20	12
SKB (Russian)	22	12
Polyisoprene (emulsion)	85–90	
Poly-2,3-dimethylbutadiene	100	
Natural rubber and balata	100	

REFERENCES

1. Anderson, J. G., *Trans. Instn. Rubber Ind.*, **14,** 266–78 (1939).
2. Anon., *Chem. Eng. News*, **23,** 1841–8 (1945).
3. Anon., *Chem. Eng. News*, **27,** 1729–30 (1949).
4. Anon., *Repts: of Meetings of the WIKAUKO and KAUTEKO,* PB 4670 (1945).
5. Anon., PB 90004 (FD 18/46).
6. Anon., PB 90005 (FD 19/46).
7. Anon., PB 90085 (FD 509/46).
8. Anon., PB 90092 (FD 517/46).
9. Anon., PB 90093 (FD 518/46).
10. Anon., PB 91260 (FD 2682/47).
11. Anon., PB 98175.
12. Anon., PB 99934.
13. Anon., PB 100083.
14. Anon., PB 101522.
15. Anon., PB 101523.
16. Anon., PB 101598.
17. Anon., PB 110027.
18. Anon., BIOS Rept. No. 124.
19. Anon., CIOS XXVI–21.
20. Anon., FD 117/47.
21. Anon., FD 508/46.
22. Anon., FD 3452/46.
23. Anon., FD 4991/47.
24. Bacon, R. G. R., *Trans. Faraday Soc.*, **42,** 140–55 (1946).
25. Barron, H., *Modern Synthetic Rubbers*, 2d ed., Van Nostrand, New York, 1943.
26. Baxendale, J. H., Evans, M. G., and Park, G. S., *Trans. Faraday Soc.*, **42,** 155–69 (1946).
27. Bebb, R. L., Private Communication, Firestone Tire & Rubber Co. to O.R.R., June 4, 1946.
28. Brazier, S. A., et al., PB 23858 (BIOS Final Rept. 349) (1945).
29. Breuer, F. W., *Rubber Age N.Y.*, **54,** 229–34 (1943); 336–40 (1944).
30. Breuer, F. W., PB 9676 (1943).
31. Bullard, R. H., PB 169 (CIOS II–10) (1945).
32. Cambron, A., and Whitby, G. S., *Can. J. Research*, **2,** 144–52 (1930).

33. Davey, W. C., et al., PB 49192 (BIOS No. 800) (1946).

34. Davey, W. C., Patterson, P. D., and Hammond, G., PB 172 (CIOS VII–8) (1945).

35. Davey, W. C., Robson, J. J., and Patterson, P. D., PB 174 (CIOS XVII–2) (1944).

36. Davies, T. L., and Williams, H. L., *Can. J. Research*, **27F,** 143–50 (1949).

37. Dawson, T. R., BIOS Overall Rept. No. 7.

38. DeBell, J. M., Goggin, W. C., and Gloor, W. E., *German Plastics Practice*, DeBell and Richardson, Springfield, Mass., 1946.

39. Downing, J., et al., PB 69121 (BIOS 1119) (1947).

40. Dunbrook, R. F., and Greer, P. S., PB 13342 (1945).

41. Ebert, G., Fries, F. A., and Garbsch, P., Ger. Pat. 532,455 (1929).

42. Ecker, R., *Kautschuk u. Gummi*, **3,** 119–26, 165–7 (1950).

43. Edwards, G. R., and Hay, N. T., PB 63621 (BIOS 966) (1946).

44. Euler, H. von, and Martius, C., *Ann.*, **505,** 73–87 (1933).

45. Fennebresque, J. D., Monrad, C. C., and Troyan, J. E., PB 512 (CIOS XXII–22) (1945).

45a. FIAT Rept. No. 618.

46. Fikentscher, H., *Cellulosechem.*, **13,** 58–64, 71–74 (1932).

47. Fordham, J. W. L., O'Neill, A. N., and Williams, H. L., *Can. J. Research*, **27F,** 119–42 (1949).

48. Fryling, C. F., U.S. Pat. 2,379,431 (1945), (Appl. 1941); 2,383,055 (1945) (Appl. 1941).

49. Garten, V., and Becker, W., *Makromol. Chem.*, **3,** 78–110 (1949).

50. Gay, G., PB 20463 (BIOS 174).

51. Goodyear Tire & Rubber Co., Private Communication to O.R.R., Oct. 9, 1946.

52. Goodyear Tire & Rubber Co., Private Communication to O.R.R., Feb. 11, 1946.

53. Goudge, M. F., BIOS Final Rept. No. 1.

54. Graulich, W., and Becker, W., *J. makromol. Chem.*, **3,** 53–77 (1949).

55. Hagen, H., *Kautschuk*, **14,** 203–10 (1938); **15,** 88–95 (1939).

56. Hale, N. H., Private Communication to O.R.R., Oct. 10, 1947.

57. Handley, E. T., et al., PB 189 (CIOS XXII–21) (1945).

58. Handley, E. T., et al., PB 193 (CIOS XXIII–4) (1945).

59. Handley, E. T., Hingeley, S. F., and Rowzee, E. R., PB 192 (CIOS XXIII–3) (1945).

60. Harries, C. D., *Untersuchungen über die natürlichen und künstlichen Kautschukarten*, Springer, Berlin, 1919.

61. Hopkinson, R., et al., PB 190 (CIOS XXIII–1) (1945).

62. Houwink, R., *Elasticity, Plasticity and the Structure of Matter*, Cambridge Univ. Press, London, 1937.

63. Houwink, R., *Chemie und Technologie der Künststoffe*, Akademische Verlagsgesellschaft, Becker and Erler, Leipzig, 1942.

64. I.G. Farbenindustrie, Brit. Pat. 469,820 (1937).

65. Johnson, P. H., and Bebb, R. L., *J. Polymer Sci.*, **3,** 389–99 (1948).

66. Keen, W. N., *India Rubber World*, **110,** 174–5 (1944).

67. Kern, W., *Makromol. Chem.*, **1,** 199–268 (1948).

68. Kline, G. M., et al., PB 949 (1945).

69. Knorr, K. E., *World Rubber and Its Regulation*, Stanford Univ. Press, 1945.

70. Koch, A., *Brit. Plastics*, **8,** 302 (1936).

71. Koch, A., *Ind. Eng. Chem.*, **32,** 464–7 (1940).

72. Koch, A., and Gartner, E., Ger. Pat. 711,568 (1941) (Appl. 1937).

73. Kolthoff, I. M., Private Communication to O.R.R., Aug. 28, 1946.

74. Kolthoff, I. M., Private Communication to O.R.R., Sept. 10, 1946.

75. Konrad, E., PB 13355 (1945).

76. Konrad, E., *Angew. Chem.*, **62,** 491–518 (1950).

77. Konrad, E., and Becker, W., *Angew. Chem.*, **62,** 423–6 (1950).

78. Konrad, E., and Becker, W., PB 100840.

79. Konrad, E., and Tschunkur, E., Ger. Pat. 658,172 (1938).
80. Lewis, J. H., PB 94311 (CIOS XXXIII-1) (1945).
81. Link, A. E., PB 554 (1945).
82. Livingston, J. W., PB 517 (CIOS XXVIII-13) (1945).
83. Livingston, J. W., PB 16714 (CIOS XXXI-76) (1945).
84. Livingston, J. W., PB 23022 (CIOS XXXI-75) (1945).
85. Livingston, J. W., PB 34722 (FIAT 607) (1945).
86. Luther, M., and Heuck, C., Ger. Pat. 543,152 (1932) (Appl. 1928).
87. Marchionna, F., Butalastic Polymers, Reinhold, New York, 1946.
88. Maron, S. H., Moore, C., Kingston, J. G., Ulevitch, I. N., Trinastic, J. C., and Bornemann, E. H., Ind. Eng. Chem., 41, 156–61 (1949).
89. Marvel, C. S., PB 11193 (1945).
90. Marvel, C. S., Deanin, R., Overberger, C. G., and Kuhn, B. M., J. Polymer Sci., 3, 128–37 (1948).
91. Meyer, K. H., Natural and Synthetic High Polymers, 1st ed., Appendix, Interscience, New York, 1942.
92. Moakes, R. C. W., PB 49193 (BIOS 792) (1946).
93. Morgan, L. B., Trans. Faraday Soc., 42, 169–83 (1946).
94. Naunton, W. J. S., et al., PB 32161 (CIOS XXXIII-19) (1945).
95. Naunton, W. J. S., Synthetic Rubber, Macmillan, London, 1937.
96. Newton, R. G., and Scott, J. R., J. Rubber Research, 13, 1–19 (1944).
97. Ozol, R. J., and Chaffee, C. C., PB 954 (CIOS XXX-10) (1945).
98. Patat, F., Z. Elektrochem., 47, 688–95 (1941).
99. Perry, G. H., and Garvey, B. S., PB 191 (CIOS XXIII-2).
100. Pfau, E. S., U.S. Pat. 2,397,201 (1946).
101. Reiner, St., Kautschuk, 16, 138–9 (1940).
102. Rubber Subcommittee, TIIC, PB 13341 (1945).
103. Rubber Subcommittee, TIIC, PB 13356 (1945).
104. Rubber Subcommittee, TIIC, PB 13358 (1945).
105. Schatzel, R. A., and White, W. L., PB 214 (CIOS XXV-34) (1945).
106. Scott, J. R., J. Rubber Research, 13, 23–6 (1944).
107. Schotz, S. P., Synthetic Rubber, Benn, London, 1926.
108. Springer, A., Kautschuk, 14, 212–8 (1938).
109. Staudinger, H., and Fischer, Kl., J. prakt. Chem. (2), 157, 158–76 (1941); Rubber Chem. and Technol., 15, 523–34 (1942).
110. Staudinger, H., and Fischer, Kl., Rubber Chem. and Technol., 15, 473–522 (1942).
111. Staudinger, H., and Staiger, F., Ber., 68, 707–26 (1935).
112. Stewart, W. D., U.S. Pat. 2,380,473–7, (1945).
113. Stewart, W. D., U.S. Pat. 2,380,710 (1945).
114. Stewart, W. D., U.S. Pat. 2,380,905 (1945).
115. Stewart, W. D., U.S. Pat. 2,383,425 (1945).
116. Stewart, W. D., U.S. Pat. 2,388,372 (1945).
117. Stewart, W. D., U.S. Pat. 2,388,373 (1945).
118. Stewart, W. D., and Zwicker, B. M. G., U.S. Pat. 2,380,617–8 (1945).
119. Stöcklin, P., Ger. Pat. 684,936 (1939) (Appl. 1933).
120. Stöcklin, P., Kautschuk, 15, 1–7 (1939).
121. Stöcklin, P., Proc. Rubber Technol. Conf., edited by T. R. Dawson, and J. R. Scott, 434–47 (1938).
122. Thompson, J. W., and Hollis, C. E., PB 85200 (BIOS 3) (1945).
123. Tschunkur, E., and Bock, W., Brit. Pat. 339,255 (1931) (Appl. 1929).
124. Tschunkur, E., and Bock, W., Ger. Pat. 542,646 (1932) (Appl. 1930).
125. Tschunkur, E., and Bock, W., Ger. Pat. 588,785 (1933) (Appl. 1930).
126. U.S. Office of Rubber Reserve, PB 13323.

127. U.S. Office of Rubber Reserve, PB 13325.
128. Van Antwerpen, F. J., *Ind. Eng. Chem.*, **35,** 126–30 (1943).
129. Weidlein, E. R., Jr., *Chem. Eng. News*, **24,** 771–4 (1946).
130. Weidlein, E. R., Jr., and KixMiller, R. W., PB 13340 (1945).
131. Whitby, G. S., U.S. Pat. 1,832,163 (1931) (Appl. 1927).
132. Whitby, G. S., and Greenberg, H., *Trans. Roy. Soc. Can.*, **23,** III, 21–4 (1929).
133. Whitby, G. S., and Katz, M., *Ind. Eng. Chem.*, **25,** 1204–11, 1338–48 (1933).
134. Williams, I., *Ind. Eng. Chem.*, **16,** 362–4 (1924).
135. Wollthan, H., and Becker, W., Ger. Pat. 753,991 (Appl. 1937).
136. Wollthan, H., and Becker, W., U.S. Pat. 2,222,967 (1940) (Appl. 1938).
137. Youker, M. A., U.S. Pat. 2,417,034 (1947).
138. Youker, M. A., and Copeland, N. A., PB 16029 (FIAT 717) (1946).

NAME INDEX

SUBJECT INDEX